Scientific Conference on Automotive Vehicles and Combustion Engines 2016 (KONMOT 2016)

IOP Conference Series: Materials Science and Engineering
Volume 148

Krakow, Poland
22 – 23 September 2016

Editors:

Marek Brzezanski
Wladyslaw Mitianiec

ISBN: 978-1-5108-3054-7
ISSN: 1757-8981

Printed from e-media with permission by:

Curran Associates, Inc.
57 Morehouse Lane
Red Hook, NY 12571

CURRAN ASSOCIATES INC.
proceedings
.com

Some format issues inherent in the e-media version may also appear in this print version.

Printed by Curran Associates, Inc. (2016)

For permission requests, please contact the Institute of Physics
at the address below.

Institute of Physics
Dirac House, Temple Back
Bristol BS1 6BE UK

Phone: 44 1 17 929 7481
Fax: 44 1 17 920 0979

techtracking@iop.org

Additional copies of this publication are available from:

Curran Associates, Inc.
57 Morehouse Lane
Red Hook, NY 12571 USA
Phone: 845-758-0400
Fax: 845-758-2633
Email: curran@proceedings.com
Web: www.proceedings.com

Table of contents

Volume 148

Scientific Conference on Automotive Vehicles and Combustion Engines (KONMOT 2016)

22–23 September 2016, Krakow, Poland

Accepted papers received: 30 August 2016
Published online: 27 September 2016

Preface

Papers

Design and reasearch of automobiles

Exploitation of automobiles

Combustion engines and emission

IOP Conf. Series: Materials Science and Engineering **148** (2016) 011001 doi:10.1088/1757-899X/148/1/011001

Preface: Scientific Conference on Automotive Vehicles and Combustion Engines KONMOT 2016, 22-23 September 2016, Cracow, Poland

The contemporary world requires ecological solutions in the area of automotive vehicles, low energy consumption, safety and easy exploitation. The role of engineers and scientists is the realization and managing of the designing and exploitation processes within that area. The growing exploitation and ecological demands set successive challenges for designers and scientists in the area of automobiles, powertrains and exploitation. Particularly the essential problem is finding solutions decreasing fuel consumption and emission of the exhaust gases as well as application of the alternative powertrains such as hybrid and electric. The cyclical automotive conference KONMOT since 1976 is a place of meeting of outstanding scientists and engineers from country and abroad, raising contemporary problems of motorization with the possibility of publishing results on the international area.

In 2016 the organized automotive conference KONMOT celebrated the 40th anniversary. It was organized by the Institute of Automobiles and Internal Combustion Engines in Cracow University of Technology (CUT). The first conference was held in Kroscienko-Szczawnica on 20-23 September 1976. The motto, that guided the deliberations, was "Progress in Research of Motor Vehicle." The conference was a nationwide and attended by many representatives of technical universities, research and development centres and plants related to the automotive industry. In the years 2000, 2004 and 2008, a conference was called KONMOT-AUTOPROGRES, the main organizer was the Cracow University of Technology. In 2002 and 2006 the main organizer of the conference was Automotive Industry Institute in Warsaw and the conference was called AUTOPROGRES-KONMOT. Papers presented on every two years conference KONMOT were published in the journal issued by the Polish Academy of Sciences “Portfolio of Science and the Problem Committee on Motorization” or published by the Technical Journal issued by CUT, as well as in the form of monographs. Some of the conferences were dedicated to the jubilee of work of outstanding scientific specialists in the automotive and engines fields.

Apart from the official agenda for the conference the gala dinners were organized for all participants, often headline performances by famous artists were important point of such meetings. The penultimate conference KONMOT was held on 25-26 September 2014 in Cracow under the slogan "Development of vehicles - construction, operation, ecology and safety." The organized scientific conference KONMOT-2016 as in previous years provided an excellent opportunity to showcase of achievements of researchers and scientists and exchange of experience for a wide range of activity in research and development centres, industrial areas and universities.

With regards to the current conference we are delighted to come up with ninety two (92) contributed research papers selected from the proceedings, as an output of the Scientific Automotive Conference KONMOT 2016 on motor-cars and engines progress held in Cracow (Poland) from September 22-23, 2016. This event marks as previous meetings in a highly successful series of conference devoted to this important technical field. Conference KONMOT gathers industrial and scientific communities in the field of design, experimental research and modelling of all scientific and industrial sectors of automotive branch. Contributions cover a wide range of topics including design and modelling of the suspension, traction and new driving systems in automobiles, method of vehicles exploitation or tribology. The new driving systems require a new design of engines for better efficiency and lower exhaust gas emission. Thirty three papers describe the research, experimental investigations and modelling of combustion processes in combustion engines.

The papers shown during KONMOT 2016 from both the oral presentation and poster sessions are included in this conference volume. They are divided into the following three topic areas: Design and Research of Automobiles, Exploitation of Automobiles and Internal Combustion Engines and Emission. More than one third of the papers are directly concerned with mathematical modelling and numerical simulation in automobiles and internal combustion engines. However, most of papers include experimental and industrial activity. Some of the papers concern the applying of special new materials and were included in the Exploitation topic.

The conference co-chairmen are grateful to the scientific committee and all the reviewers of the papers for their efficient work done in a very short period of time. These proceedings would not have been possible without the tremendous help of Mr Dr Eng. Marcin Noga and Mrs MSc Magdalena Kalarus from the Cracow University of Technology. Thanks also to CUT Publishing House for preparing of special issue of conference abstracts. Finally, we would like to thank our sponsors for their financial support, without that the conference would not be so attractive for participants, such as Development and Research Institute BOSMAL in Bielsko Biala, AVL – Poland Branch, FEV Poland, EC Engineering, Horus-Energia and Polish Scientific Society of Combustion Engines (PTNSS) and others. The logos of our main sponsors are shown below.

BOSMAL® AVL FEV

EC ENGINEERING HORUS ENERGIA

PTNSS

We invite all participants from whole the world to the next conference KONMOT 2018.

Professors of Cracow University of Technology:
Marek Brzezanski and **Wladyslaw Mitianiec**
Conference Co-chair and **Co-Editor**
E-mail: mbrzez@pk.edu.pl
wmitanie@usk.pk.edu.pl

Peer review statement

All papers published in this volume of *IOP Conference Series: Materials Science and Engineering* have been peer reviewed through processes administered by the proceedings Editors. Reviews were conducted by expert referees to the professional and scientific standards expected of a proceedings journal published by IOP Publishing.

IOP Conf. Series: Materials Science and Engineering **148** (2016) 012001 doi:10.1088/1757-899X/148/1/012001

The influence of number and values of ratios in stepped gearbox on mileage fuel consumption in NEDC test and real traffic

P Bera[1] and D Wędrychowicz[1]

[1] AGH University of Science and Technology, Faculty of Mechanical Engineering and Robotics, Department of Machine Design and Technology, Cracow, Poland

E-mail: pbera@agh.edu.pl

Abstract. The article presents the influence of number and values of ratios in stepped gearbox on mileage fuel consumption in a city passenger car. The simulations were conducted for a particular vehicle characterized by its mass, body shape, size of tires and equipped with a combustion engine for which the characteristic of fuel consumption in dynamic states was already designated on the basis of engine test bed measurements. Several designs of transmission with different number of gears and their ratios were used in virtual simulations of road traffic, particularly in the NEDC test, to calculate mileage fuel consumption. This allows for a quantitative assessment of transmission parameters in terms of both vehicle economy and dynamic properties. Also, based on obtained results, recommendations for the selection of a particular vehicle for a specific type of exploitation have been formulated.

1. Introduction

Fuel consumption of the car is for many customers one of the most important criteria for the selection of the particular model. Car manufacturers, which sell cars in Europe, shall provide in catalogues fuel consumption obtained in three homologation tests: urban (UDC), extra-urban (EUDC) and combined (NEDC). These tests are performed on a chassis dynamometer in a standardized test. The vehicle motion is described by an appropriate driving schedule, and gear-change points are strictly determined. The homologation test is the same for all passenger cars, which allows to compare their properties in the same traffic and environmental conditions. The NEDC (New European Driving Cycle) test is currently in force in Europe [1]. Unfortunately, this test significantly differs from the real traffic conditions [2]. This is one of the factors that often leads to significant discrepancies between the homologation and the real traffic fuel consumption [3]. Another reason are the gearbox ratios. They influence both car driveability and real fuel economy.

Gearbox ratios must ensure both required tractive force to put the vehicle in motion as well as reaching high top speed. To provide the best dynamic properties engine must work between the speed of the nominal torque and nominal power, which affects the theoretical number of gears. In practice, city cars usually are equipped with 5 gears regardless of the engine characteristic [4, 5]. It turns out, however, that even for vehicles with very similar technical characteristics, the gear ratios may vary significantly. This affects both vehicle dynamics and the mileage fuel consumption. The objective of the present work is to quantitative asses impact of the different gearbox ratios (in one car) on fuel consumption in NEDC test and in a few selected situations of the real traffic.

2. The use of fuel consumption characteristic in dynamic states of the engine in simulations

2.1. Measurements on engine test bed

Simulations of mileage fuel consumption for a particular vehicle are possible, only when it is equipped with an internal combustion engine, for which the fuel consumption characteristic in dynamic operating states has already been designated. Such a characteristic is created based on data collected during measurements performed on the engine test bed (Figure 1). Three basic parameters of the engine were measured: engine speed *n*, engine torque *M* and fuel mass flow *G*. Sample graph is given in figure 2.

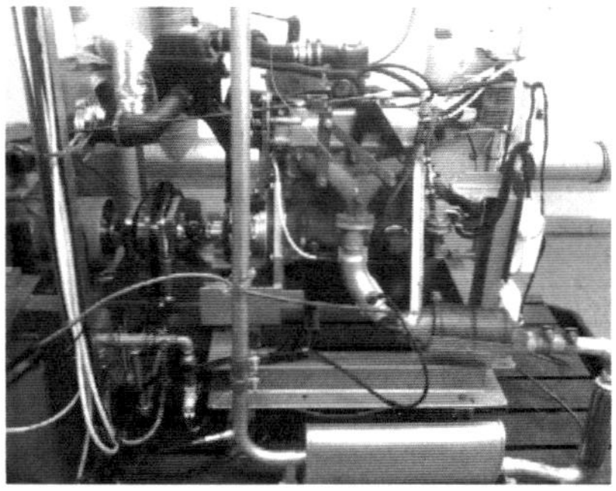

Figure 1. 899ccm SI engine for which characteristic of fuel consumption was designated.

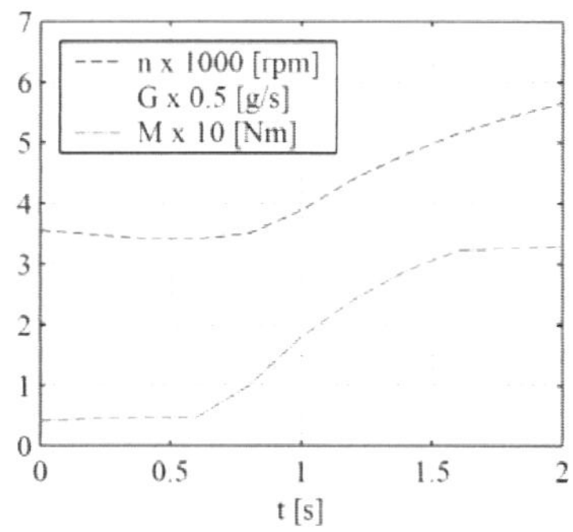

Figure 2. Engine working parameters in a single measurement.

The development of such a characteristic requires the use of artificial neural network (ANN) trained in supervised mode. The method is described in details in the works of the author [6, 7]. Architecture of ANN used to develop such a characteristics is shown in figure 3.

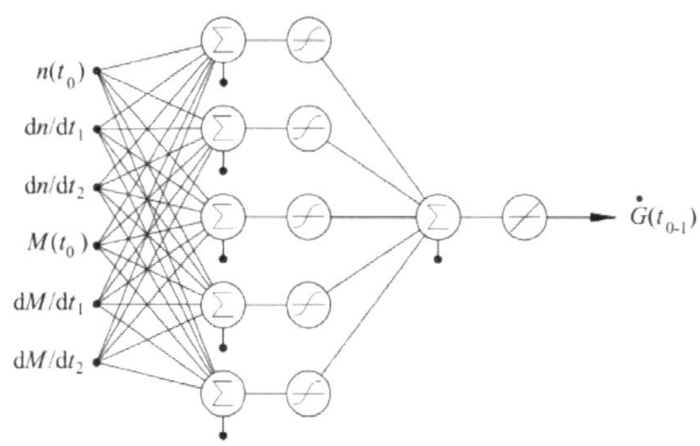

Figure 3. Artificial neural network used for calculation fuel mass flow in dynamic states.

The specificity of this method is based on analysing engine rotational speed and torque in a time window of a given length. Such an action takes into account series of phenomena occurring in the engine in dynamic operating states, e.g. changes in kinetic energy of parts in motion, changes of air-fuel mixture composition and delay in engine response to the control signal. The use of this method allows to calculate the current value of fuel mass flow *G*, both in static and dynamic working states. In comparison to other methods [8, 9] of calculation fuel consumption in dynamic states, this one is possibly the most general. In works [7, 10] it is demonstrated that in dynamic states only the use of such a characteristic allows to obtain reliable results of fuel consumption simulations. Universal characteristic can be only used in static or quasi-static operating states. The validity of the approach proposed by the author is confirmed by the regression graph shown in figure 4. The fit of ANN model to measurement data presented to the network in specific way is very good. It shows that there is a strong relationship between fuel mass flow G and three consecutive values of engine rotational speed n and torque M.

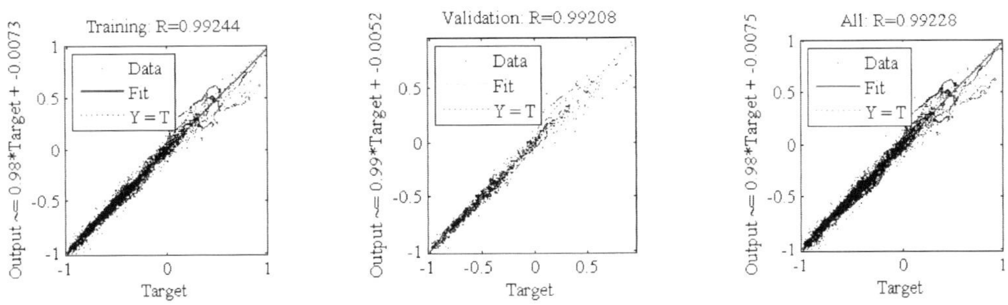

Figure 4. Regression plots received after ANN training.

Characteristic used in this article is based on measurements performed on a spark ignition engine with a nominal power 29 kW/5500 rpm and nominal torque 65 Nm/3000 rpm. Therefore it will be fitted in a small city passenger car. All further considerations will be targeted at this group of vehicles.

2.2. Application of fuel consumption characteristic for simulations of mileage fuel consumption in car

Fuel consumption characteristic in dynamic states, described in the above paragraph, can be used in simulations of mileage fuel consumption for a specific vehicle. Certainly, all vehicle key technical prameters must be known: vehicle mass m_p, wheel diameter D_k (and dynamic radius r_d), tyre rolling-resistance coefficient f, aerodynamic drag coefficient C_x, frontal area A, the total transmission efficiency η_c, coefficient of rotating masses in transmission δ, nominal engine torque and power. Based on these data, overall ratios of the transmission i_{c1}, i_{c2}, i_{c3}, i_{c4}, i_{c5} are calculated. Then with all these technical data it is possible to determine the engine rotational speed n and torque M on the basis of vehicle speed v and acceleration a. The dependence between the above parameters are described by the following relationships:

$$n = \frac{v \cdot 60 \cdot i_c}{2 \cdot 3.6 \cdot \pi \cdot r_d} \tag{1}$$

$$M = \frac{\left(m_p \cdot a \cdot \delta + F_p + F_t\right) \cdot r_d}{i_c \cdot \eta_c} \tag{2}$$

These two engine parameters are then directly used to simulate the ANN at every moment of the test. On the inputs of the network three successive values of engine rotational speed and torque are presented in time intervals 0.1 s. They are forthwith computed on the output – the value of fuel mass flow– with accordance to properly trained weights and biases.

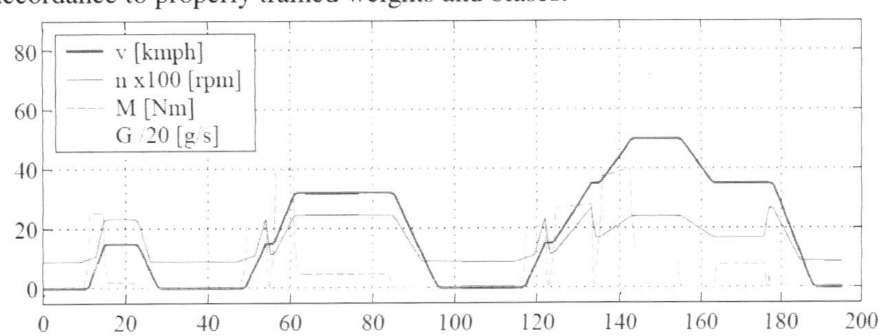

Figure 5. Vehicle A parameters in UDC test.

After completion of the test the whole amount of fuel is known, as well as the total distance of the vehicle. This allows to easily calculate mileage fuel consumption expressed in dm^3/100 km.

3. Construction parameter of city passenger cars used in further simulations

As it has been mentioned before, fuel consumption characteristic in dynamic states was designated for a small SI engine wherefore it will be used as a propulsion of a small city passenger car.
To the class of small city cars the following ones can be included: Volkswagen Up 1.0 44 kW move up! 2011 [4] Mitsubishi Space Star 1.0 Instyle 2015 [5], Mitsubishi Space Star 1.2 Intense + 2013 [5] Kia Picanto 1.0 EcoDynamics Spirit 5-d in 2015 [5] Toyota Aygo 1.0 VVT-i and x-wave Style 2015 [4], Suzuki Celerio 1.0 Comfort 2015 [4], the Hyundai i10 1.0 i-Motion Premium 2013 [5] and Fiat Panda 1.2 Pop Fire 2012 [5].
For a group of the above-mentioned cars the value of dynamic coefficient in first gear equals $D = 0.32 \div 0.39$. This means that the total value of the first gear ratio can be included within: $i_{c1śr} \pm 9\%$. It also turns out, that the theoretical vehicle speed at fifth gear, at nominal engine speed, is always higher than the vehicle top speed of 10% to 50%. This shows that designers have a wide range of possibilities when choosing a final construction solution, that will affect both vehicle dynamics and economy. Hereinafter the extreme values of the above ranges will be applied in two cars that are designed for virtual fuel consumption simulations. The technical data of both vehicles are shown in table 1.

Table 1. Basic technical parameters of vehicles A and B used in simulations of mileage fuel consumption.

Parameter	Vehicle A	Vehicle B
Vehicle mass m_p [kg]	730	
Total vehicle weight [kg]	1130	
Vehicle frontal area A [m^2]	2.0	
Aerodynamic drag coefficient C_x	0.32	
Tires size	R14 165/70	
Tyre rolling-resistance coefficient f [-]	0.01 – 0.015	
Nominal engine power [kW] / [rpm]	29 / 5500	
Nominal engine torque [N·m] / [rpm]	65 / 3000	
Top speed [kmph]	138	
1st gear overall ratio	16.709	19.919
2nd gear overall ratio	8.214	10.242
3rd gear overall ratio	5.116	6.574
4th gear overall ratio	3.732	4.891
5th gear overall ratio	3.024	4.016

Vehicle A has a long geared transmission whereas Vehicle B has a short geared transmission. Data from table 1 can be written in a different way: dynamic coeffcient in first gear for Vehicle A equals $D = 0.32$ and the top speed exceeds the theoretical maximum of 50%, however, the Vehicle B has dynamic coefficient of $D = 0.39$ and theoretical speed exceeding the top speed by 10%. Finally, the Vehicle B may be equipped with 6 gears (point 8). Figure 6 shows speed graphs for all gear ratios in each gearbox.

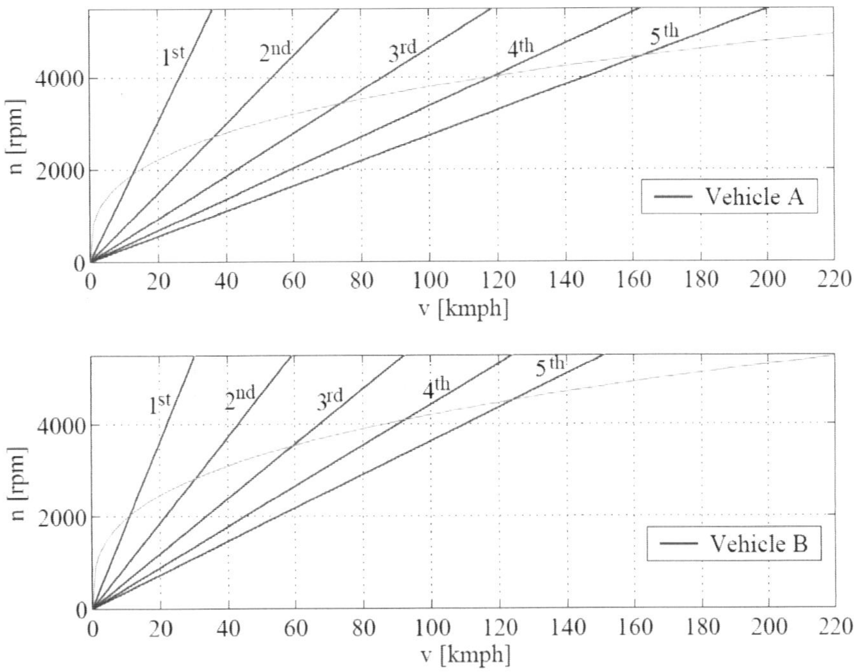

Figure 6. Engine rotational speed vs vehicle speed on each gear.

Gearbox ratios directly influence tractive force in all gears, which is shown in figure 7.

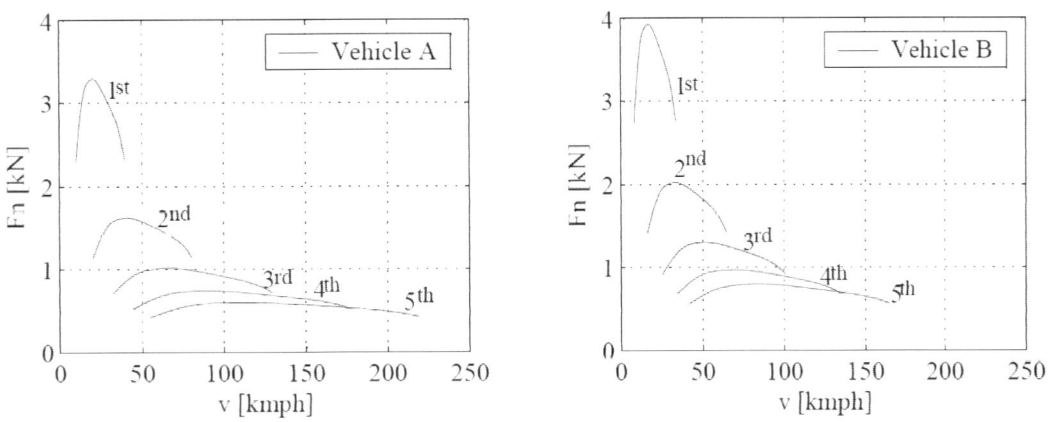

Figure 7. Tractive force on each gear for Vehicle A and B.

From the analysis of the figure 7 it results that Vehicle B has a greater tractive force on every gear than Vehicle A. The consequence of this is better dynamic, allowing the achievement of better acceleration in every gear.

4. Fuel consumption in the NEDC test

In the NEDC test vehicle speed is described by an appropriate driving schedule and the gear-changes occur at predetermined speeds. For a particular vehicle it directly determines the engine rotational speed and torque thus affecting fuel consumption in the test. Figure 8 shows the course of the vehicle speed in this test.

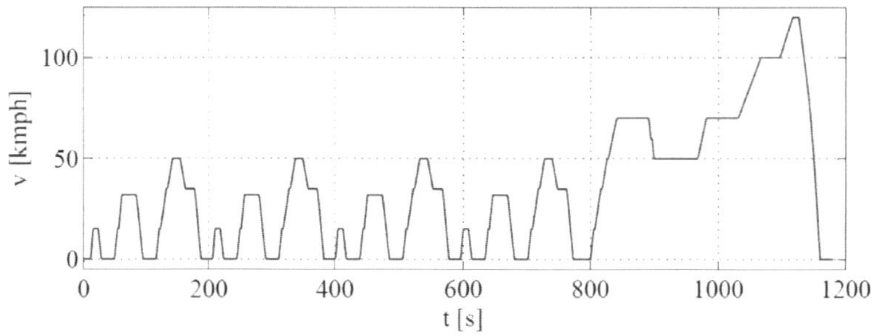

Figure 8. Vehicle speed vs time in NEDC test.

With use of the formulas described in section 2.1, engine rotational speed and torque can be easily calculated. Then these parameters are presented to network inputs, and current fuel mass flow is fortwith claculated. The above parameters for vehicles A and B are shown in figure 9.

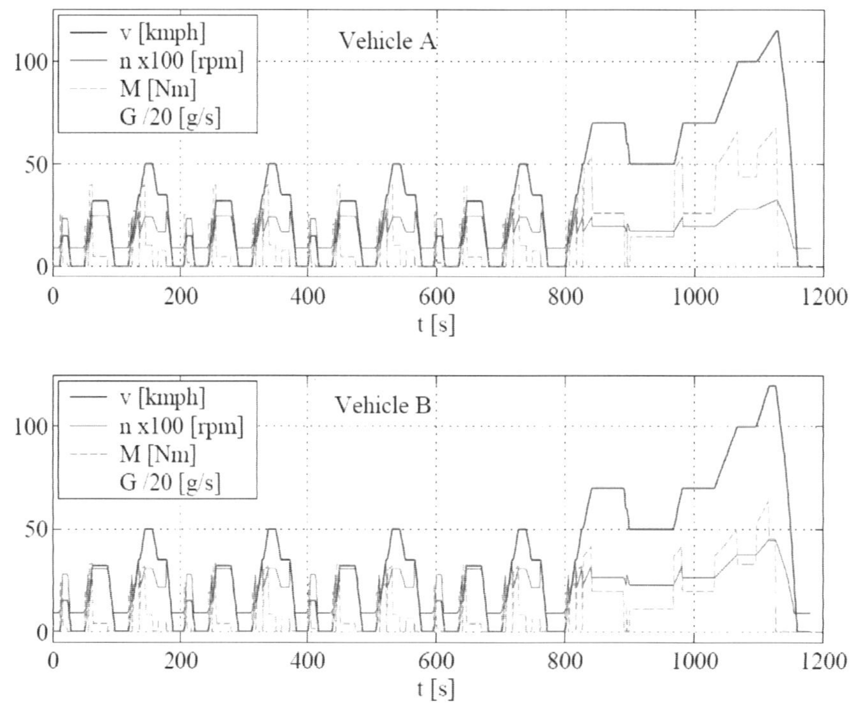

Figure 9. Vehicles speed and engine working parameters in NEDC test.

From obtained plots in figure 9 it results that in case of Vehicle A, the engine works under greater load and lower rotational speed. In such working conditions engine efficiency is possibly the highest so the fuel consumption is low, which is presented in table 2.

Table 2. Fuel consumption of Vehicle A and Vehicle B in homologation tests

Fuel consumption [dm^3/100 km]	Car A	Car B
UDC test (city)	6.8	8.0
EUDC test (highway)	4.0	4.8
NEDC test (combined)	5.2	6.1

From the performed simulations we can observe, that in case of the NEDC, long gear transmission is a significantly better solution than short gear transmission. Vehicle A has considerably lower fuel consumption than Vehicle B in all tests: UDC, EUDC and NEDC, respectively 15%, 17% and 15%. Certainly, dynamics of both vehicles, determined by the speed course in the test, is the same.

5. Fuel consumption in a real city traffic

5.1. Specificity of real city traffic

The real traffic puts greater demand on vehicle dynamics than NEDC test. Acceleration to 50 kmph takes about 14-20 s, while the 70 kmph is reached within about 20-25 s. These values are lower about 30% than in the test. Besides in the real traffic conditions driver must adjust the speed of the vehicle so that move smoothly in the stream of cars and with accordance to road traffic regulations. Moreover, gears can be shifted at any speed in contrary to NEDC test. According to the e.g. Ecodriving rules the engine should work under high load and in range of small rotational speed due to the highest efficiency in this area of work field. However, in order to ensure proper durability of the engine minimum rotation speed at which it can operate with a maximum load is 1500 rpms. The following simulations will be conducted with accordance to this rules.

5.2. Acceleration to 50 kmph

In most cases, the vehicles speed in the city is restricted to 50 kmph, similarly to UDC test. However, the acceleration time is shorter. The following simulation, shown in figure 10, will be performed for acceleration from 0 to 50kmph. Gearshift moments are different for each vehicle.

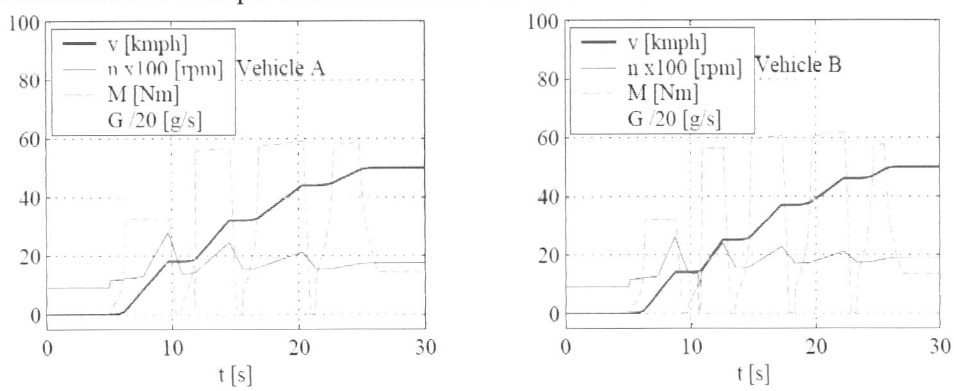

Figure 10. Acceleration to 50 kmph.

From the performed simulations it results that average fuel consumption of two vehicles is almost the same – Vehicle A – 7.2 dm^3/100, whereas Vehicle B – 7.3 dm^3/100km. It was possible due to the fact that in Vehice B gearshifts were quicker. Moreover, after reaching a speed of 50 kmph the Vehicle B can shift to 5th gear while the Vehicle A can shift only to 4th gear.

5.3. Acceleration to 70 kmph

The maximum speed which usually can be reached in a city is 70 kmph. In the next simulation, shown in figure 11, both cars will accelerate from 0 to 70 kmph in the same time. As in the previous case, both engines will work in area of maximum efficiency field, and both cars will cover the same distance.

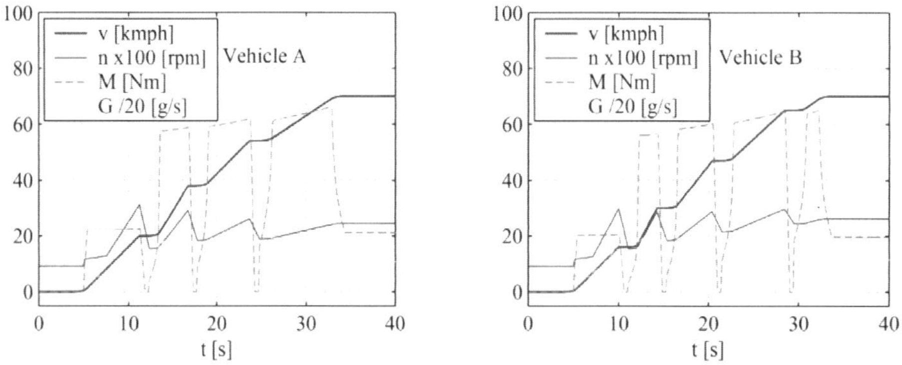

Figure 11. Acceleration to 70 kmph.

In this case average fuel consumption of Vehicle A is 7.6 dm^3/100km, and Vehicle B 7.8 dm^3/100km. This means that also in this instance, fuel consumption is similar, but requires a different gearbox shifting. Another difference is that in Vehicle B all five gears are used whereas in Vehicle A fifth gear can be shifted only after reaching final speed. Otherwise acceleration would take much more time in comparison to Vehicle B.

6. Acceleration to 100kmph with full throttle opening

Acceleration time from 0 to 100 km/h is one of the basic dynamic parameters of the vehicle. Figure 12 shows how different gearbox ratios affect acceleration process. Acceleration to 100 kmph is done so that upshifts take place after reaching nominal engine speed 5500 rpms, and the throttle is fully opened. In this case shift time is shorter than in previous simulations and lasts 1 s.

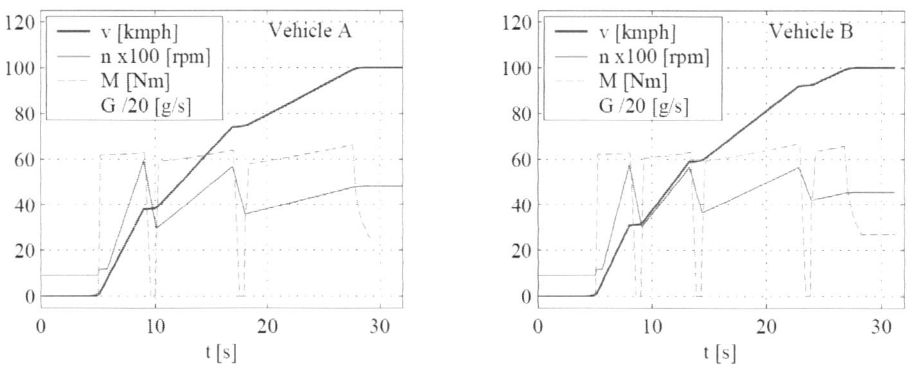

Figure 12. Vehicle parameters while accelerating to 100kmph.

The obtained results show that the acceleration time of Vehicle B is 1 s shorter than Vehicle A and the average fuel consumption equals respectively 9.3 dm^3/100km and 9.2 dm^3/100km.

7. Extra-urban / Motorway driving

At a constant highway speed of 120 kmph Vehicle A with longer gear ratios achieves a fuel consumption of 5.2 dm³/100km while vehicle B 5.6 dm³/100km. This means that the Vehicle A has 7.7% better fuel economy than Vehicle B. It's a noticeable difference, but it should be noted that in overcoming any major hill or while overtaking downshift will be needed what entails an increase in fuel consumption.

8. Increase in number of gears to 6

It is possible to increase the number of gears to six in Vehicle B. Using the method of gear graduation shown in figure 13, 6th gear in Vehicle B would have overall ratio of $i_{c6} = 3.522$.

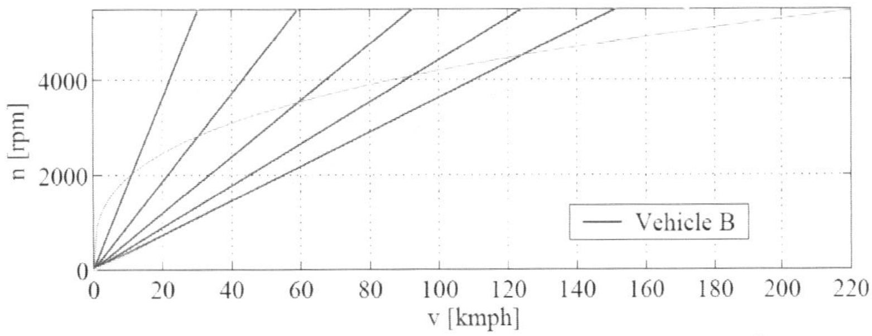

Figure 13. Vehicle B speeds on each gear including 6th.

At a constant speed of 120 kmph it would allow to decrease engine speed of 520 rpms thus reducing fuel consumption by 0.2 dm³/100km in comparison to the five-speed version. The benefits of using 6 speed gearbox are rather negligible and are achievable only in specific conditions.

9. Conclusions

The paper describes quantitative differences in mileage fuel consumption of a small city car, depending on gearbox ratios. Both solutions of vehicles A and B were within the limits applied in the construction arisen over the past five years. It turns out that the fuel consumption given by an car manufacturer and obtained in homologation test should not clearly determine the selection of a particular vehicle. Indeed - a vehicle with long gear ratios is always more economical in NEDC test (in described case it was 15% in the UDC test, 17% in extra urban and 15% in combined), however, in a real city traffic vehicle with short gearbox ratios is better. First of all it allows for full use of all five gears within speed range to 70 kmph. Secondly it provides better vehicle dynamics if needed, and finally, with proper gear handling, allows to achieve the same mileage fuel consumption as the vehicle fitted with a gearbox with long gears. The potential benefits of long geared transmission can be obtained only in extra-urban driving on the flat roads and when frequent overtaking is not required.
The use of sixth gear in the Vehicle B can reduce fuel consumption when driving at constant motorway speed of 120 kmph by 4%. In a strictly urban vehicle it is an expendable solution. The potential of the six speed transmissions can be fully used rather in vehicles with turbocharged engines, which have a high torque allowing for smooth acceleration even in sixth gear with the lowest ratio, than in small city cars.

References

[1] Official Journal of the European Communities 1991 *Council Directive amending Directive 70/220/EEC on the approximation of the laws of the Member States relating to measures to be taken against air pollution by emissions from motor vehicles (91/441/EEC)* No L 242/1
[2] *Official fuel consumption figures* Retrieved on 14.04.2016 from: http://www.theaa.com/motoring_advice/fuels-and-environment/official-fuel-consumption-figures.html

[3] Tietge U., Mock P., Zacharof N., Franco V 2015 *ICCT* Real-world Fuel Consumption of Popular European passenger car models 2015-8
[4] *Technical data of automobiles* 2016 Retrieved on 10.03.2016 from: http://www.cars-data.com/
[5] *The complete catalog of cars since 1945* 2016 Retrieved on 10.03.2016 from: http://www.automobile-catalog.com/
[6] Bera P 2013 *Combustion Engines* 4/2013 **155** p 16-25
[7] Bera P, Szybka J 2015 *Fuel consumption analysis in dynamic states of the engine with use of artificial neural network* (Cracow: AGH Press) ISBN 978-83-7464-789-2
[8] Ubysz A 2011 *Energy conservation in vehicls vol 2 General dynamic characteristics in vehicle* (Gliwice: Silesian University of Technology Press) ISBN 978-83-7335-918-5
[9] Romaniszyn K M and Wnęk H 2011 *Combustion Engines* 3/2011 **146** p 38-44
[10] Bera P 2015 *Combustion Engines* A comparison of the fuel consumption characteristic in dynamic states with the general characteristic of the combustion engine p 726-731

Analysis of limit forces on the vehicle wheels using an algorithm of Dynamic Square Method

M. Brukalski[1]

[1]MSc., Eng., PhD student, Institute of Vehicles, Warsaw University of Technology

E-mail: matbrukalski@gmail.com

Abstract. This article presents a method named as Dynamic Square Method (DSM) used for dynamic analysis of a vehicle equipped with a four wheel drive system. This method allows determination of maximum (limit) forces acting on the wheels. Here, the maximum longitudinal forces acting on the wheels are assumed and then used to predict whether they can be achieved by a specific dynamic motion or whether the actual friction forces under a given wheel is large enough to transfer lateral forces. For the analysis of DSM a four wheel vehicle model is used. On the basis of this characteristic it is possible to determine the maximum longitudinal force acting on the wheels of the given axle depending on the lateral acceleration of the vehicle. The results of this analysis may be useful in the development of a control algorithm used for example in active differentials.

1. Introduction

Analysing vehicle dynamics while cornering or while driving in a curvilinear motion, taking into account the driving forces and braking is an important issue. The analysis helps to improve active safety of the vehicle, and also in the creation of a sophisticated drive technology especially with an all-wheel drive. The Dynamic Square Method (DSM) allows us to define the maximum longitudinal force at the wheels as a function of lateral and longitudinal acceleration obtained by the vehicle while driving in a rectilinear or curvilinear motion. Knowledge of available forces (longitudinal and lateral) on wheels allows us e.g. to develop a system for the control of torque (Torque Vectoring (TV)) [5, 6].The main aim of this study was an attempt to comprehend and utilize the DSM for four-wheel vehicle model. An analysis was attempted for the change in vertical load on each wheel while driving in both a curve and on a straight track. Maximum longitudinal forces on the wheels along with lateral an longitudinal vehicle acceleration were determined as a result.

2. Analysis Dynamic Square Method for using four wheel vehicle model

The DSM firstly was presented in 1995 by Kato M., Isoda K. and Yuasa H. [2]. This method was mentioned by Klomp M. in his articles [3, 4]. This method allows to determine the actual maximum longitudinal and lateral forces which can be transferred by the wheels of the vehicle on the road. On the basis of the actual forces on the wheels longitudinal and lateral acceleration of the vehicle can be calculated. Practical use of DSM was presented by the author by a single track vehicle model [1]. This article presents the analysis of Dynamic Square Method for a two track vehicle model powered by a four wheel drive (figure 1 and 2).

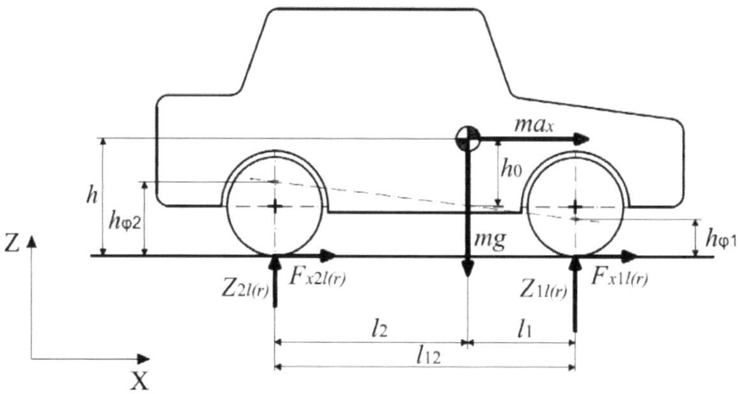

Figure1. Four-wheel vehicle model – side view.

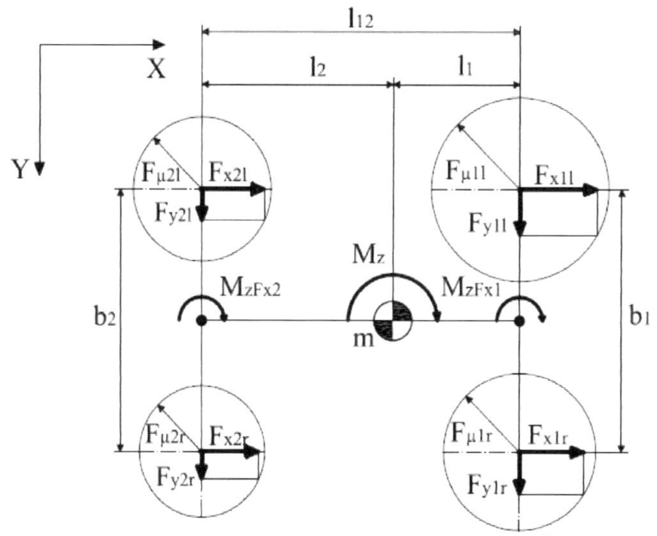

Figure 2. Four-wheel vehicle model – plane view.

The following assumptions were made for the analysis of a two track model by the DSM:

- a three dimensional model is used,
- vehicle curvilinear motions assumed as to the vehicle turning towards the right,
- traction forces $F_{n1(2)}$ and braking forces are transferred for all wheels to the ground (four wheel drive system),
- traction force or braking force on a given axis is the sum of the longitudinal forces on each of its wheels i.e. $F_{x1} = F_{x1l} + F_{x1r}$ and $F_{x2} = F_{x2l} + F_{x2r}$,
- a symmetrical distribution of driving and braking forces exists between the wheels of each axle, $F_{x1l} = F_{x1r}$ and $F_{x2l} = F_{x2r}$,
- maximal longitudinal acceleration $a_x = 9{,}81$ m/s^2,
- maximal longitudinal deceleration $a_x = 9{,}81$ m/s^2,
- account resistance forces F_p, are neglected,
- lateral load transfer between left and right wheels of the front axle ΔF_{y1} and between left and right wheel of the rear axle ΔF_{y2} for curvilinear motion is taken into account,

IOP Conf. Series: Materials Science and Engineering **148** (2016) 012002 doi:10.1088/1757-899X/148/1/012002

- front vehicle stiffness $k_{\varphi 1}$ and rear stiffness $k_{\varphi 2}$, are taken into account for roll conditions at the front,
- longitudinal load transfer between front and rear axis ΔF_x, are taken into account,
- all wheels have the same value of friction coefficient in all directions (μ_m = 1), friction coefficient of front μ_{m1} and rear μ_{m2} wheels is the same and equals μ_m.

When considering issues of classic dynamics it will possible to calculate forces on the wheels (longitudinal, lateral) knowing the acceleration (longitudinal, lateral) or velocity (longitudinal, lateral) of the vehicle.

On the basis of the assumed longitudinal forces on the wheels F_x, longitudinal acceleration a_x of the vehicle is calculated, as the quotient of the sum of the longitudinal forces acting on the wheels of the vehicle ΣF_x and its mass m.

$$a_x = \frac{\Sigma F_x}{m} = \frac{F_{x1} + F_{x2}}{m} = \frac{F_{x1l} + F_{x1r} + F_{x2l} + F_{x2r}}{m} \tag{1}$$

where:

F_{x1l} – longitudinal force on the left front wheel (during acceleration$F_{x1l} = F_{n1l}$, during deceleration $F_{x1l} = F_{h1l}$),

F_{x1r} – longitudinal force on the right front wheel (during acceleration$F_{x1r} = F_{n1r}$, during deceleration$F_{x1r} = F_{h1r}$),

F_{x2l} – longitudinal force on the left rear wheel (during acceleration$F_{x2l} = F_{n2l}$, during deceleration $F_{x2l} = F_{h2l}$),

F_{x2r} – longitudinal force on the right rear wheel (during acceleration$F_{x2r} = F_{n2r}$, during deceleration $F_{x2r} = F_{h2r}$).

During acceleration or deceleration lateral weight transfer ΔF_x is changed.

$$\Delta F_x = \frac{m a_x \boldsymbol{h}}{l_{12}} \tag{2}$$

where:

h height of gravitational centre [m],

l_{12} – wheel base [m].

Lateral load transfer at the front axle ΔF_{y1} and ΔF_{y2} at rear axle are calculated while the vehicle drives along a curvilinear path.

During curvilinear motion of the vehicle lateral load transfer front axle ΔF_{y1} and rear axle ΔF_{y2} are changing also.

$$\Delta F_{y1} = \frac{k_{\varphi 1} m a_y \boldsymbol{h_0}}{b_1 (k_{\varphi 1} + k_{\varphi 2} - m g \boldsymbol{h_0})} + \frac{m a_y l_2 \boldsymbol{h}_{\varphi 1}}{l_{12} b_1} \tag{3}$$

$$\Delta F_{y2} = \frac{k_{\varphi 2} m a_y \boldsymbol{h_0}}{b_2 (k_{\varphi 1} + k_{\varphi 2} - m g \boldsymbol{h_0})} + \frac{m a_y l_1 \boldsymbol{h}_{\varphi 2}}{l_{12} b_2} \tag{4}$$

where:

m – vehicle mass [kg],

a_y – vehicle lateral acceleration [m/s^2],

$k_{\varphi 1}$ – front vehicle roll stiffness [Nm/rad],

$k_{\varphi 2}$ – rear vehicle roll stiffness [Nm/rad],
$h_{\varphi 1}$ – height of front roll centre [m],
$h_{\varphi 2}$ – height of rear roll centre [m],
h_0 – distance between gravitational centre and roll axis [m], $\boldsymbol{h_0} = h - \frac{\boldsymbol{h_{\varphi 1}} - \boldsymbol{h_{\varphi 2}}}{l_{12}} l_1 + \boldsymbol{h_{\varphi 1}}$

$\boldsymbol{h_0} = h - \frac{\boldsymbol{h_{\varphi 1}} - \boldsymbol{h_{\varphi 2}}}{l_{12}} l_1 + \boldsymbol{h_{\varphi 1}}$,

h – height of gravitational centre[m],
b_1 –front track [m],
b_2 – rear track [m],
g – acceleration of gravity [m/s^2],
l_1 –distance between front axle and gravitational centre [m],
l_2 – distance between rear axle and gravitational centre [m],
l_{12} – wheel base [m].

To calculate the change of lateral load transfer in four-wheeled vehicle model (equations (2.3) and (2.4)) knowledge of the lateral acceleration is required. For this purpose a previously made two-wheel vehicle model [1] and on the basis of determined lateral acceleration a_y.
Knowledge of lateral load transfer between left and right wheels of the front axle $\Delta F_{y1}$and lateral load transfer between left and right wheels of the rear axle ΔF_{y2} and longitudinal load transfer between front and rear axis Δf_x makes it possible to calculate vertical loads of each wheel $Z_{1(2)l(r)}$.Signs of longitudinal and lateral load transfer depends on whether the wheel is loaded or unloaded while driving.

$$Z_{1l} = Z_{1lstat} - \frac{\Delta F_x}{2} + \Delta F_{y1} = \frac{m_1}{2} g - \frac{\Delta F_x}{2} + \Delta F_{y1} \tag{5}$$

$$Z_{1r} = Z_{1rstat} - \frac{\Delta F_x}{2} - \Delta F_{y1} = \frac{m_1}{2} g - \frac{\Delta F_x}{2} - \Delta F_{y1} \tag{6}$$

$$Z_{2l} = Z_{2lstat} + \frac{\Delta F_x}{2} + \Delta F_{y2} = \frac{m_2}{2} g + \frac{\Delta F_x}{2} + \Delta F_{y2} \tag{7}$$

$$Z_{2r} = Z_{2rstat} + \frac{\Delta F_x}{2} - \Delta F_{y2} = \frac{m_2}{2} g + \frac{\Delta F_x}{2} - \Delta F_{y2} \tag{8}$$

where:
m_1 – front mass [kg],
m_2 – rear mass [kg],
$Z_{1l(r)stat}$– static vertical load of left (right) of front vehicle wheel [N],
$Z_{2l(r)stat}$– static vertical load of left (right) of rear vehicle wheel [N].

Calculated vertical loads $Z_{1(2)l(r)}$allows to determine the friction forces under each wheel F_μ.

$$F_{\mu 1l} = \mu_{m1l} Z_{1l} \tag{9}$$

$$F_{\mu 1r} = \mu_{m1r} Z_{1r} \tag{10}$$

$$F_{\mu 2l} = \mu_{m2l} Z_{2l} \tag{11}$$

$$F_{\mu 2r} = \mu_{m2r} Z_{2r} \tag{12}$$

where:

$\mu_{m1l(r)}$– friction coefficient of left (right) front wheel,
$\mu_{m2l(r)}$ – friction coefficient of left (right) rear wheel.

An initial assumption that the longitudinal force F_x on the wheels cannot be greater than the friction force on the given wheel F_μ. This situation is represented in figure 3 by the tire friction circle [7].

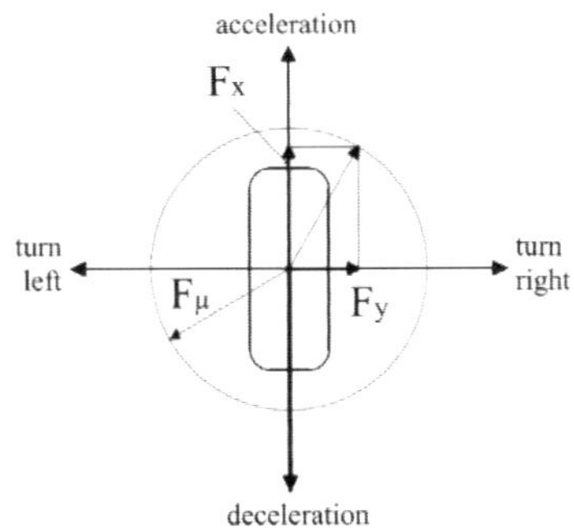

Figure 3. Tire friction circle

The figure 3 depicts tire friction circle which illustrates relationships between longitudinal forces F_x (driving F_n and braking F_h), longitudinal forces F_y and friction forces F_μ.
As it was assumed earlier that the longitudinal force on any wheel $F_x$is smaller than the friction forces on the wheel F_μ it will be possible to transfer on the road the lateral force F_y:

$$F_{y1l} = \left(F_{\mu 1l}^2 - F_{x1l}^2\right)^{\frac{1}{2}} \tag{13}$$

$$F_{y1r} = \left(F_{\mu 1r}^2 - F_{x1r}^2\right)^{\frac{1}{2}} \tag{14}$$

$$F_{y2l} = \left(F_{\mu 2l}^2 - F_{x2l}^2\right)^{\frac{1}{2}} \tag{15}$$

$$F_{y2p} = \left(F_{\mu 2r}^2 - F_{x2r}^2\right)^{\frac{1}{2}} \tag{16}$$

It can be calculated as the total lateral force which can be transferred along a given axis.

$$F_{y1} = F_{y1l} + F_{y1r} \tag{17}$$

$$F_{y2} = F_{y2l} + F_{y2r} \tag{18}$$

The longitudinal and lateral forces acting in the tire-road contact generate yaw moment M_z about the vertical axis of vehicle Z.
Yaw moment M_{zFy} depends on the lateral forces F_y, where F_{y1} is lateral force at the front and F_{y2} at the rear.

Yaw moment M_{zFy} acting around the vertical axis of vehicle Z is caused by longitudinal forces at the front axis F_{y1} and rear axis F_{y2} respectively, and are expressed by the following equations:

$$M_{zFy} = F_{y1} l_1 - F_{y2} l_2 \tag{19}$$

The difference in the friction coefficient between left and right side i.e. $\mu_{m1(2)l}$ and $\mu_{m1(2)r}$ respectively gives rise to a yaw moment M_{zFx} which leads to a difference in lateral load transfer between left and right wheels of the front axle ΔF_{y1} and lateral load transfer between left and right wheels of the rear axle ΔF_{y2}. The difference between longitudinal forces on the wheels of the front axle ΔF_{y1} and the rear axle is depicted below.

$$\Delta F_{x1} = \left(F_{\mu m1l}{}^2 - F_{y1l}{}^2\right)^{\frac{1}{2}} sgn(F_{x1l}) - \left(F_{\mu m1r}{}^2 - F_{y1r}{}^2\right)^{\frac{1}{2}} sgn(F_{x1r}) \tag{20}$$

$$\Delta F_{x2} = \left(F_{\mu m2l}{}^2 - F_{y2l}{}^2\right)^{\frac{1}{2}} sgn(F_{x2l}) - \left(F_{\mu m2r}{}^2 - F_{y2r}{}^2\right)^{\frac{1}{2}} sgn(F_{x2r}) \tag{21}$$

Orientation of the sign for the difference between longitudinal forces on the front axis ΔF_{x1} and rear axis ΔF_{x2} depends on the direction of longitudinal forces F_x on the given wheel.
Yaw moment $M_{zFx1(2)}$ acting around the vertical axis of vehicle Z is caused by the difference in longitudinal forces at the front axis ΔF_{x1} and rear axis ΔF_x respectively, and are expressed by the following equations:

$$M_{zFx1} = \Delta F_{x1lr} \frac{b_1}{2} \tag{22}$$

$$M_{zFx2} = \Delta F_{x2lr} \frac{b_2}{2} \tag{23}$$

Total yaw moment M_{zFx} caused by the difference of longitudinal forces on the wheels of each axles accepts form:

$$M_{zFx} = M_{zFx1} + M_{zFx2} \tag{24}$$

Total yaw moment M_z around the vehicle axis Z is a superposition of yaw moments which come from lateral forces M_{zFy} and yaw moment resulting from the difference between longitudinal forces on the wheels of one axle M_{zFx}.

$$M_z = M_{zFy} + M_{zFx} \tag{25}$$

Total yaw moment M_z allows us to define steering characteristics. If moment equals zero(M_z= 0),then vehicle has neutrally steered. Whereas, if M_z< 0 then vehicle is under steered and when M_z> 0 vehicle is over steered.
Previously calculated yaw moment M_z will be required to calculate the lateral acceleration a_y for four-wheel vehicle model represented by the following equations:

$$\begin{cases} a_y = \dfrac{F_{y1} + F_{y2}}{m}, & if\, M_z = 0 \\ a_y = \dfrac{F_{y1} + \frac{M_{zFx}}{l_{12}}}{m_1}, & if\, M_z < 0 \\ a_y = \dfrac{F_{y2} - \frac{M_{zFx}}{l_{12}}}{m_2}, & if\, M_z > 0 \end{cases} \tag{26}$$

The process is repeated several times (n) for calculating longitudinal forces which are a function of longitudinal and lateral acceleration of the vehicle. On the basis of the above shown scheme the algorithm is created. This was done using numerical calculation by the program Scilab. The number of iterations were 200 and as a result the lateral acceleration of the vehicle is calculated. This is done by joining points with similar values of lateral acceleration. The result was an iso-altitude lateral acceleration.10 isolines can be seen for greater readability of the graph.
Next, a principle of operation of DSM will be presented by the example of a car model, whose data is presented by Table 1. and 2.

Table1. Vehicle data

Vehicle mass m [kg]	1450
Front massm_1 [kg]	870
Rear massm_2 [kg]	580
Friction coefficient of left side of vehicle μ_{ml} [-]	1,0
Friction coefficient of right side of vehicleμ_{mr} [-]	1,0
Distance between front axle and gravitational center l_1 [m]	1,06
Distance between rear axle and gravitational center l_2 [m]	1,59
Wheel base l_{12} [m]	2,65
Front trackb_1 [m]	1,50
Rear trackb_2 [m]	1,50
Height of gravitational h [m]	0,53
Height of front roll center $h_{\varphi 1}$ [m]	0,10
Height of front roll center $h_{\varphi 2}$ [m]	0,15
Distance between t center and roll axis h_0 [m]	0,41
Front vehicle roll stiffness $k_{\varphi 1}$ [Nm/rad]	50 000
Rear vehicle roll stiffness $k_{\varphi 2}$ [Nm/rad]	30 000

Table2. Assumed driving and braking forces for the reference vehicle

Driving force on left front wheel F_{n1l} [N]	(0; 6000)
Driving force on right front wheel F_{n1r} [N]	(0; 6000)
Driving force on left rear wheel F_{n2l} [N]	(0; 6000)
Driving force on right rear wheel F_{n2r} [N]	(0;6000)
Braking force on left front wheel F_{h1l} [N]	(-6000; 0)
Braking force on right front wheel F_{h1r} [N]	(-6000; 0)
Braking force on left rear wheel F_{h2l}[N]	(- 3500; 0)
Braking force on right rear wheel F_{h2r} [N]	(-3500; 0)
Driving force on front axle F_{n1} [N]	(0; 12 000)
Driving force on rear axle F_{n2} [N]	(0; 12 000)
Braking force on front axle F_{h1} [N]	(- 12 000; 0)
Braking force on rear axle F_{h2} [N]	(-7000; 0)

Figure 4 depicts the graphical forces on the wheels of the vehicle calculated by the DSM for a four-wheel vehicle model obtained on the basis of the above calculations.

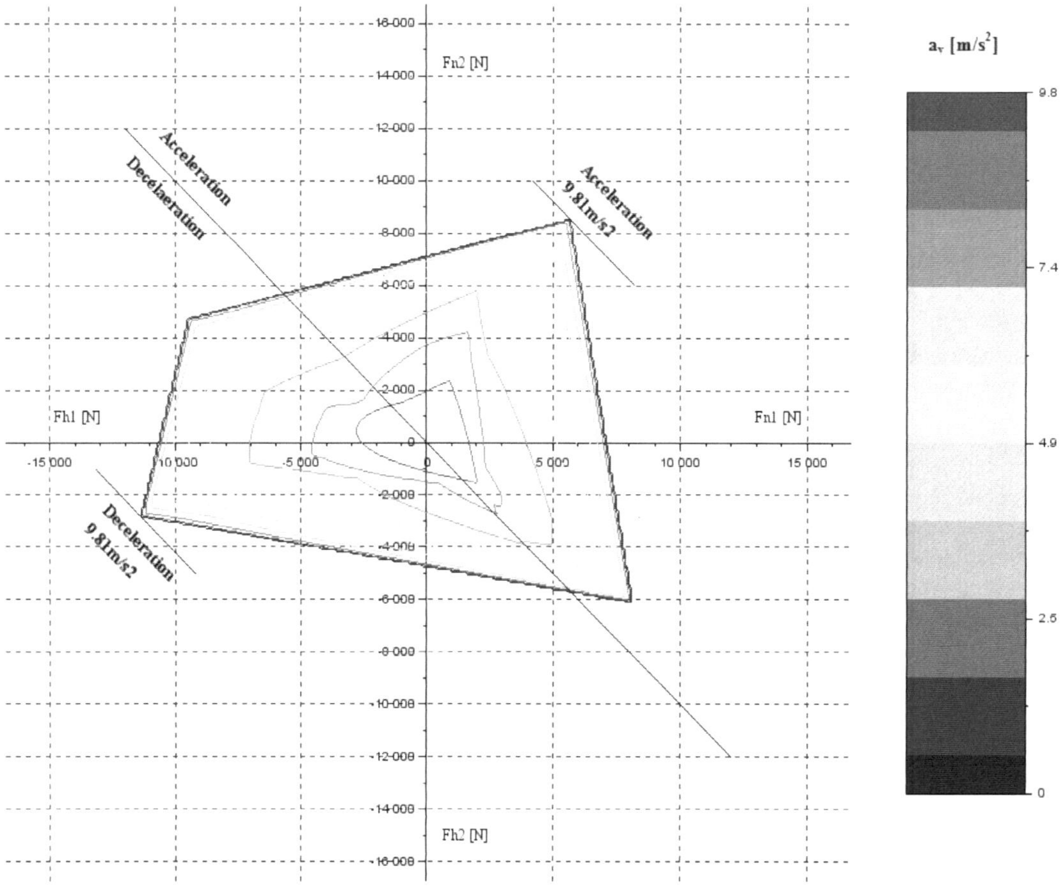

Figure 4. Limit forces on the vehicle wheels for using DSM for four-wheel vehicle model.

Figure 4 presents the limit forces acting on the wheels of the vehicle obtained via Dynamic Square Method for the four-wheel vehicle model. The analysis of the above characteristics results in understanding the relationship between the real longitudinal forces on the wheels of the vehicle, lateral acceleration a_y and longitudinal a_x. Analysing Figure 2.4 one can determine the limit of traction force acting on the axis which in turn leads us to value of lateral acceleration. The longitudinal acceleration of the vehicle results from the sum of the limit forces on the wheels front and rear axle results.

Figure 4 shows the limit forces on the wheels of the vehicle obtained via Dynamic Square Method for the four-wheel vehicle model. The analysis of the above characteristics of the resulting relationship between the real longitudinal forces on the wheels of the vehicle and achieved the lateral acceleration a_y and longitudinal acceleration a_x. Analysing Figure 2.4 can determine which limit the traction force on the axis is capable of achieving a function of the vehicle lateral acceleration. With the sum of the limit forces on the wheels front and rear axle results the longitudinal acceleration of the vehicle. So you know what is the relationship between the longitudinal and lateral acceleration of the vehicle.

It is also worth noting fact that the right upper part of the graph (for positive longitudinal forces on the wheels) and the left lower right upper part of the graph (negative longitudinal forces on the wheels) have real sense.

Also one can define the maximum driving forces on the wheels during acceleration and braking. Thus, the longitudinal forces on the wheels of the vehicle located in the upper right corner of the graph correspond to the values of the driving forces at the front and rear wheels $F_{n1(2)}$ during the rectilinear

motion when maximum longitudinal acceleration a_x equals 9,81 m/s^2. It can be stated that the maximum driving force for the front wheels and the rear (see Table 2.2) cannot be realized. The maximum value of the driving forces obtained by the car model on the front and rear wheels during the rectilinear motion of the vehicle are as follows (a_y = 0 m/s^2): $F_{n1} \approx$ 5700 N, $F_{n2} \approx$ 8500 N.
The values of longitudinal forces at the wheels contained in the lower left corner of the graph represent the vehicle driving along a rectilinear motion with a maximum longitudinal deceleration a_x = − 9,81 m/s^2.In this situation, the vehicle model gives us values of braking force at the front and rear wheels$F_{h1} \approx$ − 11 400 N and$F_{h2} \approx$ − 2800 N respectively.
The inner part of the graph (Figure 2.4) consists of iso-altitude lines depicting constant value of lateral acceleration a_y. Diagonal lines represent the relationship between longitudinal force at the front axle$F_{x1}$and the rear axle$F_{x2}$and prove that$F_{x1}+F_{x2}=$ *constant*, which means that the sum of the longitudinal forces on the front and rear wheels is always constant for the given longitudinal acceleration a_x.

3. Conclusions

The presented algorithm Dynamic Square Method has great development potential. On the basis of DSM can be determined the real lateral and longitudinal forces acting on the vehicle wheels, as well as lateral and longitudinal acceleration resulting from the action of these forces.
There are plans for the further expansion of the model for e.g. taking into account a model describing the contact of the wheel with the road and its use in creating a control system used to calculate driving and braking force (Torque Vectoring).

References

[1] Brukalski M 2015 Introduction to dynamic square method. *Zeszyty Naukowe Instytutu Pojazdów* 3(103)/2015, pp. 29-35
[2] Kato M, Isoda K and Yuasa H 1995 Study on vehicle dynamics in marginal condition usingdynamic square method, *SAE Technical Paper*958503, 1995
[3] Klomp M.: *Passenger Car All-Wheel Drive Systems Analysis*. Degree Project University of Trollhättan/Uddevalla, 2004 Sweden
[4] Klomp M 2007*On Drive Force Distribution and Road Vehicle Handling-A Study of Understeer and Lateral Grip* Chalmers University of Technology, 2007 Sweden
[5] Sawase K, Ushiroda Y and Inoue K 2007 Effect of the Right-and-left Torque Vectoring System in Various Types of Drivetrain, *SAE Technical Paper* 2007-01-3645, 2007
[6] Sawase K and Ushiroda Y 2008 Improvement of Vehicle Dynamics by Right-and-Left Torque Vectoring System in Various Drivetrains*Mitsubishi Motors Technical Review* No. 20, 2008
[7] Reński A 2011*Bezpieczeństwo czynne samochodu, Zawieszenia oraz układy kierownicze* Oficyna Wydawnicza Politechniki Warszawskiej Warszawa

A Two-Wheeled, Self-Balancing Electric Vehicle Used As an Environmentally Friendly Individual Means of Transport

D Bździuch[1] and W Grzegożek[2]

[1] Institute of Transport, Combustion Engines and Ecology, Lublin University of Technology, Poland

[2] Institute of Motor Vehicles and Combustion Engines, Cracow University of Technology, Poland

E-mail: witek@mech.pk.edu.pl

Abstract: This paper shows a concept of a model of a two-wheeled self-balancing vehicle with an electric motor drive as an environmentally-friendly personal transporter. The principle of work, modelling of construction and performing a simulation are presented and discussed. The visualization of the designed vehicle was made thanks to using Solid Works a computer-aided design program. The vehicle was modelled as an inverted pendulum. The stability of the mechanism in the equilibrium position was studied. An exemplary steering system was also subjected to the analysis that compared two controllers: PID and LQR which enabled to monitor the balance of the vehicle when the required conditions were fulfilled. Modelling of work of the controllers and the evaluation of the obtained results in required conditions were performed in the MATLAB environment.

1. Introduction

Rapid technology development and progress in civilization makes the number of cars grow constantly. These are mainly vehicles powered by combustion, whose efficiency is relatively small in comparison to efficiency of the electric engine - 3 times smaller on average [1],[2]. Their major defect is emission of toxic compounds such as CO, CO_2, NO_x, HC contained in exhaust gases. This factor, together with the increase in environmental requirements, makes producers and constructors introduce a number of changes and improvements so that their products meet progressively more restrictive emission standards. As a consequence, the produced vehicles are equipped with precise, complicated systems which are expensive to repair. Despite all their defects, combustion vehicles are still very common firstly, because of the relatively long distances they cover on one tank of petrol (it results from the ratio of power to energy obtained from 1 kg of 42MJ fuel, and to energy obtained from 1 kg of 1MJ battery[1]). In recent years electric vehicles are getting more and more popular. They offer a number of advantages, such as high efficiency, simple construction and steering, however, not enough charging stations, small distances they cover and a long time of charging in comparison to refuelling of a traditional vehicle put them at a distinct disadvantage. Nevertheless, those problems, thanks to rapid technology development, will be probably easy to tackle with in the near future. Their main advantage is the fact that the vehicles with electric engines do not produce exhaust fumes and electricity may be obtained from renewable energy sources, e.g. solar energy obtained from photovoltaic cells. Promoting electric vehicles through granting financial incentives, free parking and free charging, etc. [3], [4], are additional benefits. Despite the number of advantages, electric vehicles do not present a large share of the global automotive market. In order to meet present expectations and requirements connected to environmental protection, the idea of an urban, two-wheeled, electric and self-balancing vehicle which is environmentally friendly and has wide options for use, was drawn.

2. Principle of operation

The first self-balancing vehicles were constructed in the first half of the 20th century, but they were not very popular, mainly because they did not have automated systems enabling control over the vehicle's pitch angle. With time, new self-balancing vehicles were constructed, and the most known are Segway vehicles. The Segway's idea is based on the concept of inverted pendulum (figure 1).

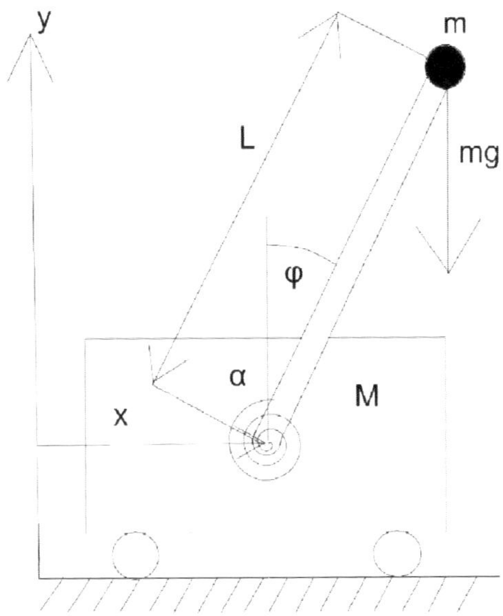

Figure 1. Inverted pendulum model.

where:
M – mass of vehicle platform,
m – point mass attached to the end of a rod,
L – length of the rod,
α – damping ratio,
φ – pivot angle,
x – coordinate describing movement of the centre of the mass of the platform,
mg – weight force applied to the point mass.

The system has a simple structure and principle of operation. By moving the vehicle platform horizontally, the rod with the point mass concentrated in its end should be balanced vertically (φ=0). The problem that needs to be solved is the control of the vehicle, which is instable. Dean Kamen Segway's inventor, created a very simple solution based on a human brain behaviour [5]. A man, when leaning forward, for example, saves himself from falling by putting his leg in the direction he is leaning to. Segway works in the same way. The horizontal position with the pitch angle equalling zero φ=0 is 'normal' for the vehicle. The moment we tilt the vehicle, it reads from the gyroscopic sensors the value of the pitch angle and then reacts correspondingly by transferring the corrected strength value to the wheels in order to prevent the construction from falling, which makes it move until the pitch is eliminated. This method is called dynamic stabilization. A mathematical model of a self-balancing vehicle may be written down in the form of differential equations of the inverted pendulum movement of the figure 1. Deriving the formulae was skipped.

IOP Conf. Series: Materials Science and Engineering **148** (2016) 012003 doi:10.1088/1757-899X/148/1/012003

$$\begin{cases} \ddot{x}(M+m)+ml(\ddot{\varphi}\cos\varphi-\dot{\varphi}^2\sin\varphi)=0 \\ ml^2\ddot{\varphi}+ml\ddot{x}\cos\varphi-mgl\sin\varphi+\alpha\dot{\varphi}=0 \end{cases} \quad (1)$$

Bringing the above system to the form of linear function, assuming small tilts, we get the following form of equations:

$$\sin\varphi\approx\varphi$$
$$\cos\varphi\approx 1$$
$$\dot{\varphi}\approx 0$$
$$\dot{\varphi}^2\approx 0$$
$$\begin{cases} \ddot{x}(M+m)+ml\ddot{\varphi}=0 \\ ml^2\ddot{\varphi}+ml\ddot{x}-mgl\varphi=0 \end{cases} \quad (2)$$

The above formula may be presented in the following way:

$$\ddot{\varphi}(\frac{ml}{M+m}-l)+g\varphi=0 \quad (3)$$

It is noticeable, that the formula in brackets comprises of constants and adopts a constant value and that is why it may be replaced with a determined constant.

$$c=\frac{ml}{M+m}-l=l(\frac{m}{M+m}-1)$$
$$c\ddot{\varphi}+g\varphi=0 \quad (4)$$

The equation (4) has to be tested by the Routh–Hurwitz stability criterion, having developed it earlier into the form of characteristic equation.
Theorem 1. Stability precondition

If a given polynomial is a Hurwitz polynomial, then all its coefficients are positive.

Theorem 2. Hurwitz stability sufficient condition

A given polynomial is a stable polynomial only if all its principal minors are positive.

Assuming the solution in the form of $\varphi(t)=e^{rt}$, calculating subsequent derivatives and putting them into the equation (4) we get the characteristic equation (5):

$$r^2+\frac{g}{c}=0 \quad (5)$$

According to the Routh–Hurwitz stability criterion, precondition of single-root polynomial coefficients was not satisfied, because the constant c assumes negative value, it results from the equations (4) and (5). The system has instable balance location with the angle $\varphi=0°$, that is with initial position for regulation by steering algorithm.

3. Concept of the vehicle

In the world, there are various solutions used in self-balancing vehicles, such as Segway or E.D.W.A.R.D, constructed by students in Adelaide, the vehicle which is an interesting object with three operating modes. A concept of a two-wheeled, self-balancing, electric vehicle with seats presented in figure 2 was based on the model in Figure 1. The basic assumption was to design a light, electric vehicle whilst, first of all, putting safety in the first place. The vehicle has unusual appearance with regular shapes and compact structure that allows two people with some small baggage to travel comfortably.

Figure 2. Concept of the self-balancing vehicle.

The frame is the basic element which, according to assumptions, is supposed to be made of light alloy material. Hub-less wheels that increase side visibility, are also worth noticing. Two electric engines that allow to steer each wheel separately, for example during rotating in place, are supposed to be driving elements. The steering system contained in a microprocessor controller and a system of sensors measuring pitch angle of the vehicle, among others, are the heart of the construction. A sliding windscreen makes it possible to travel with open passenger space on sunny, and closed on rainy days. Drive and steering elements are located between lower cover and the floor board, which allows to make better use of space. The vehicle is supposed to be easy to dismantle in order to have easy access to individual devices. At the moment the assumption is to steer the vehicle with one limb only. Another assumption is to work in a few stabilization modes and using state-of-the-art technologies to communicate with the vehicle through mobile devices, for example.

4. Control system

An outline of the control system for the simplified model of the inverted pendulum shall be described. Operation of PID and LQR regulators shall be presented. In order to analyze a simple system the MATLAB program with implemented PID and LQR regulator was used [6], [7]. A small pendulum model with the moment of inertia and the rod mass instead of the point mass attached to its end as well as friction were taken into account.

$$\begin{aligned} &(I + ml^2)\ddot{\varphi} - mgl\varphi = ml\ddot{x} \\ &(M + m)\ddot{x} + b\dot{x} - ml\ddot{\varphi} = u \end{aligned} \tag{6}$$

b- coefficient of linear dumping
I- moment of inertia of the rod
m- mass of the rod
u- forcing

The above presentation will be transformed into state-space representations. Thanks to state-space representations the relations between forcing and response are derived and they can be connected by state variables (7). In order to derive the state formula, the following state variables should be assumed:

$$\begin{aligned} x_1 &= \dot{x} \\ x_2 &= \ddot{x} \\ x_3 &= \dot{\varphi} \\ x_4 &= \dot{\varphi} \end{aligned} \tag{7}$$

After transformations we get state equations written in the matrix form

$$\begin{bmatrix} \dot{x} \\ \ddot{x} \\ \dot{\varphi} \\ \ddot{\varphi} \end{bmatrix} = \begin{bmatrix} 0 & 1 & 0 & 0 \\ 0 & \dfrac{-(I+ml^2)b}{I(M+m)+Mml^2} & \dfrac{m^2gl^2}{I(M+m)+Mml^2} & 0 \\ 0 & 0 & 0 & 1 \\ 0 & \dfrac{-mlb}{I(M+m)+ml^2} & \dfrac{mgl(M+m)}{I(M+m)+Mml^2} & 0 \end{bmatrix} \begin{bmatrix} x \\ \dot{x} \\ \varphi \\ \dot{\varphi} \end{bmatrix} + \begin{bmatrix} 0 \\ \dfrac{I+ml^2}{I(M+m)+Mml^2} \\ 0 \\ \dfrac{ml}{I(M+m)+Mml^2} \end{bmatrix} u$$

$$y = \begin{bmatrix} 1 & 0 & 0 & 0 \\ 0 & 0 & 1 & 0 \end{bmatrix} \begin{bmatrix} x \\ \dot{x} \\ \varphi \\ \dot{\varphi} \end{bmatrix} + \begin{bmatrix} 0 \\ 0 \end{bmatrix} u \tag{8}$$

Output matrix has two lines, as the response of y system consists of two elements, that is the location of the cart x and pivot point of the pendulum φ. The simulation was performed of the following data:

M=0.6[kg] – mass of the cart
m=0.1[kg] – mass of the pendulum
b=0.1[N/m/s] – coefficient of linear dumping
l=0.3 [m] – distance between the centre of the mass of the pendulum and pivot axis
I=0.003 [kg*m^2] – moment of inertia of the pendulum

The requirements for the system: achieving the wanted position in the time of up to 5 seconds, rise time for x less than 0.5s, pitch angle of the pendulum always smaller than 20° from vertical location, steady-state error smaller than 2% for x and φ. Then the simulation was performed. Below there are the results for various options of operation of the regulators. The PID regulator was analyzed as first. The value of all forcing coefficients equalling 1 was the starting point. The system did not meet the requirements, so the control parameters had to be changed appropriately. By trial and error, the control parameters that allowed to meet the requirements were selected.

The course of the response presented in figure 3 is satisfying because the basic requirements for the system were met: the pitch angle is not higher than 0.05° and, at the same time, the pendulum is stabilized after 2 seconds. Its course is sudden, but it may be optimised by selecting the right forcing coefficients.

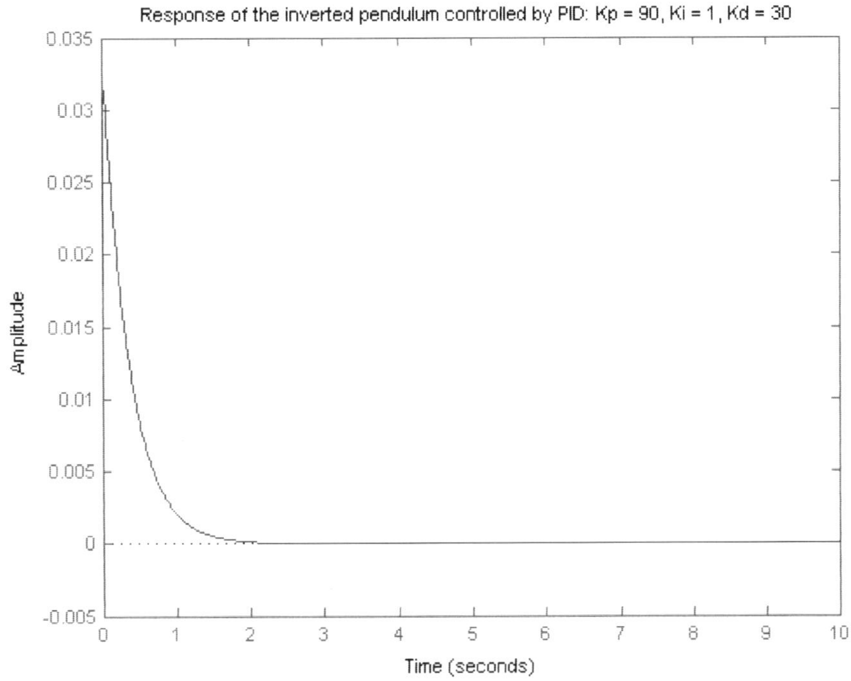

Figure 3. Response of the inverted pendulum controlled by PID.

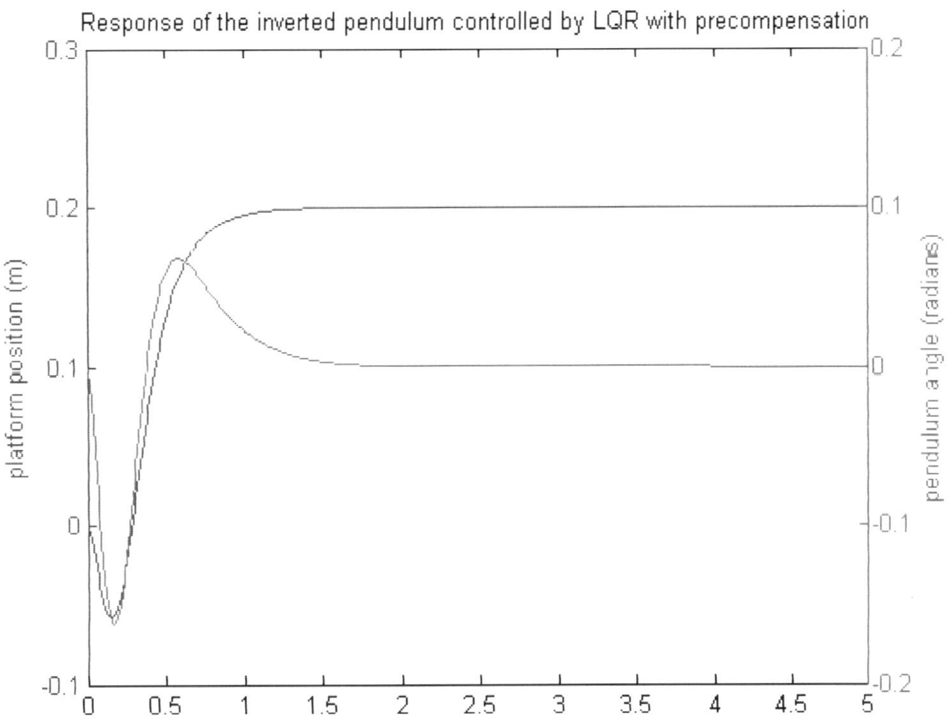

Figure 4. Response of the inverted pendulum controlled by LQR.

The LQR regulator was analysed as second. The work with LQR regulator consists in selecting proper weights and minimising costs between system errors and the effort put into controlling. First, the same weights were selected but the system did not meet the requirements. Next, the weights that reduce control (follow) errors were modified, but they increase the force put into controlling and so increase the need for energy. The pitch angle of the pendulum and the rise time presented in figure 4 are satisfying, the pendulum is stable after 1.5 seconds. Precompensation was also added. It is done to minimise steady-state error. Both PID and LQR regulators are suitable to be used in the control system of the self-balancing vehicle. In practical control systems combinations of both regulators are sometimes applied in order to improve the quality of control.

5. Conclusion

This article presents the concept of a two-wheeled, self-balancing electric vehicle as an individual means of transport. Such a vehicle could be used in the city centres, as it is environmentally friendly and easy to control. The use of electric engines helps eliminate such subassemblies as a gear box used in traditional vehicles thus decreasing the mass and increasing the accessible space. It could successfully function in the network of rental companies - one could rent the vehicle in one place and leave it in another without the need to return to the initial spot. In recent years numerous concept models of self-balancing vehicles for urban traffic have been developed. Development of the technology based on dynamic stabilization lets the driver control the vehicle of this type more effectively with minimizing the losses during controlling the vehicle. The presented concept proves that it is a construction that can be based on the inverted pendulum model and control balance with the use of simple regulators, such as PID or LQR. Despite its unusual design, the vehicle can be viewed as fully functional in everyday life.

References

[1] Kołodziejczyk L M and Zając P 2010 *Silniki Spalinowe* (Warszawa WSiP), p27 ISBN 82-02-07987-1
[2] Silniki Energooszczędne - PEMP http://www.centrum.pemp.pl/dokumenty/technika/silniki_energooszczedne/silniki_energooszczedne.pdf accessed 20-04-2016
[3] Jak Europa promuje auta hybrydowe i elektryczne – Moto.pl http://moto.pl/Ekologia/1,117005,9448532,Jak_Europa_promuje_auta_hybrydowe_i_elektryczne_.html accessed 20-04-2016
[4] Norwegia kocha auta elektryczne romans zaczyna przeszkadzać - Transport Publiczny http://www.transport-publiczny.pl/wiadomosci/norwegia-kocha-auta-elektryczne-romans-zaczyna-przeszkadzac-1211.html accessed 21-04-2016
[5] Sposób działania – Segway http://www.segway.com.pl/na-co-dzien/sposob-dzialania/ accessed 21-04-2016
[6] Inverted Pendulum: Simulink Controller Design http://ctms.engin.umich.edu/CTMS/index.php?example=InvertedPendulum&section=SimulinkControl accessed 22-04-2016
[7] Inverted Pendulum: State-Space Methods for Controller Design http://ctms.engin.umich.edu/CTMS/index.php?example=InvertedPendulum&section=ControlStateSpace accessed 22-04-2016
[8] Lauwers T B, Kantor G A, and Hollis R L 2006 *A Dynamically Stable Single-Wheeled Mobile Robot with Inverse Mouse-Ball Drive* Proc IEEE Int'l Conf. on Robotics And Automation pp 2884-2889
[9] Kaczorek T 2006 *Podstawy Teorii Sterowania* (Warszawa WNT), p32 ISBN 83-204-3251-0
[10] Rumatowski K 2004 *Podstawy automatyki* (Poznań Wydawnictwo Politechniki Poznańskiej), p168-170 ISBN 83-7143-304-2
[11] Software PID Control of an Inverted Pendulum Using the PIC16F684. Microchip Technology Inc http://ww1.microchip.com/downloads/en/AppNotes/00964A.pdf accessed 23-04-2016
[12] Grasser F, D'Arrigo A, Colombi S and Rufer A C 2002 *JOE: A Mobile, Inverted Pendulum* IEEE Transactions on Industrial Electronics vol49 no1 pp107-114
[13] Bryan P D *Designing a Self-Stabilizing Robot For Dynamic, Mobile Manipulation*
[14] Praca zbiorowa pod redakcją Morecki A and Knapczyk J 1999 *Podstawy robotyki. Teoria i elementy manipulatorów i robotów* (Warszawa WNT), p341-345 ISBN 83-204-2331-7
[15] EDWARD – Electric Diwheel With Active Rotation Damping http://sites.mecheng.adelaide.edu.au/robotics/robotics_projects.php?wpage_id=44&title=60&browsebytitle=1
[16] Jastrzębska G 2007 *Odnawialne źródła energii i pojazdy proekologiczne* (WNT Warszawa) p158-173 ISBN 978-83-204-3376-0
[17] Sieklucki G 2009 *Automatyka napędu* (AGH Kraków) p31 35 ISBN 978-83-7464-208-8
[18] Węglarz A and Pleśniak M 2011 *Samochód elektryczny* (Fundacja Instytut na rzecz Ekorozwoju Warszawa) p15-17 ISBN 978-83-89495-04-4

Parametric study on the performance of automotive MR shock absorbers

J Gołdasz[1,2] and **S Dzierżek**[3]

[1] Senior Simulation Engineer, Technical Center Kraków, BWI Group, Kraków, PL
[2] Department of Automation and Information Technology, Cracow University of Technology, Kraków, PL
[3] Engineering Manager, Technical Center Kraków, BWI Group, Kraków, PL

E-mail: janusz.goldasz@bwigroup.com

Abstract. The paper contains the results of a parametric study to explore the influence of various quantities on the performance range of semi-active automotive shock absorbers using the magnetorheological (MR) fluid under steady-state and transient excitations. The analysis was performed with simulated data and using a standard single-tube shock absorber configuration with a single-gap MR valve. Additionally, the impact of material variables and valves geometry was examined as the parameters were varied and its dynamic range studied.

1. Introduction

The so-called smart materials are capable of sensing an external stimulus and reacting to it. By definition [1], such materials are those adapting to environmental changes and manifesting their function in a organized manner. The materials can be categorized due to their responsive nature, i.e. there are materials that respond to thermal, light, electric, magnetic and stress fields, respectively. Magnetically responsive materials include magnetorheological (MR) fluids. As a suspension of micron-size solid particles in a non-conductive carrier oil the material undergoes a fast and reversible transition from a fluid to a pseudo-solid in the presence of magnetic fields. It is well-known that the so-called magnetorheological effect was first discovered by Rabinov [2] who described the changes in the rheology the material. The material's application potential was immediately recognized resulting in attempts to build a first controllable MR device (clutch). At that time several obstacles prevented the technology from being commercialized, although real-time opportunities in vibration damping and isolation were immediately recognized by both academia and the automotive industry, too [3]. In 2002 the researchers' effort resulted in the world's first semi-active vehicle suspension system based on the smart fluids [4]. The system utilized vehicle dampers based on MR fluids for real-time control of the motion of the car. Since that numerous vehicle applications of that system emerged in North America, Europe and Asia [6]. At the present moment BWI Group continues to develop the fourth generation of the system in order to implement it in future vehicle platforms. Major progress has been accomplished in the following areas: response time improvements, higher dynamic range, authority and better response at low body velocities [5, 6].

The dampers that meet specific application needs vary due to packaging constraints, load requirements and chassis configuration [7]. In general, any damper would accommodate a cylinder with MR fluid and a control valve located in the piston assembly. The control valve contains a magnetic circuit connected to an external power supply. It contains no moving mechanical components. The magnetic circuit generates field of sufficient strength to affect the rheology of the fluid while in flow through the valve. By principle, a standard automotive MR damper is a flow-mode (pressure-driven) device.

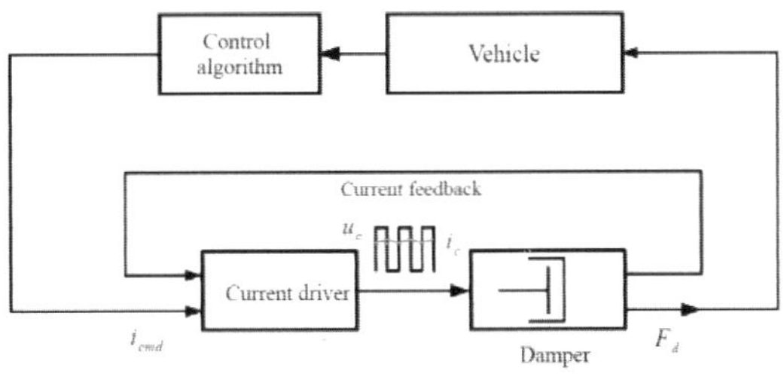

Figure 1. Simplified MR damper system layout, F_d – force, i_{cmd} – commanded current, i_c – coil current, u_c – supply voltage.

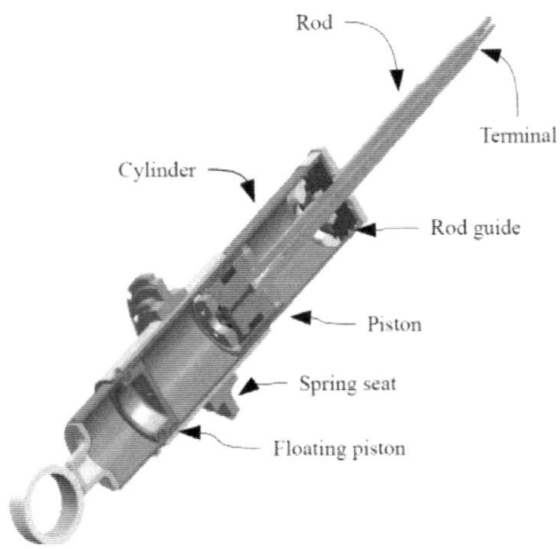

Figure 2. Automotive single-tube MR damper (courtesy of BWI Group).

In order to design an MR damper for specific application needs specification of various parameters of the electro-magnetic circuit of the actuator. Selecting the geometry of the control valve incl. flow channel sizing and the geometry of the core and coil assemblies, consideration of magnetic properties of materials constituting the magnetic circuit and their impact on the rheology of the fluid are important steps in the process. In a typical test configuration the damper's rod is driven by a prescribed displacement/force, while a current signal is applied to the coil located in the piston assembly of the damper – see figure 1. The commanded current is supplied to the coil through a pulse-width-modulated (PWM) controller. The current in the coil induces a magnetic field in the actuator in order to influence the MR fluids yield stress, and then the damping force output [8].
In principle the configuration of a single-tube gas-charged MR damper resembles the conventional monotube deCarbon damper - see figure 2. As shown in the figure, the cylinder housing incorporates the floating piston and the primary piston assembly. The primary piston divides the volume of MR fluid into the rebound (upper) fluid volume and the compression (lower) one. Next, the floating piston separates the fluid from pressurized gas. The gas volume compensates for fluid volume expansion due to temperature and prevents it from cavitation. The piston includes the control valve assembly with annular passages for the fluid to flow through.

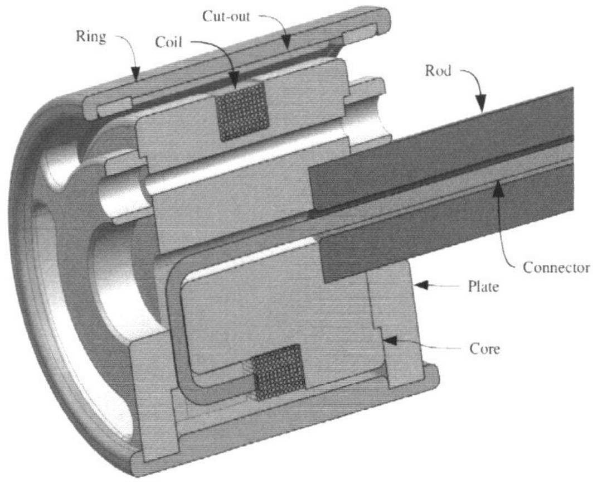

Figure 3. Single-coil piston assembly [7].

The purpose of the control valve is to alter the rheology of the fluid. In general, the valve should allow for generating low (yet optimum) damping forces in the non-energized condition (off-state) and high forces when energized (on-state). Existing control valve configurations can be classified in terms of the primary and secondary flow paths, coil arrangement, core configuration, performance enhancing features (fail-safe valve, force asymmetry, flux boosters, etc.) [7]. The most common control valve configuration is an assembly with one annular flow path and one coil assembly arranged transversely to the fluid flow direction - see figure 3. The rheology of the fluid can be altered by the magnetic field generated by a current in the coil wound onto the piston core. The magnetic flux travels through the core, the annulus, the sleeve, and then returns back into the core through the annular gap. The damping force magnitude varies with the field-induced yield stress and the flow geometry. It is, therefore, a common practice to boost the magnetic field strength in the annular gap at a design stage. As shown, the piston assembly involves a secondary flow path in the form of a thru-core passage. The feature allows for shaping the low velocity performance which would otherwise be severely degraded. Other low-velocity features involve cut-outs on either surface constituting the annulus or non-magnetic sections on the surface of the core [9]. The so-called flux (slot) bypasses degrade the field strength in the annulus. As a result, the fluid in the annulus is allowed to pass through at a lower breakaway pressure difference than in the remaining portion of the primary flow path.

During the design development process engineers usually follow the same steps that have been practiced for solenoid systems to ensure a correct flux output. That can be accomplished by selecting optimum soft magnetic materials for all components of the magnetic circuit and the hardware's geometry. Key material characteristics include high permeability, low remnant magnetisation, small hysteresis loop, high saturation field strength as well as high resistivity (low electric conductivity) [10, 11]. In general, the former ensure equivalent turn-up ratio (gain) in the energized condition whereas the latter are useful whenever an appropriate bandwidth is needed for control strategies.

The range of requirements that any damper faces is vast. Not only do they include specifying the turn-up ratio (dynamic range), but also power dissipation, response time, bandwidth, packaging, mass, external components, life span, cost and the like. With this multidisciplinary device they involve a large set of important material and geometric variables with mutual interactions between them. For example, it is well-known that the particle concentration influences both off-state viscosity of the material, its magnetization characteristics and tribology [12, 13]. Therefore, the need for development and design tools has arisen to handle the complexities in an organized and efficient fashion.

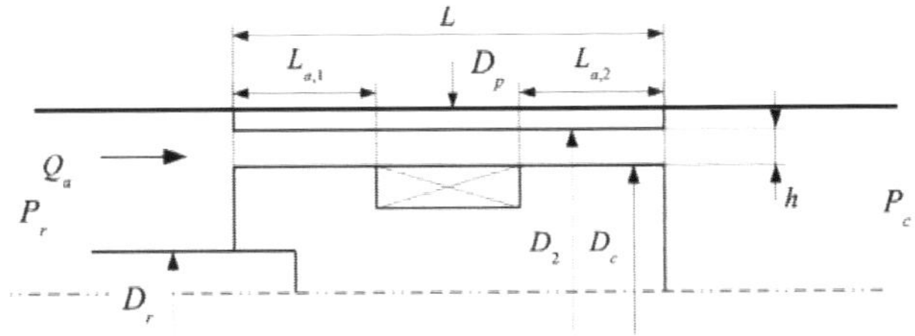

(a) Single gap piston

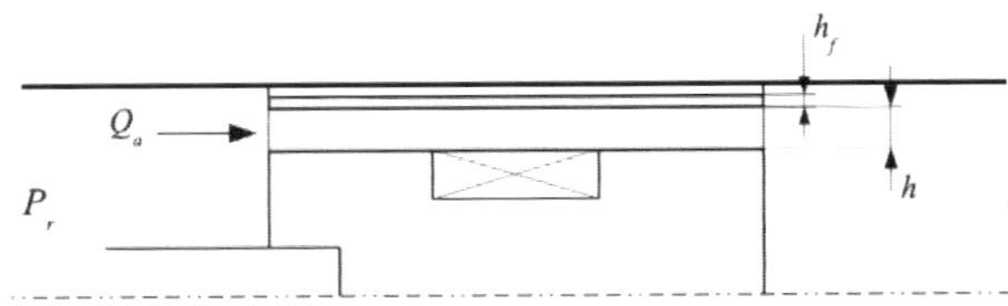

(b) Flux bypass

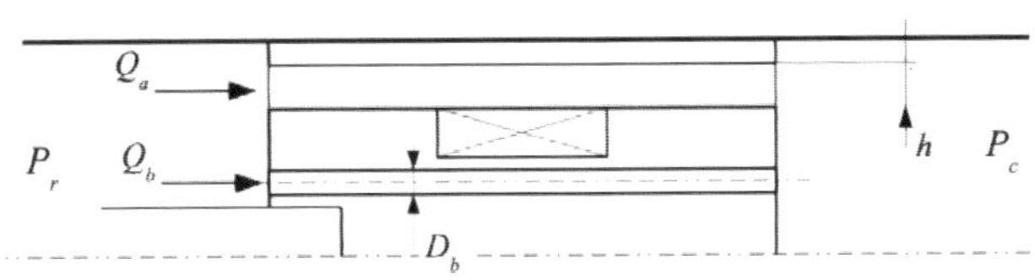

(c) Thru-core bypass

Figure 4. Piston (control valve) assembly schematic layout [7].

Therefore, in this paper we present the results of a parametric study undertaken to explore the influence of various quantities on the performance range of an exemplary monotube damper with a single annular gap in the control valve (piston). In the following sections we present key details behind the lumped parameter model of the damper system, and then examine the impact of various key geometric and material properties on the performance range of the device.

2. Damper Model

In this section we briefly highlight details of a lumped parameter model first developed by Gołdasz and Sapiński [14]. The model is based on the biplastic Bingham scheme and we use it for demonstrating the impact key engineering variables have on the damper performance. The piston assembly is presented in figure 4. Specifically, figure 4(a) reveals the single-gap piston layout, whereas figure 4(b) shows the flux bypass feature located in the annulus. Finally, the thru-core bypass location is highlighted in figure 4(c). The pressure difference across the piston is $\Delta P_{MR}=P_r-P_c$, and the total flow rate is $Q_p=v_p(A_p-A_r)=Q_a+Q_b$, and Q_a – flow rate through the annulus, Q_b – thru-core bypass flow rate, v_p – piston velocity, A_p – piston cross-section area, A_r – rod cross-section area. The annular gap height is $h=(D_2-D_c)/2$, and its length is referred to as L and the circumferential width as $w=\pi(D_2+D_c)/2$. Also, D_c is core outside diameter, whereas D_2 denotes the inner diameter of the outer ring. Moreover, the energized section length is L_a. As shown, the piston contains bypass features in the form of the flux bypass of the size h_f and the thru-core bypass whose diameter is denoted as D_b. The fluid is characterized by the field induced yield stress $\tau_0=\tau_0(B)$, B – flux density, and the viscosity μ.

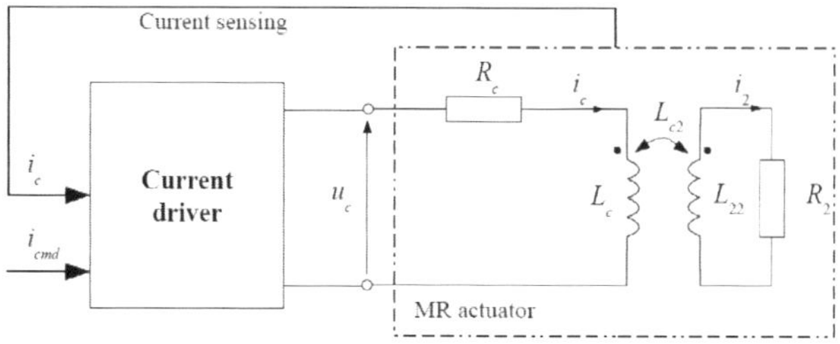

Figure 5. Controlled circuit and network model.

Using the piston layout in figure 4 and by employing the biplastic Bingham scheme the relationship between the pressure drop across the piston and the thru-core bypass and the flow rate can be expressed in the following form

$$\begin{cases} \Delta P_{MR} \approx \dfrac{2\tau_0 L_a}{h[1-\gamma(1-\delta)]} G(S) + C\dfrac{\rho Q_a^2}{A_g^2} \\ \Delta P_b - C_f \dfrac{L}{D_b} \dfrac{(Q_p - Q_a)^2}{2A_b^2} \\ G = \dfrac{h\Delta P_{MR}}{2L_a\tau_0}[1-\gamma(1-\delta)] \\ S = 12\dfrac{\mu Q_a}{wh^2\tau_0}[1-\gamma(1-\delta)] \end{cases} \tag{1}$$

The non-linear scheme for solving the pressure balance equation was highlighted in [14] and the reader should refer there for further details on handling the dimensionless variables G and S. The parameters γ and δ allow for controlling of the slope of the damper force against velocity and the intercept force at zero piston velocity. In equation 1 high velocity losses are accounted for by including the second (quadratic) term. The coefficient C (C>0) captures the effects due to the fluids's entry and exit, flow development, turbulent losses, etc. The parameters δ, γ, C can be extracted from experimental data or flow simulations.

In this study the damper model is coupled with a lumped parameter model of the electromagnetic circuit of the actuator. We use the non-linear network model [7] to incorporate time-varying effects due to eddy currents. The two-coil model of figure 5 will allow for simulating the output of the actuator when subjected to either fixed or fluctuating currents.

The model equations are as follows [7, 15]

$$\begin{cases} u_c = i_c R_c + L_c \dfrac{di_c}{dt} + L_{c2}\dfrac{di_2}{dt} \\ 0 = i_2 R_2 + L_{2c}\dfrac{di_c}{dt} + L_{22}\dfrac{di_2}{dt} \end{cases} \tag{2}$$

where i_c and i_2 are the primary and secondary circuit currents, respectively. R_c and R_2 stand for the primary and secondary circuit resistances, respectively, whereas L_c, L_{c2} and L_{2c} denote the main coil inductance, mutual inductances between the main circuit and the secondary current loop. In the model it is assumed that $L_{c2}=L_{2c}$, and the coupling coefficient between the primary inductance L_c and the secondary one L_{22} is k_c; $k_c \leq 1$. The input u_c is the supply voltage. Assuming no flux leakage the core flux is $\phi = L_c i_c = B_c A_c$. Then, the gap flux density can be computed as $B_g = B_c A_c / A_s$, where A_s is the outer surface area of the core at circumference, A_c is the core-cross section area, and B_c refers to the iron core flux density. By coupling equations 1 and 2 the impact of various material and geometric variables can be examined in detail. By itself, equation 1 can be used for predicting the steady-state output of the damper, whereas equation 2 allows for examining the dynamic behaviour of the actuator.

3. Parametric Analysis

In the sections that follow below we examine the impact of key variables on the output of the MR actuator. Specifically, Section 3.1 contains steady-state results, and Section 3.2 emphasizes the impact of some parameters that influence the damper's transient performance. The analysis was performed for the following parameter set: ρ=2680 kg/m^3, μ=0.05 Pa·s/m, D_p=46 mm, D_r=14 mm, L=27 mm, L_a=18.5 mm, D_c=37.3 mm, h=1.0 mm, h_b=2.5 mm, h_f=1.5 mm, R_c=1.15 Ω, N_c=100 (coil turns), R_2=0.0003 Ω, k_c=0.5, C=0.75 (0 A) and C=0.45 (5 A). Throughout the analysis we assumed all components of the magnetic circuit in the actuator made out of low carbon steel grades and MR fluid properties corresponding to that of a 26 % Fe vol. The magnetization characteristics were calculated based on the paper of Jolly et al. [16], and the dimensionless parameter values are those identified in [7]. The gap flux density relationship vs. coil current (B_g–I_c) was extracted from magnetostatic field calculations using the finite-element software FEMM ver. 4.2.

3.1. Steady-state

The steady-state analysis concerns the model given by equation 1. The calculated results are presented in figures 6 to 13 as steady-state maps of the damping force F_d plotted against the peak velocity of the piston V_p for a specified parameter range. The data were generated for the coil current range from I_c=0 A to I_c=5 A. We examined the impact of the following design variables: annular gap height, annulus length, thru-core bypass size, slot bypass depth and high permeability material in the piston core. Specifically, figure 6 illustrates the performance of the base (nominal) configuration of the damper. figures 7 to 9 highlight variation of turn-up ratio K, off-state damping force and on-state force with the annular gap height h. Next, figure 10 shows modification of the damping force by varying the annulus' total length. The effect the various parameters have on the actuator's performance is distinct. For example, decreasing the annular gap height improved the maximum damping force, however, the effective turn-up ratio (force gain) degraded due to increased base off-state forces. Specifically, the maximum turn-up ratio varied from 37 (h=1.0 mm) to 21 (h=0.6 mm), and at the velocity of 1 m/s it dropped from 6 to 3.5.

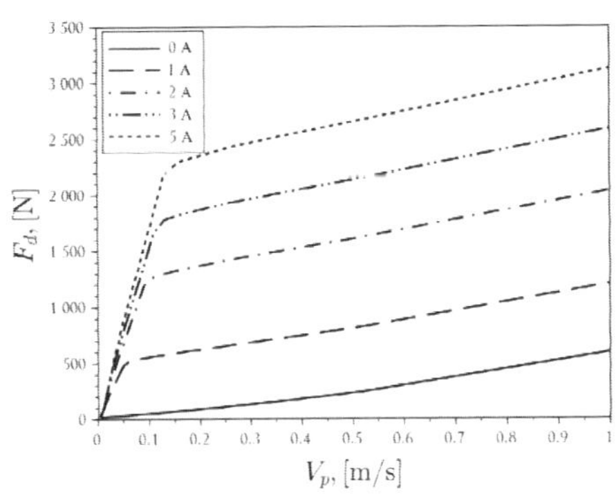

Figure 6. Base configuration: F_d vs. V_p.

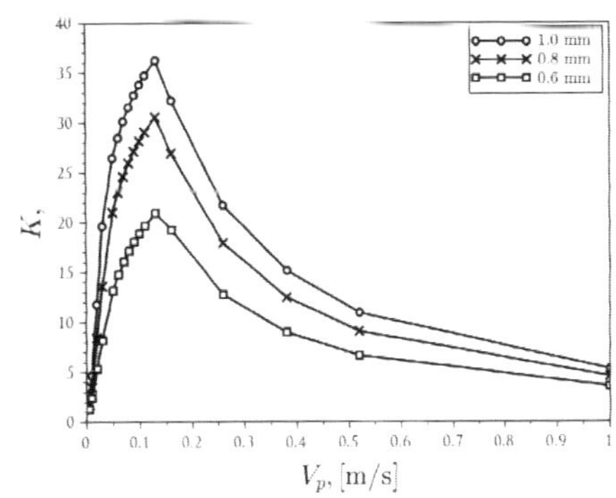

Figure 7. Turn-up ratio.

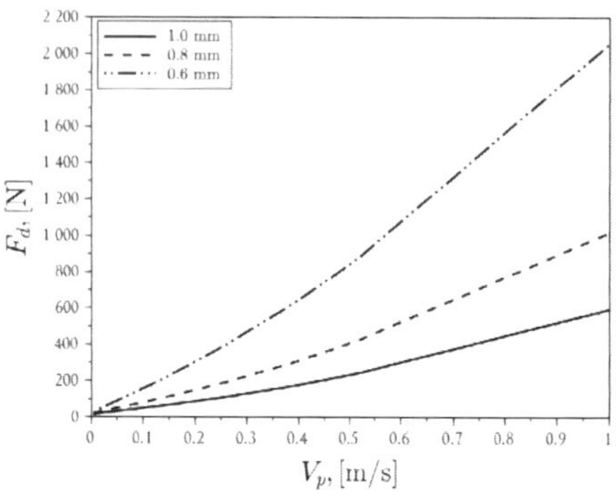

Figure 8. Gap height change: I_c=0 A.

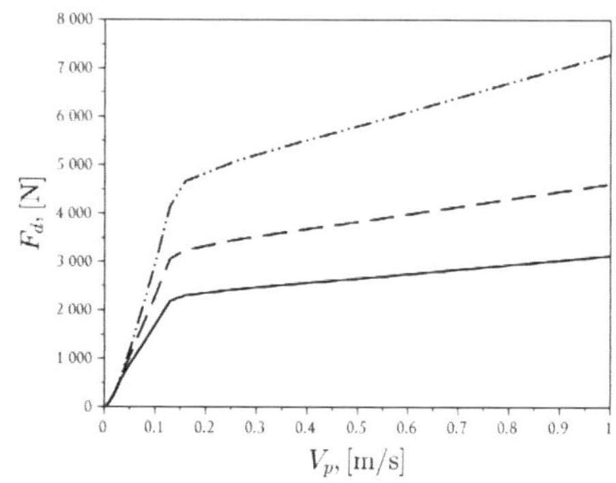

Figure 9. Gap height change: I_c=5 A.

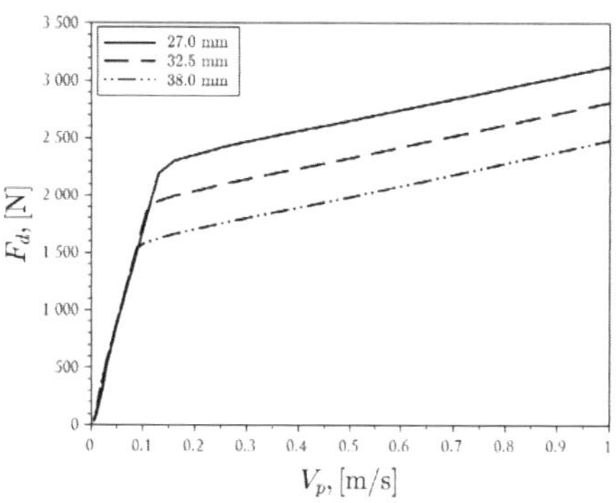

Figure 10. Gap length change: I_c=5 A.

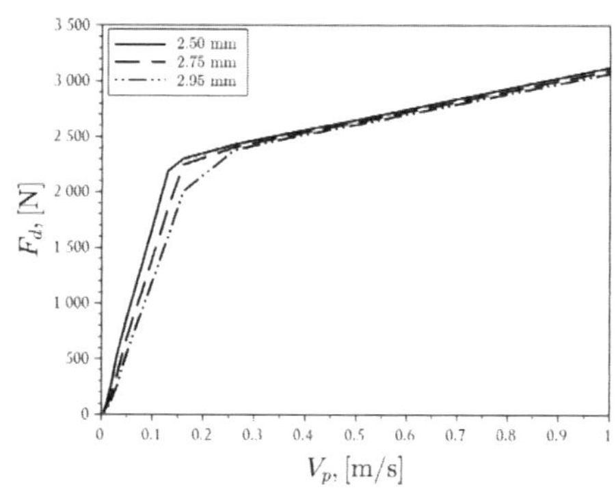

Figure 11. Bypass size change: I_c=5 A.

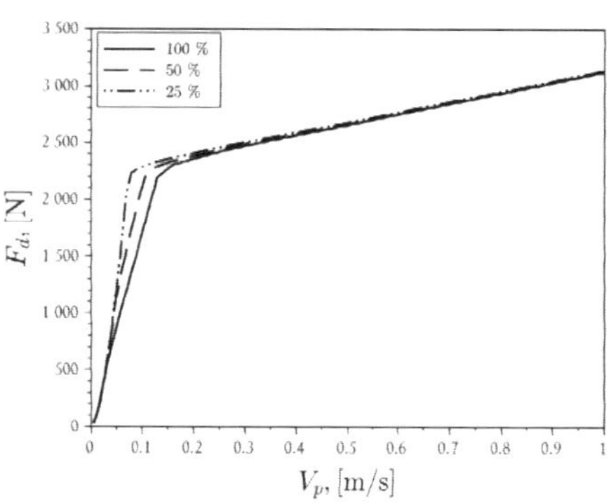

Figure 12. Bypass depth change: I_c= 5 A.

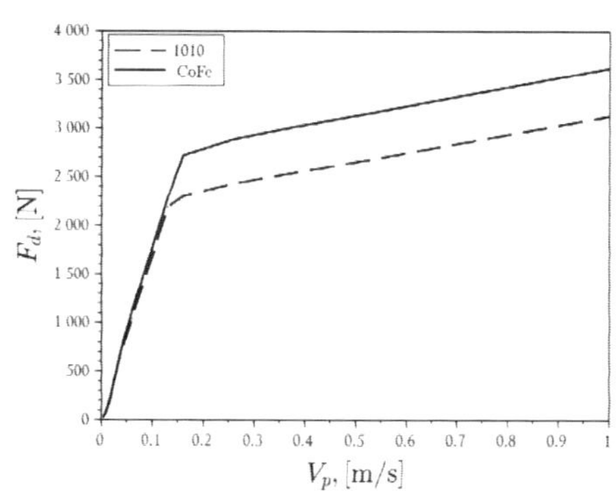

Figure 13. Impact of core material.

Next, increasing the annular length of the valve resulted in adverse data. The peak force degradation was due to the lower calculated flux density in the gap (for longer pistons) which could not be compensated for by the piston length change. Moreover, observations of figures 11 and 12 reveal that tuning the force within the low-velocity regime can be effective by varying the thru-core bypass size and the slot bypass depth. The bypass features provides engineers with two additional degrees of freedom in shaping the actuator's performance.

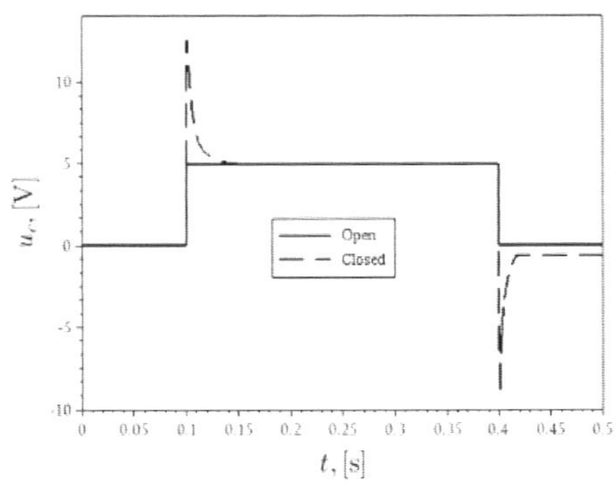

Figure 14. Supply voltage: open loop vs. controlled circuit.

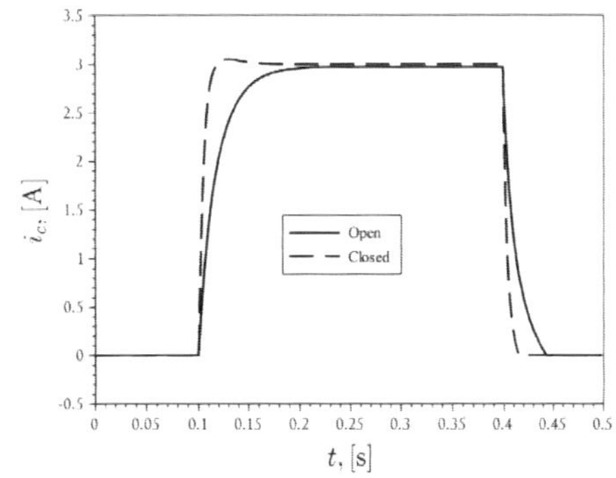

Figure 15. Coil current: open vs. controlled.

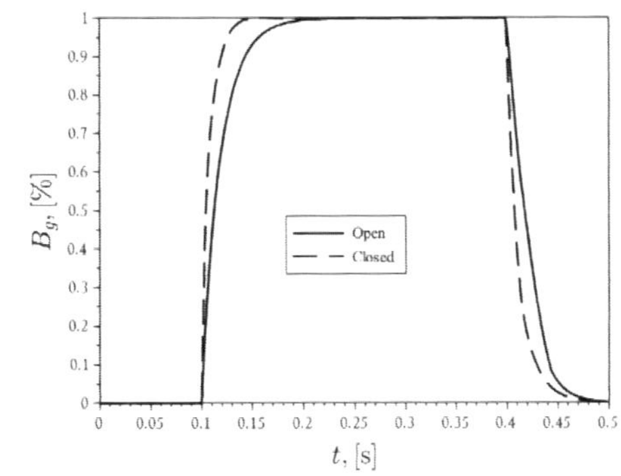

Figure 16. Flux density: open vs. controlled.

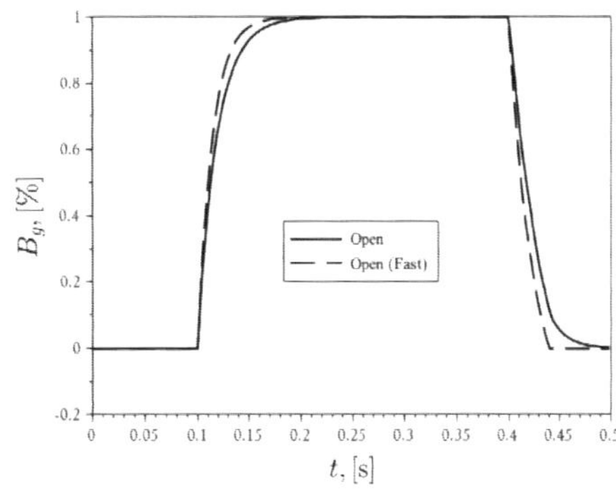

Figure 17. Flux density: open loop (base) vs. low conductivity.

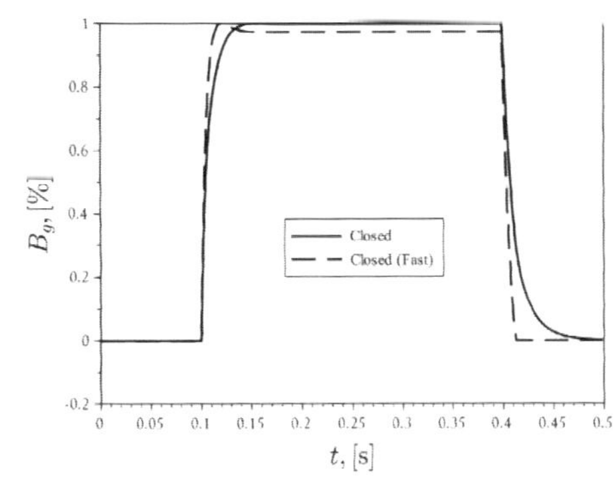

Figure 18. Flux density: controlled (based) vs. low conductivity.

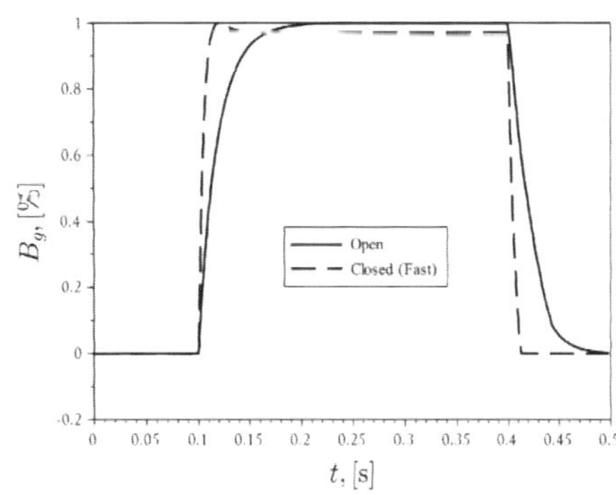

Figure 19. Flux density: open loop (base) vs. low conductivity (closed).

Note that the thru-core bypass affects the slope of the static map up to the knee point (i.e. when the pressure difference across the piston exceeds the breakaway pressure in the annulus), whereas the slot bypass allows for the slope modification within the range of force determined by the first breakaway force due to exceeding the threshold pressure difference in the slot bypass and the second breakaway force due to the annulus. Finally, a simple replacement of the core material with a material such as Vanadium Permedur having a high saturation flux density resulted in a significant performance improvement. Such materials reduce the flux density bottleneck in the area below the coil.

3.2. Transient states

The transient-state analysis concerns the model given by equation 2 coupled with a control circuit in Simulink involving a custom PWM type power supply with a PID controller model. In this section the calculated results are presented in figures 14 to 19 as time histories of the supply voltage voltage u_c, the coil current i_c and the (scaled) gap flux density B_g. We examined the impact of current control and low conductivity materials on the dynamics of the coil circuit.
In particular, figure 14 shows the comparison between the voltage step input and the regulated case with a coil current feedback loop. The controlled output is that of a PID controller featuring a large overshoot and saturated by the supply voltage. In the open loop (non-regulated) case the coil response is poor, and the flux density follows the current with a delay. For comparison, the coil current control resulted in a significant acceleration in both current and flux density's rise and decay - see figures 15 and 16. Next, figure 17 illustrates the concept of improving the core dynamics by using a low conductivity material. Here, the model was subjected to a voltage step input. In the model the effect was achieved by varying the parasitic loop resistance R_2. However, it should be observed that optimum effects can be achieved by combining the controlled circuit and the low conductivity core.

4. Summary and Conclusions

In this paper we examined a coupled lumped parameter of an MR actuator. The results that we obtained are merely an illustration of the performance range and shaping of a standard MR actuator with bypass features in the piston core. The model uses the biplastic Bingham approach for modelling the force output, and a second-order electrical network model for capturing the effect of fluctuating current inputs. We used the model to capture the influence of key design parameters on the actuator's steady-state as well as transient performance.
The parametric study that we undertook in order to explore the steady-state output concerned several geometrical as well material characteristics. For comparison, the transient-state investigation emphasized control strategies and a single material property. We illustrated that the transient output can be effectively manipulated through employing a PID current controller as well as and changes in the material of the core. For example, it was shown that the current driver improves the response time both in terms of faster coil current and flux density.
Finally, the response time acceleration seemed to provide optimum results when appropriate control strategies were accompanied by replacing the base core material with a low conductivity material for minimizing the eddy current effects.

References

[1] Takagi T 1996 *J. Int. Mat. Sys. and Struct.* **7** 346
[2] Rabinov J 1948 *AIEE Trans.* **67** 1308
[3] Gorodkin SR and Kolomentsev AV and Kordonsky WI and Kuzmin VA and Luk'ianovich AV and Prokhorov IV and Protasevich NA and Shulman ZP 1995 *US Patent 5353839*
[4] Alexandridis AA 2000 *Proc. European Conference on Vehicle Electronic Systems 2000* (Stratford-upon-Avon, England)
[5] Goldasz J 2016 *Proc. International Conference on Actuators and New Technologies 2016* (Bremen, Germany) 149
[6] Anon 2016 http://www.bwigroup.com
[7] Goldasz J and Sapinski B 2015 *Insight into magnetorheological shock absorbers* (Heidelberg: Springer)

[8] Sapinski B and Rosol M 2007 *J. Theor. and Appl. Mech.* **47** 133
[9] Foister RT and Nehl TW and Kruckemeyer WC and Raynauld O 2011 *US. Patent App. 20110100775*
[10] Farjoud Alireza and Bagherpour EA 2016 *J. Int. Mat. Sys. and Struct.* **27** 51
[11] Pawlak AM 2007 *Sensors and Actuators in Mechatronics* (New York: Taylor and Francis)
[12] Iyengar VR and Alexandridis AA and Tung SC and Rule D 2004 *Trib. Trans.* **47** 23
[13] Felt D and Hagenbuchle M and Liu J and Richard J 1996 *J. Int. Mat. Sys. and Struct.* **7** 589
[14] Goldasz J and Sapinski B 2013 *J. Int. Mat. Sys. and Struct.* **24** 1846
[15] Li W and Jeong YW and Koh ChS 2010 *IEEE Transactions on Magnetics* **46** 1859
[16] Jolly MR and Bender J and Carlson DJ *J. of Int. Mat. Sys. and Struct.* **1** 5

IOP Conf. Series: Materials Science and Engineering **148** (2016) 012005 doi:10.1088/1757-899X/148/1/012005

Experimental verification and comparison of the rubber V-belt continuously variable transmission models

W Grzegożek[1], K Dobaj[1], A Kot[1]

[1] **Institute of Automobiles and Internal Combustion Engines,** Cracow University of Technology, Al. Jana Pawla II 37, 31-864 Krakow, Poland

E-mail: witek@mech.pk.edu.pl; kdobaj@pk.edu.pl; adam.kot2@gmail.com

Abstract. The paper includes the analysis of the rubber V-belt cooperation with the CVT transmission pulleys. The analysis of the forces and torques acting in the CVT transmission was conducted basing on calculated characteristics of the centrifugal regulator and the torque regulator. The accurate estimation of the regulator surface curvature allowed for calculation of the relation between the driving wheel axial force, the engine rotational speed and the gear ratio of the CVT transmission. Simplified analytical models of the rubber V-belt- pulley cooperation are based on three basic approaches. The Dittrich model assumes two contact regions on the driven and driving wheel. The Kim-Kim model considers, in addition to the previous model, also the radial friction. The radial friction results in the lack of the developed friction area on the driving pulley. The third approach, formulated in the Cammalleri model, assumes variable sliding angle along the wrap arch and describes it as a result the belt longitudinal and cross flexibility. Theoretical torque on the driven and driving wheel was calculated on the basis of the known regulators characteristics. The calculated torque was compared to the measured loading torque. The best accordance, referring to the centrifugal regulator range of work, was obtained for the Kim-Kim model.

1. Introduction

Description of the rubber V-belt cooperation with the bevel gears assemblies is a very complex issue. The complexity is a result of the belt transverse and longitudinal flexibility, the rubber hysteresis, the V-shape cross section of the belt and changes of the forces acting on the belt in one driving cycle. Furthermore, the changes of the forces acting on the moving vehicle influence on the transmitted power and gear ratio of the CVT transmission. Further difficulties arise from the internal combustion engine vibrations (usually with one cylinder), which is a common source of motive power in this kind of vehicles.

The existing concepts of the mathematical description of the considered phenomena can be divided into three basic approaches. The first approach is the complex theoretical model, developed mostly by Göran and Gerbert [2,3]. The model considered the belt slip in radial and longitudinal direction and its elastic properties. This theory, because of its complexity, has not found wide application in practice.

Another approach is the simplified semi- experimental model, which was highly modified over the years. In the work of Dittrich [7] the two zones of the belt-bevel gear contact were assumed. The first zone was involved with the undeveloped friction area when the belt enters on the bevel gears assembly. The second zone referred to developed friction area when the changes of the belt tension forces occur. Development of this approach, introduced by Worley [5], was the introduction of the radial component of the friction force. The formulated development allowed for the assumption of the constant belt force on the whole wrap angle of the driving wheel (the single zone of contact). This assumption was explained by the occurrence of the self-locking phenomenon, preventing the radial displacement of the relaxing part of the belt. The practical applications mostly use the Kim model [5]. The model differs from the Worley approach only in the friction coefficient description. The Kim model doesn't include the difference the radial and longitudinal friction coefficient. Above conceptions assume the constant slip angle. The slip angle is defined as an angle between the friction vector and the radial direction. The value of the slip angle changes in the real CVT assembly, which is considered in the Cammalleri model [1].

The last of the three approaches is involved with the numerical models. Julió and Plante [4] introduced the discretized belt model. The model allowed for the prediction of the belt reaction according to the changes of the following parameters: load of the transmission (vehicle), angular velocity and axial force of the driving wheel. The prediction was feasible in the steady-state and transient working conditions.

The gear ratio change in the CVT transmission usually is a result of the centrifugal regulator operation. The pressing force is generated by the rollers of the regulator. The force mostly depends on the engine angular velocity and the profile of ramp curvature. In the case of the driven wheel of the transmission, the axial force is exerted by the coil spring and the torque regulator. Usually in the papers dealing with CVT operation the axial force resulting from the belt-bevel gears cooperation is not associated with the force resulting from the regulator assembly.

2. Test stand

The experiments were carried out using complete drivetrain of the TGB 101S scooter (1), mounted on the test stand frame (figure 1). The drivetrain included the CVT transmission with the centrifugal regulator. The centrifugal clutch was eliminated to allow the possibility of loading the drivetrain in the full range of the engine speed. To provide the load to the drivetrain, the water brake (2) was used. The flywheel was used to simulate the vehicle starting process.

Figure 1. Test stand.

The following parameters were measured: rotational velocity of the driving and driven pulley of the CVT transmission (3), the force exerted on the water brake (5) and the driven wheel winding radius (4). The winding radius was measured using the optical sensor. The sensors signals were recorded with the frequency of 100 Hz, using the analog-to-digital converter coupled to the PC (6). Two experimental scenarios were conducted. The first was involved with maintaining constant throttle angle and changing the load. The second scenario referred to changes of the throttle angle and maintaining constant load on the drivetrain assembly.

3. The characteristics of the regulator assembly

3.1 The centrifugal regulator

The main elements of the centrifugal regulator are: coupling plate fixed in relation to the engine crankshaft, sliding bevel gear and rollers forcing the axial movement of the bevel gear. In the range of the regulator operation, the movement of the bevel gear and thus the gear ratio change is the results of the equilibrium of the axial forces of the rollers and belt. Variable curvature of the rollers ramp provokes the difficulties in determining the relationship between the sliding movement of the bevel gear and the radial movement of the roller. On the other side, determining the actual position of the rollers is important for the regulator axial force calculation. To solve this problem, a imprint of the sliding track was made. The imprint allowed for the graphical estimation of the sliding ramp shape (curvature). Subsequently, using the CAD software, the positions of the roller for several sliding bevel gear positions were calculated (figure 2). The obtained relationships between the angle of the line tangent to the ramp (ε) and the radius of the roller centre (y), and with the axial movement of the bevel gear (x) is illustrated in figure 3. These characteristics were estimated by the third and second-order polynomial. The estimation allowed for determination of the regulator axial force for each position of the sliding bevel gear.

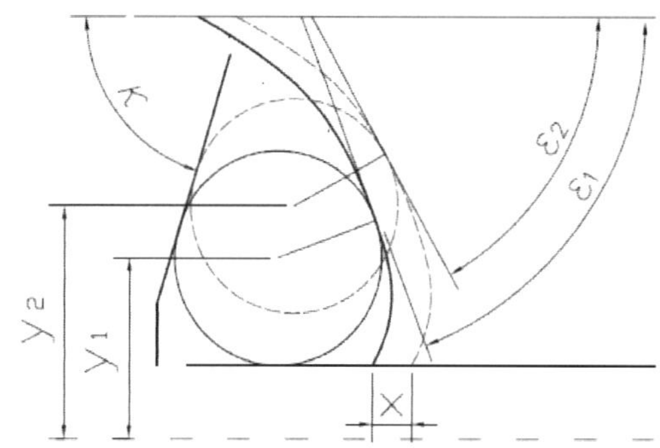

Figure 2. The centrifugal regulator scheme and example of the graphical rollers position estimation for the determined position of the sliding pulley.

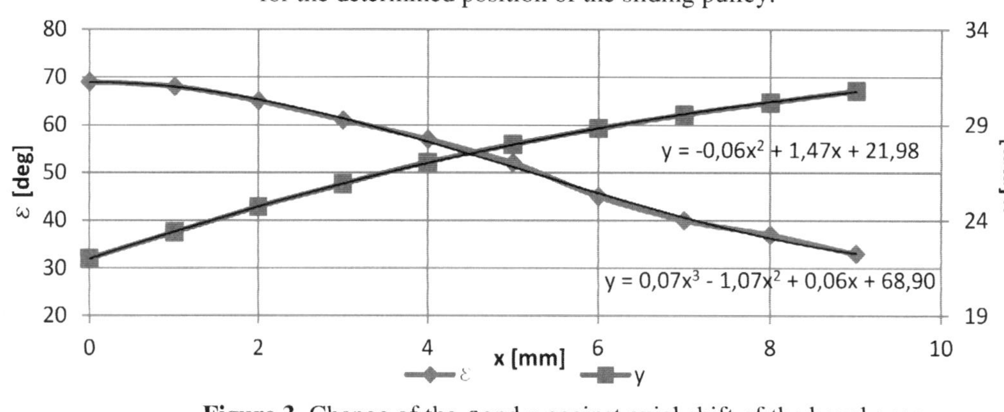

Figure 3. Change of the ε and y against axial shift of the bevel gear

The forces arrangement acting on the roller and the relationship describing the axial force are shown in [1]. The axial force is expressed as:

$$F_{zg} = \frac{zm\omega_g^2 y}{ctg(\kappa - \varphi_z) + ctg(\varepsilon - \varphi_b)}, \tag{1}$$

where:

z – number of the rollers, m - mass of the single roller, ω_g – engine angular speed, φ_z – sliding friction angle between the roller and the coupling plate, φ_b – rolling friction angle between the roller and the ramp. In the considered case, the angle of the coupling plate (κ) is always bigger than the angle of the line tangent to the ramp (ε), thus in the entire regulator working range the roller will roll on the ramp. The tangential force between the rollers and the coupling plate will have the character of the sliding friction force. By examining the next positions of the pulley with different angular velocities, the centrifugal regulator characteristics was established. The characteristics was formulated as the relation between the axial force, engine angular velocity and the transmission ratio. The roller position, involved with the ratio, has tremendous impact on the generated axial force. Wide range of the ε angle provokes that the increasing of the axial force while reducing the engine speed is possible. The increase results from the ε angle increase during the gear ratio change. This situation may occur, for example, when the load exerted on transmission increases but transmission works with low gear ratio.

3.2 The torque regulator

The driven wheel of the belt-drive CVT transmission is commonly equipped with the torque regulator of the axial force. The regulator is shown schematically in figure 4. Sliding bevel gear moves along the orienting groove, milled at a specified angle in relation with the pulley rotation axis. Thus, the axial force, acting on the belt, depends on the transmitted torque. The value of the torque regulator axial force can be calculated as follows:

$$F_{zn} = k_n(x_o + x) + \frac{2}{d_k}\left(\frac{M}{2} + k_t\xi\right) tg(\delta), \tag{2}$$

where: k_n – positioning spring stiffness, x_o – initial deformation of the spring, k_t – torsional stiffness of the spring, ξ - angle of the torsional deflection (depending on the bevel gear sliding displacement), M – torque transmitted by the CVT transmission driven wheel, d_k – diameter of the pulley guiding bushing, δ - guiding groove angle.

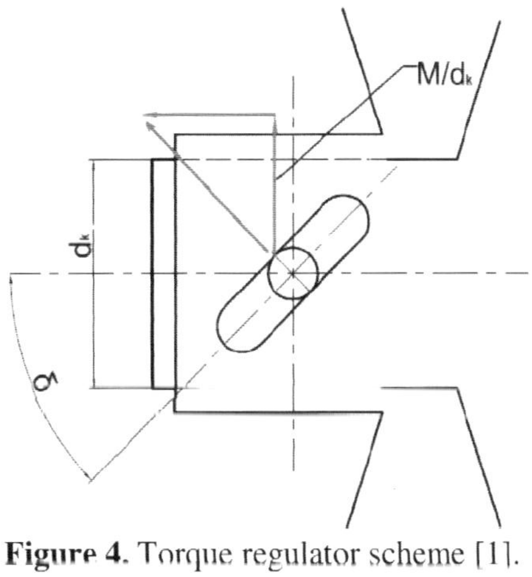

Figure 4. Torque regulator scheme [1].

3.3 General characteristics of the CVT transmission

Formulation of the general characteristics of the CVT transmission as a the torque possible to transmit in the function of the engine angular speed and transmission ratio change requires calculation of the relation between driving and driven wheel axial force and the belt force. This calculation is usually conducted using the belt cooperation model, presented in, among others, work [5] and [6]. The correctness of this approach will be justified further in this work. The sliding wheels axial forces can be expressed as:

$$F_{zg} = \frac{F_1}{2}\theta_g \frac{1 - \mu tg\frac{\alpha}{2}}{\mu + tg\frac{\alpha}{2}}, \tag{3}$$

$$F_{zn} = \frac{F_2}{2}(\theta_n - \theta_{na})\frac{1 - \mu tg\frac{\alpha}{2}}{\mu + tg\frac{\alpha}{2}} + \frac{F_1 - F_2}{2\mu}cos\frac{\alpha}{2}, \tag{4}$$

where F_{zg} and F_{zn} – axial forces of the driving ($_{zg}$) and driven ($_{zn}$) wheel, F_1 and F_2 –forces in active ($_1$) and passive ($_2$) part of the belt, θ_g i θ_n – wrap angle of the driving ($_g$) and driven ($_n$) wheel, θ_{na} – active wrap angle of the driven wheel, μ - friction coefficient between the belt and the pulley, $\alpha/2$ – belt wedge angle. The active wrap angle can be expressed as

$$\theta_{na} = \frac{1}{\mu}\left(ln\frac{F_1}{F_2}\right)sin\frac{\alpha}{2}. \quad (5)$$

When the characteristics of the centrifugal regulator is known, it is possible (basing on the equation 3) to calculate the belt force in the active part of the belt (F_1). The force F_1 can be determined in every point of the CVT transmission operation, within the range of the centrifugal regulator characteristics. Subsequently, using the equations (2) and (4) the force in the belt passive side (F_2) can be calculated. The torque of the CVT transmission driven wheel (M_n) can be calculated as follows:

$$M_n = (F_1 - F_2)\frac{d_n}{2}, \quad (6)$$

where: d_n is the belt pitch diameter in the driven wheel. The characteristics, obtained in this way is illustrated in figure 5. In addition, the curves of the constant torque value are marked. The curves were obtained by rectangular projection of the spatial characteristics on the ω-i plane. The experimentally determined parameters of CVT transmission and the assumed coefficients are established as follows: z = 6, m = 0,006kg, $\alpha/2$ = 13°, κ = 73°, μ = 0,4, μ_z = tgφ_z = 0,05, f_b = tgφ_b = 0,01, k_n = 4,8N/mm, x_o = 47,5mm, d_k = 43mm, k_t = 63Nmm/°, δ= 45°.

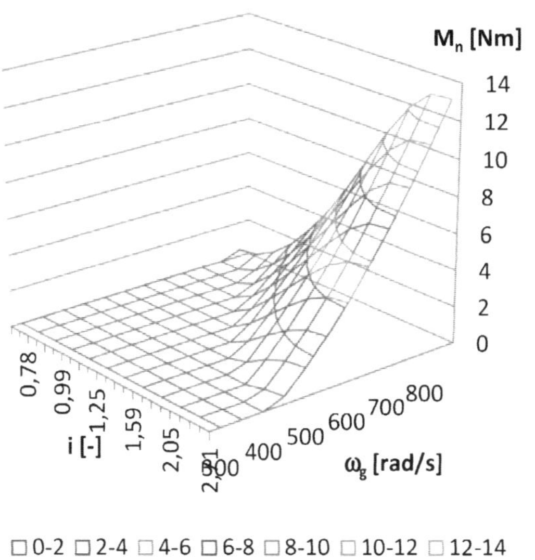

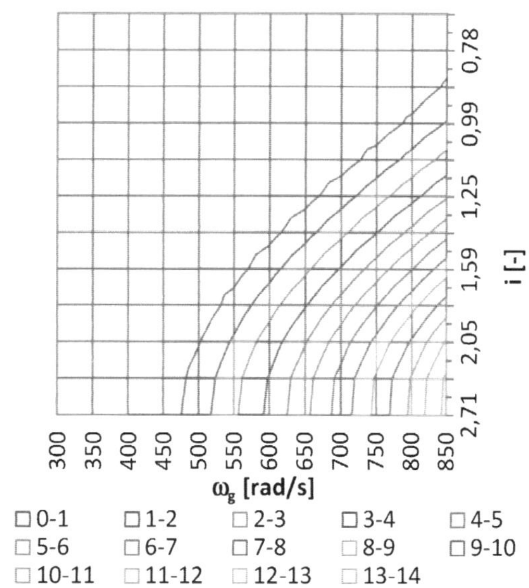

Figure 5. General characteristics of the examined CVT transmission and the constant torque curves; ω$_g$ – engine speed, i – transmission ratio, M$_n$ – driven pulley torque.

As it is shown in figure 5, the construction of the centrifugal regulator assembly provides that the torque transfer ability (in the regulator working range) is a function of the gear ratio. Whereas, the gear ratio is involved with the relative position between the rollers and the ramp. The influence of the ε angle increases with the increase of the engine rotational speed (figure 5). Hence, the ramp curvature is the essential parameter in the design of the drivetrains based on the CVT transmissions. The momentary transmission ratio value is the function the rotational velocity and the load exerted on the drivetrain. The position of the sliding driving wheel results from the equilibrium of the belt and rollers forces. Thus, the gear ratio change continues until the equilibrium of the forces described by equations (1) and (3) occurs. For the maximal ratio value, the CVT transmission works as a transmission with constant ratio. In this state, the total axial force is a combination of the regulator rollers force and the reaction force, acting on the sliding pulley. The determination of this force requires additional measurement equipment. Therefore the torque regulator, applied in the driven wheel of the transmission, allows for the extension of the torque transfer capability in the range of lower engine speeds.

4. The models of the rubber belt and bevel gear cooperation

4.1 The Dittrich model [7]

The belt circumferential forces differentiation occurs as a result of the transmitted torque. The belt force increases on the driven pulley while it decreases on the driving pulley. These changes occur only on some part of the wrap circumference, called the active wrap angle. The equilibrium conditions on the belt element on the driving wheel is shown in [7]. The conditions refer to the active wrap angle. The friction force acts only in the circumferential direction. The friction force doesn't exist in the passive area where the belt tension is constant. Hence, the axial force takes different value in this area. The total axial force is the sum of the axial forces from two belt-bevel gear contact areas and can be expressed as:

$$F_{zg} = F_{zg}^{m} + F_{zg}^{a} = \frac{F_1}{2}(\theta_g - \theta_{ga})ctg\frac{\alpha}{2} + \frac{F_1 - F_2}{2\mu}cos\frac{\alpha}{2}, \quad (7)$$

$$F_{zn} = F_{zn}^{m} + F_{zn}^{a} = \frac{F_2}{2}(\theta_n - \theta_{na})ctg\frac{\alpha}{2} + \frac{F_1 - F_2}{2\mu}cos\frac{\alpha}{2}, \quad (8)$$

where subscripts (^{a}) and (^{m}) refer to the active and passive zone (contact area). Because the belt force in the active area (F_1) is always bigger than the belt force in the passive area (F_2), the driving wheel axial force will take bigger values with the same wrap angles.

On the base of the known regulator characteristics and the registered parameters, it is possible to estimate the belt forces. Comparison of the equations (1) to (7) and (2) to (8) provides to the set of the two nonlinear equations with unknown variables F_1 and F_2. In order to verification of the obtained values, the theoretical driven wheel torque (described by equation 6) was compared with the measured load torque.

4.2 The Kim-Kim model [5]

According to the Dittrich approach, the winding radius on each of the wheels must undergo the changes as a result of the belt force differentiation. The experimental research results, described in the work [5], indicate that such belt behavior occurs only for the driven wheel. In the case of the driving wheel, the winding radius remains constant on the whole wrap angle. This behavior is explained by the self-locking phenomenon occurrence. As a consequence, it is assumed that the belt force is constant. Thus, the circumferential friction component is neglected. In the case of the active pulley (driving wheel), only the radial static friction is assumed. Thus, single contact area is considered. The total axial force for the driving pulley is expressed by equation (3). In the case of the driven pulley the two contact areas are assumed. In the passive contact area (at the entrance), the belt tension is constant. In this area the belt element forces distribution is similar to the belt element forces distribution on the driving wheel. The kinetic force in radial and circumferential direction is considered in the active contact area. The total axial force for the driven pulley is expressed by equation (4).

4.3 The Cammalleri model [1]

The main difference between the previously described models and the Cammalleri model [1] lies in the consideration of the belt flexibility and the changes of the slip angle γ along the pulley wrap angle. The axial forces of the pulleys are described in [1].

4.4 The models comparison

This chapter includes the selected results of the experimental researches for two way of the CVT transmission loading. The first loading scenario refers to the load increase with maintaining the constant, full throttle opening. The figure 6 illustrates the registered angular velocities of the pulleys and changes of the winding radiuses. In the first working stage, the CVT transmission works in the unloaded state, which corresponds with the engine speed about 8000rpm. The winding radiuses on the driving and driven wheel are then 43 and 33mm respectively. Thus the gear ratio obtained as the quotient of the radiuses has the minimal value 0,75 (figure 8). In this state, the centrifugal regulator axial force is about 400N, which corresponds to the minimal limit value of the angle ε. The real load torque and the theoretical torques obtained basing on the described approaches are shown in figure 7. The negligible theoretical capability of the torque transfer in this phase results mainly from the gear ratio value. The small driven pulley winding radii entails the large belt force value, even in the light load conditions. Furthermore, the increased wrap angle of the driving wheel provokes the necessity for increased axial force of the centrifugal regulator (equation 3). In the described CVT working area, the theoretical torque values according the Kim-Kim model are about 0,1Nm bigger than the real torque value. Similar torque difference is noticed for the Cammalleri model, but the obtained values are smaller than the real torque. In the application of the Dittrich model, the obtained torque values are about 0,3Nm smaller from the real torque values. The increase of the load (which started about 5th s.) results in the decrease of the engine speed (firstly) and change of the transmission ratio (subsequently) (figure 6 and 8). These phenomena result in the temporary loss of the CVT transmission driving capacity. This phenomenon may result from the delay operation of the regulation system. Its effect is a temporary slip increase. From the 6th to 8th second of the experiment, the decrease of the engine angular speed occurs simultaneously with an increase of the regulator axial force. The axial force increase is provoked by the gear ratio increase and thus the increase of the ε angle. Furthermore, the change of the driven wheel winding radius (associated with the belt force change) and the change of the driving wheel wrap angle influence on the reduction of the axial force, required from the regulator. Consequently, relatively small axial force change provokes the significant improvement of the theoretical torque. The best compatibility in this area was obtained for the first of the analysed models, wherein the theoretical torque is bigger than the measured torque but the difference doesn't exceed 0,8Nm. The torque values obtained using the another models are understated and the maximal difference is 2Nm for the Cammaleri model and 3,6Nm for the Dittrich model. The torque gradient obtained from the Dittrich model is twice lower than the torque gradient calculated basing on the measurements. Because the axial forces were calculated basing on the regulators characteristics, the obtained torque diagrams apply only to the centrifugal regulator working range. As it is shown in figure 6 and 8, in the range from 8th to 21th s. the sliding driving pulley was in the initial position. Therefore, the real axial force includes also the reaction force component. The reaction force determination requires additional measurement equipment. Hence, the belt forces and torques, obtained in the considered time range, achieve understate values. The forces and torques decrease (at a fixed gear ratio) is a result of the engine rotational speed decrease (figure 6). From the 21th second, the engine rotational speed increases as a result of the reduced load exerted by the water brake. During next second (to 22th s.) the gear ratio doesn't change its maximal value, which corresponds to the sudden growth of the theoretical drive capacity. Sudden decrease of the gear ratio, provoked by the centrifugal regulator operation forces the sudden decrease of the axial force (the ε angle change)

(figure 8). These changes cause the theoretical torque change (figure 7), which is also provoked by wrap angles change and the change of the belt force in the belt active part. As the result, in the range from 24th to 25th s., the theoretical drive capacity disappears which corresponds to the slip increase. From the 25th s., the CVT transmission returns to the base state.

The second loading scenario refers to the sudden throttle open at the constant load. The figure 9 shows the registered angular velocities of the pulleys and changes of the winding radiuses. In the initial stage, the CVT transmission works in the fixed load state at the partly open throttle which corresponds with the engine speed about 3000rpm. In the following conditions, the driving and the driven wheel winding radiuses are correspondingly 21mm and 55mm. The gear ratio i_r calculated as the quotient of radiuses has maximal value 2,7 (figure 11). The real loading torque and the theoretical torque obtained using the three described approaches are illustrated in figure 10. The small engine

The constant full throttle and increasing load

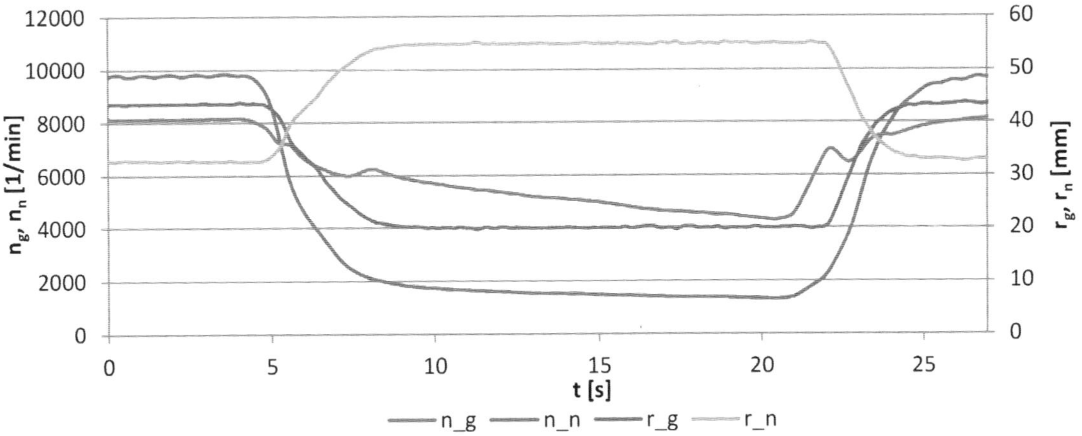

Figure 6. Change of the pulley angular speed (n) and winding radiuses (r) against time; subscript (g) – driving pulley, subscript (n) – driven pulley.

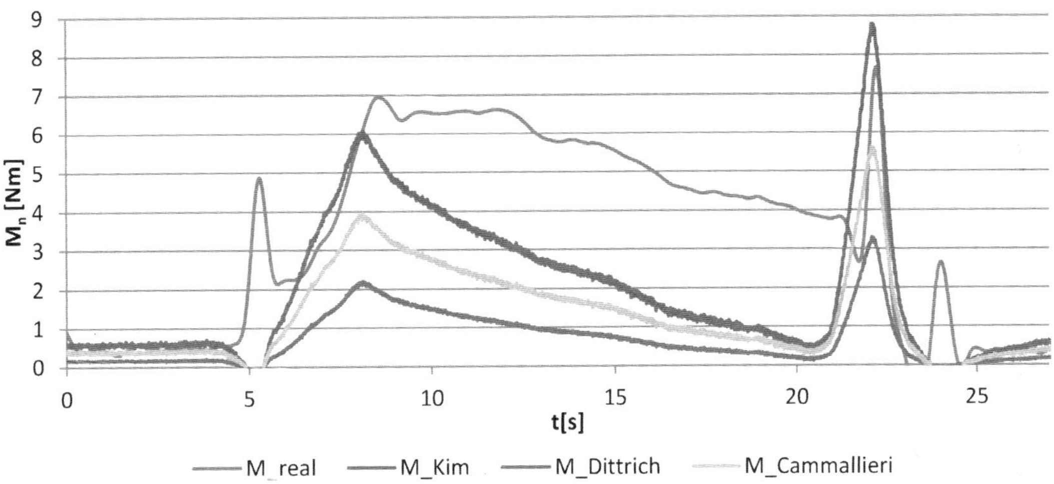

Figure 7. Real and theoretical driven pulley torque against time.

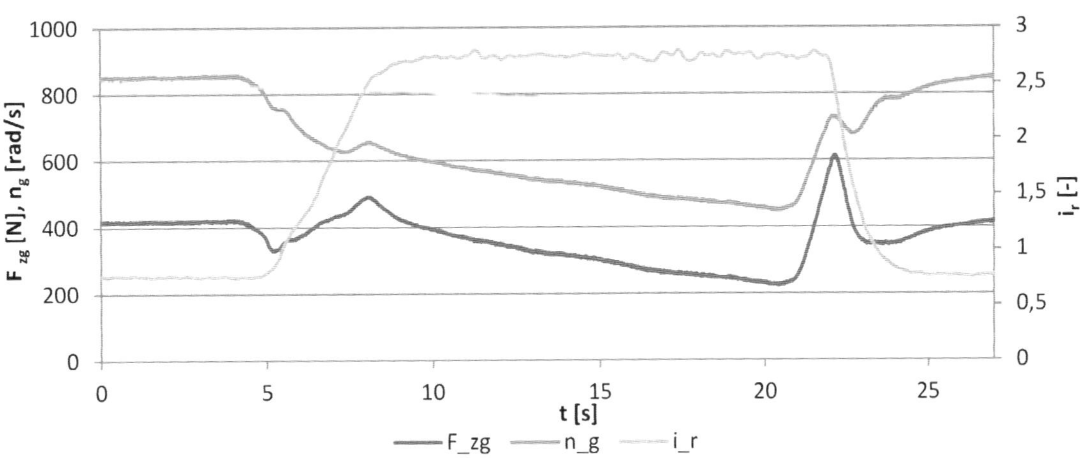

Figure 8. Centrifugal regulator axial force (F_{zg}) and transmission ratio (i_r) against time.

The constant load and rapid throttle opening

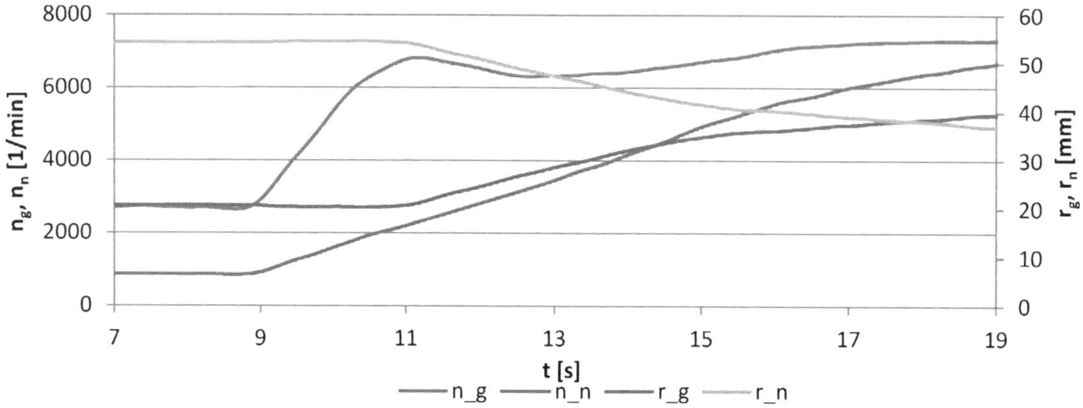

Figure 9. Change of the pulley angular speed (n) and winding radiuses (r) against time; subscript (g) – driving pulley, subscript (n) – driven pulley.

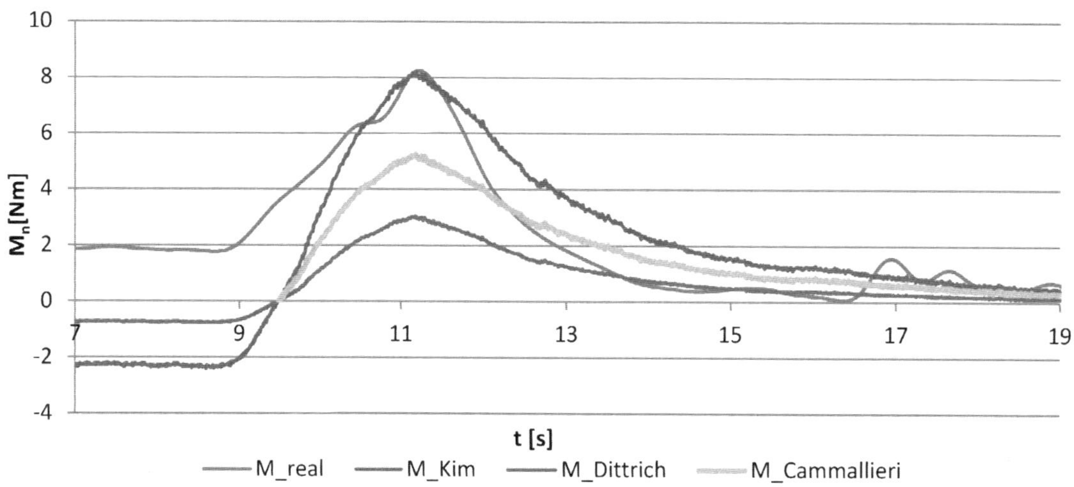

Figure 10. Real and theoretical driven pulley torque against time.

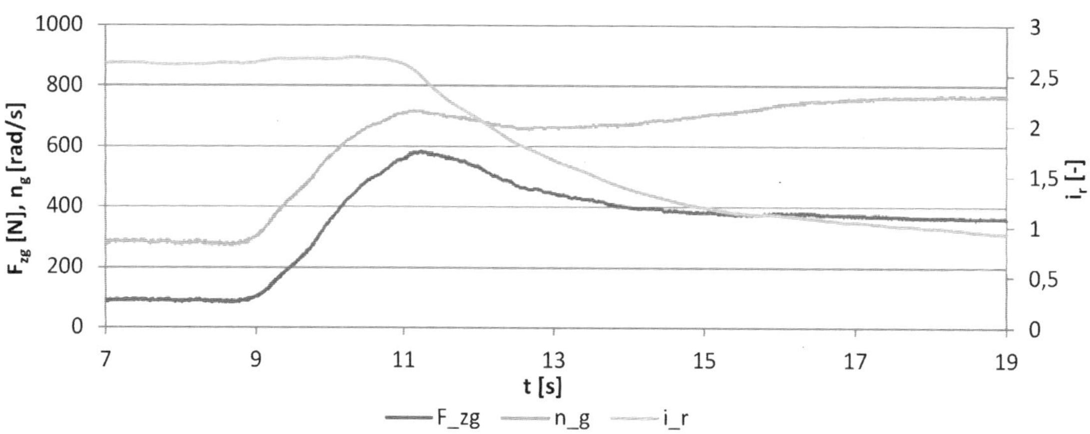

Figure 11. Centrifugal regulator axial force (F_{zg}) and transmission ratio (i_r) against time.

speed causes that the drive capacity results from the reaction force, acting on the sliding wheel of the regulator. Therefore, the theoretical torques take the negative values. The torque values in the time range from 10th to 19th s. have the physical importance. The gear ratio change begins at 11th s., at the engine speed 7000rpm (figure 10 and 11). The gear ratio change corresponds with progressive decrease of the regulator axial force and the CVT transmission drive capacity, at almost fixed engine speed. As previously the best torque compliance was obtained for the Kim-Kim model. The theoretical torque (Kim-Kim model) exceeds the real torque load and the maximal torque difference in the analysed situation is about 2Nm.

Summarizing, the best compliance between the real and modelled torque was obtained using the Kim- Kim model. The Cammalleri model, despite its complexity, gives understated torque values. However, the general trend of changes is provided. The main difficulty in the Cammalleri model is the necessity of the determination of the belt material properties, which can be the reason for the differences. Hence, the Kim-Kim model seems to be the most suitable for the practical applications. The model is characterized by simplicity and uncomplicated form. The results obtained by the Dittrich model are substantially understated, but the trend of changes is similar to another analysed models. Application of the identical belt cooperation model for the driven and driving wheel leads to values, which are not similar to the reality.

5. Conclusions

1. The experimentally obtained results indicate that the commonly cited The Kim-Kim model describes the CVT transmission working accurately.
2. The prominent difference between the real values and the parameters obtained basing on the Dittrich model justifies the occurrence of the single belt-pulley contact area on the driving wheel.
3. The Cammalleri model, including the belt elastic properties, requires the consideration of the belt material parameters. The complex form of this model limits its engineering applications.
4. The shape of the centrifugal regulator ramp provides that the temporary ε angle value and subsequently the temporary value of the transmission ratio have significant influence on the generated axial force. The engine angular speed decrease may occur correspondingly to the regulator axial force increase, provoked by the increase of the gear ratio (ε angle change).
5. The torque regulator can extend the torque transferring ability for the maximal gear ratio value, when the CVT assembly works as a transmission with a constant gear ratio.
6. The determination of the belt forces, basing on the regulators characteristics is possible only in the regulators operation ranges (excluding the maximal and minimal transmission ratio).

References

[1] Cammalleri M 2005 A new approach to the design of speed-torque-controlled Rubber V-belt Variator *Proc. ImechE* vol 219 pp 1413-27
[2] Gerbert G B 1972 Force and Slip Behaviour in V-belt Drives *Acta Politechnica Scandinavica Mechanical Engineering Series* No 67
[3] Grzegożek W 2011 *Przekładnie o ciągłej zmianie przełożenia (CVT) w układach napędowych pojazdów* Wydawnictwo PK Kraków
[4] Julió G and Plante J S 2011 An experimentally–validated model of rubber-belt CVT mechanics *Mechanism and Machine Theory* **46** pp 1037-53
[5] Kim K and Kim H 1989 Axial Forces of a V-Belt CVT *KSME Journal* vol 3 No 1 pp 56-61
[6] Kim H and Lee H and Song H and Kim H 2002 Steady State and Transient Characteristic of a Rubber Belt CVT with Mechanical Actuators *KSME International Journal* vol 16 No 5 pp 646-2002
[7] Krzyształowicz W and Zintel Z 1958 *Napęd pasowy klinowy. Obliczanie, konstrukcja i zastosowanie* PWT Warszawa

IOP Conf. Series: Materials Science and Engineering **148** (2016) 012006 doi:10.1088/1757-899X/148/1/012006

The experimentalanalysis of the slip in the rubber belt CVT

W Grzegożek[1] and A Kot[1]

[1]**Institute of Automobiles and Internal Combustion Engines,** Cracow University of Technology, Al. Jana Pawla II 37, 31-864 Krakow, Poland

email: witek@mech.pk.edu.pl; adam.kot2@gmail.com

Abstract. This work deals with the analysis of the speed losses in CVT. The bench tests have been conducted on the scooter CVT equipped with the centrifugal regulation system. This solution is typical for this type of vehicles so the conducted experiments refer to real exploitation conditions. The slip has been defined on the base of the difference between speed ratios obtained from the angular speeds and the belt pitch radii. This approach corresponds with the Dittrich model. The non-linear dependence between the slip and the transmitted torque has been obtained for the constant gear ratio. Also non-linear dependence between the slip and the gear ratio has been received for constant torque. The amount of slip value indicates that this is significant part of the total power losses as it has been described by Bertini. However it clashes with the Chen researches, where the slip corresponds with the marginal part of the overall losses.

1. Introduction

The advantages of the V-belt transmission cause this type of power train to be employed in machine drive systems from many years. The elaboration of the solutions which enable the step less gear ratio change (CVT) has contributed to the spread this type of transmission in vehicles – especially in the scooters and the snow vehicles.

A lot of works published recently concern the efficiency issue in the rubber belt continuously variable transmission. In this case the power loss mechanisms may be divided into two areas: torque losses and speed losses. The transmitted power (*P*)can be expressed as:

$$P = M(\tag{1}$$

where *M*is the torque transmitted by considered pulley while $\omega$is its angular velocity. Consequently the transmission efficiency is represented as:

$$\eta = \frac{M_n \omega_n}{M_g \omega_g}, \tag{2}$$

where subscripts ($_n$) and ($_g$) denote driven and driving pulley respectively. The fact that both M and ω are variable over time, differentiation of the above equation leads to the relation:

$$dP = d(M + dM(\tag{3}$$

Results of (1) and (3) equations show that the power losses are the sum of the torque losses and speed losses:

$$\frac{dP}{P} = \frac{dM}{M} + \frac{d\omega}{\omega}. \tag{4}$$

The first category associated with the internal friction in the belt deals with bending and compression hysteresis of the belt, its wedge-in and wedge-out at entering and exit from the pulley. The second category associated with the external friction results from the belt extension and its flexural rigidity which leads to the wrap angle decreasing and the contact area deformation. According to the research results presented in work [2] the transmission efficiency primarily depend upon the torque losses. The belt slip corresponds with only a few percent of the total power loss. This dependence is illustrated in figure1 showing changes of the CVT efficiency, torque and speed losses against the load at the constant engine speed.

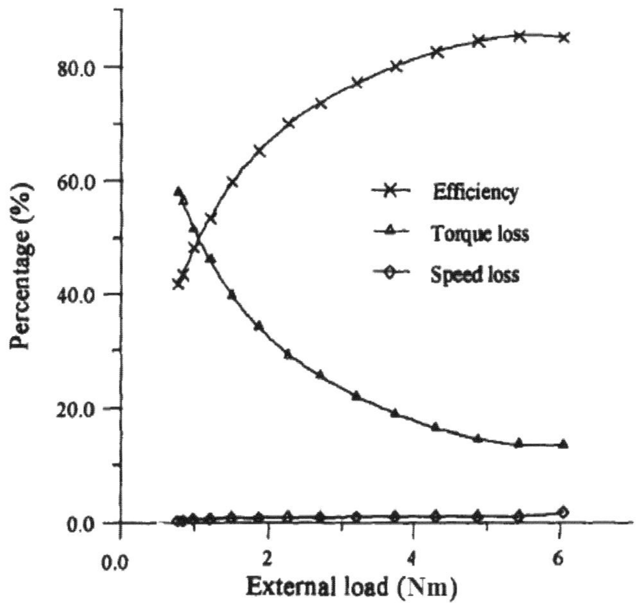

Figure 1. The CVT efficiency against load at constant engine speed (3600rpm) [2].

On the other hand the analytical power losses model in stationary conditions described in work [1] shows that the speed losses are essential for the CVT efficiency. Additionally this model has been confirmed experimentally.

The causes of the slip in belt transmissions are not fully explained. The widely accepted rubber belt transmission model proposed by Dittrich [6], assumes the presence on each of the pulleys two cooperation areas: the undeveloped friction zone at the entrance to the bevel gear and the area of kinetic friction at the exit of the wheel. In this perspective, the belt slip corresponds with the developed friction, which is accompanied by a change of the belt tension. The differentiation of the forces in tight and slack belt side causes the unequal belt extension in these sectors, so the speed of the belt needs to be changed. In the other area where the belt tension remains constant slip does not occur. In turn, in the approach described in [3] slip is related to the deflection of the belt due to the shear. Gerbert [4] however has assumed that the difference in the belt speed at the entry and exit of the pulley results from the variable winding radiuses. As a reason of this belt behaviour he has pointed out its compressibility and flexural rigidity.

2. Test stand

The experiment was realized by the test bench included complete drivetrain assembly from the TGB 101S scooter with centrifugal regulator. In order to enable the CVT load in the whole range of the engine revs the centrifugal clutch was eliminated The output bevel gear assembly was coupled with the water-supplied engine brake (dynamometer) which maximal absorbed power is 75kW. Additionally the steel flywheel coupled with the driven bevel gear too allows the simulation of the vehicle starting conditions.Figure2 shows the test stand scheme.

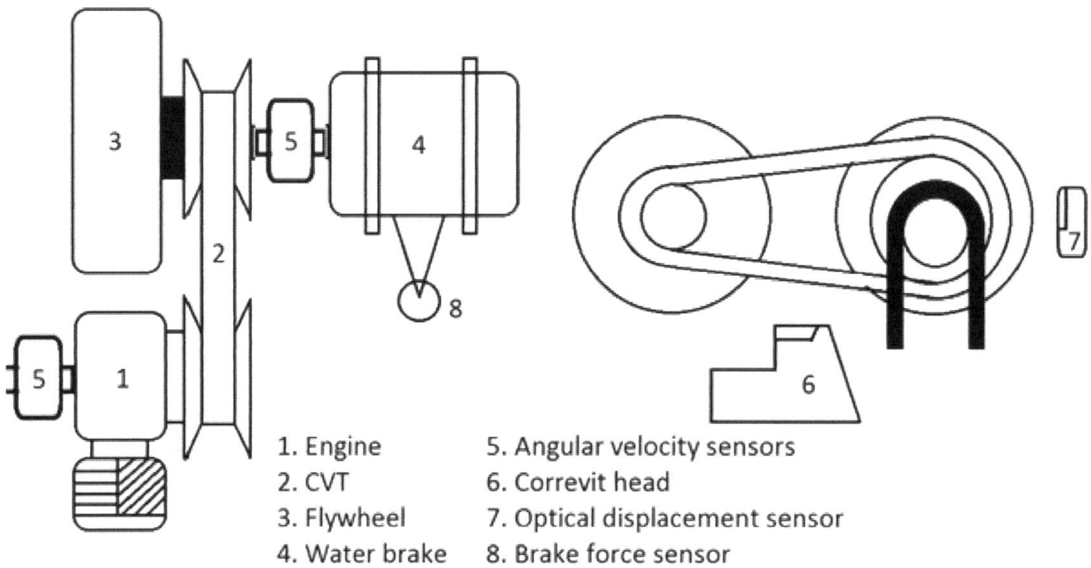

Figure2. Test stand scheme

The engine torque results in the reaction force transmitted on the engine brake lever with force sensor at its end. The measurement allows for the resistant torque calculation. Furthermore the following quantities have been measured: the angular speed on both the driving and the driven pulleys, the belt velocity and the winding radius on the driven pulley. The belt velocity and winding radius were measured by the optical sensors. The signals from the sensors were recorded by the analog-to-digital transducer connected with the computer. The registration resolution were 100Hz. The driving pulley winding radius was calculated on the base of the belt length and the pulleys centers distance. The tests were conducted with the constant throttle position and variable load as well as the variable throttle position and the constant load.

3. The belt slip

3.*1 Slip definition*

The torque load transmitted by the CVT causes the differentiation of the forces in the belt. Because of the belt compliance the strain of the tight and slack side will be different. In the effect of the mass conservation rule a tight side belt velocity has to be higher than slack side velocity in a single circuit. Therefore the belt slip can be defined as:

$$S = \frac{v_1 - v_2}{v_1}, \tag{5}$$

where v_1 and v_2 refers to the velocities in tight and slack belt side (figure3).In the basic theory of belt drives (among others in the Dittrich work [6]), the wrap angle can be divided into the reserve angle and the active (effective) angle (θ_{ga} and θ_{na} in figure 3). The change of the belt tension, and therefore slip occur in the active part of the wrap arc. In the inactive zone (at the entrance), where there is static friction, the belt velocity equals the circumferential speed of the pulley on the belt radius, hence the equation (5) can be written as:

$$S = \frac{\omega_g r_g - \omega_n r_n}{\omega_g r_g}, \tag{6}$$

where ω_g i ω_n are angular velocities whereas r_g and r_n are winding radiuses for driving ($_g$) and driven pulley ($_n$). Owing to the fact that the radius of the belt cooperation may vary along the arc of wrap, the slip determination by equation (6) requires defining winding radiuses in the area where the belt enters at the pulley.

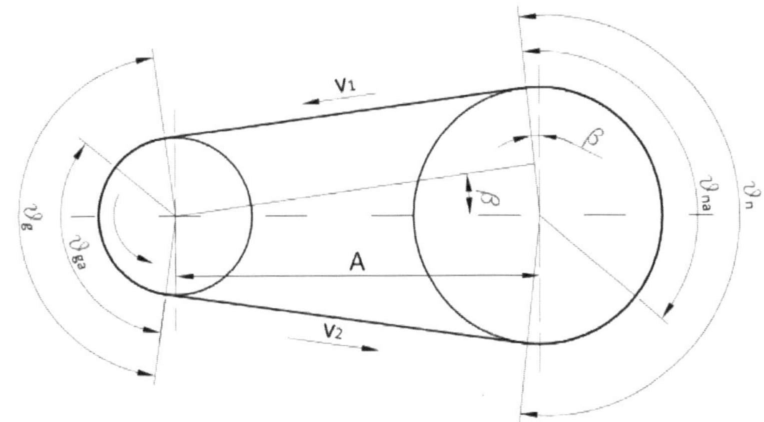

Figure3. Belt transmission scheme.

The transmission ratio can be specified by the angular velocities ratio or the winding radiuses ratio as follows:

$$i_n = \frac{\omega_g}{\omega_n}, \tag{7}$$

$$i_r = \frac{r_n}{r_g}. \tag{8}$$

Using the expressions (7) and (8) in the equation (6) the belt slip can be finally defined as:

$$S = \frac{i_n - i_r}{i_n}. \tag{9}$$

On the other hand the belt support layer made of a fabric cord prevents excessive belt extension. The experimental research presented in paper [1] shows that the currently utilized CVT rubber belts feature a very high tensile strength. The longitudinal stiffness of the belt reaches a value about 8kN/mm. For the transmission which is the subject of the present research, where the transmitted torque does not exceed 10Nm,it can be assumed that the speed of the belt in a single circuit does not change. Furthermore, widely employed in the literature models of the V-shaped rubber belt transmission [5] indicate the slip lack on the driving pulley. This belt behaviour is explained by the occurrence of the self-locking phenomenon. As a result, the winding radius and the belt tension on input pulley does not change. Accordingly, the total slip may be determined based on the difference in belt speed and the circumferential velocity of the driven wheel in the radius of the belt:

$$S' = \frac{v_l - \omega_n r_n}{v_l}, \tag{10}$$

where v_l is he measured belt speed.
As part of the examination the attempt of the non-contact belt speed measure was conducted by the Correvit L head. The real results were obtained only for the transmission ratio close to 1. The effective measurement in the whole range of the ratios was not possible because of the change in the belt slope relative to the head and occurred belt vibrations. In result the determination of the slip by equation (10) was not feasible for all the range.
Non-contact measurement of the winding radius does not give information about the location of the neutral axis of the belt. Due to the relatively small winding radii (45 mm) the consideration of this amendment is essential to determine the actual slip. In view of the above assumptions, the value of the distance between the back of the belt and the neutral axis can be determined to obtain no difference between the measured speed of the belt and the circumferential speed of the driving pulley on the diameter of the belt. For the analysed powertrain an average value of this distance, specified for the proper operation of the Correvit head, is about 3mm.
Ultimately the belt slip determined on the basis the equation (9) arises from the difference in the transmission ratios specified by the angular velocity and the winding radiuses taking into account the neutral axis of the belt.

3.2 Analysis of the research results
The samples test results for the two methods of transmission load are presented below. Using them the influence of the load torque (M_n) and the transmission ratio (i_r) on the belt slip is analysed.
The first case concerns the increasing load at a constant, full throttle. As it is shown infigure4 and 5 for the unload transmission working with minimal ratio the belt slip is about 9%. After more or less 5s.the engine speed decreasing is initiated due to the increase of load. In the next second the transmission ratio remains unchanged what is accompanied by temporary slip value peak to approx. 13%. The reason for this phenomenon is inertia of the centrifugal regulator. Instantaneous reduction of the engine angular speed entails the reduction of the driving pulley axial force. Increasing of the axial force and the traction ability appears as an effect of the change of the ramp area where the rollers act. This is associated with the transmission ratio growth. The load increase accompanied by the ratio shift leads to the belt slip rise.
Figure6 shows the changes of the slip against the driven pulley torque in the range of 9th to 21th s., when the transmission ratio is maintained at the constant value 2,7.This relationship can be approximated by a second-order polynomial with the coefficient of determination about 0,58. The slip value increases with the load torque reaching the maximum of 18% at the torque about 6Nm. This tendency arises from increasing of the active wrap angle on the driven pulley. The greater load implies the slip reduction, which may be associated with the increased axial force of the torque regulator.
The second case deals with a rapid throttle opening at a fixed load. As it is seeninfigure7 and 8 for the loaded transmission working with the maximum ratio the belt slip is about 15%. Decreasing of the ratio due to the engine speed grow then tails reduction of the slip to the value about 12%. Figure9 shows the changes of the slip against the transmission ratio in the range of 14th to 30th s., when the load torque is maintained at the constant value about 1Nm. As it was before this dependence is approximated by a second-order polynomial with the coefficient of determination about 0,64. As it can be seen the ratio increasing to the value about 1,6 causes the slip reduction to the minimum value 10%. Further the ratio growth results in the slip increasing. The maximum value in this case is 15% for the ratio 2,5. The slip growth associated with reduction of the driving pulley wrap angle (increasing ratio) follows from additional reduction of the wrap angle on account of the belt flexural rigidity. It is astonishing in this case that the minimum slip corresponds with the ratio about 1,5. The slip increase with the smaller ratios can be explained by reducing of the driven pulley wrap angle due to the belt flexural rigidity.
The experiment results clearly show a substantial influence of the transmission ratio and the transferred torque to the power loss caused by the belt slip. Thus, it confirms the test results on the

The constant full throttle and increasing load

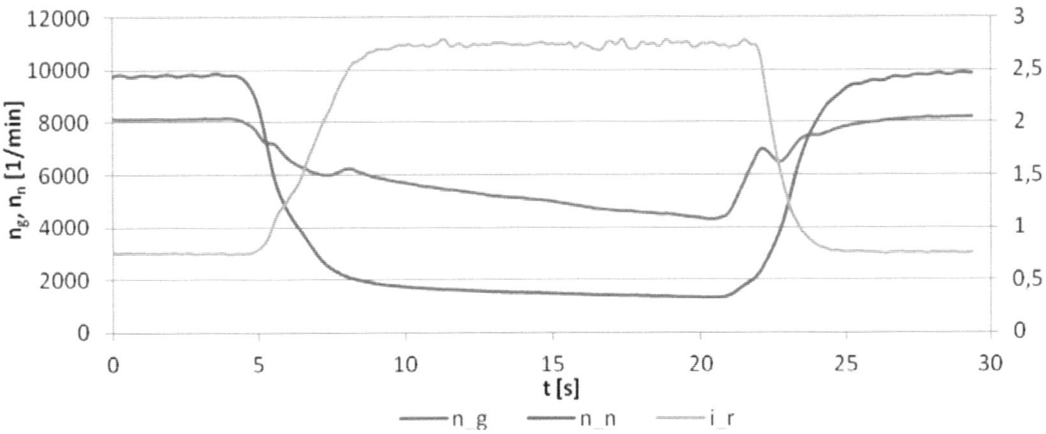

Figure 4. Change of the pulley angular speed (n) and transmission ratio (i_r) against time; subscript(g) – driving pulley, subscript (n) – driven pulley.

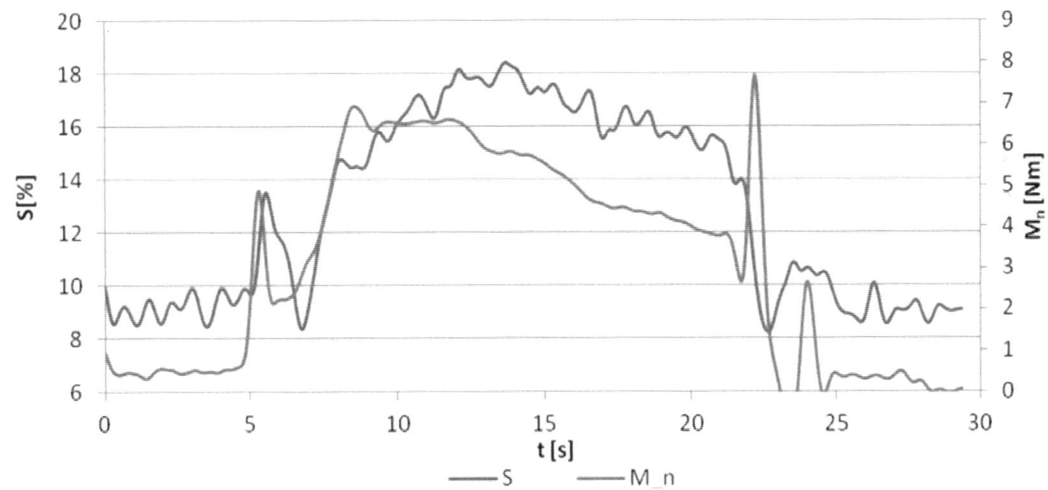

Figure 5. Change of the belt slip and driven pulley torque against time.

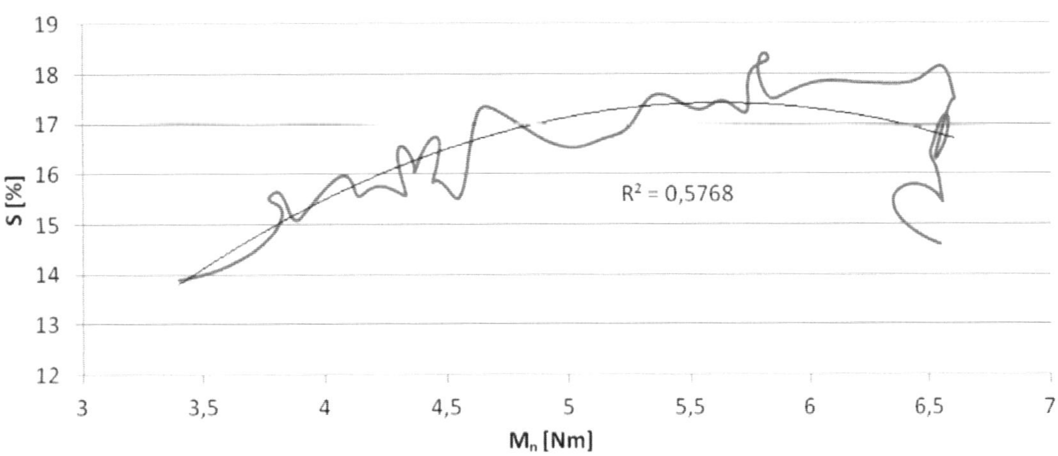

Figure 6. Belt slip against driven pulley torque for i_r =2,7 (range 9th-21th s.).

The constant load and rapid throttle opening

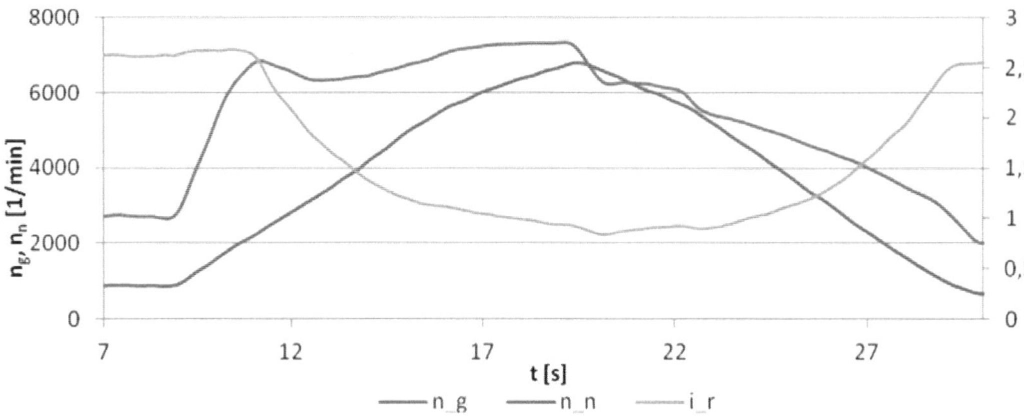

Figure 7. Change of the pulley angular speed (n) and transmission ratio (i_r) against time; subscript(g) – driving pulley, subscript(n) – driven pulley.

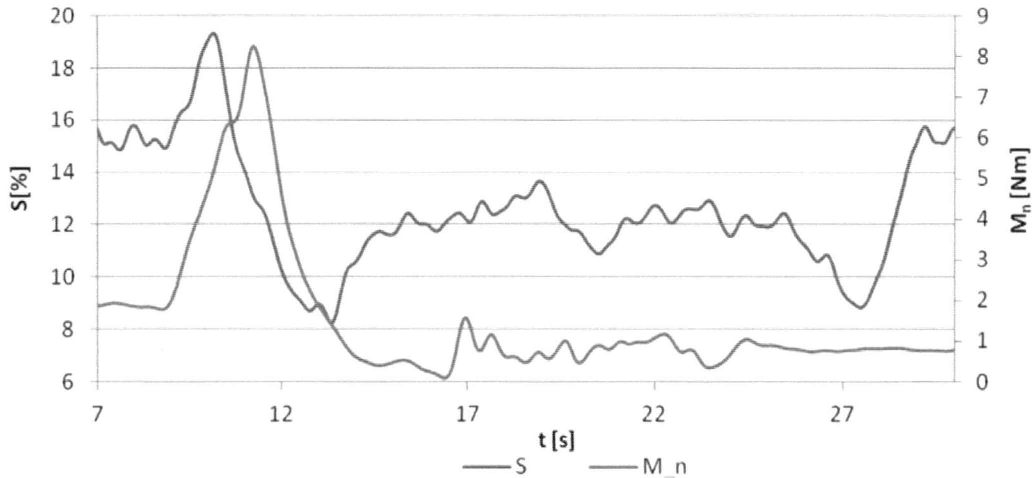

Figure 8. Change of the belt slip and driven pulley torque against time.

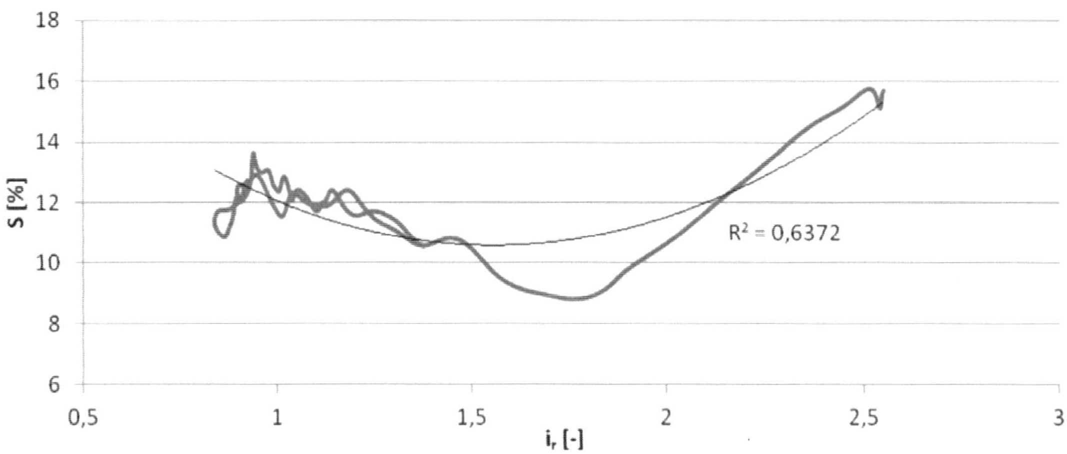

Figure 9. Belt slip against transmission ratio for $M_n \approx$ 1Nm (range 14th-30th s.).

influence of the load on the speed losses presented in [1], except for the fact that obtained relationship is nonlinear. The power loss resulting from the belt slip doesn't exceed 200W. As part of further research, it seems necessary to measure the belt temperature and to present dependences shown for a wider range of transmission ratios and loads. In each examined case, the slip is between 8% and 18%. Consequently it is a key factor in the efficiency of the CVT. That is why it does not confirm the marginal significance of the slip to the transmitted power, as Chen [2] suggested. Difficulties associated with non-contact measurement of the belt speed allow to determine the slip by the equation (10) only for the transmission ratio close to 1. Within these ranges the slip obtained by equations (9) and (10) reaches a similar value, which confirms the assumed position of the belt neutral axis.

4. Conclusions

1. The belt slip determined in the examined range of parameters reaches values from 8 to 18%, which demonstrates that the speed losses are the main or one of the main factors of the power losses.
2. The momentary belt slip value is the result of the momentary transmission conditions such as the torque and the transmission ratio. The slip appeared to be nonlinearly related to the load torque at the constant ratio. A similar trend takes place for the variable transmission ratio at the fixed torque.
3. Defining the belt neutral axis is significant for the accurate slip determination. Omitting this factor leads to the slip value overestimation.

References

[1] Bertini L, Carmignani L and Frendo F 2014Analytical model for the power losses in rubber V-belt continuously variable transmission (CVT)*Mechanism and Machine Theory***78** pp289-306

[2] Chen T F, Lee D and Wand Sung C K 1998An experimental Study on Transmission Efficiency of a Rubber V-Belt CVT*Mech. Mach. Theory*vol 33 No 4pp 351-63

[3] Firbank T C 1970 Mechanics of the belt drive *Int. J. Mech. Sci.* **12**pp 1053-63

[4] Gerbert G 1996 Belt slip-a unified approach *J. Mech. Des***118**pp 432-38

[5] Kim K and Kim H 1989 Axial Forces of a V-Belt CVT *KSME Journal* vol 3 No 1 pp 56-61

[6] Krzyształowicz W and Zintel Z 1958 *Napęd pasowy klinowy. Obliczanie, konstrukcja i zastosowanie* PWT Warszawa

Analysis of an employment of a gear ratio rate in CVT control system

WGrzegożek[1] and M Szczepka[2]

[1]Cracow University of Technology 31-864 Kraków al. Jana Pawła II 37
[2] Cracow University of Technology 31-864 Kraków al. Jana Pawła II 37

E-mail: witek@mech.pk.edu.pl

Abstract. Continuously variable transmissions(CVTs) potentially ensure the selection of such a gear ratio that scooter fuel consumption can reach minimum value. Traditionally these CVT gearboxes are mechanically controlled, causing a gear ratio to be an engine revs function. This solution does not ensure optimum gear ratio. In this paper the solution for fuel optimal control problem is presented. The results obtained during brake stand research of scooter powertrains show the significant values of brake specific fuel consumption for the velocity that is maximum for a scooter according to highway code. With the introduction of CVT gearbox in which the selection of gear ratio can be controlled according to the worked out strategy the solution for fuel consumption problem is possible. Electromechanical actuators ensure the selection of a gear ratio independently of engine revs. Such type of construction solution makes working out the suitable control strategy that ensures decreasing of scooter fuel consumption possible. Presented strategies do not use precise optimization techniques. The CVT efficiency has a strong influence on transient operation. In the paper the control strategy owing to which fuel consumption decreases by over 40% is presented. The strategy was worked out on the basis of fuel consumption map for a defined scooter exploitation model. The possibilities of realization of the worked out strategy were tested on the brake test stand.

1. Introduction

Continuously variable transmission (CVT) is more and more often used in automotive application. Its large transmission ratio coverage enables the engine to operate at more economic operating points. For every power level in internal combustion engine, there is one speed torque combination which achieves optimal fuel efficiency. Using continuously variable range of transmission ratio a line connecting these operating points can be followed for high drive unit efficiency. In spite of this, the use of CVTs in the automobile industry has remained marginal. Two-wheeler and snowmobile are major sections of automotive industry using CVTs. In this application a rubber dry belt is commonly used. Dry belts are usually used because a high friction coefficient is created between a belt and pulleys so that clamping force can be much smaller than it is in lubricated variants. Unfortunately, the problem which appears in non-lubricated belt pulley contact points results from the lack of cooling of this contact, which causes limits of the torque capacity of this type of variator. But these types of CVTs can be small and light and ideal for application in small motorbikes and scooters only. Moreover, hydraulic control of the CVTs axial thrust that is most often used

IOP Conf. Series: Materials Science and Engineering **148** (2016) 012007 doi:10.1088/1757-899X/148/1/012007

is not indispensable and that is why hydraulic losses are eliminated. Despite these advantages CVTs are perceived as the most inefficient transmission system. It is the result of the application of mechanically controlled variators. Centrifugal rollers and torque cams are widely used in rubber belt CVTs as the mechanical actuators [2]. The performance of CVT depends on the characteristics of these actuators. Using mechanical actuators it is impossible to shift the transmission ratio in such a way as to make speed-torque combination point remain on an engine optimal operating line [1].

2. Optimal Operation Line

Optimal operation line (OOL) tracking is the most fuel economical way to operate the drive line. The OOL can be calculated from engine map by minimizing the fuel consumption for a set of output power values. A large number of different control strategies exist. The most popular approaches are *speed envelope*, *single track* and *off the beaten track.* A simple approach for transient ratio shift control is to maintain the engine on the "quasi-static" pick efficiency curve which corresponds to the assumption that any arbitrary speed torque combination can be realized instantaneously. This approach is called *single track strategy* (fig. 1*)*. This strategy does not use precise optimization techniques but relies on heuristic arguments. Since the vehicle performance of the single track strategy is limited Pfiffner and all [6] have proposed another approach called *off the beaten track* strategy. Driver can select two modes: the economy and the performance modes. These two trajectories represent different driver selectable modes and they are presented in figure 2.

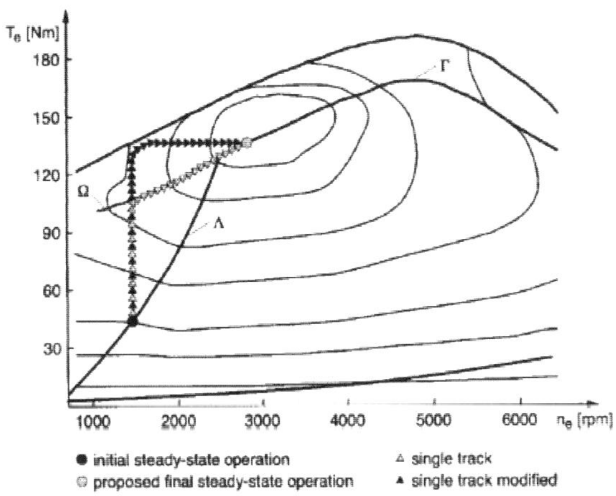

Figure 1. System trajectories of the single track and single track modified strategy. The maximum gear ratio stationary driving resistance curve Λ, the peak efficiency curve for quasi static operation Ω and the peak efficiency curve for stationary operation Γ [6].

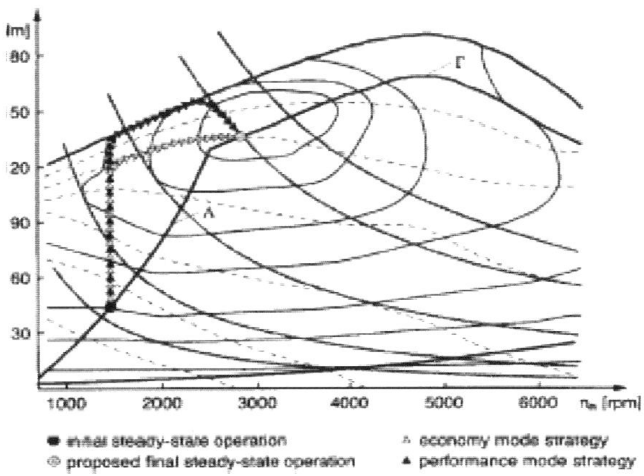

Figure 2. Engine map trajectories of the two modes *off beaten track* strategy on an accelerator step [6].

In many cases the actual engine torque is not controlled along the "quasi static" peak efficiency curve Ω but is brought directly to the final steady-state operation point. Due to this the dynamics of the vehicle increases and so is the fuel consumption.

Presented strategies do not use precise optimization techniques. The CVT efficiency has a strong influence on transient operation. Unfortunately, there are a lot of times when the demand power level is higher than engine's capability. Then engine operates at the maximum torque line until the demand power is reached. When hard acceleration is commanded, engine speed flares and corresponding torque is consumed and vehicle acceleration decreases. Due to this CVT vehicles have poor drivability. Drivability [8] is a subjective assessment and hard to measure. Figure 3 shows four objective parameters which have obvious relations with CVT drivability assessment. In figure 3 t_0 is the moment when throttle was opened. The acceleration decreased at first reached its minimum value a_{min} and began to rise. The four objective parameters related to drivability assessment were: acceleration delay $\Delta t_1=t_1-t_0$, duration of acceleration $\Delta t_2=t_2-t_0$, acceleration reduction $\Delta G=a_0-a_{min}$ and peak value of acceleration G_{max}. If the values of Δt_1, Δt_2, ΔG are very small and G_{max} is high the assessment of drivability is good. These parameters are strongly related to gear ratio rate.

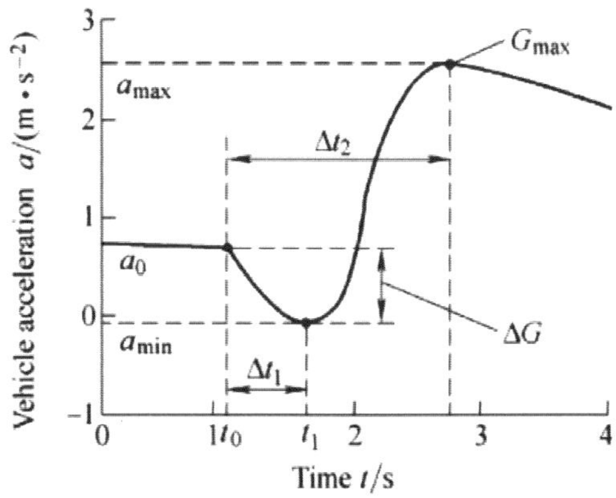

Figure 3. Objective parameters for drivability evaluation [8].

The CVT vehicles drivability is the result of gear ratio rate. It is impossible to improve the drivability under conventional control because gear shift rate depends directly on throttle position and outer resistance torque. The electromechanical actuator should be employed. It is necessary to have the CVT vehicle drive line characteristic to define the control algorithm for such actuator.

3. Test stand

Figure 4 shows the test stand. The test stand consists of: engine with CVT gearbox, flywheel, water brake, engine equipment and apparatus for measuring the following variables; engine torque, angular velocity of engine, angular velocity of inner and outer shafts of CVT and fuel consumption [4]. The flywheel is used to simulate the conditions during start of vehicle.

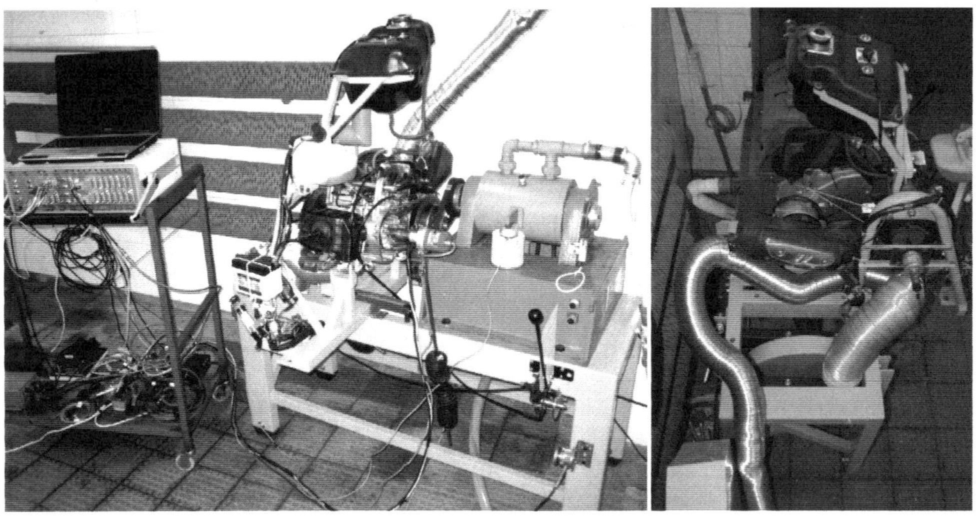

Figure 4. Test stand.

In popular motorbikes and scooters with two stroke engine very often the two minimum of brake specific fuel consumption (BSFC) were appeared. The same situation was noticed for tested engine. Due to this engine it is very difficult to approximate OOL line by using a simple function of engine rpm. From the OOL the optimal engine speed for a given throttle position can be obtained. The real operation line is beyond engine high efficiency area. The real operation line (ROL) is also presented in figure 5 (spotted line). A big difference between these lines can be observed. Maximum scooter velocity is reached by 8500 rev/min of the engine. The BSFC value at this point is very high. This big difference is the consequence of a mechanical governor employed. To ensure the tracking of a OOL the special electromechanical actuator system has to be employed. However, the tracking of OOL influences the scooter drivability due to the lack of its dynamic response. Since drivability is also important for ratio control design, the special strategy should be worked out for a good combination of fuel economy and drivability.

Unfortunately, the change of a gear ratio in order to obtain the scooter maximum velocity at lower rpm can cause noticeable diminishing of scooter performance. Moreover, there is a possibility to increase the range of typical scooter CVT by about 10% only. However, this change is still too small to obtain noticeable improvement of engine –CVT drive system efficiency. The required power of scooter drive system to reach its limited maximum velocity is 30-40% lower than the maximum power of

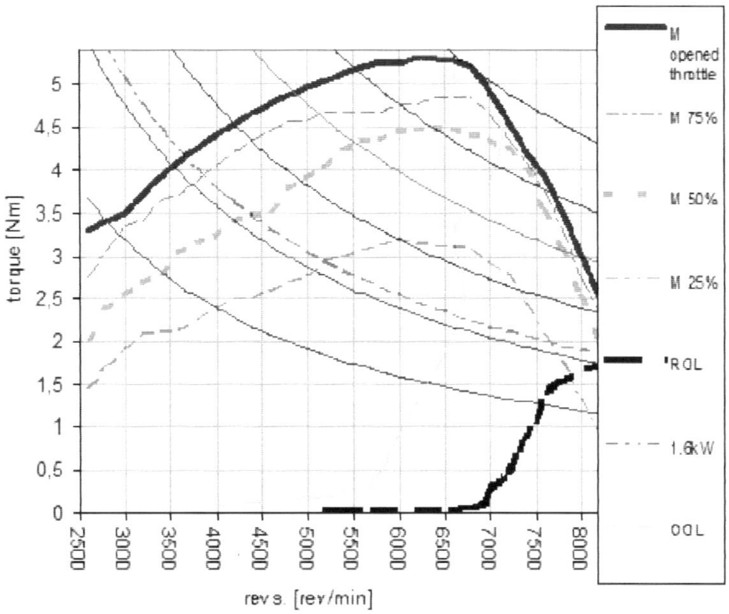

Figure 5. Optimal Operation Line [4].

the employed engine. It allows the designers of a scooter drive system to take into account the possibilities of application of the partly opened throttle for maximum scooter velocity. Introducing electromechanical actuators can improve the drive system efficiency because gear ratio is no longer a function of engine revs. To obtain noticeable decrease of fuel consumption value the change of gear ratio is necessary. The ratio control strategy of CVT should ensure slight lowering of scooter performance together with simultaneous significant lower fuel consumption. The way of the exploitation of such a vehicle is different from the exploitation of other vehicles. These vehicles move usually with wide opened throttle (WOT) from small velocity to maximum velocity. This way of scooter motion requires taking it into consideration when the ratio control strategy of CVT is evaluated. In this case to diminish fuel consumption, the engine revs corresponding to maximum scooter velocity should be lowered to 6800 revs/min.

In order to obtain this value of engine revs the gear ratio should be lowered by about 20%. The value of BSFC is the smallest one at these revs. As the power of an engine at 6800 revs/min exceeds the power calculated from scooter motion resistance the position of engine throttle has to be diminished. In other case the scooter velocity will exceed permitted velocity. The engine work with partly opened throttle will be sufficient. Partly opened throttle causes further lowering of fuel consumption. Employing an electric throttle is essential here.

In order to obtain the value of gear ratio rate under conventional control the test with rapid opened throttle was carried out. The results are shown in figures 6 and 7. The maximum value of gear ratio rate in both direction under conventional control is about 0,7 (1/s).

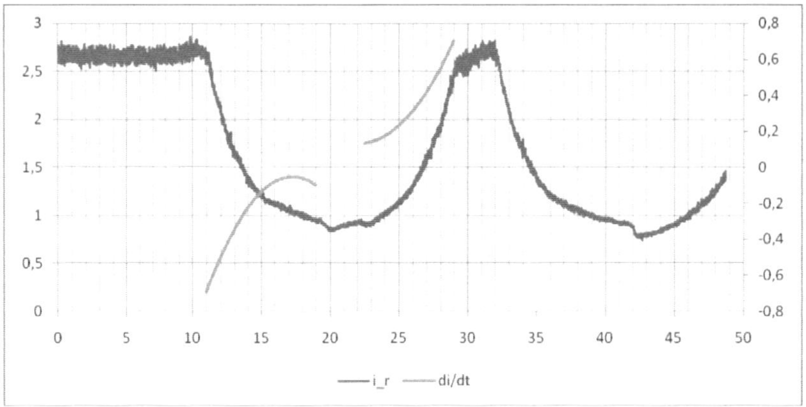

Figure 6. Gear ratio (red line) and gear ratio rate (green line).

For tested powertrain measured maximum gear ratio rate causes the diminishing of engine revs and related with engine torque (fig.7) in small quantity.

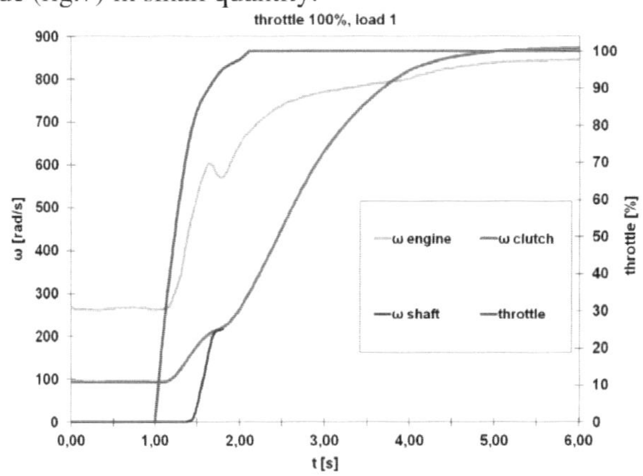

Figure 7. Throttle position (red line), revs of engine (green line), revs of clutch (rose line) revs of outer CVT shaft (blue line) during rapid acceleration.

In accordance with measured results the drivability for tested CVT vehicle is acceptable. In this case the main aim of introducing the electromechanical actuator is to diminishing of a fuel consumption. The test results can be used for limitation of gear ratio rate of electromechanical actuator.

4. System modelling

To improve the fuel consumption of considered vehicle the CVT drive line dynamics must be analysed [5,7]. The drive line is modelled only in its longitudinal behaviour and no drive train elasticities are taken into account. The efficiency of drive line is assumed as equal 1. Figure 8 shows a sketch of this system.

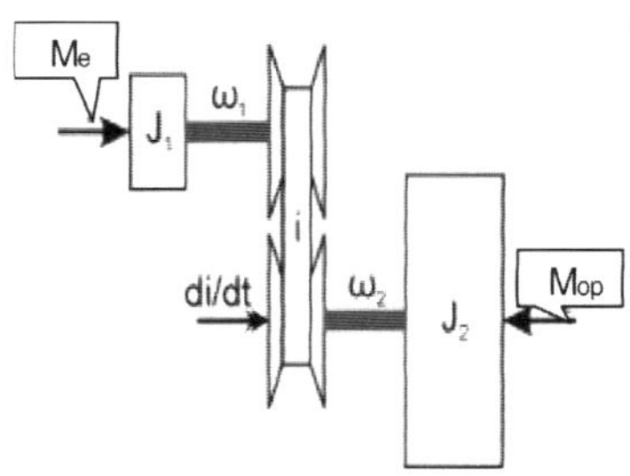

Figure 8. Drive line structure.

The gear ratio of the CVT is defined by:

$$i(t) = \omega_2 / \omega_1 \tag{1}$$

The equations of drive trains motion are as follows:

$$M_e - M_1 = J_1 \frac{d\omega_1}{dt} \tag{2}$$

$$M_2 = i(t) \cdot M_1 \tag{3}$$

$$M_2 = J_2 \frac{d\omega_2}{dt} + M_{op} \tag{4}$$

The resulting drive trains dynamic equation is as follows:

$$\frac{d\omega_1}{dt} = \frac{M_e - M_{op} \cdot i(t) - J_2 \cdot \omega_1 \cdot i(t) \cdot \frac{di}{dt}}{J_1 + J_2 \cdot i^2(t)} \tag{5}$$

where:

J_1 – inertia moment of an engine,
J_2 – equivalent inertia moment of a scooter,
M_e – engine torque,
M_1 – input moment to CVT,
M_2 – output moment out CVT,
M_{op} – resistance moment of motion,
M_R -- resistance moment related to gear ratio rate.

It can be seen from above equation that there is the resistance torque related to gear ratio rate. The torque M_R is the follows:

$$M_R = J_2 \cdot \omega_1 \cdot i(t) \cdot \frac{di}{dt} \tag{6}$$

The principle of control algorithm is that M_R is equal, during gear shifting, to M_e -M_{op} $i(t)$ to obtain the constant engine revs. When engine operates in selected point of its characteristic the decrease of fuel consumption can be obtained.

When in motion, city scooters equipped with small power engines have usually wide opened throttles. The control strategy should take into account the character of the drive which in fact is composed of extreme acceleration, drive with constant speed, breaking and reaccelerating until the maximum velocity is reached. This cycle is repeated several times; in each case the time in which the vehicle velocity is constant is different. Two values that define the run of the cycle are: the angle of opened throttle and the real gear ratio of CVT.

The start-on is usually independent of steering systems. The centrifugal clutch is used for the start on. Subsequently, the vehicle is accelerated until the velocity corresponding to engine revs slightly higher than maximum torque is reached. Then the change of a ratio in CVT begins until the minimum ratio is reached. The balance of power on a scooter wheel with power of motion resistance defines vehicle velocity. This balance appears very often in a small scooter when the revs of the engine are much higher than revs of the engine for maximum power. These revs which are much higher are caused by the necessity of achieving the maximum vehicle velocity. The difference between engine revs of maximum torque and maximum power for typical engines which is very small is the reason why engine revs should be much higher than revs of engine maximum power. It is necessary for gaining a wider range of velocity changes. Unfortunately, it results in the increase of vehicle fuel consumption. Moreover, the value of resistance power is on average 40% smaller than the maximum value of power of the engines When an actuator is mechanically steered the ratio value is the function of engine revs and thus its value is defined by a regulator structure. When electromechanical actuator is used simultaneously with electric throttle the characteristics of the engine can be better employed and the fuel consumption can be lowered. The use of electromechanical actuator should ensure the same or at least similar performance as in the traditional version. As lowering fuel consumption requires smaller revs of an engine it is indispensable to lower total gear ratio of a vehicle. In order not to change the drive power too much the vehicle acceleration should be done with possible high value of engine torque. When the rate of the gear ratio change is controlled the whole process of acceleration can be performed with constant engine revs as it is shown in revs of engine maximum torque. In such cases, even when total gear ratio is lowered, it should ensure good performance.

The control strategy is as follows: scooter accelerates with CVT low ratio till the engine reach the revs slightly higher than revs of maximum torque. Then the gear ratio starts to change with rate ensures the constant engine revs. When the maximum velocity is reached, the change of a gear ratio will be carried out simultaneously with diminishing the throttle angle to make engine power balance with the resistance power of motion. This control method was implemented in the model described above. The control run simulation is presented in figure 9.

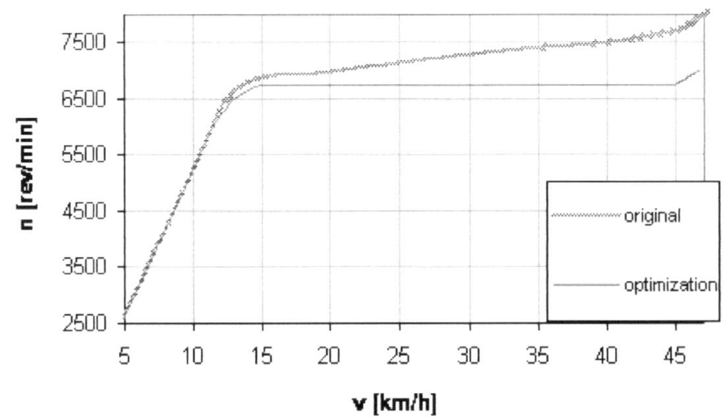

Figure 9. Control strategy.

The results of applying this control method to the mathematical model of drive train are presented in figures10 and11. The characteristics shown in figure 11 present that when the velocity is 45km/h, specific fuel consumption decrease using this control strategy is about 20% when compared with traditional method.

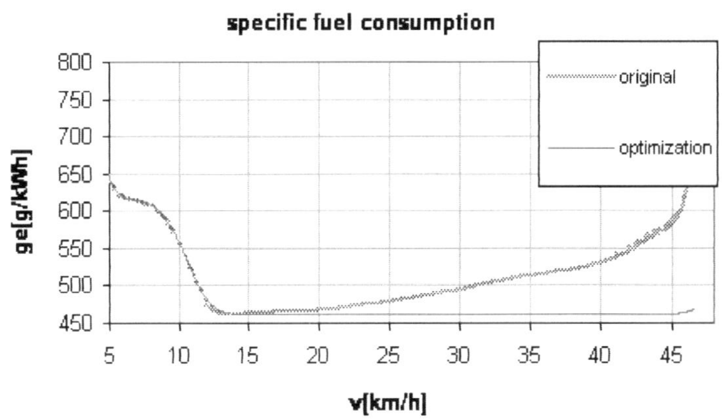

Figure 10. Specific fuel consumption.

As it was stated before the velocity of these types of vehicles is close to 45km/h, such lowering of specific fuel consumption may be also considered as fuel consumption per hour.

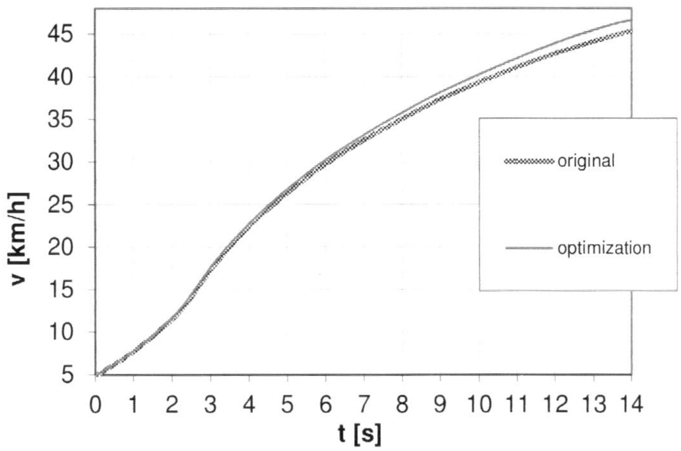

Figure 11. Performance of scooter.

As it can be seen in figure11 the performance of vehicles is almost the same.
Small scooters, motorcycles a and other vehicles of that type are driven in town in a specific way where in which extreme acceleration and braking are dominating. This kind of driving was described as Indian Drive Cycle [3] for Asian (fig. 12) countries, where the number of scooters and motorcycles exceeds 42 million.

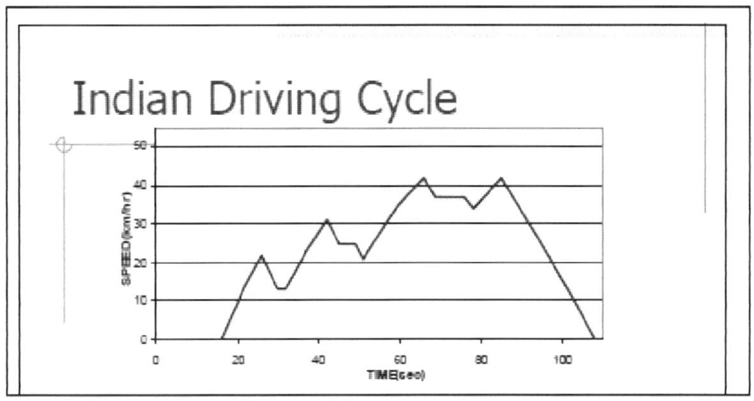

Figure 12. Indian Driving Cycle [3].

5. Theoretical analysis of fuel consumption in accordance with IDC

The mathematical model, which was worked out, was used for the analysis of fuel consumption of the scooter equipped with CVT in accordance with the suggested research procedure. The course of run of the engine torque as well as fuel consumption per hour in revs function were approximated by polynomial. Moreover, additional function courses were created to present the previously described parameters in the function of opening angle of a throttle. In figure13 the run of fuel consumption per hour for different opening angle of a throttle was presented and compared to the runs measured in test stand. These functions were used in motion simulation for the scooter equipped with original as well as optimum system of gearbox control.

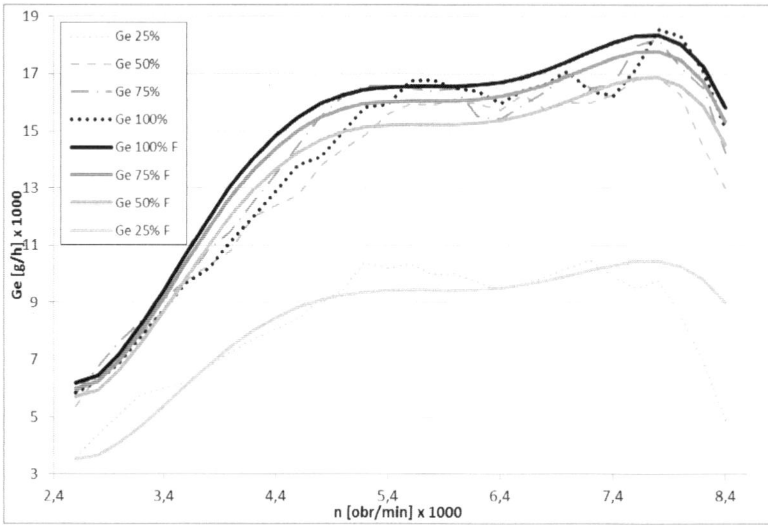

Figure 13.Test stand fuel consumption and theirs function approximations

In optimization case, the run of "u" gear ratio rate is controlled by the relevant algorithm that ensures proper engine functioning with the highest efficiency, whereas in original characteristics the gear ratio rate is changed in such a way as to present in the best way the run of the analyzed cycle. As it was in case of real vehicle research according to the road cycle, the opening angle of a throttle was manually performed.

In case of the scooter motion according to IDC cycle fuel consumption is:

a. 2,376l/100km for the scooter equipped with original control unit,

b. 2,056l/100kmfor the scooter equipped with optimum control unit.

In IDC cycle 15%reduction of fuel consumption was acquired.

Moreover, the algorithm controlling optimized gear ratio rate makes the gear ratio rate of maximum rate not larger then in original solution possible.

Applying optimum control unit gives significant savings during the scooter motion with maximum velocity (45km/h) if is compared this with fuel consumption of scooter with conventional control unit. The scooter acceleration is a little higher then acceleration with conventional unit

When a scooter is driven with maximum acceleration and velocity (range of test distance1km) its fuel consumption is:

a. 4,5 l/100km for a scooter equipped with original control unit,

b. 2,32l/100km for a scooter equipped with optimized control unit.

Fuel consumption decreases in this cycle by 50%.

When a scooter is driven with constant maximum velocity (1km) its fuel consumption is:

a.4,17l/100km for a scooter equipped with original control unit,

b.1,73l/100km for a scooter equipped with optimized control unit.

In these cases fuel consumption decrease exceeds 100%.

In the above mentioned runs the minimum velocity is 5km/h. Its due to two facts:

- control unit starts to gear change at the vehicle velocity that exceeds 10km/h,
- centrifugal clutch (which is not tested during this research) influences the vehicle driven with less than 5km/h. Applying of optimum control unit gives significant savings during the scooter motion with maximum velocity (45km/h) if compared this with fuel consumption of scooter with conventional control unit. Moreover, the scooter acceleration is a little higher then acceleration with conventional unit.

6. Conclusions

1. The gear ratio rate is one of the parameters essential for vehicle performance. So far main attention was paid to the influence of gear ratio rate on vehicle acceleration values. The use of the gear ratio rate for defining engine revs during vehicle acceleration secures optimum engine functioning considering its fuel consumption.

2. The gear ratio rate value for the realized acceleration cycle did not differ much from the gear ratio range for conventional system. In this way similar value of a belt slip and its efficiency was ensured.
3. Fuel consumption in the analyzed cycle decreases by 15% when compared to conventional control. It should be considered a great achievement.
4. The application of electromechanical control unit making gear changing as well as gear ratio rate possible allows to drive the vehicle with revs corresponding to optimum fuel consumption.

References

[1] Bonsen B, Steinbuch M and Veenhuizen P A 2005 CVT ratio control strategy optimization *Int. J. of Vehicle Design*
[2] Chen T F and others 2000 Design Considerations for Improving Transmission Efficiency of the Rubber V-belt CVT *Int. J. Vehicle Design* vol 24 No 4
[3] Chaudhari M K 2004 Motor Cycle Emission Control in India Asian Vehicle Emission Control Conference
[4] Grzegożek W and Szczepka M 2008 Analiza możliwości zmniejszenia zużycia paliwa przez pojazd jednośladowy z zastosowaniem elektromechanicznie sterowanej przekładni CVT *Czasopismo Techniczne z.6-M/2008* Wydawnictwo Politechniki Krakowskiej zeszyt 10 pp 155-167
[5] Grzegożek W and Szczepka M 2012 An Attempt of Fuel-Optimal Control of Scooter CVT Powertrains *Czasopismo Techniczne Mechanika* R.109 z.5-M pp 101-107
[6] Pfiffner R Guzzela L and Onder C H 2003 Fuel – optimal control of CVT powertrains *Control Engineering Practice* **11** pp 329-336
[7] Jantos J 2001 Control of the Transmission Ratio Derivative in Passenger Car Powertrain with CVT *SAE Technical Paper* 2001-01-1159
[8] Sun D Luo Y 2012 Integrated Control Strategy for CVT Powertrains with Consideration of Vehicle Drivability *Chinese Journal of Mechanical Engineering* vol 25 No 3

IOP Conf. Series: Materials Science and Engineering **148** (2016) 012008 doi:10.1088/1757-899X/148/1/012008

Torque vectoring for improving stability of small electric vehicles

W Grzegożek[1] and K Weigel-Milleret[1]

[1] Institute of motor Vehicles and Internal Combustion Engines, Cracow University of Technology, Al. Jana Pawla II 37, 31-864 Krakow, Poland

E-mail: witek@mech.pk.edu.pl, krzysztof.weigel-milleret@mech.pk.edu.pl

Abstract. The electric vehicles solutions based on the individually controlled electric motors propel a single wheel allow to improve the dynamic properties of the vehicle by varying the distribution of the driving torque. Most of the literature refer to the vehicles with a track typical for passenger cars. This paper examines whether the narrow vehicle (with a very small track) torque vectoring bring a noticeable change of the understeer characteristics and whether torque vectoring is possible to use in securing a narrow vehicle from roll over (roll mitigation). The paper contains road tests of the steering characteristics (steady state understeer characteristic quasi-static acceleration with a fixed steering wheel (δH = const) and on the constant radius track (R = const)) of the narrow vehicle. The vehicle understeer characteristic as a function of a power distribution is presented.

1. Introduction

Modern microcars often use electric motors to drive. That design reduces costs of driving and the negative impact of the vehicle on the environment. Common in European cities is setting a special zones with limited or total ban entry of cars equipped with combustion engines (zero emission zone), where electric vehicles can freely drive. Often the legal system is favouring microcars in relation to full size vehicles (often by lower taxes and lower insurance rates). Driver's license category AM, B1 or ID is usually sufficient to drive these vehicles. Design intent of microcars can be summarized in the following points:

1. Small size of the vehicle (length not exceeding 3 m), the body that protect against precipitation
2. Interior space should be enough for one person and a small luggage
3. Electric drive
4. The minimum range of approx. 60 km
5. Approval in L6e category (AM driver's license) or L7e (B1 driver's license)
6. Low price
7. Good maneuverability

A prototype vehicle MIST (acronym created from the first letters expressing Individual Urban Vehicle) was designed and made at the Cracow University of Technology (shown in Figure 1). MIST body is made as a welded space frame with a mounting points of the suspension. This design provides high rigidity in three axes, and high torsional rigidity of the body. Body has bilateral access to the interior and has enough space for the driver and a small luggage. Good visibility to the front and to the sides has been achieved by the large windscreen and thin front posts.

Figure 1. MIST microcar.

The front suspension of the vehicle is doubled pushed arm. The small wheelbase has allowed usage of a dependent suspension of the front wheels. Suspension movements cause the parallel movement of the front wheels. This solution increased the roll stiffness of the vehicle.
The rear suspension is semi-independent twist beam with towed arms. Semi-independent suspension solution for the rear suspension ensures high roll stiffness of the vehicle. Traction batteries container is set between rear suspension arms.
The car uses two brushless AC motors installed in the rear wheels of the vehicle to drive. An active control of the electric drive allows a regulation of the distribution of the driving torque. Hydraulic disc brakes mounted on all wheels of the vehicle.

2. Torque vectoring

The first generation of electric vehicles were build by replacing the internal combustion engine with an electric motor while keeping the same construction of the powertrain. The electric motor were positioned at the main gear of the differential. Current solutions based on the individually controlled electric motors propel a single wheel allow to improve the dynamic properties of the vehicle by varying the distribution of the driving torque (and braking torque) [3].

2.1. Steady-state torque vectoring

In order to measure a vehicle steering the gradient of steering wheel angle related to the lateral acceleration, normalized by steering system transmission ratio has been adopted. It is called the gradient of understeer, which is defined as [7]:

$$Grad_p = \frac{\partial \delta_H}{\partial a_y}\frac{1}{i_s} - \frac{\partial \delta_D}{\partial a_y} \tag{1}$$

where:

δ_H - steering wheel angle,
i_s - steering transmission ratio,
δ_D -comparative, dynamic steering wheel angle,
a_y - lateral acceleration.

$$\delta_D = \frac{L}{R} \tag{2}$$

L- wheelbase,
R- track radius of the center of mass

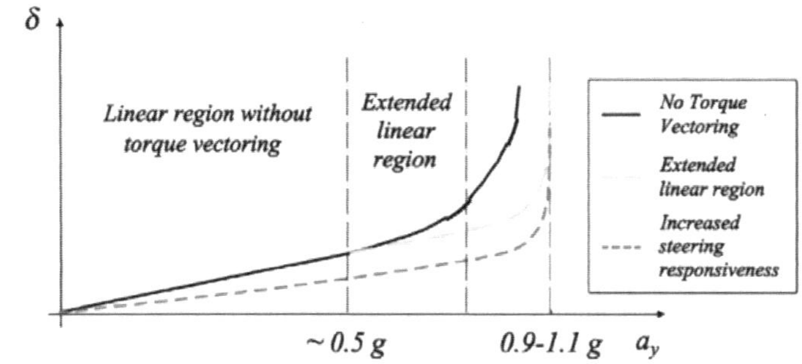

Figure 2. Modifications of the understeer characteristic achievable by torque vectoring [2].

Vehicle steering characteristic can be represented as the steering wheel angle as a function of the vehicle lateral acceleration (Figure 2). The trend of lateral acceleration at understeer vehicles is normally linear up to certain value of latera acceleration (usually approx. 0.5g) and then increases non-linearly to a maximum value. It is the maximum possible lateral acceleration to obtain in a steady state cornering. A scientific literature describes the theoretical and experimental evidence that the torque vectoring can lead to extending the linear response of the vehicle to the steering signal, increasing the maximum lateral acceleration and changing the understeer gradient [1,4,6].

2.2. Transient state torque vectoring

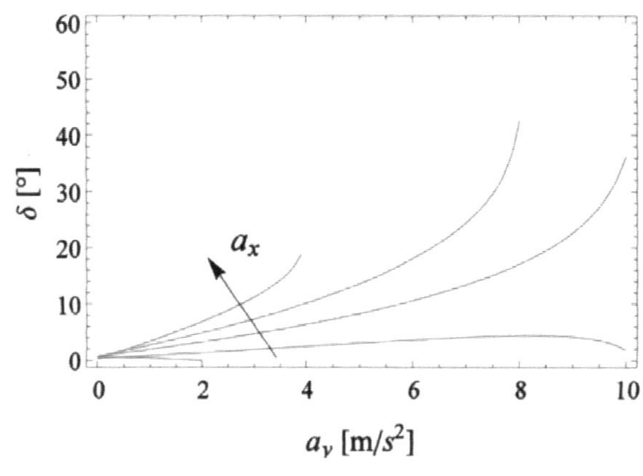

Figure 3. Vehicle understeer characteristic, steering wheel angle as a function of the lateral acceleration.

In the transient state vehicle understeer characteristic is a function of a longitudinal acceleration of the vehicle, which is highlighted in Figure 3. The increasing of the longitudinal acceleration reduces the linear response region of the vehicle to the steering signal. To improve the vehicle performance at the transient state two different torque vectoring strategies can be implemented: i) driving torque proportional to the each wheel normal (vertical) load (see Figure 4.); ii) driving torque distribution allowing achieving the same longitudinal slip ratio on each wheel of the vehicle [6]. Both strategies allow to reduce the understeer gradient difference and expand the linear respond region .

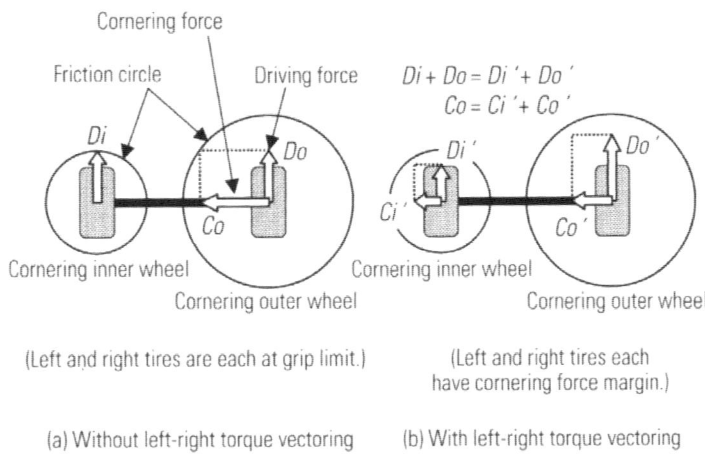

Figure 4. Driving force proportional to the friction force (b); where: **i** subscript - inside motor, **o** subscript - outside motor [5].

Torque vectoring reduces the vehicle power consumption while cornering. Simulations carried out by [2] point to a few percent reduction of the vehicle energy consumption in relation to the vehicle with the same parameters without differentiation of torque. This result shows that the torque vectoring strategies not only increases the vehicle dynamic performance, but also optimizes the usage of the battery energy.

3. Test

Most of the extensive body of literature refer to the vehicles with a track typical for passenger cars. This paper examines whether the narrow vehicle (with a very small track) torque vectoring bring a noticeable change of the understeer characteristics and whether torque vectoring is possible to use in securing a narrow vehicle from roll over (roll mitigation). The tests was conducted for a 0.7m track microcar.

3.1. Performed tests:

- acceleration and top speed test,
- steady state understeer characteristic quasi-static acceleration with a fixed steering wheel (δH = const) and on the constant radius track (R = const).

Figure 5. MIST microcar equipped with measuring set.

V-box equipment was used for measuring the speed and position of the vehicle in a fixed external coordinate system. V-box is a device for measuring the coordinates of a point associated with the vehicle, which calculate the speed, driving distance and the yaw rate, roll rate and a pitch rate. The V-box measurement is based on satellite navigation (GPS and GLONASS). Measure data are recorded on the internal memory card (Compact Flash). Furthermore, the electric motors currents were recorded. MIST equipped with measuring set is shown in Figure 5.

3.2. Steering characteristic

Tests with a constant steering angle to the left have been carried out (Delta H = const). The steering wheel has been blocked at 24° turn. Three different power distribution have been performed: i) power only to the inside wheel motor; ii) power only to the outside wheel motor; iii) constant torque split between the wheels 50%: 50%. Quasi-static acceleration from standstill start to achieve the maximum lateral acceleration or safety limit due to the possibility of rollover. The evaluation of the vehicle cornering performance has been accomplished through the analysis of the trend of the steering angle increment as a function of the lateral acceleration. The steering angle increment is defined as (see Figure 6):

$$\delta_0 - \alpha_1 + \alpha_2 = \frac{L}{R} \tag{3}$$

$$\alpha_2 - \alpha_1 = \frac{L}{R} - \frac{L}{R_0} = \frac{L}{R_0}\left(1 - \frac{R_0}{R}\right) = \delta_0\left(1 - \frac{R_0}{R}\right) \tag{4}$$

$$\Delta\delta = \delta_0\left(1 - \frac{R_0}{R}\right) \tag{5}$$

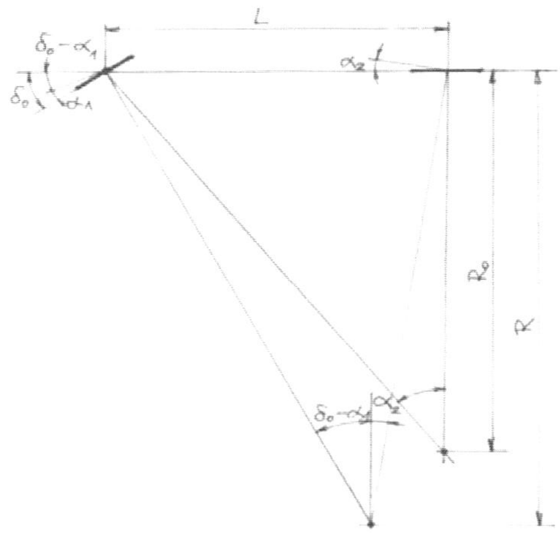

Figure 6. Constant steering wheel angle test [A Kleczkowski].

The constant radius of track R = 17m tests have been performed (turning left). Three different power distribution have been performed: i) power only to the inside wheel motor; ii) power only to the outside wheel motor; iii) constant torque split between the wheels 50%: 50%. Quasi-static acceleration from standstill start to achieve the maximum lateral acceleration or the maximum speed at which the vehicle can be maintained on the track.

4. Test result

4.1. Acceleration and top speed

The top speed of 41 km/h has been reached. This was due to a too short test track (interruption of measurement after reaching the end of the track). It differs only slightly from the top speed calculated on the data obtained from the motors manufacturer. Average acceleration in the range 0 - 25 km / h was 1 m/s^2, and in the range 0 -40 km/h was 0.7 m/s^2 (see Figure 7).

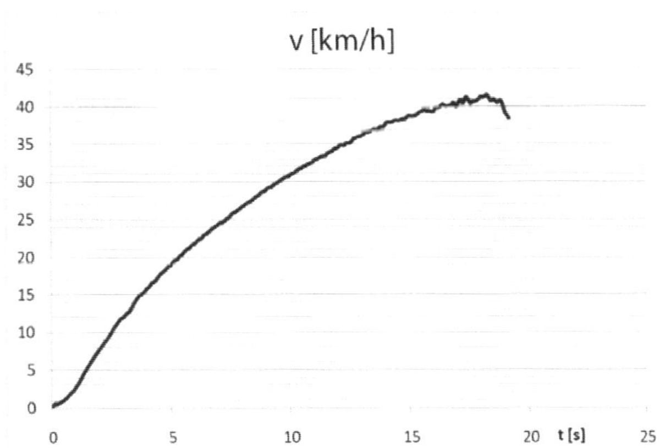

Figure 7. Speed as a function of the time during the top speed test.

4.2. Constant steering wheel angle test

The trend of the steering angle increment is dependent on the torque distribution. Supplying the power to the inside motor causes achieving the smallest linear response region and the largest understeer gradient in the non-linear region (Figure 8). The equal distribution of drive torque causes the whole tested lateral acceleration region $0 < a_y < 4m/s^2$ were almost linear respond region which understeer gradient approximately grad = 0 (Figure 9). Supplying the power to the outside motor causes the oversteer vehicle behaviour (understeer gradient <0) (Figure 10). The power distribution impact on the steering angle increment is shown in Figure 11.

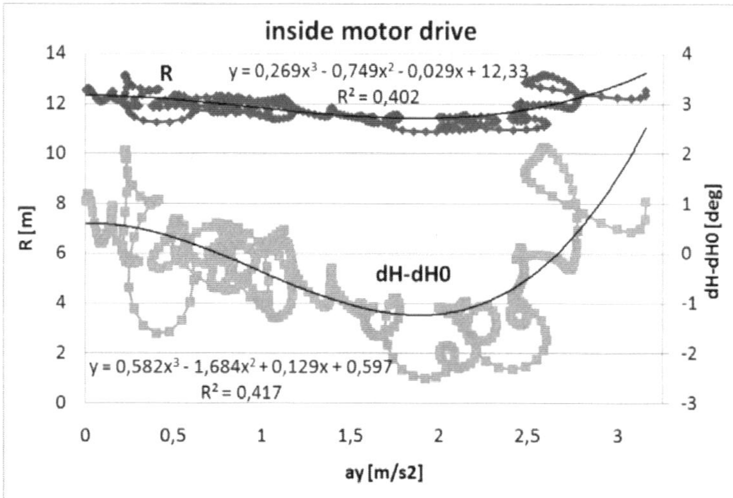

Figure 8. Radius of turn [m](black) and steering angle increment [deg](grey) as a function of lateral acceleration [m/s2] during constant steering wheel angle test - inside motor drive.

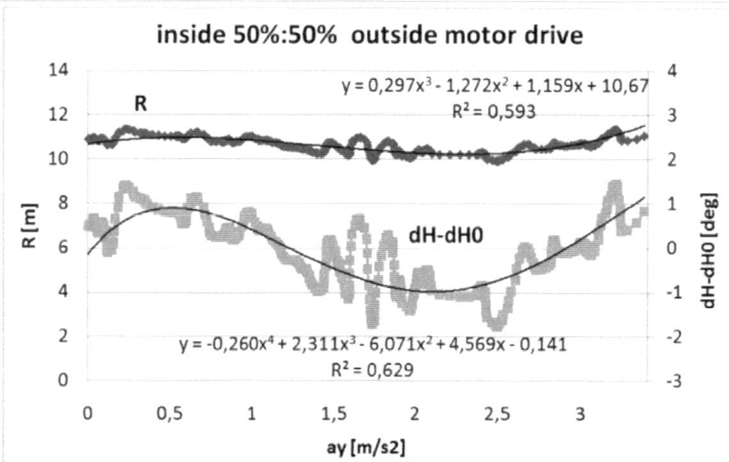

Figure 9. Radius of turn [m](black) and steering angle increment [deg](grey) as a function of lateral acceleration [m/s2] during constant steering wheel angle test - equal power distribution.

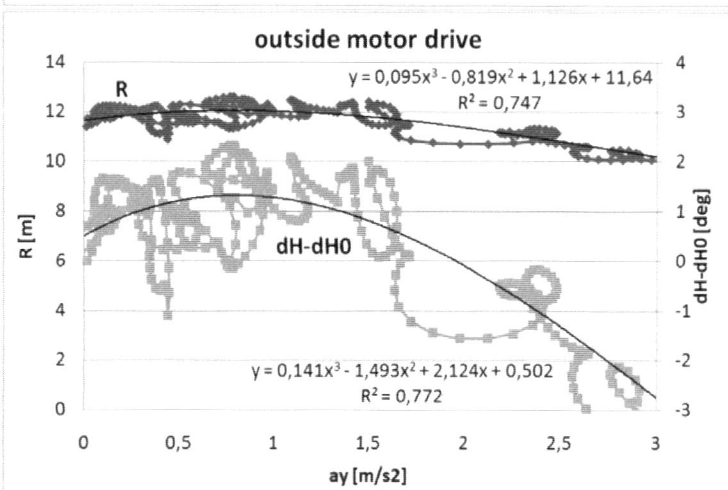

Figure 10. Radius of turn [m](black) and steering angle increment [deg](grey) as a function of lateral acceleration [m/s2] during constant steering wheel angle test - outside motor drive.

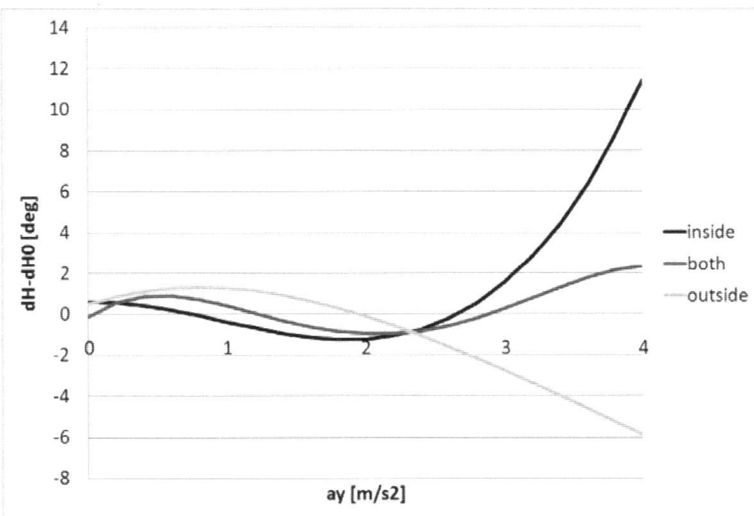

Figure 11. Steering angle increment [deg]: inside motor drive (black), both motor drive (dark grey), outside motor drive (light grey) as a function of lateral acceleration [m/s2] during constant steering wheel angle test.

4.3. Constant radius of track test

The evaluation of the vehicle cornering performance for the constant radius of track has been accomplished through the analysis of the trend of the steering angle as a function of the lateral acceleration. Supplying the power to the inside motor or equal distribution of drive torque causes understeer behaviour of the microcar (see Figure 12 and Figure 13). Supplying the power to the inside motor raises the understeer gradient. Supplying the power to the outside motor causes oversteer behaviour (see Figure 14). The understeer gradient is below 0. During the test increasing the lateral acceleration has caused decreasing the steering wheel angle.

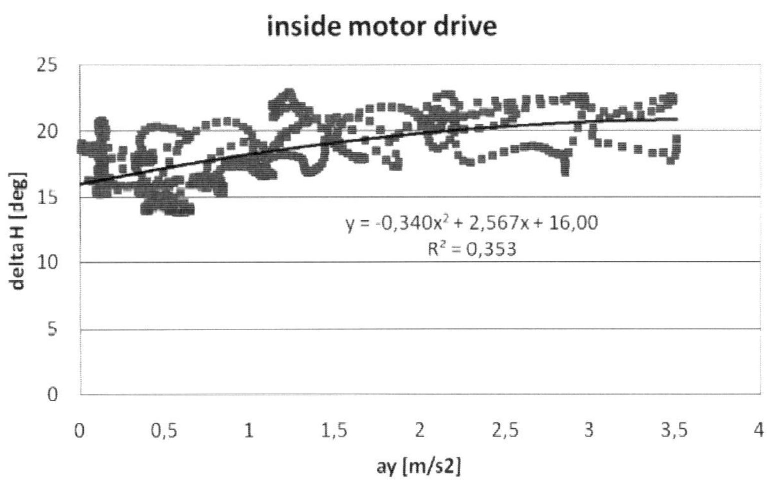

Figure 12. Steering wheel angle [deg]: as a function of lateral acceleration [m/s2] during constant radius of track test - inside motor drive.

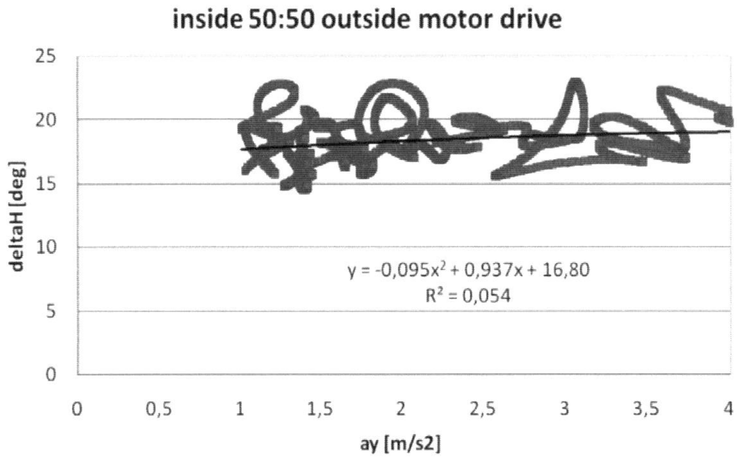

Figure 13. Steering wheel angle [deg]: as a function of lateral acceleration [m/s2] during constant radius of track test - equal power distribution.

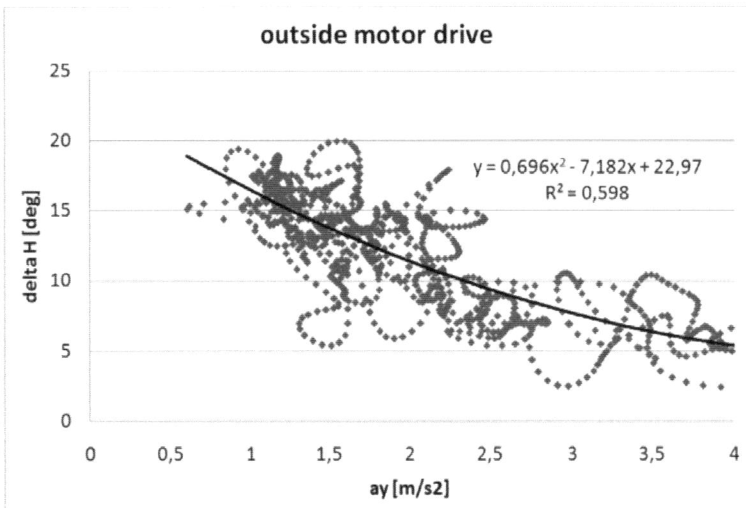

Figure 14. Steering wheel angle [deg]: as a function of lateral acceleration [m/s2] during constant radius of track test - outside motor drive.

5. Conclusion

Performed test shows that the torque vectoring can significantly affects the stability characteristics of the narrow car. A wide range of differences from understeer to oversteer and the experience gained in carrying out the tests (the detachment of two wheels from the road surface) suggests that it is possible and desirable to build a roll mitigation system based on the drive torque vectoring. Its operation is limited to acceleration of the vehicle or driving at a constant speed (with partial and full throttle). The undoubted advantage is, however, no energy losses when working, in contrast to systems based on the ESP to stabilize the vehicle's in which vehicle brakes are used. This is an advantage especially in electric vehicles. The density of energy stored in the traction batteries is many times lower than the density of energy in gasoline or diesel fuel stored in an internal combustion engine powered vehicle tank, and any energy loss dissipated in brakes significantly reduces the range of the vehicle.

References

[1] Abe M 1986 A theoretical analysis on vehicle cornering behaviors in acceleration and braking; *Vehicle System Dynamics*, **15**

[2] De Novellis L, Sorniotti A, Gruber P, Shead L, Ivanov V and Hoepping K 2012 Torque vectoring for electric vehicles with individually controlled motors: State-of-the-art and future developments; *26th International Electric Vehicle Symposium (EVS26)*

[3] Meier T, Rinderknecht S and Fietzek R 2011 Electric power train configurations with appropriate transmission systems; *SAE Technical Paper 2011-01-0942*

[4] Shibahata Y, Shimada K and Tomari T 1993 Improvement of vehicle maneuverability by direct yaw moment control, *Vehicle System Dynamics*, **22**; 465-81

[5] Sawase K, Ushiroda Y and Miura T 2006 Left-Right torque vectoring technology as the core of super all wheel control (S-AWC); *Mitsubishi Motors Technical Review*, **18**; 16-23

[6] Shimada K and Shibahata Y 1994 Comparison of three active chassis control methods for stabilizing yaw moments; *SAE Technical Paper*, **940870**; 87-97

[7] Winkler C, Aurell J 1998 Analysis and testing of the steady-state turning of multiaxle trucks; *Heavy Vehicle Weights and Dimensions, 5th international symposium*; 135-61

Proposal to use vibration analysis steering components and car body to monitor, for example, the state of unbalance wheel

R Janczur[1]

[1]Institute of Motor Vehicles and Internal Combustion Engines, Cracow University of Technology, 37 Jana Pawla II Avenue, 31-864 Krakow, Poland

E-mail: robertj@mech.pk.edu.pl

Abstract. The results of road tests of car VW Passat equipped with tires of size 195/65 R15, on the influence of the unbalancing front wheel on vibration of the parts of steering system, steering wheel and the body of the vehicle have been presented in this paper. Unbalances wheels made using weights of different masses, placed close to the outer edge of the steel rim and checked on the machine Hunter GSP 9700 for balancing wheels. The recorded waveforms vibration steering components and car body, at different constant driving speeds, subjected to spectral analysis to determine the possibility of isolating vibration caused by unbalanced wheel in various states and coming from good quality asphalt road surface. The results were discussed in terms of the possibility of identifying the state of unbalancing wheels and possible changes in radial stiffness of the tire vibration transmitted through the system driving wheel on the steering wheel. Vibration analysis steering components and car body, also in the longitudinal direction, including information from the CAN bus of the state of motion of the car, can be used to monitor the development of the state of unbalance wheel, tire damage or errors shape of brake discs or brake drums, causing pulsations braking forces.

1. Introduction

Almost every driver met with an unpleasant effect of vibration of the steering wheel angle, occurs particularly clearly in a narrow speed range, dependent on the size of the tires. The causes of the vibrations can be: dynamic unbalance wheel or tire shape errors of the rim (or both simultaneously), the variable radial stiffness of the tire at its periphery and damage the tire structure (usually visible as a local bulging of the tire) or errors shape of brake discs or brake drums, causing pulsations braking forces. Currently available apparatus for balancing wheels are equipped with special roller pressed against the rotating tire whereby, in the so-called "Road test", are defined radial force pulsation. The variable on the circumference of the radial force may be due to geometrical faults of the tire (also in case of damage to its structure), or other locally radial stiffness as a manufacturing fault. The inspiration for testing was an article by Krzysztof Prażnowski from the Department of Road Vehicles and Agricultural Opole University of Technology, presented at the conference KONMOT 2014 and published in the monograph "Research of the vehicle" [1].

2. Research object and measuring apparatus

Road tests were carried out using the car VW Passat B5 Fl Variant 1.9 TDI of the year 2003, which is in good condition and equipped with Firestone Firehawk tires 195/65R15 91T and steel wheels rim.

The air pressure in the tires was typical of partial load of the vehicle - in front of 0.23 MPa and 0.21 MPa in the rear.

The study used the following measuring devices [4]:

- unit of Racelogic VBOX 3i to measure traffic parameters of gait and self-decoding and recording linear velocity wheels of the vehicle CAN bus,
- acceleration sensors HBM B12/200 Hottinger Baldwin Messtechnik for measuring the vertical acceleration knuckle right front wheel, the body above of the tested wheel and tangential acceleration in a rotating motion of the steering wheel,
- dynamometric steering wheel CMS WA1111 Kistler for measuring the rotation angle of the steering wheel and the torque on the steering wheel,
- analog – digital converter Spider 8 Hottinger Baldwin Messtechnik,
- computer with Catman software.

Both synchronized measurement channels (VBOX and Spider 8 with a computer and acceleration sensors and wheel dynamometric) recorded the measured values with a frequency of 100 Hz. The view of where and how to place acceleration sensors and recorders measurement, are shown in figures 1 and 2. Operating wheel unbalance with additional weights (introducing deliberate unbalance of the wheel) was determined using the unit Hunter GSP 9700.

(a) (b)

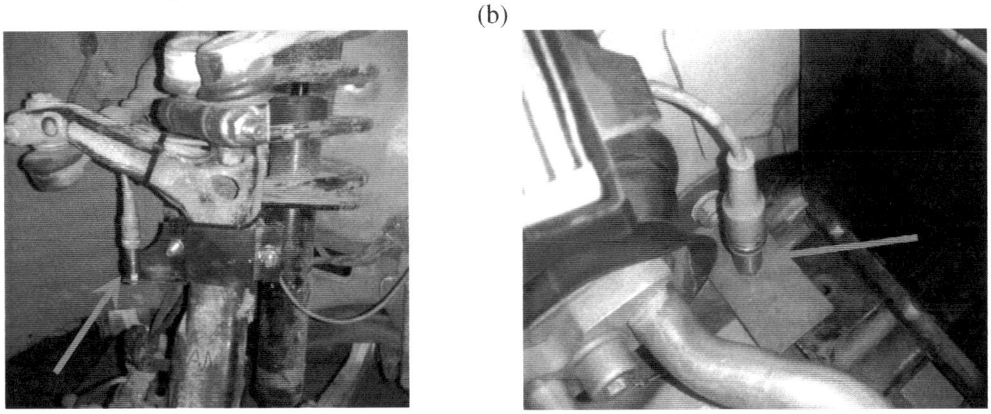

Figure 1. The view of where and how to place acceleration sensors on the front right suspension stub axle (a) and car body above this suspension (b). Sensors indicated by red arrows.

(a) (b) (c)

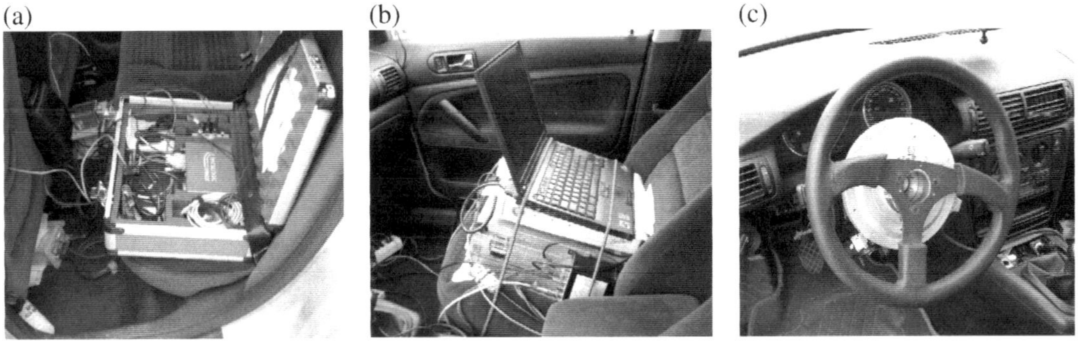

Figure 2. View recording equipment measured values: left - VBOX device (a), and the right - Spider 8 with a computer (b) and dynamometric steering wheel CMS WA1111 Kistler (c).

3. Methodology of research

Road tests the vehicle was carried out in the northern section of the ring road of Krakow, on the asphalt surface in very good condition. Measured values recorded when the vehicle standstill and the engine stopped or running, and next then very slowly increasing the velocity and the selected constant velocity. In the first stage of the study used correctly balanced all wheels and further stages of increased imbalance of the front right wheel by placing weights on the rim glued with masses respectively: 10 grams, 20 grams, 30 grams, 40 grams and 60 grams. Location and method of fixing weights shown in figure 3. For individual states unbalance wheel set "operating wheel imbalance" on the Hunter GSP 9700. Measurement results of stand unbalance the wheel for a few selected states, shown in table 1.

Figure 3. View how to place weights on the rim to the wheel unbalance and a view of the computer screen devices Hunter GSP 9700.

Table 1. Results of measurements balancing the wheels on the machine Hunter GSP 9700.

Additional weight on the wheel rim (grams)	The value of the weight balancing wheel (grams)	
	On the outside of the rim	On the inside of the rim
0 (prior to testing)	0	0
10	0	0
20	10	0
30	25	0
60	50	0
0 (after tests)	15	0

4. Presentation and analysis of results of road research

The following figures 4 and 5 show timing waveforms of vertical acceleration on the pivot of the front right wheel, the car body above this wheel and linear acceleration around the steering wheel for the car standing with the stopped engine (figure 4) and with running engine at minimum rotary speed (four-cylinder diesel engine). Running the engine with the crankshaft speed approx. 830 ÷ 840 1/min caused the vibration of the car body with vertical acceleration approx. ± 0.2 m/s^2, the stub axle vertical accelerations up to approx. ± 0.5 m/s^2, and linear peripheral steering wheel vibration - to approx. ± 0.1 m/s^2.

Figure 6 shows the frequency spectrum of acceleration measured on stub axle of front right wheel, on the car body above this wheel and linear vibration on the circumference of the steering wheel when car was stopped with the engine running at the crankshaft rotary speed approx. 830 ÷ 840 1/min. The vibration frequency of the stub axle and the steering wheel approx. 28 Hz corresponds to main frequency resulting from the combustion load in a four-cylinder four-stroke internal combustion engine, and the frequency of approx. 44 Hz is the second harmonic of the vibration. Frequency approx. 14 Hz is an additional harmonic resulting from vibrations the powertrain.

At higher speed the engine crankshaft vibrations generated by the engine and transmitted to the car body and part of the steering system are smaller. Despite this, in the following analysis includes vibration of the engine.

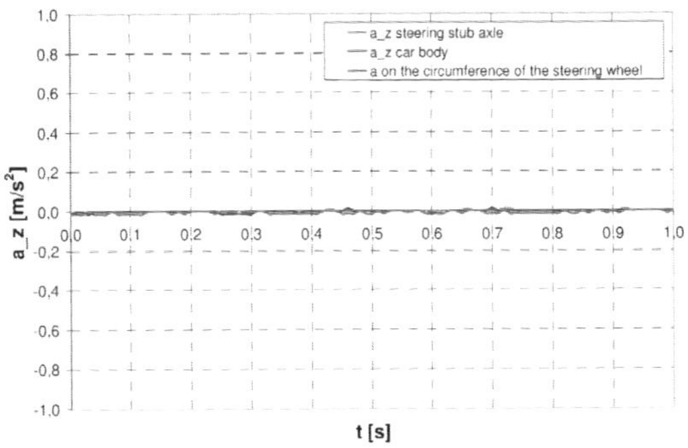

Figure 4. Timing acceleration when the car was stopped with no running engine.

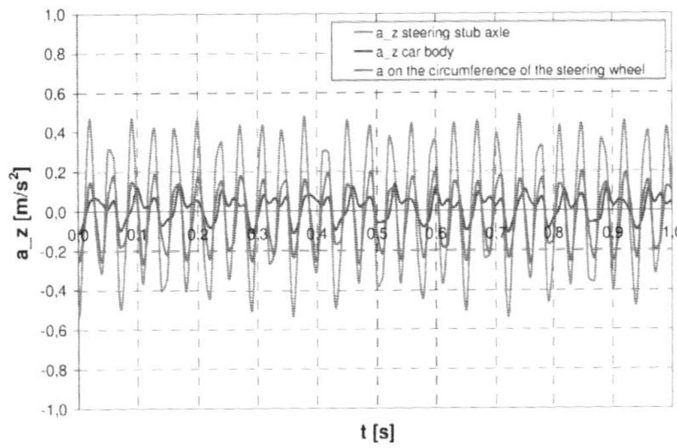

Figure 5. Timing acceleration when the car was stopped with running engine at minimum crankshaft speed (approx. 830 ÷ 840 1/min).

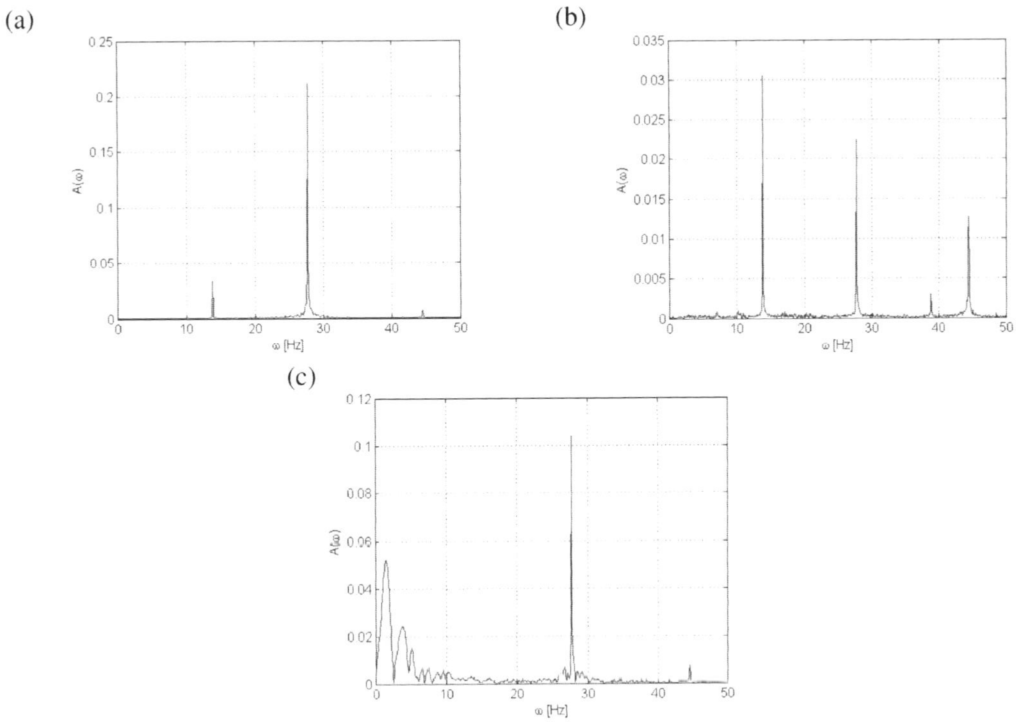

Figure 6. The frequency spectrum of acceleration measured on stub axle of front right wheel (a), on the car body above this wheel (b) and linear vibration on the circumference of the steering wheel (c) when the car was stopped with the engine running at the crankshaft speed approx. 830 ÷ 840 1/min.

Irregularities of road surface cause vibration of the elements of wheels suspension, which are transmitted to the car body and elements of the steering system, including the steering wheel. The driver can identify the wheel imbalance primarily on the basis of vibration of the steering wheel, but initial tests showed that the low mass unbalance of the front wheel, approx. 10 ÷ 20 grams, steering vibration are small enough that they cannot be observed by the driver. Vibration of stub axle of wheels make a dynamic load of the suspension and steering system and accelerate and increase their destruction, and therefore identify the occurrence of vibrations caused by wheel imbalance is very important.

For the above-described conditions of unbalance wheel, below shows the tests results at velocity approx. 100 km/h. On the figure 7 and 8 show time waveforms of the vibration elements of the steering system and the car body for the two states of the wheel unbalance resulting from the placement of the total mass of the weights 10 and 60 grams. Acceleration waveforms recorded at stub axle of the one front wheel, the car body above this wheel and on the steering wheel was subjected to spectral analysis. The results of this analysis for several unbalance state of wheel are shown in the figures 9 ÷ 12.

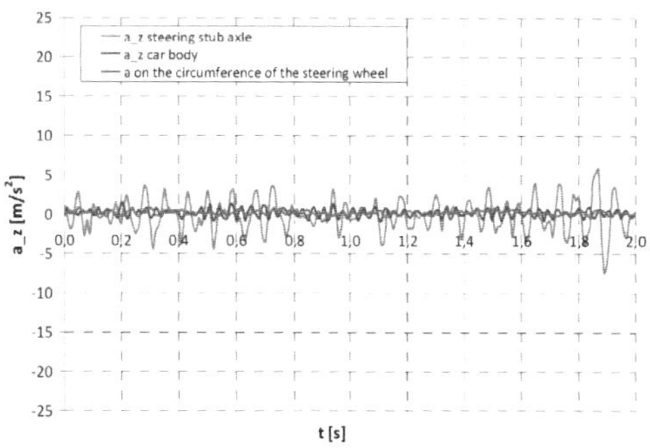

Figure 7. Timing for the acceleration of the elements of the steering system and the car body while driving at velocity approx. 100 km/h and engine crankshaft rotary speed approx. 2070 1/min, with unbalance wheel mass 10 g.

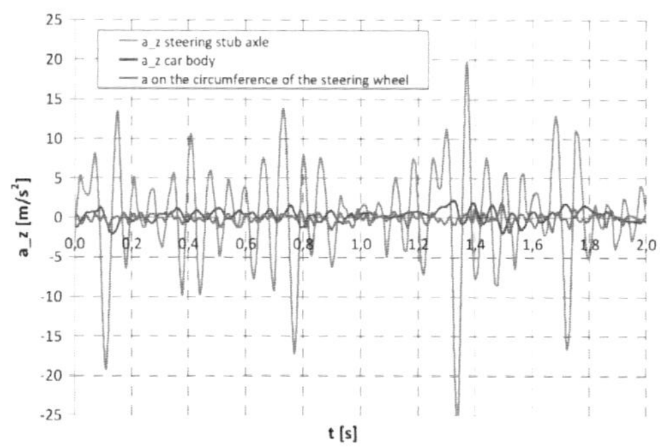

Figure 8. Timing for the acceleration of the elements of the steering system and the car body while driving at velocity approx. 100 km/h and engine crankshaft rotary speed approx. 2070 1/min, with unbalance wheel mass 60 g.

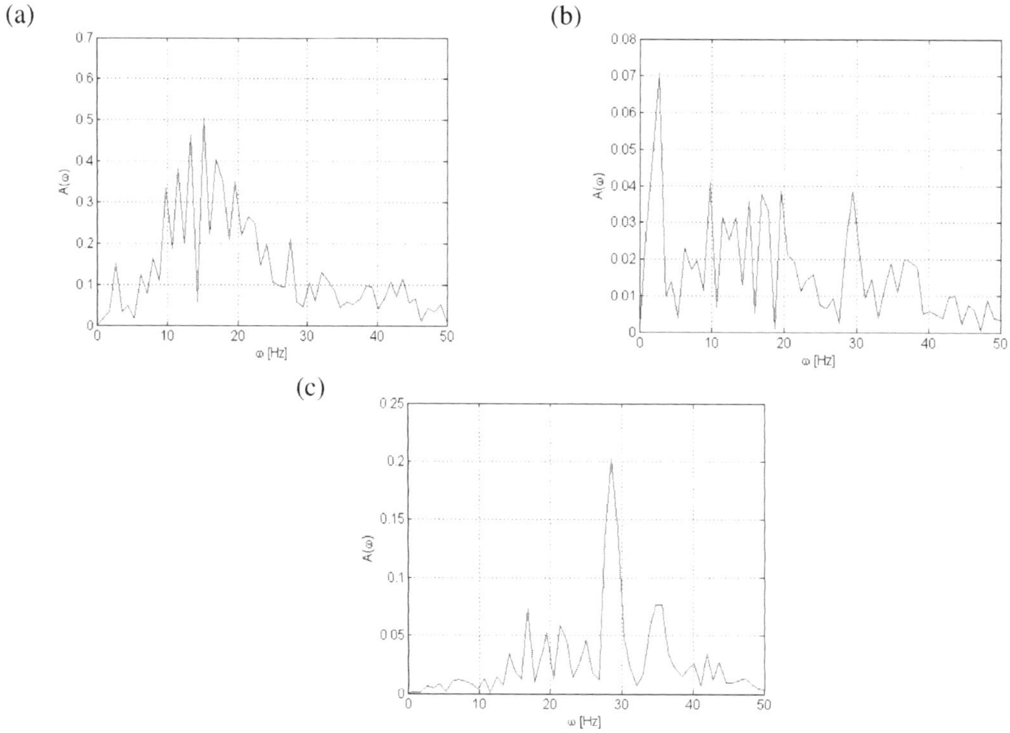

Figure 9. The frequency spectrum of acceleration measured on stub axle of front right wheel (a), on the car body above this wheel (b) and on the circumference of the steering wheel (c) while driving at velocity approx. 100 km/h and engine crankshaft rotary speed approx. 2070 1/min, with unbalance wheel mass 10 grams.

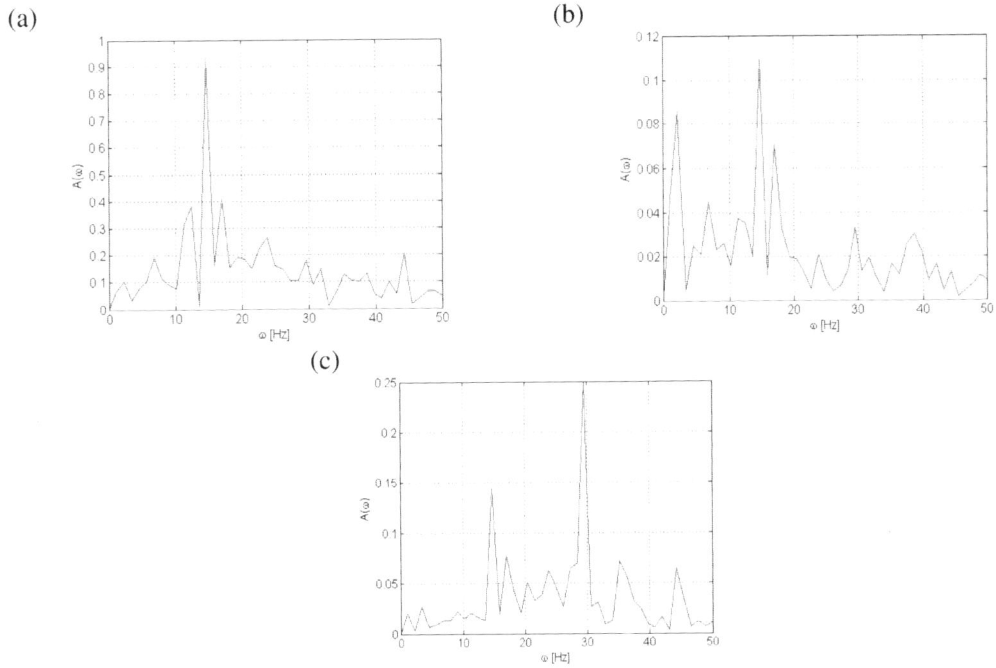

Figure 10. The frequency spectrum of acceleration measured on stub axle of front right wheel (a), on the car body above this wheel (b) and on the circumference of the steering wheel (c) while driving at velocity approx. 100 km/h and engine crankshaft rotary speed approx. 2070 1/min, with unbalance wheel mass 20 grams.

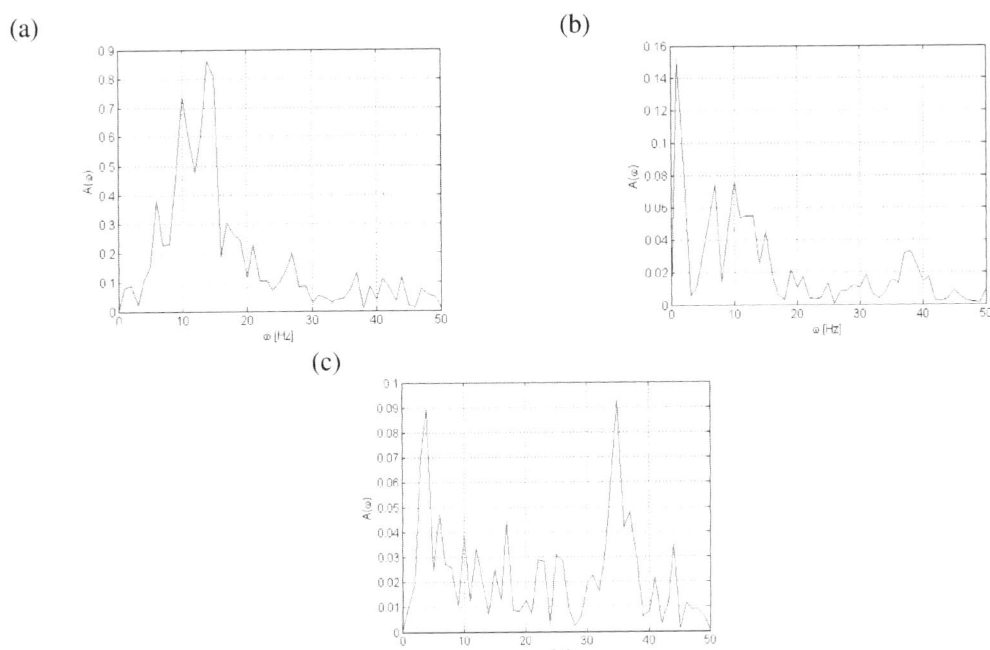

Figure 11. The frequency spectrum of acceleration measured on stub axle of front right wheel (a), on the car body above this wheel (b) and on the circumference of the steering wheel (c) while driving at velocity approx. 100 km/h and engine crankshaft rotary speed approx. 2070 1/min, with unbalance wheel mass 40 grams.

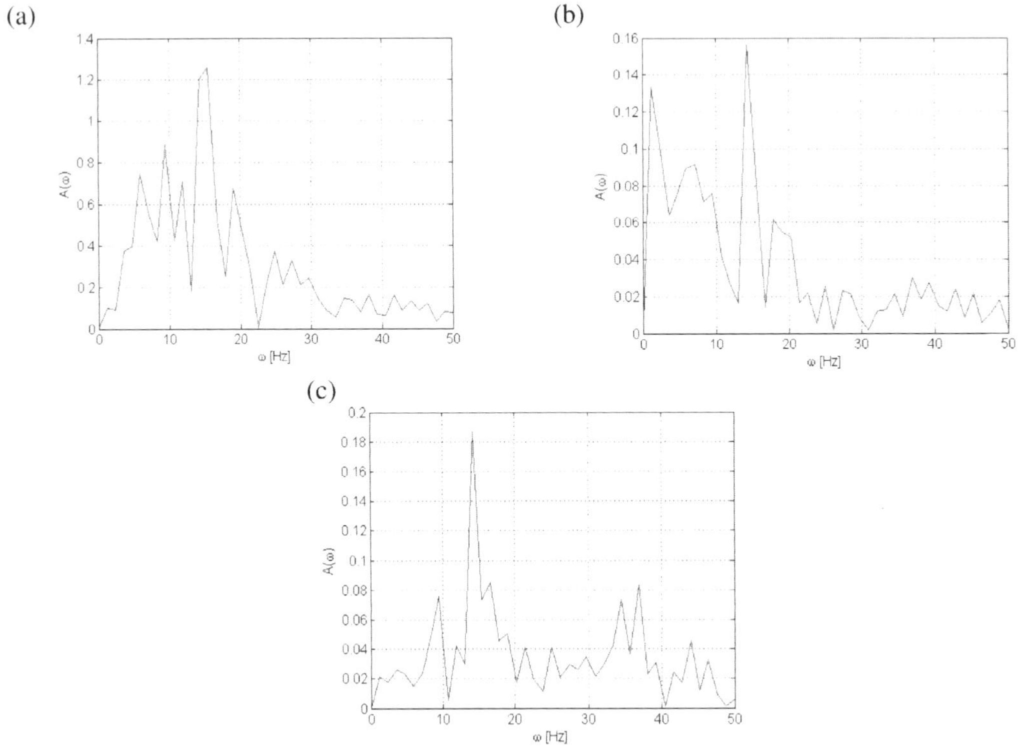

Figure 12. The frequency spectrum of acceleration measured on stub axle of front right wheel (a), on the car body above this wheel (b) and on the circumference of the steering wheel (c) while driving at velocity approx. 100 km/h and engine crankshaft rotary speed approx. 2070 1/min, with unbalance wheel mass 60 grams.

Preliminary analysis of the frequency spectrum of a wheel imbalance of 10 grams, when driving at a velocity approx. 100 km/h is indicative of the vibration of the steering wheel and a low value for a relatively broad frequency spectrum, without the dominant frequency. This points to the predominant influence of uneven road surface vibration and the wheel body [3, 5]. For comparison, figure 13 shows time waveforms of acceleration measured on stub axle of front right wheel, the car body above this wheel and on the circumference of the steering wheel and the frequency characteristics for all balanced wheels of the car. Comparison of the data shown in figure 9 and 13 indicates that the imbalance approx. 10 g cannot be correctly identified by the vibration stub axle and car body. Only the steering wheel vibration indicate the dominance of the frequency of approx. 28 Hz, which corresponds to twice the frequency resulting from the rotation of the wheels of the car traveling at approx. 100 km/h. Such domination vibrations of a similar frequency do not have road tests with a well-balanced all wheels (figure 13).

The results of spectral analysis for waveforms recorded at the wheel imbalance greater than 10 grams indicate the presence of such distinct vibration steering stub axle wheel, chassis and steering wheel, that the identification of these vibrations is possible. As part of this work introduces an imbalance of one wheel driven, but it is necessary to check identification method of vibration for the other wheels and more wheels improperly balanced. The basis for evaluation of vibration coming from the chassis of the car must be to identify the state of the road surface. It may be possible only after recognize that the vehicle is moving at a very good and flat road surface. Then it is possible to record and analyse acceleration coming from the wheels of the car. This analysis should be correlated with the speed of the car and performed in the frequency range relevant for vibration wheels. In the case of vibrations from at least two wheels, the analysis is more complex and needs to be affiliated with the identification of the change in the wheels speed and the frequency of vibration caused by driving on a curve.

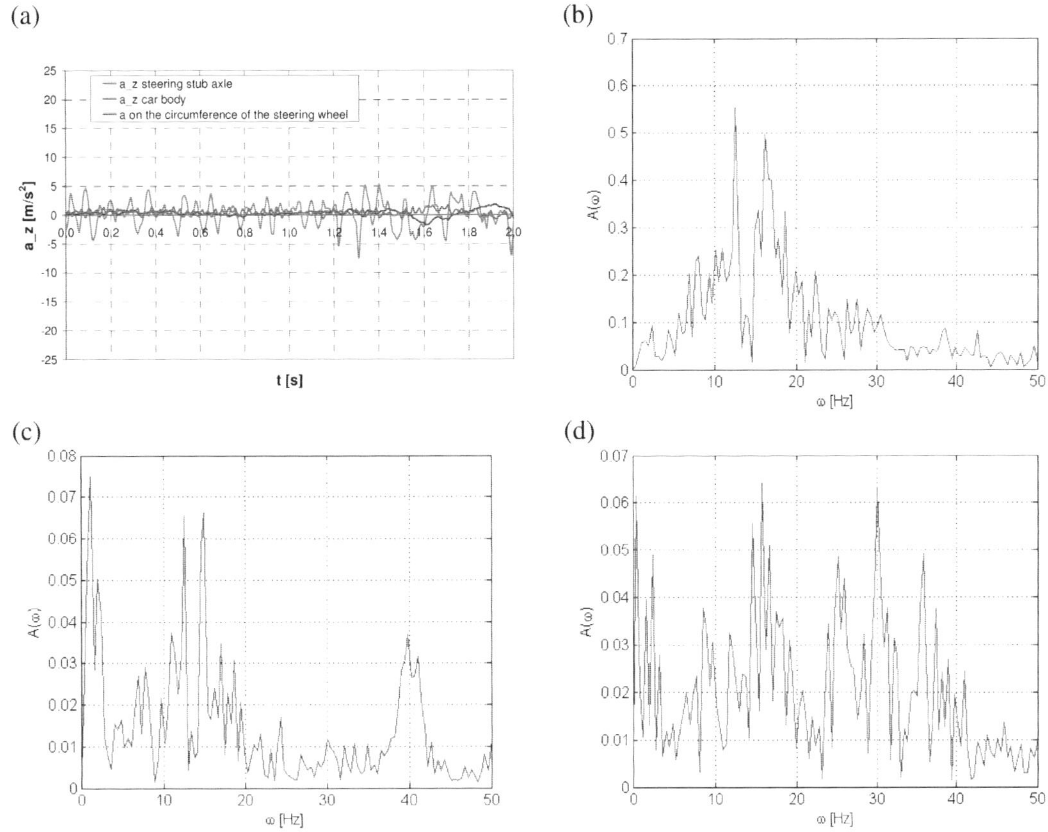

Figure 13. Timing for the acceleration of the elements of the steering system and the car body while driving at velocity approx. 100 km/h with well-unbalanced all wheels (a) and the frequency spectrum of acceleration measured on stub axle of front right wheel (b), on the car body above this wheel (c) and on the circumference of the steering wheel (d).

Installing the acceleration sensor on the stub axle wheel is technically possible, as the speed of wheel sensor ABS system. Much better working conditions can have an acceleration sensor mounted in the car body or steering wheel. Analysis of the signals recorded by means of the steering wheel dynamometer Kistler indicates that the signals of the rotation angle of the steering wheel and the torque on the steering wheel are not suitable for identification described vibration steering components, are in fact affected by manoeuvres carried out by the driver. Figures 14 and 15 show the spectral analysis of signals with an angle of rotation of the steering wheel and the torque on the steering wheel obtained when driving at a speed of approx. 100 km/h with an unbalance of the front right wheel weight 10 and 60 grams. In this signals is not a frequency described above, typical for rotating unbalance wheel at velocity approx. 100 km/h.

Therefore, in order to develop a system for measuring and analysis of vibrations caused by the wheels e.g. imbalance of wheels, should be designed the measuring system with acceleration sensors. These sensors, made in the MEMS technology, are cheap and generally available. Appropriate software will analyse the quality of the road surface on which the vehicle is currently moving and if appropriate vibration forced by rough roads are small, analyse these vibrations in order to determine whether they are caused by abnormal condition of the tires, their imbalance or abnormal shape which may indicate damage tire construction, can be made.

(a) (b)

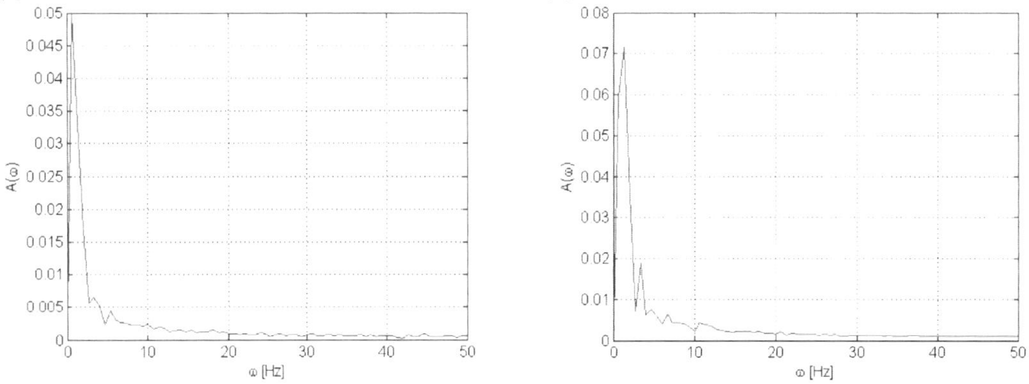

Figure 14. The frequency spectrum of the rotation angle of the steering wheel (a) and the torque on the steering wheel (b) by means of the steering wheel dynamometer Kistler while driving at velocity approx. 100 km/h and unbalance of the front right wheel mass 10 grams.

(a) (b)

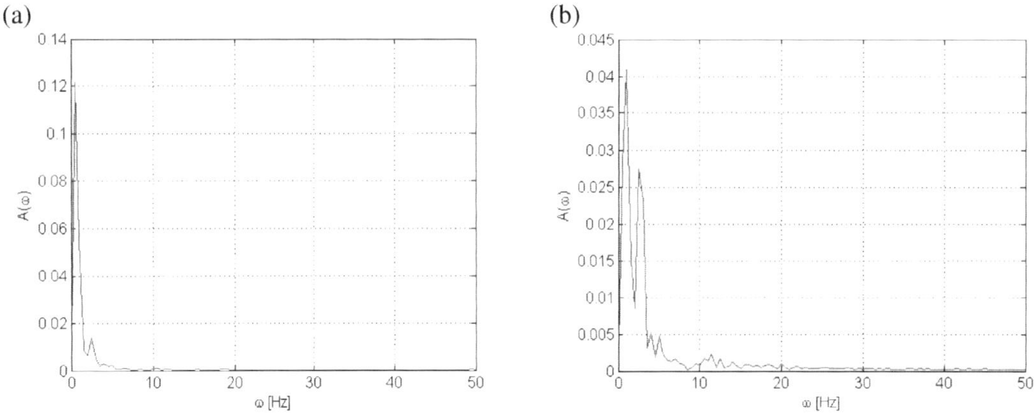

Figure 15. The frequency spectrum of the rotation angle of the steering wheel (a) and the torque on the steering wheel (b) by means of the steering wheel dynamometer Kistler while driving at velocity approx. 100 km/h and unbalance of the front right wheel mass 60 grams.

Measurements of acceleration the car body and parts of steering system in the vertical and longitudinal directions and accelerations on the circumference of the steering wheel and their spectral analysis, the knowledge of the actual speed on the basis of data from the CAN bus of the car, allow the development of a software analysing the data and forming monitor abnormal states of tires or improper effect wear of brake discs and brake drums (now rarely used). This monitor can help to increase the stability of the suspension of the car wheels and durability of tires and to maintain an optimal driving comfort.

5. Conclusion

The development of electronics in vehicles contributes to the development of surveillance systems the correct condition of the vehicle, such as the last monitors air pressure in the tires. The article proposes the use measurement and analysis acceleration of components of steering system and car body to identify the vibration caused by improper balancing wheels. Showing the acceleration waveforms recorded by

sensors placed on the steering stub axle of the front wheel of a passenger car, on the car body and on the circumference of the steering wheel for velocity approx. 100 km/h and all-wheel properly balanced and alternatively the front right wheel improperly balanced. Analysis of recorded accelerations indicates the possibility of diagnosing the state of the tires in terms of improper balance or change the shape associated with structural damage to the tires on the road with a very good and flat surface. Small vibrations can be identify although the average driver may not yet sense them on the steering wheel.
The proposed method of recording and analysing vertical acceleration including the vibrations in the longitudinal direction, can be also used to identify the pulse braking forces, e.g., caused by improper shape of disc brake. The knowledge of the actual acceleration chassis elements of vehicle and velocity on the basis of data from the CAN bus, allow the development of a software analysing the data and forming monitor abnormal states of tires or improper effect wear of discs brake. This monitor can help to increase the stability of the suspension of the car wheels and durability of tires and to maintain an optimal driving comfort.

References

[1] Prażnowski K 2014 Wykorzystanie metody STFT do identyfikacji stanu niewyrównoważenia koła ogumionego samochodu *Badania pojazdów* (Kraków: Wydawnictwo Politechniki Krakowskiej) pp 117-127
[2] Zieliński T 2007 *Cyfrowe przetwarzanie sygnałów, od teorii do zastosowań* (Warszawa: WKŁ)
[3] Lozia Z and Mazański K 2010 Typowe nierówności nawierzchni dróg *Zeszyty Naukowe Instytutu Pojazdów* (Warszawa: Politechnika Warszawska Z.1/77) pp 209-221
[4] Janczur R 2015 Vertical accelerations of the body of a motor vehicle when crossing a speed bump *The Archives of Automotive Engineering* (Warszawa: Wydawnictwo PIMOT Vol. 67, No. 1) pp 47-60
[5] Ślaski G 2012 *Studium projektowania zawieszeń samochodowych o zmiennym tłumieniu* (Poznań: Wydawnictwo Politechniki Poznańskiej, Rozprawy - Nr 481)

IOP Conf. Series: Materials Science and Engineering **148** (2016) 012010 doi:10.1088/1757-899X/148/1/012010

Calibration procedure of measuring system for vehicle wheel load estimation

M Kluziewicz[1] and M Maniowski[2]

[1, 2] Cracow University of Technology, Poland

E-mail: mmaniowski@pk.edu.pl

Abstract. The calibration procedure of wheel load measuring system is presented. Designed method allows estimation of selected wheel load components while the vehicle is in motion. Mentioned system is developed to determine friction forces between tire and road surface, basing on measured internal reaction forces in wheel suspension mechanism. Three strain gauge bridges and three-component piezoelectric load cell are responsible for internal force measurement in suspension components, two wire sensors are measuring displacements. External load is calculated via kinematic model of suspension mechanism implemented in Matlab environment. In the described calibration procedure, internal reactions are measured on a test stand while the system is loaded by a force of known direction and value.

1. Introduction and goals

Modelling of vehicle motion requires implementing characteristics of examined tires [1]. Among known methods of tire characteristics estimation the measurements can be conducted either on test benches or in the real road conditions using a test vehicle [2, 8]. The author's research is focused on a development of a tire-to-loose-surface interaction model which can be applied in a rally car dynamics analysis [3, 4]. In a view of the best representation of tire work conditions, it is crucial to conduct appropriate road tests on a real gravel surface (figure 1).

Figure 1. Test vehicle side-slipping on a gravel surface.

The most commonly used devices for road and track testing are measuring hubs. Multi-component wheel hubs for a direct load measuring are commercially accessible, but extremely expensive, not suitable for tough working conditions and result in a change of suspension system parameters. Having regard to the disadvantages of available solutions, the authorial measuring system enabling indirect tire characteristics estimation has been designed [3]. This system exploits existing front wheel suspension of Macpherson strut type. One mechanism performs various roles: wheel guiding, shock absorbing and measuring reaction forces in suspension joints (Fig. 2).

Described approach bases on a measurement of internal forces (using strain gauges and load cells) and suspension mechanism configuration coordinates (mechanical and optical displacement sensors) [Estimation of wheel]. Relations between input and output data are described by kinetostatic model of the mechanism by using jacobian matrix and spatial transformations. Computer algorithm was implemented in Matlab environment.

Featured test vehicle is a front wheel drive, fully rally-prepared Citroen Saxo VTS (Fig.1). The main advantages of using rally car are: increased crew safety, ease of sensors assembly and simplicity of setup adjustments.

This paper is focused on a calibration method and bench tests of designed measuring system. Data calculated by estimation algorithm are compared to the measured ones.

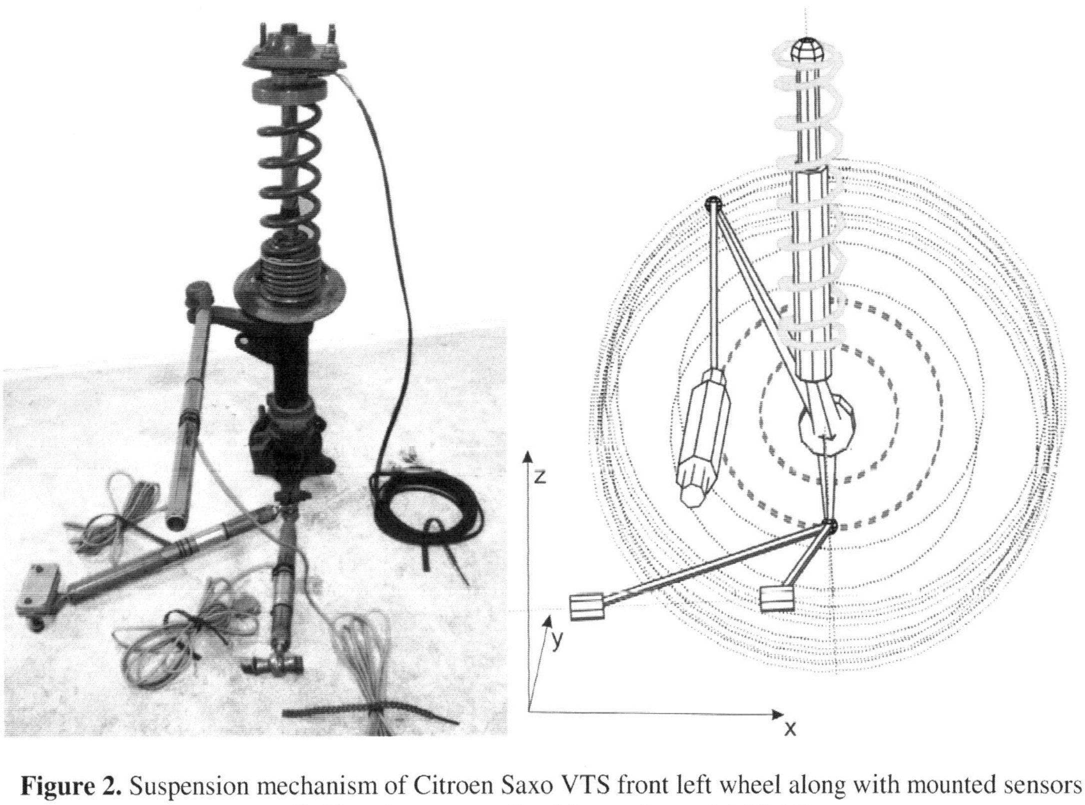

Figure 2. Suspension mechanism of Citroen Saxo VTS front left wheel along with mounted sensors (left) and corresponding kinematic model (right).

2. Estimation of tire model parameters

Tire model generates forces (F) and torques (M) in a contact patch between the tire tread and road surface in relation to different parameters, what can be expressed by the diagram (Fig.3).

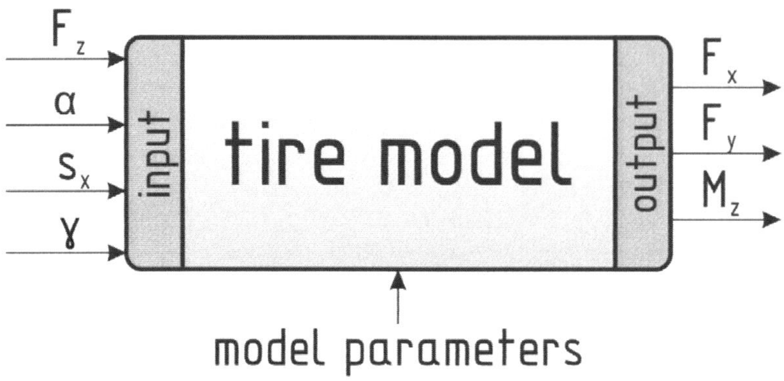

Figure 3. Diagram of tire model.

There are various tire models describing aforementioned relation by using physical and/or empirical approaches [1]. According to the model complexity, number of its parameters varies from several to hundreds. Process of model parameters estimation can be carried out when a set of inputs and outputs (Fig.3) is known from measurements, which ought to be conducted in a most similar to real work conditions.

External loads characteristics (F_x, F_y, M_z) in a contact patch between tire tread and road surface are assumed as in a following equation:

$$F_x, F_y, M_z = f(F_z, \gamma, \alpha, S_z, \beta, T_b, r_d, surface\ parameters) \quad (1)$$

where:

F_x, F_y, M_z – Tire load components estimated by internal loads of suspension mechanism in a specific suspension travel position (s) and steering rack setting (p).

F_z – Tire vertical load, measured indirectly through suspension travel and force in top mount.

γ – Wheel camber angle, estimated by: known suspension dimensions, measured suspension travel (s), steering rack setting (p) and chassis roll angle.

α – Tire slip angle, estimated via longitudinal and lateral vehicle speed (Correvit QL) and steering wheel rotation angle (cable sensor).

S_x – Longitudinal tire slip, estimated via vehicle speed, wheel rotation speed, and dynamic wheel radius.

β – Vehicle sideslip angle, calculated from longitudinal and lateral vehicle speed.

T_b – Tire tread temperature, measured with pyrometer gauge.

r_d – Dynamic wheel radius, estimated via axial force in a damper module.

Surface parameters such as loose layer depth, size distribution, moisture content, soil volume density.

For a specified mechanism setting, defined by suspension travel (s) and steering rack position (p), static task can be described by the following equation:

$$W = \boldsymbol{J}^T R \quad (2)$$

where:

$W = [F_x\ F_y\ F_z\ M_x\ M_y\ M_z]^T$ – complex load vector translated to the wheel's centre of rotation,

$R = [R_1\ R_2\ R_3\ R_4\ R_5\ R_6]^T$ – internal loads vector: R1 – reaction in a front arm rod joint; R2 – reaction in a rear arm rod joint; R3 – reaction in a steering rod; R4, R5, R6 – longitudinal, lateral and vertical reactions in a strut top mount,

$\boldsymbol{J}$ – jacobian matrix for static analysis of parallel mechanism [5].

Equation (2) describes linear relations between unknown external load (W) and measured internal load (R) in mechanism. Forces and torques in a centre of wheel rotation are related with joints reactions by the jacobian matrix. Jacobian matrix is determined for a specific suspension travel position (s) and steering rack setting (p).

In order to measure a tire load the wheel suspension mechanism was adapted. A kinematic model of MacPherson strut suspension (Fig.4) with rack-and-pinion steering system was formulated according to the following assumptions:

- the wheel knuckle is joint by spherical pairs with: lower wishbone (5, 6), tie rod (7) and top mount (8) in the strut end;
- two degrees of the wheel knuckle mobility correspond to the wheel bounce (controlled by the strut height h) and turn motions (controlled by the steering rack displacement p);
- the lower wishbone, due to a special design, can be treated as a composition of two rods (5, 6) with spherical pairs at its ends;
- direct kinematic task is solved using a vector method [5];
- joints and links are ideal;
- internal reactions in the suspension are determined under static conditions;
- following vector transformations are written instantaneously with respect to the wheel reference system.

Cartesian coordinates (**k**) of the wheel reference system are the following:

$$\boldsymbol{k} = \begin{bmatrix} b_x \\ b_y \\ b_z \\ \beta_x \\ \beta_y \\ \beta_z \end{bmatrix} \tag{3}$$

where:

$\boldsymbol{b}$ – position vector of the wheel centre [m];

$\boldsymbol{\beta}$ – vector of the wheel orientation angles [rad].

According to the considered suspension kinematic structure (Fig.2), the wheel knuckle is constrained by the following coordinates:

$$\boldsymbol{q} = \begin{bmatrix} d_1 \\ d_2 \\ d_3 \\ a_{4x} \\ a_{4y} \\ a_{4z} \end{bmatrix} \tag{4}$$

These coordinates (4) describe blocked motions, due to: constant lengths of the three rods (d_i, i=1..3) and unchangeable position of the top mount spherical pair (a_{4x}, a_{4y}, a_{4z}). Each rod (i=1..3) of the wheel suspension is described by the following vectors:

IOP Conf. Series: Materials Science and Engineering **148** (2016) 012010 doi:10.1088/1757-899X/148/1/012010

$$\boldsymbol{d}_i = \boldsymbol{b}_i - \boldsymbol{a}_i \tag{5}$$

with the length:

$$d_i = \|\boldsymbol{d}_i\|[m] \tag{6}$$

and the unit vector:

$$\widehat{\boldsymbol{d}}_i = \frac{\boldsymbol{d}_i}{d_i} \tag{7}$$

Kinematic relation between virtual changes of Cartesian (3) and constraints (4) coordinates, can be written as [5]:

$$\partial \boldsymbol{q} = \boldsymbol{J} \partial \boldsymbol{k} \tag{8}$$

Jacobian matrix (***J***) of parallel mechanism can be defined as twist coordinates (each row) of the constraints, as follows:

$$\boldsymbol{J} = \begin{bmatrix} [\widehat{\boldsymbol{d}}_1]^T & [\boldsymbol{b}_1 \times \widehat{\boldsymbol{d}}_1]^T \\ [\widehat{\boldsymbol{d}}_2]^T & [\boldsymbol{b}_1 \times \widehat{\boldsymbol{d}}_2]^T \\ [\widehat{\boldsymbol{d}}_3]^T & [\boldsymbol{b}_3 \times \widehat{\boldsymbol{d}}_3]^T \\ \begin{bmatrix} 1 & -\varepsilon_z & \varepsilon_y \\ \varepsilon_z & 1 & -\varepsilon_x \\ -\varepsilon_y & \varepsilon_x & 1 \end{bmatrix} & -[\widetilde{\boldsymbol{a}}_4] \end{bmatrix}_{6x6} \tag{9}$$

where:

$$\widetilde{\boldsymbol{a}}_4 = \begin{bmatrix} 0 & -a_{4,z} & a_{4,y} \\ a_{4,z} & 0 & -a_{4,x} \\ -a_{4,y} & a_{4,x} & 0 \end{bmatrix} \text{(skew symmetric matrix)}$$

$\varepsilon_x, \varepsilon_y, \varepsilon_z$ – small deviation angles [rad] of force sensor orientation in the top mount (around x, y and z axes accordingly).

Jacobian matrix (9) depends on the mechanism geometry and is determined for a specific suspension travel (b_z) and steering rack position (p).
Forces and torques generated in the tire tread patch are collected in a spatial load vector:

$$\boldsymbol{W}_t = [F_x\ F_y\ F_z\ M_x\ M_y\ M_z]^T \tag{10}$$

This load (8) for quasi-static conditions can be transferred to the wheel center, by the formula:

$$\boldsymbol{W} = \boldsymbol{H}\,\boldsymbol{W}_t \tag{11}$$

$$\boldsymbol{H} = \begin{bmatrix} [1]_{3x3} & [0]_{3x3} \\ [\tilde{r}]_{3x3} & [1]_{3x3} \end{bmatrix} \tag{12}$$

$$\tilde{\boldsymbol{r}} = \begin{bmatrix} 0 \\ 0 \\ -r_d \end{bmatrix} \tag{13}$$

r_d – dynamic tire radius [m].

IOP Conf. Series: Materials Science and Engineering **148** (2016) 012010 doi:10.1088/1757-899X/148/1/012010

According to the assumed constraints (4) in the suspension mechanism, there are 6 reaction forces (Fig.5), which are grouped in the vector:

$$\boldsymbol{R} = [R_1 R_2\ R_3\ R_4\ R_5\ R_6]^T \tag{14}$$

where:

R_1 – force in the front rod [N];
R_2 – force in the rear rod [N];
R_3 – force in the steering rod [N];
R_4, R_5, R_6 – longitudinal, lateral and vertical reactions in strut top mount [N].

Applying the virtual work principle, which describes balance of works done by an external and internal loads, it can be stated:

$$\boldsymbol{W}^T \partial \boldsymbol{k} = \boldsymbol{R}^T \partial \boldsymbol{q} \tag{15}$$

Transforming equation (15) by applying kinematic relations (8), the static task can be written as:

$$\boldsymbol{W} = \boldsymbol{J}^T \boldsymbol{R} \tag{16}$$

Equation (16) describes linear relations between unknown external wheel load (**W**) and the mechanism internal load (**R**), that can be measured.

3. Measuring system calibration

First step in calibration procedure was to obtain sensitivity factors for strain gages stuck to suspension arm rods and steering rod. For this purpose mentioned suspension components were coupled with hydraulic cylinder and force transducer HBM U9B inside of steel frame. While axial tensile or compression force was applied, the relation between voltage of strain gauge bridge diagonal and force measured by reference sensor was recorded. Basing on this data, slope and y-intercept of the regression line was calculated. Calibration process of three-component piezoelectric force sensor (McPherson strut top-mount) was presented in previous paper [6].

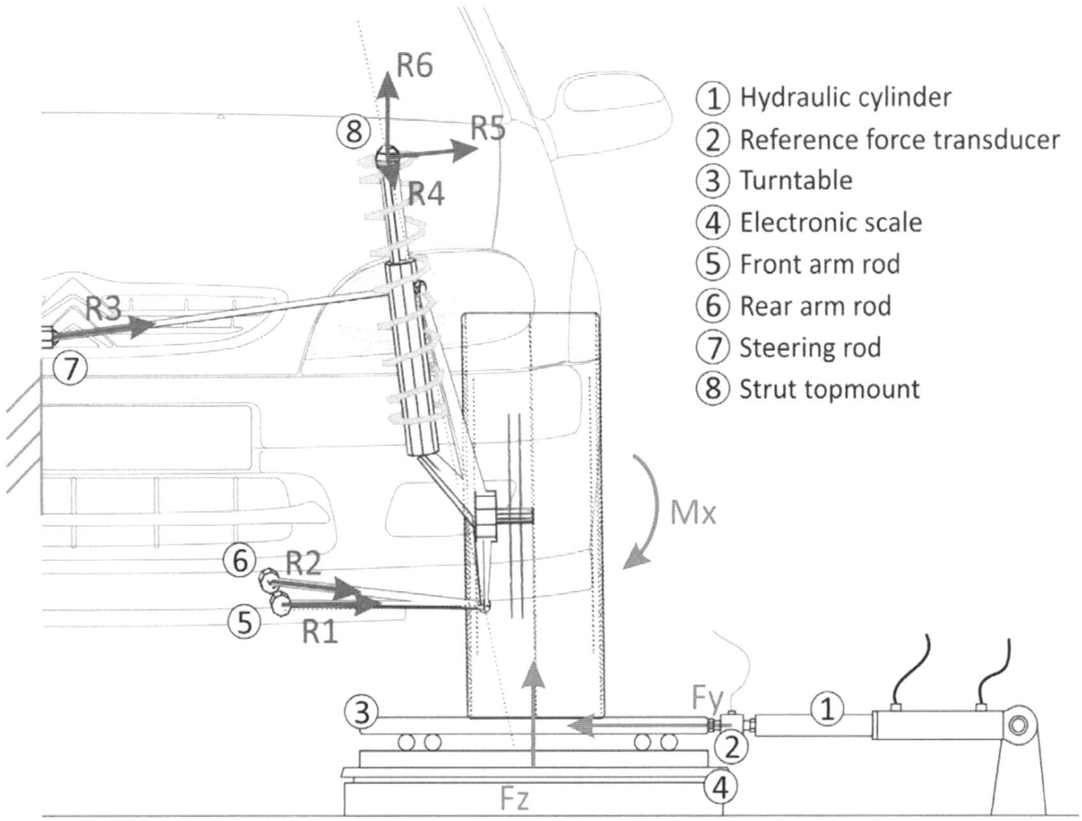

Figure 4. Scheme of test bench for side load application.

After assembly of complete measuring system vehicle was set on a test bench presented on drawing 2. Front left wheel was standing on a wheel alignment turntable, placed on an electronic scale. Turntable was attached by a force transducer to a hydraulic cylinder directed orthogonally to vehicle longitudinal axis. Relation between reaction forces in suspension mechanism (R_1 R_2 R_3 R_4 R_5 R_6) and extortion force (F_y) combined with derived torque (M_x) at a determined vertical load (F_z) was measured.
Friction force of turntable movement was modelled in accordance with Coulomb friction law. Graphical plot of reaction forces as a function of variable side load is presented on drawing 5. Thin dashed lines represent the results of three tests, whereas data obtained from calculation model are plotted as a solid lines.

Most significant values of reaction forces were measured in: McPherson strut topmount (green colour, result of the vehicle mass), front arm rod (red) and rear arm rod (orange). Relations between reaction forces and extortion force in examined range are linear.

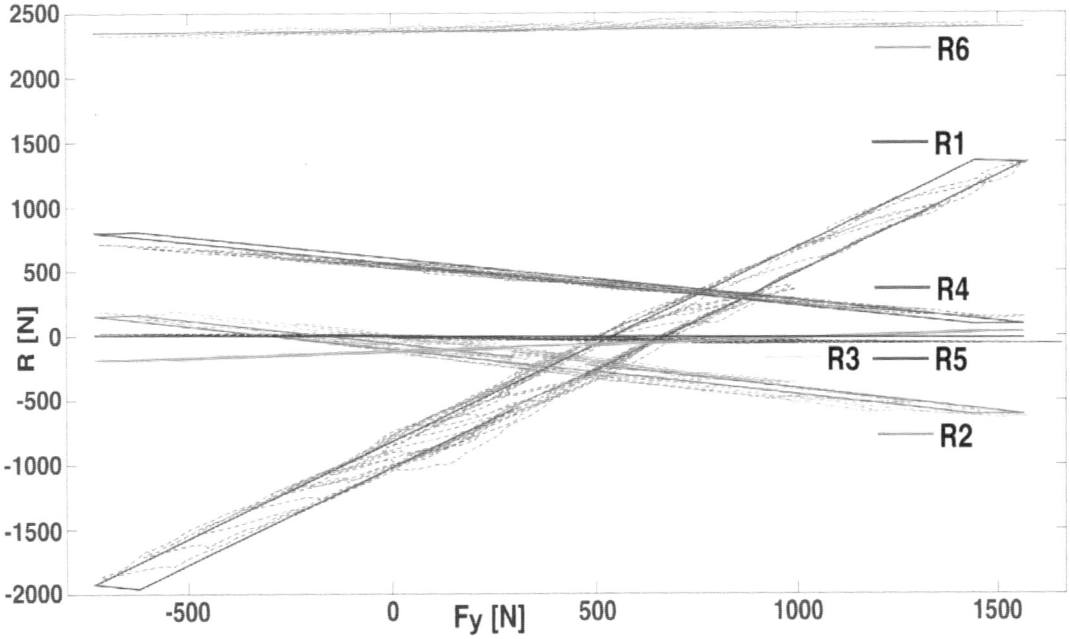

Figure 5. Plot of internal reactions as a functions of external side load.

Substantial accordance in functions graph can be a proof of proper formulation of mathematical task and valid kinematic model. Hysteresis loops are the effect of dry friction occurring in turntable under the wheel.

Figure 6. Test bench for longitudinal load application.

Similar tests were conducted with longitudinal load on a front wheel. In this case hydraulic cylinder was placed in a front of tire contact patch, parallel to vehicle longitudinal axis. Measured reaction forces (R_1 R_2 R_3 R_4 R_5 R_6) are result of longitudinal extortion force (F_x) combined with derived torque (M_y) at a determined vertical load (F_z).

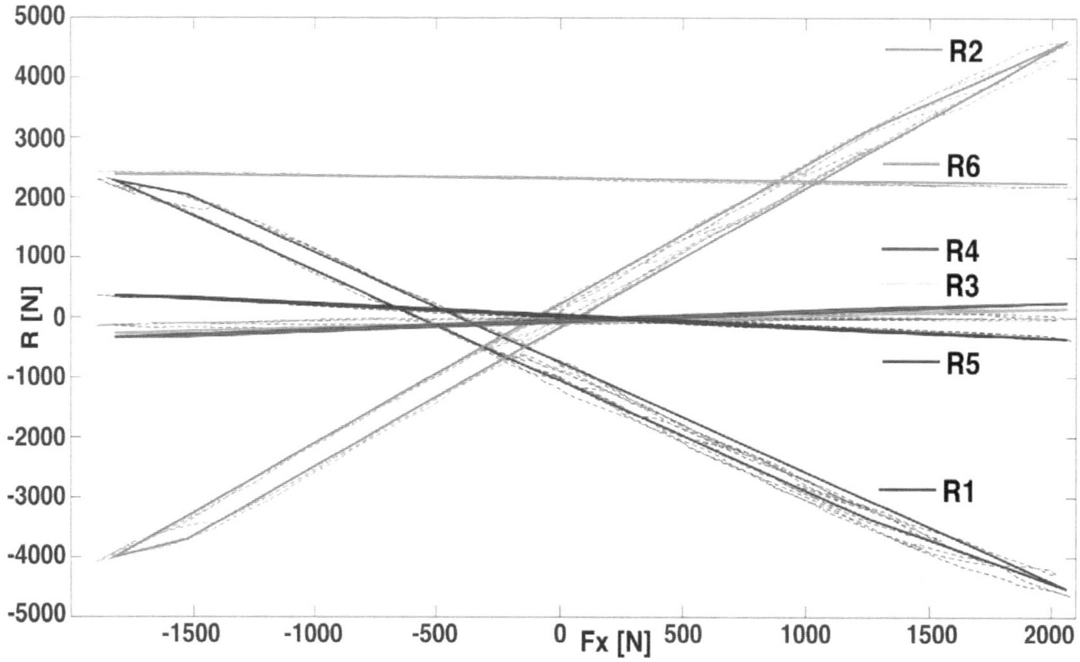

Figure 7. Plot of internal reactions as a functions of external longitudinal load.

Most significant reactions were measured in: rear arm rod (orange), front arm rod (red), McPherson strut topmount (green). Other components (R_3 R_4 R_5) are changing slightly (Fig. 7). Calculated reactions are close to the real ones. Relations between external load and internal forces are linear in the examined range.

4. Conclusions

The innovative method of measuring wheel load by using suspension mechanism is described. Forces and torques acting on a rotating wheel are estimated basing on internal forces and configuration coordinates of suspension mechanism. The relation of aforementioned input and output data featuring jacobian matrix is presented along with equation derivation and evaluation procedure.

Good correspondence between calculated and measured reactions proves the usefulness of designed system. Relations between input and output data are linear as it was expected. Thorough calibration of all measuring devices and introduction of exact mechanism dimensions to the kinematic model are crucial for the proper functioning of whole system.

An accuracy of developed estimation method is desired to be at level of 5%. Thereby system can be successfully used for tire characteristics measurement and tire model development. According to carried uncertainty analysis [7] and results of calibration, specified accuracy is very probable to be achieved.

Presented actions are just a single part of system verification. With the aim of estimating wheel load on the basis of measured internal reactions, exactly the opposite task should be solved, which requires inversion of jacobian matrix. Results of full estimation during road test will be presented in next paper.

References

[1] Pacejka H B 2002 *Tyre and Vehicle Dynamics,* Butterworth-Heinemann, (Oxford)

[2] Kluziewicz M and Maniowski M 2012 *Porównanie efektywności wyścigowej i rajdowej techniki pokonania łuku samochodem przednionapędowym*, Czasopismo techniczne 5-M/2012, Zeszyt 10, Rok 109.

[3] Kluziewicz M and Maniowski M 2014 *Estimation of wheel load state by using suspension joints reactions,* Badania Pojazdów, praca zbiorowa pod redakcją Władysława Mitiańca, Opracowanie Monograficzne, p 61 (Cracow)

[4] Kluziewicz M and Maniowski M 2010 *Stany nadsterowności podsterownego samochodu z przednim napędem* Zeszyty Naukowe Instytutu Pojazdów Politechniki Warszawskiej, Vol. 1(77), pp. 169-177

[5] Knapczyk J and Maniowski M 2006 *Elastokinematic Modeling and Study of Five-Rod Suspension with Subframe*, Mechanism and Machine Theory, Vol. 41, pp. 1031–1047,

[6] Kluziewicz M 2015 *Experimental verification of kinetostatic model of strut suspension mechanism,* Technical Transactions i. 7. Mechanics i. 2-M, Wydawnictwo PK (Cracow)

[7] Kluziewicz M and Maniowski M 2016 *Uncertainty analysis of innovative method for wheel load measurement,* Mechatronics: Ideas, Challenges, Solutions and Applications, Vol. 414, Springer

[8] Luty W 2014 *Badania eksperymentalne ogumienia w nieustalonych warunkach znoszenia bocznego*, The Archives of Automotive Engineering, Vol. 66, nr 4, p 133

IOP Conf. Series: Materials Science and Engineering **148** (2016) 012011 doi:10.1088/1757-899X/148/1/012011

Simplified planar model of a car steering system with rack and pinion and McPherson suspension

J. Knapczyk[1,2] and P Kucybała[1]

[1] Cracow University of Technology, Faculty of Mechanical Engineering, Poland
[2] State Higher Vocational School in Nowy Sącz,

E-mail: j_kn@mech.pk.edu.pl

Abstract. The paper presents the analysis and optimization of steering system with rack and pinion and McPherson suspension using spatial model and equivalent simplified planar model. The dimension of the steering linkage that give minimum steering error can be estimated using planar model. The steering error is defined as the difference between the actual angle made by the outer front wheel during steering manoeuvers and the calculated angle for the same wheel based on the Ackerman principle. For a given linear rack displacement, a specified steering arms angular displacements are determined while simultaneously ensuring best transmission angle characteristics (i) without and (ii) with imposing linear correlation between input and output. Numerical examples are used to illustrate the proposed method.

1. Introduction; review of the literature dealing with car steering system

Real steering mechanisms are complex spatial linkages. However, the kingpin inclination and caster angles that provide compliance to the steering linkage with the suspension system have little influence on the motion transmission of the steering linkage. As a result, the real rack and pinion steering linkage, which is spatial in nature, can be modelled as an equivalent planar linkage for the investigation of Ackermann condition. Therefore, by neglected the kingpin inclination and caster angle, in this study, the steering linkage of a vehicle is considered as a planar linkage with six revolute joints (*R*) and one prismatic joint (*P*) forming the kinematic chain (*R-R-R*)-*P*-(*R-R-R*) with two closed loops. Such a simplification of the steering system has been also used by other researchers [1, 2, 7].

The purpose of this research is to analyse the rack and pinion steering linkage with McPherson suspension. From this analysis, the dimension of the planar linkage model and the position of the projection plane of the steering linkage that give minimum steering error can be obtained. The steering error is defined as the difference between the actual angle made by the outer front wheel during steering manoeuvers and the calculated angle for the same wheel based on the Ackerman principle [1].
The considered slider-rocker mechanism (*P-R-R-R*) can generate, for a given slider (rack) displacement, a specified maximum rocker (steering arm) swing while simultaneously ensuring best transmission angle characteristics (i) without and (ii) with imposing linear correlation between input and output. For case (i), the multi-dimensional space of the optimization problem is inspected using partial-global-minimum plots, and some interesting properties are revealed [1]. Case (ii) is formulated as a multi-objective optimization problem that is solved using a modified Normal Boundary Intersection method. To assist practicing engineers with their designs, parametric design charts accompanied by transmission angle, input–output linearity error, and torque-to-force multiplication factor diagrams are also provided in [1].

The synthesis of the slider-rocker (*P-R-R-R*) mechanism for given limit positions can be performed graphically. However, satisfying good motion transmitting characteristics is not guaranteed, and a trial-

and-error search must be performed until acceptable mechanical-advantage or transmission angle properties are satisfied. To avoid overloading the links and joints, it is recommended that the transmission angle remains between 45°and 135°[2, 5].

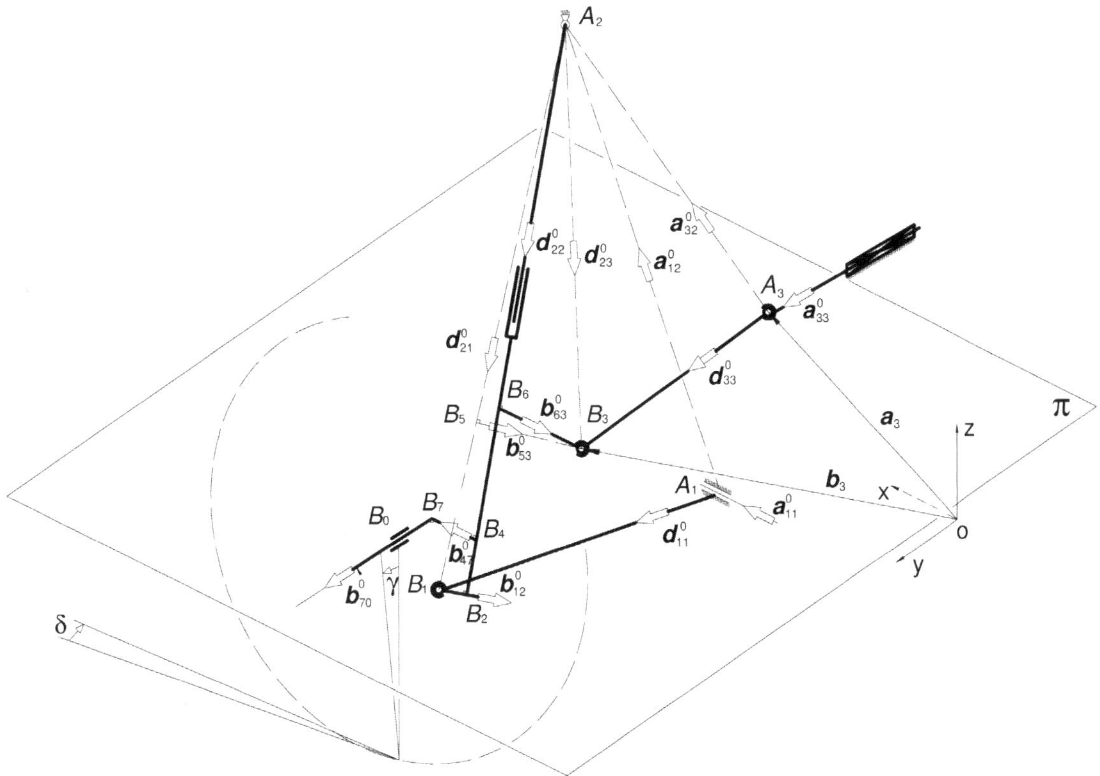

Figure 1. The scheme of the steering system with rack and pinion gear and with McPherson suspension. Notations: A_1 , B_0 – revolute joints; A_2, A_3, B_1, B_3 – spherical joints; B_6 – cylindrical joint; $\boldsymbol{a}_{ij}^{o}$ – unit vector of the line connecting points A_i and A_j ; $\boldsymbol{d}_{ij}^{o}$ – unit vector of the line connecting points A_i and B_j ; $\boldsymbol{b}_{ij}^{o}$ – unit vector of the line connecting points B_i and B_j ; δ – wheel steer (toe) angle; γ – wheel camber angle; $\boldsymbol{a}_3$ – position vector of the point A_3; $\boldsymbol{b}_3$ position vector of the point B_3 with respect to the origin of the body reference system $Oxyz$; π – plane parallel to Oxy.

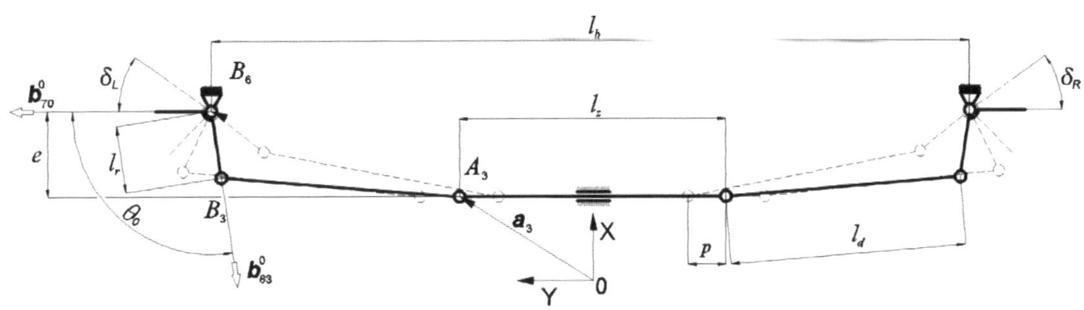

Figure 2. The simplified planar model of the steering linkage. Notations: ($A^l_3\,A^p_3 = l_z$) – the rack as input; ($A_3\,B_3 = l_d$) – the tie rod; ($B_3\,B_6 = l_r$) – the steering arm; p – displacement of the rack; δ_L – steer angle of the left wheel; δ_R – steer angle of the right wheel; e – the distance from and the rack placement to the steering axis (off-center of the rack with respect to the steering axis).

2. Objective

The work was aimed at the simplification for modelling and optimization of steering system with rack and pinion and McPherson suspension. The simplified planar model is used for analysis and design of slider-rocker (rack-steering arm) mechanisms for function generation, with transmission angle taking into account.

3. The simplified planar model

The planar model of the six-bar steering linkage (Fig.2) consists of the rack, two tie rods and two steering arms of the wheel knuckle. This side take-off configuration is more common in passenger cars. The analytical method used to solve the nonlinear problem for position and displacement analysis of this mechanism is described as follows: 1. For the assumed value of δ_L the corresponding value of p is calculated by using formulae (1)-(3); 2. To evaluate corresponding angle δ_R the formulae (4)-(11) are used.

$$\theta_1 = \pi - (\alpha_2 + \delta_L) \quad (1)$$

$$\Delta_1 = l_d^2 - (l_r sin\theta_1 + e)^2 \quad (2)$$

$$p = \sqrt{\Delta_1} + l_r cos\theta_1 - \frac{1}{2}(l_b - l_z) \quad (3)$$

$$A = l_r[p - \frac{1}{2}(l_b - l_z)] \quad (4)$$

$$B = l_r e \quad (5)$$

$$D = \frac{1}{2}\left[l_d^2 - e^2 - \left(\frac{1}{2}(l_b - l_z) - p\right)^2\right] \quad (6)$$

$$Acos\theta_2 + Bsin\theta_2 = D \quad (7)$$

$$\theta_2 = 2 \cdot \mathrm{arctg}\left[\left(-B \pm \sqrt{A^2 + B^2 - D^2}\right)/(A + D)\right] \quad (8)$$

$$\delta_R = (\alpha_2 + \theta_2) - \pi \quad (9)$$

$$\Delta_2 = l_d^2 - (l_r sin\theta_2 + e)^2 \quad (10)$$

$$p = \frac{1}{2}(l_b - l_z) - \sqrt{\Delta_2} - l_r cos\theta_2 \quad (11)$$

Note that + and – signs in equation (8) correspond to the two different assembly positions of the mechanism, where only + sign is applicable for the steering linkage.
The steering axis is determined by the centers of two spherical joints (A_3 and B_3) and its orientation is described by using: σ - kingpin inclination angle and τ - caster angle, using the formulae [3, 4, 5]

$$\sigma = arctg\left(-d_{21y}^{o}/d_{21z}^{o}\right) \quad (12)$$

$$\tau = arctg(-d_{21x}^{o}/d_{21z}^{o}) \quad (13)$$

where $\boldsymbol{d}_{21}^{o} = [d_{21x}^{o}\ d_{21y}^{o}\ d_{21z}^{o}]^T$ – unit vector of the steering axis, its coordinates are dependent on two variables: s – spring deflection of the suspension strut and p – rack displacement.

4. Numerical example

Spatial mechanism of the steering system (VW Golf) is dimensioned as follow: d_{11}= 395,8; b_{63}=126,0; d_{33}=435,7; b_{12}=57,3; b_{13}=124,4; b_{26}=51,5; $\boldsymbol{a}_1 = [245{,}7; 308{,}5; 22{,}0]^T$; $\boldsymbol{a}_2 = [271{,}9; 578; 616{,}7]^T$; $\boldsymbol{a}_3 = [147{,}0; 239{,}9; 51{,}0]^T$ [mm], $\alpha_1 = arc\ cos(-\boldsymbol{b}_{70}^{o} \cdot \boldsymbol{d}_{22}^{o}) = 81{,}5^o$; $\alpha_2 = \mathrm{arc\ cos}(\boldsymbol{b}_{70}^{o} \cdot \boldsymbol{b}_{63}^{o}) = 98{,}5^o$;

Planar mechanism considered as projection of the spatial steering linkage on the horizontal plane (π) passing through point A_3 is described by using the following dimensions: $\boldsymbol{a}_3 = [147; -239]^T$; $\boldsymbol{b}_3 = [179; -674]^T$; $\boldsymbol{b}_6=[296; -684]^T$ [mm]; $\alpha_2= arc\ cos(\boldsymbol{b}^o_{70} \cdot \boldsymbol{b}^o_{63})=98{,}5°$; $e=150$; $l_b=1370$; $l_d=435$; $l_r=117$; $l_z=480$; [mm]

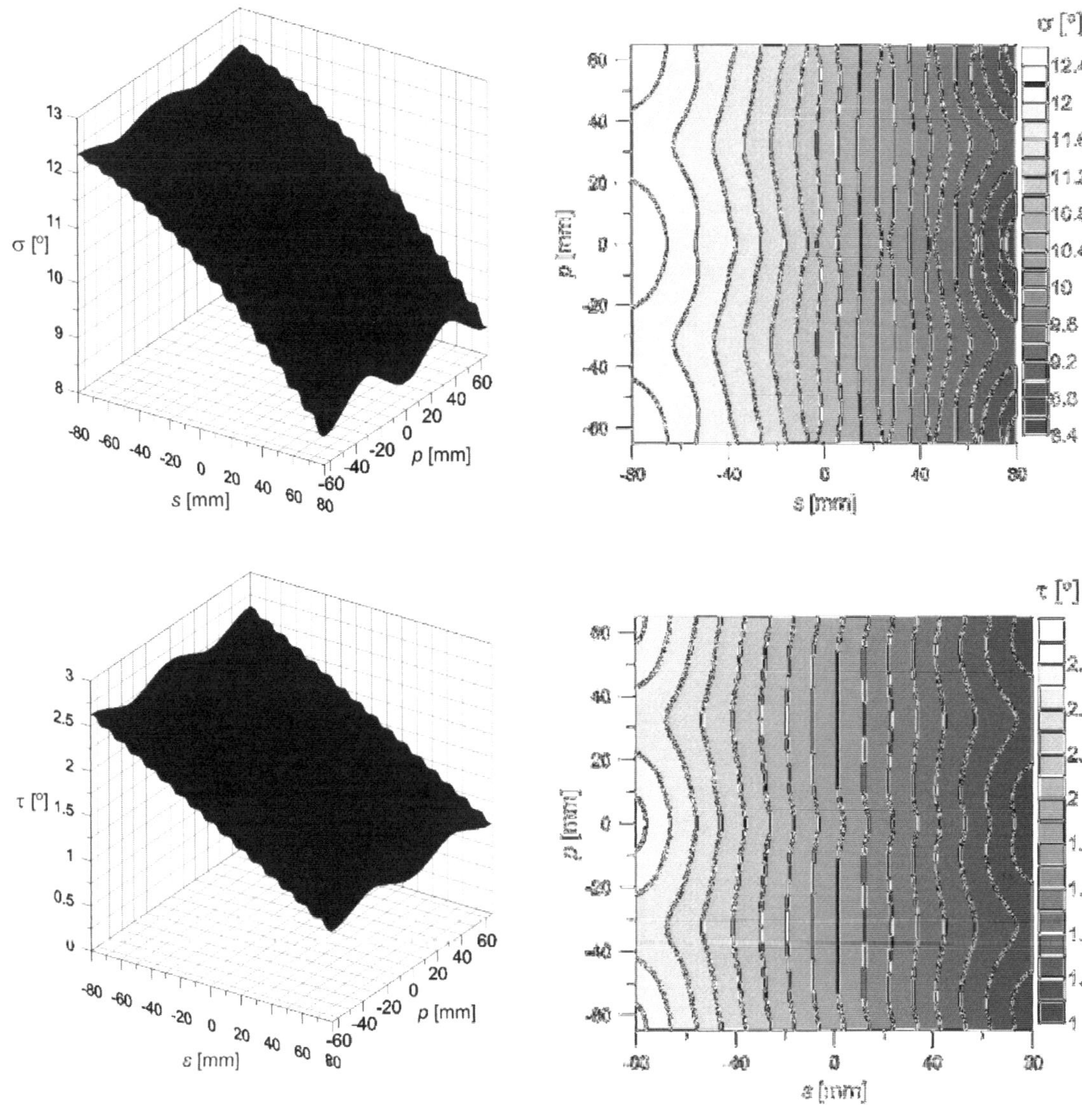

Figure 3. The surfaces describing the relations: a) kingpin inclination σ (s, p); b) caster angle τ (s, p) obtained for McPherson suspension according to data for example (VW Golf).

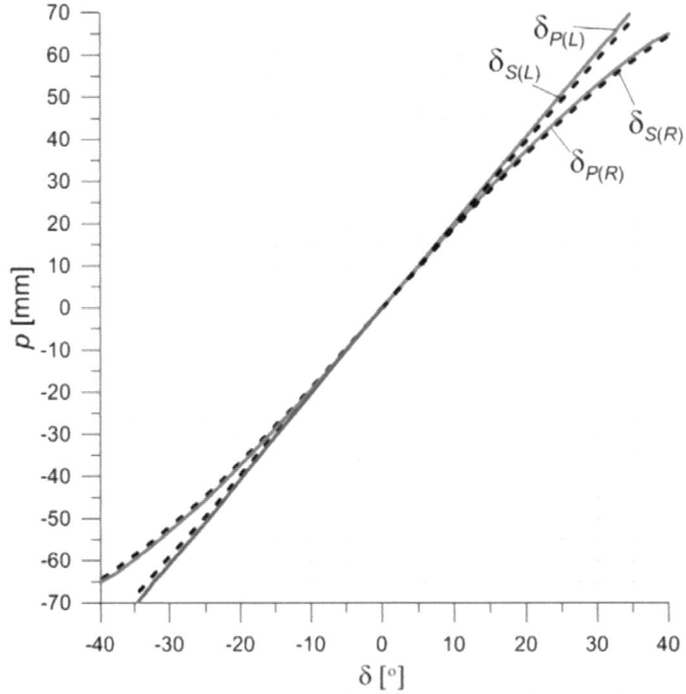

Figure 4. The relations between p - the rack displacement and the wheel steer angles: δ_L - left side and δ_R - right side, obtained using data for example 1 (VW Golf): δ_S – spatial model; δ_P – planar model.

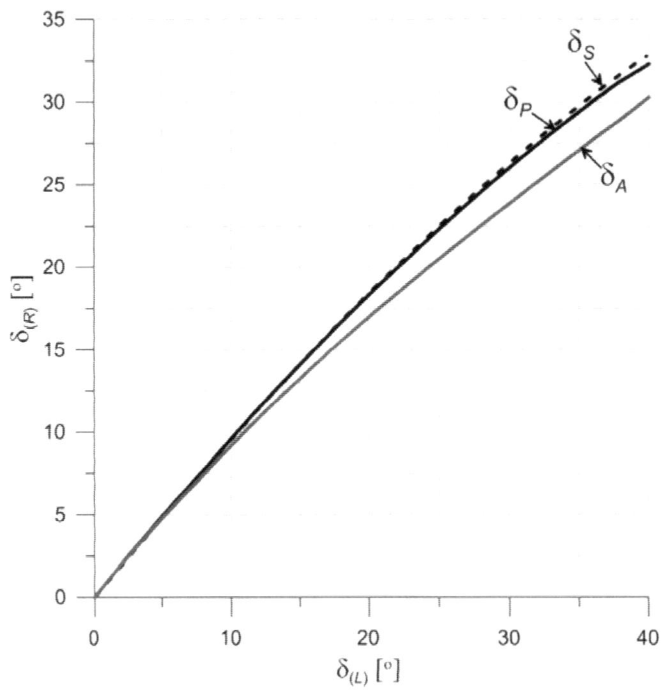

Figure 5. The relations between the wheel steer angles: δ_L - left side and δ_R - right side, obtained using data for example 1 (VW Golf): δ_S – spatial model; δ_P – planar model; δ_A – according to Ackermann.

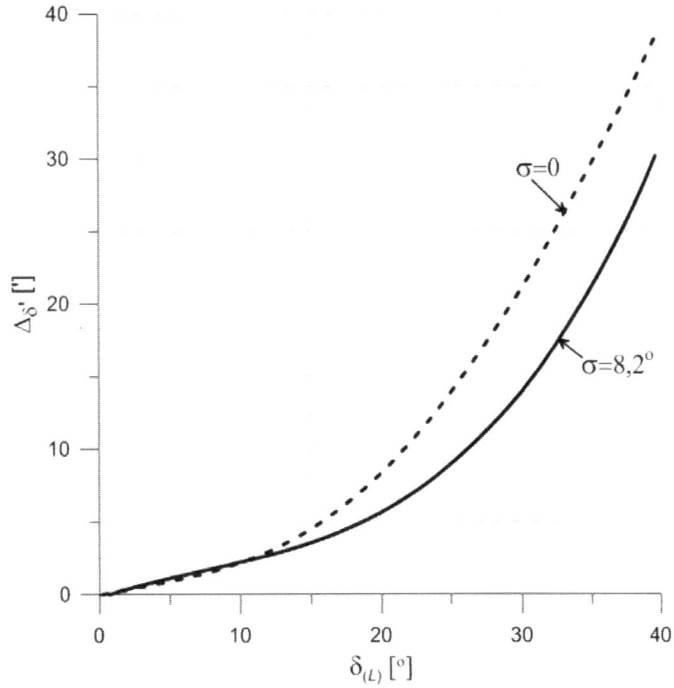

Figure 6. The relations: Δ_δ - difference between the calculated and measured right wheel steer angle, obtained using data for example 2 (Ford), with inclination angle: $\sigma = 8{,}2^{\circ}$ – for the spatial model and $\sigma = 0$ – for the planar model as functions of: δ_L - wheel steer angle of the left side [6].

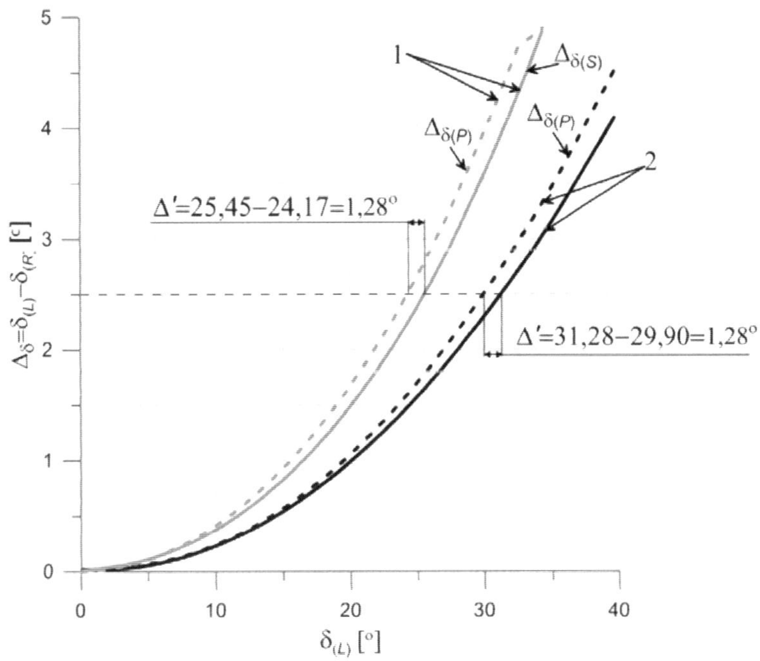

Figure 7. The relations: Δ_δ - difference of the right and left wheel steer angles, obtained using data for example 1 (VW Golf) and 2 (Ford), calculated for: $\Delta_{\delta(S)}$ – spatial model; $\Delta_{\delta(P)}$ – planar model as functions of δ_L - wheel steer angle of the left side, $\Delta' = \Delta_{\delta(S)} - \Delta_{\delta(P)}$.

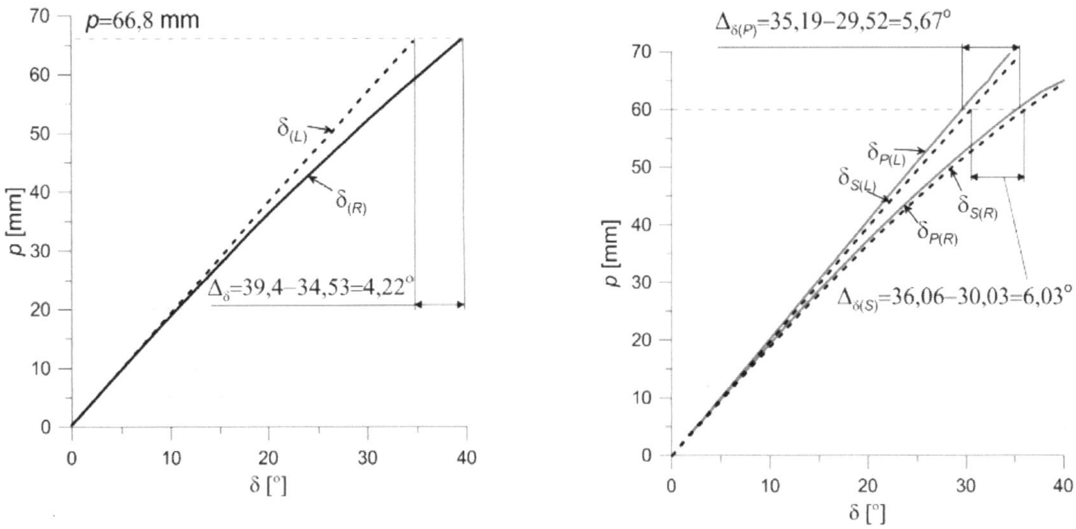

Figure 8. The relations: p - rack displacement as functions of $\delta_{(L)}$ - wheel steer angle of the left side, obtained using data for example 1 (VW Golf), calculated using: δ_S – spatial model; δ_P – planar model, $\Delta_\delta = \delta_{(L)} - \delta_{(R)}$ - difference of the right and left wheel steer angles

5. Analysis results

The differences between the results obtained for planar and spatial models arise, above all, from the relations describing the position and displacement of the wheel knuckle in planar model as functions of the rack displacement only.

For the model parameters, characteristics and the conditions as specified for planar model, the minimum differences with respect to spatial model are taken as the selection criteria for the position of projection plane.

6. Conclusions

The optimum position of the projection plane for planar model is situated near to the vertical coordinate of the rack spherical joints. The lengths of the tie rod and the steering arm can be selected as the lengths of the respective projections.

In the method presented herein, some different solutions have been obtained using a spatial model and a planar model for the calculations. In the first step, the steering linkage can be optimized using the simplified planar model. Finally, the three-dimensional model, best representing the properties of a real motor vehicle, should be used for the verifying calculations.

References

[1] Hanzaki A R at all 2008 Kinematic and sensitivity analysis and optimization of planar rack-and-pinion steering linkages. *Mechanism and Machine Theory*, v.43

[2] Habibi K H at all 2008 Roll steer minimization of McPherson-Strut suspension system using genetic algorithm method *Mechanism and Machine Theory*, vol. 43, pp. 57-67

[3] Knapczyk J and Para S 2012 Estimation of geometrical parameters of an elastokinematic model in a car McPherson suspension *Technical Trans. Mechanics* 109 (10), Cracow pp 19-34

[4] Knapczyk J and Maniowski M 2003 Dimensional synthesis of a five-rod guiding mechanism for a car front wheels *The Archive of Mechanical Engineering* vol. 50 No 1 pp 89-116.

[5] Mantaras D A, Luque P and Vera C 2004 Development and validation of a three-dimensional kinematic model for the McPherson steering and suspension mechanisms *Mechanism Mach. Theory* 39(6) pp 603-619.

[6] Reimpel J 1984 *Fahrwerktechnik: Lenkung*. Vogel Buchverlag, Wurzburg

Method of planning a reference trajectory of a single lane change manoeuver with Bezier curve

D Korzeniowski[1] and G Ślaski[2]
[1] PhD student, Poznan University of Technology, Poznan, Poland
[2] Associate professor, Poznan University of Technology, Poznan, Poland

E-mail: grzegorz.slaski@put.poznan.pl

Abstract. For a comprehensive simulation of vehicle steering process it is vital to model the decision process of planning a trajectory shape and process of the selected trajectory. A single lane change manoeuver is only slightly restricted by the road geometry. There are also other requirements of a possible trajectory of movement, such as the continuity of change (derivative) of curvature, maximizing the passenger's comfort measured with appropriate indicators based on variables of motion dynamics or parameters of motion trajectory which influence that dynamic. This article presents a suggested method of automatic generation of trajectory of single lane change manoeuver. The proposed method can be used as an integral part of driver models and is based on a combination of two symmetrical Bezier curves optionally supplemented with a straight lane connector. The method meets the requirements of a trajectory shape, which results from optimizing the value of parameters controlling Bezier curve based on minimizing the curvature and the resulting lateral acceleration while preserving the continuity of curvature derivative of the planned trajectory.

1. Introduction

The process of driving a vehicle in terms of lateral dynamics can be defined as a task of choosing and following the trajectory of movement which does not exceed the area available for the manoeuver and ensures the safety and desired comfort in terms of forces resulting from the performed manoeuver. A mathematical description of the processes realized during this tasks leads to two separate driver models:

- trajectory planning,
- following of planned trajectory.

The latter group can be further divided into compensating and anticipating models [1]. In both cases the knowledge of planned trajectory is required. This leads to the necessity to define it arbitrarily or with trajectory planning models.

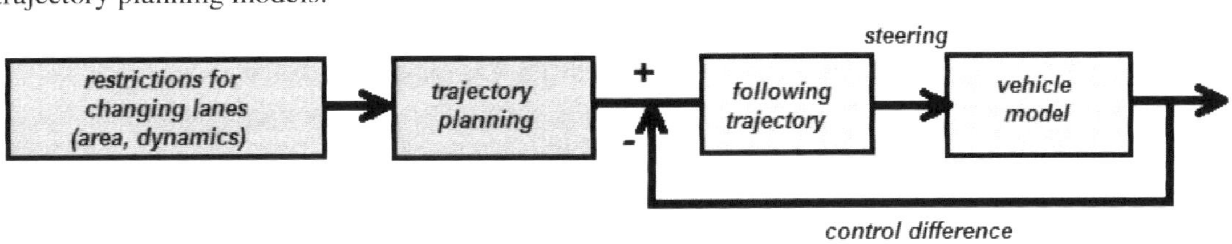

Figure 1. Simplified diagram of steering of autonomous vehicle [own study based on 1].

Besides being used in computational simulations, such models have been recently important in designing the steering systems for autonomous vehicles which perform driver tasks. Unlike computational simulations, the calculations for such vehicles have to be accurate and fast, and realized in real time.

A single line change is one of the most common elementary manoeuver in road traffic. But this does not mean that it is simple and safe one. Statistics clearly show that many collisions and accidents occur during line change manoeuvers [2]. Additional difficulty in trajectory generation for this manoeuver is result of practically infinite variants of its realization. Geometric restrictions in form of lane width and manoeuver distance are not significantly decreasing number of available movement paths.

2. Analysis of methods of generating trajectories presented in literature

The literature includes many concepts and methods to solve the problem of planning a trajectory for a single line change. The way of performing this manoeuver may be considered from two aspects. One results from an observation, according to which drivers performing this manoeuver turn the steering wheel in such a way that its angle changes in a way resembling a sine function [3]. Based on this observation the trajectory can be planned assuming a shape which resulting from a sinusoidal steering input. This method was examined in dissertation [4]. The author tested a steering input in the shape of a sine wave with different amplitudes and frequencies obtaining a range of vehicle motion trajectory, which were subjected to the optimization process to choose one with favourable lateral acceleration and geometry not exceeding the available area .

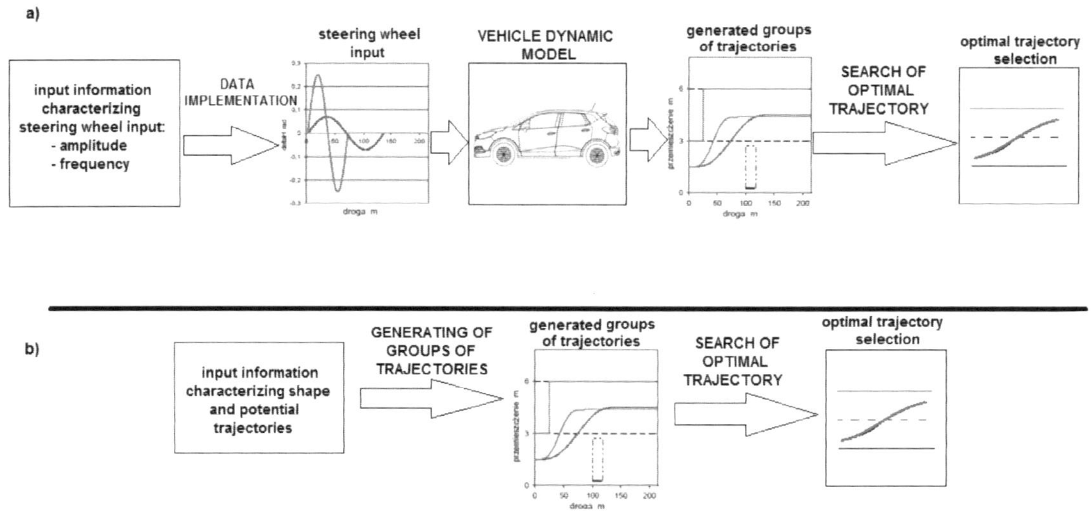

Figure 2. Methods of generating and optimizing the trajectory of a single lane change [own study].

The transition from a steering input to a trajectory requires an additional element, namely a dynamic model of vehicle (figure (2a)). Changing steering input parameters, performing calculations and simulation and finally optimizing the achieved trajectories may prolong the time of manoeuver performance to such an extent that the method cannot be used in real time steering.

The other group includes the methods which involve a generation of potential trajectories first and then optimizing without the use of dynamic vehicle models (figure (2b)). While optimizing the condition of not exceeding the designated area is met already when the trajectory is generated and lateral acceleration is calculated based on the vehicle speed and the trajectory curvature.

Since the method proposed here belongs to the latter group, the next part focuses on presenting the literature concepts which cover only the methods beginning with generation of potential trajectories.

a) circular trajectory [3]

- **method description** – vehicle trajectory is composed of two arcs of constant radii connected with a line segment;
- **number of sections** – three;
- **advantages** – a simple division into sections, utilizing only two mathematical functions, necessity of a relatively short computational time and shortest time of physical realization;
- **disadvantages** – rapid changes in momentary turning radii at points connecting the sections of the trajectory, discontinuities of the curvature, large and rapid changes of lateral acceleration values, low comfort level.

b) composition of a few arcs of different radii [3]

- **method description** – development of the previous method, both curve segments are made up by several arcs with the radii adjusted to smoother transitions and more control over the trajectory shape;
- **number of sections** – several (4-9);
- **advantages** – using only two mathematical functions, easy implementation in algorithms generating movement trajectory, smoother transition between sections than in the previous method;
- **disadvantages** – discontinuities of curvature between sections, difficult to modify the trajectory shape using only one parameter, rapid changes in lateral acceleration – lower comfort level.

c) constant radii arcs connected with clothoids [5]

- **method description** – this method involves the use of clothoid curve to connect two segments with different curvature created with other methods. The use of clothoid curve (as transition curve) helps to create smooth transition between the straight line and the arc section ;
- **number of sections** – $2n - 1$, where n is the number of sections created with the basic method;
- **advantages** – smooth changes in lateral acceleration, continuous curvature;
- **disadvantages** – additional trajectory components prolong the calculation time, necessity to numerically calculate Fresnel integral, potential problems with computing time in real time calculations.

d) 5th order polynomial trajectory [3]

- **method description** – polynomial curve of the 5th order with properly selected parameters was proposed , as a function that relatively well describes the trajectory of a single line change maneuver. What is important, the whole trajectory is created with only one curve;
- **number of sections** – one;
- **advantages** – one equation describing the whole trajectory, ensured continuity of curvature, short computational time;
- **disadvantages** – limited possibilities to modify the trajectory shape (none of input parameters is directly related to the curve shape), a long time of physical performance compared with the previous methods.

e) clothoids obtained by polynomial approximation [10]

- **method description** – a single lane change trajectory consisting of four symmetrically connected clothoids. The use of polynomial approximation helps to shorten the computational time to define curve shape ;
- **number of sections** – four;
- **advantages** – lower computational costs to determine clothoids, continuity of curvature;

- **disadvantages** – no significant disadvantages.

The review of several methods used to determine the reference trajectory of a single lane change manoeuver shows that there is no single accepted solution. The advantages and disadvantages of different concepts should be considered with consideration of the chosen optimization parameters and the purpose of performed manoeuver. A large number of possibilities in the choice of optimization parameters was indicated by G. Prokop [7]. For a single lane change manoeuver some of them are minimization of time needed for manoeuver execution, achieving the highest comfort of passengers and the related minimized lateral acceleration and minimized use of brakes or throttle controls.

The authors of this article suggest their own method to plan the trajectory based on generating a range of possible trajectories. And, in comparison with other methods, it combines a low computational complexity with the ensured continuity of curvature and possibility to optimize for a desired level of comfort measured with lateral acceleration values. The suggested method was verified in terms of achieving the assumed goals.

3. Method of determining a single lane change trajectory with symmetrically connected Bezier curves

3.1. General structure of single lane change trajectory planning

Based on an analysis of available methods the authors identified a general structure of actions useful to develop a method for trajectory planning. It assumes that the following subtasks should be performed while planning and optimizing the manoeuver trajectory:

1. **goal definition** directly related to the conditions of manoeuver execution. For example an emergency lane change to avoid collision or slow paced lane change with care for comfort,
   - definition of **optimization criteria**, constituting the basis for selecting final trajectory;
2. selecting a **method to designate a virtual desired trajectory**, which will help to achieve the defined goal most efficiently;
3. **Optimization** of the created solutions in order to indicate one movement path which best meets the defined criteria.

3.1. *Description of a method to designate trajectory and desired vehicle speed*

The suggested method to designate the virtual desired trajectory of the single lane change manoeuver is based on an assumption that a driving line can be described with two Bezier curves symmetrical in relation to the centre of the road section where the manoeuver takes place. The curves' shape is determined by the location of the so-called control points. Two control points from first curve, the initial and the final one, are synonymous with the beginning point of the manoeuver and its middle. By analogy in the case of the other curve the initial and final control points are synonymous with the middle and final point of manoeuver (figure 3). In accordance with the figures below the lane change manoeuver is divided into two parts (II and III) in the form of Bezier curves and straight line sections before and after the manoeuver (I and IV).

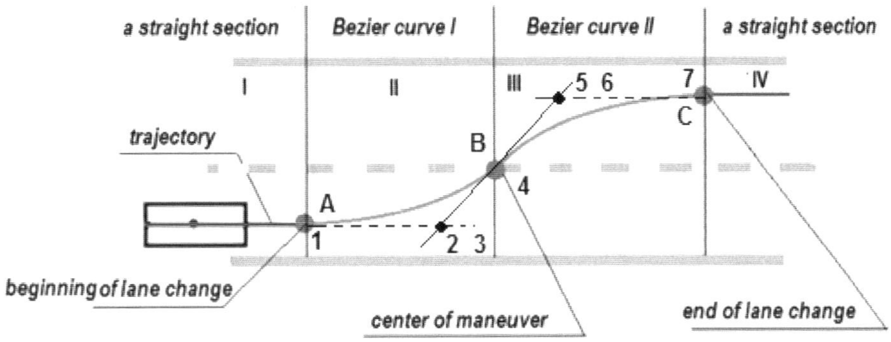

Figure 3. Trajectory of single lane change manoeuver in proposed method [own research].

Bezier curves are a type of parametric curves whose shape depends on the position of control points. The curve interpolates the beginning and ending point and approximate intermediate points [8]. An increasing number of control points can lead to an undesired shape, so it is common to use the curves of up to 3rd order combined into a spline. The coordinates of curve points are described with two mathematical functions $(B_x(t), B_y(t))$ of auxiliary parameter $t$with values range of 0-1 (domain of polynomials describing Bezier curve). Bezier curve $B(t)$ with any number of control points is described with the following equation [8]:

$$B(t) = \sum_{i=1}^{n} \binom{n}{i} (1-t)^{n-1} t^{i-1} P_i \qquad (1)$$

where:

$B(t)$ – coordinates of Bezier curve points,

t – auxiliary parameter with values range of [0, 1],

P_i – control point,

n – number of control points,

$n-1$ – order of Bezier curve,

$\binom{n}{i}$ – binomial expansion:

$$\binom{n}{i} = \frac{n!}{i!(n-i)!} \qquad (2)$$

The suggested method is based on the 3rd order Bezier curves defined by 4 control points and trajectory is assumed to be two-dimensional. If control points are defined as. $P_1 = (x_1, y_1)$, $P_2 = (x_2, y_2)$, $P_3 = (x_3, y_3)$ and $P_4 = (x_4, y_4)$, than the trajectory is described by the equations:

$$\begin{cases} B_x(t) = x_1(1-t)^3 + 3x_2t(1-t)^2 + 3x_3t^2(1-t) + x_4t^3 \\ B_y(t) = y_1(1-t)^3 + 3y_2t(1-t)^2 + 3y_3t^2(1-t) + y_4t^3 \end{cases} \qquad (3)$$
$$for\ 0 \le t \le 1$$

This method is shown in a graphic form in figure 4. Its realization is based on the following steps:

1. **Definition of input data**:
   - restrictions of available area – indication of the positions of points A, B and C (figure 3) defining the beginning, centre and ending of the lane change manoeuver,
   - definition of a range of the alpha parameter, described as an angle between the vehicle trajectory at the centre point and the axis of the road. The values of the alpha influence the shape and curvature of the trajectory by positioning control points.

2. **Determining of acceptable trajectories** – generating Bezier curves for the assumed restrictions and alpha values and rejecting the trajectories with too small alpha value result in control points positioned before the starting point of the trajectory.
3. **Selecting the optimal trajectory**:
   - calculating the parameters which describe optimization criteria, i.e. the value of maximum curvature for all the acceptable trajectories,
   - selecting the optimal trajectory: finding the trajectory with the lowest value of maximum curvature.
4. Calculating the maximum vehicle speed restricted by comfort and safety parameters for defined manoeuver.

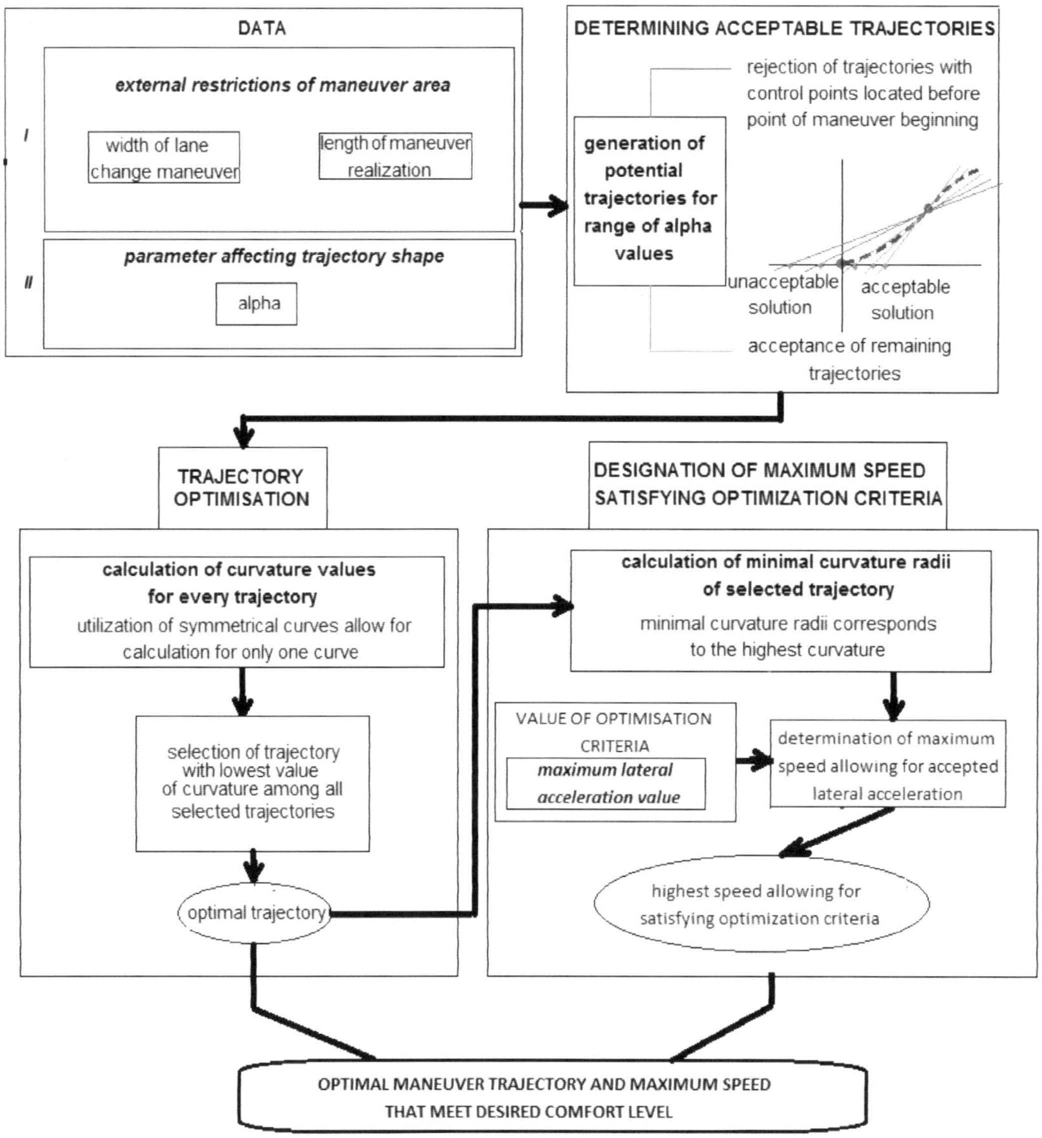

Figure 4. General idea of proposed method [own research].

4. Method implementation

The proposed method was implemented in the MATLAB environment with specifically developed procedure for determining possible trajectories and performing optimization process.

4.1. *Definition of input data*

The data which describe the restriction of the area for the manoeuver is composed of the width of lane change and the length of the route section between the beginning and end of the manoeuver. The first value results from road lane width and is easy to define. It must be noted that the generated trajectory describes the motion of gravity centre and a strict width of lane is not directly used as input information. The length of manoeuver area can be set in two ways. The first directly as the length of road available for lane change, which is results from the distance to the obstacle, for example; this approach is used in this article. The other one indirectly assumes some vehicle speed and average time of manoeuver execution, depending on conditions (emergency avoidance of an obstacle or a comfortable lane change). In the literature the average time for a single lane change manoeuver in extra urban traffic is around 5–7 s [9,10].

Alpha parameter, which influences the trajectory shape, is set as range of possible values. While verifying the method the assumed range was 100 values from 0 to 15°. This was based on a definition of this parameter as an angle between the trajectory centre and the road axis. Data from literature suggest that this angle should not exceed 45° [11].

4.2. *Method of generating potential trajectories*

Every generated manoeuver trajectory passes through three defined control points which geometrically restrict its shape. The position of other points (approximated by curve) define shape of trajectory need to be set during path generation (figure 3). Their location is determined by the alpha parameter value. One of the most important conditions applying to their position is the requirement to provide the continuity of trajectory curvature between all the manoeuver sections. Parametric continuity between two Bezier curves is ensured if tangential vectors of two sections are opposite and of equal length. The way to achieve the continuity of curvature and an example of the number and positions of control points are shown in figure 3.

A distinctive element of the method proposed by the authors is the overlapping of two control points for both curves. In figure 5 these points are indicated with the numbers 2, 3 and 5, 6. Permissible location of these control points is on an intersection of extension of the movement line before the manoeuver and line crossing centre point at an angle equal to alpha. This narrows possible trajectories and simplifies the procedure to generate trajectories.

It should be noted that low values of alpha can lead to control points placed outside the area to execute the manoeuver, i.e. left of point 1 and right of point 7 (figure 3). In this situation the trajectory is significantly deformed and should be removed as unacceptable. Examples of acceptable trajectories with 4 different values of alpha are shown in figure 5.

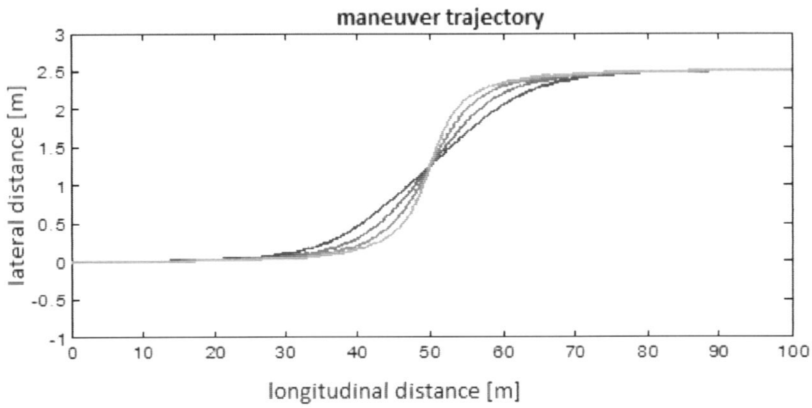

Figure 5. Example of trajectories of lane change manoeuver.

4.3. *Trajectory optimization*

The optimization of trajectory should lead to denoting one solution as best in terms of the chosen optimization criteria. In this article the criteria include the minimizing of the maximum trajectory curvature. Since the trajectory is described by parametric equations it is possible to calculate its curvature with the following formula [8]:

$$k = \frac{|y''_0 x'_0 - x''_0 y'_0|}{\left(x_0'^2 + y_0'^2\right)^{\frac{3}{2}}} \tag{4}$$

While optimizing the trajectory the stage accurate lateral acceleration value representing comfort threshold (optimization parameter). The course of curvature changes of the used Bezier curves is characterized by only one global maximum value denoting the greatest value of curvature. It is the value used to choose the optimal trajectory. The highest curvature reflects the turning radius and consequently the highest lateral acceleration. In further calculations only the maximum values of curvature of generated trajectories were taken into account. The optimal trajectory found in this way means that the of goal of selecting trajectory with lowest lateral acceleration was achieved.

4.4. *Determination of highest allowable speed for adopted indicators*

The highest acceptable vehicle speed for the following selected trajectory should not lead to exceeding the threshold of passengers comfort or threshold of safely performing manoeuver. Determining the value of such speed means achieving the next step, i.e. ensuring the fastest executing of manoeuver while maintaining comfort and safety. The procedure designating the highest acceptable speed begins with finding the minimal turning radii, which is the inverse of trajectory curvature.

To designate the speed a threshold of lateral acceleration must be determined in order to achieve the desired level of comfort or safety. Research performed on 2014 [12] indicates that while changing lanes in ideal weather conditions 85% of all lateral acceleration values were close to 1 m/s^2 with highest values around 3,6 m/s^2. The authors of this article assumed a comfort threshold of 1.6 m/s^2, an average comfort 3,6 m/s^2 and a threshold of discomfort at 5 m/s^2. These values correspond to other research results [15], where the comfort threshold was considered at around 2 m/s^2. Converting the formula for lateral acceleration allow for direct calculation of desired value of maximum speed during manoeuver realization for assumed value of acceleration $a_{y\,max}$and turning radii r:

$$v\_\max < \sqrt{a_{y\,max} r} \tag{5}$$

5. Verification of the proposed method

The correctness of the proposed method with regard to achieving the set objectives was verified in simulation, performed in MATLAB environment with the use of own procedures. The goal of experiment was to perform the process of planning movement trajectory with the use of the proposed method for a single lane change manoeuver and designating the highest acceptable speed with lateral acceleration less than threshold of comfort set at 1.6 m/s^2. The article presents the results of tests for three events with different manoeuver lengths. Input information with assumed parameters alpha and results of optimization are presented in table 1.

Table 1. Input information and results of optimization for three events with different manoeuver lengths.

No.	Lane change width (m)	manoeuver distance (m)	comfort criteria $a_{max}(m/s^2)$	range of alpha (deg)	optimal alpha (deg)	highest accepted speed (km/h)
1	2.5	30	1.6	0– 15	7.4	28
2	2.5	50	1.6	0– 15	4.6	47
3	2.5	100	1.6	0– 15	2.2	94

The acceptable trajectories from the latter case are shown in figure (6a) with the optimal trajectory highlighted in red. Figure (6b) presents the curvature values (for the first curve of every trajectory) which allow to perform optimization process.

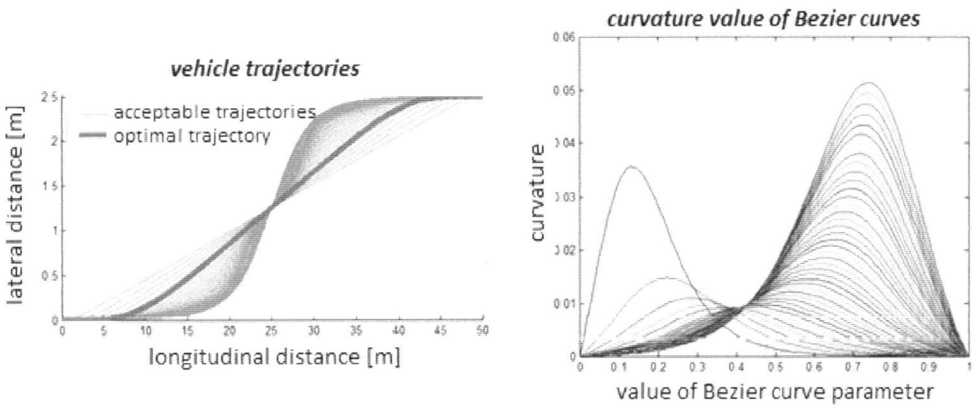

Figure 6. a) acceptable trajectories of single lane change manoeuver, and optimal trajectory highlighted in red. b) values of curvature (for first Bezier curve) of acceptable trajectories.

Based on the achieved results it can be stated that the proposed method allows to determine one optimal trajectory for the considered manoeuver. The quantitative results in the form of the highest acceptable speed and angle of vehicle deflection are within the range proposed in another research [13] and consistent with the values anticipated by authors. In the method [11] the maximum vehicle deviation (equivalent to alpha parameter in the method described in this article) for the speed around 70 km/h is around 4°, which is close to the values presented in table 1. An increase in manoeuver length with a constant level of comfort threshold (lateral acceleration) results in a decrease of alpha parameter and an increase of permissible speed to perform a lane change manoeuver.

6. Conclusions

The proposed method for generating the planned trajectory of a single lane change manoeuver was implemented and verified in MATLAB environment and proved its effectiveness for choosing the optimal trajectory for the chosen criteria. The proposed method is characterized by:

- direct determination of the planned trajectory of the manoeuver without the need for dynamic vehicle models,
- lack of need for test if the trajectory is inside the designated area due to it being initial assumption,
- single parameter optimization,
- trajectories with continuous change of curvature, consistent with physical capabilities of changing steering input and smooth trajectory,
- possibility of optimization of trajectory based on comfort or safety in terms of vehicle lateral dynamics.

With respect to other methods, presented in part 2 of this article, the proposed method is relatively computationally simple and allows to achieve a trajectory which fulfils the basic requirements of the generated trajectory, i.e. the smoothness of curvature changes and possibility to minimize curvature.

References

[1] Reński A 2011 *Bezpieczeństwo czynne samochodu. Zawieszenia oraz układy hamulcowe i kierownicze* (Warszawa: OWPW)
[2] Lee S, Olsen E and Wierwille W 2004 A Comprehensive Examination of Naturalistic Lane-Changes *Report No. DOT HS 809702* pp 1–3
[3] Chee W and Tomizuka M 1994 Vehicle Lane Change Maneuver in Automated Highway Systems California *California PATH Research Report UCB-ITS-PRR-94-22*
[4] Więckowski D 2006 *Optymalizacja wymuszenia działającego na kierownicę ze względu na kryterium poprawności wykonywania typowych manewrów w ruchu samochodu* (Kielce: Rozprawa doktorska Politechnika Świętokrzyska)
[5] Fraichard T and Scheuer A 2004 From Reeds and Shepp's to Continuous-Curvature Paths *IEEE Trans on robotics and automation* **20** 1–4
[6] Montes N and Tornero J 2004 Lane changing using s-series clothoidal approximation and dual-rate based on Bezier points to controlling vehicle pp 1–3
[7] Prokop G 2001 Modeling Human Vehicle Driving by Model Predictive Online Optimization *Vehicle System Dynamics* **35** 19–53
[8] Kiciak P 2000 *Podstawy modelowania krzywych i powierzchni* (Warszawa: WNT)
[9] Hetrick S 1997 Examination of driver lane change behavior and the potential effectiveness of warning onset rules for lane change or "side" crash avoidance systems p 35
[10] Chen J, Zhao P, Mei T and Liang H 2013 Lane Change Path Planning Based on Piecewise Bezier Curve for Autonomous *Vehicle ICVES IEEE International Conference* pp 17–22
[11] Ren D, Zhang G and Wu H 2016 Reference model od desired yaw angle for automated lane changing behavior of vehicle *J. of Harbin Institute of Technology* **23** 23–33
[12] Xu J, Yang K, Shao Y and Lu G 2015 An Experimental Study on Lateral Acceleration of Cars in Different Environments in Sichuan, Southwest China *Discrete Dynamics in Nature and Society* **2015** 1–16
[13] Satzoda R, Gunaratne P and Trivedi 2015 Drive quality analysis of lane change maneuvers for naturalistic driving studies *IEEE Intelligent Vehicles Symposium (IV)* pp 654–59

IOP Conf. Series: Materials Science and Engineering **148** (2016) 012013 doi:10.1088/1757-899X/148/1/012013

Kinematic Analysis of Four-Link Suspension of Steering Wheel by Means of Equation Sets of Geometrical Constraints with Various Structure

M S Kowalski[1]

[1] Cracow University of Technology, Faculty of Mechanical Engineering, Institute of Automobiles and Internal Combustion Engines, al. Jana Pawła II 37, 31-864 Krakow, Poland

E-mail: mskow@pobox.mech.pk.edu.pl

Abstract. In research of the kinematic and dynamic properties of complex mechanical set-ups, results of numerical experiments are used. It is required to minimize the calculation time of various problems in the domain. For the multi-link suspension of the steered wheel , sets of the equations of the geometrical constraints were presented in two structurally different forms, scalar and vector. The vector set consists of the transcendental equations. Their solution was possible after previous expanding the trigonometric functions into power series. Because of the finite amount of the computer memory for the algorithm solving the vector form, it was possible to obtain solutions consisting of three terms. The number of terms in power series of equations' solutions determines the magnitudes of increments of input parameters (degrees of freedom). In this paper it is demonstrated, that fulfilling of this demand is possible by the change of the geometrical constraint's structure of the multi-link wheel suspension system.

1. Introduction

In theoretical problems in the domain of kinematics, guidance systems of steering wheel relative to the car body are represented by the spatial mechanisms with two degrees of freedom. The structure of these mechanisms is various. Their kinematics is solved by means of several different methods [2]. In the case of McPherson suspension and with two diagonal beams the matrix analysis is used [7].

Nonlinear sets of equations describing the geometrical constraints on the relative movement of the elements of multi-link suspension system are formulated in the scalar [3] or vector [6] form. These sets are usually solved by means of numerical methods. They can also be solved by use of the perturbation method [1]. The latter one makes possible to present the solutions of nonlinear sets of geometrical constraints as a series.

In papers [4, 5] from the domain of steered wheel's suspensions usually kinematic characteristics are analysed. These characteristics constitute the intersections of the spatial characteristics. These are usually the dependences between the relative change of wheel track and wheel camber and steering angles and levels of freedom, from which one has a fixed value. Such analysis is not precise, because in the wheel suspensions movement's space the mechanism's singularity can occurred.

IOP Conf. Series: Materials Science and Engineering **148** (2016) 012013 doi:10.1088/1757-899X/148/1/012013

In problems from the domain of the analysis of suspension synthesis and car dynamics modelling, calculation time minimization is desired. Because of this, a choosing the proper solution method is very important.

The wheel suspension system synthesis consists of several stages. At first, the structure of the suspension mechanism should be defined. Next, it's geometrical parameters are determined it means the coordinates of links joining beams with the body and with the stub axle or with the wheel support. Placement of these elements in the suspension movement's space follows the designed suspension's configuration. The synthesis of the steered wheel mechanisms is performed with the simultaneous change of two parameters. Usually these are the vertical translation of the wheel's centre and translating the rack of the steering gear [6].

Examination of the dynamic properties of designed cars is possible by means of the mathematical models with many degrees of freedom. In these models, suspension systems with different structure are considered.

2. The range and purpose of the paper

In the paper there will be presented two ways of solution of the multi-link steered wheels' suspension system kinematics with two degrees of freedom using the perturbation method [1]. In the first one, the set of 17 equations representing geometrical constraints of relative displacements of the suspension's elements will be formulated in the scalar form. In the second one, 5 equations in a vector form.

The main aim of the work will be a determination of the spatial kinematic characteristics of the four-link steered wheels' suspension mechanism. It will be presented the quantitative dependence between time of effective numerical computation and the structure of the geometrical constraints of the examined mechanism.

3. Structure of multi-link steered wheels' suspension system

In Fig. 1 it is presented the diagram of the structure of the four-link steered wheels' suspension system. Points B_1, B_2, B_4, B_5 are centres of the ball-joints linking the beam with the stub axle. Point B_3 is a centre of the ball-joint linking the extreme stub-axle mechanism's rod with the stub-axle beam.

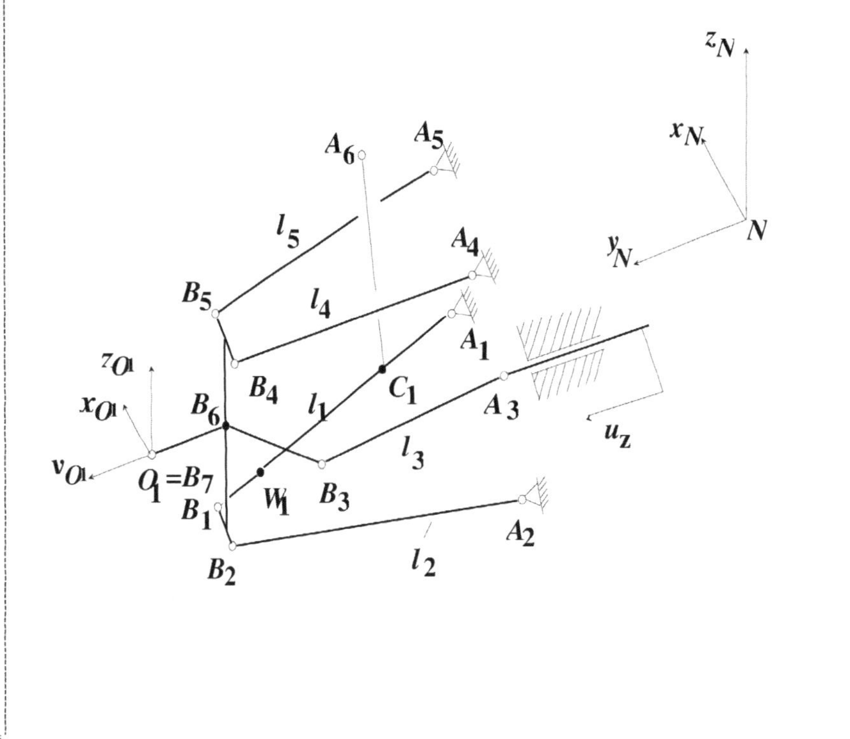

Figure 1. Scheme of the four-link steered wheel suspension mechanism .

Points A_1, A_2, A_4, A_5 are the centres of the ball-joints which replace the silent-block joints linking the beams with the body. Point A_3 is a centre of the ball-join joining the stub-axle mechanism's extreme rod with the steering gear's rack. The lower front beam, represented in the figure by the join A_1B_1, is joined with the stabilizer in the point W_1. In the point C_1 this beam is joined with the telescope column supporting the body in the point A_6. Points B_6 and B_7 represent the wheel's rotation axis.

The sets of coordinates {N}, {O_1} are connected with the body and the wheel's stub-axle respectively.

4. Equations of the mechanism's geometrical constraints

Geometrical constraints equations of the above presented suspension mechanism can be formulated as the set of 17 or 5 nonlinear algebraic equations. In the first one the equations express the squares of the distances between characteristic points of the mechanism

$$\begin{cases} \mathbf{r}_{AjBj}^{\mathbf{T}}\,\mathbf{r}_{AjBj} = l_j^2 \ , \ \text{for } j = (1)5 \\ \mathbf{r}_{BjBk}^{\mathbf{T}}\,\mathbf{r}_{BjBk} = l_{jk}^2 \ , \ \text{for} \begin{cases} j = 1 \text{ i } k = (2)5 \\ j = 2 \text{ i } k = (3)5 \\ j = (1)5 \text{ i } k = 7 \end{cases} \end{cases} \tag{1}$$

In the above set, the input parameters are coordinates q_3 of the point $O_1(q_1, q_2, q_3)$ and translation of the rack u_z, added to the coordinate y_{A3} of the point A_3 (x_{A3}, $y_{A3}+u_z$, z_{A3}). At given parameters q_3 and u_z from the set (1) the coordinates of points B_j (x_{Bj}, y_{Bj}, z_{Bj}), for j=(1)5 and coordinates q_1 and q_2 of the point O_1 are determined.

In the second method, the equations express the squares of length of vectors with origins and ends in points A_j, B_j, for j=(1)5 respectively.
They are presented in the form:

$$\left(\mathbf{r}_{NO1.N} + \mathbf{A}_{NO1}\mathbf{r}_{O1Bj.O1} - \mathbf{r}_{NAj.N}\right)^T\left(\mathbf{r}_{NO1.N} + \mathbf{A}_{NO1}\mathbf{r}_{O1Bj.O1} - \mathbf{r}_{NAj.N}\right) = l_j^2, \quad \text{for } j = (1)5. \tag{2}$$

where:

$$\mathbf{A}_{\mathrm{NO1}} = \mathbf{A}_\gamma\,\mathbf{A}_\beta\,\mathbf{A}_\alpha = \begin{bmatrix} c\gamma & -s\gamma & 0 \\ s\gamma & c\gamma & 0 \\ 0 & 0 & 1 \end{bmatrix}\begin{bmatrix} c\beta & 0 & s\beta \\ 0 & 1 & 0 \\ -s\beta & 0 & c\beta \end{bmatrix}\begin{bmatrix} 1 & 0 & 0 \\ 0 & c\alpha & -s\alpha \\ 0 & s\alpha & c\alpha \end{bmatrix} \tag{3}$$

is a rotation matrix {O_1} respectively to {N}.

However

$$l_j^2 = \mathbf{r}_{AjBj.N}^T\mathbf{r}_{AjBj.N}.$$

At given parameters q_3 and u_z from the set (1) are calculated the coordinates q_1 and q_2 of the wheel centre $O_1(q_1, q_2, q_3)$ and rotation angles {O_1} respectively to {N}: α, β, γ.

For ensuring the equivalence of the calculation range of the algorithms derived on the basis of sets (1) and (2) it is necessary to determine the rotation angles {O_1}: α, β and γ respectively to {N}. So, the set (1) must be supplemented with the calculations of the mentioned angles.

After calculation of coordinates of points O_1, and B_j(j=(1)5) it is possible to create three vectors r_{O1Bj} for j∈{1,2,3,4,5}. For every of these vector a matrix equation is fulfilled

$$\mathbf{r}_{O1Bj.O1} = \mathbf{A}_{O1N}\mathbf{r}_{O1Bj.N} \tag{4}$$

where:

$\mathbf{r}_{O1Bj.O1}$ – vector in {O₁},

$\mathbf{r}_{O1Bj.N}$– vector in {N},

$$\mathbf{A}_{O1N} = \boldsymbol{A}_{NO1}^{T} = \begin{bmatrix} c\beta c\gamma & c\beta s\gamma & -s\beta \\ s\alpha s\beta c\gamma - c\alpha s\gamma & s\alpha s\beta s\gamma + s\alpha c\gamma & s\alpha c\beta \\ c\alpha s\beta c\gamma + s\alpha s\gamma & c\alpha s\beta s\gamma - s\alpha c\gamma & c\alpha c\beta \end{bmatrix}$$

Denoting the coordinates of vectors: $\mathbf{r}_{O1Bj.O1} = [x_{bj} \quad y_{bj} \quad z_{bj}]^T$. $\mathbf{r}_{O1Bj.N} = [x_{jb} \quad y_{jb} \quad z_{jb}]^T$ and putting $j = n, m, v$; on the basis (4) we obtain:

$$\left.\begin{aligned} x_{bn} &= (x_{nb}c\gamma + y_{nb}s\gamma)c\beta - z_{nb}s\beta \\ x_{bm} &= (x_{mb}c\gamma + y_{mb}s\gamma)c\beta - z_{mb}s\beta \\ x_{bv} &= (x_{vb}c\gamma + y_{vb}s\gamma)c\beta - z_{vb}s\beta \end{aligned}\right\} \tag{5}$$

From the set (5) the rotation angles β and γ are calculated. For calculation of the rotation angle α we use equation:

$$y_{bv} = (x_{vb}s\beta c\gamma)s\alpha - (x_{vb}s\gamma)c\alpha + (y_{vb}s\beta s\gamma)s\alpha + (y_{vb}c\gamma)c\alpha + (z_{vb}c\beta)s\alpha \tag{6}$$

Next, having the rotation angles {N} respectively to {O₁}, the $\mathbf{r}_{B6B7.N}$ vector's coordinates were calculated.

$$\mathbf{r}_{B6B7.N} = \mathbf{A}_{NO1}\mathbf{r}_{B6B7.O1}\,, \tag{7}$$

The unit vector of the wheel rotation axis $\mathbf{e}_k = [e_{kx} \quad e_{ky} \quad e_{kz}]^T$ and steering and camber angles of the wheel:

$$\delta_k = -arctg\left(\frac{e_{kx}}{e_{ky}}\right), \tag{8}$$

$$\gamma_k = -arcsin(e_{kz}). \tag{9}$$

Calculation of the angles δ_k and γ_k in both algorithms is analogous.

5. Solution of the equations' sets of the suspension mechanism's geometrical constraints

The solution of sets (1) and (2) were obtained by means of the perturbation method [1].

In respect to the transcendental set (2), trigonometric functions were resolved into trigonometric series:

$$\sin(x_0 + x) = \sin x_0 + x\cos x_0 - \frac{x^2 \sin x_0}{2}, \tag{10}$$

$$\cos(x_0 + x) = \cos x_0 - x\sin x_0 - \frac{x^2 \cos x_0}{2} \tag{11}$$

The resulting set of equations can be presented in the general form:

$$f_j(q_1, q_2, \alpha, \beta, \gamma) = 0, \quad j = (1)5 \tag{12}$$

Solutions of the set (12) were split into nonlinear and linear parts:

$$f_{jN}(q_1, q_2, \alpha, \beta, \gamma) + f_{jL}(q_1, q_2, \alpha, \beta, \gamma) = 0, \quad j = (1)5 \tag{13}$$

Nonlinear parts of these equations were multiplied by the perturbation parameter ε and an auxiliary set of equations was obtained

$$g_j(\varepsilon, q_1, q_2, \alpha, \beta, \gamma) = \varepsilon f_{jN} + f_{jL}, \quad j = (1)5, \tag{14}$$

For ε=1 the sets (13) and (14) are identical, but for ε=0 the set (13) it is formed only from the linear part.
It was assumed, that the solutions of set (14) is presented in the form of the series:

$$\left.\begin{array}{c} q_1 = \sum_{i=0}^{m} \varepsilon^i q_{1i}, \quad q_2 = \sum_{i=0}^{m} \varepsilon^i q_{2i}, \\ \alpha = \sum_{i=0}^{m} \varepsilon^i \alpha_i, \quad \beta = \sum_{i=0}^{m} \varepsilon^i \beta_i, \quad \gamma = \sum_{i=0}^{m} \varepsilon^i \gamma_i, \end{array}\right\} \tag{15}$$

After substitution of (15) to (14) we obtain

$$g_j(\varepsilon, q_1(\varepsilon), q_2(\varepsilon), \alpha(\varepsilon), \beta(\varepsilon), \gamma(\varepsilon)) = 0, \quad j = (1)5, \tag{16}$$

The set (16) was expanded into series respectively to the powers of ε:

$$\sum_{i=0}^{2} \varepsilon^i g_{ji} = 0, \ j = (1)5 \tag{17}$$

Next, the linear sets of equations g_{ji} =0 were solved, at first for i = 0, next for i=1,2. The obtained solutions can be presented in the general form:

$$\begin{array}{c} q_1 = \sum_{i=0}^{2} q_{1i}, \quad q_2 = \sum_{i=0}^{2} q_{2i}, \\ \alpha = \sum_{i=0}^{2} \alpha_i, \quad \beta = \sum_{i=0}^{2} \beta_i, \quad \gamma = \sum_{i=0}^{2} \gamma_i, \end{array} \tag{18}$$

The set (1) was solved analogously.

6. Numerical example

The characteristic points were assigned to the designing placement of the suspension mechanism from Fig.1. Their values of coordinates (in mm) are as follows:

A_1(132.1, 347.4, -93.8), A_2(-245.3, 379.0, -114.3)
A_3(-88.05, 385.61, 280.92), A_4(138.74, 431.18 389.65)
A_5(6.47, 400.47, 381.42),
B_1(29.0, 697.0, -98.2), B_2(-23.7, 693.3, -132.5)
B_3(-140.3, 619.7, 284.0), B_4(9.3, 675.3, 389.0)
B_5(-24.0, 642.0, 387.0),
B_6(-0.2, 650.0, 0.5), B_7(0.0, 750.0, 0.0)

The solutions of sets (1), (2) were obtained for given input parameters q_3 and u_z with the same step equal 0.5 mm. Comparable solutions of the mentioned sets at this step were almost the same.
In Figures 2 and 3 kinematic characteristics of steering and camber angles together with the computation time are presented.

IOP Conf. Series: Materials Science and Engineering **148** (2016) 012013 doi:10.1088/1757-899X/148/1/012013

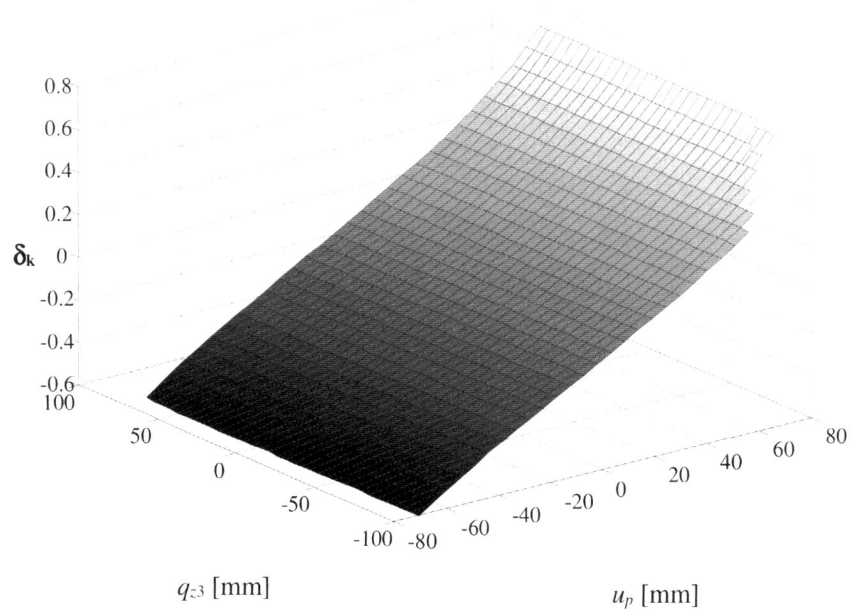

Figure 2. Dependence of steering angle – δ_k on the vertical displacement of the wheel centre – q_3 and on the displacement of the rack – u_p. Computation time for set of equation (1) = 516.02 s; Computation time for set of equation (2) = 15132.94 s.

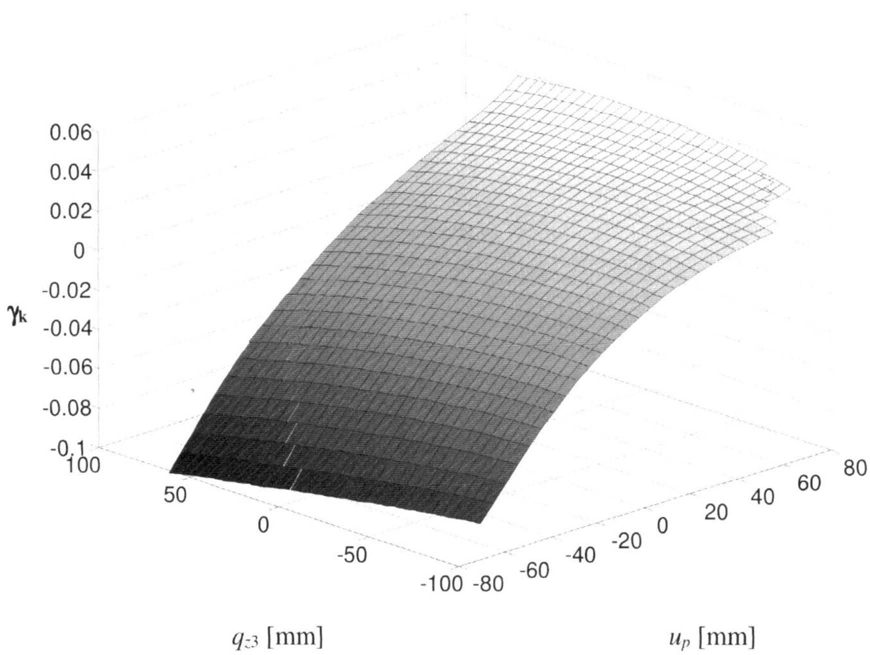

Figure 3. Dependence of camber angle – γ_k on the vertical displacement of the wheel centre – q_3 and on the displacement of the rack – u_p. Computation time for set of equation (1) = 516.02 s; Computation time for set of equation (2) = 15132.94 s.

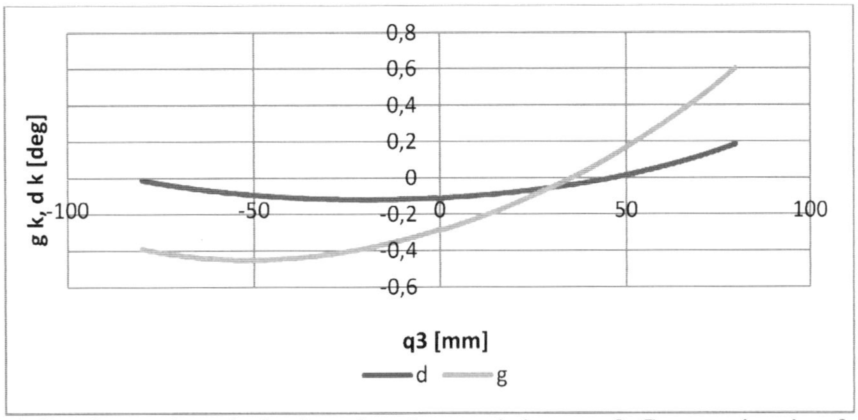

Figure 4. Steering angle and wheel camber characteristics, $u_z = 0$. Computation time for set of equation (1) = 2.276 s; Computation time for set of equation (2) = 45.216 s.

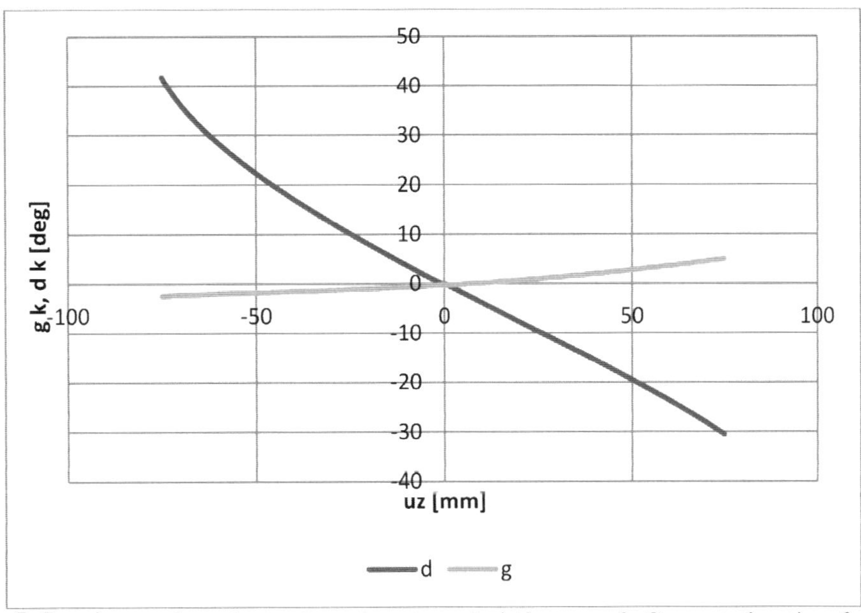

Figure 5. Steering angle and wheel camber characteristics, $q_3 = 0$. Computation time for set of equation (1) = 2.184 s; Computation time for set of equation (2) = 46.334 s.

7. Summary

For the multi-link suspension of the steered wheel , sets of the equations of the geometrical constraints were presented in two structurally different forms, scalar and vector. The vector set consists of the transcendental equations. Their solution was possible after previous expanding the trigonometric functions into power series (10) and (11).

Because of the finite amount of the computer memory for the algorithm solving the vector form, it was possible to obtain solutions consisting of three terms (18). The number of terms in power series of equations' solutions determines the magnitudes of increments of input parameters (degrees of freedom).

Solutions of the set (1) are numerical series containing 10 terms each.

The calculations were performed on a computer with Intel Core i5 M430 CPU (2.27 GHz) and 4 GB RAM. Computation times of our own programs run in the MATLAB environment, containing 5 and 17 equations for the same problems in the suspension kinematics domain considerably differ.
Computation time for the set of 5 equations is many times greater (that is about 20-30 times)_shown in all figures.

References

[1]. Grzyb A 1992 *On a perturbation method for the analysis of the kinetostatics of mechanisms* Akademie Verlag ZAMM, 2 angew. Math Mech. 72,6 T615-T618

[2]. Knapczyk J 1998 Metody analizy wielowahaczowych zawieszeń kół samochodu (Methods of analysis of car's wheels' multi-link suspensions), Interkonmot 98, *Teka Komisji Naukowo Problemowej Motoryzacji Krakowskiego Oddziału PAN*, z.**16**, 19-29, Kraków

[3]. Kowalski M S 2012 *Optymalizacja wymiarowa wybranego mechanizmu wielowahaczowego zawieszenia samochodu (Dimensional optimisation of a selected multilink suspension mechanism in vehicle suspension)* Monography 412, Ser. ISSN 0860-097X Kraków

[4]. Matschinsky W 1987 *Die Radführungen die Straßenfanrzeuge*, Verlag TUV Rheinland, Koln

[5]. Reimpell J 1997 *Podwozia samochodów. Podstawy konstrukcji (Elements of construction of car's chassis)* WKiŁ Warszawa

[6]. Struski J and Wach W 2007 Multidimensional optimization of the steered wheel multi-link suspension system, *SAE-Paper* **2007-01-0849** 2007 World Congress Detroit, Michigan

[7]. Struski J 1993 *Quasi-statyczne modelowanie sterowności samochodów (Quasi-statical modelling of car's steerability)* Monography 144, Wydawnictwa Politechniki Krakowskiej Kraków

Optimization of damping in the passive automotive suspension system with using two quarter-car models

Z Lozia[1,2] and P Zdanowicz[1]

[1] Warsaw University of Technology, Faculty of Transport, Poland
[2] Automotive Industry Institute (PIMOT), Warsaw, Poland

E-mail: lozia@wt.pw.edu.pl

Abstract. The paper presents the optimization of damping in the passive suspension system of a motor vehicle moving rectilinearly with a constant speed on a road with rough surface of random irregularities, described according to the ISO classification. Two quarter-car 2DoF models, linear and non-linear, were used; in the latter, nonlinearities of spring characteristics of the suspension system and pneumatic tyres, sliding friction in the suspension system, and wheel lift-off were taken into account. The smoothing properties of vehicle tyres were represented in both models. The calculations were carried out for three roads of different quality, with simulating four vehicle speeds. Statistical measures of vertical vehicle body vibrations and of changes in the vertical tyre/road contact force were used as the criteria of system optimization and model comparison. The design suspension displacement limit was also taken into account. The optimum suspension damping coefficient was determined and the impact of undesirable sliding friction in the suspension system on the calculation results was estimated. The results obtained make it possible to evaluate the impact of the structure and complexity of the model used on the results of the optimization.

1. Introduction; review of the literature dealing with quarter-car models and their applications

The quarter-car model (figure 2) is a system with two degrees of freedom (2DoF), describing the vertical vibrations of a "sprung mass" (a part of the vehicle body solid) and an "unsprung mass" (connected with the road wheel). The two masses are linked by a spring-damper system representing the spring-damping properties of the suspension of the specific wheel. A spring-damper (or spring-only) element, situated between the unsprung mass and the ground, represents the road wheel properties in the radial direction. The models of this type appeared in 1970s (e.g. [11, 19, 27, 28]) and were employed in many publications of 1980s (e.g. [10, 15, 20, 30, 31]) and 1990s (e.g. [5, 12, 21]); they are useful even today, e.g. for the works that provide guides for vehicle designers (such as [2, 6, 7, 13, 17, 26, 29, 31, 32, 33, 34, 35]). They are both linear and non-linear, describing the passive, semi-active, and active suspension systems. In some of the works, the model analyses were supplemented with experimental examinations of systems of the quarter-car type [13, 26, 33, 36, 37]. The quarter-car models were used in sophisticated optimization algorithms, where Pareto-optimal solutions were sought [14], with random nature of selected model parameters (sprung mass determined by hardly-predictable vehicle load and tyre stiffness depending on inflation pressure, e.g. [6]) being taken into account, and at the assessment of designs of variable-damping, semi-active, and active suspension systems (e.g. [2, 4, 5, 6, 7, 29, 30, 31, 33, 34, 35]). Most authors of the publications where the optimization of suspension system parameters is addressed have enumerated three major assessment criteria, which are related to the minimization of measures of the driver and passengers' ride discomfort and of changes in the normal reaction at the tyre/road contact. A reduction in the range of working displacements of the suspension system has also been considered [4, 5, 6, 7, 11, 12, 17, 19, 20, 21, 26, 27, 28, 29, 30, 31, 33, 34, 35]. A contradiction between these requirements has been highlighted in [30]: increased damping in the suspension system impairs the comfort but improves the safety. It has been shown in [17] that optimum levels of damping in suspension systems can be determined as a compromise between the two assessment criteria. In most of the works reported in the literature, the excitation was assumed as a Gaussian stationary random process characterized by the power spectral density of road profile height. For the non-linear models, sinusoidal waveforms or random process realizations were applied as the input. The wavelength values ranged from 0.1 m to 100 m [19, 20, 27, 30]. The smoothing properties of vehicle tyres were taken into account [3, 4, 15, 16, 17]. The vehicle speed range was 10-50 m/s [7]. In the publications quoted, frequencies from the 0-80 Hz band were taken for the analyses [2, 19, 20, 22, 28, 33, 34, 35]. This range is also mentioned in the ISO standards applicable to the comfort of vibrations (e.g. [8]).

2. Objective

The work was aimed at the optimization of damping in the passive suspension system of a motor vehicle moving rectilinearly with a constant speed on a road with rough surface. The impact of the structure and complexity of the quarter-car model used on the optimization results obtained was assessed within this work.

3. The quarter-car models and their equations of motion

Figure 1a shows a linear quarter-car model. The model consists of a "sprung mass" m_1 and an "unsprung mass" m_2. The suspension stiffness and the radial stiffness of the pneumatic tyre have been denoted by k_1 and k_2, respectively. The symbols c_1 and c_2 have the meaning of the suspension damping coefficient and the coefficient of radial damping of the tyre, respectively. The symbol ζ represents the time-dependent kinematic excitation (input) from the uneven road surface, measured along the $O\zeta$ axis. The vertical displacements of individual model masses from the static equilibrium position for zero road profile height are denoted by z_1 and z_2, respectively. The motor vehicle, as well as the model under consideration, moves rectilinearly with a constant speed V [km/h] (or v [m/s]).

Figures 1b and 1c show a non-linear quarter-car model. The symbols m_1, m_2, z_1, and z_2 have the same meanings as they have in the case of the linear model. Symbols ζ_1 and ζ_2 represent the vertical coordinates of the centres of model masses along the $O\zeta$ axis, which is fixed to the road. Fs_1 is the suspension spring force, approximated by degree 5 polynomials; Fts_1 is the sliding friction force in the suspension system, represented by a broken line [15, 16, 36]; Ft_1 is the viscous damping force in the suspension system, represented by a broken line [16]; Fs_2 is the radial force of tyre elasticity, approximated by a degree 3 polynomial [15, 16, 36]; Ft_2 is the damping force in the pneumatic tyre, represented by a linear function of the rate of radial deflection of the tyre [1, 10, 11, 12, 15, 16, 20, 27, 33]; $g = 9.81$ m/s^2; ζ is the excitation (input) from the uneven road surface.

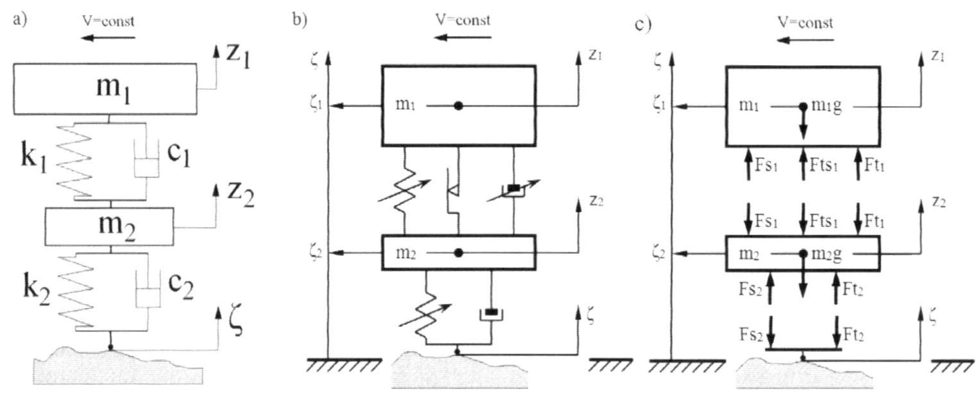

Figure 1. Linear (a) and non-linear (b, c) quarter-car model (for the notation used, see the text); a, b – general structure of the models; c – system of forces in the non-linear model.

The equations of motion for the linear model have been derived from the principle of dynamic force analysis. Their matrix form is shown in relation (1), where the symbols of the matrices of inertia **M**, viscous damping **C**, stiffness **K**, excitation influences transmitted by damping in the pneumatic tyre $\mathbf{C}_\zeta$, and excitation influences transmitted by radial stiffness of the pneumatic tyre $\mathbf{K}_\zeta$ have been pointed out. The vectors of generalized coordinates (displacements), velocities, and accelerations have been denoted by $\mathbf{q}, \dot{\mathbf{q}}, \ddot{\mathbf{q}}$, respectively. This notation is used in the concise form (2) of this relation.

$$\underbrace{\begin{bmatrix} m_1 & 0 \\ 0 & m_2 \end{bmatrix}}_{\mathbf{M}} \cdot \underbrace{\begin{bmatrix} \ddot{z}_1 \\ \ddot{z}_2 \end{bmatrix}}_{\ddot{\mathbf{q}}} + \underbrace{\begin{bmatrix} c_1 & -c_1 \\ -c_1 & c_1 + c_2 \end{bmatrix}}_{\mathbf{C}} \cdot \underbrace{\begin{bmatrix} \dot{z}_1 \\ \dot{z}_2 \end{bmatrix}}_{\dot{\mathbf{q}}} + \underbrace{\begin{bmatrix} k_1 & -k_1 \\ -k_1 & k_1 + k_2 \end{bmatrix}}_{\mathbf{K}} \cdot \underbrace{\begin{bmatrix} z_1 \\ z_2 \end{bmatrix}}_{\mathbf{q}} = \underbrace{\begin{bmatrix} 0 \\ c_2 \end{bmatrix}}_{\mathbf{C}_\zeta} \cdot \dot{\zeta} + \underbrace{\begin{bmatrix} 0 \\ k_2 \end{bmatrix}}_{\mathbf{K}_\zeta} \cdot \zeta \tag{1}$$

$$\mathbf{M} \cdot \ddot{\mathbf{q}} + \mathbf{C} \cdot \dot{\mathbf{q}} + \mathbf{K} \cdot \mathbf{q} = \mathbf{C}_\zeta \cdot \dot{\zeta} + \mathbf{K}_\zeta \cdot \zeta \tag{2}$$

For equation (2), the Laplace transform has been formulated, at zero initial conditions. After transformations, a solution in the form of equation (3) has been obtained, where the domain $s = r + i \cdot \omega$ has a real part r and an imaginary part ω, while $i^2 = -1$ (ω [rad/s] is the radian frequency):

$$\mathbf{q}(s) = \mathbf{H}_\mathbf{q}(s) \cdot \zeta(s) \tag{3}$$

The operational transmittance (transfer function) for displacements $H_q(s)$ has the form (4):

$$\mathbf{H}_\mathbf{q}(s) = \begin{bmatrix} H_{q_1}(s) \\ H_{q_2}(s) \end{bmatrix} = \frac{\mathbf{q}(s)}{\zeta(s)} = \left(\mathbf{M} \cdot s^2 + \mathbf{C} \cdot s + \mathbf{K}\right)^{-1} \cdot \left(\mathbf{C}_\zeta \cdot s + \mathbf{K}_\zeta\right) \tag{4}$$

We can easily pass from the Laplace transform to the Fourier transform. The operational transmittances will then become spectral transmittances. In formal terms, this is expressed in passing from domain s to argument $i \cdot \omega$, by assuming the real part r of the expression $s = r + i \cdot \omega$ as zero. With such a substitution, the relations (3) and (4) still hold.

The equations of motion for the non-linear model (figures 1b and 1c) have also been derived with the use of the principle of dynamic force analysis. They have the form presented by equations (5):

$$\begin{cases} \ddot{z}_1 = \dfrac{Fs_1 + Fts_1 + Ft_1}{m_1} - g \\ \ddot{z}_2 = \dfrac{-Fs_1 - Fts_1 - Ft_1 + Fs_2 + Ft_2}{m_2} - g \end{cases} \tag{5}$$

Their solutions are obtained in the time domain with the use of approximation techniques, by numerical integration. The results presented herein have been obtained with the use of an authorial program built in the Matlab Simulink environment [36].

4. Qualitative differences between the linear and non-linear quarter-car models

The differences between the two models arise, above all, from the form of the functions describing the forces in the suspension system and pneumatic tyre. The description of the suspension spring force in the non-linear model reflects the impact of suspension travel limiters in the rebound and compression phase (beginning and end of the characteristic curve), noticeable in experimental tests [15, 16, 36]. In the non-linear model, the sliding friction force is taken into account (while it is ignored in the linear model) and the asymmetry of shock absorber characteristics can be represented. The approximation of changes in the force of tyre elasticity by a degree 3 polynomial [15, 16, 36] can reflect the nonlinearities of this force at small and large tyre deflections and the possibility of wheel lift-off.

5. Random excitation from an uneven road surface

An assumption has been made here that the road surface is undeformable and its vertical irregularities are a realization of a stationary Gaussian random process. In the relevant ISO standard [9], road surfaces have been classified from A to H, i.e. from very good (A) through good (B), average (C), and poor (D) to very poor (E and further F, G, and H). The road is described in the ISO standard [9] by the function of power spectral density (PSD) $S_d(\Omega)$ [m^3/rad] of one track parallel to the road centreline:

$$S_d(\Omega) = S_d(\Omega_0)\cdot(\Omega/\Omega_0)^{-w} \tag{6}$$

where: $\Omega = 2\pi/L$ [rad/m] is angular frequency of the longitudinal road profile; L [m] is road roughness wavelength; Ω_0 [1/m] is reference angular frequency (in most cases, $\Omega_0 = 1.0$); $S_d(\Omega_0)$ [m^3/rad] is road roughness indicator, defining the general road surface condition (whether it is good or poor); w [–] is road waviness indicator, informing whether long or short waves predominate in the road profile. Roads of different classes differ from each other in the $S_d(\Omega_0)$ values. The exponent w has a constant value of w = 2. The road roughness wavelength L values have been assumed as 0.1–100 m. This is the "micro-profile" of road surface irregularities [10, 19, 20]. Based on the known function $S_d(\Omega)$ and with the use of subprograms (taken from publication [10]), where the FFT (fast Fourier transform) techniques were employed, a discrete realization of the road profile was generated, which was then approximated by cubic splines [15, 16]. Such a form enables the use of realizations with high wavelength values, exceeding the maximum micro-profile wavelength (100 m), thanks to the periodical form of the approximating function, which has been described in publications [15, 16].

In the pneumatic tyre model, the "point contact" model is normally used for the radial direction of vibrations [3, 4, 15, 16]. Therefore, the road roughness wavelength that is taken into account must be bounded from below for unnatural overestimation of the damping forces in the tyre model to be avoided. To overcome this difficulty, the smoothing properties of vehicle tyres may be simulated by adopting a "fixed footprint tyre model" [3, 4, 15, 17] (with averaging the road profile height over the tyre/road contact patch, whose length is $2\cdot l_{op}$) and by filtering the road roughness spectra. To do this, a filter was used with the absolute value of its transmittance being as specified in [15]. The inputs thus filtered make it possible to employ the one-point tyre/road contact model [3, 4, 15, 17].

6. Automotive suspension damping selection criteria (optimization criteria)

6.1. Basic optimization criteria

As in publications [2, 4, 5, 6, 7, 11, 12, 17, 19, 20, 21, 27, 28, 29, 30, 31, 32, 33, 34, 35], three criteria have been adopted to assess the correctness of selection of suspension damping coefficient c_1:

- minimization of the measure of vehicle occupants' discomfort, i.e. the standard deviation of sprung mass acceleration, σ_a [m/s^2];
- minimization of the safety hazard, measured by the standard deviation of the vertical component of the normal reaction at the tyre/road contact, σ_F [N];
- reduction in the working displacements (range of changes in the deflection) of the suspension system to a value lower than the suspension displacement limit r_{zg} [m].

They may be described as functions of suspension damping coefficient c_1 and vehicle speed V:

$$\sigma_a(c_1, V) \Rightarrow \text{minimum} \qquad \sigma_F(c_1, V) \Rightarrow \text{minimum} \qquad 6\cdot\sigma_{uz}(c_1, V) \le r_{zg} \tag{7}$$

where: $\sigma_a(c_1, V)$ and $\sigma_F(c_1, V)$ [m/s^2] are the objective functions subject to minimization and $\sigma_{uz}(c_1, V)$ [m/s^2] is the standard deviation of suspension deflection; r_{zg} [m] is the suspension displacement limit.

The above criteria were adopted for each of all the road surface quality classes taken into consideration. Factor "6" in formula (7) comes from the Gaussian distribution of the suspension deflection, but for the linear model. For such a model, the standard deviations σ_a, σ_F, and σ_{uz} were calculated from the known formulas [1, 10, 11, 12, 17, 18, 19, 20, 22, 23, 24, 25, 27, 35] as definite integrals of power spectral densities of sprung mass accelerations, vertical component of the normal road reaction, and suspension deflection. These densities were calculated as products of the power spectral density of the excitation (road profile height) and squared absolute values of the transmittances of the quantities mentioned above [1, 10, 11, 12, 17, 18, 19, 20, 22, 23, 24, 25, 27, 35].

In the case of the non-linear model, σ_a, σ_F, and σ_{uz} were calculated from time histories of the vertical acceleration of the sprung mass, $\ddot{z}_1(c_1, V, t)$, vertical component of the normal road reaction, $F(c_1, V, t)$, and suspension deflection, $u_1(c_1, V, t)$, for the assumed values of parameters c_1 and V.

6.2. Modified optimization criteria

The criterion that is easiest to use is the one related to reducing the range of working travel of the suspension system. The other two criteria (see (7)) give rise to more difficulties. Due to different units of measure, it is not easy to formulate a joint synthesized measure of discomfort and safety hazard. A way to solve this problem has been shown in [17]. The proposed modification of the criteria should be started with normalizing the criterial quantities, i.e. σ_a and σ_F, by dividing each of the values in an individual curve (plotted for a specific vehicle speed V) by the maximum value of the curve within the c_1 domain. The symbol "max" means here "the highest value" instead of a formal maximum.

$$\sigma_{au}(c_1, V) = 100\ \% \cdot \sigma_a(c_1, V)/\max\{\sigma_a(c_1, V)\} \text{ (max. relative to } c_1\text{, for a specific V value)} \quad (8)$$

$$\sigma_{Fu}(c_1, V) = 100\ \% \cdot \sigma_F(c_1, V)/\max\{\sigma_F(c_1, V)\} \text{ (max. relative to } c_1\text{, for a specific V value)} \quad (9)$$

The minimums of the curves are better visible than they were before the normalization [17].

The next step in the modification process is to change the reference system, with maintaining the variability range from zero to 100 %. Thus, the comfort indicator WP1 and the safety indicator WP2 are determined [17]. Here, the symbol "min" has the meaning of a formal minimum.

$$WP1(c_1) = -\sigma_{au}(c_1, V) + \min\{\sigma_{au}(c_1, V)\} + 100\ \% \text{ (min. relative to } c_1\text{, for all V values)} \quad (10)$$

$$WP2(c_1) = -\sigma_{Fu}(c_1, V) + \min\{\sigma_{Fu}(c_1, V)\} + 100\ \% \text{ (min. relative to } c_1\text{, for all V values)} \quad (11)$$

The modified criteria of selection of suspension damping coefficient c_1 (i.e. the criteria of optimizing the selection of suspension damping in respect of ride comfort and safety) have the form [17]:

$$Q_z(c_1) = w_k \cdot WP1(c_1) + w_b \cdot WP2(c_1) \quad \Rightarrow \text{maximum} \quad (12)$$

$$6 \cdot \sigma_{uz}(c_1, V) \leq r_{zg} \quad (13)$$

where the new symbols are defined as follows: $Q_z(c_1)$ [%] is the modified objective function subject to maximization; w_k [–], w_b [–] are weighting factors for comfort and safety, from within a range of <0, 1>; WP1 [%],WP2 [%] are comfort and safety indicator, respectively (see relations (10) and (11); r_{zg} [m] is suspension displacement limit. In publication [17], attention has been directed to a change in the terminology used: the terms "discomfort" and "safety hazard" have been replaced with the notions "comfort indicator WP1" and "safety indicator WP2", because of their different monotonicity as functions of c_1. Now, the maximum of the new objective function $Q_z(c_1)$ is sought.

7. Calculation data: parameters of the model and of the test conditions

The model parameters taken as an example corresponded to the data of the front suspension system of the Isuzu D-max motor vehicle: $m_1 = 578$ kg, $m_2 = 69.5$ kg. The coefficients of stiffness for the linear model were $k_1 = 42\,520$ N/m, $k_2 = 220\,000$ N/m. The coefficient of viscous damping of the tyre was $c_2 = 150$ N·s/m and c_1 was a variable in the calculations. The length of the tyre/road contact patch in static conditions was $2 \cdot l_{op} = 0.185$ m. Three roads were chosen for the calculations, i.e. road of class B (good), with $S_d(\Omega_0) = 0.000004$, $\Omega_0 = 1.0$, and $w = 2$; road of class C (average), with $S_d(\Omega_0) = 0.000016$, $\Omega_0 = 1.0$, and $w = 2$; and road of class D (poor), $S_d(\Omega_0) = 0.000064$, $\Omega_0 = 1.0$, and $w = 2$ (see [9]). The wavelength values varied between 0.1 m and 100 m. In consideration of the ISO standard provisions that concern the vibration comfort (e.g. [8]), the frequency band of 0-80 Hz (0-502.65 rad/s) was adopted for the analyses. The analyses were carried out for 4 constant vehicle speeds V [km/h], i.e. 30 km/h, 60 km/h, 90 km/h, and 120 km/h. The vehicle speed values expressed as v [m/s] and V [km/h] are connected with each other by the generally known relation $v = V/3.6$. The damping coefficient c_1 [N·s/m] was changed and its values were indirectly expressed by means of a relative damping coefficient [1], i.e. by values γ [–] defined as follows for the linear model:

$$\gamma = c_1/c_{1kr}, \quad \text{where:} \quad c_{1kr} = 2 \cdot m_1 \cdot \omega_{01}; \quad (14)$$

c_{1kr} [N·s/m] is critical damping coefficient; $\omega_{01} = 2 \cdot \pi \cdot f_{01}$ [rad/s] is the first (lower) natural radian frequency of the undamped system; f_{01} [Hz] is the first (lower) natural Hertz frequency of the undamped system. For the system under analysis, these values were $\omega_{01} = 7.839$ rad/s, $f_{01} = 1.248$ Hz, $c_{1kr} = 9\,062$ N·s/m. The analyses were carried out for 26 values of the relative damping coefficient γ [–], changed from 0.1 to 0.6 in steps of 0.02. The damping coefficient in the suspension system was:

$$c_l = \gamma \cdot c_{lkr} = \gamma \cdot 2 \cdot m_l \cdot \omega_{0l} \quad [\mathrm{N \cdot s/m}] \tag{15}$$

The working displacements of the suspension system were limited to $r_{zg} = 0.12$ m, corresponding to the real linear range of operation of the suspension system of the vehicle under analysis.

8. Calculation results

For the model parameters and characteristics and the test conditions as specified, the quantities taken as the criteria of selection of damping in passive automotive suspension systems (optimization criteria) were calculated in accordance with the algorithm presented in a previous part of this paper.

8.1. Determining the optimum values of the relative suspension damping coefficient γ

The optimization criterion related to reduction in the working displacements of the suspension system can be most easily applied. In figure 2, standard deviations of suspension deflection σ_{uz} have been presented as functions of relative damping coefficient γ for the linear and non-linear model, for roads of class B, C, and D, and for vehicle speed values of 30 km/h, 60 km/h, 90 km/h, and 120 km/h. In qualitative terms, the curves are similar to each other, but their values are considerably lower for the non-linear model. For the linear model, the requirements expressed by (7) and (13) were met for the road of class B (good). For the road of class C (average), they were also met, but only for the relative damping coefficient γ higher than 0.159. On the road of class D (poor), the requirements of relations (7) and (13) could only be met when the vehicle speed did not exceed 60 km/h and the relative damping coefficient γ was higher than 0.30. For the non-linear model, the requirements expressed by (7) and (13) were met for the road of class B (good) and C (average) for all the vehicle speeds and the whole range of changes in the relative damping coefficient under analysis. For the requirements to be met on the road of class D (poor), the vehicle speed had to be limited to 60 km/h and the relative damping coefficient γ had to be higher than 0.25. Actually, it is hard to expect that any driver would decide to travel with a high speed on a road being in poor technical condition (of class D). It should be stressed that the nonlinearities of the real vehicle suspension system and tyres could be reproduced in the latter model and this explains the fact that the criterion related to reduction in the working displacements of the suspension system, expressed by relations (7) and (13), could be easier satisfied.

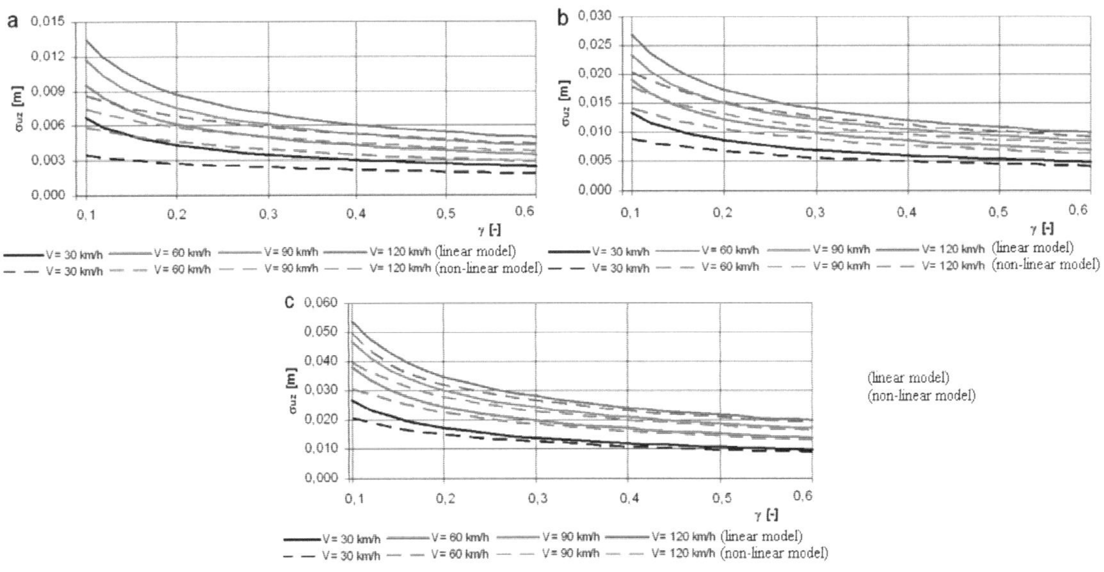

Figure 2. Standard deviations of suspension deflection σ_{uz} for the linear model (solid lines) and non-linear model (dashed lines) vs. relative damping coefficient γ, for roads of class B (good, figure a), C (average, figure b), and D (poor, figure c) according to ISO [9] and for four vehicle speed values.

Figure 3 shows standard deviations of sprung mass acceleration σ_a and of the vertical component of the normal road reaction σ_F as functions of relative damping coefficient γ for the linear and non-linear model, for road of class C (average), and for the four vehicle speed values. The curves show similar monotonicity, but their minimums for both models occurred at different locations. It can be noticed that for the non-linear model, the minimum (i.e. optimum) values are distinctly shifted towards lower γ values. The results obtained for roads of class B and D are similar, in qualitative terms.

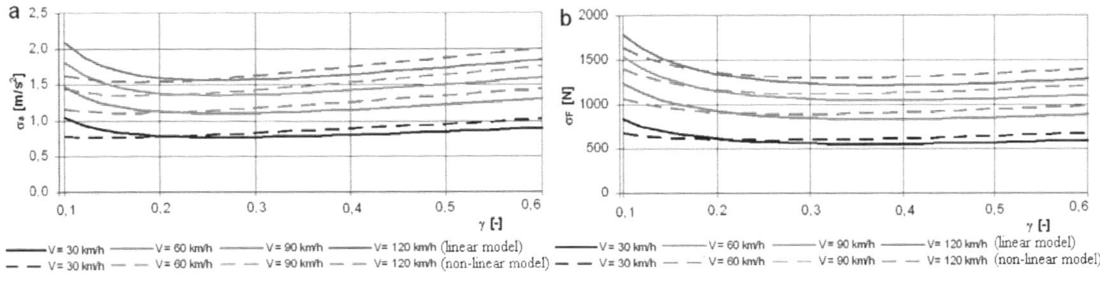

Figure 3. Standard deviations of sprung mass acceleration σ_a (figure a) and of the vertical component of the normal road reaction σ_F (figure b) vs. relative damping coefficient γ for the linear model (solid lines) and non-linear model (dashed lines), for road of class C (average) according to ISO [9], and for the four vehicle speed values.

In result of the procedure described by relations (8)-(21), the criteria of assessment were modified, i.e. they were normalized and the reference system was changed, with their variability range from zero to 100 % having been maintained. Thus, a modified objective function $Q_z(c_l)$ has been created, which describes the quality of tuning the suspension damping, c_l, in respect of comfort and safety of vehicle ride. The independent variable is now the relative damping coefficient γ of the linear model, related to c_l according to equation (15). The motor vehicle ride comfort and safety may be treated equally ($w_k = w_b$) or with different preferences ($w_k \neq w_b$).
Figure 4 shows modified objective function Q_z as a function of relative damping coefficient γ for the linear and non-linear model, for roads of class B, C, and D, for four vehicle speed values, and for two sets of the weighting factors for comfort and safety: set 1 (solid lines in the graph) with $w_k = w_b = 0.5$ and set 2 (dashed lines in the graph) with $w_k = 0.4$ and $w_b = 0.6$. Set 2 represents the case that safety is deemed 1.5 times as important as comfort ($1.5 = 0.6/0.4$). For this set, the optimum values of the relative damping coefficient γ (at the maximums of Q_z) are somewhat higher than those for comfort and safety being equally treated (set 1), whether the linear or non-linear model is used.

The results obtained from the linear model are similar for different roads in terms of curve shape and monotonicity. They show identical optimum damping values and similar Q_z function values for the corresponding damping. For the non-linear model, the results largely depend on the road type. The worse the road (changing from B to C and D), the higher the optimum damping is for both sets of the weighting factors. In the extreme case, the optimum damping coefficient value is even doubled (from 0.12 for road B to 0.24 for road D, at a vehicle speed of 30 km/h). For each of the roads and vehicle speeds considered, the optimum damping determined with the use of the non-linear model is lower than that obtained from the linear model. The better the road, the bigger differences are observed.

8.2. Discussion on the reasons for the differences in the optimum suspension damping values for the non-linear model compared with the linear model

The authors have formulated a hypothesis that the shift of the optimum suspension damping value obtained from the nonlinear model in relation to that determined with the use of the linear model chiefly arises from the fact that sliding friction in the suspension system is represented in the non-linear model. Figure 5 shows results analogous to those presented in figure 4, but obtained without taking into account the sliding friction in the non-linear model. As it can be seen, the optimum values of the suspension damping values came considerably nearer to those obtained from the linear model. An exception was the case where the vehicle moved on road D (poor) with a speed of 120 km/h, but this may only serve information purposes, as a vehicle drive in such conditions is hardly imaginable.

The results presented confirm the hypothesis having been formulated and highlight the significant importance of sliding friction for vibration damping and safety measures. Although friction, which is inevitable in the construction of motor vehicle suspension systems, is connected with the known disadvantageous properties of suspension systems [36, 37], it softens the requirements for the necessary level of viscous damping in shock absorbers.

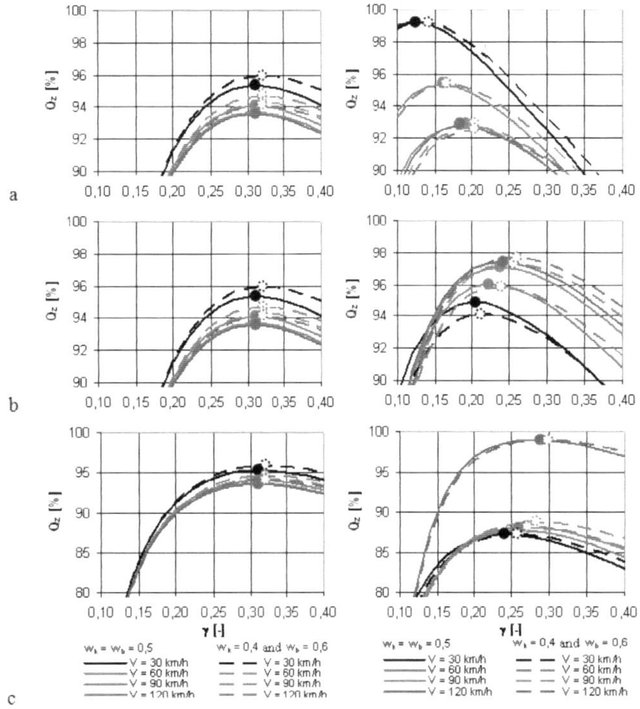

Figure 4. Objective function Q_z vs. relative damping coefficient γ for the linear model (left) and non-linear model (right), for a road of class B (good, figure a), C (average, figure b), and D (poor, figure c), for four vehicle speed values, and for two sets of the weighting factors for comfort and safety: set 1 (solid lines) with $w_k = w_b = 0.5$ and set 2 (dashed lines) with $w_k = 0.4$ and $w_b = 0.6$.

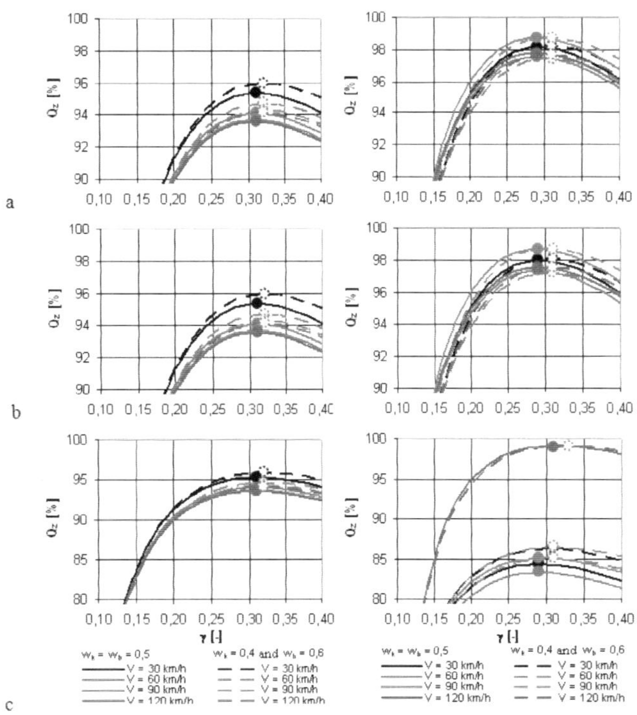

Figure 5. Objective function Q_z vs. relative damping coefficient γ for the linear model (left) and non-linear model (right), with the sliding friction being disregarded in the latter model. Graphs plotted for road of class B (good, figure a), C (average, figure b), and D (poor, figure c), for four vehicle speed values, and for two sets of the weighting factors for comfort and safety: set 1 (solid lines) with $w_k = w_b = 0.5$ and set 2 (dashed lines) with $w_k = 0.4$ and $w_b = 0.6$.

8.3. Determining the optimum damping values for the two sets of the weighting factors for comfort and safety

The results obtained from the linear model (figure 4) show that the optimum value of the relative damping coefficient γ for set 1 of the weighting factors for comfort and safety is 0.31; for set 2, it is 0.32. Based on equation (15) and the model data, the optimum values of the suspension damping coefficient c_l may be calculated:

– for set 1: $c_l = 0.31 \cdot c_{lkr} = 0.31 \cdot 9\,062$ N·s/m ≈ 2 810 N·s/m (16)

– for set 2: $c_l = 0.32 \cdot c_{lkr} = 0.32 \cdot 9\,062$ N·s/m ≈ 2 900 N·s/m (17)

The difference between the c_l values is not too big, in spite of significant differences between the weighting factors adopted. The results obtained from the non-linear model, but with disregarding the sliding friction (see figure 5), show that the optimum value of the relative damping coefficient γ for set 1 of the weighting factors for comfort and safety is about 0.29 and for set 2, it is about 0.31 (except for the case with the vehicle moving on road D with a speed of 120 km/h). These values are close to those obtained from the linear model (compare with (16) and (17): the differences are about 6.5 % and 3.1 %, respectively). Therefore, it is reasonable to adopt the results obtained from the linear model and represented by relations (16) and (17). In spite of efforts made to reduce the sliding friction in suspension systems, this phenomenon will always occur in real systems, which will result in stronger suspension damping. The values of the friction impact are not only small; they will also lower the said percentage differences between the calculation results obtained from the non-linear and linear models.

9. Conclusion

The optimum values of the suspension damping coefficient determined by means of the non-linear model are markedly lower than those obtained from the linear one. For the road of class B (good) and the vehicle speed being low enough (e.g. 30 km/h), the optimum values of coefficient γ calculated from the more complex model are lower by even more than a half than those obtained with the use of the linear representation (γ falling in the ranges of 0.12-0.14 and 0.31-0.32, respectively). This is chiefly due to the fact that sliding friction, always taking place in suspension systems but not taken into account in the linear approximation, is represented in the non-linear model. The disregarding of this undesirable phenomenon leads to a very important conclusion that the calculation results obtained from the linear and non-linear model are close to each other (γ of about 0.31-0.32 and 0.29-0.31, respectively, if the unrealistic case of the vehicle being driven on the road of class D with a speed of 120 km/h is ignored). The next stage of this work will be dedicated to optimization of the shock absorber asymmetry coefficient. Afterwards, the authors will focus on the verifying calculations, carried out with the use of a three-dimensional vehicle motion model, which most accurately represents the properties of a real motor vehicle.

Acknowledgements

This work was done within Research Project PBS3/B6/27/2015, sponsored by the National Centre for Research and Development. Part carried out by the team from the Warsaw University of Technology.

References

[1] Arczyński S 1993 *Mechanika Ruchu Samochodu* (Warszawa: WNT)

[2] Cao D, Song X and Ahmadian M 2011 Editors' perspectives: road vehicle suspension design, dynamics, and control *Vehicle System Dynamics* **49 (1-2)** pp 3-28

[3] Captain K M, Boghani A B and Wormley D N 1979 Analytical tire models for dynamic vehicle simulation *Vehicle System Dynamics* **8** pp 1–32

[4] Crolla D A 1996 Vehicle dynamics – theory into practice J. of Autom. Eng. 210 pp 83–94

[5] Firth G R June 1991 *The Performance of Vehicle Suspensions Fitted with Controllable Dampers* (The University of Leeds, Department of Mechanical Engineering, Ph.D. thesis)

[6] Gobbi M, Levi F and Mastinu G 2006 Multi-objective stochastic optimization of the suspension system of road vehicles *J. of Sound and Vibration* **298** pp 1055–72

[7] Gobbi M and Mastinu G 2001 Analytical description and optimization of the dynamic behaviour of passively suspended vehicles *J. of Sound and Vibration* **245(3)** pp 457–81
[8] ISO 2631–1: 1985 and 1997 *Mechanical vibration and shock – Evaluation of human exposure to whole-body vibration – Part 1: General requirements*
[9] ISO 8608:1995 *Mechanical vibration – Road surface profiles – Reporting of measured data*
[10] Kamiński E and Pokorski J 1983 *Teoria samochodu. Dynamika zawieszeń i układów napędowych pojazdów samochodowych* (Warszawa: WKŁ)
[11] Kasprzyk T Prochowski L and Szurkowski Z 1974 Optymalizacja własności sprężystych i dobór konstrukcji ogumienia samochodu osobowego dla różnych warunków eksploatacji Techn. Mot. (Part I 10/1974 pp 10–12, Part II 11/1974 pp 14-19)
[12] Kasprzyk T and Prochowski L 1990 *Teoria samochodu. Obciążenia dynamiczne zawieszeń* (Warszawa: WKŁ)
[13] Konieczny J 2011 Laboratory tests of active suspension system *J. of KONES Powertrain and Transport* **18(1)** pp 263–272
[14] Kwarciński T 2007 Sprawiedliwość czy efektywność? Analiza wykorzystująca ekonometryczny model wzrostu gospodarczego z historycznie optymalnym zróżnicowaniem płac *Acta Universitatis Lodziensis, Folia Oeconomica* **213** pp 109–124
[15] Lozia Z 1985 *Wybrane zagadnienia symulacji cyfrowej procesu hamowania samochodu dwuosiowego na nierównej nawierzchni drogi* (Warsaw Univ. of Technology, Faculty of Automotive and Construction Machinery Engineering, doctoral thesis)
[16] Lozia Z 1998 Analiza ruchu samochodu dwuosiowego na tle modelowania jego dynamiki *Prace Naukowe Politechniki Warszawskiej. Transport* vol **41** (a monograph)
[17] Lozia Z 2016 *The use of a linear quarter-car model to optimize the damping in a passive automotive suspension system – a follow-on from many authors' works of the recent 40 years.* (The Archives of Automotive Engineering vol 71(1) pp 39-71)
[18] Meirovitch L 1975 *Elements of Vibration Analysis* (McGraw-Hill Kogakusha)
[19] Mitschke M 1989 *Teoria samochodu. Dynamika samochodu* (Warszawa: WKŁ)
[20] Mitschke M 1989 *Dynamika samochodu (Drgania vol 2)* (Warszawa: WKŁ)
[21] Muluka V May 1998 *Optimal Suspension Damping and Axle Vibration Absorber for Reduction of Dynamic Tire Loads* (Montreal: Concordia Univ., Dep. of Mech. Eng., M.A.Sc. thesis)
[22] Newland D E 1984 *An Introduction to Random Vibrations and Spectral Analysis* (Longman, 2nd edition)
[23] Osiecki J 1979 *Podstawy analizy drgań mechanicznych* (Kielce Univ. of Technology, Poland)
[24] Osiecki J 1994 *Dynamika maszyn* (Warszawa: Military Univ. of Technology)
[25] Osiecki J, Gromadowski T and Stępiński B 2006 *Badania pojazdów samochodowych i ich zespołów na symulacyjnych stanowiskach badawczych* (Radom, Poland: ITE–PIB) ISBN 83-7204-510-0
[26] Patil S A and Joshi S G 2014 Experimental analysis of 2 DOF quarter-car passive and hydraulic active suspension systems for ride comfort *Systems Science & Control Engineering:* An Open Access Journal vol **2** pp 621–631
[27] Rotenberg R W 1974 *Zawieszenie samochodu.* (Warszawa: WKŁ)
[28] Ryba D 1973 Improvements of dynamic characteristics of automobile suspension systems (Part I. Two-mass systems) *Vehicle System Dynamics* **3** pp 17-46
[29] Sekulić D and Devidović V 2011 The effect of stiffness and damping of the suspension system elements on the optimization of the vibrational behavior of a bus *Int. J. for Traffic and Transport Eng.* **1(4)** pp 231–244
[30] Sharp R S and Crolla D A 1987 Road vehicle suspension system design – a review. *Vehicle System Dynamics* **16(1987)** pp 167–192
[31] Sharp R S and Hassan S A 1986 An evaluation of passive automotive suspension systems with variable stiffness and damping parameters *Vehicle System Dynamics* **15(1986)** pp 335–350
[32] Smith M C and Swift S J 2016 Design of passive vehicle suspensions for maximal least damping ratio *Vehicle System Dynamics,* **54 (5)** pp 568-584

[33] Ślaski G 2012 Studium projektowania zawieszeń samochodowych o zmiennym tłumieniu *Wydawnictwo Politechniki Poznańskiej. Rozprawy* **no 481**

[34] Verros G, Natsiavas S and Papadimitriou C 2005 Design optimization of quarter-car models with passive and semi-active suspensions under random excitation *J. of Vibration and Control* **no 11** pp 581–606

[35] Wong J Y 2001 *Theory of ground vehicles* (Canada-USA: John Wiley & Sons, Inc.) ISBN 0-471-35461-9

[36] Zdanowicz P 2012 *Ocena stanu amortyzatorów pojazdu z uwzględnieniem tarcia suchego w zawieszeniu* (Warsaw Univ. of Technology, Faculty of Transport, doctoral thesis)

[37] Zdanowicz P 2010 Problematyka oceny stanu amortyzatorów na stanowisku EUSAMA. Logistyka **4(2010)** CD1 ISSN 1231-5478

IOP Conf. Series: Materials Science and Engineering **148** (2016) 012015 doi:10.1088/1757-899X/148/1/012015

Simulation research of the tire Basic Relaxation Model in conditions of the wheel cornering angle oscillations

W Luty[1]

[1]Warsaw University of Technology
The Faculty of Transport, Koszykowa 75, 00-662 Warsaw, Poland

E-mail: wluty@wt.pw.edu.pl

Abstract. A description of the tire Basic Relaxation Model (BRM) is presented in this paper. Simulation research of the tire BRM model in conditions of oscillatory changes of the wheel cornering angle were performed. During the simulation tests the courses of changes in the value of lateral reaction force, transmitted by the wheel, as a response to the sinusoidal changes in the value of the wheel cornering angle have been presented. There have been compared the simulation results obtained for the model of tire-road interaction in two modes: including and not including the BRM. The simulation results allowed to verify prepared BRM and also to determine the influence of the tire relaxation process on the tire behavior in conditions of dynamic changes of the wheel cornering angle.

1. Introduction

There are several reasons for changing the lateral reaction force transmitted by the tire, during the oscillatory dynamic changes in a value of the wheel cornering angle. The first reason is the shape of a tire cornering characteristics (figure 1a). The cornering angle oscillations can change the effective value of the lateral reaction force F_y especially in a scope of the wheel cornering angle values, where the tire cornering characteristics is strongly nonlinear (figure 1a).

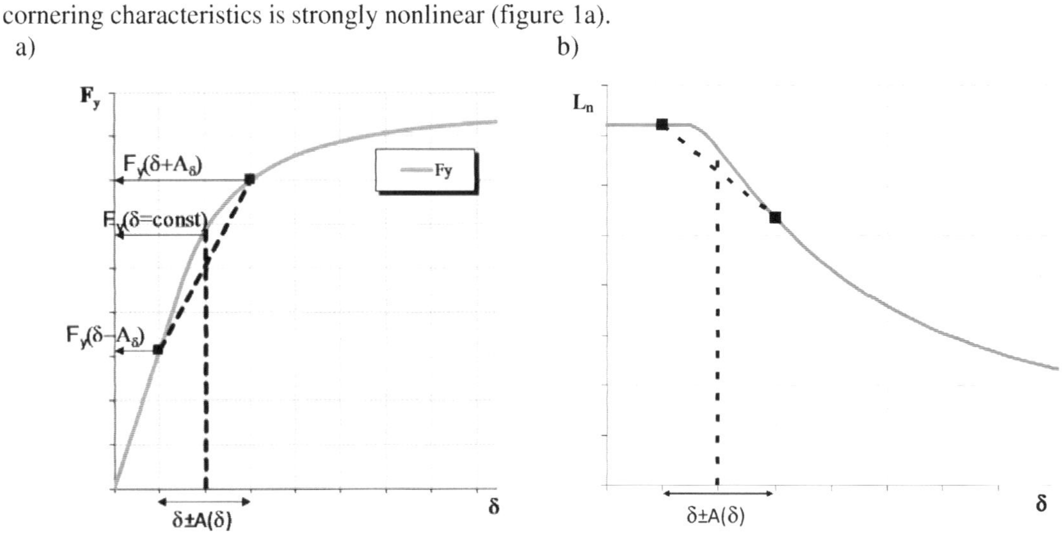

Figure 1. Example of the tire cornering characteristics - possible change in effective value of the lateral reaction force F_y (a) and relaxation length L_n (b) as a result of the cornering angle oscillation around its certain constant value.

IOP Conf. Series: Materials Science and Engineering **148** (2016) 012015 doi:10.1088/1757-899X/148/1/012015

The effective value of the lateral reaction force F_y decreases when the cornering angle oscillates in the scope of values where the tire cornering characteristics is nonlinear and convex. But it is also possible that the effective value of the lateral reaction force F_y does not change or even increases, when the tire cornering characteristics is linear or nonlinear but concave (e.g. in the scope of small values of the wheel cornering angle). In simulation research, when calculating the value of the lateral reaction force, that tire performance can be achieved using anyone of the available tire-road interaction model for steady-state wheel running conditions [7,1].

The second reason for changing in the effective value of the lateral reaction force F_y transmitted by the tire during oscillatory changes of the wheel cornering angle can be the phenomenon of the tire relaxation. This is because the relaxation length L_n which depends nonlinearly on the wheel cornering angle (figure 1b) [3,4]. Therefore, the growth rate of the lateral reaction force is higher than its decrease rate even than the cornering angle oscillates with constant amplitude. In addition, it is worth noting that due to the tire relaxation, the amplitude of the lateral reaction force transmitted by the tire can strongly decrease with increasing the cornering angle oscillation frequency. In simulation research, when calculating the value of lateral reaction force, such tire performance can be achieved using appropriate tire relaxation model [3,4,5,10].

Presented on the Figure 1 tire properties may be in opposition. It means that nonlinear tire cornering characteristics may decrease and the tire relaxation process may increase the effective value of the lateral reaction force transmitted by the tire during wheel cornering with oscillations of the wheel cornering angle. So when building the tire-road interaction model, as a module of the vehicle dynamics model, it is worth to take care of mentioned above tire properties. Especially the tire properties in transient cornering conditions may be very important when performing the vehicle simulation research in aspect of its safety or lateral dynamics. The achievement of tire properties in transient cornering conditions requires use of appropriate tire relaxation model and its properly matched coefficients.

The main purpose of this work is to assess the behavior of prepared tire Basis Relaxation Model in conditions of dynamic changes in the wheel cornering angle.

2. The mathematical description of the Basic Tire Relaxation Model

During work the tire relaxation model has been prepared. Its name is the Basic Relaxation Model (BRM) because it is based on the elementary tire properties, which are related to the lateral dynamics of the tire envelope. The essence of the model are both mathematical description and its coefficients which have been determined based on the experimental research results of real tires. Mathematical description of the BRM model is presented in the following expressions:

The tire model for steady-state wheel cornering conditions

$$F_{yu} = S_y \cdot k_\delta \tag{1}$$

enter to the BRM

The BRM

$$m_{tr} \cdot \ddot{u}_y + \left(\frac{k_\delta}{v_x} + k_{yd}\right) \dot{u}_y + k_{ys} \cdot u_y = F_{yu} \tag{2}$$

$$F_{ys} = k_{ys} \cdot u_y \tag{3}$$

$$F_{yd} = k_{yd} \cdot \dot{u}_y \tag{4}$$

$$F_y = F_{ys} + F_{yd} \tag{5}$$

exit from the BRM

Where particular markings mean:

F_{yu} – the lateral reaction force transmitted in the contact area between the tire tread and the road surface in steady state cornering conditions – the value obtained from the tire-road interaction model without the tire relaxation (it is input to the BRM model),

δ – the wheel cornering angle,
s_y – the wheel lateral slip $s_y = tan(\delta)$,
v_x – the longitudinal component of the wheel speed,
u_y – the tire lateral deflection,
$\dot{u}_y$ – the speed of the tire lateral deflection,
m_{tr} – the mass of the tire envelope part deflected in lateral direction, reduced to its tread staying in contact with the road surface,
k_{yd} – the tire envelope lateral damping coefficient,
k_{ys} – the tire lateral stiffness,
F_{yd} – the lateral viscous damping force in the tire envelope,
F_{ys} – the lateral spring force in the tire envelope,
F_y – the lateral reaction force transmitted into the wheel rim, and further into the vehicle chassis (output from the BRM model).

Prepared model does not include the resistance of the tire envelope to deflection under the influence of its gyroscopic moment during wheel roll. The main differential equation of the model (2) is linear, but the model coefficients (k_δ, k_{ys}), based on the tire experimental research results are under nonlinear influence of the wheel running conditions. Values of some model coefficients (m_{tr}, k_{yd}) are established basing on the own experimental research results of the truck tire and also confirmed basing on information from the literature references [3,5,4,8].
The advantage of the prepared BRM is that there is no need to modify the main part of the model used for calculating the lateral reaction force F_{yu} in steady-state wheel cornering conditions. In this case, as the main part of the model, the DFS (Dugoff, Fancher, Segel) model have been used [1]. Prepared the BRM has been successfully verified by comparing its behavior with the real truck tire behavior, observed in different transient cornering conditions during the experimental laboratory tests. Therefore, it is possible to test the BRM in extended or extreme transient cornering conditions of the wheel, which are difficult to achieve in the laboratory.

3. Simulation based tests of the tire Basic Relaxation Model in extreme transient cornering conditions

In this case, the BRM model was tested by sinusoidal changing the wheel cornering angle about its constant non-zero value. During the tests the cornering angle oscillation frequency were increasing monotonously from 0 to 10 Hz. Whereas the oscillation amplitude were constant (Figure 2).

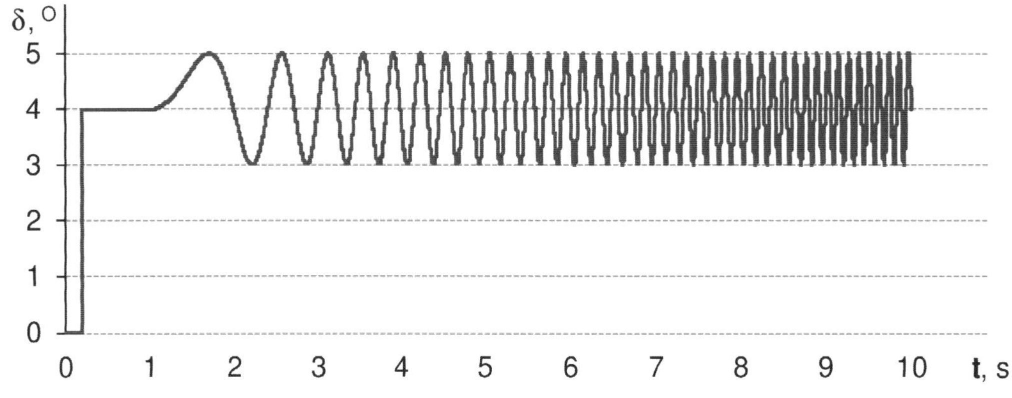

Figure 2. Example of the wheel cornering angle waveform, entered into the BRM model as an input signal (the constant component of the cornering angle $\delta = 4^O$, the amplitude of cornering angle oscillation $A(\delta) = 2^O$, the oscillation frequency range $f = 0 \div 10\ Hz$).

Particular test conditions were as follows:
- steady state value of the wheel cornering angle $\delta=4^{o}$,
- values of the cornering angle oscillations amplitudes $A(\delta)=1^{o}$ or $A(\delta)=2^{o}$,
- the wheel rolling speeds $v=30$, 60 or $90\ km/h$.

Applied test method was based on a comparison between waveforms and values of the lateral reaction force F_y obtained for the tire-road interaction model in two modes: with BRM or without BRM. So the tests results were described as obtained from the model with or without the tire relaxation.

4. Simulation tests results of the tire BRM

The example of the simulation research result is presented on the figure 3.

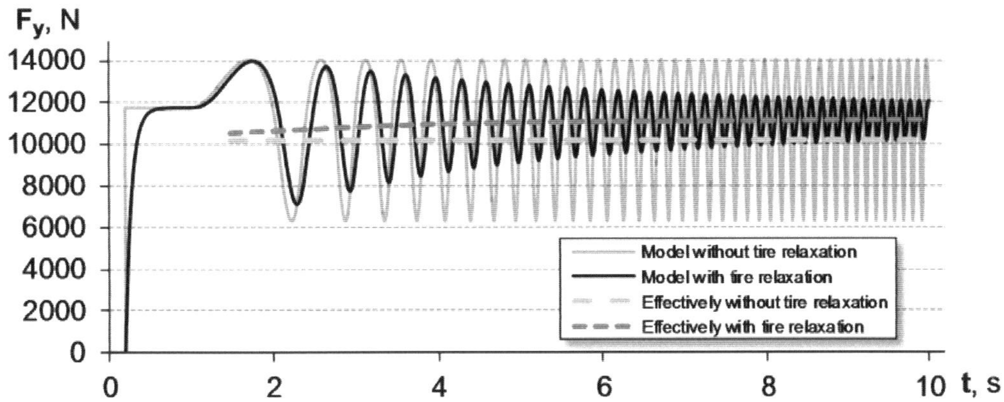

Figure 3. A single set of the test results obtained for the tire-road interaction model with and without the BRM (with and without the tire relaxation) ($v=30km/h$, $\delta=4^{o}$, $A(\delta)=2^{o}$).

The figure 3 shows clear differences between waveforms of the lateral reaction force F_y obtained for the model without and with the tire relaxation. For both tire model modes a decrease of the effective value of the lateral reaction force, after starting the cornering angle oscillation, is visible. However comparing the tire model with and without relaxation, there have been achieved:
- lower value of the lateral reaction force amplitude, decreasing with the oscillation frequency increase,
- time lag of the lateral reaction force oscillation,
- higher effective value of the lateral reaction force, increasing with the frequency increase.

Indicated differences between two tire model modes change in other wheel rolling conditions (figure 4). It means that the tire BRM reacts to changes in the wheel running conditions. Basing on presented simulation tests results, there have been calculated oscillation amplitudes $A(F_y)$ and effective values F_{ye} of the lateral reaction force F_y. That values obtained for the different wheel running conditions have been presented due to the cornering angle oscillation frequency f (Figure 5).

For the model without tire relaxation there have been obtained a permanent reduction of the lateral reaction force value F_y as a result of the wheel cornering angle oscillation. That reduction does not depend on the speed and the oscillation frequency but is directly greater for the greater amplitude of the cornering angle oscillation. Whereas for model with the tire relaxation, the maximum reduction of the lateral reaction force F_y effective value have been obtained from the beginning of the wheel cornering angle oscillation (at minimum frequency). However, that reaction force effective value clearly increases with the cornering angle oscillation frequency. Therefore, the tire model with the BRM behaves like getting stiffer with increasing frequency of the cornering angle oscillation.

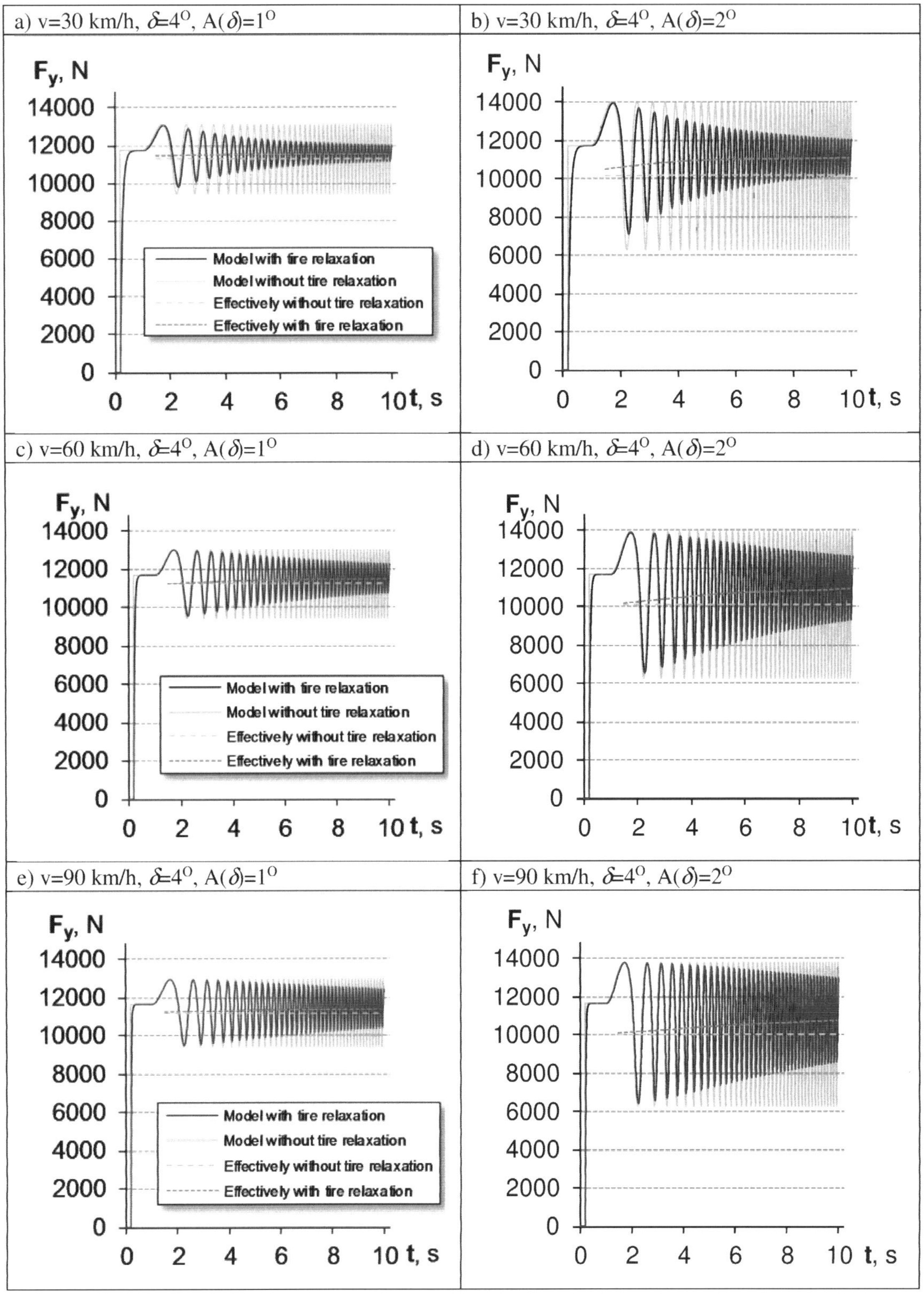

Figure 4. Complete simulation research results of the tire model in two modes: without and with BRM model.

Observed reduction of the lateral reaction force effective value is greater for higher wheel rolling speeds and amplitude of cornering angle oscillation (figure 5c, d).
For the tire model with the BRM, there have been obtained decreasing amplitude of lateral reaction force F_y oscillation with increasing frequency of the cornering angle oscillation (figure 5a, b,). That visible decrease of the lateral reaction force amplitude gets less with increasing wheel speed. Observed tire model behavior with BRM is quite similar to the real tires behavior in conditions of the cornering angle oscillation [2,6,9].

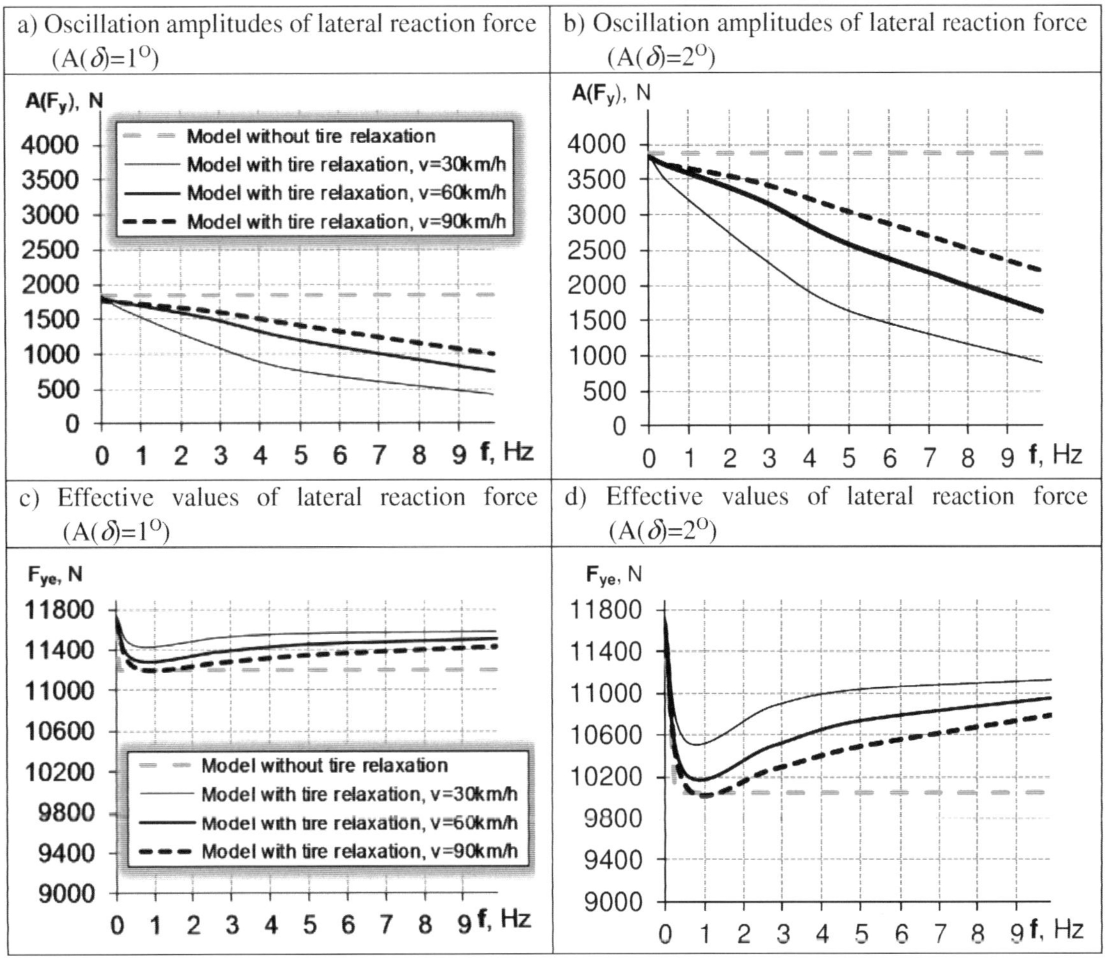

Figure 5. Influence of the wheel running conditions on the oscillation amplitude A(F_y) and effective values F_{ye} of the lateral reaction force F_y calculated using the tire model in two modes: without and with the BRM model.

4. Conclusions

The simulation tests results show that including the tire relaxation model into a model of the tire-road interaction may significantly change the model properties in conditions of oscillatory changes in the wheel cornering angle values. However, it is necessary to achieve the required properties of the relaxation model by the appropriate mathematical description and data for modeling.

Prepared tire Basic Relaxation Model despite expanded mathematical description has many advantages. Important feature of the tire BRM is that it can be used as an addition module connected in series with the main part of the model established earlier for the tire-road interaction in steady-state wheel running conditions, without need of its any modification. The tire BRM is specifically related by its coefficients with elementary properties of a tire. The tire BRM model features allow to achieve its very close behavior to behavior of real tire in transient states of the wheel side cornering. This gives the possibility of simulating the behavior of the tires in extreme wheel running conditions, which are difficult to achieve in laboratory but which may occur in the traffic conditions.

The tire BRM ensured the achievement of its expected behaviors, which are typical for the real tires. In this case of simulation research, by using the tire model with the BRM, special tire behavior in conditions of the wheel cornering angle oscillations have been achieved e.g.:

- expected reduction in the effective value of the lateral reaction force,
- clear relationship of the reduction in effective value of the lateral reaction force with the wheel speed, the amplitude and frequency of cornering angle oscillations,
- expected decrease of the amplitude of lateral reaction force due to frequency of the cornering angle oscillations and also due to the wheel running speed.

References

[1] Dugoff H, Fancher P S and Segel L 1970 *An analysis of tire traction properties and their influence on vehicle dynamic performance,* SAE Transactions 700377

[2] Gansiniec O 2001 *Wpływ zmian kąta znoszenia i obciążenia koła samochodu na zdolność przenoszenia sił stycznych.* Praca doktorska. (Politechnika Krakowska)

[3] Loeb J S and others *1990 Lateral stiffness, cornering stiffness and relaxation length of the pneumatic tire.* SAE Paper 900129

[4] Luty W 2009 *Analiza nabiegania ogumienia nowych konstrukcji podczas toczenia ze znoszeniem bocznym w quasi-statycznych warunkach ruchu* Warszawa ISBN978-83-61486-21-3 (WAT)

[5] Luty W 2009 *Analiza właściwości nowych konstrukcji ogumienia kół jezdnych samochodu ciężarowego średniej ładowności* Warszawa ISBN978-83-61486-21-3 (WAT)

[6] Meier-Dörnberg K and Strackerjan B 1977 *Prüfstandsversuche und Berechnungen zur Querdynamik von Luftreifen.* Automobil-Industrie 22 (4)

[7] Mitschke M., *Dynamika samochodu.* WKŁ Warszawa 1977

[8] Pacejka H B 2002 *Tyre and Vehicle Dynamics* (Oxford Butterworth-Heinemann)

[9] BRAESS, H H and others 1982 *Reifeneigenschaften und Pehrsicherheit* F. Porsche AG (Entwicklungszentrum Weissach)

[10] Von Yong Q W, Gnadler R and Schieschke R 1994 *Einlaufverhalten und relaxationslänge von automobilreifen* ATZ 96/1994

IOP Conf. Series: Materials Science and Engineering **148** (2016) 012016 doi:10.1088/1757-899X/148/1/012016

Does an active adjustment of aerodynamic drag make sense?

Marek Maciejewski

Politechnika Poznańska, Instytut Maszyn Roboczych i Pojazdów Samochodowych, 60-965 Poznań, ul. Piotrowo 3

E-mail: marek.maciejewski@put.poznan.pl

Abstract. The article concerns evaluation of the possible impact of the gap between the tractor and semitrailer on the aerodynamic drag coefficient. The aim here is not to adjust this distance depending on the geometrical shape of the tractor and trailer, but depending solely on the speed of articulated vehicle. All the tests have form of numerical simulations. The method of simulation is briefly explained in the article. It considers various issues such as the range and objects of tests as well as the test conditions. The initial (pre-adaptive) and final (after adaptation process) computational meshes have been presented as illustrations. Some of the results have been presented in the form of run chart showing the change of value of aerodynamic drag coefficients in time, for different geometric configurations defined by a clearance gap between the tractor and semitrailer. The basis for a detailed analysis and conclusions were the averaged (in time) aerodynamic drag coefficients as a function of the clearance gap.

1. Introduction

A characteristic feature of articulated vehicles, unlike conventional trucks, is additional free space between the rear wall of the cab and the front wall of the trailer. In the case of a small clearance gap between both units, it can be assumed that the aerodynamic effect will not be significant. It appears that in this situation the pressure acting on the rearward-facing surface of the tractor and the forward-facing surface of the trailer are more or less similar, and therefore will not lead to a significant change in the aggregated drag. On the other hand, in case of a large distance between the vehicle units, the pressure difference on both surfaces may become large enough to result in clear increase of the aerodynamic drag for the entire vehicle. Therefore, it can be assumed that increasing the distance between the vehicle units will increase the aerodynamic drag. This finding is consistent with the principle that the drag coefficient of the tractor-trailer system depends solely on the geometry of individual units and the distance between them.

Recently, the author of this publication made a contact with a company engaged in transport operations involving articulated vehicles. The company representative observed that their drivers report a noticeable dependency of aerodynamic drag on the vehicle speed and the distance between units. We confirmed this observation and explained that in general:

- aerodynamic drag coefficient depends only on the relative position of the two units of the articulated vehicle, and is speed independent,
- aerodynamic drag of the vehicle (with given geometric configuration, i.e. with specific drag coefficient) depends solely on speed.

The above statements have been confirmed by many aerodynamic tests and simulations that can be found in literature relating to the aerodynamics of articulated vehicles and the available ways to reduce their aerodynamic drag forces [1, 2, 3, 4, 5, 6, 7, 8, 9].

The company representative did not agree with above statements and was convinced that speed also affects the drag coefficient. He believed, that in order to minimize the aerodynamic drag, the gap between the tractor and the trailer should be adjusted during the drive according to the current speed of the vehicle. This difference of opinion led the company to turn to the author of the publication with the proposal to carry out fast numerical simulations to finally resolve the matter. Thus, the subject of carried out simulations was not only finding a certain optimum amount of space between the tractor and trailer, but also to examine whether the optimal value of the gap depends on the speed. Although the dependence of the drag coefficient on speed seemed doubtful due to the aerodynamic interference phenomena occurring in the space between the vehicle units, it was considered to be worth checking. If it turned out to be true, an active adjustment of the gap magnitude depending on the speed would be a simple and convenient method to reduce aerodynamic drag.

2. Method of the numerical simulation

The numerical simulations of aerodynamic phenomena around the articulated vehicle were carried out on the basis of the Navier–Stokes equations for incompressible fluid, because for typical speeds for road vehicles, relative errors made due to the assumption of flow incompressibility are negligibly small. For incompressible formulation, in the continuity equation is missing an evolutionary term, and this equation should be treated as divergence-free (solenoidal) condition for velocity field. To circumvent this limitation, an additional term containing the time derivative of pressure has been introduced into the continuity equation. This introduced artifical compressibility, that ensured a sufficient coupling between the continuity and momentum equations.

For computational simulations, it is necessary to replace the initial continuous problem with its discrete counterpart. The discretization process consists in replacing a continuous system (with infinite degrees of freedom) by its discrete representation (with a finite number of degrees of freedom) expressed through nodal points defined usually on a computational mesh. This requires a simultaneous definition of behaviour of the points lying inside each mesh cell, depending on behaviour of nodal points of given cell. For this purpose, the polynomial approximation methods are used most often. The polynomials are simultaneously applied to reconstruct the effect of operators included in mathematical description. Such form of discretization (in combination with appropriate approximation) is applied to both, space and time, in our case, sequentially.

The space discretization of the problem is realised with the use of approximation corresponding to the finite volume method, where the solution change (in a time step) for single computational cell is resulted from current balance conditions on the cell walls, i.e. from the flux balance established on the basis of the Riemann problem solution for reconstructed fields of primitive variables. In this resolving, the upwind numerical scheme referred to the signs of local eigenvalues is used. Determination of numerical fluxes through the cell walls is the basis to carry out a time integration (the explicit Runge–Kutta method) and determine variable values inside particular cells in next time step of automatically selected length.

Due to the significant impact of computational mesh configuration (the resolution and form) on the accuracy of the achieved results, and inability to predict a priori the appropriate mesh arrangement, the computational program was developed in the adaptive version to ensure required accuracy of simulations. It carries out standard computations on initial mesh and then, on the basis of assessment of the solution quality, determines subdomains where the solution is not sufficiently accurate. The mesh density gets appropriately increased in indicated subdomains. The cyclical procedure for successive mesh refinement is continued until assumed criteria for the accuracy have been fulfilled. The adaptive program that has been used includes, unlike the standard one, two new elements: a procedure to estimate the solution quality (on the basis of solution error estimation, or estimation of the intercellular variable gradients) in particular cells, and a procedure for the computational mesh refinement based on increasing the local mesh density as a result of successive bisection of the longest cell edges in chosen mesh subdomain.

3. Range and conditions of simulations

The scope of considerations was to determine the aerodynamic properties of articulated vehicles depending on following two factors: the clearance gap between the tractor and semitrailer, and the vehicle speed. These properties are determined mainly by the aerodynamic drag coefficient, and to the lesser degree by distributions of pressure and velocity components.

All numerical simulations have been performed in the two-dimensional (2D) space. Admittedly, for precise reconstruction of aerodynamic phenomena around vehicles, it would be appropriate to use 3D simulations, but for comparative studies of vehicles with similar outlines, the 2D simulations are sufficient. It is even more justified when comparing aggregate properties, e.g. drag coefficients. Other reasons for 2D simulations were:

- the easier identification of main factors influencing the aerodynamic drag,
- simpler and clearer forms of aerodynamic phenomena,
- the simpler and faster creation of vehicle models,
- far shorter times of simulations.

Conducted simulations tried to reconstruct typical traffic conditions, i.e. the vehicles moving in open space, without artificial, physical or computational, limitations related to insufficient size of computational domain. This size was equal for all simulations. It is worth noting that assumed boundary conditions did not reconstruct aerodynamic circumstances from wind-tunnels, but reflected real, windless conditions from road traffic.

Objects of aerodynamic simulation were the articulated vehicle models with simplified bodywork forms achieved by straightening out the bodywork lines and omission of local curvatures. This had an undoubtful impact on the increase of aerodynamic drag but, at the same time, caused their greater diversity. The simplification of bodywork forms had no impact on simulation processes, but enabled much more to emphasize some forms of aerodynamic phenomena, in comparison to more streamlined models.

Figure 1 presents an exemplary articulated vehicle model with the clearance gap indicated by the letter B. The total number of considered tractor-trailer gap sizes is equal to 16: from 0.0 m (a one-body or 'monospace' form) to 1.5 m, with a step of 0.1 m. The geometrical shape and position of the tractor are the same in all of computational models. The trailer shape and sizes (the length: 15 m, the height: 4 m, and the ground clearance: 0.4 m) are identical, but the trailer position is variable, appropriate to the gap size B.

All the created vehicle models were placed in the duly geometrically defined outer outline of assumed flow domain. The height of computational domains was always equal to 24 metres and the length was equal to 72 meters (from -12 m to 60 m). Such defined flow domains were used to generate the initial computational meshes for all considered models. A sample initial mesh, created for the gap size B equal to 1 m, is presented in figure 2.

The boundary conditions for each computational domain have been defined according to the following scheme:

- on the vehicle profile: the motionless solid boundary,
- on the left- and right-side of the 'tunnel': inflow and outflow boundaries, respectively,
- at the top of flow domains: the symmetric boundary conditions,

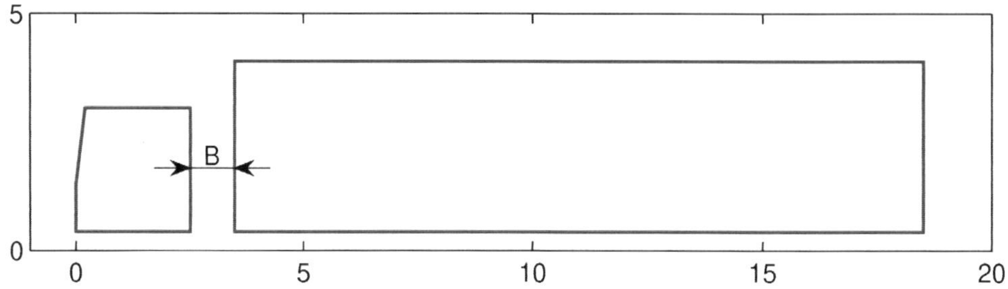

Figure 1. The articulated vehicle model, where the clearance gap size is indicated by the letter B (all dimensions are in metres).

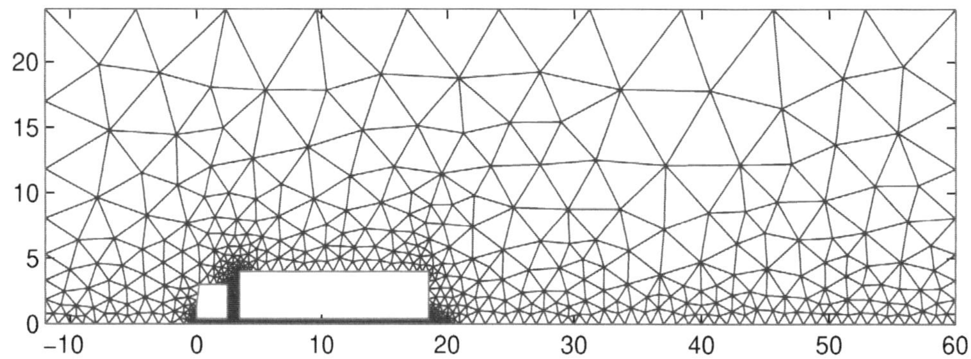

Figure 2. Exemplary initial mesh for the computational configuration with the tractor-trailer gap size equal to 1 m (all dimensions are in metres).

- at the bottom of 'wind-tunnel': the moving (with inflow speed) solid boundary.

The considered vehicle speeds, or the air inlet velocities, have been assumed as 15, 20, 25 and 30 m/s. The control parameters for each of computational processes were identical.

For description of the computational domain configurations, the uniform notation has been introduced according to the following pattern:

AB-C

where particular symbols (A, B and C) denote:

A – v (always), as distinctive sign of all the simulations considered,
B – the clearance gap (in decimetres) between the tractor and semitrailer,
C – the vehicle speed (in metres per second).

Total number of computational simulations was 64.

4. Direct results of simulations

Results of carried out numerical simulations of articulated vehicle aerodynamics may take one of two forms: direct or indirect. Direct results of aerodynamic simulations are relevant to:

- temporary distributions of variables (i.e. pressure, and horizontal and vertical components of velocity) in each flow domain, for predefined points in simulation time,

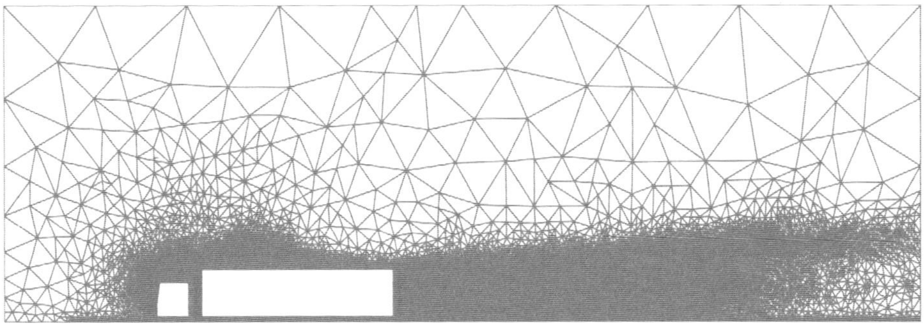

Figure 3. Exemplary final mesh for the computational configuration v10-25 and 20th second of simulation.

- temporary configurations of adaptive computational meshes (for predefined time points), indicating the different forms of aerodynamic phenomena around vehicle.

Direct results of simulations, and in particular those referring to variable distributions, tell a lot about aerodynamic phenomena around road vehicles. The consideration of variable distribution in a flow field is justified only if we want to explain certain specific questions related to local flow forms. Such analysis of direct results is not too effective in the context of overall and concise evaluation of the aerodynamic properties of vehicles.

The exemplary post-adaptive mesh form, from the final time point (20 s) of the simulation, for the model v10-25 (the gap is equal to 1 m, and the velocity – 25 m/s), is presented in figure 3. The distributions of pressure and both components (horizontal and vertical) of velocity, for the time point equal to 20 s, can be found in figure 4.

5. Aggregate indirect results

Unlike presentation of direct results, which always concerns only one specific case of computational configurations and only one of primitive variables (pressure or velocity components) distributed over a computational mesh, the presentation of indirect results, understood as overall aerodynamic properties, may refer simultaneously to one or many different computational configurations dynamically, in a function of time. Typical indirect and comprehensive measures with regard to description of aerodynamic features of vehicles are the aggregate aerodynamic drag and lift forces. These quantities are however inconvenient to compare the aerodynamics of different vehicles. For this purpose the dimensionless drag and lift coefficients are more suitable. Therefore, in our comparative studies on aerodynamic resistance, we mainly use the aerodynamic drag coefficient determined typically on the assumption that the overall aerodynamic drag has a fully convective (non-diffusive) character.

For a specific aerodynamic simulation, the aerodynamic drag coefficient is not a fixed value; it changes in time according to temporary distributions of primitive variables. Moreover, the temporary changes in the drag coefficient can be directly compared to other ones from different vehicles. It allows comparing many such coefficients on a single timeline chart. Examples of the drag coefficient comparisons can be found in figures 5 and 6. The first one shows temporary changes of drag coefficients for exemplary clearance gaps between the tractor and semitrailer (for constant vehicle speed), and the second one presents temporary changes of drag coefficients for exemplary vehicle speeds at a constant gap between the vehicle units.

As already noted above, the aerodynamic drag coefficient usually evolves in time, which is related to variable forms of aerodynamic phenomena around the vehicle. It is usually not

(*a*)
(*b*)
(*c*)

Figure 4. Exemplary variable distributions for the model v10-25 (20th second): (*a*) pressure, (*b*) horizontal velocity, and (*c*) vertical velocity, where the field values are locally defined by colours (red – maximum, blue – minimum).

noticeable in wind-tunnel tests because of a high inertia of vehicles, and also, to a limited extent, an elasticity of the weighting system with force-measuring transducers. Because it is widely accepted that the aerodynamic drag coefficient is defined as a constant value, the averaging of that coefficient was carried out based on the values obtained for various time steps. Due to the fact, that the first few seconds of each simulation are characterised by dissimilarity or fluctuation of results, the averaging was carried out in relation to the recurrent and stable part of results. These large fluctuations result from assumption of approximate, and in general spurious, initial conditions for the flow domain. Hence, the averaging was carried out for the results from the second half of the time steps, which was sufficient enough to achieve a reliable value of the constant drag coefficient.

Moreover, observing the velocity component distributions at different vehicle speeds, it can be noted that the aerodynamic phenomena, e.g. detaching air vortices behind a vehicle, occur at a rate dependent on the vehicle speed. The greater the running speed, the greater the vortices

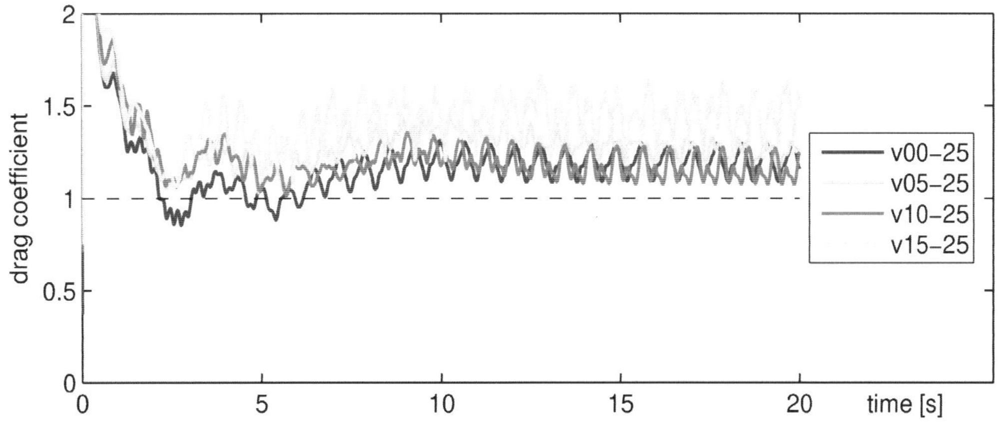

Figure 5. Drag coefficients as a function of time, for some computational configurations with various gap sizes (in dm) and the constant speed (25 m/s).

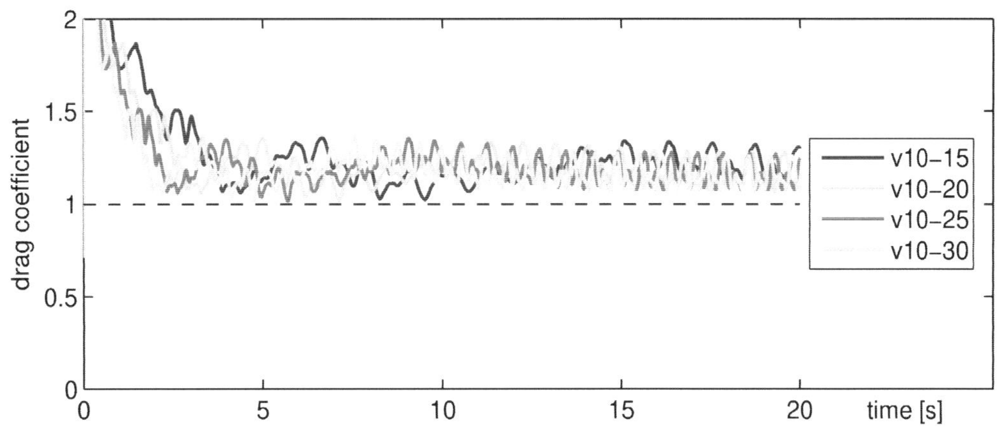

Figure 6. Drag coefficients as a function of time, for some computational configurations with various running speeds and the constant gap size (10 dm, i.e. 1 m).

creation rate behind a vehicle. Consequences of this can be noticed in relationships between an aerodynamic drag coefficient and a vehicle speed, e.g. in figure 6. At higher vehicle speeds, the correct periodic forms of aerodynamic phenomena are achieved in less time. Stabilization of these phenomena normally takes place after a certain number of time steps, i.e. after propagation of boundary conditions to the whole flow domain.

Because of automatic adjustment of size of each time step, its length depends on maximal, local flow velocities from all computational cells. With some approximation it can be claimed that the time step size is inversely proportionate to a vehicle speed. This dependence may be distorted by process of automatic computational mesh adaptation, but all parameters of that process were uniform for all simulations.

Because the end time of simulations for each vehicle speed was equal to 20 s, and the sizes of time steps are different and dependent on vehicle speeds, the resulting numbers of time steps are accordingly different. At high speeds, the number of short time steps was large, and at low

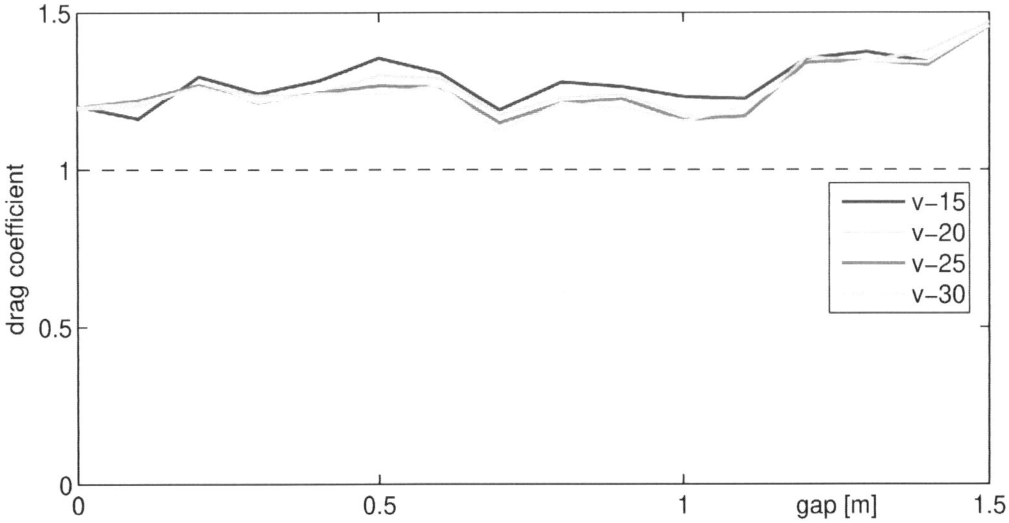

Figure 7. Averaged drag coefficients in a function of the tractor-trailer gap – trajectories for various speeds.

speeds, the number of long time steps was small. As a result, a flow stabilisation was ensued in a short time (quickly) at high speeds, and in a long time (slowly) at low speeds. Implications of this can be observed in figure 6, where subsequent stabilization of aerodynamic drag coefficients at low vehicle speeds is clearly visible.

After calculating the constant, aggregated aerodynamic drag coefficients for each of carried out simulations, it was possible to determine general relationships for aerodynamic drag coefficients in terms of the clearance gap between tractor and trailer, and the speed of articulated vehicle. Figure 7 presents the curves of drag coefficients versus gap size, for all considered vehicle speeds: 15, 20, 25 and 30 m/s. These relationships are the basis for the answer to the question from article title: Does an active adjustment of aerodynamic drag make sense?

6. Conclusions

Fundamental conclusions from the carried out simulations have been outlined in the following two points.

(1) In the range of studied gaps between tractor and semitrailer (from 0.0 m to 1.5 m, with the step 0.1 m), it was found that the lowest values of aerodynamic drag coefficient took place in the case of no gap and for the gaps equal to 0.7 and 1.1 m, regardless of the vehicle speed.

(2) There is no need to control a tractor-trailer gap size depending on vehicle speed in order to actively reduce the aerodynamic drag, because the value of the drag coefficient is not influenced by vehicle speed. This conclusion is consistent with our initial assumption presented in the introduction to this article.

Finally, it would be advisable to draw the attention to two other issues. The first concerns little diversity of results, i.e. different values of drag coefficients for different speeds. That issue does not apply to zero gap, i.e. a one-body or 'monospace' case, but to the cases with real (from 0.1 m to 1.5 m) gaps between the tractor and semitrailer. In these cases, some, considerably insignificant, differences in the level of drag coefficient can be noticed. This is due to the differences in aerodynamic interference at different speeds.

The second issue raises the question: why the aerodynamic drag coefficient does not grow successively with the increase of the clearance gap? In fact, it stems from the nature of 2D simulations, where the exchange of air between the tractor and trailer occurs only vertically, whereas in reality and 3D simulations, this exchange occurs in both horizontal and vertical directions. It does not affect, in any way, the above presented conclusion regarding no effect of running speed on the level of aerodynamic drag coefficient for each considered configuration of articulated vehicles.

References

[1] Hammache M and Browand F 2004 On the aerodynamics of tractor-trailers *Lecture Notes in Applied and Computational Mechanics* 19 (Berlin: Springer) pp 185–205

[2] Roy C, Payne J, McWherter-Payne M and Salari K 2004 RANS simulations of a simplified tractor/trailer geometry *Lecture Notes in Applied and Computational Mechanics* 19 (Berlin: Springer) pp 207–18

[3] Nayeri C N, Haff J, Greenblatt D, Loefdahl L and Paschereit C O 2009 Drag reduction on a generic tractor-trailer using active flow control in combination with solid flaps *Lecture Notes in Applied and Computational Mechanics* 41 (Berlin: Springer) pp 179–91

[4] Leuschen J and Cooper K R 2009 Summary of full-scale wind tunnel tests of aerodynamic drag-reducing devices for tractor-trailers *Lecture Notes in Applied and Computational Mechanics* 41 (Berlin: Springer) pp 451–462

[5] van Raemdonck G M R and van Tooren M J L 2016 Numerical and wind tunnel analysis together with road test of aerodynamic add-ons for trailers *Lecture Notes in Applied and Computational Mechanics* 79 (Switzerland: Springer International) pp 237–52

[6] El-Alti M, Chernoray V, Kjellgren P, Hjelm L and Davidson L 2016 Computations and full-scale tests of active flow control applied on a VOLVO truck-trailer *Lecture Notes in Applied and Computational Mechanics* 79 (Switzerland: Springer International) pp 253–67

[7] Devesa A and Indinger T 2012 Fuel consumption reduction by geometry variations on a generic tractor-trailer configuration *SAE Int. J. Commer. Veh.* **5** 18–28

[8] Hirz M and Stadler S 2013 A new approach for the reduction of aerodynamic drag of long-distance transportation vehicles *SAE Int. J. Commer. Veh.* **6** 453–8

[9] Anbarci K, Acikgoz B, Aslan R A, Arslan O and Icke R O 2013 Development of an aerodynamic analysis methodology for tractor-trailer class heavy commercial vehicles *SAE Int. J. Commer. Veh.* **6** 441–52

On the impact of size and position of semi-trailer on the aerodynamic drag of an articulated vehicle

M Maciejewski
Politechnika Poznańska, Instytut Maszyn Roboczych i Pojazdów Samochodowych, 60-965 Poznań, ul. Piotrowo 3
E-mail: marek.maciejewski@put.poznan.pl

Abstract. The paper relates to the numerical aerodynamic simulation of an articulated vehicle. All simulations were comparative, where variables were only three geometric parameters: the height and length of the semi-trailer, and the gap between the tractor and semi-trailer. Simulation results have been investigated and compared in terms of observed values of aerodynamic drag coefficients. These coefficients have been presented in time (for example only), but above all, as constant, averaged magnitudes which are more suitable for comparisons. Total number of simulations and thereby geometrical configurations was 125, i.e. for 5 sizes of the height and length of trailer, and 5 clearance gaps between the tractor and trailer. Results have been presented graphically and discussed.

1. Introduction
In the conventional meaning, reducing the aerodynamic drag is associated with appropriate shaping of the front and rear of a vehicle, in conjunction with change of the bluff bodied shapes to streamlined ones. In case of a truck, where a cargo box is located directly behind the cab, there is only one overpressure zone covering the front face of the cab and an upper part (above the cab) of the cargo box, and only one zone of underpressure located behind the truck rear face. In case of an articulated vehicle there is an additional free clearance gap between the rear face of tractor and the front face of semi-trailer, resulting in underpressure at rear face of the tractor, and overpressure at front face of the trailer. The pressure levels cannot be determined without carrying out relevant studies [1, 2, 3, 4, 5], and in our case – without numerical simulations. In these simulations the main focus was put on determining the effect of aerodynamic interference processes taking place between tractor and semi-trailer, and on the aerodynamic drag level (as a function of the clearance gap between both units of an articulated vehicle). Moreover, the impact of characteristic trailer parameters, i.e. height and width, on the results was investigated.

In this study, only primary, rectilinear geometric forms of the tractor and trailer were considered. The models did not have any roundings or aerodynamic drag reduction devices. The study was concerned, above all, with estimating the impact on aggregated level of aerodynamic drag coefficient for an articulated vehicle, depending on two primary covariates, i.e. trailer height and length, under different conditions defined by the clearance gap between the tractor and trailer. Determining the most advantageous geometric parameters of trailer as a gap function, will allow to continue studies on developing more complex geometric forms of trailers in order to further reduce the aerodynamic drag by rounding of corners and edges, and introducing other, more complex geometric forms, to be realised in future studies.

The results presented here are focused mainly on changes in aerodynamic drag coefficients as a function of the trailer height and length, for different tractor-trailer gaps.

2. The adaptive solution method

Numerical simulations of aerodynamic processes around moving vehicles were carried out on the basis of the incompressible Navier–Stokes equations. The lack of an evolutionary term in the continuity equations made such a description of a problem less convenient, as the equation could only be taken as a divergence-free constraint for velocity. The simple solution which allows to circumvent the difficulties connected therewith and which has been used in these simulations, is an introduction to the continuity equation of a new term comprising a time derivative of pressure. This means an introduction of artificial compressibility, which causes convenient, in computational context, coupling of the continuity equation and momentum equations.

The idea of the theoretical description of air (as a fluid) behaviour presented above relates to the problem description on the continuous level. In order to solve this problem with the use of a computer, the continuous level of description should be replaced with the discrete one. The discretization process consists in replacing an infinite number of material points by a certain finite number of 'nodal' points, usually characteristic points of a computational mesh. Through these nodal points and related flow parameters, it is possible to describe the state of flow at any point in the domain under consideration. The (usually polynomial) approximation is also used to express the effect of differential operators from our problem description.

The discretization process applies to both, the space and time. This may take place simultaneously for both dimensions, or sequentially (alternately). For the vehicle aerodynamics, i.e. for a subsonic problem, it is natural to assume the sequential approach. The space discretization may be in general implemented through the finite differences, finite elements or finite volumes. Here, the last approach has been applied, while the time discretization is based on finite differences and the explicit integration is used. The solution in individual cells is determined from temporary equilibrium conditions for fluxes on local cell boundaries, in the basis of an solution of the Riemann problem. The calculateded values of numerical fluxes are used to integrate with respect to time, and to determine the primitive variables in cells on the next time step.

The flow around any vehicle can be categorised as a laminar or turbulent, but if the flow is turbulent only in a subdomain, the entire flow is already considered as turbulent. For typical vehicle speeds, the flow at the rear of a vehicle is always turbulent – superimposed on the main flow. The turbulence is difficult to define in a universal way. It can be characterised by the irregularity, distinct and effective mixing, the dissipation, high Reynolds numbers, and the wide range of length and time scales. It is assumed that the largest eddies present in the flow are of the same order as the characteristic flow size. Besides that, there is also the whole spectrum of eddies with much smaller sizes, dissipated due to the viscosity.

Each flow around a vehicle, regardless of whether it is laminar or turbulent, is correctly described using the Navier–Stokes equations. In order to take these small scales of turbulent flow into account, an adequate spatial resolution (density) of the computational mesh should be ensured. This may, however, result in a huge number of degrees of freedom (unknowns) in solved problem, and can make a direct solution of the Navier–Stokes equations a very demanding and challenging task, even for parallel computing. As a remedy, a turbulence modelling can be used, i.e. an eddy parametrization, either all or only those with the smallest sizes. This makes sense especially when the purpose of the simulation is to determine the forms of phenomena, their effects and parameters characterizing them, such as the drag coefficient. The turbulence modelling consists essentially in modification of the Navier–Stokes equations through direct reference only to the averaged flow and simultaneous parametrization of mixing processes. Three general approaches to compute turbulent flows can be distinguished:

- the Reynolds-averaged Navier–Stokes equations (RANS), where almost all of the eddies are subject to modelling,
- the large eddy simulation (LES), where the parametrization concerns only the smallest eddies (due to mesh resolution),
- the direct numerical simulation (DNS), where no parametrization is carried out.

To explain the classification above, it should be noted that the eddy parametrization is not related to any real reconstruction of aerodynamic phenomena on a computational mesh, but rather only with the description of their effects, mainly the smallest ones.

In case of comparative studies, the purpose is to estimate the effects of modifications introduced to the original computational configuration, or just to compare similar geometric forms of vehicle body to search for the optimal solution. In such cases, it can be possible and perfectly reasonable to ignore the reconstruction of small eddy effects because they are characterised by rather large homogeneity, and this means that comitted errors in the flow simulations, for geometrically similar objects, are almost equal, and thus it does not affect the correctness of comparative simulations.

The proper local resolution (density) of computational mesh is always of primary importance for correct reconstruction of aerodynamic phenomena. Generally, the smaller the specific phenomenon (e.g. an eddy) is, the finer the mesh should be. Users are not able to properly arrange the unstructured mesh, according to the temporary flow conditions. It requires a criterion to indicate a mesh subdomain, and a procedure to carry out the suitable modification of this mesh subdomain. Such an algorithm is build into the program used to realise all simulations. The software was written by the author under a grant from the European Commission: Highly efficient parallel 3D codes for industrial applications (CP-94-01239). The version of the program used for our aerodynamic simulations is marked with the identification code ll2-35.

3. Scope of simulations

The purpose of simulations was to determine the impact of geometric parameters characterising the trailer, on the aerodynamics of an articulated vehicle. The study concerned a simple shape of the tractor and trailer, without any roundings and additional geometric forms which are usually made to reduce aerodynamic drag, such as side skirts, boat tails, gap reducers, vortex generators and so on. The reason for such an assumption was to focus on two fundamental trailer parameters: the length and height. The simple shape of tractor remained identical in all simulations. The determination of the most profitable length–height relationships will be the basis for further studies in the near future, taking more sophisticated geometric trailer shapes into consideration.

Moreover, due to a significant impact of the tractor-trailer gap size on the aerodynamic drag, the simulations were carried out for different clearance gaps. This allows to show the impact of aerodynamic interference between the tractor and trailer on aerodynamic properties of the entire vehicle. Thus, our studies included 125 independent simulations carried out for 5 different trailer heights, 5 different trailer lengths, and 5 tractor-trailer gaps.

Although the direct simulation results allow to observe changes in the flow field and its parameters, i.e. the pressure and velocity components, further analyses have been mainly related to the aerodynamic drag coefficient as a consolidated gauge allowing the unambiguous and convenient evaluation of the tractor-trailer system. However, rather than using real, varying in time, values of drag coefficients, the analyses were conducted based on the averaged values characteristic for particular computational configurations. The values of these coefficients, due to the simple and austere shape of vehicle models, will be very high in relation to real vehicles, but at the same time their diversity will be greater.

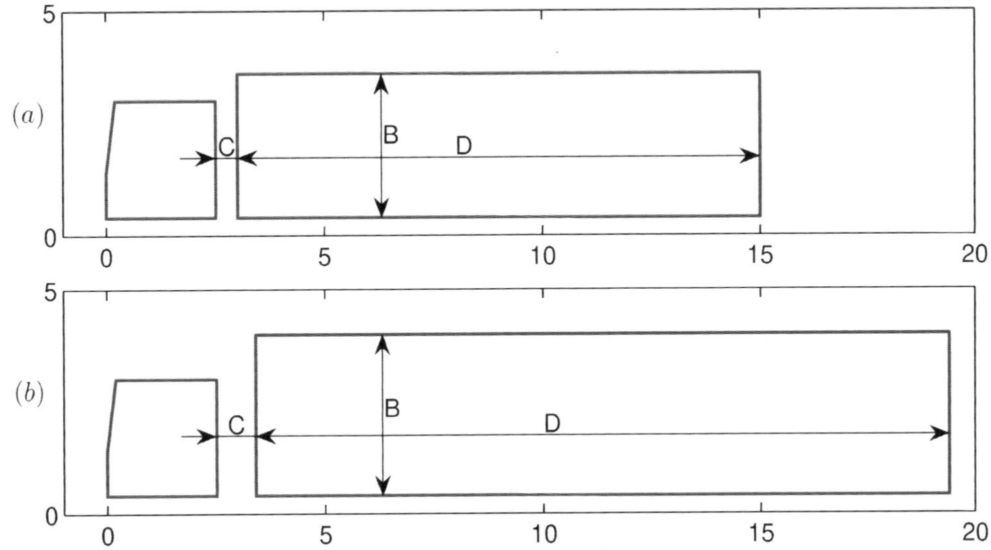

Figure 1. The articulated vehicle models: (*a*) p36-5-120, where B=3.6 m, C=0.5 m and D=12 m, and (*b*) p40-9-160, where B=4.0 m, C=0.9 m and D=16 m.

All of 125 numerical simulations were performed in the two-dimensional (2D) space. Admittedly, for precise reconstruction of aerodynamic phenomena around vehicles, it would be appropriate to use the 3D simulations, but for comparative studies of vehicles with similar outlines, the 2D simulations are sufficient. It is even more justified when aggregate properties (e.g. the drag coefficient) are compared. Other reasons for 2D simulations were: the easier identification of main factors influencing the aerodynamic drag, simpler and clearer forms of aerodynamic phenomena, the simpler and faster creation of vehicle models, and far shorter times of simulations.

Numerical simulations use the opposite flow conditions, i.e. with the stationary vehicle and the moving road and air. It is an analogy with the situation encountered in wind-tunnels, however without any physical restriction for the flow. The introduced limitations for computational flow domains had no effect on the results, and assumed forms of boundary conditions did not reconstruct those from wind-tunnels.

4. Vehicle models, flow conditions and meshes

In accordance with the purpose of this work and reasons already given, the objects of aerodynamic simulation were two-dimensional models of articulated vehicles with simplified bodywork forms achieved by straightening out the bodywork lines and omission of local curvatures. Two extreme (with the smallest and largest trailer and clearance gap) examples of these vehicle models are presented in the figure 1. The trailer height, trailer length and clearance gap are denoted by the capital letters B, D and C, respectively. Each of these quantities includes 5 specific values:

- the height B: 3.6, 3.7, 3.8, 3.9 and 4.0 metres,
- the gap C: 0.5, 0.6, 0.7, 0.8 and 0.9 metres, and
- the length D: 12, 13, 14, 15 i 16 metres.

All created vehicle models were placed in duly geometrically defined outer outline of assumed flow domain. The height of computational domains was always equal to 24 metres and the length

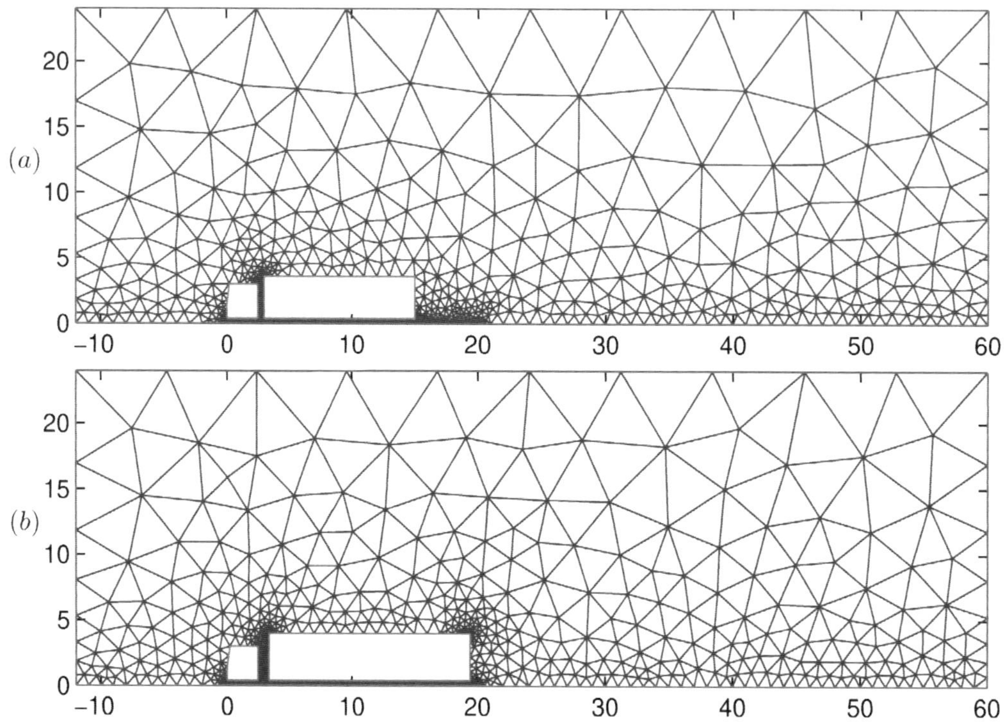

Figure 2. The sample initial meshes for the models: (*a*) p36-5-120 and (*b*) p40-9-160 (the coordinates are scaled in metres).

was equal to 72 meters (from -12 m to 60 m). Such defined flow domains were used to generate the initial computational meshes for all concerned models. The sample initial meshes, created for the smallest and largest values of the geometric parameters B, C and D, are presented in figure 2.

The boundary conditions for each computational domain were defined according to the following scheme:

- on the vehicle outline: the motionless solid boundary,
- on the left- and right-side of the pseudo-tunnel: inflow and outflow boundaries, respectively,
- at the top of flow domains: the symmetric boundary conditions, and
- at the bottom of wind-tunnel: the moving (with inflow speed) solid boundary.

The vehicle speed, or the air inlet velocity, was always equal to 20 m/s. The physical parameters, like the air density and kinematic viscosity, were equal in each computational case. The control parameters for each of computational and adaptive processes were identical. The simulation time was always equal to 15 s.

The names of particular computational configurations has been defined in accordance with the following scheme:

AB-C-D

where particular symbols (A, B, C and D) denote:

A – the letter 'p', as the distinctive sign for carried out simulations,

B – the trailer height (in decimetres),

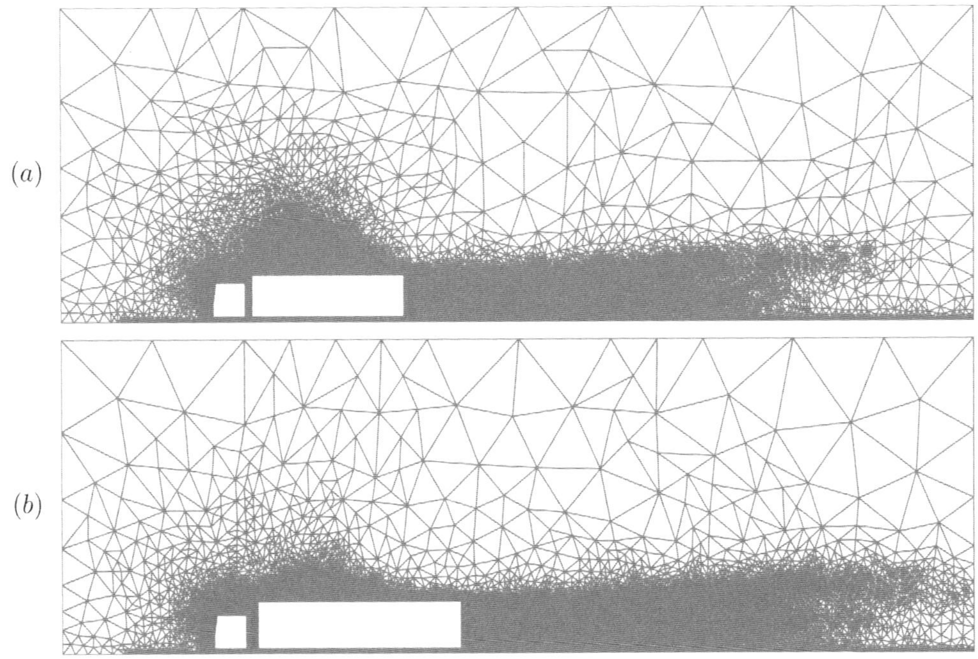

Figure 3. The sample final meshes for the models: (*a*) p36-5-120 and (*b*) p40-9-160, in 15th second of simulation.

C – the clearance gap (in decimetres),
D – the trailer length (in decimetres).

5. Aerodynamic drag coefficients

As previously mentioned, the presentation of simulation results is limited only to aerodynamic drag coefficients, because issues related to forming the flow field described among others through distributions of pressure and velocity components, are not subject to these considerations. However, to illustrate the outcome of adaptation processes, the figure 3 presents sample final forms of computational meshes, which had the forms as in figure 2 at the beginning of computations.

In general, for unsteady problems such as the vehicle aerodynamics, the magnitude of drag coefficient is variable in time. For the sample forms of computational configurations: p36-5-120 and p40-9-160 (in the figures 2 and 3), the temporal variation of drag coefficients is presented in the figure 4. Such form of aerodynamic drag coefficients is, however, unsuitable for comparisons. Therefore for further considerations on obtained results, the constant, consolidated values of these coefficients were defined for each computational configuration. The constant drag coefficients have been determined based on the results from second half of time steps because the initial results (from the first seconds) were disturbed by the inadequate form of initial conditions. Thus the determined (constant) values of drag coefficients are reliable.

In order to illustrate the formation of aerodynamic drag coefficients depending on varying parameters characterised by the tractor-trailer gap, the trailer height and the trailer length, for each clearance gap following have been determined:

- the dependence of the drag coefficient on the trailer length, in figure 5, and
- the dependence of the drag coefficient on the trailer height, in figure 6.

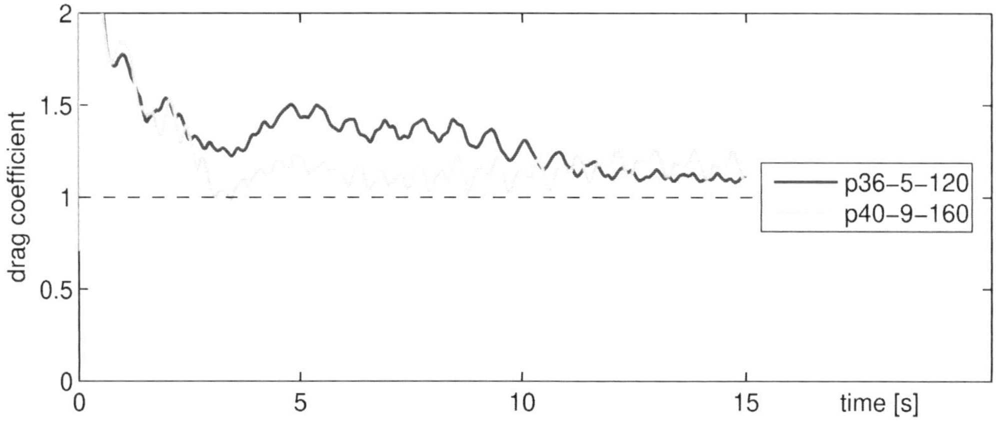

Figure 4. The drag coefficients as a function of time, for the models p36-5-120 and p40-9-160.

Moreover, for each tractor-trailer gap all drag coefficients have been averaged, and percentage differences in relation to these average values have been determined. These percentage differences have been shown in the figure 7, where the coordinates are the trailer length (D) and the trailer height (B). The last three figures have given rise to conclusions.

6. Conclusions

Based on the analysis of the aerodynamic drag coefficient levels presented in the figures 5, 6 and 7, the following can be concluded.

(1) There is a general correlation between the drag coefficient magnitude and the trailer length: the longer trailer, the higher drag coefficient.

(2) There is no clear correlation between the drag coefficient and the trailer height. Nevertheless, for smaller trailer heights, the scatter in the drag coefficients is larger.

(3) Because of the aerodynamic interference between the tractor and the trailer, some combinations of the trailer height and length can cause particular (beneficial or negative) consequences for the drag coefficient magnitude. For such cases it is recommended to carry out the 3D simulations, which can describe the aerodynamic interference and determine the aerodynamic drag coefficient in more reliable way.

References

[1] Hakansson C and Lenngren M J 2010 *CFD Analysis of Aerodynamic Trailer Devices for Drag Reduction of Heavy Duty Trucks* (Göteborg: Chalmers University of Technology, Master thesis)

[2] Patten J, McAuliffe B, Mayda W and Tanguay B 2012 *Review of Aerodynamic Drag Reduction Devices for Heavy Trucks and Buses* (Ottawa: National Research Council Canada, CSTT-HVC-TR-205)

[3] Choi H, Lee J and Park H 2014 Aerodynamics of Heavy Vehicles *Annual Review of Fluid Mechanics* **46** 441–68

[4] McAuliffe B R 2015 *Improving the Aerodynamic Efficiency of Heavy Duty Vehicles: Wind Tunnel Test Results of Trailer-Based Drag-Reduction Technologies* (Ottawa: National Research Council Canada, Aerodynamics Laboratory, LTR-AL-2015-0272)

[5] Devesa A and Indinger T 2012 Fuel consumption reduction by geometry variations on a generic tractor-trailer configuration *SAE Int. J. Commer. Veh.* **5** 18–28

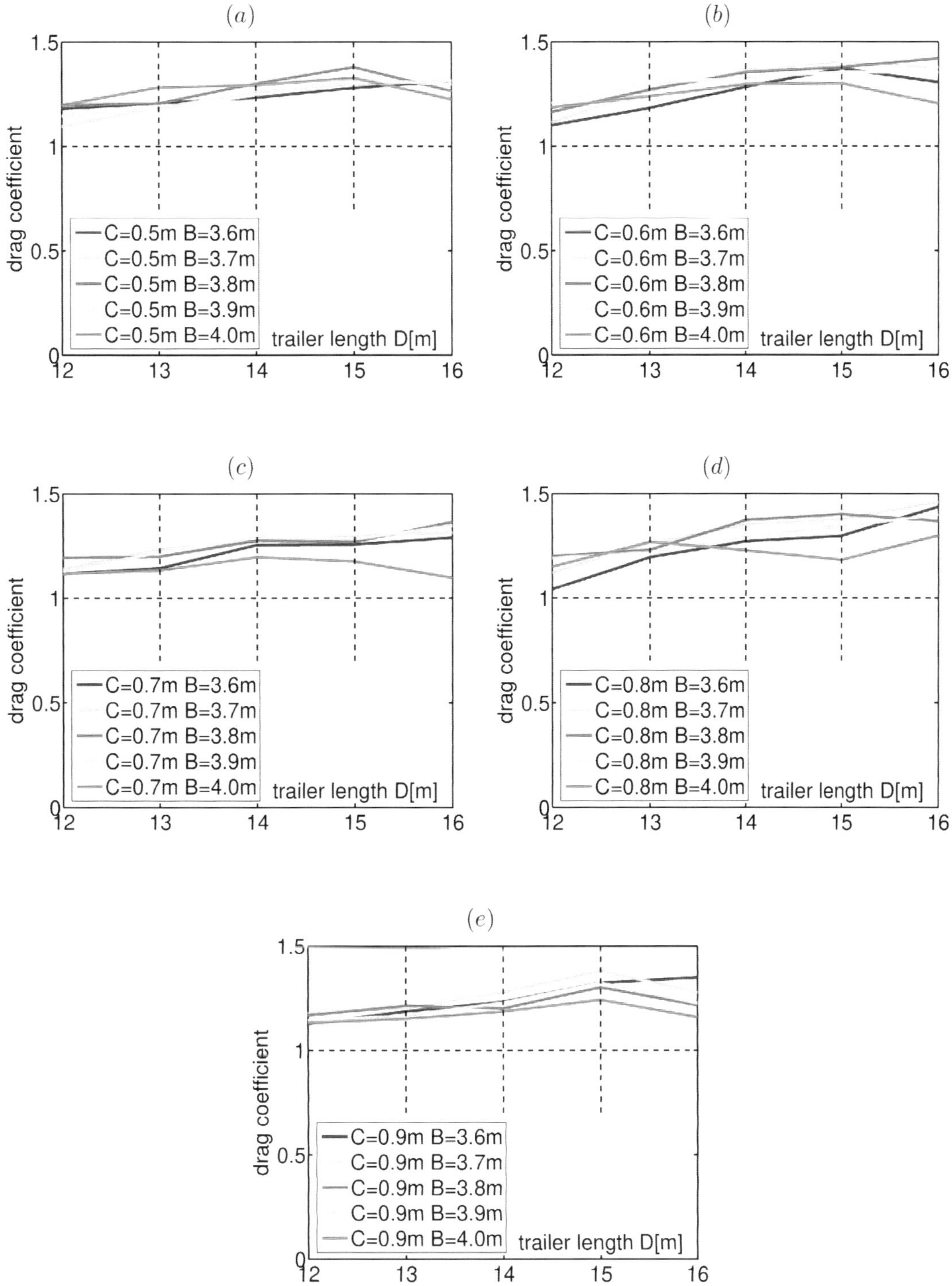

Figure 5. The averaged drag coefficients in a function of the trailer length, for different tractor-trailer gaps: (*a*) C=0.5 m, (*b*) C=0.6 m, (*c*) C=0.7 m, (*d*) C=0.8 m and (*e*) C=0.9 m.

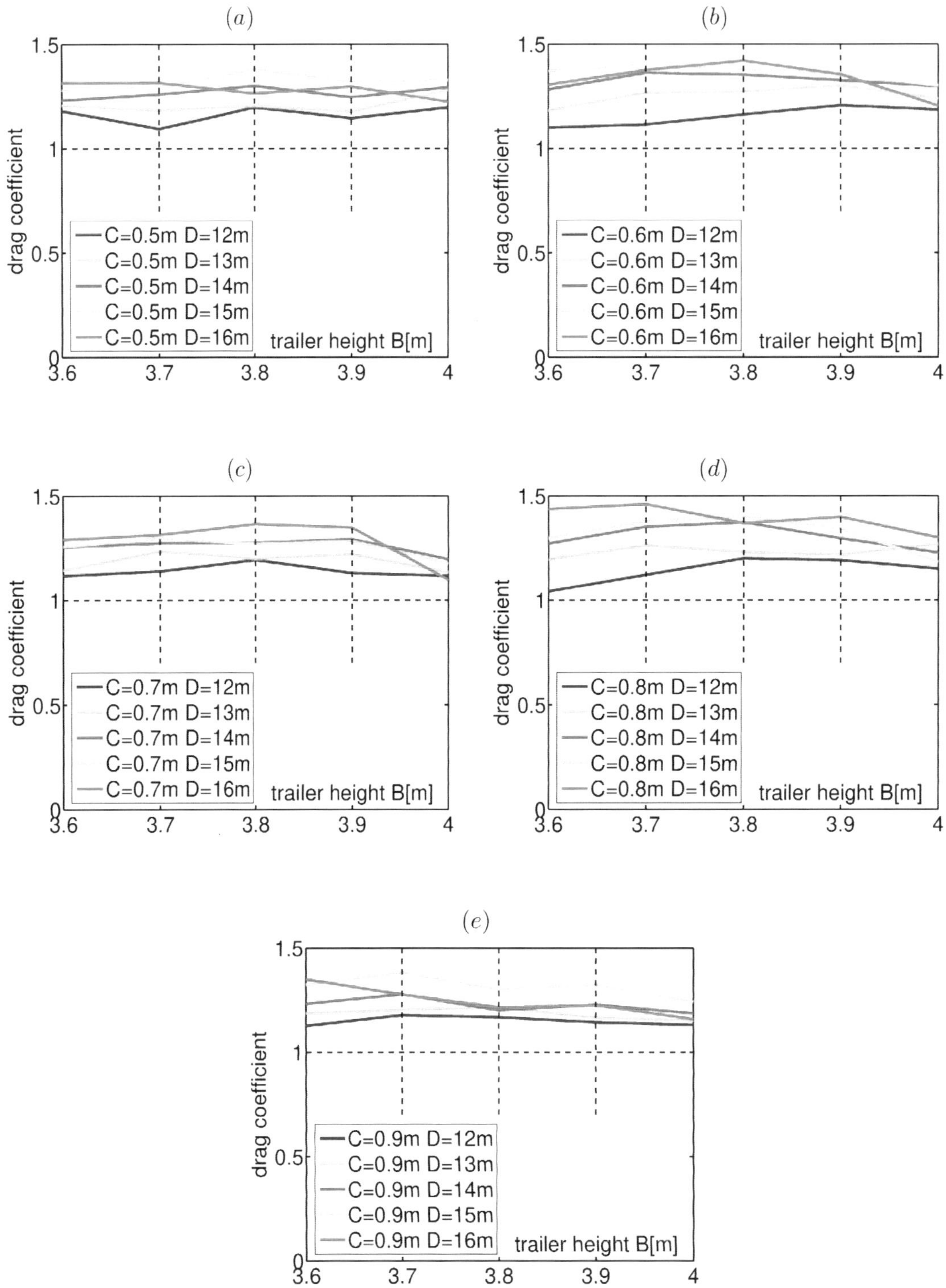

Figure 6. The averaged drag coefficients in a function of the trailer height, for different tractor-trailer gaps: (*a*) C=0.5 m, (*b*) C=0.6 m, (*c*) C=0.7 m, (*d*) C=0.8 m and (*e*) C=0.9 m.

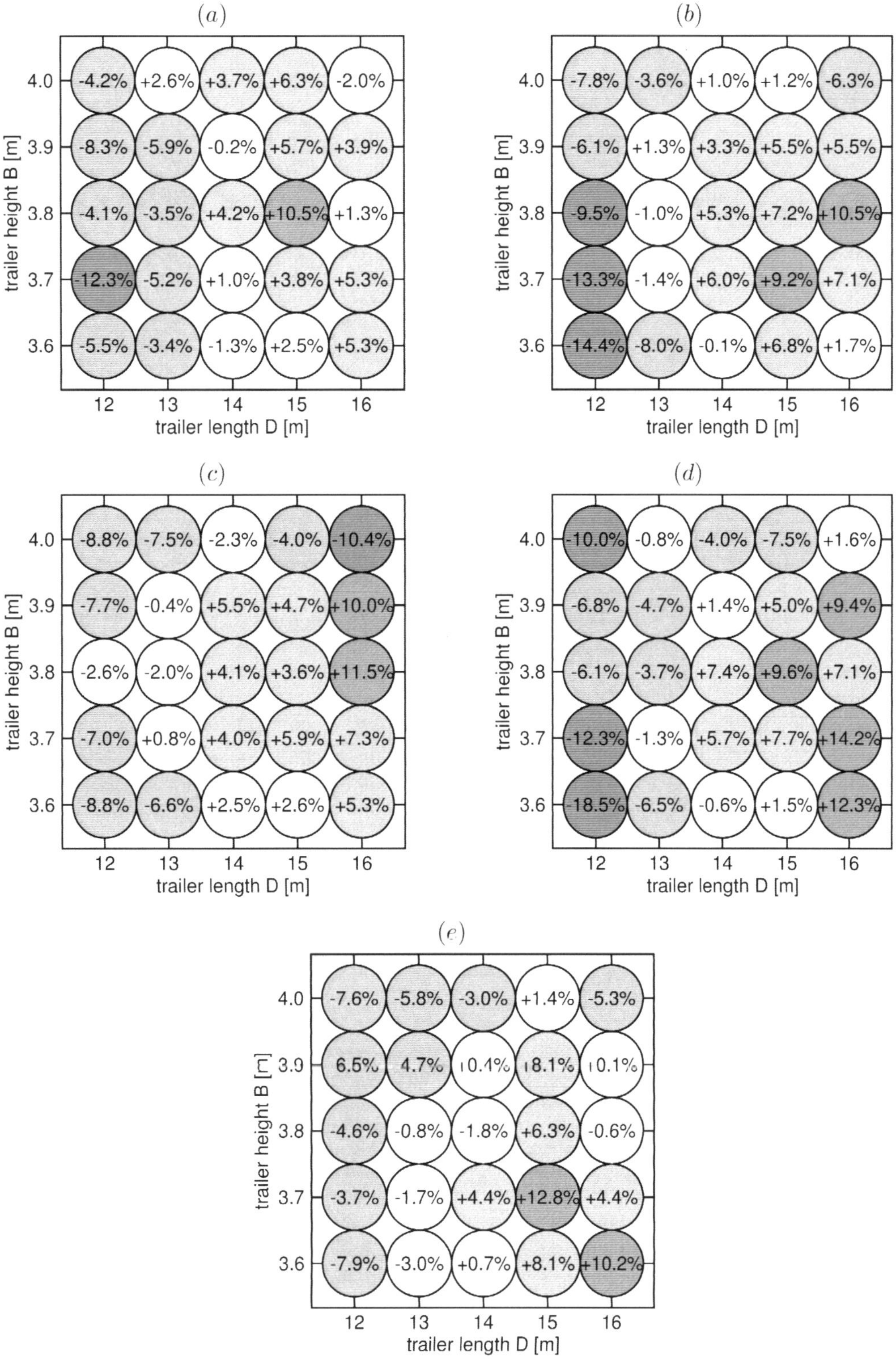

Figure 7. Percentage differences of particular drag coefficients in relation to their mean value, for the gaps: (*a*) C=0.5 m, (*b*) C=0.6 m, (*c*) C=0.7 m, (*d*) C=0.8 m and (*e*) C=0.9 m.

The test bench for testing torsional stiffness of active anti-roll bar made of extended profiles with rectangular cross-section

K R Macikowski[1] and S Kaszuba[1]

[1] Department of Vehicles and Fundamentals of Mechanical Engineering, Faculty of Mechanical, Lodz University of Technology, 116 Żeromskiego St, 90-924 Lodz, Poland, tel. + 48 42 631 23 93.

E-mail: krzysztof.macikowski@p.lodz.pl

Abstract: The article describes the test bench constructed to determine the characteristics of torsional stiffness of extended rod elements, which can be used, for example, in cars as anti-roll bars. The bench has been designed so as to allow an examination of the samples with variable length and variable cross-sectional dimensions. It is possible to perform tests for different materials. The article contains a detailed description of the mentioned test bench and presentation of the results obtained from preliminary tests.

1. Introduction

Anti-roll bars improve driving comfort, steerability and safety. They limit the tilting of the body during driving around a curve. Tilting is a consequence of the centrifugal force. Little body roll is beneficial for accurate and safe driving. Too much of it can cause loss of steerability and, thus, might lead to a side slip and often a roll-over of a vehicle [1].

Classic (rod) anti-roll bar is an appropriately formed torsion bar characterized by constant stiffness, selected by the vehicle manufacturer[2]. Its value is described by the following formula:

$$K_s = \frac{M_s}{\alpha} \tag{1}$$

where:

M_s [Nm] – torsion torque
α [deg] – torsion angle

In an anti-roll bar called active, stiffness can vary in certain range, fluently or gradually. The devices changing stiffness are generally costly constructions, with a complex structure. They require additional actuators, for instance: hydraulic, pneumatic, mechanical or mixed. An example of the mentioned device is BMW mechanism, shown in Figure 1.

Listed drawbacks of described equipment influenced the creation of the concept of fluent adjustment of torsional stiffness. For the object in question, which normally acts on the torsion, additional bending stresses are introduced. They change a resultant stress state and, as a consequence, also expected torsional stiffness. The developed concept assumes that the middle part of the anti-roll bar consists of a packet (two or four elements) of extended profiles with rectangular cross section each. Hence, bending-torsional stiffness is considered.

The described test bench has been built in order to determine the characteristics of torsional stiffness of extended rod elements. The results will be used to define a range of variation of this stiffness, nature of the change and an impact of individual factors (samples dimensions, mechanical properties of the material type) on this change.

Figure 1. The mechanism of an active stabilization installed in a BMW vehicle.

2. Test bench for testing torsion stiffness of extended rectangular cross-section profiles intended for the middle part of anti-roll bar

The test stand for determining stiffness characteristics of twisted elements should provide reproducible, easy to write and edit, research results. A multitude of samples in different sizes, packetized in a variety of configurations shall enable a quick disassembly and reassembly of the unit in which tests are conducted. In the laboratory of the Department of Vehicles and Fundamentals of Mechanical Engineering in the Lodz University of Technology was built the bench that meets these requirements. Its block diagram is shown in Figure 2.

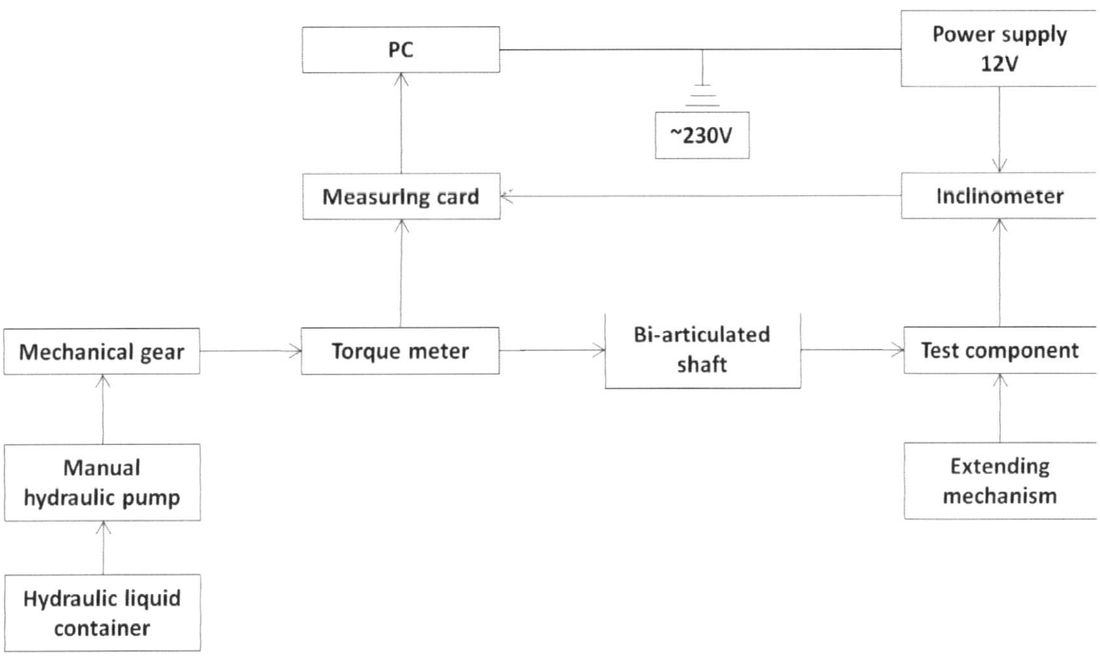

Figure 2. Block diagram of the stand for testing torsional stiffness.

The test bench was equipped with two sensors: torque and incline. Obtained from these analog signals are directed to the measuring card National Instrument NI 9215 BNC type, which forwards them in a digital form to the recording unit (PC in Figure 2). Special software allows to record measurement results on PC and their subsequent processing for the purpose of presentation.
Figure 3 shows the test bench. Number (1) indicates a computer set for signal recording and data processing. The hydraulic drive unit (2), equipped with a manual hydraulic pump, is used to twist the profiles. The hydraulic actuator causes the rotation of particular elements of the bench. They are connected with each other in a series and they are respectively: torque meter MI50 (3), bi-articulated shaft and a set of testing profiles (minimum of two) which are mounted in a special test unit (5). Number (4) indicates an inclination sensor (which is part of the test unit).
The test unit is shown in Figure 4. It has two supports: fixed (3) and movable (2). The fixed support is placed in an extreme position of a bench and aims to permanently immobilize the studied profiles (1). The movable support acts as a base on which is mounted an inclinometer (7) Posital Fraba type ACS0802SV20HE2PM (tilt sensor). Apart from its basic function of profiles rolling, the support also allows their axial shortening and lengthening depending on a degree of the distance between the profiles and their twist. Extending mechanism (4) consists of two halves. Each of them is mounted on a corresponding profile in a way which prevents movement. Test elements are extended in the middle of their active length, by distancing both halves from each other. The distance is adjusted by wrenching two screws M12x1 (5) into the upper part of mechanism. These screws press against two pins (6) which are fixedly mounted in the lower part of mechanism. In this way, both halves move away from each other by an equal distance in relation to the longitudinal axis of the examined profile.

Figure 3. Test bench for testing torsion stiffness. 1 – recording unit. 2 – hydraulic drive unit. 3 – torque meter. 4 – inclinometer. 5 – test unit.

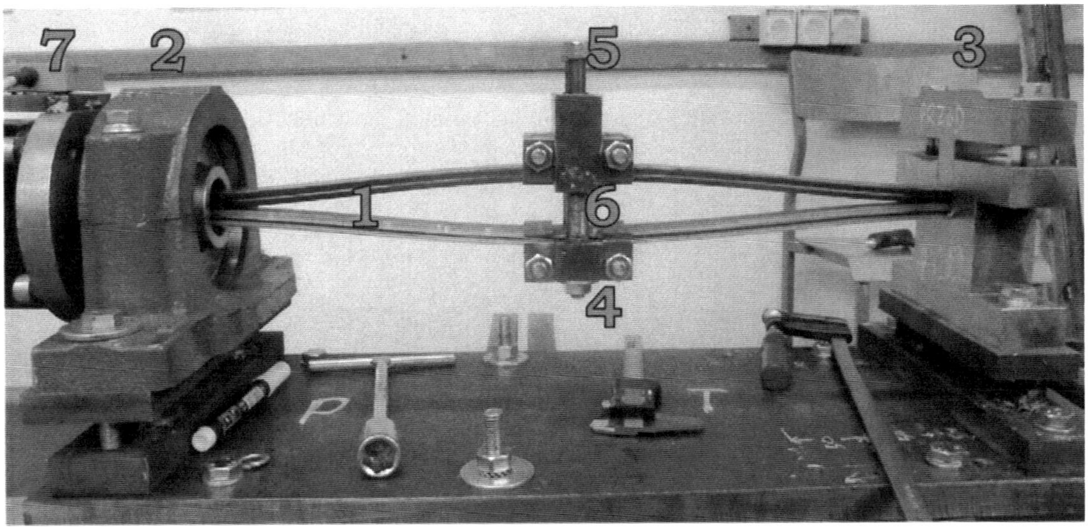

Figure 4. Test unit in position of work. 1 – a packet of profiles. 2 – movable support. 3 – fixed support. 4 – extending mechanism. 5 – two screws M12x1. 6 – two guide pins. 7 – inclinometer.

It is expected to carry out a series of studies of profiles with variable amounts in various length, width and height. Figure 5 shows an example of 4 profiles with beveled edges (they were prepared for construction reasons to allow an installation of an extending mechanism) with dimensions 5x22,5x650mm. They were made of spring steel 50HF and then hardened throughout and tempered.

Figure 5. Research profiles before hardening process.

3. The course of research and description of preliminary results

Profiles presented in Figure 5 were installed on the bench and there were conducted 7 series of tests. In each series there were recorded torsion angle and torque generated for the 9 following angle positions : 0°, 3°, 6°, 9°, 12°, 15°, 18°, 20° and a maximum angle for which allowed a stroke of hydraulic actuator (in this case 21,22°). In each series, the distance between both halves of extending mechanism increased from 0mm up to 30mm with the 5mm stroke. Figure 6 shows the profiles in three characteristic positions: ready for testing, after extending both parts of the mechanism by 30mm (without twisting) and after extending both parts of the mechanism by 30mm and twisting by 21.22°. An effective length of the test pieces was 528mm. These values resulted from the initial assumptions which had been made beforehand.

(a)

(b)

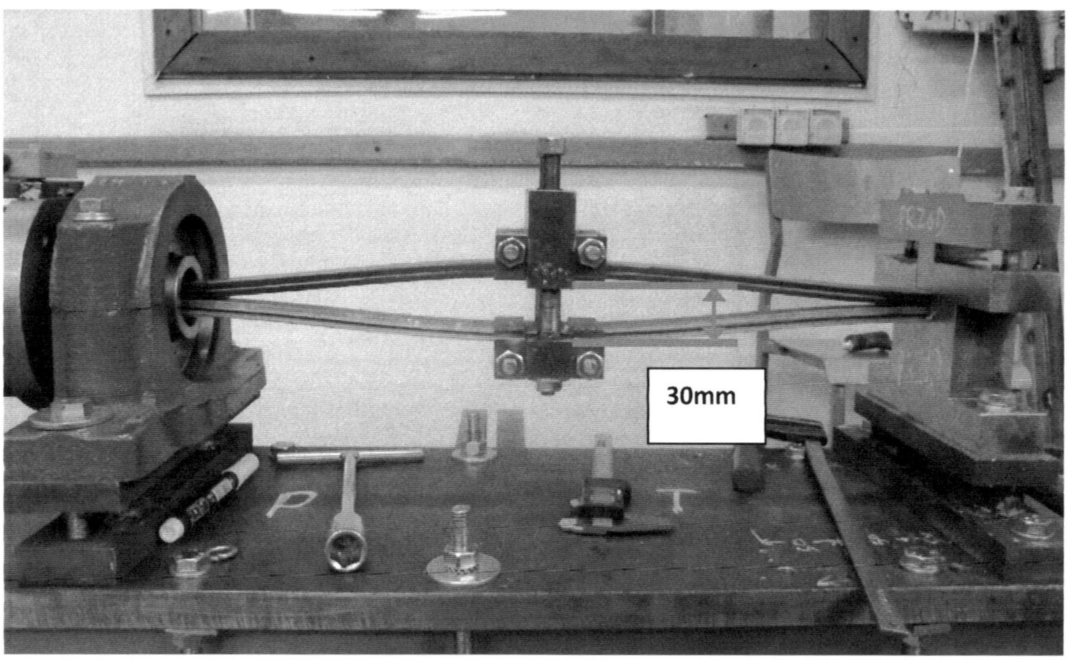

(c)

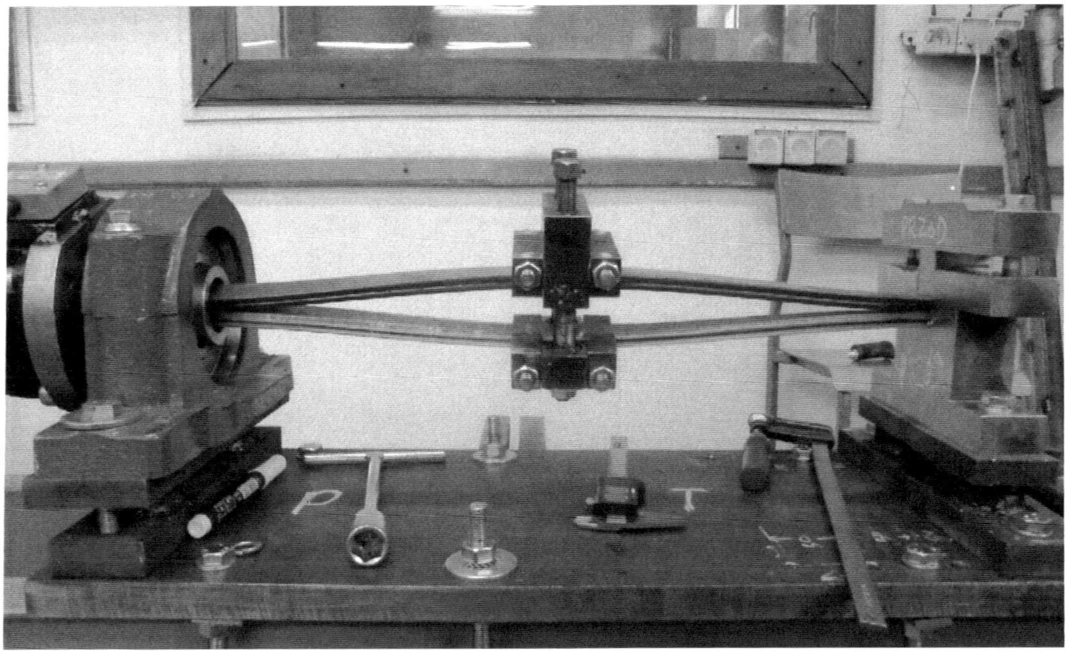

Figure 6. Research profiles in characteristic positions. a) ready for testing (the first measurement was without extending). b) after extending both parts of the mechanism by 30mm (without twisting). c) after extending both parts of the mechanism by 30mm and twisting by 21.22 °.

Obtained results were subject to an electronic processing and after this phase changing stiffness graphs were prepared (they included a generated torque [Nm] in a function of rotation angle [deg]). Received curves were similar in shape to straight lines, so the linear trend line was used. The graph referring to the profiles from Figure 5 was placed below (in Figure 7). It includes test results from the 1st and 7th series. The variable stiffness area is placed between both curves. For each series (between loading and unloading) we observe hysteresis. It is the effect of the friction appearing between profiles and in particular parts of the test bench.

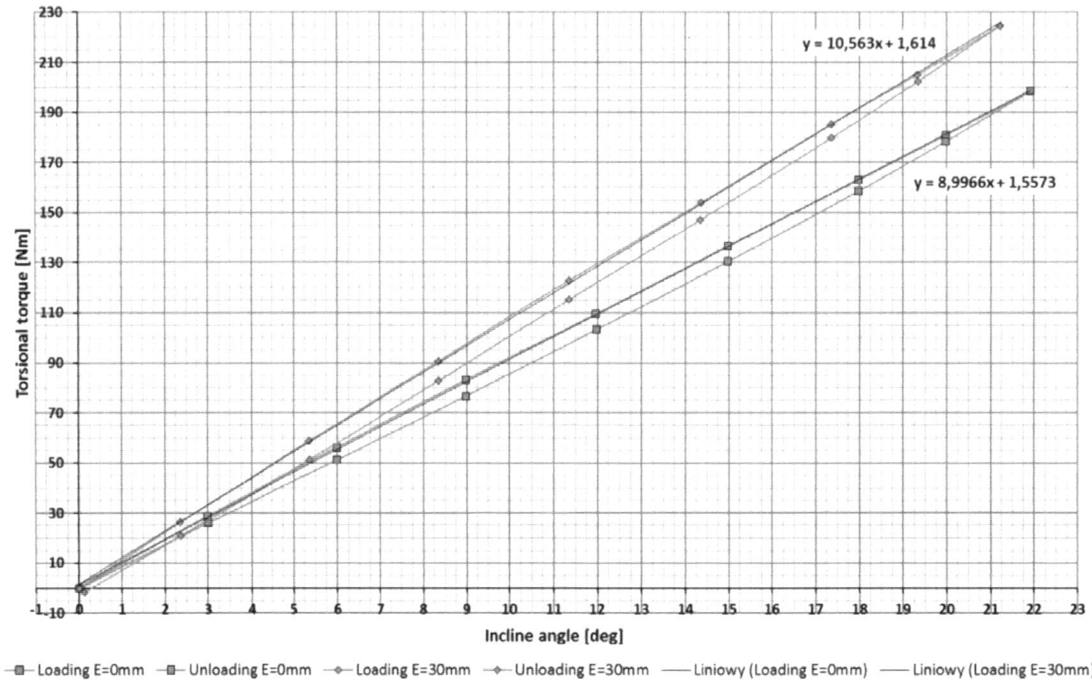

Figure 7. The graph showing changing stiffness in a set of 4 twisted profiles with following dimensions 5x22,5x650mm (active length equal 528mm for each sample).

On the basis of known functions of trend curves from series 1 and 7 there were determined the maximum and minimum torque acting for twisting 20°. Then, using the formula (2), strengthening of profiles (increase of stiffness) was calculated and it amounted to 17.3%.

$$K_{20} = \left(\frac{M_{max}}{M_{min}} * 100\right) - 100 \quad (2)$$

where:

K_{20} [%] – increase stiffness in the profile twisted by 20°

M_{max}, M_{min} [Nm] – maximum torque, minimum torque

In the next step there was conducted research of the same elements. Their active length was shortened consequently by 50mm to 478mm and 428mm. Figure 8 shows them in the position "before initiating the test".

Table 1. Summary of stiffness increase for 4 profiles with following dimensions 5x22,5x650mm while shortening their active length.

Pos	No of profiles	Profile height [mm]	Profile width [mm]	Profile active length [mm]	Stiffness increase in profile [%]	Stiffness increase resulting from profile shortening [%]	Maximum torque read at 7th test series [Nm]
1	4	5	22,5	528	17,3	-----	213
2	4	5	22,5	478	20,4	18,0	255
3	4	5	22,5	428	30,8	77,9	328

A change of profiles strengthening for 20° torsion angle was recognized in a tabular form and presented in Table 1. An analysis of the data it contains indicates an upward trend in the strengthening of the samples by shortening their active length (strengthening from 17.3% at the length of 528mm to 30,8% at the length of 428mm). During the studies, an increase in the generated torque while extending profiles was observed (data for not extended profiles have not been included in the table). The mentioned increase is explained by theory of strength of materials [3], where in the rod-like elements, a torsional angle varies linearly according to the change of their length. So, to twist the same sample with a shorter length it must be ordered to a greater torque. This theory results from a well-known formula for torsion bars with a rectangular cross-section:

$$\varphi = \frac{M_s l}{\eta a b^3 G} \tag{3}$$

where:

φ [°]	–	torsional angle of the rod with rectangular cross section
M_s [Nm]	–	torque acting on profile
l [m]	–	profile length
η	–	calculation factor
a [m]	–	longer edge of rectangle
b [m]	–	shorter edge of rectangle
G [Pa]	–	Kirchhoff modulus

The above considerations do not allow to clarify the increase in stiffness during extending profiles taking into account the change in their active length. Further studies and analysis will have to clarify what factors are responsible for and how they affect the stiffness on extended and twisted profiles.

(a)

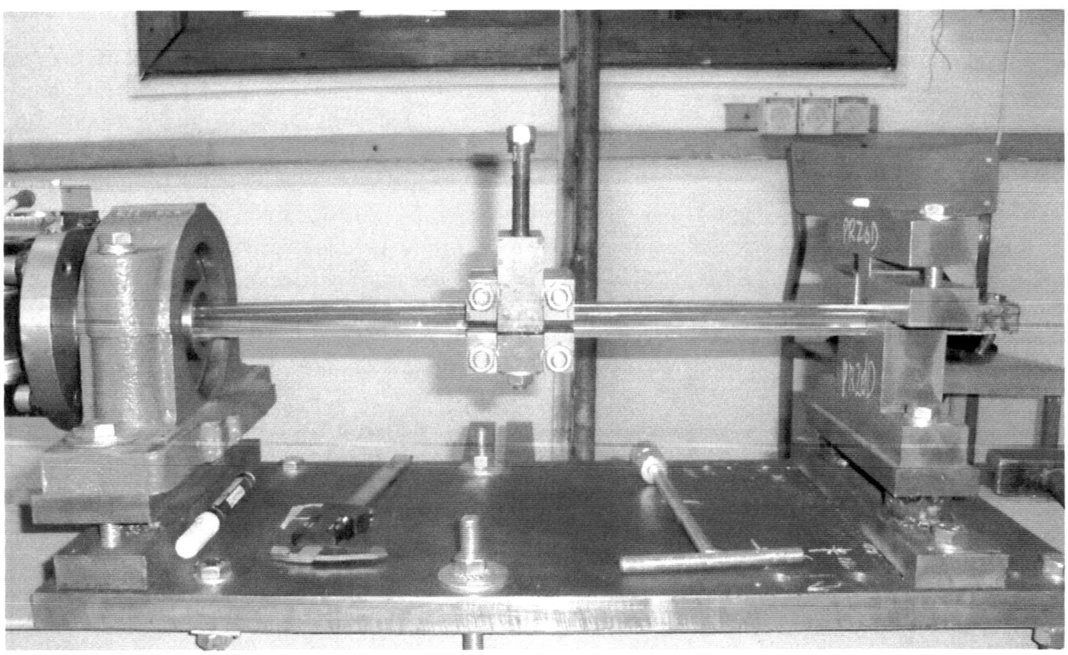

(b)

Figure 8. Profiles ready for testing. a) with active length equal 478mm. b) with active length equal 428mm.

4. Conclusions

The aim of the conducted research is to explain a phenomenon of increasing stiffness in twisted profiles which are extended in the middle of their active length. An assessment of the design parameters and a degree of their impact on the stiffening profiles constitute the result of the study. Preliminary studies lead to the following conclusions:

1. It is possible to increase the torsional stiffness of the elements with rectangular cross-section by extending them in the middle of their active length. It was verified by investigation of twisted and extended profiles (each time up to 30mm) with following lengths 528, 478 and 428mm. The following strengthening (showed in mentioned sequence) was obtained: 17,3%, 20,4%, 30,8%.
2. The increase of torsional stiffness (in profiles made of spring steel) depends on their length and cross-sectional dimensions.

Further work will include the investigation of more experimental profiles so that the obtained results enable the development of an empirical relationship describing the change of torsional stiffness of extended elements. Simultaneously, an attempt on an analysis of examined cases will be made. It is also planned to create a digital model and test it by using finite element method in the ANSYS 16.2 program. Properly constructed model will allow the analysis of more samples in various combinations with selected dimensions. It will reduce the cost of its preformation, implementation and reduce the duration of the study.

References

[1] Werner J 1966 *Budowa samochodów. Konstruowanie podwozi,* Warszawa: WKiŁ p 477
[2] Wittek M 2010 Stabilizer Bars: Part 1. Calculations and Constructions. *Transport Problems* **5**
[3] Jakubowicz A and Orłoś Z 1972 *Wytrzymałość materiałów* Warszawa: WNT

Multi-criteria optimization of chassis parameters of Nissan 200 SX for drifting competitions

M Maniowski[1]

[1] Institute of Automobiles and Internal Combustion Engines, Cracow University of Technology, Poland

E-mail: mmaniowski@pk.edu.pl

Abstract. The work objective is to increase performance of Nissan 200sx S13 prepared for a quasi-static state of drifting on a circular path with given constant radius (R=15 m) and tyre-road friction coefficient (μ=0.9). First, a high fidelity "miMA" multibody model of the vehicle is formulated. Then, a multicriteria optimization problem is solved with one of the goals to maximize a stable drift angle (β) of the vehicle. The decision variables contain 11 parameters of the vehicle chassis (describing the wheel suspension stiffness and geometry) and 2 parameters responsible for a driver steering and accelerator actions, that control this extreme closed-loop manoeuvre. The optimized chassis setup results in the drift angle increase by 14% from 35 to 40 deg.

1. Introduction

The paper deals with chassis parameters optimization of Nissan 200sx S13 (figure 1) with Rear Wheel Drive and 300hp engine prepared for a drifting competition [1], where not a lap time, like in most of motorsport disciplines, but the biggest drift (side slip) angles are evaluated.

Figure 1. Nissan 200sx S13 while drifting on R-const test.

An oversteering without a suitable driver intervention is unstable mode of a vehicle motion. There are a few great papers [2, 3, 4, 5] describing dynamic behaviour of drifting vehicles, but focused on a control theory standpoint and with rather basic vehicle-driver models utilized. This kind of extreme motion states are not covered in classical vehicle dynamics [6, 7, 8].

Possibilities of effective initiation, sustaining and decay of extreme drift angles depend on a drift-vehicle chassis setup. Typically, this setup is selected based on many track tests with trial and error methods. There are, to the author's knowledge, no published papers dealing with a computer aided selection method of the best drift vehicle parameters.

In order to improve this expensive process a model (called "miMA") of driver-vehicle-road system adapted for multi-criteria optimization problems has been formulated by the author in Matlab environment (figure 2).

"miMA" software is characterized by:

- multibody spatial model of a vehicle with hundreds of discrete parameters specialized in motorsport, like off-road [9], racing [10, 11], rally [12], drifting [13] or karting;
- effective code yielding proper balance between model accuracy and computation time;
- combined optimization of driver actions for closed-loop manoeuvres, vehicle chassis and motion trajectory;
- implemented genetic algorithm which enable global search of nonlinear and multi-criteria tasks;
- substitution of actual driver actions by additional optimization variables, what emulates driver adaptation without using oversimplified driver models.

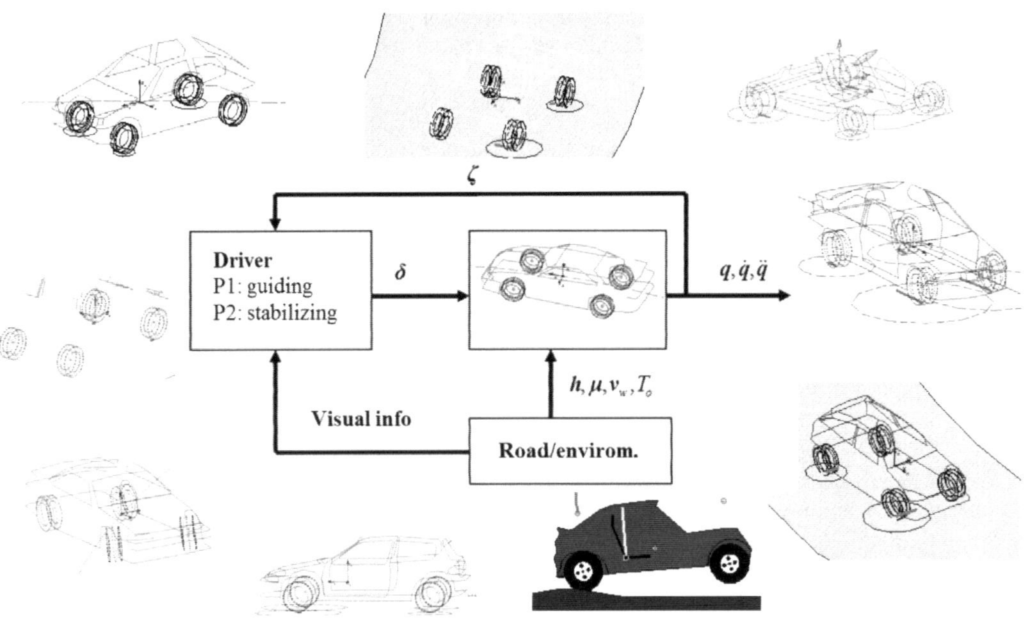

Figure 2. „miMA" model of driver-vehicle-road system used for optimisation tasks in motorsport.

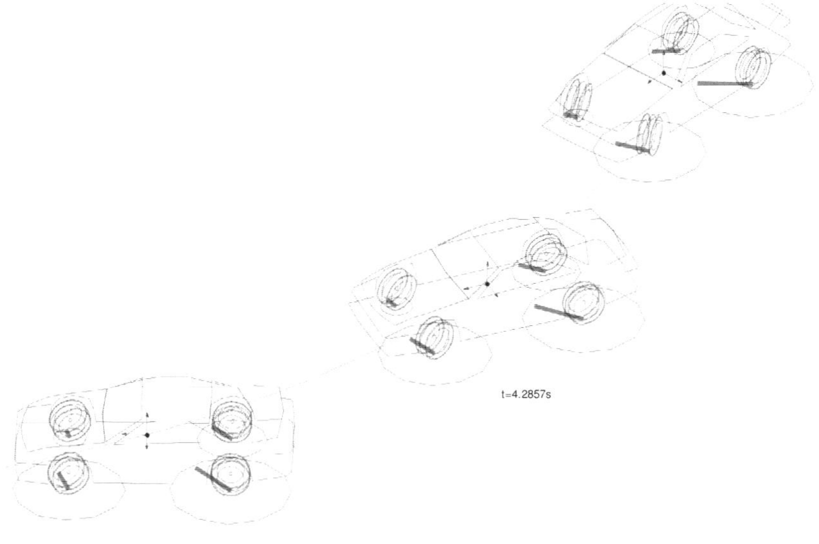

Figure 3. Nissan 200sx S13 cornering (R = 15 m, μ = 0.9) with maximal stable drift angle (β_{max} = 35.3 deg) obtained for the base chassis setup.

In this work the given drift-vehicle (figure 1) has to follow R-const (R=15m) curve (figure 3), in a quasi-static fashion, with the biggest possible drift angle (β) on even and isotropic road with tyre-road friction coefficient μ=0.9.The best setup of the drift-vehicle chassis is sought, together with the driver actions needed for stabilization of unstable oversteering motion. This problem is solved by using an innovative definition of driver-vehicle system optimization (miMA).

2. Definition of "miMA" model

The model of driver-vehicle-road system (figure 3) adapted for optimisation problems is formulated by the author in Matlab environment. "miMA" is characterized by deep specialization in motorsport applications, e.g. rally [8], race [13], drifting [11] and off-road [12]. The vehicle model relates its design parameters ($\boldsymbol{p}$) and driver actions ($\boldsymbol{\delta}$) with the vehicle dynamic characteristics, which can be derived based on its motion states ($q, \dot{q}, \ddot{q}$). A multibody dynamic model with discrete parameters is described by system of nonlinear ordinary differential equations (ODE). Special formulation of closed form solutions of the wheel suspensions makes it possible to avoid time consuming differential algebraic equations (DAE).

The virtual racing driver (figure 3) has to guide a vehicle through a desired path, stabilize it and seek a maximum performance from a vehicle on a specific road surface. The driver actions are described by:

$$\boldsymbol{\delta} = [\delta_h\ \delta_b\ \delta_a\ \delta_c\ \delta_g\ \delta_e]^{\mathrm{T}} \tag{1}$$

where: δ_h – steering wheel angle [deg]; δ_b – normalized position of brake pedal ($\delta_b \in$ <0,1>); δ_a – norm. position of accelerator pedal ($\delta_a \in$ <0,1>); δ_c – norm. position of clutch ($\delta_c \in$ <0,1>); δ_g – gear shift position ($\delta_g \in \{-1,0,1,2,\dots\}$); δ_e – normalized position of e-brake ($\delta_e \in$ <0,1>).

The driver actions (1) are based on observations (figure 3) of selected vehicle motion states (ζ) and visual perception. The road-environment model includes descriptions of road profiles ($\boldsymbol{h}$), friction potential ($\boldsymbol{\mu}$), wind velocity (v_w) and ambient air temperature (T_a) and road surface temperature (T_r).

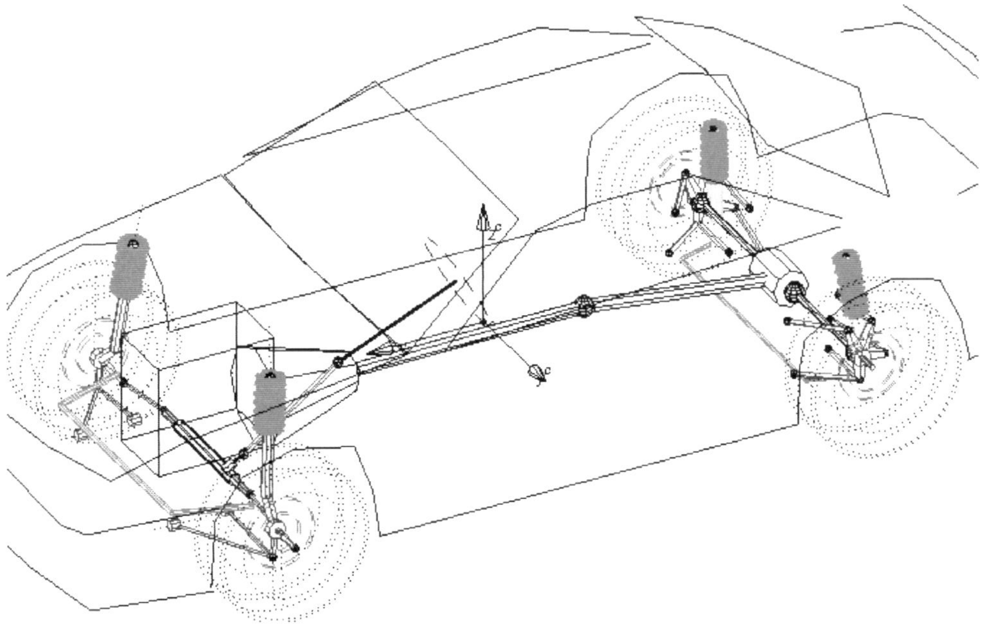

Figure 4. "miMA" multibody model of Nissan 200sx S13 with 26 gen. coordinates.

3."miMA" model of Nissan 200 sx S13

The numerical example considers Nissan 200sx S13 (figure 1) with: Rear Wheel Drive, 300 horsepower engine, 1250 kg curb weight (50/50%), limited slip differential and performance tires [1].

The nonlinear multibody vehicle model is presented in figure 4. Main components of "miMA" model are defined in Table 1. The model, described by 26 generalized coordinates (***q***) and 440 discrete parameters, is suitable for dynamic analyses in a frequency range up to 20 Hz.

The wheel suspension mechanisms, i.e. a McPherson strut with trailing rack&pinion steering system in the front axle and a multi-link in the rear axle, are described by spatial kineto-static models []. Magic Formula with the first order lag dynamics [5] is implemented as tyre model.

Table 1. Description of „miMA" model components for dynamic analysis of Nissan 200sx S13.

Model parts	Generalized coordinates (q)	Description
Car body	6 (x, y, z, φ, θ, ψ)	Position and orientation of rigid body
Wheels	4 (φ_1, φ_2, φ_3, φ_4)	Rotation about wheel bearings
Suspensions	4 (z_1, … z_4)	Bounce motion
Tyres	4+4 (F_{x1}...F_{x4}, F_{y1}...F_{y4})	First-order dynamics of tire horizontal forces
Steering sys.	1 (φ_k)	Compliance of steering shaft
Powertrain	1+1+1 (M_s, φ_m,φ_w)	Engine torque + differential + shafts compliance
SUM	26	

Figure 5. Front axle of Nissan 200sx S13 on test rig for kinematic characteristics determination by using wire based platform.

In order to limit the paper scope, it is assumed that: the road is even and isotropic ($\boldsymbol{h}$ = 0, $\boldsymbol{\mu}$ = const), temperature effects of tyres, dampers and brakes are neglected. More on tyre temperature influence on a race car performance one may find in [10].

Most of "miMA" model parameters are estimated on the basis of in&outdoor experiments by utilisation of the Institute test rigs and measuring apparatus.

Characteristics of the wheels suspension springs (coil springs, bumpers, anti-roll bars) and dampers (figure 4) are determined on the basis of test rig measurements and described as passive elements with nonlinear characteristics.

Major dimensions of the actual suspension mechanisms in Nissan 200sx S13 are measured directly. Those difficult to measure are identified by comparison of kinematic characteristics obtained from measurements and simulations [14]. The test rig utilising a wire based platform [15] for determination of wheel suspensions kinematic characteristics is presented in figure 5.

The considered vehicle was instrumented (figure 6) in order to verify the simulation model on the basis of the measured motion states.

Figure 6. Acquisition apparatus in the vehicle cockpit and optical measurement of the vehicle horizontal velocity.

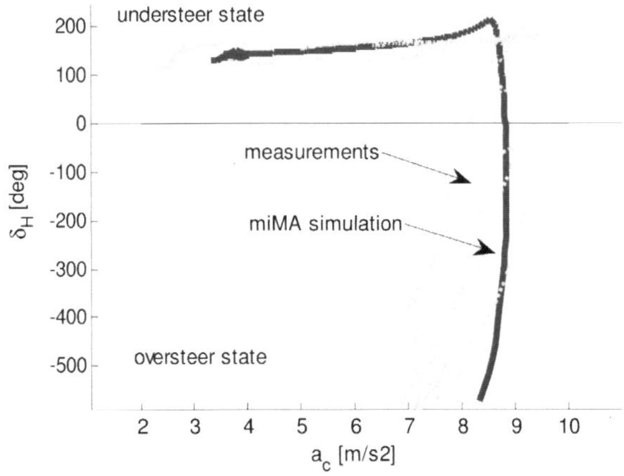

Figure 7. Measurement and simulation results of R-const test (R =15 m, μ = 0.9).

Sample results of measurements and simulations for the considered R-const (figure 3, R =15 m, μ = 0.9) test are presented on a handling diagram in figure 7. An expert drift driver starts in an understeer state, where the steering wheel angle (δ_h) increases in proportion to the central acceleration (a_c), until the front axle saturation at a_c= 9 m/s^2. Next, an oversteer state is generated by excessive application of the engine power. The drift angle (β) increases to 35 deg, what requires a countersteer of δ_h=-550 deg to continue R-const trajectory as semi-steady state (figure 3), with slight modulations on the steering wheel and the accelerator (not shown). High adequacy of the formulated "miMa" model may be confirmed by comparing the motion states from the road measurements and the simulations. The positively verified model may be further incorporated in an optimization algorithm.

The following 11 chassis parameters (***p***) of Nissan "miMA" model (figure 4) are selected as important for an optimization. Its definitions, initial values and ranges are given in Table 2.

Table 2. Definitions of Nissan chassis parameters (p_i) and driver actions (δ_i).

No	Definition	Bounds	Initial	After optimization
p_1	FR steering axis inclination angle [deg]	10÷16	13	15
p_2	FR steering axis caster angle [deg]	3÷9	6	4
p_3	Steering system ratio [-]	12÷20	17	12
p_3	Rate of steering Ackerman [-]	-1÷1	0.5	0
p_5	RR wheel camber angle [deg]	-5÷-1	-3	4
p_6	FR antiroll bar (linear) stiffness [N/mm]	5÷50	25	20
p_7	RR antiroll bar (linear) stiffness [N/mm]	0÷20	10	0
p_8	Preload of limited slip differential [Nm]	0÷200	100	100
p_9	Rate of 2-way limited slip differential [-]	0÷1	1	1
p_{10}	Maximal engine torque [Nm]	355÷400	355	0
p_{11}	Mass distribution to FR axle [%]	45÷55	50	48
δ_h	Steering wheel angle [deg]	-700÷0	-550	-560
δ_a	Accelerator position [-]	0÷1	0.5	0.7

4. Vehicle chassis optimization for R-const drifting

The numerical example is based on the following assumptions:

- the vehicle (figure 4) has to: (I) generate the biggest possible drift angle (β), (II) in a quasi-static fashion (when $\dot{\beta} \rightarrow 0$), (III) moving on R-const (R=15m, μ=0.9) curve with a minimal deviation (ΔR), and at (IV) minimal cost of the vehicle setup changes;
- this closed-loop manoeuvre is controlled be an virtual driver by changing only the steering wheel angle (δ_h) and the accelerator position (δ_a) on given ($\delta_g = 2$) gear;
- the manoeuvre takes 8 seconds and starts with a drift angle defined by initial conditions;
- steady-state drifting, i.e. state of motion when drift angle velocity vanishes ($\dot{\beta} = 0$), has to be obtained during the last 4 s.

In that case, an optimization algorithm of the driver actions and the vehicle chassis parameters is defined as follows:

- goal functions: $\boldsymbol{w} = [\max(\beta)\ \ \min(\dot{\beta})\ \ \min(\Delta R)\ \ \min(c)]_{1\text{x}4}$ (2)
- decision variables: $\boldsymbol{d} = [\boldsymbol{\delta}\ \ \boldsymbol{p}]_{1\text{x}13}$ (3)
- under constraints: $\boldsymbol{d}_{min} < \boldsymbol{d} < \boldsymbol{d}_{max}$, (4)

where: $\boldsymbol{\delta} = [\delta_h\ \delta_a]_{1\text{x}2}$ driver actions,
$\boldsymbol{p} = [p_1 \ldots p_{11}]_{1\text{x}11}$ vehicle chassis parameters ,
ΔR deviation from R-const trajectory,
$c = \sum_{i=1}^{11} p_i$ vehicle reassembly cost.

The problem is solved by using multicriteria Genetic Algorithm after 30k iterations (ca 8hrs on PC).

5. Results of Nissan 300sx chassis optimization

The obtained optimization results for Nissan 200sx (R = 15 m, μ = 0.9) are presented on a drift phase plane (figure 8) defined by the first two (w_1 and w_2) criteria (2). The obtained V-like profile of each solutions family has the following interpretation. The best solution is described by a maximal drift angle (β) and a minimal drift angle rate ($\dot{\beta}$), that approaches zero. This state of the drifting vehicle motion fulfils a stability condition from a technical point of view.
In case of the base chassis setup, the vehicle can be driven (figure 8, left) with a steady-state power-slide at β = 35.3 deg drift angle (figure 3). Other solutions are unstable under the defined work assumptions. In case of the optimized chassis setup (figure 8, right), the steady-state drift angle is increased by 14% to ca. 40 deg. The optimized chassis parameters are given in Table 2.

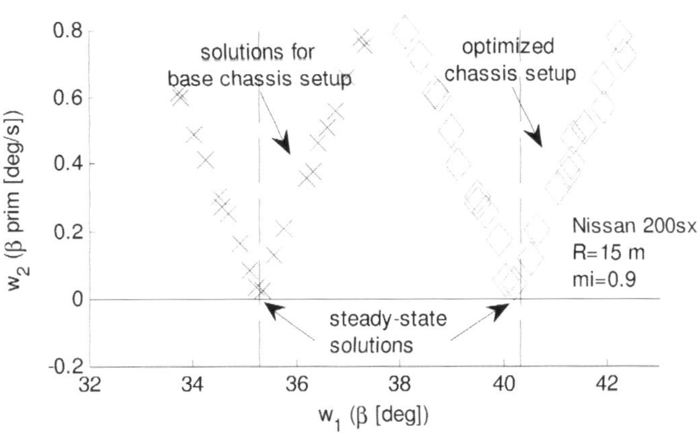

Figure 8. Optimization results of Nissan 200sx (R = 15 m, μ = 0.9) on drift phase plane ($\dot{\beta}$ vs β).

6. Concluding remarks and further steps

In the paper an innovative method of RWD vehicle chassis optimization for quasi-static drifting on defined path with given constant radius and tyre-road grip, is presented.

The high fidelity vehicle model, called "miMA", with 26 state variables and more than 400 parameters is formulated for the optimisation purposes. "miMA" enables to describe details of a vehicle chassis components (relevant in motorsport) at a reasonable computational time (3 times shorter than real time). Similar results are rarely achievable with commercial general purpose software. The developed model of Nissan 200sx S13 was successfully verified on the basis of track tests with an expert drift-driver.

Control problem of oversteering mode stabilization was reduced to a static optimization problem with virtual driver actions included as decision variables, in addition to chassis parameters.

The considered Nissan 200sx S13 with optimized chassis setup (for R = 15 m, μ=0.9) can stably sustain a 14% bigger drift angle, i.e. β = 40 deg instead 35 deg. It was mainly achieved by changing the front axle inclination angle and anti-roll bar stiffness.

The subsequent step of the drift vehicle preparation usually considers an optimization in cases of dynamic manoeuvres, like a slalom or a drift initiation.

References

[1] Jugowski A 2014 *Research on controlled oversteer states in Nissan 200SX* Bachelor thesis (promoter Maniowski M) Mechanical Department Cracow University of Technology

[2] Abdulrahim M 2006 *On the Dynamics of Automobile Drifting* SAE paper **01-1019** pp. 169–178

[3] Edelmann J Plöchl M 2009 *Handling characteristics and stability of the steady-state power slide motion of an automobile* Regular&Chaotic Dynamics **14** pp.682–692

[4] Velenis E et al. 2011 *Steady-state drifting stabilization of RWD vehicles* Control Eng. Practice **19**

[5] Hindiyeh Y Gerdes Ch 2011 Driftkeeping: *path tracking at the friction limits using high sideslip cornering* International Association of Vehicle System Dynamics Manchester

[6] Mitschke M 2004 *Dynamik der Kraftfahrzeuge* Gabler Wissenschaftsverlage

[7] Pacejka H 2012 *Tire and Vehicle dynamics* Butterworth – Heinemann

[8] Milliken W Milliken D 1995 *Race Car Vehicle Dynamics* Warrendale PA USA SAE International

[9] Maniowski M 2014 *Optimization of wheel suspension dampers in off-road rally car* Proceedings of the Institute of Vehicles **4 (100)** pp. 57-69

[10] Maniowski M 2016 *Optimization of driver actions in RWD race car including tyre thermodynamics* Vehicle System Dynamics 54 **4**

[11] Maniowski M 2016 *Optimization of driver and chassis of FWD racing car for faster cornering* The Dynamics of Vehicles on Roads and Tracks pp. 77-86

[12] Kluziewicz M Maniowski M 2012 *Comparison of race and rally driving techniques efficiency in front-wheel drive car cornering* Technical Transactions Mechanics **10** pp. 51-62

[13] Maniowski M (2016?) *Bifurcation states of rear-wheel-drive car motion* Journal of Theoretical and Applied Mechanics (after review)

[14] Knapczyk J Maniowski M 2006 *Synthesis of a five-rod suspension for given load-displacement characteristics* Proc. Instn. Mech. Engrs Part D J. Automobile Eng **220** pp.879-889

[15] Góra M Maniowski M 2011 *Verification of 6dof platform with wire-based sensors for spatial tracking* The Archive of Mechanical Engineering **LVIII** pp.157-174

IOP Conf. Series: Materials Science and Engineering **148** (2016) 012020 doi:10.1088/1757-899X/148/1/012020

Modernisation of a test rig for determination of vehicle shock absorber characteristics by considering vehicle suspension elements and unsprung masses

M Maniowski[1], S Para[1] and M Knapczyk[2]

[1]Institute of Motor Vehicles and Internal Combustion Engines, Cracow University of Technology, 37 Jana Pawla II Avenue, 31-864 Krakow, Poland
[2] BWI Group S.A., Cracow, PL

E-mail: mmaniowski@pk.edu.pl

Abstract. This paper presents a modernization approach of a standard test bench for determination of damping characteristics of automotive shock absorbers. It is known that the real-life work conditions of wheel-suspension dampers are not easy to reproduce in laboratory conditions, for example considering a high frequency damper response or a noise emission. The proposed test bench consists of many elements from a real vehicle suspension. Namely, an original tyre-wheel with additional unsprung mass, a suspension spring, an elastic top mount, damper bushings and a simplified wheel guiding mechanism. Each component was tested separately in order to identify its mechanical characteristics. The measured data serve as input parameters for a numerical simulation of the test bench behaviour by using a vibratory model with 3 degrees of freedom. Study on the simulation results and the measurements are needed for further development of the proposed test bench.

1. Introduction

Properties of automotive dampers are fit to the customer's requirements, by adapting the valve configuration, choosing the right damper elements dimensions, the hydraulic oil properties, the gas pressure and some other parameters [8]. There is no simple method of evaluating overall damper performance on the bench with good correlation to vehicle test results [9 Bench rig like in Figure 1 is current automotive industry standard for damper performance evaluation. The test is usually conducted with a sine-wave excitation at constant stroke and different frequencies. Peak forces measured during subsequent tests are used to plot force vs. velocity diagrams used to describe the damper.

The test bench consists of a frame and a force sensor which is mounted to the beam. In the lower part a hydraulic ram is fixed to the bottom. Its cylinder movement is numerically controlled which allows to move the test object with piston displacements and piston velocities defined by the user. The limiting values of operation are given by the utilized hydraulic pumps. Another option to force a displacement of the tested object is to use a driving shaft which can realize different piston velocities but with quasi-sinusoidal profile and constant stoke.

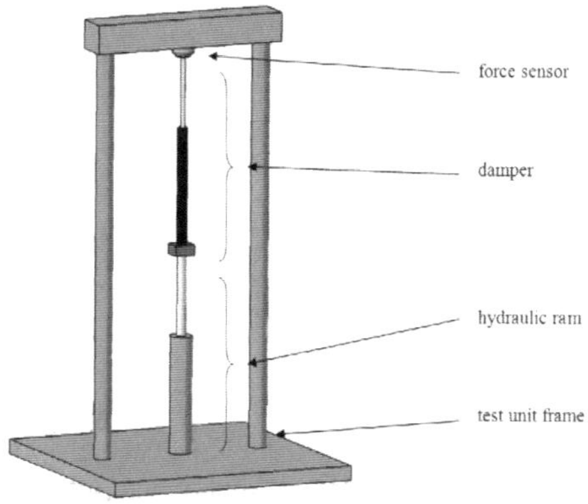

Figure 1. Typical test rig for determination of damping force characteristics of an isolated shock absorber.

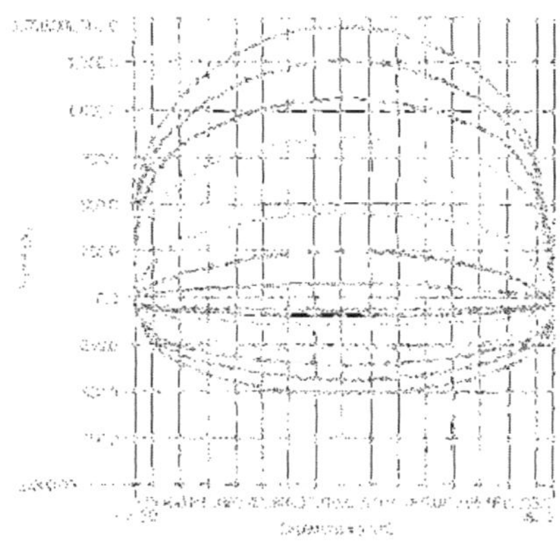

Figure 2. Damper characteristics as damping forces vs. displacement.

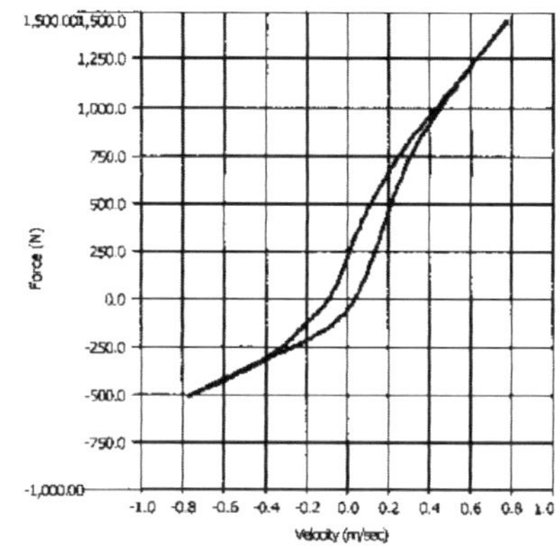

Figure 3. Damping forces vs. velocity.

Equipped with a measurement system, in detail the mentioned force sensor, displacement sensor and/or velocity sensor, damper characteristics can be obtained, similar to these presented within Figure 2 and Figure 3.

Figure 2 shows the damping force as a function of the piston displacement. The maximum force during the compression phase is about 500 N whereas the rebound force is nearly three times greater at about 1450 N at the greatest stroke velocity. An alternative is shown in Figure 3 where the damping force is presented as a function of the piston velocity.

This testing procedure only describes the properties of the dampers, but installed in a vehicle the whole suspension system together with unsprung masses let dampers work in a different way [2]. Especially on rough roads where the piston displacement is small and the damper can generate a noise which deteriorates the ride comfort.

The function of the shock absorber is to damp both sprung and unsprung resonance [5,6]. In terms of vehicle characteristics, its function is to damp body roll motion when the vehicle is rapidly steered, to damp body pitch motion when the vehicle is rapidly accelerated or decelerated, and to damp unsprung mass vibrations for ensuring the tyre-ground contact while driving on undulated roads. On the other hand, smaller damping forces can improve ride comfort while driving on a smooth road and reducing ride harshness while passing over a road bump. For instance, driving on a rough road insists high damping forces within the compression phase and tension respectively. In contrast, handling manoeuvres, like a lane change, do not require such high damping forces.

The main idea is extending the existing damper test bench (Figure 1) with a tyre-wheel and a guiding arm, imitating a vehicle suspension, joined with a damper module composed of a shock absorber, a coil spring, a spring aid, elastic top mount and bushings. The designed test bench should fulfil additional criteria, like easy change of the suspension parameters, simplicity, limited space taken.
In order to better understand the suspension behaviour at the test rig and to provide information about boundary conditions for the real life tests, a dynamic model of the vibrating system with 3 degrees of freedom is formulated by using MATLAB/SIMULINK.

2. The considered wheel suspension system

The rear, undriven suspension of a Mini Morris Cooper S was considered as the research object. It consists of a trailing arm consisting the wheel hub and the attachment for the spring-damper module. The trailing arm is constrained by two lateral links. This kind of suspension mechanism belongs to the kinematic group of spherical mechanisms, where the sphere centre is located at the trailing arm front [7].

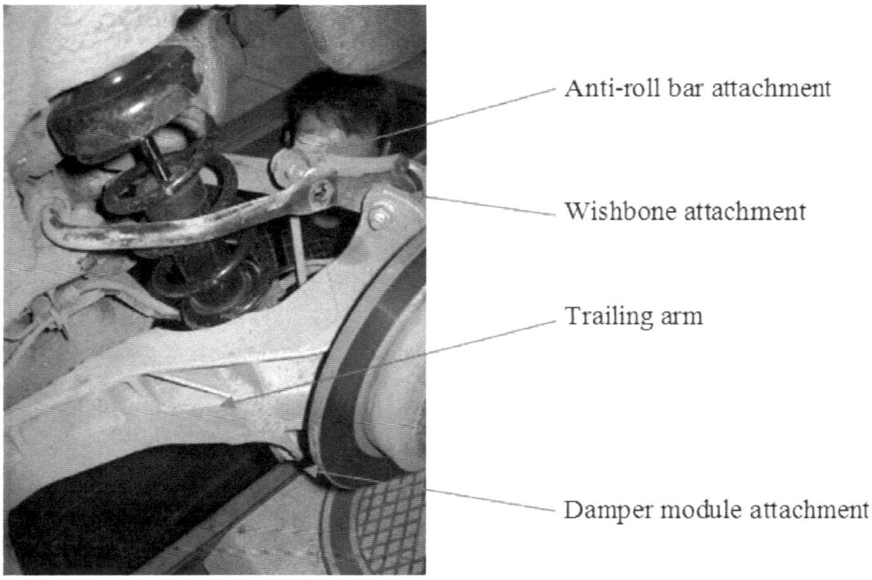

Figure 4. Measurement of unsprung masses on the rear wheel suspension of a Mini Morris Cooper S.

An overview of the measured elements, which were carried out at experimental test benches, is given in Figure 5.

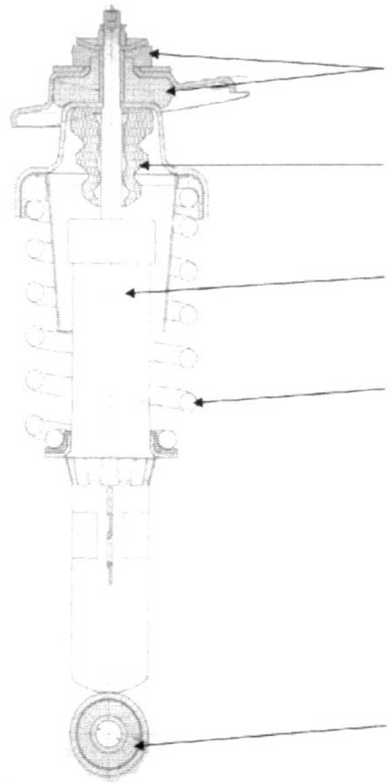

Figure 5. Assembly of the considered coil over shock module.

Starting at the top of Figure 5 there are two elastic elements (top mount bushings) which are located at the upper and lower side of a disk and have an initial preload caused by the assembly. The disk is mounted to the car body. The upper element springs into action during a rebound and the lower element during a bump event. The next considered part of the damper module is the so-called spring aid. If the damper deflection in compression stroke is greater than a given value the spring aid comes into action. It works as a parallel spring to the main suspension spring increasing the net suspension stiffness. Next, the twin-tube damper and the coil spring. The last element used within the damping module is the lower bushing. All these described elements have been disassembled and measured at special test rigs. The obtained readings give information about the reaction forces as result of a quasi-static deflection.

Until now, only the elements which are in connection with the damper module have been considered. To imitate a vehicle suspension some more aspects are important. Also the pneumatic tyre of the 205/45R17 84V (Bridgestone Potenza with Run Flat), inflated with 2.6 bar, is taken into account.

The tests were carried out in the lab [4] at the Institute of Automobiles and Internal Combustion Engines at Cracow University of Technology.

The measured mass on single wheel (250 kg) can be divided into the unsprung and the unsprung masses. To differentiate between them the suspension was prepared for separate measurements like presented in Figure 4. One step was to detach the anti-roll bar. Next, the damper module was disconnected from the trailing arm. This allowed to measure the unsprung suspension mass which consists of: brake disk, screws, wheel bearing, moving part mass of the trailing arm, break calliper which includes the brake fluid, the brake pads and the hydraulic pipes. Separately the wheel mass was measured, which consists of: rim and tyre.

To sum up, the goal was to obtain information about elements installed within the tested car suspension. Different methods have been considered. The easiest and quickest way is to make use of data sheets, but often the accessibility is poor which does mean that tests have to be carried out at test rigs. In this paper, readings were made on machines which are quasi-static or by making use of dynamic test rigs. The stiffness is described as a function of displacement or velocity respectively. An additional possibility to obtain information is to make use of non-commercial instruments like a scale, weights and a measuring instrument. This is an easy and cheap way to gain information about elements but is connected with a long preparation time.

All obtained characteristics of each car suspension component will be used within a numerical calculation. The next chapter will describe approaches of modelling a quarter-car model to obtain more realistic information about an extended and the newly designed guiding arm.

3. Mechanical models for vertical vibrations analysis

A numerical model has been created which takes into account suspension masses and characteristics of elastic and damping elements. The following models were assumed.

4-DOF quarter car model with longitudinal arm

A so called quarter car model with four degrees of freedom for analysis of vertical vibrations in the wheel suspension system, is presented in Figure 6.

Due to the fact that the wheel and the damper module are not assembled in one axis, which does mean that that there is a ratio between this elements at the trailing arm, this has to be taken into account within the calculation.

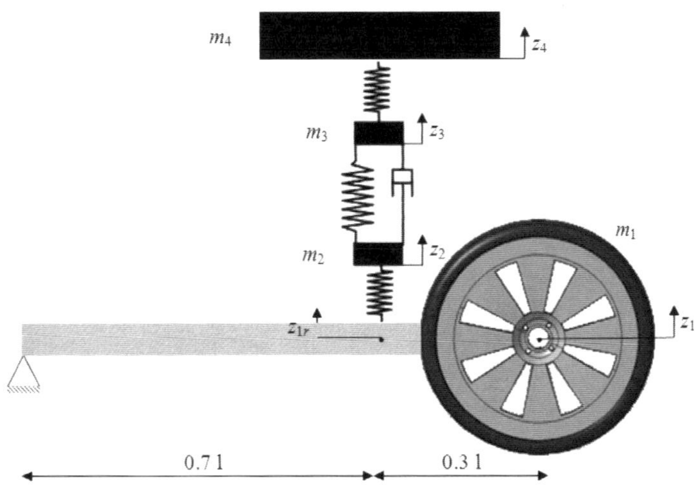

Figure 6. Geometry of the wheel suspension with longitudinal arm.

The suspension ratio, defined as relation between the wheel vertical motion and the spring-damper deflection, is equal to 0.7. This does mean that the trajectory is different which effects that the vertical displacement or the difference (z_2-z_1) is not correct within the calculation. Respecting the ratio, the proper displacement (z_{1r}) has to be taken:

$$z_{1r} = 0.7 \cdot l \cdot \sin(\alpha) \quad (1)$$

4-DOF quarter car model

The 4 DOF model allows to execute a more precise analysis by considering all characteristics and elements, as well as to estimate the error made by testing on a test rig without a sprung mass, which is not possible within a rigid frame at the test bench due to the available space. This simulation can be carried out by adapting the vibratory system as shown in Figure 7.

IOP Conf. Series: Materials Science and Engineering **148** (2016) 012020 doi:10.1088/1757-899X/148/1/012020

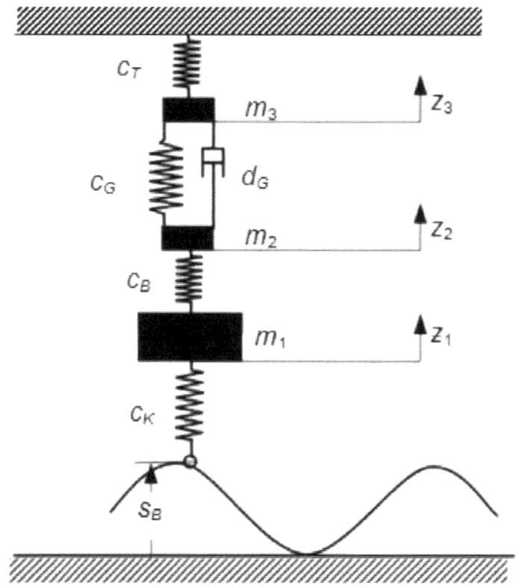

Figure 7. 3-DOF system with unmovable car body.

3-DOF quarter car model

The absence of the sprung mass and attaching the top mount to the force sensor changes the system. Thereby z4 equals zero. The equations of the oscillatory system are as follows:

$$\ddot{z}_1 = \frac{c_B}{m_1}(z_2 - z_{1r}) - \frac{c_K}{m_1}(z_1 - s_B) \tag{2}$$

$$\ddot{z}_2 = -\frac{c_B}{m_2}(z_2 - z_{1r}) + \frac{c_G}{m_2}(z_3 - z_2) + \frac{d_G}{m_2}(\dot{z}_3 - \dot{z}_2) \tag{3}$$

$$\ddot{z}_3 = \frac{c_T}{m_3}(-z_3) - \frac{c_G}{m_3}(z_3 - z_2) - \frac{d_G}{m_3}(\dot{z}_3 - \dot{z}_2) \tag{4}$$

These equations describe the movement of the suspension. Solving the equations the forces of each structural element can be obtained, which is an input for designing the test rig elements.

Table 1. Mass parameters of 4-DOF quarter car model.

m_1	37 kg	Unsprung mass
m_2	2 kg	Lower valve, inner and outer tube, lower bushing
m_3	1 kg	Piston rod, piston, valve
m_4	213 kg	Sprung mass (car body part)

4. Design of the proposed test rig

By knowing the masses and dimensions of all components, as well as the movement obtained by the simulation, the new elements of the modernised test rig have been designed. Overviews are shown in the following figures.

An existing test rig (Figure 9) which consists of a frame, a force sensor at the top and a hydraulic ram with a base plate at the bottom has been extended by a guiding arm, shown in Figure 8. This simplified suspension considers all geometrical ratios, all masses, real elements of the suspension strut and the real wheel of the vehicle.

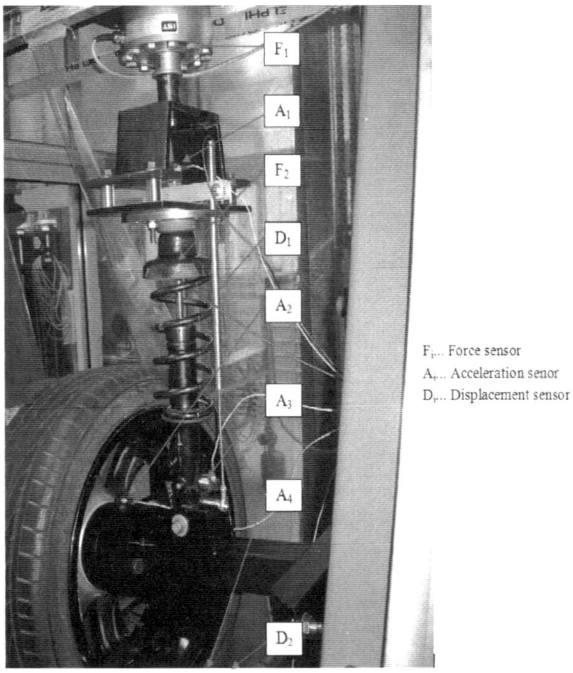

Figure 8. Test rig in initial position and sensors layout for extended damper testing.

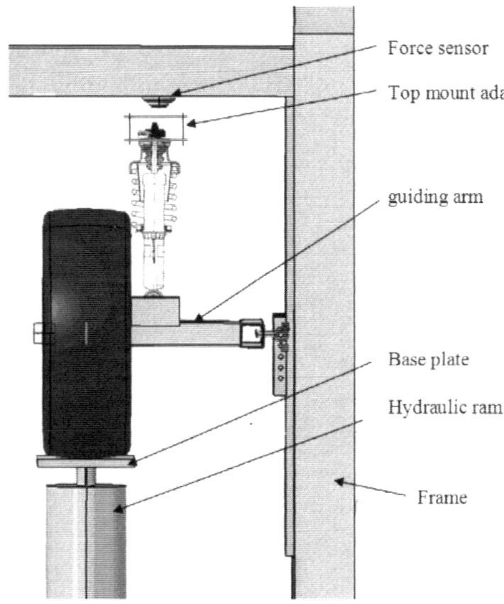

Figure 9. Final test rig assembly.

Before a test run was started, the test stand had to be equipped with measuring equipment. The amount and the types of available sensors have been:

- 6 acceleration sensors,
- 2 displacement sensors (at damper module and hydraulic ram),
- 1 force sensor (at the damper rod),
- 1 force sensor (at the test rig).

The signal of the base plane accelerometer A4 (Figure 8) provided information about the excitation frequency. The frequency was also obtained by reading the displacement measured by the test rig controller but to make a cross-check this information was also gained by acceleration sensor A4.

To obtain the damper characteristic, or more precisely the damping force as a function of the rod velocity, which is dependent on the frequency and excitation displacement, sensors D1 and F2 were used. The velocity is not measured directly but calculated by numerically differentiating the displacement signal.

Signals gained with sensors F1 and D2 are used to compute the transfer function from the excitation displacement up to the net module force, which is transferred up to the imitated car body attachment.

The dependency between signals gained by sensors A2 and A3 provided information about the suspension ratio. Accelerations measured with sensors A1 and A2 are used for NVH analysis of the damper module.

5. Damper characteristics obtained from measurements

The measured characteristics of the mean piston displacement as a function of excitation frequencies for different amplitudes (2 , 5 and 10 mm) of sinusoidal excitation, are presented in Figure 10. Due to the fact that the car body is substituted by the fixed test rig frame, the unsprung mass can move only. The obtained characteristics is typical for a nonlinear system, where the unsprung resonance frequency depends on the excitation amplitude. In conditions of higher frequencies and amplitudes the tyre starts to loss contact with the exited base. This phenomenon can be used to compare different characteristics of the dampers.

The values of the piston displacement multiplied by the suspension ratio give approximated the mean displacement of the oscillating unsprung mass. Due to the unsymmetrical damper characteristic (the force during rebound is nearly four times greater than the force during bump motion) the displacement drifts from the initial position at 0.

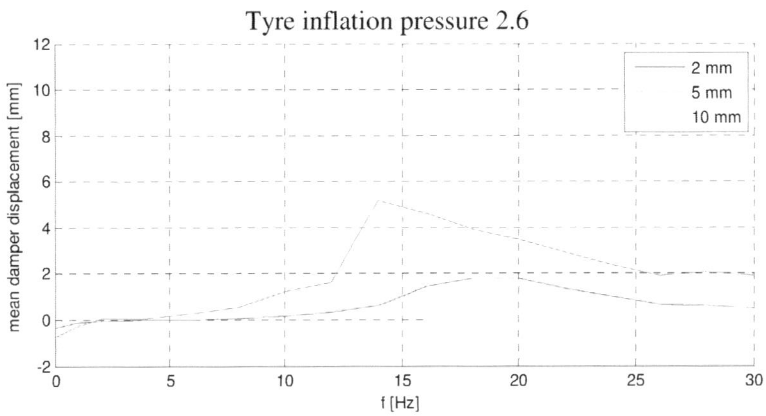

Figure 10. Mean damper displacement D1 vs. excitation frequency.

A comparison of the standard deviations of the two measured reaction forces (F1, F2) can also be suggested for evaluation purposes (Figure 11). Recognizable is the fact that a deviation of the net force (F1) at the test rig differs from a deviation of the damping force (F2), especially at the resonance frequency.

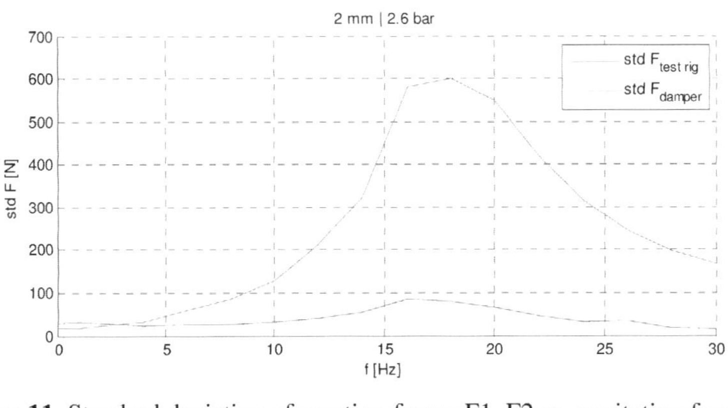

Figure 11. Standard deviation of reaction forces F1, F2 vs. excitation frequency.

6. Validation of the numerical model

Based on measured data the numerical model (3-DOF, chapter 3) can be validated. For a first comparison a frequency of 1 Hz, a peak amplitude of 2 mm and the nominal tyre pressure is defined. The data obtained by the measurement is printed blue whereas the simulated results are plot in red colour within Figure 12.

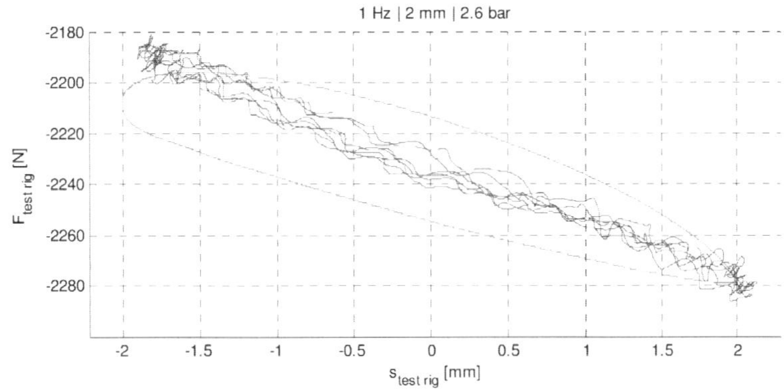

Figure 12. Measurement (blue) vs. simulation (red); Force F1 vs. Displacement D2.

At low excitation frequencies the controller of the hydraulic ram generates overlaid oscillations with a higher frequency. The irregularity within the measured signal is recognizable. Comparing the two signals similar displacements and reaction forces at the test rig sensor can be observed.

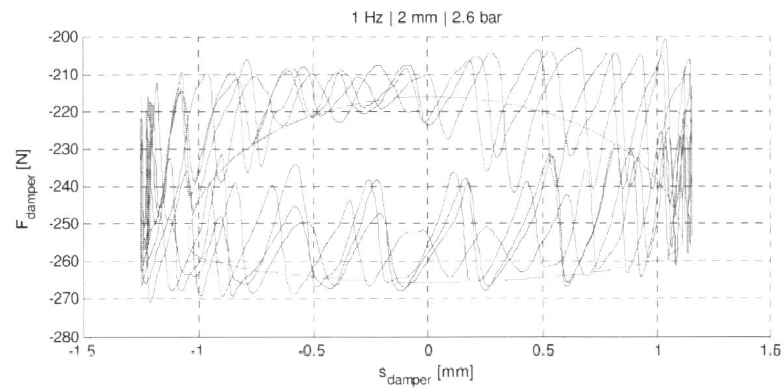

Figure 13. Measurement (blue) vs. simulation (red); Force F2 vs. Displacement.

The simulation provides a similar damper characteristic (damper force as a function of damper displacement) as presented in Figure 13.

Unfortunately the simulation model provides differing results at higher frequencies than 4 Hz. A reason is that the characteristics of the compliant elements have been estimated with a quasi-static method. The high dynamic of this oscillatory system would require dynamic characteristics like reaction forces as a function of displacement and frequency. Additionally a more advanced damper model should be used.

7. Concluding remarks and further steps

This paper presents an approach of modernizing a standard shock absorber test bench. The test mechanism consists of a guiding arm with an additional unsprung mass, the wheel and other compliant suspension components fixed to a simplified suspension. Measurements of car suspension elements were carried out which serve as input parameters for a numerical simulation and for the design process, respectively. With the help of this enhanced test rig NVH analyses can be carried out which can be compared to standard damper tests and real life application or to compare dampers of the same type but with different characteristics.

Initial information about the system response was captured using formulated multi body model analysis. The simulation results were based on solution of three, nonlinear, inhomogeneous, ordinary differential equations, which were implemented in Simulink/Matlab. Effects like change of mean position of the unsprung body mass caused by the unsymmetrical damper characteristic can be observed already before the real test mechanism is built.

Designing a test mechanism, which should fit to an existing damper test rig, must fulfil some criteria. For instance, the friction between the moving parts should be as less as possible. Standard components should be used to reduce the amount of parts which have to be newly machined. The assembling of the mechanism should be as quick and simple, as possible. And finally, the geometric ratio of the suspension mechanism should be considered. During assembling and attaching the test mechanism to the test rig it turned out that all parts are easy to install and give good possibilities of adjustment. Additionally, the guiding arm withstood all the excitations, even those near to the resonance frequencies.

The carried out measurements required lot of preparation time. The most complicated measuring element to install was the displacement sensor, but combined with the additional force sensor, it provided very important information about the damper characteristics and the oscillating system.

The measurement gave additional information about the ability of the test rig hydraulic system how it realizes different excitation frequencies and displacements. In the range of higher frequencies and lower amplitudes of excitation, special focus should be laid on a background noise coming from the power hydraulic unit. Lateral slip and vibrations of the tire with respect to base plate, and collisions of tire side wall with suspension spring, occurred at the highest amplitudes, what led to additional disturbances.

For future analyses, based on the gained data, some parameters can be defined to evaluate the damper behaviour by comparing different shock absorbers at this test mechanism. As refinement for future projects a compensation of the unnatural unsprung mass movement, which already leads to tyre contact loss at higher frequencies and amplitudes, can be considered. Additionally, emphasis should be laid on the elastic bushing behaviour. The bushing reactions at higher frequencies are different to the quasi static measurements and herewith also the modelled system behaviour. Additional measurements should be carried out to obtain a better correlation between measurement and simulation results. Measurements with microphone should also be planned to find correlation between the system mechanical response and its noise emission.

References

[1] Brandon P and White N 2006 *Sensitivity of damper valve architecture to changes in mount stiffness/damping/hysteresis and subsequent combined damping characteristics.* Kingston University Lotus Engineering Vehicle Dynamics Expo

[2] Fukushima N 1983 *Optimum characteristics of automotive shock absorbers under various driving conditions and road surfaces* Int. J. Vehicle Design vol 4 no 5

[4] Knapczyk J 1993 *Skrypt do laboratorium z budowy samochodów.* PK

[5] Mitschke M 2004 *Dynamik der Kraftfahrzeuge* Gabler Wissenschaftsverlage

[6] Prochowski L 2008 *Mechanika ruchu. Pojazdy samochodowe* WKiŁ

[7] Reimpell J 1997 *Podwozia samochodów. Podstawy konstrukcji* WKiŁ

[8] Dixon J 1999 *The Shock Absorber Handbook. SAE Warrendale Pa*

[9] Duym S *2000 Simulation Tools, Modelling and Identification, for an Automotive Shock Absorber in the Context of Vehicle Dynamics* Vehicle System Dynamics vol 33

IOP Conf. Series: Materials Science and Engineering **148** (2016) 012021 doi:10.1088/1757-899X/148/1/012021

Merging into platoons

A Modi[1]

[1]Warsaw University of Technology, SIMR, ul. Narbutta 84 , Warsaw ,Poland

E-mail: Abhishek.Modi@simr.pw.edu.pl

Abstract. Series of vehicles interacting with each other while driving in a straight line or while maneuvering is termed as a platoon of vehicles and known as platooning. This paper describes platooning and merging into platoons in general. The first part of this paper focuses on developing control strategies for automated driving. In this part a controller for longitudinal driving is designed and analysed. The second part of the paper focuses on the analysis of merging into a platoon on highways. This is done by validating experimental results by integrating a driver model with a single track bicycle model resulting in the calculation of merging path.

1. Introduction

Vehicle platooning or platooning in general can be described as a group of vehicles driving together where the interaction between these vehicles leads the driver to create a stable environment for driving. Platoon formation is the term used for a situation where two platoons (or a combination of a platoon and a vehicle or two vehicles) are combined into one platoon. An example of such a situation is illustrated in Figure 1 where vehicle number 4 accelerates in order to join the platoon (i.e. vehicles 1 to 3) [1].

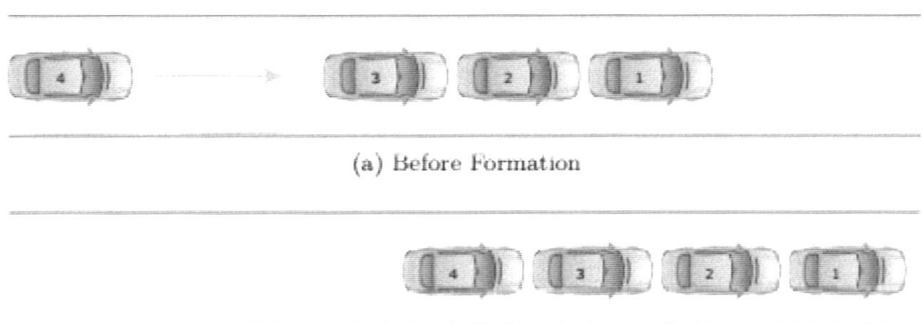

(b) After Formation

Figure 1. Platoon formation [1]

Traffic congestion is easily identified as a personal annoyance; however, the problem accumulates on a global scale. In 2015, the United States population alone spent 42 billion hours in traffic, wasting 10.7 billion gallons of gas and$78.2 billion (USD) [11]. There are over 600 million cars in operation in the world today, and by 2030, that number is projected to double [12]. With this influx of vehicles and the potential for increase in traffic congestion, it is important to look to a variety of social and technological solutions to help mitigate the problem. One unique solution to traffic management is the vehicle platoon.

A platoon of driverless vehicles moves at high speeds along a dedicated lane of a highway. Each vehicle operates autonomously, accelerating or decelerating in response to local sensor measurements of velocity and position, as well as inter-vehicle communication, to maintain the platoon formation. Because the vehicles are automatically controlled, they can safely operate at higher speeds and closer inter-vehicle distances than human-driven vehicles, thus increasing the safety [14].

Parallel computing and a wide range of sensor networks are widely used and look to be the answer to the problem of vehicle platooning. Vehicle platoons help in the reduction of traffic congestion and have huge environmental benefits. A deep understanding of sensor accuracy will be required to perform intricate analysis which can lead us to the solution of vehicle platooning. From radars to magnets to radio communication systems, vehicle platooning systems have been found to be more efficient than most human drivers. Typical platooning systems consist of a longitudinal and lateral control system. These control systems are explained below.

2. Control strategies for platooning

For automated driving in a platoon, design of longitudinal and a lateral controller is imperative and hence there is the need for a robust control methodology to suitably deal with uncertain non-linear time-varying systems. Therefore, introduced is a new driving assistance algorithm in order to make the automated vehicles capable of cooperating and navigating through urban traffic. This leads eventually to a more effective organization of the urban transport. The key functionality is that a platoon, which is composed of a leader followed by vehicles, should be able to travel at a close distance in a single line. Hence, a longitudinal control scheme is proposed for the platooning functionality. The proposed control schemes are based on the use of cameras and laser sensors which are mounted on the front of every car preceding the leading car. There are 2 control strategies proposed, a first order sliding model and a second order sliding model. But only the first order sliding model (FOSM) will be discussed. This FOSM is particularly suitable to deal with the uncertain nonlinear time-varying systems. Yet, FOSM controllers generate a discontinuous control action which has the drawback of producing high frequency chattering, with the possibly excessive mechanical wear and passengers' discomfort, due to the propagated vibrations throughout the different subsystems of the controlled vehicle. This will be proven in the following discussion of the system.

2.1. Vehicle model

Consider a platoon comprising of n+ 1 number of identical vehicles, where n is the total number of vehicles. The model of this platoon is represented by a single track bicycle model as shown below [2].

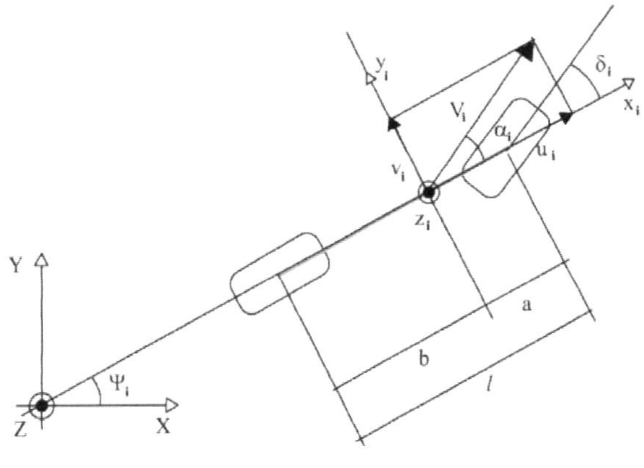

Figure 2. Bicycle model for sliding mode control [2]

Then, the model of the i-th vehicle is represented by the following equations [2],

$$\dot{u}_i = \frac{1}{m}\left[mv_ir_i - mfg + u^2{}_i(fK_1 - K_2) + C_f\left(\frac{v_i+ar_i}{u_i}\right)\delta_i + T_i\right] \quad (1)$$

$$\dot{v}_i = \frac{1}{m}\left[mu_ir_i - (c_f + c_r)\frac{v_i}{u_i} - (bc_r + ac_f)\frac{r_i}{u_i} + c_f\delta_i + T_i\delta_i\right] \quad (2)$$

$$\dot{r}_i = \frac{1}{Jz}\left[(bc_r - ac_f)\frac{v_i}{u_i} - (b^2c_r + a^2c_f)\frac{r_i}{u_i} + ac_f\delta_i + aT_i\delta_i\right] \quad (3)$$

$$\dot{x}_i = u_i \cos\varphi_i - v_i \sin\varphi_i \quad (4)$$

$$\dot{y}_i = u_i \sin\varphi_i + v_i \cos\varphi_i \quad (5)$$

$$\dot{\varphi}_i = r_i \quad (6)$$

Where, 'u_i', 'v_i' and 'r_i' are the longitudinal velocity, lateral velocity and the yaw rate respectively; 'x_i' and 'y_i' are the position coordinates in the centre of gravity. The 2 control signals are 'δ_i' the steering wheel angle and 'T_i' the traction force at the contact point between tire and road. 'm' is the mass of the vehicle and 'a', 'b' and 'l' are the distance from the front axle to the centre of gravity, distance of rear axle to centre of gravity and wheel base respectively. 'K_1 ' and 'K_2 ' are the control gains for front and rear respectively. For the control design a linearized decoupled model obtained from equation (1) to equation (6) is used.

The longitudinal dynamics is given by equation (7) [3],

$$u_i = 2u_{0i}(fK_1 - K_2)\frac{u_i}{m} + \frac{T_i}{m} \quad (7)$$

While the handling dynamics are given by equation (8),

$$\begin{bmatrix}\dot{v}_i\\ \dot{r}_i\end{bmatrix} = \begin{bmatrix}a_{11} & a_{12}\\ a_{21} & a_{22}\end{bmatrix}\begin{bmatrix}v_i\\ r_i\end{bmatrix} + \begin{bmatrix}b_1\\ b_2\end{bmatrix}\delta_i \quad (8)$$

Where,

$$a_{11} = -\frac{(C_f + C_r)}{m \cdot u_0}; \quad a_{12} = \frac{(bC_r - aC_f)}{(m \cdot u_0) - u_0}; \quad a_{21} = -\frac{(bC_r - aC_f)}{J_z \cdot u_0}; \quad a_{22} = -\frac{(b^2C_r - a^2C_f)}{J_z \cdot u_0}$$

$$b_1 = \frac{C_f}{m}; \quad b_2 = \frac{aC_f}{J_z}$$

'C_f' and 'C_r' are the cornering stiffness at the front and the rear tyres and 'J_z', the polar moment of inertia.

2.2. *Longitudinal control design*

To understand the design of the controller we must first look into the leader of the platoon. The leader of the platoon has an arbitrary speed and acceleration [4]. The chosen sliding variable for the 0th vehicle is the error between the reference speed $u_{ref}\ (t)$ and the actual speed $u_o\ (t)$ of the vehicle represented by equation (9) [5].

$$S_o(t) = u_{ref}(t) - u_o(t) \tag{9}$$

The FOSM is given by equation (10),

$$T_o(t) = m\left[\eta_0 sign(S_0(t)) + \dot{u_{ref}}(t)\right] - 2u_{oo}(fK_1 - K_2)u_o(t) \tag{10}$$

The objective of this longitudinal control of the ith vehicle is to maintain a safe distance from the preceding vehicle. Here the Constant Time-Headway (CTH) is used, which is used primarily to describe the following vehicles implying the vehicles behind the leader and also used frequently to design the adaptive cruise control (ACC) [6].

CTH policy is explained by equation (11),

$$S_{d_i}(u_i(t)) = S_{d_0} + hu_i(t) \tag{11}$$

Where S_{d0} is the distance between stopped vehicles, and h is the headway time. The spacing error is expressed by equation (12) [6],

$$e_i(t) = S_{di}(u_i(t)) - d_i(t) = S_{d0} + hu_i(t) - x_{(i-1)}(t) + x_i(t) \tag{12}$$

Where, $d_i(t)$ is the longitudinal distance between the ith vehicle and the (i-1)th vehicle. This distance can be measured by a laser sensor mounted on the front of the vehicle. So in this case the sliding variable is the spacing error, i.e. $S_i\ (t) = e_i(t)$. Then the control law to make the sliding variable vanish in finite time is given by equation (13).

$$T_i(t) = \frac{m}{h}\left[-\eta_i \sin(S_i(t)) - u_i(t) - u_{i-1}(t)\right] - 2u_{0i}(fK_1 - K_2)u_i(t) \tag{13}$$

This control law guarantees that the vehicles correctly keep the desired distance and that the system is string stable i.e., the spacing errors never amplify along the platoon [7].

2.3. *Simulation with Simulink*

The considered platoon is composed by a leader vehicle and 2 follower vehicles. For the simulation realistic parameter are used. The leader of the platoon starts travelling with u_{ref} .We can see the longitudinal error in Figure 3 and we can note that the error is steered to zero in finite time. From Figure 4 we can observe that the safe distance is remained and adapted with variation of velocities. That becomes even clearer in Figure 5 where the distance of the follower vehicles is depicted with respect to the leader. So at a velocity of 14 m/s we see the gap is 16 m that implies a safety distance at standstill S_{d0}= 2 m and a headway time h = 1s, 2 m + 1 s . 14 m/s = 16 m. Finally, the traction input is reported in Figure 6, it's obvious to see the signal is not a smooth one. Hence, it is concluded that the sliding mode control has very robust properties against disturbances, model inaccuracies and parameter variations. The FOSM generates a discontinuous signal which causes a high actuator stress resulting in vibrations induced through the vehicle. Therefore this FOSM should be improved.

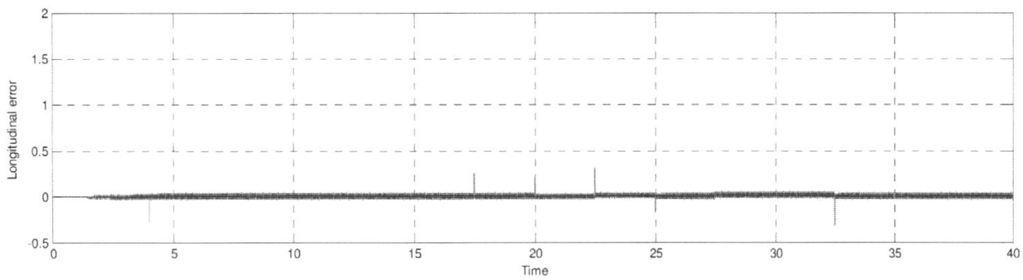

Figure 3. Longitudinal error FOSM.

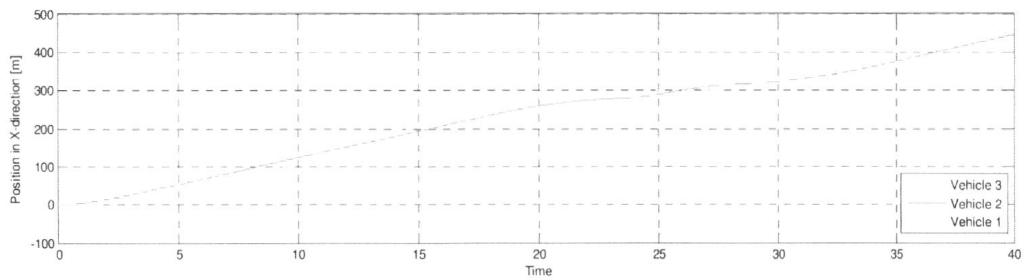

Figure 4. Position in X-direction FOSM.

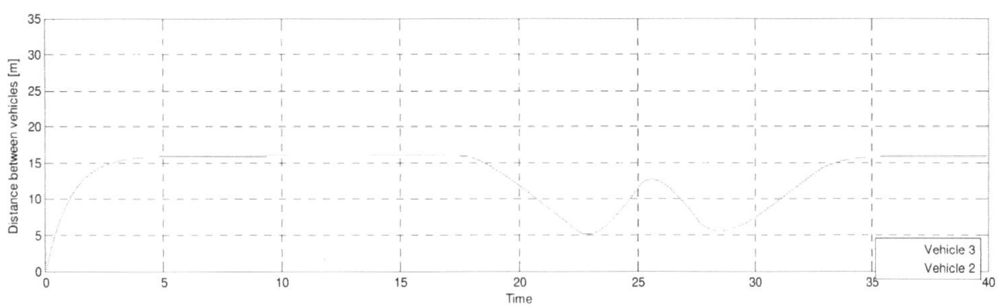

Figure 5. Distance between vehicles FOSM.

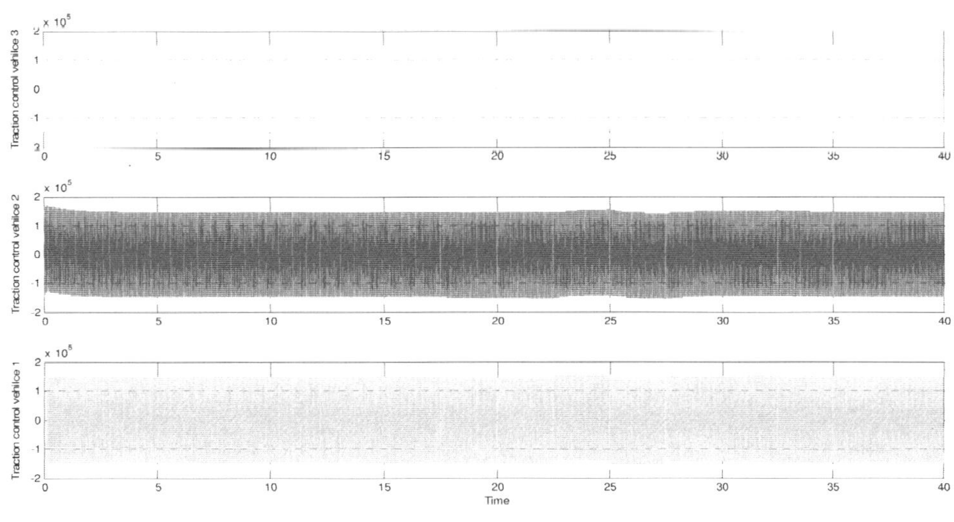

Figure 6. Traction control FOSM.

3. Experimental Validation of vehicle behaviour

The main objective of the experiment is to reconstruct a merging trajectory, more specific the merging that occurs when entering the highway. The car comes into the traffic from the right side, merging in between 2 cars driving in the lane. The merging manoeuvre was performed with three different longitudinal speeds and three different gap sizes to analyse the merging behaviour. The longitudinal speed during the merging process was maintained constant. Based on safety reasons and constant time-headway (CTH) policy the gap sizes were chosen as 25m. The desired constant speed was maintained by the front vehicle at 50km/hr. The second vehicle maintained the desired gap between the vehicles with the assistance of the view distance restriction on the front window. This distance was pre-set when standing still before each measurement and maintained constant during whole manoeuvre. For the objective evaluation and comparison of the results two different drivers were performing the merging manoeuvre. Both drivers repeated the measurement twice for each speed and each gap. For the purpose of validation a basic vehicle single track bicycle model integrated with a driver model was used which will be analysed in section 3.2.

3.1. Description of the experiment

The testing vehicle is BMW 320i equipped with the measurement system as described below. The coordinate system represented by axis X, Y and Z is pictured in Figure 1. The Z-axis points upwards. Acceleration sensors are mounted in the center of gravity and measure in X, Y and Z direction. Yaw rate sensor mounted is mounted at the Center of Gravity and measures angular velocity of the vehicle in degrees per second. Steering wheel sensor measures the steering angle and the steering moment which a driver applies.

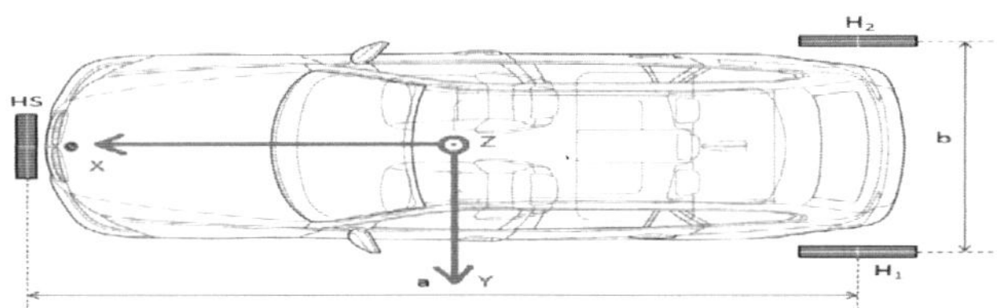

Figure 7. Sensor Position W.R.T Co-ordinate System [9].

The experiment was performed on Lelystad testing track consisted of two curves with constant radius R = 160 m and two straight parts with the length L=720 m. The speeds for merging were set to 50, 60 and 70 km/h in order not to exceed certain values of lateral acceleration since the tire behaviour is linear under $a_y < 0.4g$.The three sensors (HS, H1 and H2) measure the differences in height therefore the pitch and roll movements can be determined. The front sensor measures besides the longitudinal and lateral speeds. Distance of the sensor HS from the centre of gravity was measured as 2416 mm. The testing car was during test equipped with Vredestein Quatrac 3 225/45/17 tires. The Cornering stiffness on the front and rear axle C1 and C2 respectively is, 74000 N/deg and 128 000 N/deg respectively. The mass of the vehicle 'm' is 1650 kg and the wheelbase 'l' is 2.76 m. Yaw inertia of the vehicle 'J_z'is calculated by equation (14) which provides sufficiently accurate results for the simulation model.

$$J_z = m \times a \times b \qquad (14)$$

Where, 'a' is the distance from front axle to centre of gravity and 'b' is the distance from rear axle to centre of gravity.

3.2. Simulation of merging using an integrated single-track vehicle - driver model

For the simulation of vehicle behaviour, a simplified linear bicycle model and a driver model has been used which is represented in Figure 8 and Figure 9 respectively. This model provides relatively accurate results under non-extreme conditions. The following equations have been used to build the simulation model [9].

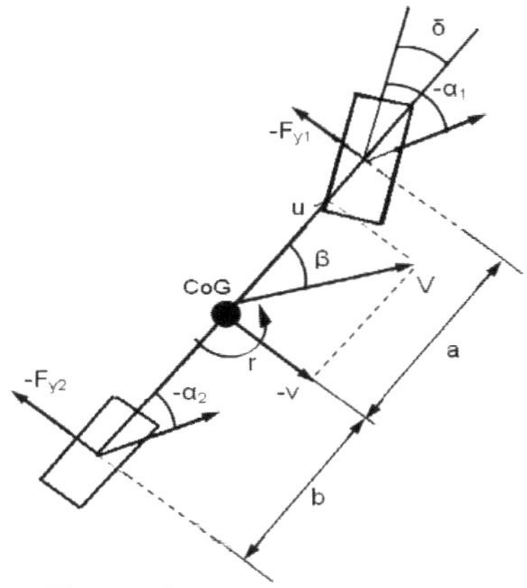

Figure 8. Bicycle Model [10]

$$m.a_y = m.(v + u.r) = F_{y1} + F_{y2} = -C_1.\alpha_1 - C_2.\alpha_2 \tag{15}$$

$$J_z.r = F_{y1}.a - F_{y2}.b \tag{16}$$

$$\alpha_1 = (v + ra)/u - \delta \tag{17}$$

$$\alpha_2 = (v - r.b)/u - \delta \tag{18}$$

Where 'm' is vehicle mass, 'a_y' is lateral acceleration, 'v' is lateral speed, 'u' is longitudinal speed, 'r' is yaw rate, 'F_{y1}' and 'F_{y2}' are lateral forces, 'C_1' and 'C_2' are cornering stiffness's at the front and rear axle respectively, 'a' and 'b' are the distances from the front and rear axle to the centre of gravity , 'α_1' and 'α_2' are slip angles at front and rear respectively, δ is steering angle and J_z is yaw moment of inertia [9].

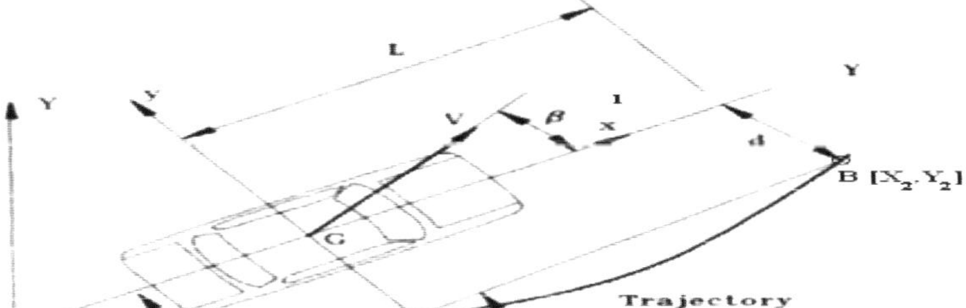

Figure 9. Driver Model [11].

For simulation a driver model represented by Figure 9 with a sort of a predictive behaviour from was used [10]. This driver model describes the human behaviour in a very simplified way but it gives satisfactory results in many cases. In the time domain the model is described by equation (19).

$$\tau\dot{\delta}(t)+\delta(t)=-K_d\left[\psi(t)-\psi_1(t)+\frac{y(t)}{L}\right] \tag{19}$$

Where, 't'- time delay of the driver, '$\delta(t)$' is the steering angle, 'K_d' is the gain of the driver, 'φ' is the yaw angle of the vehicle explained as the angle between X-axis and a line passing through two points of the trajectory at a preview distance 'L', 'y' is the lateral displacement of the vehicle .

For the simulation of the merging behaviour the measurement data with the velocity 50 km/h, the gap of 25 m and the driver 1was considered. Before the simulation the driver parameters were chosen. The common values for the time delay, the gain and the preview distance are 0.2 s, 0.3 and 25 m respectively [10]. For our simulation we assumed the value of the time delay of the driver to be 0.2 s the value of the gain to be 0.45 and the value of the preview distance to be 15 m because the driver has to look ahead in to merging into a gap of 25 m. Figure 10 depicts three curves representing the path. The red curve is the measured data of the position of the vehicle obtained from the GPS system, the blue is the input to the driver model which is definition of the path which the driver should follow and the green curve is the path from the vehicle model.

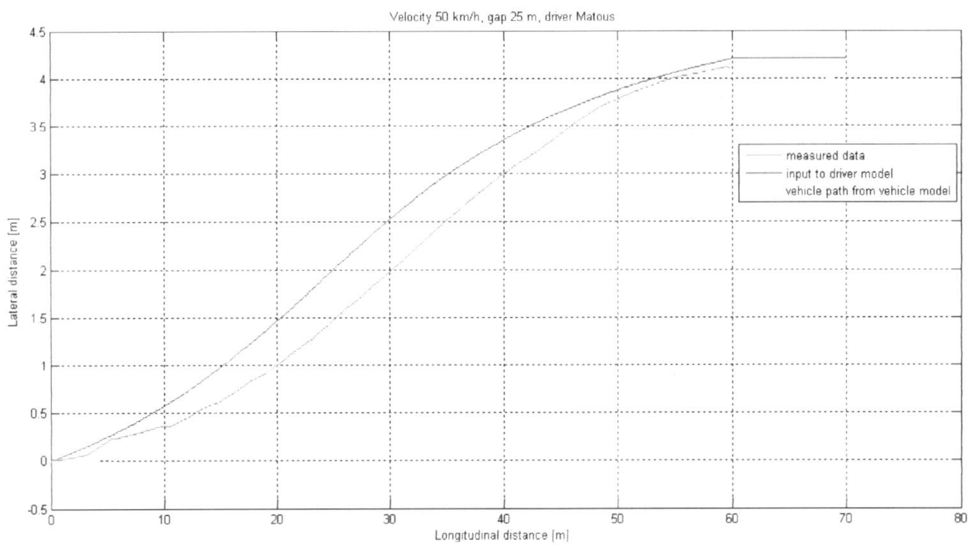

Figure 10. Simulation of the merging task using the driver model.

The results show that the driver model follows the input path with a certain delay, which is significant at the Beginning of the merging. In the middle part the simulated path is very similar to the driven one. In the last part the simulated path is differing more from the real one and it is not that quickly approaching the straight part of the input curve.

4. Conclusion

Our understanding and by comparing theoretical results and performance results we can gain a lot of valuable information into the development of control strategies for vehicle platoon systems. Firstly we conclude that a control strategy was developed which led us to the understanding of the error and therefore improvements can be made on the design by including a lateral controller alongside the longitudinal controller. Secondly, the understanding of the integrated vehicle–driver model led us to understanding merging behaviour at different speeds for different driver types. By varying the gain and time delay one can comprehend the age of the driver and the reaction times as well. There are many other issues that must be addressed to develop a complete platoon solution, including emergency actions such as obstacle avoidance and accommodation for vehicles that exit from the platoon. The investigation of strategies for these events is the subject of future work.

Appendix A

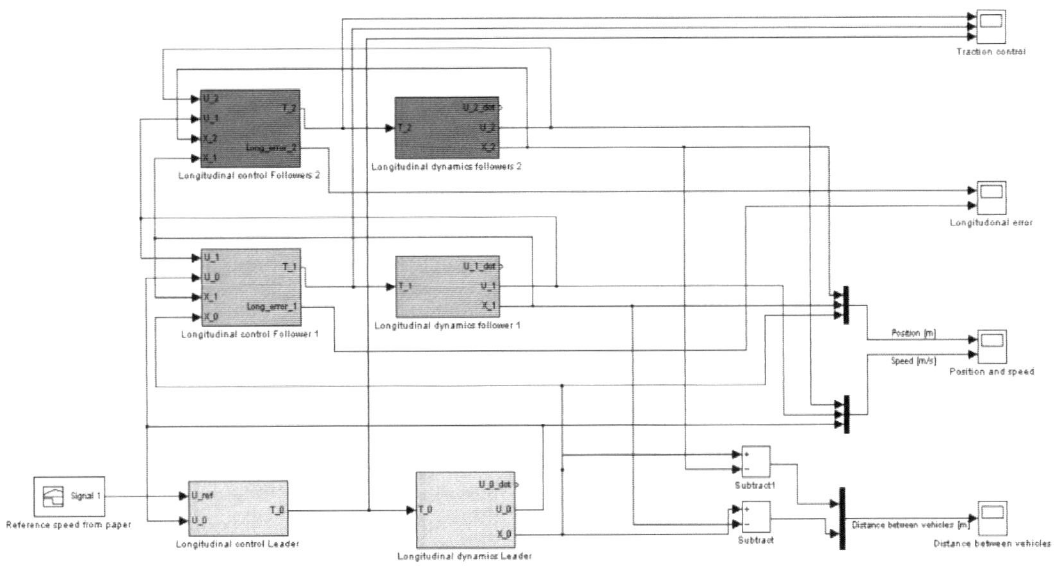

Figure A 1. Simulink Model of the longitudinal controller.

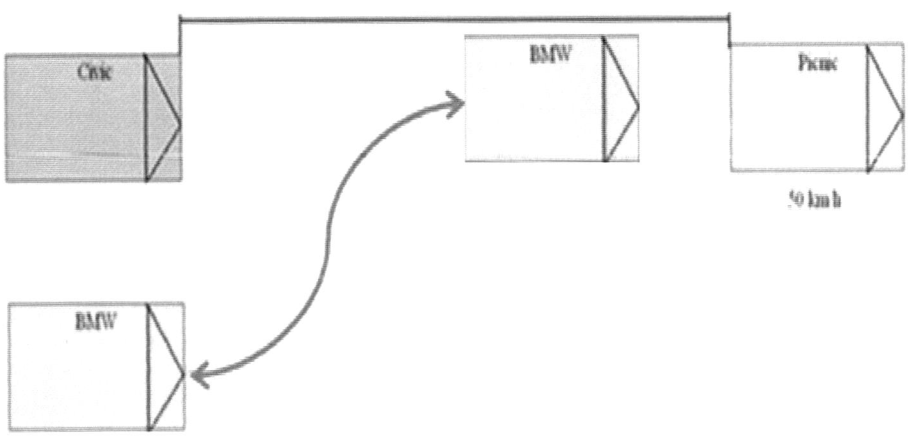

Figure A 2. Outlay of merging experiment.

References

[1] Hobert L 2012 A study on platoon formations and reliable communication in vehicle platoon 2012 *MSc.dissertation,*Dept .Mechanical Engineering, University of Twente, Twente ,The Netherlands
[2] Ali A, Garcia G and Martinet P 2014 *The flatbed platoon towing model for safe and dense platooning on highways* 2014, *IEEE Intelligent Transportation systems Magazine*
[3] Ali A, Garcia G and Martinet P 2014 *Minimizing the inter-vehicle distances of the time headway*

policy for platoons control in highways", 2014 10th International Conference on Informatics in Control, Automation and Robotics (ICINCO13), pp. 417-424
[4] McMahon D , Hedrick J K and Shladover S E 1990 *Vehicle Modelling and Control for Automated Highway Systems Proceedings of the 1990 ACC*
[5] J. Zhou J and H. Peng H 2004 *String stability conditions of adaptive cruise control algorithms* IFAC Symp. on Advances in Automotive Control Salerno, Italy
[6] Liang C Y 2000 *Traffic-friendly adaptive cruise control design* 2000 Ph.D. dissertation Dept. Mech. Eng., Univ. of Michigan, Ann Arbor, MI
[7] Xu Q, Hedrick K, Sengupta R and VanderWerf J 2002 *Effects of vehicle-vehicle/roadside-vehicle communication on adaptive cruise controlled highway systems* Proc. IEEE Vehicular Technology Conf. pp.1249-53
[8] Nielsen, Uwekiencke & lars. 2005 *Automotive Control Systems* Springer-Verlag Berlin Heildelberg,. 978-3-540-26484-2
[9] Pauwelussen J 2014 *Essentials of Vehicle Dynamics* Butterworth-Heinemann
[10] Giancarlo, G. 1997 *Motor Vehicle Dynamics -Modelling and Simulation* World Scientific.
[11] *Urban mobility report* 2008,Texas Transportation Institute , Texas A&MUniversity System. 2007.http://mobility.tamu.edu/ums/, 2007. Online
[12] *World Business Council for Sustainable Development Mobility* 2030: Meeting the challenge sto. sustainability, Nov 2007
[13] Jadbabaie A, Lin J and A. S. Morse Coordination of groups of mobile autonomous agents using nearest neighbor *rules IEEE Transactions on AutomaticControl*, **48**(6):988–1001
[14] Bassam B 2005 "*On the Feasibility of Large-Scale Automated Highways* Department of mechanical engineering University of California, Santa Barbra

IOP Conf. Series: Materials Science and Engineering **148** (2016) 012022 doi:10.1088/1757-899X/148/1/012022

Classification of the road surface condition on the basis of vibrations of the sprung mass in a passenger car

K Prażnowski[1] and J Mamala[2]

[1,2] Opole University of Technology, Faculty of Mechanical Engineering,
[1.] PhD [2.] DSc., DEng - Professor in Opole University of Technology

E-mail: j.mamala@po.opole.pl

Abstract. In order to identify the state of the wheel balance in a passenger car on the basis of vibrations of the car body under actual conditions of its operation, it is necessary to determine the impact of random interferences resulting from a changing environment. For this purpose, the criterion for the evaluation of the road surface condition was developed on the basis of longitudinal vibrations of the car body of the tested car in the speed range from 50 km/h to 110 km/h. Selected functions such as: probability distribution and methods in the frequency domain: short-time Fourier transform (STFT) and power spectral density (PSD) were used to analyse recorded signals.

1. Introduction

Kinematic inputs caused by random road irregularities are the main source of vibrations of the while driving (h_w - figure 1). Other sources may include dynamic inputs connected with the rotation of wheels and suspension system (e.g. those connected with the heterogeneous nature of tires, improper shape or balance of wheels) and resulting inertial forces affecting the car body. As a result of the impact of the kinematic input (h_w)on the wheel (unsprung mass), the following variables change: (zm_1) and (zm_2) which determine the suspension and wheel deflection, acceleration of both of these masses, and contact force between the tire and the surface $F_u(\omega)$, presented in figure 1 as the so-called quarter car model.

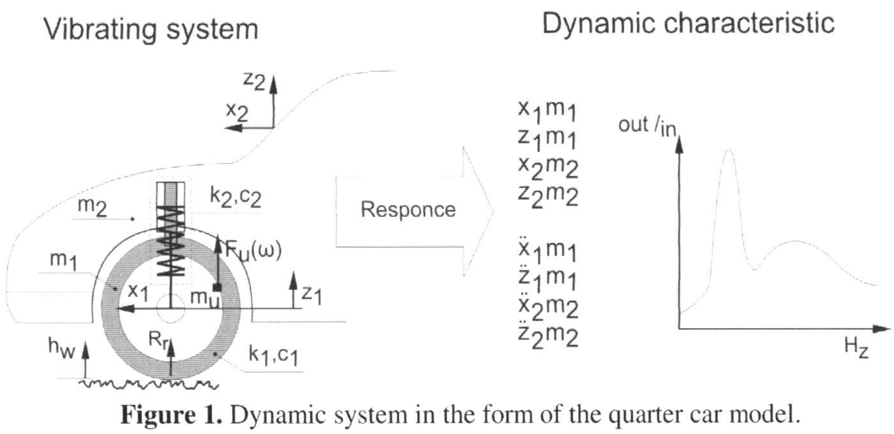

Figure 1. Dynamic system in the form of the quarter car model.

In case of the occurrence of the wheel malfunction in the form of unbalance $F_u(\omega)$or heterogeneous nature of the tire, the described mass transfers (m_1) will increase due to the rotation of the wheel. It is particularly disadvantageous in the range of the wheel natural frequency, i.e. in the wheel rotation frequency range of approx. 12 Hz. The resonance phenomenon leading to a significant increase in the amplitude of mass accelerations $\ddot{z}m_2$ is the result of such a situation. The relationship between the input and variables is described by the transfer (amplification) function presented in figure 1. It is very important to understand this function from the point of view of the evaluation of comfort and safety. Comfort is evaluated by means of the function of amplification of sprung mass accelerations whereas safety is evaluated by means of the analysis of the function of amplification of the dynamic load [1]. During the operation of the passenger car, ratios (k_1, k_2, c_1, c_2) are subject to change (damage, ageing of the material, material loss), which, in turn, results in the change in the range of natural frequencies, acceleration values and amplification function. It results in an increased amplitude of mass vibrations (m_2–car body) in relation to the amplitude of the road irregularities causing vibrations [1,2,3,4,5]. While maintaining constant linear velocity should not be instantaneous longitudinal acceleration of the vehicle body. Dominant amplitude resulting from irregular work power train (engine, transmission), have a frequency range different from the frequency of rotation of the wheel. The occurrence of instantaneous longitudinal acceleration may therefore indicate the presence of significant unevenness in the road.

2. Measurement condition (SNR)

In each real measuring signal there are interferences in the form of noise that is defined as a stochastic process independent of the transmitted information [5,6]. The SNR ratio (signal-to-noise ratio) described by the below given equation (1) is used for the evaluation of stochastic disturbances in the signal. The SNR ratio also determines the useful signal power value in the predetermined frequency band in relation to the noise (interference) power occurring in the same frequency band or as a ratio of the amplitude of these signals which are often presented in the logarithmic form for dB unit to emphasise the difference from the SNR ratio.

$$SNR = 20\ log_{10}\left(\frac{A_{signal}}{A_{noise}}\right) = 10\ log_{10}\left[\left(\frac{A_{signal}}{A_{noise}}\right)^2\right] \quad (1)$$

where:

A_{signal}- signal amplitude,
A_{noise}- noise amplitude.

The relationship above can be used to present examples of changes in the values of the SNR ratio of the analysed signal of acceleration of the car body (mass m_2), obtained in test-stand measurements and road tests in relation to the speed profile (figure 2). The SNR ratio for standard tests carried out under conditions free from road irregularity interferences reaches values in the range from 15 dB to 170 dB (figure 2a) whereas the maximum values are obtained at 85 dB and unsprung mass resonance frequency (12Hz) and then they decrease. Such vibrations may result from the contact of the tire tread face with the roller of the chassis dynamometer. As a result of the predetermined unbalance of the (front right) wheel with the mass m_{nw}= 0.08 kg, the value of the SNR ratio increases (figure 2b), reaching a significant value of 350 dB (at 17.2 Hz), as the wheel rotation frequency increases. In this case, the SNR value reaches 200 dB at 12 Hz.

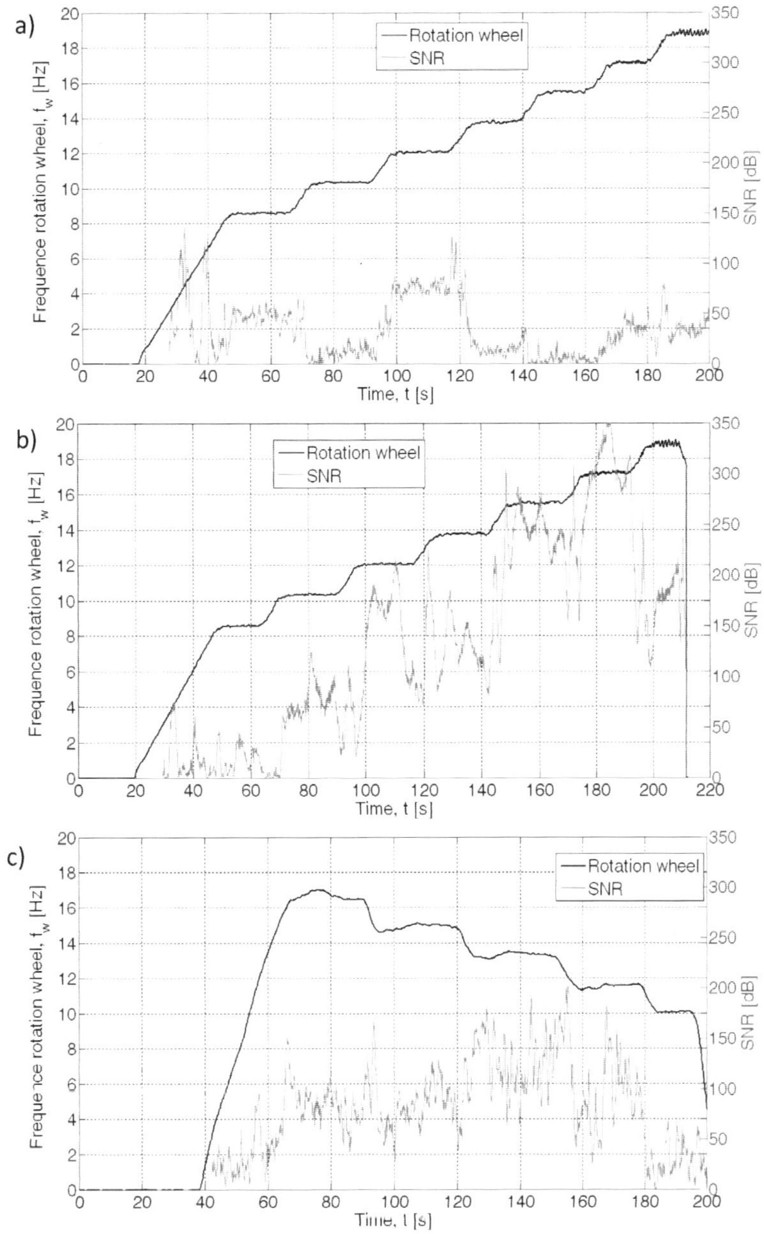

Figure 2. Changes in the SNR of the acceleration signal a_z: a) standard test-stand measurements, b) test-stand measurements, unbalanced wheel with the mass m_{nw} =0.08 kg, c) road tests, unbalanced wheel with the mass m_{nw} =0.08 kg.

Road irregularity interferences (figure 2c) led to the decrease in the SNR values the maximum value of which did not exceed 175 dB at 14 Hz. The mass acceleration signal(m_2) is useful in the diagnostics of selected elements of wheels and suspension system for frequencies exceeding 10 Hz.

3. Inputs caused by road irregularities

The road profile is described in literature[1,2,7,8] as the sum of waves of different lengths which form road macro profiles (waves of length exceeding 100 m) and micro profiles (waves of length from 0.1 to 100 m). Road parameters are determined on the basis of the autocorrelation and power spectral density (PSD) function under the assumption of the stationary function of random road surface irregularities. According to ISO 8606 of year 1995, roads were divided into 8 classes (marked with letters: A, B, C, D, E, F, G, H) to the description of which the PSD function was used. Graphs of functions $S_g(\Omega)$ are determined for the frequency Ω determined in the coordinate domain of horizontal movement of the vehicle on section sand bound by the relationship $\Omega=1/\lambda$, (λ- length of the wave of the harmonic component of the road irregularity signal). Indicated functions are described by the following relationship (2):

$$S_g(\Omega)=A_g\Omega^{-m} \qquad (2)$$

where: A_g - road roughness index, m- function slope (with a value close to 2).

The quality of the road surface is then evaluated on the basis of the value $S_0=S_g(\Omega_0)$ determined for the harmonics with specified wave length λ (for asphalt surfaces $\Omega_0=1/2\pi\approx0.16$ c/m [cycle/metre]). The identification of the specific road surface class requires specialised equipment; it is not used in the conditions of normal operation of the vehicle. However, damage to the road surface, such as: wheel tracks, transverse technological welds, cracks and spalling cracks, affects the acceleration of the unsprung and sprung mass of the vehicle.

It was assumed in the analysed matter that the detection of the unbalanced wheel during the road test is possible provided that the value of the amplitude caused by the unbalanced wheel is higher than the noise (interference) amplitude in the analysed range of frequency band from 7 Hz to15 Hz. The SNR ratio(signal-to-noise ratio) determining the useful signal amplitude value in the predetermined frequency band to the noise power in the same band is used for the evaluation of stochastic disturbances of the signal [5].

4. Measuring equipment

The measuring system used (figure 3) consists of the acceleration sensor 3DM-GX3-25 with the measurement range of ±50 m/s^2, and initial bias error of $\pm$ 0.002 m/s^2, head for the non-contact car body speed measurementL-350 AQUA (measurement range: 0.3- 250 km/h) connected to the card NI USB 6212. The acceleration measurement sensor was installed on the windshield inside the cabin of the vehicle.

A mid-size passenger car with a petrol engine of capacity 1.2 16V was used for the road tests. Experiments were carried out on roads with bituminous surface. Tire pressure was 0.2 MPa for each of the wheels. Vehicle weight during tests was 1160 kg (kerb weight 990 kg plus two persons 170kg).

The car used in road tests has front suspension type McPherson with a single-control arm and a telescopic shock absorber as a steering column with a coil spring which is a spring element. In order to reduce the impact of vibrations generated by rolling wheels on the car body and the impact of the irregularity of roads with small wavelengths, apart from the vertical flexibility of suspension, its flexibility in the longitudinal direction is also required. It is ensured by the damping element in the form of the rubber and steel bushing. Suspension systems with steering columns show greater sensitivity to unbalanced wheels, improper shape of tires or other structures, e.g. multi-arm suspension [2,3].

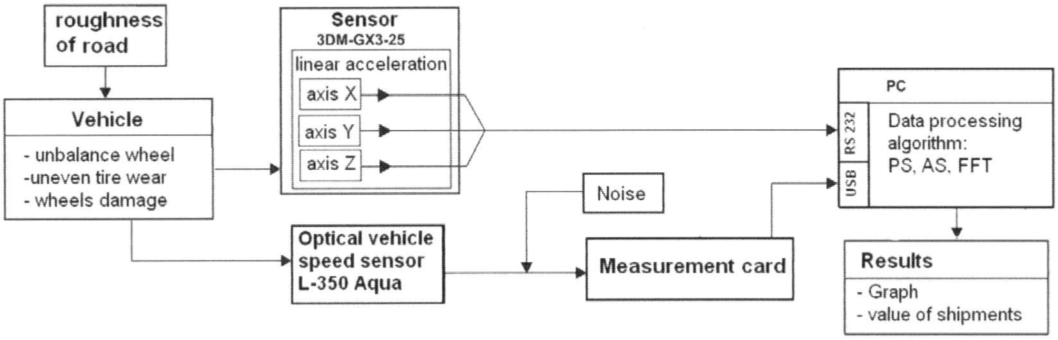

Figure 3. Flow chart of the measurement system.

It was assumed on the basis of the analysis of the kinematics of the front suspension of the tested front vehicle that road surface irregularities lead to the increase in undamped vibrations both in the longitudinal direction (x axis) and in the vertical direction (z axis).

5. Determination of the criterion for the evaluation of road surface condition

The influence of the type of road surface on the value of the amplitude of the acceleration of the tested car body was determined by means of tests carried out with the use of the chassis dynamometer and on selected sections of road with bituminous surface which are presented in figure 4. Due to the surface defects, the selected sections of the road were defined as: road **A** without visible surface defects, road **B** with visible surface defects, road **C** with considerable bituminous surface defects, transverse cracks and transversely deformed surface (folds).

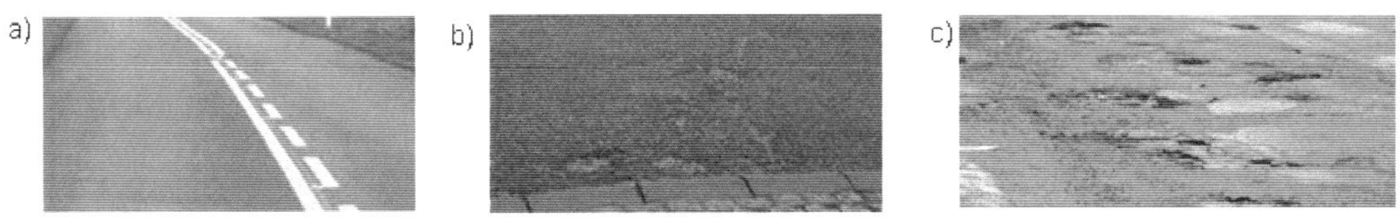

Figure 4. Selected types of road surface: a) road A, b) road B, c) road C.

During the tests, the authors tried to maintain a constant linear velocity with accuracy (+/- 5 km/h) for selected speed ranges (from 50 km/h to 90 km/h). The car body acceleration recording time was established in such a way that it included at least 20 wheel rotations. The autocorrelation function, power spectral density, short-time Fourier transform (STFT) and probability distribution were used for the analysis of data obtained under stationary conditions (using the chassis dynamometer) and in road tests for bituminous surfaces defined as road A, B and C.

The influence of the road surface condition on the value of the power spectral density was presented in figure 5 As a result of test-stand measurements carried out by means of the chassis dynamometer MAHA MSR 500, the values of the vertical acceleration of the sprung mass m_2 and unsprung mass m_1 free from interferences caused by road irregularities were obtained. It was stated that in case of a good road surface (road A and chassis dynamometer), in the wheel rotation frequency range (11.3 Hz) the values were similar and amounted to 0.042 $m^2/(s^4Hz)$ whereas in case of road C, the PSD value was 0.2 $m^2/(s^4Hz)$.

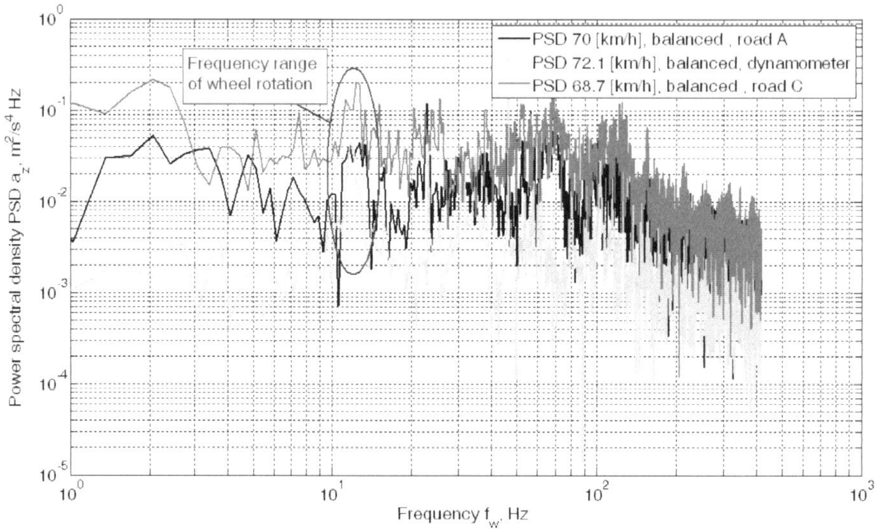

Figure 5. PSD of the component of the acceleration of vibrations of the car body in the vertical direction, caused by the technical condition of the road surface on which a wheel rolls.

The use of the STFT function, the information on the change in the spectrum of the analysed signal was obtained by monitoring its characteristics in the time and frequency domain. Irregularities of roads B and C result in an increase in the amplitude of vibrations of the car body in the vertical direction (a_z) in the frequency band range from 1 Hz to 30 Hz. Contrary to the spectrum obtained for road A, there are no evident amplitudes corresponding to the rotational speed of wheels of the vehicle. It is connected with the occurrence of significant road surface irregularities. It results in an apparent increase in background amplitudes in the frequency range up to 60 Hz from 0.02 m/s^2 for road A, up to 0.12 m/s^2 for road B and 0.18 m/s^2 for road C. Local increases in amplitudes can also be seen in obtained spectrograms which prove that there are significant irregularities and defects in the road surface.

In amplitude spectra obtained for tests carried out with balanced wheels there is a multiple increase in the amplitude of vibrations of the car body in the longitudinal direction (x axis) and in the vertical direction (z axis) for tests carried out on roads B and C in relation to road A and test-stand measurements.

Dominating amplitudes for speed from 90 km/h to 92 km/h occur in the frequency range from 5 Hz to 25 Hz (figure 6). In the longitudinal direction there is a significant increase in the average amplitude value for road B (0.02731 m/s^2) and road C (0.04573 m/s^2) as compared to road A (0.007627 m/s^2). Therefore, an attempt was made to determine the limit value of acceleration in the direction of the horizontal axis a_x, above which the measurement aiming at the identification of the unbalanced wheel under road test conditions cannot be carried out. For this purpose, the average value of the filtered acceleration signal of the longitudinal axis (x axis) in the frequency range from 0 to 100 Hz was used for each flexible time window of the STFT function. In order to evaluate the distribution of amplitudes of acceleration of the sprung mass m_2 in the longitudinal direction, the statistical inference was carried out for the assumed measurement range (3s) on the basis of the probability distribution function (figure 7).

a)

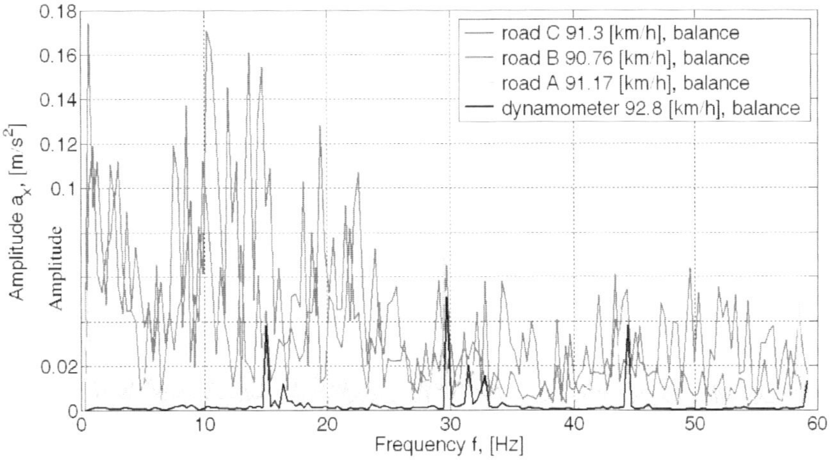

b)

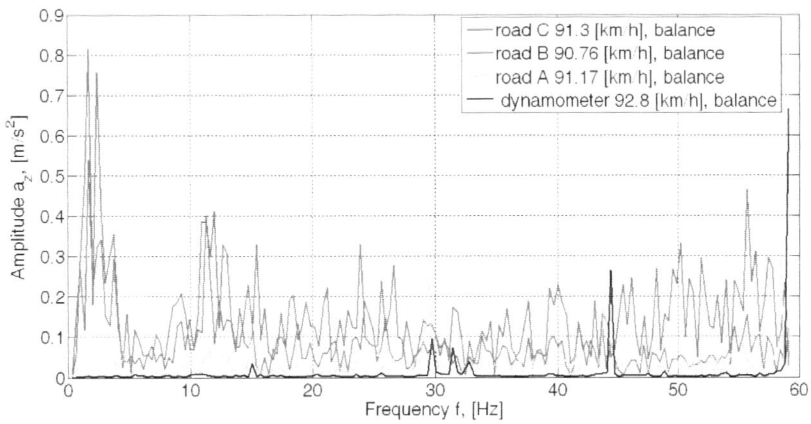

Figure 6. Amplitude of vibrations of the car body for selected types of road surfaces A, B and C: a) in the longitudinal direction (x axis), b) in the vertical direction (z axis).

It can be seen from the graph of the probability distribution function for road A (figure 7a) that the distribution of amplitudes in the longitudinal direction for frequencies from 9 Hz to 16 Hz is included in the range from 0.08 m/s^2 to 0.017 m/s^2. For frequencies: 10 Hz, 11 Hz and 13Hz, the amplitude ranges from 0.069 m/s^2 to 0.0125 m/s^2 and takes a more uniform shape with two apparent modes. In case of 16 Hz, there is a shift of the maximum towards greater values of acceleration amplitudes amounting to 0.0165 m/s^2. The impact of speed on the average value of acceleration in the longitudinal direction can be observed at 16 Hz.

Dominating amplitudes for road B for wheel rotation frequency range from 9 Hz to 15 Hz (figure 7b) occur in the range from 0.069 m/s^2 to 0.0754 m/s^2. The graph shows a more even distribution of amplitudes at a frequency of 15 Hz. Despite the increase in the wheel rotation frequency, all dominating modes of vibrations of mass m_2 fall within the above-mentioned range, which proves that there is no significant impact of speed of the vehicle on the amplitude value. However, the difference in the amplitude value can be observed at 17 Hz; it amounts to 0.0983 m/s^2.

In case of road C (figure 7c), there is no apparent impact of speed on the value of the dominating mode as its value for the wheel rotation frequency 16 Hz is lower than for 9 Hz. However, there is an apparent increase in the value of amplitudes for 14 Hz and 16 Hz as compared to the results obtained for road B. There is also a significant increase in the value of amplitudes (higher by an order of magnitude) as compared to road A.

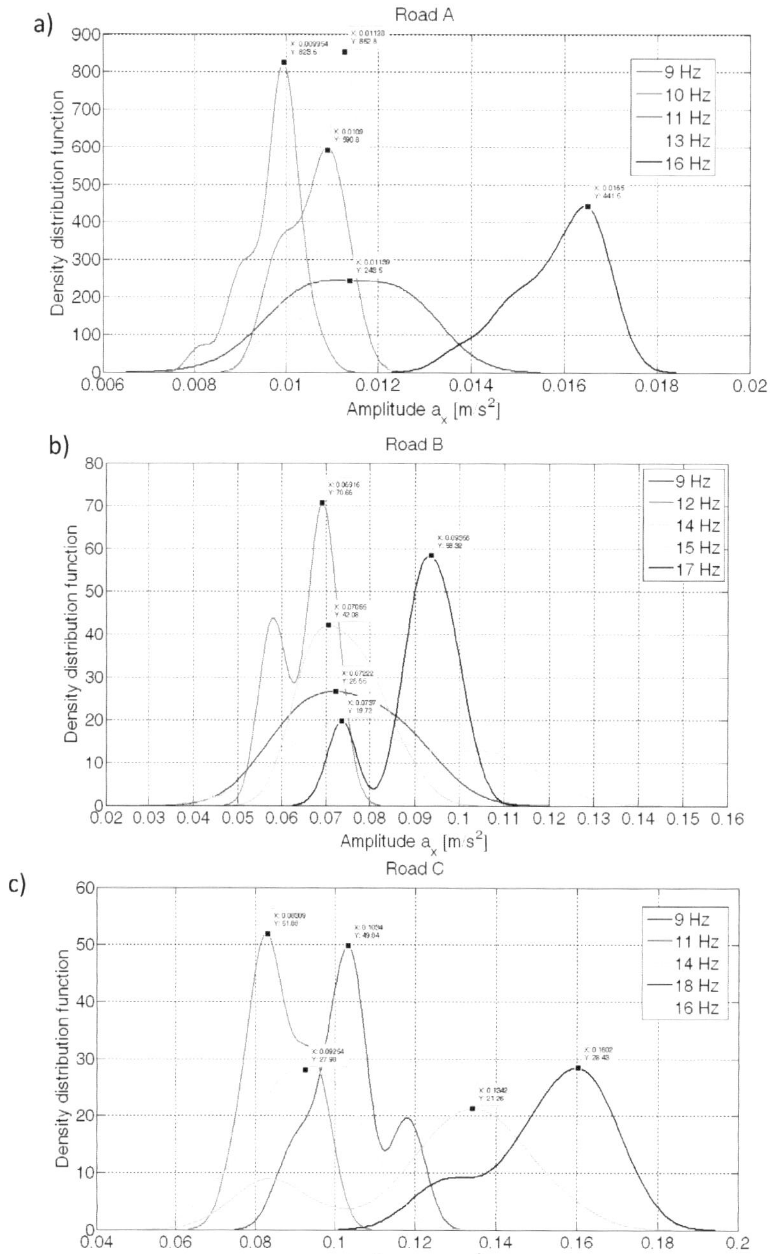

Figure 7. Probability distribution function of the variable a_x for: a) road A, b) road B, c) road C.

Thus, the information about the value of amplitudes dominating in the probability distribution function was obtained for selected ranges of frequency of rotation of the wheel rolling on roads A, B and C. The parameters of longitudinal accelerations of the car body for constant speed were calculated on the basis of such information in the wheel rotation frequency function (figure 8). On this basis the limit value of acceleration in the longitudinal direction was assumed at the level of 0.02 m/s^2, above which the measurements of wheel unbalance under road test conditions cannot be carried out.

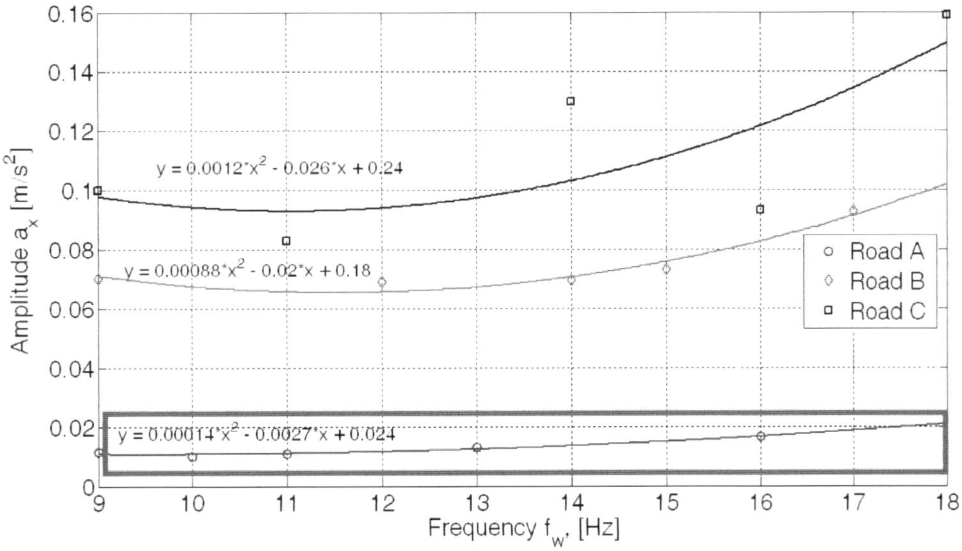

Figure 8. Classification of road surface on the basis of the probability distribution function of the variable a_x.

The results obtained by means of the spectral method using the average value of the tested signal window were the basis for the development of the algorithm of the influence of the road surface condition in terms of diagnostics of wheel unbalance under road test conditions (KSND – road surface condition classifier).

6. Conclusion

As a result of the analysis, test-stand measurements and road tests, it was stated that interferences caused by road surface irregularities have a significant influence on vibrations of the sprung mass of the car body. Therefore, an early detection of unbalanced wheel by means of the measurement of the sprung mass of the car body may be subject to interferences caused by road surface irregularities. Moreover, it was found that road surface irregularities lead to accelerations both in the vertical direction as well as in the longitudinal direction. Bituminous surface irregularities for arbitrarily selected road A with a good surface result in the decrease in the SNR ratio as compared to standard results obtained during test-stand measurements. The most preferred values of the SNR ratio for road tests were obtained at 13.8 Hz.

The criterion for the evaluation of the road surface condition under road test conditions was determined on the basis of the analysis based on the probability distribution; this criterion clearly indicates the limit value of the amplitude of longitudinal vibrations of the car body for a balanced wheel in the range of the tested wheel rotation frequency f_k (from 8.5 Hz to 18 Hz), which did not exceed 0.018 m/s^2. The range for which it is possible to carry out measurements to identify wheel balance was thus determined. It was determined as a result of the experiment of the tested vehicle up to the value of 0.02 m/s^2 for longitudinal acceleration at constant linear velocity of the vehicle.

References

[1] Ślaski G 2012 *Study design automotive suspension with variable damping* (in Polish). Wydawnictwo Politechniki Poznańskiej, Rozprawy - Nr 481, Poznań
[2] Mitschke M 1989 *Dynamic car* (In Polish), tom 2 Drgania. Wydawnictwa Komunikacji i Łączności, Warsaw
[3] Prażnowski K, Brol S and Augustynowicz A 2014 *Identification of static unbalance wheel of passenger car carried out on a road.* Solid State Phenomena, Trans Tech Publications, vol. 214, pp 48-57, Switzerland

[4] Lozia Z and Mazański K 2010 *Typical uneven road surfaces* (in Polish). Zeszyty Naukowe Instytutu Pojazdów, Z.1/77, pp 209 - 221, Politechnika Warszawska, Warsaw
[5] Mamala J, Brol S and Jantos J 2008 *The estimation of the characteristics of the drive system based on the measured acceleration of the vehicle* (in Polish). Studia i Monografie, z. 235, Oficyna Wydawnicza Politechniki Opolskiej, Opole
[6] Brol S 2013 *Analysis of the possibilities of using direct measurement of the acceleration to determine the characteristics of the traction car* (in Polish). Studia i Monografie, z. 350, Wydawnictwo Politechniki Opolskiej, Opole
[7] Tyan F, Hyong Y F, Tu S H et al. 2009 *Generation of random road profiles* Journal of Advanced Engineering vol 4 no 2 pp 1373-1378
[8] Snamina J, Kowal J and Wzorek T 2011 *Analysis of energy dissipation of vibration in active vehicle suspension for the selected control algorithms* (in Polish) Mechanika, R. 108, z. 1-M, pp 233-240, Wydawnictwo Politechniki Krakowskiej, 2011

IOP Conf. Series: Materials Science and Engineering **148** (2016) 012023 doi:10.1088/1757-899X/148/1/012023

The influence of the pressure force control signal on selected parameters of the vehicle continuously variable transmission

A. Bieniek[1], M. Graba[2] and K. Prażnowski[3]

[1,2,3] Opole University of Technology, Faculty of Mechanical Engineering,
[3] PhD student, [1,2] DSc., DEng - Professor in Opole University of Technology

E-mail: a.bieniek@po.opole.pl

Abstract. The paper presents results of research on the effect of frequency control signal on the course selected operating parameters of the continuously variable transmission CVT. The study used a gear Fuji Hyper M6 with electro-hydraulic control system and proprietary software for control and data acquisition developed in LabView environment.

1. Introduction

Modern cars are often equipped with gears, in which the shifting is carried out automatically. From year to year notes increased interest in new vehicles equipped with automatic transmission. Among the various solutions used in automotive vehicles, we can distinguish transmissions with continuously variable shifting

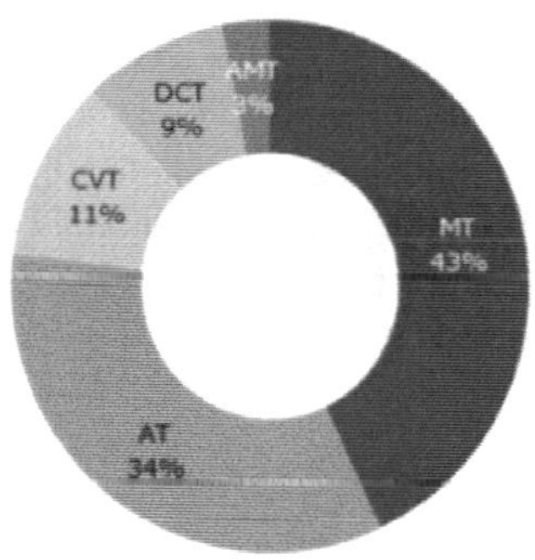

Figure.1. The share of different types of gear in the global market for automotive powertrain (AMT – Automatized Manual Transmission, DCT- Double clutch transmission, CVT- Continuously Variable Transmission, AT – Automatic Transmission, MT- manual transmission) [16].

The share of continuously variable transmission in the global market is significant amounting to approx. 11%, and among transmissions with automatic gear larger share have only traditional solutions based on the torque converter (figure 1). Also, the forecast for 2017 years provide a significant increase in the production of propulsion systems equipped with continuously variable transmission (figure. 2). These types of transmissions are used mainly in vehicles, municipal compact and equipped with combustion engines with little power class, although they are used also in the vehicle middle and upper classes and SUVs equipped with diesel engines that achieve significant power and generate high torque values.

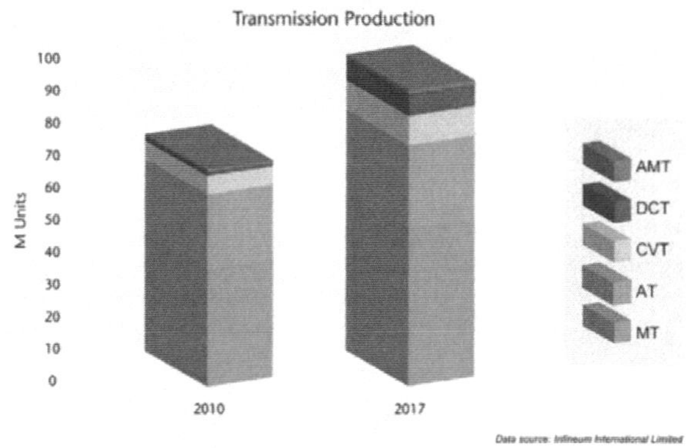

Figure 2. Number produced different transmission types in 2010 and 2017 (AMT - automatized Manual Transmission, DCT- Double clutch transmission, Continuously Variable CVT- Transmission, AT - Automatic Transmission, manual transmission MT-) [15].

In the first case, the most commonly used transmission with a metal belt segment, which segments combined package of steel belts cooperate with treadmills bevel gears and the continuously variable transmission combines with torque converter forming a major part of the drive train to the wheels. In the latter case the construction of the transmission linkage is based on a metal chain, which bolts compressed between the raceways gears move by the effect of friction torque (Audi - Multitronic). Continuously variable transmission chains combined with a friction clutch usually carriage wet controlled automatically. The use of variable speed drive systems belt segment cooperating with the torque converter cause a considerable comfort of use, on the other hand contributes to the occurrence of significant slippage between the pump and the turbine, and thereby decrease efficiency of the entire drive system.

2. Tension control of the continuously variable transmission

Continuously shifting CVT equipped with a metal strand takes place under the control of oil pressure in the hydraulic system. The conical gear pulleys are mounted on shafts the gear unit in such a way that one of the conical discs of each pulley is mounted on a fixed, while the other is axially movable. Chambers sliding discs from each of the transmission pulleys is supplied oil under pressure, giving rise to an axial thrust force, which compresses the tension member cooperating with transmission pulley. The pressure in each thrust cylinder moving targets is the result of the valves installed in the hydraulic circuits of each of the chambers. The most commonly used are the valves, which are controlled by PWM, the different parameters such as the filling degree of the PWM signals directly affect the pressure in the hydraulic circuit. Control of the oil pressure value in the individual circuits, ie. A circle around the primary and secondary pulley will work continuously variable transmission with specific indicators. The pressure in the cylinders of thrust sliding transmission pulleys affects the ability of the torque transmisison, hence the thrust force required for primary transmission pulley.

$$A_{1\min} = \frac{T_o \pi (i_p + 1)}{2\mu_T (l - 2a)\cos\beta} \tag{1}$$

where:

T_0- load (torque) input, Nm

i_p –transmission ratio

μ_T- friction coefficient between the transmission pulley and the belt

L - length of the belt, m

a- transmission pulley base, m

β- angle of inclination of the conical surface of the transmission pulley.

The thrust force depends on the pressure and thrust surface (A_{dyn}) associated with the centrifugal forces on the fluid in the cylinder pressure of [6].

$$A_1 = S_1 \cdot p_{stat1} + A_{dyn1} \tag{3}$$

where:

S_1 - the surface pressure of the primary transmission pulley cylinder [m^2],
p_{stat1} - oil pressure (static) in a primary cylinder, [kPa],
A_{dyn1} - component of the thrust of which is the result of the dynamic oil pressure on the pressure cylinder.

This component can be determined from the relationship:

$$A_{dyni} = \frac{\pi \omega_i^2 \rho_{oil}}{4}\left(R_{oci}^4 - R_{sci}^4\right) \tag{4}$$

where:

ω_i - angular velocity at the i transmission pulley [rad / s],
ρ_{oil} - density of hydraulic oil, [kg / m^3],
R_{oci} - outer radius of the cylinder pressure of the transmission pulley "i", [m],
R_{sci} - the inner radius of the cylinder pressure the transmission pulley "i", [m],
i - marking transmission pulleys i= 1 – primary pulley, and i= 2 – secondary pulley.

On the other hand, introducing a safety factor gear kp obtain the expression for the required thrust, for the primary pulley:

$$A_{1dem} = \frac{k_p M_o \pi (i_p + 1)}{2\mu_T (l - 2a)\cos\beta} \tag{5}$$

The thrust force depends on the pressure and the pressure of the surface and the effect of centrifugal force, [6], and the quotient of the cylinder pressures ζP, could be described with

$$\zeta p = \frac{p_p}{p_s} \tag{6}$$

where:

P_p pressure in the primary pulley [Pa],
P_s-pressure at the secondary pulley [Pa].

and in fact the thrust forces ratio:

$$\zeta F = \frac{F_p}{F_s} \tag{7}$$

where:

F_p-thrust force at primary pulley [N],
F_s- thrust force at secondary pulley [N]

is associated with obtaining a particular transmission speed ratio [3, 4]. The curve quotient of the pressure ratio depends on the given transmission safety factor kp.
Such knowledge on the impact of signal parameters of control valves and the inclusion of specific CVT transmission design features such as thrust areas for each pulley is important for the design of the transmission control algorithm.
Transmission control must primarily take into account the safety of its operation, ie. such a choice of pressure in the thrust cylinder of each pulley and taking into account that there is no slip linkage between belt and pulleys surface [2, 5, 7, 11, 12, 14]. On the other hand, it is very important to ensure adequate

dynamics of changes in operating parameters of the transmission [8, 11, 12, 13] and to enable optimum operation of the power unit [10, 13] resulting in the improvement of properties of dynamic, economic and environmental reasons [1, 7, 9].

3. Experimental research

The research was carried out on the automotive continuously variable transmission mounted on the bench. During the test transmission worked without a load in the stationary states. There was transmission pulleys rotational speeds, and the pressure in the individual hydraulic circuits recorded. Used for this purpose factory rotational pulley speed sensors and pressure sensor in the secondary pulley, and additionally installed a pressure sensor in the primary pulley. Tests were conducted at a constant rotational speed of the electric motor, respectively 800, 1000, 1200, 1400 rev / min.

As a research facility used automotive continuously variable transmission Fuji Hyper M6 fitted with a metal belt pushed. Continuously variable transmission consists of two components: a continuously variable transmission with a metal belt tensioning and cooperating with its hydrokinetic torque converter. Table 1 shows the basic properties characterizing the test transmission.

Table 1. The Fuji Hyper M6 main properties.

No	Properties	Value
1.	Transmission ratio range D, DE	0,442 – 2,432
2.	Transmission ratio range S1	2,432 - 1,890
3.	Transmission ratio range S2	1,473 - 1,300
4.	Transmission ratio value S3	0,980
5.	Transmission ratio value S4	0,742
6.	Transmission ratio value S5	0,602
7.	Transmission ratio value S6	0,520
8.	Transmission ratio value R	1,931
9.	Control system type	el.-hydraulic
10.	Hydrokinetic torque converter	3 part + lock
11.	Hydrokinetic conv. torque ratio	1,71
12.	Hydrokinetic conv. diameter	200 mm
13.	Belt type	metal belt
14.	Hydraulic pressure range	1,80 – 5,35 MPa
15.	Primary wheel thrust area, S_1	0,0207 m^2
16.	Primary wheel thrust area, S_2	0,0097 m^2
17.	Wheel notch angle, β	11°
19.	Wheel center distance, a	0,155 m
20.	Belt length, l	0,648 m
21.	Construction index, κ	0,3871

The test transmission was mounted on a test bench equipped with an electric motor is used to drive the test facility coupled with Optidrive inverter, which enables realization of the selected scenarios of the transmission pulleys rotational speed changes (Fig. 3).

IOP Conf. Series: Materials Science and Engineering **148** (2016) 012023 doi:10.1088/1757-899X/148/1/012023

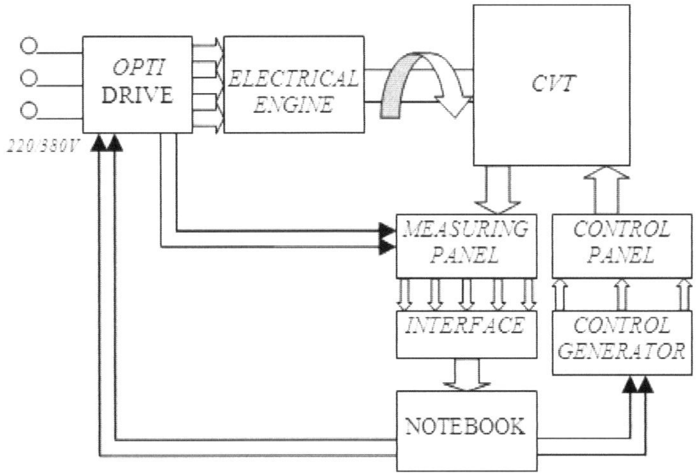

Figure 3. Scheme of the CVT Test Bench.

Transmission control was carried out using valves installed in a hydraulic unit of primary and secondary pulleys. There was controled using the author's application control and the measurement made in the package Lab-View via PWM control signals of different frequencies and different degrees of filling. The block diagram of the control the measurement is shown in figure 4.

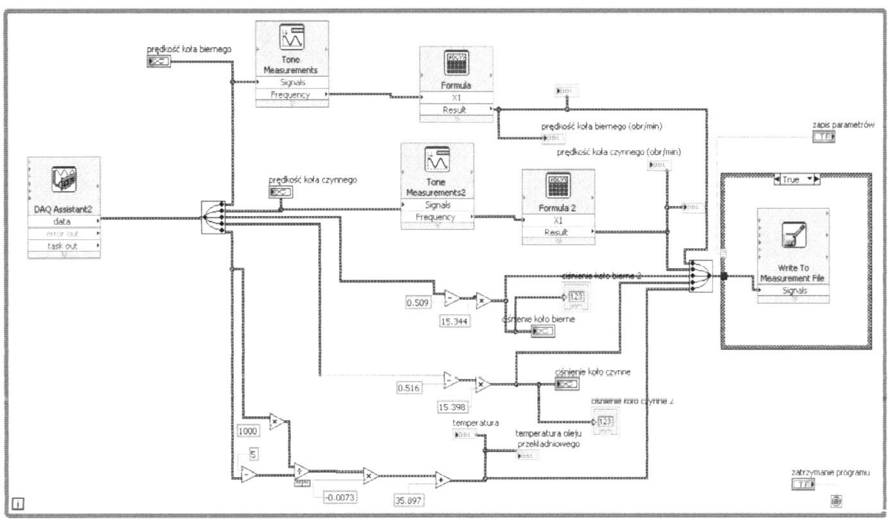

Figure 4. Block diagram of LabView measuring application.

Control and measurement hardware was NI interface which allows simultaneous recording of selected operating parameters of the transmission and control of solenoid valves placed in the hydraulic system. The frequency control signal varied in the range of 10 to 400 Hz and the degree of filling of the PWM signal was in the range of 0-100%.

4. Analysis of the results

4.1. The influence of the frequency of the PWM control signal to transmission operation

The results of stand tests found that the stabilization of the operating parameters of the transmission after crossing the frequency of approx. 50 Hz. Further increasing the frequency does not significantly affect the achieved operating parameters such as rotational speed. This applies to the control valves of both transmission pulleys. Noticeable is the only occurrence of unevenness larger secondary pulley

IOP Conf. Series: Materials Science and Engineering **148** (2016) 012023 doi:10.1088/1757-899X/148/1/012023

rotational speed. The smallest value is obtained from the control signal frequency of approx. 250, 350 Hz, nevertheless considered in the entire frequency range for both the standard deviation of the transmission pulley rotational speed is small. In the case of a secondary pulley, but the situation is reversed, than the primary pulley. A statement of the value of the average velocity and the standard deviation for the selected frequency ranges are shown in figure 5.

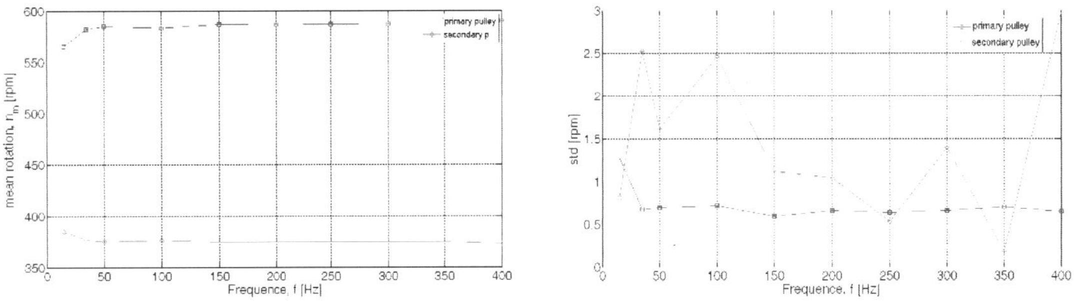

Figure 5. The impact of frequency on the average wheel speed and the standard deviation of the speed.

Simultaneously with increasing frequency control signal, an increase in pressure in the both primary and secondary cylinder pulley, and for the control signal frequency of 15 Hz the value of the average pressure in the primary pulley (P_{cz}) is 13,6 bar (σ = 0,12), whereas for a secondary pulley P_b = 22,58 bar (σ = 9,1). Increase the frequency to 35 Hz increased the pressure at primary pulley P_{cz} = 19,22 bar (σ = 0,11) and secondary pulley P_b = 34,28 bar (σ = 1,35). A further change in frequency control signal caused an increase in pressure and a decrease in the value of the standard deviation.

4.2. The impact of filling the PWM control signal on the transmission operation

The analysis of experimental results allowed us to assess the impact of filling the PWM signal on the selected parameters of the transmission. There was analyzed the following parameters of the transmission operating indexes: the rotational speed of the primary and secondary pulley, primary and secondary pulley cylinder pressure and transmission ratio i_T, and the pressure thrust ratio ÇP. The evaluation was conducted in transmission for given constant values of electrical motor rotational speed, driving torque converter coupled with the primary pulley. The results showed that with increasing duty cycle slightly increases the rotational speed of the secondary pulley. Exceeding characterizing the PWM filling degree of 60% followed by a sudden change in speed and then after crossing the 70% duty cycle control to re-stabilize the speed (Fig. 6).

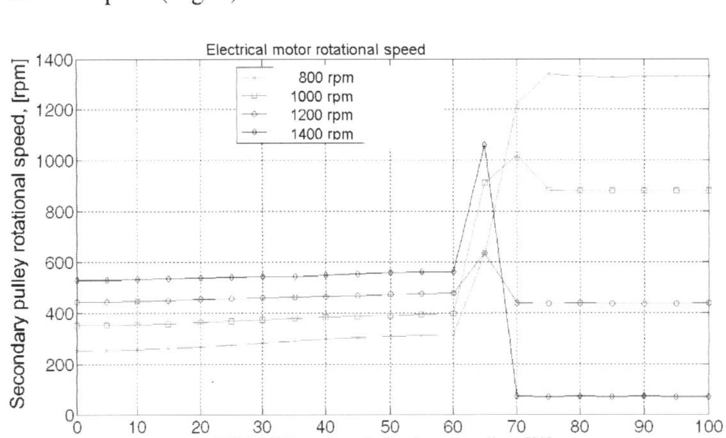

Figure.6. The influence of the fulfilment degree of the PWM signal for the secondary pulley rotational speed

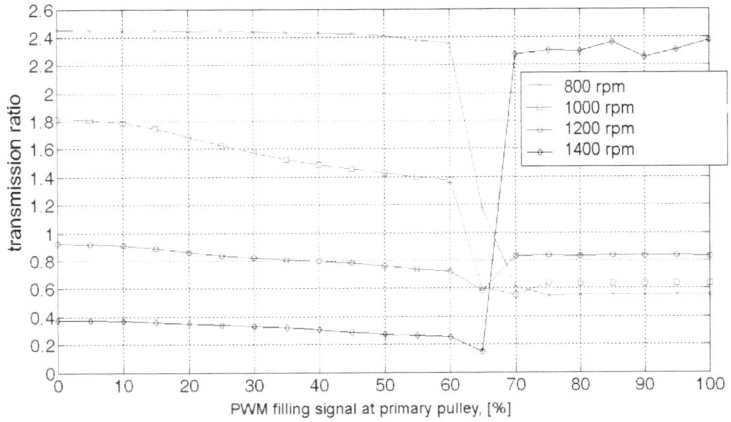

Figure. 7. The influence of the fulfilment degree of the PWM signal for the transmission ratio

The course of the continuously variable transmission ratio understood as the quotient of the rotational speeds of the transmission pulleys, depending on the degree of filling of the PWM signal shown in figure 7. The duty cycle of the PWM range from 0-60% transmission ratio is kept constant values. In terms of a preset duty cycle of 60-70% suffer from rapid change and stabilize the PWM value above 70%. In turn, the results of primary pressure indicate that the pressure in the cylinder pulleys decreases with a PWM wheel active in the range of 0-60% PWM, and then stabilized. The highest pressure values are for the engine speed of 800 rev / min, while the lowest for a speed of 1200 rev / min. The differences in pressure in the primary pulley for different engine speeds are particularly evident for low PWM (0-30%). All observed changes in pressure are uninterrupted, and there were no sudden abrupt pressure changes (Figure 8, 9).

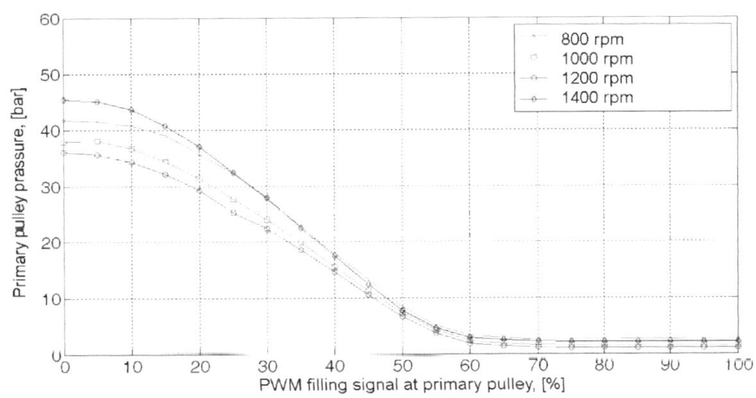

Figure. 8. The influence of the fulfilment degree of the PWM signal for the primary pulley pressure

The course of the pressure in both cylinder pulleys are similar (Figure 8, 9). With the increase of the PWM pressure it decreased to the PWM value of about 60% and then plateaued. Pressure values for the individual engine speeds were very close to each other in practically the whole range of changes in the degree of fulfilment of the PWM signal. Only for low PWM (0-10%) you will notice small differences.

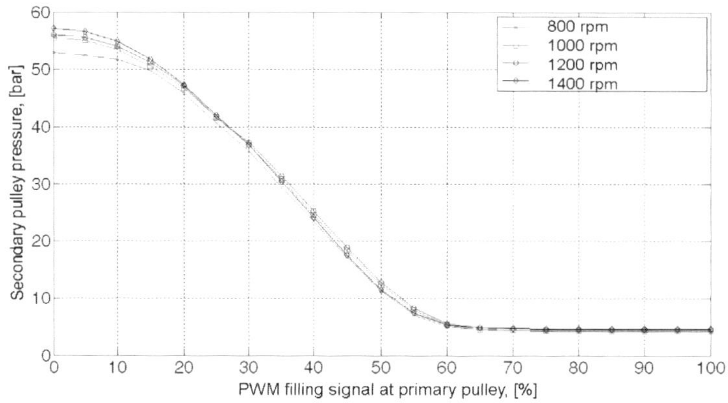

Figure 9. The influence of the fulfilment degree of the PWM signal for the secondary pulley pressure

The quotient of the pressure thrust ζP, Figure 10 shows an example of waveforms for control valves of both wheels with the same parameters of the control signal f = 100 Hz and the degree of PWM varying between 0 and 100%.

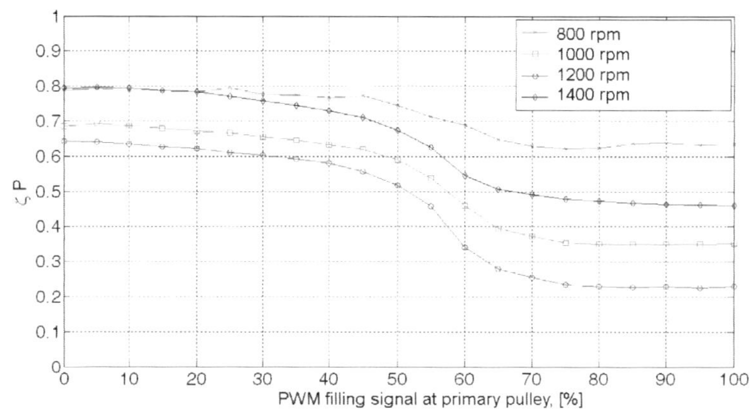

Figure 10. Sample waveforms using the pressure ratio control valves simultaneously with the same parameters of the control signal for the valves of both transmission pulleys

Theoretically, it is possible to obtain the quotient of the pressure of a very wide range, however, transmission operating under load by a defined, in practice, considerably narrows the range of variation of up to approx. 0,8 – 2,6. Such a range of changes quotient of the pressure, however, is sufficient to produce a full range of gear changes.

5. Conclusion

The use of solenoid-controlled PWM signal allows for a smooth change of pressure in the cylinder pressure of the transmission pulleys. Research has shown that in the low speed range used real degree of fulfilment of the PWM signal is about 60%. However, the use of higher speed, and thus a proportional increase in the efficiency of the hydraulic pump supply hydraulic transmission may require the use of the entire range of the degree of fulfilment of the PWM signal. Controlled by a signal from a different degree of filling enables smooth change of the pressure in each cylinder pulley which translates into a wide range of transmission ratio change and affect its ability to carry the load and the properties such as the safety and overall efficiency. The studies also indicate that in the case of stationary conditions sufficient stabilization of the operating parameters of the transmission is obtained at a frequency greater than approx. 50 - 100 Hz.

References

[1] Behrenroth J and Gueter C 2002 *Aspekte zur Systemarchitektur, Triebstrangregelung und Abstimmung – wie haben Kundenerwartung die Entwicklung des CVT für den neuen Mini beeinflusst*, CVT 2002 Congress Munich, October 2002 VDI-Berichte 1709 pp 195-210

[2] Bieniek A 2006 *Neue Methode zur Bestimmung von Reibungsabwicklungsgrad des Stufenloses Umschlingungsgetriebe*, International Symposium „Advances in Mechanical Engineering", Gdańsk 2006 pp 27-38

[3] Bieniek A 2006 *Analitisches Modelbildung der Stufenloses Umschlingungsgetriebe*, International Symposium „Advances in Mechanical Engineering" Gdańsk 2006, pp 39-50

[4] Bieniek A and Jantos J 2006 *Analytical model of continuously variable transmission*, International Congress Motor Vehicles & Motors, Kragujevac 2006, paper nr 20060074, s. 1-6

[5] Bieniek A 2006 *Analysis of transmission load and frictional conditions in continuously variable transmission*, International Congress Motor Vehicles & Motors, Kragujevac 2006 paper No 20060064, pp1-6.

[6] Bieniek A and Augustynowicz A 2012 *Zjawiska hydrodynamiczne w układzie zacisku cięgna samochodowej przekładni CVT a wskaźniki jej pracy*, Technical Transactions Mechanics (Czasopismo Techniczne, Mechanika) 5-M/2012, Is. 10, Y 109 pp 109-118

[7] Ide T. 1999 *Effect of Power Losses of Metal V-Belt CVT Components on the Fuel Economy*, International Congress on Continously Variable Power Transmissions CVT'99 Eindhoven 1999 pp 93-98

[8] Jantos J 2001 *Transmission ratio derivative value and driveability of the vehicle with CVT*, Archives of transport 2001 vol 13 pp 47-59

[9] Jantos J, Mamala J and Bieniek A. 2006 *Estimation of Friction Degree in the continuously variable transmission*, International Conference "Power Transmission '06" pp 377- 380

[10] Nowatschin K, Hommes G, Fleischmann H-P, Faust H, Gleich T and Friedmann O. 2000 *Multitronic- Das neue Automatikgetriebe von Audi -Teil I* Automoblitechnische Zeitschrift 102(2000) nr 7/8, pp 548-553

[11] Renius K T and Sauer G 1996 *Optimale Ketteneinspannung stufenloser Umschlingungsgetriebe*, 1996 Sonderforschungsbericht 365 Teilprojekt C1 pp. 261-287

[12] Sauer G 1996 *Grundlagen und Betriebsvehalten eines Zugketten-Umschlingungsgetriebes*, VDI Verlag 1996 nr 293

[13] Serrarens A and Veldpaus F 2002 *Driveability Assessment of a CVT-powertrain with mechanical Torque Assist*, FISITA World Automotive Congress, Helsinki 2002 No F02V067

[14] Stöckl B 2002 *Development of a Pressure-Controlled clamping system for Continously Variable Belt and Chain Transmissions*, CVT 2002 Congress Munich, 7th/8th October 2002 VDI-Berichte 1709 pp 571-582

[15] Infineum International Limited – 2015

[16] China market share statistics - 2015

Evaluation of the influence of velocity on dynamic passenger loads during a frontal minibus impact against an obstacle

L Prochowski[1], A Dębowski[1], A Żuchowski[1], K Zielonka[2]

[1] Military University of Technology, Institute of Motor Vehicles and Transport, Warsaw, Poland
[2] Automotive Industry Institute (PIMOT), Warsaw, Poland

E-mail: lprochowski@wat.edu.pl

Abstract. The safety of people travelling by minibus is a very complex issue, in which the decisive role is played by load-bearing vehicle structure, passenger seats, and personal protection means. In order to maximize the number of people transported, the seats are spaced very closely to each other and this may pose a hazard to the passengers. Based on an analysis of experimental test results, a computer model representing a system composed of a minibus floor segment, seats, and dummies was built. For the analysis, seats integrated with seat belts were adopted. A seat of this type was based on a high-rigidity frame necessary to bear, *inter alia*, the strong force exerted (during a collision) by passenger's torso on the shoulder seat belt and transmitted to the upper seat belt anchorage point on the seat backrest. Within this work, the frontal minibus impact against an obstacle with velocities ranging from 20 km/h to 70 km/h was considered.

The analysis covered the motion of, and dynamic loads on, a test dummy representing a 50th percentile adult male (Hybrid III dummy). Within the analysis, realizations of dynamic loads caused by inertial forces and reactions exerted by a three-point seat belt were taken into account. Special attention was paid to the extreme values of the loads that acted on dummy's head, neck, and torso when the head hit the backrest of the preceding seat in the culminating phase of the vehicle impact against an obstacle. The values of biomechanical indicators *HIC*, *ThAC*, N_{ij}, and *FAC* and of the joint injury risk indicator were calculated.

1. Introduction

The safety of minibus passengers, especially in the situation of ongoing growth in travelling speeds, is a very important issue, in which the decisive role is played by load-bearing vehicle structure, passenger seats, and personal protection means. The close spacing of vehicle seats adversely affect passengers' comfort and may pose a hazard to the passengers [2, 5, 6], in particular when the vehicle frontally hits an obstacle. The exploration of the hazard was started from an analysis of the trajectory of passenger's body during an accident; the analysis has been presented in [3]. Due to small space available, different passenger's position, and limited possibilities of using standard airbags in minibuses, the results of typical research on the motion of motor car driver and passenger's bodies during an accident are hardly applicable to this case [1, 11, 14]. The significant mass of an adult (a 50th percentile adult male as used in this work) often results in big deformation of the seat to which the seat belt is anchored and, in consequence, in a growth in the displacement of passenger's body in the direction of inertial forces, which translates into an additional hazard to the occupant of the preceding seat [7, 9].

The work was undertaken to identify some characteristics of the process of passenger head's impact against the backrest of the preceding seat during a minibus collision with an obstacle. This identification includes determining the instant when the impact begins, impact duration time, location of the place of the head's impact against the backrest, values of the impact velocity and force, as well as position of the impact velocity vector. A separate stage of the calculations was dedicated to determining the influence of the velocity v_0 of vehicle impact against the obstacle on the extreme values of the loads generated during the collision. Results of the analysis of passenger's motion in the space between the seats and of the loads generated during the collision will provide a basis for improving the dynamic properties of the system consisting of seat with its anchorage to the floor, seat occupant, and seat belts as an important part of the minibus safety system. One of the important numerical indicators of the effectiveness of the protection provided by the safety system is the risk of injury to vehicle passengers during a road accident.

At present, vehicle seats are assessed by checking their conformity with UN ECE Regulations Nos. 14, 17, and 80 [10]. According to these Regulations, the seats are subjected to dynamic tests at $v_0 = 30$ km/h. However, such a scope of the assessment is far from satisfactory because it does not reflect the current knowledge of the threats that arise during road accidents at impact velocities v_0 exceeding 30 km/h.

2. Modelling and validation

Based on an analysis of results of experimental tests carried out at the Automotive Industry Institute (PIMOT) in Warsaw [3, 16], a computer model was built, which represented a road accident with a frontal minibus impact against an obstacle. The calculations were based on a model of the dynamics of a system consisting of a vehicle floor segment with three seats anchored to it. In this model, built in the PC-Crash 10 program, the test dummies representing vehicle passengers are restrained with three-point seat belts. The dummies and seats' positions have been shown in Figure 1. The Hybrid III test dummy representing a 50th percentile adult male occupying the third seat has been denoted by H3.

The model of a human passenger [12, 13] was built of 20 ellipsoids connected with each other by ball joints. The passenger is placed on a seat, which is a multibody system consisting of 11 elements. Each of the elements that constitute the models of passengers and seats has properties described by defining e.g. geometrical parameters of the ellipsoid, mass, tensors of inertia, linear velocity, and angular velocity of the centre of mass of the body, as well as the elasticity characteristics necessary to calculate the contact forces and friction between individual model bodies or between the bodies and the interfacing surfaces. The values of the contact forces between the bodies are calculated on the grounds of a predefined non-linear function depending on the current interpenetration depth of the bodies. The global and local coordinate systems make it possible to define the positions and orientations of the bodies in the space.

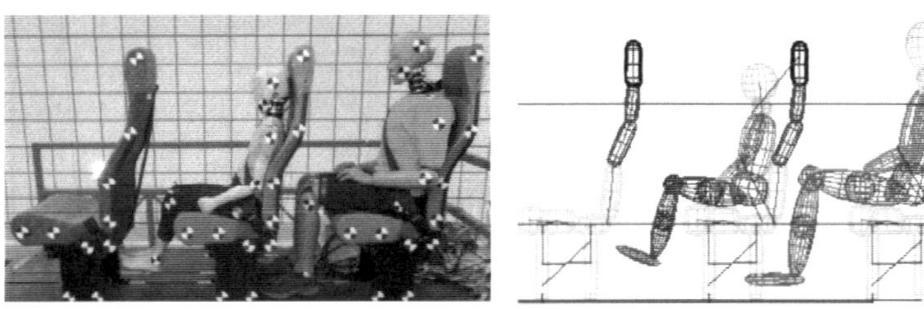

Figure 1. Physical vehicle model for the experimental tests and test dummies on vehicle seats in the computer model.

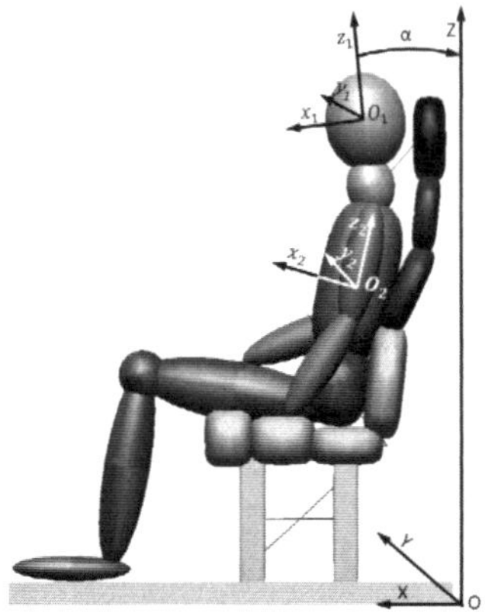

Figure 2. Global and local coordinate systems (e.g. OXYZ and $O_ix_iy_iz_i$, respectively).

The model having been built was subjected to a validation process. With this end in view, an impact of the computer minibus model against an obstacle was simulated and the test conditions adopted for the simulation corresponded to those of the experimental test. The kinematic input applied as a time history of the minibus floor acceleration in the OX direction (see Figure 2) was in conformity with the excitation applied during the test with the physical model, i.e. during the experiment (see Figure 1).
The validation process was carried out in two stages:

- assessment of conformity between the acceleration vs. time curves recorded for the computer model and the experiment (Figure 3);
- comparison of the joint injury risk indicator values determined from the model and the experiment (presented in Section 5 herein).

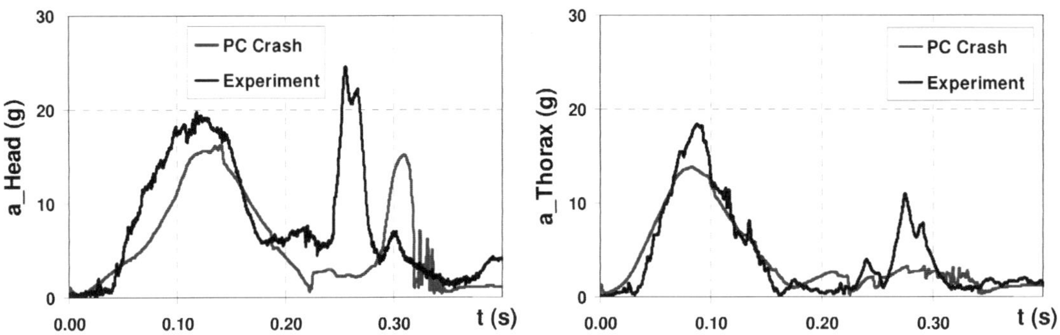

Figure 3. Acceleration vs. time curves for the centre of mass of the head and torso (thorax) of dummy H3.

During the model tests, the physical quantities that characterized the possible motion of, and dynamic loads on, minibus passengers were determined as functions of time. The tests were carried out with taking into account realizations of the dynamic loads generated by reactions in the areas of contact between the dummy and the surrounding objects, reactions at the belt-to-seat anchorage points, and inertial forces. Special attention was paid to the extreme values of the loads that occurred during the impact of dummy's head against the backrest of the preceding seat. Based on the load vs. time curves, the values of biomechanical indicators *HIC*, *ThAC*, N_{ij}, and *FAC* and of the joint injury risk indicator were calculated.
The simulation tests were carried out for vehicle impact velocity values ranging from 20 km/h to 70 km/h. In this velocity range, the head of the child occupying the seat in the second row does not hit the backrest of the first seat and the seat belt forces do not cause any considerable deformation of the child's seat. Therefore, the child's motion was not analysed in this work.

3. Kinematics of passenger's motion between the seats

The calculation results were used to identify the place and time of the beginning of impact of dummy's head against the backrest and to determine the impact duration time and the impact velocity value. A contact of dummy's head with the backrest of the preceding seat was found to occur at velocities of $v_0 \geq 40$ km/h (Table 1).

Table 1. Contour of the silhouette of dummy H3 immediately before hitting the backrest of the preceding seat.

v_0 [km/h]	20	30 (experiment)	30	40	50	60	70
Dummy's position	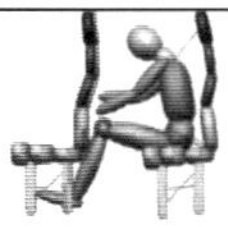		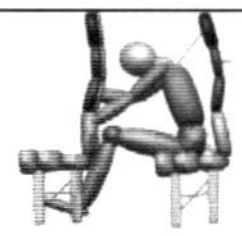	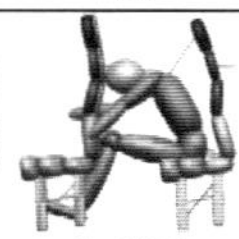	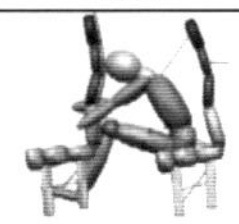	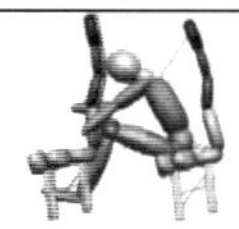	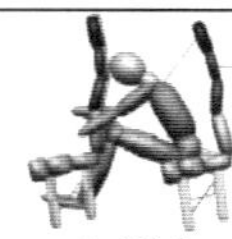
Time* [s]	0.161	0.125-0.130	0.155	0.132	0.109	0.100	0.093

*) Time to the instant of maximum tilt of dummy's head or to the beginning of the impact of dummy's head against the backrest of the preceding seat

Figure 4 shows values of the following quantities as functions of vehicle impact velocity v_0:

- T_{U1} – time of the beginning of the impact of dummy's head against the backrest;
- T_{FK} – time of duration of the phase of contact between dummy's head and the backrest;
- H_{PU} – height of the point of impact of dummy's head against the backrest;
- V_{XYZ} – value of the resultant velocity of impact of dummy's head against the backrest at the instant T_{U1};
- Alfa – angle defining the position of the head impact velocity vector at the instant of impact (cf. Figure 6).

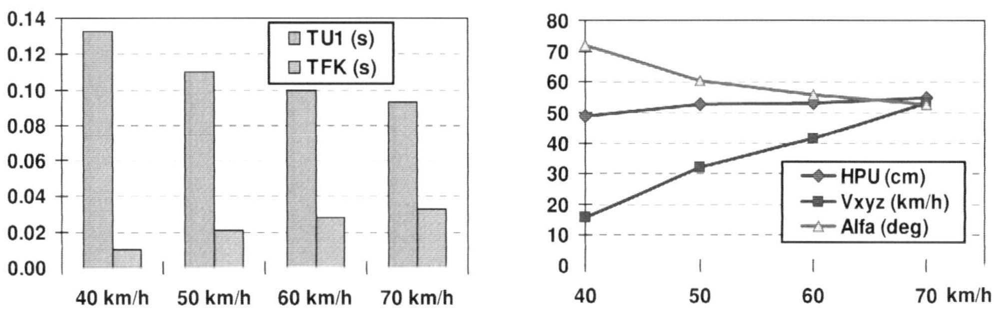

Figure 4. Summary of values of the quantities that characterize an impact of dummy's head against the backrest of the preceding seat.

The results obtained provide grounds for formulating the following findings:

- When the values of velocity v_0 grew from 40 km/h to 70 km/h then the impact of dummy's head against the backrest began more and more early and the value of T_{U1} decreased from 0.132 s to 0.093 s.
- In result of shortening of time T_{U1}, the value of the head impact velocity V_{XYZ} became closer to v_0, i.e. a growth from 39 % at $v_0 = 40$ km/h to 76 % at $v_0 = 70$ km/h could be observed in the proportion U:

$$U = \frac{V_{xyz}}{v_0} 100\% \tag{1}$$

- Time T_{FK} of duration of the phase of contact between dummy's head and the backrest increased with growing v_0 to reach a level of 0.033 s at $v_0 = 70$ km/h.

- The height of the centre of the area of contact between dummy's head and the backrest slightly rose with growing v_0.

The quantities referred to above considerably affect the state of loads on the neck, e.g.:

- the value of time T_{FK}, which is taken into account in the neck shear criterion (maximum shearing force) F_{XY} (F_X, F_Y; Figure 5) and the neck tension criterion (maximum tensile force, F_T; Figure 5) in UN ECE Regulation No. 94;
- the angle of impact Alfa, which affects the resolution of the impact force vector into components F_X, F_Y, F_T, F_C, which in turn have an influence on the values of moments M_F and M_E, loading the neck as shown in Figure 5 [4].

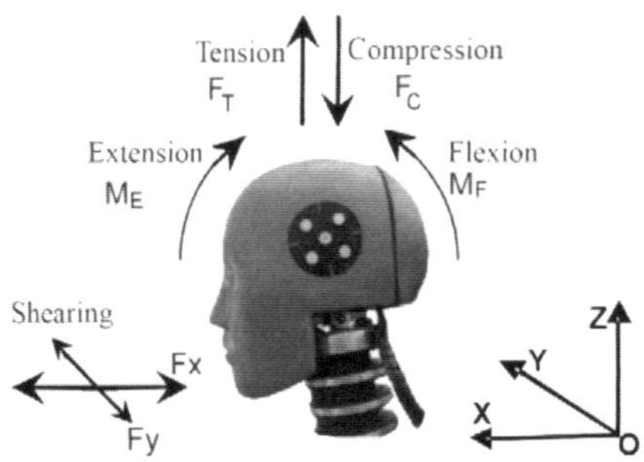

Figure 5. Schematic diagram of the loading of dummy's neck in the sagittal plane.

The position of the head impact velocity vector (Figure 6) shows that with growing v_0, an unfavourable change takes place in the direction of this vector, which results in an increase in its component parallel to the O_1z_1 axis. This causes dangerous neck extension combined with compression of the cervical spine.

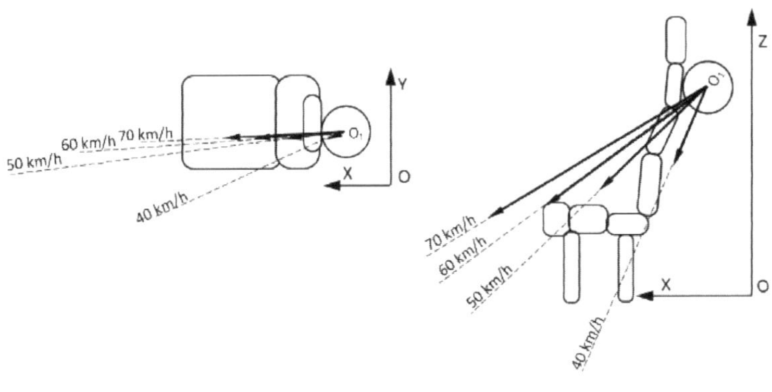

Figure 6. Position of the head's centre of mass velocity vector in the OXZ and OXY planes at the instant of impact.

4. Dummy loads calculation results

Below is shown the influence of velocity v_0 on the time histories of some dynamic loads and on the extreme values of the loads. Figure 7 shows the curves representing the head and neck loads during the head impact against the backrest of the preceding seat and afterwards against the headrest of the seat occupied by the dummy. For the vehicle velocity range of $v_0 \geq 40$ km/h, the areas of predominating load values can be seen, representing the head impact against the backrest of the preceding seat (t = 0.09-0.14 s) and against the headrest (t = 0.22-0.30 s).

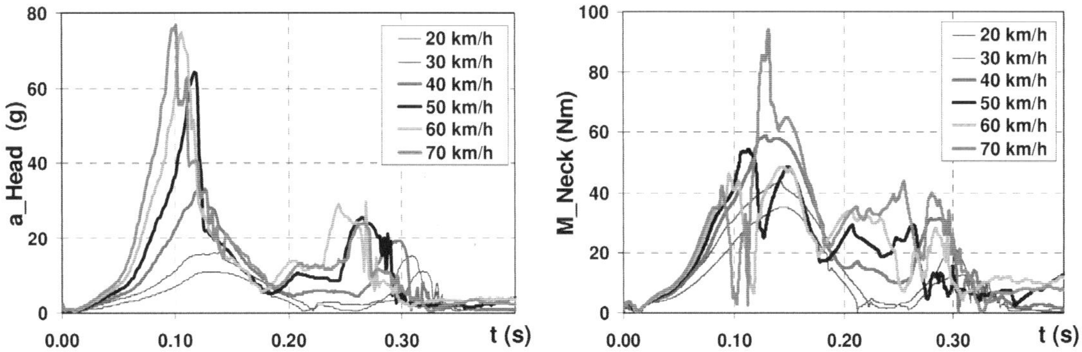

Figure 7. Resultant head acceleration and resultant moment on the force acting on the neck.

The extreme values of the head and thorax accelerations and of the force and moment of the reactions acting on the neck and developed in the lap and shoulder belt straps (F_LB and F_SB, respectively) as well as of the forces in the left and right femurs of the dummy have been presented together in the form of bar graphs in Figure 8. The extreme values displayed were calculated as the highest values of the loads acting for at least 3 ms, 5 ms, and 10 ms, determined from signals having been subjected to low-pass filtration CFC180. Such a method of load presentation connects the maximum values of the loads with their time of duration, while the load duration time is of decisive importance for the assessment of the load effects on the human body. Two ranges of the load values can be discerned in the drawing: a range of relatively low loads at velocities of 20 km/h and 30 km/h (at which dummy's head did not hit the backrest of the preceding seat) and a range of loads rapidly growing with increasing values of the vehicle impact velocities ($v_0 \geq 40$ km/h).

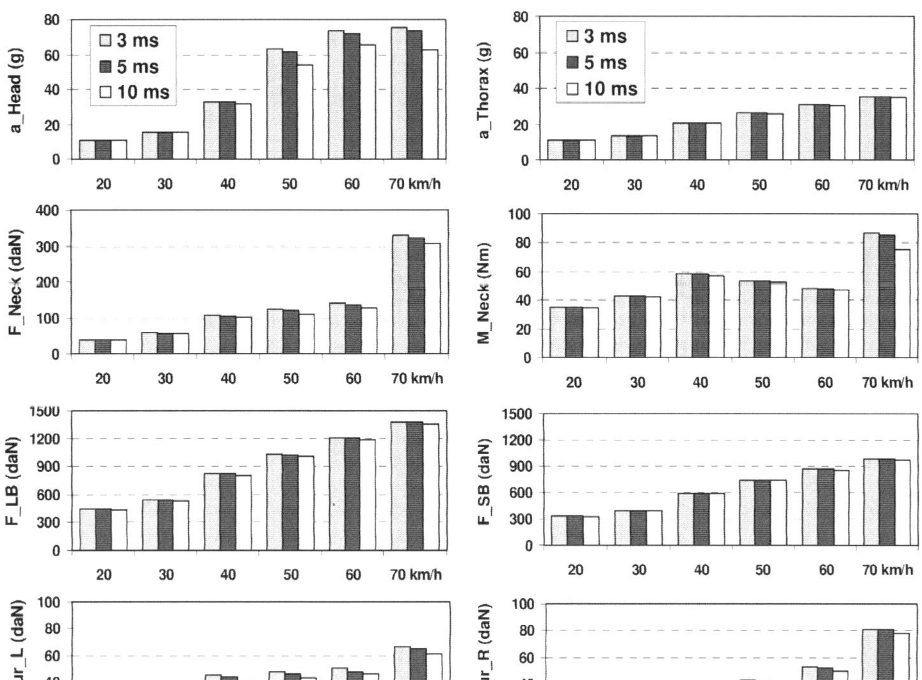

Figure 8. Extreme load values, calculated for load duration times of 3 ms, 5 ms, and 10 ms.

Time histories of the force of head impact against the backrest of the preceding seat (resultant force F_{xyz} and its component F_x in the local head coordinate system) have been shown in Figure 9. The extreme values of the head impact force at vehicle impact velocities of 60 km/h and 70 km/h reached a level of 50-80 % percent of the force destructive to the seat backrest. Under a force of 7-8 kN, the backrest is tilted by an angle of 5-8 °, depending on the seat frame stiffness. The experimental tests carried out within the work described in [7] revealed that the seat backrest struck by the dummy placed on the seat behind was deformed to a significant extent when the vehicle impact velocity was 55 km/h.

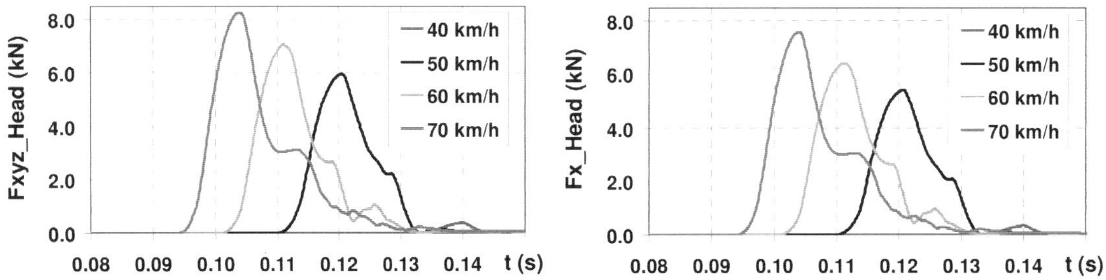

Figure 9. Time histories of the force of head impact against the backrest of the preceding seat.

Some characteristic values of the force of head impact against the backrest of the preceding seat have been specified in Table 2, where the force symbols used have the following meaning:

F_{xyzMAX} – maximum value of the resultant force;

F_{xyz3ms}, F_{xyz5ms}, $F_{xyz10ms}$ – the highest values of resultant force F_{xyz} acting for at least 3 ms, 5 ms, and 10 ms, respectively;

F_{x3ms}, F_{x5ms}, F_{x10ms} – the highest values of component force F_x acting for at least 3 ms, 5 ms, and 10 ms, respectively.

Table 2. Summary of values of the force of head impact against the backrest of the preceding seat (dummy H3).

v, km/h	T_{FK} t [s]	F_{Pmax} F [N]	$F_{ŚR\ 3}$ F [N]	$F_{ŚR\ 5}$ F [N]	$F_{ŚR\ 10}$ F [N]	$F_{XŚR\ 3}$ F_X [N]	$F_{XŚR\ 5}$ F_X [N]	$F_{XŚR\ 10}$ F_X [N]
40	0,017	401,6	369,8	328,9	193,5	-326,7	-289,4	-169,8
50	0,075	6001,6	5844,8	5615,5	4684,7	-5322,9	-5108,3	-4287,2
60	0,084	7110,9	6913,5	6700,9	5641,7	-6300,8	-6112,1	-5159,1
70	0,102	8340,7	8107,1	7800,6	6291,8	-7435,9	-7136,7	-5777,1

F_{Pmax} - value of the extreme forces during the impact phase of the head restraint,
$F_{ŚR}$ - average value of force , calculated for the interval 3, 5 and 10 ms,
$F_{XŚR}$ – average value of the component F_X, calculated for the interval 3, 5 i 10 ms.

The values given in Table 2 have been calculated identically as the extreme values shown in Figure 8.

5. Synthetic injury indicators and the risk of the corresponding injuries

The test results provided grounds for determining the values of the following biomechanical indicators of bodily injury caused by impact loads:

- *HIC* – Head Injury Criterion;
- *ThAC* – Thorax Acceptability Criterion;
- N_{ij} – Neck Injury Criterion;
- *FAC* – Femur Acceptability Criterion.

The analytical relations necessary to calculate the above indicators have been given in publications [4, 8, 16] and results of the calculations carried out to determine the indicator values have been brought together in Figure 10. The calculations were done with taking into account the loads caused by the impact against the backrest of the preceding seat and, during the backward motion in the final phase of the collision, against the headrest of the seat occupied by the dummy.

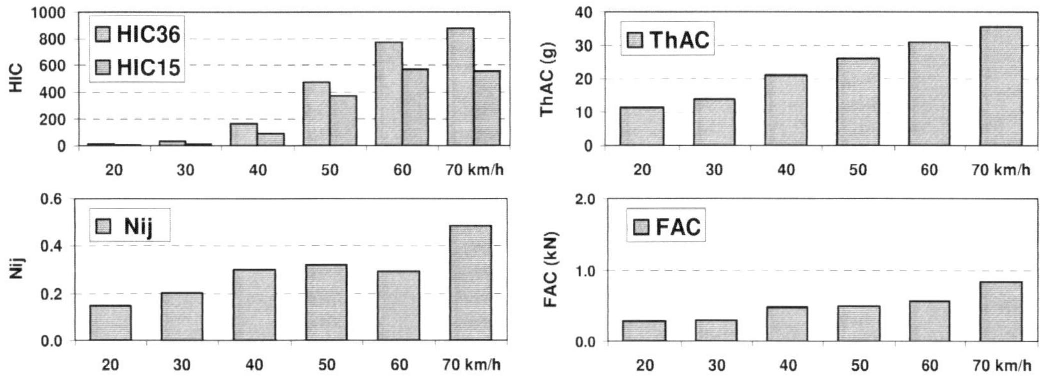

Figure 10. Values of the biomechanical indices calculated from time histories of the loads on dummy's head (HIC), neck (N_{ij}), thorax (ThAC), and femurs (FAC).

A road accident can result in many different injuries. Therefore, the joint injury risk indicator is determined with taking into account the joint effect of impact loads on passenger's head, neck, torso, and legs. The joint injury risk indicator P_{Joint} is expressed in percentage terms, in the form as follows [8]:

$$P_{Joint} = [1 - (1 - P_{Head}) \cdot (1 - P_{Neck}) \cdot (1 - P_{Chest}) \cdot (1 - P_{Femur})] \cdot 100\ \% \qquad (2)$$

where: P_{Head} – head injury risk (calculated on the grounds of HIC_{36});
P_{Neck} – neck injury risk (calculated on the grounds of N_{ij});
P_{Chest} – thorax injury risk (calculated on the grounds of $ThAC$);
P_{Femur} – leg injury risk (calculated on the grounds of FAC).

The hazard to vehicle passengers was considered in relation to the injury risk classified in the Abbreviated Injury Scale as "moderate" (AIS2). The HIC_{36} and HIC_{15} indicator values were calculated with taking into account different lengths of the time interval for which the highest acceleration values were recorded. The N_{ij} indicator values resulted from joint effect of force F_z (F_T, F_C) and moment M_y (M_E, M_F) on the neck [15, 16].
Results of calculations of the P_{Joint} risk indicator and the risk of injury to head, neck, thorax (torso), and legs resulting from the dynamic loads that acted on a vehicle passenger in the space between the seats have been presented in Figure 11.

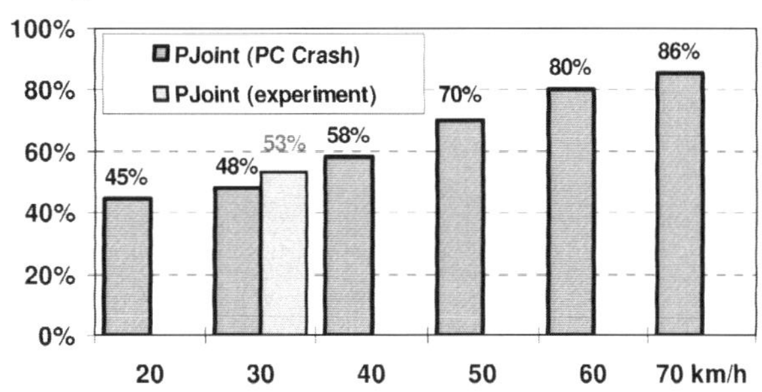

Figure 11. Results of calculations of the P_{Joint} risk indicator; the bar marked green represents the risk indicator value obtained from an experiment with v_0 = 30 km/h.

6. Recapitulation and conclusions

The tests carried out confirmed the fact that the vehicle impact velocity has a strong influence on the risk of severe injuries to minibus passengers and that minibus seats should also be tested at vehicle impact velocities v_0 higher than 40 km/h. An important argument is provided by the injury risk values, which rise from 47.9 % at v_0 = 40 km/h to 85.5 % at v_0 = 70 km/h. Such a growth in the vehicle impact velocity results in an increase in the values of all the injury-generating factors, as specified below:

- head acceleration from 33.2 g to 75.6 g;
- force in the neck from 1.06 kN to 3.31 kN;
- torso acceleration from 21.0 g to 35.7 g;
- force in the lap belt strap from 8.26 kN to 13.81 kN;
- force in the shoulder belt strap from 5.89 kN to 9.83 kN;
- force of head impact against the seat backrest from 0.38 kN to 8.26 kN.

Simultaneously, the following consequences of the growth in the velocity of vehicle impact against an obstacle can be observed:

- The value of time T_{U1} considerably drops, which results in a definite reduction in the time available for a protective action of the passenger safety system, if provided.
- With declining time T_{U1}, the velocity V_{XYZ} of head impact against the backrest is growing.
- An increase in the pre-impact vehicle velocity v_0 causes a growth in the time of duration of the dynamic interaction between the head and the backrest.

The factors described above and the impact load values specified (clearly presented in Figure 8) indicate the necessity of extending the present scope of type approval tests of passenger seats to a vehicle impact velocity of 50 km/h. At such an impact velocity, the values of dynamic loads on the head and neck as observed at present constitute a serious risk of injury to vehicle passengers and this risk should be reduced by improvements in the passengers' personal protection system.

References

[1] Araszewski M, Roenitz E and Toor A 1999 *Maximum Head Displacement of Vehicle Occupants Restrained by Lap and Torso Seat Belts in Frontal Impacts*. SAE **1999-01-0443**

[2] Chimba D, Sando T and Kwigizile V 2010 *Effect of bus size and operation to crash occurrences*. Accident Analysis and Prevention **42** (2010) 2063–2067

[3] Dziewiecki K, Prochowski L and Zielonka K 2015 *Modelling and experimental investigation of the motion of a microbus passenger in the space between seat rows during a road accident*. The 3rd Polish Congress of Mechanics and the 21st International Conference on Computer Methods in Mechanics, Short Papers Vol. **2**, Krakow

[4] Eppinger R, et al, 1999 *Development of Improved Injury Criteria for the Assessment of Advanced Automotive Restraint Systems – II*. NHTSA

[5] Haddadin S, Albu-Schäffer A and Hirzinger G 2007 *Dummy crash-tests for the evaluation of rigid human-robot impacts*. IARP International Workshop on Technical Challenges and for Dependable Robots in Human Environments (IARP2007), Rome

[6] Lawrence G J L 2001 *Study of Improved Safety for Minibuses by Better Seat and Occupant Retention*. The 17th International Technical Conference on the Enhanced Safety of Vehicles, Amsterdam

[7] Martínez L, Vicente T, García A, Alcalá E and Aparicio F 2005 *Analysis of Coaches Rows Seats Distance Influence on the Passengers Comfort and Safety*. The 19th International Technical Conference on the Enhanced Safety of Vehicles, Washington

[8] Mitsuishi H, Sukegawa Y, Okano S, Nagase H 2003 *Frontal Collision Safety of Bus Passengers in Japan*. The 18th International Technical Conference on the Enhanced Safety of Vehicles, Nagoya

[9] National Highway Traffic Safety Administration 2008 *New Car Assessment Program.* Federal Register Vol. **73**, No. 134

[10] Prochowski L, Fitas M and Zielonka K 2014 *Analysing the results of research on the influence of the position of a seat belt on the movements and dynamic loads of child's head and torso during a bus impact against an obstacle.* The Archives of Automotive Engineering – Archiwum Motoryzacji, Vol. **66**, Warsaw

[11] Prochowski L and Zielonka K 2010 *Metody badań i wyniki pomiarów obciążeń dynamicznych pasażerów podczas uderzenia autobusu w przeszkodę (Test methods and results of measurements of dynamic occupant loads during a bus in frontal collision).* Zeszyty Naukowe Instytutu Pojazdów, Vol. **1/2010**, Politechnika Warszawska (Warsaw University of Technology), Warsaw

[12] Prochowski L and Żuchowski A 2014 *Analysis of the influence of passenger position in a car on a risk of injuries during a car accident.* Eksploatacja i Niezawodnosc – Maintenance and Reliability, Vol. **16**, no. 3

[13] Steffan H 2000 *Validation of the Coupled PC-CRASH – MADYMO Occupant Simulation Model.* SAE **2000-01-0471**

[14] Wach W 2009 *Symulacja wypadków drogowych w programie PC Crash (Simulation of Road Accidents in the PC Crash Program).* Instytut Ekspertyz Sądowych (Institute of Forensic Research), Krakow

[15] Witte M and Mousel, T, Bauer H 2008 *Car Interior Sensing Environment for Adaptive Individual Safety.* Congress FISITA, F2008-08-117

[16] Żuchowski A, Prochowski L, Muszyński A and Zielonka K 2015 *Obciążenia dynamiczne i ryzyko obrażeń głowy i szyi osób uczestniczących w wypadkach drogowych (Dynamic loads and head and neck injury risk in persons involved in road accidents).* In a monograph by Torbicz, W. (ed.): Inżynieria Biomedyczna – Podstawy i Zastosowania (Biomedical Engineering – Fundamentals and Applications), Polish Academy of Sciences, Vol. **3**, Akademicka Oficyna Wydawnicza EXIT

[17] Żuchowski A, Prochowski L and Zielonka K 2014 *Ocena ryzyka obrażeń pasażerów podczas wypadku drogowego mikrobusu jako kryterium doskonalenia konstrukcji foteli i ich rozmieszczenia (Evaluation of the risk of passenger injuries during a road accident of a minibus as a criterion of improvement of the design and deployment of bus seats).* Logistyka **3/2014**, 7166-7174

IOP Conf. Series: Materials Science and Engineering **148** (2016) 012025 doi:10.1088/1757-899X/148/1/012025

A portable wheel tester for tyre-road friction and rolling resistance determination

J Pytka[1], P Budzyński[1], P Tarkowski[1] and M Piaskowski[1]

[1]Lublin University of Technology, Institute of Automotive Engineering, Nadbystrzycka 36, 20-618 Lublin, Poland

E-mail: j.pytka@pollub.pl

Abstract. The paper describes theory of operation, design and construction as well as results from primarily experiments with a portable wheel tester that has been developed by the authors as a device for on-site determination of tyre-road braking/driving friction and rolling resistance. The paper includes schematics, drawings, descriptions as well as graphical results form early tests with the presented device. It is expected that the tester can be useful in road accident reconstruction applications as well as in vehicle dynamics research.

1. Introduction

Knowledge of the wheel-surface forces is essential for vehicle dynamics studies, vehicle design, accident reconstruction, road construction and maintenance. The μ parameter, a non-dimensional longitudinal surface reaction on a driving wheel, is a basic factor describing vehicle handling dynamics and, consequently, road safety. Braking distance, forward acceleration and lateral dynamics measures are directly dependent upon the μ parameter. Similarly, the f_{RR}, coefficient of rolling resistance plays a significant role in vehicle dynamics analysis as a measure that describes energy consumption by a rolling wheel. Both the parameters, μ and f_{RR} can be determined experimentally with the use of several devices.

Methods and devices for wheel-surface forces determination include the following [2 – 7]:

- devices for sideway force coefficient determination, in which a test wheel is inclined and rolls freely, without slip, for example SCRIM Tester, α=20deg, s=34%, Mu-Meter, α=7.5deg, s = 13%;
- devices for fixed-slip, longitudinal wheel force determination, the test wheel is straight positioned and the slip percentage is set up for a given value, for example: Griptester, s=14%, DWW Trailer, s=86%, Skiddometer BV-11, s=18%, Norsemeter, s=20% ;
- devices for variable-slip measurements of wheel-road friction, examples: Norsemeter, s=5–95 %, Petra, s=0–100 %;
- devices for locked wheel measurements, s=100 %, LPCP Adhera, Skiddometer BV-8, Stuttgarter Reibungstester, ASTM E-274, Skid Resistance Tester SRT-3 [13].

Besides the over-mentioned methods and devices there are numerous laboratory or field test stands, as for example multi sensor measuring head for three-dimensional wheel-surface forces determination, drum testers or the so called dynamometric trailer for wheel force determination in road or off-road condition. One method utilizes an instrumented vehicle with a rotating wheel force transducer that enables three-dimensional wheel force measurements. The measuring device, rotating wheel dynamometer was built with strain gage force transducers and utilizes a RF, non-contact data transfer. The device was used in multiple off-road tests.

Figure 1. STR-3 trailer-type wheel tester.

Another type of measuring equipment for surface friction determination are portable friction measuring devices, which are used mainly by road construction or maintenance engineers. They include:

- British Pendulum Number device, which enables to determine the BPN factor by simple drop test of a rubber–coated test skidder, supported on a bearing;
- Dynamic Friction Tester with a rotating skidder to determine friction at a speed between 5 and 90km/h;
- VTI Grip Tester, a device in a form of a portable wheel trailer, operated by a walking technician, enables to determine surface friction at 25% wheel slip (see figure 2);
- Drag-Sled Tester, handy operated, enables to determine surface friction in a simple pull or push test, they utilize either external dynamometers or built-in electronic force measuring instruments (TORTUS 3).

Figure 2. VTI Grip Tester.

The aim of this study was to develop a portable wheel tester that will be easy in operation and allow to determine both braking friction and rolling resistance coefficients. Our goal was also to develop a RF non-contact microprocessor based device for data acquisition and transfer to any available portable device (a smartphone or tablet computer). Finally, we also have developed a custom specialized software which supports the field measurements and helps in determining μ and f_{RR} values online.

2. Design of the portable wheel tester

2.1. Mechanical design of the tester

The tester is a three-wheeled portable device designed for easy and quick operation over a variety of surfaces, road (asphalt, bitumen, Belgian pavement, etc.) and off-road (grass, soft soil of smooth surface). In the design, two strain gage type transducers have been used to measure horizontal force acting on the axis of a test wheel. The front test wheel together with two rear wheels are of 150mm diameter, they have pneumatic tyres with air valves in order to maintain inflation pressure. The main frame of the tester supports rear wheels, which rolls freely and a suspension of the test wheel. The suspension (shown in figure 3) integrates the test wheel and its axis, brake disc bearings and their housings, which are mounted to the strain gage transducers. The suspension functions so that any action (a force or a torque moment) on the test wheel in horizontal plane is reduced to pure longitudinal forces acting 50/50% on the two transducers. The bearings that support the test wheel axis are rocker type to compensate any nonlinearities that could be the result of assembly as well as the measuring action.

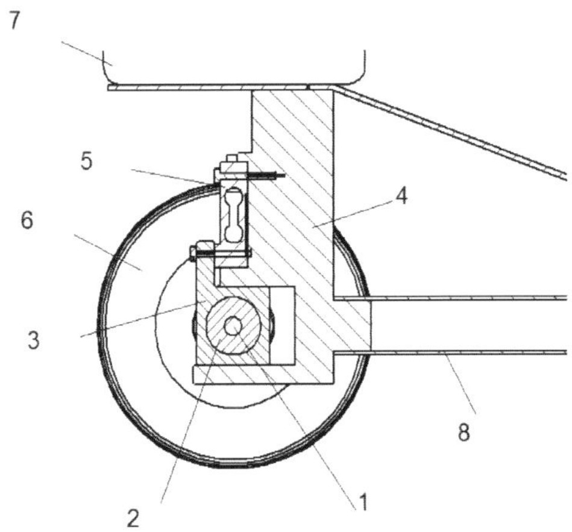

Figure 3. A schematic of the test wheel suspension in the portable wheel tester. 1 – wheel axis, 2 – ball bearing, 3 – bearing housing, 4 – suspension frame, 5 – strain gage sensor, 6 – test wheel, 7 – ballast.

2.1.1. Tyre size effect

In order to reconstruct wheel-surface interactions, an effect of tyre size (width, diameter) has to be considered. Regarding the contact pressure as a parameter describing the wheel-surface interactions and disregarding the tread effects, contact area and contact pressure for the test wheel under various loads has been determined at a normal inflation pressure. Wheel load – contact pressure pairs have been taken to construct a relationship which is a nomogram to calculate an additional weight that has to be added to load the test wheel for a given tyre. The net weight on the test wheel (without additional ballast) is 3kg. Sport weight-lifting round weights have been used and this solution enables to apply the additional load in steps: 5, 10, 15, 20, 25, 30 and 35kg. Ballast heavier than 50kg can damage the mechanical structure of the tester and may have significant effects on load cells readings. The relationship has shown included in figure 4. Based on this simple model, it is possible to determine additional weight for a given tyre for which we want to determine the μ or f_{RR} value. As an example, for a tyre that generate contact pressure 100 kPa an additional weight of 30kg is needed to reconstruct tyre – surface interactions and determine the values of μ and f_{RR}.

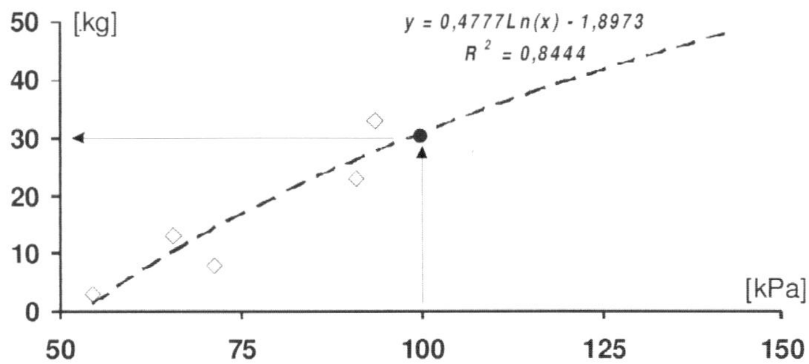

Figure 4. Wheel load – contact pressure relationship for the test wheel for determining the additional ballast weight.

2.2. Microprocessor device for measurements control and data acquisition

The measuring system of the tester has been built with a scale wheel, supported on two load cells, which give readings of both horizontal and vertical forces acting on the wheel. Based on the measured data, rolling and braking friction can be determined. The portable wheel tester wirelessly transfers measured data—horizontal force acting on the test wheel—to a mobile device (a phone or a tablet), and custom data analysis software determines friction coefficients just after the measurements. A RF wireless data transfer system utilizes the Bluetooth IIs technology and it is integrated with a microprocessor control device into a small, battery operated system (shown in figure 5). One can perform measurements wirelessly, with no physical contact with the device. All those features facilitate the operation of the tester and moreover, such a design idea is related to the philosophy of the internet of things.

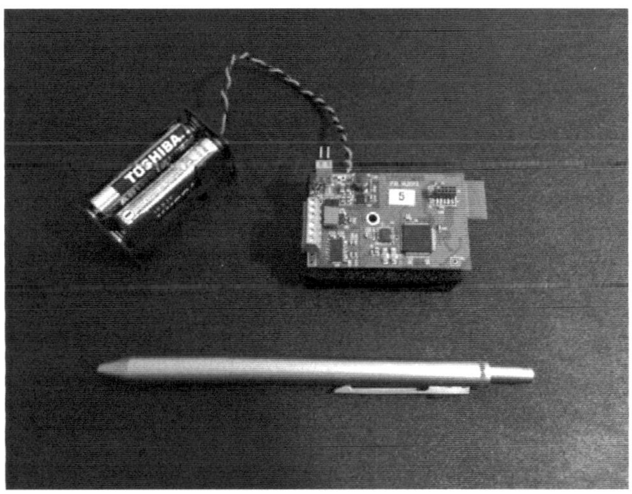

Figure 5. A microprocessor-based electronic control unit for acquisition and wireless transfer of measured data.

2.3. Data analysis software

The software uses an algorithm that allows the choice of a range of My data for averaging and determining the f_{RR}. This is performed by selecting the starting and ending point of the chosen range.

The algorithm also provides a smart tool for searching for My data required for braking friction μ_B determination. The final results, the f_{RR} and μ_B are readily presented on the screen. Figure 6 shows sample readings from the tester with the software activated for surface friction determination.

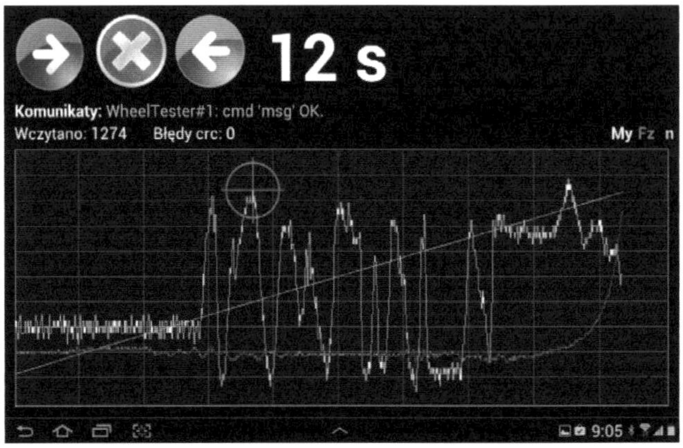

Figure 6. Sample readings from the tester as visualized by the software, together with markings for determining the μ and f_{RR} values online

2.4. *Operating the tester*

The wheel tester operation is very simple. An operator pushes the arm to move the tester forward at a low speed, as seen in figure 7. The push arm is mounted to the main frame of the tester and is movable in the vertical plane in order to fit the level of the grips with the height of the technician for ease operation. During rolling resistance readings, the test wheel rolls freely; for braking friction measurements, a hand-operated brake is activated to block the test wheel. An engineer, standing nearby in the range of the RF device can see the readings online. Using a touch screen of a portable device with installed data analysis software and active Bluetooth RF transceiver, the engineer chooses a portion of data obtained during a free rolling test for determination of f_{RR} value and the result appears on the screen. There is a possibility of recording the complete data set. For a braking friction test, the two persons have to correlate their procedures. First, the operator of the tester begins to push the tester with a typical forward speed. At a moment known for the analysis engineer, the operator brakes the test wheel and continues to push the tester for 2-3 seconds. The engineer can read the data online and has to choose a peak value of the longitudinal force on the touch screen. The software performs calculations and the result appears on the screen.

Figure 7. Operating the wheel tester

2.5. Early tests and calibration of the tester

The device has been examined in several test runs, performed on three different surfaces: bitumen, Belgian pavement and grass. Figure 8 includes raw data in form of time courses of horizontal force measured by the tester. This data has been captured onto the memory stick card and this function of the tester is useful in any research oriented measurements. Simply, while not determining the actual values of f_{RR} and μ_B , one may obtain the complete data an use it for further analysis.

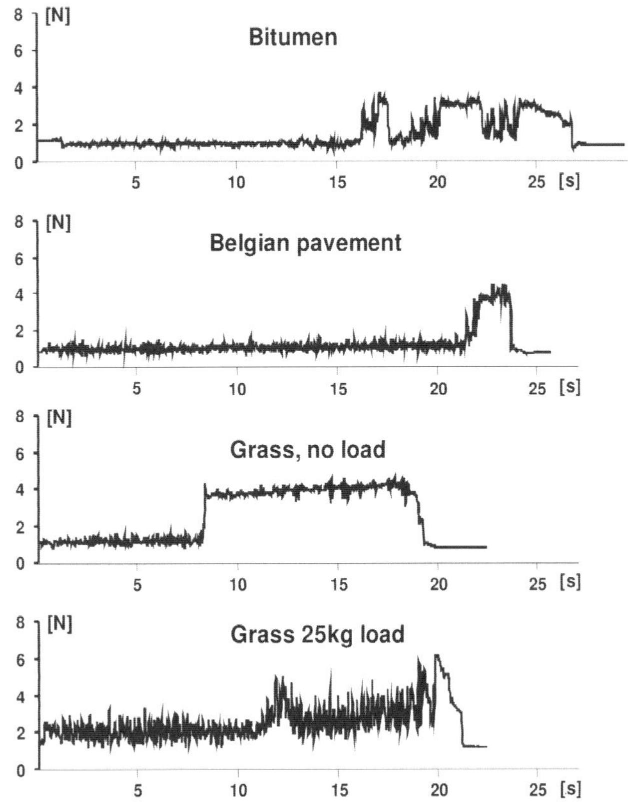

Figure 8. Time courses of the horizontal force obtained for three different surfaces: bitumen, Belgian pavement, grass – no load on the tester, grass – 25kg load (top to bottom)

3. Applications of the tester

Generally, the portable wheel tester can be useful as a supporting device in on-road and off-road vehicle dynamics studies, where knowing an actual value of μ and f_{RR} is important. Applications include vehicle braking and acceleration tests, tyre testing, lateral dynamics and vehicle handling dynamics experiments, etc. Those applications are typical for academia and research institutions and there the ability of determination and recording of raw data (horizontal forces versus time) will be of interest. Another subject, where such a device can be of help is road accident reconstruction. There a quick yet precise determination of actual values of μ is crucial for the procedures. The tester can be transported even in a compact passenger car, so bringing the device to the accident place will not be a problem. The data analysis software can be used in such application as a toll that provides with immediate results. Besides the on-road application, the presented device can be used in terrain conditions to determine μ and f_{RR} coefficients which corresponds to wheel-soil interactions. The general concept and mechanical design of the tester enable to determine braking/driving friction as well as rolling resistance on soft, deformable tractive surfaces such as arable soils, grass, snow, etc. The tester has already been used as a evaluation tool in a method for testing grassy runway surfaces.

For a given grassy airfield, determining of actual μ and f_{RR} values is important to make decision on performing take-offs or landing on this runway, especially after rain, when the grass and surface is wet. The tester has proved its functionality. However, after the tests on a grassy surface it was concluded that for off-road applications the mechanical design of the tester have to be changed. Modification have to be made to the wheel brake, as the prototype disc brake may collide with uneven, rough surface of grass blades during test runs. This drawback can be easily eliminated by the use of a rim brake.

4. Conclusion

We have designed and developed a prototype of a portable wheel tester for determining the μ and f_{RR} values in field experiments. The tester utilizes strain gage type sensor technology together with RF and microprocessor based electronic for data acquisition. There is a possibility to determine wheel friction and rolling resistance online and a specialized custom software has been developed to support this functionality. Application of the tester include vehicle dynamics testing, both on- and off-road, road accident reconstruction and testing and evaluation of grassy runway surface. At the current stage of the project, the tester can reconstruct wheel-surface conditions for tyres inflated to approx. 120kPa. For higher inflation pressures, it is needed to modify the mechanical design of the device. Further development of the tester will also include modification of the brake system for off-road conditions.

References

[1] Currey N S 1992 *Aircraft Landing Gear Design: Principles and Practices* (Washington, USA)
[2] Grzesikiewicz W., Pokorski J., Szwabik B: Modelowanie i badania eksperymentalne przyczepności hamowanego koła *Przegląd Mechaniczny*, Rok Wyd. LXII, Zeszyt 10/2003
[3] Grzesikiewicz W and Pokorski J 2005 *Modelowanie i symulacja cyfrowa zestawu pomiarowego do wyznaczania charakterystyk przyczepności opon samochodowych i nawierzchni drogowych.*X Międzynarodowa Konferencja Naukowa nt.: Badania Symulacyjne w Technice Samochodowej. Kazimierz Dolny
[4] Luty W, Prochowski L and Z. Szurkowski 1999 *Stanowisko do badań ogumienia dużego rozmiaru.* VII Międzynarodowe Sympozjum IPM. Warszawa-Rynia, 8-10.12.1999
[5] Luty W and Wysocki T 1999 *Przyczepa dynamometryczna do badań dynamicznych ogumienia.* VII Międzynarodowe Sympozjum IPM Warszawa-Rynia, 8-10.12.1999
[6] Luty W and Prochowski L 2002 Modelowanie charakterystyk przyczepności ogumienia samochodów ciężarowych. *Zeszyty Naukowe Instytutu Pojazdów* 1(44)/2002 Politechnika Warszawska, 2002r
[7] Prochowski L and Żuchowski A 2001 Wyznaczanie charakterystyk przyczepności ogumienia podczas krótkotrwałego hamowania koła samochodu ciężarowego. PWN *Archiwum Motoryzacji* nr 4/2001
[8] Pytka J, Tarkowski P, Fijałkowski S, Budzyński P, Dąbrowski J, Kupicz W and Pytka P 2011 *J of Terramechanics* **48** 384-395
[9] Pytka J, Tarkowski P, Kaznowski Ł and Piaskowski M 2014 Metoda i urządzenia do oceny stanu nawierzchni lotniska trawiastego *PAK Pomiary Automatyka Kontrola*
[10] Pytka J 2014 *J of Aircraft*, **51**, 353-360
[11] Rafalski L, Godlewski D and Czarnecki K 1993 Porównanie metod pomiarowych współczynnika przyczepności i tekstury nawierzchni - eksperyment międzynarodowy. *Drogownictwo* 5/1993
[12] Szwabik B and Pokorski J 1996 Przesłanki teoretyczne a techniczne środki pomiaru współczynnika przyczepności nawierzchni drogowych. *Konstrukcja, badania, eksploatacja, technologia pojazdów samochodowych i silników spalinowych*, Zeszyt 8, Polska Akademia Nauk, Kraków
[13] Szwabik B, Pokorski J and Hanula W 1997 *Zestawy pomiarowe SRT-3. Podstawowe urządzenia do badania przyczepności nawierzchni drogowych w Polsce.* Seminarium naukowo-

techniczne „Diagnostyka i ocena stanu dróg". Polska Akademia Nauk, Politechnika Szczecińska, 1997

[14] Tarkowski P, Pytka J, Budzyński P and Kaznowski Ł *Maintenance &Reliability (Eksploatacja i Niezawodnosc)* **15** 272-277

IOP Conf. Series: Materials Science and Engineering **148** (2016) 012026 doi:10.1088/1757-899X/148/1/012026

Design and Calibration of Rolling Resistance Test Trailer R^2Mk.2

G Ronowski[1]

[1]Technical University of Gdańsk, Gdańsk, Poland

E-mail: gronowsk@pg.gda.pl

Abstract. The paper describes construction and calibration procedures of the rolling resistance test trailer R2 Mk.2. The trailer is design to measure rolling resistance of passenger car tyres in various road conditions on trafficked roads. The trailer utilizes so called angle method also known as vertical arm method. The paper presents also calibration procedures that are necessary to ascertain good precision of the measurements.

1. Introduction

For many years different research institutions and companies carry out research focused on savings in energy consumption. This also applies to a public road transport. To reduce energy consumption of road vehicles it is necessary to reduce restrictive forces acting on vehicles during their motion. One of such a force is tyre rolling resistance that is especially important during driving with constant, moderate speed. It is common practice to describe rolling resistance as Coefficient of Rolling Resistance (C_{RR}) defined as ration of the Rolling Resistance Force F_{RR} and Vertical Load acting on the tyre (Q):

$$C_{RR} = F_{RR}/Q \tag{1}$$

Figure 1. Roadwheel facility at the Technical University of Gdańsk equipped with replica road surfaces.

As the Rolling Resistance Force is directly proportional to the Coefficient of Rolling Resistance and tyre Vertical Load there are two ways to reduce this force - making the vehicle lighter and reducing CRR. Reduction of the tyre load is out of the scope of this paper so it will concentrate on the measurements of the Coefficient of Rolling Resistance, as the ability to measure is a basic problem in each optimization process. CRR is related to many factors like properties of the tyre, road surface, load, inflation pressure and speed. Road surface characteristics play very important role in tyre energy losses [1, 2] thus it is necessary to account for them. For many years laboratory methods of rolling resistance measurements are utilized especially by tyre manufacturers, but usually they can't account for road pavement properties. The reason is that on outer drums it is not possible to mount real road surfaces due to their inadequate strength. At the best some kind of epoxy resin or polyurethane based replica road surfaces may be used - see Fig. 1. One may say that laboratory methods of rolling resistance measurements are already well established although they are lacking good representability. In contrast, the measurements of rolling resistance in road conditions poses many problems. Methods of measuring rolling resistance by towing car or coast-down are not accurate enough, and for practical reasons not suitable for use in road conditions.

For many years several research centres in Europe work on the development of effective methods of measuring rolling resistance in real road conditions. One of these centres is the Technical University of Gdańsk (TUG) that developed and constructed test trailer R^2 Mk.2 designed to measure CRR on the road. It is one of five known trailers specially designed for this type of measurements, wherein two of them were constructed and used in TUG.

In Poland, in late seventies, successful test trailer SRT-3 for skid resistance measurements was developed. Although the trailer was designed to measure tyre traction (skid resistance), it was possible also to use it for rough estimation of tyre rolling resistance. Unfortunately in order to precisely measure tyre rolling resistance it is necessary to construct specialized facilities that are immune to air drag and inertia forces as well as road grade so use of "universal" trailers is very limited.

2. Design of the test trailer

During 2012 and 2013 TUG's Faculty of Mechanical Engineering has built an enhanced version of the R^2 trailer (that was built a few years earlier) to study the rolling resistance. The trailer was named R^2Mk.2 (see Fig. 2). The idea behind the design was retained from the first trailer R^2, but a lot of improvements in design and equipment were introduced to the Mk. 2 version [3].

Trailer R^2 Mk.2 has three wheels. Two front wheels are used to stabilize it on the road while the third wheel is a test wheel and runs between wheel tracks of the front wheels. The trailer may accomodate test wheels of diameter 570 - 730 mm and width up to 245 mm.

Figure 2. Rolling Resistance test trailer R^2 Mk.2.

Both trailers of R^2 series use "*vertical arm principle*" described in Fig. 3. Rolling resistance force acting on the test wheel is transferred to the wheel hub located on a vertical swinging arm supported at the bottom end on the axle connected to the main frame of the trailer. The force acting on the arm produces momentum that deflects the arm in backward direction. The angle of deflection is measured by a laser system. To compensate for the influence of acceleration and longitudinal slope of the road surface a special counterbalance system is used (see Fig. 4). This system is patented by TUG. The trailer is loaded by a certain mass supported on a spring and shock absorber. As the angle of the vertical arm deflection is measured in relation to the frame being the reference (not to true vertical direction), it is necessary to provide information about the frame position related to the pavement and horizontal plane. To do so, two laser sensors measure and compare distance from the pavement at the front and rear of the frame. The trailer is also equipped with a system evaluating road grade operating on barometric principle. Interior of the trailer is presented in Fig. 5.

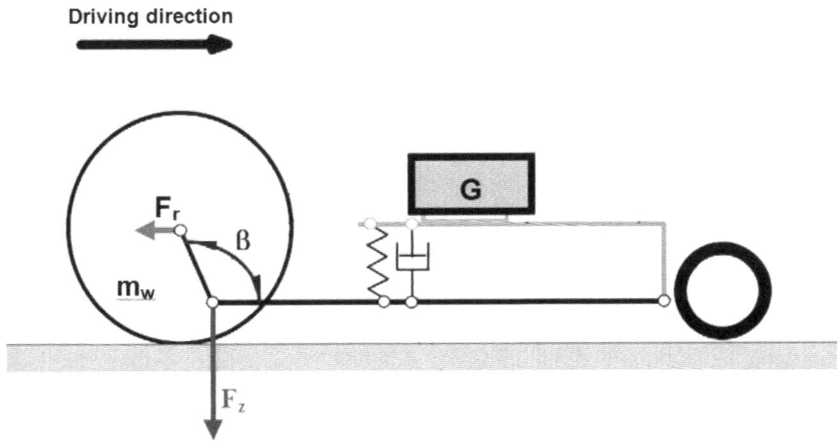

Figure 3. Principle of the vertical arm measuring method.

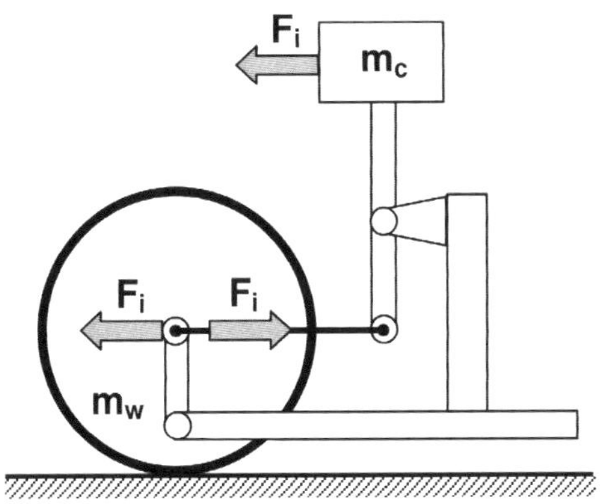

Figure 4. The counterbalance system.

Figure 5. Interior of the R^2 Mk.2 trailer.

3. Calibration of the test trailer R^2 Mk.2

In order to ascertain that test trailer measures rolling resistance accurately it is necessary to perform many calibration procedures. Some of the procedures must be performed only once just after finishing the construction, some others must be performed frequently to check condition of the systems and properly set up the measurements. Most important calibration procedures are described below.

3.1. Tyre load test

Trailer R^2 Mk.2 is equipped with gravity based tyre loading system thus calibration was necessary only during initial testing and after major modifications. During load calibration the test wheel is placed on legalized weighting scale class III with measuring range of 6000 N and accuracy ±3 N - see Fig. 6. In order to ascertain that the trailer is properly levelled two pads are placed under front wheels of the trailer. In years 2014-2015 the test tyre load was 4002 N while in early 2016 it increased to 4150 N due to the trailer's modernization. TUG uses test wheel with tyre P1 [4] weighting 212 N and test wheel with tyre H1 weighting 186 N.

Figure 6. Calibration of R2 Mk.2 trailer - testing tyre load.

3.2. Road grade influence

Trailer R^2 Mk.2 is equipped with patented system that compensates the road grade influence on tyre rolling resistance measurement results. The calibration is performed in order to adjust this system for optimal work. During calibration, the test wheel of the trailer is replaced by the hardened steel disc of 620 mm diameter and weight of 200 N (see Fig. 7.A) in order to eliminate hysteresis attributed to the pneumatic tyre. During the whole calibration the measuring system is subjected to periodical oscillations induced by the rotating shaker presented in Fig. 8. At first the calibration of the system is performed on levelled steel surface plane. Level of the surface is checked with accuracy of 0.5 mm / 1m. Backward acting force of 40 N is applied to the test disc spindle in direction parallel to the road plane. The force is applied via flexible rope and the tilt angle of the measuring arm (g_o) is recorded. After this the graded steel blocks are placed under the steel disc and the front wheels of the trailer. The geometry of the blocks is such that they simulate road grade of 3% (see Fig. 7-B). Trailer is calibrated both for up-hill grade and down-hill grade. During the calibration, for each grade the test disc spindle is subjective to force of 40 N acting parallel to the road surface and corresponding (g) angle is recorded. It should be the same like in the case of no grade conditions (g_o). If there is difference between the tilt on the graded and horizontal road surface the adjustments of the compensation mechanism are made and the calibration is repeated.

a) b)

Figure 7. Calibration of R^2 Mk.2 trailer - testing grade influence; (a) - steel disc resting on tilted steel block, (b) - pads under front wheels to control the tilt.

Figure 8. Rotating shaker and angle *g*.

3.3. Acceleration/deceleration influence

TUG's trailer R^2 Mk.2 is equipped with a mechanical system compensating for road grade and acceleration/deceleration. Unfortunately, with this system it is not possible to fully correct the road grade and acceleration/deceleration at the same time as grade acts only on mass, while acceleration acts on mass and rotational inertia (as a moment of inertia). In the case of R^2 Mk.2 trailer it was decided to compensate for the road grade as much as possible and accept certain uncompensated influence of acceleration/deceleration. Calibration of the R^2 Mk.2 trailer in respect of the acceleration/deceleration is based on the rolling resistance measurements performed on uniform, levelled, good quality road surface when driving with intentional acceleration and deceleration. The tests are composed with three phases. During the first phase test vehicle is driving with fairly constant acceleration as close to 0.4 m/s^2 as possible starting with speed of 40 km/h. When speed of 50 km/h is reached the vehicle drives with constant speed for at least 200 m (phase two) and after this the phase three (deceleration as close to -0.4 m/s^2 as possible) is performed.

In Fig. 9 results of calibration performed for R^2 Mk.2 trailer are presented. In the shown example the regression line CRR=0.0034*a+0.0088 describes influence of acceleration/deceleration on the measured results. Based on the slope and intercept of this line it may be calculated that for acceleration of 0.02 m/s2 for tested tyre and road surface (SMA8) the measured CRR = 0.008868 while for deceleration -0.02 m/s2 the measured CRR=0.008732. This corresponds to 1.5% difference in CRR results.

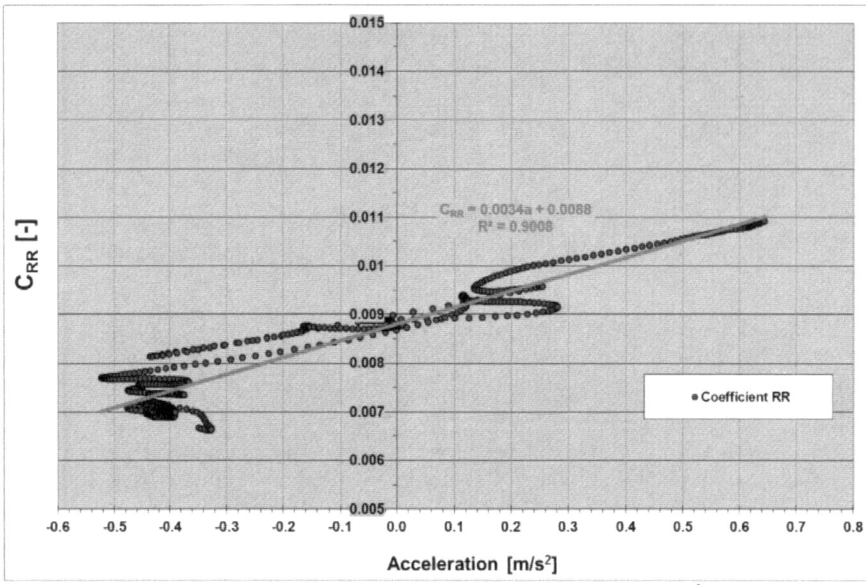

Figure 9. Results of acceleration/deceleration calibration of R^2 Mk.2 trailer.

3.4. Calibration of the temperature sensors

In the trailer R^2 Mk.2 air temperature is measured by the thermistor sensor Pt100 while road and tyre temperatures are measured by IR sensors. Calibration involves comparing the indications of the sensors with reference temperature measured by system with thermocouple (Fluke 52 II, accuracy 0.3°C) - see Fig. 10.

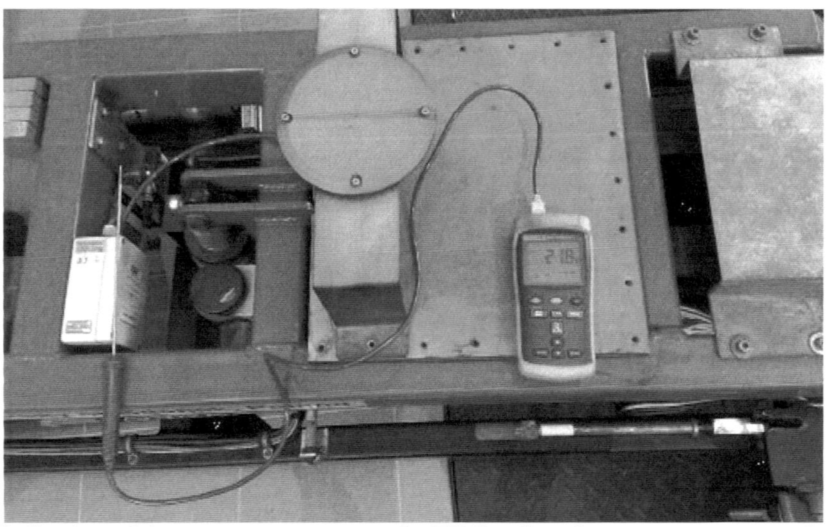

Figure 10. Testing of the temperature sensors (in TUG trailer air temperature sensor is mounted outside the wind protective chamber).

During periodic calibrations the trailer is stored for at least 5 hours in the room with constant temperature of 20±2°C and the comparison of indications is made. To check road temperature sensor mounted at the trailer a piece of asphalt concrete surface having ambient temperature is placed in the beam of the sensor. This ascertains that calibration accounts for typical surface emissivity. If necessary the sensors are readjusted. Checking of calibration at 10±2°C and 30±2°C was done only once, when the system was initially tested.

3.5. Calibration of tyre inflation sensor

TUG's trailer R^2 Mk.2 is equipped with two inflation pressure sensors. The first sensor sends data to the data recorder and control display while the second one is used by the system automatically adjusting the inflation pressure, thus the measurements may be performed in two modes - capped and regulated inflation pressure. Calibration of the system is performed by comparing readings from the first sensor with readings from certified manometer of accuracy 0.6%. During initial calibration the readings were compared for inflation pressures from 150 kPa to 250 kPa (with increment of 10 kPa). During periodical checks calibration is made only for 210 kPa.

3.6. Tyre size influence

During calibration the test wheel of the TUG's trailer R^2 Mk.2 is replaced by a hardened steel disc having sectors of following radiuses: 300, 310 and 320 mm (see Fig. 11). The calibration plane is determined by three steel plates (positioned under the test and front wheels) set with an accuracy of 0.5 mm per 1 m. The steel on steel interface eliminates hysteresis of the tyre and ascertains good accuracy of the calibration. During whole calibration the measuring system is subjected to periodical oscillations induced by the rotating shaker presented in Fig. 8.

After levelling of the trailer the backward acting force of 40 N is applied to the test disc spindle in direction parallel to the road plane. The force is applied via flexible rope and the tilt angle of the measuring arm (g_o) is recorded. The measurements are repeated for all three radiuses of the steel disc. During the measurements the distance between test surface and front and rear of the trailer is also recorded by on-board laser sensors. Resulting data are used to calculate influence of the test wheel radius on the measured rolling resistance force. The obtained characteristic is used later during measurements to account for influence of test wheel radius.

Figure 11. Calibration of R^2 Mk.2 trailer - radius of the test wheel influence.

3.7. Calibration of speed measuring equipment

Information about trailer speed is NOT used in TUG's system for corrections of inertia forces thus standard GPS based system is adequate. This system operates without calibration. If there is a drop of satellite signal, the information about speed is taken from sensors evaluating rotational speed of the trailer's supporting wheels. This system self-calibrates when GPS signal is available. The calibration is performed on straight road section 1000 m long.

3.8. Calibration of the rolling resistance force sensor

During calibration the test wheel of the TUG's trailer R^2 Mk.2 is replaced by hardened steel disc of 620 mm diameter (see Fig. 11). The calibration plane is determined by three steel plates (positioned under the test and front wheels) set with an accuracy of 0.5 mm per 1 m. The steel on steel interface eliminates hysteresis of the tyre and ascertains good accuracy of the calibration. During whole calibration the measuring system is subjected to periodical oscillations induced by the rotating shaker presented in Fig. 8.

After levelling of the trailer the backward acting force is applied to the test disc spindle in direction parallel to the road plane. The force is applied via flexible rope. The force increases from zero up to 110 N and then decreases back to zero. Readings of the measuring system are recorded and compared with known forces applied to the test wheel spindle via gravitationally loaded system (see Fig. 12). In Fig. 13 relation between tilt of the measuring arm and horizontal load applied to the test wheel spindle is presented. In Fig. 14 relation between measured and calibration force is presented together with the hysteresis loop.

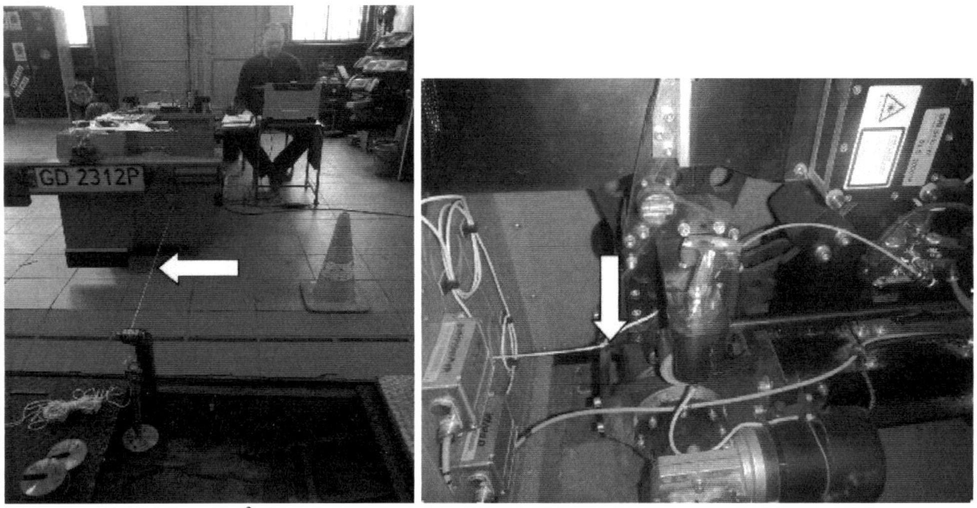

Figure 12. Calibration of R^2 Mk.2 trailer - force calibration (arrows indicate flexible rope pulling the test wheel spindle).

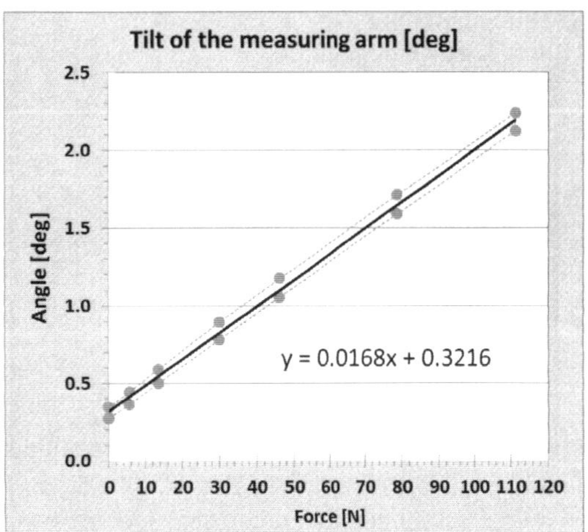

Figure 13. Calibration of R^2 Mk.2 trailer - relation between tilt of the measuring arm and calibration force for steel disc of 620 mm diameter.

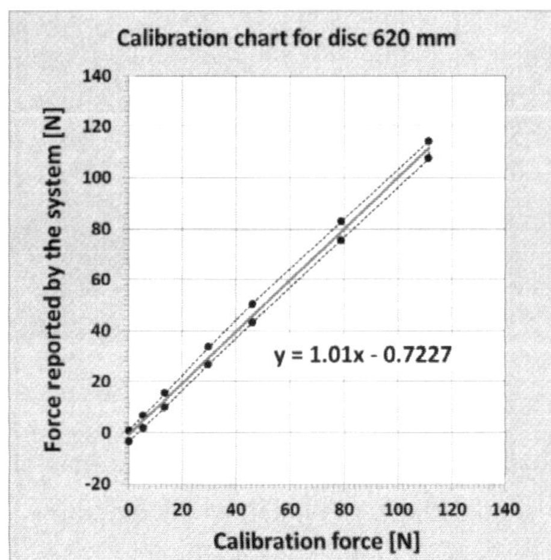

Figure 14. Calibration of R^2 Mk.2 trailer - relation between measured force and calibration force.

3.9. Influence of the rutting

During calibration the test wheel of the TUG's trailer R^2 Mk.2 is replaced by a hardened steel disc of 620 mm diameter (see Fig. 11). The calibration plane is determined by three steel plates (positioned under the test and front wheels). The steel on steel interface eliminates hysteresis of the tyre and ascertains good accuracy of the calibration. During whole calibration the measuring system is subjected to periodical oscillations induced by the rotating shaker presented in Fig. 8. Calibration is perform for situations where steel plate under the test wheel has different heights than plates positioned under the supporting wheels. Following heights differences are tested: -30, -20, -10, 0, 10, 20, 30 mm - see Fig. 15.

The backward acting force of 0, 10, 20, 40, 60, 80 N is applied to the test disc spindle in direction parallel to the road plane. The force is applied via flexible rope. The tilt of the test arm is recorded and on the base of those results the influence of rutting (trailer tilt) is established. During standard measurements tilting of the trailer is constantly monitored by two laser sensors and is used to correct the results.

Figure 15. Calibration of R2 Mk.2 trailer - rutting influence.

4. Example of results obtained by the R^2 Mk.2 trailer

Fig. 16 shows an example of the results obtained by the R^2 Mk.2 trailer. These measurements were performed at a speed of 80 km/h, the tyre load 4150 N and inflation pressure 210 kPa. Test section 1200 m long was divided into two sub-sections: the first covered by road surface SMA 8 and the second covered by experimental surface SMA 6.

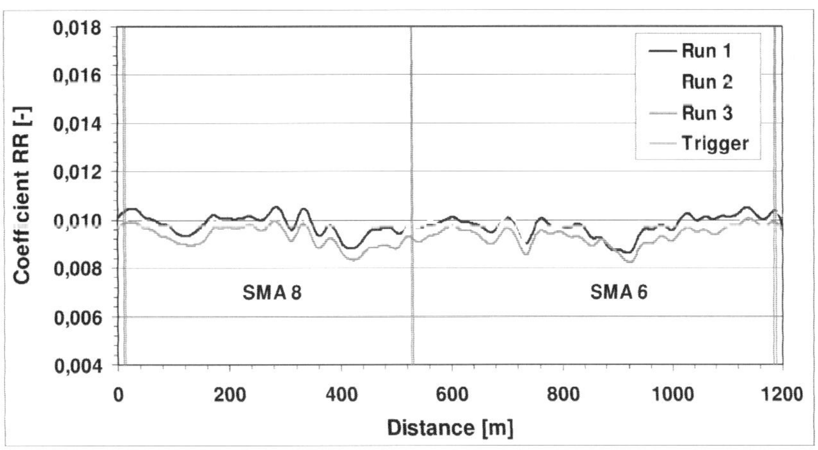

Figure 16. Example of measurement results.

Test tyre P1 (SRTT - UNIROYAL P225 / 60R16 97S M + S Tiger Paw) was used during measurements. Temperatures were as follows: air 20.2°C, road surface 32.9°C and the tyre sidewall 36.3°C. The measurements were repeated three times, one after another. Each run is marked with different colour in Fig. 16. The figure shows very good correlation between each run and the same "details" of the rolling resistance changes along the test sections that are caused by the pavement non-uniformity.

5. Conclusions

The coefficient of rolling resistance of tyres is a very important parameter because of its influence on energy consumption. Equally important is having the right tools to measure this factor in road conditions. Therefore, Technical University of Gdańsk has built a test trailer to study the rolling resistance of tyres in real road conditions. After successful introduction of the first R^2 trailer in 2003 its modified version was developed in 2012-2013. This trailer was a base for rolling resistance measurements in several national and international projects including ROLRES [5], ROSANNE [6] and LEO [7].

Due to the severity and complexity of the trailer, calibration and testing procedures are numerous, complex and time-consuming. Nevertheless, all procedures performed by TUG proofed high quality of the R^2 Mk.2 trailer. The calibration procedures described in this paper are also used in pre-standard developed by the ROSANNE project that will be used in further standardization activities of international bodies. At present R^2 Mk. 2 trailer owned by TUG is the most advanced trailer of this type in the world and is extensively utilized in many European countries as well as in USA.

Acknowledgements

Research work reported in this article was sponsored by the European Commission within 7th FP 'ROSANNE' [Grant Agreement 605368] and Polish National Centre for Research and Development (NCBiR) within Polish-Norwegian Research Programme CORE, project LEO (Grant Agreement 196195 /2013) and SPB project ROLRES (Grant Agreement PBS1/A6/1/2012).

References

[1] Ejsmont J, Ronowski G, Taryma S and Świeczko-Żurek B 2015 *The influence of road surface unevenness on tyre rolling resistance* The Archives of Automotive Engineering vol 70 No. 4

[2] Ejsmont J, Ronowski G, Świeczko-Żurek B and Sommer S 2016 *Road texture influence on tyre rolling resistance* Road Materials and Pavement Design 03/2016 DOI: 10.1080/14680629.2016.1160835

[3] Świeczko-Żurek B, Jaskula P, Ejsmont J, Kędzierska A and Czajkowski P 2016 *Rolling resistance and tyre/road noise on rubberised asphalt pavement in Poland* Road Materials and Pavement Design 03/2016 DOI: 10.1080/14680629.2016.1159245

[4] ISO/TS 11919-3 *Acoustics - Measurement of the influence of road surfaces on traffic noise - Part 3* Reference tyres, the International Organization for Standardization

[5] ROLRES *Wpływ tekstury i równości nawierzchni drogowych na straty energetyczne w oponach i zawieszeniach pojazdów samochodowych* Project PBS1/A6/1/2012, NCBiR, Poland

[6] ROSANNE *Rolling resistance, skid resistance and noise emission measurement standards for road surfaces* European Union's Seventh Framework Programme (FP7/2008–2013) under grant agreement No 605368. Retrieved from www.rosanne-project.eu

[7] LEO *Low Emission Optimised tyres and road surfaces for electric and hybrid vehicles* Polish-Norwegian Research Programme CORE, project "LEO" No. 196195/2013. Retrieved from http://www.leo.mech.pg.gda.pl/

[8] Pokorski J, Szwabik B 2001 *Doświadczalne i obliczeniowe charakterystyki przyczepności opon samochodowych i nawierzchni drogowych* Zeszyty Naukowe Instytutu Pojazdów 3(42)/2001 s.73-84. Politechnika Warszawska

[9] Szwabik B, Pokorski J, Hanula J 1997 *Zestawy pomiarowe SRT-3 – podstawowe urządzenia do badania przyczepności nawierzchni drogowych w Polsce* Seminarium Naukowo-Techniczne „Diagnostyka i ocen stanu dróg". Materiały Konferencyjne s.339-348. Politechnika Szczecińska. Szczecin 3-4 kwietnia 1997

IOP Conf. Series: Materials Science and Engineering **148** (2016) 012027 doi:10.1088/1757-899X/148/1/012027

A statistical analysis of energy and power demand for the tractive purposes of an electric vehicle in urban traffic – an analysis of a short and long observation period

G Slaski[1] and B Ohde[2]
[1] PhD, DSc, Assistant Professor, Poznan University of Technology, Poznan, Poland
[2] PhD student, Poznan University of Technology, Poznan, Poland

E-mail: Grzegorz.Slaski@put.poznan.pl

Abstract. The article presents the results of a statistical dispersion analysis of an energy and power demand for tractive purposes of a battery electric vehicle. The authors compare data distribution for different values of an average speed in two approaches, namely a short and long period of observation. The short period of observation (generally around several hundred meters) results from a previously proposed macroscopic energy consumption model based on an average speed per road section. This approach yielded high values of standard deviation and coefficient of variation (the ratio between standard deviation and the mean) around 0.7-1.2. The long period of observation (about several kilometers long) is similar in length to standardized speed cycles used in testing a vehicle energy consumption and available range. The data were analysed to determine the impact of observation length on the energy and power demand variation. The analysis was based on a simulation of electric power and energy consumption performed with speed profiles data recorded in Poznan agglomeration.

1. Introduction

Nowadays attention is paid to minimizing fuel and energy consumption of vehicles. For battery electric vehicles the question of finding the route optimized for the lowest energy consumption is especially important due to their low range and the resulting range anxiety. Vehicle power and energy demand for tractive purposes depends on speed and resistance forces. The values of both these variables are random by nature and hence a wide dispersion is possible. They are affected by a random speed profile due to urban traffic nature caused by various interactions between vehicles in traffic, the infrastructure, driving style and traffic lights.

As the energy is an integral of a power history over time, and when the time is quite long, the variability of energy is much smaller than the variability of the value of instantaneous power demand. Due to random traffic conditions special standardized speed profiles are usually applied to estimate the average energy consumption for ICE and EV vehicles to enable comparison between various vehicles. Such standardized profiles simulate in laboratory conditions a type of driving in a city traffic and are several kilometers long.

The authors of this article previously suggested a macroscopic model for estimating the average power demand and energy consumption ("Tank-to-Wheel") over a short road section (several dozen meters to a hundred meters) based on the average speed of a vehicle over this road section [1]. The model was

IOP Conf. Series: Materials Science and Engineering **148** (2016) 012027 doi:10.1088/1757-899X/148/1/012027

based on statistical data obtained from real speed data profiles and the microscopic energy consumption model calculated for short periods of time and distance traveled.

The fact that the same average speed value can result from different speed profiles poses w problem while building a model in this manner. It also influences the average power demand and average energy consumption values and one can find that the same average speed results on different levels of energy consumption (figure 1) or average power demand.

With this data it is possible to find equations with adequate parameter values to approximate these statistical data [2, 3]. If only aggregated statistical data is used, e.g. the average speed and average power demand or energy consumption, the models extrapolating this data may yield highly inaccurate results as it is difficult to comply with speed variation inside the observation period (in the case of traffic flow simulation: simulation step). Thus, due to this random nature of power demand and energy consumption data, questions arise if also power demand and energy consumption values variability can be taken into account. The authors assumed that it will be useful to consider an additional value (for example a standard deviation) to asses and model energy consumption with different levels of probability.

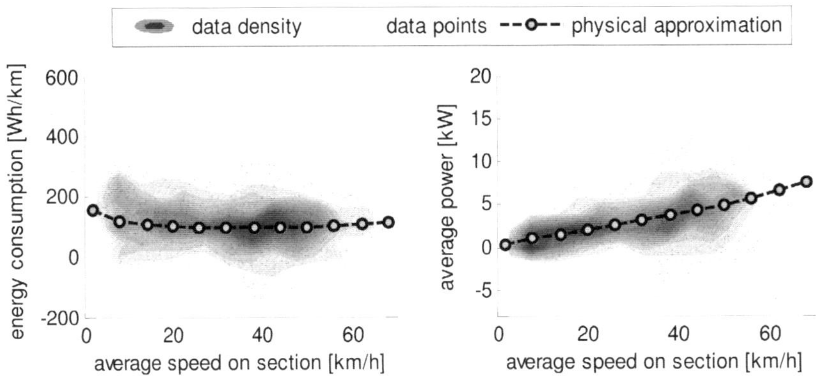

Figure 1. Specific energy consumption per route section calculated for speed profiles recorded in Poznan agglomeration.

2. Power and energy consumption data and models

The models the authors suggested in previous works were an approximation of the data presented in figure 1. The data of power demand and energy consumption was obtained with the simulation of Backward-Facing Model of longitudinal dynamics of battery electric vehicle Nissan LEAF with speed profiles recorded during everyday traffic used as an input. Speed profiles were recorded in Poznan agglomeration during normal, everyday driving performed by four drivers. The recording was carried out with a GPS receiver and data logger. Vehicle parameters were adjusted to represent the first generation of Nissan LEAF battery electric vehicle and energy consumption values observed during standardized tests [4, 5]and reported by users [6].

It is important to note that by energy consumption and power demand the authors mean the energy and power supplied from the battery to the inverter for traction purposes (Fig. 2).

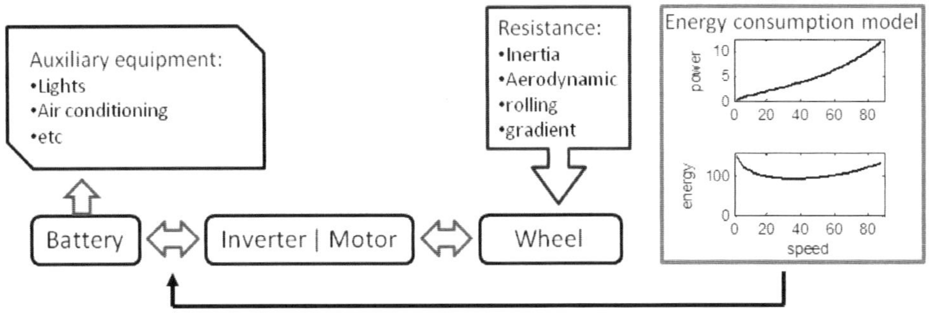

Figure 2. Flow of energy in a battery electric vehicle and a point of interest of this article supplemented with the presented model of energy consumption.

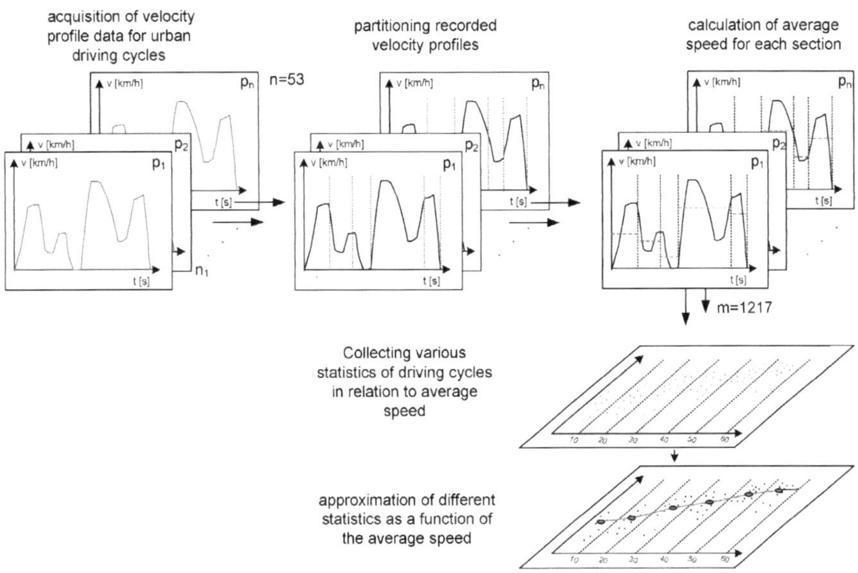

Figure 3. Driving cycle data processing procedure.

The following methodology was used to obtain the average energy consumption data in relation to average speed (Fig. 3):

a) power demand data and energy consumption data were modelled with a microscopic simulation of Backward-Facing Model [7]
b) the trips resulting in a several kilometers long speed, power and energy consumptions time histories were divided into much shorter route sections (links between nodes) adequately to time points at which the car was in a road network node (all junctions),
c) statistical parameters: the average speed, average acceleration and deceleration, average power, and energy consumption were automatically calculated for every section with specially programmed procedures in Matlab,
d) the average values obtained were ranked according to the speed intervals they belonged to, which were established with a constant width of 5 km/h from 0 to 70 km/h
e) for every average speed interval one mean average speed and mean average variable were calculated with the points of relationship between the average speed over the section and the average power demand and average energy consumption per unit distance
f) the obtained relationship for power demand was approximated with a function based on a sum of motion resistance:

$$P_{el} = \frac{P_{roll}+P_{aero}+P_{inertia}}{\mu} \quad (1)$$

where: P_{el} is electric power supplied from battery for tractive purposes, P_{roll} is rolling resistance power, P_{aero} is aerodynamic resistance power, $P_{inertia}$ is inertia resistance power and μ is battery to wheel efficiency of electric drive.

The user of this model then needs to know the average speed over every route section (link). Then it is possible to estimate the average power demand and later to calculate the energy used over the section by multiplying the average power demand and time spent in this section (fig. 4). It is a good method also to include the energy consumed by different auxiliaries. The consumed energy can also be calculated by multiplying energy consumption and the distance unit and the section distance.

The model obtained after the approximation is deterministic and predicts the energy consumption with 50% of probability that a real energy consumption will be lower than or equal to the one predicted. To provide tools to comply with higher values of probability the authors analyzed the energy consumption variability.

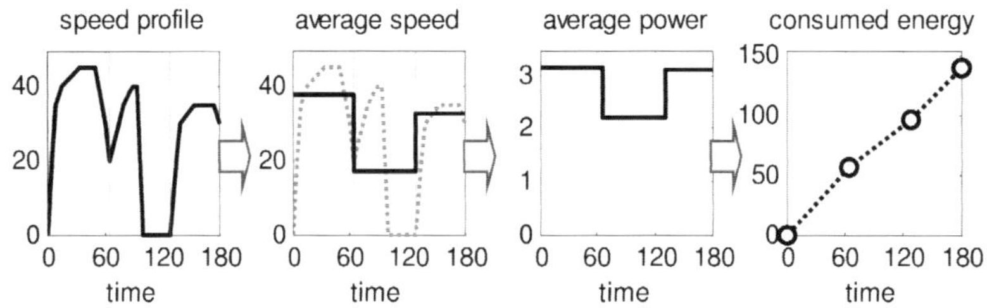

Figure 4. The idea of proposed macroscopic energy consumption model. Speed data are available only as average speed over section of road, based on this information average tractive power is estimated. Value of average power is multiplied by time spend on section to calculate energy consumed on this section.

2.1. *Long and short observation period*

As the nature of road infrastructure results in big differences in section (link) lengths, the authors decided to compare the average power demand and energy consumption per distance unit obtained from the whole trip analysis and from the proposed model.

2.1.1.Long observation period data.

In the article a long observation period was represented by the length of the whole trip. The average length of a single trip recorded in Poznan agglomeration was around 7 km, the shortest trip was only 1.44 km long, while the longest was 13.54 km long. The average number of sections per trip was 23.4, the lowest was 6, and the highest was 61. The mean average speed of all trips was 24.55 km/h, with a standard deviation of average speed equal 6.06 km/h.

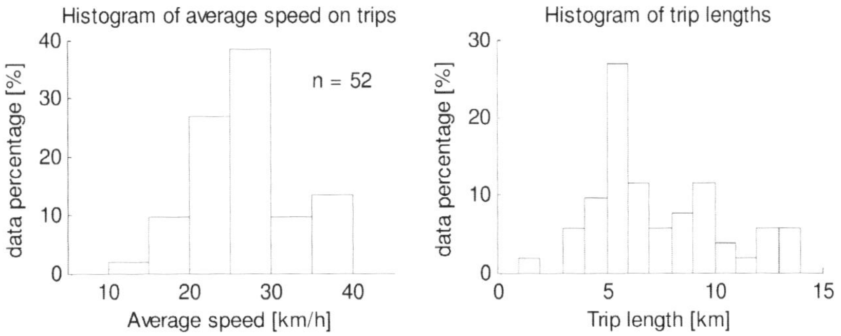

Figure 5. Composition of data with long observation period. Each bar represents data within speed interval 5 km/h wide, or trip length interval 1 km wide.

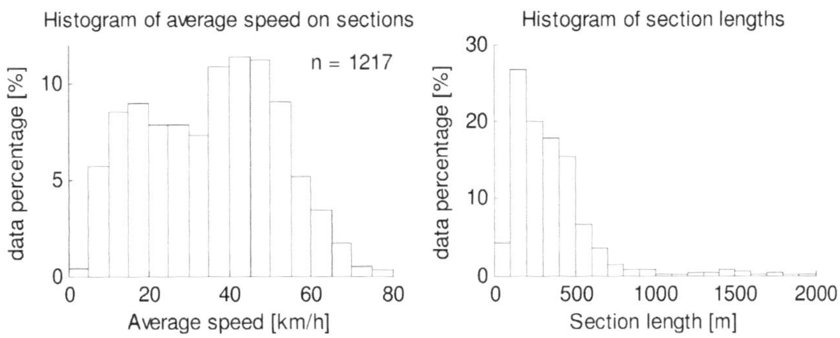

Figure 6. Composition of data with short observation period. Each bar represents data within speed interval 5 km/h wide, or trip length interval 100 m wide.

2.1.2.Short observation period data.

In this article a short period of observation is represented by the length of road section between two intersections. The average length was around 300 m and over 92% of sections were no longer that 600 m. The longest recorded section is slightly over 2.5 km long and the median length was 243 m. Over 30% of sections were shorter than 150 m.

The mean average speed of all sections was 32.7 km/h, with 84% of sections having average speed lower than 50 km/h (legal speed limit for urban areas in Poland).

2.2. Methodology of analysis and interpretation of chosen variables

Speed profiles gathered in Poznan agglomeration were used to calculate consumed energy and average power demand of electric vehicle using methodology presented in section 2 of this paper. The authors analysed the relationship between the average speed of vehicle over a short or long observation period and the average value of power demand or energy consumed in these periods. To analyse data for a short observation period for every section of the route the authors calculated statistical parameters of energy consumption per distance unit and average power demand. In the next step the obtained values were ranked according to average speed intervals they belonged to, established with a constant width of 5 km/h from 0 to 70 km/h. Then statistical parameters were calculated :

g) for every section:

- average value – was calculated as the mean value of the continuous variable $(f(t))$ over time (T):
  - $$X_{avg} = \frac{1}{T}\int_0^T f(t)\,dt \quad (2)$$

h) standard deviation – was calculated with the formula:

- $$s = \left(\frac{1}{T}\int_0^T \left(f(t) - X_{avg}\right)^2 dt\right)^{\frac{1}{2}} \quad (3)$$

i) for every average speed interval:

j) mean value – was calculated as the arithmetic mean of a discrete number (N) of samples (x_i):

- $$X_{mean} = \frac{1}{N}\sum_{i=1}^{N} x_i \quad (4)$$

k) standard deviation – was calculated using the formula:

$$s = \left(\frac{1}{N-1}\sum_{i=1}^{N}(x_i - X_{mean})^2\right)^{\frac{1}{2}} \quad (5)$$

l) Coefficient of variation (CV) – was calculated as a ratio of a standard deviation and the mean or average value. It allows to asses data dispersion. For normal distribution it means that:

m) 68,3, 95,4 or 99,7 % of probability that a sample value is between $mean(1 \pm CV), mean(1 \pm 2\,CV)$ or $mean(1 \pm 3\,CV)$ respectively.

- It also means an 84% or 98% probability that the sample value is lower than $mean(1 + CV)$ or $mean(1 + 2\,CV)$ respectively,

n) Cumulative Distribution Function (CDF) – a function describing probability that any data point (d_i) has value equal or lower than argument (x) of function:

$$CDF(x) = P(d_i \leq x) \quad (6)$$

or it can be interpreted as a percentage of samples having value equal to or lower than x.

3. Results of analysis o statistical dispersion of power and energy variables

Calculated statistical parameters of average energy consumptions per unit distance and average power demand are presented in fig. 7. The level of a mean value of average energy consumptions per unit distance is similar in whole range of examined average speed, in contrast to increasing mean average power demand. Because of this the article mainly focuses on analysis of statistical dispersion of an average energy consumptions per unit distance, while the statistical dispersion of an average power demand was treated as supplementary data.

The values of standard deviation, both for the average energy consumption per unit distance and the average power demand - are quite high, almost equal to the mean value of the analysed variables. For average energy consumption per unit it is between 90 and 160 Wh/km, and for power demand the value rises linearly with speed, which results from power definition as the force multiplied by speed in the direction of the force.

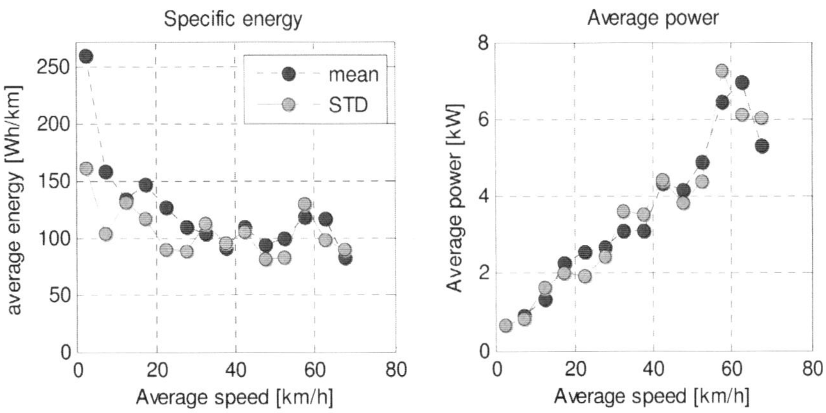

Figure 7. Mean value and standard deviation of average energy consumption per distance unit and average power relative to average speed on section.

3.1. Statistical dispersion of average speed and average energy consumption per distance unit

Statistical dispersion of the average energy consumption per distance unit for a short observation period is very high, with coefficient of variation values for individual speed intervals between 0.6 and 1.1, and the overall value of 0.91 - fig. 8. There is a slight inclination, probably resulting from decreasing mean values with rising speed. The variation coefficient of speed within a section falls from the top value of almost 2 to about 0,1 at speeds higher than 45 km/h.

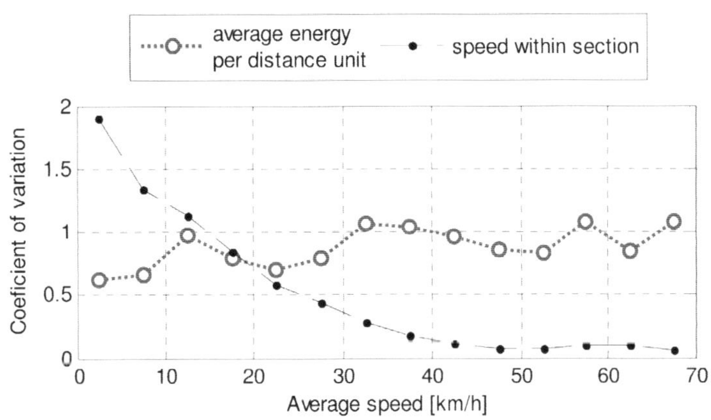

Figure 8. Coefficient of variation of average energy consumption per distance unit energy for short observation period data, supplemented with mean coefficient of variation of speed within section.

Contrary to preliminary hypothesis a statistical dispersion of the average energy consumption per distance unit and power demand (Fig. 10) does not decrease as the average speed increases despite a decreasing coefficient of speed variation on section (i.e. the mean variation -of speed inside an individual section, not the variation between data samples). A statistical dispersion of average energy consumption per distance unit for a long observation period is significantly smaller, with coefficient of variation around 0.15.

Figure 9. shows a significant disproportion between the distribution of data from individual sections and the data for whole trips. Due to the number of data points for individual sections, the distribution is shown by cumulative distribution function (CDF) values calculated for each speed interval.

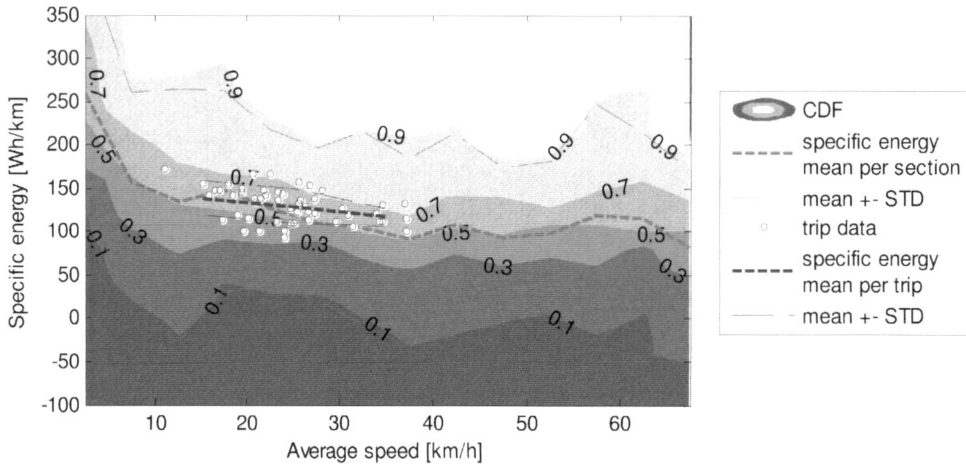

Figure 9. Mean average energy consumption per distance unit for short (red) and long observation period (blue). Graph is supplemented with visualisation of CDF of specific energy for short period of observation and data points from long period of observation.

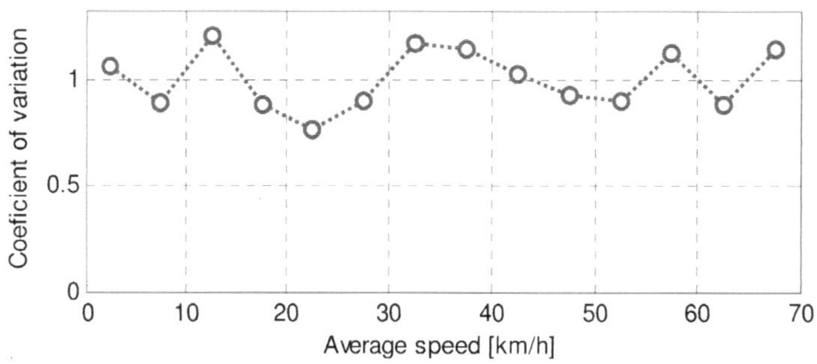

Figure 10. Coefficient of variation of average power depending on average speed.

A statistical dispersion of average power demand shows similar trends to the dispersion of average energy consumption per distance unit, but without a visible trend (increasing or decreasing) and with slightly higher values of coefficient of variation, ranging from 0.76 to 1.2.

3.2. Impact of section length on statistical dispersion of average energy consumption and average speed

As there was big difference observed between coefficients of variation for short and long periods of observations additional analysis was performed to examine the impact of section (link) length on statistical dispersion on calculated average energy consumption per distance unit. In order to perform this analysis all sections were ranked according to their lengths to analyze a section length frequency distribution. For a section length the constant interval width of 100 m was used from 0 to 700 m and over this value the width of length interval was increased to avoid creating data intervals with less than 13 samples (fig. 11).

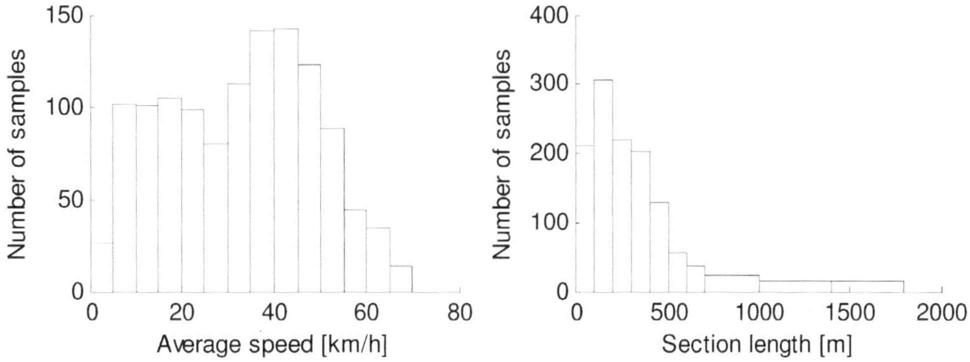

Figure 11. Average speed and section length intervals with sample count.

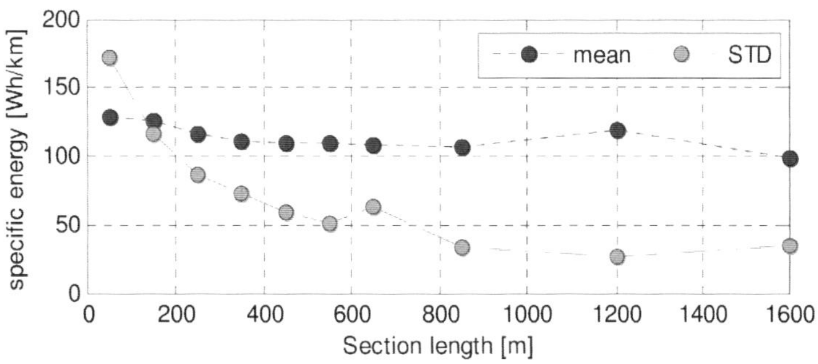

Figure 12. Mean value and standard deviation of specific energy relative to section length.

Mean of average energy consumption per distance unit calculated for length interval changes only slightly with the increasing section length, but the standard deviation noticeably decreases (Fig. 12). As a result, coefficient of variation of average energy consumption per distance unit decreases with the increasing section length (also period of observation)(Fig. 11), from value over 1.3 for sections shorter than 100 m to the lowest value of 0.22 for sections over 1 km long. The local peak of CV for the length of 600 m to 700 m may be the result of changes in average consumed energy relative to section length, which in turn results from the randomness of the collected data (more data could help to overcome this problem). CV of specific energy calculated for whole trips is around 0.15. For section lengths over 700 meters CV has a value over 0.3, but this could be the result of data scarcity and drop in the mean value of average energy.

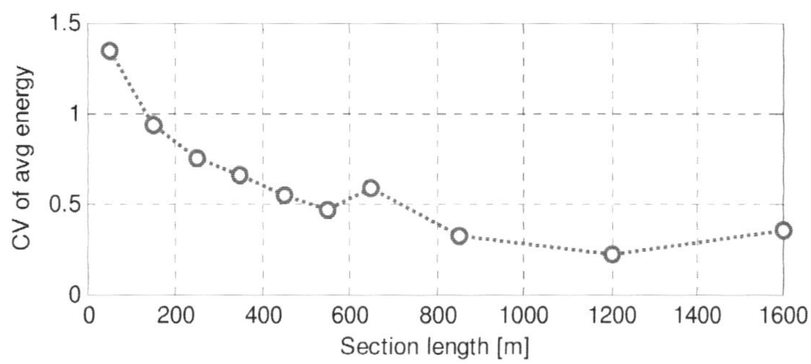

Figure 13. Coefficient of variation of specific energy

In order to test how CV of average energy consumption per distance unit behaves for longer periods of observation individual trips were merged to form routes with length over 20 km. The result was 15 trips with mean length of 24 km and CV of specific energy about 0.12. Aggregation to trips with length over 30 km resulted in only 10 trips with CV around 0.095.

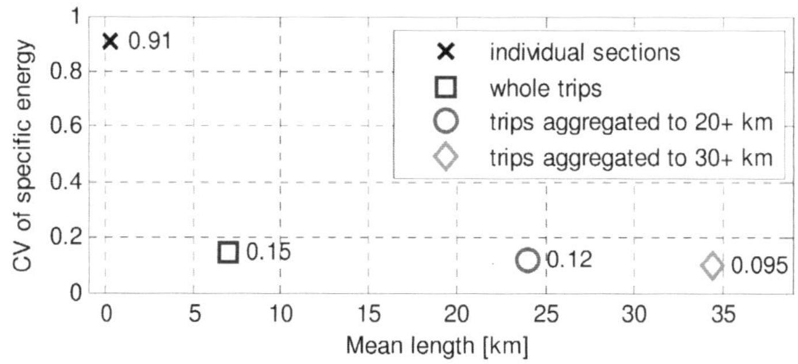

Figure 14. Coefficient of variation of specific energy for short and long period of observation.

In order to test how CV of specific energy behaves for much longer periods of observation the individual trips recorded were then merged to form routes over 20 km long. The result were 15 trips with the mean length of 24 km and CV of specific energy about 0.12. Aggregation to trips with length over 30 km resulted in only 10 trips with CV around 0.095 - figure. 13.

The graph in figure 13 shows the results of the analysis of the impact of the section length on average speed. This result shows that longer sections generate data with higher average speed, which can be associated with freer flow traffic (fewer obstacles and turns which slow down a vehicle). The standard deviation of average speed maintains a similar level regardless of the section length.

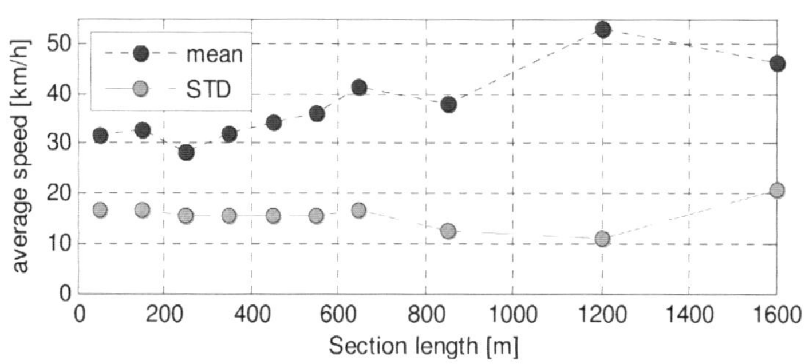

Figure 15. Impact of section length on average speed on section.

4. Summary

The concept of a model of average energy consumption for short sections was adopted because this method of representing roads infrastructure is used in traffic simulation software along with a similar way of obtaining and presenting data in on-line navigation systems. The model is expected to improve the accuracy of energy consumption predictions by using the average speed on each section data in comparison to the models based on the average speed for the whole route. However, an analysis of energy consumption distribution presented in the article indicates a high level of uncertainty for predictions for individual route sections based only on average speed information.

The key findings from obtained results are:

- for short observation period a statistical dispersion of specific energy is roughly independent of average speed. With an increasing length of observation period data dispersion decreases.
- the value of coefficient of variation for energy consumption is higher than 0.4 for section lengths shorter than 700 m.
- while predicting energy consumption it means a high inaccuracy for individual route sections but with increasing travel length the accuracy will improve.

In future the authors will develop the statistical model which helps to aggregate energy consumption estimates from a number of sections and aggregate its uncertainty, thus enabling to obtain much lower values of it for the whole route.

References

[1] Ślaski G, Ohde B and Maciejewski M 2015 Makroskopowy model zużycia energii i jego walidacja dla testowych cykli jezdnych *Logistyka* **4** 1025-36

[2] 1999 *Meet: Methodology for calculating transport emissions and energy consumption,* (Luxembourg: Office for Official Publications of the European Communities)

[3] Yao E, Yang Z, Sond Y and Zuo T 2013 Comparison of electric vehicle's energy consumption factors for different road types *Discrete Dynamics in Nature and Society*

[4] Lohse-Busch H 2012 *Advanced Powertrain Research Facility AVTA Nissan Leaf testing and analysis* (Argonne IL: Argonne National Laboratory)

[5] Geringer B and Tober W K 2012 *Battery Electric Vehicles in Practice* (Vienna: Austrian Society of Automotive Engineers), 34-9

[6] My Nissan LEAF internet forum (2016-05-13): http://www.mynissanleaf.com/viewtopic.php?f=31&t=2523

[7] Ślaski G, Ohde B and Pikosz H 2014 Modelowanie energochłonności eksploatacji samochodu elektrycznego w warunkach ruchu miejskiego dla potrzeb symulacji zużycia energii przez flotę taksówek *Logistyka*, **3** 4777-86

Method and algorithm of automatic estimation of road surface type for variable damping control

K Dąbrowski[1] and G Ślaski[2]

[1] PhD student, Poznan University of Technology, Poznan, Poland

[2] PhD, DSc, Assistant Professor, Poznan University of Technology, Poznan, Poland

E-mail: Krzysztof.dabrowski.put@gmail.com

Abstract. In this paper authors presented an idea of road surface estimation (recognition) on a base of suspension dynamic response signals statistical analysis. For preliminary analysis cumulated distribution function (CDF) was used, and some conclusion that various roads have responses values in a different ranges of limits for the same percentage of samples or for the same limits different percentages of samples are located within the range between limit values. That was the base for developed and presented algorithm which was tested using suspension response signals recorded during road test riding over various surfaces. Proposed algorithm can be essential part of adaptive damping control algorithm for a vehicle suspension or adaptive control strategy for suspension damping control.

1. Introduction

The validity of idea of using the recognition of road kinematic excitation in front of a vehicle as way of improving passenger comfort has been confirmed in many publications since 1960s [1-3]. The recognition of road unevenness ahead of vehicle is technically reasonable if vehicle's suspension is able to be actively adjusted to ride over incoming obstacles on the road surface. It does not mean that the implementation of this idea for vehicles with only an adaptive suspension (instead of an active one) is not possible. Such implementation can be based on the assumption that next to ride road section will generate a similar kinematic excitation as previously driven.

The classification of road surfaces according to generated kinematic excitation is defined by ISO 8606 standard [4]. It is based on amplitude-frequency analysis of Fourier decomposition of road surface. Some publications [5, 7] describe differences between frequency distribution for various road types (asphalt, sett). This publications confirm that it is possible to identify a road surface type using an analysis of vehicle vertical dynamic response.

When it is possible to recognize the road surface (magnitude of kinematic excitation) the vehicle is just riding on, it is also possible to adaptively adjust the suspension damping to this type of road.

Publication [5] confirms that the amplification function for frequencies of 0–2 Hz has the lowest value for a high dumping level. For a road surface with dominant frequencies in this range (highway, expressway) it is most favourable to use the highest dumping setting. For frequencies higher than 2 Hz the lowest value of amplification is achieved with the lowest dumping level, which means that for road surface with a dominant frequency higher than 2 Hz the best insulating properties are obtained with lowest dumping setting [6].

2. Literature analysis of the possibility of kinematic excitation from road surface recognition

The idea of controlling the dumping of suspension widely used nowadays in automotive applications are often focused on increasing comfort by minimizing the effect of road imperfections that the vehicle has already ridden over. An ideal possibility would be to know about the forthcoming surface irregularities for an earlier reaction and preparation of suspension parameters for every bump or particular surface type. There are 3 approaches for predictive control:

- scanning road surface ahead of vehicle,
- controlling the rear axle suspension damping based on front axle excitation or dynamic answer (wheelbase predictive control),
- assuming that the next road segment will be similar to the previous one.

An example of the first approach is presented in 2008 Mercedes F700 concept vehicle with recognition of road irregularities in front of the car. The vehicle was equipped with two laser scanners that measured the surface 15 m ahead of the vehicle. Such a distance was required by Active Body Control system to overcome its time delay, as it takes 0.66 s to switch the actuator from one end position to the other one [8]. This prototype vehicle was thoroughly described in Schindler's doctoral dissertation [9] from 2009. The real implementation of this concept of predictive control was extensively presented. Schindler proposed an idea of measurement, selection of sensors, its properties and assessment of usefulness in this application.

An example of second approach of controlling vehicle rear axle damping an a base of front axle excitation was described in 1985 by Frühlauf [10]. He tried to use vehicle motion model with 7 degrees of freedom (body bounce, roll and pitch movement and vertical movement of unsprung masses) and linear regulator to control rear axle based on front axle excitation. To asses time delay between excitation on front and rear axle 4th degree Padé approximant was used. One of the most interesting approach of wheelbase prediction control idea was presented in Huisman's doctoral dissertation [11] in 1994. It used a half vehicle model of truck with adjustable shock absorbers on rear axle. Huisman created both the control model and the state observer that estimates physical parameters used in control model but not measured directly by sensors.

The assumption that the next segment of the road is the same type as actual or previous with using statistical analysis of surface type is the third approach to predictive control of suspension damping. Estimation of road irregularities (road excitation) in front of vehicle is based on classifying it into particular types of road surfaces (road excitations values) not on recognition of single road unevenness. This idea doesn't utilize an additional road irregularities scanning system, because it assumes that the nearest section to ride over road is statistically similar to the one behind the vehicle. The knowledge of road surface type allows to choose suitable suspension setting for every particular road type [5-6]. For example in [12] authors propose to modify sky-hook strategy in such way that suspension dumping coefficient d_{sky} is selected on base of Fourier distribution of road surface irregularities. Input signal for road surface recognizing algorithm was vehicle body acceleration. With knowledge of body acceleration and vehicle body acceleration transfer function it is possible to estimate excitation of road surface. To ensure algorithm stability authors used low pass filter with frequency cutoff at 20 Hz. The authors pointed out that real road surface does have some band of frequencies with dominant amplitude.

Other authors [13] presented similar adaptive sky-hook strategy based on different coefficient of dumping d_{sky} for each surface type. Authors in their idea of suspension damping control proposed for road of good quality high level of dumping for better comfort, and low level of suspension dumping for roads of poor quality.

Strategy of dumping control can be also based on an analysis of frequency of passing excitation of road surface. The authors of publications [14-15] presented a frequency band detector, which with use of simple mathematical function allows to determine band of frequency of passing excitation. Detector is equipped with parameter α that defines switching frequency between extreme dumping settings. Formula for two-state control based on detection of range of frequencies is presented in [14]:

$$\begin{cases} c_{max}\ for\ (\ddot{x}_M^2 - \alpha^2\ \dot{x}_M^2\ ) < 0 \\ c_{min}\ for\ rest\ cases \end{cases}, \qquad (1)$$

where: α – value of switching value of frequency to select between levels of dumping, $\dot{x}_M$ – sprung mass velocity.

Frequency range detector enables to adapt a passive setting of suspension to passing frequency of excitation from surface. It appears that the use of only a frequency range detector for passive setting selection of suspension damping allows to achieve more favourable shape and values of vehicle body acceleration transfer function than the use of two-state sky-hook steering [16].

3. Method and algorithm of assessing road surface class based on statistical analysis of suspension response

The authors of this paper propose a method to estimate road surface type based on statistical analysis of dynamic suspension response to kinematic excitation. The following assumptions were adopted:

- the quality of road surface can be determined by statistical analysis of a response of vehicle vertical dynamic to kinematic excitation from surface,
- the method should allow for reliably assessing type of road surface based on signals from real sensors and account for interference,
- the method should provide resistance against changes in signal values from different suspension settings.

3.1. Method to analyse the selected suspension response signals

To statistically analyse dynamic response signals cumulated distribution function (CDF) was used, with assumption that data distribution is normal [17]:

$$F(u) = \int_{-\infty}^{u} \frac{1}{\sigma\sqrt{2\pi}} e^{-\frac{(u-\mu)^2}{2\sigma^2}} du \text{ where } \sigma = \sqrt{\frac{\sum_{i=1}^{N}(u_i-\mu)^2}{N}} \quad (2)$$

where: u – variable, μ – mean value, N – number of measurements, σ – standard deviation.

Example of obtained CDF and analysis of boundary values is shown in figure 1. To categorize excitation based on suspension response analysis two methods can be used:

- determination of limit values of the range that some percentage (for example 90%) of measurement samples are located, fig. 1a,
- determine percentage of samples outside of established indicating range (fig. 1b).

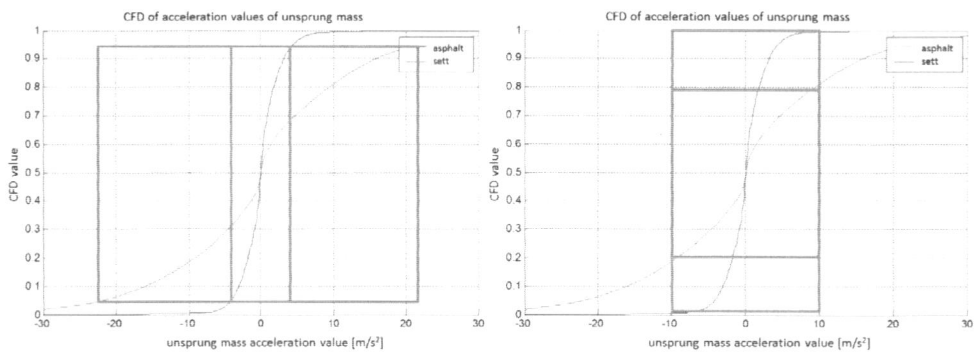

Figure 1. Comparison of CDF of measurements of acceleration of unsprung mass for asphalt and sett surface.

In the first case CFD 90% range and resulting limit values of response signal for asphalt surface type is denoted with blue rectangle (acceleration range limit values ±4 m/s^2) and for cobblestone surface is denoted with green rectangle (acceleration values ±22 m/s^2). In practice with short time available for analysis it is easier to utilize second approach, that is based on counting samples with values outside selected indication range of analysed suspension response signal. Analysing CDF it was determined that limit values should be selected to be between values obtained for asphalt and cobblestone surface.

Based on the previous analysis the example where limit value was assumed as ±10m/s^2 - Fig - 1.b it was found out that probability that acceleration of unsprung mass is outside this range is 2% for asphalt surface and 40% for cobblestone surface, what is corresponding with a percentage of analysis time.

3.2. Algorithm for road surface type estimation

In accordance with the method described the algorithm for surface type estimation counts how many samples are outside of the assumed range during a 0.5 s period and based on this with use of road type indicator asses probability that individual sample is outside of that value range. Period of 0.5 s was chosen to ensure short time of surface categorization and with sampling rate of 0.005 s allows to obtain 100 samples per period. Diagram of algorithm is shown in figure 2. For the purposes of algorithm road type indicator P was defined:

$$P = \frac{n}{100}k \tag{3}$$

where: n – number of samples outside of ±10 m/s^2 range, k – correction factor related with suspension current dumping setting.

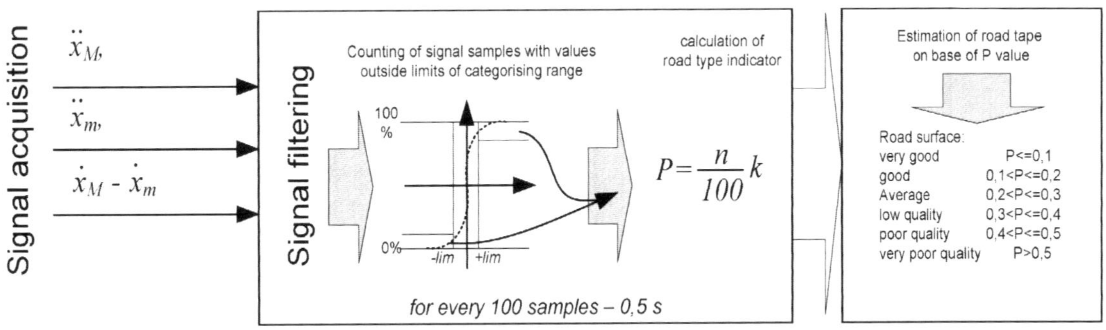

Figure 2. The structure of algorithm for road surface type estimation.

To make algorithm working it was necessary to:

- determine limit values $\pm lim$ for various surface types taking also into account various damping level,
- possible selection of one suspension dynamic response signal, optimal for surface categorization,
- determining correction factor k to take into consideration variation of damping levels and its impact on dynamic suspension response.

3.3. Determining the limit values of criteria ranges

The first task could be done in a few ways – for example simulation tests or analysis of road test data acquired while riding over various road surfaces types with different suspension damping settings. The authors have chosen another approach; they analysed the data acquired during road tests described in section 4.1. It was assumed that limit values for criteria range would be determined in such a way that 90% of samples would be below limit value. After performing analysis limit values were empirically determined for various surface types (table 1.)

Table 1. Limit values (for 90% of samples) determined based on road tests with different suspension damping settings

		body acceleration [m/s2]		wheel acceleration [m/s2]		suspension deflection velocity[m/s]	
speed [km/h]	road surface	damping max	damping min	damping max	damping min	damping max	damping min
140	highway asphalt	1,18	1,25	5,5	7	0,05	0,07
90		1,13	1,05	5	5,8	0,045	0,055
80	extra urban road surface	1,8	1,6	8,5	12	0,08	0,12
60	Asphalt 1	1,1	1,03	3,7	5,4	0,06	0,08
60	Asphalt 2	1,7	1,2	5	7	0,09	0,12
30	jointed plain concrete pavement	1,7	1,25	5,5	7,5	0,07	0,11
40		2,3	1,8	9	12	0,1	0,15
50		3	2,5	12	16	0,12	0,18
20	cobblestone	2,6	2,2	12	16,5	0,12	0,19
30		3,2	2,8	17	22,5	0,15	0,24
40		4,1	3,6	23,5	30	0,17	0,28

3.4. Determination of optimal suspension dynamic response signal for road type estimation

The second task was to select of most appropriate signal to perform road surface categorization for roads of extremely different values of kinematic excitation and generated dynamic suspension response. Among three available signals of suspension response one was selected on a basis of the following parameters:

- The mean difference of limit values lim of criteria range due to change of damping settings from minimal T_{min} to maximal T_{max}. The data to calculate these parameters were obtained from table 1. Differences in limit values for each of 6 surface type were calculated with following formula:

$$\frac{lim_{Tmin} - lim_{Tmax}}{lim_{Tmax}} = \Delta lim$$

The mean difference of limit values between extreme settings of suspension dumping for 6 types of surface was determined with the following formula:

$$\overline{\Delta lim} = \frac{1}{6}\sum_{i=1}^{6}\left(\frac{lim_{Tmin} - lim_{Tmax}}{lim_{Tmax}}\right)$$

- standard deviation of differences in limit values was calculated with following formula:

$$\sigma = \sqrt{\frac{\sum_{i=1}^{6}(\Delta lim - \overline{\Delta lim})^2}{6}}$$

- difference of limit values between extreme types of surface (asphalt and cobblestone), named interval of limit values were calculated with following formula:

$$Z = \frac{lim_{very\ good} - lim_{very\ poor}}{lim_{very\ good}}$$

These parameter values for three dynamic response signals are presented in table 2.

Table 2. Impact of damping setting on limit values of criteria range (consisting of 90% of samples).

	body acceleration [m/s²]	wheel acceleration [m/s²]	suspension deflection velocity[m/s]
$\overline{\Delta lim}$	-0,097	0.29	0.39
σ	0,1	0.14	0.2
Z	4	6	6,2

After an analysis of the obtained values limit values were assumed to correspond to extra urban road surface with an appropriate operating speed. The limit values for respective signal are as follows:

- suspension travel rate: 0.1 m/s,
- acceleration of sprung mass: 1.7 m/s^2,
- acceleration of unsprung mass: 10 m/s^2.

With these limit values road type indicator *P* for highway is between 0 and 0.1, for extra urban road it is between 0.1 and 0.2, and for roads of poor quality it has higher values. For example for cobblestone surface value of road type indicator *P* is shown in figure 1b where 40% of samples is outside boundary range of ±10 m/s^2. For other types of surface range of road type indicator *P* was divided linearly into ranges as shown in figure 3.

In addition to the impact of damping setting on ranges limit, the same importance has interval of limit values *Z* between extreme types of surface. This interval value defines width of ranges for each of 6 surface types. If this width for each signal of suspension response is significantly higher than even higher standard deviation σ of mean difference of limit values due to damping setting it will result in higher accuracy of road surface evaluation by proposed road type indicator *P*. Figure 3 explains need to consider difference in interval of limit values *Z* for each dynamic response signal. Limit values are shown on the blue axis and are divided into 6 main ranges defining 6 types of road surface, red color denotes standard deviation σ of mean difference of limit values due to damping setting. Figure 3 shows that a similar standard deviation in case of a small width of categorization ranges of particular types of road surfaces can lead to incorrect estimation of surface type that vehicle rides over.

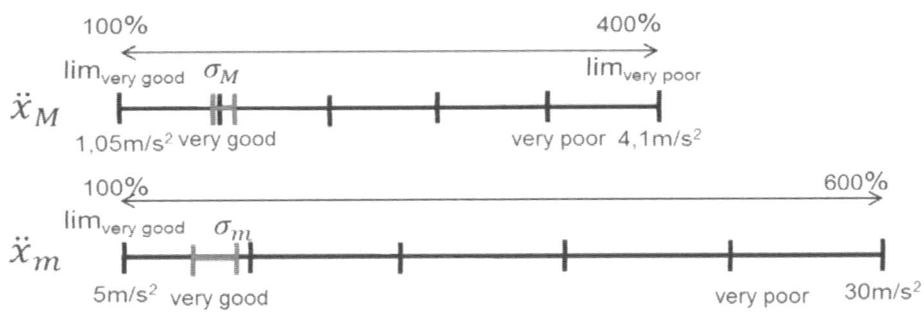

Figure 2. Illustration of methodology of selecting input signal for algorithm.

The results presented in table 2 show that the increase of the limit value between $lim_{very\ good}$ and $lim_{very\ poor}$ for wheel acceleration and suspension travel rate is around 600% which means that limit value for worst surface is 6 times bigger than for best surface. This Z value for body acceleration is significantly lower - around 400%. That means that seemingly the best measured signal is the least favorable one. Therefore acceleration of unsprung mass was selected for use in algorithm estimating the type of road surface.

3.5. Selection of correction factor k

Results of analysis of road tests data presented in table 2 confirm that impact of suspension damping setting on acceleration of vehicle body for different surface types can be contrary. Analysis of limits of criteria ranges (consisting of 90% of samples) shows that change of suspension damping setting from maximal to minimal increases limit value for expressway roads and decreases limit value for other road types. Result of this gives correction factors for different road types that are contrary to each other. This in turn results in lack of possibility to use single correction factor for each road type. In table 2 mean value and standard deviation σ of mean difference of limit values due to damping setting are determined. It appears that suspension damping level has strongest impact on suspension travel rate and weakest on body acceleration. standard deviation σ values show similar tendency – for body acceleration it is 10%, and for travel rate it is 20%. During simulation tests acceleration of wheel and constant correction factor for highest damping setting were assumed. Used correction factor value was $k = 1.29$ which complies with mean value of impact of damping change $\overline{\Delta lim}$ for each operational travel speed values over analysed road surfaces.

4. Study of efficiency of proposed method of assessing road surface class

After selection of limit values for algorithm tests were carried out in order to:

- test of repeatability of surface type recognition (estimation) by proposed algorithm,
- determination witch of vertical dynamic response signal allows to achieve most reliable results,
- determination of value of correction factor associated with damping setting impact,
- check if it is required to apply correction factor associated with vehicle speed change for operational speed for each road surface.

The suggested algorithm of road surface type estimation (recognition) was verified in simulation tests using suspension dynamic response signals acquired during road tests. These signals - acceleration of sprung and unsprung mass and suspension travel - had been changing with various excitation from road surface irregularities. Based on this measured signals target input signal for algorithm estimating road surface type was prepared and the influence of correction factor related with current suspension setting on output signal - road type indicator P - was analysed. The developed algorithm was tested during simulation and repeatability of estimating road surface type was verified. Figure 4 presents diagram of used test procedure.

Test examination was performed to represent operational conditions, suspension response signals measurements were carried over different surfaces with set of vehicle speed values similar to values reached in real life travels, not with whole range of speeds achievable by vehicle. Seven routes with diverse surface were chosen in Poznan agglomeration. Road tests were divided according to vehicle speed:

- 90, 120, 140 km/h for expressways,
- 60, 80 km/h for extra urban roads and main urban roads,
- 20, 30, 40, 50 km/h for urban roads (asphalt, jointed plain concrete pavement and cobblestone surface).

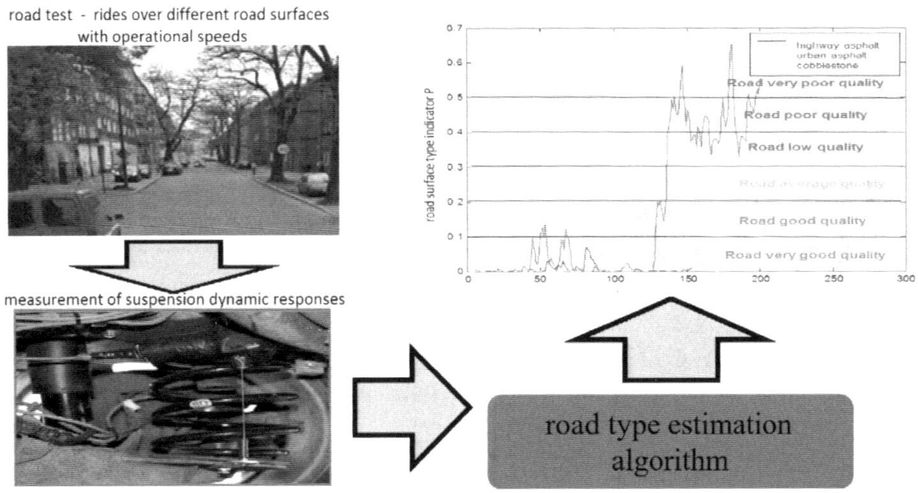

Figure 3. Methodology of algorithm testing.

Simulation tests for signals from rides over asphalt surface shown that for individual periods of road surface type estimation - 0,5 s - value of surface coefficient *P* can sporadically be higher than 0.3, which indicates poor quality surface. Crossing of this limit value occurred only momentary for single observation period. That short term exceeding of limit values causes that unambiguous estimation of road surface type can be successfully made only for three surface classes: good quality road (highway, expressway), moderate quality road (urban asphalt road) and poor quality road (cobblestone). Behaviour of indicator *P* for different road surface type is shown in figure 5. During tests it was found that the impact of a rear axle load on estimation results based on vehicle front left quarter suspension responses is negligible in comparison with the impact of vehicle speed and damping setting.

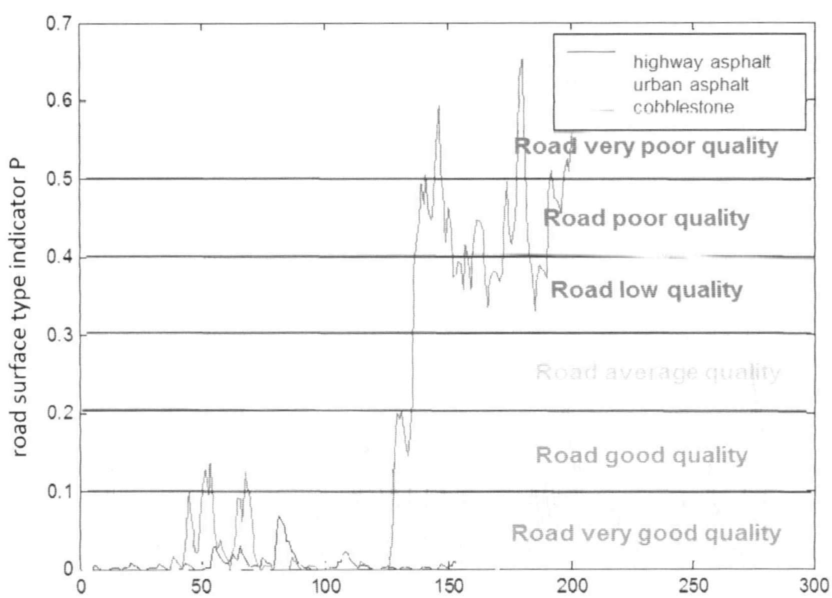

Figure 4. Behaviour of road surface type indicator P calculated on base of wheel acceleration.

5. Conclusions

The presented algorithm estimating road surface type was suggested on the basis of a study of the variation range of measured signals of suspension dynamic response to excitation from road surface in various road conditions and operational speeds. The verification of the algorithm for multiple passes confirmed that it is possible to assess a road surface type based on the suspension dynamic response signal. This concept disregards the influence of manoeuvers of acceleration, deceleration or time when vehicle is being stationary, when signals are treated as signals from a surface with the lowest kinematic excitation levels (very smooth road). This poses a challenge for future research and the results acquired while algorithm testing:

- introduction of correction factor related with impact of speed on dynamic response,
- utilization of estimation of sprung mass for considered vehicle quarter,
- extension of observation time by decision making with consideration of data from few recent estimations, with different weights factors.

The developed algorithm allows to select suspension damping setting, or damping control strategy, according to recognized road surface. For highway ride suspension the damping can be set to maximum, which was proven to provide the highest comfort, while for a cobblestone surface the suspension can be set to minimum damping. The road surface recognition system allows to utilize various control strategies for suspension damping described in literature, and to switch between them adaptively. Determination of strategy with most benefits regarding passengers comfort is considered as subject for further study and way to increase knowledge about suspension control.

References

[1] Bender E K 1968 Optimum linear preview control with application to vehicle suspension. *Journal of Basic Engineering 100* (), pp 213–221
[2] Tomizuka M 1976 Optimal linear preview control with application to vehicle suspension – revisited, *Trans. of ASME, Journal of Dynamic Systems, Measurement and Control*, pp 309-315
[3] Sheridan T B 1966 Three models of preview control, *IEEE Transaction on human factors in electronics* pp 91– 02
[4] ISO 8608:1995. Mechanical vibration - Road surface profiles - Reporting of measured data
[5] Ślaski G 2013 Symulacyjne i eksperymentalne badania adaptacyjnego sterowania tłumieniem zawieszenia w zależności od częstotliwości sinusoidalnego wymuszenia kinematycznego *Archiwum Motoryzacji*
[6] Ślaski G 2011 The Influence of adaptive damping level on vehicle vibration comfort-passenger car experimental tests results, *Konferencja TRANSBALTICA 2011* Wilno
[7] Unger A 2012 Serientaugliche quadratisch optimale Regelung für semiaktive Pkw-Fahrwerke. (Cuvillier Verlag Göttingen, Technische Universität München)
[8] Streiter R 2008 ABC Pre-Scan im F700 - Das vorausschauende aktive Fahrwerk von MercedesBenz *Automobiltechnische Zeitschrift (ATZ)* pp 388–397
[9] Schindler A 2009 Neue Konzeption und erstmalige Realisierung eines aktiven Fahrwerks mit Preview-Strategie, (Karlsruher Institut für Technologie)
[10] Frühauf F *et al* 1986 Design of an Active Suspension for a Passenger Vehicle Model using Input Processes with Time Delay, *Vehicle System Dynamics*, pp 126-138

[11] Husiman. R.G.M. *et al* 1993, Preview estimation and control for (semi) active suspensions *Vehicle System Dynamics Vol. 22* pp 335 – 346
[12] Hong K S, Sohn H C and Hedrick J K 2002 Modified skyhook control of semi-active suspensions: A new model, gain scheduling, and hardware-in-the-loop tuning. *Journal of Dynamic Systems Measurement and Control 124* pp158–167
[13] Kashem S B A and Chowdhury M A 2015 Study and Review on Vehicle Suspension Control Theories and Introduction of Novel Adaptive Skyhook Control System, *Australian Journal of Basic and Applied Sciences*
[14] Savaresi S, Spelta C, 2009 A single sensor control strategy for semi-active suspension, *IEEE Transaction on Control System Technology, 17(1)*pp 143-152
[15] Savaresi S M, Poussot-Vassal C, Spelta C, Sename O and Dugard L 2010 Semi-active suspension control design for vehicles. *Access Online via Elsevier*
[16] Poussot-Vassal C and Spelta C *et al* 2011 Survey on Some Automotive Semi-Active Suspension Control Methods: a Comparative Study on a Single-Corner Model, *18th IFAC World Congress,*
[17] Rill G 2012 Road vehicle Dynamics, (New York CRC Press)

IOP Conf. Series: Materials Science and Engineering **148** (2016) 012029 doi:10.1088/1757-899X/148/1/012029

The theoretical analysis of an instrument for linear and angular displacements of the steered wheel measuring

K Wach[1]

[1]Cracow University of Technology
Al. Jana Pawla II 37, 31-864 Krakow, Poland

E-mail: krzysztof.wach@mech.pk.edu.pl

Abstract. In the paper the theoretical analysis of the measuring instrument for determination of translation and rotation of the stub axle with the steered wheel against car body was presented. The instrument is made of nine links with elongation sensors embedded in it. One of several possible structures of instrument of this kind was presented. Basing on solution of the geometrical constraints system of equations of the device, the numerical analysis of the measurement accuracy was conducted.

1. Introduction

The analysis of the car movement parameters on the curved track is one of the fundamental issues of stability and steerability. The change of these parameters is always caused by the change of the external forces acting on the vehicle.

The car comparative studies carried out so far show that even at the same kinematic extortion, realized by the change of steering angle, the change of the forces generated at the wheel- road contact patch is dependent on many factors associated with tire, suspension and steering system construction.

Steering wheels are carried out against a car body through the spatial mechanisms with flexible constraints [1]. The flexibility is the reason that during a car ride along the same path, with different speeds, the real kinematic steering ratio is changing; a significant difference between the real and theoretical steering angles appears. A measurement of real steering and camber angles in experimental car studies has a significant value. The results of these measurements are used to work out the relationships between car movement parameters as well as for stability and steerability valuation [2, 3,4].

In case of independent suspensions, wheel vertical movements caused by unevenness of the road surface cause a track change. It leads to wheels drifting and adversely affects the straight driving [5, 6]. The measurement of the position and orientation of the wheel relative to the car body is very difficult, only a few studies on this topic, mainly related to the dynamic measurement of the steering angle, can be found in the literature [2,7, 8]. A Datron RV-4 instrument [9] allows the measurement of the position and orientation of the steering wheel relative to the car body, the measurement, however, is complicated and the measured values are not obtained directly but as a result of complex calculations. This instrument has large dimensions and considerable weight in proportion to the weight
of the wheel. The durability of the instrument is being reduced under the influence of the dynamic loads generated while driving the car over the road unevenness.

2. The goal and scope of the work

The main goal of the work is an introduction of the indirect measurement method of the stub axle with steering wheel translation and rotation.

The scope of this work concerns problems of solving the kinematics of four- link suspension and the proposed instrument mechanisms as well as determination of the steering and camber angles and the characteristics of lateral displacements of the wheel centre. The analysis of the measurement accuracy were conducted.

3. Multi- link steered wheels suspension mechanism system

In Figure 1 the four- link steering wheel suspension mechanism scheme is shown. Points: B_1, B_2, B_4 and B_5 are centres of ball joints connecting links with the stub axle. Point B_3 is a centre of the ball joint connecting steering linkage with the stub axle arm. Points: A_1, A_2, A_4 and A_5 are centres of ball joints, which replace a metal- rubber joints connecting links with the car body. Point A_3 is centres of the ball joint connecting steering linkage with a rack. A lower front link, represented in the figure by link A_1B_1, is connected with anti- roll bar at the point W. A telescopic column is connected with this link at the point C. Frames {N} and {K} are associated, respectively, with the car body and the stub axle.

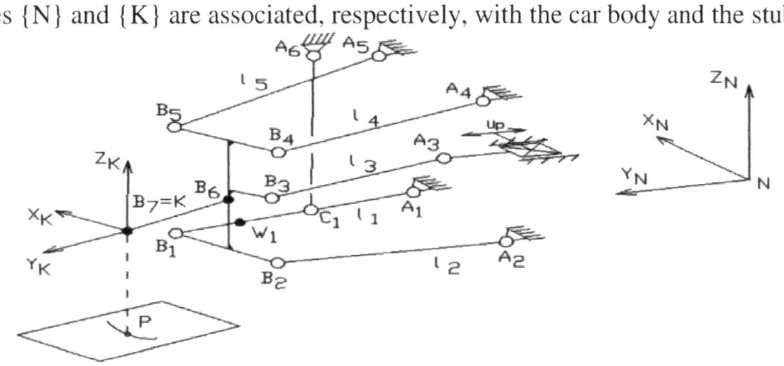

Figure 1. Scheme of four- link steered wheels suspension mechanism.

4. Equations of geometric constraints of the suspension mechanism

The equations of geometric constraints of the mechanism shown in Figure 1 were written in the form of 14 nonlinear algebraic equations. These equations express squares of distances between characteristic points of the suspension:

$$\begin{cases} \vec{r}^{\,T}_{A_jB_j}\cdot\vec{r}_{A_jB_j} = l_j^2, & \text{for } \ j = 1(5), \\ \vec{r}^{\,T}_{B_jB_k}\cdot\vec{r}_{B_jB_k} = l_{jk}^2, & \text{for } \begin{cases} \mathrm{j} = 1 \text{ and } \mathrm{k} = 2(5), \\ \mathrm{j} = 2 \text{ and } \mathrm{k} = 3(5), \\ \mathrm{j} = 3 \text{ and } \mathrm{k} = 4(5). \end{cases} \end{cases} \tag{1}$$

In the above system of equations given parameters are coordinate z_{B1} of the point $B_1(x_{B1}, y_{B1}, z_{B1})$ and steering rack displacement- u_p, added to coordinate y_{A3} of the point $A_3(x_{A3}, y_{A3}+u_p, z_{A3})$. At the given parameters z_{B1} and u_p, coordinates of the point $B_j(x_{Bj}, y_{Bj}, z_{Bk})$, for j=(1)5 and k=(2)5 were determined from the system (1) using the perturbation method [10]. A constructional position of the points B_6 and B_7 is given, therefore determination of their coordinates in the movements space of the suspension {N} is possible from the following systems of equations:

for point B_6:

$$\vec{r}^{\,T}_{B_kB_j}\cdot\vec{r}_{B_kB_j} = l_{kj}^2, \qquad \text{for } \begin{cases} \mathrm{k} = 6 \text{ and } \mathrm{j} = 1, \\ \mathrm{k} = 6 \text{ and } \mathrm{j} = 3, \\ \mathrm{k} = 6 \text{ and } \mathrm{j} = 5, \end{cases} \tag{2}$$

for point B_7:

$$\vec{r}^{T}_{B_kB_j}\cdot\vec{r}_{B_kB_j} = l^2_{kj}, \qquad \text{for} \begin{cases} k = 7 \text{ and } j = 1, \\ k = 7 \text{ and } j = 2, \\ k = 7 \text{ and } j = 4. \end{cases} \tag{3}$$

Next, a rotation angles: α, β, γ of {K} against {N} were determined. After calculating the coordinates of points K and Bj (j=(1)5) a three vectors for $\vec{r}_{KB_j}$ can be created. For each of these vectors a matrix equation is satisfied:

$$\vec{r}_{KB_j.K} = A_{KN}\cdot\vec{r}_{KB_j.N}, \tag{4}$$

where: $\vec{r}_{KB_j.K}$ - vector in {K}, $\vec{r}_{KB_j.N}$ - vector in {N},

$$A_{KN} = A^T_{NK} = \begin{bmatrix} c\beta\cdot c\gamma & c\beta\cdot s\gamma & -s\beta \\ s\alpha\cdot s\beta\cdot c\gamma - c\alpha\cdot s\gamma & s\alpha\cdot s\beta\cdot s\gamma + s\alpha\cdot c\gamma & s\alpha\cdot c\beta \\ c\alpha\cdot s\beta\cdot c\gamma + s\alpha\cdot s\gamma & c\alpha\cdot s\beta\cdot s\gamma - s\alpha\cdot c\gamma & c\alpha\cdot c\beta \end{bmatrix}. \tag{5}$$

Denoting coordinates of vectors: $\vec{r}_{KB_j.K} = [x_{bj}, y_{bj}, z_{bj}]^T$, $\vec{r}_{KB_j.N} = [x_{jb}, y_{jb}, z_{jb}]^T$ and assuming j = n, m, v on the basis of (4) we obtain:

$$\left.\begin{aligned} x_{bn} &= (x_{nb}\cdot c\gamma + y_{nb}\cdot s\gamma)c\beta - z_{nb}\cdot s\beta \\ x_{bm} &= (x_{mb}\cdot c\gamma + y_{mb}\cdot s\gamma)c\beta - z_{mb}\cdot s\beta \\ x_{bv} &= (x_{vb}\cdot c\gamma + y_{vb}\cdot s\gamma)c\beta - z_{vb}\cdot s\beta \end{aligned}\right\} \tag{6}$$

From the system of equations (6) the rotation angles α and β are calculated. In order to calculate the angle γ equation (7) is used:

$$y_{bv} = (x_{vb}\cdot s\beta\cdot c\gamma)s\alpha - (x_{vb}\cdot s\gamma)c\alpha + (y_{vb}\cdot s\beta\cdot c\gamma)s\alpha + (y_{vb}\cdot c\gamma)c\alpha + (z_{vb}\cdot c\beta)s\alpha. \tag{7}$$

Then, having rotation angles of system $\{K\}$ against system $\{N\}$, coordinates of vector $\vec{r}_{B_6B_7.N}$:

$$\vec{r}_{B_6B_7.N} = A_{KN}\cdot\vec{r}_{B_6B_7.K}, \tag{8}$$

unit vector $\vec{e}_k = [e_{kx}, e_{ky}, e_{kz}]^T$, lying on the wheel rotation axis as well as steering and camber angles:

$$\delta_k = -arctg\left(\frac{e_{kx}}{e_{ky}}\right), \tag{9}$$

$$\gamma_k = -arcsin(e_{kz}), \tag{10}$$

were calculated.

5. Coordinates of characteristic points of the suspension mechanism

A constructional location of the mechanism in the suspension movements space {N} is defined by the coordinates of ball joints connecting links with the car body and the stub axle [11]: A_1 (144.1,345.2,-92.2); A_2 (-229.2,362.2,-101.7); A_3 (-99.7,400.0,306.2); A_4 (-69.0,396.3,413.5); A_5 (134.6,428.5,408.9;); B_1 (28.7,690.9,-98.0); B_2 (-24.4,687.0,-131.6); B_3 (-135.7,617.1, 286.9); B_4 (-18.1,639.8,388.4); B_5 (15.4,673.3,389.5); B_6 (0.5,647.0,1.1); B_7 (1.0,747.0,0.6).
Coordinates values were given in mm.

IOP Conf. Series: Materials Science and Engineering **148** (2016) 012029 doi:10.1088/1757-899X/148/1/012029

6. Suspension characteristics

In Figure 2 characteristics of the analysed mechanism, obtained on the basis of its kinematics solution are shown.

a)

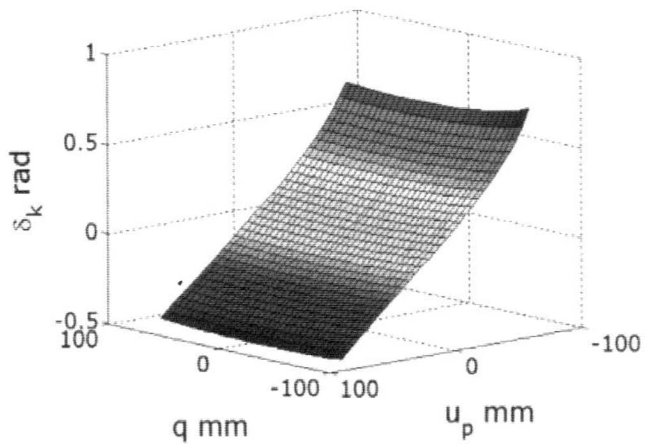

b)

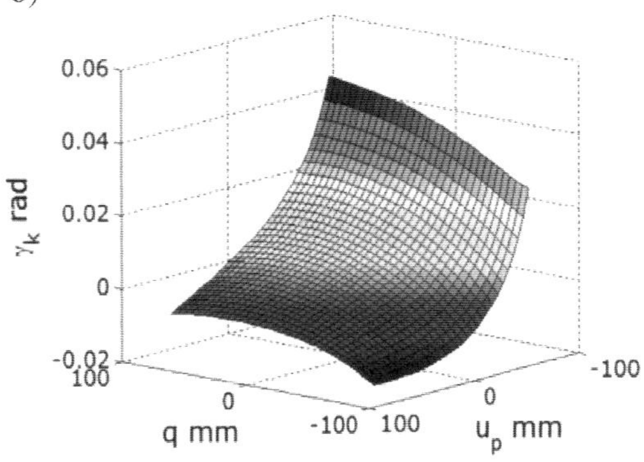

c)

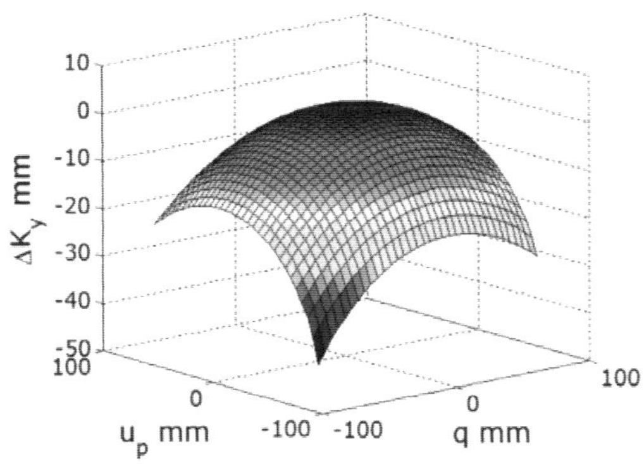

Figure 2. Dependences of steering and camber angles: δ_k, γ_k- respectively and relative lateral displacement of the wheel centre ΔK_y on suspension deflection- q and steering rack displacement-u_p, determined on the basis of the suspension kinematics solution.

7. Structure and mobility of the measuring instrument mechanism

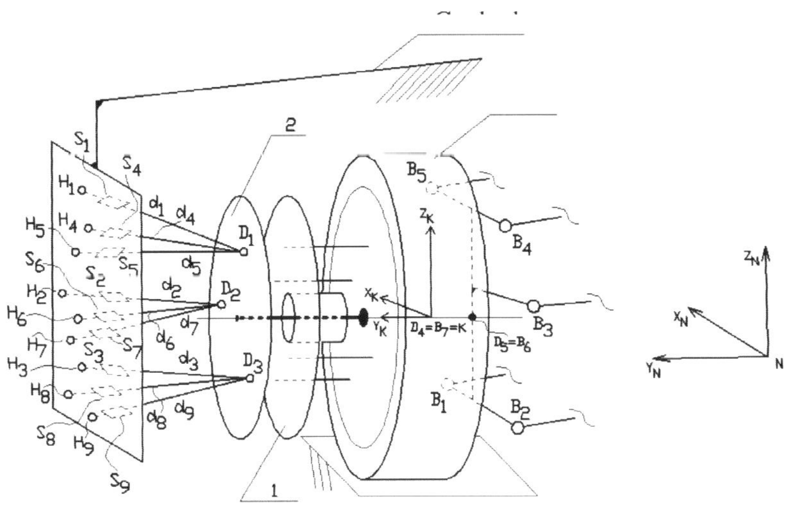

Figure 3. Scheme of the measuring instrument for determination of the translation and rotation of the steered wheel stub axle. Points D_i, i=1(3) are the centres of three concentric ball joints. 1- disc attached to a rim, 2- disc immobilized against a stub axle.

The mechanism of the measuring instrument, shown in Figure 3, is composed of nine links d_j, j=(1)9 connected by rotary- sliding kinematic pairs s_j, j=(1)9. At the points: D_i, i=(1)3, which are the ball joints centres, the links are attached to the bearing disc, which is connected to the stub axle and immobilized against it. By points: H_j, j=(1)9, centres of the ball joints connecting links with the disc connected with

IOP Conf. Series: Materials Science and Engineering **148** (2016) 012029 doi:10.1088/1757-899X/148/1/012029

the car body were noted in the figure. Characteristic for this mechanism is that each point marked as D_i, i=(1)3 is a common centre of three concentric ball joints.
The instrument mechanism has 6 mobility degrees- using the formula from the of mechanisms and machines theory [12] :

$$R = R_t - R_p, \tag{11}$$

where:

$$R = 6(n-1) - \sum_{i=1}^{5} i \cdot p_i, \tag{12}$$

R - real mobility of the mechanism,
R_t - theoretical mobility of the mechanism,
R_p - apparent mobility of the mechanism,
p_i - kinematic pair of the i- class,
n - number of links creating mechanism.

With regard to the considered mechanism: n = 20, p_4 = 9, p_3 = 18, p_5 = p_2 = p_1 = 0, R_t = 24. After subtracting the apparent degrees of mobility R_p = 18 from theoretical mobility, the real mobility R=6; is equal to the stub axle degrees of freedom in {N}.

8. Kinematics of the measuring instrument mechanism

Centres of the ball joints: D_i, i=(1)3 belong to the wheel stub axle. So the distances of these points from points Bj, j=1(6) can be calculated. However, the coordinates of points: D_i, i=(1)3 in {N} can be determined from systems of equations:

$$\vec{r}^{\,T}_{B_jD_j} \cdot \vec{r}_{B_jD_j} = l^2_{B_jD_j}, \qquad \text{for } D_1 \begin{cases} j = 1, \\ j = 2, \\ j = 3, \end{cases} \text{for } D_2 \begin{cases} j = 1, \\ j = 2, \\ j = 4, \end{cases} \text{for } D_3 \begin{cases} j = 2, \\ j = 3, \\ j = 4. \end{cases} \tag{13}$$

Thus it becomes possible to calculate the relative elongations s_j of links d_j, j=(1)9, relative to their constructional distances. Elongations s_i depend on given parameters q and up: s_j(q and u_p).
In the practical application of the measuring instrument for determining the rotation and translation of the steered wheel stub axle it is needed to measure the coordinates of points D_i, i=(1)3 as well as H_j, j=(1)9, for constructional suspension configuration; links elongations s_j(q and u_p), measured by the sensors are also needed.
Determination of the steered wheel stub axle rotation and translation using a measuring instrument boils down to solving of the inverse problem. In this case, at given elongations s_i(q and u_p) of the links d_j, j=(1)9 and constructional positions of the points D_i, i=(1)3; as well as the additional points, eg. $D_4 \equiv B_7$ and $D_5 \equiv B_6$, the coordinates of these points, as the functions of parameters q and up, should be determined from the systems of equations:
for point D_1:

$$\vec{r}^{\,T}_{D_1H_j} \cdot \vec{r}_{D_1H_j} = (l_{D_1H_j} + s_j)^2, \qquad \text{for} \quad \begin{cases} j = 1, \\ j = 4, \\ j = 5, \end{cases} \tag{14}$$

for point D_2:

$$\vec{r}^{\,T}_{D_2H_j} \cdot \vec{r}_{D_2H_j} = (l_{D_2H_j} + s_j)^2, \qquad \text{for} \quad \begin{cases} j = 2, \\ j = 6, \\ j = 7, \end{cases} \tag{15}$$

for point D_3:

$$\vec{r}^{\,T}_{D_3H_j} \cdot \vec{r}_{D_3H_j} = (l_{D_3H_j} + s_j)^2, \qquad \text{for} \quad \begin{cases} j = 3, \\ j = 8, \\ j = 9. \end{cases} \tag{16}$$

The coordinates of point D_4 are obtained by solving the system of equations:

$$\begin{cases} \vec{r}_{D_4D_1}^{T}\cdot\vec{r}_{D_4D_1} = l_{D_4D_1}^2, \\ \vec{r}_{D_4D_2}^{T}\cdot\vec{r}_{D_4D_2} = l_{D_4D_2}^2, \\ \vec{r}_{D_4D_3}^{T}\cdot\vec{r}_{D_4D_3} = l_{D_4D_3}^2. \end{cases} \tag{17}$$

The coordinates of the point $D_5 \equiv B_6$ are calculated similarly as the point D_4. The above systems of equations were solved using triangulation method. The steering and camber angles (δ_d and γ_d respectively), as well as the relative changes in lateral position of the wheel ΔK_{yd} were calculated in the same way as it is shown in paragraph 4.

9. Constructional coordinates of the characteristic points of the measuring instrument mechanism

The coordinates of centres of the joints: D_i, i=(1)3 as well as coordinates of the points D_4, D_5 belonging to the wheel rotation axis assigned to the constructional configuration of the suspension are given below in millimetres: D_1(50.0,815.0,30.0); D_2(25.0,835.0,-90.0); D_3(-50.0,855.0,200.0); H_1(90.0,1165.0,-40.0); H_2(10.0,1164.0,-70.0); H_3(-40.0,1169.0,270.0); H_4(45.0,1171.0,-90.0); H_5(110.0,1162.0,10.0); H_6(30.0,1167.0,-95.0); H_7(30.0,1166.0,-20.0); H_8(-85.0,1163.0,220.0); H_9(-20.0,1164.0,320.0); D_4(1.0,747.0,0.6); D_5(0.5,647.0,1.1).

10. Elongations of the measuring instrument links

The following graphs show sample dependences of the elongations of links of the measuring instrument on the steering rack displacement- u_p and the suspension deflection- q.

a) b)

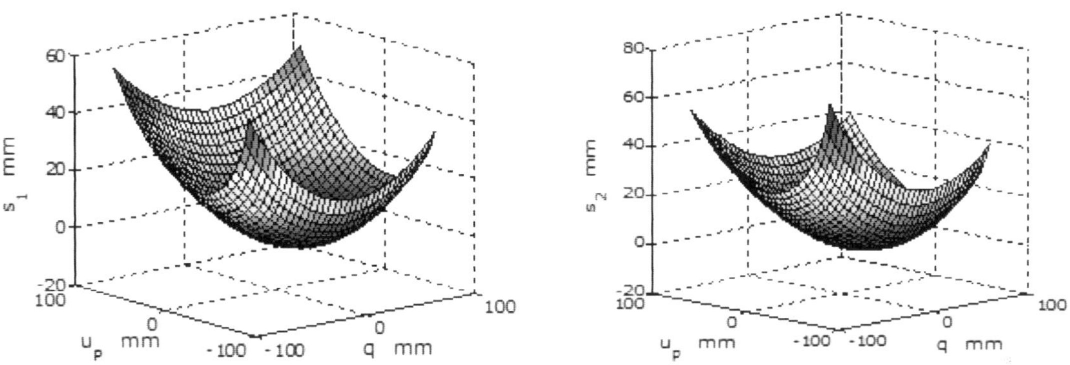

c) d)

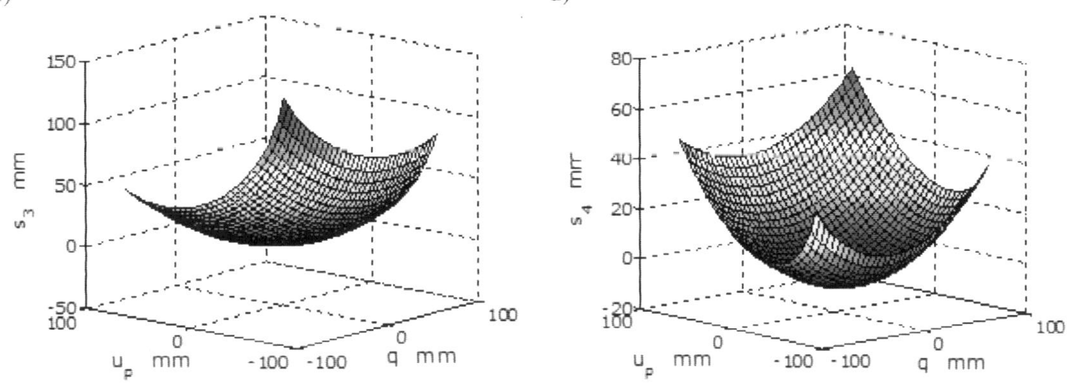

e) f)

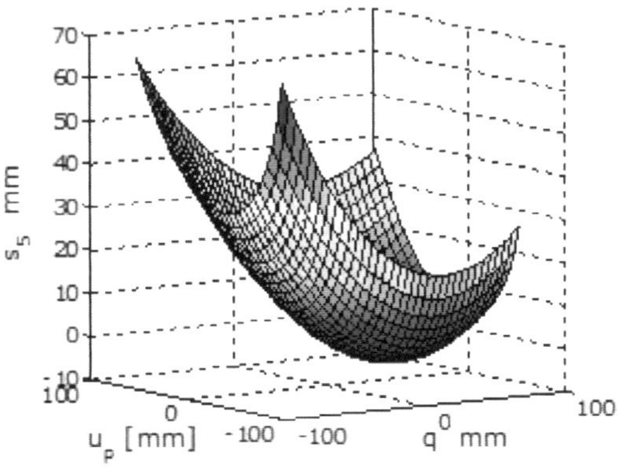

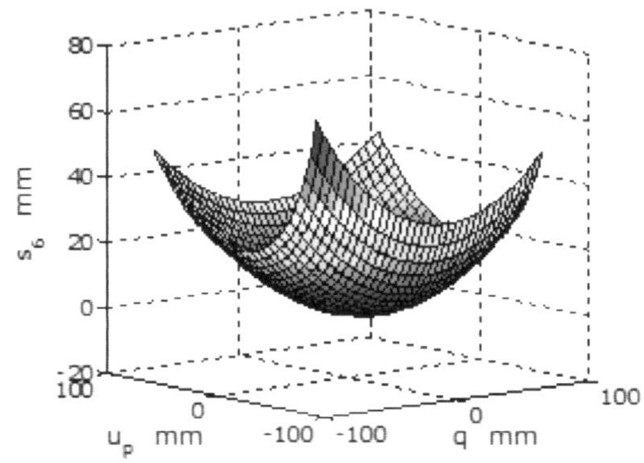

i)

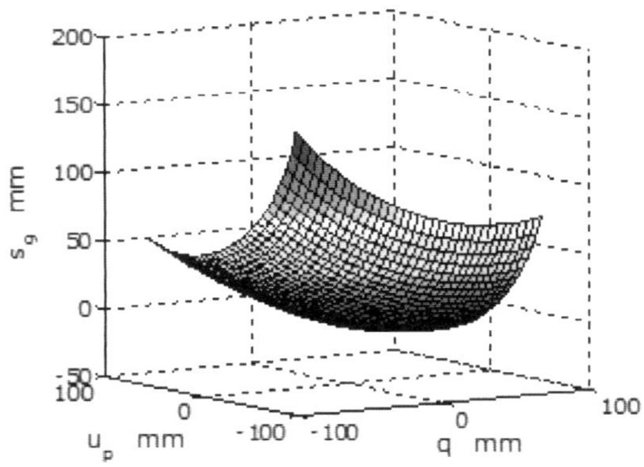

Figure 4. Sample dependences of the measuring instrument links elongations s_j on suspension deflection- q and steering rack displacement- u_p.

11. Suspension characteristics

Suspension characteristics obtained via numerical simulation using the proposed measuring instrument are show in the Figure 4.

a)

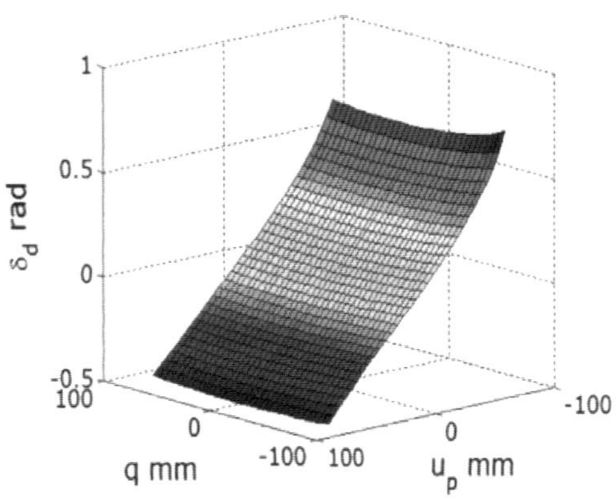

b)

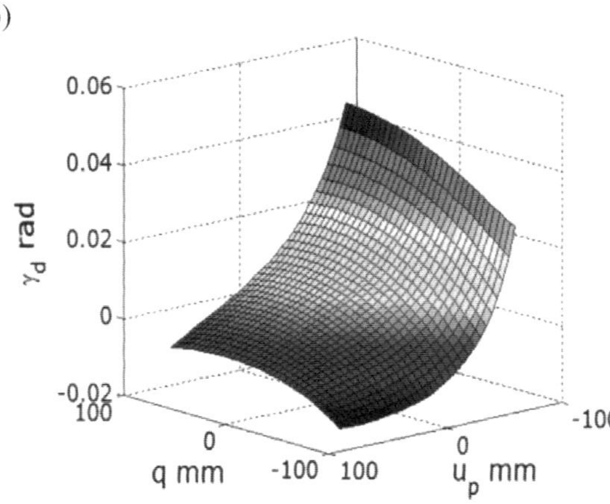

c)

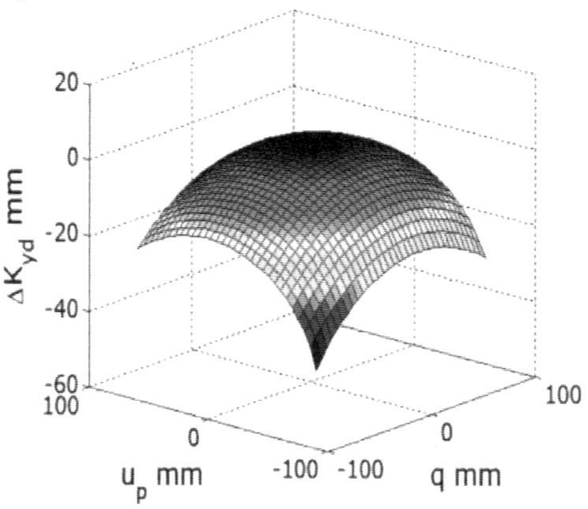

Figure 5. Dependences of steering and camber angles: δ_d, γ_d- respectively and relative lateral displacement of the wheel centre ΔK_{yd} on suspension deflection- q and steering rack displacement- u_p, determined on the basis of the suspension kinematics solution.

12. Measurement uncertainty analysis

Standard uncertainty of the measurement was determined on the basis of numerical simulation using formula shown in equation (17):

$$\Delta f = \pm[\sum_{j=1}^{9}\left(\frac{\partial f}{\partial s_j}\cdot\Delta s_j\right)^2]^{1/2} \quad (17)$$

where:

f - respectively:
- steering angle $\delta_d(s_1,\ldots,s_9)$;
- camber angle $\gamma_d(s_1,\ldots,s_9)$;
- relative lateral displacement of the wheel centre $\Delta K_{yd}(s_1,\ldots,s_9)$;

s_j - j-th displacement sensor elongation

Δs_j - an accuracy of j-th displacement sensor, $\Delta s_j = 0{,}02$ mm was assumed for j=1,…,9.

The sample values of measurement uncertainty were shown below in table 1.

Table 1. The sample maximum values of the measurement uncertainty for the steering δ_d and camber γ_d angels, as well as for the relative lateral displacement of the wheel centre ΔK_{yd}.

	Δδ [°]	Δγ [°]	Δ**ΔK**$_y$ [mm]
Maximum value of measurement uncertainty in case of the maximum left turn and constructional load of the vehicle	± 0,09	± 0,03	± 0,11
Maximum value of measurement uncertainty in case of the return of steering wheels to straight ahead position and constructional load of the vehicle	± 0,10	± 0,02	± 0,02
Maximum value of measurement uncertainty in case of the maximum right turn and constructional load of the vehicle	± 0,04	± 0,01	± 0,11

13. Conclusions

The mechanism of the proposed instrument consists of nine links with optical elongation sensors built-in. The structure of it enables its connection to the stub axle via three joints. Each of them is a conjunction of three joints with a common centre, each with three degrees of freedom.

A four link suspension and rack- and- pinion steering system kinematics problem was solved in the paper. The basic kinematic characteristics of steering and camber angles and the relative lateral displacements of the wheel centre, as well as the links lengths changes of the measuring mechanism were determined. Changes of links lengths were taken into account in the equations of geometric constraints of the instrument, the quantities needed to make up analogical kinematic characteristics of suspension and steering system were calculated. The susceptibility of coupling measuring instrument and car's bod was not taken into account during the presented analysis, while the ideal stiff coupling was assumed.

The analysis of kinematics of steering wheels suspensions mechanisms should be done using spatial characteristics, which are described by functions, which independent variables are the suspension deflection and steering rack displacement. Cross sections of these characteristics can be used for comparison with experimentally determined characteristics.

Basing on the results of the computer simulation and the determined measurement uncertainty, it was found, that- considering practical use- the developed device enables an accurate determination of suspension characteristics. Moreover, there is always a solution of the system of equations of the geometrical constraints of the measuring instrument mechanism, in contrast to an instrument based on a structure of the Stewart platform [13].

References

[1] Reimpell J and Betzler W 2002 *Podwozia samochodów. Podstawy konstrukcji* WKiŁ (Warszawa)

[2] Janczur R 2002 *Analityczno- eksperymentalna metoda badań sterowności samochodu.* Politechnika Krakowska. Rozprawa doktorska. (Kraków)

[3] Struski J 1993 *Quasi- statyczne modelowanie sterowności samochodu* Wydawnictwo Politechniki Krakowskiej, Monografia **144** (Kraków)

[4] Struski J *Przyrząd do pomiaru dynamicznego kąta skrętu koła kierowanego* Patent nr P-267693

[5] Rill G 1986 *Steady State Cornering on Uneven Roadways* SEA Technical Paper No.860575 (Warrendale PA)

[6] Walczak S 2003 *Analiza dynamicznych obciążeń różnych typów niezależnego zawieszenia kół samochodu.* Politechnika Krakowska. Rozprawa doktorska. (Kraków)

[7] Blumenfeld W and Schneider W 1985 *Opto- elektronisches Verfahren zur Spur- und Sturz-winkelmessung am fahrenden Farzeug* ATZ **87** 1 pp 17-21

[8] Struski J and Kowalski M 2008 *Podstawy teoretyczne uogólnionych zagadnień z zakresu parametryzacji układów prowadzenia kół względem nadwozia* Czasopismo Techniczne PK 6-M, pp 119-29

[9] CORSYS- DATRON RV-4 *Wheel Vector Sensor for Simultaneous Measurement of all Wheel Positions and Orientations in 5 Axes* 2008 User manual volume I, (Wetzlar, Germany)

[10] Grzyb A 1992 *On a perturbation method for the analisys of the kinetostatics of mechanisms* Akademie Verlag ZAMN Z Angew. Math. Mech **72**, T615-T618

[11] Góra M 2002 *Analiza kinematyczna wielowahaczowych mechanizmów zawieszeń samochodów* Politechnika Krakowska. Rozprawa doktorska (Kraków)

[12] Młynarski T Listwan A and Pazderski E 1987 Teoria maszyn i mechanizmów Cz. 1 *Synteza i analiza strukturalna mechanizmów* Wydawnictwo Politechniki Krakowskiej (Kraków)

[13] Struski J and Wach K 2012 *Analiza mechanizmu przyrządu pomiarowego do wyznaczania translacji i rotacji zwrotnicy z kołem kierowanym* Czasopismo Techniczne Politechniki Krakowskiej

[14] Struski J and Wach K 2012 *Analiza mechanizmu przyrządu pomiarowego do wyznaczania translacji i rotacji zwrotnicy z kołem kierowanym Engeenering Transactions, 1(**60**)/2012 (Warszawa) pp 40-54*

IOP Conf. Series: Materials Science and Engineering **148** (2016) 012030 doi:10.1088/1757-899X/148/1/012030

Analysis of vehicle dynamics under sadden cross wind

S Walczak[1]

[1]Lecturer, Faculty of Mechanical Engineering, Cracow University of Technology, Cracow, Poland

E-mail: swalczak@mech.pk.edu.pl

Abstract. In this paper, the way of calculating aerodynamic forces acting on a vehicle passing in the region of sadden cross wind was presented. The CarDyn, a vehicle dynamics simulation program, developed by the author was used. The effects of the cross wind were studied with a fixed steering wheel simulation. On the base of computer simulations the car cross wind sensitivity were determined, and vehicle responses such as lateral offset, side acceleration and yaw angular velocity are presented.

1. Introduction

During the drive, car vehicles are particularly exposed to a series of dangerous situations. Those situations may result from the technical condition of the vehicle, the driver extortion and the weather conditions. One of such a dangerous situation, resulting from the weather conditions, can be impact of crosswind on vehicle.

Road test results and computational fluid dynamics numerical simulations (CFD) show that crosswind-vehicle interaction affects both the frontal aerodynamic drag, as well as lateral drag and the aerodynamic lift force.

However, in a fundamental way, the crosswind cause occurrence of the significant component of aerodynamic reaction on lateral axis direction and yawing moment around vertical axis of vehicle. The last one results from the fact that centre of wind pressure of coachwork is situated usually out of the vehicle centre of mass. Influence of crosswind on the vehicle motion can have a significant impact in case of a vehicle driving into open area, e.g. from forested area or during a overtaking manoeuver of truck tractor with trailer. Experimental tests of the car vehicle behaviour under the crosswinds impact are performed in aerodynamic tunnel [4], or during road tests with the use of a wind generator, in accordance with ISO 12021 standard [8]. Road tests should be carried out on dry, paved roads with optimal case of asphalt surface. In special cases, test are allowed to be performed on wet surface, providing the measurement of the water layer depth on the pavement. The preferred length of the wind zone during the test is equal to 25 meters. The velocity of the natural wind during the experimental procedure should be minimized and should not exceed 3 m/s in each direction. The wind generator should produce wind with its velocity equal to 20 ±3 m/s. The direction of generated wind should be perpendicular to the direction of the car vehicle motion. The researchers are conducted for cars, cars with trailers and light truck. Another possibility of the influence of crosswind on car vehicle is conducting the computer numerical simulations using a methods of the Computational Fluid Dynamics (CFD) [3].

2. Mathematical model of vehicle

The vehicle, in the author's program CarDyn, is modelled as a system of rigid bodies described by state parameters, which are divided into two groups: generalized coordinates (y) and generalized velocities (z). In order to describe the motion of rigid bodies with respect to the car body configuration coordinates are chosen, to give a the minimum number of differential equations of motion, since some of these degrees of freedom are eliminated through geometric constraints.

Differential equations of motion have been divided into two groups, the kinematic equations and the dynamic equations. The kinematic equations are expressed by the relationship between generalized coordinates and generalized velocities. The form of these equations depends on the choice of the generalized coordinates and the generalized velocities.

In general case, the kinematic equations have the form:

$$\frac{d}{dt}[\mathbf{y}] = \mathbf{K}^{-1} \cdot [\mathbf{z}] \tag{1}$$

where $\mathbf{K}$ – kinematic matrix.

The dynamic equations take the form:

$$\mathbf{M}_{pq}\dot{\mathbf{z}}_{p} = \mathbf{Q}_{p} \tag{2}$$

General form of symmetric mass matrix $\mathbf{M}_{pq}$ and generalized forces vector $\mathbf{Q}_p$ take following form:

$$\mathbf{M}_{pq} = \sum_{i=1}^{k} \left\{ \frac{\partial \mathbf{v}_{Oi.N}^{T}}{\partial z_p} m_i \frac{\partial \mathbf{v}_{Oi.N}}{\partial z_q} + \frac{\partial \boldsymbol{\omega}_{Oi.N}^{T}}{\partial z_p} T_{Si.N} \frac{\partial \boldsymbol{\omega}_{Oi.N}}{\partial z_q} \right\}_{f \times f} \tag{3}$$

$$\mathbf{Q}_{p} = \sum_{i=1}^{k} \left\{ \begin{array}{l} \frac{\partial \mathbf{v}_{Oi.N}^{T}}{\delta z_p} \left[\mathbf{F}_{i.N}^{z} - m_i \mathbf{a}_{Oi.N}^{R} \right] + \\ + \frac{\partial \boldsymbol{\omega}_{Oi.N}^{T}}{\partial z_p} \left[\mathbf{M}_{i.N}^{z} - \mathbf{T}_{Si.N} \dot{\boldsymbol{\omega}}_{Oi.N}^{R} - \boldsymbol{\omega}_{Oi.N} \times \mathbf{T}_{Si.N} \boldsymbol{\omega}_{Oi.N} \right] \end{array} \right\}_{f \times 1} \tag{4}$$

where:

m_i – the mass the ith body,

$\mathbf{T}_{Si.N}$ – the inertia tensor of the ith body with respect to the vehicle-fixed reference frame {N},

$\mathbf{F}_{i.N}^{z}$ and $\mathbf{M}_{i.N}^{z}$ – external forces and moments,

$\mathbf{a}_{Oi.N}^{R}$, $\dot{\boldsymbol{\omega}}_{Oi.N}^{R}$ - „residual" accelerations [9] of the ith body centre of mass.

The vehicle is modelled as a system 9 rigid bodies with 17 degrees of freedom:

- vehicle body – 6 degrees of freedom,
- front suspension – 5 degrees of freedom,
- rear suspension – 2 degrees of freedom,
- each of the wheels – 1 degree of freedom.

The simulation program adopted nonlinear characteristics of elastic-damping elements of suspensions, designated on the basis of experimental research.

In order to describe the interaction of wheel tires with the road surface, the *TM-Easy* tires model [9] is used, modified in papers [11], [12]. The model takes into account the elastic-damping properties of tires in vertical, longitudinal and transverse direction.

The transient states of each tire are described in two differential equations of the first order [9], [12].

3. Aerodynamic forces and moments acting on the vehicle

Taking advantage of mentioned in research methods, one can assign aerodynamic forces and moments acting on vehicle for given car motion condition and wind velocity and direction. If car vehicle in driving with velocity $\mathbf{v}_{0,0}$, whereas air is moving with velocity $\mathbf{v}_{w,0}$, then the velocity of relative motion of moving air to vehicle is equal to:

$$\mathbf{v}_{w0,0} = \mathbf{v}_{w,0} - \mathbf{v}_{0,0} \tag{1}$$

In the system of vehicle body, this velocity is equal to:

$$\mathbf{v}_{r,N} = \mathbf{A}_{ON}^{T} \cdot \mathbf{v}_{w0,0} \tag{2}$$

where:

$\mathbf{A}_{ON}$ – the matrix of the transformation from the system associated with the body to the absolute system,

ψ – the yaw angle,
θ – the pitch angle,
φ – the roll angle.

$$\mathbf{A}_{ON} = \mathbf{A}_{\psi}\mathbf{A}_{\theta}\mathbf{A}_{\varphi} = \begin{bmatrix} \cos\psi & -\sin\psi & 0 \\ \sin\psi & \cos\psi & 0 \\ 0 & 0 & 1 \end{bmatrix} \begin{bmatrix} \cos\theta & 0 & \sin\theta \\ 0 & 1 & 0 \\ -\sin\theta & 0 & \cos\theta \end{bmatrix} \begin{bmatrix} 1 & 0 & 0 \\ 0 & \cos\varphi & -\sin\varphi \\ 0 & \sin\varphi & \cos\varphi \end{bmatrix}$$

While the length of the velocity vector is equal to:

$$v_r = \left|\mathbf{v}_{r,N}\right| = \sqrt{v_{rN,x}^2 + v_{rN,y}^2 + v_{rN,z}^2} \tag{3}$$

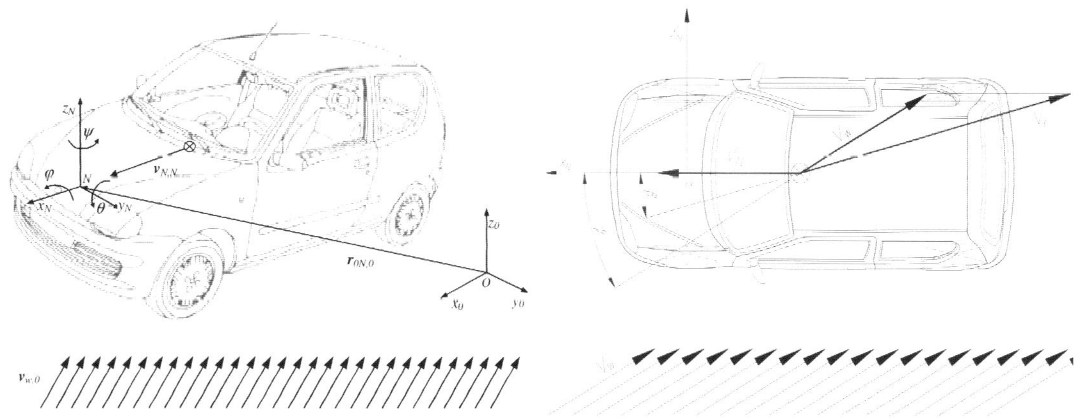

Figure 1. The view of vehicle model with definitions the coordinate system applied in this study.

In case analysed plane motion, the velocity of the relative motion of air and vehicle, and the angle of deviation of relative velocity from the longitudinal axis of the vehicle may be expressed in following relationship:

$$\begin{aligned} v_r^2 &= \left(v_N + v_w \cos\psi\right)^2 + v_w^2 \sin^2\psi \\ \beta &= \frac{v_w \sin\psi}{v_N + v_w \cos\psi} \end{aligned} \tag{4}$$

Aerodynamic forces and moments acting of the vehicle, which are a consequence of cross-interaction of car and air, are the resultant of normal surface forces which come from pressure distribution on car surface and tangential forces which are a consequence of air viscosity. This are viscosity causes a complex flow around the vehicle, thus the assignation of the aerodynamic forces and moments is conducted by the use of empirical dependences:

$$\mathbf{F}_{Air} = \begin{bmatrix} F_{Air,x} \\ F_{Air,y} \\ F_{Air,z} \end{bmatrix} = A \cdot \frac{\rho_p \cdot v_r^2}{2} \begin{bmatrix} c_{Fx} \\ c_{Fy} \\ c_{Fz} \end{bmatrix}$$
$$\mathbf{M}_{Air} = \begin{bmatrix} M_{Air,x} \\ M_{Air,y} \\ M_{Air,z} \end{bmatrix} = A \cdot L_O \cdot \frac{\rho_p \cdot v_r^2}{2} \begin{bmatrix} c_{Mx} \\ c_{My} \\ c_{Mz} \end{bmatrix} \tag{5}$$

where:

A – the frontal area,

ρ_p – the air density,

L_O – the vehicle wheel base,

c_x, c_y, c_z, c_{Mx}, c_{My}, c_{Mz} – dimensionless coefficients of the drag force, the lateral force, the lift force, the pitching moment, the rolling moment and the yawing moment.

A typical characteristics of the aerodynamic dimensionless coefficients for a car vehicle [1] are presented in figure 2.

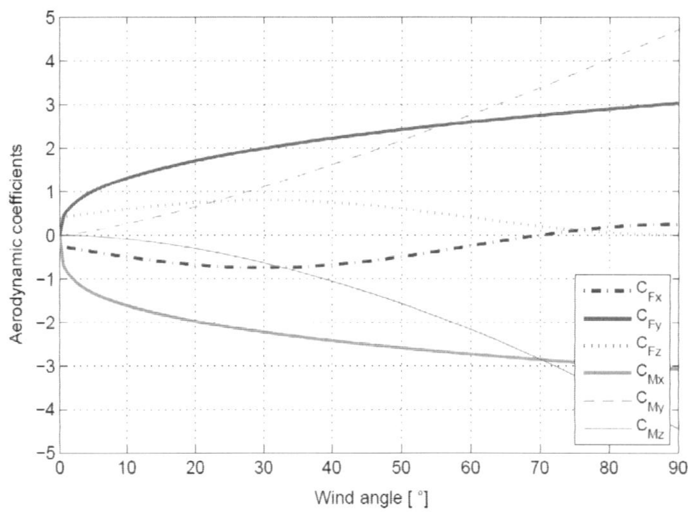

Figure 2. Typical characteristics of the aerodynamic dimensionless coefficients for a passenger car according to [1].

As it follows from the presented characteristics, during the symmetrical flow around the vehicle ($\psi = 0$), a car is only under the load of the drag force ($F_{Air,x}$) and the lift force ($F_{Air,z}$). The stability and the steerability of the car and the road safety under the cross wind is under the greatest influence of the lateral force ($F_{Air,y}$) and the yawing moment ($M_{Air,y}$). Due to this fact, in the simulation program, following aerodynamic forces and moments are included:

- the drag force ($F_{Air,x}$),
- the lateral force ($F_{Air,y}$),
- the yawing moment ($M_{Air,z}$).

The lateral force is acting on the car body in point called the centre of pressure (COP), which usually is not coincident with the centre of mass of the vehicle. Due to this fact, during the car drive through the area of cross wind, the yawing moment acting on the car is produced, thus it leads to the change of the drive direction.
Location of the centre of pressure (COP) with respect to the centre of mass which is used to assign aerodynamic moments acting on the vehicle can be determined by the comparison of dependence describing the lateral force multiplied by the arm of its action with the relationship of the yawing moment. It leads to formula:

$$r_{Air,x} = \frac{c_{Mz}}{c_y} \cdot L_O \tag{6}$$

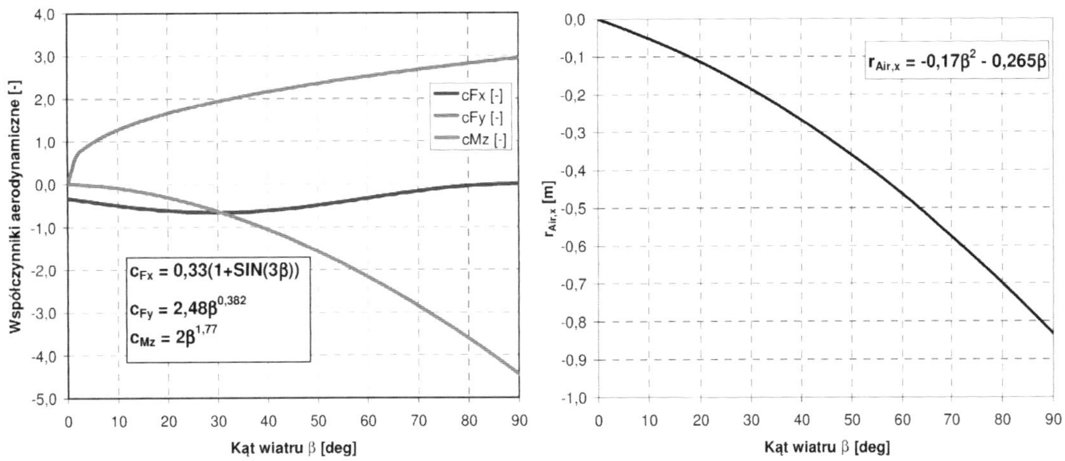

Figure 3. Typical characteristics of the aerodynamic dimensionless coefficients for a modelled vehicle, taken from the figure (2) and calculated on the base of formula (2).

Dimensionless coefficients of aerodynamic forces and moments in steady state flow conduction depend primarily on the angle of deviation of the relative velocity from the longitudinal axis of the vehicle, while the frontal area occurs in dependencies which define the aerodynamic forces and moments.
In presented paper the analysis of the car drive through the 20m long area of cross wind which flows in direction perpendicular to the initial direction of the vehicle motion. In such case, in the initial section when the vehicle drives in a cross wind zone, and in the end section when it drives out of this zone, the air does not act directly on the entire lateral surface of the vehicle but only on this part which is in the zone of air flow. This causes a change of the character of the flow around the vehicle body in comparison with the flow conditions for which dimensionless coefficients of aerodynamic forces and moments are determined. It is assumed that in this driving areas value of the aerodynamic lateral force is proportional to the lateral surface of the car directly exposed to the wind. The lateral aerodynamic force is calculated from the modified Formula:

$$F_{Air,y} = \frac{A_y(x)}{A_{y0}} A \cdot \frac{\rho_p \cdot v_r^2}{2} c_{Fy} \tag{7}$$

where:

A_y – the lateral area,

$A_y(x)$ – lateral surface of the car body directly exposed to the wind (fig. 4-5),

For this reason, the lateral surface of the car has been divide into section, thereby defining the relationship describing the lateral surface of the car directly exposed to the cross wind. It is assumed, that the yawing moment is caused by the aerodynamic lateral force applied at the point called the centre of pressures and it is also proportional to car lateral surface which is directly exposed to the cross wind.

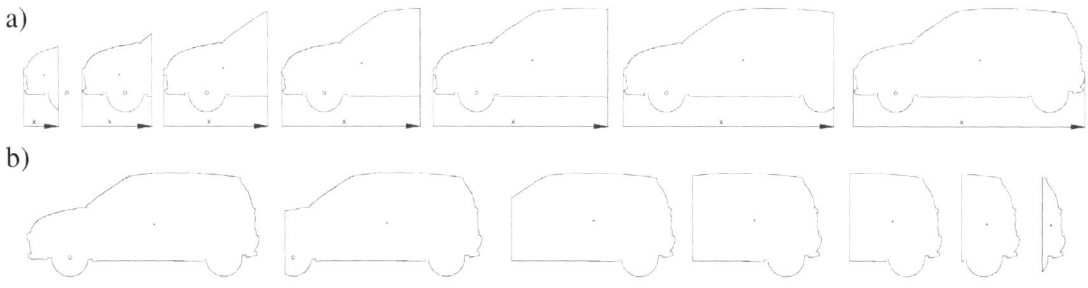

Figure 4. The lateral surface of the car body directly exposed to the wind a) in the stage when car is driving in a cross wind area, b) in the stage when car is driving out of this a cross wind area.

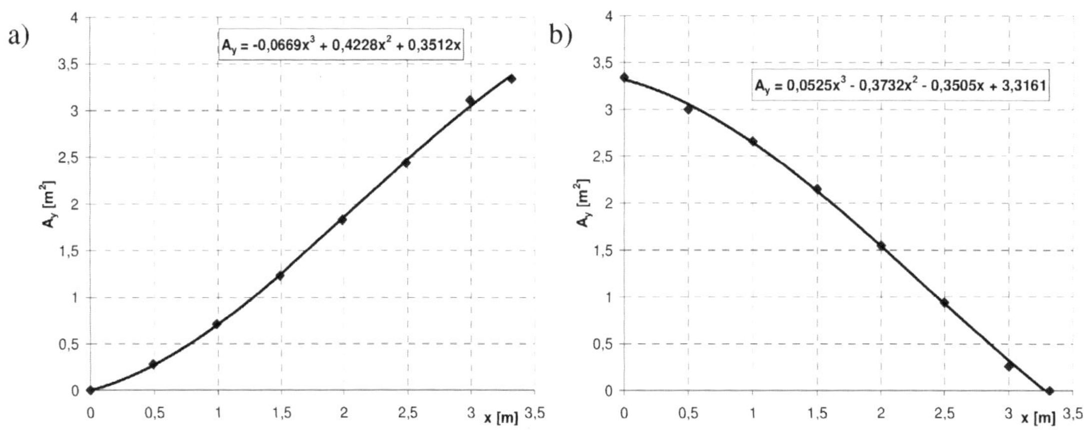

Figure 5. The relationship describing the lateral surface of the car directly exposed to the wind: a) in the stage of driving in the cross wind area in the function of the distance of the front of the car to the starting point of the wind area, b) in the stage of driving out of the cross wind area in the function of the distance of the front of the car to the ending point of the wind area.

Relationships describing aerodynamic coefficients used in this research are developed on the basis of the formula given in paper [2] and characteristics of dimensionless aerodynamic coefficients for a small car, illustrated in figure 2, taken from [7]. Aerodynamic coefficients used in the simulation are determined from the following relationships:

$$\begin{aligned} c_{Fx} &= 0{,}33\cdot\left(1+\sin\left(3\beta\right)\right) \\ c_{Fy} &= 2{,}48\cdot\beta^{0{,}382} \\ c_{Mz} &= 2\cdot\beta^{1{,}77} \end{aligned} \tag{8}$$

4. Simulation tests

In the analysis, a small car is chosen with the wheel base equal to 2,2 m and total mass equal to 940 kg. Simulation research have been carried out for three different driving speeds corresponding to the maximum speed limits for different type of roads – 50 km/h, 90 km/h and 140 km/h respectively. In the simulation the kinematic input coming from unevenness of the road surface is omitted. Simulations have been performed with the steering wheel hold down in toward direction. It is assumed that the car is driving through the 20 m long area of cross wind which flows with three different velocities equal to 20 m/s, 25 m/s and 30 m/s respectively (see in figure 6).

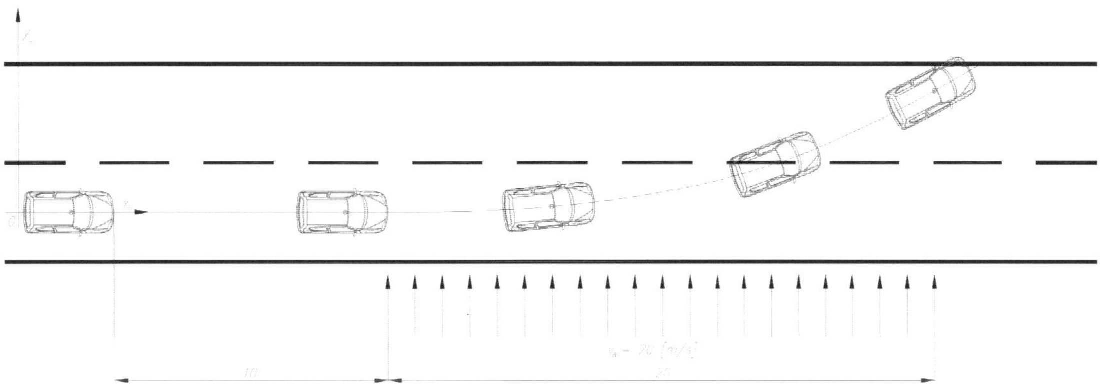

Figure 6. The geometry of driving manoeuver in cross wind area according to [7]

Results of selected simulations are presented in graphs (figures 7-9). Results present the lateral displacement of the car in function of time. Safety of driving car is dependent on this parameter. The maximum allowable lateral displacement is assumed as a distance equal to 1 m at time of 1 s form the driving the car in the cross wind area. Adopted time is equal to the time of psychophysical response of the driver to identified abnormal behaviour of the car.
Allowable displacement results from the average width of the car equal to 1,5 m, which driving around the lane axis of its width equal to 3,5 m, has approximately 1m distance to the lane border [5].

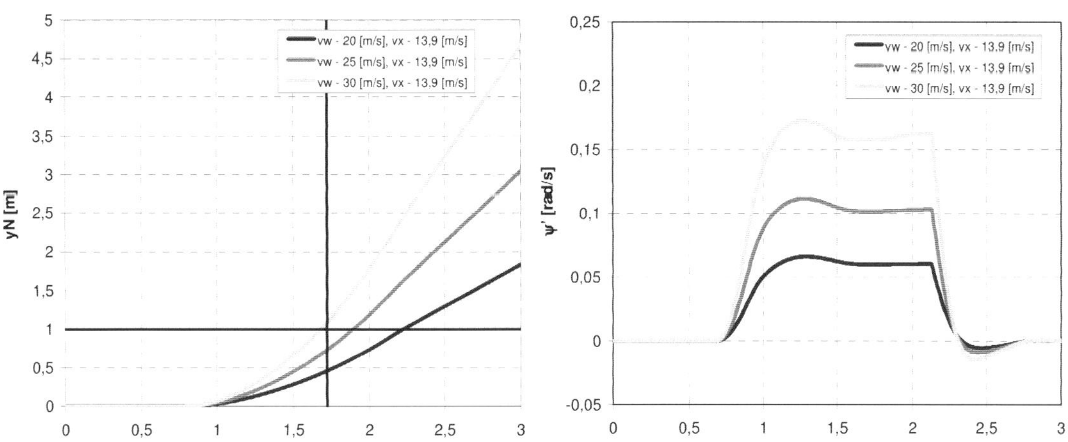

Figure 7. Driving of the small car with speed of 50 km/h through the cross wind area blowing with velocity equal to 20 m/s, 25 m/s and 30 m/s a) lateral displacement of the vehicle, b) yaw velocity.

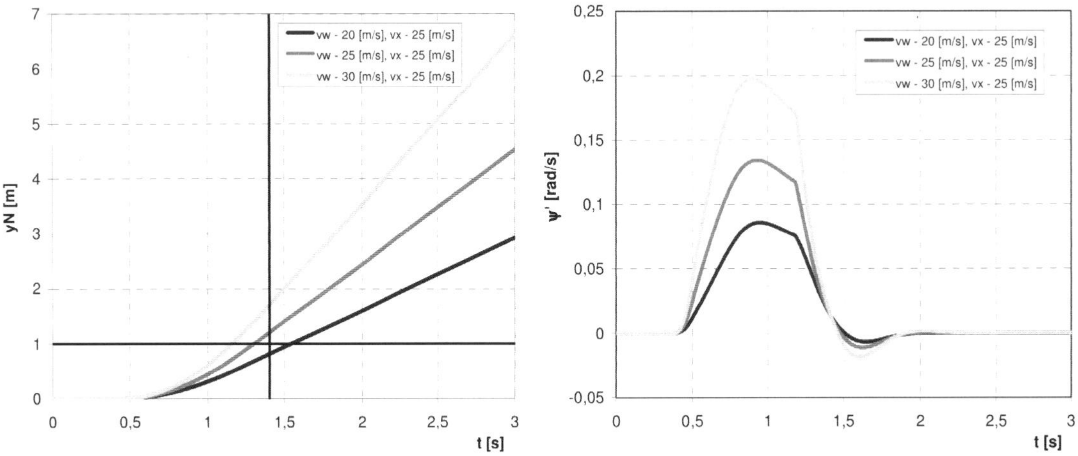

Figure 8. Driving of the small car with speed of 90 km/h through the cross wind area blowing with velocity equal to 20 m/s, 25 m/s and 30 m/s a) lateral displacement of the vehicle, b yaw velocity.

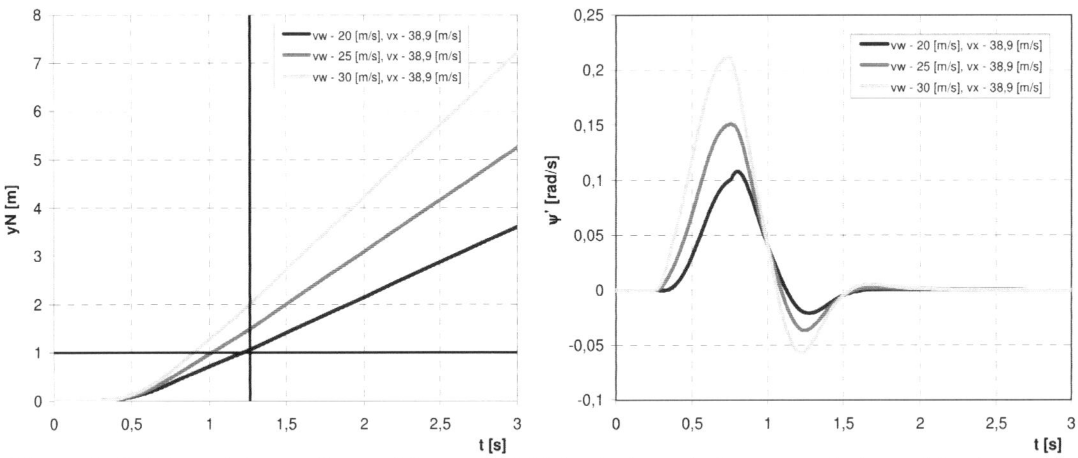

Figure 9. Driving of the small car with speed of 90 km/h through the cross wind area blowing with velocity equal to 20 m/s, 25 m/s and 30 m/s a) lateral displacement of the vehicle, b yaw velocity.

5. Summary

Presented simulation results indicates that for adopted date relating to the aerodynamic coefficients, driving the analysed car with speed of 50 km/h through the cross wind area, the safety criterion is satisfied for the cross wind blowing at speed of approximately 30 m/s. For the driving with speed of 90 km/h through the cross wind area, the safety criterion is satisfied for the cross wind blowing at speed of approximately 20 m/, whereas for a driving speed of 140 km/h the safety criterion is satisfied for neither of analysed wind speed.

The developed mathematical model of the car, implemented in the author's program CarDyn gives wide opportunity to test the behaviour of car vehicles exposed to the cross wind action. It can be useful in reconstruction of road accidents, for which the sudden change of trajectory of the car motion could be caused by the drive through the area of cross wind, which can take place on bridges, viaducts, during driving out of the forested areas, or during overtaking large trucks.

Verification of adopted aerodynamic dimensionless coefficients for the analysed car is planned through the numerical simulation of air flow around a vehicle with the use of Computational Fluid Dynamics method (CFD).

References

[1] Baker C 1987 Measures to control vehicle movement at exposed sites during windy periods. Journal of Wind Engineering and Industrial Aerodynamics, 25(2), pp. 151–161.
[2] Batista M and Perkovič M 2014 A simple static analysis of moving road vehicle under crosswind. Journal of Wind Engineering and Industrial Aerodynamics 128(2014) pp. 105-113.
[3] Bin Wang, You-Lin Xu, Le-Dong Zhu and Yong-Le Li 2014 Crosswind Effect Studies on Road Vehicle Passing by Bridge Tower using Computational Fluid Dynamics. Engineering Applications of Computational Fluid Mechanics, 2014, 8:3, pp. 330-344
[4] Cheli F, Corradi R, Sabbioni E and Tomasini G 2011 Wind tunnel tests on heavy road vehicles: Cross wind induced loads - part 1. Journal of Wind Engineering and Industrial Aerodynamics, 99(10):1000–1010, 2011.
[5] Gajek A and Walczak S (2002) Analiza wpływu nierównomierności sił hamowania na stateczność ruchu samochodu. Zeszyty Naukowe Politechniki Świętokrzyskiej. Mechanika 76, str. 169-177, Politechnika Świętokrzyska, Kielce 2002.
[6] Gillespie T.D 1992 Fundamentals of Vehicle Dynamics. SAE Inc., Warrendale, PA.
[7] Hanke O, Bertram T, Hiller M (2001) Analysis and Control of Vehicle Dynamics Under Crosswind Conditions. IEE/ASME 6–12 July 2001, Como, Italy.
[8] ISO 12021 (2010) *Road vehicles - Sensitivity to lateral wind - Open-loop test method using wind generator input.*
[9] Rill G 1994 Simulation von Kraftfahrzeugen. Vieweg & Sohn Verlagsgesellschaft, Braunschweig/Wiesbaden.
[10] Snæbjörnsson J Th, Baker C J and Sigbjörnsson R 2007 Probabilistic assessment of road vehicle safety in windy environments. Journal of Wind Engineering and Industrial Aerodynamics 95 (2007) pp. 1445–1462.
[11] Walczak S 2003 Analiza dynamicznych obciążeń elementów niezależnych zawieszeń kół samochodu. Politechnika Krakowska. Rozprawa doktorska, Kraków.
[12] Walczak S 2012 Wpływ właściwości dynamicznych modelu ogumienia na dynamikę poprzeczną samochodu. Czasopismo Techniczne z. 3-M/2012, str. 101-113.

Influence of the vertical load exerted by the trailer on the coupling device on towing vehicle's steerability and stability

D Więckowski[1], T Pusty[1] and P Jędrys[1]

[1] Automotive Industry Institute (PIMOT), ul. Jagiellońska 55, 03-301 Warszawa

E-mail: d.wieckowski@pimot.eu, t.pusty@pimot.eu, p.jedrys@pimot.eu

Abstract. The paper concerns an analysis of the steerability and stability of a motor vehicle towing a trailer during the double lane change manoeuvre. As results of experimental road tests carried out on an airfield runway, time histories of the following parameters were determined: steering angle, lateral acceleration, and drift angle of the towing vehicle, as well as angular velocity of towing vehicle yaw. The tests were performed for various states of the vertical load exerted on towing vehicle's coupling device.

1. Introduction

For a trailer to be safely towed, the coupling device components installed on the towing vehicle and on the trailer should be correctly selected and matched up to each other so that adequate cooperation between the vehicles combined together might be ensured. When a tractor-trailer unit is moving, the following forces act on the vehicle coupling device:

1) Tractive force necessary to overcome trailer's rolling resistance, air resistance, and grade resistance and to accelerate the trailer;
2) Braking force:
   a) for unbraked trailers, the total braking force is transmitted by the coupling device;
   b) for trailers with brakes, a braking force may occur in the coupling device due to imperfect synchronization between the operation of brakes of both vehicles;
3) Vertical forces applied by the drawbar in the case of a centre-axle trailer, pressing down the coupling device.

The coupling devices should meet the following requirements [3, 4]:

- The coupling device components installed on the towing vehicle and on the trailer should be compatible with each other in terms of their dimensional and strength characteristics.
- They shall prevent separation of the vehicles from each other when in motion.
- They shall attenuate any shocks and jerks.
- They shall ensure articulation of the vehicle coupling and enable angular displacement of the longitudinal vehicle centrelines, especially on road bends and uneven ground.
- Their safe coupling and uncoupling shall be possible by one person without the use of tools.
- All the couplings must be made as mechanical connections.
- The longitudinal centreline of the coupling device components should coincide with the longitudinal vertical symmetry plane of the towing vehicle during rectilinear forward motion of the tractor-trailer unit.

- All the couplings shall have positive mechanical engagement and the closed position shall be locked at least once by further positive mechanical engagement.
- The effects of vibration must not cause component parts of the coupling device to move to positions where the device may open or disengage.
- The coupling device on the motor vehicle shall be designed to support the vertical load that might be exerted on it by a centre-axle trailer.

The above shows that the coupling devices must meet various requirements arising from complex issues related to strength of materials, dynamics, and kinematics.

At present, the coupling devices are subject to type-approval requirements laid down in UN ECE Regulation No. 55.01 [4]. It should be stressed here that the type-approval tests do not include road tests, i.e. they are limited to rig tests only. The advisability of the conducting of analyses concerning the dynamics of the mechanical connection between the trailer and the towing vehicle has been shown in publications [6, 7, 8]. In this connection, it has been considered reasonable to undertake the road testing of the coupling devices, with carrying out dynamic measurements in real (road) conditions. It should be stressed here that the complex construction and operation of the coupling devices on the one hand, and their critical importance related to the ensuring of reliable mechanical connection between two (or more) vehicles on the other hand, must be kept in mind. Moreover, the impact of the trailer on the steerability and stability of the towing vehicle must also be taken into consideration.

2. Experimental road tests

Road tests are very important from the explorative and research work point of view. They should not be skipped when the issues related to the tractor-trailer interactions are explored. The mechanical coupling devices and their parts are defined as all the elements of vehicle frame as well as all the load-carrying parts of the motor vehicle and trailer substructure (undercarriage) and superstructure (body) with the use of which the vehicles are coupled together to form a tractor-trailer unit or an articulated vehicle. They also include the permanently fixed or detachable components intended for the fastening or operation of the coupling devices and parts mentioned above [4, 5].

2.1. Vehicle testing during the double lane change manoeuvre [1]

The tests of a tractor-trailer unit during the double lane change manoeuvre according to [1] are carried out in a closed-loop system, i.e. with the driver-vehicle-driver feedback, and they are dedicated to passenger cars and light trucks with up to 3 500 kg GWM (Gross Vehicle Mass). During the test, the vehicle is driven along a prescribed test road section (within a predefined corridor), schematically presented in Figure 1, with a constant speed of V = 80 km/h and with a maximum possible speed. If the vehicle cannot achieve the speed of V = 80 km/h, only the latter test option is performed.

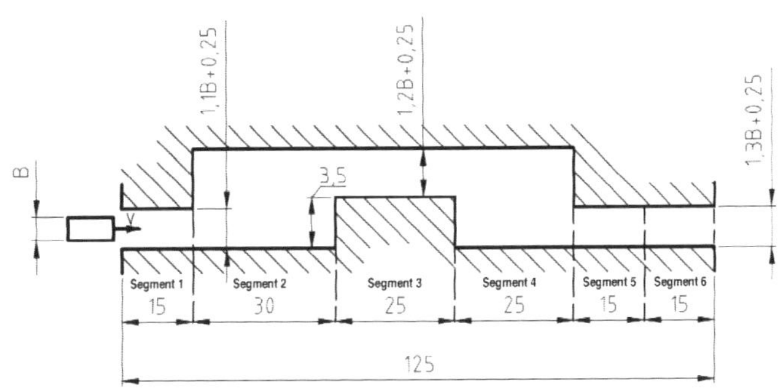

Figure 1. Schematic outline of the test road section (test corridor) during the double lane change manoeuvre: dimensions in [m]; B – vehicle width [1].

Based on the test, time histories of the following parameters can be experimentally determined [1]:
- lateral acceleration $a_y(t)$;
- angular velocity $\dot{\psi}(t)$ of towing vehicle yaw (rotation around the vertical axis);
- roll angle $\varphi(t)$ of the towing vehicle (rotation around the longitudinal axis);
- steering torque $M_h(t)$;
- drift angle $\beta = \text{arctg}\dfrac{V_y}{V_x}$,

  where V_x and V_y are longitudinal and lateral components of the vehicle velocity, respectively.

In the tractor-trailer unit, the distribution of the load carried on the trailer was varied with the use of two tanks filled with water, which represented the real trailer loading conditions; the towing vehicle load remained unchanged (the vehicle was loaded to its Gross Vehicle Mass). The trailer load was arranged in two ways, referred to as option I and option II, in order to obtain two different values of the load on the device that coupled the trailer with the motor truck used as the towing vehicle. The values of the coupling device load were selected on the grounds of the data given in the normative documents [2, 5].

The technically permissible maximum load on the coupling point (defined as the mass corresponding to the maximum permissible static vertical load to be transferred by the trailer to the towing vehicle at the coupling point) must be at least equal to 4 % of the Gross Trailer Mass or 25 kg, whichever is the greater (for the trailer under test, the calculated value of the load on the coupling point was 100 kg); the force P necessary to pull the ball out of its socket should meet the inequality below:

$$P > 20\ W_s \tag{1}$$

where: W_s – maximum acceptable vertical force applied to the ball socket [N]; W_s = 300 N for trailers with a mass of $m_R \leq 430$ kg; $W_s = 7\ \%\ (m_R * g)$ for trailers with a mass m_R from 431 kg to 1 700 kg; W_s = 1 200 N for trailers with a mass m_R from 1 701 kg to 3 500 kg; m_R – actual trailer mass [kg].

For the vehicles under test, the value of force P was 24 000 N.

Option I (denoted by "PS_NOK"). The arrangement of the test load placed on the trailer corresponded to incorrect distribution of real trailer load in result of which the load on the coupling device was directed opposite to the acceleration of gravity g (i.e. the pressure of the rear axle wheels of the motor vehicle on the road was reduced). In static conditions, the value of the vertical load on the coupling device was PS = −1118 N (the load was directed opposite to the acceleration of gravity g, i.e. vertically upwards). The distance between the ball of the coupling device and the front plane of the tanks filled with water was 3 200 mm.

Option II (denoted by "PS_OK"). The arrangement of the test load placed on the trailer corresponded to correct distribution of real trailer load in result of which the load on the coupling device was directed like the acceleration of gravity g (i.e. the pressure of the rear axle wheels of the motor vehicle on the road was increased). In static conditions, the value of the vertical load on the coupling device was PS = 1746 N (the load was directed like the acceleration of gravity g, i.e. vertically downwards). The distance between the ball of the coupling device and the front plane of the tanks filled with water was 2 800 mm.

The view of the trailer prepared for testing has been shown in Figure 2.

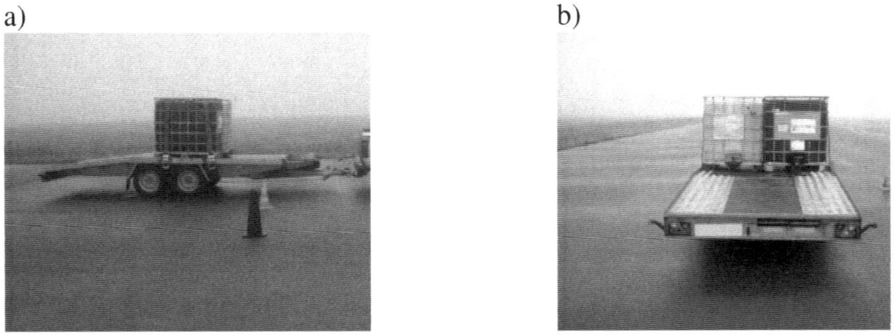

Figure 2. The trailer coupled with the motor vehicle during the tests:
a) side view; b) rear view.

2.2. Analysis of the test results

The maximum speed that could be achieved for the manoeuvre to be safely completed (i.e. for the vehicle to go through the whole predefined corridor) was 65 km/h. When the vehicle speed exceeded this limit, trailer wheels were lifted off the road surface. The tests were carried out for option I and option II of trailer load distribution. Time histories of the steering wheel angle δ_H, recorded during the double lane change manoeuvre, have been presented in Figure 3. The following symbols have been adopted: "DMC_65" represents the test drive of the motor vehicle without the trailer, "PS_NOK_65" represents the test drive of the motor vehicle with the trailer loaded to option I, and "PS_OK_65" represents the test drive of the motor vehicle with the trailer loaded to option II.

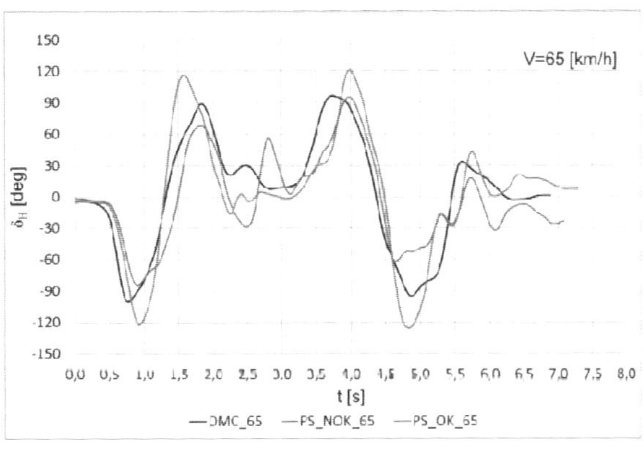

Figure 3. Time histories of the steering wheel angle δ_H.

The extreme absolute values of the steering wheel angle δ_H in individual sectors (see Figure 3, cf. Figure 1) were recorded for option II, i.e. PS_OK_65. In the tests of the tractor-trailer unit, lower amplitudes of δ_H were recorded for option I (PS_NOK_65); however, distinct corrections of the steering wheel angle occurred when the tractor-trailer unit went through sectors 3, 5, and 6 of the test corridor. The extreme absolute values of the steering wheel angle δ_H in individual sectors of the test corridor have been given in Table 1.

Table 1. Extreme absolute values of the steering wheel angle δ_H in individual sectors of the test corridor.

Sector of the test corridor	Absolute value of δ_H [deg]		
	DMC_65	PS_NOK_65	PS_OK_65
Sector 1	100.7	84.6	121.7
Sector 2	88.7	67.8	116.4
Sector 3	96.3	95.2	121.2
Sector 4	102	62.5	126.1

For further analysis, the values denoted by DMC_65, recorded in the test during which no force was applied to the coupling device, were adopted as a reference and compared with the extreme absolute values of the steering wheel angle δ_H in individual sectors of the test corridor. In the test denoted by PS_NOK_65, the extreme absolute values of the steering wheel angle δ_H were lower by about 40 % than the corresponding reference values. Conversely, the extreme absolute values of δ_H recorded during the test denoted by PS_OK_65 in individual sectors of the test corridor exceeded the reference values by 20-30 %.

Figure 4 shows time histories of the lateral acceleration a_y, recorded during the manoeuvre under analysis.

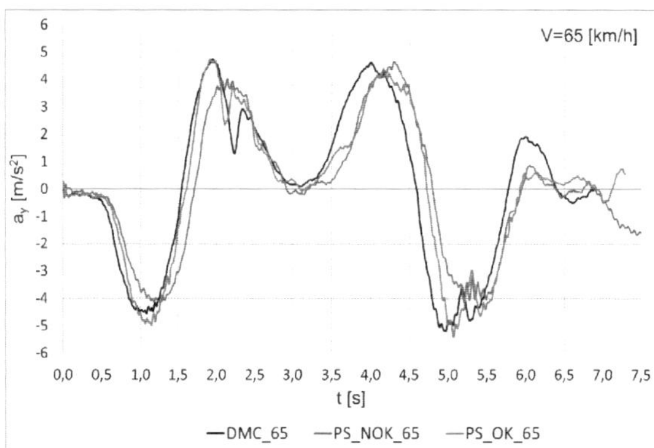

Figure 4. Time histories of the lateral acceleration a_y.

The lateral acceleration vs. time curves ($a_y = f(t)$) having been recorded are similar to each other in qualitative and quantitative terms, but a phase shift in the a_y amplitudes occurred when the tractor-trailer unit went through sectors 4 and 5 of the test corridor. The extreme absolute values of the lateral acceleration a_y in individual sectors of the test corridor have been given in Table 2. In the test denoted by PS_NOK_65, the absolute values of the lateral acceleration a_y were lower by about 10-20 % in sectors 1, 2 and 4 of the test corridor than the corresponding reference values. Conversely, the absolute values of the lateral acceleration a_y, recorded during the test denoted by PS_OK_65, were higher by about 10 % in sectors 1 and 4 and lower by up to 10 % in sectors 2 and 3 as against the corresponding reference values.

Table 2. Extreme absolute values of the lateral acceleration a_y

Sector of the test corridor	Absolute value of a_y [m/s^2]		
	DMC_65	PS_NOK_65	PS_OK_65
Sector 1	4.6	4.1	5.0
Sector 2	4.7	4.0	4.7
Sector 3	4.6	4.7	4.2
Sector 4	5.2	4.1	5.4

Time histories of the drift angle β have been presented in Figure 5. The extreme absolute values of the drift angle β in individual sectors of the test corridor have been given in Table 3. The extreme absolute values of the drift angle β recorded in individual sectors of the test corridor were highest for the PS_OK_65 option and lowest for the DMC_65 option. It should be simultaneously stressed that growths in the β values occurred in the PS_NOK_65 test in sectors 3, 5, and 6 of the test corridor. In the PS_NOK_65 test and the PS_OK_65 test, the absolute values of the drift angle β were higher by 5-160 % and by 40-180 %, respectively, than the corresponding reference values.

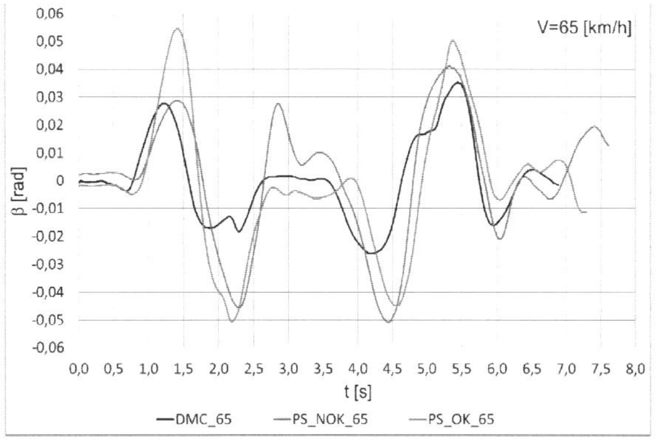

Figure 5. Time histories of the drift angle β.

Table 3. Extreme absolute values of the drift angle β

Sector of the test corridor	Absolute value of β [rad]		
	DMC_65	PS_NOK_65	PS_OK_65
Sector 1	0.028	0.029	0.054
Sector 2	0.018	0.046	0.051
Sector 3	0.026	0.051	0.045
Sector 4	0.035	0.041	0.050

Figure 6 shows time histories of the angular velocity $\dot{\psi}$ of motor vehicle yaw.

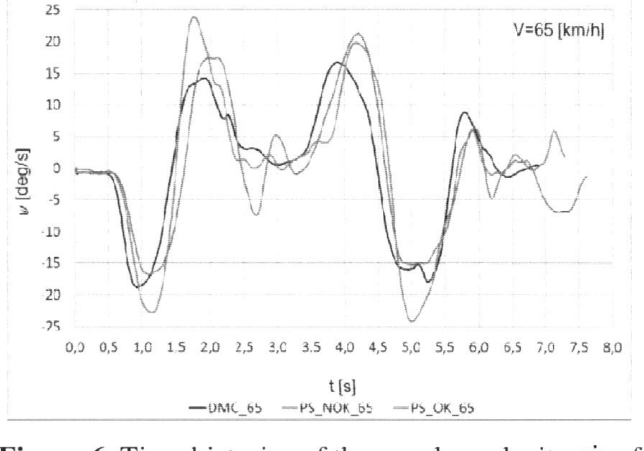

Figure 6. Time histories of the angular velocity $\dot{\psi}$ of motor vehicle yaw.

The extreme absolute values of the angular velocity $\dot{\psi}$ of motor vehicle yaw as recorded in individual sectors of the test corridor were highest for the PS_OK_65 option and lowest for the DMC_65 option. It should be simultaneously stressed that growths in the $\dot{\psi}$ values occurred in the PS_NOK_65 test in sectors 3, 5, and 6 of the test corridor. The extreme absolute values of the angular velocity $\dot{\psi}$ of motor vehicle yaw in individual sectors of the test corridor have been given in Table 4. During the test denoted by PS_NOK_65, the absolute values of the angular velocity $\dot{\psi}$ of motor vehicle yaw were lower by 10-15 % in sectors 1 and 4 of the test corridor and higher by about 20-30 % in sectors 2 and 3 than the corresponding reference values. In the test denoted by PS_OK_65, in turn, the absolute values of the angular velocity $\dot{\psi}$ of motor vehicle yaw were higher by 20-70 % as against the corresponding reference values during the whole manoeuvre.

Table 4. Extreme absolute values of the angular velocity $\dot{\psi}$ of motor vehicle yaw

Sector of the test corridor	Absolute value of $\dot{\psi}$ [deg/s]		
	DMC_65	PS_NOK_65	PS_OK_65
Sector 1	18.89	16.82	22.80
Sector 2	14.25	17.53	23.91
Sector 3	16.78	21.42	19.79
Sector 4	17.94	15.22	24.20

Figure 7 shows time histories of the motor vehicle roll angle φ.

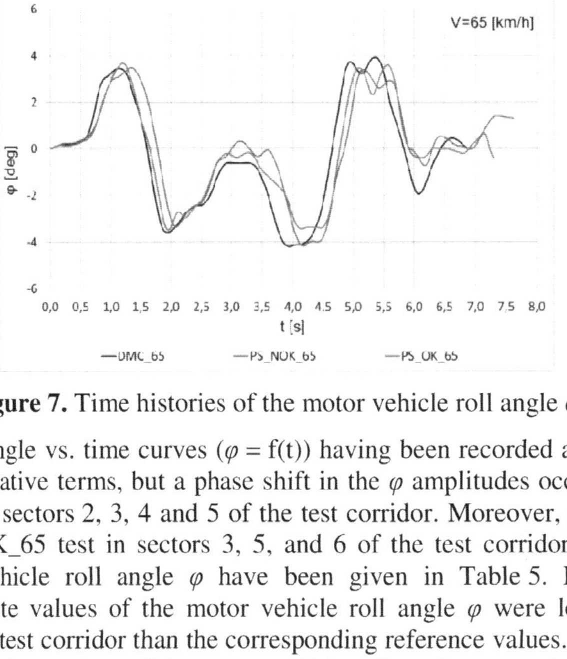

Figure 7. Time histories of the motor vehicle roll angle φ.

The motor vehicle roll angle vs. time curves (φ = f(t)) having been recorded are similar to each other in qualitative and quantitative terms, but a phase shift in the φ amplitudes occurred when the tractor-trailer unit went through sectors 2, 3, 4 and 5 of the test corridor. Moreover, growths in the φ values occurred in the PS_NOK_65 test in sectors 3, 5, and 6 of the test corridor. The extreme absolute values of the motor vehicle roll angle φ have been given in Table 5. In the test denoted by PS_NOK_65, the absolute values of the motor vehicle roll angle φ were lower by up to 10 % in sectors 2, 3, and 4 of the test corridor than the corresponding reference values. During the test denoted by PS_NOK_65, the absolute values of the motor vehicle roll angle φ were lower by 3-20 % than the corresponding reference values in the same sectors (2, 3, and 4) of the test corridor. Based on the recorded time histories of the values of the steering wheel angle δ_H, lateral acceleration a_y, drift angle β, angular velocity of motor vehicle yaw $\dot{\psi}$, and motor vehicle roll angle φ, a statement may be made that differences existed between dynamic characteristics of the vehicles subjected to the DMC_65, PS_NOK_65, and PS_OK_65 test options. The qualitative and quantitative differences show that:

- the impact on lateral dynamic behaviour of vehicle articulated with trailer;
- the value of the static load on the coupling device affects the steerability and stability of the towing vehicle.

Table 5. Extreme absolute values of the motor vehicle roll angle φ

Sector of the test corridor	Absolute value of φ [deg]		
	DMC_65	PS_NOK_65	PS_OK_65
Sector 1	3.46	3.49	3.67
Sector 2	3.59	3.35	3.48
Sector 3	4.18	4.17	3.43
Sector 4	3.94	3.42	3.61

3. Conclusions

When a tractor-trailer unit is moving, the towing vehicle is subjected to the impact of significant forces applied by the trailer to the mechanical coupling device. They chiefly consist of time-varying longitudinal forces occurring when the vehicle is braked or accelerated, lateral forces generated when the vehicle is turning or moving on a road bend, and vertical forces. The values of all of them depend on the vertical load exerted by the trailer drawbar. Due to the fact that the type-approval tests of coupling devices do not include road tests (they are limited to rig tests only), the knowledge of trailer's influence exerted through the coupling device on towing vehicle's steerability and stability is rather inadequate. Therefore, it is recommendable to carry out dynamic road tests to explore this issue.

The results obtained from this analysis show the reasonability of conducting research on the influence of the vertical load exerted by the trailer on the coupling device on towing vehicle's steerability and stability.

Based on the presented test results, a statement may be made that the maximum values of the parameters examined were not always associated with the case that incorrect load was applied to the coupling device. In such cases, however, the driver had to correct additionally the vehicle trajectory with the steering wheel for the tractor-trailer unit to be kept within the predefined corridor. The necessity to make the additional corrections indicates, therefore, a reduction in the intrinsic system capability to attenuate the oscillations of the transient process and deterioration in the stability of the tractor-trailer unit.

By extending the scope of the road testing to other tests, such as circular driving test or test with a jerk applied to the steering wheel, other information may be obtained, e.g. regarding the oversteering or understeering of the towing vehicle.

References

[1] ISO 3888-1 1999 *Passenger car-test track for a severe lane-change manoeuvre. Part 1: Double lane-change*

[2] PN–S–48024: *Zaczepy kulowe urządzeń sprzęgających. Wymagania i badania (Tow balls of coupling devices)*

[3] Prochowski L and Żuchowski A 2004 *Samochody ciężarowe i autobusy (Motor trucks and buses).* WKŁ Issue 1 Warszawa

[4] UN ECE Regulation No. 55.01.: *Uniform provisions concerning the approval of mechanical coupling components of combinations of vehicles*

[5] *Commission Regulation (EU) No. 1230/2012 of 12 December 2012 implementing Regulation (EC) No. 661/2009 of the European Parliament and of the Council with regard to type-approval requirements for masses and dimensions of motor vehicles and their trailers and amending Directive 2007/46/EC of the European Parliament and of the Council.*

[6] Luijten M F J 2010 *Lateral Dynamic Behaviour of Articulated Commercial Vehicles.* Eindhoven

[7] Więckowski D 2014 *Analiza dynamiki kulowego urządzenia sprzęgającego podczas hamowania (Analysis of the dynamics of a ball-type coupling device during braking).* In: Mitianiec, W. (ed): *Badania pojazdów (Vehicle testing)*, a monograph issued by The Cracow University of Technology, Kraków 2014, ISBN 978-83-7242-761-8, pp. 217-226

[8] Więckowski D and others 2012 *Analiza kinematyki i dynamiki elementów mechanizmów sprzęgających na przykładzie kulowych urządzeń sprzęgających (Analysis of the kinematics and dynamics of coupling mechanisms, with ball-type coupling devices taken as an example).* Problem Study Report No. BLY.001.12N Automotive Industry Institute, Warszawa

Research methodology simplification for teaching purposes illustrated by clutch automatic control device testing

J Wojs[1]

[1] Cracow University of Technology, Faculty of Mechanical Engineering, Institute of Automobiles and Internal Combustion Engines, al. Jana Pawła II 37, 31-864 Krakow, Poland

E-mail: wojs@mech.pk.edu.pl

Abstract. The paper proves that simplified, shorter examination of an object, feasible in laboratory classes, can produce results similar to those reached in scientific investigation of the device using extensive equipment. A thorough investigation of an object, an automatic clutch device in this case, enabled identifying the magnitudes that most significantly affect its operation. The knowledge of these most sensitive magnitudes allows focusing in the teaching process on simplified measurement of only selected magnitudes and verifying the given object in the positive or negative.

1. Introduction

To determine the characteristics of the operation of a given car mechanical subassembly it is infrequent to employ multichannel measurements. Their purpose is to properly characterise and describe the subassembly operation, determine its time and power characteristics to finally connect the magnitudes and parameters that identify the object. To achieve this it is necessary to design a relevant plan of tests and experiments, often time-consuming and requiring advanced apparatus for the measurement and recording of signals, and eventually processing the results with error analysis. Once the detailed investigations identify the parameters of the subassembly operation and its functionality (including merits and weaknesses of its operation), it is possible to distinguish the properties and physical magnitudes that are essential for the role of this subassembly in the car. This enables, within the framework of teaching process, select the physical magnitudes most important for the proper functioning of the subassembly as well as simplify tests performed in time-limited laboratory classes in order to – without neglecting the aim of tests– measure and analyse only the parameters that will enable (even only in terms of quality and roughly) verifying (positively or negatively) the test object during the laboratory classes.

2. Test object – Polish clutch automatic control device CEBRON

The automatic clutch [1] is a device enabling driving a car without using the pedal. It is designed for disabled persons after amputation of the left or both legs or suffering dysfunction of these.
The working element of the self-engaging clutch is a vacuum servo-motor, shown in figure.1, interacting with three or four sensors and valves. It is divided into two chambers, left –vacuum chamber and right – compression chamber. The vacuum chamber is connected with vacuum tank by an electromagnetic valve, while the compression chamber is connected with the atmosphere by:
- return valve delivering air to the compression chamber,

- release control valve responsible for removing air from the compression chamber only in the first phase of the piston movement,
- release mechanical valve controlled by throttling valve,
- release electromagnetic valve controlled electronically.

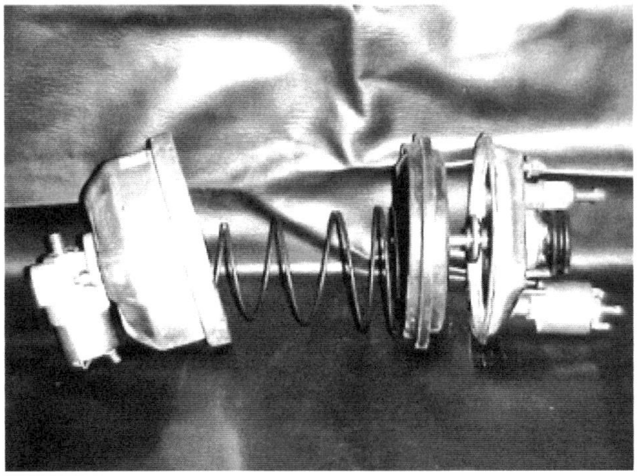

Figure 1. Vacuum servo-motor of automatic clutch [2].

The working medium is negative pressure in the suction manifold.
For the tests a relatively small, cheap passenger car shown in Fig. 2 was used.

Figure 2. Test vehicle with automatic clutch [2].

Specification of the test car [2] :

- VIN	KLY3S11BDWC491576
- engine cubic capacity	796cm3
- maximum power	30kW (41KM)
- rotational speed at maximum power	5500 rpm
- maximum torque	59 Nm
- rotational speed at maximum torque	2500 rpm

- friction, single-plate, dry clutch with compression spring and torsion damper.

The tests on the automatic clutch were done by the author of this paper within the framework of his doctoral research and the details are given in [2]. The results are quoted in the paper for comparison.

The testing equipment used for advanced investigations [2] included:

- a telemetric set with bridge foil strain gauges (measurement of torque on half-shafts)
- potentiometric sensor (measurement of throttle angular displacement and measurement of clutch lever angular displacement), Figs 4 and5,
- CORREVIT contactless head (measurement vehicle velocity), Fig.3,
- rotational speed impulse converter (measurement of wheels rotational speed) Fig.5,
- all the data were processed and recorded using a processing system shown in Fig 7.

Figure 3. Correvit head [2].

Figure 4. Clutch lever angular displacement [2].

Figure 5. Throttle angular displacement converter [2].

Figure 6. Wheel rotational speed converter [2].

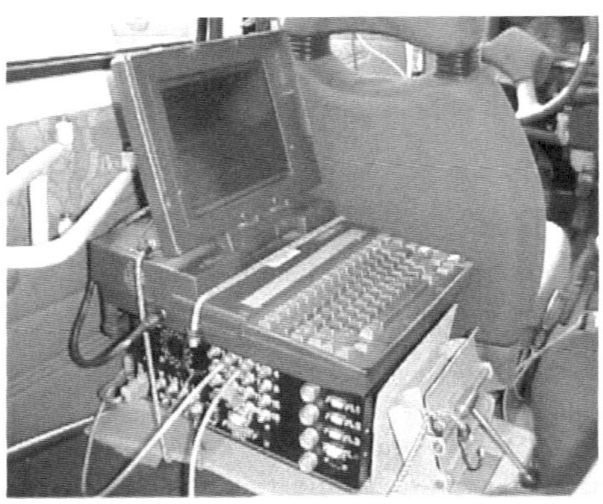

Figure 7. Data processing system [2].

Figure 8 presents the results of tests on vehicle running-up with automatic clutch disengaged. The aim was to compare the behaviour of the vehicle during running-up from rest with automatic clutch engaged and disengaged. The driver operates the clutch in such a way as to start from rest fast and smoothly, avoiding sharp jerks.

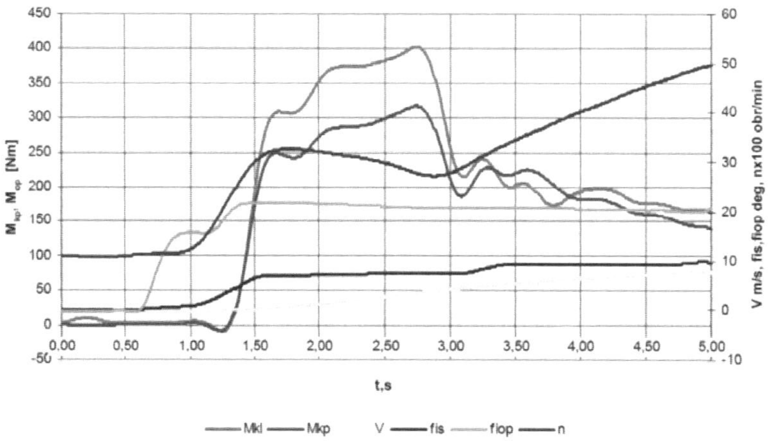

Figure 8. Results of running-up a vehicle without automatic clutch [2].

Notation in the diagram:

M_{kl} – torque on the left half-shaft [Nm],
M_{kp} – torque on the right half-shaft [Nm],
φ_s – clutch lever angular displacement [°],
φ_{op} – throttle opening angle [°],
V – vehicle velocity [m/s],
n – engine speed.

The highest values of driving torque on half-shafts occur at clutch full engagement the duration of which is ca. 3 seconds. The engine rotational speed drop is then observed. As can be seen in the diagram, there is a time relationship between the throttle opening angle and clutch lever angular displacement, and torque drop on half-shafts typical of vehicle running-up from rest, after engine speed and clutch shaft speed become equal.

Figure 9 presents the results of tests on running-up a vehicle with automatic clutch engaged.

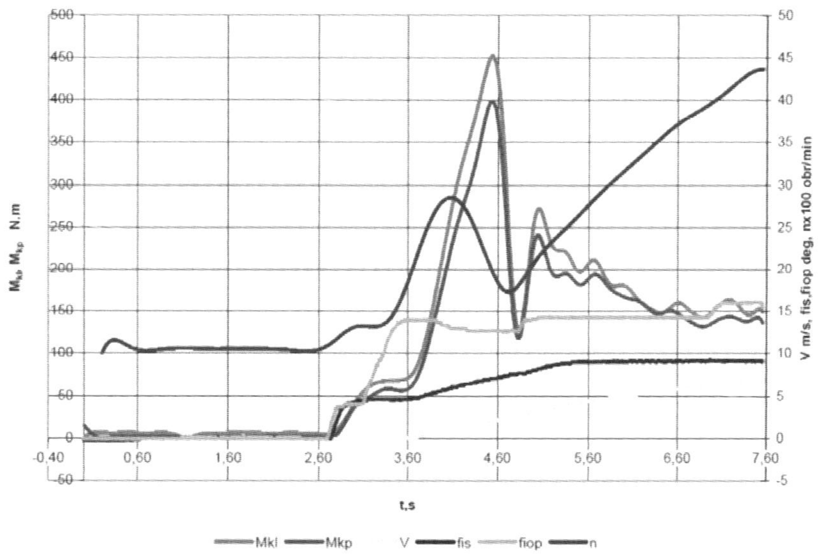

Figure 9. Results of tests on running-up a vehicle with automatic clutch engaged [2].

Notation in the diagram:

- M_{kl} – torque on the left half-shaft [Nm],
- M_{kp} – torque on the right half-shaft [Nm],
- φ_s – clutch lever angular displacement [°],
- φ_{op} – throttle opening angle [°],
- V – vehicle velocity [m/s],
- N – engine speed.

With the automatic clutch engaged higher o values of torque are observed than in the case of running-up the vehicle with the clutch disengaged. The same is true for engine speed. The total time of clutch engagement was reduced compared with the previous test.
The analysis of errors indicate the occurrence of standard deviation in the measurements of the random variables, which are the measurement results biased with error. Mean value was calculated.

Table 1. Measurement devices errors [2].

Value measured	Measurement device	Unit	Measurement accuracy
Vehicle velocity	CORREVIT contactless head	m/s	0,1 m/s
Clutch lever angular displacement	Resistance converter	1°	0,5 °
Throttle opening angle	Resistance converter	1°	0,25 °
Half-shaft torque	a telemetric set with strain gauges system	Nm	0,1 Nm
Rotational speed	Impulse converter	rpm	1/60 rpm
Range (measurement section)	Ruler	mm	1 mm

The road tests were run on a dry road surface. The measurements of the parameters shown in table 1 were recorded by means of processing and digital recording apparatus. For each parameter standard deviation and mean value were calculated. The diagrams below show the values of torque on half-shafts with standard deviation and clutch lever angular displacement with standard deviation.
Notation in the diagrams:

M_{kl} – torque on left half-shaft [Nm],
M_{kp} – torque on right half-shaft [Nm],
φ_s – clutch lever angular displacement [°],
σ – standard deviation,
$w_{śr}$ – mean value.

The left half-shaft torque curve (figure 10) shows the standard deviation to be minor since its value does not exceed 6% at the highest torque. It can be concluded that the accuracy of the tests was satisfactory.

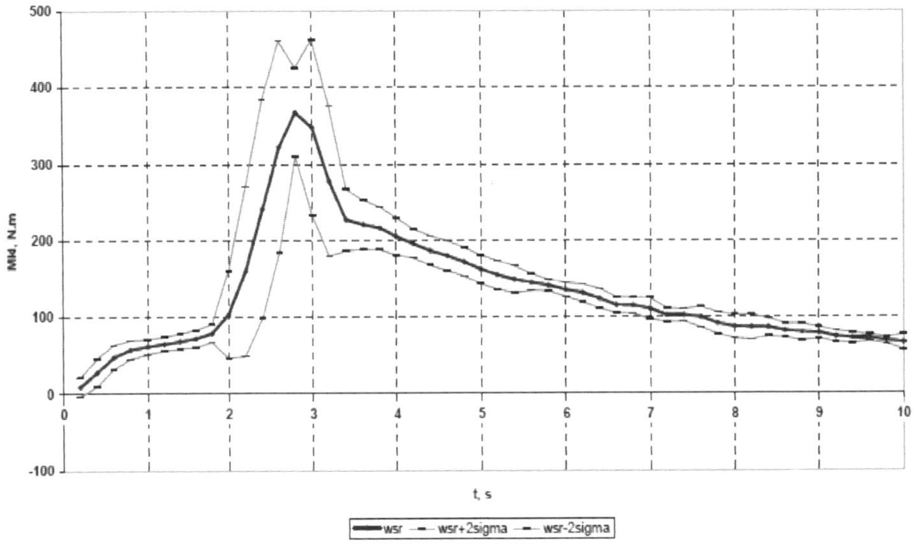

Figure 10. Torque on left half-shaft with standard deviation when vehicle starts moving.[2].

The curve for the left half-shaft is similar right shown in figure 11. The value of standard deviation does not exceed 7% of the mean value of torque.

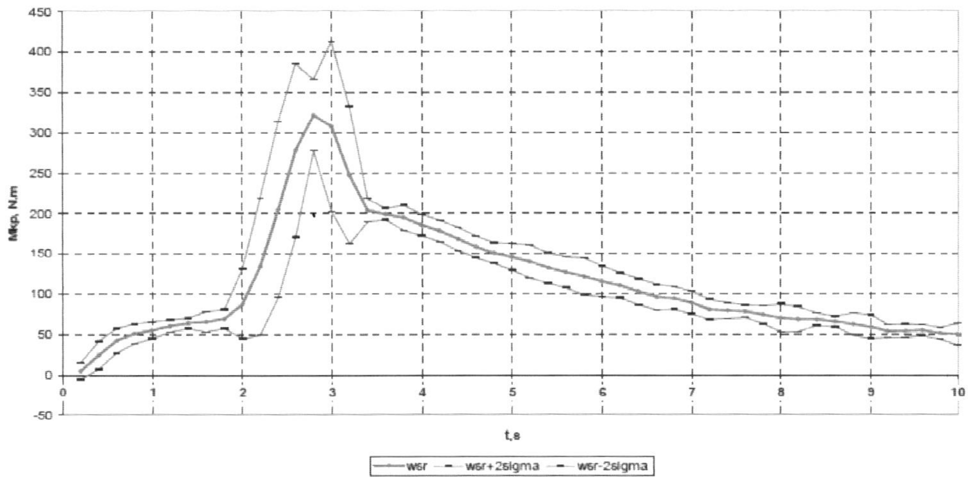

Figure 11. Torque on right half-shaft with standard deviation when vehicle starts moving [2].

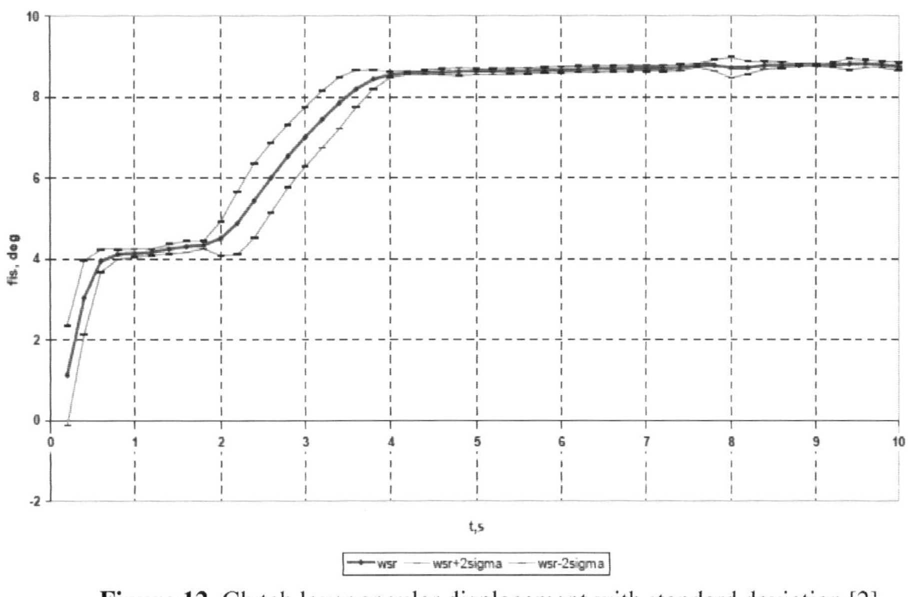

Figure 12. Clutch lever angular displacement with standard deviation [2]

The clutch lever displacement angle curve shown in figure 12 also indicates a minor percent of standard deviation from the mean value, which is a satisfactory result. To sum up, in the performed tests the values of standard deviation were close to one another.

The analysis of error for both simplified and thorough tests proves that the tests were performed correctly, and their results are close and relatively easy to interpret. They confirm the proper execution of the intentions of the designer of the device. The device operates correctly.

3. Simplified tests

The tests were performed on a dry asphalt surface on the test section of 50m. The task of the driver was to run up the test car from rest fast and smoothly on the test section, in other words to operate both the accelerator and clutch pedals in such a way as to limit the phenomena (e.g. jerking) adversely affecting the measurements. In total 24 time measurements were taken, eight for each configuration, that is:

- with disengaged clutch,

- with engaged clutch in the city mode,
- with engaged clutch in the road mode.

The results of time measurements of the ride were recorded with a mechanical stop-watch. The watch was switched on at the moment of starting from rest and switched off when the front wheels of the car crossed the measurement section end line.

Table 2. Running-up time of vehicle with automatic disengaged and engaged clutch [5].

Measurement number	RUNNING-UP TIME [s]		
	Clutch DISENGAGED	Clutch ENGAGED ROAD MODE	Clutch ENGAGED CITY MODE
1	8,3	9,0	18,0
2	10,6	11,0	14,8
3	10,5	10,1	15,1
4	9,6	9,4	16,2
5	10,6	10,9	14,6
6	9,8	10,1	15,4
7	8,6	8,8	17,7
8	8,6	8,9	16,9
$t_{śr}$	9,575	9,775	16,0875

As can be seen, the automatic clutch "imitates" the driver in the road mode, while in the city mode it works similarly but with the application of "driving on half-shaft"[5].

When the automatic clutch is disengaged and when engaged in the road mode the running-up times are similar, almost identical. This indicates that the clutch works correctly, there is no adverse effect on the clutch operation and comfort. Its operation is different in the city mode. The running-up times for the city mode are significantly longer due to due to slower engaging the clutch. Adjusting the automatic clutch in the city mode results in slower running-up time due to reduced jerking movements while starting, which is also an intended effect for the city traffic.

Below (Table 3) vehicle velocities calculated on the running-up time are tabulated.

Table 3. Vehicle velocities calculations [5]

Measurement number	Velocities [m/s]		
	Clutch DISENGAGED	Clutch ENGAGED ROAD MODE	Clutch ENGAGED CITY MODE
`	6,02	5,56	2,78
2	4,72	4,55	3,38
3	4,76	4,95	3,31
4	5,21	5,32	3,09
5	4,72	4,59	3,42
6	5,10	4,95	3,25
7	5,81	5,68	2,82
8	5,81	5,62	2,96
$V_{śr}$	5,27	5,15	3,13

4. Conclusions

In the road tests of a vehicle for disabled persons, equipped with an automatic clutch, during starting from rest the running-up time was determined, during the specific tests – also the time characteristics of velocity and torque on half-shafts.

The tested automatic clutch fixed in cars correctly controls the clutch operation. The test results indicate how close to each other are the running-up times of a car with automatic clutch in the road mode. This confirms the correct operation of the clutch controlling device, no adverse effect of the device on clutch operation was observed.

In the city mode, however, the device behaves differently. The running-up times in the city mode differ from such values during running-up times of a vehicle without the automatic clutch. This is caused by slower engaging the clutch by the device to avoid jerking movements during start (different regulation adjustment is employed than for the road mode). It is a favourable and intended effect improving the comfort of starting from rest.

During vehicle work with the automatic device engaged a larger torque on half-shafts and higher rotational speeds are observed.

References

[1] Instrukcja obsługi http://www.autoadaptacje.com/

[2] Wojs J 2007 *Kształtowanie charakterystyk automatu sprzęgłowego ze względu na drgania wzdłużne samochodu dla niepełnosprawnych* Doctor Thesis Cracow University of Technology Krakow

[3] Walczak S and Wojs J 2000 *Specyfika diagnostyki samochodów przystosowanych dla osób niepełnosprawnych,* Materiały Konferencji Diagnostyka Pojazdów Samochodowych, Politechnika Śląska, Katowice

[4] Articles http://www.cebron.com.pl

[5] Grabda K 2015 Engineer Thesis supervisor Jacek Wojs: Rozpędzanie samochodu z wykorzystaniem automatu sprzęgłowego Cracow University of Technology Mechanical Faculty

IOP Conf. Series: Materials Science and Engineering **148** (2016) 012033 doi:10.1088/1757-899X/148/1/012033

Impact of multisource VOC emission on in-vehicle air quality: test chamber simulation

K Brodzik, J Faber, A Gołda-Kopek and D Łomankiewicz

BOSMAL Automotive Research and Development Institute Ltd, Sarni Stok 93, 43-300 Bielsko-Biała, Poland

E-mail: krzysztof.brodzik@bosmal.com.pl

Abstract. Air quality inside vehicle may be strongly influenced by the presence of volatile organic compounds (VOC). The sources of these compounds may be different. In case of new vehicles VOC mainly originate from off-gassing of interior materials, while in used cars exterior pollution, like exhaust gases, starts to dominate. The aim of this work was to check the influence of multiple VOC sources on concentration of volatile organic compounds emitted from car interior parts. For this purpose material emission tests were performed in 1 m^3 emission testing chamber (WKE 1000, Weiss, Germany) at 65 °C, 5% RH and with air exchange. Three different car parts were studied: sun visor, headlining, and handbrake lever cover. It was stated that volatile organic compounds concentration inside test chamber during the test performed with three different parts inside was significantly lower than those being result of addition of the results obtained for parts tested separately. Presented results indicate interactions between different materials and their emissions as well as prove that some of materials acts like sorbents.

1. Introduction

Interior air quality in vehicle's cabin is gaining an increased attention because car is frequently used mean of transport. People spend much of their time indoors, inter alia in cars, commuting to work and travelling for business or private purposes [1-4].

The presence of organic compounds, among them volatile organic compounds (VOC), very volatile organic compounds (VVOC) and semi-volatile organic compounds (SVOC), is the main factor influencing air quality and its regarding by people. Organic compounds present in air have or may have a negative impact on human health, but may also influence driving safety. Selected compounds are responsible for:

- drowsiness (e.g. saturated and unsaturated aliphatic hydrocarbons, cyclohexane, ethylbenzene, 1-butanol);
- headache and dizziness (e.g. hexane, heptane, cyclohexane, toluene, naphthalene, trimethylbenzene);
- eye irritation (e.g. 2-ethylhexanol, benzaldehyde, styrene, propylbenzene);
- respiratory problems and throat irritation (e.g. styrene, methycyclopentane, decalin);
- fatigue, nausea, skin irritation, toxic, carcinogenic or mutagenic effect, and many others [5,6].

The other action of organic compounds, which has not direct influence on human health, but on driving safety, is a phenomenon called fogging. Semi-volatile organic compounds with boiling temperatures above 240-260°C [7], mainly phthalates, condensate on interior surface of windscreen, reducing the transparency and causing light reflections [8-10].

Some of organic compounds has also odorous character. Human nose is a very sensitive odour detector, capable to detect odour on very low concentration level [11]. On the other hand, people can recognize differently the same odour, because odour perception is a very subjective task. Moreover, even if specific odour is recognized as a pleasant, in long-term exposure may lead to odour annoyance and discomfort [12].

Organic compounds in new car cabin originate mainly from materials' off-gassing [3,13-15], as in vehicle interior multiplicity of different materials is used [9], among them polystyrene, polyethylene, polypropylene, polyamide, acrylonitrile-butadiene-styrene, rubber, natural or artificial leather, polyurethane foam etc. Each of this material is a potential source of VOCs emission. It was found that ethylbenzene and styrene are specific compounds for polystyrene presence, disregarding air temperature. In ambient temperature polyurethane foam emits inter alia toluene, phenol, decane, nonanal, benzaldehyde, diethyl phthalate etc. Benzaldehyde, nonanal, tridecane, decanal, and diethyl phthalate can be emitted in 23 °C from polyethylene, while in higher temperature (70 °C) 2-ethyl-1-hexanol was additionally detected. The main compound, which is usually emitted from polypropylene, is 2,2,4,6,6-pentamethylheptane [16]. Additionally, plastics are often enriched by different additives to achieve desired properties. Among those additives plasticizers, stabilizers or flame retardants can be found. Many of this additives are regarded to be hazardous, and for that reason controlling the emission of those compounds is significant [8]. The toxicity of organic compounds mixtures in new passenger vehicles was described elsewhere [17].

Concentrations of organic compounds depend on atmospheric conditions, vehicle model, it's age and mileage and user's habits [14]. Within the same car model the concentration of selected compounds may differ significantly, depending on materials and equipment used in vehicle interior [4,18]. Along with the increase of vehicle age and mileage, when proper ventilation is applied, the concentration of VOCs emitted from materials usually decreases, but compounds originating from external sources may additionally appear [14]. Those sources include fuel combustion, fuel leakage and others, and concentration of pollutants depends inter alia on air exchange efficiency, season of the year, time of travelling, traffic and travelling routes [19,20]. Car exhaust gases are the main source of human exposure to particulate matter and soot [21], aldehydes, including formaldehyde and acetaldehyde [22], and naphthalene [23]. The main compounds under study in used vehicles were so called 'emission markers', that is: benzene, toluene, ethylbenzene and xylenes (BTEX) [14,20,24]. Interior air quality in used vehicles was widely investigated and quite great knowledge was gathered. In case of new vehicles still little is known about interior air quality, as a result of variety of models, types of equipment and materials used in vehicle interior.

The emission of VOCs from materials is affected by air temperature and relative humidity (RH), depending on type of material and type of compound. It was found that in typical ambient temperature (23-30°C) concentration of VOCs was independent on temperature, and the impact of RH was negligible [25]. Materials which are used in indoor environments are usually tested in temperature of 23°C [8]. Although in indoor environment the temperature can reach approximately 30°C in maximum, in vehicle interior in direct sunlight (i.e. in summer months) the temperature may reach an excessive maximum (over 65°C) [26]. The increased temperature in vehicle interior would result in higher emission of different organic compounds, and therefore higher human exposure to harmful pollutants [25]. For that reason materials for automotive industry are tested in temperature of 65°C.

The tendencynowadays is to reduce the use of toxic materials indoors in order to minimalize occupants' exposure to harmful chemicals emitted from those materials [27]. Similarly, in automotive industry (in vehicle production and furnishings), producers intend to reduce harmful materials presence.

There is no clear requirements regarding interior air quality in new vehicles [15], however concentration of organic compounds may be at significantly high level. Total volatile organic compounds (TVOC) concentration inside car cabin may vary from less than 1 mg/m^3 to over 10 mg/m^3. Such a high VOC concentration may have effect in driver's and passenger's discomfort and irritation feelings. For that reason monitoring of air composition in new vehicles and measuring of VOCs emitted from materials is essential to ensure a high quality of produced cars and to take care of human health and comfort.

The aim of this work was to compare results obtained during separate car interior parts emission testing with the results of tests performed with multiple parts. Gathered information should expand knowledge regarding in-vehicle air quality and possibility to asses quality control check based on separate parts emission testing.

2. Experimental methods

2.1. Materials under study

Three different car interior parts were chosen for this study, namely: handbrake lever cover, sun visor and headlining. All parts were made of synthetic materials and in brand new condition. Handbrake lever cover was made of only one material (polyvinyl chloride – PVC). Soft PVC was used as covering decorating layer of sun visor with some hard plastic used as inside support material while headlining was made of a wafer material of synthetic fibres/polyurethane foam/synthetic fibres (SF/PUR/SF). Emission tests were performed for separate parts as well as for all of them together.

2.2. Emission tests

Emission tests of car interior parts were performed in airtight, specially constructed emission testing chamber of ca. 1 m^3 internal volume (WKE 1000, Weiss, Germany). In case of handbrake lever cover and sun visor tests two pieces of each part were tested while in case of headlining only half of a part was placed inside the chamber. Each set of parts was placed into the preconditioned chamber at 65 °C supplied with humidified (5% RH) purified air (air exchange set to 0.4). After 5 h of test item conditioning air samples were collected with the use of Tenax TA sorbent in stainless steel tubes (Markes, UK) for VOC analysis and cartridges filled with 2,4-dinitrophenylhydrazine (2,4-DNPH) on silica gel (Sigma-Aldrich, Poland) for carbonyl compounds analysis. Cartridges with sorbents were connected to heated distribution tree of WKE 1000, SKC PCMTX-8 (SKC,UK) pumps were used for sample collection with the air flow rate set to 100 cm^3/min in case of VOC sampling on Tenax TA and to 750 cm^3/min in case of carbonyl compounds sampling on 2,4-DNPH cartridges.The accuracy of flow rates was provided by calibration with DRYCAL flowmeter (BIOS, USA). Duplicate samples were collected for 20 min allowing collection of 2 dm^3 air sample for VOC analysis, and 15 dm^3 sample for analysis of carbonyl compounds. After sampling sorbent tubes were closed and stored in refrigerator prior the analysis.

2.3. Analytical methods

Analysis of volatile organic compounds emitted from tested car interior parts was performed as described previously [13,14,18] The air samples collected on Tenax TA were desorbed with the use of a thermal desorption system Unity2 (Markes, UK) equipped with Series2 Ultra TD autosampler (Markes, UK). Samples were desorbed at 300 °C for 15 minutes and transferred via a heated transfer line to a gas chromatograph equipped with a flame ionization detector (7890A, Agilent, USA) and a mass spectrometer (MSD 5975C, Agilent, USA). The gas chromatograph was equipped with a column splitter allowing to perform simultaneous analysis with use of both detectors. Detailed TD-GC/FID-MS analysis parameters are presented in Table 1. Identification of volatile organic compounds was based on retention times (RT) and mass spectra of more than 140 VOCs standards (Sigma-Aldrich, Poland) and National Institute of Standards and Technology (NIST08) spectra library. The concentrations of emitted volatile organic compounds were calculated in toluene equivalent (µg/m^3) i.e. were calculated against the standard toluene curve.

Table 1. Details of chromatographic analysis.

Thermal desorber	Tube: 280ºC (15 min); Purge: 1 min;

	Cryotrap: -30°C to 300°C at ca. 60°C/min Column pressure: 23 psi; Split ratio: (5:1)
Gas chromatograph	Capillary column: DB-5MS UI (60 m x 1μm x 0,32 mm) Carrier gas: He (6.0) Temperature program: from 40°C (2 min), at 3°C/min to 92°C, at 5°C/min to 160°C, 10°C/min to 290°C (12 min)
Mass spectrometer	Transfer line to MS: 300°C; Ion source: electron impact (EI) Ion source temperature: 230°C Scan mode: SCAN; *m/z*: 35-260 amu; NIST08 library
HPLC	Column: ZorbaxStablebond C-18 (4.6m x 250mm, 5μm) Column temperature: 40 °C; Flow rate: 1 cm^3/min Mobile phase: A = acetonitrile, B = water Gradient program: 0 min 60%A, 10 min 60%A, 20 min 70%A, 30 min 100%A, 31 min 60%A, 36 min 60%A DAD detector; VWD: 365nm

Carbonyl compounds collected on DNPH cartridges were extracted with 1 cm^3 of acetonitrile (Sigma-Aldrich, Poland) what was followed by liquid chromatography (HPLC) analysis with use of Infinity 1260 system (Agilent, USA) equipped with ZorbaxStablebond C-18 collumn. In case of HPLC analysis identification and quantification of carbonyl compounds was based on retention times and calibration curves of 16 carbonyl compounds (formaldehyde, acetaldehyde, acrolein, acetone, propionaldehyde, crotonaldehyde, methacrolein, 2-butanone, butyraldehyde, pentanal, cycloheksanon, *o*-tolualdehyd, *m,p*-tolualdehyde, hexaldehyde, 2,5 –dimethylbenzaldehyde, all from Sigma-Aldrich, Poland).

3. Results and discussion

Altogether 377 volatile compounds were detected by TD-GC/FID-MS in collected air samples where: 142 compounds were detected only in samples collected from single part testing, 92 compounds were detected in samples collected from 2 separate test, 74 compounds in samples collected from 3 tests. It was stated that 69 volatile organic compounds were present in air during all the tests i.e. 3 tests of separate parts and 1 test of 3 parts together. Chromatograms of collected VOC samples are compared in Fig. 1. Even brief analysis of obtained chromatograms allows noticing that concentration of volatile organic compounds emitted during the "3 parts together" emission test is not a simple sum of VOCs emitted when parts were tested separately. There are visible differences not only in quantity, in this case signal intensity, but also in compound's nature. Some compounds are present only in case of separate parts, however other ones were detected only while all, three parts, were tested simultaneously.

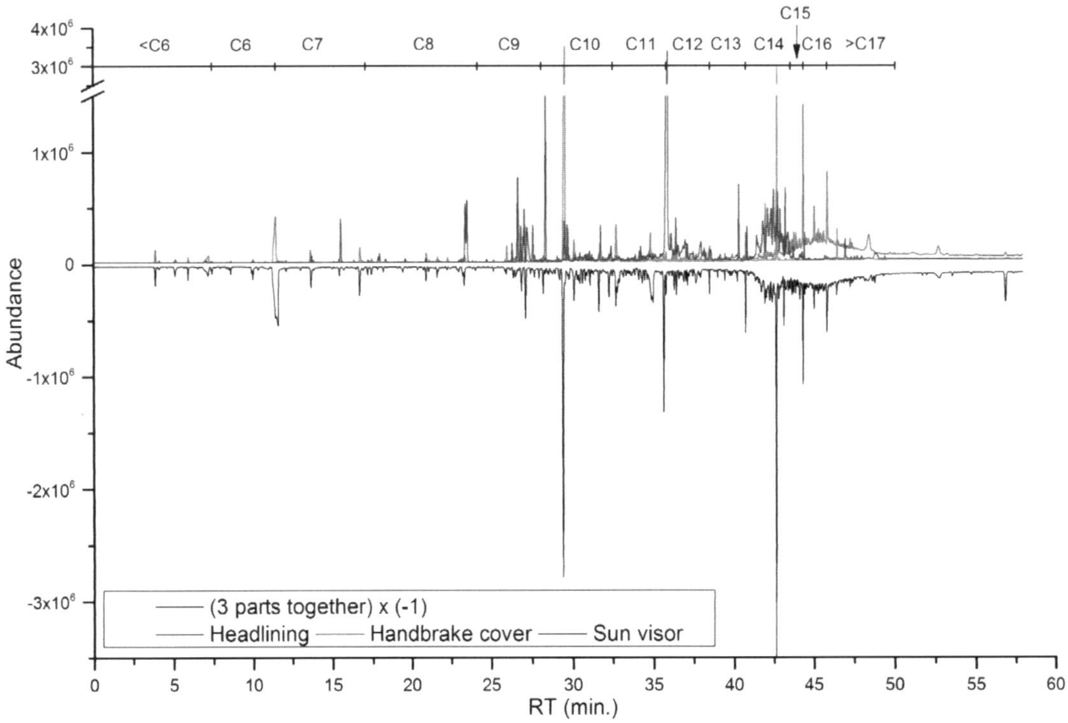

Figure 1. Comparison of chromatograms collected during the emission tests of car interior parts.

Volatile organic compounds analyzed by TD-GC/FID-MS were divided into 13 groups, "RT groups", in reference to their retention times (RT) (Fig. 1), where <C6 stands for all compounds with RT lower than RT of hexane, C6 stands for hexane and all compounds with RT lower than RT of heptane, C7 stands for heptane and all compounds with RT lower than RT of octane etc. For each compounds' group concentrations were summarized and values were compared (Fig. 2). It can be seen that for three first groups (<C6…C7) VOCs emitted during the testing of 3 parts may be close to sum of the same VOCs emitted from separate parts, however in all other cases no such dependence can be noticed. For example handbrake lever cover emitted ca. two times more C10 compounds than three parts together and even bigger differences are observed in case of sun visor. This simple approach clearly indicates that there are important differences between concentration of emitted VOCs from car interior parts when different items are in contact/proximity with one other.

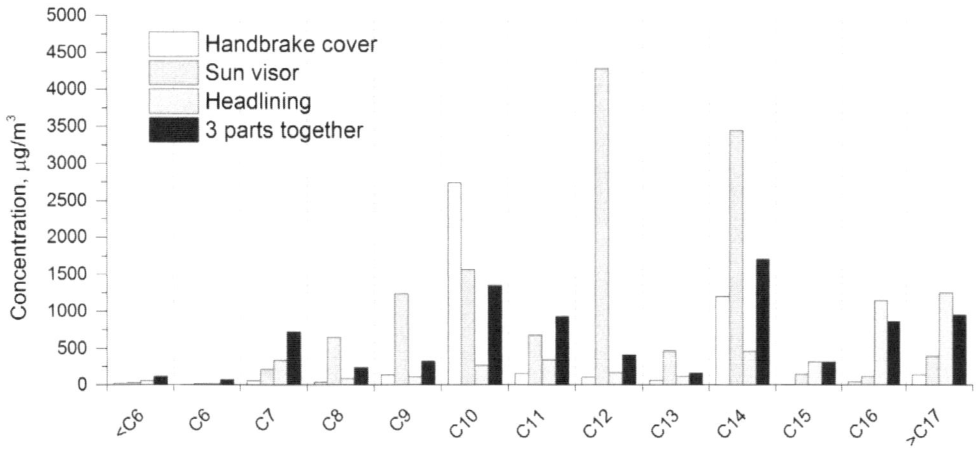

Figure 2. Comparison of VOC RT groups concentration.

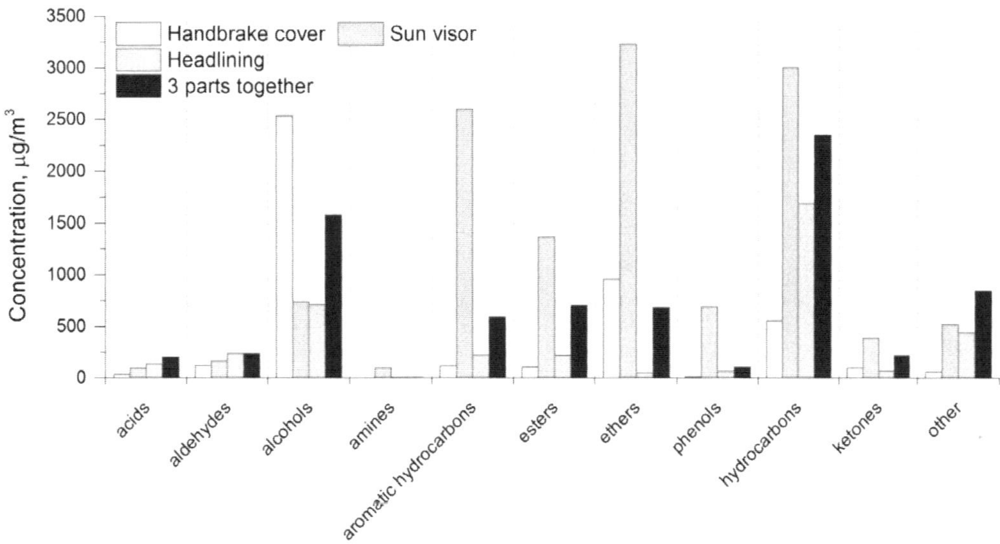

Figure 3. Comparison of VOC concentration.

Similarly, in case of comparison of different types of volatile organic compounds concentrations there are no clear dependence between substances emitted from separate parts and from test performed for parts together (Fig. 3). Large differences are observed mainly in case of ethers, aromatic hydrocarbons, phenols and alcohols. Taking into account that total concentration of volatile organic compounds (Table 2) was the highest for sun visor and the lowest for headlining, which was also the largest tested partone can assume that in case of simultaneous testing of different items some of VOCs emitted from one item may adsorb onto another's item surface. This assumption seems to be confirmed when emission from headlining is deeper analyzed. In this case VOC lowers than C15 "RT groups" have significantly lower summary concentration than those from other separate parts. These groups emitted from 3 parts together are significantly lower than from dominating separate part, other than headlining. On the other hand when C15, C16 and >C17 RT groups are taken into account, results are in opposite to the above.

Headlining emission is larger than from two other parts and only slightly differs from emission detected from simultaneous testing of three parts.

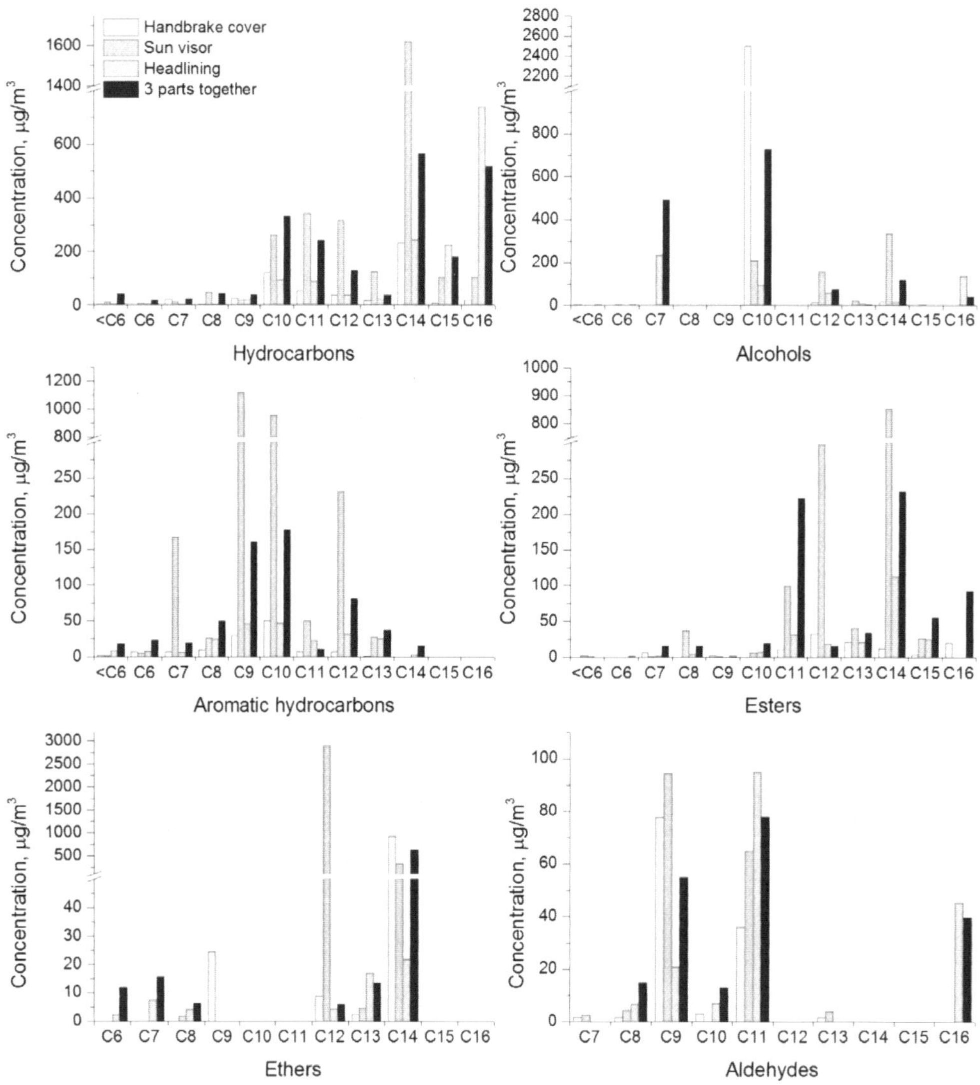

Figure 4. Comparison of VOC type and RT groups, based on TD-GC/FID-MS results.

Some additional information provides comparison of VOC type and RT (Fig. 4). Also in this case the most significant differences between concentration of emitted organic volatile compounds from separate parts and parts tested together are observed when headlining is not important source of selected VOC group. This is especially noticeable in case of hydrocarbons: RT C14 group is dominated by the emission from sun visor, while RT C16 is dominated by the emission from headlining, but similar concentrations levels are observed from 3 parts testing. However in case of RT C14 group concentration of VOCs emitted from sun visor is nearly three times higher than concentration of these compounds emitted from all parts together. On the other hand, concentration of RT C16 group emitted from headlining is only ca. 40% higher than from 3 parts. These seems to further confirm the thesis that presence of large surface headlining, which additionally is the most porous material from three tested, have the deciding effect on total VOC concentration in case of simultaneous testing of different parts. When separate emission test performed for headlining yield in significant concentration of VOC RT or type group the same effect is observed in case of multiple parts testing. When considered groups of compounds are not emitted in large amount by headlining their concentration in case of 3 parts testing is significantly lower than from their dominant donor. Therefore conclusion seems to be justified that parts made of different materials, which additionally differs in porosity, may as well emit and adsorb volatile organic compounds at elevated temperatures when in proximity.

Table 2. Summary of TD-GC/MS-FID and HPLC results.

	Handbrake lever cover	Sun visor	Headlining	3 parts together
TD-GC/FID-MS				
Total VOC concentration, µg/m^3	4 751	13 281	4 720	8 210
HPLC				
Formaldehyde, µg/m^3	3.4	63.9	437.0	497.6
Acetaldehyde, µg/m^3		40.8	103.3	100.1
Acetone, µg/m^3	2.1	3.4	11.3	10.3
Propionaldehyde, µg/m^3			13.3	10.5
2-Butanone, µg/m^3			10.8	5.5
Benzaldehyde, µg/m^3	15.6	35.2	20.4	30.8
Cyclohexanone, µg/m^3		90.8	1.6	7.6
Hexaldehyde, µg/m^3			8.7	8.4

Results obtained with use of HPLC method are presented in Table 2. In case of carbonyl compounds analysis previously noticed dependence is confirmed. As a proof of headlining importance in multisource emission one can notice that in case of acetaldehyde theoretically possible concentration of this compound during three parts test may be ca. 140 µg/m^3 (emissions from sun visor and headlining) but the real value from testing all of the parts is 100 µg/m^3. On the other hand in case of cyclohexanone theoretically possible concentration for 3 parts is ca. 92 µg/m^3 but the real value is only 7,6 µg/m^3 probably due to previously mentioned adsorption of compound on headlining surface.

Presented results confirms previously suggested interactions between VOC emitted from different materials [27, 28] what in case of our results can be noticed by differences between theoretically possible total concentration of volatile organic compounds and the real values. Deeper look at the presented results allowed conclusion that some of materials, like headlining which can be characterized by significant porosity, acts like sorbents what leads to decrease of specific compounds or group of compounds concentrations.

4. Summary

The results presented highlight the issues in predicting air composition on the basis of single emission source measurements, thus make car quality control based on in-vehicle VOC concentration very difficult to achieve while only single parts results are available. Taking into account that in new car VOCs emitted from interior materials are main source of air contamination [13], and may have negative effect on users comfort, health and safety, interaction of emitted substances with other materials may lead to elongation of exposure, due to their adsorption and expected desorption in future.

References

[1] WHO 2006*Development of WHO Guidelines for Indoor Air Quality*
[2] WHO 2010 *Selected pollutants. Guidelines for Indoor Air Quality*
[3] Yoshida T and Matsunaga I 2006 *Environ. Int.***32** 58-79
[4] Zhang G-S, Li T-T, Luo M, Liu J-F, Liu Z-R and Bai Y-H 2008 *Build. Environ.* **43** 315-19
[5] Salthammer T and Uhde E 2009*Organic indoor air pollutants*(Weinheim:Wiley-VCH Verlag GmbH)

[6] Sarbu I and Sebarchievici C 2011 Olfactory comfort assurance in buildings *Chemistry, emission control, radioactive pollution and indoor air quality* (InTech)
[7] ISO 16000-6:2011*Indoor air. Part 6: Determination of volatile organic compounds in indoor and test chamber air by active sampling on Tenax TA sorbent, thermal desorption and gas chromatography using MS or MS-FID*
[8] Wensing M, Uhde E and Salthammer T 2005 *Sci. Total Environ.* **339** 19-40
[9] Loock F, Lampe T and Bahadir M 1993 Fresenius *J. Anal. Chem.* **347** 280-85
[10] Kalman E-L, Löfvendahl, Winquist F and Lundström I 2000 *Anal. Chim. Acta***403** 31-38
[11] Delahunty C M, Eyres G and Dufour J-P 2006 *J. Sep. Sci.* **29** 2107-25
[12] Bartsch J, Uhde E and Salthammer T 2016 *Anal. Chim. Acta***904** 98-106
[13] Faber J, Brodzik K, Gołda-Kopek A and Łomankiewicz D 2013 *Pol J. Environ. Stud.* **22** (6) 1701-09
[14] Faber J, Brodzik K, Gołda-Kopek A and Łomankiewicz D 2013 *J. Environ. Sci.***25** (11) 2324-30
[15] Buters J T M et al. 2007 *Environ. Sci. Technol.***41** 2622-29
[16] Mitchell G, Higgit C and Gibson L T 2014 *Polym. Degrad. Stabil.* **107** 328-40
[17] Janicka A 2013 *Proc. ECOpole***7** (1) 93-99
[18] Brodzik K, Faber J, Łomankiewicz D and Gołda-Kopek A 2014 *J. Environ. Sci.***26** (5) 1052-61
[19] Jo W-K and Park K-H 1999 *Atmos. Environ.* **33** 397-404
[20] Rahman Md M and Kim K-H 2012 *J. Hazard. Mater.* **241-242** 267-78
[21] Lee K, Sohn H and Putti K 2010 *J. Air Waste Manage.* **60** 130-36
[22] Jo W-K and Lee J-W 2002 *Environ. Res.* A **88** 44-51
[23] Jo W K and Lee J H 2011 *Int. J. Environ. An. Ch.* **91** 1412-24
[24] Som D, Dutta C, Chatterjee A, Mallick D, Jana T K and Sen S 2007 *Sci. Total Environ.* **372** 426-32
[25] Lin Ch-Ch, Yu K-P, Zhao P and Lee G W-M 2009 *Build. Environ.* **44** 525-33
[26] Mandalakis M, Stephanou E G, Horii Y and Kannan K 2008 *Environ. Sci. Technol.* **42** 6431-36
[27] Setnikova I 2014 *Mat. Sci. Eng.* C **36** 1-6
[28] Guo H, Murray F and Wilkinson S 2000 *J. Air Waste Manage. Assoc.* **50** 199-206

IOP Conf. Series: Materials Science and Engineering **148** (2016) 012034 doi:10.1088/1757-899X/148/1/012034

Simulation analysis of the EUSAMA Plus suspension testing method including the impact of the vehicle untested side

K Dobaj[1]

[1] **Institute of Automobiles and Internal Combustion Engines,** Cracow University of Technology, Al. Jana Pawla II 37, 31-864 Krakow, Poland;

E-mail: kdobaj@ pk.edu.pl

Abstract. The work deals with the simulation analysis of the half car vehicle model parameters on the suspension testing results. The Matlab simulation software was used. The considered model parameters are involved with the shock absorber damping coefficient, the tire radial stiffness, the car width and the rocker arm length. The consistent vibrations of both test plates were considered. Both wheels of the car were subjected to identical vibration, with frequency changed similar to the EUSAMA Plus principle. The shock absorber damping coefficient (for several values of the car width and rocker arm length) was changed on one and both sides of the vehicle. The obtained results are essential for the new suspension testing algorithm (basing on the EUSAMA Plus principle), which will be the aim of the further author's work.

1. Introduction

Increasing number of vehicles in use requires continuous improvement of the car diagnostics methods, especially related to active safety. The car suspension is one of the most important assemblies in relation not only with active safety but also with ride comfort. The suspension diagnostic methods must follow the specified requirements. The diagnosis must be relatively fast and easy to perform. Furthermore, differences in the car suspension design must not influence on the formulated suspension condition indicators. Indicators must give clear values for one state of the car suspension damage. These requirements must be able to application in the test stand which corresponds to in-road driving as far as it is possible.

Suspension quality, referring to active safety and ride comfort, results in the vehicle behaviour under the influence of the road unevenness. The test stand methods analyse the vehicle response under the provoked regular vibrations. Several testing methods are developed [1,2,3]. These methods vary in the amplitude of the excitation. The exciter (steel plate on which the car wheel is placed) in supplied with constant frequency with value over the two vehicle resonance frequencies (from sprung and unsprung mass). The excitation frequency is maintained for few seconds and then the idle run takes place. The behaviour of the test stand- vehicle assembly during the phase of damping is analysed. For the determination of the suspension, several testing indicators were formulated. Some of them are related with the value of the exciter amplitude while passing through the resonance frequencies [2,3], some refer to the damping ratio calculation (in ex. according to Lehr) [3]. Another indicators are related with the force of the load exerted on the exciter plate from the wheel and mass of the quarter of the vehicle.

The suspension test stands vary in the construction but all of them consider only one wheel of the vehicle and its loading mass during single testing cycle. Thus, the effects of interaction between two sides of the vehicle are neglected. These effects include car transverse moves and influence of

damping qualities of the side of the vehicle, on which the suspension criteria are not calculated. The author conducted simulation analysis the modification of the EUSAMA Plus suspension testing method. The modification introduced consistent vibrations of both sides of the vehicle.

2. Suspension testing methods development

The issues of determination of the vehicle parameters influence on active safety and comfort criteria indicated on the test stands were considered by many authors. Innovative conceptions of the EUSAMA Plus suspension testing method were introduced in work [1]. The author of that paper formulated automatic diagnostic method based on the neural network system. The neural network was used to conduct the image analysis involved with the EUSAMA rate (WE). The analysed images were formulated as a figures of the EUSAMA rate change during the time of the measurement. The measurement was performed according to the EUSAMA plus testing principle. The excitation frequency was changed from 30 Hz to 1 Hz. Each value of the frequency, with step of 1 Hz, was maintained in 5 seconds. Damping qualities of the passive shock absorber were analysed. Simulation model was formulated [1]. The model included characteristics of the shock absorber with the level of hydraulic fluid varied from 100 % to 25 % with the step of 25 %. The characteristics of shock absorbers with different levels of hydraulic fluid included in the simulation model were obtained experimentally [1]. Different levels of the hydraulic fluid resulted in different values of the EUSAMA rate and provoked shift of the range of maximal EUSAMA rate (minimum of the test plate loading force). Formulated automatic classifier (based on neural network system) analysed the figures and immediately matched considered state of damage in shock absorber (loss of hydraulic fluid) into one of four states of the suspension (mostly shock absorber) condition. The results proved that the neural network application is an appropriate way of EUSAMA suspension testing improvement.
Another approach of the suspension testing method development is involved with the vibroacoustics vibrations analysis [2]. The analysis is focused on determination of the kind of the shock absorber damage. Measurements were conducted on the EUSAMA test stand, but additional transmitters were mounted on the vehicle body and the rocker arm. It allowed for the measurement of the relative acceleration between the sprung and unsprung mass (joined by, among other, shock absorber). The measured acceleration values underwent the discretization process. After this process the signals were processed by following two-dimensional methods: Fast Fourier Transformation, Wavelet Transform, Wigner- Ville Distribution [2]. The processed signals analysis was oriented on application of the signal processing methods application to determination of the kind of the shock absorber failure. The time- frequency spectra of the relative shock absorber accelerations were created. The created spectra were involved with standard shock absorber and shock absorber with 25 % hydraulic fluid volume loss. Relatively small volume losses (25 %) provoke increases of the acceleration in the range of excitation frequencies over the resonance. Furthermore vibroacoustics method of the shock absorbers diagnostics was proposed [2]. The method was based on the wavelet transform, which allows the adjustment of the observation range and analysis of the time- frequency characteristics in the short, selected range. The new diagnostic estimator was created. The estimator allowed for detection of the kind of the shock absorber failure.
In the work [3], the comparison of suspension (referenced mostly to shock absorber) diagnostic methods was conducted. The authors described following methods: BOGE, EUSAMA, modified BOGE method, phase shift principle and Theta principle. The experimental analysis was performed. The modified BOGE method was introduced by MAHA company. The car wheel is placed on the testing plate joined with crank mechanism of the exciter and also supported on the ground by the coil spring [3].

The measurement system allowed for measurement of exciter plate amplitude and also the force exerted on the plate during the testing cycle. The suspension efficiency percentage indicator in calculated according to the algorithm specified by MAHA company. The phase shift method consists in determining the phase angle between the signals amplitude and force acting on the exciter plate.

The suspension criterion is defined as the minimal value of the angle in the range of frequency between the resonance of the sprung and unsprung mass. Development of the suspension testing is now strongly concerned with Theta principle. The principle uses the test stand similar to BOGE method. The footprint plate is excited to the frequency of 10 Hz. Then the propulsion is disconnected and the idle run occurs [3]. On the basis of the measured parameters, the damping ratio according to Lehr is calculated. It is defined as a ratio of linear suspension damping to the critical value of damping. The definition is correct for the mechanical assembly with 1 DOF. The quarter of car- test stand assembly has 3 DOF. Thus the manufacturers apply simplifying algorithms in the Lehr damping ratio calculations. The optimal Lehr damping ratio for the car with new shock absorber is determined in the range from 0.25 to 0.35. In the work [3] the experimental analysis was conducted. The analysis was involved with verification of the influence of the unsprung mass and tyre pressure change on results of the suspension testing according to EUSAMA and Theta testing principles. The Theta principle is much less sensitive to the considered parameters change in comparison with the EUSAMA principle. The shock absorber parameters, obtained during the tester plate excitation can be also used for the vehicle suspension modelling [6]. The tested suspension parameters are measured while the test stand excitation run performed is. The accelerometers, strain gauges and race transducers, mounted on the shock absorber, are used. The another approach of the additional sensors usage is introduced in the work [7] and [9]. The sensors which are attached to the shock absorbers are used. The micro-mechanical sensors measure the acceleration and the internal pressure of the shock absorber [9]. Furthermore, the oil temperature measured is [7,9]. The continuous monitoring, during complete driving cycles, allows for the shock absorber replacement before its damping qualities decrease rapidly. For the shock absorbers diagnosis, the artificial intelligence was also adapted. The work [8] describes the application of the neural network with its training and testing process.

3. Simulation analysis

3.1. The model formulation

To conduct the vibrational analysis, a numerical model was formulated. The Matlab simulation software was used. The half- car vehicle model was developed as the assembly of two quarter- car vehicle models. The sprung masses were modelled as concentrated points, joined by the massless lever (the half of the lever with the length x). The lever was joined to another lever which corresponded to the anti- roll bar of the car (length of the half of the anti- roll bar described by y). The sprung masses were joined to the rocker arms with dimensions a and b (Fig.1) via shock absorber and coil spring with the replacement damping and stiffness. To the endings of the rocker arms the unsprung masses were joined. The unsprung masses were supported on the test stand plates with masses m_p. The plates are coupled with the crank mechanisms, with radiuses r, joined to the frequency sources.

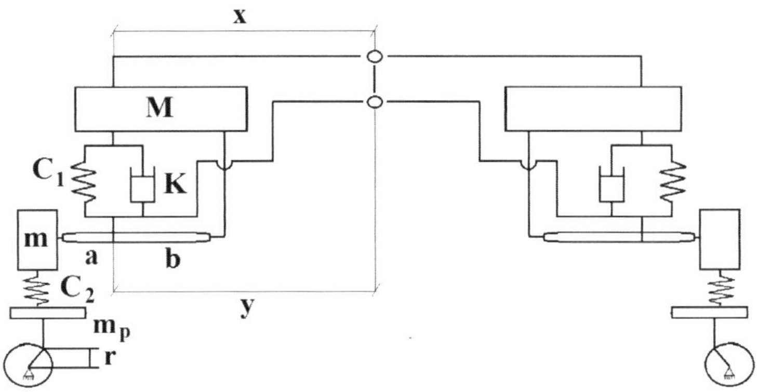

Figure 1. The scheme of the simulation model.

Table 1. The simulation model parameters

	Mass parameters			Stiffness parameters:	
M	270	kg	**C_1**	45000	N/m
m	40	kg			
m_p	15	kg	**C_2**	200000	N/m

	Shock absorber damping K			Dimensional parameters	
K	1748	Ns/m	**x**	0.844	m
a	0.1	m	**y**	0.744	m
b	0.4	m	**r**	3	mm

The car wheel suspension was simplified as an assembly of the rocker arm, perpendicularly joined to the spring and the damper. Thus, the inclination of the suspension column was neglected. Further simplifying assumptions were involved with the replacement parameters. The nonlinear passive shock absorber was modelled as a linear damper with the replacement damping coefficient K. The base value of the damping coefficient was calculated correspondingly to the car mass parameters, in the formula described in [4]. The suspension spring element (a coil spring) was described by replacement stiffness coefficient C_1. The tyre was modelled by one parameter- the radial stiffness C_2. The model parameters are included in Tab.1 The mass and dimensional parameters refer to the Fiat Grande Punto, with the parameters obtained in [5].

3.3. Results of the analysis

The simulation was conducted basing on the half car vehicle model, subjected to uniform vibrations of the tester plates. The frequency applied to the test stand crank mechanisms was changed from 1 to 30 Hz with the step of 1 Hz per second (similar to the EUSAMA Plus testing principle). During every simulation attempt, the tester plate dynamic force was measured. The measurement was conducted on the one side of the test stand- vehicle assembly. The maximum of this force occurred in different frequency ranges according to considered simulation scenario. The simulation scenarios were involved with applying the same frequency, changed in the same time periods, to both tester plates. This approach was conducted as a step for the EUSAMA Plus method modification. The modification takes into account the influence of the suspension condition of the vehicle untested side. The knowledge of the system properties in terms of the vibrations transmission from the one to another side of the vehicle will be essential for the new testing algorithm formulation (as a modification of the EUSAMA Plus method). The algorithm will be the object of the author's further work.

3.3.1 Influence of shock absorber damping coefficient

The first simulation scenario was involved with the change of the shock absorber damping coefficient. The analysis results are included in Tab. 2. Each value of the damping coefficient was changed on the only one vehicle side (where the plate dynamic force was measured) and on both sides. The EUSAMA coefficient values for the parameter change on one and both sides are labelled "EUS., tested" and "EUS., both" respectively (Tab. 2). Furthermore the ranges of frequency, where the maximal dynamic force occurred are also included in the way that "f. range" label is placed above and under the corresponding "EUS., tested" and "EUS., both" table field.

Table 2. Influence of the shock absorber damping coefficient on the EUSAMA coeff.

damp. coeff.	1748 [Ns/m]	1448 [Ns/m]	1148 [Ns/m]	848 [Ns/m]	548 [Ns/m]	448 [Ns/m]
f. range	15-25 Hz	15-25 Hz	15-25 Hz	15-25 Hz	15-25 Hz	15-25 Hz
EUS., tested	0.69	0.68	0.66	0.65	0.62	0.61
EUS., both	0.69	0.66	0.62	0.55	0.33	0.22
f. range	15-25 Hz	15-25 Hz	15-25 Hz	5-15 Hz	5-15 Hz	5-15 Hz
damp. coeff.	348 [Ns/m]	148 [Ns/m]	98 [Ns/m]	48 [Ns/m]	18 [Ns/m]	8 [Ns/m]
f. range	15-25 Hz	15-25 Hz	15-25 Hz	5-15 Hz	5-15 Hz	5-15 Hz
EUS., tested	0.61	0.59	0.58	0.57	0.56	0.56
EUS., both	0.03	0	0	0	0	0
f. range	5-15 Hz	5-15 Hz	5-15 Hz	5-15 Hz	5-15 Hz	5-15 Hz

The obtained results confirmed the commonly known disadvantage of the suspension diagnostics methods. The methods react only to prominent decreases of the shock absorber damping force. The conditions of the simultaneous excitation of the plates with the same frequency provoked the behaviour of the parallel coupling of the shock absorber. In the case of the damping coefficient decrease only in the "tested" model side, the EUSAMA coefficient maintained its value corresponding to the very good damping conditions (exceeding 0,6) in the range of damping coefficient from 1748 to 348 Ns/m. From 148 to 8 Ns/m the EUSAMA coefficient value referred to good damping condition range (from 0,6 to 0,4). The damping coefficient 1748 Ns/m maintained on the untested side of the model supported the poor damping qualities of the tested side and provoked limited decrease of the EUSAMA coefficient.

In the case of the damping coefficient change on both sides, the model response is more significant. The damping coefficient deterioration from 1748 to 548 Ns/m provides the change of the EUSAMA Coefficient from 0.69 to 0.33. This change corresponds to descriptively suspension damping change from very good (0,69) to satisfactory (0,33). The model reaction is involved with the uniform damping qualities deterioration of both shock absorbers, mounted in parallel. The change damping coefficient from 348 to 8 Ns/m results in the complete lack of the vibrations damping, resulting in the achievement of the 0 EUSAMA coefficient. This value refers to the detachment between the tire and the tester plate.

IOP Conf. Series: Materials Science and Engineering **148** (2016) 012034 doi:10.1088/1757-899X/148/1/012034

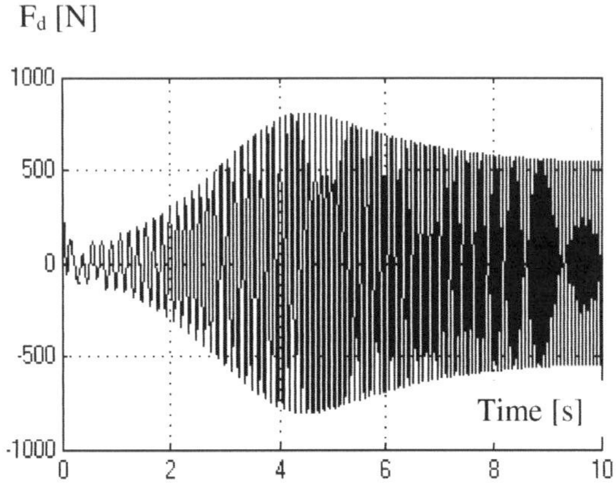

Figure 2. Time change of the force F_d, 1448 Ns/m on both sides, f. range 5-15 Hz.

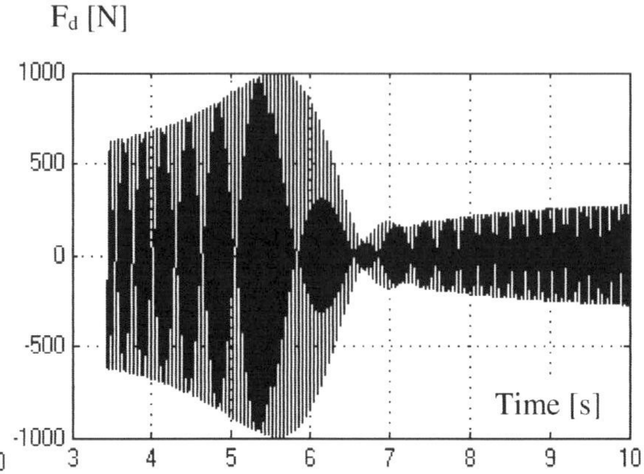

Figure 3. Time change of the force F_d, 1448 Ns/m on one side, f. range 15-25 Hz.

The time change of the plate dynamic force (F_d) in the selected simulation cases is shown in every figure of the article. The Fig. 2 and 3 show the testing process in the range from 5 to 25 Hz, with the frequency change speed 1_Hz per second. The process took 20 seconds and was divided in the two simulation scenarios. During the frequency change, the resonances of sprung and unsprung mass occur. The smaller frequency value, obtained about 4 second (Fig. 2) corresponds to the resonance of the sprung mass. Whereas, the frequency value obtained about 5,5 second in the process of frequency change from 15 to 25 Hz (Fig.3) corresponds to the resonance of the sprung mass. The maximal dynamic force value is obtained in the resonance frequencies. It is noticeable that decrease of the global damping coefficient of the half car model assembly provides that the frequency range of the maximal damping force occurrence is postponed from bigger to lower values. Due to the decreased global damping, the assembly vibrations are increased. In these conditions, the inertia of a large sprung mass vibrating with a large resonance amplitude is responsible for the maximal dynamic force value. This phenomenon is shown in Fig 4 and 5, where the low damping conditions are presented. In the situation in Fig.5, the plate dynamic force is about 7000 N, which exceeds the wheel static load force for more than 50 %.

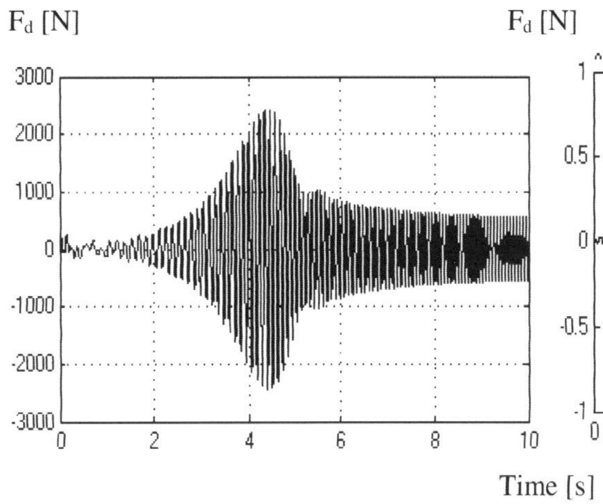

Figure 4. Time change of the force $F_{d,}$ 448 Ns/m on both sides, f. range 5-15 Hz.

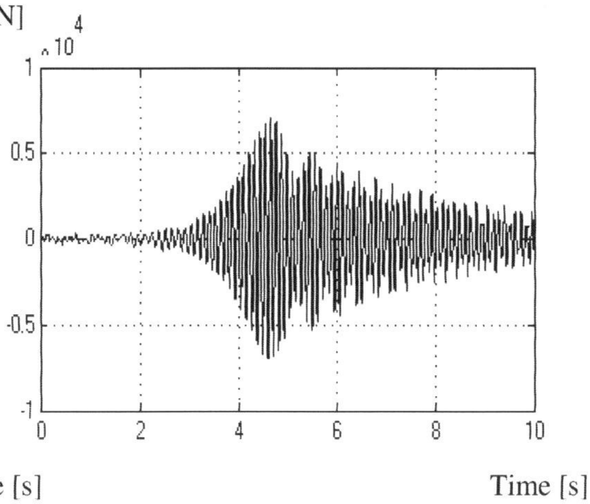

Figure 5 Time change of the force F_d, 48 Ns/m on both sides, f. range 15-25 Hz.

3.4.2 Influence of tire radial stiffness

Another simulation scenario was involved with the tire radial stiffness change. The change range of the considered parameter was selected to analyse the influence of different tire pressure and application of the tires, varied in their profiles.

Table 3. The analysis of the tire radial stiffness influence

radial stiff.	50000 [N/m]	100000 [N/m]	150000 [N/m]	200000 [N/m]	250000 [N/m]	300000 [N/m]
f. range	15-25 Hz	15-25Hz	15-25 Hz	15-25 Hz	15-25 Hz	15-25 Hz
EUS., tested	0.85	0.8	0.74	0.69	0.64	0.59
EUS., both	0.77	0.74	0.78	0.69	0.66	0.65
f. range	15-25 Hz	5-15 Hz	5-15 Hz	15-25 Hz	15-25 Hz	20-30 Hz

Results of the tire radial stiffness influence are included in Tab. 3. The table fields "EUS., tested" and "EUS., both" contain the EUSAMA coefficient values obtained in the previously described manner. The "f. range" captions above and under the EUSAMA coefficient values refer to the "EUS., tested" and "EUS., both" table field respectively. The increase of the tire radial stiffness provokes the deterioration of the EUSAMA coefficient value (Tab. 3). However the parameter change in the considered range, allows for maintaining the EUSAMA coefficient in the range corresponding to very good damping conditions. The value 50000 N/m of the radial stiffness allows for the achievement of very beneficial (according to ride comfort) value of the EUSAMA coefficient, but such low radial stiffness may provoke the loss of the vehicle stability. The Fig 6 shows the change of the dynamic force of the tester plate in the considered conditions. In addition to low value of its maximum, the regularity (small changes) of the force is noticeable. The regularities result in high comfort, perceptible by the car passengers. On the other hand, when the tyre radial stiffness is high (Fig. 7), the tester plate dynamic force shows rapid decrease in the considered frequency range. The stiffer tyre undergo smaller displacements in the specified excitation conditions, which results in higher test plate dynamic forces. The decrease influences both active safety and ride comfort of the vehicle. The road vibrations (resulting in the force changes) may cause that the car reaction, provoked by the driver's steering is different than expected. The influence of the one side of the vehicle to another is also noticed. When the global value (as a parallel springs assembly) of tyre radial stiffness coefficient is lower, the EUSAMA coefficient takes more beneficial values (Tab. 3).

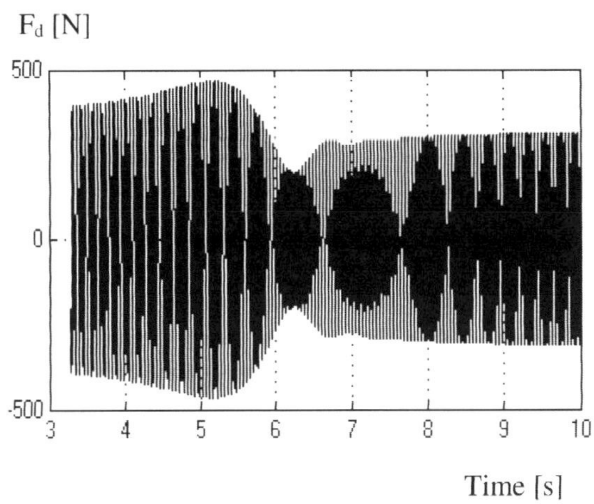

Figure 6. Time change of the force F_d, 50000 N/m on one side, f. range 15–25 Hz.

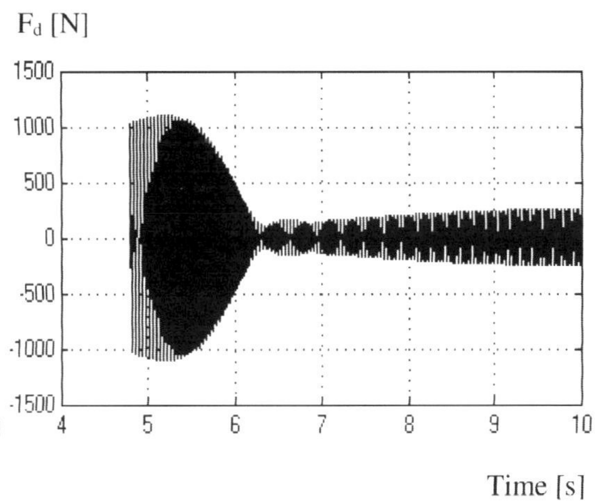

Figure 7. Time change of the force F_d, 300000 N/m on both sides, f. range 5-25 Hz.

3.4.3 Influence of the car width

To analyse the testing method sensitivity on the parameters of various vehicles, the influence of the car width was examined. Three width values were considered (Tab. 4). The simulation attempts were performed in the conditions of high (1748 Ns/m), medium (1000 Ns/m) and low (20 Ns/m) shock absorber damping coefficients, changed on the "tested" and both model sides.

Table 4. The results of the car width influence analysis.

Car width / frequency range	EUS., tested/ both 1748 [Ns/m]	EUS., tested 1000 [Ns/m]	EUS., both 1000 [Ns/m]]	EUS., tested 20 [Ns/m]	EUS., both 20 [Ns/m]
fr. range	15-25 Hz	15-25 Hz	15-25 Hz	5-15 Hz	5-15 Hz
1,588 m	0.69	0,66	0.6	0.56	0
fr. range	15-25 Hz	15-25 Hz	15-25 Hz	15-25 Hz	5-15 Hz
1,688 m	0.69	0.65	0.6	0.57	0
fr. range	15-25 Hz	15-25 Hz	15-25 Hz	5-15 Hz	5-15 Hz
1,788 m	0,69	0.65	0.6	0.56	0

The model response proved very small sensitivity on the car width change. As in the previous simulation scenarios the poor damping condition of one side of the vehicle provide that the influence of vibrations meets the resistance from the other side of the vehicle, considering the leverage between the suspension elements. The increased vehicle width must provide that the influence of the damping of one side to another decreased is. However, in the considered width range, this decrease is negligibly small (Tab.4). The conditions of medium damping coefficient values are shown in Fig. 8. Whereas, the conditions of poor damping are shown in Fig. 9. The indicated in these figures changes of the test plate dynamic force are similar to the previously obtained results, in similar model damping qualities.

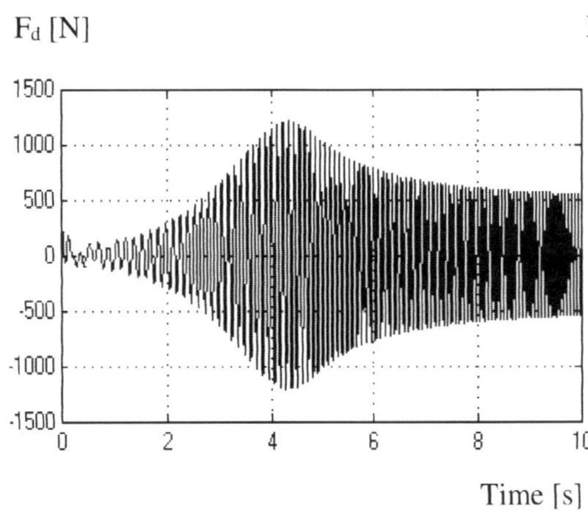

Figure 8. Change of the force F_d, 1000 Ns/m on both sides, width 1,588 m, f. 15–25 Hz.

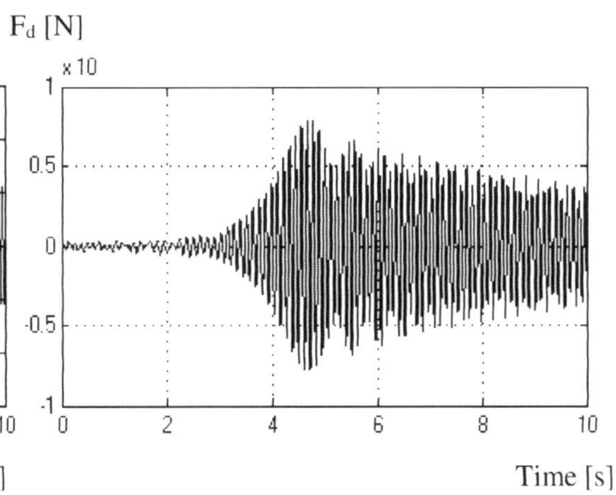

Figure 9. Change of the force F_d, 20 Ns/ m on both sides, width 1,788 m f. 5-25 Hz.

3.4.3 Influence of the rocker arm length

Another simulation scenario referred to the suspension leverage change. It was modelled by adjusting the rocker arm length. The distance between the rocker support and the point of the suspension-

IOP Conf. Series: Materials Science and Engineering **148** (2016) 012034 doi:10.1088/1757-899X/148/1/012034

damping column was changed (distance "b", look at chapter 3.1). The situations referred to the high, medium and poor damping conditions, for the tested and both sides of the vehicle.

Table 5. The rocker arm length change influence on the EUSAMA coeff.

distance b / frequency range	EUSAMA tested/ untested 1748 [Ns/m]	EUS., tested 1000 [Ns/m]	EUS., both 1000 [Ns/m]]	EUS., tested 20 [Ns/m]	EUS., both 20 [Ns/m]
fr. range	15-25 Hz	15-25 Hz	5- 15 Hz	5- 15 Hz	5- 15 Hz
30 cm	0.67	0.62	0.56	0.51	0
fr. range	15-25 Hz	15-25 Hz	15-25 Hz	15-25 Hz	5-15 Hz
40 cm	**0.69**	0.65	0.6	0.57	0
fr. range	15-25 Hz	15-25 Hz	15-25 Hz	15-25 Hz	5- 15 Hz
50 cm	0.7	0.67	0.61	0.59	0

The rocker arm length has small impact on the EUSAMA coefficient value (Tab.5), but the influence is more significant than the car width changes. The increased distance between the rocker arm support and the suspension- damping column mounting position results in the increase of the EUSAMA coefficient value. This phenomenon can be explained by the decreased relative displacement of the parts of the shock absorber, under the specified excitation conditions, when the considered distance is extended. Decrease of the relative displacement provokes the smaller values of the tester plate dynamic force. The comparison of the dynamic force change under the same excitation conditions, exerted on the two suspension assemblies, varied in the rocker lengths, is shown in Fig. 10 and 11. The figures refer to the simulation scenarios marked in white font in the Tab. 6. The increase of the length from 30 (Fig. 10) to 50 cm (Fig. 11) in the considered conditions, results in the improvement of the EUSAMA coefficient from 0.62 to 0.67 (Tab. 6).

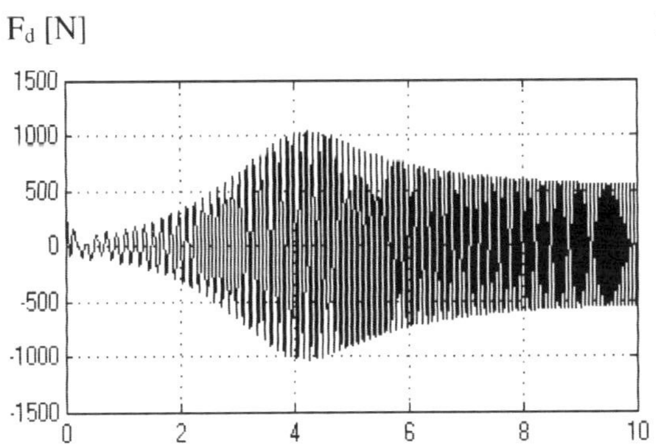

Figure 10. Time change of the force F_d, 1000 Ns/m on one side, b= 30 cm f. range 15–25 Hz.

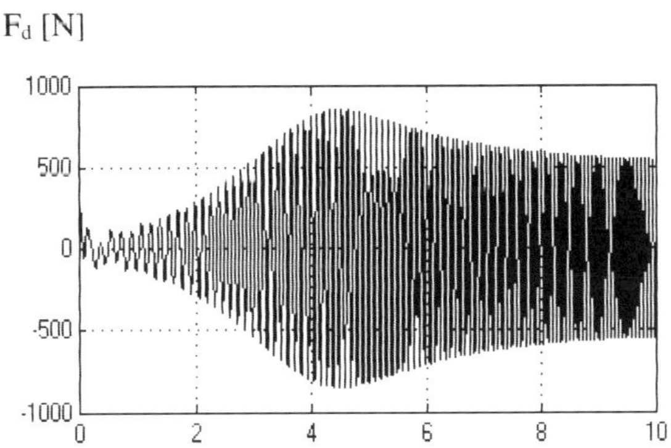

Figure 11. Change of the force F_d, 1000 Ns/ m on one side, b= 50 cm f. range 5-25 Hz.

4. Conclusions

The first simulation scenario considered the influence of the shock absorber damping coefficient. Change of the considered parameter in only one side of the vehicle provokes that the damping qualities of the non-changed side strongly affect the changed side. The second simulation scenario referred to the tire radial stiffness change. Increase of the parameter results in the deterioration of the EUSAMA coefficient values. The third scenario was involved with the analysis of the car width influence. The extended width (from 1.588 m to 1.788 cm) provided that the influence of the one side damping qualities to another was deteriorated but the changes were very small. The last scenario referred to the rocker arm length changes. The increase of the length provokes smaller relative displacements of the suspension- damping column and provides increased EUSAMA coefficient values. The influence (in the considered range from 30 to 50 cm) provides small changes. But they are more significant from the case of the car width changes. The conducted analysis corresponds to the conditions of passing through the perpendicular, symmetrical road unevenness, acting simultaneously on the both sides of the car. Real example of these conditions is involved with passing through the speed bump. The obtained results constitute the step for the formulation of the new suspension testing algorithm (as a modification of the EUSAMA Plus principle), which will be the object of the Author's further work.

References

[1] Cempiel D 2013 *Automatic Classifier of the Kind of Car Shock Absorber Damage* Combustion Engines 154(3) pp 1067-1075

[2] Gardulski J 2006 *Badania Diagnosyczne Amortyzaotrów* DIAGNOSTYKA'2 (38) pp 187-198

[3] Stańczyk T L and Jurecki R 2014 *Analiza Porównawncza Metod Badania Amortyzatorów Hydraulicznych* ZESZYTY NAUKOWE INSTYTUTU POJAZDÓW 4(100) pp 25-45

[4] Mitschke M. and Wallentowitz H 2004 *Dynamik der Kraftfahrzeuge, 4. Auflage*, Springer Verlag, Berlin/Heidelberg

[5] Ślaski G and Grześkowiak P 2011 *Masy Nieresorowane- Analiza Mas Elementów Konstrukcyjnych Zawieszeń Samochodów Osobowych* Logistyka 6/2011 pp 3743- 3750

[6] Popescu M O S and Mastorkakis N E 2009 *Testing and Simulation of a Motor Vehicle Suspension* INTERNATIONAL JOURNAL OF SYSTEMS APPLICATIONS, ENGINEERING & DEVELOPMENT Issue 2, 3/2009 pp 74- 83

[7] Ferreira C Ventura P Grinde C Morais R Valente A Neves C and Reis M 2010 *Characterization and testing of a shock absorber embedded sensor* Procedia Engineering 5/2010 pp 319- 322

[8] Sincebaugh P Green W and Rinkus G 1996 *A neural network based diagnostic test system for armored vehicle shock absorbers* Expert System with Applications vol. 11 issue 2 pp 237- 244

[9] Ferreira C Ventura P Grinde C Morais R Valente A Neves C and Reis M 2009 *A novel monolithic silicon sensor for measuring acceleration, pressure and temperature on a shock absorber* Procedia Chemistry 1/2009 pp 88-91

Energy balance in motor vehicles

M Dziubiński, A Drozd, M Adamiec and E Siemionek

Department of Automotive Vehicles, Lublin University of Technology

E-mail: m.dziubinski@pollub.pl

Abstract. This paper present the concept of testing energy balance. The test was conducted on the test bench equipped with the alternator, battery and standard mounted current receivers. The course of measurements consisted in recording the indications of three ammeters and a tachometer. On the basis of the recorded current values, it was possible to determine: energy received from the battery, consumed by the receivers and the energy drawn from the alternator.

1. Introduction

Energy balance includes the analysis of energy consumed by the receivers, supplied by the alternator and received by the battery; on this basis it can be determined whether a given alternator has enough power to ensure the proper functioning of the receivers and the appropriate charging of the battery [1-3].

The way energy balance is tested is determined by the industry standard. It shows that the verification of balance consists in comparing battery capacity before a test drive with its capacity after the appropriate drive cycle has been carried out. This standard specifies that the battery capacity at 25 ^{0}C should not be less than 80% of the capacity at the beginning of the test. The method of checking balance consists in calculating the energy consumed by the receivers and calculating the energy returned by the alternator. This method, however, is a theoretical one; it does not take battery work into consideration, and it also takes into account a lot of correction factors, e.g. the time in which the alternator returns energy, in relation to the duration of the whole drive cycle.

The concept of testing balance presented in this paper eliminates the drawbacks of the method described in literature. The presented method is accurate and makes it possible to:

- prepare the accurate balance sheet,
- take into account changes in the current drawn by the battery,
- take into account and analyse transient states of receivers work,
- take into account additional receivers which are not the standard equipment of the car,
- take into account the actual power consumed by the receivers,
- control the charging process of the battery at any time; it can be determined how long it will take the alternator to charge the battery to its rated capacity.

The method of determining balance consists in mounting three ammeters and a tachometer in the car. The ammeters measure the following currents:

- battery current
- alternator current
- current of the receivers

The essence of the method is based on the basic balance equation, which is determined by the following formula:

$$A_{alt} - A_{odb} - \frac{A_{ak}}{\eta_Q} = 0 \quad (1)$$

A_{alt} - energy returned by the alternator,
A_{odb} - energy consumed by the receivers,
A_{ak} - energy charging of the battery,
η_Q - battery charging efficiency.
Testing balance consists in registering, by means of ammeters, the three currents and the rotational speed at regular intervals.
On the basis of the obtained results, graphs of currents measured in time and the chart of alternator rotational speed (after taking ratio into consideration) can be drawn up. Then, the obtained results are multiplied by the mean value of the voltage in the system during the preparation of the balance sheet (approximately 13.5 V) according to the formula:

$$P(t) = U \cdot I(t) \quad (2)$$

and the graphs of power similar to the graphs of current are obtained. Then, the obtained graphs of power are integrated according to the following formula:

$$A(t) = A(t) = \int_0^t N(t)dt \quad (3)$$

in time from t = 0 to the end of measurement duration, and what is obtained are the graphs of energy:
- consumed by the receivers.
- returned by the alternator,
- either returned or collected by the battery.

Analysing the charts the energy consumed by the receivers and returned by the alternator makes it possible to decide:
- whether the power consumed by the receivers will be fully covered directly by the alternator or partly by the battery,
- whether there are possibilities to install additional receivers in the car.

The analysis of energy collected or returned by the battery leads to some conclusions about the state of battery charge at any time of conducting the balance [4-6,12,13]. On the basis of the graph, it can be stated after how much time of the test duration the battery is charged to its rated capacity. On the grounds of the graph of the power returned or collected by the battery, the chart of energy stored in the battery can be drawn up.
The initial energy A_0 can be determined from the formula:

$$A_0 = I \cdot U_{śr} \cdot t = Q \cdot U_{śr} \quad (4)$$

A_0 - energy stored in the battery [Wh]
I - current discharge amperage [A]
$U_{śr}$ - mean discharge voltage [V]
t - discharge time [h]
Q - battery capacity [Ah]
Energy stored in the battery as a percentage is calculated from the equation.

$$A_{\%} = \frac{A_{Wh}}{A_{100\%}} 100\% \quad (5)$$

$A_{\%}$ - energy expressed in [%]
A_{Wh} - energy expressed in [Wh]

$A_{100\%}$ - energy expressed in [Wh] equivalent to 100% .

On the basis of the drawn up current-speed characteristics of the alternator, the value of current which the alternator can provide at a given rotational speed can be determined for the tested vehicle. For complete energy balance while driving in different conditions, one should adopt coefficients of the simultaneous use of particular receivers, taking their power into consideration [8-11].

2. The description of the test bench and the course of measurements

The test was conducted on the test bench , equipped with the alternator type A115-34B, 45 Ah battery, air conditioning system (figure 1), ESP system (figure 2) and standard mounted current receivers. On the experimental bench, the ammeters and a voltmeter were mounted. The course of measurements consisted in recording the indications of three ammeters and a tachometer every 5 seconds, for a randomly selected drive cycle which lasted 1800s. This cycle was simulated for a particular rotational speed of the alternator. During the cycle, engine rotational speed was changed and different current receivers were switched on.

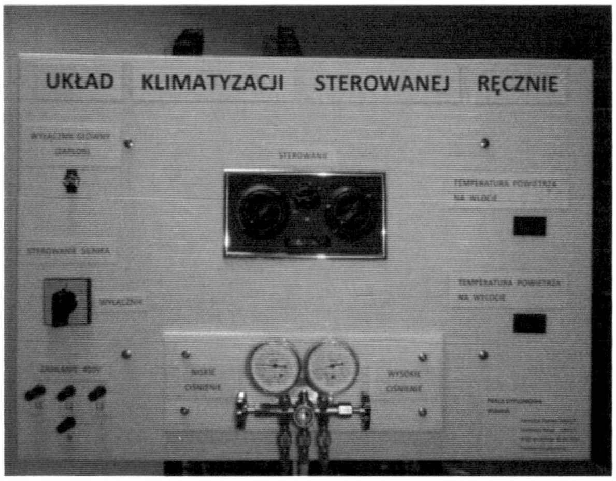

Figure 1. Air conditioning test bench.

Figure 2. ESP test bench.

3. Test results

On the basis of the recorded measurement results, it was possible to determine: current drawn from the battery (Fig. 3), current in the circuit of the receivers (Fig. 4) and the current flowing from the alternator (Fig.5). On the basis of the recorded current values, it was possible to determine: energy received from the battery (Fig. 6), consumed by the receivers (Fig. 7) and the energy drawn from the alternator (Fig. 8). Figure 9 shows the external characteristics of the alternator A115-34B.

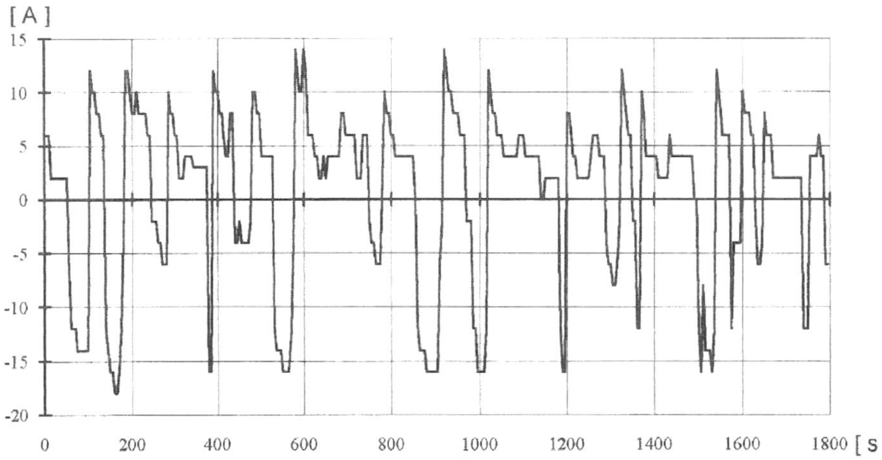

Figure 3. Current drawn from the battery.

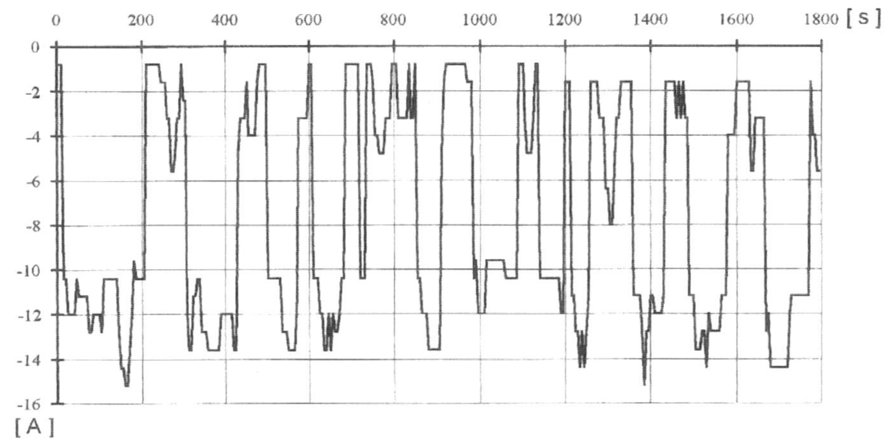

Figure 4. Current consumed by the receivers.

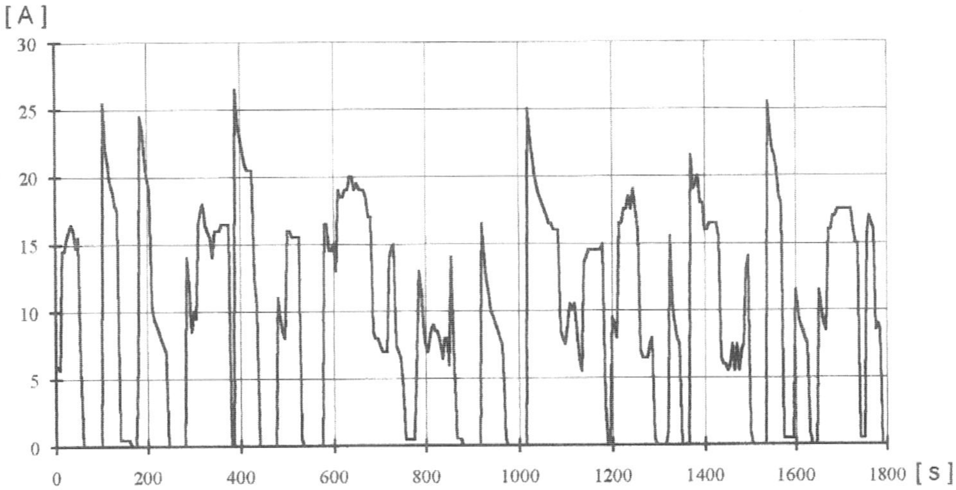

Figure 5. Current flowing from the alternator.

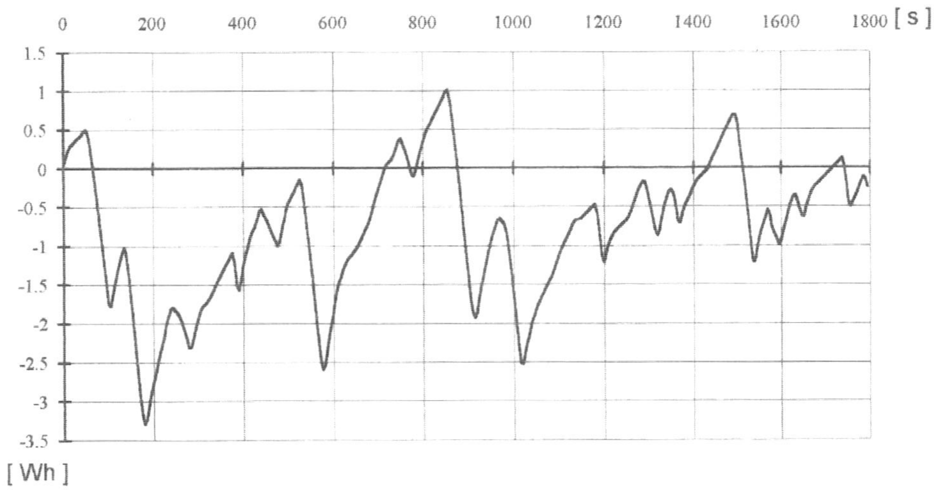

Figure 6. Energy delivered to the battery.

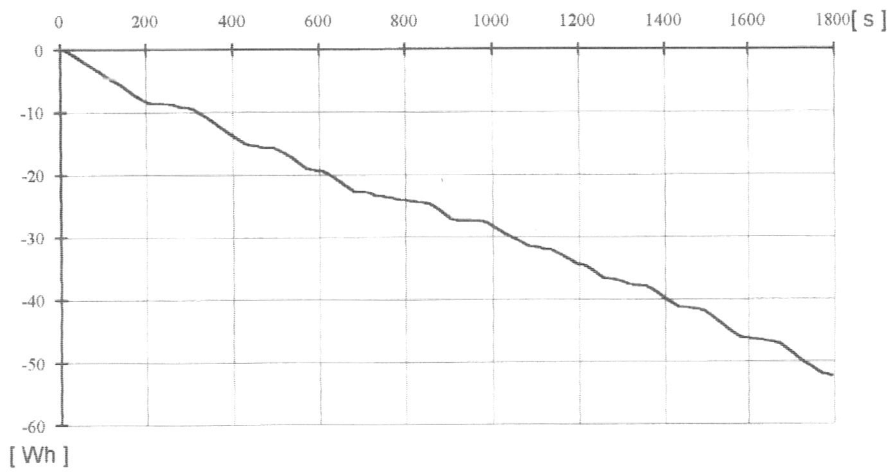

Figure 7. Energy consumed by the receivers.

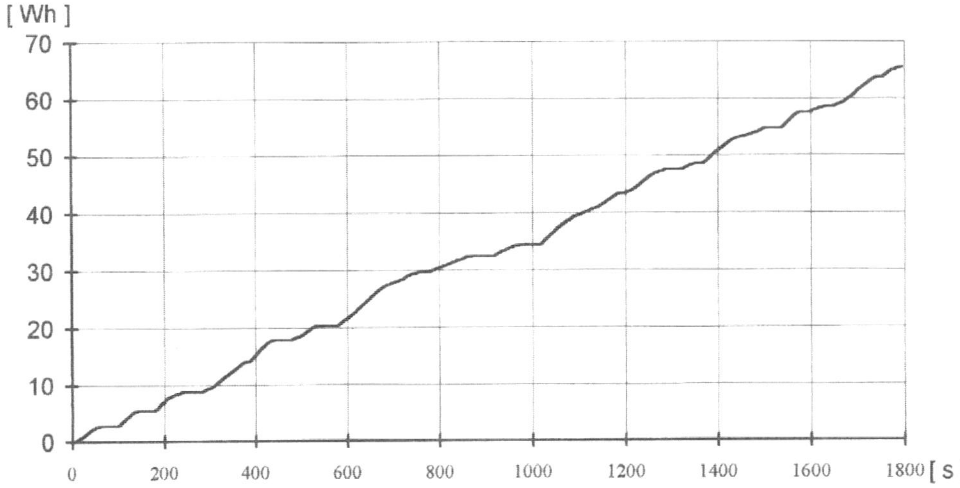

Figure 8. Energy returned by the alternator.

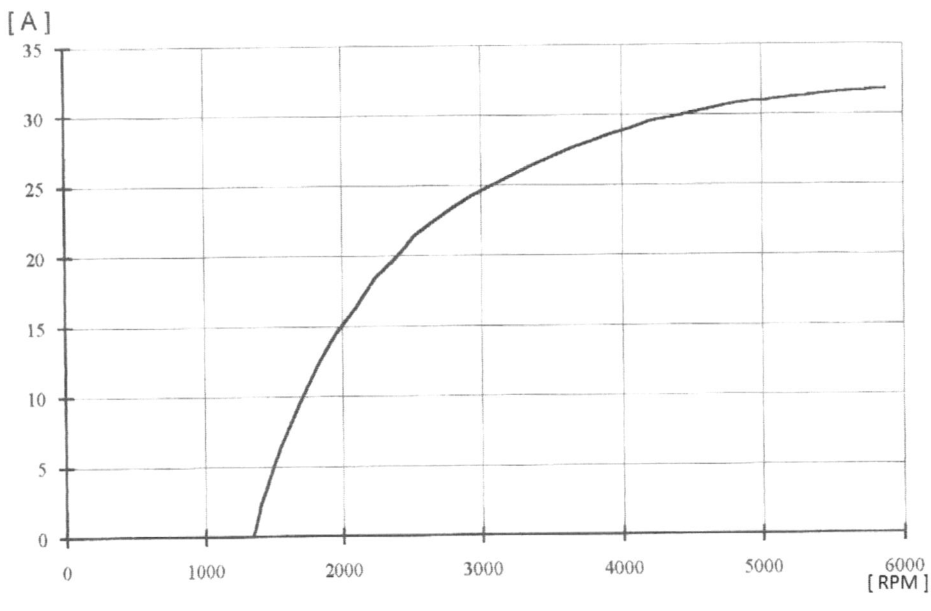

Figure 9. The characteristics of the alternator load.

Further in the study, the analysis of co-operation between the power supply system, the air-conditioning system (Fig. 10) and the ABS system (Fig. 13) was carried out. Measurements were conducted for four different air-conditioning fan speeds and variable rotational speed of the alternator. The measurement results have been shown in Fig. 11 and Fig.12.

Testing energy balance of the wiring system in co-operation with the ABS system was realized by changing the rotational speed of the alternator. The aim of the study was to determine the voltage value of the power supply system at which the ABS system works properly, and the voltage at which a failure will be recognized.

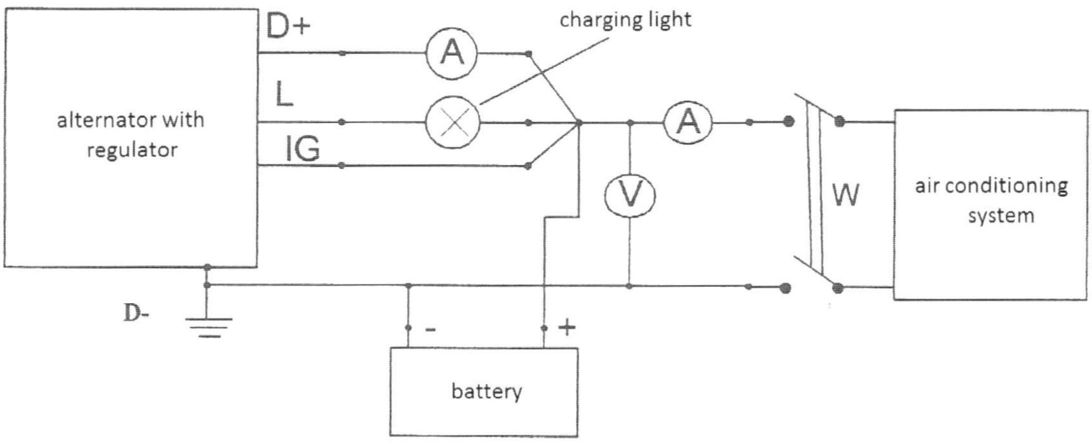

Figure 10. Scheme for examining the load imposed on the power supply system by air-conditioning. A - ammeter, V - voltmeter, W – switch.

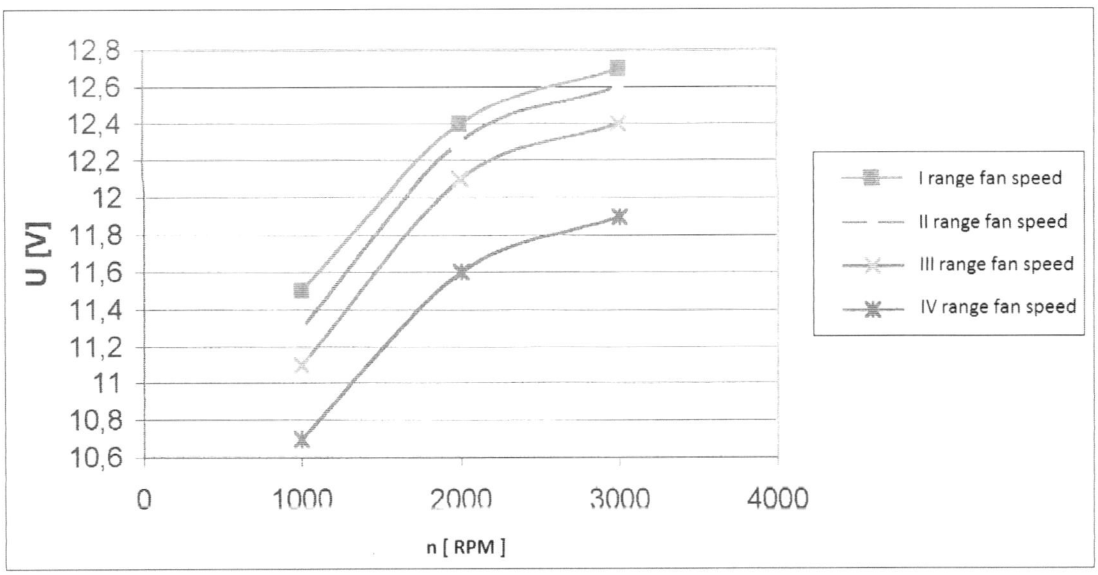

Figure 11. Voltage - speed characteristics of the alternator load.

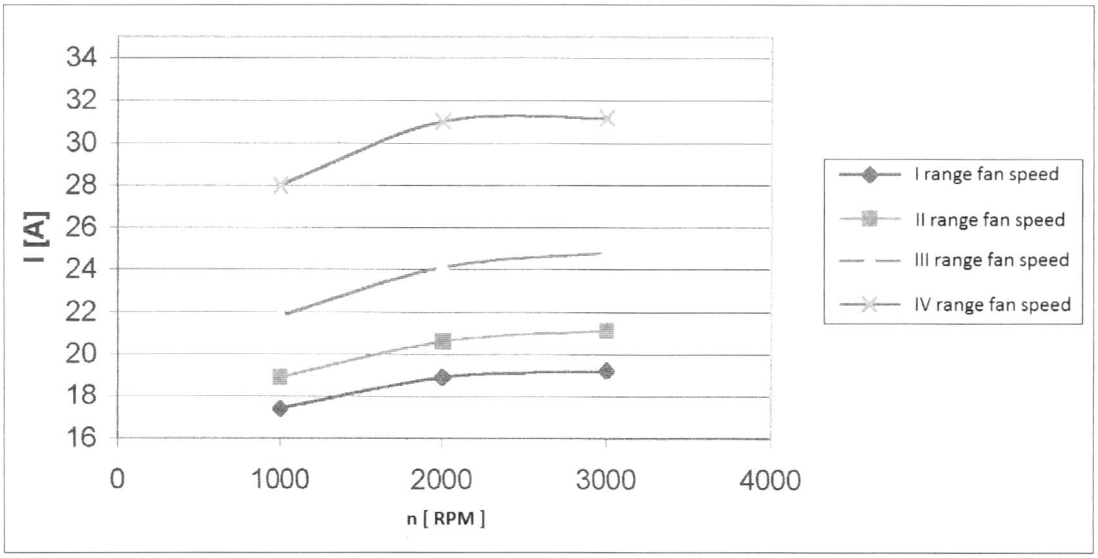

Figure 12. Current - speed characteristics of the alternator load.

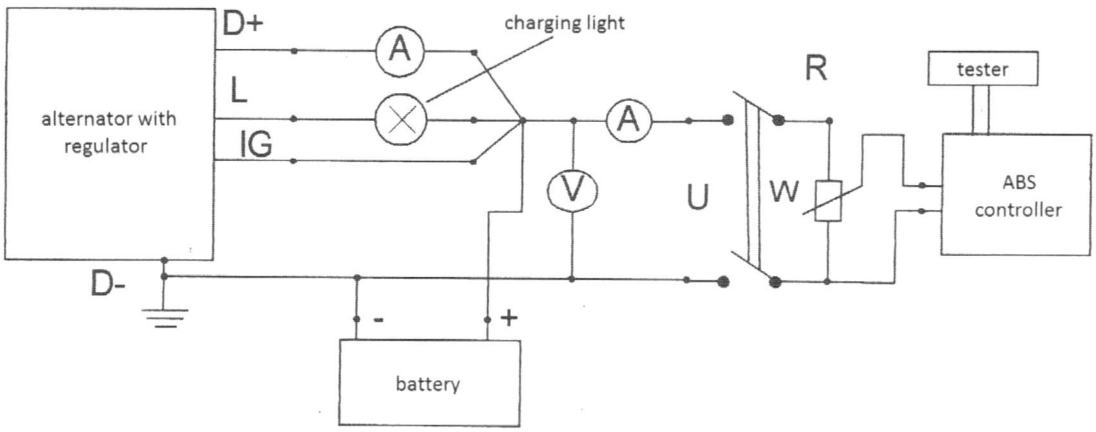

Figure 13. Diagram of supplying the system for testing the ABS controller
D-, D +, L, IG – alternator connectors, A - ammeter, V - voltmeter, R -resistor, W – switch.

4. Conclusions

The continuous improvement in safety and driving comfort led to the increase in the demand for alternator power. As a result of excessive load in the system, the supply voltage decreases, which leads to the malfunction of electronic devices and is indicated by control lights.

The measurement results of three characteristic currents: drawn from the alternator, consumed by the receivers and the current in the battery circuit are used in the method of drawing up energy balance of the car wiring. The balance method provides valuable information about the process of charging or discharging the battery. The indicator lamp indicated charging status in the supply system does not provide information about the distribution of currents in the battery circuit.

The carried out tests of power balance, in which the alternator and air-conditioning co-operated, confirmed the increase in values of current drawn from the alternator and voltage decrease in the wiring system for particular ranges of fan rotational speeds.

As the supply system and the ABS system cooperated, signalling a failure in the ABS circuit by lighting the ABS warning lamp in the case of voltage decrease was confirmed.

References

[1] Khaligh A and Zhihao L 2010 *Battery, Ultracapacitor, Fuel Cell, and Hybrid Energy Storage Systems for Electric, Hybrid Electric, Fuel Cell, and Plug-In Hybrid Electric Vehicles: State of the Art* Vehicular Technology, IEEE Transactions on , vol.59, no.6, pp.2806,2814, July 2010

[2] Agarwal V and Dev M 2013 *Introduction to hybrid electric vehicles: State of art* Engineering and Systems (SCES), 2013 Students Conference on, pp 1-6 12-14 April 2013

[3] Chan C C 2007 *The State of the Art of Electric, Hybrid, and Fuel Cell Vehicles*, Proceedings of the IEEE, vol.95, no.4, pp.704,718, April 2007

[4] Dziubiński M 2001 *The parameter analysis of selcted multifunctional voltage regulators co-operating whith compact alternator*, Teka - Kostrukcja, Badania, Eksploatacja, Technologia, Pojazdów Samochodowych i Silników Spalinowych, Polish Academy of Sciences , Zeszyt Nr 22, 2001, s.77-83

[5] Dziubiński M and Walusiak S *Simulation study of reliability car electronic wires* Teka Polish Academy of Sciences, ISSN 1642-1639, Zeszyt 29-30, 2005 s.123-130

[6] Dziubiński M and Walusiak S 2000 *The Characteristics of Compact Alternator Collaborating with the Multifuunctional Voltage Regulator*, 26th International Scientific Conference On Combustion Engines KONES 2000, ISBN 83-910906-7-1, 2000, s. 49-55

[7] Horyński M 2013 *The application of dispersed processing networks in order to optimize the energy consumption in contemporary buildings* Przegląd Elektrotechniczny, ISSN 0033-2097, R. 89 Nr 4 2013 , s. 293-295

[8] Larminie J, Lowry J, 2012 *Electric Vehicle Technology Explained* Second Edition United Kingdom ISBN 9781119942733

[9] Leitman S and Brant B 2009 *Bulid Your Own Electric Vehicle" Second Edition*, United States of America ISBN 978-0-07-154373-6

[10] Maggetto G and Van Mierlo J *Electric and electric hybrid vehicle technology: a survey Electric, Hybrid and Fuel Cell Vehicles (Ref. No. 2000/050)*, IEE Seminar

[11] Crist P 2012 *Electric Vehicles Revisited – Costs, Subsidies and Prospects*, Discussion Paper 2012 (03), International Transport Forum, OECD, Paris 2012, p. 6.

[12] Siemionek E and Dziubiński M 2015 Testing energy consumption in the trolleybus and the bus on a chosen public transport line in Lublin, *Advances in science and technology research journal* - 2015, nr 26, vol. 9, s. 152-153

[13] Siemionek E 2013 *Performance tests of a one-truck vehicle with an electric drive system, Selected issues from machine building and exploitation*, Edition by Leszek Gardyński, Mirosław Szala, ISBN 978-83-63569-85-3, Lublin 2013, p. 61-75

[14] Siłka W 2002 *Teoria ruchu samochodu*, Wydawnictwo Naukowo-Techniczne, ISBN 83-204-2748-7, Warszawa

[15] Feng-chun S, Liu B and Zhen-po W., *Analysis of Energy Consumption Characteristics of Dual-source Trolleybus*, Transportation Electrification Asia-Pacific (ITEC Asia-Pacific), 2014 IEEE Conference and Expo Beijing

Electromagnetic interference in electrical systems of motor vehicles

M Dziubiński[1], A Drozd[1], M Adamiec[1] and E Siemionek[1]

Department of Automotive Vehicles, Lublin University of Technology

e-mail: m.dziubinski@pollub.pl

Abstract. Electronic ignition system affects the electronic equipment of the vehicle by electric and magnetic fields. The measurement of radio electromagnetic interference originating from the ignition system affecting the audiovisual test bench was carried out with a variable speed of the ignition system. The paper presents measurements of radio electromagnetic interference in automobiles. In order to determine the level of electromagnetic interference, the audiovisual test bench was equipped with a set of meters for power consumption and assessment of the level of electromagnetic interference. Measurements of the electromagnetic interference level within the audiovisual system were performed on an experimental test bench consisting of the ignition system, starting system and charging system with an alternator and regulator.

1. Introduction

The source of electromagnetic interference in vehicles is the electrical system, which has a range of devices whose natural operate (spark plugs, sparking contacts) is associated with the generation of radio electromagnetic interference. The source of electromagnetic interference in automotive vehicles is the ignition system, which can affect the operate of radio communication equipment, even within a few hundred meters from the vehicle [8,9,11]. Devices that generate electromagnetic interference inside vehicles include: elements of injection systems, vibrating voltage regulators, blower motors, wipers, fans, washers, as well as bimetal sensors and relays.

Electromagnetic interference can travel through galvanic isolation in the form of voltage drops on metallic pieces and lead resistance and trough inductive way via system capacity [1,7,12]. The existence of inductive coupling between parallel wires results in rapid changes of current, which interfere with the operate of electronic devices, and as a result may even damage the input transistor. Power surge occurs during a power failure.

By interrupting the flow of electric current by means of a mechanical switch, the spark that occurs between the contacts causes a reduction in the height and steepness of the induced pulse. There are many sources of power surge in the vehicle electrical system. It should be noted that the battery that operates with the electrical system effectively suppresses power surge which originates from the alternator, ignition system and induced in other circuits with inductance. Disconnection of the battery results in deprivation of the natural power surge protection. That is why additional electronic equipment is used more frequently to protect the system of a vehicle.

The power surge generated by the ignition system arises upon the delivery of power supply through the ignition switch. An unfavourable situation may occur if the circuit of the battery gets interrupted during charging, when the other receivers are disconnected. This results in an increased power supply to the ignition system, which was created in the winding of the alternator at the moment of burden reduction. Dangerous power surges can occur during normal operation of the whole system, it disables the ignition switch, which in effect disconnects the battery from the receivers. Power surges can also arise in the case of welding of vehicle subassemblies using an electric arc. For this reason, manufacturers of alternators and electronic voltage regulators recommend to detach them during any electric arc welding.

An electromagnetic interference meter is used to measure radio electromagnetic interference. This is a special measurement receiver that allows, using accepted methods of measurement, to clearly assess the distortive effect in the radio or television reception, regardless of the nature and waveform of electromagnetic interference.

Electromagnetic interference measurements, with sinusoidal waveform, can be made with conventional measurement receivers, such as field strength meter or selective voltmeter. For electromagnetic interference, whose waveform is sinusoidal, there is a relationship between the amplitude of the electromagnetic interference and the effect of electromagnetic interference for a given measurement frequency.

Considering the electromagnetic interference of pulse type, which is generated by the majority of interfering devices in a vehicle, the knowledge of the pulse shape and its peak value is not enough to predict the effect of electromagnetic interference. The parameter to be considered is the repetition frequency of pulses per unit of time. It was found that the effect that interferes with radio or television reception depends largely on: the shape and the absolute value of the pulse and the pulse repetition frequency. For this reason, to measure a series of pulsed electromagnetic interference one uses an electromagnetic interference meter, instead of using a selective micro-voltmeter.

2. The analysis of anti-electromagnetic interference protection systems

We can distinguish three types of protection systems due to the different effectiveness:

- normal protection systems: reducing radio electromagnetic interference to the N level,
- tightened protection systems: reducing radio electromagnetic interference to the O level,
- special protection systems: reducing radio electromagnetic interference to the S level,
- the W level (high) is used on separated areas [2-4].

In order to reduce the level of electromagnetic interference intensity of a continuous nature, arising in the vehicle electrical system, one uses measures that suppress the radiation of electromagnetic interference along the wires. The radiation of electromagnetic interference can be reduced by shielding. The condition for this protection to be effective is a good electrical connection between the metal shield enclosing the source of electromagnetic interference and the mass of a vehicle.

Capacitors, inductors and LC filters are used on the low voltage side. In order to reduce the level of radio electromagnetic interference at a frequency of long- and medium-wave ratio one attaches capacitors in parallel to the source of electromagnetic interference (figure 1.). Feed through capacitors, conducting high currents (up to 60 A) enable good suppression of electromagnetic interference at higher frequencies of FM.

Reactors are used to improve the effectiveness of suppression in the range of UHF waves. The inclusion of a reactor in the power cord of an electrical equipment prevents electromagnetic interference from the power supply. In other cases one can use LC filters and shielding. In order to suppress electromagnetic interference from the high-voltage side, one uses special energy absorbing resistors. Such a solution may be fully effective only if suppressing resistors are embedded as closely as possible to the source of electromagnetic interference, for instance in a spark plug or high-voltage wires.

IOP Conf. Series: Materials Science and Engineering **148** (2016) 012036 doi:10.1088/1757-899X/148/1/012036

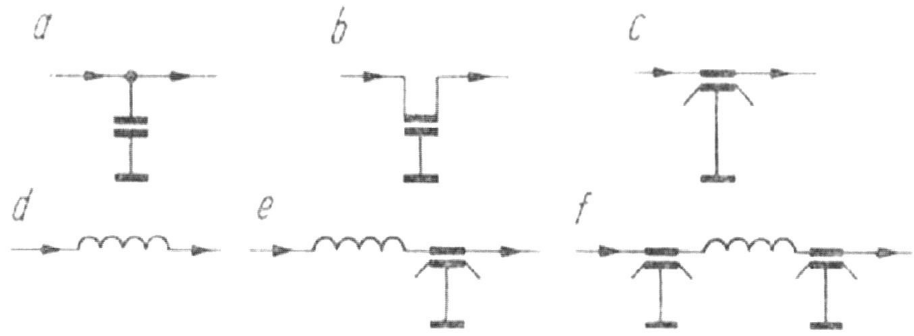

Figure. 1. Typical electromagnetic interference suppression components: *a* – a parallel capacitor, *b* – a bypass capacitor, *c* – feedtrough capacitor, *d* – inductance (reactor) *e* – a simple LC filter, *f* – LC filter with increased effectiveness.

A spark plug can be depicted as a capacitor made of metal body divided by an insulator and an inner electrode (figure 2a). After placing a resistor in place of the center electrode, the equivalent circuit is presented by figure 2b. Placement of the resistor closer to the spark plug reduces the system capacity. More than half of the energy stored in the capacity of the spark plug is forced to go through the R resistance. This causes a reduction in the amplitude of the current pulse in the capacitive phase of spark discharge. Ignition systems with a lower input impedance are more sensitive to the inclusion of suppressing resistance.

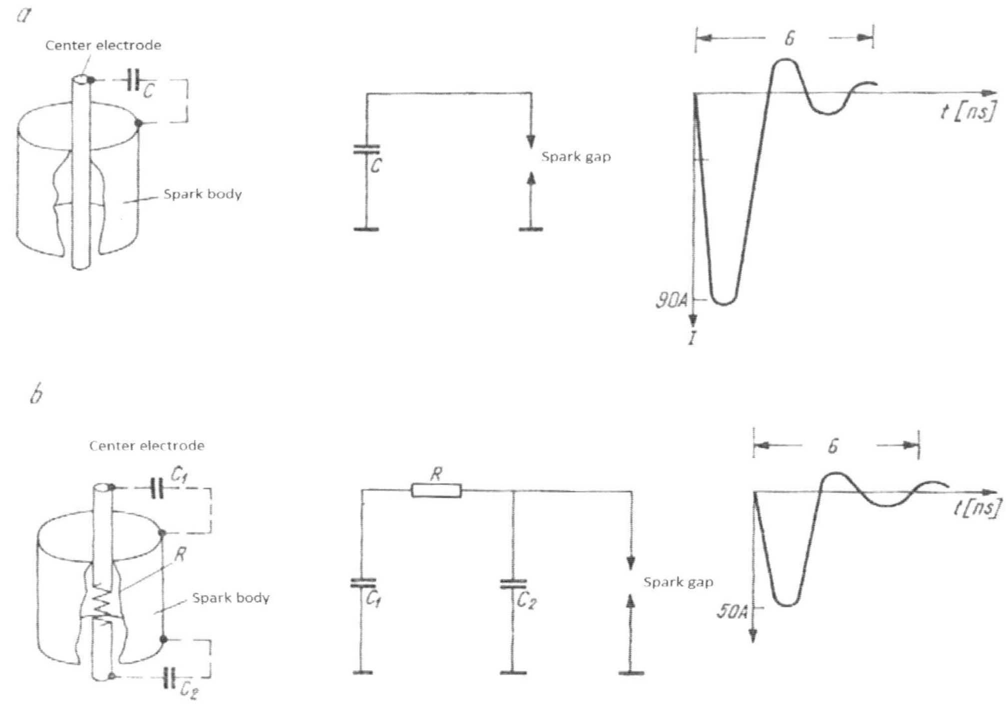

Figure 2. Anti-suppression protection systems in the form of a series resistance: *a* – scheme of a conventional spark plug, *b* – scheme of a spark plug with a built-in suppression resistor.

A very effective way of limiting the level of electromagnetic electromagnetic interference emitted from the ignition system is shielding of its individual elements. These results in increased capacity of the entire circuit of the high voltage, which in effect reduces the voltage across the electrodes of a spark plug, thus contributing to ignition misses. The lack of connection between the engine, the body and the weight of a vehicle can help to reduce the effectiveness of the suppression equipment. Ground connection should be as short as possible and its cross section must be at least 16 mm^2.

3. Measurements of radio electromagnetic interference in automobiles

Measurements of the intensity of any electromagnetic interference field are performed using an electromagnetic interference meter equipped with a special measurement antenna, enabling determination of the intensity of an electromagnetic interference field. The voltage at the antenna terminals is proportional to the electromotive force (SEM) induced in the antenna and to the intensity of the electromagnetic interference field, in which the antenna is located. The ratio of the intensity of the electromagnetic interference field *(E)* to the voltage measured *(Uz)* at the terminals of the antenna attached to the electromagnetic interference meter is determined by the manufacturer of a given electromagnetic interference meter, and is known as a calibration coefficient of the antenna *(kf)* in a function of frequency. The intensity of the electromagnetic interference field is determined from the following formula:

$$E = k_f + U_z \tag{1}$$

where:

E – the intensity of the electromagnetic interference field [dB], (0 dB = 1μV/m),

U_z – the voltage level measured by an electromagnetic interference meter [dB], (0 dB = 1μV/m),

k_f – calibration coefficient of the antenna [dB].

In the short-, medium- and long-wave ratio, that is from 0.15 to 30 MHz, one applies the rod antenna, whose length is 1 m. It allows to measure the electrical component of the field strength. The magnetic component is also measured in this frequency ratio using a tuned dipole antenna (requires a change in the length of dipole arms for a particular frequency), or a wide-band antenna which does not require tuning for a particular frequency. With these antennas one can measure the electric component of the electromagnetic interference field strength for the vertical and horizontal polarization.

4. Experimental tests

The experimental audiovisual test bench has full radio and monitor equipment (two-way). In order to determine the level of electromagnetic interference, the audiovisual test bench was equipped with a set of meters for power consumption and assessment of the level of electromagnetic interference. The block diagram of audiovisual test bench is presented by figure 3.

The experimental test bench consists of a two-channel stereo amplifier, two liquid crystal panels and a set of speakers. The system requires power supply from the battery or alternator. The system is protected by a serially switched fuse (15 A). The liquid crystal panels *DVD052 Dual Screen Player* are equipped with screens with a diagonal measurement of 7", built-in stereo speakers and the ability to play multimedia files from portable storage media (DVD, VCD, MP3, MPEG4, JPG). The amplifier *Voice Kraft C-238* has a nominal effective power of 2 x 20W, supplies power to two, two-way coaxial speakers *Houston Acoustic HA-402* with a maximum power of 120W.

The battery is connected to two terminals Z1 and Z2 of the test bench. Between the Z1 terminal and the W1 switch one can find a serially switched fuse that protects the system. The Z1 terminal supplies voltage to the power amplifier and the car lighter socket, into which the M1 monitor is connected. The audio-video signal (A/V) of the M1 monitor is supplied to the M2 monitor with a mini-jack cable. The power supply of the M2 monitor is implemented with the M1 monitor. A stereo audio signal of the M1 monitor is led out of the mini-jack headphone socket to the input of the phono amplifier. A memory stick with sample media files, that one can play using the system, is connected to the USB port of the M1 monitor. The system has two ways to adjust the volume, the first at the M1 monitor and the second at the stereo power amplifier. The view of the test bench is shown in Fig. 4.

IOP Conf. Series: Materials Science and Engineering **148** (2016) 012036 doi:10.1088/1757-899X/148/1/012036

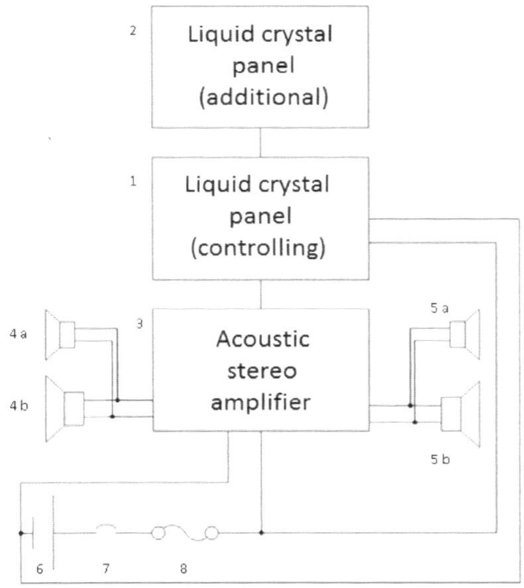

Figure 3. The block diagram of test bench: 1 – liquid crystal panel (controlling), 2 – liquid crystal panel (additional), 3 – acoustic stereo amplifier, 4 – two-way speaker, left *a* – tweeter, *b* – woofer, 5 – two-way speaker, right *a* – tweeter, *b* – woofer, 6 – battery/alternator, 7 – power switch, 8 – fuse (15 A).

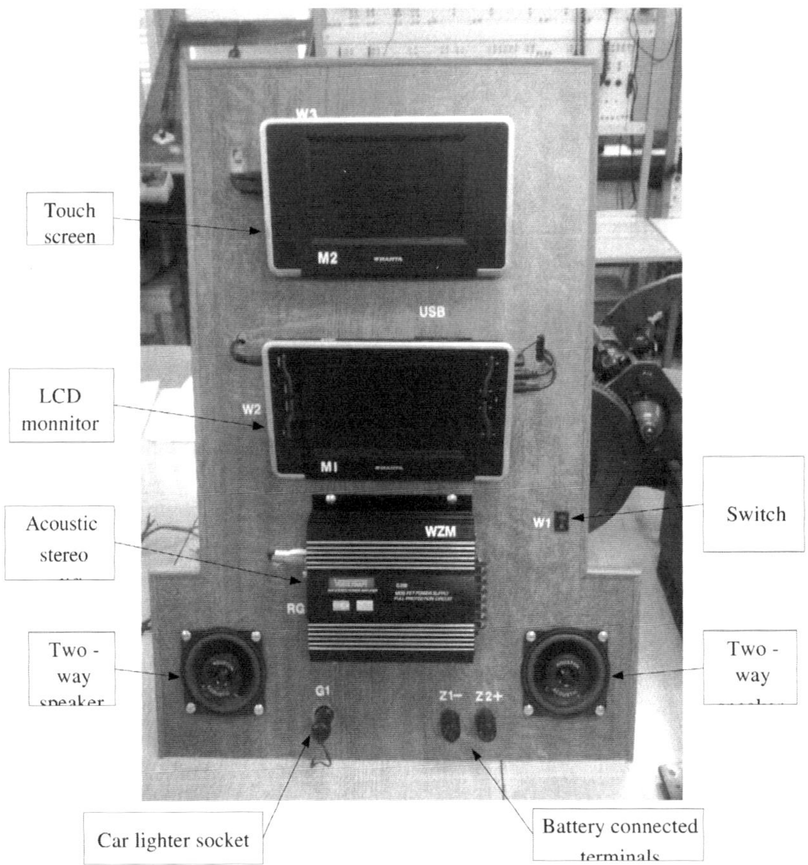

Figure 4. The view of the audiovisual test bench.

IOP Conf. Series: Materials Science and Engineering **148** (2016) 012036 doi:10.1088/1757-899X/148/1/012036

The research program conducted on the audiovisual test bench. The test equipment used to determine the active power of the system and the level of electromagnetic interference included: two analog ammeters to measure the direct current, magnetoelectric having isolation equal to 3 kV, accuracy class 0.5 and two analog voltmeters to measure the DC voltage, magnetoelectric having isolation equal to 3 kV, accuracy class 0.5, finder of electromagnetic interference type ZSS-1 (figure 6), speed meter DT-2234C+ and the universal diagnostic tool from Bosch model MOT-150.
Measurements of the electromagnetic interference level within the audiovisual system were performed on an experimental test bench consisting of the ignition system, starting system and charging system with an alternator and regulator. The tests were conducted by registering:

- the level of radio electromagnetic interference originating from the ignition system,
- the measurement of current and voltage intensity generated by the alternator and the level of radio electromagnetic interference resulting from its operate,
- the measurement of power consumption and voltage on the audiovisual test bench,
- the measurement of current and voltage during switching of the starter.

The block diagram of the complete measurement system is presented by figure 5, and the view of the complete measurement system is presented by figure 6.

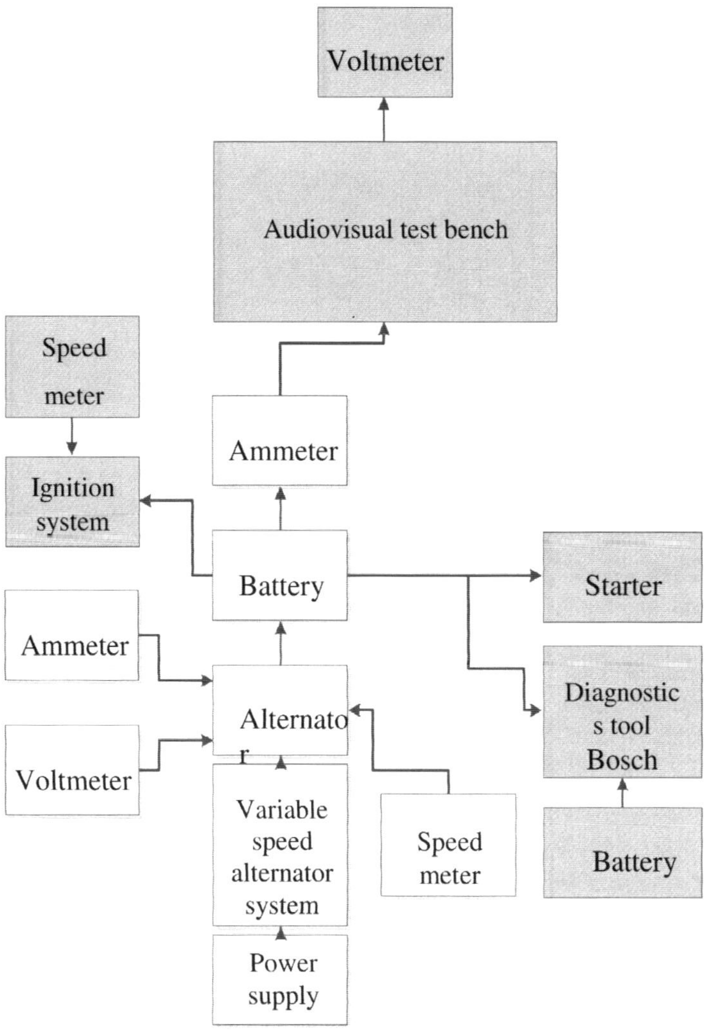

Figure 5. The block diagram of the measurement system.

Figure 6. The view of the complete measurement system.

The finder of electromagnetic interference and the diagnostic tester from Bosch MOT 150 were used in the study (presented in Fig. 7).

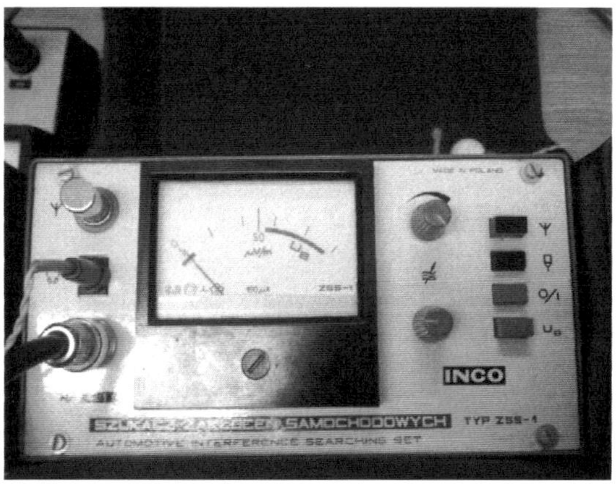

Figure 7. The finder of electromagnetic interference type ZSS-1.

5. The results of experimental tests

The measurement of radio electromagnetic interference originating from the ignition system affecting the audiovisual test bench was carried out with a variable speed of the ignition system. The electromagnetic interference level in a function of rotational speed of the ignition device is presented by figure 8.

The measurement of current and voltage generated by the alternator and the measurement of electromagnetic interference resulting from its operate was carried out with variable speed alternator. The measurement results are shown in figure 9 and figure 10.

The measurement of current and voltage consumption was performed by changing the volume. The results of the measurement of intensity of the current and active power consumption are presented by figure 11 and figure 12.

The measurement of current and voltage during the operate of a starter was carried out on neutral gear and for a short circuit condition.

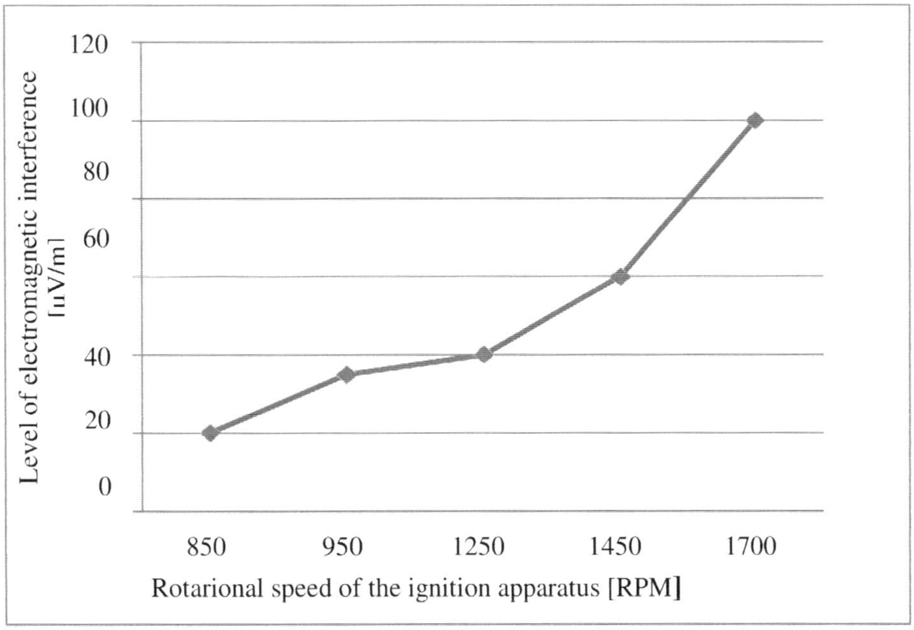

Figure 8. The level of electromagnetic interference in a rotational speed of the ignition apparatus.

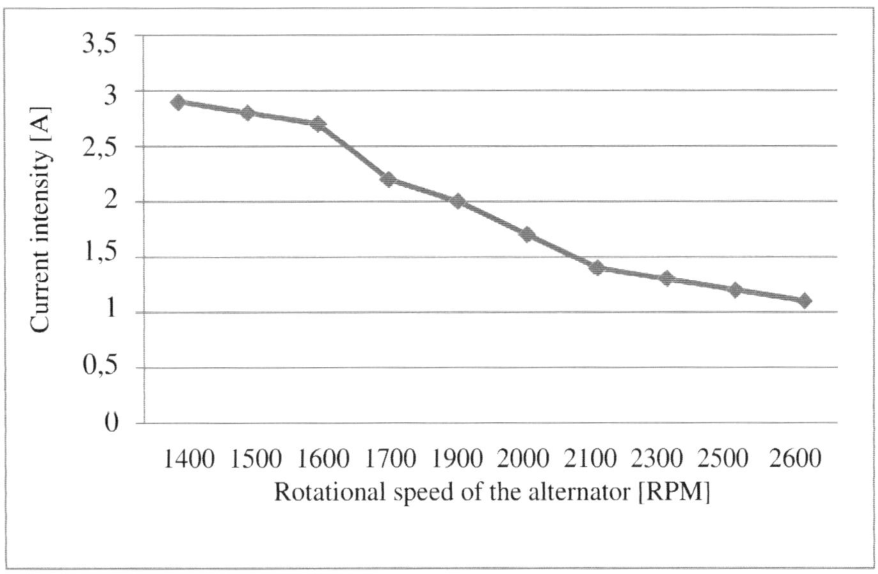

Figure 9. The current intensity in a rotational speed of the alternator.

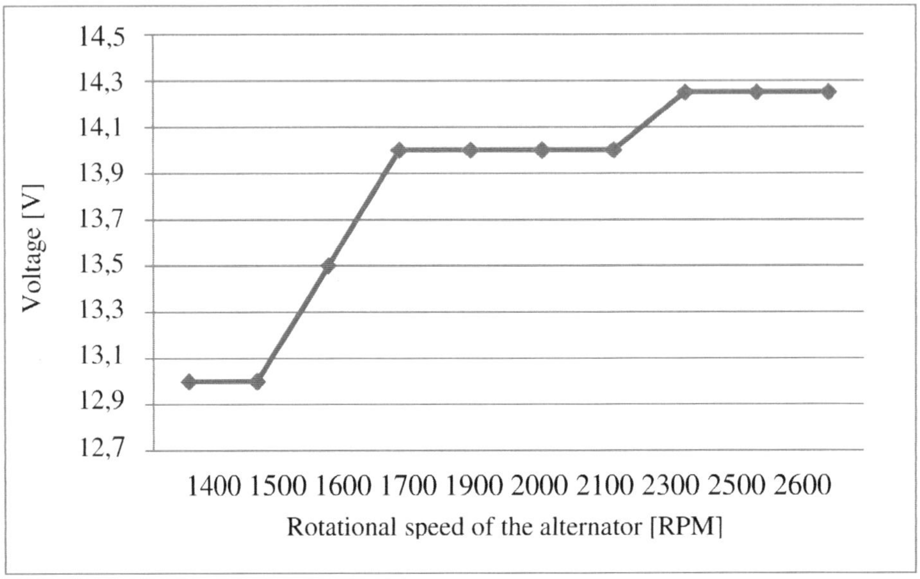

Figure 10. The voltage in a rotational speed of the alternator.

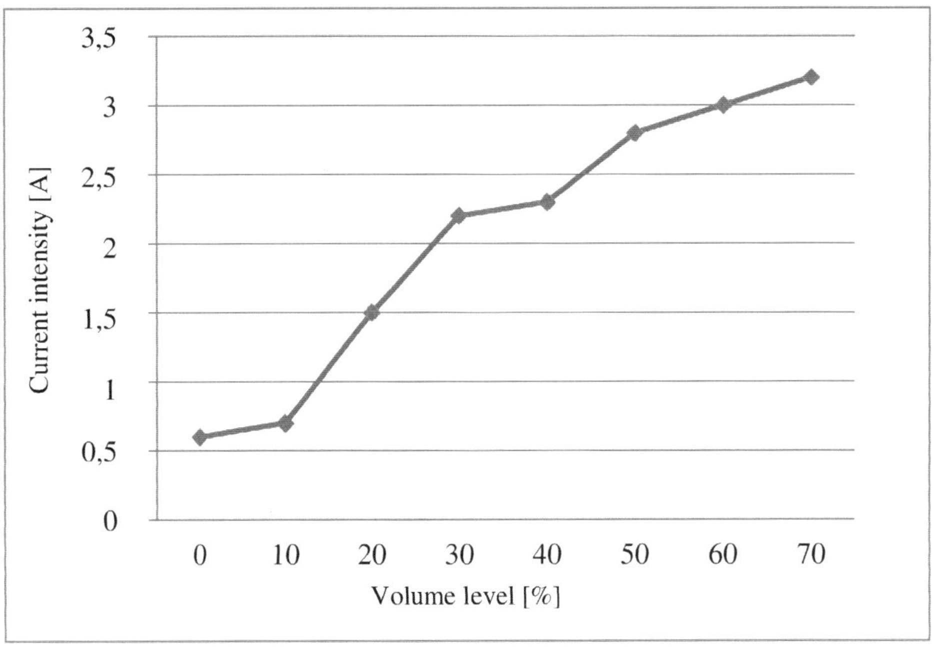

Figure 11. The current intensity in the volume level.

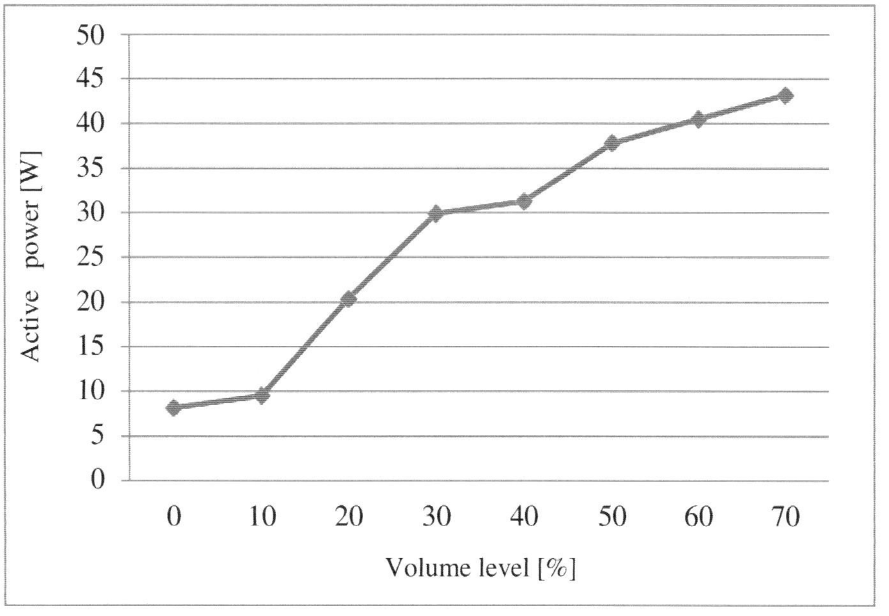

Figure 12. The active power in the volume level of the audiovisual test bench.

5. Conclusions

The analysis of tests results as well as visual and listening observations shows that despite the exceeding values of radio electromagnetic interference, the operation of the monitor and amplifier was not disturbed by the operation of the ignition system and alternator.

The current on the neutral gear 37 A was recorded when the starter was engaged, while the current 295 A was recorded during the braking torque, at which there was a momentary drop in voltage at the experimental test bench.

The measurements of current and voltage of the alternator at variable speeds up to 2600 rev/min showed an electromagnetic interference level of 30 μV/m.

References

[1] Cieśla A, Kraszewski W, Skowron M and Syrek P., 2015, Analysis of the distribution of the magnetic field generated by the physiotherapy device, Przegląd Elektrotechniczny, ISSN 0033-2097, R. 91, Nr 2, s.162-165
[2] CISPR-25: 2002 - Limits and methods of measurement of radio disturbance characteristics for the protection of receivers used on board vehicle
[3] ISO 10605, 2008, Road vehicles - electrical disturbances from electrostatic discharges.
[4] ISO 11452, Road vehicles - electrical disturbances by narrowband radiated electromagnetic energy - component test methods.
[5] Jarzyna W, Augustyniak M and Bochenski M 2012, et al., PD and LQR controllers applied to vibration damping of an active composite beam. Przegląd Elektrotechniczny, Volume: 88 Issue: 10B s. 128-131
[6] Jarzyna W, Augustyniak M and Wojcik A 2012 Application of a fuzzy PI controller in regulation of active piezoelectric composite structure. Przegląd Elektrotechniczny Volume: 88 Issue: 10A , s. 298-301

[7] Karpowicz J and Gryz K 2003 The methods of assessment of exposure to complex electromagnetic fields in office rooms, *Przegląd Elektrotechniczny* , ISSN 0033-2097, R.79, Nr 12, s. 923-926
[8] Mazurek P 2009 Measurements of high frequency electromagnetic field in urban environment, *Przegląd Elektrotechniczny*, ISSN 0033-2097, R 85, Nr 12, s.118-120
[9] Mazurek P 2013 Selected of legal and technical aspects of the automotive EMC *Przegląd Elektrotechniczny* ISSN 0033-2097, R. 89 Nr 12 pp 328-331
[10] Ogłodek E and Mo D 2009 The effect of extremely low frequency magnetic fields on osteogenesis, Borgis - New Medicine, Vol. 3, p. 58-60
[11] Prauzner T 2012 The electromagnetic electromagnetic interferences in the electronic alarm systems *Przegląd Elektrotechniczny* ISSN 0033-2097, R. 88, Nr 12b, s. 205-208.
[12] Zeńczak M 2008 Approximate relationships for calculation of electric field and magnetic field intensities under power transmission lines, *Przegląd elektrotechniczny*, ISSN 0033-2097, R.84, Nr 4, s. 117-120
[13] Żurek Z 2001 Application of magnetic methods in investigations of rail vehicle elements, *Zeszyty Naukowe Politechniki Śląskiej* Seria Transport z. 42, s. 169,175

IOP Conf. Series: Materials Science and Engineering **148** (2016) 012037 doi:10.1088/1757-899X/148/1/012037

The analysis of the accuracy of the wheel alignment inspection method on the side-slip plate stand

A Gajek[1] and P Strzępek[1]

[1] Cracow University of Technology, Institute of Automobiles and Internal Combustion Engines PL 31-364 Kraków, Al. Jana Pawła II 37

E-mail: gajeka@mech.pk.edu.pl, piotrs@mech.pk.edu.pl

Abstract. The article presents the theoretical basis and the results of the examination of the wheel alignment inspection method on the slide slip plate stand. It is obligatory test during periodic technical inspection of the vehicle. The measurement is executed in the dynamic conditions. The dependence between the lateral displacement of the plate and toe-in of the tested wheels has been shown. If the diameter of the wheel rim is known then the value of the toe-in can be calculated. The comparison of the toe-in measurements on the plate stand and on the four heads device for the wheel alignment inspection has been carried out. The accuracy of the measurements and the influence of the conditions of the tests on the plate stand (the way of passing through the plate) were estimated. The conclusions about the accuracy of this method are presented.

1. Introduction

The vehicle inspection stations are equipped with the slide surface plate stands for the wheels toe-in control in dynamic conditions (figure 1). Test is realized during crossing the plate by the vehicle with small constant speed. The plate is crosswise bearing. The lateral displacement of the plate is proportional to the toe-in of the car wheels.

Figure 1. Side slip tester for control of the car wheels toe-in [2].

The article presents theoretical basis and the results of the examination of that method. The dependence between the lateral displacement of the plate and toe-in of the tested wheels has been shown. The comparison of the toe-in measurements on the plate stand and on the four heads device for the wheel alignment inspection with laser measurement system has been carried out (figure 2). The accuracy of measurements and the influence of the conditions of the investigation on results of tests on the plate stand were estimated. The conclusions about the accuracy of this method are presented.

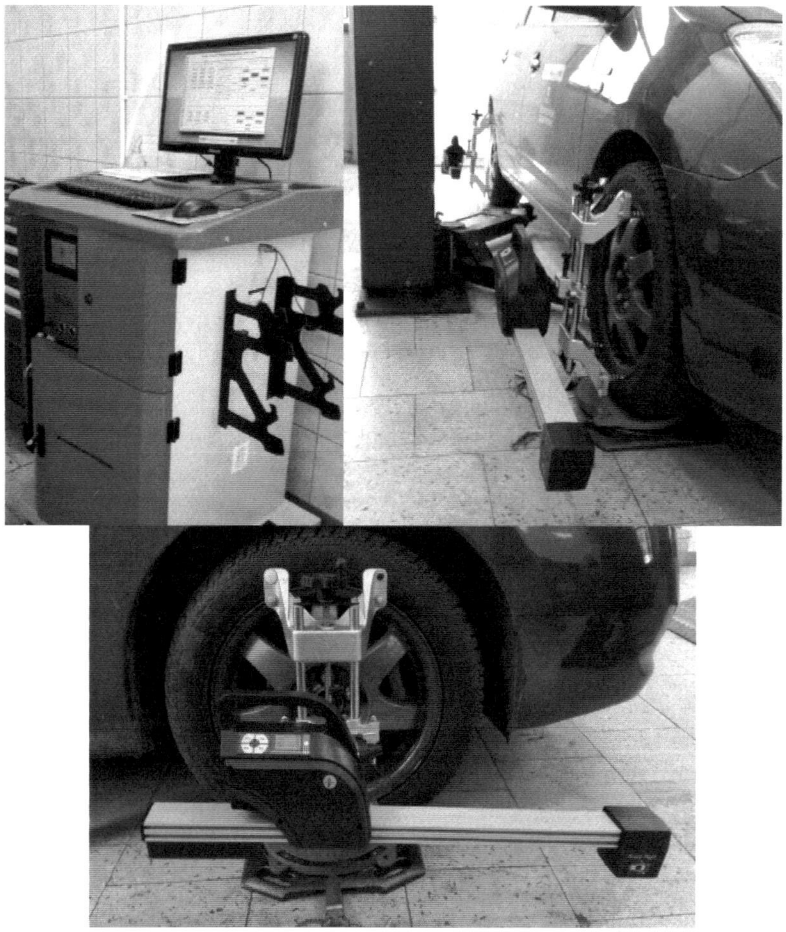

Figure 2. Four heads device with CCD camera measurement system.

2. Base of the measurement – theoretical analysis

The plate length 1 m can shift in the lateral direction x, but doesn't shift in the longitudinal direction y (figure 3). Friction resistances during lateral movement of the plate are minimalized by special bearing. The car wheel with toe-in or toe-out regulation crosses the plate with the constant small speed. Second wheel of the axle rolls on the stiffness surface of the stand floor. The direction of the right wheel setting and reaction between this wheel and stiff surface of the floor causes the lateral displacement of axle wheels axle and reaction on the left wheel. This wheel together with the crosswise bearing plate moves lateral in relation to the surface of the stand floor. The plate displacement is proportional to the toe-in or toe-out and is measured. The value of toe-in/out is calculated on the base of the plate displacement and evaluated. The analysis was done for few cases of the wheels alignment.

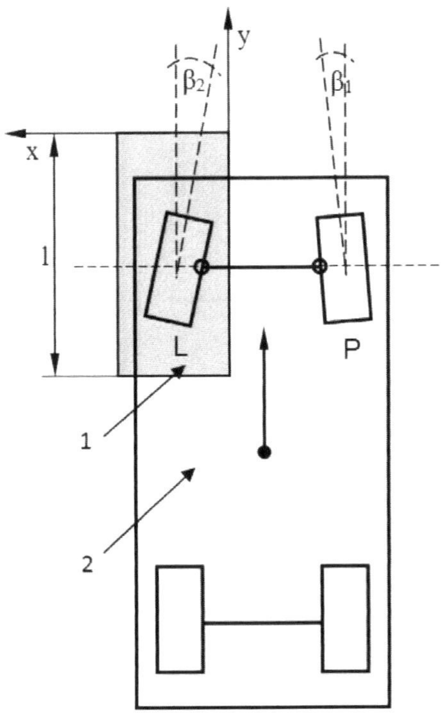

Figure 3. Scheme of the vehicle position in relation to the side slip plate:
1 – slide plate, 2 – vehicle, β_1, β_2 – toe-in angles.

I case

The right wheel setting with toe-in angle, the left wheel setting directly (straight ahead), (figure 4).

Right wheel forces the direction of the wheel axle in relation to the stand floor. The right wheel displacement in the y direction on the length l causes lateral displacement the centre of this wheel by line segment Δx_1. If the wheel slip will be omitted (the rolling resistance has very small value) then the centre of the right wheel displacement on the length l is:

$$\Delta x_1 = l \cdot tg\beta_1 \tag{1}$$

β_1 – toe-in angle of the right wheel,
l – length of the plate

This displacement reacts on the car body and the left wheel. If the steering wheel will be positioned directly (straight ahead), the reaction form the right wheel causes the lateral displacement of the left wheel centre by line segment Δx_2 together with the plate. Ignoring the friction resistances in the plate bearings and slip of the tire, the lateral displacement of the plate will be equal lateral displacement of the right wheel centre:

$$\Delta x_2 = \Delta x_1 \tag{2}$$

For this case (right wheel with toe-in angle β_1, left wheel with zero toe-in angle) the toe-in of the wheels z_1 (figure 4) is:

$$z_1 = d \cdot \sin \beta_1 \tag{3}$$

d – diameter of the wheel rim

On the base of figure 4 we can write:

$$\frac{z_1}{d} = \frac{\Delta x_1}{\sqrt{\Delta x_1^{\,2} + l^2}} \tag{4}$$

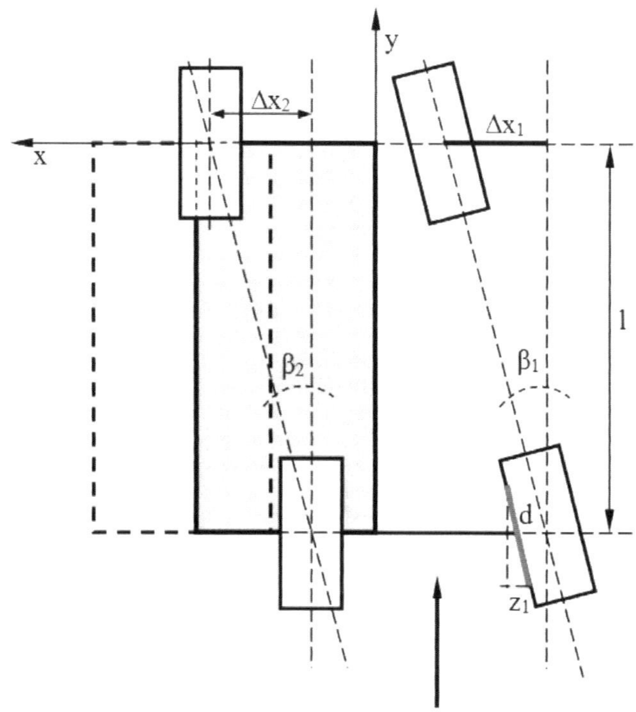

Figure 4. The position of the front wheels of the car at the start and end of the measurement on the side slip plate.

On the stand the displacement of the plate Δx_2 is measured. Because $\Delta x_2 = \Delta x_1$ then toe-in of the wheels equals:

$$z = z_1 = \frac{\Delta x_2 \cdot d}{\sqrt{\Delta x_2^{\,2} + l^2}} \tag{5}$$

II case

Right wheel P setting directly (straight ahead), left wheel L with toe-in angle β_2 (figure 5)

Left wheel rolls at the angle β_2 to the longitudinal axis of the plate. Right wheel objects to this direction. It rolls on the stiff surface of the stand floor paralelly to the longitudinal axis of the slide plate. This wheel forces the direction of the motion of the axle of the wheels. It causes reaction between left wheel and plate and displacement of the plate on left in relation to the left wheel. The lateral slip of the left wheel in relation to the plate is compensated (balanced) by lateral displacement of the plate.

Component v_x (figure 5) is the speed of the wheel in relation to the slide plate in the x direction. At the same time the plate shifts in the opposite direction with the speed **-** $\mathbf{v_x}$. Value of the v_x is:

$$v_x = v_y tg\beta_2 \tag{6}$$

During uniform motion of the vehicle on the l distance the lateral displacement of the plate is:

$$\Delta x_2 = v_x \cdot t = v_y tg\beta_2 \cdot t$$
$$t = \frac{l}{v_y} \tag{7}$$

$$\Delta x_2 = l \cdot tg\beta_2 \tag{8}$$

As in previous case the toe-in of the wheels is:

$$z = z_2 = \frac{\Delta x_2 \cdot d}{\sqrt{\Delta x_2^{\ 2} + l^2}} \tag{9}$$

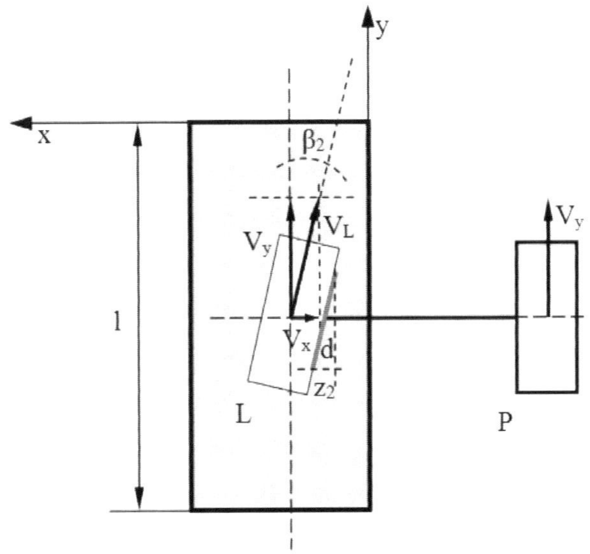

Figure 5. The orientation of the front left wheel L on the plate and the front right wheel P on the floor of the stand.

III case

The right wheel setting with toe-in angle β_1, left wheel with toe-in angle β_2 (figure 6)

In this case the sum of the reactions from two wheels effects on the plate. The right wheel rolls on the stiff surface and hold its direction of motion at the angle β_1. This motion on the length l causes the lateral displacement of the left wheel together with the plate on the left by section $\Delta x_1 = ltg\beta$. Left wheel tries to keep its direction of rolling β_2. Right wheel sets against of this direction. It rolls on the stiff surface. It causes reaction between left wheel and plate and displacement of the plate in relation to the wheel on left by Δx_2. Displacements Δx_1 i Δx_2 have the same directions.

Vectors analysis of the wheels speed allows to set sum of the plate displacement. Axle of the left and right wheels displacements in relation to the stand floor with speed $\mathbf{v_P}$. Component $\mathbf{v_{x1}}$ of this speed is the speed of the left wheel and plate in x direction. Left wheel is positioned in β_2 direction and rolls with speed $\mathbf{v_L}$ (figure 6). Value of the speed $\mathbf{v_L}$ results from value of the component v_y of the speed $\mathbf{v_P}$. Component $\mathbf{v_{x2}}$ of the speed $\mathbf{v_L}$ is the speed of the left wheel in relation to the plate. On the contrary the plate moves in the reverse direction with speed $\mathbf{-v_{x2}}$. The speed of the plate displacement in relation to the stand floor is the sum of the vectors $\mathbf{v_{x1}}$ and $\mathbf{-v_{x2}}$. If the slip between the wheel and the plate is omitted then the speed value of the plate in relation to the floor of the stand can be described as:

$$v_x = v_{x1} + v_{x2} = v_y tg\beta_1 + v_y tg\beta_2 \tag{10}$$

v_{x1} – component of the speed v_p in x direction, it causes shift of the left wheel together with plate on the left,
v_{x2} –speed of the plate in x direction in relation to the left wheel

During uniform motion of the vehicle the lateral displacement of the plate is:

$$\Delta x_2 = v_x \cdot t = v_y (tg\beta_1 + tg\beta_2) \cdot t$$
$$t = \frac{l}{v_y} \tag{11}$$

On the base of these dependences the displacement of the slide plate is:

$$\Delta x = l \cdot (tg\beta_1 + tg\beta_2) = \Delta x_1 + \Delta x_2 \tag{12}$$

Toe-in of the wheels is:

$$z = z_1 + z_2 = \frac{\Delta x_1 \cdot d}{\sqrt{\Delta x_1^{\;2} + l^2}} + \frac{\Delta x_2 \cdot d}{\sqrt{\Delta x_2 + l^2}} \tag{13}$$

Because $\Delta x_1 << l$ and $\Delta x_2 << l$, after simplification we can describe:

$$z = z_1 + z_2 = \frac{\Delta x_1 \cdot d}{l} + \frac{\Delta x_2 \cdot d}{l} = \Delta x \frac{d}{l} \tag{14}$$

If the wheels are toe-out the sign of the sum $z_1 + z_2$ will be reversed. Displacement of the plate will be on right. If one of the wheels will be toe-in positioned and second toe-out then the displacement of the plate will be the difference between Δx_1 and Δx_2.

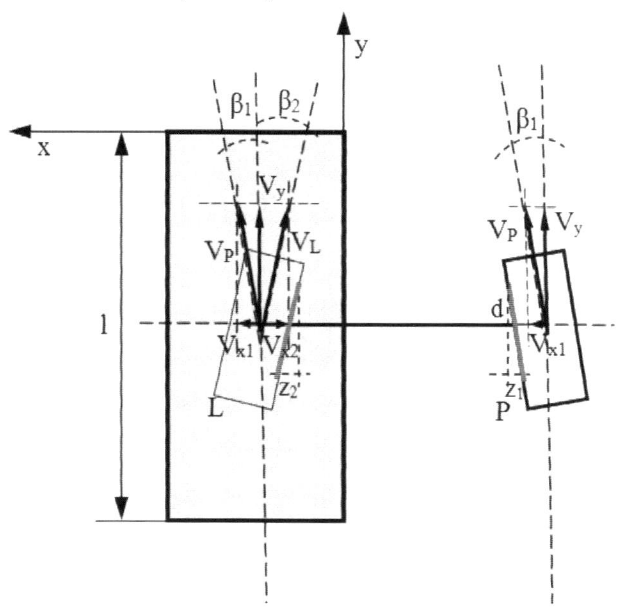

Figure 6. The orientation and speed of the left wheel on the plate and right wheel on the floor of the stand.

The analysis shows that measurements of the plate displacement allows to determine the sum of the toe-in for the data wheel rim diameter. Analysis was performed for the ideal conditions of measurement and state of the plate stand: the friction in bearings was disregarded, the slip of the wheels was ignored, during crossing the plate the driving or braking forces doesn't act, the crossing the plate is parallel to the plate axis. During real tests these factors can influence the measurements of the plate displacement. The influence of these factors and the examination the accuracy of that method the were realized by verification tests.

3. Influence of the exploitation parameters on the results of the toe-in measurement

The parallel direct during crossing the plate can be not maintained (figure 7). It causes the change of the angle of the left wheel during crossing the plate by angle α in relation to parallel drive. Also right wheel crosses the floor with angle α. For the wheels with toe-in regulation as in figure 7 and for small angle α can be described:

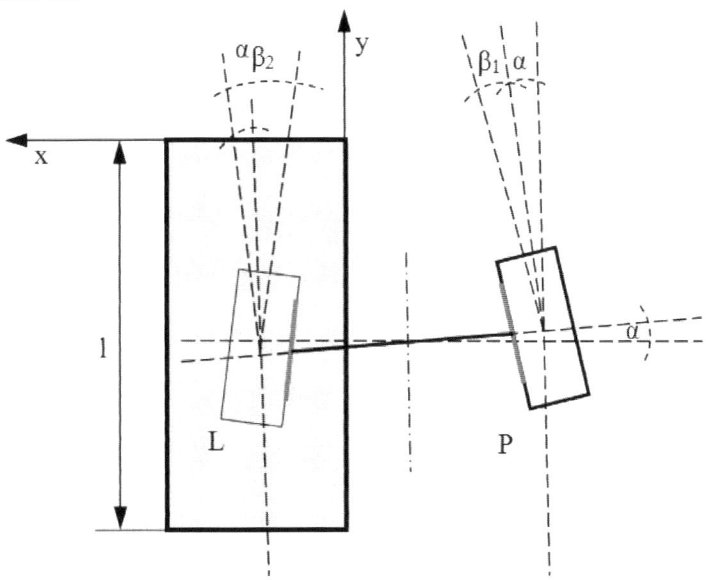

Figure 7. The orientation front axle during not parallel drive in on the plate.

$$\Delta x_1 = l \cdot tg(\beta_1 + \alpha)$$
$$\Delta x_2 = l \cdot tg(\beta_2 - \alpha) \tag{15}$$

$$\Delta x = \Delta x_1 + \Delta x_2 = l \cdot (tg(\beta_1 + \alpha) + tg(\beta_2 - \alpha)) = l \cdot (\frac{tg\beta_1 + tg\alpha}{1 - tg\beta_1 tg\alpha} + \frac{tg\beta_2 - tg\alpha}{1 + tg\beta_2 tg\alpha}) \tag{16}$$

For the small angles α the product $tg\beta_i \cdot tg\alpha$ is near zero. If we take it into consideration, it can be described:

$$\Delta x = \Delta x_1 + \Delta x_2 = l \cdot (tg\beta_1 + tg\beta_2) \tag{17}$$

It shows that for the small angles α the result of the toe-in measurement isn't disturbed by not parallel drive by the plate. The tests confirmed above theoretical analysis (table 1). During tests the toe-in of the vehicle was regulated up to manufacturing data 0°0" (0mm).

Table 1. Results of the toe-in measurements on the plate stand depending on the angle α of the drive in the plate

$\alpha = 3°$ right		
	nr of the test	toe-in value [mm]
	1	0
	2	0
	3	0
$\alpha = -3°$ left		
	nr of the test	toe-in value [mm]
	1	0
	2	0
	3	0

Another factor which can influence the result of the measurement is the change of the car speed during crossing the plate. If the vehicle accelerates or brakes the tangent forces act between plate and tire. The influence of this parameter was analysed in the small range of speed changes possible in practise ($\Delta v \approx 5$km/h). Before the tests the the toe-in of the wheels was regulated up to manufacturing data 0°0" (0mm). Tests didn't show the influence of the speed changes on the results of toe-in measurements (table 2).

Table 2. The results of the toe-in measurements for the front wheels on the plate during acceleration and braking on the plate

nr of the test	toe-in for acceleration [mm]	toe-in for braking [mm]
1	0	0
2	0	0
3	0	0

The next controlled factor was changing of the drive direction during crossing the plate. The driver turned the steering wheel about ± 15°. Before the tests the toe in value was regulated up to manufacturing data 0°0" (0mm). The results are shown in table 3.

Table 3. The results of the toe-in measurements for the front wheels on the plate. Driver turned the steering wheel about ± 15°

nr of the test	toe-in value [mm]
1	-0.8
2	-0.8
3	-0.4
4	-1.1

Scatter of measurements was due to the inaccuracy of setting the steering wheel angle.
The measurements showed the clear influence of this parameter in the results of toe-in measurements. It means that the steering wheel ought to be held immovable during crossing the plate.

4. Accuracy of the toe-in measurement on the slide slip plate stand

The investigation were done by comparison of the results on the slide slip plate stand MAHA to the results obtained by four heads device Custor HWA G58 (figure 2).

Accuracy of the plate lateral displacement measurement was 1 mm. Accuracy of the toe-in calculation depends on the diameter of the wheel rim. For the wheel rim 15" the toe-in accuracy is 0.4 mm and for rim 18" the accuracy is 0.5 mm. It corresponds to the angle of toe-in 3' for rim 15"and about 3.5' for rim 18". Four heads device makes possible to measure toe-in with accuracy ±1' (table 4).

Table 4. Range and accuracy measurements by Custor HWA G59 device [1]

	Parameter	Range	Accuracy
Front axle	Toe-in	±24°	±1'
	Camber angle	±10°	±1'
	Castor axis angle	±20°	±2'
	King pin axis angle	±20°	±2'
	Axis out of parallel	±10°	±1'
Rear axle	toe-in	±24°	±1'
	Camber angle	±10°	±1'
	Angle deviation body axis of symmetry from the geometric axis of drive	±10°	±1'

The measurements on the plate stand were done for wheels toe-in 0°0' ±10' up to manufacturing data and for toe-in -0°20'. The toe-in angles for right and left wheel were regulated symmetrical and next asymmetrical. The tests were repeated three times for each toe-in setting. The results of the tests are shown in table 5. The measurements on the slide plate show less toe-in value than controlled by four heads device. It can result from friction in the bearings of the plate which can increase during exploitation of the stand.

Table 5. The results of the toe-in measurements for front wheels on the plate stand and measured four heads device

wheels toe-in – four heads device		left wheel toe-in - four heads device	right wheel toe-in - four heads device	toe-in - the slide plate
0°0'	0 mm	0°0'	0°0'	0 mm 0 mm 0 mm
-0°20'	-2,5 mm	-0°20'	0°0'	-1.5 mm -1.5 mm -1.9 mm
-0°20'	-2,5 mm	0°0'	-0°20'	-1.9 mm -1.9 mm -1.9 mm
-0°20'	-2,5 mm	-0°10'	-0°10'	-1.5 mm -1.5 mm -1.9 mm

5. Conclusions

The investigations have shown the usefulness of the side-slip plate stand for control and estimation the wheels toe-in during periodical technical inspection of the vehicle. The side-slip plate stand in all cases of the wheels alignment have shown the proper sign and value of the toe-in.

The accuracy of the toe-in calculation depends on the diameter of the wheel rim. For the rim diameter 15" it is 0,4 mm, for rim 18" it is 0,5 mm. This accuracy is sufficient for the periodic technical inspections but too low for the accuracy of the regulation. The accuracy of measurements on the side-slip plate can be increased by use more precise sensors of the lateral displacement of the plate.

The estimated method isn't sensitive on the not parallel crossing the plate and on the small changes of the car speed. However changes of the direction by turning the steering wheel during crossing the plate cause the errors which make impossible proper estimation of the wheels toe-in.

The side-slip plate stand can measure less value of toe-in than four heads device. It can be caused by friction resistance in plate bearing and lateral stiffness of the tires. During stand exploitation friction resistance in plate bearing can increase.

References

[1] *User manual CUSTOR HWA G58*, Anwa-Tech Sp.z o.o. Sulejówek 2011
[2] *User manual Side-Slip Tester MAHA,* Maschinenbau Haldenwang GmbH 2016
[3] Bocheński C 2000 *Badania kontrolne samochodów* WKŁ Warszawa
[4] Trzeciak K 2008 *Diagnostyka samochodów osobowych* WKŁ Warszawa
[5] Sitek K and Syta S 2011 *Badania stanowiskowe i diagnostyka* WKŁ Warszawa
[6] *Linia diagnostyczna Uniline Quantum* Materiały firmy UNIMETAL Złotów 2015

Diagnostics monitor of the braking efficiency in the on board diagnostics system for the motor vehicles

Andrzej Gajek

Cracow University of Technology
Institute of Automobiles and Internal Combustion Engines, PL

E-mail: gajeka@mech.pk.edu.pl

Abstract. The article presents diagnostics monitor for control of the efficiency of brakes in various road conditions in cars equipped with pressure sensor in brake (ESP) system. Now the brake efficiency of the vehicles is estimated periodically in the stand conditions on the base of brake forces measurement or in the road conditions on the base of the brake deceleration. The presented method allows to complete the stand – periodical tests of the brakes by current on board diagnostics system OBD for brakes. First part of the article presents theoretical dependences between deceleration of the vehicle and brake pressure. The influence of the vehicle mass, initial speed of braking, temperature of brakes, aerodynamic drag, rolling resistance, engine resistance, state of the road surface, angle of the road sloping on the deceleration have been analysed. The manner of the appointed of these parameters has been analysed. The results of the initial investigation have been presented. At the end of the article the strategy of the estimation and signalization of the irregular value of the deceleration are presented.

1. Introduction

On board diagnostics program for the ABS/ESP system tests the state of the electronic and electric elements of this system. Mechanical and hydraulic elements are not tested, although quality of working the whole system depends on the efficiency of the electronic, as well as mechanical elements. Inspection of the mechanical elements needs to build more advanced diagnostic monitors. Example of these monitors are procedures for control of the brake lining wear and monitor of the unequal brake forces on the same axle of the vehicle [2, 4, 8].

The paper presents new diagnostics monitor for control of the efficiency of brakes in various road conditions in cars equipped with pressure sensor in brake (ESP) system. This procedure together with monitor of the unequal brake forces on the same axe of the vehicle allow to test currently these parameters which are controlled during obligatory periodical technical inspection of the vehicles on the roller stands. Presented method allows to replace or to complete the stand periodical tests by current on board diagnostics system OBD. It is the important problem for systems responsible for active safety in the road traffic.

The brake efficiency of the vehicles during periodical technical inspection is tested in the stand conditions on the base of brake forces measurement on the roller stands or in the road conditions on the base of the brake deceleration [1]. The conditions of these tests are established by special regulations [1]. The efficiency of braking tested in the stand conditions is calculated on the base of dependence:

$$z = \frac{\sum F_h}{G_c} \cdot 100\% \tag{1}$$

$\sum F_h$ – sum of the maximum brake forces measured on the stand,
G_c – vehicle total weight from vehicle total mass *dmc*

The efficiency of braking in the road conditions is calculated on the base of dependence:

$$z_{dr} = \frac{b_m}{g} \cdot 100\% \tag{2}$$

b_m – mean brake deceleration, measured in road conditions,
g – earth acceleration, $g=9{,}81 m/s^2$
The conditions of the road tests are established by regulations: the measurement of the deceleration during intensive braking on the flat surface of the road in good state, the initial speed about 30 km/h. For example, the coefficient of the brake efficiency calculated on the base of the stand or road tests for the passenger cars ought to be not less than 58% (for cars registered from 07/2010).

2. Diagnostics monitor of the brake efficiency

The state of the brakes and the conditions of braking change during car exploitation and influence on the brake efficiency. For these reasons the special procedure for on board diagnostics system has been worked. It allows to estimate if the deceleration measured in current road conditions performs the requirements of the regulations.
Dependence between vehicle deceleration and forces acting on the braked car in the longitudinal direction is following:

$$m\ddot{x} = \sum F_{br} - \frac{\sum M_{iw}}{r_d} + F_{rol} + F_{air} + F_{eng} \pm F_{hill} \tag{3}$$

where:
m – vehicle mass,
$\ddot{x}$ - vehicle deceleration,
$\sum F_{br} = \sum M_{br}/r_d$ - sum of braking forces caused by the braking moments,
$M_{iw} = J_k \ddot{\varphi}_k = \frac{J_k}{r_t} \ddot{x}$ - inertia moment of the wheel,

$F_{air} = Ac_x \frac{\rho}{2} \dot{x}^2$ - aerodynamic drag,
$F_{rol} = Gf\cos\alpha$ - rolling resistance,
$F_{eng} = M_{eng} i / r_d$ - the engine resistance reduced to the vehicle wheels,
i – overall drivetrain ratio
$F_{hill} = G\sin\alpha$ - the climbing resistance
G – vehicle weight,
α – angle of the road slope,
r_d – dynamic radius of the wheel,
r_t – rolling radius of the wheel,

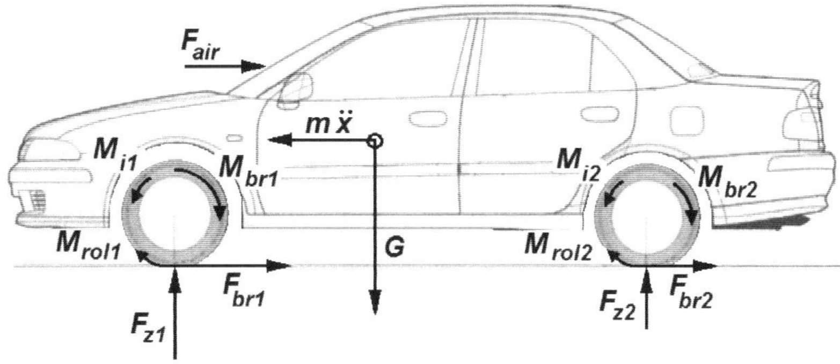

Figure 1. The system of the forces and moments acting on the braking vehicle on the horizontal surface of the road, $M_{rol1} + M_{rol2} = Gfr_d$.

For the brakes with ABS/ESP system with pressure sensor, dependence between sum of braking forces and pressure in the hydraulic system, for the brakes without braking force regulator, can be calculated on the base of dependence:

$$\sum F_{br} = f(p) = (k_p + k_t)p \tag{4}$$

p – pressure in the brake master cylinder,
k_p, k_t – proportional coefficients for front and rear wheels calculated experimentally
Dependence (4) is actual to the moment of ABS operation and it can change with the brakes temperature and initial speed of braking. These parameters can be taken into consideration as the coefficients k_{temp} and k_v established experimentally:

$$\sum F_{br} = f(p) = (k_p + k_t)p \cdot k_{temp} \cdot k_v \tag{5}$$

On the base of equation (3) and (5) we can establish relation between deceleration of the vehicle and pressure in the brake system:

$$\ddot{x}(p) = \frac{1}{m + \sum \frac{J_k}{r_d r_t}} \cdot [(k_p + k_t)p \cdot k_{temp} \cdot k_v + mgf\cos\alpha + Ac_x \frac{\rho}{2}\dot{x}^2 \pm mg\sin\alpha + F_{eng}] \tag{6}$$

During working out of this diagnostic monitor, the dependence between deceleration and brake pressure $\ddot{x}(p)_{dmc}$ for the vehicle with the efficient brakes, vehicle total mass *dmc* and for the established conditions of the braking, ought to be measured, fig. 2. This characteristic ought to be written to ROM of the brake controller. It is the base for following analyses and estimations. The base characteristics ought to be made for few ranges of the initial speed of braking. The load of the vehicle changes during exploitation of the car and for the same pressure p the brake deceleration depends on the current mass m. Quantity estimation of the brake efficiency depends on the comparison of the measured deceleration $\ddot{x}_m$ for the current mass m of the vehicle to the deceleration $\ddot{x}_{dmc}$ for the *dmc* mass and the same pressure p in the hydraulic system and similar conditions of braking.
During braking the vehicle with the mass m and mass *dmc*, on the road with the same hill slope and rolling resistance, the characteristic $\ddot{x}(p)_m$ ought to run over the characteristic $\ddot{x}(p)_{dmc}$ (fig. 2). It means that if the tested car with mass m reaches deceleration $\ddot{x}_{gr}$ required by regulations, then brake pressure p for this deceleration $\ddot{x}_{gr}$ ought to be less than p_{gr} for $\ddot{x}_{gr}$ for mass *dmc*, fig. 2.

The value of the pressure p_{gr} which causes the required deceleration $\ddot{x}_{gr}$ for the mass *dmc* and efficient brakes ought to be established in time of the worked out of this diagnostic monitor.

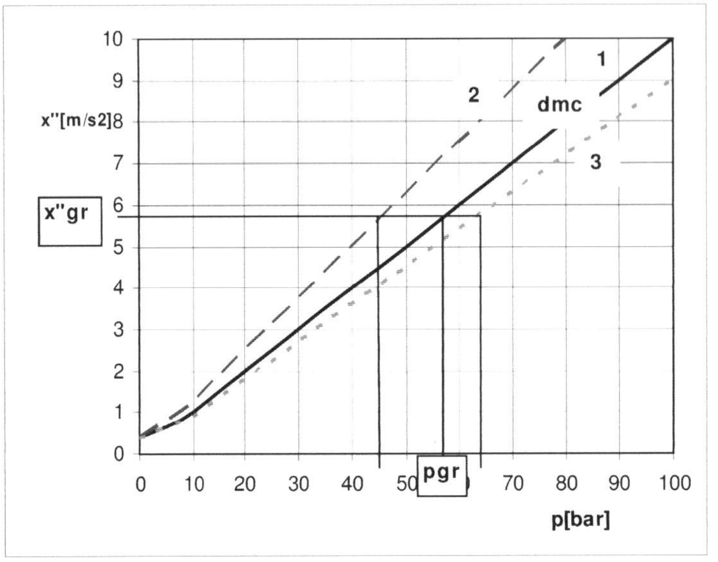

Figure 2. Scheme of the dependence of the brake deceleration from the braking pressure: 1 – for mass *dmc*, 2 – for current mass *m<dmc,* proper state, 3 – for mass *m<dmc*, too small braking efficiency

During braking on the horizontal road with small rolling resistance, for the deceleration about 4 m/s^2 and more, the inert resistance of wheels and rolling resistance of wheels are similar and have opposite signs [9]. For these condition, the equation (3) can be described as:

$$m\ddot{x} = \sum F_{br} + F_{air} + F_{eng} \quad (7)$$

The air resistance F_{air} doesn't depend from the mass of the vehicle but from the speed. For the same conditions of braking, on the base of the equations (4) and (7) can be described:

$$\ddot{x}_m(p) = \frac{1}{m}[(k_p + k_t)p + F_{air} + F_{eng,m}] \quad 8)$$

and $$\ddot{x}_{dmc}(p) = \frac{1}{dmc}[(k_p + k_t)p + F_{air} + F_{eng,dmc}] \quad (8a)$$

From the equation (8) and (8a) results that during braking in the same road conditions for the same pressure p and initial speed, with clutch uncoupled, the ratio of the decelerations for mass m of the vehicle and mass dmc ought to be:

$$\ddot{x}_m(p)/\ddot{x}_{dmc}(p) = dmc/m \quad (9)$$

Because of the small difference between deceleration $\ddot{x}_m(p)$ and $\ddot{x}_{dmc}(p)$, the engine resistance values for the same transmission ratio, initial speed and clutch coupled are similar for the mass m and dmc: $F_{eng,m} \approx F_{eng,dmc}$. For this assumption the dependence (9) is actual during braking with transmission engaged.

Above conditions oblige in the range of braking without ABS acting.

If the diagnostic program contains procedure for estimation of the current mass of the vehicle then the diagnostic monitor compares the measured deceleration $\ddot{x}_m(p)$ to the deceleration $\ddot{x}_{dmc}(p)$.

Estimation if the vehicle achieved deceleration $\ddot{x}_{gr}$ required by regulations (for example 5,8 m/s^2 for the passenger cars) is seldom performed during intensive braking. On the slippery roads deceleration required by regulations can be not reached. For these situations we can enlarge the evaluation of the brakes on the wide range of braking efficiency by calculation of the coefficient γ described as:

$$\gamma = \frac{\ddot{x}_m(p)}{\ddot{x}_{dmc}(p)} \quad (10)$$

$\ddot{x}_m(p)$ - measured brake deceleration for pressure p and current mass m,

$\ddot{x}_{dmc}(p)$ - deceleration for the mass dmc and pressure p according to characteristic written in ROM

In the all range of the braking pressure, it ought to be $\gamma>1$ for $m<dmc$.

3. Conditions of the on board diagnostics

Monitor of the brake efficiency requires particular conditions for performing the measurements. The initial investigation confirmed the necessity of the performing few conditions for the trustworthy assessment of the current measured deceleration $\ddot{x}_m(p)$. From dependence (6) results that the brake deceleration depends from many parameters: design parameters (J_k, r_t, m, A, c_x) and current conditions of braking (vehicle speed, engine resistance, brake temperature, grade of hill resistance, rolling resistance, road coefficient of friction). These factors ought to be taken into consideration in the diagnostic algorithm.

Construction parameters can be written to the diagnostics program as constant coefficients. External parameters ought to be analysed.

3.1. Vehicle speed and deceleration

Characteristics $\ddot{x}_{dmc}(p)$ ought to be tested and set in the program for a few values of the initial speeds of braking. Then the measured brake deceleration $\ddot{x}_m(p)$ for the real mass ought to be compared to the characteristic $\ddot{x}_{dmc}(p)$ for the similar initial speed.

The deceleration of the vehicle is calculated on the base of wheel rotational speed measured by ABS/ESP sensors and differentiation of that speed. Exact calculation of the vehicle speed $\dot{x}$ requires to take into consideration the wheel slip s. The vehicle speed and deceleration are:

$$\dot{x} = \frac{\dot{\varphi}_k r_t}{1-s}, \qquad \ddot{x} = \frac{r_t}{1-s}\frac{d\dot{\varphi}_k}{dt} \tag{11}$$

$\dot{\varphi}_k$ – wheel rotational speed,

s – wheel slip

The monitor of the brake efficiency will be realised for the slip range when the ABS doesn't work. It results from the lack of dependences between pressure in master cylinder and deceleration when slip exceeds the boundary value and ABS is working. In the range of braking without ABS acting the wheel slip has small value and can be neglected. The optimal situation is if the ESP system has longitudinal deceleration sensor.

The trustworthy calculation of the mean deceleration requires appropriate long time of braking and establishing the moment of the start and end of the calculation [6]. During breaking up to stop, the start of calculation is in the moment when the speed of the vehicle is 0,8 of the initial speed of braking and the end is when the car speed is 0,2 of the beginning speed of braking. The investigation shows that the calculation $\ddot{x}_m(p)$ ought to be performed for initial speed of the car and braking pressure bigger than minimum values. For the tested car it was: $\ddot{x}_{\min} \approx 2m/s^2, p_{\min} \approx 20bar$. For the less values disturbances caused the excessive scatter of results.

3.2. Engine resistance and rolling resistance

Engine resistance has real influence on the brake deceleration and ought to be taken into consideration in the diagnostic algorithm. The base characteristics $\ddot{x}_{dmc}(p)$ ought to be made for the transmission ratios, bigger then 1. Measured deceleration for the mass m will be compared to the deceleration $\ddot{x}_{dmc}(p)$ for the same transmission ratio.

The influence of the rolling resistance on the road in the good technical state is minimal in relation to the disperse of the results of measuring of the deceleration and can be neglected. Bad state of road can be recognized using the sensor of the vibrations of the car body or by analysis of unequal wheel speed as in the OBB system [7]. If the level of vibrations established in the program is exceeded, the monitor isn't realised. Also when the tire pressure is too small the monitor isn't realised.

3.3. Brakes temperature

This parameter isn't measured in the on board diagnostics system. It can be taken into consideration indirectly. If the successive intensive braking are one after second in the short time period, then the result of next deceleration isn't analysed.

It is necessary to pay attention that efficiency of the heated brakes also ought to be sufficient. In connection to it if the efficiency of the heated brakes reduces under required value, then the driver ought to be informed about this situation, although the formal conditions of the test are not in accordance with the proper regulations.

3.4. Vehicle mass

The mass of the vehicle can be identified by measuring the suspension deflection or the pressure in the pneumatic suspension of the vehicle. The point of dynamic method is that the deceleration measured during overrun strip with power transmission engaged for the current mass of the vehicle is compared with the deceleration for the overrun strip for the *dmc* mass and the same transmission ratio. If these measurements will be realised in the same road conditions and speed, then on the base of the dependence (8) and (8a) is:

$$m\ddot{x}_m = F_{air} + F_{eng,m} \tag{12}$$

$$\text{and} \quad m_{dmc}\ddot{x}_{dmc} = F_{air} + F_{eng,dmc} \tag{12a}$$

After side subtraction of these equations the mass m can be calculated:

$$m = \frac{m_{dmc} \cdot \ddot{x}_{dmc} + \Delta F_{eng}}{\ddot{x}_m} \tag{13}$$

$$\Delta F_{eng} = F_{eng,m} - F_{eng,dmc}$$

Value ΔF_{eng} is connected with difference between engine moment of inertia for decelerations $\ddot{x}_m$ and $\ddot{x}_{dmc}$. It can be estimated during working out of the program or neglected. If it is neglected, then the equation (13) can be described as:

$$m = m_{dmc} \frac{\ddot{x}_{dmc,eng}}{\ddot{x}_{m,eng}} \tag{14}$$

The values of the deceleration $\ddot{x}_{dmc,eng}$ for mass *dmc*, transmission ratios higher than 1 and for a few speeds of the car ought to be measured and written to the diagnostic program. Deceleration $\ddot{x}_{m,eng}$ is measured currently during driving. Calculations of the mean value of deceleration ought to be realised many times during driving. If that algorithm will be introduced to diagnostics program, the analyse of the braking efficiency may be extended on the wide range of decelerations.

3.5. Angle of the road sloping

During upgrade braking (α>0) the deceleration $\ddot{x}(p)$ and the brake efficiency $z_{dr}(p)$ will be bigger than on the horizontal surface for the same pressure p, fig. 3. During braking on the slope surface (α<0) the deceleration $\ddot{x}(p)$ will be less and the estimation of the efficiency can be negative, despite of the correct state of brakes. Diagnostics program ought to distinguish these situations.

For the estimation of the slope resistance we can use the inertial sensor of the longitudinal acceleration used in the hill holder system, programmed for the measurement angle of road sloping. The second possibility is calculation on the base of the difference between the deceleration during overrun strip on the horizontal road (written in the diagnostic program) and on the road before current braking.

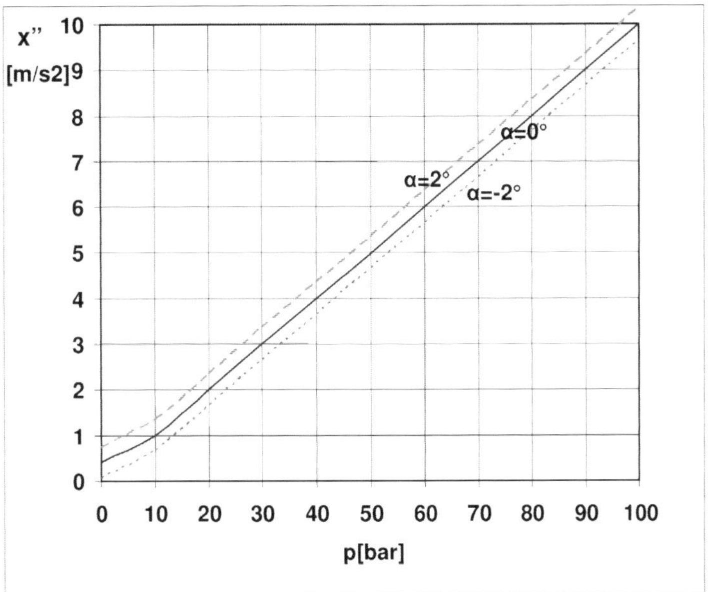

Figure 3. Changes of the deceleration characteristics $\ddot{x}(p)$ considering the angle of the road sloping α.

3.6. Conditions of the road surface

Road conditions have decisive influence on the maximum brake deceleration. On the slippery surfaces deceleration doesn't achieve the value required by regulations. Protection against the improper estimation of the brakes is control if ABS is in action in time of braking. If the ABS system operates before the achievement of the boundary deceleration $\ddot{x}_{gr}(p)$, then this measurement isn't taken into consideration to brake efficiency estimation. For conditions of braking on the slippery surface or during gentle braking, the coefficient γ can be calculated according to dependence (10). That coefficient ought to satisfy the condition $\gamma>1$ which obliges in all range of the decelerations.

Irregularities of the road surface influence on the changes of the wheel speed measured by ABS sensors and next on the calculation of the deceleration. The course of the wheel speed ought to be filtered before differentiate [5].

4. Results of the investigation

Fig. 4 presents the results of the brake deceleration measurements as the function of the brake pressure for the passenger car with mass m_1 = 1585 kg and m_2 = 1910 kg. The tests were performed on the horizontal, dry asphalt road, with clutch uncoupled, for the initial speed 50 – 60 km/h. The results were approximated by linear function. The ratio m_1/m_2 was 0,83, the ratio $\ddot{x}_{m2}/\ddot{x}_{m1}$ was 0,81. That difference results from the spread (scatter) of the deceleration value about ±0,5 m/s^2 for the same brake pressure in current conditions of braking. This fact ought to be taken into consideration for determination of the conditions and criterion of the brake efficiency estimation.

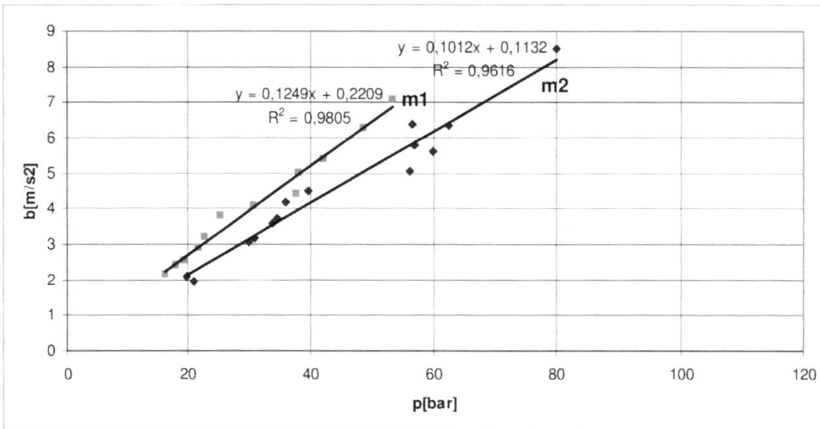

Figure 4. Dependences of the brake deceleration and brake pressure measured in the road conditions and functions of regression, *m1* = 1585 kg, *m2* = 1910 kg.

On the base of the equation (9) for the pressure p and braking on the horizontal surface, the deceleration of the vehicle with mass m ought to reach value:

$$\ddot{x}_m(p) \geq \ddot{x}_{dmc}(p)\frac{dmc}{m} \tag{15}$$

For example, for the results presented in fig. 4 the deceleration for pressure p=50 bar and mass m_1 calculated form dependence (15) as $\ddot{x}_{m1} = \ddot{x}_{m2} \cdot \frac{m_2}{m_1}$ is 6,2 m/s^2. Measured deceleration for mass m_1 was $\ddot{x}_{m1} = 6{,}4 \pm 0{,}5 m/s^2$.

The braking efficiency z, according to dependence (2), for the vehicle with mass m ought to be:

$$z_m(p) \geq z_{dmc}(p)\frac{dmc}{m} \tag{16}$$

Braking efficiency for mass m_1 calculated from the dependence (16) is 63% and calculated on the base of the measurements and dependence (2) was 65%±5%.

Coefficient γ as $\gamma = \ddot{x}_{m1}(p)/\ddot{x}_{m2}(p)$ was 1,24.

The measurements and calculations have shown the efficiency of the brake system.

Estimation of the vehicle mass on the base of overrun strip

Measurements were made for the car with mass m_1=1585 kg and mass m_2=1910 kg, with engine 1,9 TDI, with transmission ratio 2 and initial speed 50 km/h. For 5 repeats, the mean value of the deceleration for mass m_1 was 0,85 m/s^2, for mass m_2 was 0,72 m/s^2. Standard deviation was 0,035 m/s^2. For these values the mass m_1 calculated from the dependence (14) was 1618 kg±14 kg. Real mass was 1585 kg. It means that the percentage error was 2%. Because of scatter of the measurements, the value of the mass ought to be calculated as the mean value of the minimum 5 measurements of the deceleration. Fig. 5 shows the example of the deceleration during overrun strip with power transmission engaged for mass m_1 and mass m_2.

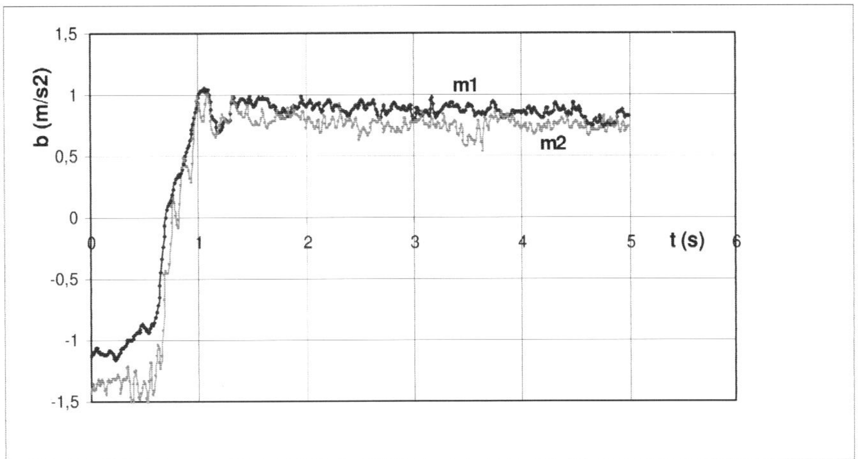

Figure 5. Example of the car deceleration during overrun strip with power transmission engaged for mass m_1 = 1585 kg and m_2 = 1910 kg. Engine 1,9 TDI.

5. Signalling of the braking efficiency state

Failure of the brake system the driver can notice immediately by identification of big reduction of the brake deceleration without additional signals from the on board diagnostics system. However progressive, by steps, wear of the brakes causes the reduction of deceleration in the long period and these changes can be unnoticed. For that growing inefficiency the special strategy of the estimation and signalization of the boundary state of the diagnostics parameters ought to be adapted. The statistical analysis or moving weighted mean value can be used.

5.1. Statistical analysis

For the *n* realized measurements of the deceleration $\ddot{x}_m(p)$ and calculations of the coefficient γ, the probability *P* that brake system doesn't perform the condition of efficiency ($\gamma < \gamma_{gr}$) is calculated as:

$$P(\gamma < \gamma_{gr}) = \frac{n_i}{n}[\%] \tag{17}$$

n_i – quantity successive measurements for which is $\gamma<\gamma_{gr}$

γ – value of current coefficient $\gamma = \ddot{x}_m / \ddot{x}_{dmc}$

γ_{gr} – boundary value of coefficient γ for mass *m*, $\gamma_{gr} = dmc / m$

If $P(\gamma<\gamma_{gr}) \geq P_{gr}$ then the driver is informed that braking system is inefficient.

Small boundary of probability P_{gr} is profitable for safety but can cause faults type α - false alarm. Acceptation of more value boundary probability P_{gr} causes that estimation of the state of brakes will be with the bigger certainty, but can cause too late signalization of the state of inefficiency.

5.2. Moving weighted mean value

Wear of brake system in the long time, by steps, causes that in the successive braking the state of this system can changes form an efficient to inefficient and on the contrary. It is profitable for these systems to take the strategy of estimation of the moving weighted mean value [7]:

$$\bar{\ddot{x}}_n = \ddot{x}_n \cdot w + \bar{\ddot{x}}_{n-1}(1-w) \tag{18}$$

or

$$\bar{\gamma}_n = \gamma_n \cdot w + \bar{\gamma}_{n-1}(1-w) \tag{19}$$

$\ddot{x}_n$, γ_n –value of the deceleration or coefficient γ in the *n* current measurement,

$\bar{\ddot{x}}_{n-1}$, $\bar{\gamma}_{n-1}$ – mean value of the deceleration or coefficient γ from n-1 previous measurements

w - the weight of the current measurement, assumed value

When the vehicle started the tour after parking, the value of the mean deceleration from the last driving cycle is accepted. Then it is compared to the first measurement in current driving cycle. Next the process of the calculation is repeated.
Signaling of the incorrect state of brakes follows if the mean value of deceleration is less than boundary value: $\bar{\ddot{x}}_n < \ddot{x}_{gr}$ or $\bar{\gamma}_n < \gamma_{gr}$

6. Conclusions

Presented algorithm of the vehicle deceleration calculation requires to store in the controller memory the maps of the deceleration value in dependence of the brake pressure, initial speed of braking and engine resistance for *dmc* mass. This algorithm allows to calculate and estimate the deceleration for vehicles with pressure sensor in ESP system without assembling additional sensors.
Proposed diagnostic monitor allows to extend the range of the on board diagnostics of brakes on the important parameter of the efficiency of brakes. It is the conditional monitor. It means that it requires to perform the following basic conditions:
- start of the braking,
- exceed the minimum initial speed and deceleration of brake,
- the braking during the longitudinal direction of movement and without ABS acting,
- braking on the good state of the road surface,
- determined time of braking

On the base of the presented conditions the program for the braking efficiency evaluation acting in the on board diagnostic system can be worked out.
Presented diagnostics monitor of the brake efficiency together with the monitor of the unequal brake forces control [4] complete the obligatory periodical technical inspection of the vehicles. In the future the on board diagnostics system OBD-Brakes can replace the system of the stand periodical tests of the brake forces.

References

[1] *Rozporządzenie Min. Transportu, Budownictwa i Gospodarki Morskiej w sprawie zakresu i sposobu przeprowadzania badań technicznych pojazdów oraz wzorów dokumentów stosowanych przy tych badaniach.* Dz. U. z 7.09.2012 r., Pos. 996

[2] Gajek A 2014 Prognozowanie zużycia okładzin ciernych hamulców tarczowych w procesie eksploatacji pojazdów samochodowych *Konferencja naukowa DIAGNOSTYKA 2014.* UTP Bydgoszcz

[3] Gajek A and Walczak St 2011 Dynamic identification of the vehicle mass. *Proceedings of the 7th International Scientific Conference TRANSBALTICA 2011*, p. 36 – 39. Technical University Wilno. O5/2011. ISSN 2029-2376

[4] Gajek A and Świder P 2009 Ocena nierówności momentów hamujących kół w pokładowym systemie diagnostycznym pojazdu samochodowego. *Archiwum Motoryzacji, Polskie Towarzystwo Naukowe Motoryzacji.* Wyd. PWN W-wa, nr 3/2009.

[5] Gajek A 2002 Modelowanie i analiza układu samochód – stanowisko bębnowe do badań i diagnostyki hamulców. *Monografia. Politechnika Krakowska,* Kraków

[6] Gajek A and Kuranowski Al. 2000 Analiza wyników badań skuteczności hamowania samochodów osobowych w warunkach stanowiskowych i drogowych, *V Konferencja Naukowo Techniczna „Diagnostyka Pojazdów Samochodowych"*, Instytut Transportu Politechniki Śląskiej, Katowice, XI/2000

[7] Merkisz J and Mazurek St 2007 Pokładowe systemy diagnostyczne pojazdów samochodowych OBD. *WKŁ,* W-wa

[8] Gajek A and Walczak St 2008 Monitor diagnostyczny stanu okładzin ciernych hamulców pojazdów samochodowych. *Czasopismo Techniczne. Mechanika.* Wyd. Politechniki Krakowskiej, z. 8 – M/2008

[9] Świder P 2012 Teoria ruchu samochodu. cz. I. *Skrypt, Wyd. Politechnika Krakowska,* Kraków

IOP Conf. Series: Materials Science and Engineering **148** (2016) 012039 doi:10.1088/1757-899X/148/1/012039

Analysis of motion of the body of a motor car hit on its side by another passenger car

M Gidlewski[1,3] and L Prochowski[2,1]

[1]Automotive Industry Institute, 03-301 Warszawa, ul. Jagiellońska 55, Poland
[2]Military University of Technology, 00-908 Warszawa, ul. Gen. S. Kaliskiego 2, Poland
[3]University of Technology and Humanities, 26-600 Radom, ul. Malczewskiego 29, Poland

E-mail: m.gidlewski@pimot.eu

Abstract. Based on an analysis of the course of a few experimental crash tests, a physical model and afterwards a mathematical model were prepared to describe the motion of bodies of the vehicles involved during the phase of impact. The motion was analysed in a global coordinate system attached to the road surface. Local coordinate systems were also adopted with their origins being placed at the centres of mass of the vehicles. Equations of motion of the model were derived. The calculation results enabled defining the influence of the location of the point of impact against the vehicle side on e.g. the following:

- time history of the impact force exerted by the impacting car (A) on the impacted car (B) as well as characteristic values of this force and of the impulse of the impact force;
- time histories showing changes in the velocity of the centre of vehicle mass and in the angle of deviation of the velocity vector from the direction of motion of the impacted vehicle before the collision;
- trajectory of the centre of mass and angle of rotation of the body of the impacted vehicle.

The calculations were focused on the initial period of motion of the body of the impacted vehicle, up to the instant of 200 ms from the start of the collision process. After this time, the vehicles separate from each other and move independently. The results obtained from the calculations covering this initial period make it possible to determine the starting-point values of the parameters to be taken for further calculations of the free post-impact motion of the cars.

1. Introduction

Right-angle collisions of motor vehicles amount to more than 25 % of road accidents in Poland. Knowledge of the course of accidents of this kind and, first of all, exploration of the course of the dynamic interactions that take place during the right-angle collision of two motor vehicles will enable further improvement of traffic incident modelling and reconstruction methods. The identification of the dynamic interactions between the vehicles in the culminating phase of the collision process will also make it possible to analyse the processes of injury formation in vehicle occupants and will provide a basis for the creation of new personal protection and safety means.

Many articles dealing with side collisions of motor vehicles have been published. Predominantly, they present analyses of the deformation of the side of a motor vehicle body and of the distribution of forces on the barrier face [2, 3] during a right-angle and oblique impact [7]. The results of such research works are oriented at improving the vehicle occupant protection system and designing a Mobile Deformable Barrier (MDB) [5, 6, 9]. In publication [1], the displacement of a motor car hit on its side has been presented, with highlighting the influence of the stiffness of the load-bearing vehicle structure on this process. Interesting are the results of an analysis of road accidents in Japan, where the occurrence

of injuries was linked with the speed and type of the vehicle hit on its side [8]. There is a definite lack of works where the issue of impact-caused interactions between vehicles involved in a right-angle collision would be addressed. An attempt to analyse this problem has been presented in [4].
The objective of this work was to determine the influence of the place of a frontal impact of a motor car against the side of another car on the course of the impact-caused interaction between the two vehicles and on the change in the velocity vector and trajectory of the body of the impacted vehicle. In this work, experimental testing and mathematical modelling were combined together. The results presented have been obtained for the time interval 0-0.2 s. This is the period after which the vehicles separate from each other and move independently in accordance with their individual kinetic energy. Based on the experimental test results having been collected, a physical model and afterwards a mathematical model were prepared to describe the motion of bodies of the vehicles involved during the phase of impact. The motion was analysed in a global coordinate system attached to the road surface. Local coordinate systems were also adopted, with their origins being located at the centres of mass of the vehicles. The local systems were used for presenting the measurement results and for deriving the equations of motion of the model.

2. Preparation and carrying out of experimental tests

Six crash tests were performed at the Automotive Industry Institute (PIMOT) in Warsaw with the use of 12 passenger cars of the same make and model. In each test, the front of car A crashed into the left side of car B (Fig. 1). The pre-impact speed of car A was about 50 km/h and it was twice as high as that of car B. The crash tests were carried out on a test yard with dry concrete surface, in good weather conditions. During the tests, the steering wheel of each car was left free and car wheels were not braked.

The relative positions of cars A and B were changed in successive crash tests. The location of the point of impact on the side of vehicle B was defined by the distance L_{AB} between the longitudinal plane of symmetry of car A and the front wheel axis of car B (see Fig. 1). The values of the characteristic parameters of successive crash tests have been specified in Table 1.

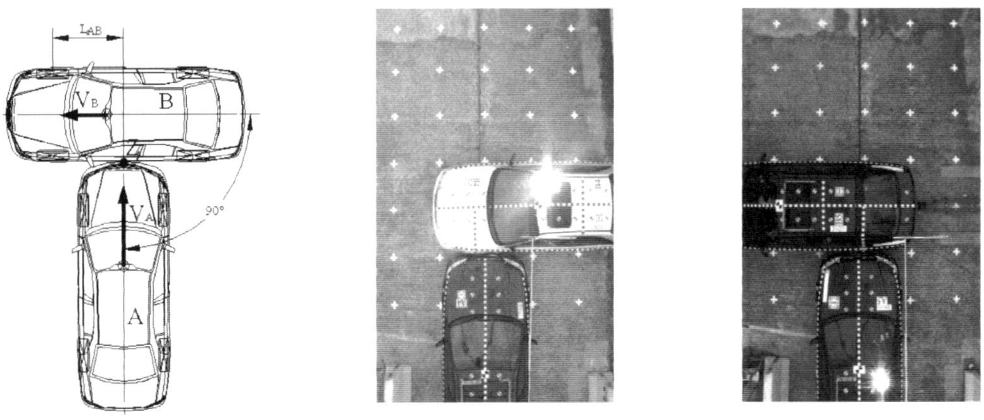

Figure 1. Examples of the relative positions of cars A and B during the crash tests

Table 1. Velocity of vehicle A at the instant of the collision

L_{AB} [m]	0.04	0.17	1.25	1.34	2.57	2.71
Velocity V_A [m/s]	13.6	13.1	15.1	12.7	14.2	14.7

The tests were carried out with the use of Honda Accord cars manufactured in 2000-2002. The cars were in good technical condition and had undamaged and non-corroded bodies, which had not been previously repaired. At the centre of mass of each car, there was a three-axial acceleration sensor, installed together with sensors measuring the angular velocity of the vehicle body in relation to the three coordinate axes, whose directions and senses were consistent with those of the local coordinate systems attached to the cars.

3. Calculation of the impact force applied to the side of vehicle B

To analyse the processes that took place during the vehicle collision, a global coordinate system and local coordinate systems were adopted.

- A global coordinate system $O_GX_GY_GZ_G$, attached to the road. The $O_GX_GY_G$ plane of this system is situated at the road surface level and the O_GZ_G axis is pointing vertically upwards. The O_GX_G axis is parallel to the vector of pre-impact velocity of vehicle A and the O_GY_G axis is parallel to the vector of pre-impact velocity of vehicle B.
- Local levelled coordinate systems, attached to the centres of mass of vehicle A and B. The local coordinate system $O_Ax_{PA}y_{PA}z_{PA}$ has its origin O_A situated at the centre of mass of vehicle A, the $O_Ax_{PA}y_{PA}$ plane is parallel to the road surface plane, and the O_Ax_{PA} axis is parallel to the longitudinal vehicle symmetry plane. The local coordinate system $O_Bx_{PB}y_{PB}z_{PB}$ has its origin O_B situated at the centre of mass of vehicle B, the $O_Bx_{PB}y_{PB}$ plane is parallel to the road surface plane, and the O_Bx_{PB} axis is parallel to the longitudinal vehicle symmetry plane.
- Local coordinate systems, attached to the bodies of vehicles A and B. The local coordinate system $O_Ax_Ay_Az_A$ has its origin O_A situated at the centre of mass of vehicle A and the O_Ax_A axis is parallel to the longitudinal vehicle centre line. The local coordinate system $O_Bx_By_Bz_B$ is fixed to the body of vehicle B. The origin O_B of this system is situated at the centre of mass of vehicle B and the O_Bx_B axis is parallel to the longitudinal vehicle centre line.

Before the collision, the axes of the local coordinate systems $O_Ax_{PA}y_{PA}z_{PA}$ and $O_Ax_Ay_Az_A$ are parallel to the corresponding axes of the global coordinate system and the local coordinate systems $O_Bx_{PB}y_{PB}z_{PB}$ and $O_Bx_By_Bz_B$ are rotated in relation to the global system by an angle of 90 deg around the vertical axis. The positions of the local coordinate systems relative to the global system are defined by angles Φ_i – vehicle roll angle (angle of rotation around the O_ix_i axis), Θ_i – vehicle pitch angle (angle of rotation around the O_iy_i axis), and Ψ_i – vehicle yaw angle (angle of rotation around the O_iz_i axis). Subscript "i" indicates vehicle A or B.

The system of equations (1), which constitutes a basis for calculating the impact force acting between the vehicles, represents the state of equilibrium of the model reduced to the $O_GX_GY_G$ plane. This model with the system of forces under consideration has been shown in Fig. 2.

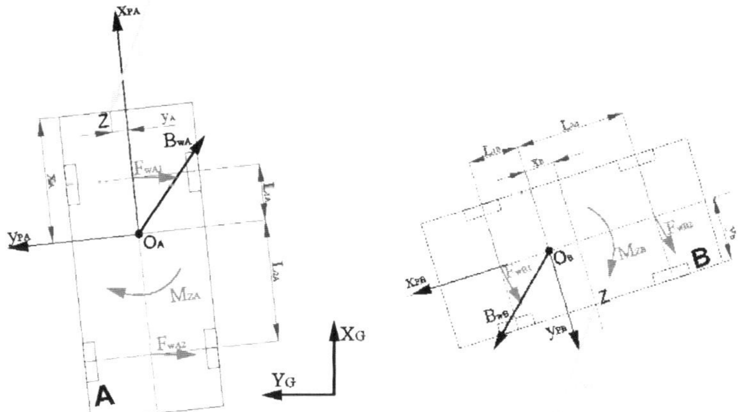

Figure 2. Coordinate systems adopted, basic dimensions used, and position of the impact force vector.

$$\begin{cases} m_A \cdot a_{xPA} = F_{nA} + F_{xA1} + F_{xA2} \\ m_A \cdot a_{yPA} = F_{sA} + F_{yA1} + F_{yA2} \\ I_A \cdot \ddot{\Psi}_A = F_{sA} \cdot x_A - F_{nA} \cdot y_A + F_{yA1} \cdot L_{1A} + F_{yA2} \cdot L_{2A} \\ m_B \cdot a_{xPB} = F_{sB} + F_{xB1} + F_{xB2} \\ m_B \cdot a_{yPB} = F_{nB} + F_{yB1} + F_{yB2} \\ I_B \cdot \ddot{\Psi}_B = F_{nB} \cdot x_B - F_{sB} \cdot y_B + F_{yB1} \cdot L_{1B} + F_{yB2} \cdot L_{2B} \\ F_{nB} = F_{nA} \cdot \cos\gamma_{AB} - F_{sA} \cdot \sin\gamma_{AB} \\ F_{sB} = F_{nA} \cdot \sin\gamma_{AB} + F_{sA} \cdot \cos\gamma_{AB} \end{cases} \tag{1}$$

where:

B_i (B_{ix}, B_{iy}) – inertial force acting on the ith vehicle; i = A or i = B;

M_{zi} – moment of inertia acting on the vehicle;

m_i, I_i – vehicle mass and mass moment of inertia of the vehicle relative to the vertical axis;

a_{xPi}, a_{yPi} – longitudinal and lateral components of the vector of acceleration of the centre of vehicle mass, calculated from (3);

F_{i1} (F_{xi1}, F_{yi1}) – tangent road reaction acting on the wheels of the front vehicle axle and the components of this reaction force (longitudinal and lateral, respectively);

F_{i2} (F_{xi2}, F_{yi2}) – tangent road reaction acting on the wheels of the rear vehicle axle and the components of this reaction force;

F_{wi} (F_{ni}, F_{si}) – impact force acting on the vehicle and the longitudinal and lateral components of this force;

x_i, y_i – coordinates of point Z, i.e. the point of application of the impact force, in the levelled coordinate system (Fig. 2);

γ_{AB} – angle of rotation of vehicle B relative to vehicle A around the vertical axis during the collision.

The quantities recorded during the crash tests included components of the vector of acceleration of the centre of vehicle mass (a_{xi}, a_{yi}, a_{zi}) and components of the vector of angular velocity of each vehicle (P_i, Q_i, R_i), measured in relation to the local coordinate system fixed to vehicle A or B. Example measurement results in the form of time histories of acceleration a_{yB} and angular velocity R_B have been presented in Fig. 3.

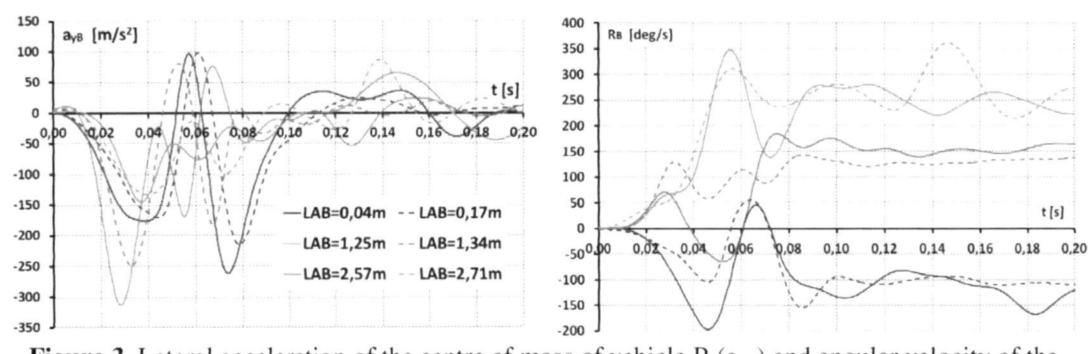

Figure 3. Lateral acceleration of the centre of mass of vehicle B (a_{yB}) and angular velocity of the body of the same vehicle relative to axis O_Bz_B (R_B) in the local coordinate system $O_Bx_By_Bz_B$ (low-pass filtered with a cut-off frequency of 50 Hz).

The values of the quantities used in equations (1), i.e. the angular accelerations of the vehicle body relative to the vertical axis and the components of the acceleration of the centre of vehicle mass, were calculated with using the results obtained from the experiment and the relations between the coordinate systems. The kinematic calculations, the results of which were used at the analysis of the post-impact motion of the vehicles, were carried out with taking into account a transformation of the measurement results from local coordinate systems of vehicles A and B to the global coordinate system. The transformation matrices (2, 3, and 4) have been presented below:

$$\begin{bmatrix} \dot{\Phi}_i \\ \dot{\Theta}_i \\ \dot{\Psi}_i \end{bmatrix} = \begin{bmatrix} 1 & \sin\Phi_i \,\mathrm{tg}\Theta_i & \cos\Phi_i \,\mathrm{tg}\Theta_i \\ 0 & \cos\Phi_i & -\sin\Phi_i \\ 0 & \dfrac{\sin\Phi_i}{\cos\Theta_i} & \dfrac{\cos\Phi_i}{\cos\Theta_i} \end{bmatrix} \cdot \begin{bmatrix} P_i \\ Q_i \\ R_i \end{bmatrix} \tag{2}$$

$$\begin{bmatrix} a_{xPi} \\ a_{yPi} \\ a_{zPi} \end{bmatrix} = \begin{bmatrix} \cos\Theta_i & \sin\Theta_i \sin\Phi_i & \sin\Theta_i \cos\Phi_i \\ 0 & \cos\Phi_i & -\sin\Phi_i \\ -\sin\Theta_i & \cos\Theta_i \sin\Phi_i & \cos\Theta_i \cos\Phi_i \end{bmatrix} \begin{bmatrix} a_{xi} \\ a_{yi} \\ a_{zi} \end{bmatrix} \tag{3}$$

To determine the velocities and coordinates of the centre of mass of vehicle B in the global coordinate system, the following relation was used:

$$\begin{bmatrix} \ddot{X}_{GB} \\ \ddot{Y}_{GB} \\ \ddot{Z}_{GB} \end{bmatrix} = \begin{bmatrix} \cos(\frac{\pi}{2}+\Psi_B) & -\sin(\frac{\pi}{2}+\Psi_B) & 0 \\ \sin(\frac{\pi}{2}+\Psi_B) & \cos(\frac{\pi}{2}+\Psi_B) & 0 \\ 0 & 0 & 1 \end{bmatrix} \begin{bmatrix} a_{xPB} \\ a_{yPB} \\ a_{zPB} \end{bmatrix} \tag{4}$$

4. Results of the calculation of the impact force applied to vehicle B

Results of the calculation of the impact force F_{wB} applied to car B and its normal component F_{nBXG} in the global coordinate system have been shown in Fig. 4.

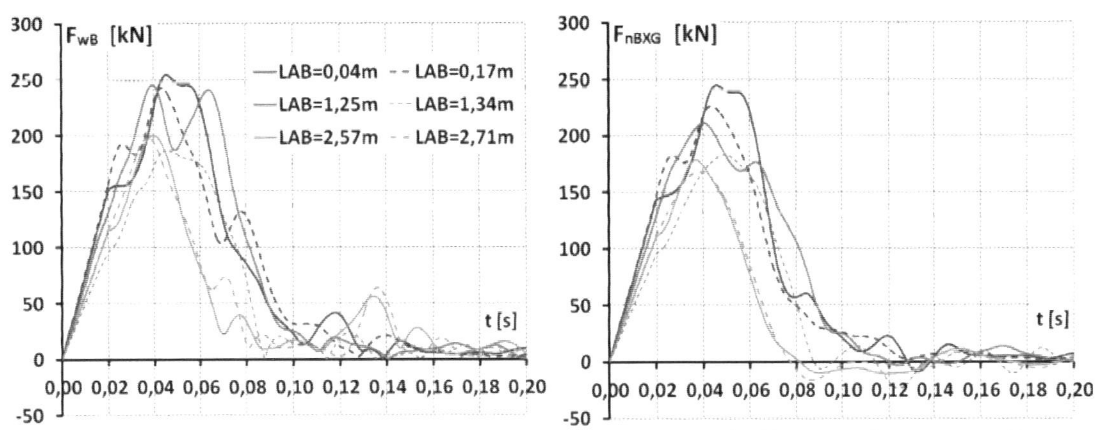

Figure 4. Time histories of the impact force F_{wB} applied to car B and of its normal component F_{nBXG} in the global coordinate system for different L_{AB} values.

The impact force vs. time curves having been determined were used for the calculations of several characteristic values, i.e.:

- extreme values of the impact force, calculated for time intervals of 3 ms, 5 ms, and 10 ms;
- impulse of the impact force, lasting during the period from 0.03 s to 0.06 s.

The time history of the impact force has the nature of a quick-changing process. Therefore, the extreme (instantaneous) values of this force do not have a decisive influence on the post-impact motion of vehicle B; they are strongly dependent upon the low-pass filter used. In this consideration, the maximum values obtained after averaging the function F_{nBXG} (t) over time intervals of 3 ms, 5 ms, and 10 ms have been given in Table 2. Such a method of load presentation links the maximum values of the impact force with the time of duration of the resulting load. The load duration time is of decisive importance for the assessment of the effects of a side impact. Table 3 shows values of the impulse of the impact force and average values of the impact force, calculated for the time interval in which the highest values of the force F_{nBXG} were recorded.

Table 2. Extreme values of the impact force F_{nBXG} [kN], calculated over time intervals of 3 ms, 5 ms, and 10 ms, based on the time histories presented in Fig. 4

L_{AB} [m]	$\Delta t = 0.003$ s	$\Delta t = 0.005$ s	$\Delta t = 0.010$ s
0.04	244	243	241
0.17	225	225	222
1.25	211	210	207
1.34	183	182	181
2.57	178	177	174
2.71	168	167	166

Table 3. Characteristic values describing the force input applied to vehicle B (Values determined on the grounds of the time history of the normal force F_{nBXG} (t))

L_{AB} [m]	Impulse of the impact force, for the period from 0.03 s to 0.06 s [Ns]	Average value of the impact force, for the period from 0.03 s to 0.06 s [kN]
0.04	6549	218
0.17	6021	201
1.25	5718	191
1.34	5047	168
2.57	4361	145
2.71	4363	145

5. Vehicle motion caused by the impact force

Below are shown the primary effects caused by the impact force, i.e.:

- time histories of the velocity of the centre of mass of vehicle B and of the angle of deflection of the velocity vector from its pre-impact direction (Fig. 5);
- trajectories of the centre of mass of vehicle B and changes in the vehicle body yaw angle Ψ_B, i.e. the angle of rotation of the vehicle body around the vertical axis (Fig. 6).

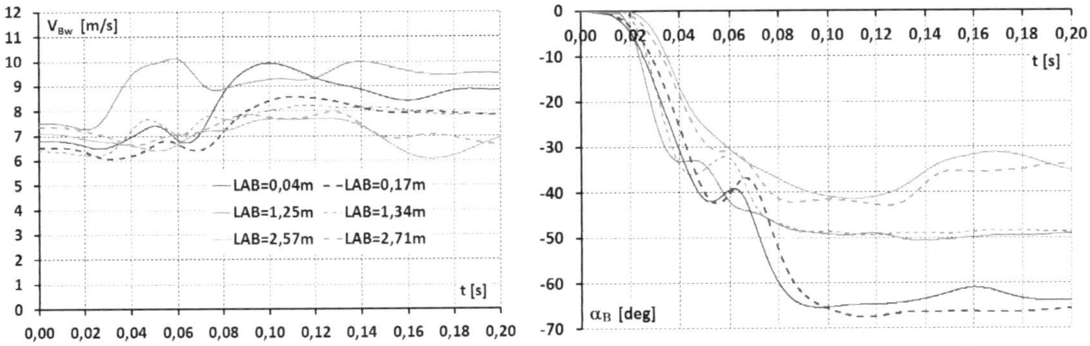

Figure 5. Velocity of the centre of mass of vehicle B and angle of deflection of the velocity vector from its pre-impact direction

The curves presented in Fig. 6 make it possible to determine the position of vehicle B on the road during the collision and at the instant of beginning of the free vehicle motion. The points marked with "x" in the graphs show the position of the centre of vehicle mass and the angle of vehicle body rotation around the vertical axis at the instant of beginning of the free vehicle motion.

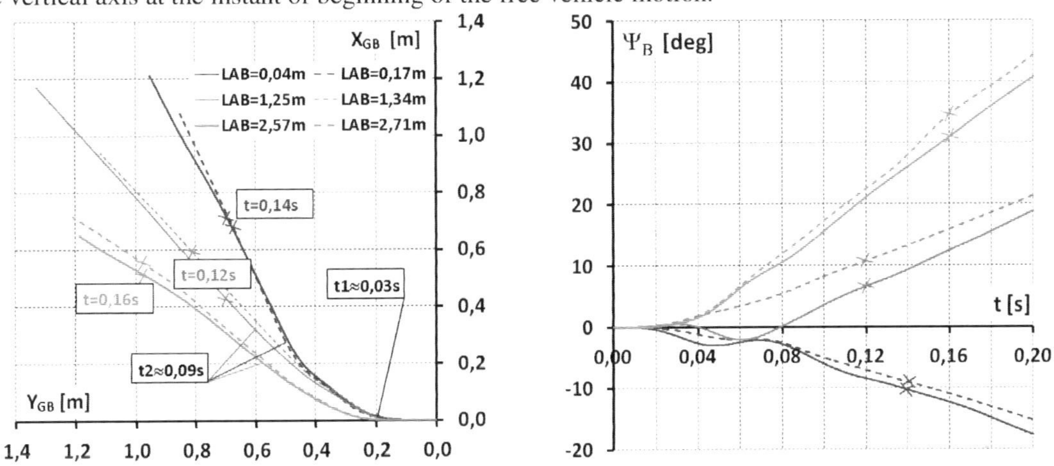

Figure 6. Trajectories of the centre of mass and changes in the body yaw angle of vehicle B

6. Vehicle body tilt versus reactions on wheels

The side impact force is balanced by the force of inertia of the impacted vehicle and the tangential reactions developing at the contact between the vehicle wheels and the road surface. Of particular interest are the lateral tangential reactions. The course of changes in their values strongly depends on the roll angle Φ_B resulting from the side impact. Fig. 7 shows the lateral tilt (roll) of the body of vehicle B at several instants in one of the crash tests (L_{AB} = 1.34 m). A time history of the roll angle Φ_B has been presented in Fig. 8, where the curves plotted have been based on measurement results obtained from six crash tests.

Noteworthy are the very high values of the vehicle body roll angle, observed not only in the contact phase of the collision process but also after the vehicle separation. A vehicle body roll angle of about 7-8 deg causes the wheel suspension on one of the vehicle sides to be fully compressed. Since the roll angle values are observed to exceed this limit, they show that the vehicle body not only tilts by rotating around the longitudinal axis O_Bx_B but also rotates around an instantaneous axis going through the centres of the areas of contact between the wheels and the road on one of the vehicle sides (left or right), as it can be seen in Fig. 7. In the case illustrated there, such a rotation caused the right wheels of vehicle B to be lifted off the road surface.

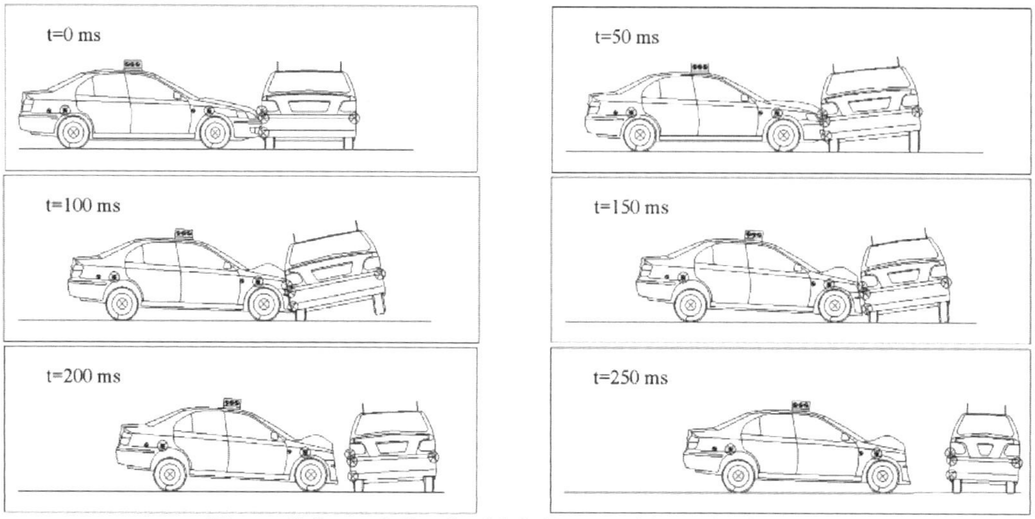

Figure 7. Lateral tilt of vehicle B caused by a side impact

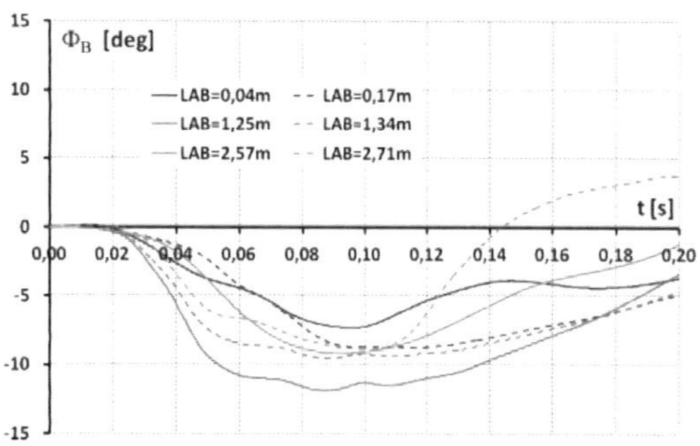

a) Roll angle $\mathbf{\Phi_B}$ (t = 0-0.2 s)

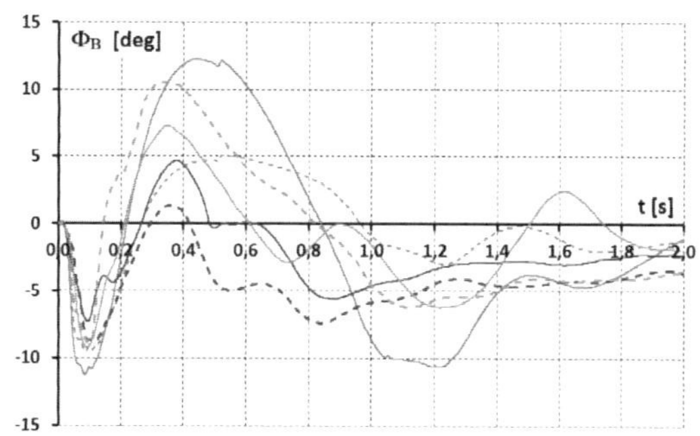

b) Roll angle $\mathbf{\Phi_B}$ (t = 0-2 s)

Figure 8. Time histories of the vehicle B body angle of rotation around axis O_GX_G of the global coordinate system

Fig. 9 shows time histories of the lateral tangential road reactions on the wheels of the front and rear axle of vehicle B. The very high values of the lateral road reactions at the contact with the wheels of vehicle B were caused by very high values of the vehicle body roll angle during the collision.

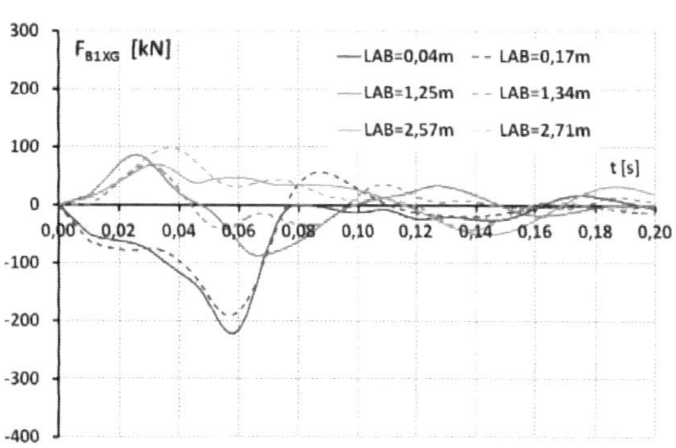

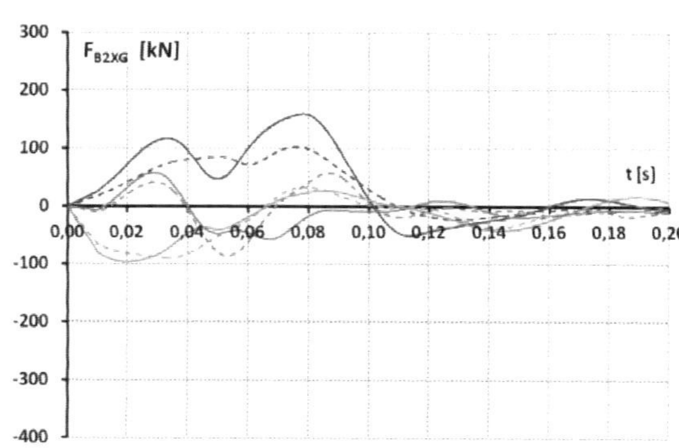

Figure 9. Lateral tangential road reactions F_{B1XG} and F_{B2XG} exerted on the wheels of the front and rear axle of vehicle B, respectively, in the global coordinate system

7. Recapitulation and conclusions

The test results obtained and the test result analysis carried out show, first of all, characteristics of the process of generating dynamic effects on the body of a motor car being struck on its side. The impact-caused interaction between vehicles (impact force) chiefly depends on the vehicle deformation process, development of the inertial resistance force exerted by the impacted vehicle, and development of the tangential reactions in the area of contact between the vehicle wheels and the road. Each of these factors has a complex influence on the time history of the impact force.

The calculations carried out were focused on the initial period of vehicle motion, up to the instant of 0.2 s from the start of the collision process. After this time, the vehicles separate from each other and move independently. The results obtained from the calculations covering this initial period make it possible to determine the starting-point values of the parameters to be taken for further calculations of the free post-impact motion of the cars.

The measurement and calculation results enabled defining the influence of the location of the point of impact against the vehicle side on e.g. the following:

- time history of the impact force exerted by car A on car B as well as characteristic values of this force and of the impulse of the impact force;
- time histories of the angular velocity of the vehicle body, translational velocity of the centre of vehicle mass, and angle of deviation of the vector of this velocity from the initial direction of motion of car B;
- trajectory of the centre of mass and time history of the angle of rotation of the body of car B, i.e. changes in the position of car B on the road.

An analysis of the measurement and calculation results provided grounds for formulating the following conclusions:

1. The values that characterize the impact force and the impulse of the impact force (Tables 2 and 3) decrease with increasing distance L_{AB} between the place of impact and the front axle of car B.
2. When car B is hit on a place close to its front axle, its pre-impact speed is raised by about 30 % (Fig. 5) and the direction of its velocity vector is changed by about 65 deg; when the point of impact is situated close to the rear axle of car B, the pre-impact speed of the impacted vehicle is decreased by about 7 % and the direction of the vehicle velocity vector is changed by about 35 deg. The changes in the magnitude and direction of the vector of velocity of the centre of mass of a car hit on its side decrease with increasing distance L_{AB}.
3. At the end of the phase of contact between the vehicles, the lateral and longitudinal displacement of the centre of mass of car B does not exceed 1 m (Fig. 6) for all the values of distance L_{AB}. At this instant, the angle of rotation of the vehicle body around the vertical axis is within a range from −10 deg to 30 deg, increasing with a growth in distance L_{AB}. After the separation of the vehicles, the direction of motion of the centre of mass of car B is approximately rectilinear and deflected from the direction of the pre-impact motion by an angle, the value of which decreases with increasing distance L_{AB}.
4. The lateral reactions at the contact between the wheels of vehicle B and the road reach very high values, of up to 80 % of the value of the impact force.

The tests reported herein were carried out and their results were analysed within an authors' own research project No. N N509 559440.

References

[1] Bumjin K, Hyunwoo K, Dongho P 2013 Research on the AE-MDB CAE analysis for the improved Euro NCAP side impact test procedure. *Proceedings of the 23rd International Technical Conference on the Enhanced Safety of Vehicles (ESV)*

[2] Digges K, Eigen A 2001 Measurements of vehicle compatibility in front-to-side crashes. *Proceedings of the 2001 International IRCOBI Conference on the Biomechanics of Impact, Isle of Man*

[3] Digges K and Eigen A 2001 Measurements of stiffness and geometric compatibility in front-to-side crashes.

[4] Gidlewski M, Prochowski L and Zielonka K 2015 Analysis of the Influence of Motor Cars' Relative Positions During A Right-Angle Crash on the Dynamic Loads Acting on Car Occupants and the Resulting Injuries. *Paper Number 15-0107. Proceedings of the 24th International Technical Conference on the Enhanced Safety of Vehicles (ESV)*

[5] Malak A, Poisson P and Touratier D 2013 Evaluation of the AFL advanced European mobile deformable barrier for side impact versus the ECE-R95 in dynamic load cell wall test. *Proceedings of the 23rd International Technical Conference on the Enhanced Safety of Vehicles (ESV)*

[6] Roberts A K and van Ratingen M R 2003 Progress on the development of the advanced European mobile deformable barrier face (AE-MDB). *Proceedings of the 18th International Technical Conference on the Enhanced Safety of Vehicles (ESV)*

[7] Welsh K J and Struble D E 1999 Crush energy and structural characterization. *SAE Technical Paper* No. 1999-01-0099

[8] Yonezawa H, Harigae T, Ezaka Y 2001 Japanese research activity on future side impact test procedures. *Proceedings of the 17th International Technical Conference on the Enhanced Safety of Vehicles (ESV)*

[9] Yonezawa H, Minda H, Harigae T, Sakurai M, Nishimoto T 2003 Investigation of new side impact test procedures in Japan. *Proceedings of the 18th International Technical Conference on the Enhanced Safety of Vehicles (ESV)*

IOP Conf. Series: Materials Science and Engineering **148** (2016) 012040 doi:10.1088/1757-899X/148/1/012040

Use of certain alternative fuels in road transport in Poland

W Gis[1], J Pielecha[2], J Waśkiewicz[1], M Gis[1] and M Menes[1]
[1]Motor Transport Institute, 80 Jagiellońska Steet, 03-301Warsaw, Poland
[2]Poznan University of Technology, Institute Combustion Engines and Transport 3 Piotrowo Street, 60-965 Poznan, Poland

E-mail: wojciech.gis@its.waw.pl

Abstract. The development of biomethane and hydrogen technology in the road transport in the EU countries is recommended, among the others, in the Directive of the European Parliament and of the Council 2014/94/EU of 22 October 2014. Under the provisions of the said Directive, it is recommended to EU countries to use biomethane and progressively ensure accessibility to hydrogen cars on their territories, and above all to ensure the possibility of driving hydrogen vehicles between the member States.
The territorial accessibility for biomethane vehicles is determined by the availability of biomethane refuelling infrastructure in the first place in cities and then on the road network distances recommended in this directive. The territorial accessibility for hydrogen vehicles is determined by the availability of hydrogen refuelling infrastructure, in the first place along the TEN-T network. The article presents the possibilities of using these alternative fuels in Poland, presenting some of the results of research and analysis in this area.

1. Introduction

Modern alternative fuels such as biomethane and hydrogen are finding increasing use in road transport. The first action in this area in Poland is very important taking into account the existing air pollution. Biomethane is currently used mainly in urban transport. Hydrogen now allows the movement of cars with fuel cells for a distance of about 700 km. A very important activity is the construction in this field of the relevant infrastructure, which in Poland is lacking. Below are some aspects of the application of both alternative fuels.

2. Biomethane consumption

The road transport in Europe almost fully depends on fossil fuel. Diversification of the road transport fuels will be a key attribute for road transport in the coming years. Biogas is one of alternative renewable fuels. At present in Poland biogas is used for generating electricity and heat, but it built the first biomethane refuelling station (Figure 1). The installation is located in a landfill in Niepołomice. It is a city in the southern Polish, near Krakow, the former capital of Poland.

IOP Conf. Series: Materials Science and Engineering **148** (2016) 012040 doi:10.1088/1757-899X/148/1/012040

Figure 1. The view of the mobile installation for biogas upgrading, compression, storage, and distribution of biomethane in Poland.

In some countries (for example in Sweden), upgraded biogas to natural gas quality (biomethane) is used as a vehicle fuel and the European Project Baltic Biogas Bus (BBB) contributed to this. This project stimulated the use of biogas (biomethane) as fuel of city buses aiming to reduce environmental impact.

The use of biomethane to power engines of buses (vehicles) also gives significant benefits in the form of a particulate matter numbers, emissions of aldehydes, PAH (polycyclic aromatic hydrocarbons) and direct NO_2 emissions. Motor vehicles powered by methane engines run quieter than vehicles powered by gasoline or diesel [1]. The above mentioned advantages of vehicles with engines supplied with methane are confirmed by the tests shown below.

The tests on the emissions of pollutants and the consumption of fuel were conducted on buses: bus I – fuelled with CNG – the tests conducted, using the bus Jelcz M121 M/4, equipped with a SI combustion engine, meeting the Euro III emission standard, with the cylinder capacity of 11967 cm^3 and maximum power 180 kW. Bus II was equipped with a CI combustion engine – fuelled with diesel (bus - Jelcz M121 MB, engine's cylinder capacity 11970 cm^3, maximum power 180 kW). The tests were performed on the same driving distance and comparable road conditions (tests were performed simultaneously on two buses, which were driven directly one after the other). The Figure 2 shows tested buses. The Figure 3 shows the testing device type PEMS for the measurement of exhaust emission. Two such devices were used, one for each bus.

Figure 2. View of tested buses (bus with engine powered by CNG– left, bus with engine powered by diesel – right).

Route (between Niepołomice and Krakow outward and return) characteristics: (i) distance: 41 km, (ii) change in altitude: minimum 190 m, maximum 202 m, (iii) maximum slope: 3.1%, minimum slope: -3.1% (iv) average positive slope: 0.2%, average negative slope: -0.2% (Figure 4).

Figure 3. View of the testing device type PEMS.

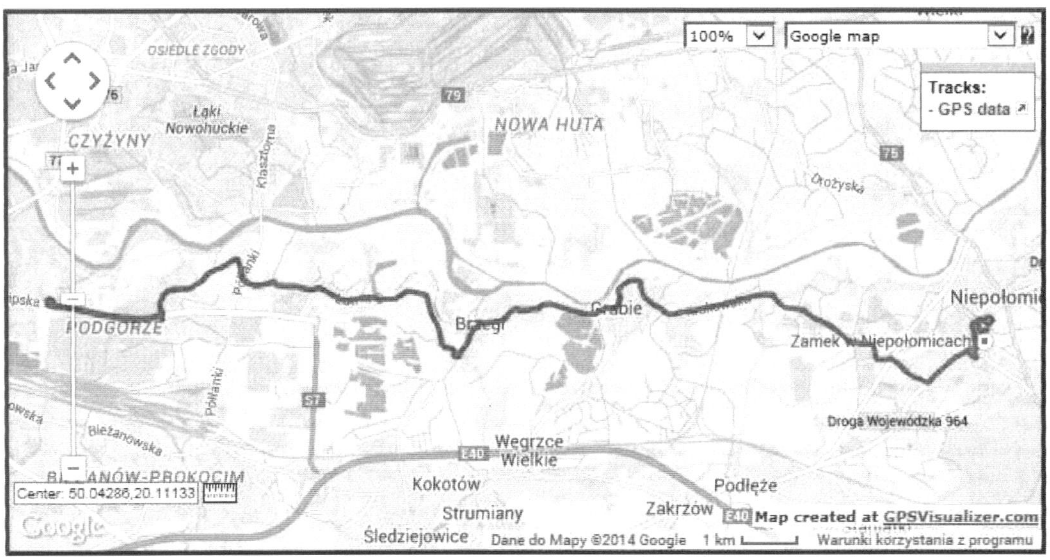

Figure 4. Characteristics of the test route.

The Figure 5 shows comparison of the mileage consumption of fuel and the road emission of pollutants, obtained during testing.

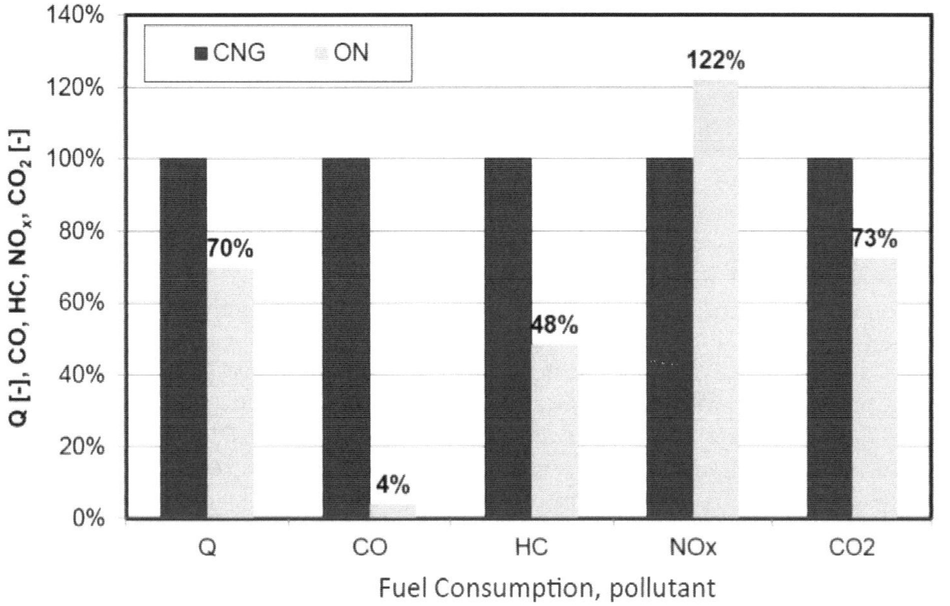

Figure 5. Comparison of the mileage consumption of fuel and the road emission of pollutants, obtained during testing.

The conducted tests indicate, that the road emission of nitric oxides (NO_x) of the bus with an engine refuelled with CNG is lower than the road emission of NO_x of the bus with the compression ignition engine (CI). Considering the fact, that excess emission of nitric oxides from CI engines is a serious problem, especially in urban traffic, the benefit of utilizing buses with engines refuelled with CNG is obvious here.

The emission of particulate matter, the second most significant component of the exhaust gases in the CI engines (in the presented tests it was not measured due to lack of such possibility, using the PEMS type measuring devices), in the case of using engines refuelled with CNG is lower than the one occurring in the CI engines (without utilizing expensive purifying systems).

In the case of emission of the carbon dioxide (CO_2) by the bus with the engine fuelled with CNG, therefore with the spark-ignition engine (SI), it is greater than the one occurring in the bus with the engine fuelled with diesel. When using biomethane, though, which is nearly entirely renewable, during life cycle (from well to wheel), the bus with the engine fuelled with biomethane will be a better solution than utilizing a bus with the CI engine for the reason of the CO_2 emission, which is lower during the life cycle of the CNG engine, comparing to the CI engine.

There is an advantage in the bus with the CI engine over the one with the SI engine, fuelled with CNG (biomethane) in relation to the emission of carbon monoxide (CO), and in the case of the emission of the hydrocarbons (HC). The measurements, performed with the PEMS devices, allowed to define their total value (non-methane hydrocarbons and methane), and indicate the higher values for the bus with the SI engine. It needs to be emphasized, that methane is not toxic, although it is a greenhouse gas. The utilization of the more efficient catalytic systems will allow the reduction of the road emission of CO and HC.

2. Hydrogen consumption

The development of hydrogen technology in the road transport brings a lot of hope for the operation of zero emission vehicles (Fuel Cell Electric Vehicle (FCEV)).

This technology is supported primarily by [2]:

- no pollutants emissions,
- reduction in noise emissions,
- reduction in fossil fuel imports,

- increasing the export and services potential and new jobs.

However, there are some disadvantages of using this technology e.g. among the others [2]:
- difficult shift from paradigm - fossil fuels to the paradigm - renewable fuels (hydrogen),
- investments related to the construction of hydrogen refuelling stations network,
- limited, in most countries of the world, political support to hydrogen technology with the envisaged foreground development of the road transport methanization and focusing on EVs,
- lack of up-to-date knowledge on modern hydrogen technology.

The pre-commercial phase (2020 – 2030), involves construction in Europe, of about 200 - 300 hydrogen refuelling stations located primarily in large urban agglomerations and along the TEN-T network. It is to handle about 5,000 passenger cars and 500 buses equipped with fuel cells. The early commercial phase of hydrogen propulsion technology should appear in Europe around 2020 and ensure the creation of hydrogen infrastructure including approximately 2,000 hydrogen filling stations / min 1000 / along the strategic transport routes, serving 500 thousand passenger cars and 1,000 buses equipped with fuel cells. The commercial development phase (2040 – 2050) of the hydrogen power technology should begin at the end of the third decade of the twenty-first century. These forecasts are burdened however with a relatively high degree of uncertainty, particularly with regard to the numbers of fleet of hydrogen-powered FCEVs [3].

The current global fleet of such cars can be estimated at 2-3 thousand, while yet recently it was assumed that only in Europe it will reached a level of 5,000 passenger cars and 1,000 buses [3].

In Europe, the most advanced program of the transport hydrogenization, with a budget of 1.4 billion Euro has Germany. The program entitled -"National Innovation Programme Hydrogen and Fuel Cell Technology" is coordinated by the Nationale Organisation Wasserstoff – und Brennstoftzellentechnologie (NOW GmbH) [3].

Other automotive and technologically advanced countries, began at the end of the first decade of the twenty-first century to create long-term programs of work on the use of hydrogen in road transport [3-9].

At present, there are over 60 models of cars powered by fuel cells in various stages of technical and market development, prepared by practically all major automotive companies [2-3].

In practice, so far, on the relatively mass scale, Toyota launched the production of fuel cell cars in the autumn of 2014. In 2014 of the Mirai model produced in 700 units, and the plan for 2015 envisages production of 3500 cars. Cars with fuel cells are also manufactured by Hyundai [3]. In Poland, there are practically no vehicles with fuel cell. There is also lack refuelling stations for hydrogen vehicles equipped with these cells.

2.1. The future network of hydrogen refuelling stations on the Polish territory

Development of hydrogen stations infrastructure is a key factor in the development of hydrogen technology in the road transport by the year 2050. In March 2015 there were 184 hydrogen refuelling stations operating in the world (in this e.g. 82 in Europe, 63 in North America). Unfortunately only 40% hydrogen refuelling stations (74 stations) were of a public character. The remaining ones frequently functioned within various types of research centres, industrial facilities and energy units, or as private use stations [3].

Despite the strategic importance of developing hydrogen filling stations infrastructure, in the available materials, including various national programs for the hydrogen propulsion technology developments, the explicitly formulated programming methodology for the development of these stations, has not been encountered [2].

An attempt to formulate the methodology has been made in the framework of the European HIT-to-Corridors project within the work: Circumstances of the national plan for hydrogenization of road transport in Poland [3].

The methodology developed is of multi-stage character. Individual steps leading to the designation of the location of hydrogen refuelling stations in Poland (as the methodology alone seems to be of universal character) are as follows [3]:

Stage I: Method allowing to identify regions in which the hydrogen refuelling stations should be located in the first place.

Stage II: Method allowing to identify urban centres, in which should be located the said stations.

Stage III: Method for determining the area of the station location.

Stage IV: Method used to indicate a specific location of hydrogen refuelling station.

Stage V: Method indicating the preferred order of building investments in creating future network of hydrogen filling stations on the Polish territory. In any of the said stages, the group of 3-5 basic characteristics was adopted that determines, according to the experts, the potential future demand for hydrogen fuel, whose likely impact strength was determined by giving them the appropriate rank on a scale of 1 to 5.

In stage I, were taken into account : the average value of GDP / capita, the average population density (inhab./km^2) or the average traffic flow of cars on the roads of international importance running through the province (cars/day). With the criteria adopted in the II stage, the cities predisposed for the location on their territory of the refuelling hydrogen stations, are in the order of ranking: Warsaw, Katowice (group of towns – Górnośląskie conurbation), Kraków, Tri-City (Gdańsk, Gdynia, Sopot), Łódź, Wroclaw, Poznań. The indication in the III stage of the hydrogen refuelling station locations in the selected cities or in groups of cities (Warsaw, Katowice (group of towns - Górnośląskie conurbation), Krakow, Tri-City (Gdańsk, Gdynia, Sopot), Łódź, Wrocław, Poznań and Szczecin) was based on the measurement results of the average passenger cars traffic intensity on the roads leading to these cities or on selected road junctions located in the vicinity of these cities and also takes into account the possibility of refuelling hydrogen vehicles transiting through Poland through the trans-European transport (TEN-T) network, including the direction between Germany and the Baltic countries[3].

The location criteria for the hydrogen refuelling stations specified in stages I - III are of course general in nature. It is necessary to consider by the experts the issues of specific site location, such as the existing or planned service station, nearby existing petrol stations, or the plot, on which in accordance with applicable local development plan the said location is possible and there is an interest of potential investor of such a project (stage IV). Pointing to the proposed order of the investments in the construction of hydrogen refuelling stations in Poland (Stage V) and taking into account the above-mentioned reasons, the preliminary aspect of locations in the cities or urban areas selected according to the rankings stages I to III, was considered.

In the first place taken into account were [3]:

- already existing refuelling opportunities in the neighbouring countries,
- the expected future hydrogen refuelling station locations in the Baltic countries,
- gradually increasing the area available for hydrogen-powered cars as a result of the subsequent location of new stations at distances up to 300 km from the existing or sequentially from the newly-opened ones.

With the above criteria, the order of preliminary proposals to build hydrogen refuelling stations in Poland are as follows: 1 - **Poznań** 2 - **Warsaw**, 3 - **Białystok**, 4 - **Szczecin**, 5 – **Łódź area**, 6 - **Tri-City area**, 7 - **Wrocław**, 8 - **Katowice region**, 9 - **Kraków** (Figure 6-7).

Figure 6. Map of Poland with marked sites of the proposed public hydrogen refuelling station locations. Source: ITS own compilation.

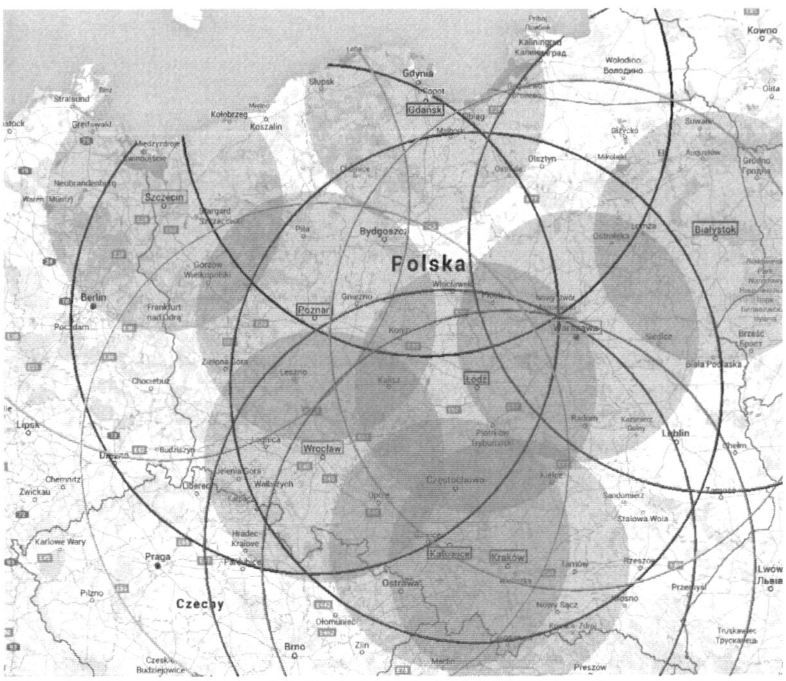

Figure 7. Penetration areas of cars with fuel cells (within 300 km of areas of hydrogen refuelling stations), of the stations proposed to be built by 2030.

3. Summary

The advantages of biomethane as an automotive fuel is primarily lower emissions of nitrogen oxides, particulate matter, and carbon dioxide in (life-cycle) as compared to the emissions of these components emitted from a diesel engines. The use of biomethane in the road transport brings about independence from the import of crude oil and crude oil derived fuels, as is it in case of hydrogen.
The advantages of hydrogen as an automotive fuel is the lack of pollutants emission from motor vehicles' engines, which is especially important in crowded city centres and with the possibilities of its local production.
As a result of verification, under Polish conditions, of the original method developed for determining the initial location of the hydrogen refuelling station in Poland, in the pre-commercial phase, the said location has been indicated along with the order of investment, taking into account above all the freedom to move around Poland of cars powered by hydrogen visiting Poland and transiting our country between other EU countries.
The places where it is proposed to be built hydrogen refuelling stations should be (in order of their creation): **1 - Poznań 2 - Warsaw, 3 - Białystok, 4 - Szczecin, 5 –Lódź region, 6 - Tri-City area, 7 - Wrocław, 8 - Katowice region, 9 - Kraków.**

References

[1] Nylund Nils-Olof 2010 *Environmental benefits with biogas buses.* Nordic Biogas Conference. Oslo 2010.3.10
[2] Wallmark C, Schaap G, Mohseni F, 2015.*Infrastruktura wodorowa dla transportu. Fakty i plan koncepcyjny dla Szwecji* SWECO
[3] Gis W, Menes E, Waśkiewicz J i inni, 2015. *Przesłanki narodowego planu wodoryzacji transportu samochodowego w Polsce.* Instytut Transportu Samochodowego, Warszawa, listopad 2015
[4] *UK Hydrogen Energy Strategic Framework – Analysis*, ww.dti.gov.uk/energy/sepn/hydrogen
[5] Dunwoody E 2011 *Hydrogen and renewables in California*, November 2011
[6] Hirose K 2015 *Toyota's Approach to Sustainability Mobility and Fuel Cell Vehicle Development*, Toyota Motor Corporation, 2013
[7] Pier-Etienne F 2011 *Fuel Cell and Hydrogen Sector on the road to markets: Key issue and early take aways*, November 2011.
[8] Byung K A 2011 *Development of Fuel Cell Electric Vehicle in Hyundai Motor*, Hyundai Motor Group, November 2011
[9] Karlson T 2013 *Hydrogen and Fuel Cell in Canada County Update*, 20 IPHE Steering Committee Meeting City of Fukuoka, Japan November 20-21, 2013

IOP Conf. Series: Materials Science and Engineering **148** (2016) 012041 doi:10.1088/1757-899X/148/1/012041

Torque measurement issues

J Goszczak

Department of Vehicles and Fundamentals of Machine Design, Lodz University of Technology, Żeromskiego 116, 90-924 Łódź

E-mail: jaroslaw.goszczak@p.lodz.pl

Abstract. Problems with torque measurement in operational tests are considered. Introduction with torque definition is included. Short overview of different types of torque meters is presented. Own results and remarks about torque measurement and torque meters are quoted. Author takes into account such problems as: electromagnetic and mechanical noise (from componentry, e.g. clutches). Different ways of averaging and their results are discussed. Conclusions based on test results are included in the summary.

1. Introduction

Reliable and repeatable measurement is a crucial matter during every research and in every kind of system which operation is based on torque value. Everybody who have tried to conduct any test knows that it is usually very difficult to achieve pure test result. In the majority of cases it is mandatory to perform some result processing, e.g. hardware or software filtration or to use different statistic functions. This paper is about torque measurement, which is a very important technical parameter, necessary to determine e.g. mechanical efficiency of tested object.

Torque measurement is therefore of fundamental importance in all rotating bodies and applies to the rotation of shafts in many things like pumps, rotational cutting equipment, gearbox shafts, vehicle axles, and electric motors. Torque measurement is also a necessary part of measuring the power transmitted by rotating shafts [1].

The moment of force $\mathbf{M}_O$ about a point O is defined as a cross product of vectors **r**and **F**, where **r** is the position vector from **O** to the point of force application and **F** is the force vector [2].

$$\mathbf{M}_O = \mathbf{r} \times \mathbf{F} \quad (1)$$

2. Different ways of torque measurement

It is not the main aim of this paper to present a detailed list and description of different types of torque meters which are available today. Nevertheless a short overview will be done. Wider description of different types of torque meters is available in [3].

In general [3], we can share torque measurement into two groups: direct and indirect measurement of the torque. Direct methods consist in measurement of some physical values (usually elastic strain) which varies according to the torque value. After calibration detector gives directly the value of torque (example of direct measurement is described in [4]). Indirect methods consist in measurement of physical values which change can be calculated into torque. It could be for example measurement of force acting on arm, which length is known or measurement of current and rotational speed of electric

motor. Often this kind of indirect torque measurement could be faster, easier and enough precisely in industrial conditions in comparison with direct methods.
Figure 1 presents the classification of torque measurements methods.

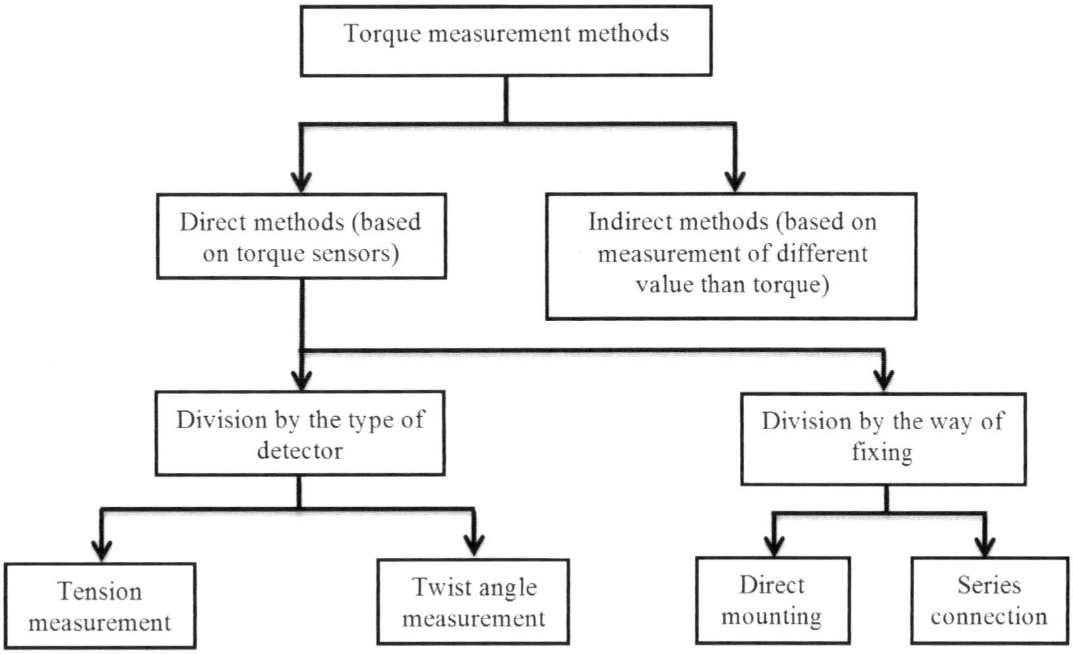

Figure 1. Classification of torque measurement methods [3].

Two ways of torque mounting can be distinguished. The first one, series connection is when the sensor transmit the whole value of torque in the mechanical circuit, i.e. that two ends of shafts are connected by torque meter. It is by far the most popular solution. The second one occurs when the sensor is mounted directly on twisted element which transmits torque (usually on the shaft). In this case it is necessary to design a non-series detector and calibrate it.

Nowadays, we can distinguish such types of torque meters:

- Strain gauges type, which based on changing of strain gauges resistance under the elongation.
- Surface Acoustic Wave (SAW)- this technology based on the sensors resonant frequency changing due to deformation.
- Magnetoelastic- in this case torque is measured by changing of magnetic permeability of ferromagnetic material.
- Piezoelectric reaction torque sensors- twisting of quartz discs positioned between metal plates cause a strain. This phenomena enable to generate electric charges on the surfaces.
- Optical- two optical elements (discs) are placed at the opposite ends of the torsional element. Due to the twisting relatively to each other, amount of light which goes through both discs changes and it is measured by detector.
- Inductive- two cores' coils are twisting relatively to each other. The change of inductance is connected with measured signal.

Figure 2 shows the classification of torque measurement sensors. You can read more about torque meters principles of operation for example in [3], [5], [6], [7], [8]. [9].
When applying directly the torque definition it is possible to design an original solution of torque sensor, for example developed in the existing machine, like in [4] or [10].

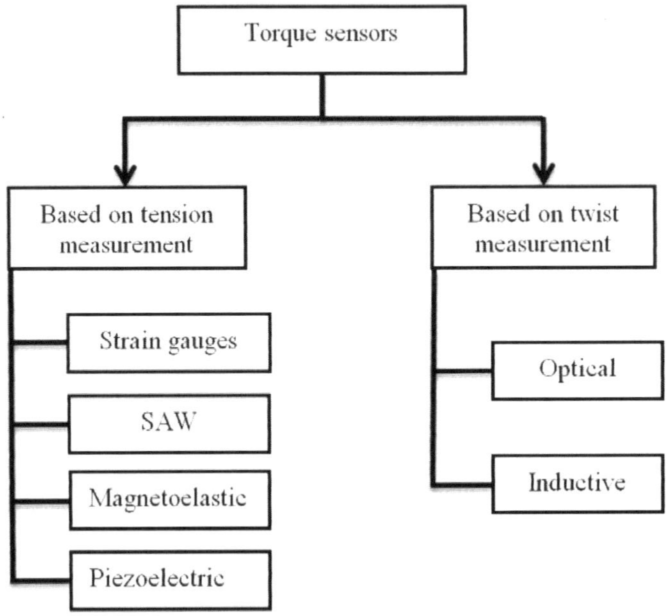

Figure 2. Classification of torque measurement sensors [3].

3. Own remarks about torque measurement

3.1 Torque meter mounting

In case of mounting torque meters in series connection, what is by far the most popular solution due to freely available different torque transducers from many manufacturers it is important how the torque meter will be installed at the test bench. In case of the modern sensors "T" series (figure 3) from HBM company it is possible the only one way of mounting. Rotating part (rotor) is installed in series connection, the second part (stator) functions as the antenna and as the power supplier for the stator. Between the antenna and the stator it is a gap, 2÷3mm big. This original solution, called flange torque meter allows to reduce significantly the noising phenomena of bending, tensile, and compressive forces which are unavoidable due to a geometric errors in the shaft train [11].

Figure 3. T series torque meter from HBM [12].

In conventional solutions with twisted shaft, the length of the torque meter is significantly bigger than the diameter (height) of the sensor. This fact contributes to the necessity of elimination the forces caused by e.g. mounting errors because those forces disturb the torque measurement. What is more, in this type of torque meters it is mandatory to remember about locking a housing of the sensor to prevent from turning.

In terms of this type of torque meters we can distinguish at least three possibilities of mounting:

- The first of two most popular ways of fixing the torque meter is to catch the housing of the sensor 3 by the handle, stiffly connected with the ground as it is shown in the figure 4. Near clutches 2 are used additional bearings 1, which bear driven and braking shafts. This solution aims to not load additionally the bearings inside torque meter by forces from geometric errors in the shaft train and weight of the both parts of the shaft. Drawbacks of this method are the necessity of expansion the test bench, deterioration of dynamic characteristics of sensors and occurrence of hysteresis phenomena due to the application the additional bearings near the sensor.

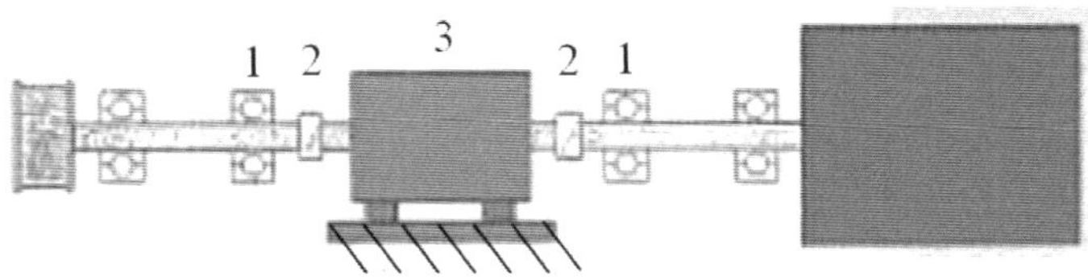

Figure4. Torque meter montage with additional bearings[11].

- The second popular way of torque meter installation is to suspend the sensor by two clutches as it is shown in the figure 5. This solution is mainly used for small rotational speed values[13] and torque meters which weight is small in comparison to the stiffness of both parts of the shaft. This solution is simpler than the previous one (it is not necessary to use a handle and an additional bearings) but we have to prevent the housing against turning and try to diminish the geometry errors in the test bench to minimum because every imperfections are reduced only by clutches.

Figure 5. Torque meter suspended by two clutches.

- The next possibility of torque meter mounting is shown in the figure 6. This solution is similar to the first one described in this paper but the main difference is that additional bearing are not before clutches like in the figure 4 but behind. This way of torque meter fixing is rather rare and can be used when the errors in the shaft train are relatively big. In this case those additional forces (mainly bending forces) will be taken by additional bearings and will not acting on the transducer with inner bearings. In this solution a deterioration of dynamic characteristic and hysteresis phenomenon may be observed but if the influence of those disadvantages are less than the effect of forces from mechanical noise in the test bench, this design could be beneficial.

Figure 6. Torque meter suspended by two bearings.

3.2 Bench mechanical noise

In the previous subsection it was pointed out the issue of geometry errors which are one of the reasons of the mechanical noise. The author of this paper called by mechanical noise every mechanical origin phenomena which disturb the value of the torque measured by torque meter.

Figure 7 depicts the basic type of assembly errors. The real error is a superposition of different forms of offsets.

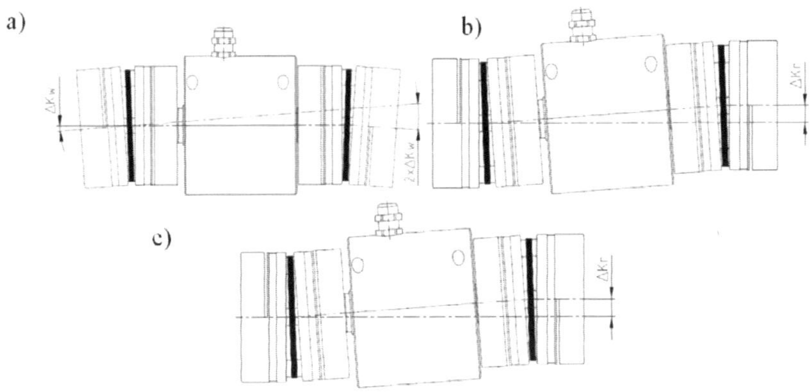

Figure 7. Different types of assembly errors: a) radial angular offset, b) radial parallel offset, c) axial offset [14]

To confirm the thesis of the mechanical noise caused by above mentioned assembly errors the test was perform. The shaft of the big test bench with powertrain of 160 kW was driven evenly by a rope which was wound on the driven shaft of the test bench. In the powertrain the torque meter with range 15 Nm was applied.

This way of driving (instead of typical use of electric motor from the test bench) was performed to omit the additional pulsation of torque value caused by rotational speed control mode of the inverter, which keeps in this mode constant rotational speed instead of current. Figure 8 presents the test results which confirm the assumption. One grid equals to 1,5 Nm so the mechanical noise equals to almost 1,5 Nm.

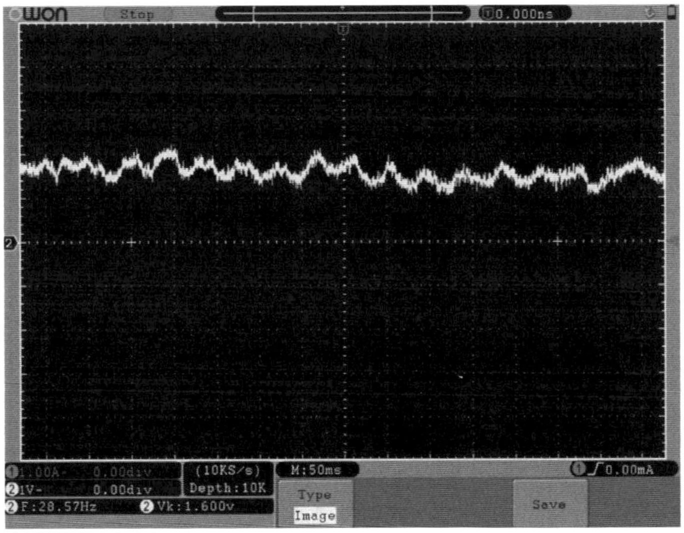

Figure 8. Mechanical noise test results

3.3 Electrical noise

Beyond mechanical noise measurement values could be disturbed by an electrical noise. We can distinguish noise caused by incorrect power supply tension, electromagnetic noise or potential difference on different test bench elements. When conducting a research it is vital to pay attention to noise of mains frequency 50Hz.

Popular source of noise in electrical grid are inverters. The basics rules to reduce the electrical noise caused by them are:

- Using electrical filters in a wiring, both before and behind the inverter.
- Putting the inverter into Faraday cage.
- Using shielded cables.
- Good ground of all devices.

If the electrical noise is caused by the supply voltage it is likely that beneficial will be:

- Using mains filters.
- Checking the ground of all devices at the test bench.
- Using a uninterruptible power supply with possibly big capacity of the battery (in this case we have to remember about protection against electromagnetic noise) or if it is possible use a battery as a source of power, during the tests.

To reduce the electromagnetic noise it is advisable to:

- Put all electronic devices into Faraday cage.
- Use possibly short shielded cables.

Another way to reduce the electrical noise is to apply galvanic isolation and to use e.g. opto-isolators. What is more, every device at the test bench should have the same ground to avoid any differences of electric potential.

Next crucial issue is to use digital transmission of data if it is available. If the torque meter has both the analog and digital output it is definitely more beneficial to use digital one. If the data processing is being done in the housing of the sensor, the housing acts as a Faraday cage. From the housing comes out the digital signal which is much more resistant to a noise.

3.4 Data averaging

Due to an occurrence of the unavoidable phenomena causing equally by mechanical and electrical noise but also other phenomena (e.g. due to the accuracy of the sensor) it is at least sometimes a need to average the test data in order to reduce the oscillations of the torque value to some assumed range. It could be necessary i.e. if we want to use a digital indicator. Fast changing values in a big range makes it impossible to read the value.

The reducing of oscillations is possible by using different filters but in the static measurement often could be enough a simple averaging by function implemented in digital oscilloscopes what is shown in the figure 9,10 and 11. Those data were recorded directly one after another in the same conditions at the test bench. N-fold averaging means that currently displayed curve is the average of n curves. In all screens (figures 9 to 11) one grid corresponds to 0,75 Nm.

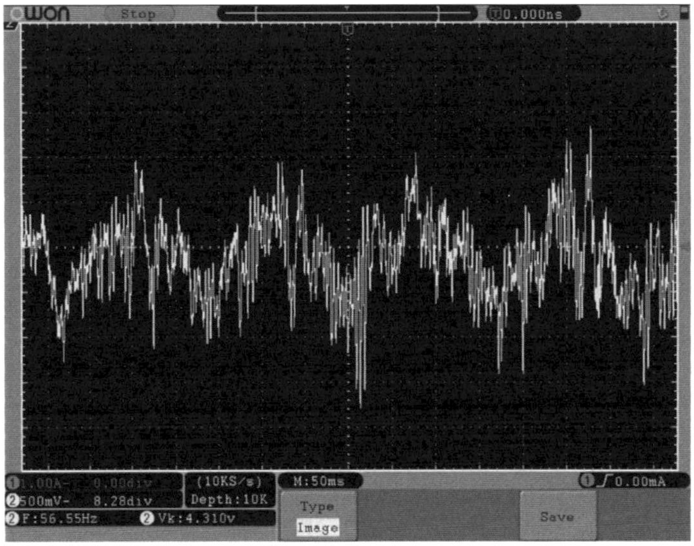

Figure 9. The curve of the torque value without any averaging.

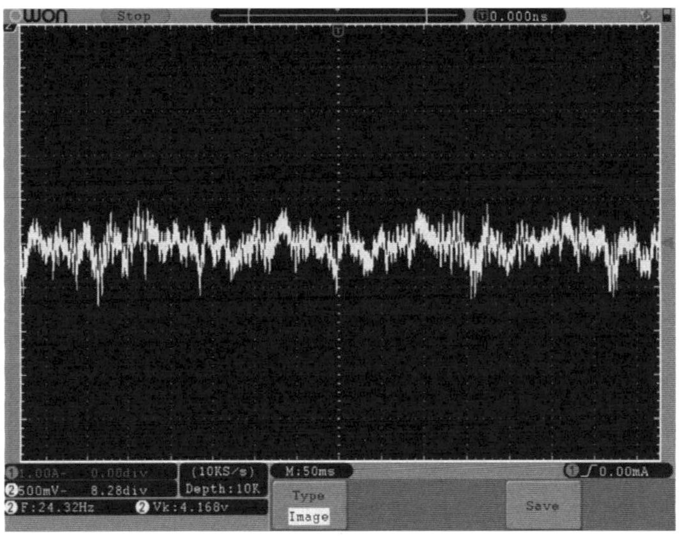

Figure 10. The curve of the torque with fourfold averaging.

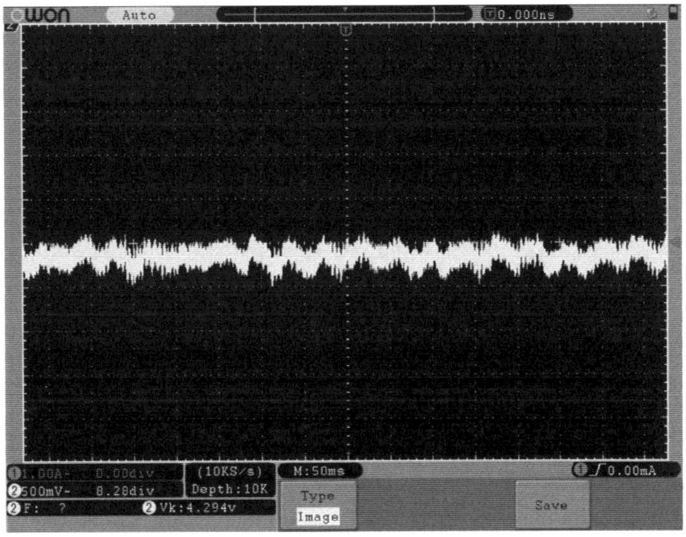

Figure 11. The curve of the torque with 64-fold averaging.

This method of averaging is not suitable for the dynamic process. In this case we cannot average the data from even about one minute.

During the tests it was noticed that the main impact for torque value has the noise proportional to the rotational speed of the shaft. This assumption is confirmed by the data recording presented in the figure 12, where is visible the repetitive shape of two revolutions of the shaft disturbed by mechanical noise (caused by e.g. assembly errors).

Based on these observations it appeared an idea to average the shape of torque value in each revolution. In such a solution it is necessary to measure simultaneously the rotational speed to change the averaging period, preferably directly in the torque meter.

The realization of this idea needs to elaborate a new algorithm which should be executed in the sensor. The work on the algorithm and then testing the proposal are currently being developed.

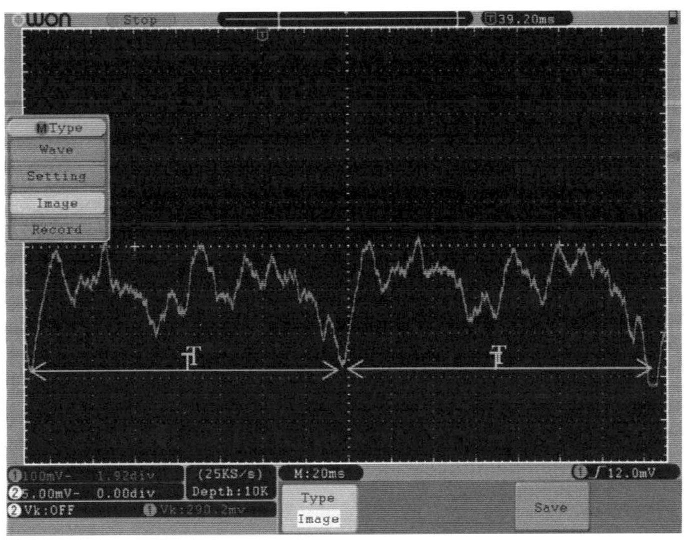

Figure 12. The screen shot from oscilloscope with shape of torque from two shaft revolutions.

4. Summary

Nowadays on the market there is a big choice of different torque meters. It is possible to select the most suitable design and measurement method of the sensor but none of the torque meters is prevented from noise. Every user of torque meters is responsible to provide a good conditions for torque measurement. It is vital to minimize the mechanical noise in the powertrain caused inter alia by assembly errors, prevent all the devices against electrical noise and to choose correct method of torque meter installation including proper type of clutches which enable to compensate the mounting offsets.

References

[1] Morris A S and Langari R 2015 *Measurement and Instrumentation: Theory and Application* (United States of America: Hayton J.) chapter 18 p 557
[2] Latash M L and Zatsiorsy O 2015 *Biomechanics and Motor Control : Defining Central Concepts* (United States of America: Nikki Levy) chapter 1 p 4
[3] Kamiński M 2012 Czujniki momentu obrotowego *Pomiary Automatyka Robotyka* **5** 20-27
[4] Decner A and Iskierski L 2015 Torque measurement in industrial conditions *Napędy i Sterowanie* **7/8**139-143
[5] Sadowski A andŻółtowski B 2012Non-contact measurement methods of torque *Logistyka* **6**915-924
[6] Skala B 2004 The torque measurement based on various principles *Prace Instytutu Elektrotechniki* **220** 17-27
[7] Naruszewicz W 2007 *Automatyka B2B*Accessed: 30.04.2016 http://automatykab2b.pl/technika/760-pomiary-momentu-obrotowego#.V0GL3uR34hS
[8] BOSCH 2014 *Czujniki w pojazdach samochodowych* (Warszawa: WKiŁ) pp 70-77
[9] Honeywell International Inc. 2005 Accessed 29.04.2016 https://www.element14.com/community/servlet/JiveServlet/previewBody/16410-102-1-50374/White%20Paper_SAW%20Torque%20Sensor%20Technology.pdf
[10] Dudziak M, Malujda I and Talaśka K 2014 Torque measurement in worm aglomeration machine *Zeszyty Naukowe. Transport / Politechnika Śląska* **82** 67-73
[11] Zajączkowski J Accessed 3.05.2016 http://www.hbm.com.pl/art11.html
[12] HBM Accessed 4.05.2016 http://www.hbm.com/en/3004/t40b-torque-transducer-with-a-rotational-speed-measuring-system/
[13] Pomianowski R Accesed 5.05.2016 http://www.sensorat.alpha.pl/momentomierze.html
[14] KTR Accesed 7.05.2016
https://www.ktr.com/fileadmin/ktr/media/Manuals/49011pl000000.pdf

The influence of battery degradation level on the selected traction parameters of a light-duty electric vehicle

Z Juda[1] and M Noga[1]

[1]Institute of Motor Vehicles and Internal Combustion Engines, Cracow University of Technology, 37 Jana Pawla II Avenue, 31-864 Krakow, Poland

email: zjuda@pk.edu.pl; noga@pk.edu.pl

Abstract. The article describes results of an analysis of the impact of degradation level of battery made in lead-acid technology on selected traction parameters of an electric light duty vehicle. Lead-acid batteries are still used in these types of vehicles. They do not require complex systems of performance management and monitoring and are easy to maintaining. Despite the basic disadvantage, which is the low value of energy density, low price is a decisive factor for their use in low-speed electric vehicles. The process of aging of the battery related with an increase in internal resistance of the cells and the loss of electric capacity of the battery was considered. A simplified model of cooperation of the DC electric motor with the battery assuming increased internal resistance was presented. In the paper the results of comparative traction research of the light-duty vehicle equipped with a set of new batteries and set of batteries having a significant degradation level were showed. The analysis of obtained results showed that the correct exploitation of the battery can slow down the processes of degradation and, thus, extend battery life cycle.

1. Introduction

Traction batteries in electric vehicles have a limited lifespan. Battery life depends on many factors, and the end result is the increasing of an internal resistance of a cell and the loss of a battery capacity, which results in limiting the range of the vehicle. Another consequence is a reduction of the stiffness of discharge characteristics, which results in deterioration of operating parameters of the electrical drive. After a long use the degradation processes are accelerated. Another factor degrading the lead acid battery is also the sulfation of plates of battery cells. The result is a reduction in capacity and increase in internal resistance. Charging of such a battery requires increased charging voltage. Losses of the active material due to mechanical damage reduce the surface of cell plates and also cause a loss of capacity. The objective of the study is to evaluate the effect of degradation processes in a long-term operation in a relation to the important battery parameters such as capacity and the internal resistance of the battery.

2. The object of research and data acquisition path

The research which are the subject of this study were conducted on four-wheeled, light-duty electric vehicle Melex 945 with two seating positions for driver and passenger equipped with a loading platform that allows transport of loads up to 150 kg. The source of propulsion of the vehicle is a separately excited DC motor [1]. Reducing of rotational speed in the drive system is implemented through a two-stage cylindrical gear, which was joined with an electric motor and with a rigid rear axle suspended on leaf springs. Front suspension is based on transverse-wishbones and a single transverse leaf-spring. Damping of vibrations coming from the road surface roughness is carried out using twin-tube oil shock absorbers located at each wheel. The vehicle uses hydraulic dual-circuit braking system. The fluid pressure produced by the pump acts on the drum brakes of all four wheels. The vehicle also has a set of lighting required to travel on public roads. Basic technical data of the vehicle are shown in Table 1.

Table 1. Basic technical data of the test vehicle

Parameter	Value
Length x width, Wheelbase	2660 x 1230 mm, 1660 mm
Curb weight	700 kg
Load	2 persons + 150 kg load
Motor type	DV3-4006AA, DC, separately excited
Supply voltage	48 V
Armature current	100 A
Nominal motor power	3,9 kW at 4300 rpm
Nominal motor torque	8,2 Nm
Motor efficiency	75%
Overall transmission ratio	16
Maximum speed	29 km/h
Range	About 40 km

2.1. Traction batteries

The electrical energy storage system consist of the 8 lead-acid batteries located in the middle of the vehicle under the driver and passenger seats. Serial connection of batteries with a voltage of 6V each gives the nominal voltage value of 48V for the set. The used batteries are characterized by an increased durability to remain in a state of deep discharge - so called "deep cycle". Technical data of the battery packs used for the tests are shown in Table 2. New batteries have a slightly lower nominal capacity in relation to the previously used traction batteries T-145, and are characterized by a lower mass. The both sets of batteries are of a "deep-cycle" type have thicker plates (the alloy of lead and antimony) and a special type of separator. In turn, the active mass of the plates has a higher density. As a result, the battery life increases. The life of "deep-cycle" battery type declared by the manufacturer in relation to conventional battery shows an increase of nearly 100%.

Table 2. Technical data of energy storage systems used in the tests [2], [3]

Parameter	Old set	New set
Battery type	Trojan T-145, „deep cycle"	Trojan T-125, „deep cycle"
Nominal voltage, V_{bat}	48 V (8x6V)	48 V (8x6V)
Capacity (10h), C_{10}	239 Ah	221 Ah
Stored energy, E_{bat}	11.47 kWh	10.61 kWh
Mass of the battery set, m_{bat}	264 kg	240 kg
Charging system	External charger 230V AC/ 48V 30 A DC	

For the Depth of Discharge (DoD) equal to 20% the Life Cycle reaches 3000 deep charge/discharge cycles. For DoD equal to 50% - Life Cycle is 1200, and for DOD = 100% - Life Cycle is only 600. It should be noted that in real operating conditions the DoD over 80% is not allowed. Reducing of the operating costs caused by increasing of the battery lifetime is mainly related to the extension of time to change the battery pack [4]. Batteries of a "deep-cycle" type are designed for the DoD from 50% to 80%, which qualifies this type of battery for vehicles with purely electric drive. An important factor limiting Life Cycle value is the battery-charging-current in a mode of regenerative braking. High values of the charging current for lead-acid batteries have a degrading impact on the battery plates and the active mass.

2.2. Control system of a motor

Management of the operation of the electric motor is carried out by a programmable electronic control system Curtis 1266 designed to operate with a nominal voltage of 36V or 48V, and a maximum continuous current of the armature of 140A.

The controller is based on the MOSFET technology and adjusts the voltage, the armature- and excitation current by pulse width modulation method (PWM). The frequency of the PWM equals 16 kHz. In the field control subsystem the transistors are connected in a H-bridge circuit [5], so that it was possible to change the polarity of the supply voltage, what allows to change the direction of rotation of the motor. The circuit of the motor armature is powered by the half-bridge transistors, which allows current to flow in both directions to and from the battery, respectively, while the machine works as a motor during acceleration and driving the vehicle at a constant speed and as a generator during regenerative braking. The controller has the ability to change a number of configuration parameters, such as e.g. speed limit, maximum armature- and field current of the motor, and delaying the onset and intensity of the braking process with the recovery of energy to the battery pack. Defining a separate set of controller settings, it is possible for the four driving modes, where the selection is made using two bi-stable switches. Changing the controller settings is carried out using a dedicated programmer Curtis 1311.

2.3. Data acquisition path

To record measured values a portable PC computer and data acquisition card of NI USB-6008 type were used. The data acquisition card has 8 analogue inputs with 12-bit resolution and sampling rate up to 10 kS/s. During the tests the waveforms of voltage V_a and I_a armature current, field current I_f and the pulse signal of motor speed Vs from the Hall sensor were recorded at a frequency of 1 kS/s.

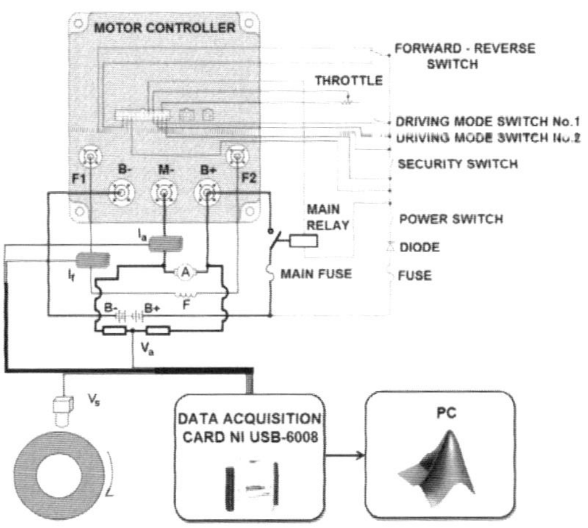

Figure 1. Wiring diagram of the vehicle and scheme of data acquisition path; A - motor armature, F – field wound of the motor, B+, B- - battery connections, M- - negative connection of the motor armature, F1, F2 – connections of the filed wound, I_a – sensor of the armature current, I_f – sensor of the field current, V_a – voltage divider for measuring the armature voltage, V_s – sensor of the rotational speed of the motor.

Registration of signals was carried out using a script running in Matlab. The data acquisition path is shown in Figure 1. Besides recording of the voltage signals to the file, the application running in the

Matlab environment allows to convert the stored data to the actual values of the measured parameters. In the case of the armature- and field current measurements, the sensors with Hall effect were used.

Armature voltage is calculated taking into account the ratio of the resistive voltage divider used for measurement, what equals to 6. Application of the divider of voltage armature allows to adjust the level of the measured signal to the input range of data acquisition card.

The modernized system of motor rotational speed measurement, which replaced the original sensor cooperating with the speedometer, was built on the basis of a Hall sensor and a trigger of the motor shaft position made of a ferromagnetic plate. This sensor generates 5 periods of the rectangular signal 0-5V for each revolution of the motor shaft.

The calculation of the speed of the vehicle consisted of counting of falling edges of the rectangular motor speed signal at the time of 200ms using an application developed in MATLAB and converting this indication to the value of speed, taking into account the dynamic radius of the drive wheels and overall transmission ratio. An accurate calibration of the vehicle speed measurement system was performed using a GPS device with a frequency of 100 Hz. In a Figure 2 the courses of vehicle speed obtained from the GPS100Hz-device and newly developed measurement system used in the vehicle were presented. The modernized measurement system allows for the much more precise determination of speed compared to the previously used method.

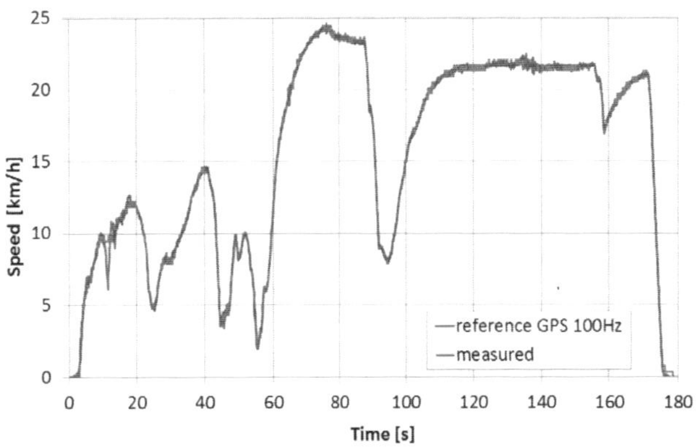

Figure 2. The courses of vehicle speed obtained from the GPS100Hz-device and newly developed measurement system used in the vehicle.

3. Methodology and the scope of research

In September 2015, a set of batteries used since six years in the test vehicle has been replaced by a new one. Batteries used previously have been naturally degrade, which resulted in a drastic reduction of electric capacity and increased internal resistance.

To assess the impact of the application of new batteries on selected traction parameters of Melex vehicle, just before the planned replacement of the battery the research of the vehicle equipped with the old, fully charged, battery pack was carried out. Then, after replacing the batteries with new ones and finished the full charging exactly the same set of tests was performed.

The research of the vehicle was conducted on straight and flat road track with a dry asphalt surface. The test rides to the registration of the results were carried out only in one direction so as to avoid the influence of the inclination of the road on the obtained results. For analogous reasons, during the research the vehicle ran one driver. Before testing, the batteries were fully charged. Vehicle load was about 165 kg (driver, passenger and data acquisition system). Research for both sets of batteries were held in windless and dry weather, at an ambient temperature of 17 °C.

During each of the test runs, after the sudden push the accelerator pedal the vehicle was accelerated to a certain speed, followed by a quick release of the accelerator pedal and the vehicle slowed to a stop with the generator operation of the electric machine and without using the brakes [6]. The vehicle was accelerated to a speed of 26 km/h, which is close to the maximum speed of the vehicle of a fully charged battery pack. This procedure aimed at eliminating the impact of the ongoing battery discharge on the value of maximum speed obtained by the vehicle.

4. The results of research and their analysis

To protect the battery pack and the vehicle electrical circuits against overload the maximum permissible armature current equal to 300A was defined in the motor control system. Due to the short rise time of the voltage to the maximum and the low rotational speed of the motor the armature current reaches the limit value, what sets the controller to a mode of current limiting. This state lasts for about 0.8 seconds and then the motor reaches maximum power. Then the reducing the field current occurs, what allows the motor to operate at a higher rotational speed when adjusting for constant power. This is a classic way to control this type of electrical machines.

4.1. Estimation of characteristic parameters of the two battery sets

An analysis of the obtained results was preceded by estimating of the increase of internal resistance and the actual capacity of the old battery set. The theoretical - experimental method was used for this purpose the details of which will be presented below.

In the first step the internal resistance of a new set of batteries was calculated. For lead-acid batteries the following relationship is true (4.1) [7]:

$$R_{int} = N \cdot \frac{0{,}022}{C_{10}}, \tag{4.1}$$

where:
R_{int} – internal resistance of the battery, [Ω]
N – number of cells in series connection, -
C_{10} – 10-hour battery capacity, [Ah]

In both cases the number of cells in the battery pack amounts 24, while in the case of a new battery, the 10-hour capacity is 221 Ah. In this situation, an internal resistance of the new battery set is $R_{int\_new} = 0.0024$ [Ω].

Estimating the internal resistance of the old set of batteries was based on a comparison of experimental results obtained in similar conditions for both sets of batteries. The comparison was carried out for the first tests of acceleration of the vehicle with fully charged batteries for the both sets of batteries.

An equivalent circuit of the electric drive (Figure 3) was used in order to calculate the resistance of the old battery set. The model of the battery impedance represented only by a real part (resistance) was adopted for this analysis.

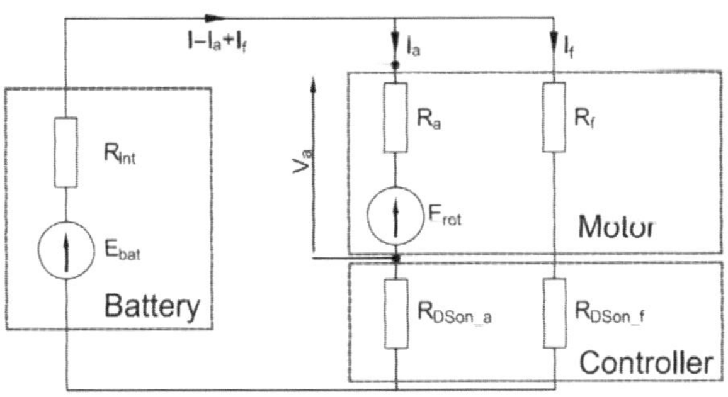

Figure 3. Equivalent circuit of the electric drive.

In the scheme above the following symbols and abbreviations are used:
E_{bat} – electromotive force of the battery, [V]
E_{rot} – electromotive force generated in the armature, [V]
I - current taken from the battery, [A]
I_a – armature current, [A]
I_f – field current, [A]
R_a – armature resistance, [Ω]
R_f – field winding resistance, [Ω]

R_{int} – internal resistance of the battery, [Ω]
$R_{DSon\_a/f}$ – on-resistance of MOSFETs in the armature/field circuit of motor controller, [Ω]
V_a – armature voltage, [V]

To determine the internal resistance of the old battery the waveforms of armature current, field current and the voltage measured at the terminals of the armature of the electric motor registered during acceleration of the vehicle with the new and old sets of batteries were analysed.

For a range of speed close to the maximum the same value of the armature current in the waveforms recorded for the both batteries was selected (4.2).

$$I_a = I_{a\_old} = I_{a\_new} = 96.27\ [A], \tag{4.2}$$

The controller provides a flow of field current proportional to the armature current value, what makes that for the both selected points in the waveforms of the armature current the field current also reaches the same values (4.3).

$$I_f = I_{f\_old} = I_{f\_new} = 8.20\ [A], \tag{4.3}$$

Ensuring for both sets of batteries a fully charged state and the same speed of the vehicle causes that the increase in internal resistance of the old battery set is directly proportional to the difference in voltage measured in the analysed conditions on the armature of the motor supplied from new and old batteries, and other circuit parameters remain unchanged. In this case, the internal resistance of the old battery set can be calculated using the formula (4.4).

$$R_{int\_old} = R_{int\_new} + \frac{V_{a\_new} - V_{a\_old}}{I_a + I_f}, \tag{4.4}$$

The armature voltage values registered for the new battery set and the old one amount to I_{a_new} = 48.36 [V] and I_{a_old} = 45.72 [V], respectively, so eventually the estimated value of resistance of the old battery is R_{int_old} = 0.0277 [Ω].

Having the calculated value of the internal resistance of the old battery set it is possible to estimate the capacity of the storage system. For this purpose appropriately transformed formula (4.1) was used. Inserting the calculated value of the internal resistance R_{int_old} gives as a result C_{10_old} = 19.1 [Ah].

The calculated capacity of the old, exploited battery set is approximately 8% of the original value, what is reflected in the results of research presented in the next section of work.

4.2. Simulation of the behaviour of new and used set of batteries

Simulations were performed using Implement Battery Model from Electrical Sources, Electric Drives / Extra Sources Matlab / SimPowerSystems library [8]. Discharge curves of constant current values 50, 100 and 150 A show the characteristics of rigidity to the entire area of the battery not showing signs of wear – Figure 4, and large voltage drops to the exploited battery, what is clearly visible in the Figure 5.

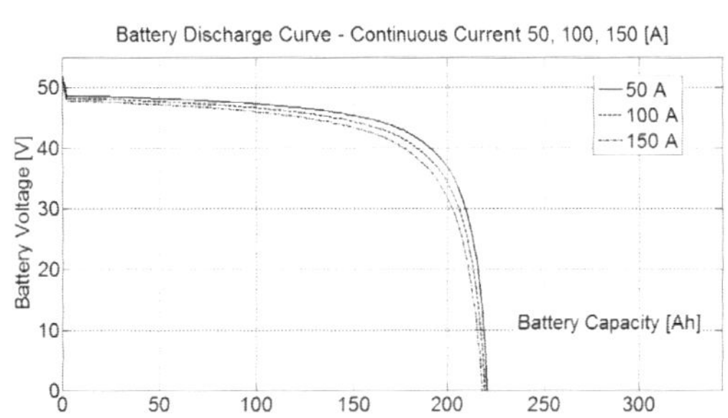

Figure 4. Battery discharge curve – capacity 220 Ah – new set.

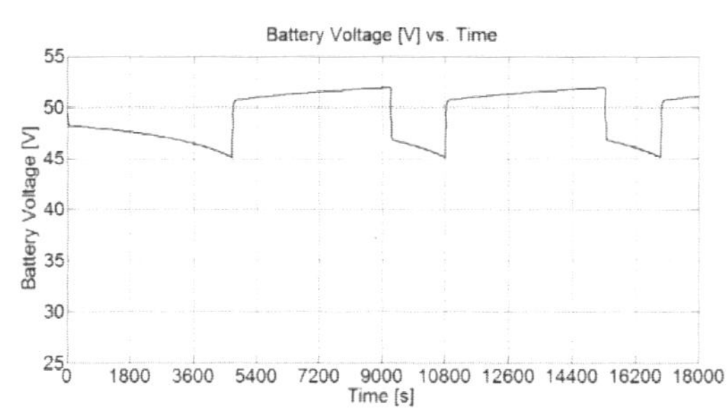

Figure 5. Battery discharge curve – capacity after degradation - 20Ah.

Figure 6 shows the result of simulation of a constant current discharging the battery 100 A from the state of full charge (SOC = 1) to SOC = 0.4, and constant current charging to achieve the SOC value = 0.6. Discharge of the battery is made within 1.31 hour. Recharging to the SOC = 0.6 required a time of 1.26 hour. Additional discharged to the SOC = 0.4 occurred after 0.43 hour. During the process of charging and discharging a terminal voltage of the battery has changed in the range of 45.12 V to 51.93 V.

A similar process for the exploited battery – Figure 7 - showed different behaviour under identical conditions. Battery discharge occurred after the time of 8.55 min. and to recharge to the SOC = 0.6 time 9.51 min was needed. Another discharged to the SOC = 0.4 occurred at 2.65 min. In the process of charging and discharging terminal voltage of the battery has changed in the range of from 27.99 V to 55.27 V.

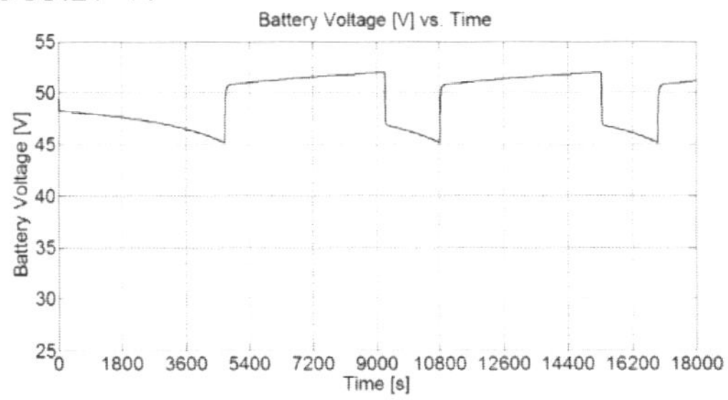

Figure 6. Battery voltage during charging/ discharging – 220 Ah – new set.

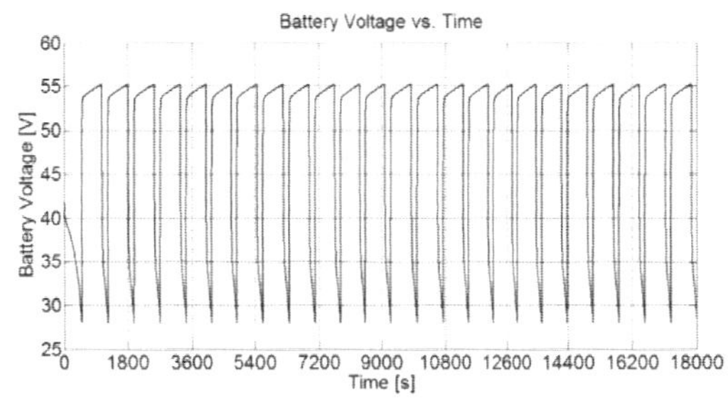

Figure 7. Battery voltage during charging/ discharging – 20 Ah – after degradation.

The loss of battery capacity due to progressive degradation leads to faster discharge, thus increasing the frequency of charge / discharge cycle which has a direct impact on battery life [9].

4.3. Selected issues of the process of acceleration of the electric vehicle with new and old energy storage systems

Figure 8 presents a comparison of the maximum power taken from the battery pack during the process of acceleration of the electric vehicle with the old and new set of batteries as a function of the parameter Acceleration rate, i.e. the elapsed time of increasing of the armature voltage from zero to the maximum value.

Analysis of the content of Figure 8 indicates that the course of changes of maximum power taken from battery is similar in nature for the new and old batteries. Lower values obtained for the old batteries are the result of the increased voltage drop across the internal resistance of the battery, which increases with the degree of discharge of battery and with its degradation.

Figure 9 shows a comparison of the maximum armature current registered in the same conditions as before. The maximum values of the armature current obtained for successive values of the parameter Acceleration rate are practically the same for both sets of batteries. This is due to the fact that the controller enters in the phase of current limitation when starting the motor regardless of the state of the battery. This is caused by low armature resistance and the low value of the electromotive force induced in the rotor at a low rotational speed.

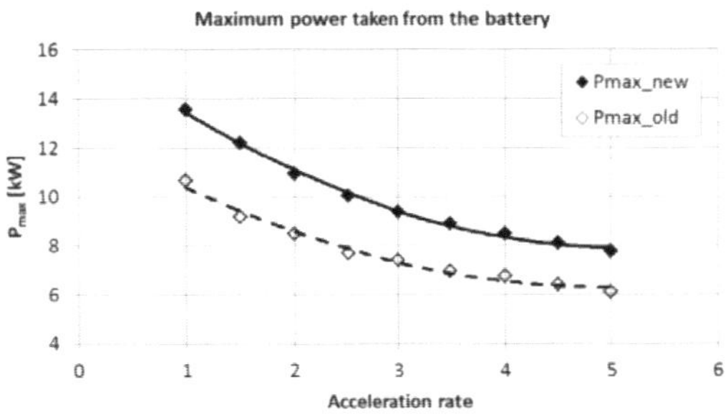

Figure 8. The maximum power taken from the battery during acceleration of the electric vehicle with the old (Pmax_old) and new (Pmax_new) set of batteries as a function of Acceleration rate.

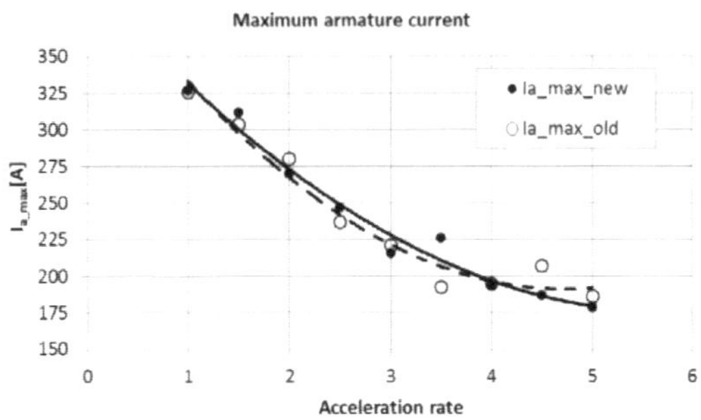

Figure 9. The maximum armature current during acceleration of the electric vehicle with the old and new set of batteries as a function of Acceleration rate.

The maximum value of the voltage measured at the motor armature is observed when the vehicle reaches maximum speed. This value depends directly on the internal resistance of the battery. Figure 10 shows the maximum armature voltage recorded in the successive tests for the both sets of batteries.

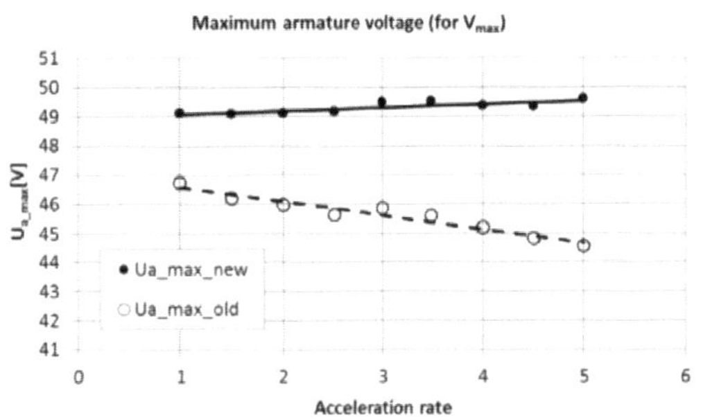

Figure 10. The maximum armature voltage as a function of Acceleration rate for the both set of batteries.

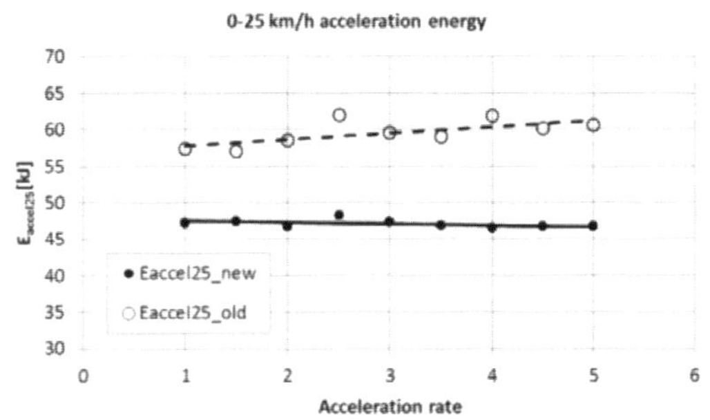

Figure 11. Electric energy taken from the battery pack during acceleration of the vehicle to a speed of 25 km/h for different values of the parameter Acceleration rate.

The maximum value of the armature voltage in the terms of measurements represents the condition of the battery pack. A curve with a slightly increasing trend in subsequent tests was obtained for the set of new batteries. This phenomenon can be explained by the fact that in subsequent attempts the time of voltage rise to the nominal value was gradually increasing, what gave lower values of armature current observed in the first phase of acceleration. If so, the lower load of the power supply resulted in lower voltage drop during the first phase of acceleration. In this situation the armature voltage at the end of acceleration reached gradually increasing values. In the successive tests significantly decreasing maximum values of the armature voltage were obtained for the set of old batteries. This is due to a very rapid process of discharging of the old batteries caused by their sulfation what lowers electric capacity.

The dependence of electric energy taken from the battery pack during acceleration of the vehicle to a speed of 25 km/h for different values of the parameter Acceleration rate is presented in Figure 11. For the new set of batteries the energy of the vehicle acceleration to desired speed is practically independent on the intensity of the first phase of the process of acceleration. For the old battery set, the energy of acceleration is significantly higher than for the new battery set - to about 25%. In addition, the energy rises in the successive test drives. This results from the large energy losses on the increased internal resistance of the old battery set, additionally increasing in subsequent test drives, as described above.

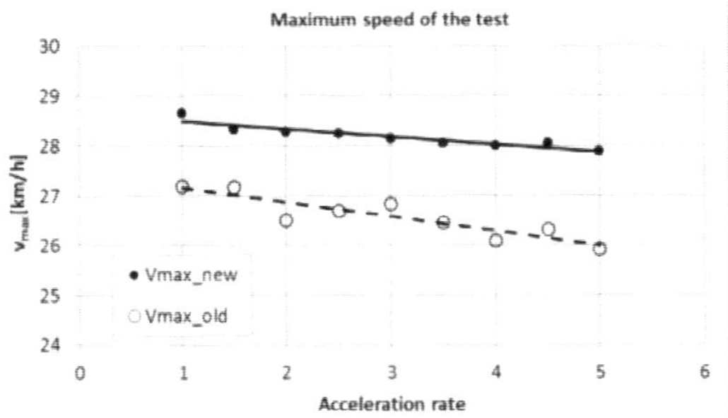

Figure 12. Maximum speed reached in the test track as a function of the parameter Acceleration rate.

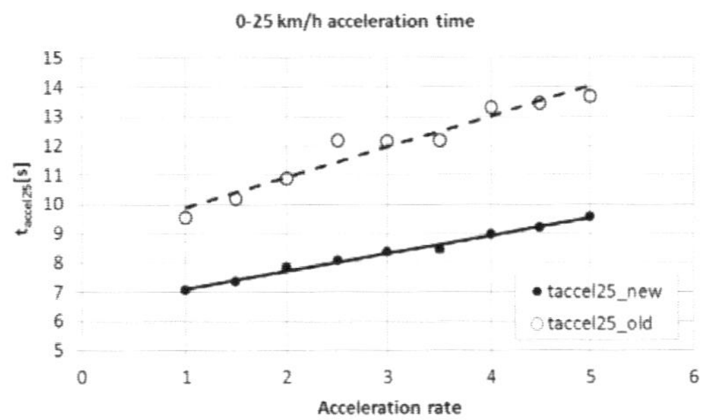

Figure 13. Time of acceleration to a speed of 25 km/h as a function of the parameter Acceleration rate obtained for the vehicle fitted with a set of old and new batteries.

A graph of the maximum speed of the vehicle obtained on the length of the test track is shown in Figure 12. For the both battery sets new and old the maximum speed decreases with increasing value of the parameter Acceleration rate. This is caused by the fact that the acceleration takes place all the time on the track of the same length, so if accelerating is milder, the speed obtained by the vehicle will have a gradually decreasing value. In the case of old set of batteries this effect is compounded by a gradual discharge of the battery, what, as mentioned above, can be seen analysing the successive test results.

Graphs of time of acceleration to a speed of 25 km/h as a function of the parameter Acceleration rate obtained for the vehicle fitted with a set of old and new batteries were presented in Figure 13. The acceleration time of the vehicle to a speed of 25 km/h depending on the intensity of increase of the voltage to the maximum value is rising in both cases. For the new set of the batteries the time is increased from about 7 seconds for the Acceleration rate equalled 1.0 s to about 9.5 s for the setting Acceleration rate = 5.0 s. For the old set of batteries, the values being 9.5 and 13.8 s, respectively, what indicates a significant decrease in maximum performance of the vehicle.

Figure 14 shows waveforms of changes in the value of the average acceleration to a speed of 25 km/h depending on the value of Acceleration rate parameter obtained for the vehicle fitted with a set of new and old batteries.

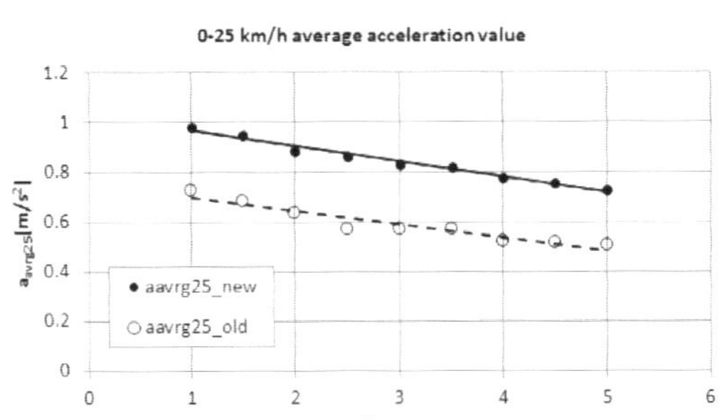

Figure 14. The average acceleration to a speed of 25 km/h depending on the value of Acceleration rate parameter obtained for the vehicle fitted with a set of new and old batteries.

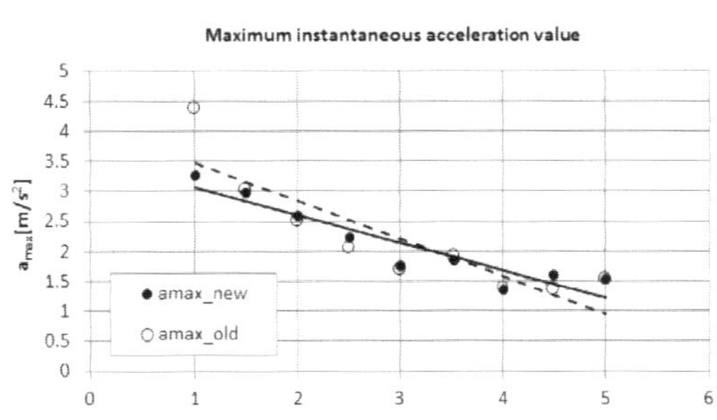

Figure 15. The maximum instantaneous acceleration of the vehicle in a function of Acceleration rate parameter for the vehicle with the old and new battery pack.

The curves of changes in the value of the average acceleration to a speed of 25 km/h are inverted to the waveforms of acceleration time shown in the previous figure. Figure 15 shows a comparison of the maximum instantaneous acceleration of the vehicle in a function of Acceleration rate parameter for the vehicle with the old and new battery pack. The maximum instantaneous acceleration value is recorded in the first phase of acceleration of the vehicle from a standing start. The acceleration depends on the torque to the wheels of the vehicle, and in turn, this value is derived from the torque of the motor which is directly proportional to the armature current. It should be remembered that the maximum value of the armature current in both cases was similar. If so, the instantaneous maximum torque underlying the maximum acceleration of the vehicle will obtain similar waveforms, regardless of the state of charge of the battery in the vehicle.

5. Conclusions

Basing on the results of carried out research described in the paper, the following conclusions were formulated:

- In the case of electric drive the intensity of the process of acceleration to the desired speed has no significant impact on energy consumption,
- The operation of an electric vehicle with a worn out battery pack is associated with a drastic deterioration of performance, such as maximum speed, acceleration and most of all range,
- Increased internal resistance of the battery exhibiting a high degree of wear causes high energy losses, in the case of the test vehicle. This value reached a level of about 25%. In the same way drops the overall efficiency of the electric power source,
- The relationship similar to the above mentioned will be valid also in the process of charging of the worn out batteries. A part of the energy provided by a rectifier will dissipate to the heat on the internal resistance of individual cells,
- The efficiency of the recovery of kinetic energy in the process of regenerative braking will decrease for the same reasons in the case of operation with worn out batteries.

Acknowledgements

Authors would like to thank Dr. Robert Janczur for the lending of the 100Hz GPS device used to calibrate speed measurement system.

References

[1] Khajepour A, Fallah S and Goodarzi A 2014 *Electric and Hybrid Vehicles* (Chichester: John Wiley & Sons)

[2] Trojan T-145 Battery Data Sheet 2015 (Santa Fe Springs: Trojan Battery Company)

[3] Guo J, Wang J and Cao B 2009 *Regenerative Braking Strategy for Electric Vehicles* Intelligent Vehicles Symposium (Xi'an: IEEE) p 864-868

[4] Eshani M, Gao Y, Gay S E and Emadi A 2005 *Modern Electric, Hybrid Electric and Fuel Cell Vehicles, Fundamentals, Theory and Design* (Boca Raton: CRC Press)

[5] Trojan T-125 Battery Data Sheet 2016 (Santa Fe Springs: Trojan Battery Company)

[6] Juda Z 2014 *Hamowanie odzyskowe pojazdów z napędem elektrycznym – strategie sprawności odzysku i komfortu jazdy* Badania Pojazdów (Cracow: Politechnika Krakowska) p 49-60

[7] Larminie J and Lowry J 2003 *Electric Vehicle Technology Explained* (Chichester: John Wiley & Sons)

[8] Tremblay O, Dessaint L A and Dekkiche A I 2007 *A Generic Battery Model for the Dynamic Simulation of Hybrid Electric Vehicles*, IEEE Vehicle Power and Propulsion Conference, (Arlington: IEEE)

[9] Juda Z 2001 *Simulation of Energy Conversion in Advanced Automotive Vehicles* SAE Technical Paper 2001-01-3341, (Warrendale: SAE)

Technical and legal considerations and solutions in the area of battery charging for electric vehicles

Z Juda[1]

[1]Institute of Motor Vehicles and Internal Combustion Engines, Cracow University of Technology, 37 Jana Pawla II Avenue, 31-864 Krakow, Poland

E-mail: zjuda@pk.edu.pl

Abstract. The issue of protecting health of residents of urbanized areas from the effect of excessive particulate matter and toxic components of car exhaust gases imposes the need of introduction of clean electric vehicles to the market. The increasing market availability of electric vehicles, especially in the segment of short-range (neighborhood) vehicles is followed by development of new and advanced infrastructure solutions. This also applies to the increasingly popular hybrid vehicles PHEV (Plug-in Hybrid Electric Vehicles). However, problems with the existing designs are primarily associated with limited driving range on a single battery charge, the density of charging stations in urban and suburban area, energy system efficiency due to increased electricity demand and the unification of solutions for charging stations, on-board chargers and the necessary accessories. Technical solutions are dependent on many factors, including the type and size of battery in the vehicle and access to power grid with increased load capacity. The article discusses the legal and technical actions outlined in the above directions. It shows the available and planned solutions in this area.

1. Introduction

The need for the wider popularity of electric vehicles seems obvious and is primarily due to the necessity of reducing the impact of transport on environmental pollution and warming the earth's atmosphere. The rapid development of this type of vehicles, however, faces difficulties both of a technological nature, as well as from the legislation. The main concern of potential buyers of electric vehicles refers the uncertain range of EVs. This problem is also related to the development of both home and public charging stations, especially fast charging stations. A dense network of quick charging stations could reduce the current imbalance of replenishment of energy time in conventional vehicle (refueling) and electric vehicle (charging). Another issue is associated with unification of standards for charging, so that the users can be sure the possibility of recharging their vehicles, regardless of the brand.

2. The European Standards of EV battery charging equipment

IEC 62196-1 (PN EN 62196) *Plugs, socket-outlets, vehicle connectors and vehicle inlets - Conductive charging of electric vehicles - Part 3: Dimensional compatibility and interchangeability requirements for d.c. and a.c./d.c. pin and contact-tube vehicle couplers* [1].

The standard is applicable to vehicle computer, plugs, socket-outlets, connectors, inlets and cable assemblies for electric vehicles, charged via conductive connection. The operating voltage of such a systems cannot exceed 690 V AC and current to 250A, and 600 V DC and current to 400 A.

IEC 62851 (PN EN 61851) *Plugs, socket-outlets, vehicle couplers and vehicle inlets – Conductive charging of electric vehicles* [2].
The standard defines four modes of charging batteries:
Mode 1: Slow charging from a household-type socket-outlet. In Europe, this means single-phase 16 A 230 V, 3 kW socket. EV can be equipped with a cable for connecting the socket and the vehicle.
Mode 2: Slow charging from a household-type socket-outlet with an in-cable protection device.
Mode 3: Slow or fast charging using a specific EV socket-outlet with control and protection function installed.
Mode 4: Fast charging using an external charger.
IEC 61851-1 also describes the pilot signal which communicates the charging requirements by means of PWM method [3].
The EU and US road signs indicating a charging station are shown in Figure 1. Table 1 lists the standard charging parameters.

Figure 1. Road signs indicating a charging station- EU left, US right

Table 1 Details of charging modes

Charging Mode	AC grid type (phases)	Maximum current [A]	Maximum power [kW]	Estimated charging time [h]	EV connector type
Mode 1	1 or 3	16	3.7	5.5	A, B
Mode 2	1 or 3	32	22	1.0	A, B
Mode 3	1 or 3	63	44	0.5	A, B, C
Mode 4	DC	400	up to 200	0.1 (CC phase only)	C

Assuming the energy demand of about 0.2 kWh / km energy capacity of 20 kWh is enough to cover the distance about 100 km [4]. EV's can be connected to power supply by one of three ways:

A- Connection EV – grid using a fixed connection of supply cable and plug to EV,
B- Connection EV – grid using a releasable connection between EV and AC supply device,

C- Connection EV – grid using the supply cable and EV connector permanently attached to the supply equipment; this is only solution for Mode 4 charging.

In all charging systems is mandatory PE (protective earth) connection, while the connection CP (pilot connection) is required in systems mode 2, 3.4.

IEC 61980 *Electric vehicle wireless power transfer (WPT) systems - Part 1: General requirements*
The regulations included in IEC 61980 standard apply to equipment the wireless electric power transfer (WPT) to the Electric Vehicle energy storage. The technologies considered here include:

- inductive power transfer (MF-WPT, electric energy transferred through the magnetic field),
- capacitive power transfer (EF-WPT, electric energy transferred through the electric field),
- microwave power transfer (MW-WPT, electric energy transferred through electromagnetic waves in range 1 GHz to 300 GHz),
- infrared power transfer, (IR-WPT, energy is transferred through electromagnetic waves in range 300 GHz – 400 THz).

The standard describes the following issues:

- the characteristics and operating conditions,
- the specification for the required level of electrical safety,
- requirements regarding basic communication for safety and process matters if required by a WPT system,
- requirements for basic positioning of the primary and secondary devices, efficiency and process matters if required by a WPT system
- specific EMC requirements for WPT systems
- as well as others.

3. Pilot signal

Electric Vehicle Charging Equipment (EVSE) transmits input PWM signal with a frequency of 1 kHz and a voltage of 12 V into the EV controller. The EV driver to pin CP is accompanied by a system of diodes and two resistors: 2.74 kΩ and 1.3 kΩ. When the charger plug is powered and current loop PWM signal CP-PE is closed permanently by a 2.74 kΩ resistor, this results in a decrease in the voltage to 9 V. After the procedure is initiated by EV driver charging, accompanied by the parallel resistor 1.3 kΩ, which gives the result of the resistance of 880 Ω and fall voltage is 6 V. EVSE controller by changing the PWM duty cycle informs the EV controller about the maximum charge current. Pilot signal circuit is shown in Figure 2. Table 1 lists the PWM pilot signal meanings.

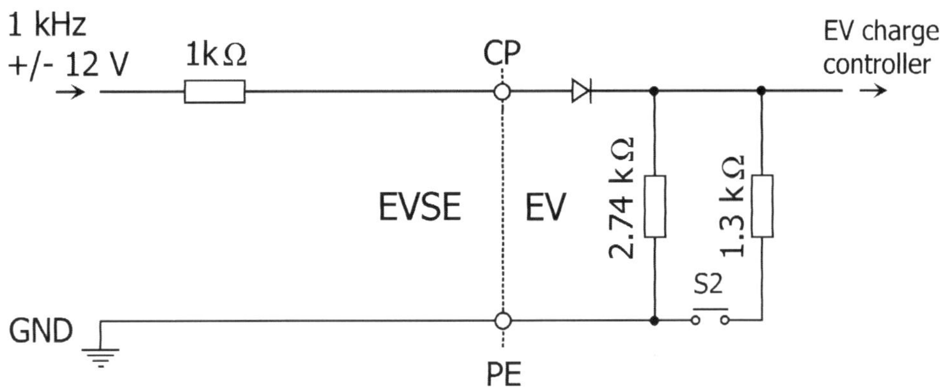

Figure 2. Pilot signal connection.

Table 2. PWM Pilot signal meaning.

PWM Duty Cycle	Maximum charge current
16%	10 A max
25%	16 A max
50%	32 A max
90%	Fast Charging Mode

4. Fast Charging Systems

The development of rapid-charging is related to the appearance of the electrochemical battery technology which can provide high charge current values (above 1C) [5]. The use of such batteries in mass-produced vehicles (Mitsubishi, Toyota, Nissan) enabled the Japanese EV's manufacturers to introduce DC fast charging system called CHAdeMO. The fast charging system was developed by SAE and later adopted by several European producers called CCS (Combo Charging System). The system forms a combination of connector DC fast charging systems and charging AC mode 1 and 2. The manufacturer of popular EV called Tesla has initiated and is successfully developing a worldwide network of fast charging stations Superchargers dedicated to Tesla electric vehicles. China is developing one of the largest network of EV quick charging stations in cooperation with ABB. The network is based on the Terra 53Z DC fast chargers.

4.1. CHAdeMO fast charging system

CHAdeMO quick charger is designed to feed an input power of 62.5 kW and can charge the battery at a voltage of 50 V to 500V. The system uses a dedicated connector charger for each specific vehicle. For digital communication it applies CAN protocol. Fast charge is based on the CC (Constant Current) principle in a first charging phase (up to 80% of target SOC) and the controllers provide CV (Constant Voltage) loop to complete the process (80% to 100% SOC). View of the CHAdeMO connection is shown in Figure 3.

Figure 3. CHAdeMO charging interface.

CHAdeMO interface pins designations:

5, 6 - (-) charging current, (+) charging current,
8, 9 - CAN Bus
1 - GND

2, 10 - charge sequence signal 1, charge sequence signal 2
4, 7 - EV charge permission, connector proximity detection

4.1.1.The procedure for digital communication
After a connector is used to couple the charger plug with the EV socket, the charger transmits the initiation of the charging process signal to the computer in the EV. EV controller connects via CAN and transmits input about battery parameters in the EV to the computer in the charging station, including: required battery voltage, battery capacity, maximum charging time, etc. The computer in the charging station checks the compatibility of EVCS/EV connection. Then EVCS computer transmits the following EVCS parameters to the vehicle controller: the possible values of voltage and current, and the EV controller checks and closes the connection when it has completed the process of coupling settings. This signal is transmitted to EVCS. EVCS approves the connection and performs insulation test, then sends the signal confirming the completion of the preparatory procedure. After receiving this signal, the computer turns on the main contactors in EV, and then calculates the optimum battery charging parameters, taking into account the battery current state of charge and transmits a signal (every 100 ms) with the desired value of the charging current. The charging current is continuously monitored.

4.2. Tesla Superchargers
Each Tesla Electric Vehicle (EV) is equipped with an on-board charger, capable of single-phase socket charge rated at 40 A or 3 x 16 A three-phase current input. Such a garage should be equipped with Wall Connector device [6], which is compatible with Tesla EVs and is capable of power input of at a rate of 48 A or 72 A through a dedicated connector cable. In the first case, charging 60 minute charge cycle offers vehicle range of 48 km to 55 km (depending on the vehicle model), and in the second case from 74 km to 84 km. Of course, this requires an adequate electrical installation and layout of the garage. In EV's, Tesla uses advanced Lithium-Ion batteries with a maximum charging current of 1.75 C, which means that the battery with energy capacity of 60 kWh could be charged by device of power 105 kW.
Tesla made a significant step forward in the field of fast charging EV batteries. This was possible as a result of introducing a system of free life fast charging batteries for all Tesla vehicle owners. The total cost of charging the battery is included in the vehicle price. The current network of Tesla quick charging stations includes 628 locations equipped with 3738 chargers called Supercharger. Each station is typically equipped with 4 to 12 superchargers. Stations are located mainly in the US, Canada and Europe, as well as in Japan, China and Australia. The geographical distributions of the stations is such that another one is within the range of the vehicle after the battery is fully charged. Of course, in case of the lack of supercharging station in a given location, it is possible to charge traditionally through an on-board charger if a single-phase grid slot is available. Of course, such a charge cycle is much more time consuming. In Poland, there is currently only one station with 4 superchargers in operation, whereas in the nearest future 5 further new stations are planned, at such distances from each other that they will be within the Tesla vehicle range. Stations are open 24 hours a day. The maximum capacity of a single supercharger is 120 kW. The design of a supercharger is such that it is design by analogy to on-board chargers operating in parallel. Thus, a 120 kW supercharger includes 12 standard chargers with a capacity of 10 kW connected in parallel. Tesla Supercharger example performance is shown in Fig. 4.

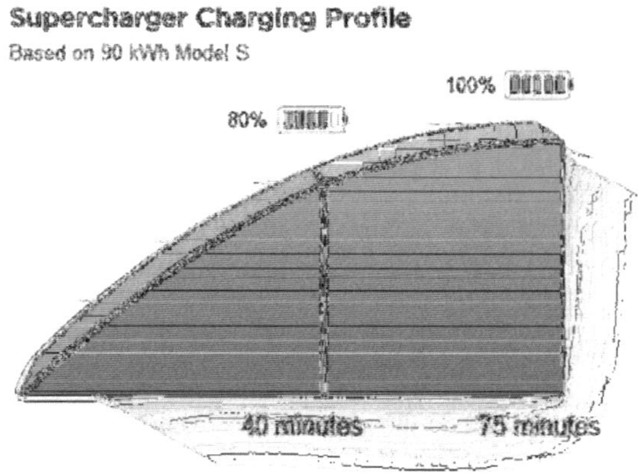

Figure 4. Tesla Supercharger performance.

Tesla EVs are equipped with a Lithium-Ion battery with the rated power capacity of 60 kWh, 70 kWh, 85 kWh or 90 kWh, which naturally affects the vehicle range varies from 370 to 555 km.
The procedure for charging the battery from the state 10 % SOC to 80 % SOC is fast. 80 % SOC gives the opportunity to get you to the next charging station. Further loading (from 80 % SOC to 100 % SOC requires a much longer time due to the reduced charge current in this range. This is due to the need to protect the battery from early wear. Therefore, the actual charging time is variable. Computer installed in the Tesla vehicle communicates the message to start charging with the station controller, and the latter decides on the appropriate charging parameters of the state of battery charge current load charging station and the current climate. Due to the fact that charging is faster for a low SOC than for a higher state of SOC, the selection regarding the charging parameters has to take into account the driving range. It might be the case that the complete (and longer) charging is not necessary in given circumstances. Many external factors can affect the actual charging parameters: outdoor temperature, grid limitations, the number of simultaneously loaded vehicles at the station and many others. The loading procedure is simple. A vehicle is parked in a designated spot, after connecting the cable to the vehicle, and station controllers and the vehicle communicate and check the compatibility of the connection. Subsequently, charging starts and the progress of this process can be monitored by a special application on a smartphone. Most of the stations use between 20 to 40 minutes for this, and thus the potential waiting time is not long.

4.3. CCS (Combined Charging Solution)
The idea of CCS was demonstrated during the 15 International VDI Congress in 2011 by European car manufacturers (Audi, BMW, Daimler, Ford, GM, Porsche and VW). This system (also called a combo) is a universal solution, which allows both low-power (1 or 3 phase for the home grid) as well as a fast DC charging . This standard applies PLC (Power Line Communication) protocol to control the charging process and the exchange of information between the EV and EVSE. The design of the CCS connection is shown in Figure 5.

Figure 5. View of CCS charging interface.

4.4. China Fast Charging System

Stations Terra 53Z 50 kW can recharge the batteries with current up to 125 A with a voltage of 220 V to 550 V DC. The battery charge is completed within 20 min and can ensure the increase of the vehicle range of about 100 km. Digital communication based on GSM and Ethernet. Table 3 gives an example with parameters of EV charging. The design of the China GB standard connection is shown in Figure 6.

Table 3. Examples of EV charging performance.

Electric Vehicle	Battery energy capacity [kWh]	On-Board Charger Power [kW]	Range [km]	Range after 30 min. Fast charging [km] (est.)	Specific Energy Consumption [kWh/km]	Recharging Time (230 V AC) [h] (est.)	Remarks
Nissan Leaf	24	3.3	161	120	0.149	8.0	
i-MiEV	16	3.3	100	73	0.160	7.0	
Tesla 60	60	10	370	270	0.162	7.5	40 [A]
Tesla 70	70	10	432	315	0.162	9.1	40 [A]
Tesla 85	85	10	510	370	0.166	10.0	40 [A]
Tesla 90	90	10	555	400	0.162	10.5	40 [A]
Toyota Prius PI (model 2017)	8.8	3.3	35		0.25		Electric Mode

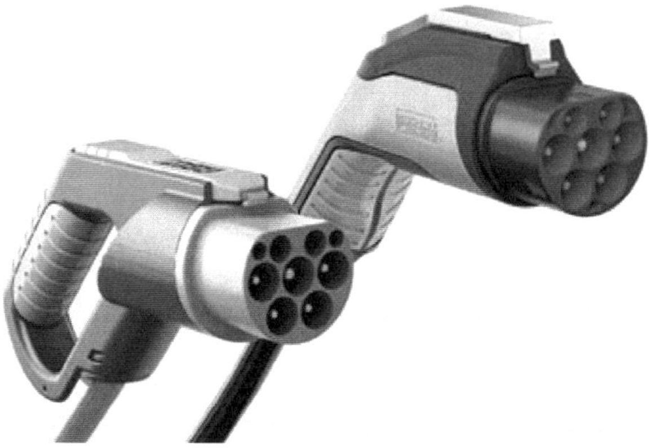

Figure 6. The view of China GB charging interface.

5. Wireless Charging Systems

Qualcomm system [7] is composed of a wall power supply, transmitter pad, receiver pad, on-board charging controller and a battery. The transfer of energy between the transmitter and receiver pad is carried out wirelessly. These systems are designed to match the power of 3.3 kW to 20 kW. A single induction coil has a rectangular shape with rounded corners (Circular Rectangular) or double bipolar (Bipolar or DD). The DD model provides more focused magnetic flux and more compact transmission and reception coil as well as demonstrates 50 % higher power density. CR coils require more precise positioning. The position tolerance for the transfer of full power to the coil CR is 75 mm in the lateral direction and 100 mm in the longitudinal direction, while the 100 mm coil DD in the lateral direction and 150 mm in the longitudinal direction. The vertical spacing between DD coils in the range of 160 to 220 mm provides adequate power transfer. In the case of CR coils, after 220 mm spacing is exceeded , it is followed by a significant decrease. The Halo System operates at a frequency of 85 kHz. This ensures EMC compatibility. The development of Halo technology was utilized as a technique of charging EV battery during normal drive, and called Dynamic Electric Vehicle Charging (DEVC). For this purpose, special street lines are equipped with transmission pads, which are longitudinally spaced every 2 m [8]. This is designed for slow moving vehicles like taxis or buses.

6. Conclusions

The increase in popularity of electric vehicles is limited by the low accessibility of public charging stations. In addition, the lack of universal, global standards, which would normally offer users confidence in the possibility of charging EVs, discourages potential customers from buying such cars. One solution to the problem of charging stations could be the application of more than one standard. However, this makes the design of energy conversion and control system considerably more complex. The result is a higher cost of such multi-system solutions. The appearance of fast charging and technologically advanced batteries on the market reduces the charging time of conventional and electrical vehicles. The range of today's electric vehicles gradually approaches that of the conventional vehicles. The strategy of building networks and developing brand dedicated fast battery charging stations adopted by some vehicle manufacturers can also increase the attractiveness of buying an electric vehicle.

References

[1] International Electrotechnical Commission. International Standard IEC 62196-1, *Plugs, socket-outlets, vehicle couplers and vehicle inlets – Conductive charging of electric vehicles – Part 1: Charging of electric vehicles up to 250 A a.c. and 400 A d.c.* ISBN 2-8318-6975-7, Geneva, Swiss

[2] International Electrotechnical Commission. International Standard IEC 61851-1, *Electric vehicle conductive charging system – Part 1: General requirements ISBN 978-2-88912-222-6,* Geneva, Swiss
[3] U.S. Department of Energy April 2012 *Plug-In Electric Vehicle Handbook for Public Charging Station Hosts* DOE/GO-102012-3275
[4] Garcia-Valle R, Pecas Lopes J A 2013 *Electric Vehicle Integration into Modern Power Networks,* ISBN: 978-1-4614-0133-9, Springer
[5] Kettles D February 2015 *Electric Vehicle Charging Technology Analysis And Standards*, Florida Solar Energy Center, FSEC Report Number: FSEC-CR-1996-15,
[6] Tesla Motors Web Page, https://www.teslamotors.com
[7] Qualcomm Web Page 2011 *Wireless charging for electric vehicles* https://www.qualcomm.com/products/halo
[8] Williams R, N January 2016 *Everybody hertz*, Electric & Hybrid Vehicle Technology International, UKIPM&E

The influence of nitrogen ion implantation on the tribological properties of piston rings made of Hardox and Raex steels

P Budzyński[1], M Kamiński[1] and K Pyszniak[2]

[1] Faculty of Mechanical Engineering, Lublin University of Technology, Nadbystrzycka 36, 20-618 Lublin, Poland
[2] Institute of Physics, M.Curie-Sklodowska University, PlacM.Curie-Sklodowskiej 1, 20-031 Lublin, Poland

E-mail: p.budzynski@pollub.pl

Abstract. The implantation of nitrogen, carbon, and oxygen can be used for enhancing the tribological properties of critical components for internal combustion engines. Hardox and Raex steels have very similar strength parameters as for steel used for piston rings in internal combustion engines. An essential criterion when selecting material for the production of piston rings is a low friction factor and a low wear index. The aim of this study was to determine the extent to which these parameters can be enhanced by nitrogen ion implantation. Samples were implanted with nitrogen ions with 65 keV energy and the fluence of implanted ions set to $1 \cdot 10^{17}$ N^+/cm^2. Friction and wear measurements were performed on a pin-on disc stand. The results demonstrate that implantation with nitrogen ions significantly reduces the friction factor and wear of Hardox 450 and Raex 400 steels. Implantation can and should be used for enhancing the tribological properties of steel used for friction elements in internal combustion engines, particularly when heat treatment is excluded. Final elements can be subjected to implantation, as the process does not change their dimensions.

1.Introduction

Advanced combustion engines must meet very strict standards regarding exhaust purity, fuel consumption, and power output per engine capacity. Given such requirements, manufacturers must use materials which are lightweight and highly resistant to operating conditions at the same time. The group of elements that are most exposed to the effect of high pressure, temperature and mechanical loads includes piston rings [1].

Piston rings make part of the piston-rings-cylinder system. Besides sealing the combustion chamber, piston rings also transfer heat from the piston to the cylinder sleeve, maintain the piston in the axis of the cylinder, scrape oil excess from the cylinder surface, and distribute oil over the cylinder surface [2]. Given the number of functions they perform, piston rings can have various designs. Individual rings can play different roles, too. Top compression rings ensure sealing of the combustion chamber; compression-scraper rings help maintain the required sealing of the combustion chamber and scrape oil excess from the cylinder bearing surface. Finally, there are oil control rings which not only remove oil excess but also distribute oil over the cylinder bearing surface. The design of modern combustion engines is based on a combination of the three types of rings.

Depending on their type and ring function, ring pistons for combustion engines can be made of various materials, the most popular being cast irons. Nonetheless, given the increasing effort of engine design, numerous solutions use steel rings, including top compression rings, oil control rings, and compression-

scraper rings [3]. To enhance their strength properties, steel rings are often subjected to nitriding or coating, e.g. with molybdenic or chromic coatings [4].
Besides strength, emphasis is put on reducing mechanical losses resulting from friction between the active surfaces of the rings and the surface of the cylinder sleeve. It is estimated that losses caused by friction in the rings-cylinder system amount to approx. 24% of the total mechanical losses due to friction in the combustion engine [5].
This paper investigates the effect of nitrogen ion implantation of the tribological properties of two wear-resistant steels: Hardox and Raex. The properties of these steels are similar to those of steels used for piston rings. After confirming the improvement of the tribological properties of these steels by ion implantation, we will continue our work to check their suitability as a material for the piston rings.In many cases, ion implantation leads to enhancing the tribological wear resistance of a given material, particularly under poor lubrication. In addition, it leads to a significant reduction in the friction factor of the implanted material. The objective of this study is to determine the effect of nitrogen ion implantation on enhancing the tribological properties of steel ring pistons.

2. Materials and methods

Given the design of combustion engine pistons and the arrangement of piston rings in such engines, top compression rings located in the vicinity of the piston crown are the most exposed to tribological wear. They also carry the highest thermal loads and operate at reduced lubrication, which means that they often operate under boundary friction. As a result, their surface should be resistant to tribological wear. The steel used for such pistons should have high mechanical strength. Based on this criterion, we selected two steels for tribological tests: Hardox 450 and Raex 400. The properties of MS 064 steel used for ring pistons by the MAHLE company [6] and those of the tested steels are compared in Table 1, while their chemical compositions are given in Table 2.

Table 1. Properties of the tested steels.

	Hardness (HRC)	Tensile strength (MPa)	Young modulus (GPa)	Density (g/cm^3)
MS 064	48-54	1125-1325	210	7.80
Hardox 450	48	1400	210	7.85
Raex 400	41	1250	210	7.85

Table 2. Chemical composition of the tested steels.

	C, %	Si, %	Mn, %	P, %	S, %	Cr, %
MS 064	0.50	1.20	0.80	0.035	0.040	0.80
Hardox 450	0.26	0.70	1.60	0.025	0.010	1.40
Raex 400	0.20	0.70	1.70	0.030	0.015	1.50

To enhance their tribological properties, the tested steels were subjected to ion implantation. This physical process consists in “bombing” the workpiece surface with ions of other elements. Atoms of the implanted elements impact the surface of the material with such a high kinetic energy that they form a structure with different tribological and chemical properties. Due to the permanent location of the atoms in the structure, the material is not exposed to the risk of delamination, as is the case with coatings applied to the surface. This process proceeds in low temperatures (<200°C), so the crystal structure of

IOP Conf. Series: Materials Science and Engineering **148** (2016) 012044 doi:10.1088/1757-899X/148/1/012044

material does not change, and outside dimensions of the workpiece do not change due to ion implantation. Since there are many reported cases of enhanced tribological properties of materials due to ion implantation [7, 8], it is necessary to investigate the effect of this process on the tested steels, Hardox 450 and Raex 400.

The samples were implanted using nitrogen ions with an energy of 65 keV. The dose of implanted ions was set to $1 \cdot 10^{17}$ N^{+}/cm^{2}. After that, the samples were subjected to tribological tests on a pin-on-disc stand using the Nano Tribometer (NTR2) manufactured by Anton Paar. As counter sample, we used a ball with a diameter 0.5 mm made of WC (wolfram carbide). A load of 0.5 N was applied to the ball. Each sample made 5000 work cycles (revolutions) on the test stand. As a result, we could determine variations in the friction factor for individual samples under dry friction conditions.

Next, the wear trace was measured using the Taylor Hobson Form Talysurf 50mm Intra profile measurement gauge. The measurements provided profilograms enabling determination the mean wear and the depth of wear trace. The wear was defined as the mean surface of the cross section of contact between the sample and the counter-sample.

In order to compare the effect of ion implantation on the tribological properties of the tested samples, we performed tests on unimplanted samples according to the above design.

3. Results and discussion

Nitrogen ion implantation leads to changes in the surface layer properties of the tested samples, as clearly shown in Figure 1. The wear trace of the implanted samples indicates a change in the character of wear. The unimplanted Hardox 450 sample (Figure 1a) shows the dominance of adhesive-abrasive wear. A dominant trend to be observed after implantation is oxidizing wear (Figure 1b). The surface of the trace is covered with a thick layer of oxides. A similar trend can be observed regarding the variations in wear of the implanted Raex 400 sample. As can be seen in Figure. 1c, the trace bottom of the unimplanted sample is covered with fragments of peeled metal. Following the implantation, the trace bottom is covered with a uniform layer of oxides.

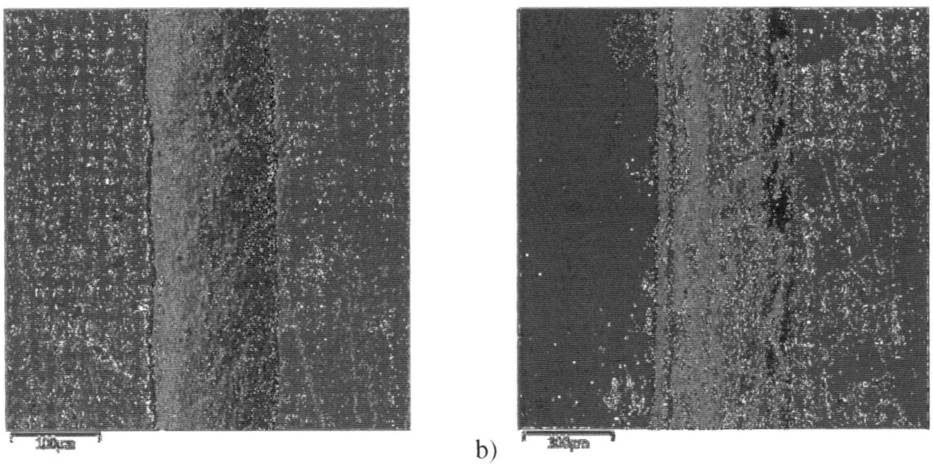

a) b)

IOP Conf. Series: Materials Science and Engineering **148** (2016) 012044 doi:10.1088/1757-899X/148/1/012044

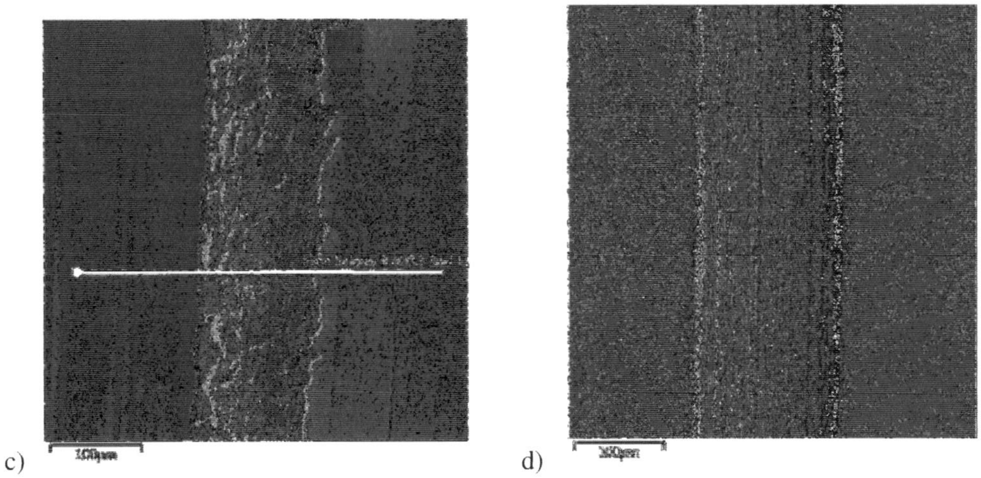

Figure 1. Microphotography of wear trace of Hardox 450 sample – a) before implantation, b) after nitrogen ion implantation and Raex sample - c) before implantation and d) after nitrogen ion implantation.

The measurements provide data about variations in the friction coefficient for individual samples. It was observed that after implantation the friction factor was lower for both Hardox 450 and Raex 400 than was the case with the unimplanted samples.
The friction coefficient of the unimplantedHardox 450 steel sample reaches 1.0 already after~700 cycles, and then increases to 1.4 after ~3500 cycles. In contrast, the friction factor of the implanted sample does not exceed 0.6 until ~2900 cycles. Importantly, the friction coefficient of the implanted sample does not exceed the value of 1.0, as can be seen in Figure 2.

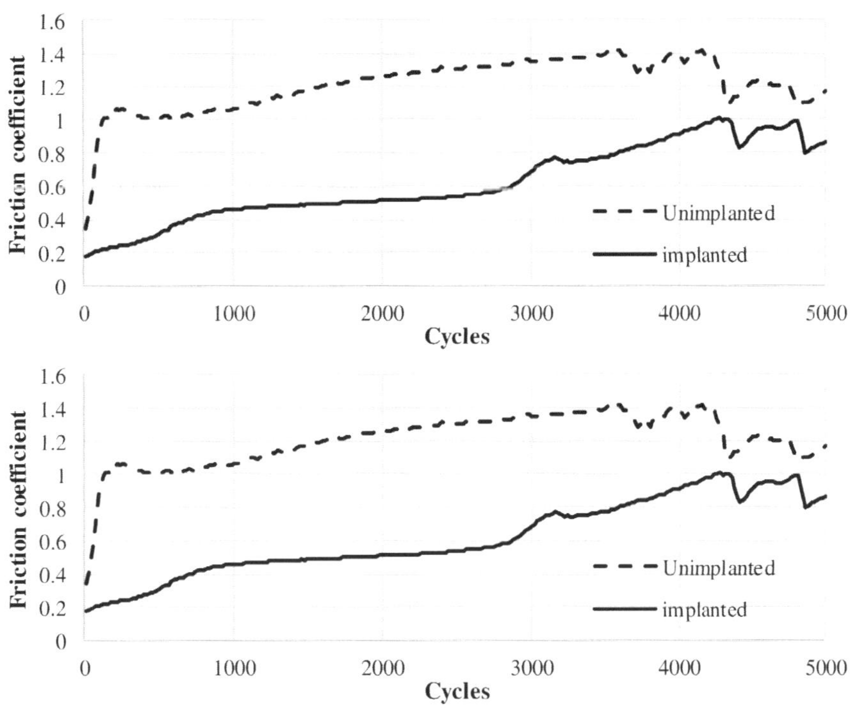

Figure 2. Variations in friction coefficient for Hardox 450 steel samples.

The unimplanted Raex 400 steel sample has a much higher friction factor than the implanted sample. The friction coefficient is 0.6 after ~350 cycles. Another characteristic point is the value of 0.75; after exceeding this value, the friction coefficient rapidly increases reaching 1.2 after ~3800 cycles. Until the end of the test, the friction coefficient does not decrease below 1.2. The friction coefficient of the implanted sample is not higher than 0.45 in the entire measuring range, as can be seen in Figure 3.

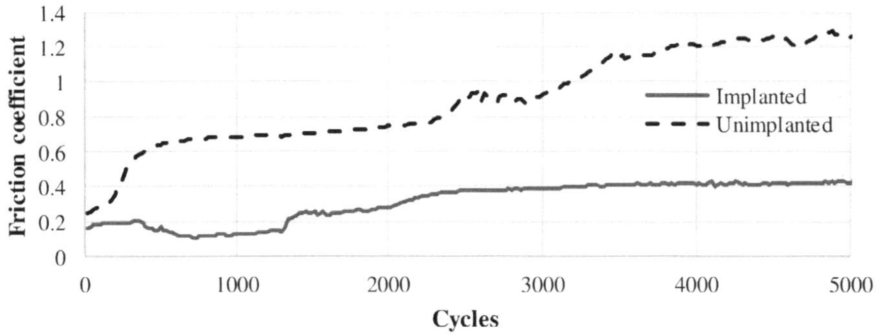

Figure 3. Variations in friction factor for Raex 400 steel samples.

The curve illustrating variations in the friction coefficient of the implanted sample reveals that the friction coefficient decreases after ~350 work cycles. This is due to the nature of ion implantation, as the layer with the highest concentration of implanted ions is not located on the surface of the sample but at a certain depth depending on the energy and type of implanted ions [9].

The significant decrease in friction factor is connected with reduced friction. Reduced friction has a positive effect on mechanical efficiency of the engine. It is estimated that the friction of a top compression ring consumes 1.6% of the fuel of a vehicle [5]. An 80% decrease in the friction coefficient will lead to reducing fuel consumption by 1.28%.

Following the tribological test, the wear of individual samples was assessed. The assessment was done based on the mean surface area of wear trace and the mean wear depth (Figure 4.). The results are given in Table 3.

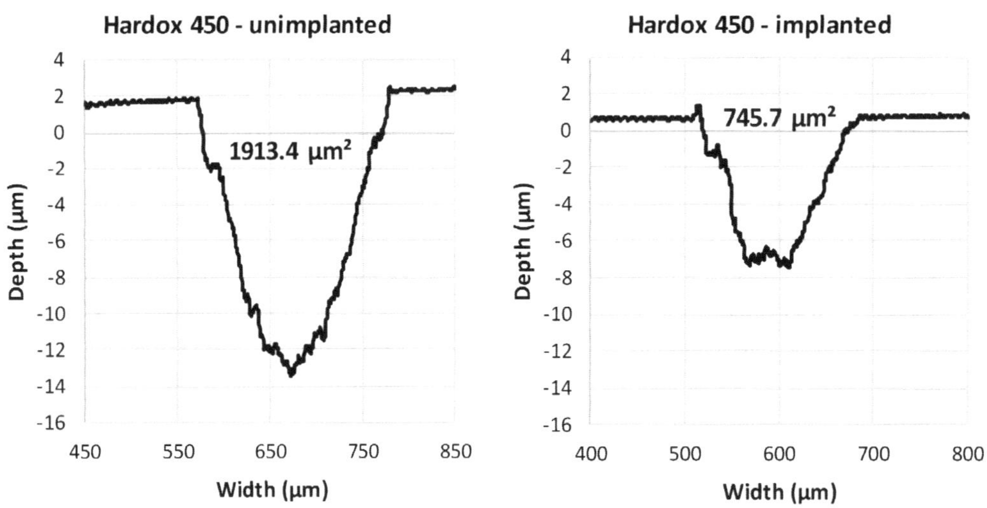

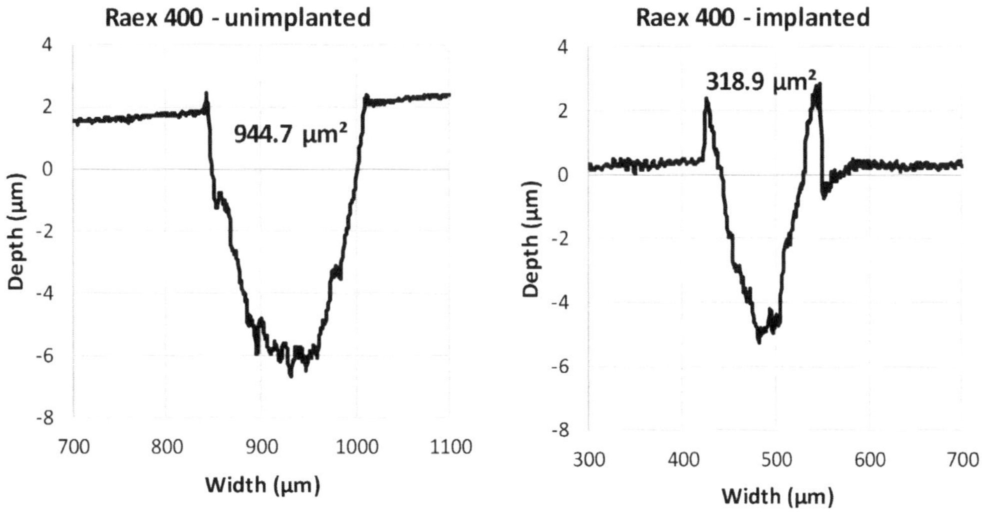

Figure 4. Exemplary profilograms of the wear tracks left by the balls on Hardox 450 and Raex 400 samples.

Table 3. Wear of the tested steels

Material		Mean wear (μm^2)	Mean wear depth (μm)
Hardox 450	unimplanted	1911.74	15.51
	implanted	739.54	7.99
Raex 400	unimplanted	894.57	8.23
	implanted	312.97	5.26

The results of wear demonstrate that ion implantation led to reducing the wear of Hardox 450 steel sample by over 60%. Moreover, the wear depth is almost two times lower than that of the unimplanted sample.
Similar results were obtained for Raex 400 steel samples. The mean friction wear decreased by nearly 64% and the mean wear trace depth decreased by 36%.

4. Conclusions

The results of the study investigating the effect of nitrogen ion implantation on the tribological properties of Hardox 450 and Raex 400 steels demonstrate that the process can be used for machining piston rings. The tribological results show a significant drop in the friction coefficient for both Hardox 450 and Raex 400. The applied nitrogen ion implantation clearly increased the tribological wear resistance of the tested materials. Given its characteristics, this process can be used as finish treatment. The tribological properties of the tested steels obtained due to ion implantation remain unchanged despite the mechanical deformation of rings during assembly. The results of tribological tests entitle us to prepare the piston rings made of implanted Hardox 450 and Raex 400 steels and verify their suitability to work in a real engine.

References

[1] Wajand J A 2005 *Tłokowe silniki spalinowe średnio- i szybkoobrotowe* (WNT, Warszawa)

236-43
[2] Andersson P, Tamminen J and Sandström C E 2002 *Piston ring tribology A literature survey* (Finland) 9-13
[3] Motor Service 2013 *Pierścienie tłokowe do silników spalinowych* (Germany) p 5, 17, 19-22
[4] Napadłek W, Balicki P 2010 *Teksturowanie laserowe warstwy wierzchniej pierścieni tłokowych Silnika spalinowego* Transcomp XIV 2341-55
[5] Kennedy M, Hoppe S, Esser J 2012 *Piston Ring Coating Reduces Gasoline Engine Friction*MTZ worldwide**73** 40-43
[6] MahleGmbh 2016 *Piston rings* (German/ Springer Fachmedien Wiesbaden) p 22
[7] Budzynski P, Kara L, Küçükömeroğlu T and Kaminski M 2015 *The influence of nitrogen implantation on tribological properties of AISI H11 steel* Vacuum**122** 230-35
[8] Budzynski P 2015 *Long-range effect of nitrogen ion-implanted AISI 316L stainless steel* Nuclear Instruments and Methods in Physics Research Section B: Beam Interactions with Materials andAtoms**342** 1-6
[9] Budzynski P 2010 *Problematyka tarcia i zużycia tworzyw metalowych implantowanych jonowo w technologii maszyn* (WPL, Lublin) 15-24

Long-range effect of ion implantation of Raex and Hardox steels

P Budzyński[1], M Kamiński[1], A Droździel[2], M Wiertel[2]

[1]Faculty of Mechanical Engineering, Lublin University of Technology, Nadbystrzycka 36, 20-618 Lublin, Poland
[2]Institute of Physics, M. Curie-Sklodowska University, Place M. Curie-Sklodowskiej 1, 20-031 Lublin, Poland

E-mail: p.budzynski@pollub.pl

Abstract. Ion implantation involves introduction of ionized atoms of any element (nitrogen) to metals thanks to the high kinetic energy that they acquired in the electric field. The distribution of nitrogen ions implanted at E = 65 keV energy and D = $1 \cdot 10^{17}$ N^{+}/cm^{2} fluence in the steel sample and vacancies produced by them was calculated using the SRIM program. This result was confirmed by RBS measurements. The initial maximum range of the implanted nitrogen ions is ~0.17 μm. This value is relatively small compared to the influence of nitriding on the thickness surface layer of modified steel piston rings. Measurements of the friction coefficient during the pin-on-disc tribological test were performed under dry friction conditions. The friction coefficient of the implanted sample increased to values characteristic of an unimplanted sample after ca. 1500 measurement cycles. The depth of wear trace is ca. 2.4 μm. This implies that the thickness of the layer modified by the implantation process is ~2.4 μm and exceeds the initial range of the implanted ions by an order of magnitude. This effect, referred to as a long-range implantation effect, is caused by migration of vacancies and nitrogen atoms into the sample. This phenomenon makes ion implantation a legitimate process of modification of the surface layer in order to enhance the tribological properties of critical components of internal combustion engines such as steel piston rings.

1. Introduction

A rapid development of automotive technology requires designers to use modern materials. They must have high strength and low specific weight in order to reduce the overall weight of the vehicle. The materials used to construct the most loaded components such as the components of internal combustion engines are often subjected to various treatments to enhance their mechanical properties [1]. There is a widespread application of various coatings, for example nitriding or chrome-plated piston rings. Another example is the doping with other materials, for example by welding as is the case with the surfacing of stellite valve face and valve seat [1].
One of the known methods for material modification in terms of tribological, mechanical and chemical properties is ion implantation. It is a physical process which comprises implanting atoms of one element into the structure of another material. In the absence of specific requirements for the implantable element and the workpiece, it is possible to implant atoms of any element in any structure of the material [6].

IOP Conf. Series: Materials Science and Engineering **148** (2016) 012045 doi:10.1088/1757-899X/148/1/012045

Ion implantation is performed in a device called ion implanter. Atoms of an element are ionized and then accelerated along the electric field to reach the high-speed power from tens to hundreds of keV and to impact the surface of the workpiece, thereby permanently digging into its structure. This process is performed under vacuum and at relatively low temperatures (<200° C), so that heat causes no changes to the material structure. Another feature of ion implantation is that it does not affect the changes in external dimensions of the workpiece, so it can be applied as the final step in the production of components.

The influence of ion implantation on the mechanical, tribological [9] and chemical properties of the material is well known. There are a lot of works on the impact of the implantation process with respect to enhancing properties of different materials, such as [4] and [5]. Nitrogen atoms are most often used for implantation. In addition, carbon, titanium, boron and phosphorus are used, too. The use of ion implantation of nitrogen results in higher micro-hardness, lower friction coefficient, reduced tribological wear and higher corrosion resistance of materials [7].

In spite of so many advantages of ion implantation, the process is not often used in the engineering industry due to the fact that the atoms implanted into the element during implantation penetrate the structure of the material to a depth of ~ 0.2μm. However, when analysing the results of studies on the effect of implantation on the properties of materials, one can often encounter information that the impact of implanted ions is visible at much greater depths than the depth resulting from the implantation process [10]. This effect is called the long-range effect and it extends the life of the implanted element as a result of ion implantation. There is no definitive cause of the long-range effect. It is believed that this effect occurs due to migration of atoms implanted into the material during the friction process, as found during the tribological test reported in [3].

Since the application of ion implantation seems to result in a much longer service life than initially expected, this process appears suitable for producing machine parts. This paper presents the long-range effect of ion implantation on the tribological properties of Hardox 450 and Raex 400 steels. These steels are similar to the steel used for piston rings in an internal combustion engine, so they are used to investigate the suitability of using ion implantation in the construction of vehicle engines.

2. Material and methods

To study the long-range effect of ion implantation on the tribological properties of the material, we performed tests on Raex 400 and Hardox 450 steels. They have been selected due to the fact that their physical properties are very similar to those of the steel used in the production of piston rings for internal combustion engines. The mechanical properties of the selected steels and those of the steel used for producing rings [8] are given in Table 1, while Table 2 shows the chemical composition of the tested steels.

Table 1. Properties of the tested steels.

	Hardness (HRC)	Tensile strength (MPa)	Young modulus (GPa)	Density (g/cm^3)
MS 064	48-54	1125-1325	210	7.80
Hardox 450	48	1400	210	7.85
Raex 400	41	1250	210	7.85

Table 2. Chemical composition of the tested steels.

	C, %	Si, %	Mn, %	P, %	S, %	Cr, %	Ni, %	Mo, %	B, %
Hardox 450	0.26	0.70	1.60	0.025	0.010	1.40	1.50	0.60	0.005
Raex 400	0.20	0.70	1.70	0.030	0.015	1.50	0.40	0.50	0.004

Test samples were made of a steel sheet metal. They were cut by a high pressure water jet in order to avoid structural changes induced by high temperature typical of waste machining methods. The samples had the shape of discs with a diameter φ = 25 mm. The sample surface was polished to Ra = 0.01. The samples were implanted using nitrogen ions. The implantation process was run with the power E = 65 keV, and the dose of implanted ions was set to $D=1\cdot10^{17}$ N^+/cm^2.

After that, the samples were subjected to tribological tests on a pin-on-disc stand using the Nano Tribometer (NTR2) manufactured by Anton Paar. As counter sample, we used a ball with a diameter 0.5 mm made of WC (tungsten carbide). A load of 0.5 N was applied to the ball. As a result, we could determine variations in the friction coefficient for individual samples under dry friction conditions.

Next, the wear trace was measured using the Taylor Hobson Form Talysurf 50mm Intra profile measurement gauge. The measurements provided profilograms enabling determination of the mean wear and the depth of wear trace. The wear was defined as the mean surface of the cross section of contact between the sample and the counter-sample.

For the Raex 400 steel sample, the number of cycles (revolutions) was determined based on the value of friction coefficient. The threshold friction coefficient characteristic of the unimplanted sample was set. After reaching this value by the implanted sample, the test was stopped and the average cross-sectional area and the depth of wear truck were measured. The same number of cycles was performed for both samples of Hardox 450.

3. Results and discussion

For the purpose of preliminary determination of ion-implantation, a numerical simulation was performed. The distribution of implanted ions and point defects (gaps) at various depths can be calculated with the use of numerical programs. SRIM is the most commonly used program for this type of calculations [11]. Usually, numerical results correspond well with experimental findings obtained with e.g. RBS or SIMS methods.

Figure 1 shows the calculated distribution of nitrogen ions implanted at energy E = 65 keV and fluence $D = 1\cdot10^{17}$ N^+/cm^2 into a Raex 400 sample and the vacancies generated by them. It is clear that the calculated range of the implanted nitrogen ions does not exceed 0.17 μm, and the maximum concentration of gaps is located at a depth of ca. 0.05 μm.

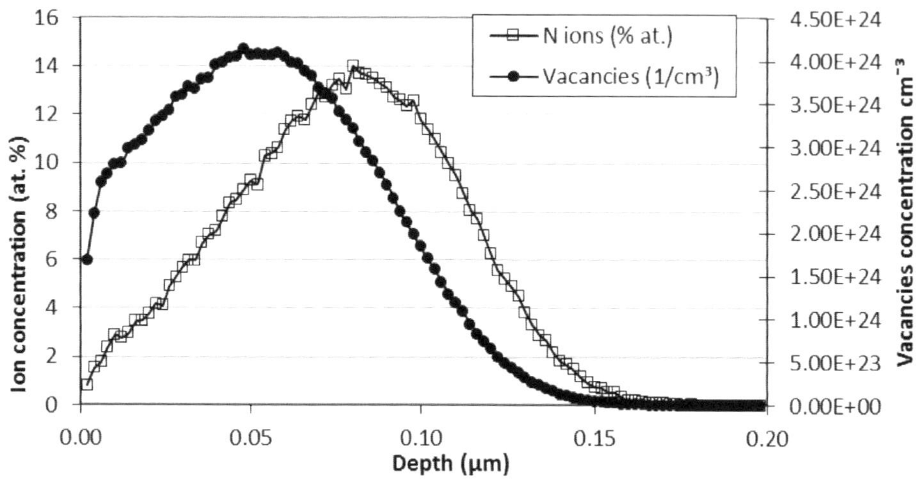

Figure 1. Calculated distribution of nitrogen ions implanted at energy E = 65 keV and fluence $D = 1 \cdot 10^{17}$ N+/cm^2 into Raex steel and the generated vacancies.

The results of friction coefficient for the Raex 400 sample are shown in Figure 2. The implantation of nitrogen leads to a significant reduction in the coefficient of friction.

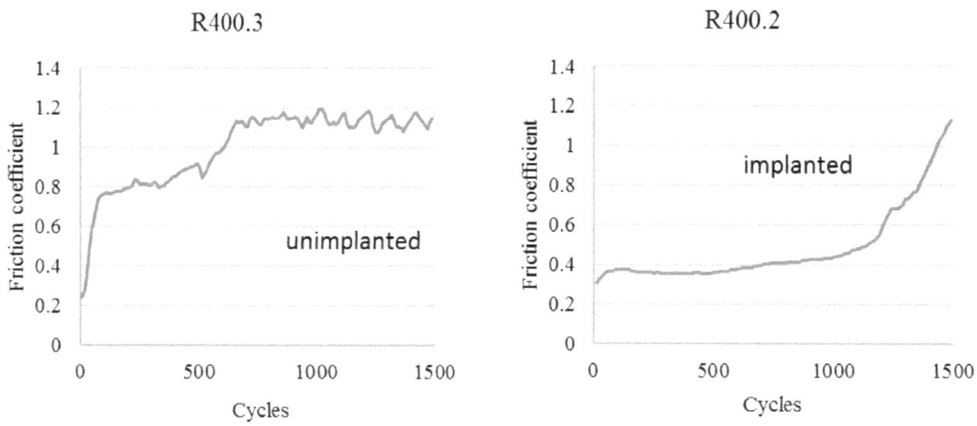

Figure 2. Friction coefficient of Raex 400 sample before and after implantation of nitrogen ions.

This means that the thickness of the modified layer in the implantation depth can be determined from the track produced during the test when the friction coefficient is lower than that of the unimplanted sample. The testing of the samples implanted was stopped when the value of the friction coefficient reached the value characteristic of the unimplanted sample, that is $\mu = 1.15$. The cross section of the track is measured using a profilometer (Figure 3).

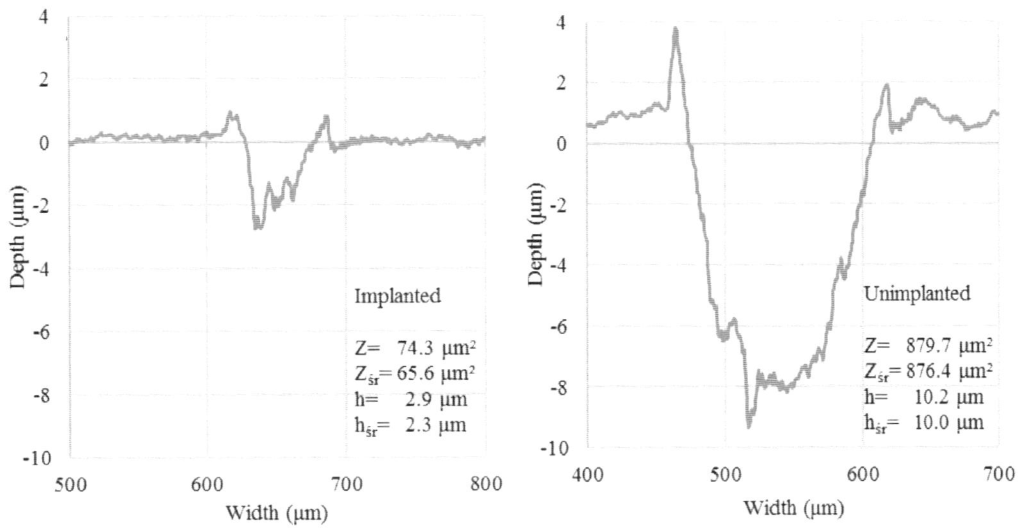

Figure 3. Profilogram of a track left by the ball on Raex 400 sample in a 1500-cycle test.

As shown in Figure 2, the value of the friction coefficient of the implanted sample is equal to the values of the friction coefficient of the unimplanted sample after ca. 1500 cycles. The depth of the wear track is approx. 2.3 µm. This indicates that the thickness of the layer modified during the implantation of the Raex 400 sample is ca. 2.3 µm and it exceeds the initial range of the implanted ions by an order of magnitude. This effect is known as the long-range implantation. A similar effect of implantation of nitrogen ions with an energy of 65 keV and a fluence of $1 \cdot 10^{17}$ N^+/cm^2 leading to reduction in the friction coefficient was observed for the Hardox 450 sample (Figure 4).

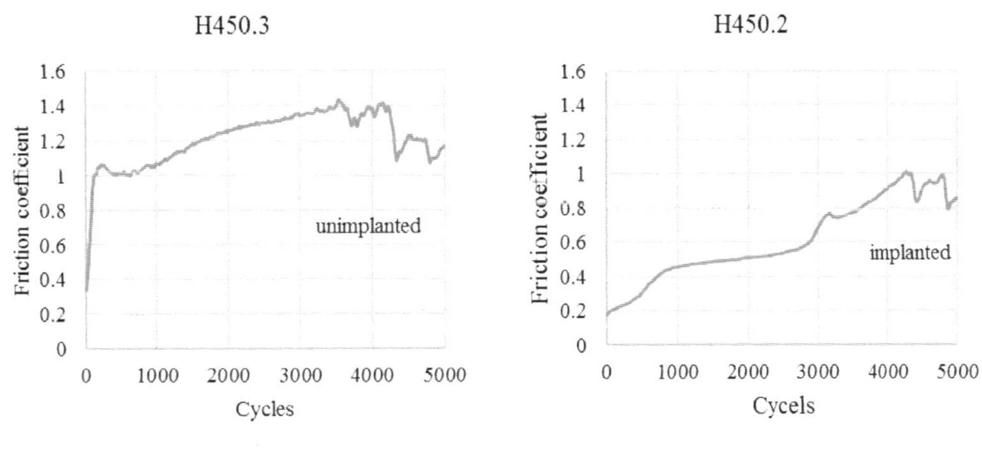

Figure 4. Friction coefficient of Hardox 450 steel, unimplanted and implanted with nitrogen ions at energy E = 65 keV and fluence D = $1 \cdot 10^{17}$ N^+/cm^2.

The reduction in the friction coefficient after nitrogen implantation in the Hardox 450 sample occurs within ca. 4000 measurement cycles. The depth of wear track is 2.1 µm. In the case of the Hardox 450 sample, the thickness of the layer with tribological properties enhanced by an order of magnitude exceeds the range of the implanted nitrogen ions.

The long-range effect is caused by the migration of vacancies and nitrogen ions into the sample. The migration of vacancies deeper inside the sample is caused by the pressure generated by a high concentration thereof at a depth of ca. 0.05 μm, which is only slightly lower than the maximum concentration of the implanted nitrogen ions. In turn, the migration of nitrogen atoms is caused by diffusion driven by the gradient of the chemical potential, which depends on the temperature, pressure, and concentration [6].

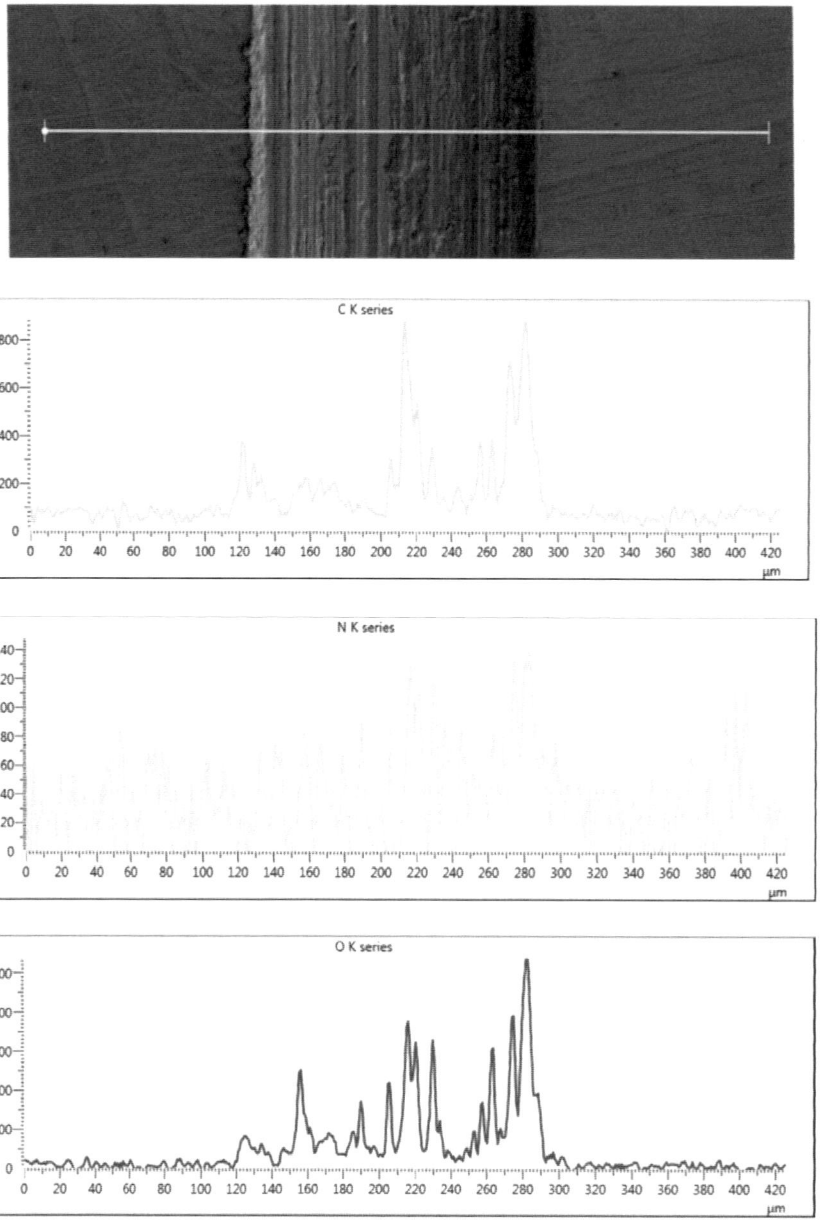

Figure 5. Microphotograph of a fragment of implanted Raex 400 sample and the scan line and distribution of carbon, nitrogen, and oxygen ions.

The distribution of oxygen, carbon, and nitrogen atoms on the sample surface and on the track bottom was examined by EDS method. As shown in Figures 5 and 6, the content of carbon and nitrogen on the track bottom introduced during the implantation process is greater than that on the sample surface. This implies that the migration of nitrogen and carbon atoms into the sample facilitates radiation damage and an increase in the diffusion coefficient caused by a local temperature increase during the tribological test. The increased numbers of carbon atoms in the implanted sample requires a comment. The carbon originated from vacuum oil vapours deposits on the sample during implantation although the pressure in the collector chamber was p = $2 \cdot 10^{-6}$ Pa. The temperature of the sample during the implantation process did not exceed 50°C to prevent implantation heat effects, which could have contributed to deposition of oil vapours on the relatively cool sample.

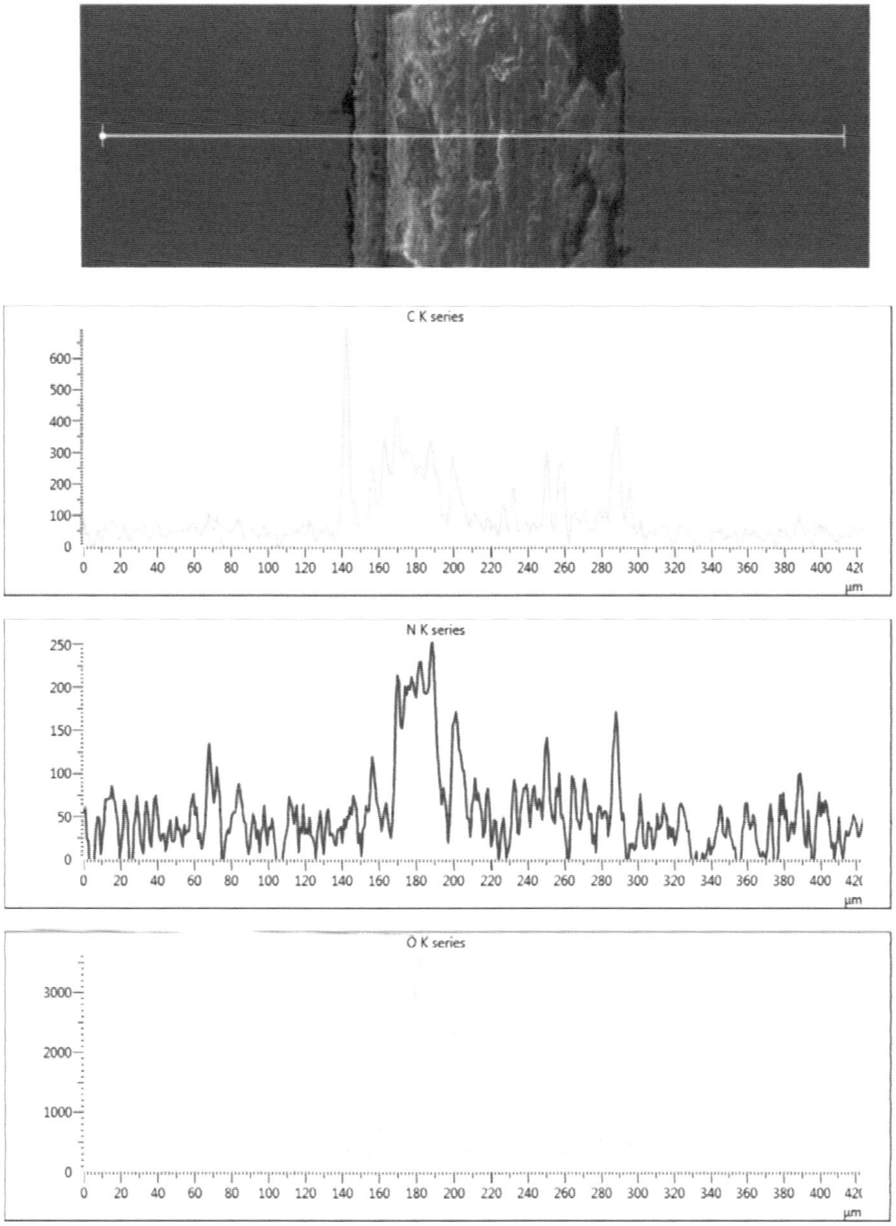

Figure 6. Microphotograph of a fragment of implanted Hardox 450 sample and the scan line and distribution of carbon, nitrogen, and oxygen ions.

4. Conclusions

The results have demonstrated that the long-range effect is caused by the movement of both the implanted nitrogen and dopant atoms in the vicinity of the friction zone. The study is the first to document the diffusion of carbon atoms. In addition, the increased content of oxygen in the bottom trace shows the dominant oxidizing wear of the Raex 400 and Hardox 450 samples after the implantation of nitrogen at an energy of 65 keV and a dose of $1 \cdot 10^{17}$ N^{+}/cm^{2}. The results of the friction coefficient have revealed that the effect of ion implantation is several times greater than assumed. With the ion implantation of nitrogen, it is possible to extend the life of the implanted part of the machine. The implantation can be used to enhance the mechanical properties and chemical combustion of vehicle engine components.

References

[1] Adamiec P and Dziubiński J 2005*Wytwarzanie i właściwości warstw wierzchnich elementów maszyn transportowych* (WPŚ, Gliwice) 90, 174
[2] Budzynski P 2010 *Problematyka tarcia i zużycia tworzyw metalowych implantowanych jonowo w technologii maszyn* (WPL, Lublin) 15-24
[3] Budzynski P 2015 Long-range effect of nitrogen ion-implanted AISI 316L stainless steel *Nuclear Instruments and Methods in Physics Research Section B: Beam Interactions with Materials andAtoms* **342** 1-6
[4] Budzynski P, Kara L, Küçükömeroğlu T and Kaminski M 2015 The influence of nitrogen implantation on tribological properties of AISI H11 steel *Vacuum***122** 230-35
[5] Budzyński P, Tarkowski P and Penkała P 2001 Influence of nitrogen ion implantation on tribological properties of tool steel NC10 *Vacuum***63** 731-736
[6] Burakowski T and Wierzchoń T 1995 *Inżynieria powierzchni metali* (WNT, Warszawa) 106-12, 392-448
[7] Guo Z, Pang X, Yan Y, Gao K, Volinsky A A and Zhang T Y 2015 CoCrMo alloy for orthopedic Implant application enhanced corrosion and tribocorrosion properties by nitrogen ion implantation *Applied surface science* **347** 23-34
[8] Mahle Gmbh 2016 *Piston rings* (German/ Springer Fachmedien Wiesbaden) p 22
[9] Oñate J I, Alonso F and Garcîa A 1998 Improvement of tribological properties by ion implantation *Thin Slid Films* **317** 471-76
[10] Sharkeev Y P and Kozlov E V 2002 The long-range effect in ion implanted metallic materials: dislocationstructures, properties, stresses, mechanisms*Surface and Coatings Technology* **158-159** 219-24
[11] Ziegler J F 2004 SRIM-2003 steel *Nuclear Instruments and Methods in Physics Research Section B: Beam Interactions with Materials and Atoms* **219-20** 1027-36

IOP Conf. Series: Materials Science and Engineering **148** (2016) 012046 doi:10.1088/1757-899X/148/1/012046

The influence of nitrogen ion implantation on microhardness of the Stellite 6 alloy

P Budzyński[1], M Kamiński[1], K Pałka[1], A Droździel[2] and M Wiertel[2]

[1]Faculty of Mechanical Engineering, Lublin University of Technology, Nadbystrzycka str. 36, 20-618 Lublin, Poland
[2]Institute of Physics, M.Curie-Skłodowska University, M.Curie-Sklodowska sq. 1, 20-031 Lublin, Poland

E-mail: p.budzynski@pollub.pl

Abstract. Cobalt alloys known as Stellite used to produce or surfacing machine elements subjected to combustion gases and heat. They are used a currently in the manufacture of valves and valve seats in internal combustion engines. Because of the small thermal conductivity, stellite may not be subjected heat treatment. In order to improve the mechanical properties of cobalt alloys, samples were implanted with nitrogen ions with 65 keV energy and ion dose of $1{\cdot}10^{16}$, $5{\cdot}10^{16}$, $1{\cdot}10^{17}$ N^+/cm^2. The influence of ion implantation on properties of strength was determined by measuring microhardness using a Vickers hardness test. The measurement results allowed to determine the increase in the microhardness of 20% with dose $5{\cdot}10^{16}$ N^+/cm^2 compared to the sample not implanted. Implantation of nitrogen ions can increase the strength of the valves and the valve seats having Stellite without changing the external dimensions of the final element, and without interfering with its inner structure by low-temperature of modification the surface layer.

1. Introduction

Current internal combustion engines competing with electric motors must be characterised by high efficiency and performance. Manufacturers strive to reduce engine displacement in order to lower the weight of the engine, simultaneously increasing their generated power. This necessitates application of light and extremely durable materials for construction of engines. The efficiency of internal combustion engines depends on their capability to generate and maintain high pressure in the combustion chamber. Besides the piston-ring-cylinder system, valves and valve seats are largely responsible for the tightness of the combustion chamber. They are elements of the valve timing in four-stroke engines designed to facilitate supply of fresh air or the air-fuel mixture to the cylinder; next, they have to ensure tightness during compression and combustion of the load and facilitate removal of exhaust gas from the cylinder [12].

Valves and valve seats are exposed to direct contact with high temperature while sealing the combustion chamber and fuel flow around the valves and valve seats in the exhaust stroke, in particular in the case of exhaust valves [9]. This results inconsiderable thermal loads during heat exhaust through the valves and valve seats into the cylinder head. Also, the work of the valves causes high mechanical loads associated with the impact of the surface of the valve heads against the surface of valve seats.

Additionally, exhaust valves should exhibit high thermal conductivity, heat resistance, a low coefficient of thermal expansion, and resistance to tribological wear and corrosion [11].
Damage to the contact surface between the valve face and the valve seat face induces loss of tightness of the combustion chamber, which is associated with reduction of the efficiency of the engine. To prevent their rapid wear, valves and valve seats are made of materials that are resistant to substantial changes in temperature and to cyclic impact loads related to valve opening and closure. The intake valves are usually made of low alloy steels or chromium steels in the case of highly loaded engines. In turn, exhaust valves are produced from various types of steel, e.g. ferritic-pearlitic, austenitic, and silchromesteels. Given the relatively low hardness of valves and valve faces made of austenitic steels, their strength is improved by hard facing [2], with satellite as the main welding material [1].
The name "stellite" denotes cobalt alloys with chromium, tungsten, carbon, and others. These alloys have a multiphase structure. In the microstructure, the metallographic sections of these alloys exhibit star-like regions (stella –the Latin name for "star"), hence their name. These alloys are resistant to abrasion and activity of organic acids; they are also creep-resistant and heat-resistant materials [2]. They are widely used in aviation, space, and automotive technologies as well as biomedical engineering as implant materials. Stellites are used for production of the highest quality tools and parts of combustion engines and other elements operating in extreme heat conditions. The alloy was developed in the early 20th century and its mechanical properties are still being improved.
The strength properties of stellite-welded valve faces and valve seats can be improved using ion implantation. This process facilitates implantation of ions of any element into the structure of materials, which changes their tribological, chemical, and mechanical properties. There are many well-known examples of improvement of material properties by application of ion implantation described e.g. in publications [4] and [5]. A great advantage is the negligible effect of ion implantation on the dimensions of the treated element; hence, the process can be applied in the final stage of manufacturing of products that already have their final dimensions [3].
Besides improvement of tribological properties, ion implantation contributes to an increase in mechanical strength. This is associated with an increase in the microhardness of the implanted element. The implantation process is accompanied with appearance of compressive stresses and inclusions of nitrides, carbides, and borides. The implantation-induced hardening process depends on the type and dose of implanted ions and the temperature of the implanted material [7].It has been proven that the increase in the microhardness of the implanted material is a permanent change, retained even in the heating process [10].
This paper presents investigations of the effect of nitrogen implantation on the microhardness of a selected stellite type.

2. Characteristics of the material and research methodology

Given the multiple applications of stellite as a material for construction of valves and valve faces, this material was chosen for analysis and assessment. A commercial variety Stellite 6 was chosen. The alloy comprises 27-32 % Cr, 4-6% W, and 0.9-1.4% C. The samples for the analyses were taken from a ϕ25 mm-diameter rod. Sample surfaces intended for the analysis were polished in order to achieve a roughness value of Ra=0.01.

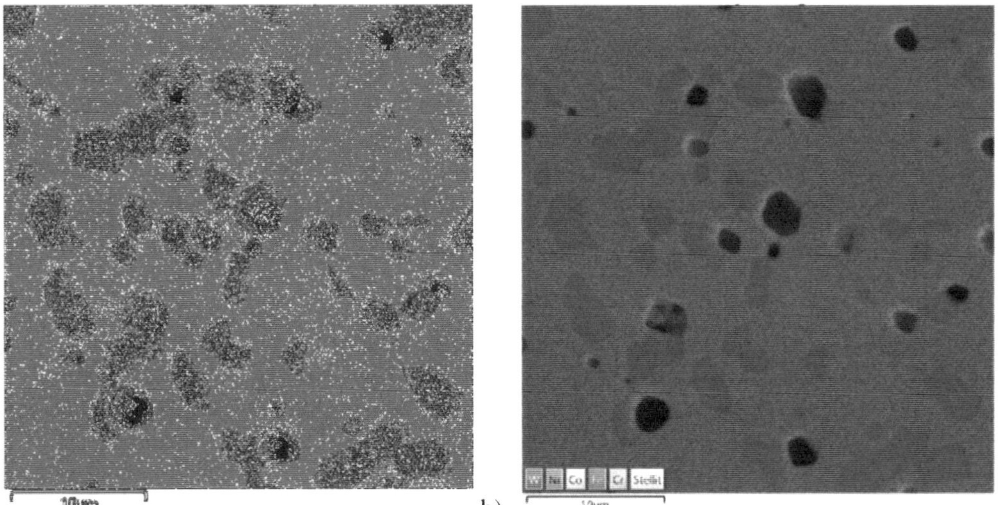

a) b)

Figure 1. Microphotograph of the sample surface (a); analysis of the content of the main elements on the sample surface(b).

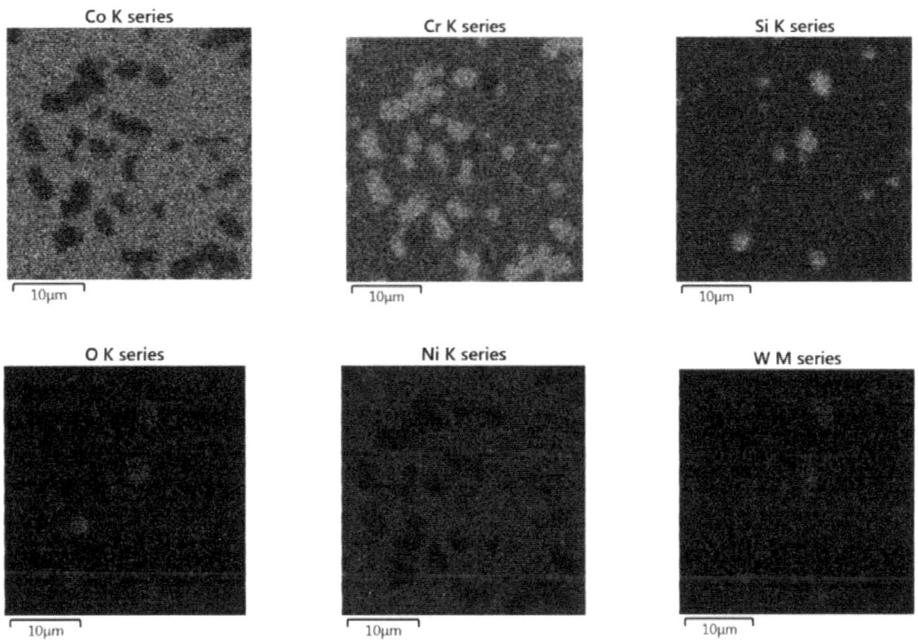

Figure 2. Contents of elements exhibiting the highest heterogeneity of distribution per unit area of the sample surface.

After preparation of the samples, microphotographs of their surfaces were taken with a scanning electron microscope TESCAN Vega 3LMU.The sample surface topography is shown in Figure 1a, and Figure 1b shows the content of the major alloy components. The microphotographs reveal a mosaic and heterogeneous structure characteristic for stellite. The heterogeneous distribution of the alloy components is shown in Figure 2.

IOP Conf. Series: Materials Science and Engineering **148** (2016) 012046 doi:10.1088/1757-899X/148/1/012046

The samples were subjected to the ion implantation process. Nitrogen was the implanted element. The implantation was carried out evenly over the entire surface of the sample at energy 65 keV. The dose of the implanted ions varied between the samples; the values of the fluences of the implanted nitrogen ions are presented in Table 1.

Table 1. Fluences of the implanted ions for individual samples.

Sample	Fluences of implanted ions (N^+/cm^2)
S6.1	$1 \cdot 10^{16}$
S6.2	$5 \cdot 10^{16}$
S6.3	$1 \cdot 10^{17}$
S6.4	unimplanted

After ion implantation, the samples were again analysed under the scanning electron microscope and next microhardness was assessed. The measurement was performed using a microhardness meter FM-800 from Future-Tech. Hardness was assessed with the Vickers method, and the load of 5g (0.049 N) was applied. A microhardness value of $HV_{0.005}$was determined based on the measurements. The implanted samples and the unimplanted sample were analysed in order to compare the microhardness results.

3. Results and discussion

The reassessment of the surface structure revealed that the distribution of the implanted nitrogen ions was uniform over the entire surface (Figure 3), in contrast to the other alloy components.

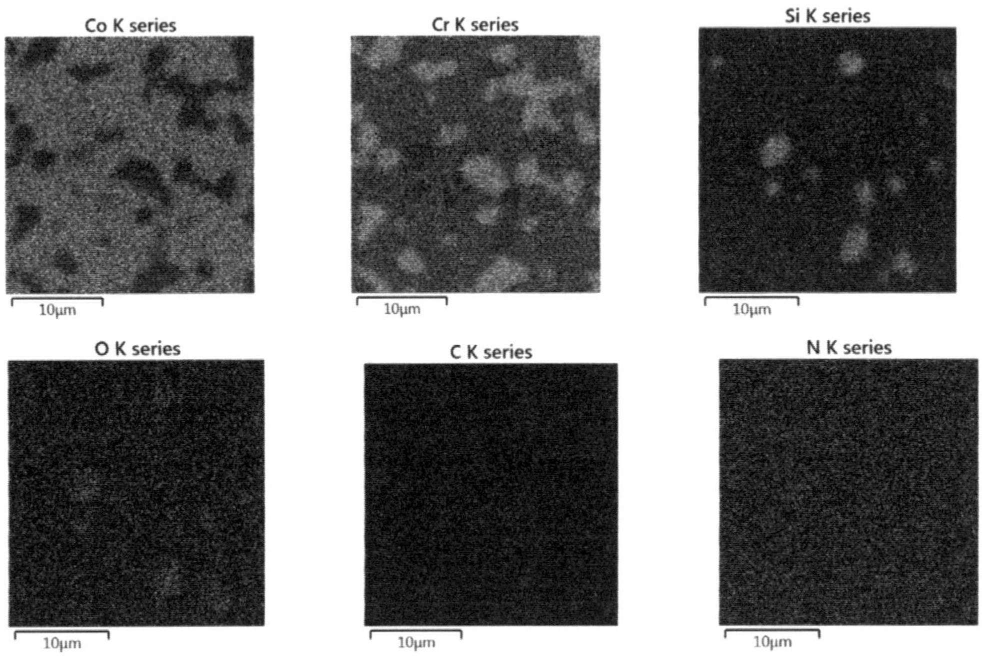

Figure 3. Density of elements per unit area of the analysed sample after nitrogen ion implantation at energy 60 keV and a fluence of $5 \cdot 10^{16}$ ions/cm^2.

The implanted samples were analysed for microhardness. Several measurements were carried out to obtain a reliable result. The results of each measurement are shown in Figure 4.

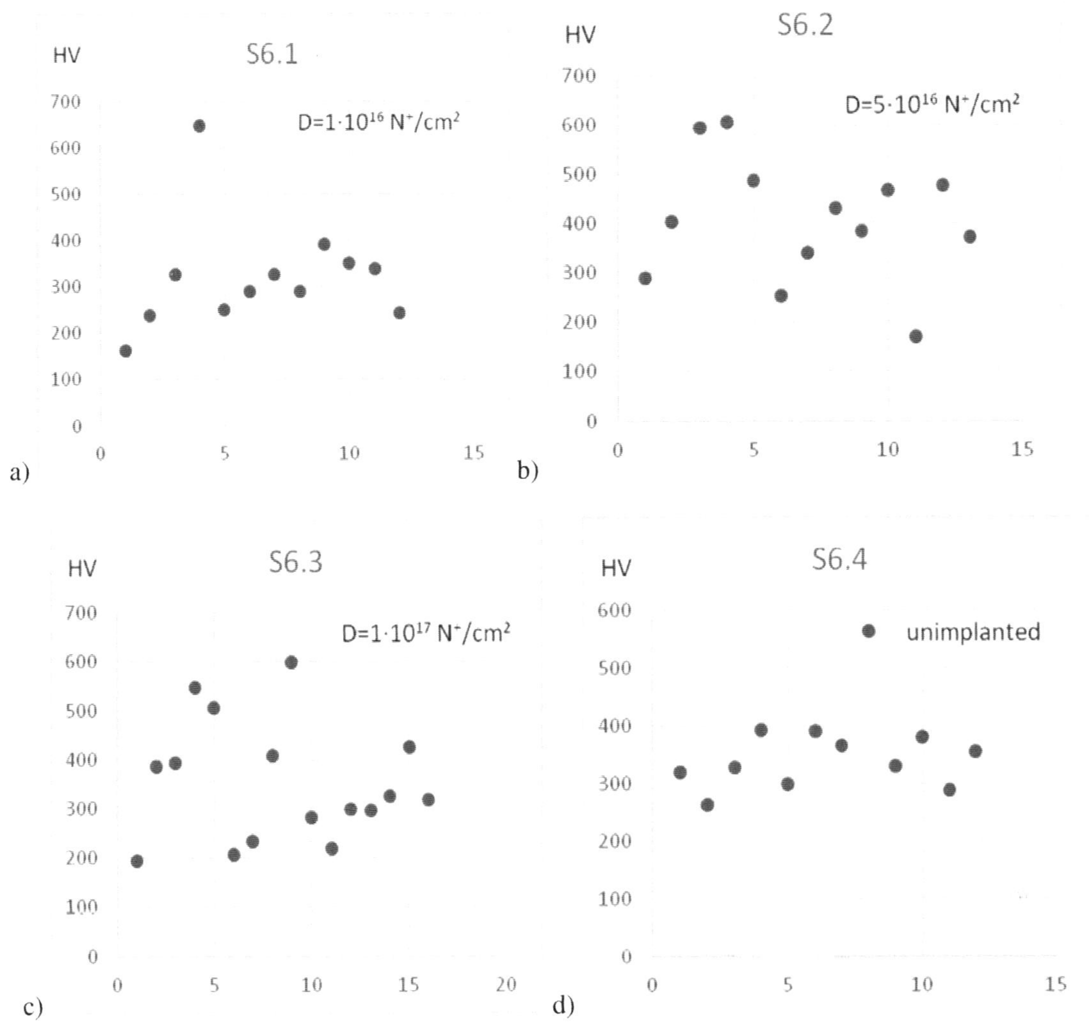

Figure 4. Results of measurements of microhardness for individual samples

There is a noticeable dispersion of the values of microhardness assessed at different sites on the sample surface (Figure 4). This is related to e.g. the different contents of elements contained in the alloy in the different fragments of the sample. The varying microhardness is a result of the differences in the contents of alloy and structural components in the measured areas. In the analysed samples, the chromium, molybdenum, or tungsten solid solution in cobalt enhances the dispersion of hard carbides (primarily Cr_7C_3), which determine the (macroscopic) hardness of the alloy.

During the implantation process, the microhardness of surface sections, which exhibited high hardness in the basic unimplanted material, was increased. The microhardness of weaker fragments exhibited a lower increase, hence the increasing coefficient of variation of the results.

The analysis of the results obtained indicated that some measurements for samples S6.1 and S6.4 (Figure 1a, 1d) apparently deviated from the other values. To determine whether these measurement points were significant for the results of the investigations, the results were analysed taking into account the3S_x criterion. To this end, a mean value of the measurement results was established using equation (1), disregarding the overestimated value.

$$\bar{x} = \frac{1}{n}\sum_{k-1}^{n} x_k \tag{1}$$

In the equation, $\bar{x}$ denotes the mean value and k=1…n are values of each measurement in the series. After calculation of the mean value from all the measurements for each sample, equation (2) was used to calculate the standard deviation $S_{\bar{x}}$. Since the number of the measurements was greater than 10, a formula for calculation of the standard deviation from the mean value of Gaussian distribution was used.

$$S_{\bar{x}} = \left(\frac{\sum_{k-1}^{n}(x_k - \bar{x})^2}{n(n-1)}\right)^{\frac{1}{2}} \tag{2}$$

To meet the 3S_x criterion the analysed measurement results must be within the range defined as (3)

$$\bar{x} - 3S_{\bar{x}} < k < \bar{x} + 3S_{\bar{x}} \tag{3}$$

Based on (3), it was found that the 647 HV value for sampleS6.1 was not within the acceptable range. Similarly, the 563 HV value for sample S6.4 was discarded. The mean values of microhardness measurements for each sample as well as standard deviations and coefficients of variation are shown in Table 2.

Table 2. Averaged results of microhardness measurements for each sample

	Implanted ion fluence (N^+/cm^2)	Mean HV value	Standard deviation	Coefficient of variation (%)
S6.1	$1 \cdot 10^{16}$	292.02	64.02	22
S6.2	$5 \cdot 10^{16}$	405.75	125.33	31
S6.3	$1 \cdot 10^{17}$	353.13	122.13	35
S6.4	unimplanted	337.15	43.31	13

The highest coefficient of variation, i.e. 35%, which corresponds to an average variation, is noted in the sample implanted with the highest ion fluence. The lowest coefficient of variation was obtained for the unimplanted sample. There is a clear correlation between the coefficient of variations calculated for the measurements and the fluence of the implanted ions. The coefficient of variation increased with the increase in the ion dose.

As indicated by the results presented in Table 2, the ion implantation process had a positive effect on the value of microhardness in the analysed samples. The microhardness of the sample implanted with the $1 \cdot 10^{16}$ N^+fluence was lower than that of the unimplanted sample. In the case of ion implantation at fluences of $5 \cdot 10^{16}$and $1 \cdot 10^{17}$, the microhardness value increased by 20% and 5%, respectively.

4. Conclusions

Cobalt-based alloy steels, referred to as stellites, are widely used for improvement of the strength of valve faces and valve seats in internal combustion engines. They exhibit high resistance to mechanical damage as well as acid- and heat-resistance. Application of ion implantation contributes to an increase in the average microhardness of stellite. Nitrogen implantation at a fluence of $5 \cdot 10^{16}$ N^+/cm^2proved the most effective, as it increased the mean microhardness of the sample by 20%, compared with the unimplanted sample. The structure of the stellite alloy largely differentiates the values of surface microhardness depending on where measurements are performed. This phenomenon is enhanced by ion implantation. Besides enhanced microhardness, ion implantation improves the tribological [6] and chemical [8] properties of implanted materials.

References

[1] Adamiec P and Dziubiński J 1998 *Wybrane zagadnienia materiałów kontrukcyjnych i technologii wytwarzania pojazdów* (WPŚ, Gliwice) 65, 282-3

[2] Adamiec P and Dziubiński J 2005 *Wytwarzanie i właściwości warstw wierzchnich elementów maszyn transportowych* (WPŚ, Gliwice) 90-1

[3] Budzyński P 2010 *Problematyka tarcia i zużycia tworzyw implantowanych jonowo w technologii maszyn* (WPL, Lublin) 15-24

[4] Budzyński P 2015 Long-range effect of nitrogen ion-implanted AISI 316L stainless steel *Nuclear Instruments and Methods in Physics Research Section B: Beam Interactions with Materials and Atoms* **342** 1-6

[5] Budzyński P, Kara L, Küçükömeroğlu T and Kamiński M 2015 *The influence of nitrogen implantation on tribological properties of AISI H11 steel* Vacuum **122** 230-35

[6] Budzyński P, Tarkowski P and Penkała P 2001 *Influence of nitrogen ion implantation on tribological properties of tool steel NC10* Vacuum **63** 731-736

[7] Burakowski T and Wierzchoń T 1995 *Inżynieria powierzchni metali* (WNT, Warszawa) 392-448

[8] Guo Z, Pang X, Yan Y, Gao K, Volinsky A A and Zhang T Y 2015 *CoCrMo alloy for orthopedic Implant application enhanced corrosion and tribocorrosion properties by nitrogen ion implantation* Applied surface science **347** 23-34

[9] Lewis R and Dwyer-Joyce R S 2002 *Wear of diesel engine inlet valves and seat inserts* Proceedings of the Institution of Mechanical Engineers, Part D: Journal of Automobile Engineering **216** 205-16

[10] Moćko W 2014 *Analiza wytężenia zaworu silnikowego wykonanego ze stopu tytanu Ti6AL4V obciążonego mechanicznie i cieplnie* Transport samochodowy **4** 55-71

[11] Picard S, Memet J B, Sabot R, Grosseau-Poussard J L, Rivière J P and Meilland R 2001 *Corrosion behaviour, microhardness and surface characterisation of low energy, high current on implanted austenitic stainless steel* Materials science and engineering **303** 163-72

[12] Wajand J A and Wajand J T 2005 *Tłokowe silnlki spalinowe średnio i szybkoobrotowe* (WNT Warszawa) 440-3

IOP Conf. Series: Materials Science and Engineering **148** (2016) 012047 doi:10.1088/1757-899X/148/1/012047

ZeroFlow – new, environmentally friendly method of controlled gas nitriding used for selected car parts

J Kowalska[1] and L Małdziński[1]

[1]Institute of Machines and Motor Vehicles, Poznan University of Technology, Poznan, ul. Piotrowo 3, 60-965, Poland

E-mail: jagoda.daria.kowalska@gmail.com

Abstract. This article presents new method of controlled gas nitriding called ZeroFlow, which is used for nitriding of selected car parts, such as crankshafts, camshafts, piston rings, poppet valve springs and discs, piston pins or nozzles for unit injectors. This article will discuss the essence of controlled gas nitriding process, with an emphasis on the influence of process parameters on results of nitriding process. This information are the basis to understand the issue of the kinetics of nitrided layer growth, and as it follows – for its practical application in designing, regulation and control of nitriding processes using simulation models (simulator of the kinetics of nitrided layer growth). This article will also present the simulator of the kinetics of nitrided layer growth, which supports nitriding using ZeroFlow method – through the use of simulator layers are obtained in the shortest possible time, which is connected with the lowest energy consumption; therefore, nitriding process using ZeroFlow method and simulator of the kinetics of nitrided layer growth is both economical and environmentally friendly.

1. Introduction

ZeroFlow is a new method of controlled gas nitriding, developed at the Poznan University of Technology, which enables precise forming of nitrided layers using nitriding kinetics. It is characterized by much lower consumption of gases, as well as simplification of the nitriding furnace and of the process itself, while full control over the kinetics of nitrided layer growth is still maintained. The same as in the traditional process, ZeroFlow method assumes the use of atmosphere consisting only raw ammonia. However, unlike to the traditional method, kinetics can be controlled by adjusting the chemical composition of the atmosphere in the furnace retort (by adjusting the nitriding potential) through the regulation of ammonia flow rate, or more precisely – through the regulation of ammonia inflow rate by stopping and reactivating ammonia feeding into the furnace retort. It is significant that in ZeroFlow method ammonia inflow rate is temporarily reduced to zero, which makes much easier to control the chemical composition of the atmosphere. To sum up, using a unary atmosphere makes the ZeroFlow method simpler than currently popular methods based on binary atmospheres, but simultaneously it allows to control the kinetics of nitrided layer growth by regulating of ammonia inflow rate [1]. Until now, ZeroFlow nitriding method has been introduced in 26 industrial plants in Poland and many other countries worldwide, such as: Italy, Great Britain, Canada, Sweden, Singapore, South Korea, Germany, Czech Republic, Belarus, Russia, India, Pakistan, Switzerland. 36 industrial installations have been constructed and implemented so far, another 4 are launched. Several thousands of nitriding processes for various parts of machines and vehicles have been conducted with positive results, which is the best verification of this method rightness.

Gathered experience shows that ZeroFlow method enables forming of nitrided layers with respect to the required phase structure, zone thicknesses, and hardness distribution. Through the use of phenomenological models layers are produced in the shortest possible time, which is connected with the lowest energy consumption. Mathematical models of the kinetics of nitrided layer growth additionally allow determining nitrided layer phase structure and thickness of each zone that occurs in it as a function of process parameters: time, temperature and nitriding potential of the atmosphere. Layer obtained this way does not need costly and time-consuming grinding of iron nitrides, which leads to lower consumption of energy and materials. Several times lower consumption of gases in comparison with its consumption in another currently used nitriding processes, and as a result – much lower emission of post-process gases, also indicates on economical and environmentally friendly character of ZeroFlow method.

Thus, ZeroFlow is a new, rapidly evolving method of controlled gas nitriding, which fits into currently prevalent trends in the world: minimization of raw materials and energy consumption, as well as minimization of the negative impact on the environment [1]. However, it should be noted that ZeroFlow method is only general technological and structural solution, which enables carrying out processes in different, strictly selected parameters, such as temperature, nitriding potential of the atmosphere, time, speed of heating, reduction of potential etc. Therefore, each element requires the development of specially dedicated process with specific parameters; this task is supported by simulation models, which considerably speed up the development of new processes [2,3]. It is also worth mentioned that simulations of heat treatment processes are a really important issue and they allow observing various phenomena, which usually are not perceived in industrial practice (sometimes even in experimental research). They can also be a part of regulation and control of the processes [4].

Forming of nitrided layers occurs on steels that are most commonly used in transport machines engineering, like vehicles and aircrafts, or in technological machines and tools (used for example for the wood industry) engineering; therefore, ZeroFlow nitriding is used for parts of machines and vehicles such as: toothed wheels for wind power plants, plates for casting glass bulbs for lamps, inlet sleeves with the pushing piston, used in molds for aluminium pressure die casting, etc. Also many processes were conducted for parts of car engines: crankshafts, camshafts, piston rings, poppet valve springs and discs, piston pins or nozzles for unit injectors. Aim of nitriding of this elements is to obtain layer with higher surface hardness, improved fatigue life and corrosion resistance, increased wear resistant and antigallic properties. According to that, this layer increase durability of nitrided parts, and as a result – durability of machines and vehicles.

2. The essence of nitriding process using ZeroFlow method

Controlled gas nitriding ZeroFlow is carried out in a heightened temperature (usually from range between 490°C÷590°C), using an atmosphere consisting of ammonia. The low durability of the ammonia molecules (at average nitriding temperature ammonia is in unstable thermodynamic state) results in their disintegration after contact with a metal surface, as shown in the following catalytic reaction of dissociation:

$$NH_3 \overset{Fe}{\Leftrightarrow} N + {}^3/_2 H_2 \qquad (1)$$

As a result of atomic nitrogen diffusion into the steel surface, layer with different properties than original material (material of the core) is formed. Its phase structure, zone thicknesses, and as a consequence – its properties – depend both on the type of the steel and the parameters of the nitriding process: time, temperature and nitriding potential of the atmosphere. As a nitriding potential of the atmosphere is understood a ratio of partial pressures of the atmosphere active ingredients. Atomic nitrogen and hydrogen, obtained by dissociation of ammonia, tends to recombine into diatomic molecules as soon as possible, whereas nitrogen diffusion occurs only when it is in atomic state – that means it is necessary to feed fresh amount of ammonia continuously into the furnace retort in order to carry out the process. Therefore, atmosphere in furnace retort during the process consist of nitriding gas NH_3 and denitriding gas H_2, which are the active ingredients of the atmosphere, and inert gas N_2 [1,5,6]. Thus, we can describe nitriding potential of the atmosphere with a following formula:

$$N_p = \frac{p_{NH_3}}{p_{H_2}{}^{3/2}} \tag{2}$$

Depending on the activity of nitrogen in the atmosphere (which is connected with nitriding potential), nitriding of steel may be accompanied by growth of the single-, double- or triple-phase layer (α, γ'+α, ε+γ'+α). The connection between phase structure and parameters such as nitriding potential and temperature is shown on T-N_p phase equilibrium diagram, also known as a Lehrer's diagram. The boundaries of phases occurrence (α/γ, α/γ', γ/ε, γ'/ε) are presented at this diagram as a function of temperature and nitriding potential. It is worth noting that the chemical potentials of nitrogen in particular phases are equal at boundaries between them, and at equilibrium conditions with atmosphere NH_3+H_2 they are equal to chemical potential of nitrogen in this atmosphere.

As far as Lehrer's diagram predict in some measure phase structure of the nitrided layer, it does not deliver any information about the concentration of nitrogen in α, γ' and ε phases as a function of temperature and nitriding potential, however, these data are the basis for description of the kinetics of nitrided layer growth. L. Małdziński, on the basis on his own analysis and research, had determined lines and curves of iso-concentration (constant values of nitrogen concentration) in T-Np phase equilibrium diagram, which allowed for the development of a modified T-N_p-N phase equilibrium diagram (figure 1). It delivers essential information required to develop models of the kinetics of nitrided layer growth [1,4].

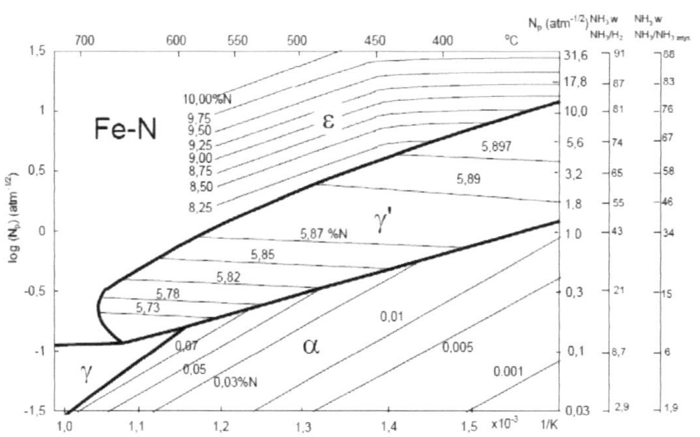

Figure 1. T-N_p-N phase equilibrium diagram.

From T-N_p-N phase equilibrium diagram we can conclude that depending on the nitriding potential of the atmosphere (if process temperature remains the same), nitrided layer can have three different forms. At the nitriding potential N_p^{α}, located in the graph on the range of α phase occurrence, only diffusion zone will increase – that means we get single-phase layer α (figure 2), which does not contain any iron nitrides (γ' and ε). This is the simplest case of nitrided layer growth, analogous to the growth of carburized layer. Furthermore, because of the low activity of nitrogen in α phase (low concentration of nitrogen), growth of this layer is not accompanied by reaction of nitrogen recombination, which is an additional simplification [4].

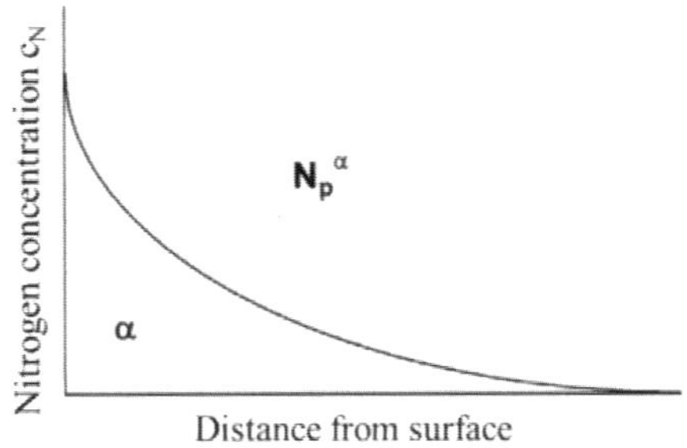

Figure 2. Single-phase layer α structure.

At the higher nitriding potential $N_p^{\gamma'}$, located in the graph on the range of γ' phase occurrence, double-phase layer γ'+α (figure 3) will increase. This is a case much more complex than single-phase layer growth, and it proceeds in several stages. At first stage α phase increases until saturation of the ferrite surface with nitrogen. Then second stage – nucleation of γ' phase – occurs, and immediately after second stage starts third stage – coalescence of γ' crystals, which lasts until a continuous γ' phase is formed on the iron surface. At fourth stage the γ' phase increases with a variable (or more precisely – with an increasing) nitrogen concentration at the surface $c_N^{s/\gamma'}$. After nitrogen at the surface reaches the local equilibrium with an atmosphere, starts the last, fifth stage - γ' phase growth on ferrite (γ'+αlayer growth) with a fixed nitrogen concentration at the surface $c_N^{s/\gamma'}$ [4].

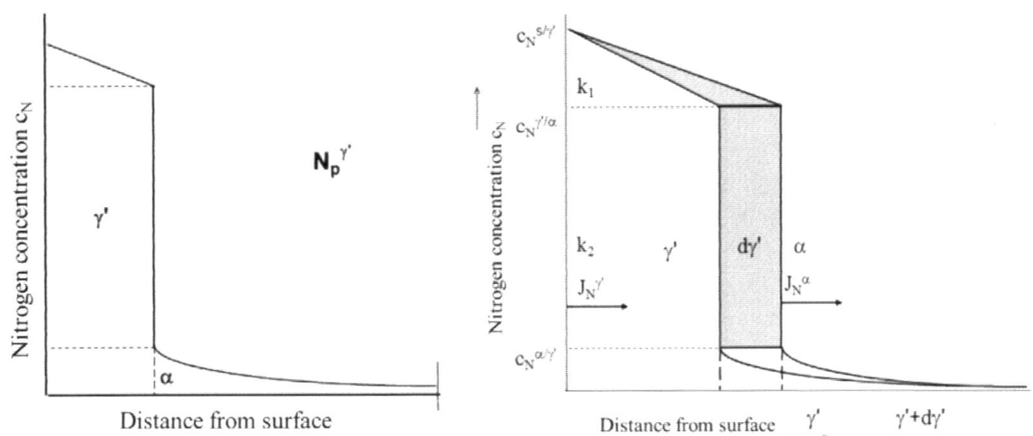

Figure 3. Double-phase layer γ'+α structure.

Growth of the triple-phase layer ε+γ'+α (figure 4) occurs at the nitriding potential N_p^ε, located in the graph on the range of ε phase occurrence, or in other words – at the nitriding potential higher than border potential $N_p^{\gamma'/\varepsilon}$. Triple-phase layer growth in initial stages is the same as the double-phase layer growth, but this case is even more complex. After the γ'+α layer is formed, at the γ' layer surface occurs the nucleation of ε phase; the next stage is coalescence of ε phase crystals until a continuous ε phase is formed. Then starts its growth – first with a variable, and next with a fixed concentration at the surface $c_N^{s/\varepsilon}$. Therefore, ε phase growth is analogous to the γ' phase growth, however, because of accompanying phenomena this case is more difficult to mathematical description. In fact, ε phase growth can occur not only on continuous γ' phase with maximum, border nitrogen concentration at the surface, but also on apexes of that phase embryos before they form a continuous γ' phase. Complications are also connected with nitrogen concentration at the surface $c_N^{s/\varepsilon}$, which does not reach equilibrium value due to the strong desorption of atomic nitrogen to atmosphere [4].

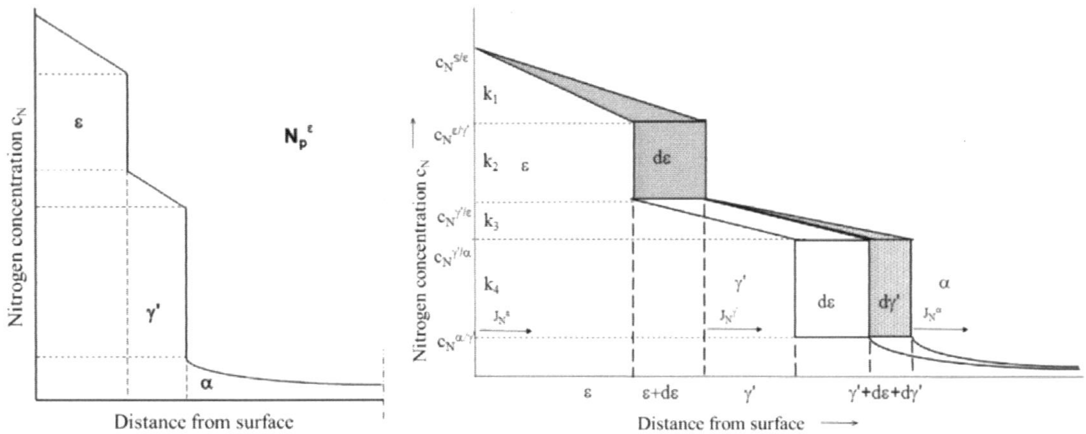

Figure 4. Triple-phase layer ε+γ'+α structure.

Assuming a fixed nitrogen concentration at the surface c_N (which means local equilibrium or stationary state between nitrogen and atmosphere at the surface) and local equilibrium of nitrogen in ε, γ' and α phases on borders γ'/ε and α/γ', the growth of iron nitrides is determined by streams of nitrogen atoms J_N. The values of that streams are connected with gradient of nitrogen concentration at the surface c_N on the border of the individual phases [4,7]. Influence of nitrogen concentration at the surface c_N on the growth of phases has been shown on figure 5.At the at the nitriding potential $N_p^ε(c_N^{s/ε}>c_N^{ε/γ'})$, located in the graph T-N_p-N on the range of ε phase occurrence, both ε and γ' phases will grow. With increasing nitrogen concentration at the surface $c_N^{s/ε}$, ε phase will grow faster and faster; if the stream of nitrogen atoms$J_N^{γ'}$ will be bigger than the stream of nitrogen atoms $J_N^α$, then the γ' phase will be consumed towards the ε phase. The reverse situation occurs at nitriding potential of the atmosphere equal to the border nitriding potential $N_p^{ε/γ'}$, when the gradient of nitrogen concentrations at the surface $c_N^{s/ε}$ and $c_N^{ε/γ'}$ will be equal to 0 ($c_N^{s/ε} = c_N^{ε/γ'}$); then the stream of nitrogen atoms$J_N^{γ'}$ will be smaller than the stream of nitrogen atoms $J_N^α$ and the ε phase will be consumed towards the γ' phase [7].

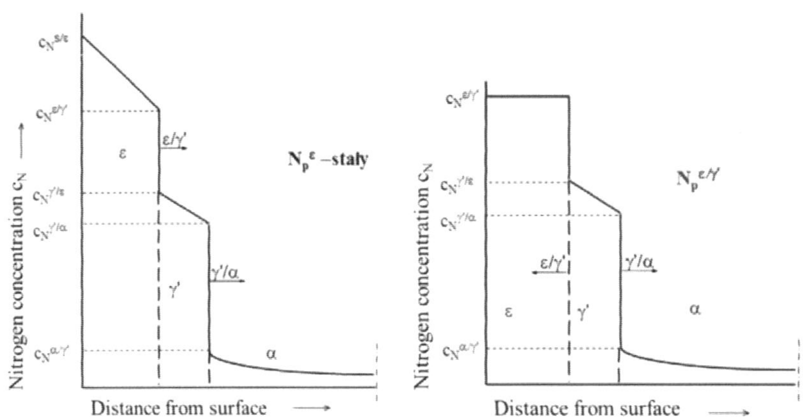

Figure 5. Influence of nitrogen concentration at the surface c_N on the growth of phases.

As seen from above, nitriding potential of the atmosphere is one of the most important parameters determining kinetics of nitrided layer growth. Another important parameter, also included on T-N_p-N phase equilibrium diagram, is temperature of nitriding process. From T-N_p-N phase equilibrium diagram we can conclude that depending on the process temperature (if nitriding potential of the atmosphere remains the same), we can obtain layers with different phase structure. For example, if we lower process temperature from 500°C to 450°C (with unchanged nitriding potential of the atmosphere N_p=3,2), we get double-phase layer instead of triple-phase layer. Moreover, in multistage processes, higher temperature after the first stage of process lowers the case hardness and simultaneously increases the case depth; it also may lower the core hardness depending on the prior tempering and total nitriding cycle or the apparent effective case depth because of the loss of core hardness, depending on how effective case depth is defined.

It is worth noting that on the basis of T-N_p-N phase equilibrium diagram we cannot draw conclusions about kinetics (speed) of nitrided layer growth and each phase that occurs in it. Phase structure and particular zone thickness depends on time of the process as well. The influence of time has been taken into consideration in phenomenological models of the kinetics of nitrided layer growth, developed by L. Małdziński and his co-workers [1]. Phenomenological description of nitrided layer growth assumes that this growth is caused by diffusion of nitrogen atoms through the interstitial positions of nitrogen atoms subnets; iron atoms participation in diffusion is negligible in comparison with participation of nitrogen in diffusion. Thus, general model of the kinetics of nitrided layer growth can be described using Fick's laws of diffusion. First Fick's law describes diffusion stream of nitrogen atomsj_N^φin φ phase, in the distance of x from the surface:

$$j_N^\varphi = -D_N^\varphi \frac{\partial c_N^\varphi}{\partial x} \tag{3}$$

In the formula above, c_N^φ means nitrogen concentration in φ phase, and D_N^φ is an actual coefficient of nitrogen diffusion in φ phase, connected with coefficient of self-diffusion $D_N^{\varphi *}$, according to the following formula:

$$D_N^\varphi = D_N^{\varphi *} \frac{dlna_N^\varphi}{dlnc_N^\varphi} \tag{4}$$

$dlna_N^\varphi/dlnac_N^\varphi$ quotient is a thermodynamic factor of nitrogen diffusion in φ phase (where a_N^φ is a nitrogen activity in φ phase, corresponding to concentration c_N^φ), which can be related with nitriding potential of the atmosphere:

$$\frac{\partial lna_N^\varphi}{\partial lnc_N^\varphi} = \frac{\partial lnN_p}{\partial lnc_N^\varphi} \tag{5}$$

As mentioned before, nitriding potential of the atmosphere is one of the most important parameters determining kinetics of nitrided layer growth, and its influence on phase structure of layer is shown on T-N_p-N phase equilibrium diagram. Thus, through the nitriding potential of the atmosphere, we can connect a thermodynamic factor of nitrogen diffusion in particular phases of Fe-N phase equilibrium diagram with T-N_p-N phase equilibrium diagram. Distribution of nitrogen concentration as a function of distance from surface in particular phases of Fe-N phase equilibrium diagram can be described by second Fick's law:

$$\frac{\partial c_N^\varphi}{\partial t} = \frac{\partial}{\partial x}\left(D_N^\varphi \frac{\partial c_N^\varphi}{\partial x}\right) = \frac{\partial D_N^\varphi}{\partial c_N^\varphi}\left(\frac{\partial c_N^\varphi}{\partial x}\right)^2 + D_N^\varphi \frac{\partial^2 c_N^\varphi}{\partial x^2} \tag{6}$$

Models developed on the basis of this relations together with determined experimental data describe the growth of nitrided layer in the field controlled by nitrogen diffusion in nitrided layer phases on iron. They are used in the simulations of nitriding processes, where through the adjusting process parameters they enable forming phase structure of nitrided layer, i.e. types of formed phases and their thicknesses, nitrogen concentrations at the surface, profiles of nitrogen concentration in particular phases. Moreover, they allow watching the growth of ε and γ' phases as a function of temperature and nitriding potential of the atmosphere not only in single-stage processes, but also in multistage processes. Unfortunately, they have also some limitations, resulting from complicated physico-chemical nature of nitriding. In industrial practice the biggest limitation is that this models can be used only for pure iron, whereas in fact the carbon and alloy steels are nitrided. The last one are the most problematic, because kinetics of nitrided layer growth on alloy steels is determined by many factors, sometimes contrary interacting, which is a major obstacle in the development of a satisfactory mathematical models and the determination of the necessary physico-chemical data. L. Małdziński proved that T-N_p-N phase equilibrium diagram can be used only for alloy steels containing Cr, Mn, Mo, V, because content of this elements does not change its shape [1,4].

L. Małdziński with collaboration with E. Mittemeijer and M. Sommers attempted to describe quantitatively the kinetics of γ' and ε+γ' phases on steels. They used the growth models of this phases on pure iron, and then modified them using correction factors kJ_N, which take into account the cumulative effect of alloying elements and carbon on diffusion streams of nitrogen atoms in ε, γ' and α phases, and as a result – on speed of growth of this phases. Thus, improved models has been received, which are supported by experimental verification; they are based on the phenomenological models of mass flow (nitrogen flow) and diffusion laws mentioned before [4]. Further research of L. Małdziński and his co-workers allows development models of the kinetics of nitrided layer growth. On the basis of this models, which enable precise forming of nitriding layer phase structure, simulator of the kinetics of nitrided layer growth has been created.

3. Simulator of the kinetics of nitrided layer growth

Simulator of the kinetics of nitrided layer growth (figure 6) is a computer programme which enables graphical and computational prediction of the growth of nitrided layer thickness and particular phases that will occur in it (ε, γ', ε+γ', α), effective case depth and hardness distribution as a function of process parameters: time, temperature and nitriding potential of the atmosphere. In practice it means that by using a simulator of the kinetics of nitrided layer growth, it is possible to form the nitrided layer precisely, and as a result – to obtain nitrided layer with strictly defined, required properties.

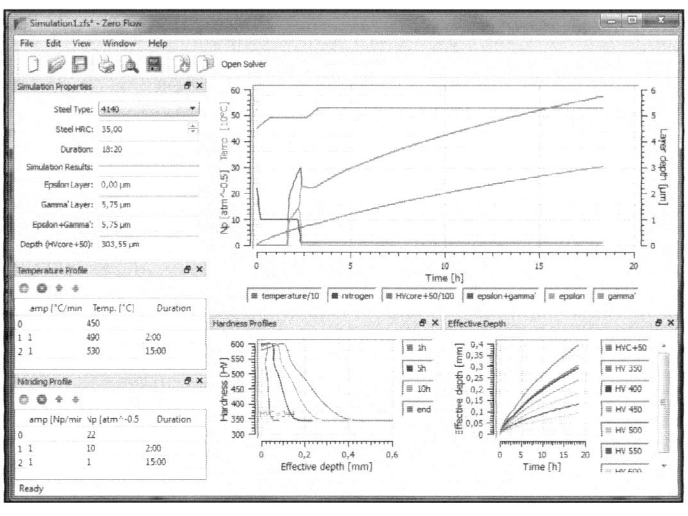

Figure 6. Simulator of the kinetics of nitrided layer growth.

Main simulator window consists of two major parts; on the left side we can recognize a number of cells, which are intended for nitriding parameters input (figure 7), and on the right side – graphs showing the results of conducted simulation (figure 8). Simulation is carried out automatically after entering data about the simulated process, such as simulation properties, temperature profile or nitriding profile. Simulation properties allows choosing the type of steel or determining core hardness (steel HRC). Currently simulator contains models for Armco iron and 12 types of steel, including: 1020 and 1050, low-alloy steels 4140, 5140, 4340, N135M, 18HGT, 8620, medium-alloy steels 32CVD13, H11, H13

and D2. Range of core hardness was based on hardness of steels used in experimental research and it is about 10HRC, however the minimum and maximum values are different for each type of steel. Moreover, in simulation properties we have also "Duration" cell, which shows overall time of nitriding process, computed on the basis of temperature and nitriding profiles defined by user.
Temperature and nitriding profile have the form of tables, where each row represents one section of the profile. Every section can be edited, moved or added/deleted. One section corresponds to one stage of process and it is described by the following quantities: the duration of the section, its target value and a ramp (speed at which the target value is reached). Ramp is connected with actual conditions of heating and nitriding potential reduction, resulting, among others, from the size of the furnace (or more precisely – from the size of its retort) or batch size. Each of these values are entered into another column.

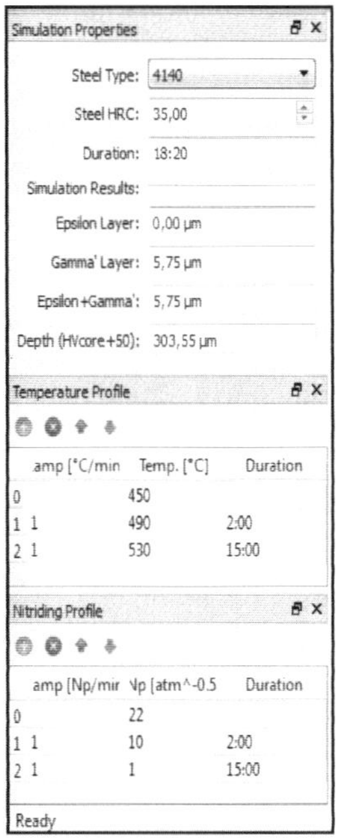

Figure 7. Simulator of the kinetics of nitrided layer growth – cells intended for nitriding parameters input.

Any change in parameters (time, temperature, nitriding potential of the atmosphere, as well as the properties of the steel) will restart the simulation and automatically perform calculations for the new input parameters. Results of the simulation are presented both in numerical and graphical form. Selected results in numerical formare presented under simulation properties and they contains following values:

- epsilon layer thickness,
- gamma' layer thickness,
- compound zone thickness (epsilon + gamma' layer), also known as a white layer,
- layer effective thickness with core hardness of +50HV (HVC+50).

As well as values in numerical form show only the final result of simulated nitriding process, graphs allow following the whole course of process and determining on its basis specific values during the process. Main part of simulator window contains graph ofε, γ', ε+γ'layers thickness and effective case depth HVC+50 growth as a function of nitriding time, which also includes curves presenting changes of temperature and nitriding potential during nitriding process. Below are another two graphs. First of them shows hardness profiles on the cross-section of nitrided layer after different times of nitriding. Second graph shows growth of the effective case depth for specific values of hardness (e.g. HVC+50, HVC400, HVC500, HVC600) as a function of nitriding time.

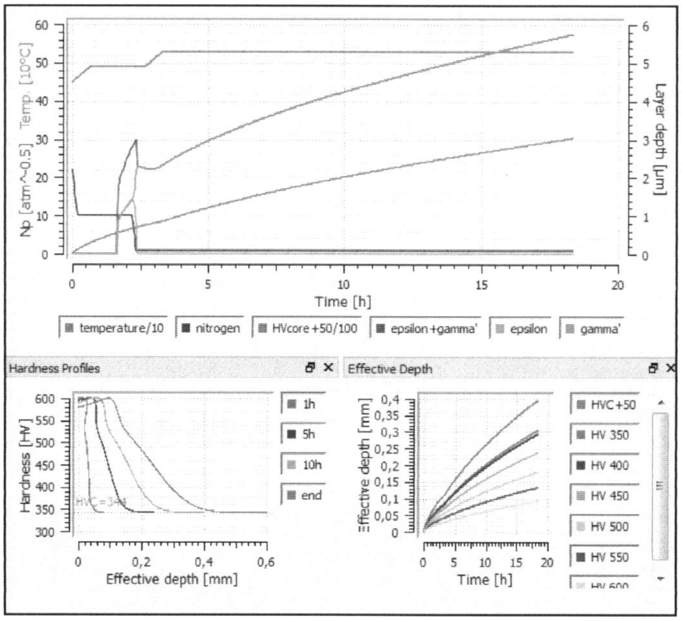

Figure 8. Simulator of the kinetics of nitrided layer growth - graphs showing the results of conducted simulation.

To sum up, simulator of the kinetics of nitrided layer growth enables precise simulations of nitriding process as a function of process parameters: time, temperature and nitriding potential of the atmosphere. Parameters can be adjusted to suit actual conditions of nitriding process, which delivers detailed information about the growth of nitrided layer. Thus, simulator can function both as a regulatory and control. By regulatory is understood the ability to determine the process parameters in order to obtain specific properties of the layer; by control – the ability to compare layers obtained after nitriding process with simulation results. However, the most important advantage of simulator is the ability to optimize process parameters, and as a result – forming required nitrided layer in the shortest possible time and with the lowest energy and gases consumption. It is possible by using a simulation solver (figure 9).

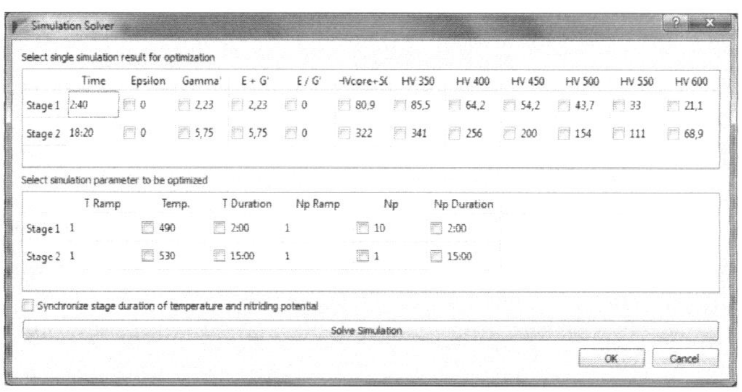

Figure 9. Simulator of the kinetics of nitrided layer growth – simulation solver.

Main solver window consist of two groups of cells. First of them, described as a "Select single simulation result for optimization", contains results of simulation which is currently carried out. Among them we can distinguish: total duration of each stage of the process, ε, γ', ε+γ' layers thickness, ε/γ' ratio, effective case depth HVC+50 and others. In second group, described as a "Select simulation parameter to be optimized", we have process parameters – time, temperature, nitriding potential of the atmosphere – which are optimized in order to obtain required layer structure and properties. To optimize the parameters, we have to select single cell from the first group and set its target value, and then select parameter from the second group, which will be changed. After pressing the button "Solve simulation", new value of selected parameter will be shown instead of previous one. Without no doubt, solver is one of the most important functions of simulator of the kinetics of nitrided layer growth, because it is an essential factor determining economical and environmentally friendly character of nitriding using ZeroFlow method.

Gathered industrial experience not only confirmed the effectiveness of ZeroFlow method, but they also confirmed the effectiveness of using a simulator of the kinetics of nitrided layer growth as an application that supports precise forming of nitrided layer with respect to the required phase structure, zone thicknesses and hardness distribution.

4. Summary

The simulator of the kinetics of nitrided layer growth is an application which supports nitriding using ZeroFlow method. Despite of many advantages and proven effectiveness it still requires further development and research – like verification and modification of the existing models or development of models for new types of steel. However, it should be noted that limitations of the simulator does not preclude from using them, both in experimental and industrial scale. Gathered experience has shown that simulator of the kinetics of nitrided layer growth allows precise forming of nitrided layers with respect to the required phase structure, zone thicknesses and hardness distribution. Furthermore, through the use of mathematical models these layers are produced in the shortest possible time and with the lowest energy and gases consumption.

References

[1] Małdziński L 2011 *Niskoemisyjne, ekonomiczne regulowane azotowanie gazowe ZeroFlow (Technologie zero-emisji)*ed J Jabłoński (Poznań: Zeszyty Naukowe Politechniki Poznańskiej) pp 189–229

[2] Kowalska J 2015 *Regulacja i kontrola kinetyki wzrostu warstwy azotowanej na stalach w procesie regulowanego azotowania gazowego ZeroFlow (Badania i Rozwój Młodych Naukowców w Polsce. Nauki Techniczne i Inżynieryjne vol 12)* (Poznań: Młodzi Naukowcy) pp 120–6

[3] Małdziński L, Bazel M, Korecki M, Miliszewski A and Przygoński T 2009 Controlled nitriding using a Zero Flow metod *HeatTreating Progress***9/4** 19–22

[4] Małdziński L 2002 *Termodynamiczne, kinetyczne i technologiczne aspekty wytwarzania warstwy azotowanej na żelazie i stalach w procesach azotowania gazowego* (Poznań: Wydawnictwo Politechniki Poznańskiej)

[5] Campbell F C 2008 *Elements of metallurgy and engineering alloys* (Ohio: ASM International)

[6] Pye D 2003 *Practical nitriding and ferritic nitrocarburazing* (Ohio: ASM International)

[7] Kowalska J 2016 *Wpływ parametrów procesu azotowania na budowę fazową warstwy (Badania i Rozwój Młodych Naukowców w Polsce. Nauki Techniczne i Inżynieryjne vol 23)* (Poznań: Młodzi Naukowcy) pp70–5

IOP Conf. Series: Materials Science and Engineering **148** (2016) 012048 doi:10.1088/1757-899X/148/1/012048

Test and assessment method of Automotive Safety Systems (SSB) particularly to monitor traffic incidents

B Pijanowski, S Łukjanow and R Burliński

Automotive Industry Institute)PIMOT), Warsaw, Poland

E-mail: b.pijanowski@pimot.eu

Abstract. The rapid development of telematics, particularly mobile telephony (GSM), wireless data transmission (GPRS) and satellite positioning (GPS) noticeable in the last decade, resulted in an almost unlimited growth of the possibilities for monitoring of mobile objects. These solutions are already widely used in the so-called "Intelligent Transport Systems" – ITS and affect a significant increase for road safety. The article describes a method of testing and evaluation of Car Safety Systems (Polish abbreviation - SSB) especially for monitoring traffic incidents, such as collisions and accidents. The algorithm of SSB testing process is also presented. Tests are performed on the dynamic test bench, part of which is movable platform with car security system mounted on it. Crash tests with a rigid obstacle are carried out instead of destructive attempts to crash test of the entire vehicle which is expensive. The tested system, depending on the simulated traffic conditions, is mounted in such a position and with the use of components, indicated by the manufacturer for the automotive safety system installation in a vehicle, for which it is intended. Then, the tests and assessments are carried out.

1. Introduction

The telematics development allows to generally equip vehicles with different types of safety systems with an option of monitoring. Standardisation documents, which are recognised by law, define minimum technical and functional requirements for the above-mentioned systems. However, the test and assessment methods of systems, which are particularly specialised in monitoring traffic incidents, are not determined. There is no particular test method of the operation quality of devices (systems) of this kind. There is also no method of conducting a comparative assessment of the systems.

The method described in this paper allows the manufacturers of these systems (and vehicles equipped with such systems) to carry out a comparative assessment and check of various technical solutions for safety systems. In case of the vehicles' users, it should contribute to a significant increase of the road safety level. It is especially important in Polish conditions, where the number of fatal and serious accidents is still high (in spite of a significant decrease), which contributes to huge financial outlays associated with the loss of life or health of drivers, passengers and pedestrians, as well as the damage to the road infrastructure and vehicles.

On the basis of the conducted single crash tests of the entire vehicles, it is difficult to assess and compare the operation quality of individual systems, because a vehicle should be examined in a single test at the predetermined speed. The speed increase in another test may be limited by the test station structure (catapult). The standardised tests for the entire vehicles are carried out at a speed of approx. 50 km/h. The impact speed increase is not always possible due to technical reasons. In addition, a big disadvantage of testing the safety systems installed in complete vehicles includes the necessity of purchasing the vehicles, which are damaged as a result of the test.

2. The proposed test method of automotive safety systems

The tests of automotive safety systems according to the proposed method are carried out by a simulation of traffic collisions (10...20 km/h) and a simulation of the traffic accident (50...120 km/h). In case of the tested safety systems, for tests simulating road conditions, the tests at 3 degrees of freedom, as the sufficient approximation of actual conditions, were adopted. In case of comparative tests within the above-mentioned ranges, the constant test conduct speed, e.g.: 15, 50, 80, 110 km/h, is determined.

The diagram of the test station for the proposed method implementation was presented in figure 1.

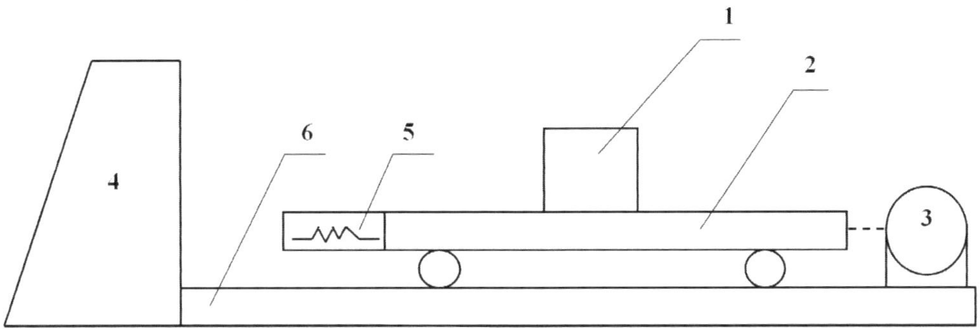

Figure 1. Station for checking the automotive safety systems.

The essence of the developed method for testing and assessment of automotive safety systems, particularly for monitoring traffic incidents, includes the fact that each tested safety system (1) is mounted on the test station that has a mobile platform (2) which is driven by the propulsion system (3) and is led to a collision with a rigid barrier (4) by means of a damping element (5) that is fixed to the platform or rigid barrier (4). The tested system, depending on the simulated traffic conditions, is mounted in such a position and with the use of components, indicated by the manufacturer for the automotive safety system installation in a vehicle, for which it is intended. Then, the tests and assessments are carried out according to the below-described stages:

In the first stage, a frontal impact of a vehicle with the installed automotive safety system into a rigid barrier at the speed corresponding to the traffic collision, selected from the range of 10... 20 km/h, was enforced on the test station, then, the automotive safety system is rotated on the platform by 90°, and the test is repeated for the next (simulated in this way) "sides of the vehicle". During the tests, in this stage, monitoring and recording (and transmission) devices transmit data on the incident and transmission parameters to the operator's panel.

In the second stage, a frontal impact of a vehicle with the installed automotive safety system into a rigid barrier at the speed corresponding to the traffic accident, selected from the range of: 50...120 km/h was enforced on the test station, then, the automotive safety system is rotated by 90°, and the test is repeated for the next sides of the examined system. During the tests, in this stage, monitoring and recording (and transmission) devices transmit data on the incident and transmission parameters to the operator's panel; but in this stage, the safety systems, which met the test positive results on the basis of the adopted assessment criterion, are checked.

According to the specified method, the test results can be used at the automotive safety system comprehensive assessment described in section 3 (and in the examples of section 4, as mechanical parameters). The algorithm of the motor vehicle's safety monitoring system is presented in Fig. 2 [13].

Table 1. Functional Classes for typical systems/devices (automotive safety systems).

Equipment symbol	Description of the system or unit	Manufacturers' markings	Functional class
S1	Safety belts and belt pre-tensioners	SRS	A
S2	Airbags	other	A
S3	Anti-lock and four-wheel anti-lock braking systems	ABS, ASR, EBO, EBS and other	A
S4	Track stabilisation	ESP	A
S5	Adaptive Cruise Control	ACC	B
S6	Automatic braking to 50 km/h, detection of obstructions	other	A
S7	Control of: the driver's fatigue, lane	other	A
S8	Illumination and visibility	other	B
S9	Monitoring of traffic incidents	e-Call	A
S10	Surrounding observation cameras	other	C

3. The proposed assessment method of automotive safety systems

In order to develop the assessment, the criteria of the research level assessment (amplitude) and functional status, which were specified in the ISO 7637-2 standard, were used. Among 5 listed classes describing the functional status: A, B, C, D and E, the use of the first three of them is proposed:

A class: all functions of the tested device, during the test and after the resistance test, are implemented according to the assumptions within the specified tolerance.

B class: all functions of the tested device, during the resistance test, are implemented according to the assumptions; it is accepted that one or more functions are implemented outside the specified tolerance limits. After the tests, all the functions are implemented in accordance with the specified tolerance limits.

C class: one or more functions of the tested device are not implemented during the test in accordance with the assumptions, but after the test, there is automatic return to the device normal operation.

The range of systems and motor vehicles' safety devices is very large and constantly growing. The most typical automotive safety systems in the current state of the technology development and in the currently produced vehicles were presented in table 1. In case of these automotive safety systems, the classes of the above-mentioned ones describing the functional status of the systems were offered (column 3 table 1).

The class for other systems should be determined prior to the automotive safety system tests. In a given automotive safety system, it is possible to define the tolerance limits for each function implemented by the system. The information flow analysis in the automotive safety system, according to the algorithm determined on the basis of technical specifications, is very helpful. The example specifications for a model of the automotive safety system information flow with the e-call function was shown in table 2. In the model, it is important to specify "validity" (weight) of a signal, e.g. 0, 1, 2, 3, 4 given in column 4 of table 2, with the following markings:

0 – a signal may be wrong,
1 – a certain signal,
3, 4 – e.g. priorities.

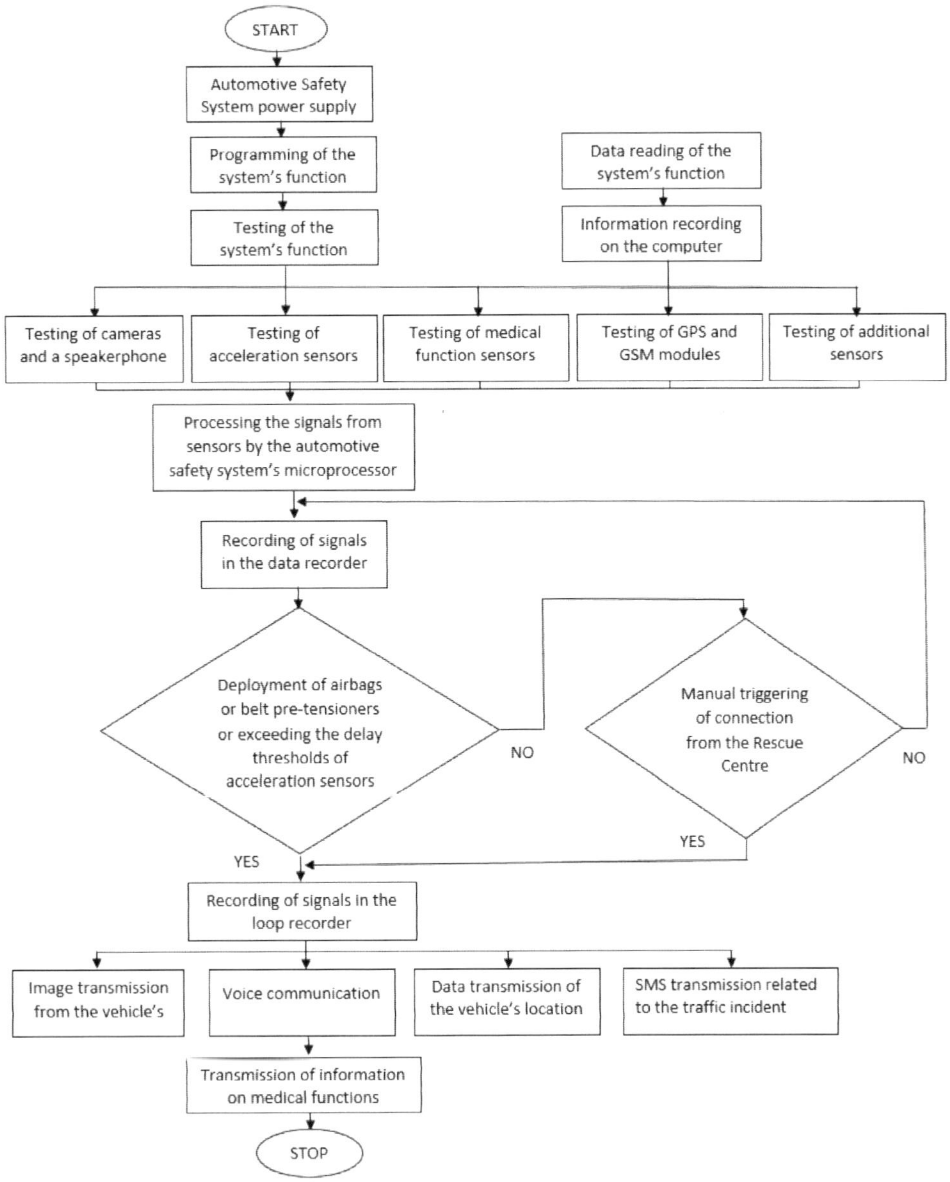

Figure 2. Algorithm of the vehicle's safety monitoring system.

In case of a given type of the automotive safety system, apart from the class resulting from the functional assessment, it is important to limit the speeds (or their range), at which they should operate efficiently. It is possible to determine it on the basis of the tests of different classes (functional assessment) depending on the speeds (or their range). Due to the multi-functional nature of systems and devices, it is also possible to determine the classes that describe the functional status for individual functions and essential operating parameters.

Table 2. Specification of the automotive safety system's information flow module signals.

Item	Name	Definition	Signal status
1	MAX1	Longitudinal deceleration of the 1 module's sensor	0, 1, 2, 3, 4
2	MAX2	Longitudinal deceleration of the 2 module's sensor	0, 1, 2, 3, 4
3	MAY1	Lateral deceleration of the 1 module's sensor	0, 1, 2, 3, 4
4	MAY2	Lateral deceleration of the 2 module's sensor	0, 1, 2, 3, 4
5	MAZ1	Vertical deceleration of the 1 module's sensor	0, 1, 2, 3, 4
6	MAZ2	Vertical deceleration of the 2 module's sensor	0, 1, 2, 3, 4
7	MB1	Fire signal of the 1 module's sensor	0, 1
8	MB2	Fire signal of the 2 module's sensor	0, 1
9	MC1	Smoke signal of the 1 module's sensor	0, 1
10	MC2	Smoke signal of the 2 module's sensor	0, 1
11	MD1	Sink signal of the 1 module's sensor	0, 1
12	MD2	Sink signal of the 2 module's sensor	0, 1
12	PLP	Airbag signal left front	0, 1
13	PPP	Airbag signal right front	0, 1
14	PLT	Airbag signal left back	0, 1
15	PPT	Airbag signal right back	0, 1
16	ZLP	Car body crumple signal left front	0, 1, 2
17	ZPP	Car body crumple signal right front	0, 1, 2
18	ZLB	Car body crumple signal left side	0, 1, 2
19	ZPB	Car body crumple signal right side	0, 1, 2
20	ZLT	Car body crumple signal left back	0, 1, 2
21	ZPT	Car body crumple signal right back	0, 1, 2
22	V	Linear speed before the accident	0, 1, 2, 3, 4
23	OMEGA	Deviation angular velocity before the accident	0, 1, 2, 3, 4
24	H	Brake pressing signal	0, 1

4. Assessment examples of automotive safety systems

Two examples for determining the assessment of the SSB-1 and SSB-2 safety systems, which are used to monitor traffic incidents, were presented below.

SSB-1. System (automotive safety system): the system is designed to automatically inform the receiving centre about the traffic incident occurrence based on acceleration in three axes of coordinates x, *y*, *z* when forward and backward. This system is designed for installation in cars. It enables the voice contact with the receiving centre. In the radio transmission, among others, information on the vehicle identification, its location and speed are transmitted.

Table 3 shows the results of the SSB1 assessment.

The tests on the station defined in section 2 were carried out using the technology demonstrator made within the framework of the research project described in the paper [13].

The tested automotive safety system is a multifunctional device, and therefore, a single functional class cannot be determined.

The tests at simulation of the collision (15 km/h) and the accident (60 km/h) showed compliance with the above-mentioned functional classes.

SSB-2. The system (automotive safety system) is intended to automatically inform the receiving centre about the traffic incident occurrence. It has the same functions as the above-presented SSB-1, and in addition, it is equipped with a camera and control systems of the travellers' "vital functions" by checking the blood pulse (e.g. by means of a sensor placed on the hand wrist).

IOP Conf. Series: Materials Science and Engineering **148** (2016) 012048 doi:10.1088/1757-899X/148/1/012048

Table 3. SSB1 assessment results.

Item	Function name	Class
1	The voice communication using the GSM telephone with the Public-Safety Answering Point in the event of the traffic incident occurrence	A
2	Location	A
3	Movement direction	A
4	Speed	A
5	Acceleration values measured in three axes	A
6	The vehicle's position in relation to the reference plane	B
7	Engine speed	C
8	Sending a message about the traffic incident occurrence containing the data package from a data recorder to the Public-Safety Answering Point	A

The SSB2 assessment results were presented in table 4.
The test included the same range as in case of SSB-1, and in addition:

Table 4. SSB2 assessment results.

Item		**Class**
9	The status of the driver's pulse sensor (presence of the pulse signal, battery status, the communication status of the pulse sensor and the receiver)	A
10	The transmission of images from the cameras placed in the vehicle	C

The tests at simulation of the collision (15 km/h) and the accident (60 km/h) showed compliance with the above-mentioned functional classes.

5. Validation of the test method of automotive safety systems

According to the EN/PN 17025:2001 standard, the validation is to confirm, by testing and presenting the objective evidence, that the requirements on the intended use were met. It is a process which confirms that the analytical procedure (tested method), used for the implementation of a concrete specified marking/test, is suitable for the intended use (fitness for purpose).
The method should be validated when it is important to demonstrate that its parameters are suitable for a particular analytical problem. This can include the following situations:

- introduction of a new method,
- improvement of the method or adjusting it to its new problem,
- if the quality control reveals that the determined method changes over time,
- if a given method is to be used by other personnel, on a different apparatus.

The validation of the described test method of the automotive safety system with the use of the test station with a mobile platform was carried out by comparison and correction of parameters based on the crash test results of the FORD Focus car. The course of the test with the installed automotive safety system was described in the publication [13].

6. Conclusion

The rapid telematics development, noticed in the last decade, owing to the technology of: mobile telephony (GSM), data transmission (GPRS) and satellite location (GPS), resulted in almost an unlimited increase in opportunities for monitoring mobile objects. These solutions are already widely used in the so-called intelligent transport systems – ITS, and affect a significant increase of the road safety. In this paper, the applied test method of the station Automotive Safety Systems (SSB), which are designed especially to monitor traffic incidents, such as collisions and accidents, was presented.

The algorithm of the vehicle's safety monitoring system was demonstrated. Moreover, the automotive safety system assessment method based on the functional status (classes: A, B and C) was presented.

7.References

[1] http://dlakierowcow.policia.pl/dk/statystyka/47493.dok.html 2012 2011 2010 2009 2008 - Polish Police Headquarters; Traffic Office Prevention and Analysis Team *Wypadki drogowe w Polsce [Car accidents in Poland]*

[2] PN-EN 15722:2011 *Transport drogowy i telematyka ruchu-E Bezpieczeństwo- e-Call minimalne bazy danych [Road transport and traffic telematics E-safety e-Call minimum databases]*

[3] FprEN 16072:2011 *Intelligent transport systems - eSafety - Pan-European e-Call operating requirements*

[4] FprEN 16102:2011 *Intelligent transport systems - e-Call - Operating requirements for third party support*

[5] PN-EN ISO 24978:2010 *Systemy inteligentne w transporcie - Systemowe informacje bezpieczeństwa i komunikaty alarmowe przekazywane w każdych dostępnych radiowych środkach masowego przekazu - Procedury rejestrowania danych [Intelligent transport systems – System safety information and alarms transmitted in all available radio media – Data recording procedures]*

[6] BOSCH 2010 *Elektrotechnika i Elektronika Samochodowa, Układy bezpieczeństwa i komfortu jazdy [Automotive Electrotechnics and Electronics, Safety and driving comfort systems,* Technical Information Guide

[7] Burliński R, Łukjanow S and Pijanowski B 2013 *System monitorowania tętna osób zwłaszcza w pojeździe samochodowym po zdarzeniu drogowym [Systems for monitoring the pulse of people especially in a vehicle after the traffic incident]* - a registered patent in the Patent Office of the Republic of Poland No. P.406745

[8] Haddabin - S, Aibu - Schaffer A, Hirzinger G Dumny *Crash - Tests for the Evaluation of Rigid Human - Robot Impacts*; Institute of Robotics and Mechatronitcs DLR – German Aerspace Center

[9] Insurance Institute for Highway Safety 2009 *Guidelines for Rating Injury Measures 1005 N*; Glebę Road, Ariington, VA 22201

[10] KERATRONIK; Wasikowski A 2012 *Inteligentny System Ratunkowy ISR [Intelligent Emergency System] – IES* (company materials)

[11] Łukjanow S and Pijanowski B 2012 *Układ monitorowania bezpieczeństwa w pojeździe samochodowym, zwłaszcza w czasie zdarzeń drogowych [Motor vehicle's safety monitoring system especially during traffic incidents*] - a registered patent in the Patent Office of the Republic of Poland No. P.402200

[12] Pijanowski B and Żardecki D 2013 "*Selected Vehicle monitoring problems with the "e-Call" system taken as an example*. The Archives of Automotive Engineer - ing, Vol 61, No 3

[13] Pijanowski B and Łukjanow S 2014 *Monitorowanie bezpieczeństwa pojazdów samochodowych szczególnie w czasie zdarzeń drogowych [Monitoring the safety of vehicles especially during traffic incidents].* International Automotive Conference KONMOT 2014, Kraków

[14] UN/ECE Regulation No. 10 Series of amendments 03 2008 *Jednolite przepisy dotyczące homologacji pojazdów w odniesieniu do kompatybilności elektromagnetycznej [Uniform provisions concerning the approval of vehicles with regard to electromagnetic compatibility]* (EMC)

[15] Żardecki D and Pijanowski B 2012 *Information in the motor vehicle safety system with the e- call function*. The Archives of Automotive Engineering, Vol. 58, No 4

IOP Conf. Series: Materials Science and Engineering **148** (2016) 012049 doi:10.1088/1757-899X/148/1/012049

Some control problems of continuously variable belt transmission

B Radzymiński[1]

[1]Department of Vehicles and Fundamentals of Machine Design, Lodz University of Technology, Żeromskiego 116, 90-924 Łódź

E-mail: bartosz.radzyminski@p.lodz.pl

Abstract. Control problems of continuously variable belt transmission used in passenger cars have been discussed. Pulley adjustment solutions and choice of control and feedback signals are the main topics. Intention to use such a transmission as part of a complex system containing mechanical energy storage caused that the adjustment transition time become crucial problem.

1. Introduction

Continuously variable belt transmission are increasingly used in conventional drivetrains of passenger cars. They allow to adjust the ratio of drivetrain to the currently prevailing traffic conditions kipping the operating point of the internal combustion engine in the area of greatest efficiency. In currently used solutions of gearboxes equipped with continuously variable transmissions produced by one of the leading manufacturers - JATCO company, high-pressure hydraulic system provides the necessary clamping force and adjusts the ratio of variator. Control of all functions of the system performs control valve assembly [6]. In this system, input pulley is responsible for establishing ratio and output pulley is responsible for generating clamping force causing belt tension caused in minimum level avoiding belt slipping during torque transmitting. This division is due to the fact of placing axial movement limiters of the movable conic wheel on the input pulley. To obtain maximum efficiency of continuously variable belt transmission and minimum power consumption by the hydraulic system, belt should be all the time tension to an extent no greater than is required for the safe transfer of the instantaneous torque without slipping. In the hybrid drivetrain, the secondary power source must be connected to the drive wheels by variable speed system for providing energy transfer to and from the drive wheels.

2. Discussed gearbox

Figure 1shows a schematic of JATCO CVT7 gearbox used in Nissan Micra K12 - the gearbox was used as a basis for consideration of the application in a hybrid drive system. More details on the study of such a transmission and analysis of control valve assembly can be found in [1] and [2].

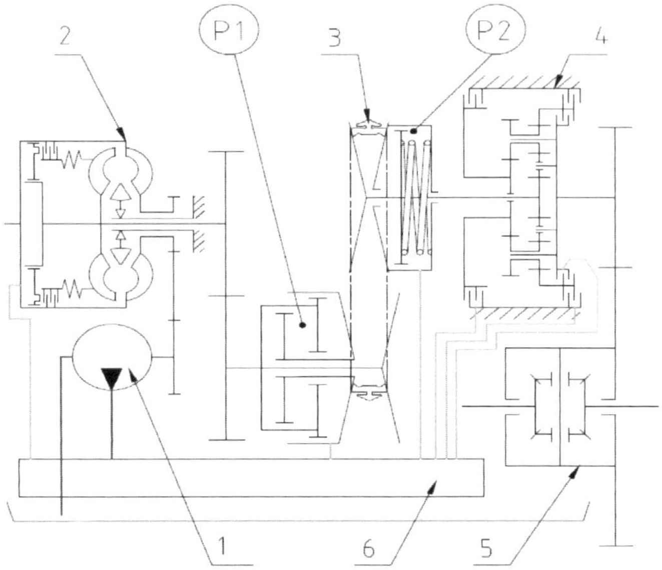

Figure 1. Mechanical scheme of JATCO CVT7 gearbox; 1 – oil pump; 2 – torque converter; 3 – pushbelt; 4 – Ravigneaux planetary transmission; 5 – final drive; 6 – control valve assembly; P1 – pressure in primary (input) pulley actuator; P2 - pressure in secondary (output) pulley actuator.

The diagram portion of control valve assembly in part responsible for the development of the pressure in the pulleys actuators and the preparation of lubricating oil is shown in Figure 2.

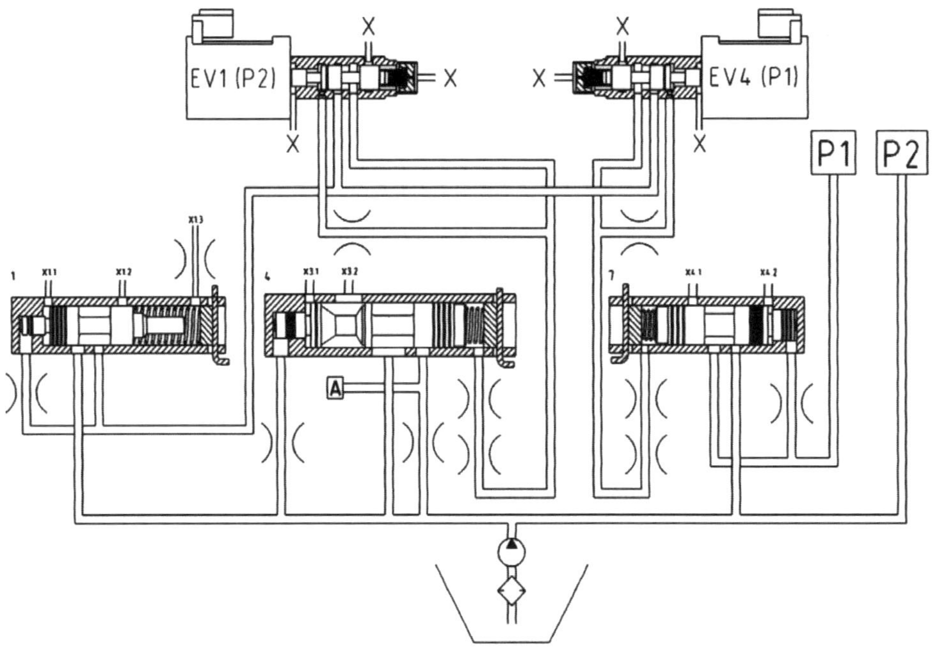

Figure 2. Portion of hydraulic scheme of TOSOK control valve assembly in part responsible for generating control pressures for the CVT and controlling lube oil pressure (part no 6 in figure 1); 1, 4, 7 – spool valves; EV – electrovalves; A – receivers responsible for pushbelt lubrication and oil cooling.

3. Disturbance signals

Control systems with open loop-scan make the control process faster than control systems with closed loops. Moreover open loop systems can be cheaper because of the absence of some sensors.

3.1. Oil temperature influence on the process of generating oil pressure[3] and [4]

Based on the study it was found that in the control system used by TOSOK, temperature have big impact on the process of pressure generating in pulleys actuators. This impact is visible on electro valves characteristics.

$$P2 = f(PWM_1, t) \tag{1}$$

$$P1 = f(PWM_4, t) \tag{2}$$

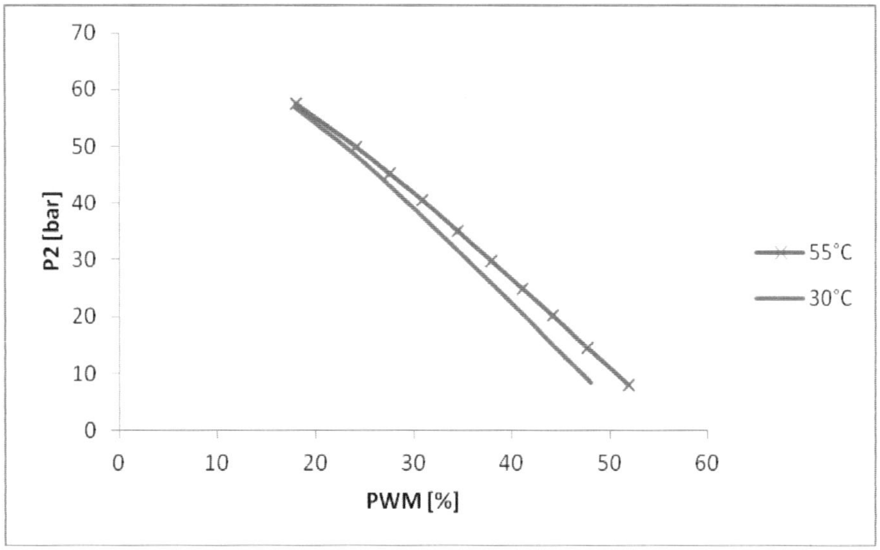

Figure 3. Influence of oil temperature on pressure in pulley actuator with PWM as a input signal.

The graphs placed in Figure 3, shows that with increasing temperature of the oil in order to achieve the same pressure, PWM signal should increase.

However, the studies and observations found that phenomenon presented in Figure 3 is not related with an increase of the oil temperature but it is related with the rises of temperature of the coil of solenoid resulting from the operation of the flowing current. In fact, coil changes resistance with temperature. Solenoid coil actuated with maximum PWM heat up the coil to a temperature of 115°C, so it can be stated that the oil flow causes cooling of solenoid coil. Another important aspect is oscillating within some limits voltage in vehicle electrical system. In order to disambiguate the input signal responsible for generating the force in the coil, and therefore the pressure, it is necessary to introduce the concept of average current supply solenoid valve.

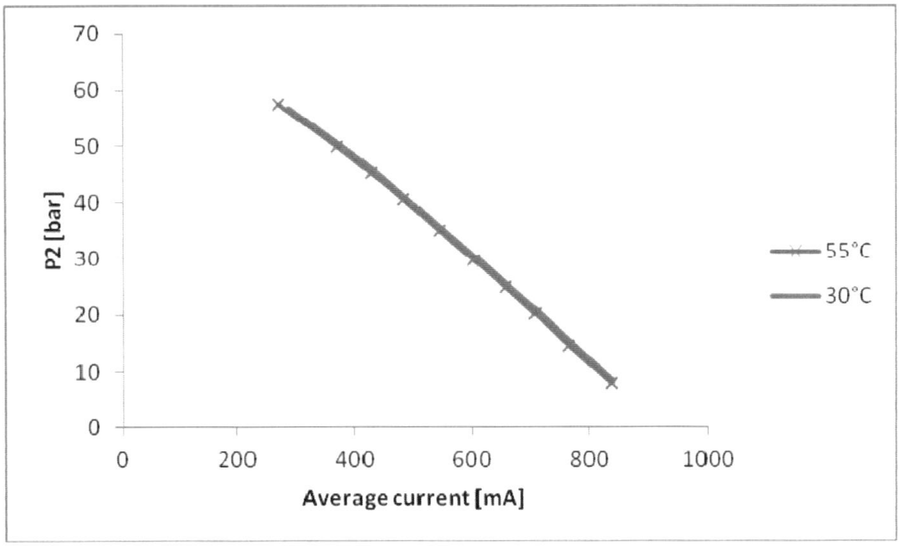

Figure 4. Influence of oil temperature on pressure in pulley actuator with average current as a input signal.

The graphs placed inFigure 4, shows that by controlling the average current of solenoid valve,it is possible to be completely independent in the algorithm responsible for the development of the pressure in the pulleys actuators of temperature. Then electro valves characteristic presented in equation 3 is not required.

$$P = f(PWM, t) \tag{3}$$

3.2. Torque influence on the P1 pressure selection

In this study, the continuously variable transmission ratio was calculated on the basis of information provided by pulleys speed sensors.

Continuously variable transmission ratio is defined as:

$$i_{CVT} = \frac{\omega_2}{\omega_1} \tag{4}$$

List of pressures required to obtain requested ratio of continuously variable transmission without torque load with constant pressure in secondary pulley are shown in Figure 5. Those graphs are open loop control maps.

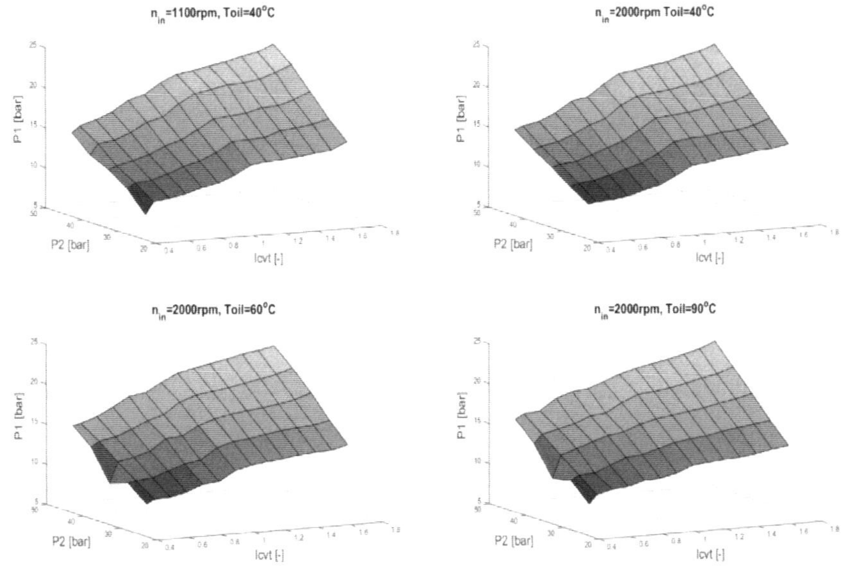

Figure 5. Listing of P1 and P2 pressures for CVT ratio for different input rotational speed and oil temperature.

The use of open-loop in the given form does not give grounds to change the belt tension depending on the value of the input torque, which, as it turns out the study, at steady pressures in both actuators definitely affects the value of continuously variable transmission ratio.
It is important to keep in mind that three of the four signals $P1$, $P2$, l_{CVT}, T_{in} determine fourth of them:

$$P1=f(l_{CVT}, P2, T_{in}) \quad (5)$$

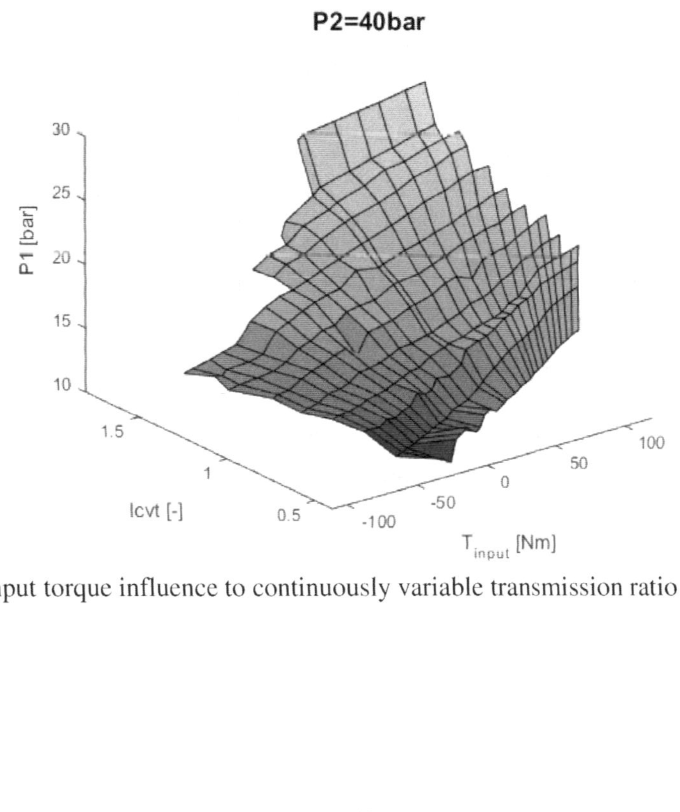

Figure 6. Input torque influence to continuously variable transmission ratio for P2=40 bar.

IOP Conf. Series: Materials Science and Engineering **148** (2016) 012049 doi:10.1088/1757-899X/148/1/012049

This should be understood in such a way that with increasing of input torque, in order to prevent the occurrence of belt slippage, the pressure in the output pulley actuator should be increase. Then, in order to maintain the same ratio, the pressure in the input pulley actuator should increase.

4. The combination of a continuously variable belt transmission with flywheel

Using the continuously variable belt transmission in a conventional drivetrain is to considerthe transferred torque in control algorithm in order to maximize system efficiency. Information on the value of the transferred torque can be derived from the characteristics of the engine in the form of:

$$Torque=f(throttle_{position},\ engine_{speed}) \tag{6}$$

When joining in hybrid drivetrain the secondary energy source in the form of mechanical energy accumulator (flywheel), both of these sources are connected to the drive wheels through the continuously variable transmission. Realising the potential of the hybrid drive is expressed when:

- The driver demands braking torque (resulting charging of the flywheel)
- The driver request drive torque (resulting discharging of the flywheel)

Scheme of such a hybrid drive system shown in Figure 7[5].

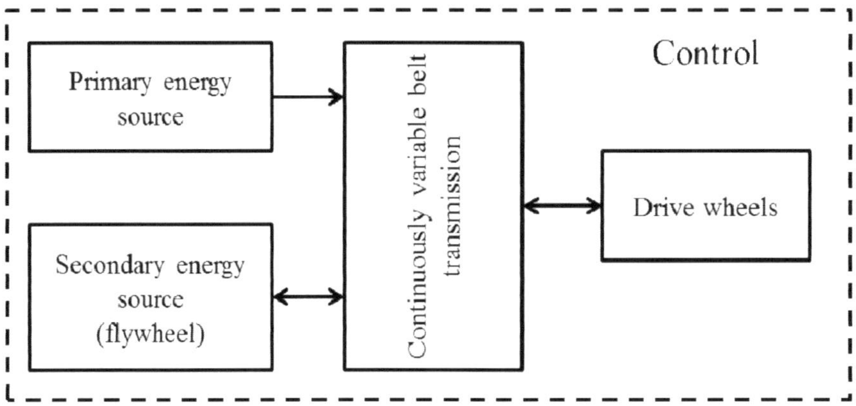

Figure 7. Diagram of a prototype hybrid powertrain.

In this system, during deceleration of the vehicle, its kinetic energy is converted into kinetic energy of the rotating flywheel. In this situation, during reduction of the vehicle speed the rotational speed of the flywheel should increase. To do this the ratio of continuously variable transmission should continuously increase - simplified example is shown in the graphs in Figure 8, where driver request is to achieve braking torque (vehicle deceleration). Similarly, while putting the kinetic energy of the flywheel to increase the kinetic energy of the vehicle, continuously variable transmission ratio should be reduced continuously. In both cases it is necessary to ensure the occurrence of angular acceleration or deceleration of the flywheel to form a braking or driving torque.

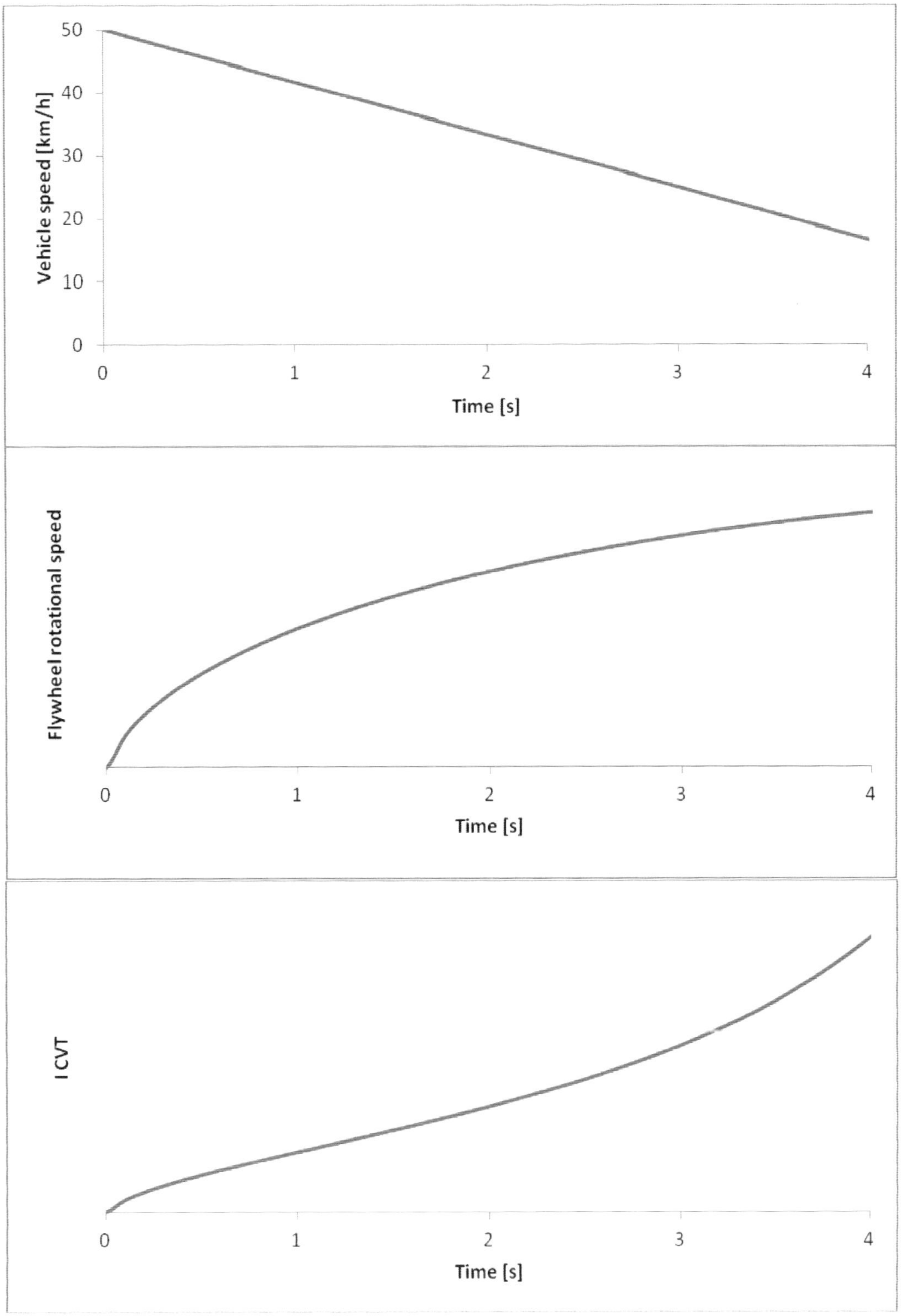

Figure 8. Sample curves of vehicle and flywheel speed and the course of ratio of CVT.

5. Conclusions

- To become independent from the influence of temperature in pressure generation process, average current as a input signal should be used
- For continuously variable belt transmission transmitted torque is important disturbing signal
- It is necessary to design an algorithm that allows, for a requested deceleration process of the vehicle, calculate required ratio of continuously variable transmission
- The instantaneous values of the continuously variable transmission ratio (so the rate of continuously variable transmission ratio change) depend on the driver's desired braking and driving torque
- Rate of ratio change of continuously variable belt transmission is critical from flywheel point of view
- To check the dynamic properties of the continuously variable belt transmission currently used in conventional drivetrains is crucial

References

[1] Radzymiński B and Pawelski Z 2015 Wpływ parametrów zakłócających na pracę skrzyni biegów wyposażonej w przekładnię CVT *Problemy rozwoju maszyn roboczych* ed Kosecki A pp 286-296

[2] Radzymiński B, Goszczak J, Werner A and Pawelski Z 2015 Analiza bloku elektrohydraulicznego samochodu Nissan Micra w części dotyczącej wypracowania ciśnień do siłowników przekładni CVT*Logistyka* **3** pp 4109-4117

[3] Goszczak J, Radzymiński B, Werner A and Pawelski Z 2015 Badania samochodowego elektrozaworu sterowanego sygnałem PWM*Logistyka* **3**pp 1562-1572

[4] Radzymiński B 2016Wpływ temperatury na ciśnienie generowane przez elektrozawór sterownika bezstopniowej skrzyni biegów*Zagadnienia aktualnie poruszane przez młodych naukowców* **5** pp 431-438

[5] Radzymiński B 2016 Koncepcja hybrydowego układu napędowego samochodu osobowego wyposażonego w mechaniczny akumulator energii*Zagadnienia aktualnie poruszane przez młodych naukowców* **5** pp 427-430

[6] BOSCH 2005 *Elektroniczne sterowanie skrzynką biegów EGS* (Warszawa: WKiŁ)

Acknowledgments

The author of this paper would like to express the greatest gratitude to Mr Andrzej Werner and Mr. Zbigniew Pawelski for their invaluable help during preparing this article. Thanks to his knowledge, experience and patience this paper is much more valuable.

IOP Conf. Series: Materials Science and Engineering **148** (2016) 012050 doi:10.1088/1757-899X/148/1/012050

The impact of European Registers of Road Transport Undertakings on security and enforcement of the system of digital tachograph

M Rychter[1] and R Rychter[2]
[1] University of Life Sciences in Lublin, Faculty of Production Engineering 28 Głęboka St., 20-612 Lublin, Poland
[2] Łazarski University, 43 Świeradowska St., 02-662 Warsaw, Poland

E-mail: [1]rychter@poczta.fm and [2]Renata.Rychter@interia.pl

Abstract. Secure digital tachograph system is indisputably a means having the great influence on road safety, enforcement of the social conditions for drivers in road transport as well as supporting the fair competition between road transport undertakings. In order to better monitor the compliance of road transport with the rules in force, inter alia the provisions on usage the digital tachographs, the European Union regulation introduces the European Registers of Road Transport Undertakings (ERRU), a system that allows a better exchange of information between Member States. The main purpose of this analysis is to present the measures for enforcement the execution of the provisions on obligation of installation and usage of digital tachographs in road transport and for improving the security of data recorded within digital tachograph system especially through exchange of the information on the most serious infringements committed by hauliers in any Member State, which may lead to the loss of good repute and the e-document engineering method used by the ERRU system.

1. Introduction

The success of the digital tachograph system, introduced in member states of the European Union by *the Council Regulation (EC) No 2135/98 of 24 September 1998 amending Regulation (EEC) No 3821/85 on recording equipment in road transport and Directive 88/599/EEC concerning the application of Regulations (EEC) No 3820/84 and (EEC) No 3821/85*[1] since 1 of May, 2006, is the guarantee of fair competition and enforcement of legal provisions on common rules on admission to the occupation of road operator and social rules of drivers. The usage of digital tachograph has undoubtedly a crucial role to play in improvement of road safety, because the main target of the tachograph is the observance of driving and rest times and speed limits. The effects of road accidents involving motor vehicles or set of vehicles for the actual total laden weight of over 3,5 tones used in the carriage of goods by roads; often vehicles carrying dangerous goods or vehicles used for passenger carriages, are much more harmful in terms of both economic and social than accidents with the participation of so called: light vehicles (the

[1] Official Journal L 274, 09.10.1998, p. 1

total laden weight of less than 3,5 tones). For that reasons, the development of the safe digital tachograph system shall be one of the main purposes of the enforcement policies of the member states in the range of road transport, *inter alia* because of the permanent and fast technical progress and high risk of installation of unpermitted devices used for manipulation of the records of such equipment. European Register of Road Transport Undertakings (ERRU) performs both auxiliary and monitoring function regarding the digital tachograph system. This linked-up database, operating since 1 January 2013, allows better monitoring of the compliance of road transport undertakings with the rules of using digital tachographs. Undertakings that do not respect the provisions on digital tachographs will face the consequences in the Member State where they are based. They can loss their good repute and access to the road transport market. The well balanced and parallel system based on digital tachograph system and ERRU is a kind of "squeeze–out mechanism" for unfair road transport operators, from the market of professional road carriages.

2. The safety of digital tachograph system

Undoubtedly, the tools for the prevention of accidents in the course of commercial road transport of heavy vehicles should be created and developed. Digital tachograph system is one of such tools, giving the opportunity to the completed and successful monitoring of periods of drivers' activity and speed of driving during conducting road carriage. The technical progress undermines the possibility to build a model of digital tachograph system fully secure from the point of view of data recorded by such system. The natural consequence is the need for recommendations on technical requirements for digital recording equipment in road transport, which would prevent manipulating of data recorded by this device. Security of digital tachograph system is unquestionably one of the most important elements affecting effectiveness of this system. The methodology of prevention and technical solutions in order to eliminate and decrease the reaction of prohibited devices, manipulating data, recorded in register equipment, has a crucial role.

To ensure the safety of data stored and processed within the framework of the digital tachograph system it is also necessary to use the knowledge and technology of cryptology. Data security has a long tradition in the automotive industry. However, the issue of data security and protection against attempts to falsify or delete is still little known in the environment of manufacturers and suppliers of motor vehicles, which are used for road transport and in which recording devices should be installed. Cryptography plays a significant role for the security of the digital tachograph system. Mechanisms and importance of cryptography should be respected by each participant of digital tachograph system. Data encryption mechanisms are correspondingly important in connection with the introduction of the obligation to apply the intelligent digital tachographs and the ability to copy data from the recording equipment using the interface and electronic communication.

Fitters and workshops play an important role in the security of tachographs. Important role in the process of testing, periodic checks and calibration of digital recording equipment plays a suitable quality level of training of technicians in workshops and the quality of tests and examinations carried out by workshops.

3. The smart tachograph

Bearing in mind the significant meaning of control functions of digital recording equipment, it was necessary to develop technical solutions, enabling remote control of digital recording equipment, using intelligent transport systems, including the global GPS location system (Global Positioning System). Changes in European legislation, according to the provisions of regulation (EU) No 165/2014 of the European Parliament and of the Council of 4 February 2014 on tachographs in road transport, repealing Council Regulation (EEC) No 3821/85 on recording equipment in road transport and amending Regulation (EC) No 561/2006 of the European Parliament and of the Council on the harmonisation of certain social legislation relating to road transport[2], confirmed the legitimacy and desirability of

[2] OJ L 60, 28.2.2014, p. 1.

researches conducted by and obtained as a consequence of their results, including, in particular, in the scope of use of recording equipment, connected to the global satellite navigation system, as a means of enabling the automatic registration of the vehicle's position and supporting the control function of the system.

With regard to the communication on the distance between the recording equipment, digital tachograph system plays a significant role in enforcement instrument for control services, authorized to the road side checks and inspections at the premises of undertakings.

Consideration for the fair competition and conducting the commercial activity regarding to the regulations of existing Community law require the uniform application of common rules. The scale of conducted carriages by road at the global level of the economic market of the European Union brings the need of closer monitoring of compliance with the regulation on road transport throughout imposing on the transport operators and drivers the obligation of the implementation of the digital recording devices and making records of drivers' activities. Because of the safety reasons in road transport the obligation of registration of rests and driving periods as well as records of journey and speed of driving have the special meaning. Experience with the use of analogue recording devices has presented the serious unreliableness of the monitoring system of working time of drivers in road transport, based on such analogue devices. Digital recording equipment and multi-element, completed system based on digital data gives the opportunity to monitor the compliance with legal provisions on driving and rest periods, as well as it allows to reconstruct the history of carriages conducted by the driver.

In 2019, a new generation of digital tachograph (the smart tachograph) will be introduced on the market, and will be mandatory for all newly registered commercial vehicles. Within next fifteen years, the smart tachograph will be mandatory for all commercial vehicles in the scope of the tachograph regulation, operating within the territory of European Union. Once operational, one of the functions of the smart tachograph will be to automatically record the exact geo-positioning of vehicle and driver at the start and the end of the driver's working day, as well as every three hours of cumulated driving time. This information will be stored in a secure way both on-board vehicle and at the company site, as requested by the EU rules, and thus will be downloadable both during roadside and company checks. This certain tachograph function will be essential in controlling the period of time over which vehicle and driver operated on a territory of a given Member State.

Hence, suitable worked out and developed, allowing work in on-line mode, the system of intelligent digital tachograph shall eliminate the infringements in road transport in the range of rules on driving and rest periods for drivers. This only will be achieved under the condition of compatible functioning of ERRU and registering of all cases of infringement against tachograph rules, which cause the loss of good repute of transport manager of road transport operator. Furthermore, to provide the security of data recorded and stored in digital tachograph system, it must be ensured the permanent monitoring of the level compliance with the provisions of European Union law on road carriage and social rules for drivers. Intelligent tachograph system will be the necessary and effective method for control of compliance with the good repute condition in road transport and for risk estimation of the infringements concerning the provisions on working time rules for drivers. Consequently, strengthening the effectiveness of the certain elements of digital tachograph system and providing its interoperability, as well as the data security against its loss and change, will create the good instrument, countering the road accidents, collisions and improving social conditions for the drivers. It directs to the statement on the same resonance of the digital tachograph system and ERRU.

4. Legal bases and obligation of arrangement of the national registers and their interconnection

4.1 National registers as enforcement tool of common rules on access to the road transport market

Provided by every member state in European Union, national electronic registers should perform subsidiary role for the enforcement[3] of *regulation (EC) No 1071/2009 of the European Parliament and*

[3] Strachowska R., *Komentarz do art. 16 rozporządzenia nr 1071/2009 ustanawiającego wspólne zasady dotyczące warunków wykonywania zawodu przewoźnika drogowego*, Lex/el. 2013.

of the Council of 21 October 2009 establishing common rules concerning the conditions to be complied with to pursue the occupation of road transport operator and repealing Council Directive 96/26/EC. [4] Pursuant to article 16 paragraph 1 of regulation 1071/2009, for the purposes of the implementation of this regulation, in particular articles 11 to 14 and Article 26, each Member State shall keep a national electronic register of road transport undertakings which have been authorised by a competent authority designated by it to engage in the occupation of road transport operator. The data contained in that register shall be processed under the supervision of a public authority designated for that purpose. The relevant data contained in the national electronic register shall be accessible to all the competent authorities of the Member State in question.

The obligations of the member states regarding the national electronic registers covers the task of recording in the register the data relating to undertakings which it authorises; in order to improve transparency and to allow the client of a transport undertaking to verify whether that undertaking is in possession of the appropriate authorisation, certain data contained in the national electronic register are publicly accessible.

According to article 16 paragraph 2 of regulation 1071/2009 national electronic registers of road transport undertaking shall contain, inter alia, the number, category and type of serious infringements, which have resulted in a conviction or penalty during the last 2 years.

Even though the obligation of introducing the national register of road transport undertakings has no derogations, Poland has not implemented in national law the appropriate provisions on such register.[5]

5. Infringements regarding usage of digital tachographs as the most serious infringements in road transport

5.1. Definition of serious infringements

The regulation 1071/2009 does not define the term "serious infringement." Nevertheless, pursuant to article 2 point 7 of *regulation (EC) No 1072/2009 of the European Parliament and of the Council of 21 October 2009 on common rules for access to the international road haulage market Text with EEA relevance*[6] "serious infringement of Community road transport legislation" means an infringement which may lead to the loss of good repute in accordance with article 6 of regulation 1071/2009 or to the temporary or permanent withdrawal of a Community licence.

Furthermore, in the absence of EU specific provisions in the regard, which serious infringements of European law applies to the transport manager and which to the undertaking, it is a matter of national (criminal or administrative) legislation to determine the active subject of infringements to the road transport legislation.[7] Article 6 of regulation 1071/2009 provides that when assessing the good repute of an undertaking the Member State should asses the undertaking itself, the transport manager and any other relevant person as may be determined by the Member State. Member States have got the delegation to determine the conditions to be met by undertakings and transport managers in order to satisfy the requirement of good repute, one of the conditions of the access to the occupation of the road transport operator. Serious infringements include convictions, penalties or infringements of national rules in force

[4] OJ L 300, 14.11.2009, p. 51 – 71.

[5] According to the project of amendments of law on road transport of, accepted by the Cabinet of Ministers on 12 of April, 2016 and addressed to the Polish Parliament, in Poland until April 2017 the National Electronical Register of Road Transport Undertakings will be created and connected to ERRU. The Polish register will be administrated by the General Inspector of Road Transport and will consists of three data bases: register of road transport operators, possessing permission on occupation of the road transport operator, the evidence of persons without skills on management of road transport operations and register of serious infringements in road transport.

[6] Official Journal L 300, 14.11.2009, p. 72 – 87.

[7] European Commission, *FREQUENTLY ASKED QUESTIONS RELATING TO THE IMPLEMENTATION OF ERRU (Electronic Register of Road Transport Undertakings), Regulation (EC) No 1071/2009*, 2013, http://ec.europa.eu/transport/modes/road/access/doc/faq-on-erru.pdf.

in the fields of: commercial law, insolvency law, pay and employment conditions in the profession, road traffic, professional liability, trafficking in human beings or drugs.
Additionally, the term "serious infringements in road transport" means that the transport manager or the transport undertaking have not in one or more Member States been convicted of a serious criminal offence or incurred a penalty for a serious infringement of Community rules relating in particular to: the driving time and rest periods of drivers, working time and the installation and use of recording equipment, the maximum weights and dimensions of commercial vehicles used in international traffic, the initial qualification and continuous training of drivers, the roadworthiness of commercial vehicles, including the compulsory technical inspection of motor vehicles, access to the market in international road haulage or, as appropriate, access to the market in road passenger transport, safety in the carriage of dangerous goods by road, the installation and use of speed-limiting devices in certain categories of vehicle, driving licences, admission to the occupation, animal transport.
Most infringements listed in Annex IV are by their nature committed by drivers. For instance, in the case of breach of EU rules on driving times, breaks and rest periods established by Regulation (EC) No 561/2006, the transport undertaking is liable for infringements committed by drivers (article 10 paragraph 3) unless competent authorities in Member State establish that there is an evidence that the undertaking cannot be reasonably held responsible (the transport undertaking complied with requirements of article 10 paragraph 2 of this regulation to organise the work of drivers, to properly instruct them, to check compliance with the provisions of the regulation). Such a principle of responsibility of the undertaking for the illegal acts of the driver could be applied in the same way to other infringements referred to in Annex IV when assessing the good repute of the undertaking. Such serious infringement, which was mentioned in point 2 of the annex IV to regulation 1071/2009, is referred to the digital tachograph system – not having a tachograph or using a fraudulent device able to modify the records of the recording equipment or falsifying record sheets or data downloaded from the tachograph and/or the driver card.[8]

5.2. Categories and seriousness of serious infringements
Pursuant to the article 6 paragraph 2 letter b of regulation 1071/2009, the European Commission is obliged to draw up a list of categories, types and degrees of seriousness of serious infringements of Community rules which, in addition to those set out in Annex IV of regulation 1071/2009, may lead to the loss of good repute. A common list of categories, types and degrees of serious infringements of the European Union rules in the commercial road transport, in addition to those set out in Annex IV to regulation 1071/2009, which may lead to the loss of good repute of a road transport operator, was pointed by *the European Commission regulation 2016/403 of 18 March 2016 supplementing Regulation (EC) No 1071/2009 of the European Parliament and of the Council with regard to the classification of serious infringements of the Union rules, which may lead to the loss of good repute by the road transport operator, and amending Annex III to Directive 2006/22/EC of the European Parliament and of the Council.*[9] Categorisation of serious infringements is presented in annex I to the regulation 2016/403 in division into twelve tables, containing categories and types of serious infringements against the EU rules in commercial road transport, separated into three categories of seriousness (MSI = most serious infringements/VSI = very serious infringement/SI = serious infringement), according to their potential to create a risk of fatalities or serious injuries.
Groups of infringements regarding digital tachograph system, against *regulation (EU) No 165/2014 of the European Parliament and of the Council of 4 February 2014 on tachographs in road transport, repealing Council Regulation (EEC) No 3821/85 on recording equipment in road transport and*

[8] Regarding to frauds and manipulations on digital tachographs, see more: M Rychter, 2011 2012 Budowa i zastosowanie systemu tachografii cyfrowej Warszawa; M Rychter, SAE Technical Paper, Directions of Counteracting Manipulations of Digital Recorders in the Light of the Safety Functioning of the Digital Tachographs System, 2013, No 1, p. 9.
[9] OJ L 74, 19.03.2016, p. 8.

amending Regulation (EC) No 561/2006 of the European Parliament and of the Council on the harmonisation of certain social legislation relating to road transport[10] are categorised in table 2 of annex I to regulation 2016/403. As most serious infringement (MSI), concerning the installation of tachograph, was defined: "not having type-approved tachograph installed and used (e.g.: not having a tachograph installed by fitters, workshops or vehicle manufacturers approved by the competent authorities of the Member States, using a tachograph without the necessary seals placed or replaced by an approved fitter, workshop or vehicle manufacturer or using a tachograph without the installation plaque)." Furthermore, the most serious infringements, regarding use of digital tachograph and driver card, are the following five infringements: driving with a driver card that has been falsified (considered as driving without driver card), driving with a driver card of which the driver is not the holder (considered as driving without driver card), driving with a driver card which has been obtained on the basis of false declarations and/or forged documents (considered as driving without driver card), using a fraudulent device able to modify the records of the tachograph, falsifying, concealing, supressing or destroying data recorded on the record sheets or stored and downloaded from the tachograph and/or the driver card.

The second category of seriousness of infringements, regarding digital tachograph, are very serious infringement (VSI), such as: using a tachograph not inspected by an approved workshop, driver holding and/or using more than one own driver card, tachograph not correctly functioning (e.g.: tachograph not properly inspected, calibrated and sealed), tachograph improperly used (e.g.: deliberate, voluntary or imposed misuse, lack of instructions on correct use, etc.), undertaking not keeping printouts and downloaded data, recorded and stored data not available for at least a year, incorrect use of driver card, unauthorised withdrawal of driver card which has an impact on the record of relevant data, driver card used to cover a period longer than that for which it is intended and data is lost, use drivers card and data not legible, not using manual input when required to do so, incorrect use of switch mechanism, refusing to be checked, unable to produce records of current day and the previous 28 days, unable to produce records of the driver card if the driver holds one, unable to produce manual records and printouts made during the current day and the previous 28 days, unable to produce a driver card, if the driver holds one, tachograph not repaired by an approved fitter or workshop, driver not marking all required information for the periods of time which are no longer recorded while tachograph is unserviceable or malfunctioning.

Only one infringement regarding the digital tachograph was considered in regulation 2016/403 as very serious, that is: driver card not in the correct slot (multi-manning).

The same categorisation of serious infringements dealing with digital tachograph system was introduced in *directive 2006/22/EC of the European Parliament and of the Council of 15 March 2006 on minimum conditions for the implementation of Council Regulations (EEC) No 3820/85 and (EEC) No 3821/85 concerning social legislation relating to road transport activities and repealing Council Directive 88/599/EEC*[11] and the risk assessment system on road transport undertakings for the purposes of road side checks and checks at the premises.

Very important in in terms of procedure on analysing the condition of good repute are the provisions of annex II to the regulation 2016/403 on frequency of occurrence of serious infringements. The serious (SI) and very serious (VSI) infringements regarding digital tachographs, when committed repeatedly shall be regarded as more serious by the competent authority of a Member State of establishment. When calculating the frequency of occurrence of repeated infringements Member States shall take into account factors such as: seriousness of infringement (SI or VSI), time (at least one rolling year from the date of a control), number of drivers used for the transport activities managed by the transport manager (average per year). According to the potential of creating a risk to road safety the maximum frequency of serious infringements beyond which they should be considered as more serious shall be established as: 3 SI/per driver/per year = 1 VSI; 3 VSI/per driver/per year = launch of a national procedure on good repute.

[10] OJ L 60, 28.2.2014, p. 1.
[11] OJ L 300, 14.11.2009, p. 51.

The number of infringements per driver per year is an average figure calculated by dividing the total number of all infringements of the same level of seriousness (SI or VSI) by the average number of drivers employed during the year. The frequency formula provides for a maximum threshold for occurrence of serious infringements beyond which they shall be considered more serious. It must be underlined that Member States may establish stricter thresholds if envisaged in their national administrative procedure for assessing good repute. Poland and most of the EU member states have not introduced such national rules. However it must be noted that regulation 2016/403 is the example of new UE legislation and it shall apply from 1 January 2017.

5.3. The level of infringements concerning digital tachographs

According to European Commission Staff Working Paper Summary of the Impact Assessment, the biggest part of controlled vehicles is found breaching the social rules. The number of infringements by drivers and road transport operators against the social rules is dangerously high: 3.9 million offences were detected in the period 2011-2012 (4.5 million offences reported in 2009-2010 and 3.3 million in 2007-2008).[12] Approximately one fourth of these are found breaching tachograph rules. On average around 45 000 vehicles are in breach of EU tachograph rules. Such persistent non-compliance with obligations for minimum rest periods and the resulting driver fatigue can be estimated to produce an increase in the societal cost of accidents of close to €2.8 billion.[13]

For that reason, one of the operational objective of the European Commission is to eliminate the "most serious infringements" against tachograph rules by 2020.[14]

Road inspections performed by the Polish Road Transport Inspection presented that in first half of the year 2015 there were 55 869 infringements in road transport, according to the law on road transport. Infringements regarding the usage of tachographs stated 30% of these infringements. Almost 10% of the infringements regarding usage of tachograph is manipulating this recording device.[15] Conversely the structure of infringements indicated as the result of checks at the premises of undertakings in Poland presents approximately 72% of infringements, regarding the improper usage of tachographs.[16] According to the data gathered by the Polish Road Transport Inspections in the whole year 2015, 78% of infringements occurred as the result of checks at the premises of the road transport undertakings were violations concerning usage of tachographs in road transport (147579 such cases).[17]

It must be also underlined that type and sanctions for the same infringement against the rules on tachographs can vary in Member States of EU. Hence, it has got a very negative impact on security of digital tachograph system and the level of road safety. For instance, the infringement "Driving with no type approved recording equipment installed and used" is punished in chosen Member States of EU, within administrative procedure, on very various level of punishment. The level of sanction for this infringement in Germany is very high: fine up to € 15,000 (for undertaking) and up to € 5,000 for a driver. In Austria it is fine up to 2.180 € or up to 3.600 € (for the employer) and fine up to 5.000 € or

[12] European Commission, DG MOVE/D3, *Evaluation of social legislation in road transport (Regulation (EC) No 561/2006 on driving times and rest periods, Directive 2002/15/EC on working time of mobile workers, Directive 2006/22/EC on enforcement)*, Brussels 2015.

[13] European Commission, *Commission Staff Working Paper Summary of the Impact Assessment, SEC/2011/0947 final - COD 2011/0196, 52011SC0947*, eur-lex.europa.

[14] Commission Staff Working Paper, Impact Assessment on measures enhancing the effectiveness and efficiency of the tachograph system Revision of Council Regulation (EEC) No 3821/85. Accompanying the document Proposal for a Regulation of the European Parliament and of the Council amending Council Regulation (EEC) No 3821/85 to enhance the security, effectiveness and efficiency of the tachograph in road transport, Brussels, 2011, p. 28.

[15] Główny Inspektorat Transportu Drogowego, *Informacja dotycząca działalności wojewódzkich inspektoratów transportu drogowego w pierwszym półroczu 2015 roku*, Warszawa 2015, p. 21.

[16] *Ibidem*, p. 59.

[17] Central Evidence of Infringements, 2015.

imprisonment in case of recurrence (for driver and transport operator), whereas in Lithuania the head of undertaking will pay fine up to €869 and the driver - up to €290.[18]

6. Conclusions – The legal proposals for strengthening the linkage between ERRU and digital tachograph system

The perspective *de lege ferenda* concerns the question of procedural unified rules on inspections in road transport under the European Union law and common sanction system for the infringement in road transport regarding the usage of digital tachographs and social rules of drivers. Only the equivalent reaction on the infringement against the union law and the same obligations, rights and rules under the procedural provisions will ensure the necessary countermeasures for frauds and infringements on digital tachograph system. Present European Union regulation 1071/2009 allows Members States to determine the conditions of fulfilling the requirement of good repute. The impact of ERRU on the safety of digital tachograph system and, consequently, the safety in road transport and fair competition will be increased by the unified system of sanctions and procedural provisions of controls. These issues are necessary for elimination of the key harms of the road transport market. Pursuant to the preamble of regulation 1071/2009 this problems are still vital: "distortion of competition and a lack of market transparency and of uniform monitoring, as well as the risk that undertakings employing staff with a low level of professional qualification may be negligent in respect of, or less compliant with, the rules on road safety and social welfare, which may harm the image of the sector."

References

[1] European Commission, Commission Staff Working Paper Summary of the Impact Assessment, SEC/2011/0947 final - COD 2011/0196, 52011SC0947, eur-lex.europa
[2] European Commission Directorate-General for Mobility and Transport 2013 Study on sanctions in the field of commercial road transport, CONTRACT NO 2011/MOVE/D3/ETU/514-2011/SI2.612723 – FINAL REPORT – PART I Brussels p. 36
[3] Główny Inspektorat Transportu Drogowego 2015 Informacja dotycząca działalności wojewódzkich inspektoratów transportu drogowego w pierwszym półroczu 2015 roku Warszawa.
[4] Rychter M 2011 2012 Budowa i zastosowanie systemu tachografii cyfrowej Warszawa
[5] Rychter M, SAE Technical Paper, Directions of Counteracting Manipulations of Digital Recorders in the Light of the Safety Functioning of the Digital Tachographs System, 2013, No 1, p. 9
[6] Strachowska R 2013 Komentarz do art. 16 rozporządzenia nr 1071/2009 ustanawiającego wspólne zasady dotyczące warunków wykonywania zawodu przewoźnika drogowego Lex/el

[18] European Commission Directorate-General for Mobility and Transport, *Study on sanctions in the field of commercial road transport, CONTRACT NO 2011/MOVE/D3/ETU/514-2011/SI2.612723 - FINAL REPORT – PART I,* Brussels 2013, p. 36.

Modelling and optimization of rotary parking system

A Skrzyniowski

Faculty of Mechanical Engineering
Cracow University of Technology
John Paul II Avenue 37, Cracow, Poland

E-mail: jendrek@mech.pk.edu.pl

Abstract. The increasing number of vehicles in cities is a cause of traffic congestion which interrupts the smooth traffic flow. The established EU policy underlines the importance of restoring spaces for pedestrian traffic and public communication. The overall vehicle parking process in some parts of a city takes so much time that it has a negative impact on the environment. This article presents different kinds of solution with special focus on the rotary parking system (PO). This article is based on a project realized at the Faculty of Mechanical Engineering of Cracow University of Technology.

1. Introduction

The rapid development of motorization in Poland in the second part of the 1990s generated difficulties with commuting and parking motor cars at the workplace, schools, universities, market places and also in the place of residence, both in old and modern parts of the city.

Therefore it is necessary to ensure a sufficient amount of parking space for motor cars in the close vicinity of newly erected offices, malls, edifices, residential areas and multi-storey buildings.

Much more difficult problems arise when ensuring adequate quantity of parking space for cars in the existing building is concerned [3].

The idea of construing multi-storey parks has progressively developed all over the world and its scale can be measured by, e.g. the obligation to provide parking places (1.5 parking lots per one newly built household) in some of the West European countries [2].

Nowadays the mechanized and automated multi-storey parks are adjusted to the applied system and are equipped with such elements as slide plates or ramps, transport carriages, mobile floors, lifts, elevators, turntables, etc., and also the complicated control and automation system. Actually they do not differ too much from former mechanized systems, e.g. 'Kent', 'Dolly', 'Autosilos', 'Buranelli' or 'Pigeon', which used to be manufactured in the interwar period.

The mechanization and automation of moving vehicles within the car park area worked out in the western countries in the interwar period only partly contributed to the lowering of cubature index of the presently built objects without reducing the index of total space per one car, because the driving lanes were not eliminated. On the other hand these changes were associated with a high capital and exploitation cost, which in turn was not outbalanced by the substitution of reinforced concrete walls and floors with light steel constructions, typical of the presently used solutions [1].

Car parks offered by specialist companies, e.g. Wöhr, Doppelmeyer, Klaus or Parktec have about 35% surface loss (and so also cubature loss) for driving lanes. As a result the cost of one parking lot in these modern objects and in traditional ones is similar, and the only advantage is partially lowered cubature because of the lowered height of particular storeys (vehicles are moved without clients' participation) [1].

Car parks and garages are a frequent problem encountered by the designers and investors, especially when the designed and modernized objects are planned in the city centre area. The lack of sufficient space, its high cost, limitations imposed by the water and soil conditions as well as the high cost of underground cubature frequently hinder the smooth placement and designing of offices, hotels, as well as residential and sometimes production buildings. The modernization of the existing buildings also meets problems with providing necessary garage or parking lots. These problems known in developed countries for years are now more and more frequently touching Polish designers, mostly in big agglomerations [1].

2. Analysis of car parking solutions

A comparative analysis of the preferred option, i.e. rotary (optimum) parking, with elements of other parking systems, was based on a review of literature on multi-storey mechanical car parks with special emphasis on solutions involving inter-storey rotary transport of cars, i.e. rotary car parks. The comparison was supplemented by a review of characteristics of traditional multi-storey (especially indoor ones) as typical construction objects. For the simplicity's sake, the following abbreviations have been used throughout this paper:

PA – automatic car park (generally),
PP – automatic car park with inter-storey progressive transport of cars,
PO – automatic car park with inter-storey rotary transport of cars (rotary parking),
PTW – traditional indoor multi-storey car park in a building object,
OPO – optimum rotary car park, preferred solution.

3. Justification of optimum type of car park OPO as a solution to the car parking problem

The OPO type car park (for a PO 10 model, i.e. for 10 cars as the smallest for this type of unit) was selected because of the following properties and parametric comparison with before mentioned car park types:

1) Coaxial, modular, increasing of the parking volume (up to 120 or more, thanks to which comparison analyses could be performed); the assumed direction of car entry/exit is rectangular to the drum,
2) No waste gases emission, making PTW (according to calculations in point 4) an 'environmental benefit' as compared to the OPO and all PA car,
3) about 50 percent energy savings as compared to the parking discussed in point 3 thanks to the smaller resistance in the rotational movement of the drum than in the progressive movement of the platforms, thanks to which the installed capacity can be increased twice,
4) about 60 percent shorter time of car retrieval than in a parking presented in point 3 (90 instead of 230 sec); shorter inner driving lanes and a bigger number of entry/exit stands,
5) object surface index 2.8 m^2/l lot generating about 22% and 86% saving as compared to car parks listed in points 3 and 4, respectively,
6) object volume index 48 m^3/l lot generating about 17% and about 47% saving on space as compared to car parks mentioned in points 3 and 4, respectively,
7) unit investment cost of 30,000 PLN/l lot generates about 60% and about 75% saving on capital cost in serial production as compared with car parks listed in points 3 and 4, respectively,
8) shortcomings of car parks PO and MP-750, presented in point 2.2, were eliminated by making the following design assumptions;
   - leveling of platforms with two prismatic guides (shape of a circle section), giving the openwork structure some lightness and possibility to support the drum on both sides,
   - the drive is shifted from the axis to the circumference of the side of the drum with the use of a cycloidal chain gear, which apart from minimizing the size of the drum bearing also provides full self-attenuation of its movement,
   - lowered depth of the trench to about 1/3 of the surface part, i.e. parameter conditioning rational assumption of self-acting (gravity-based) onward motion of parking cars (car enters on one side and leaves on the other),
   - the way in which cars enter and leave the parking eliminates driving lanes and access paths, thus reducing the terrain and space losses,

- modular multiplication of the parking space, limiting the consequences of potential break-downs of the system to one module only.

4. Description of rotary parking

The presented technical variant of a multi-storey automated car park is an avant garde solution meeting the goals discussed in this paper. And so, [1]:

- The capital cost of a parking lot in this solution is several times lower;
- The total space index is lowered;
- The average time of parking in and out is shortened;
- Its reliability increases;
- The cost of exploitation of the parking is lowered.

Generally, this car park, as visualized in Figure 1, resembles the Vienna 'Riesenrad or Chorzów 'Devil's Wheel' (Poland), where the carriages for people were substituted with parking platforms for cars. This car park rotates around the horizontal axis. There are no internal driving lanes or paths because each car goes directly to its parking lot. There are no roofs, carrying walls, poles and the gravitational transport of vehicles makes use of the existing slope of the drive-in way and does not require any mechanical aid or drives. This system is simple, cheap and reliable [3].

Depending on the size of the car park, the platform with hangers, installed rotationally on the sides of the drum, can be single or multi-storey, and each of the levels comprises two cars side by side. The size of the drum also depends on the number of arms with platform hangers [3].

The drum, fixed on two supports, rotates whereas the drive in and out is realized in the plane perpendicular to the rotation axis, at half of the drum's height. Accordingly, the car park is partly located in the ground and part of it is exposed on the surface [3]. An exemplary rotary car park solution has been presented in Figure 1.

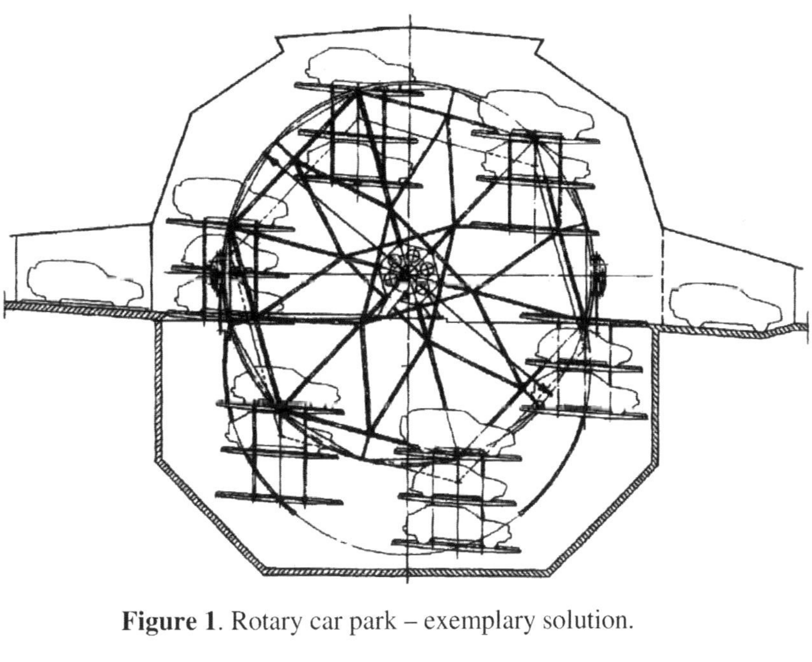

Figure 1. Rotary car park – exemplary solution.

5. Model of a rotary car park – computer simulation and optimization of parking parameters

The car park was modelled at two stages: control automation and automation of mechanical part.

5.1. Model optimization of control automation – computer simulation

The automation of the drive in-park-drive away parking was based on a modelled control presented in Figure 2.

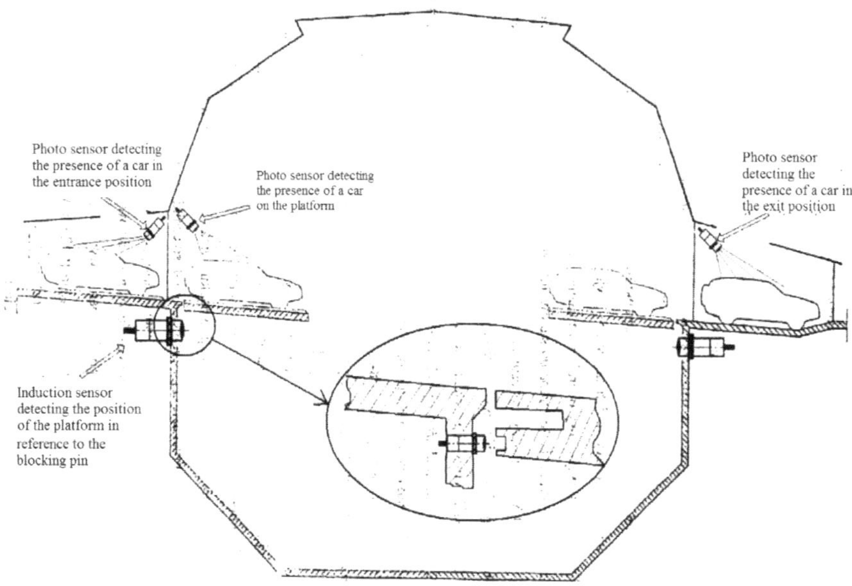

Figure 2. Distribution of sensors for automated control: drive in-park-drive away.

The automation tasks were realized with the use of a controller, e.g.: sensor Omron CJ1, providing 2560 signals of digital ins and outs, which in the case of the extended car park should extension of controller capacities. The software package Omron-CX-Programmer V4.0 was used for writing the program (Figure 2).

The program was divided into sections, which was facilitated by programming and easier use of the program.

Each section is responsible for particular elements of car park control, as in the table below:

Section 1	Entry to platform no. 1
Section 2	Entry to platform no. 2
Section 3	Exit from platform no. 1
Section 4	Exit from platform no. 2
Section 5	Initial conditions
Section 6	Major problem
Section 7	Activate sub-programs
Section 8	Check out the parking for free lots
Section 9	Calculate weights
Section 10	Select position on platform no. 1
Section 11	Select position on platform no. 2
Section 12	Calculate angle of parking rotation on entry
Section 13	Realization of parking rotation
Section 14	Start counting the parking time
Section 15	Select position of exit
Section 16	Calculate angle of parking rotation on exit
Section 17	Realization of parking rotation
Section 18	Stop counting the parking time.

5.2. Description of optimization program

The assumed optimization program could be used more intensely, therefore authors presented fragments of the program controlling the operation of the car park responsible for finding an optimum solution for work: cost, access lane and efficiency. An exemplary solution of a drive-in and drive-away from the platform have been presented in Figure 3 and in Figure 4, respectively.

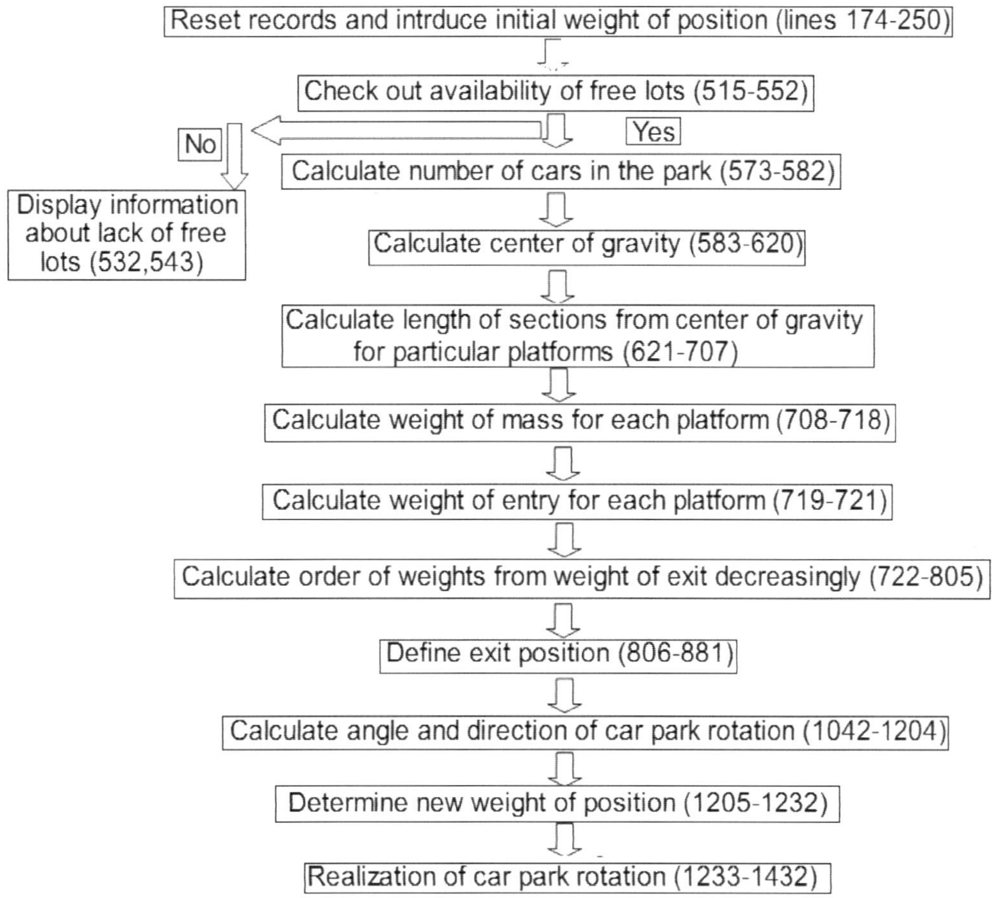

Figure 3. Algorithm of determining 'enter' position [2].

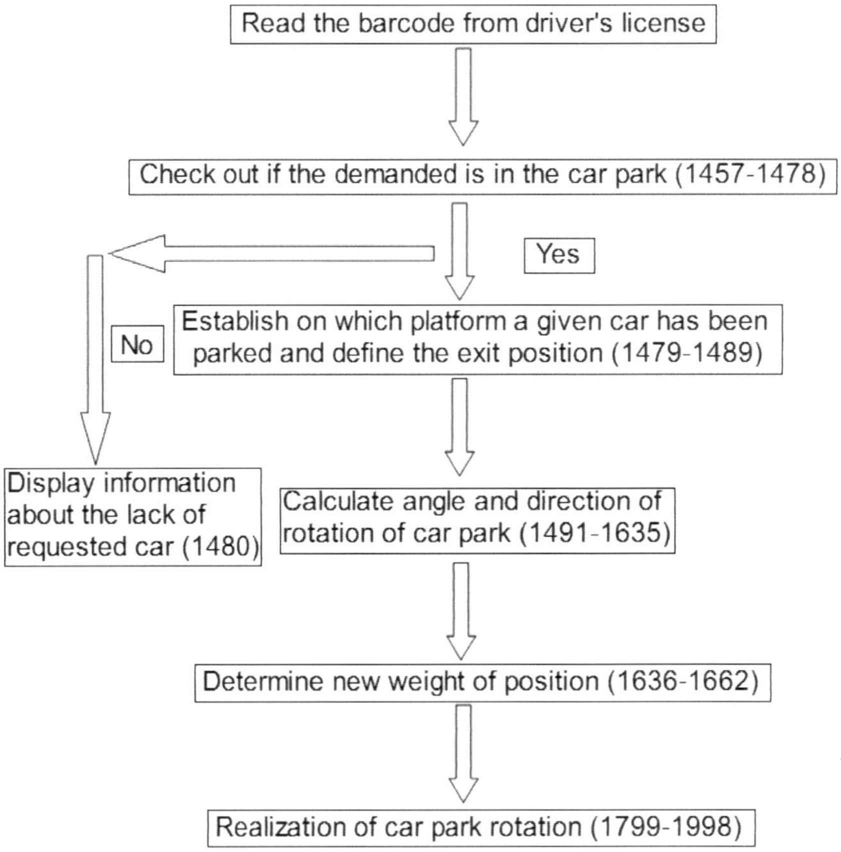

Figure 4. Algorithm of determining 'exit' position [2].

Knowing the number of cars and the distance of platforms from the centre of the car park, one can calculate the coordinates of the centre of gravity x_{cc} y_{cc}. The sub-program is used for calculating lengths of sections as a function of centre of gravity for each platform. Then the weights of mass of each of the platforms are defined on the basis of these lengths. After calculating the length of sections connecting the centre of gravity with each platform, the program sorts them out decreasingly, ascribing weights to each of them, where 1 corresponds to the platform staying closest to the center of gravity and 10 to the most distant one. In this way the number of the platform, onto which a given car should drive in to make the centre gravity closest to the centre of the car park, is established.

5.3. Verification of InTouch program

Now the process visualization systems are a very powerful and dynamically developing branch of automatic control. The visualization in the project have been realized with the versions of the InTouch 8.0 program. InTouch consists of three major programs: Application Manager, WindowMaker and WindowViewer [10].

The Application Manager of InTouch is used for managing created applications. It is also used for configuring WindowViewer as an NT service, configuring Network Application Development (NAD), both for client and server architectures, configuring Dynamic Resolution Conversion (DRC) and/or dispersed alarming. From the level of the Application Manager we can also activate database tools DBDump and DBLoad.

The WindowMaker program is an editing environment with object-oriented graphics used for creating anonymous windows and keys. These windows can be connected with industrial IN/OUT systems and also other applications of Microsoft Windows [10]. The WindowViewer is an Environment, where windows created with WindowMaker are displayed. WindowViewer performs scripts of InTouch. QuickScript collects archival data, processes alarm signals, generates reports, may work as a client or server, both for DDE communication protocol or for SuiteLink[10].

6. Model optimization of construction part – computer simulation

The computer simulation of the rotary car park was performed with the use of program C Robot BAT, thanks to which the construction nodes could be defined and the static and dynamic strength calculated. The forces acting on the prototype in real conditions could be determined. The static models create possibility of determining elements and nodes of the parking.

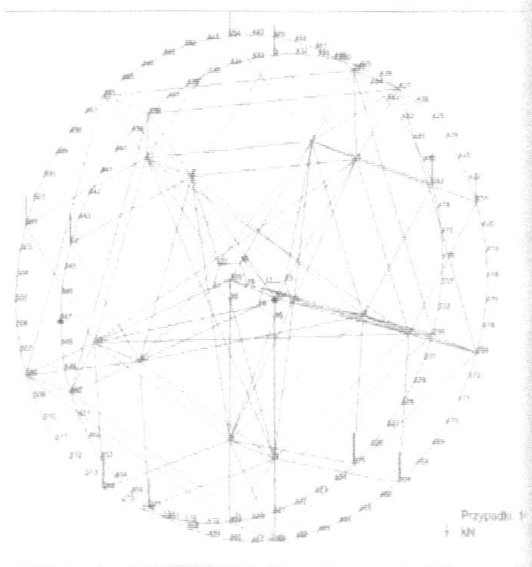

Figure 5. Exemplary numeration of nodes.

The results were presented in the form of plots. The curve of probability that a given momentum G was exceeded (M > Mk) has been presented at the assumed probability of drive-in/drive-out (Mode). The considered value: Mode={17, 33, 50, 67, 83}*100%. The number of trials in simulation MaxTrial = 10 000 was assumed on the basis of a comparison of results presented in Figure 6.

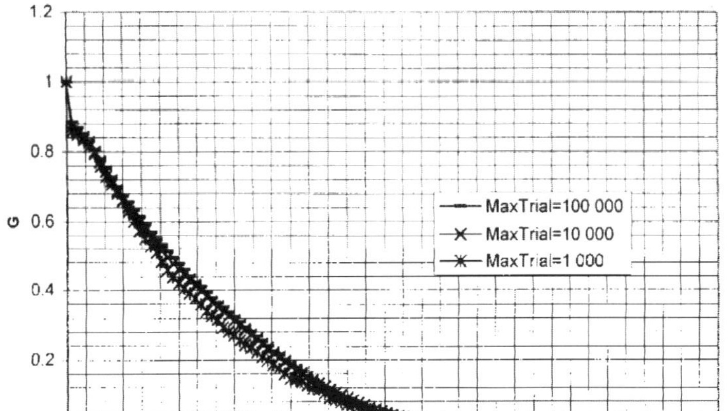

Figure 6. Plots G(M>M$_k$) obtained for different number of trials MaxTrial, Mode=50%.

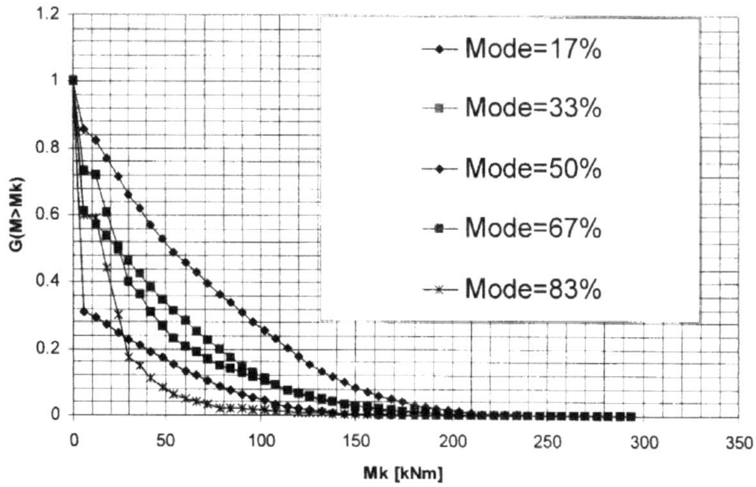

Figure 7. Curves of probability that M$_k$ is exceeded by momentum M for various frequencies of drive-ins/drive-outs.

The computer simulation of the car park allowed for defining significant construction nodes. By assuming real conditions of the car park in the ROBOBAT program, attempts could be made to perform strength calculations (exemplary calculations provided). This extends the scope of the discussed problem and creates possibilities of further realization, i.e. construction of a prototype.

7. Concluding remarks

A car park model was worked out and some constituent systems selected on the basis of the project. The rotary car park model PO – 10 was designed as a reduction model in scale 1 ÷ 12.5 (to fit in the available car models in this scale). Prior to the building of the model, computer simulations of the nominal model, as well as the construction and exploitation analyses were realized. Then the model was construed. It consisted of a general description, function and limitations, design project with documentation, electronic project with documentation, design project with documentation of the model. The model consists of a mobile part with platforms where the cars are parked, electronic part, i.e. drive of the drum with regulated rotational velocity, automatic part allowing the cars to automatically drive-in to the parking lot and then exit. An original and characteristic way of entering and leaving the parking lot is based on the gravity motion generated by the inclination of the parking lanes, being the parking places.

The experiments on the gravity-based car movement (entering and leaving the parking) were performed.

The model experiments revealed that a 2° inclination of the parking lanes on exit and entrance was sufficient. This inclination also suffices to stop the car before entering the platform and then after entering the platform in the place of parking.

The tests were performed in real conditions (on 5 cars of different weight) for the 2°- 4° angle of inclination of the drive lane revealed that the inclination of 2° was sufficient.

The drive of the model drum was solved by a gear with toothed belt and electronic change control of rotation velocity (In real conditions – figure: 'Drive of drum in a prototype of rotary car park PO-10" – further in the text). The rotation of the drum is activated by the drive system, where the motor drives the chain gear through the Nord flexible reducer. The pin chain is fixed in many points on the outer side of the drum. The total drivetrain i = 2550, which gives rotations of the drum n = 0.57 rpm.

Moreover by using the inverter to the electrical motor the rotations of the drum can be accelerated or slowed down.

The analysis of the reduction model, especially the reliability of the car movement within the car park area, necessitated seeking for technical solutions to the blocking problems. A construction solution of the drive in and out from the parking platform was also provided.

Acknowledgment
My hearty words of appreciation go to the late Professor Bolesław Stolarski, whose work has been presented in this paper. He introduced me to the field of mechanics, motor cars and modern technology. He is enthusiasm in finding new engineering solutions has had an immense impact on new generations of scientists.

References
[1] Borek J Mruk A Stolarski B 2005 Studium obrotowego parkingu samochodów *Mechanics and Mechanical Engineering* Vol.**9**, no.1
[2] Stolarski B et al. 2011 Modelowanie i optymalizacja obrotowego parkingu samochodowego, research project PK-WM
[3] Borek J Stolarski B 2005 Wielopoziomowy, wolnostojący, obrotowy parking samochodowy – studium Not Publisher work Kraków
[4] Korzeniowski W 2000 Parkingi i garaże dla samochodów osobowych, WC**0113**, Warszawa
[5] Kwaśniewski J 1999 Programowalne sterowniki przemysłowe w systemie sterowania J Kwaśniewski & Fundacja Dobrej Książki Kraków
[6] Urbanowicz H 1979 Napęd elektryczny maszyn roboczych. *WNT* Warszawa
[7] Miedziński B 2000 Elektrotechnika. Podstawy i instalacje elektryczne. *PWN* Warszawa
[8] Przeździecki F 1982 Elektrotechnika i elektronika. *PWN* Warszawa
[9] Niderliński A 1985 Systemy komputerowe automatyki przemysłowej t. **2**. *WNT* Warszawa
[10] Wonderware: InTouch 1997 Podręcznik użytkownika. Kraków
[11] Stolarski B 1990 Metody taksonomiczne w technologii samochodów – Monografia *PK* **97**
[12] Taguchi O 1996 Tokio Introduction to Quality Engineering Design Quality Engineering design Quality into Products and Process. Translation: Sekkeisha
[13] Kacznia J Polański Ł Praca dyplomowa (graduate work)
[14] http://www.woehr.de/-internetowy
[15] http://www.morski.com.pl/-internetowy
[16] http://www.sels.com.pl/index2php-internetowy
[17] https://www.youtube.com/watch?v=q3Kju88On7E

IOP Conf. Series: Materials Science and Engineering **148** (2016) 012052 doi:10.1088/1757-899X/148/1/012052

Study of the various factors influencing deposit formation and operation of gasoline engine injection systems

Z Stepien

Oil and Gas Institute – National Research Institute, 31-503 Krakow, Lubicz 25 A, POLAND

E-mail: zbigniew.stepien@inig.pl

Abstract. Generally, ethanol fuel emits less pollutants than gasoline, it is completely renewable product and has the potential to reduce greenhouse gases emission but, at the same time can present a multitude of technical challenges to engine operation conditions including creation of very adverse engine deposits. These deposits increasing fuel consumption and cause higher exhaust emissions as well as poor performance in drivability. This paper describes results of research and determination the various factors influencing injector deposits build-up of ethanol-gasoline blends operated engine. The relationship between ethanol-gasoline fuel blends composition, their treatment, engine construction as well as its operation conditions and fuel injectors deposit formation has been investigated. Simulation studies of the deposit formation endanger proper functioning of fuel injection system were carried out at dynamometer engine testing. As a result various, important factors influencing the deposit creation process and speed formation were determined. The ability to control of injector deposits by multifunctional detergent-dispersant additives package fit for ethanol-gasoline blends requirements was also investigated.

1. Introduction

Emission control and energy conservation policies are the main drivers for change in engine and fuel technology. Modern spark ignition engines are designed to use high octane fuel as for example ethanol-gasoline fuels for optimal fuel efficiency. Moreover, ethanol heat of evaporation is about 280% higher versus gasoline and effective evaporative cooling of air-fuel mixture allows the compression ratio to be raised, providing a higher level of thermodynamic efficiency. Ethanol fuel emits less pollutants than gasoline, it is completely renewable product, allows to increase economic and energy independence state and has the potential to reduce greenhouse gases (GHG) emission. However ethanol-gasoline blends can present a multitude of technical challenges to engine operation including creation of very adverse deposits. It is well known that harmful deposits would build up inside an engine (intake ports, intake valves, combustion chambers) and inside as well as outside fuel injectors if commercial gasoline or the most alternative fuels (as for example ethanol fuels) – Fig. 1, did not contain effective deposit control additives [1-5]. Deposits in the fuel intake system can have detrimental influence on preparation of air-fuel charge leading to suboptimal engine performance. The challenge for modern additives is not only to keep vital engine and fuel systems parts clean, but they should also remove existing deposits [1, 2, 6 - 8]. Fuel injectors are highly sensitive even to small deposits in the regions where fuel is metered and atomized. A period of hot soak when engine is fully warmed up and next switched off when the injector tip temperature has been measured to reach about

100 °C is essential for PFI (Port Fuel Injection) deposit formation [9 - 11]. During the hot soak period, the lighter hydrocarbons, from small amount of fuel left at the injector tip, evaporate, leaving a thin film rich in heavy hydrocarbons on the orifice surfaces. These hydrocarbons oxidize to form gums and resins, and this sticky varnish can bake into hard deposit given sufficient time [9 - 14]. These deposits, mostly from chemical (incompatible additives with base fuel) or thermal degraded fuel (high temperature pyrolysis) and its poor stability can restrict fuel flow and alter the spray pattern, which can deteriorate drive ability, decrease power and fuel economy as well increase exhaust emissions [1, 12 - 18]. Therefore if a fuel injector becomes clogged due to gasoline deposits, the injector will not be able to supply the proper amount of fuel to air ratio, leading to a sluggish, inefficient motor.

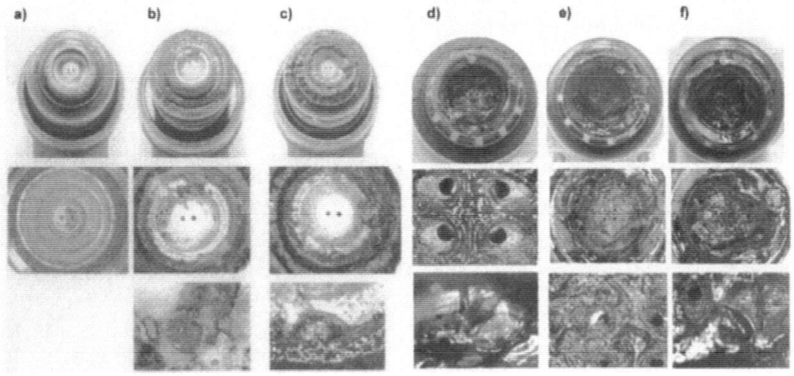

Figure 1. Injector appearance of Mercedes M111 PFI system (a – clean injector), deposits after 60 hours test with unadditized E10 fuel (b, c) and FORD 1.8L Duratec-HE PFI FFV injector deposits after 100 hours tests of various unadditized E85 fuel (d, e, f).

The most serious consequences of injector fouling, especially in case of direct injection, are pre-ignition, and engine misfiring and malfunction [14 - 16].
Furthermore, deposits in spark ignited engines are formed not only by the fuel influence but also by engine lubricating oil flow and especially its high consumption quantity, blow-by gases (positive crankcase ventilation - PCV) and combustion gases (exhaust gas recirculation – EGR). Particulates such as airborne dirt and solids from combustion products brought into the intake system by EGR and PCV systems may have also considerably affect on deposit formation. A great significance for deposit formation have also engine operation conditions in connection with fuel composition [1, 8, 9].
Regardless of factors causing both engines as well injectors harmful deposits it appears that fuel treatment with efficient detergent additives compatible and fitted for fuel composition are the most effective method in controlling and counteracting deposit formation.
The objective of this paper is presentation of study results evaluating the influence of various causative factors of deposit formation and operation of gasoline engine PFI system. Dynamometer engine simulation tests were crucial for deposit reproducing and special test bench including fully adjustable actual engine fuel injection system was used for assessment of generated injector deposits on this injection system operation.

2. Materials and methods

2.1. Test fuels

The E20 and E85 fuels used in dynamometer engine tests were splash blends consisting of fuel grade ethanol and base (unadditized) E0, RON 98 unleaded gasoline – Table 1. In one test, conventional gasoline - RON 95 complying with EN 228, from the gasoline station, was used to the preparation of the E85 fuel. This fuel was composed of the 15 % v/v commercial RON 95 gasoline blended with 85

IOP Conf. Series: Materials Science and Engineering **148** (2016) 012052 doi:10.1088/1757-899X/148/1/012052

%v/v fuel grade ethanol. In addition in one test E85 base fuel was treated with developed in Oil and Gas Institute – National Research Institute detergent-dispersant additive dedicated for high ethanol-gasoline blends. The additive containing a detergent-dispersant DEM1 of N-alkylated benzoxazine derivative structure, possessing hydrogenated heterocyclic ring and aromatic ring substituted by an alkyl and a synthetic carrier oil of oxyalkylated alkyl phenol structure and an aromatic solvent.

Table 1. Physicochemical properties of the E0, E20 and E85 fuels

Properties	Base Gasoline E0	Fuel E20	Fuel E85	Test Methods
Research Octane Number	98,1	100,3	108,2	PN-EN ISO 5164
Motor Octane Number	89,5	90,8	93,7	PN-EN ISO 5163
Lead, mg/l	<2,5	-	-	PN-EN 237
Density in 15^0C, kg/m^3	725,6	748,0	784,0	PN-EN ISO 12185
Sulphur, mg/kg	<3	<2	<0,5	PN-EN 20846
Copper, mg/kg	-	-	<0,05	EN 15837
Phosphorus, mg/l	-	-	<0,15	EN 15487
Induction period, min.	>480	>360	>360	PN-ISO 7536
Gums content, mg/100ml:				PN-EN ISO 6246
- unwashed		5,4		
- washed	<1	0,3	1,0	
Copper corrosion	1A	1A	1A	PN-EN ISO 2160
Benzene, %(*V/V*)	0,25	-	-	PN-EN 238
Oxygen, %(*m/m*)	0,0	7,4	-	PN-EN 1601
Ethanol, %(*V/V*)	<0,1	20	70,5	PN-EN 1601
Methanol, %(*V/V*)	-	-	0,3	EN 1601
Inorganic chloride mg/kg	-	-	<4,0	EN 15492
Vapor pressure (DVPE), kPa	57,8	67,0	45,2	ASTM D 4953
Distillation characteristics:				PN-EN ISO 3405
IBP, °C	33,4	35,7	-	
- up to 70^0C distilled volume, %(*V/V*)	21,1	48,2	-	
- up to 100^0C distilled volume, %(*V/V*)	52,5	71,2	-	
- up to 150^0C distilled volume, %(*V/V*)	93,1	91,2	-	
- end of distillation, ^{0}C	181,9	184,6	-	
- residues, %(*V/V*)	1,0	1,0	-	

2.2. Test benches and procedures

A special engine test bed and in-house test procedure were developed to study the effect of fuel composition, fuel additives, engine systems design, engine operating parameters and quantity of lubricating oil consumption on PFI system injector deposits. As a test engine on the test bench was used FORD 1.8L Duratec-HE PFI FFV (125PS) – Table 2.

The engine was operated on the engine test bench according one of the two newly (in-house) established 100 hours tests cycles which were prone to build up injector deposits. Both tests cycles included 4 stages differ in engine speed, load and time of duration. The test was automatically performed up to total cycles time of 100 hours. The first quasi static operation pattern test assumed mild operation condition, and the second one more severe operation conditions by more dynamic pattern, higher engine load and speed. The speed / load profiles for both test cycles are shown graphically in Fig. 2.

In order impact assessment of injector deposits formation on injector operation, directly after engine simulation test, not only visual rating of the external injector tips coking deposit was performed but also checking and assessment fuel injectors flow and spray pattern degradation on the special in-house flow bench. The main part of the flow bench was complete PFI system of FORD 1.8L Duratec-HE PFI FFV (125PS) engine connected to the control block allowing for variable test fluid (instead gasoline) pressure adjustment range (from 0 to 0,6 MPa) as well provides constant control over fuel injection variables such as injector pulse width (from 0,02 ms to 50 ms), injector frequency opening (from 15 Hz to 150 Hz) and operating time (from 1 s to 3 h) per test – Fig. 3.

Table 2. Specification of the test engine FORD 1.8L Duratec-HE PFI FFV (125PS)

Engine operating cycle	Four-stroke, spark ignition engine
Fuel delivery	Indirect fuel injection
Configuration	Straight 4
Aspiration	Naturally aspirated
Firing order	1-3-4-2
Valvetrain	DOHC (chain driven) /4 VPC
Cylinder bore	83,0 mm
Piston stroke	83,1 mm
Displacement	1798 cm^3
Power	125 bhp (92 kW) @ 6000 rpm
Torque	165 Nm @ 4000 rpm
Compression ratio	10,8 : 1
Medium fuel consumption (E85)	10,5 l/100 km
Tappet clearance	Hydraulically adjustment
Oil capacity with filter	4,3 dm^3
Engine emission level	Euro IV

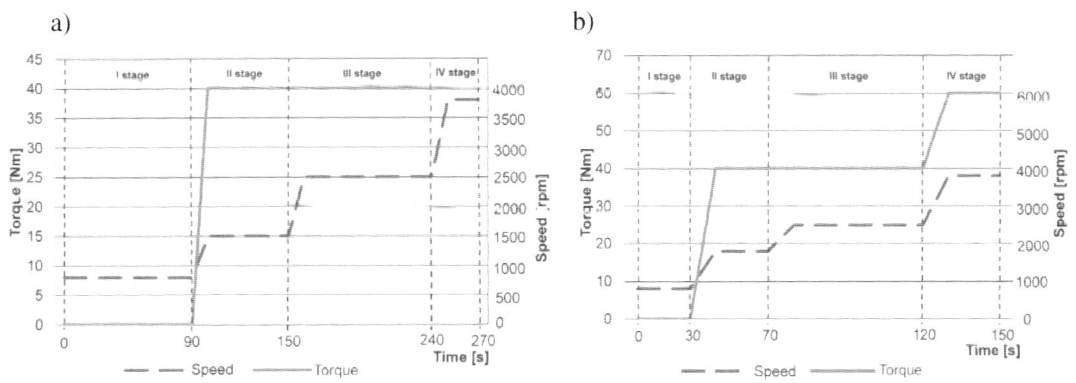

Figure 2. Engine test cycles. a) mild operation conditions, b) more severe operation conditions.

3. Tests results

Within the frame of the project a research study of various factors influencing injectors deposit formation and their operation in a PFI gasoline FFV engine were carried out. Since high ethanol-gasoline fuel blends are viewed as a potential alternative fuels to decrease fossil CO2 emissions and dependence on crude oil, such fuels formed the basis of the study. As in the case of fossil fuels, FFV fueled by ethanol-gasoline fuels deliver optimal performance when the sensitive areas in the intake system and in the fuel injectors are kept clean that is free of deposits. No standard test is available to assess the tendency of ethanol-gasoline blends to form intake system and injector deposits. So, a proprietary engine test bench and devised in-house test method in two versions was used to provoke by various factors and next to evaluate tendency of deposits formation in and on the fuel injectors. Directly, after engine test, measurements of fuel quantity delivery by particular injectors for various parameters of their operation, and evaluation spray pattern degradation were performed.

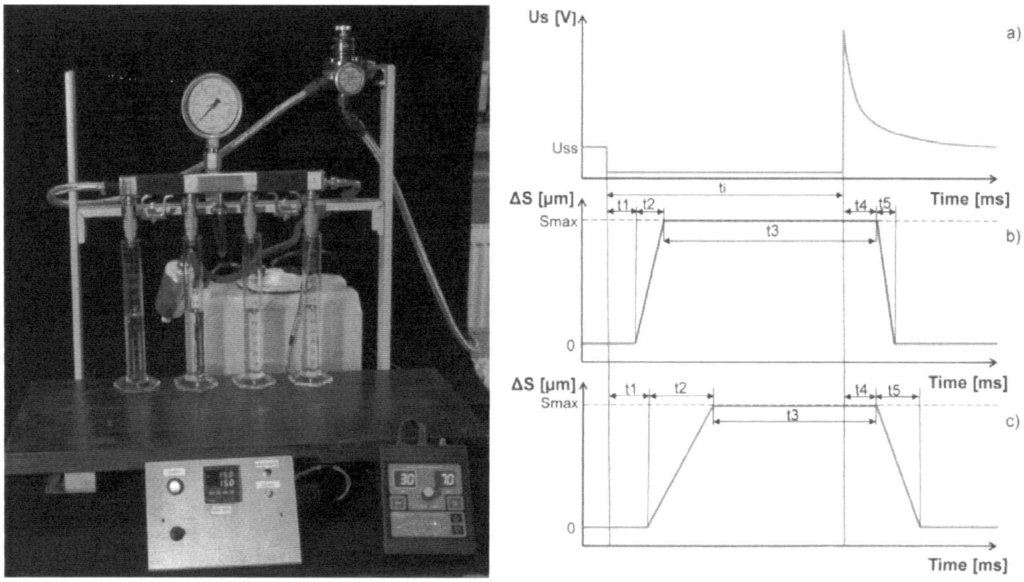

Figure 3. Flow bench for assessment of injectors operation and electronic pulse oscillation of injector control (a), correct course of raising and lowering injector metering needle (b), delayed by internal deposits course of raising and lowering injector metering needle (c)

- Evaluation of base (unadditized) E85 fuel impact for injector deposit formation of PFI FFV engine without EGR and blow-by in mild operation conditions (Fig. 2a) – Fig. 4a.

Port fuel injector deposits form especially during the hot soak period after an engine has been turned off. Fuel trapped in an injector tips is exposed to a higher temperature for a longer time than fuel that flowed through the injector when the engine was running, and forms both external coking deposits as well internal deposits especially in the absence of efficient deposit control additive. Therefore injector flow assessments were performed 24 h after engine bench test. Since injectors are designed for pulsed operation they were tested by pulsing. It is well known that engine and fuel injector deposits are formed not only by the fuel and its treatment but also by engine lubricating oil, blow-by gases (positive crankcase ventilation – PCV) and EGR. Therefore in the first test was used base E85 fuel and engine without EGR and PCV.

After the test was observed negligible external injector deposits, and asymmetrical control liquid spray pattern for injector No. 4. In this case droplets settling on the surfaces of intake manifolds is possible and as a result deterioration HC emissions due to increased wall wetting and worse air-fuel mixing. There is also evidence injector-to-injector diversification of fuel quantity delivery for rail pressure 0,38 MPa – Fig 4a.

- Evaluation of base (unadditized) E85 fuel impact for injector deposit formation of PFI FFV engine without EGR and blow-by in more severe operation conditions (Fig. 2b) – Fig. 4b.

The only difference between first and the second test are more severe engine operation conditions. This change led to a spray pattern degradation in the case second and fourth fuel injectors. Taking into account lack of external injector deposits, the reason of observed spray pattern degradation is probably due to internal injector fuel channels deposits. This increase possibilities of droplets setting on the surfaces of intake manifolds and early described negative results.

The diversification of fuel quantity delivery by particular injectors are smaller, but it should be remembered that new set of fuel injectors usually flowing within 2 – 3% of one another. However following injector deposit formation, the flow rate varies increasingly from injector to injector and the ECU continually correcting fuel rate up to about 10% by adjusting individual injector pulse width to balance fuel delivery. If the flow rate varies by too much from injector to injector, the engine will cause drive-ability problems.

a) b)

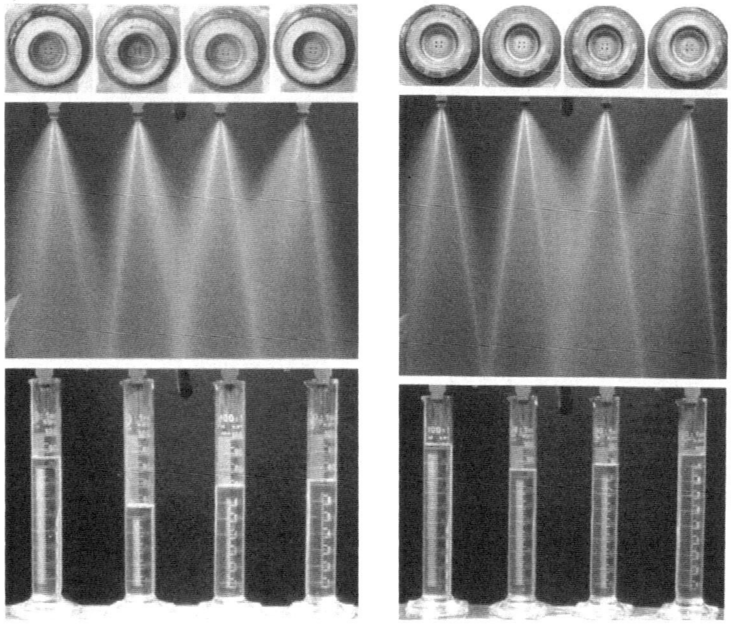

Figure 4. View of injector tips deposits, spray patterns and quantity of delivered fuel by injectors after bench test of PFI FFV engine

a) without EGR and blow-by run with base E85 fuel in mild operation conditions
b) without EGR and blow-by run with base E85 fuel in more severe operation conditions

- Evaluation of base (unadditized) E85 fuel impact for injector deposit formation of PFI FFV engine with EGR and blow-by in mild operation conditions (Fig. 2a) – Fig. 5a.

The results in Fig. 5a confirm that EGR and blow-by gases which are directed into intake system with E85 base fuel injected by PFI system, as a result each other interaction, to form significant injector deposits. Spray pattern degradation due to external injector deposits in combination with diversification in spray penetration resulting in poorer engine performance and increased HC emission. It is clear also significant injector-to-injector difference of fuel quantity delivery for rail pressure 0,38 MPa, but also for lower pressure - 0,12 MPa, Table 3.

- Evaluation of base (unadditized) E20 fuel impact for injector deposit formation of PFI FFV engine with EGR and blow-by in mild operation conditions (Fig. 2a) – Fig. 5b.

Reduction of ethanol content in the fuel to 20%v/v (E20) resulted in significantly less injectors tips deposits for comparison to E85 fuel – Fig. 5a and 5b. But despite that offset spray pattern especially third injector indicate the possibility of droplets setting on the surface of intake manifolds and increasing HC emission. Moreover poor injector-to-injector metering consistency has been observed probably as a result of deposit build-up inside internal injector flow holes and channels – Fig. 5b.

- Evaluation of base (unadditized) E85 fuel (including 15%v/v commercial standard gasoline) impact for injector deposit formation of PFI FFV engine without EGR and blow-by in mild operation conditions (Fig. 2a) – Fig. 6a.

Detergent additives for EN 228 gasoline application are compatible with an ethanol content up to 10%v/v but not always for higher ethanol content because of poor solubility of some conventional additives in ethanol. So, higher additive treat rates or different detergent-dispersant additive chemistry may be necessary for optimal detergent performance of ethanol-gasoline fuel.

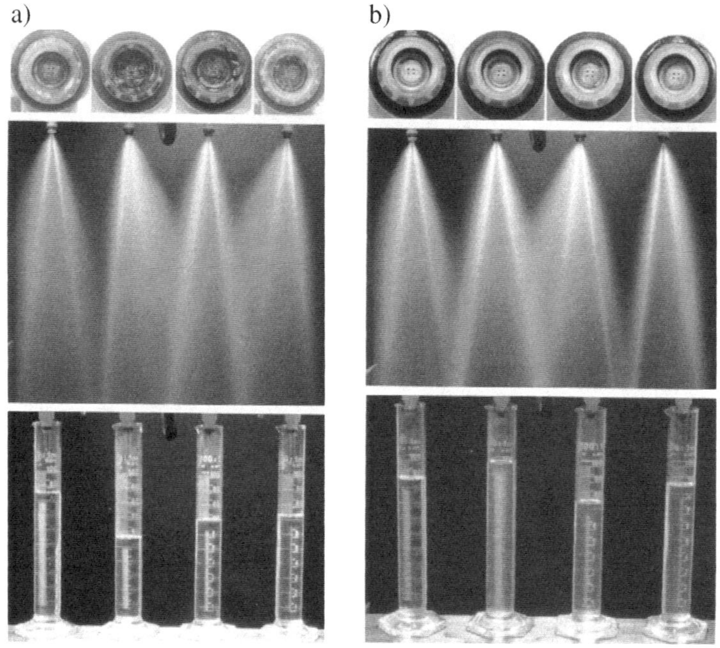

Figure 5. View of injector tips deposits, spray patterns and quantity of delivered fuel by injectors after bench test of PFI FFV engine
a) with EGR and blow-by run with base E85 fuel in mild operation conditions
b) with EGR and blow-by run with base E20 fuel in mild operation conditions

Presented results of the test – Fig. 6a, represent example where traditional detergent additives used to treatment of commercial gasoline, constituting 15%v/v tested E85 fuel, have solubility problem in the high ethanol fuel and through the incompatibility leading to external and internal injector deposit formation. Significant injector deposits caused considerable diversification in spray pattern and cone angle between injectors. This leads to deterioration HC emissions due to increased wall wetting and worse air-fuel mixing – Fig. 6a.

Injectors flow measurements both for rail pressure 0,38 MPa as well as for 0,12 MPa showed very poor injector-to-injector metering consistency and very significant differences of fuel delivery between injectors – Fig. 6a, Table 3. Increasing rail pressure to 0,5 MPa for injector pulse width 0,1 ms led to switch off two injectors operation (No. 2 and 3) through injector internal deposits build up. Internal deposits can slow the injector response for control electric signal and change dynamic response characteristics resulting in a loss of control of injection event timing and/or amount of fuel delivered. Injector deposits, both external-coking as well as internal will increase engine-out emissions and deteriorate engine performance.

- Evaluation of base (unadditized) E85 fuel impact for injector deposit formation of PFI FFV engine with EGR and blow-by and increased engine lubricating oil consumption in mild operation conditions (Fig. 2a) – Fig. 6b.

Increased engine lubricating oil consumption was achieved through application of used intake valves stem seals. Lubricating oil consumption was at the level 0,8 liter per 100 h of engine operation in mild conditions. 24 h after engine bench test (injector soaking period) has been stated that a few injectors fuel holes are blocked by heavy, black, sticky deposits. This had an impact on diversification in spray patterns and fuel atomization quality injector-to- injector. Reduced test fluid flow by fouled injectors holes has resulted in significant diversification of fuel quantity delivery – Fig. 6b especially for rail pressure 0,38 MPa – Table. 3.

- Evaluation of treated E85 fuel impact for injector deposit formation of PFI FFV engine with EGR and blow-by in mild operation conditions (Fig. 2a) – Fig. 6c.

E85 base fuel was treated with developed in Oil and Gas Institute – National Research Institute detergent-dispersant efficient additive specifically designed and optimized for high ethanol-gasoline blends. The results of a research test proved that specifically developed for high ethanol-gasoline blends detergent-dispersant additives can be effective in preventing formation of internal and external injector deposits (keep clean) and remove already formed deposits (clean up) – Fig. 6c.

After engine test on the injectors tips have not been found any deposits formation, and diversification of spray pattern was very small. Very good injector-to-injector metering consistency has been ascertained both for rail pressure 0,38 MPa as well as for 0,12 MPa – Fig. 6c. This contributes to the notion that deposit control additives provide means to effectively and efficiently modify fuel properties counteracting injector deposit formation.

Considering fuel injector design, improved injector deposit resistance can be achieved due to laser drilling and smoother surface in flow holes.

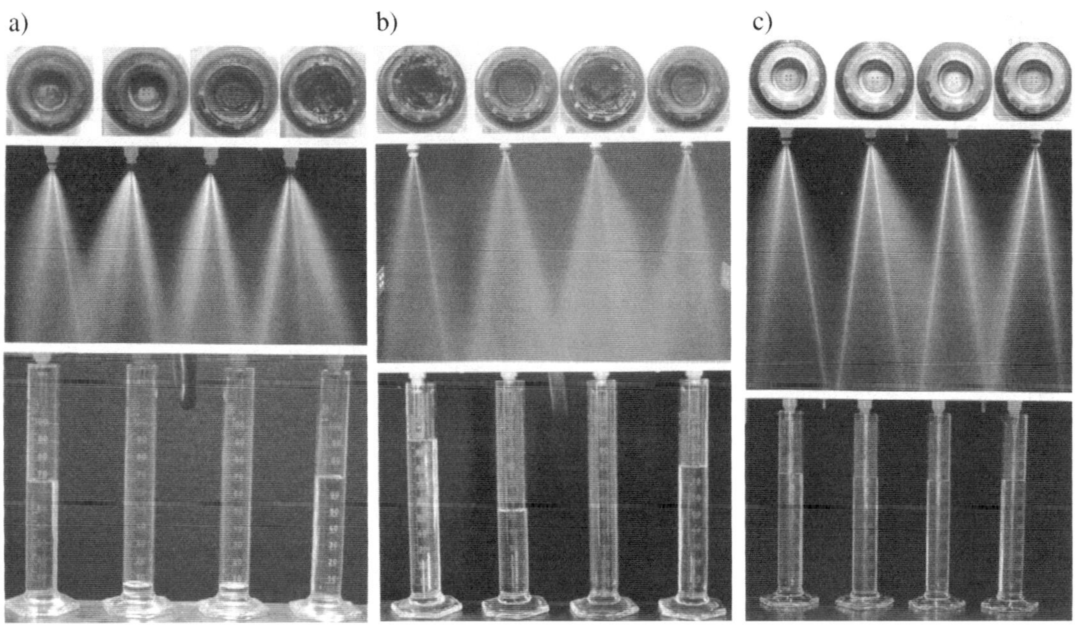

Figure 6. View of injector tips deposits, spray patterns and quantity of delivered fuel by injectors after bench test of PFI FFV engine: a) without EGR and blow-by run with base E85 fuel (including 15%v/v commercial standard gasoline) in mild operation conditions, b) with EGR and blow-by and increased engine lubricating oil consumption run with base E85 fuel in mild operation conditions, c) with EGR and blow-by run with treated E85 fuel in mild operation conditions.

All tests and results have been summarized in Table 3 below.

Table 3. Breakdown of all carried out tests and obtained results.

Fuel and assessing conditions	Figure No.	Quantity of injected fuel [ml]		External injector tips deposits	Quality of fuel atomization
		Rail pressure 0,12 [MPa]	Rail pressure 0,38 [MPa]		
Unadditized E85 without EGR and blow-by – low engine load	5	68 52 60 63	87 58 71 74	Negligible deposits	For injector No 4, asymmetrical control liquid spray pattern, too broad on its bottom. Possibility of droplets settling on the surfaces of intake manifolds.
Unadditized E85 without EGR and blow-by – higher engine load	6	80 71 74 82	103 85 90 98	Negligible deposits	For injector No 2 and 4, asymmetrical control liquid spray pattern. For injector No. 4 spray too broad on its bottom. Possibility of droplets settling on the surfaces of intake manifolds.
Unadditized E85 with EGR and blow-by – low engine load	7	68 52 60 63	87 58 71 74	Significant resin and lacquered deposits especially on the injectors of the second and fourth cylinder	For injector No 2 and 4, asymmetrical control liquid spray pattern, too broad on its bottom. Possibility of droplets settling on the surfaces of intake manifolds.
Unadditized E20 with EGR and blow-by – low engine load	8	71 81 66 70	88 99 76 87	Negligible deposits	For injector No 3, asymmetrical control liquid spray pattern, too broad on its bottom. Possibility of droplets settling on the surfaces of intake manifolds.
Unadditized E85 including 15% commercial standard gasoline, without EGR and blow-by – low engine load	9	67 64 50 69	67 7 6 71	Heavy, black, sticky deposits especially on the injectors of the third and fourth cylinder	For injector No 2 and 4, asymmetrical control liquid spray pattern. For injector No. 2 and 4 spray too broad on its bottom. Possibility of droplets settling on the surfaces of intake manifolds.
Unadditized E85 with EGR and blow-by – low engine load (increased engine lubricating oil consumption)	10	77 61 59 71	100 54 0 76	Heavy, black, sticky deposits especially on the injectors of the first and third cylinder	For injector No 3, asymmetrical control liquid spray pattern, too broad on its bottom. Possibility of droplets settling on the surfaces of intake manifolds. Quality of the control liquid spray atomization varies between the injectors.
Treated by additive dedicated for E85 with EGR and blow-by – low engine load	11	72 70 71 71	88 87 86 87	Clean injectors	The correct shape of the control liquid spray pattern, and good quality of spray atomization in the injectors.

The work reported here is resulting from the project "Influence of bioethanol fuels treatment on the operational performance, ecological properties and GHG emissions of spark ignition engines (BIOTRETH). The project was financed by the European Economic Area (EEA) Grants / Norway Grant, contract No. Pol-Nor / 199100/6/2013.

4. Conclusions

- Evolving engine construction and fuel technology defines additive requirements including cleanliness challenge for counteract fuel injector external and internal deposits
- Reduced injector deposit build-up improves performance over entire injector lifetime
- Injector deposits, both external – coking as internal will increase engine-out emissions and deteriorate engine performance
- Properly designed fuel additives required to control injector deposits
- Detergent-dispersant additive system will be critical in controlling deposit forming tendencies
- Fuel deposit control additives enable the safe and efficient utilization of fuel
- Regardless of the reasons injector deposits build-up and their location the most effective means to prevent these harmful phenomenon are fit for fuel composition detergent-dispersant additives

References

[1] Stępień Z 2015 Multidirectional investigations of high-ethanol fuels on deposit formation in spark ignition engines; *Combustion Engines*. 3/2015, **162(3)**, p. 608-618. ISSN 2300-9896, PTNSS-2015-3418
[2] Stępień, Z 2015 Types of internal Diesel injector deposits and counteracting their formation; *Combustion Engines*. 4/2015, **163(4)**, 79-91, ISSN2300-9896, PTNSS-2015-410
[3] Stępień Z 2014 Intake valve and combustion chamber deposits formation – the engine and fuel related factors that impacts their growth; *Nafta-Gaz*, No. **4/2014** p. 28 – 34. Wydawnictwo Instytutu Nafty I Gazu ISSN 0867-8871
[4] Bergström K, Melin S-A and Coleman J. 2007 The New ECOTEC Turbo BioPower Engine from GM Powertrain – Utilizing the Power of Nature's resources; *28th Internationales Wiener Motorensymposium 2007*
[5] Hauet B, Grand JG, Jouron C and Tran-Dinh C. 2010 E85: Impact of the fuel properties on „flex fuel" engine tuning; *31 Internationales Wiener Motorensymposium 2010*
[6] Stępień Z 2015 Zagrożenia eksploatacyjne stwarzane przez paliwa etanolowe dla silników o zapłonie iskrowym; *Prace Naukowe Instytutu Nafty i Gazu - Państwowego Instytutu Badawczego* No **204**, Paliwa alkoholowe dla transportu – uwarunkowania, badania i rozwój ISSN 2353-2718, DOI: 10.18668/PN2015.204
[7] Żak G, Ziemiański L, Markowski J, Krasodomski W, Wojtasik M, Bujas C and Stępień Z. 2015 Ocena właściwości detergentowych mieszanek benzyny z wysoką zawartością etanolu w zależności od zastosowanych formulacji pakietów dodatków; *Prace Naukowe Instytutu Nafty i Gazu – Państwowego Instytutu Badawczego* No **204**, Paliwa alkoholowe dla transportu – uwarunkowania, badania i rozwój ISSN 2353-2718, DOI: 10.18668/PN2015.204
[8] Schwahn H and Lutz U. 2010 Deposit Formation of Flex Fuel Engines Operated on Ethanol and Gasoline Blends *SAE Paper* No **2010-01-1464**
[9] Kalghatgi G 2013 Fuel/Engine Interactions; *SAE International*, ISBN 978-0-7680-6458-2
[10] Taniguchi B Y, Peyla R J, Parsons G M, Hoekman S K and Voss D A 1986 Injector Deposits-the Tip of Intake System Deposit Problems; *SAE Paper* No. **861534**
[11] Abramo G P, Horowitz A M, Trwella J C 1986 Port Fuel Injector Cleanliness Studies *SAE Paper* No **861535**
[12] Benson J D and Yaccarino P 1986 Fuel and Additive Effects on Multiport Fuel Injector Deposits; *SAE Paper* No. **861533**
[13] Tupa R and Koehler D E 1986 Gasoline Port Fuel Injectors – Keep Clean/Clean Up with Additives *SAE Paper* No. **861536**
[14] Richardson C B, Gyorog D A and Beard L K 1989 A Laboratory Test for Fuel Injector Studies; *SAE Paper* No. **892116**
[15] Hongming Xu, Chongming Wang, Xiao Ma, Asish K and Adam Weall, Jens Krueger-Venus 2015 Fuel injector deposits in direct-injection spark-ignition engines; *Progress in Energy and*

IOP Conf. Series: Materials Science and Engineering **148** (2016) 012052 doi:10.1088/1757-899X/148/1/012052

Combustion Science **50 (2015)** p. 63 – 80
[16] Aradi A A, Hotchkiss A, Imoehl B, Sayar H and Avery N.L. 1999 The effect of fuel composition, engine operating parameters and additives on injector deposits in a high-pressure direct injection gasoline (DIG) research engine. *SAE Technical Paper*, No. **1999-01-3690**
[17] Rosner D and Seshadri K. 1990 Experimental and theoretical studies of the laws governing condensate deposition from combustion gases, *18th Symposium on Combustion, Waterloo, Canada 1990*
[18] Davis J 1983 A New Theory of Aerosol Deposition From Turbulent Fluids; *Chemical Engineering Science*, **Tg. 38 (1)**

The use of surface layer with boron in friction pairs lubricated by engine oils

W Szczypiński-Sala[1] and J Lubas[2]
[1] Cracow University of Technology
[2] Dept of Combustion Engines and Transport, Rzeszow University of Technology

E-mail: ws@mech.pk.edu.pl

Abstract. The aim of the present work is to determine the influence of surface layers with boron and engine oil on the processes of friction and wear in friction pairs. The ring samples with borided surface layer cooperated under test conditions with counterparts made with CuPb30 and AlSn20 bearing alloys. During the tests, the friction pairs were lubricated with 15W/40 Lotos mineral oil and 5W/40 Lotos synthetic oil. The lubrication of friction area with Lotos mineral oil causes the reduction of the friction force, the temperature in the friction area and the wear of the bearing alloys under study, whereas the lubrication with Lotos synthetic oil reduces the changes in the geometrical structure of the cooperating friction pair elements. Lubrication of the friction area in the start-up phase of the friction pair by mineral oil causes faster stabilization of the friction conditions in the contact area than in the cause of lubrication of the friction pair by synthetic oil. The intensity of wear of the AlSn20 bearing alloy cooperating with the borided surface layer is three times smaller than the intensity of use of the CuPb30 alloy bearing.

1. Introduction

Design and modernization of the means of transport includes problems with the reliability and the durability of the construction. The reliability and durability of the construction is the result of the selection of appropriate forming technology of elements and components working in kinematics systems of the means of transport. Modern technology aims to create the elements working under maximum unit pressure, at high relative speeds of movable elements and high temperature of operation of kinematics systems. The implementation of these processes is possible by the application of surface layers with suitable tribological properties, such as wear resistance, low friction coefficient, corrosion resistance and seize resistance. The development of production processes of surface layers of kinematic elements of friction pairs and the knowledge of forming the selected parameters of these elements allows for the use surface layers with boron in a friction pair as an element affecting the performance characteristics. The modification of the surface layers with boron can be performed using various technologies: diffusion of boron in the surface layer of materials and their alloys, spraying coating containing boron and as an added element of metal alloy in metallurgical processes [8].
Boronizing is a thermochemical process that is widely used for boride-type coating. Borides formed in steel surfaces significantly increases their hardness (about 2,000 HV), wear resistance, and corrosion resistance [4, 7]. The tribological properties of these layers depend on the physical state of boride source used, boronizing temperature, treatment time, properties of the boronized materials, as well as any possible thermal treatment [10]. Boronizing is very effective, especially on grey and ductile iron

or low-alloy steel with chromium [1, 2]. Current boronizing processes allow for obtaining surface layers with high hardness, improved wear resistance, and low friction coefficient [6, 8], with low brittleness and no tendencies toward cracking [2].

These properties cause the increasing interest in the use of these surface layers on the parts working under wearing conditions. However, the main interest in the literature is in the use of boron to form surface layer in the friction pairs operating under abrasive wear without lubrication.

The paper presents the influence of the type of engine oils on the conditions of friction and wear in the friction pairs with the elements containing boron in the surface layers. The tests were carried out under limited lubrication with the use of engine oil is order to determine the impact of the surface layer with boron and the lubricant on processes of friction and wear.

2. Experimental

The aim of the study is to determine the influence of the technological surface layer with boron on the friction parameters in sliding pairs under lubricated friction conditions. The research involves the determination of work parameters of friction pairs as a function of bearing material, engine oil and changeable parameters of the load of friction pairs. In the research, the friction pairs were composed of a ring sample and a counterpart made from bushing bearings (figure 1).

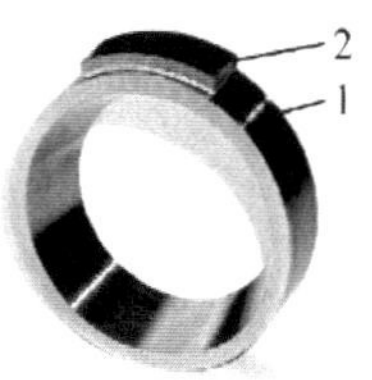

Figure 1. Friction pair; 1-ring sample, 2 counterpart

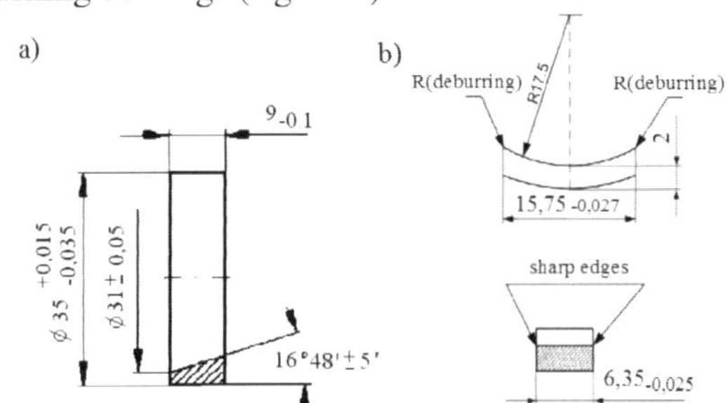

Figure 2. Shape and dimension of the elements of friction pair; a) ring sample, b) counterpart.

Figure 3. Microstructure of borided layer (x200).

Ring samples (figure 2a) were made of 46Cr2 steel and were borided in powder at a temperature of 950°C for 8 hours, and then were austempered. In the boronizing process, the powder of the following composition was used: B4C (30%), Al2O3 (68%), NH4Cl, and NaF. The borided layers obtained the characteristic structure ("tooth-shaped") with a hardness of 2000 HV and a thickness of 40 µm (figure 3). The ring samples with borided layer cooperated during the test with the counterpart made from AlSn20 and CuPb30 bearing alloys (figure 2b) and the area of friction was lubricated with the engine oils: Lotos mineral 15W/40 and Lotos synthetic 5W/40.

Stand test was conducted on a T-05 block on ring tester. The tests were executed by following a specified algorithm, which included the initial running-in of the samples and the correct co-operation process at the pre-determined load parameters. The running-in was executed on a test site at a pressure of 5 MPa until a complete adhesion of the ring specimen and the counterpart was achieved. At the starting phase, it was assumed that that the pair would be accelerated from a speed of 0 to 500 rpm in 30 s and the friction coefficient was measured as a function of pressure. The friction force, the temperature in the friction area and the wear of materials of the elements of friction pairs were tested under pre-determined conditions: the ring specimen rotational speed of 100 rpm and the unit pressure of 5, 10, 15 and 20 MPa.

3. Results of research

The work of friction pairs during start-up stage was characterized by high dynamics of tribological processes as a consequence of external loads which had an important influence on lubrication conditions. Assessment of changes is possible by registering the friction coefficient as a function of variable velocity and configuration of friction pairs (bearing alloys and the type of engine oils)

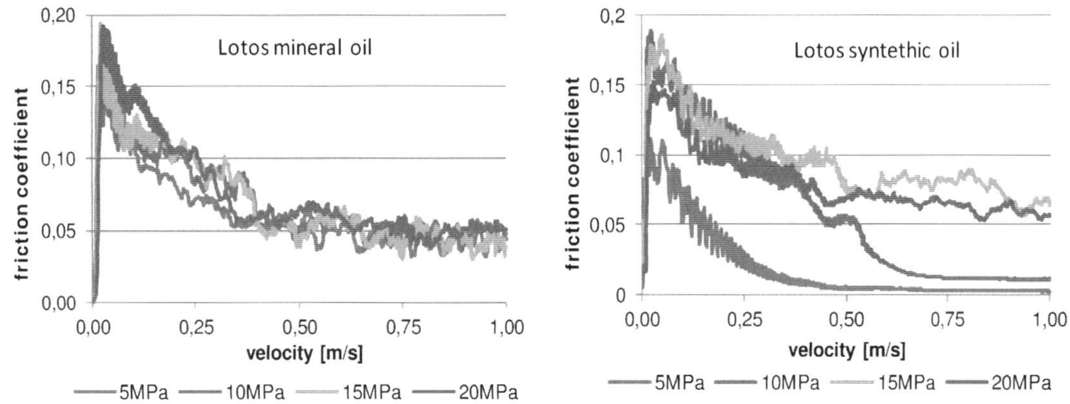

Figure 4. Friction coefficient in the friction pairs with the AlSn20 alloy counterpart lubricated by mineral oil and synthetic oil.

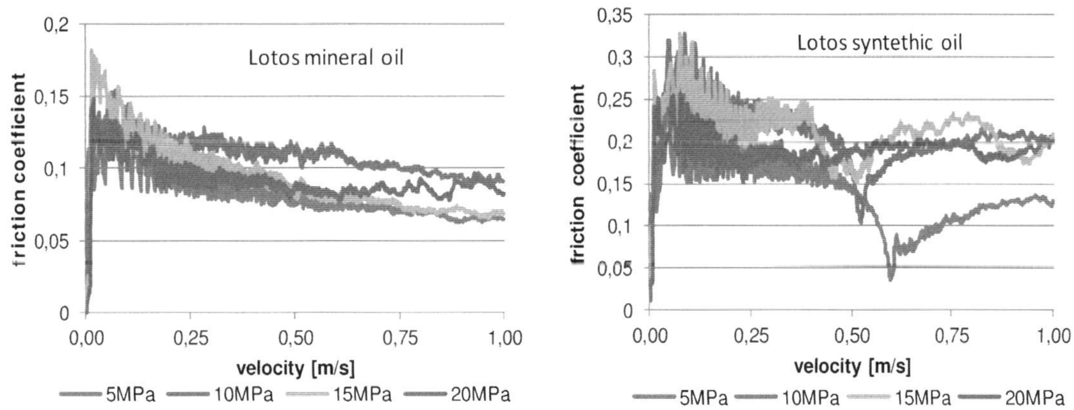

Figure 5. Friction coefficient in the friction pairs with the CuPb30 alloy counterpart lubricated by mineral oil and synthetic oil.

The record of change of friction coefficient in the start-up phase of friction pairs showed a fast growth of friction resistance up to the point of achieving a maximum value and then a decrease of friction resistance was observed with the growth of velocity (figure 4, 5). In the friction pairs lubricated with the mineral oil a significant decrease of the friction coefficient were observed. In the friction pairs with the AlSn20 bearing alloy counterparts the leveling of the compensation friction coefficient at the level μ~0.05 from speed of 0.5 m/s was observed (figure 4). But, in the case of friction pairs with the CuPb30 bearing alloy counterparts the scattering of the friction coefficient from μ=0.06 to μ=0.095 was observed. The lowest friction coefficient was recorded at the pressure of 5MPa, the maximum friction coefficient occurred at the pressure 10 MPa (figure 5). When used to lubricate the friction area the synthetic oil was observed to have had important change in the course of friction coefficient with reference to the registered in friction pairs lubricated with mineral oil. In the friction pairs with the AlSn20 bearing alloy counterparts stabilization of frictional resistance was observed, but there are two groups of pairs. At low unit pressures of 5-10 MPa there significant decrease of the friction coefficient below μ=0.02, and at high unit pressures of 15-20 MPa, the friction coefficient is stabilized at a level of μ=0.06 (figure 4). Important changes in the friction processes occurred in the friction pair with the CuPb30 bearing alloy counterparts which had high coefficient of friction; μ~0.2 for unit pressure from 10 to 20 MPa. In friction pairs at the pressure of 5 MPa, the friction coefficient stabilizes oat the level of μ=0.13 (figure 5). In these pairs, at the unit pressure of 5-10 MPa, a significant decrease of the friction resistance was observed at the velocity 0.5-0.6 m/s, then the friction coefficient increased and then it stabilized itself.

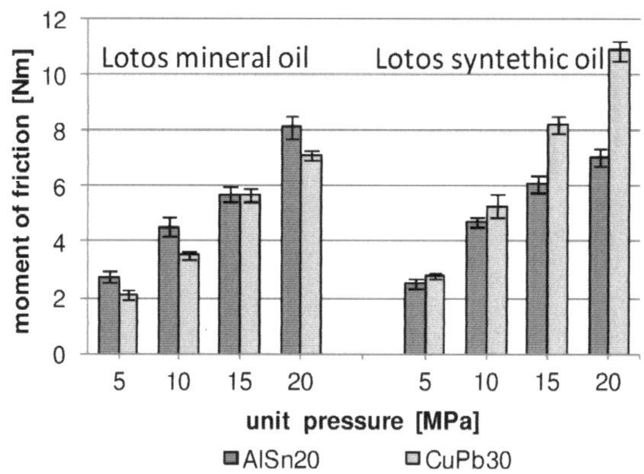

Figure 6. Influence of the bearing alloy and the type of engine oil on the start-up moment in function of the load of a friction pair.

Another significant aspect pertaining to friction pairs is determination of the value of the start-up moment. The results of the measurements of the maximum the start-up moment differed significantly. The differences depended on the type of bearing alloys used and the type of engine oils applied to lubricate the friction area (figure 5). The use of the Lotus mineral oil in the lubrication of friction area generated a lower start-up moment in pairs with the CuPb30 bearing alloy counterparts as compared to the pairs with the AlSn20 bearing alloy counterparts. The use of the Lotos synthetic oil in friction pairs changed the friction conditions. In the pairs with the AlSn20 bearing alloy counterparts a lower start-up moment was observed than in the friction pairs with the CuPb30 bearing alloy counterparts. By analysing the value of the moment of friction significant differences can also be seen between the pairs with CuPb30 and AlSn20 bearing alloy counterparts. Under the conditions of lubrication by Lotos mineral oil, the difference in the start-up moment between the researched pairs is ~24% at the pressure of 5MPa and it decreased with the increase unit pressure and at the unit pressure of 15 MPa, the start-up moment was at the similar level. However, at the pressure of 20 MPa, the start-up moment was again higher in the pairs with AlSn20 counterparts by about 12% than in the pairs with CuPb30 counterparts. The use of the Lotos synthetic oil for lubrication generated lower start-up moment in the pairs with AlSn20 counterparts by about 10 % (at the pressure 5 MPa) than in pairs with CuPb30 counterparts. The increase of the unit pressure also caused the increase of the start-up moment and at the pressure of 20 MPa, the difference between the researched pairs was more than 35%. The recorded maximum start-up moment can be the basis for the design of the friction pairs in terms of providing an adequate level of load at the time of start-up or creating adequate friction conditions. This allows one to determine the energy necessary to properly start the machine and allows for the design of friction pairs in terms of heat load. The differences between the start-up moment in the researched pairs are mainly the effect of the impact of the surface layer in the friction elements, the lubricant and the intensity of the formation of boundary layers which reduce the frictional resistance [9].

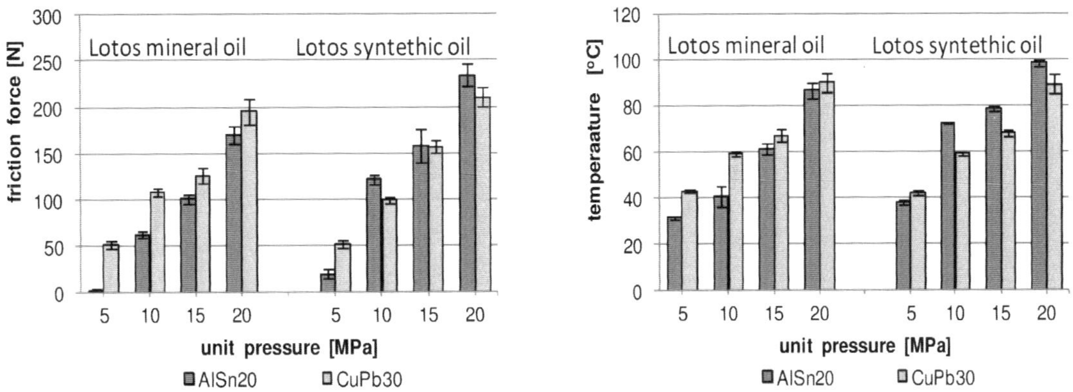

Figure 7. Influence bearing alloy and type of engine oil on friction forces and temperature depending on unit pressure (at 100 rpm and after 500 s).

The influence of engine oils on the friction conditions in the friction pairs with elements including boron in surface layers was tested under pre-determined conditions, at the ring specimen rotational speed of 100 rpm and at the jump unit pressure of 5, 10, 15 and 20 MPa. With the lubrication of the friction area by Lotus mineral oil, higher values of friction force were observed in friction pairs with the CuPb30 bearing alloy counterparts than in friction pairs with the AlSn20 bearing alloy counterparts. These changes were particularly important at lower unit pressures. An increase of the unit pressure caused a decrease of the percentage difference and when the unit pressures accounted for 20 MPa, the difference was 12%. A similar character of changes is observed during the recording of temperature in the friction area. The measurements showed higher temperature in friction pairs with the CuPb30 bearing alloy counterparts, compared to the friction pairs with the AlSn20 bearing alloy counterparts. Also, the difference of temperature between the researched friction pairs decreased with the increase of unit pressure (from 26% at the pressure of 5 MPa to 4% at the pressures of 20MPa). The lubrication by Lotos synthetic oil generated higher friction force and temperature in friction pairs with the AlSn20 bearing alloy counterparts. The trend of these changes was observed at the unit pressure of above 10 MPa. At low unit pressures of 5 MPa, the friction forces and temperatures in fiction pairs with the AlSn20 bearing alloy counterparts were lower than in the pairs with the CuPb30 bearing alloy counterparts. Under these conditions of lubrication in the friction pairs, the percentage difference between the friction pairs with the AlSn20 bearing alloy counterparts and the ones with the CuPb30 bearing alloy counterparts was 10% for friction force and 9% for temperature (at the pressures of 20 MPa).

In the friction contact area of the cooperating surface layers, the removal processes of the material occurred, caused by a relative movement of the friction elements. The results of the stand test have shown no measurable weight loss of the material of ring samples. Significant changes were observed only in the topography of the surface layer (Table 2). Instead, the measurement of wear of the bearing alloys has shown that the wear intensity of the AlSn20 alloy is significantly lower compared to the wear intensity of CuPb30 alloy (figure 5). Also, the application of the engine oils to the friction pairs has exerted an important influence on the wear of the bearing alloys - the use of mineral oil reduced the wear intensity of researched bearing alloys. In the conditions of lubrication by the mineral oil, the wear intensity of the CuPb30 alloy is almost three times higher than that measured for the AlSn20 alloy. On the other hand, under the lubrication by the synthetic oil, the maximum difference in the value of wear intensity of the researched bearing alloys is ~ 250%.

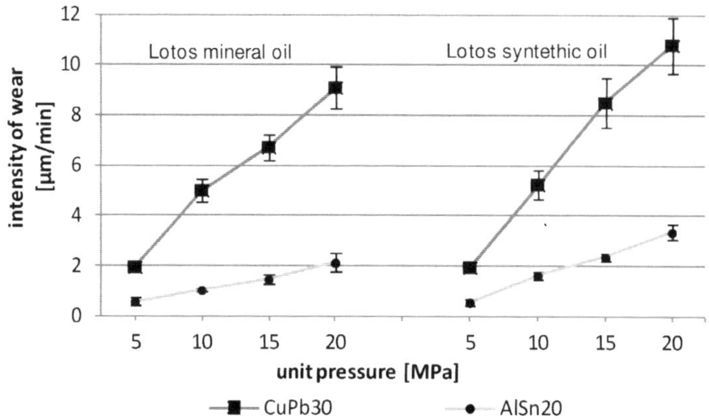

Figure 8. Wear intensity of bearing alloy material vs. unit pressure and type of engine oil used for lubrication of friction pairs.

In the friction contact area of the cooperating surface layers, the removal processes of the material occurred, caused by a relative movement of the friction elements. The results of the stand test have shown no measurable weight loss of the material of ring samples. Significant changes were observed only in the topography of the surface layer (Table 2). Instead, the measurement of wear of the bearing alloys has shown that the wear intensity of the AlSn20 alloy is significantly lower compared to the wear intensity of CuPb30 alloy (figure 5). Also, the application of the engine oils to the friction pairs has exerted an important influence on the wear of the bearing alloys - the use of mineral oil reduced the wear intensity of researched bearing alloys. In the conditions of lubrication by the mineral oil, the wear intensity of the CuPb30 alloy is almost three times higher than that measured for the AlSn20 alloy. On the other hand, under the lubrication by the synthetic oil, the maximum difference in the value of wear intensity of the researched bearing alloys is ~ 250%

Table 1. Surface roughness of ring samples and counterparts before the test [μm]

Parameter of surface roughness	Ring sample borided in powder	Counterpart	
		CuPb30	AlSn20
Ra	0.42	0.48	0.34
Rz	4.9	3.0	3.2
Ry	5.3	3.4	4.9
Sm	88	48	70

The measurements of surface roughness of the friction pair elements after the tests revealed significant changes in the Ra, Rz and Sm parameters of roughness, as compared to the values of those parameters before the tests (Table 1). The measurements of the Ra parameter of ring sample showed that in the friction pair lubricated synthetic oil the Ra parameter was much smaller than in the friction pairs lubricated mineral oil (Table 2). Important change of Ra parameter was observed in the friction pairs with CuPb30 bearing alloy counterparts when were lubricated mineral oil - the Ra parameter increased by 37% compared to the surface roughness before the tests and when the friction pair was lubricated synthetic oil the Ra parameter increased by ~ 11%. Similar changes were observed when the Ry parameter was measured/while measuring the Ry parameter. During the co-operation of the friction pairs with the AlSn20 bearing alloy counterparts lubricated by the synthetic oil, a reduction/decrease of 5%. in the value of the Ry roughness parameter was observed. In the case of Rz parameter, a reduction/decrease of surface roughness was observed, except for friction pairs with the CuPb30 bearing alloy counterparts lubricated by the mineral oil, where the Rz parameter of surface roughness increased by 12%. Measurements of Sm parameter have shown its growth in the case of lubrication by mineral oil, and in the case of synthetic oil, a minimal decrease of stabilization of the Sm parameter roughness was observed.

Table 2. Surface roughness of ring samples after test.

Parameter of surface roughness	Counterpart CuPb30				Counterpart AlSn20			
	Mineral oil		Synthetic oil		Mineral oil		Synthetic oil	
	Value (μm)	Change (%)	Value (μm)	Change (%)	Value (μm)	Change (%)	Value (μm)	Change (%)
Ra	0.58	37	0.47	11	0.59	39	0.54	29
Rz	5.5	12	4.9	-1	4.2	-15	3.4	-32
Ry	7.7	44	6.8	27	6.4	21	5.0	-5
Sm	92	4	86	-3	97	10	88	0

Table 3. Surface roughness of counterpart after test.

Parameter of surface roughness	Mineral oil				Synthetic oil			
	CuPb30		AlSn20		CuPb30		AlSn20	
	Value (μm)	Change (%)	Value (μm)	Change (%)	Value (μm)	Change (%)	Value (μm)	Change (%)
Ra	1.26	152	0.58	71	1.04	108	0.54	59
Rz	8.9	197	3.9	11	8.0	167	3.6	3
Ry	12.9	279	5.1	4	11.3	232	5.0	2
Sm	150	213	135	93	113	135	122	74

The measurements of counterparts made of bearing alloys showed an increase of their surface roughness (Table 3). The surface layer measurements of the counterparts indicated more substantial changes than in the case of ring samples, the largest of them exceeding even several dozen percent. The biggest changes were recorded for the parameter Ry, which in the case of the Lotos mineral oil increased by 279%, and in the conditions of lubrication by the Lotos synthetic oil - by 232%. Also, significant changes were recorded for parameters Sm, Rm and Ra. The changes of parameters measured in the friction pairs with the AlSn20 bearing alloy counterparts were at a lower level and did not exceed 93%, while in the case of the Sm and Ra parameters – they reached 71%. The lowest change observed in the case of the Rz and Ry parameters amounted to a few percent (a maximum 11% in the conditions of lubrication by Lotos mineral oil). In the case of both materials, higher values of the surface roughness could be observed in the friction pairs lubricated by the Lotos mineral oil, in relation to the pairs lubricated by the Lotos synthetic oil.

4. Discussion

The start-up phase of sliding pairs indicates the behaviour of the system during its further work until the modified layer is used up. The most favourable operating conditions are presented in sliding pairs in which the friction coefficient increases in the initial stage of start-up, and then decreases significantly and stabilises itself at a constant level. Those sliding pairs which exhibited the tribochemical equilibrium within the shortest time generate optimal conditions for their further operation. The physiochemical processes and the changes in the surface microgeometry caused the adaptation of the system to the conditions of friction. The registered change of the friction force reveal the ability of the sliding pairs to adapt to the friction conditions in the extension of the pair's operation time. The changes occurring in the reaction of the pair to the stabilization forces upon the start-up, and to the time flow, explain whether the system allows for a long-term and reliable operation or not. The stabilisation of the friction resistances indicates the adaptation of the pair to the existing forces and the generation of stable anti-wear and anti-seizure layers. The created surface layers ensure a separation of the co-operating elements of friction pairs and a reduced rate of direct adhesion between the surface irregularities. These conditions create a state of equilibrium between the processes of layer destruction and creation within the tribochemical processes occurring in the friction pair [9].

The use of AlSn20 bearing alloy in friction pairs had an impact on the reduction of wear. This is mainly due to the continuous and homogeneous construction of the sliding layer of the AlSn20 bearing alloy, which enables formation of a hard, thin, continuous layer of aluminium oxide. However, in the case of the CuPb30 alloy, the high temperature in the contact area promotes the formation of brittle cracks in the layer of friction, and the organic acids in the oil cause leaching of lead, exposing the copper structure of the bearing material. In this situation, the products of wear caused by friction may accumulate in the resulting discontinuity areas of the friction layer and interact with the processes on the surface layer of the second element, which causes its cutting and enhances wear [11].

The differences in the wear of the bearing alloys and the absence of measurable wear of the surface layer ring specimen were the effect of the interaction between the cooperating surface layers, as well as of the physiochemical changes of their surfaces that were induced by external forces [9]. These phenomena result from the elementary wear processes that occurred within the contact area of the sliding pair on the elementary surfaces of the cooperating layers. The lubrication factor is crucial for these processes, because it creates favourable or unfavourable friction conditions depending on its transformation. These changes contribute to the generation of boundary layers on the surface layers of the cooperating elements that are either highly resistant to ruptures or are quickly destroyed under variable operating conditions [5]. The high wear of bearing alloy observed in pairs is explained by the increased initial surface roughness and the load of the system. Due to the influence of the hard areas on the areas of the second material, a stress concentration occurs, which leads to the interaction between the two surface layers and a more intense abrasion of the softer material [11]. These changes may lead to the smoothening of the surface and removal of its irregularities, which eliminates the potential sources of further material transfer and stabilizes the wear process. However, the hard wear products created in the friction process induce chipping, slicing, and grinding, which intensify the wear process [11].

The surface roughness parameters measured indicated the intensity of friction and its influence on the shaping of the sliding pair's geometric structure. As an effect of the processes occurring within the friction area under the external forces, the system processes the preexisting geometric structures of both elements into a system with a structure that ensures the most favourable friction conditions [3]. As a result of these changes, a structure was created that reflected the changes, ensuring the given association of a certain optimal functionality; that is, an operating surface layer was generated. In pairs where this kind of relationship does not exist and the load conditions as well as pair composition cannot create a state of equilibrium, the effect is the destruction of a kinematic sliding pair [11].

5. Conclusions

The following conclusions may be drawn on the basis of the experimental test performed and the analysis of their results:

1. Lubrication of the friction area by Lotos mineral oil reduces start-up moment, friction force, temperature in the friction area and wear of tested bearing alloys;
2. Lubrication of the friction area in the start-up phase of the friction pair by mineral oil caused a faster stabilization of the friction conditions in the contact area than when synthetic oil was used for lubrication of the friction pair.
3. The intensity of wear of the AlSn20 bearing alloy cooperating with the borided surface layer is three times smaller than the intensity of wear of the CuPb30 bearing alloy;
4. The surface roughness of the ring samples and bearing alloy counterparts showed less change in the geometric structure of the friction pair elements lubricated by Lotus synthetic oil in relation to the friction pair elements lubricated by Lotus mineral oil;

References

[1] Balusamy T, Sankara Narayanan T S N, Ravichandran K, Song Park II and Ho Lee M 2013 Pack boronizing of AISI H11 tool steel: Role of surface mechanical attrition treatment *Vacuum* vol 97 pp 36-43

[2] Çavdar U, Sadık Ünlü B, Pinar A M and Atik E 2015 Mechanical properties of heat treated iron based compacts *Materials & Design* vol 65 pp 312-73

[3] Guha D and Chowdhuri R 1996 The effect of surface roughness on the temperature at the contact between sliding bodies *Wear* vol 197 pp 63-73

[4] Gunes I 2013 Wear behaviour of plasma paste boronized of AISI 8620 steel with borax and B_2O_3 paste mixtures *Journal of Materials Science & Technology* vol 29 pp 662-668

[5] Komvopoulos K, Pernama S A, Ma J, Yamaguchi E S and Ryason P R 2005 Synergistic effects of boron-, sulfur-, and phosphorus-containing lubricants in boundary lubrication of steel surfaces *Tribology Transaction* vol 48 pp 218-229

[6] Lubas J 2010 Practical application of boron-modified sliding pairs in I.C. Engine *Tribology International* vol 43 pp 2046-2050

[7] Mao D, Wang X, Wang W and Wei X 2012 Effect of boronizing on the dry sliding wear behavior of DC53/0.45 mass% C steel pairs *Surface and Coatings Technology* vol 207 pp 190-195

[8] Taktak S 2006 Tribological behaviour of borided bearing steels at elevated temperatures *Surface Coating Technology* vol 201 pp 2230-2239

[9] Nehme G, Mourhatch R and Aswath P B 2010 Effect of contact load and lubricant volume on the properties of tribofilms formed under boundary lubrication in a fully formulated oil under extreme load conditions *Wear* vol 268 pp 1129–1147

[10] Pranay H, Liang M, Usta and Ucisik A H 2006 Wear and surface characterization of boronized pure iron *Journal of Tribology* vol 129 pp 1-10

[11] Schouwenaars R, Jacobo V H and Ortiz A 2007 Microstructure aspect of wear in soft tribological alloys *Wear* vol 263 pp 727-735

IOP Conf. Series: Materials Science and Engineering **148** (2016) 012054 doi:10.1088/1757-899X/148/1/012054

The analysis of diagnostics possibilities of the Dual- Drive electric power steering system using diagnostics scanner and computer method

W Szczypiński-Sala and K Dobaj[1]

[1]Institute of Motor Vehicles and Internal Combustion Engines, Cracow University of Technology, 37 Jana Pawla II Avenue, 31-864 Krakow, Poland

E-mail: ws@mech.pk.edu.pl, kdobaj@pk.edu.pl

Abstract. The article presents the analysis of diagnostics possibilities of electric power steering system using computer diagnostics scanner. Several testing attempts were performed. There were analyzed the changes of torque moment exerted on steering wheel by the driver and the changes of the angle of rotation steering wheel accompanying them. The tests were conducted in variable conditions comprising wheel load and the friction coefficient of tyre road interaction. Obtained results enabled the analysis of the influence of changeable operations conditions, possible to acquire in diagnostics scanners of chosen parameters of electric power steering system. Moreover, simulation model of operation, electric drive power steering system with the use of the Matlab simulation software was created. The results of the measurements obtained in road conditions served to verify this model. Subsequently, model response to inputs change of the device was analyzed and its reaction to various constructional and exploitative parameters was checked. The entirety of conducted work constitutes a step to create a diagnostic monitor possible to use in self-diagnosis of electric power steering system.

1. Introduction

Power steering system is at present used in all vehicle size classes. The aim of its use is to diminish the torque, the driver needs to turn the steering wheel in order to do a desirable maneuver. In some trucks, the application of the steering systems is enforced by legal regulations.

Hydraulic and electric power steering system solutions used contemporarily ensure the change of the value of steering feedback torque, what is particularly desirable during maneuvering at small travel speed and especially while parking the vehicle.

However, in some road situations the change of the level of vehicle steering effort tends to be seen negatively by the driver. Subjective feelings of the driver are of vital significance here, since based on them and the driver's earlier experience, they are able to predict the behavior of the driven vehicle in response to the steering wheel movements planned by them. At present, one can see increasingly stronger tendencies to broader use of electric power steering systems and moving from hydraulic power steering. That is why, with regard to the increasing number of vehicles with electric systems, the issues associated with the possibilities of proper diagnostics of work of these devices take on significance. In the paper, the attempt was undertaken to analyze the range of diagnostics possibilities of electric power steering system using computer diagnostics scanner. It was also evaluated how changeable operational conditions can influence the parameters of electric power steering system obtained in car diagnostic scanner.

2. Electric power steering systems

Electric power steering systems have a lot of advantages. They are characterized by considerably smaller demand for power consumption in comparison to hydraulic power steering system and they are also smaller in size. The systems can lower steering wheel torque and improve the returnability [1,2,3]. These assets of electric power steering systems are the reason why they are used in almost all contemporary motor vehicles. Electric power steering systems began to be introduced to motor vehicles in the 90-s of the last century. Since then, they have been the subject of many studies and theoretical research papers. In systems of this type, electric motor is used, creating necessary torque and ipso facto fulfilling supporting functions [4,5,6]. The torque, the driver turns the steering wheel is registered by the torque sensor. On this basis, taking signals also from other sensors into account, assisting force is calculated, what in turn constitutes the basis for proper regulation of the current supplying the assistance motor [7,8,9,10]. In addition, one should notice that the steering stiffness resulting from the stiffness of the elements fixing the electric power steering device to the steering column has the influence on car steerability [10,11].

3. Driving and steering the vehicle

Both the driver's comfort and their safety depend on the operation and the regulation of controlling the steering system assistance. While running the vehicle, the driver bases the decisions taken by them on the steering feel, that is why, the degree this feel is considered will decide on the further acting strategy [11]. Thus, the operation of the steering system decides on the driver's feelings [12]. Some authors [11] are inclined to assume that the steering feel depends only on the steering wheel torque. Also in the publication [13], the authors indicated the torque on the steering wheel and steering wheel rotation angle as factors which are most clearly felt by the driver, therefore they are the most significant stimuli from the point of view of the steering feel. There are various ways of influence the steering feel. This can be the choice of shifting the steering gear, change of the length of the steering arm or change of the diameter of the steering wheel [14].
The steering torque is associated with steering effort [2,3,12]. Friction between tire surface and ground influences the steering wheel torque. The intensity of this influence depends on the concrete solution of constructional steering system [15]. During parking manoeuver or turning back, higher values of shifting in the steering system are beneficial. The value of steering effort is influenced by the type of tires, their stiffness coefficient and the tire pressure. It is assumed, that two factors influence in the most significant way the subjective evaluation of the facility of running the vehicle. Namely, the steering effort or steering wheel torque and the time of vehicle reaction to the steering wheel. Beneficial steering feel influences the reassurance of safe and convenient driving comfort [6,17,21].
The dependence of the vehicle reaction in the form of vehicle yaw rate and its lateral acceleration in response to the action of the driver expressed with the change of steering wheel rotation angle and the speed of this change has to be predictable for the driver and should not undergo too violent hesitations. Only then, the driver has the feeling of full control over the vehicle [18,19,22,23].

4. Experimental procedure

On the first stage of the experimental analysis, the essential torque applied by the driver of the vehicle during turns of the steering wheel to the right and left side range was determined. It was made in a maximum range, possible to attain for steering system construction reasons. The manoeuver was performed three times. In turn, with power steering system turned off and subsequently on both levels of reinforcement – i.e. in normal mode and 'city' mode. The measurement was made on dry asphalt surface. The courses of the changes of the steering wheel rotation angle and the torque obtained from diagnostic tester were juxtaposed in figure 1.

IOP Conf. Series: Materials Science and Engineering **148** (2016) 012054 doi:10.1088/1757-899X/148/1/012054

As one can observe from the obtained courses during the manoeuvers performed with power steering system turned off, the maximum registered torque applied by the driver exceeded 28 Nm. This value was achieved very quickly after starting the manoeuver and maintained during it, despite increasing the angle of wheel turn. This is a maximum value, possible to be read from system control unit by SolusPro tester. With power assistance steering system turned off, the total steering wheel torque required to realize a specific turn is applied by the driver. With power steering system turned on, based on the torque measured on the torsion bar, control unit of the system controls the torque of the electric motor, supporting the movement of the wheels of the vehicle. The torque of the electric motor is transmitted to the lower part of the torsion bar, allowing to diminish the torque on the steering column, resulting from the turn of the steering wheel. The torque recorded by the tester is lower. The controlled torque of the electric motor is higher for the city mode. Only after reaching the extreme angular position of vehicle wheels one can observe the increase of the torque on the steering wheel.

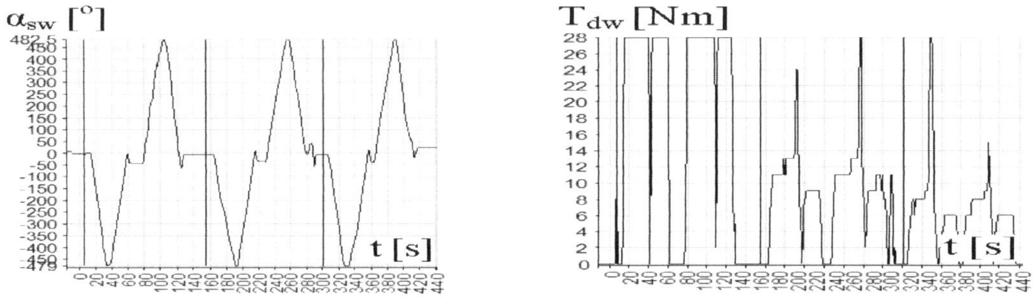

Figure 1. The course of the changes of the torque on the steering wheel during turns to the right and left on dry asphalt.

This can be seen on both modes i.e. normal mode and city mode. The controlled level of the assistance of the power steering system is corrected with the signal from the torsion bar. The torque on the rotor of electric motor helps to obtain the angle of wheel turn indispensable to perform a manoeuver. The torque on the wheels constitutes the sum of the components of the electric motor torque and the one applied by the driver. On the presented diagram we can observe the course of the changes of the torque on the steering wheel. It is of vital significance from the point of view of the rate of the realization of driver's intentions. As one can observe, after turning on city mode, the torque is diminished by about 5 Nm.

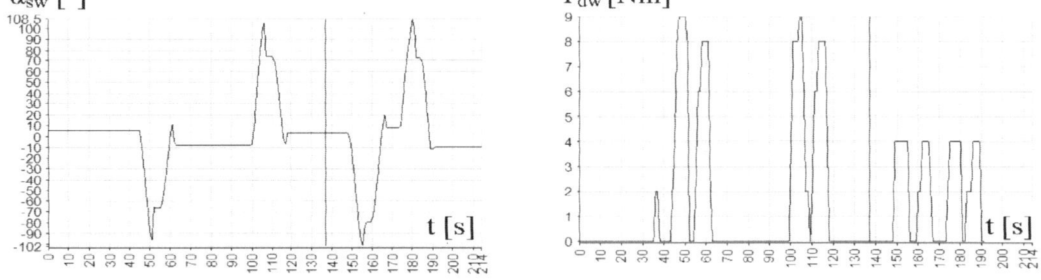

Figure 2. The changes of the torque on the steering wheel during turns to the right and left, manoeuvers performed on the asphalt covered with 5 cm fresh snow.

In order to evaluate the friction between tire surface and ground, tests were carried out on the asphalt covered with the layer of 5 cm of snow. Figure 2 shows the courses of the torque on the steering wheel registered on the diagnostic tester. The following manoeuvers were performed: turn left, the return of the wheels to straight – ahead position, turn right, the return of the wheels to straight-ahead position. With the aim of precise determination of the influence of the conditions of interaction of the wheel with snowy surface, the test was repeated many times. Power steering system efficiency was compared on both levels of reinforcement i.e. normal mode and in "city" mode, which is characterized by higher assistance.

In real operating conditions, flexibility of suspension and the steering system as well as the coefficient of friction between tire surface and the ground affect the efficiency of control of the electric motor in the power steering system. In the analysed case, the highest attainable value of the torque amounted to 9 Nm. It was achieved during the phase of the steady turn of the wheels. The increase of the torque on the registered course is steep. This is why at every moment during the manoeuver, an optimal assistance of power steering system has to be ensured. The difference between "city" mode and normal one in the assistance torque, which can be seen in the figure 2 amounted to about 5 Nm i.e. about 50 %.

An analogous test, as described above was also made on wet asphalt. The obtained courses of the torque on the steering wheel were shown in figure 3. In comparison to the manoeuvers performed on snow, one can notice that the values of the torque are in this case higher by about 2 Nm, both in normal mode and in "city" mode.

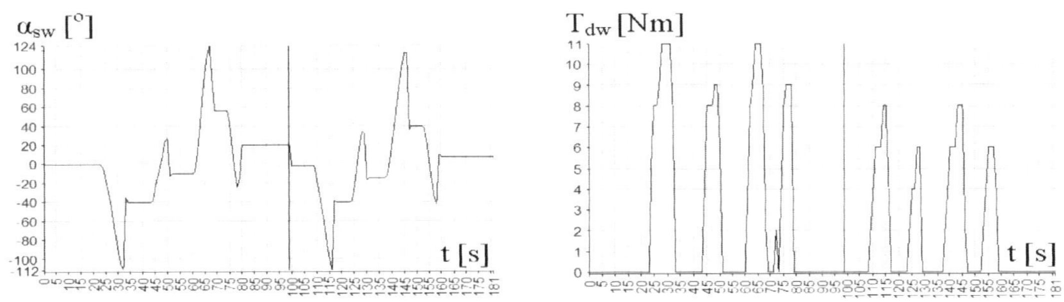

Figure 3. The changes of the torque on the steering wheel during turns to the right, left, manoeuvers performed on wet asphalt.

In order to check the reaction of the control unit of electric power steering system on any changes of the state of contact of the vehicle wheel with the surface, manoeuvers with the wheels were performed with changing the front axle load. The courses of the torque on the steering wheel with the load on the front axle of 7200 N, obtained in the tests were shown in figure 4, whereas the figure 5 shows the changes of the torque on the steering wheel with the load on the front axle of 4100 N. Comparing these courses one can notice that unloading the steering wheels of the vehicle influences considerably the resistance of the friction between tire and ground, what in turn decides on the torque on the steering wheel. This is visible with power steering system turned off. However, in such case, the level of controlling the assistance of power steering is only slight changed. We can observe this comparing the course of the torque both in normal mode and in city mode. The torque on the steering wheel changes maximally by 1 Nm with power steering system turned on.

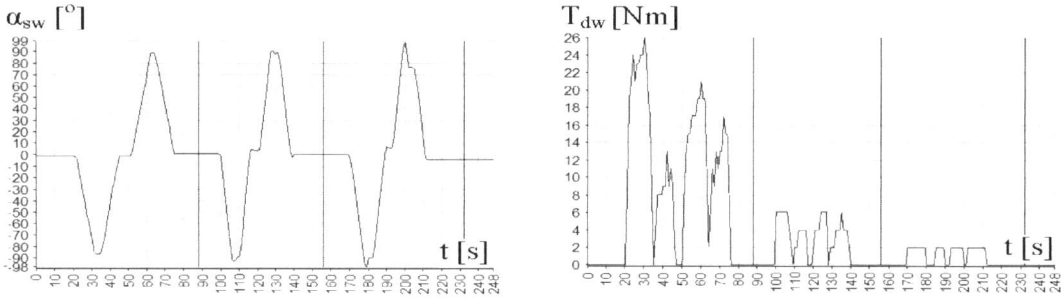

Figure 4. The changes of the torque on the steering wheel during turn right, left, manoeuvers performed with loading the front axle with 7200 N, subsequent manoeuvers were performed: with power steering system turned off, normal mode and city mode.

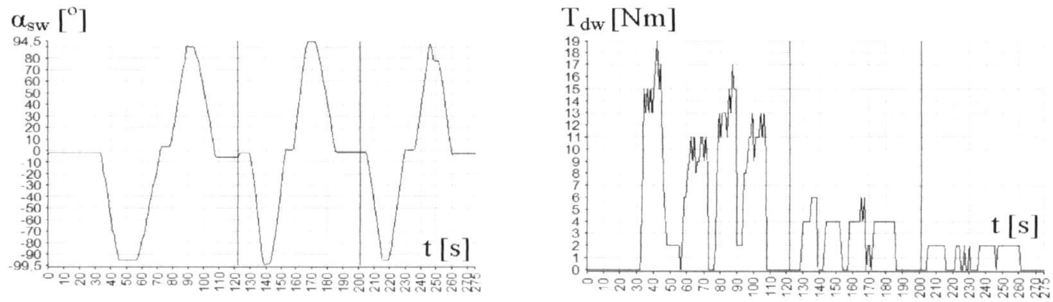

Figure 5. The changes of the torque on the steering wheel during turn right, left, manoeuvers performed with loading the front axle with 4100 N, subsequent manoeuvers were performed: with power steering system turned off, normal mode and "city" mode.

Apart from the manoeuvers described above, test rides were carried out on the same short section of asphalt surface at two various speeds. During each of these rides, manoeuver of the turn by 90° was performed twice. In given drive conditions, the test was made with normal mode and in "city" mode. The courses of the changes of the turning angle and the torque on the steering wheel during the ride on dry asphalt surface were presented in figure 6. On the basis of the obtained course of the change of the angle of turning the steering wheel versus time, one can single out particular phases of the test ride, i.e. starting to turn, then the return to straight drive and subsequently rapid turn.

In the phase of starting to turn, the torque on the steering wheel reached 10 Nm. During starting the second turn, the correction of the running track is visible, which was followed by a maximum turn of the wheels. From here, the change of the torque on the steering wheel is visible preceded by the achievement of extreme position. In case of "city" mode on the same ride route with analogous speeds, the torque on the steering wheel was about 40% lower.

Test rides, described above with the turns of the vehicle by 90° were repeated on the asphalt covered with snow. In case of these rides, the values of the torque on the steering wheel were slightly higher and reached from 8 to 10 Nm. However, in this case, higher range of the fluctuations of the torque was observed what could be the consequence of the changes of the resistance of rolling resulting from the presence of snow cap, over which the wheels were rolling. On the one hand, the coefficient of friction between the wheel and compacted snow was lower than on dry asphalt, on the other hand, as it was mentioned; the snow layer could be the cause of the increase of the rolling resistance of straight rolling.

IOP Conf. Series: Materials Science and Engineering **148** (2016) 012054 doi:10.1088/1757-899X/148/1/012054

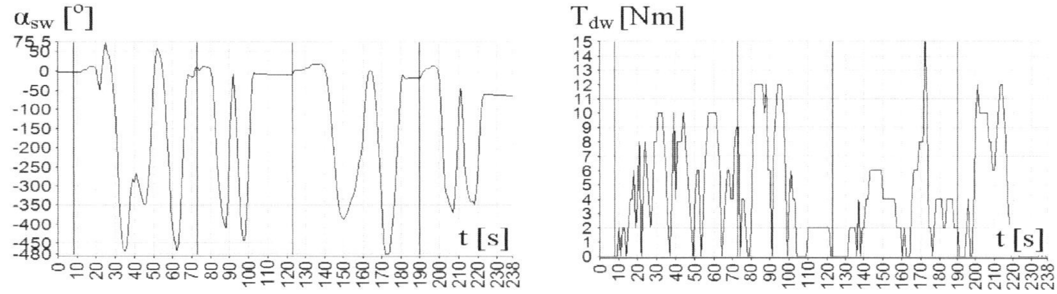

Figure 6. The changes of the torque on the steering wheel during the ride on dry asphalt; normal mode and "city" mode.

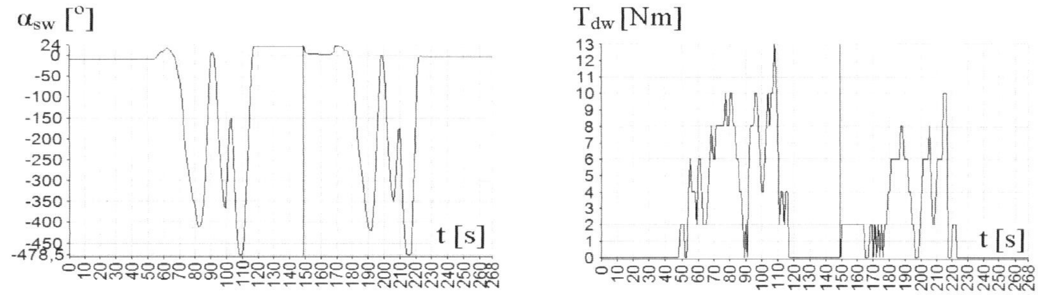

Figure 7. The changes of the torque on the steering wheel during manoeuvers; asphalt surface covered with snow, normal mode and "city" mode.

5. Model

The driver exerts the torque (T_{dw}) on the driving wheel (with the moment of inertia J_{sw}). The steering angle (α_{sw}), which is the result of the exerted torque, is measured with the use of the specified measuring equipment. The torque is transmitted from the steering wheel to the upper part of the steering column (with neglected stiffness and damping) and to the torsion bar with torsional stiffness c_{tb} and damping D_{tb}. The torsion bar allows to generate the signal for the control unit. The applied torque is calculated based on the signal. The torsion bar angular displacement is mechanically limited to a few degrees [4]. The power steering electric motor is coupled with the steering column via a worm gear. The worm wheel is joined to the steering column via elastic sleeve coupling (with stiffness c_{cl} and damping D_{cl}). The lower part of the steering column (with stiffness c_{lps} and damping D_{lps}) is placed under the worm gear assembly. The lower part of the steering column is coupled with the pinion of the rack and pinion mechanism (with neglected friction).

The formulated model doesn't include the steering system kinematics changes. The tire- road contact surface was modelled as the wheel with determined surface s. The surface s was constant during the simulation attempts. Thus, the tire's pliability on the tire- road contact area was neglected. The influence of the wheel suspension was also omitted. Normal load force F_l and the friction coefficient μ, included in the model are the tire-road contact parameters.

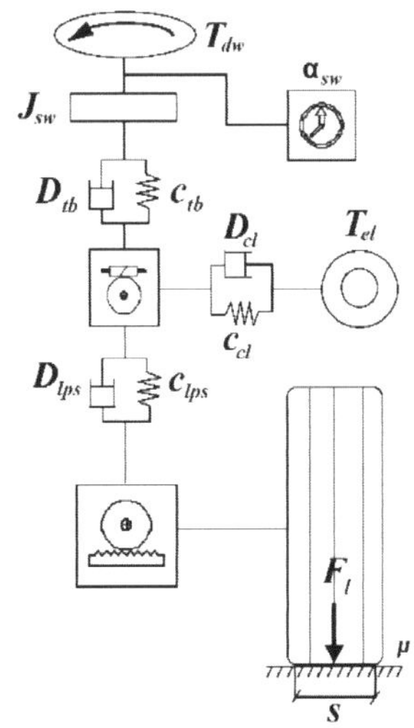

Figure 8. The formulated Matlab model.

6. Simulation analysis

6.1. Verification

The analysis of the results of computational calculations and the verification of a computational model were the next step of the work. The courses of the changes of the angle of the turn of the steering wheel, obtained from the calculations were related to those obtained during the measures made on the vehicle and registered with a diagnostic tester. Real technical data of the steering system were adopted for calculations. This allowed to obtain compatibility between the results registered during experimental research and computational calculations. Adopted parameters correspond to those established in the paper [4]. The values of the torque of the electric motor adopted in the simulation model both for "city" and normal mode of power steering system amounted to about 0.5 Nm.

Due to the flexibility of the collection of components of the steering system, the torque on the steering wheel showed fluctuations to some extent during repeated tests. The range of this variability was designated for obtained results both as a response of the real system and the results of computational calculations obtained based on the steering system mathematical model. So that the obtained results can be compared more easily, all the juxtapositions were examined as manoeuvers performed in two seconds' intervals of time. On the presented graphs one can see irregularities of the change of the angle of the turn of the steering wheel in the initial phase of the work of the system, what is the result of the slack in the steering system as well as steering stiffness and steering damping. Such a situation can be observed at the beginning of each of the manoeuver, since a constant torque of electric motor was adopted in calculations.

6.2. Influence of the change of the axles load

Both the operating conditions and malfunction which may occur can influence the work of the system and ipso facto its evaluation based on the read-outs registered by diagnostic tester. Simulation analysis was conducted in order to assess the influence of changeable operating conditions on the torque on the steering wheel.

The load on the wheels of the vehicle was the operating factor influencing the operation of the steering system, which underwent evaluation. The range of the change of the load by 1000 N, from 3922 N to 4922 N was analysed. The lowest load corresponded to unladen kerb mass of the vehicle burdened by a driver and one passenger with a joint mass of 150 kg. However, the biggest load corresponded to admissible gross vehicle weight rating. The analysis was made in conditions of interaction of tire with wet asphalt and normal mode of power steering system.

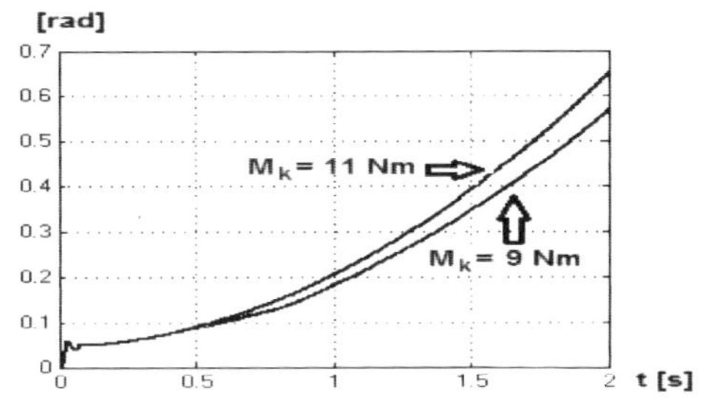

Figure 9. The range of fluctuation results obtained from computational calculations and measurement response of the real system.

The increase of the pressure causes in an obvious way the increase of the resistance of the movement during manoeuvers with the wheels of the vehicle. As one can observe on the courses shown on figure 10, in case of the lowest load, at a constant torque on the steering wheel and the constant torque of electric motor of the steering system, after two seconds of performing the manoeuver of the turn, the angle of the turn of 0,65 rad was obtained. However, in case of the highest load, due to higher resistance of the friction between tire surface and ground, the steering wheel was turned by 0,6 rad. The difference between these values is relatively small in view of the range adopted for the analysis of the change of the load. The change of the load by 1000 N causes the difference of the turn angle by 7%. This confirms a good optimization of the analysed solution.

6.3. Influence of the change of the friction coefficient between tire surface and the ground

Friction coefficient between tire surface and the ground was the next operating factor which was evaluated. The influence of its changes within the range from 0,3 to 0,9 was analysed. This corresponds to the range we can deal with on various surfaces; from the case of ice-covered and wet carriageway to dry, very adhesive asphalt. The results obtained from model simulations were shown on the figure 11.

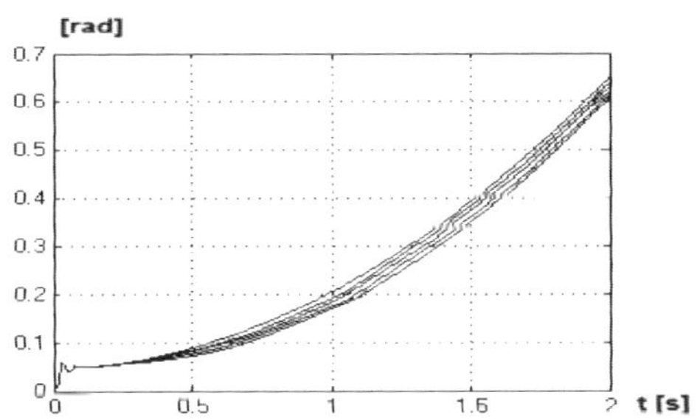

Figure 10. The influence of the change of the vehicle axles loads on rotation steering wheel angle at assumed constant torque on the steering wheel.

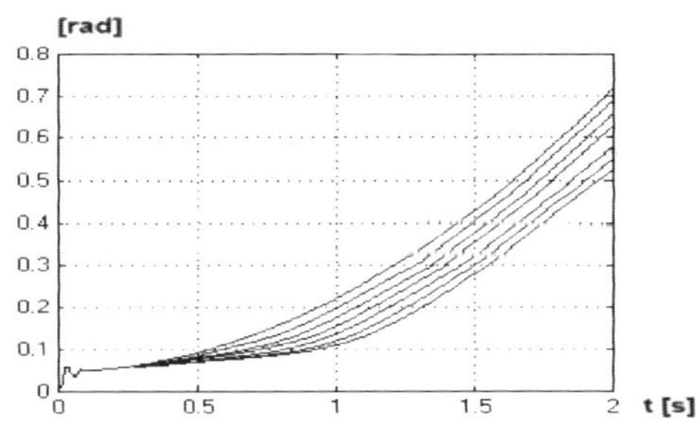

Figure 11. The influence of the change of the friction coefficient between tire surface and the ground on rotation steering wheel angle at assumed constant torque on the steering wheel.

For the adopted friction coefficient amounting to 0,3, i.e. the lowest examined resistance of the turn, after 2 seconds of the duration of the manoeuver, the steering wheel is turned by about 0,72 rad. However, in case of the friction coefficient amounting to 0,9, the steering wheel is turned by about 0,9 rad. The difference observed in the angle of the turn amounts to about 26 %. This shows considerably higher susceptibility of the system on the changes of the friction coefficient in comparison to the changes of the load on the vehicle axle. The abovementioned observations confirm the necessity to pay more attention to this factor during tests and diagnostic research of the system.

7. Conclusions

With regard to the increasing number of vehicles with electric systems, the issues associated with the possibilities of proper diagnostics of work of these devices take on significance. The following conclusions may be drawn on the basis of the experimental test performed and the analysis of their results:

1. Based on the formulated Matlab model of electric power assistance system, comparative analysis of the parameters of electric power steering system is possible.
2. Conducted measures enabled the evaluation of statistical dispersion of the values of the parameters of work of electric power steering system recorded by diagnostic tester.
3. The values of the angle of the steering wheel turn and the torque on the steering wheel decide about the steering effort, therefore they should be the parameters examined by diagnostic testers during electric power steering system evaluation;
4. The conducted evaluation of the influence of measurement conditions enables the determination of the range of variability of the records from the tester, which should be the basis to specify the guidelines of the range of allowed fluctuation of test conditions necessary for a reliable and valid measurement;
5. The entirety of conducted work constitutes a step to create a diagnostic monitor possible to be used in self-diagnosis of electric power steering system.

References

[1] Liu Z, Yang J and Liao D 2003 *The optimization of electric power assisted steering to improve vehicle performance* Proceedings of the Institution of Mechanical Engineers, Part D: Journal of Automobile Engineering August 1 2003 217 pp 639-646

[2] Parka T J, Hanb Ch S and Leec SH *2005 Development of the electronic control unit for the rack-actuating steer-by-wire using the hardware-in-the-loop simulation system* Mechatronics vol 15, Issue 8, October 2005, pp 899–918

[3] Nogowczyk P 2009 *Kształtowanie charakterystyk układu kierowniczego, przy wykorzystaniu przyspieszenia poprzecznego pojazdu* PhD Thesis Kraków

[4] Janczur R and Kuranowski A 2014 *Doświadczalna identyfikacja wybranych parametrów elektrycznego urządzenia wspomagającego układ kierowniczy* Logistyka Nauka nr 3 pp 2501-2508

[5] Kuranowski A 2012 *Effect of the Variable Input Function Parameters on the Characteristics of an Electric Power Steering System* Automotive Safety Problems vol 1 Politechnika Świętokrzyska Kielce pp 53-61

[6] Jaksch F 1979 *Driver-Vehicle Interaction with Respect to Steering Controllability* SAE Technical Paper 790740 doi: 10.4271/790740

[7] Grzegożek W and Nogowczyk P 2008 *Analiza możliwości kształtowania charakterystyk wspomagania układu kierowniczego z wykorzystaniem wysiłku kierowania* Czasopismo techniczne z.8-M/2008 pp 55-62

[8] Kuranowski A 2012 *Wpływ parametrów zmiennego wymuszenia wejściowego na charakterystyki elektrycznego urządzenia wspomagającego układ kierowniczy* VIII Międzynarodowa Konferencja Naukowo – Techniczna Problemy Bezpieczeństwa w Pojazdach Samochodowych Kielce Politechnika Świętokrzyska pp 175-187

[9] Kuranowski A 2012 *Effect of Variable Input and Output on the Characteristics of a Steering System with EPS (Electric Power System)* Journal of KONES Powertrain and Transport vol 19 no 1 European Science Society of Powertrain and Transport Publication Warsaw 2012 pp 215-226

[10] Kim J-H and Song J-B 2002 *Control logic for an electric power steering system using assist motor* Mechatronics vol 12, Issue 3 April 2002 pp 447–459

[11] Yao Y 2006 *Vehicle Steer-by-Wire System Control* SAE Technical Paper Series 2006-01-1175 SAE World Congress 2006 Detroit Michigan April 3-6

[12] Data S, Pesce M and Reccia L 2004 *Identification of steering system parameters by experimental measurements processing* Proceedings of the Institution of Mechanical Engineers Part D Journal of Automobile Engineering 08/2004 218(8) pp 783-792

[13] Newberry A C, Griffin MJ and Dowson M 2007 *Driver perception of steering feel* Proceedings of the Institution of Mechanical Engineers Part D Journal of Automobile Engineering 04/2007 221(4) pp 405-415

[14] Essma S 2000 Steering Effort Analysis of an Oval Racing Track Setup Champ Car *Newman Haas Racing 2000* International ADAMS User Conference

[15] Toffin D, Reymond G, Kemeny A and Droulez J 2003 *Influence of steering wheel torque feedback in a dynamic driving simulator* DSC North America 2003 *Proceedings* Dearborn Michigan October 8-10 2003 (ISSN 1546 5071)

[16] Nogowczyk P and Grzegożek W 2010 *Power steering adjustment considering lateral acceleration influence* Archiwum Motoryzacji Wydawnictwo Naukowe Przemysłowego Instytutu Motoryzacji no 2 pp 127-144

[17] Janczur R and Kuranowski A 2014 *Obciążenia układu kierowniczego w granicznych stanach ruchu samochodu osobowego* Logistyka Nauka nr 3 pp 2509-2516

[18] Amberkar S, Kushion M, Eschtruth K and Bolourchi F 2000 *Diagnostic Development for an Electric Power Steering System* SAE Technical Paper 2000-01-0819

[19] Badawy A, Zuraski J, Bolourchi F and Chandy A 1999 *Modeling and Analysis of an Electric Power Steering System* SAE Technical Paper 1999-01-0399 doi:10.4271/1999-01-0399

[20] Janczur R, Kuranowski A, Nogowczyk P and Pieniążek W 2015 *Badania eksperymentalne sił w drążkach kierowniczych* Archiwum Motoryzacji Wydawnictwo Naukowe Przemysłowego Instytutu Motoryzacji vol 68 no 2 pp 171-180

[21] Bertollini G and Hogan R 1999 *Applying Driving Simulation to Quantify Steering Effort Preference as a Function of Vehicle Speed* SAE Technical Paper 1999-01-0394 doi:10.4271/1999-01-0394

[22] Hai Chenguang Ji Peng and Yu Haifeng 2010 *Analysis of Influence of Steering System on Vehicle Straight Line* WASE International Conference on Information Engineering (ICIE) Beidaihe, Hebei 14-15 August 2010 vol 3 pp 336-339

[23] Reinelt W, Klier W, Reimann G, Schuster W et al 2004 *Active Front Steering (Part 2): Safety and Functionality* SAE Technical Paper 2004-01-1101 doi: 10.4271/2004-01-1101

Evaluation the course of the vehicle braking process in case of hydraulic circuit malfunction

W Szczypiński-Sala[1] and J Lubas[2]

[1]Institute of Motor Vehicles and Internal Combustion Engines, Cracow University of Technology, 37 Jana Pawla II Avenue, 31-864 Krakow, Poland
[2] Dept of Combustion Engines and Transport, Rzeszow University of Technology, Poland

E-mail: ws@mech.pk.edu.pl

Abstract. In the paper, the results of the research were discussed, the aim of which was the evaluation of the vehicle braking performance efficiency and the course of this process with regard to the dysfunction which may occur in braking hydraulic circuit. As part of the research, on-road tests were conducted. During the research, the delay of the vehicle when braking was measured with the use of the set of sensors placed in the parallel and the perpendicular axis of the vehicle. All the tests were conducted on the same flat section of asphalt road with wet surface. Conditions of diminished tire-to-road adhesion were chosen in order to force the activity of anti-lock braking system. The research was conducted comparatively for the vehicle with acting anti-lock braking system and subsequently for the vehicle without the system. In both cases, there was a subsequent evaluation of the course of braking with efficient braking system and with the dysfunction of hydraulic circuit.

1. Introduction

Currently, vehicles are equipped with many kinds of devices that assist drivers in emergency situations. Vehicle ride control depends on the vehicle design and also on the condition of the vehicle braking hydraulic circuit components. Modern passenger cars are equipped with anti-lock braking systems (ABS). Anti-lock braking systems are one of the basic construction solutions which have the influence on active safety of the vehicles. The system prevents the wheels from locking during braking. Anti-lock braking systems ensure vehicle steerability even in conditions of emergency stop. The system also attempts to keep the wheel slip within the range where maximum braking force can be generated, thereby optimizing braking performance [6, 10]. From this point of view, their research and functional on-roads testing gain particular significance [3, 4, 9, 10].

We can distinguish three basic units in anti-lock braking systems, namely: wheel speed sensors, vehicle ABS controller and central hydraulics block with valves. When braking, the evaluation of the longitudinal slip is based on the measurement of wheel rotary speed. On the basis of the information on the speed it is possible to control hydraulics valve, which regulates the pressure in braking hydraulic circuit of particular wheels. In anti-lock braking system, vehicle velocity and the rotational speed of each wheel are assumed as input signals constantly sent to the controller, whereas control signal of brake pressure modulator is the output signal from the system. Thanks to this, change of the pressure of the brake fluid in brake caliper pistons of particular wheels is possible. However, what is the exerted value in this system is the pressure on brake pedal, ipso facto pressure in the master cylinder [4,6,8,11].

The operation of the control system of anti-lock braking system can be based on various strategies. The assumed strategy of the division of braking forces on particular wheels decides on the maintenance of steerability and stability of vehicle movement. The issues are the subject of theoretical considerations [1], which constitute the basis for working out the logic of the activity of anti-lock braking system [2,11]. Operation strategies have high significance for the ABS controller's ability to adapt to changing slip–force characteristics of tyres and provide estimates of the magnitude of the effects of different tyre and road operating conditions. Vehicle ABS control algorithm allows to compensate eventual dysfunctions of particular components of the braking system, too. Usually, the algorithms of controlling anti-lock braking system are realized with full efficiency of the remaining part of the braking system. However, during vehicle operation, confounding factors may occur. The obsolescence of brake fluid and the absorption of humidity from the surroundings by it may considerably lower its boiling point. During intensive braking, when the elements of wheel brakes heat considerably, vapour of brake fluid may be created in brake caliper pistons. Likewise, leakiness in the system or its air bubbles in the lines result in the restriction of the pressure possible to obtain in lines, since in such cases the change of compressibility of the brake fluid filling the system occurs. The most unfavourable situation is the significant diversification of braking forces on wheels on the right and left side of the vehicle. In case of the occurrence of the mentioned dysfunctions, the evaluation of the effectiveness of the operation of anti-lock braking system is most reliable in operation conditions, in road tests.

The aim of the paper was the evaluation of the influence of restricted efficiency of the hydraulic system on the operation of anti-lock braking system during braking. Since the restriction of maximum pressure in brake caliper pistons influences the change of the division of braking forces on particular wheels and extorts the correction of them by anti-lock braking system, road tests were conducted, which evaluated obtained delay in braking on various levels of braking system dysfunction. During braking maneuver, parallel and perpendicular acceleration of the vehicle was recorded with the aim of the evaluation of car body movement dynamics during braking process.

2. Experimental procedure

As part of the research, on-road tests were conducted. During the research, the delay of the vehicle when braking was measured with the use of the set of sensors placed in the parallel and the perpendicular axis of the vehicle. On-road tests were conducted on the vehicle – B Segment in Euro Market Segment car classification with independent suspension. On the front, the vehicle was equipped with multi-link suspension, whereas at the back of the vehicle – trailing-arm suspension. Telescopic shock absorbers were installed both on the front and at the back. All the tests were conducted on the same flat section of asphalt road with wet surface. Conditions of diminished tire-to-road adhesion were chosen in order to force easier activity of anti-lock braking system. The research was conducted comparatively for the vehicle with acting anti-lock braking system and subsequently for the vehicle without the system. Turning off the possibility of the operation of anti-lock braking system was achieved by the removal of a circuit breaker fuse. In both cases, there was a subsequent evaluation of the course of braking with efficient braking system and with the dysfunction of hydraulic circuit. All the tests with the same conditions of their realization were four times repeated. The tests were conducted with air-locked hydraulic system and with fully blocked flow of brake fluid to particular brake calipers. During road research, the acceleration in the parallel and perpendicular axis of the vehicle was measured with the use of acceleration sensors of the company Hottinger Baldwin Messtechnik GmbH of B12/200 type. Measurement sensors were connected to an analog-to digital converter also produced by HBM. The courses of acceleration obtained during test rides were recorded on the computer disc.

IOP Conf. Series: Materials Science and Engineering **148** (2016) 012055 doi:10.1088/1757-899X/148/1/012055

3. Results of research

After conducting road tests, the courses of the signals obtained from acceleration sensors were evaluated in the first place. The figure 1 depicts the courses of acceleration in the parallel and perpendicular axis of the vehicle when braking from the speed of 14 m/s, recorded by the set of sensors installed in the vehicle. Obtained courses of acceleration changes illustrate the behaviour of car body during intensive braking. The delay in the perpendicular axis of the vehicle obtained in road conditions, during the test with the efficient braking system and with the operation of anti-lock braking system reached peak values exceeding 8 m s^{-2} - Figure 1.

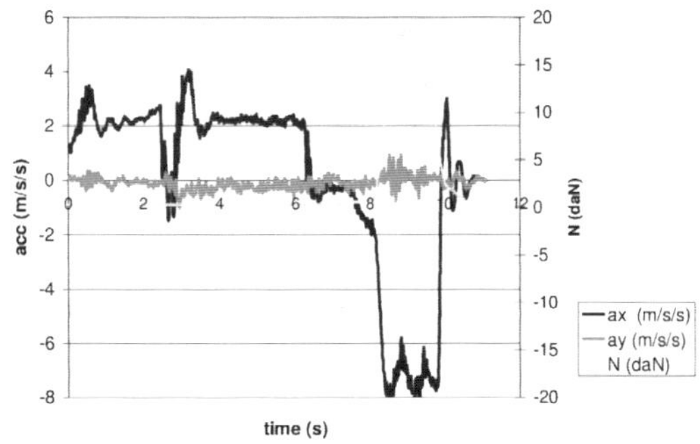

a)

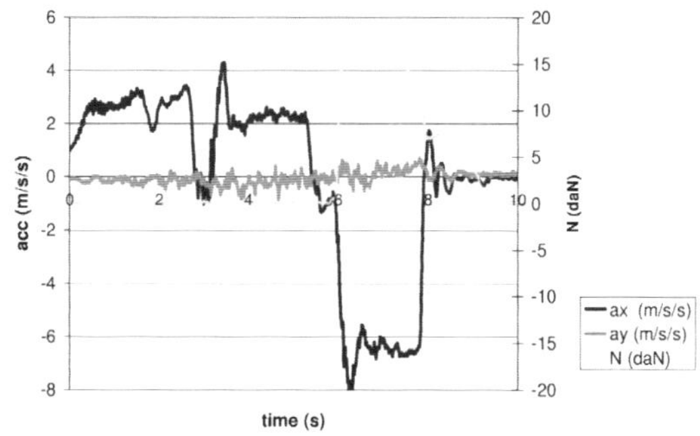

b)

Figure 1. The course of the changes of the delay during braking test in perpendicular (a_x) and parallel (a_y) axis of the vehicle; a – braking with the efficient hydraulic circuit and operating anti-lock braking system, b – braking with the efficient hydraulic circuit without anti-lock braking system.

The introduction of the air into braking system, the consequence of which is the decrease of the pressure of brake fluid in brake caliper piston of the right front wheel with the same pressure force on the brake pedal was the reason why obtained peak values of braking delay were half lower. In such conditions, due to the occurring differences of the developing braking force in the right and left wheel of the vehicle, pressure hydraulic valves were started by anti-lock braking system, what is very clearly visible on the courses presented on figure 2. The constraint of the possibility of increasing pressure in hydraulic caliper cylinder of one of the wheels resulting from the presence of air bubbles was the reason for the occurrence of the tendency of blocking and skidding the second wheel, the anti-lock braking system needed to counteract. However, its consequence was considerable decrease of the obtained braking delay – in maximum values up to 4 m s^{-2}. Analogous relationships can be observed by comparing the courses presented on subsequent figures 1b and 1 d respectively, which illustrate braking attempts for the vehicle without anti-lock braking system. Also in this case, the presence of vapour bubbles of the hydraulic caliper cylinder caused lowering the maximum obtainable braking delay by 50 %, and additionally it can be noticed that in the initial braking phase i.e. in the phase of the increase of longitudinal delay, the increase of the value of parallel acceleration of the vehicle is also recorded. Such a course of parallel acceleration can be explained with the increasing difference of braking forces on the wheels on the right and left side of the vehicle, not corrected by the operation of anti-block system. The changes of the braking force on the vehicle wheels are visible on all acceleration courses recorded by the sensor in parallel axis of the vehicle. The course of their changes when braking during rectilinear vehicle movement and with fully efficient braking system is considerably smaller both during the operation of the anti-lock system and without the system.

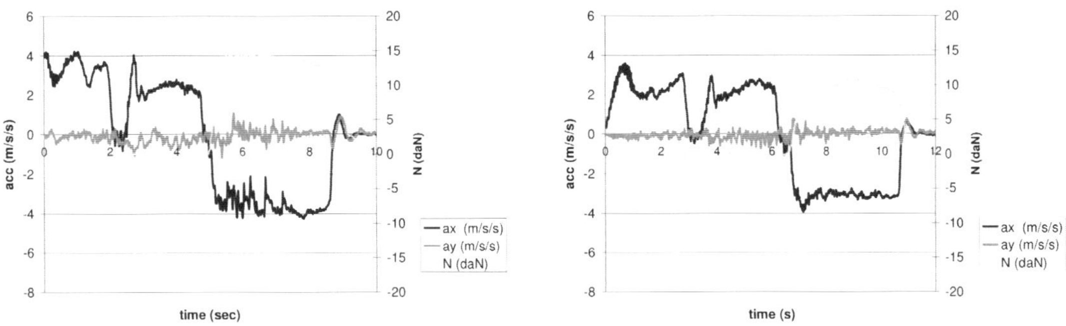

a) b)

Figure 2. The course of the changes of the delay during braking test in perpendicular (a_x) and parallel (a_y) axis of the vehicle; a – braking with air-locked braking wires and operating anti-lock braking system, b – braking with air-locked braking wires and without the anti-lock braking system.

The figures 3a and 3b present exemplary results obtained during braking test conducted with the fully blocked flow of brake fluid to brake caliper cylinder to the right back wheel of the vehicle. In such case, the obtained values of perpendicular delay decreased by about 9% in comparison to the values obtained with correct functioning brake calipers of all the wheels of the vehicle. The lack of the possibility of correcting braking forces on the remaining wheels after turning off anti-lock braking system was the reason for even larger decrease of the obtained perpendicular delays. This was when the difference exceeded 10 %. Higher values of maximum parallel acceleration can also be observed during these tests. The recorded values of parallel acceleration were more than 50 % higher. Fully blocked flow of brake fluid to the brake caliper cylinder of the right front wheel of the vehicle caused even larger difficulties with the maintenance of the track of vehicle movement. Exemplary results obtained during such a test were presented on figure 4. We can observe here very high values of parallel acceleration, particularly with turned off anti-lock braking system. Average values obtained from the courses of acceleration of 5 subsequent ride tests were presented on figure 5.

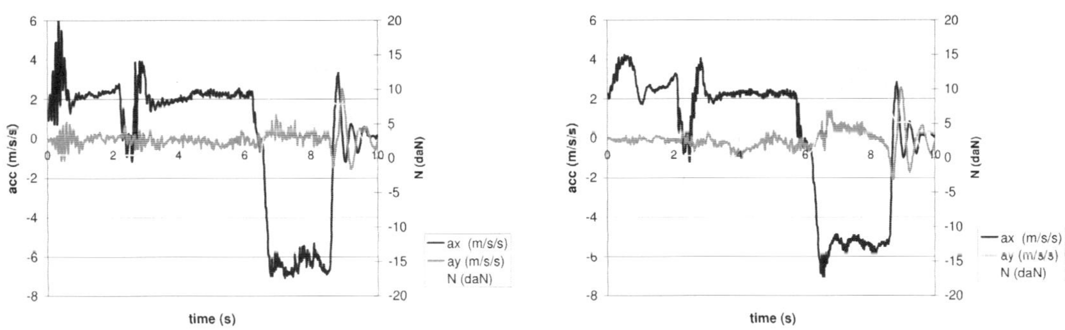

a) b)

Figure 3. The course of the changes of the delay during braking test in perpendicular (a_x) and parallel (a_y) axis of the vehicle; a – braking with the blocked brake caliper cylinder of the right back wheel and with operating anti-lock braking system, b – braking with the blocked brake caliper cylinder of the right back wheel and without anti-lock braking system

The average values of delay obtained in the same road conditions in the vehicle with the diminished braking force, resulting from the air pocket of brake caliper piston, reached similar values both in case of the operation of anti-lock system and with the lack of it. However, anti-lock braking system influences the restriction of the acceleration fluctuations registered in the parallel axis of the vehicle. This is because, it allows for more profitable division of braking forces on both sides of the vehicle. During rectilinear braking with fully efficient braking system, maximum acceleration registered in parallel axis reached 1.5 m/s^{-2} , whereas in case of the lack of anti-lock system it was 20% higher. Also in case of fully blocked brake caliper cylinder of one of the wheels of the vehicle, anti-lock braking system can considerably influence the course of parallel acceleration of car body.

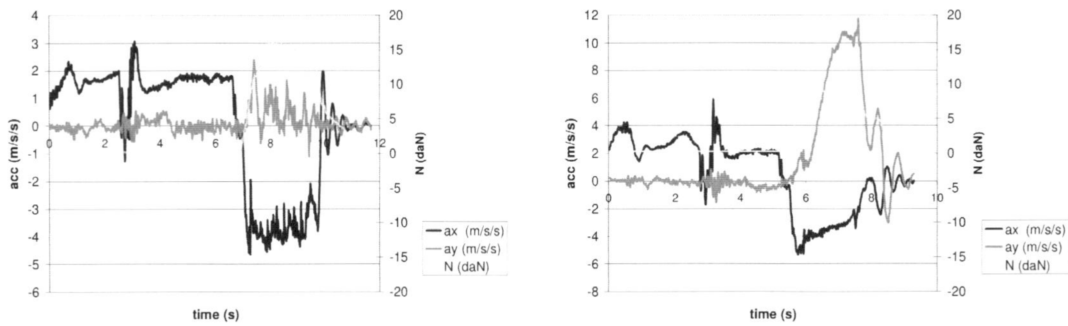

a) b)

Figure 4. The course of the changes of the delay during braking test in perpendicular (a_x) and parallel (a_y) axis of the vehicle; a - braking with the blocked brake caliper cylinder of the right front wheel and with operating anti-lock braking system, b – braking with the blocked brake caliper cylinder of the right front wheel and without anti-lock braking system.

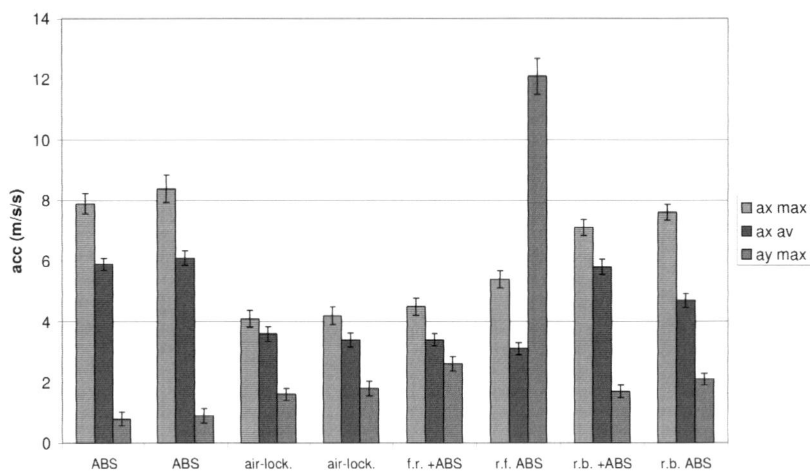

Figure 5. The juxtaposition of average acceleration values in perpendicular and parallel axis of the vehicle obtained during braking tests.

4. Conclusions

Anti-lock braking systems are one of the basic applied constructional solutions which have the influence on vehicle active safety. Conducted road tests of the course of the vehicle braking process allowed to compare the efficiency of their operation with the disturbances occurring in functioning in hydraulic circuit. The measurement of the acceleration in the parallel and perpendicular axis of the vehicle, applied during the tests gave the possibility of the evaluation of the changes of dynamics movement as a result of occurring disturbances.

On the basis of the conducted research we can draw particularly the following conclusions:

1. The change of the compressibility of the brake fluid caused by the presence of vapour bubbles in hydraulic braking system, the consequence of which is the decrease of the brake fluid pressure in brake caliper piston, can lead to lowering the maximum obtainable braking delay by 50 %;
2. The initial braking phase, the phase of increasing pressure constitutes the neuralgic part of the process – recorded parallel acceleration reach the highest values in this phase;
3. With the lack of anti-lock braking system, when braking during straight drive, the parallel acceleration is several times larger than with the operation of the system;
4. The anti-lock braking system is able to correct the dysfunctions of the hydraulic system. Even in case of blocking the flow of brake fluid into the brake caliper cylinder at one of the wheels, anti-lock braking system gives the possibility of the revision of braking course.

References

[1] Botero J C, Gobbi M, Mastinu G, Di Piazza N and Martorana R 2007 *On the reformulation of the ABS logic by sensing forces and moments at the wheels*, Proceedings of the Fifth IFAC Symposium on Advances in Automotive Control

[2] Day T.D Roberts G 2002 *A simulation model for vehicle braking systems fitted with ABS*, SAE Paper (2002-01-0559)

[3] Hegeduš H and Ilić D 2006 *Comparison of measurement methods for vehicle acceleration* XVIII *IMEKO World Congres* Metrology for a Sustainable Development Rio de Janeiro, Brazil

[4] Kudarauskas N 2007 Analysis of emergency braking of a vehicle, *Transport* vol. XXII no 3

[5] Norma ISO/DTR 13487 F: *Braking of road vehicles - Definition of mean fully developed deceleration*

[6] Roger L 2000 Vehicle sensitivity to brake torque differences – test and simulation results, Int. Conf. on Aut. Braking, Leeds

[7] Rozporządzenie Min. Infrastruktury z 18 września 2009 r. w sprawie zakresu i sposobu przeprowadzania badań technicznych pojazdów oraz wzorów dokumentów stosowanych przy tych badaniach. Dz. U. nr 155/2009 poz. 1232

[8] Sokolovskij E 2005 Experimental investigation of the braking process of automobiles *Transport* vol. XX no 3

[9] Szczypiński-Sala W 2014 Wykorzystanie trójosiowego pomiaru przyspieszeń w drogowych badaniach opóźnienia hamowania pojazdu, *Logistyka*, 2014, nr 6

[10] Yasui Y, Nitta H, Yoshida T, Hosome T and Kawamura K 2000 *Experimental approach for evaluating tire characteristics and ABS performance* World Congress SAE

IOP Conf. Series: Materials Science and Engineering **148** (2016) 012056 doi:10.1088/1757-899X/148/1/012056

Kinetic energy recovery systems in motor vehicles

C Śliwiński[1]

[1]Military Institute of Engineer Technology, Wrocław, Poland

E-mail: sliwinski.c@witi.wroc.pl

Abstract. The article draws attention to the increasing environmental pollution caused by the development of vehicle transport and motorization. Different types of design solutions used in vehicles for the reduction of fuel consumption, and thereby emission of toxic gasses into the atmosphere, were specified. Historical design solutions concerning energy recovery devices in mechanical vehicles which used flywheels to accumulate kinetic energy were shown. Developmental tendencies in the area of vehicle manufacturing in the form of hybrid electric and electric devices were discussed. Furthermore, designs of energy recovery devices with electrical energy storage from the vehicle braking and shock absorbing systems were presented. A mechanical energy storing device using a flywheel operating under vacuum was presented, as were advantages and disadvantages of both systems, the limitations they impose on individual constructions and safety issues. The paper also discusses a design concept of an energy recovery device in mechanical vehicles which uses torsion springs as the main components of energy accumulation during braking. The desirability of a cooperation of both the mechanical- and electrical energy recovery devices was indicated.

1. Introduction

The increasing environmental pollution causes the tightening of emission standards. In order to meet their requirements, car manufacturers seek various methods of reducing emissions of harmful and toxic substances, which are produced during the process of fuel combustion in internal combustion engines. This is achieved via improvement of the power unit's efficiency and fuel consumption reduction, the use of innovative systems of exhaust gas recirculation, particle filters and other systems. The most popular methods of reducing fuel consumption include:

- "downsizing" the internal combustion engines by reduction of the stroke volume while maintaining or improving its performance;
- "economy" modes of automatic transmissions, which allow the engine to work in the most economic rotation speed range;
- "start-stop" systems, which allow the engines to shutdown automatically, for the time of stoppage, and then start-up automatically;
- hybrid material frames, built with composite and aluminium materials, to reduce the vehicle's weight;
- hybrid drive systems, which use internal combustion engines and electric motors to power vehicles;
- improvement of vehicle's aerodynamics;
- the use of "low rolling resistance" tires.

Recently, car manufacturers increasingly use hybrid drives with energy recovery systems to lower the vehicle fuel consumption. These systems most often use the braking phase to recover the momentum from the crankshaft, drive shaft or the car's axles and then store it in the form of different types of energy (electric one, kinetic one or compressed air). The most popular energy recovery methods include the use of[1]:

- electric motors, converting the kinetic- to electric energy and storing it in battery packs,
- flywheels, storing kinetic energy.
- compressors, converting the kinetic- to compressed air energy and storing it in hydraulic accumulators[2].

The recovery of kinetic energy in motor vehicles is achieved mostly in three phases. In the first, the energy from the moving vehicle is converted, in the second – it is stored, and in the third, the stored energy is used to power the vehicle.

2. Historical design solutions of energy recovery systems

In the past, most interests in terms of energy accumulating were focused on kinetic accumulators. Devices using flywheels to store kinetic energy were used in hybrid buses, trams and cars. In 1946 a hybrid drive system was developed, consisted of electric motors and a flywheel module, allowing short distance travels. It was used in the industry and public transport. The concept was to use a centrally placed flywheel module, with a flywheel weighing 1.5 t and of a 1.6 m diameter. The hermetic flywheel module was filled with hydrogen, for the air friction lowering, allowing the flywheel to rotate at a speed of up to 3000 rpm.

The Gyrobus power principle (Figure 1) involves powering the electric motor 5, from overhead power lines. The motor spins-up the flywheel 6, on bus stop. It causes the flywheel to accumulate kinetic energy. During the later stage, that is the acceleration and travel, the energy accumulated in the flywheel is transferred through the shaft to the electric motor 5 acting as a generator, converting the kinetic- to the electric energy. The electric energy produced charges the electric motor drive 1 which accelerates the bus. To sum up, the Gyrobus's energy recovery system's purpose was to recover the kinetic energy from the flywheel and convert it to electric energy.

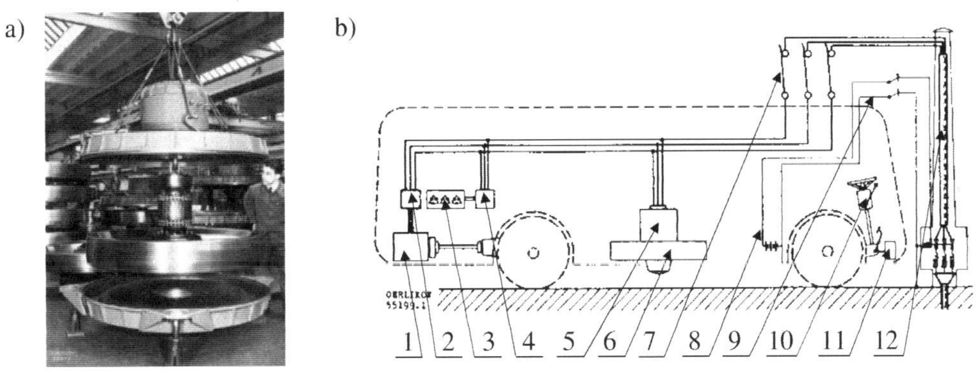

Figure 1. Gyrobus's flywheel module (a) and Gyrobus's system's components (b) [www.photo.proaktiva.eu/digest/2008_gyrobus.html]
1 – electric motor drive, 2 – speed control, 3 – capacitors, 4 – traction control, 5 – electric motor, 6 – flywheel module, 7 – pantograph, 8 – batteries, 9 – on-switch and grounding, 10 – traction control steering, 11 – speed control steering, 12 – charge mast.

The Gyrobus bus line had charging stations every 5 km. There, a 5 minute spinning-up of the flywheel took place. The flywheel spin-up from standstill to operational rotation speed took about 40 minutes. The kinetic energy accumulated in the flywheel allowed the acceleration of a bus up to 55 km/h. Limitations of technology, large dimensions and weight of the flywheels used in that epoch caused the project shutdown and the design did not to come into everyday use.

In the 70s in the United States of America a number of car concepts were created equipped with hybrid drives with mechanical energy recovery systems. They consisted of internal combustion engines and kinetic accumulators. The use of a flywheel module for energy recovery purposes requires a complicated powertrain control system with an automatic transmission. The reason for this is the need to ensure smooth torque transfer between the system's flywheel and the drive shaft, during both acceleration and braking, when the car's kinetic energy is being recovered. An example of a hybrid drive with a flywheel acting as an kinetic accumulator and an automatic gearbox is shown in Figure 2[3].

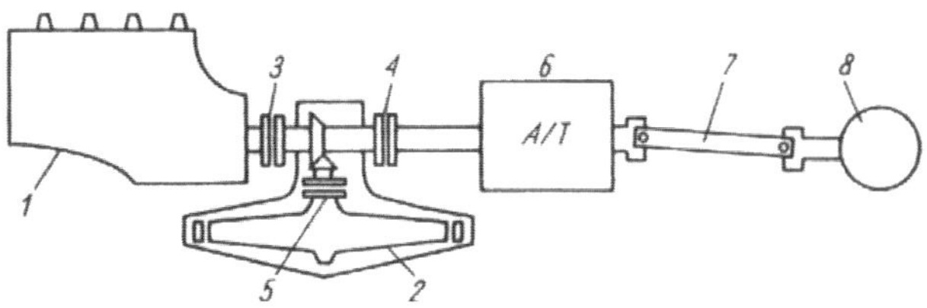

Figure 2. Hybrid drive with a mechanical energy recovery system [3]
1 – internal combustion engine, 2 – flywheel, 3, 4, 5 – clutches,
6 – automatic transmission, 7 – Cardan drive shaft, 8 – driveline.

During travel the internal combustion engine 1, through clutches 3 and 4 and automatic transmission 6 powers the vehicles driveline 8. Clutch 5 is disengaged and the flywheel 2 is in standstill. While braking, in kinetic energy recovery phase, clutch 3 is disengaged and clutch 5 engaged which causes the flywheel 2 to spin-up. Afterwards, during the vehicle's acceleration the torque gathered in the flywheel 2 is being transferred to the driveline 8.
Currently similar kinematic systems of flywheel based energy recovery are used. The difference, when compared to the historical ones, lies in the use of advanced electronic control systems, the high-strength multi-material design and the complexity of the construction. A schematic of a 2013 developed system for energy recovery and storage during braking motion with intermittent rotary velocity for e.g. hybrid vehicles is shown in Figure 3.

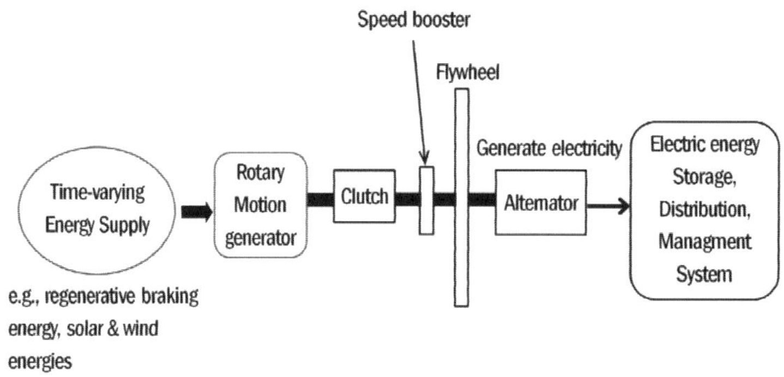

Figure 3. Major components of a regenerative braking system[4]

3. The principle of operation of energy recovery systems

Kinetic energy recovery from a moving vehicle is being initiated mostly during a vehicle's braking phase. In order to do so, the shaft of an energy recovery system must be included in the powertrain, most often connected to a crankshaft or to one of the axles. The first phase of the energy recovery process is the conversion of the energy from the moving vehicle. During the vehicle's slowing down, energy recovery devices (electric motor, flywheel or compressor) are being coupled through transmission to the driveline. This results, dependent on the system, in spinning-up of the flywheels, carrying out work by the electric motors or compressors in order to convert the kinetic- to electric- or hydraulic energy. The next phase is the accumulating of the energy, which is achieved with the use of flywheels, battery packs or hydraulic accumulators. The third and the last phase of the energy recovery process is the utilization of the energy gathered to accelerate the vehicle or to power the vehicle's electric consumers. By that

means the energy, which in normal conditions would be lost in the braking process in the form of generated heat, is being recovered and utilized.
The principle of operation of an energy recovery system with the use of an electric motor, converting the kinetic- to electric energy and storing it in a battery pack is shown in Figure 4. The arrows in the diagrams a), b) and c) show the flow of the energy between the vehicle's engine, driveline and the energy recovery system's components.

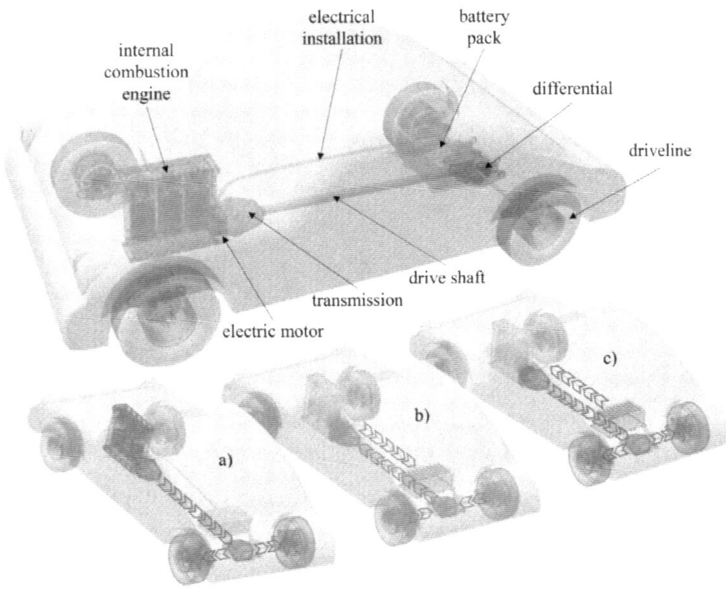

Figure 4. Principle of operation of an electric energy recovery system [own work]
(a)accelerating of the vehicle with the use of an internal combustion engine,
(b) energy recovery during the vehicle's braking and its storing in the battery pack,
(c) accelerating of the vehicle with an electric motor.

4. Contemporary design solutions of energy recovery systems

Currently, electric energy recovery systems are most often used in the motor industry. It is due to the World's development trends and the fact that electronics plays a dominant role in motorization for the sake of increasing the user's comfort. Hybrid electric, electric devices and fuel cells are more often used by the car manufacturers. Table 1 shows the selected experimental car models and the mass produced ones, with hybrid drives installed.

Table 1. Makes and models of vehicles that have forms of energy recovery systems installed.

Manufacturer	Model
Citroen	XsaraDynactive
Dodge	Durango Hybrid
Fiat	Multipla Hybrid Power
Ford	Escape Hybrid SUV
Honda	Civic Hybrid, Insight
Nissan	Leaf
Toyota	Auris Hybrid, Prius
Volvo	V60 Plug-In Hybrid

4.1. Electric energy recovery systems

As was already mentioned, the motor industry largely prefers the electrical energy recovery systems. These systems are frequently based on electrical motors coupled with internal combustion engines through crankshafts with the battery packs placed in the rear part of the vehicle (Figure 5).
This allows the electric motor to work as a "generator" during the vehicle's slowing down and a "starter" during the vehicle's acceleration/travel. The energy stored is utilized automatically to power both the car during acceleration and the electric consumers during stoppage when the engine's shut down (through the start-stop system). In the case of low battery power, the internal combustion engine, through the electric engine, powers the battery pack to ensure their suitable efficiency.

a) b) c)

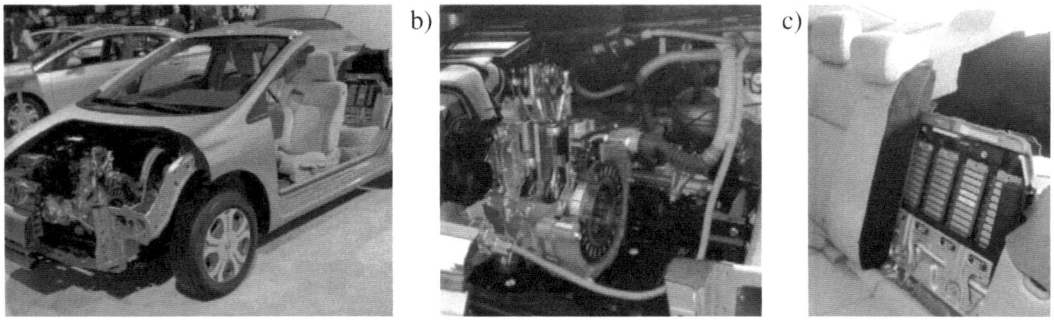

Figure 5. Honda Civic hybrid (a) drive components cut through (b) and battery pack placed in the rear seat's backrest (c) [www.theexpensivecars.com/2012-honda-civic-hybrid.html].

Among the electric energy recovery systems not all devices use the braking phase of the vehicle to recover energy. An example of such a system is the Bose's electromagnetic suspension (Figure 6) [www.bose.com/controller?url=/automotive/bose_suspension/index.jsp]. This device converts the reciprocating motion, accompanying the vehicle's shock absorbing during travel through uneven terrain, into electric energy. This system uses the obtained energy to raise the user's comfort by the horizontal stabilizing of the vehicle. The results of tests with vehicles equipped with conventional and Bose's suspension show that the Bose system is more effective with the car displaying lesser tilt and better shock absorbing (Figure 7).

Figure 6. Bose suspension

a) b)

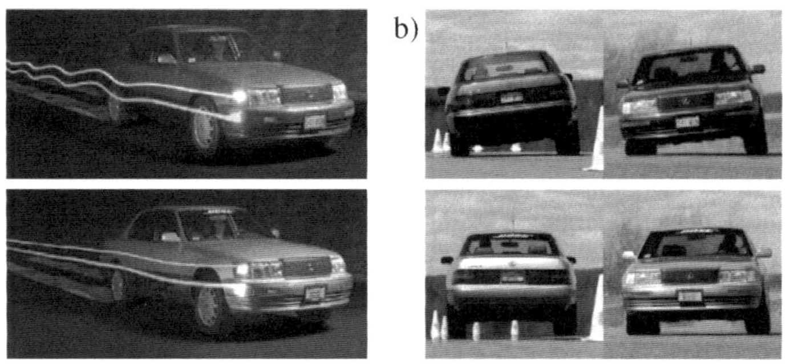

Figure 7. Motion (b) and turning (c) of vehicles equipped with conventional (upper figures) and Bose suspension (lower figures)

The advantage of an electric energy recovery system's construction is that it's reliable and maintenance free. When in use, it reduces the wear of the conventional braking system's components. Tests performed on a hybrid-electric car resulted with an energy recovery ratio ranged from about 16 % to 45 %[5]depending on the degree of use of the conventional braking system. Electric energy recovery systems are also relatively safe with installed current protections and battery packs sealed off.
However, it should be noted that there is a potential danger from the battery packs, for the health of the user and the environment, especially during road accidents. Lifetime of battery packs still depends on

the continuity of the user's vehicle usage. The lack of the vehicle's regular use causes the lowering of the battery pack power and efficiency. The lifetime of a battery pack varies from 3 to10 years.

4.2. Kinetic energy recovery systems

It's in the past that kinetic energy recovery systems were equipped with flywheels of great weight and large diameter to ensure a proper moment of inertia. In 2009 Flybrid Systems designed a hybrid kinetic accumulator for the Formula 1 race cars allowing the flywheel to work in the speed ranges of up to 60.000 rpm. It's steel and carbon-fiber construction weighed 25 kg and was rated at 60 kW.

Since then Flybrid worked with Jaguar and Volvo on developing a reliable system for everyday use. Applications of Flybrid's system for passenger cars feature long life and maintenance free work with the automated process of recovering and utilizing of the gathered kinetic energy. It's design consists of a flywheel locked in an airtight housing (Figure 8.b). By creating a vacuum it was possible to minimize friction losses.

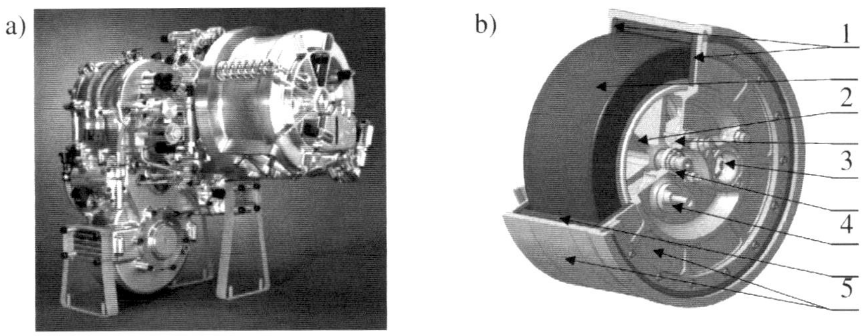

Figure 8. Volvo-Flybrid kinetic energy recovery system with transmission (a)and Flybrid flywheel module cut through (b):1 – Containment Disks, 2 – Flywheel rim,3 – Flywheel hub, 4 – Vacuum seals,5 – Vacuum port, 6 – Flywheel bearing, 7– Speed sensor,8– Containment Ring,9– Flywheel housing [www.media.volvocars.com/global/en-gb/media/pressreleases/48800].

The advantage of the presented kinetic energy recovery system's construction is its small size and weight. Kinetic accumulators have a longer life cycle, relative to electric devices, because of the lack of the negative impact of low temperatures on its lifetime. The production and disposal process of such a system is not as potentially harmful for the environment as the one of an electric system, which includes the need to use electrolytes for batteries.

The disadvantage of kinetic energy recovery systems based on flywheels, is the need to ensure high rotating speeds of its working unit. It poses a health threat for the user and passers-by in the case of its damage and detachment of its components. In the 90s, safety issues caused the shutdown of a Chrysler hybrid race car project, code-named "Patriot". Its kinetic accumulator system included a 61 kg flywheel, rotating at up to 58.000 rpm[5].

5. Design concept of a kinetic energy recovery system

The Military Institute of Engineer Technology in Wroclaw submitted a patent application for "A device for recovering energy of a motor vehicle" to the Polish Patent Office (application no. P.409589) and to the European Patent Office (application no. 15169299.3) [7]. The concept comprises torsion springs as the main components of kinetic energy storage during a vehicle's slowing down phase. The system's increased lifetime will mainly be due to a simple design and a decreased amount of electronic components for steering purposes only. Ultimately the device should be installed on the vehicle's rear axle, to balance its centre of gravity, but it's possible to install the spring-modules alongside of the vehicle's axis of symmetry in a circular or rectangular pattern. These modules could be engaged in the system with additional gears or clutches to allow for sequential utilization of the stored energy. The principle of operation of this concept relies on the operation of five clutches: a friction clutch (1), two claw clutches (2), two one-directional clutches (5) and two locking mechanisms (10) (Figure 9). The figure shows a concept of an energy recovery system with the turned off view of housings, to illustrate the connections between clutches, torsion springs and gears.

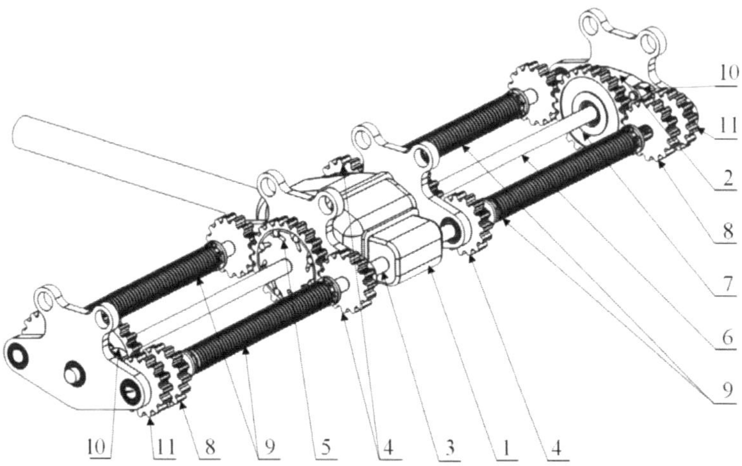

Figure 9. A device for recovering kinetic energy of a motor vehicle [7],
1 –friction clutch housing,2 – claw clutch engagement mechanism housing,
3 – fourth shaft, 4 – fourth gear, 5 – one-directional clutch, 6 – drive shaft, 7 – first gear,
8 – third gear, 9 – torsion spring, 10 – locking mechanism, 11 – second gear.

The friction clutch (1) stops the fourth shaft (3) alongside with fourth gears (4) and the cooperating one-directional clutches (5). The claw clutches (2) transfer the torque from drive shafts (6) to first gears (7) causing the rotation of third gears (8). The difference in rotation speeds between the fourth gears (4) and the third gears (8) causes the twisting motion of the torsion springs (9). When the desired angle of rotation of torsion spring fronts (9) is reached, the claw clutches (2) disengage, and the locking mechanisms (10) engage, leaving the cooperating second gears (11) third gears (8) and first gears (7) in standstill. The disengagement of the friction clutch (1) causes the utilization of the stored in the torsion springs (9) torque, through the fourth gears (4) and the one-directional clutches (5) to the drive shafts (6). When the angle between the torsion spring fronts (9) reaches zero degrees, meaning all of the stored torque is used, the locking mechanisms (10) are being disengaged leaving the device in standstill.
Using the torsion springs to recover energy allows for smooth torque transfer between the driveline and the energy recovery device, as well during the vehicle's braking and acceleration without the use of complicated transmissions. This is mainly due to the work characteristics of torsion springs. During a vehicle's slowing down phase, a steady rise of torque occurs, from near zero to maximum, limited with the vehicle's wheel friction to the road surface (Figure 10).

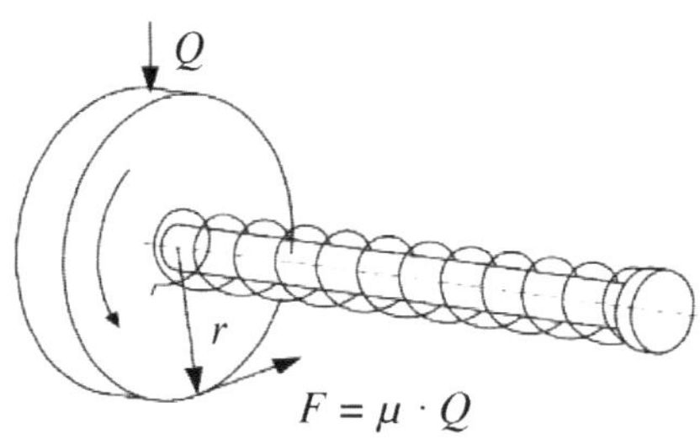

Figure 10. Distribution of forces on a vehicle's wheel during braking while recovering energy through a torsion spring.

The dependence between the torsion spring's maximum torque T_{max}, the road surface's friction coefficientμ, the wheel's load Q and radius r is shown in equation (1),

$$T_{max} = \mu \cdot Q \cdot r \tag{1}$$

Figure 11 shows a graph of torsion springs characteristics in form of the correlation of a spring's stored torque T and the spring's angular deflection. The torsion spring's smooth torque's transfer is suitable for the vehicle's energy demands. During braking, when a smooth speed decrease is desired for the user's comfort, the torque increase should be linear; from a minimal T_{min} to a maximal T_{max} value. During start-up, the energy needed to overcome the car's rolling resistance, mostly related with the vehicle's weight, is high. Therefore the applied torque should also be linear from a maximal T_{max} to a minimal value T_{min}, such as that of the torsion spring.

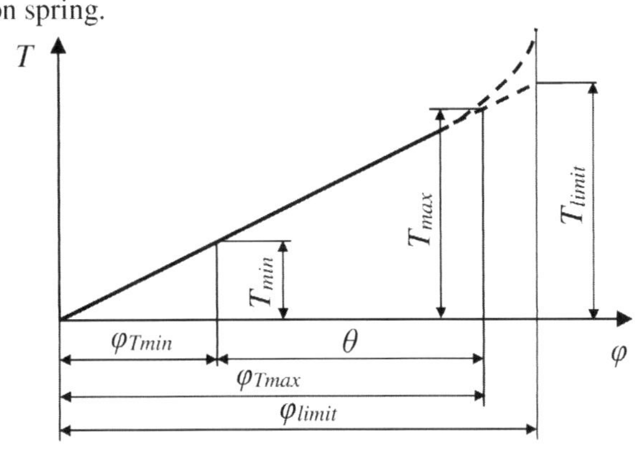

Figure 11. Graph of torsion spring's characteristics in form of torque value T versus angular deflection φ[8].

When choosing springs for an application, an important parameter is the torsion spring index. It is dependent on spring's wire diameter d and the mean spring diameter D, and it is calculated as in equation (2). The lower the factor, the more exposed to the stress damage the spring's fibres closest to the axis of the spring will be. Therefore using springs of index not lower than 3 or 4 is recommended.

$$C = \frac{D}{d} \tag{2}$$

What is important for the torsion spring's maximum torque T, is the correlation of the wire's cross-sectional shape and the curvature of the spring's coil which are known as the K factor. It is defined differently, dependent of the literary source. For example literary source [8] for calculation purposes of the K factor provides the following equation (3). However, the factor's value is similar, regardless of the formula used.

$$\mathrm{K} = \frac{4C\text{-}1}{4C\text{-}4} \tag{3}$$

For design assumptions, an exemplary 1H18N9 steel will be used of tensile strength rating R_m = 1500 N/mm^2 = 1500 $\cdot 10^6$ N/m^2 and allowable stress in bending as in equation (4) [9]. The torsion spring's maximal torque as in [8] is dependent on the spring's material, in the form of the allowable stress in bending k_g [N/m^2], the spring's wire diameter d [mm], and the K factor. For a spring with mean diameter of D = 30 mm, the wire's diameter of d = 6 mm and n = 35 number of coils, the maximal achievable torque can be calculated from an equation (5).

$$k_g = \frac{R_m}{x_{Rm}} = 0.7 \cdot R_m = 0.7 \cdot 1500 \cdot 10^6 = 1050 \cdot 10^6 \frac{\mathrm{N}}{\mathrm{m}^2} \tag{4}$$

$$T_{max} = k_g \cdot \frac{\pi \cdot d^3}{32 \cdot K} = 1050 \cdot 10^6 \cdot \frac{3.14 \cdot 0.006^3}{32 \cdot 1.19} = 18.7 \text{ N·m} \quad (5)$$

The maximal torque can be achieved safe only with an optimal twist angle. The spring's twist angle φ_{Tmax}, dependent of the maximal torque, the spring's material, in form of the Young's modulus E, the spring's wire diameter d, the K factor and the length of the spring l can be calculated as in equation (6).

$$\varphi_{Tmax} = 114.5 \cdot \frac{k_g \cdot l}{K \cdot d \cdot E} = 114.5 \cdot \frac{k_g \cdot \pi \cdot D \cdot n}{K \cdot d \cdot E} = 114.5 \cdot \frac{1050 \cdot 10^6 \cdot 3.14 \cdot 0.03 \cdot 35}{1.19 \cdot 0.006 \cdot 190 \cdot 10^9} = 292.2° \quad (6)$$

The spring's twist angle φ_{Tmax}, which is dependent of the maximal torque T_{max}, should be checked for buckling. The spring's limit twist angle φ_{limit}, given safety factor's value 2, can be calculated as in equation (7).

$$\varphi_{limit} = 123.1 \cdot (n)^{1/4} = 299.4° \quad (7)$$

The lesser value of the spring's twist angle φ_{Tmax}, in comparison with the limit twist angle value φ_{limit}, means that the chosen torsion spring can safely produce 18.7 Nm with the spring's twist angle of 292.2°. It should be reviewed whether the spring's project length was correctly assumed and what correlations are there between the spring's length, diameter and twist angle. Therefore, new factors will be introduced:

- the ΔD factor, the difference between the spring's mean diameter D, and the decreased diameter D' while in twist angle φ_{Tmax}, shown in equation (8).
- the φ' factor, a percentage use of the limit twist angle, a quotient of the spring's twist angle φ_{Tmax} and the spring's limit twist angle value φ_{limit} is shown in equation (9).

$$\Delta D = D - D' \quad (8)$$

$$\varphi' = \frac{\varphi_{Tmax}}{\varphi_{limit}} \cdot 100\ \% \quad (9)$$

The correlations of the spring's length and the ΔD and φ' factors for a spring of 30 mm mean diameter, of 6 mm wire diameter, under the load of 18.7 Nm, for 4 sample springs were presented in table 2 and graphically in Figure 12.

Table 2. The correlation of the spring's length and the ΔD and φ' factors

Parameter	Sample spring number			
	#1	#2	#3	#4
l, mm	37	99	161	223
φ_{Tmax}, °	42	125	209	292
φ_{limit}, °	184	242	275	299
φ', %	23	52	76	98
ΔD, mm	2.78	1.29	0.89	0.70

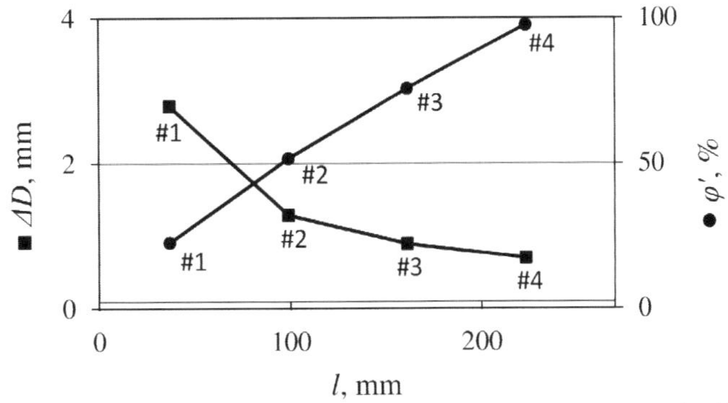

Figure 12. The correlation of the spring's length and the ΔD and φ' factors

The results from table 2 show that with the increase of spring's length the percentage use of the spring's limit twist angle φ_{limit}, increases. However, the ΔD factor decreases, which means that the longer the spring, the lower is the diameter reduction during the twist.

Analysing the formula from equation 6 and the results from table 2, the torque obtained is not dependent on the spring's length, which means that the desired torque T_{max} could be achieved with a shorter spring with lower twist angle, but with a greater mean spring diameter reduction. However, this would result in earlier deterioration of the spring's characteristics. For the kinetic energy recovery system's reliability purposes, a rather long spring should be used, because of the lower decrease in the spring's mean diameter D while in twist angle φ_{Tmax}.

Analysing the results of the above performed calculations, it can be assumed that the proposed concept of the kinetic energy recovery system comprising four torsion springs, would allow great recovery of the vehicle's kinetic energy, which is normally lost during the braking process. It should be noted that the described concept's parameters, such as: dimensions, weight, production costs, energy recovery effectiveness and savings resulting from the system's use are unknown. Nonetheless, the device's advantages, such as low maintenance cost, reliability and no efficiency losses due to the lack of energy conversion, allows the author to believe that it is reasonable to develop a test model of such a solution. Additionally, because of the aforementioned work characteristics of torsion springs, this system would ideally complement the electric energy recovery systems used nowadays. The use of both kinetic- and electric energy recovery systems would greatly raise the effectiveness of energy recovery.

6. Conclusions

Recovery of a motor vehicle's kinetic energy, as the most developmental method of reducing fuel consumption, attracts the attention of both car manufacturers and vehicle users. Contemporarily, many car manufacturers use hybrid electric drives, which recover energy during a vehicle's braking phase and utilize it for acceleration- or powering of electric consumers purposes. Kinetic energy recovery systems, e.g. using flywheels as means of energy storage, draw little interest. The proposed idea of a kinetic energy recovery device using torsion springs to store energy, would complement the currently used electric energy recovery systems perfectly. It requires little time to store kinetic energy and it is safe, when comparing to kinetic energy accumulators using flywheels as means of kinetic energy storage. The use of both kinetic- and electric systems alongside each other in order to reduce fuel consumption and thereby lower emissions of harmful and toxic substances in the atmosphere seems very promising.

References

[1] Kapoor R and Parveen C M 2013 Comparative study on various KERS,*Proceedings of the World Congress on Engineering* Vol III, London, U.K.

[2] Siadkowska K and Skiba K 2014 Pneumatyczno-hydrauliczny układ odzyskiwania energii hamowania w pojazdach ciężarowych,*Czasopismo Logistyka*, Poland

[3] Szumanowski A 1984 Teoria samochodu, akumulacja energii w pojazdach, Wydawnictwa Komunikacji i Łączności, Warsaw, Poland
[4] Hsu T R 2013 On a flywheel-based regenerative braking system for regenerative energy recovery, *Proceedings of Green Energy and Systems Conference*, CA, USA
[5] Gajek A and Strzępek P 2015 Examining the braking energy recovery in a vehicle with a hybrid drive system, *The Archives of Automotive Engineering*,
[6] Cibulka J 2009 Kinetic energy recovery system by means of flywheel energy storage,*Advanced Engineering3*, ISSN 1846-5900
[7] Śliwiński C 2015 A device for recovering energy of a motor vehicle,application no. P.409589 (Polish Patent Office), application no. 15169299.3, European Patent Office
[8] Żukowski S 1955 Sprężyny, Państwowe Wydawnictwa Techniczne, Warsaw, Poland
[9] Baranowski B 1997 Sprężyny metalowe, Wydawnictwa Naukowe PWN, Warsaw, Poland

Comparison of spectral analysis of vibration using commercial knock sensor and 3-axis acceleration sensor

Ł Zieliński[1], D Walczak[2], K Szczurowski[3] and S Radkowski[4]
[1]PhD student, Warsaw University of Technology Institute of Vehicles, Warsaw, PL
[2]PhD student, Warsaw University of Technology Institute of Vehicles, Warsaw, PL
[3]Doctor, Warsaw University of Technology Institute of Vehicles, Warsaw, PL
[4]Professor, Warsaw University of Technology Institute of Vehicles, Warsaw, PL

E-mail: lzielinski@mechatronika.net.pl

Abstract: With the development of internal combustion engines, engineers attempt to reduce the noise and vibration generated. Due to the high cost of fuel, are increasingly looking for new sources of power in order to reduce costs. In diesel engines, an increasingly popular method is the admixture of propane-butane. This follows because of the price of the fuel as well as to improve the efficiency of combustion. With the development of this type of dual fuel power seems to be a reasonable study of the effects of LPG to generate noise and vibration, as well as an attempt to evaluate the combustion process. Unfortunately, too much addition of LPG causes a phenomenon called knock consisting in abnormal, uneven, explosive combustion of fuels in reciprocating engines. This phenomenon may lead to a reduction in engine performance and permanent damage. Control of the knock detection uses vibration acceleration sensors recording the high frequency ranges. Within the framework of the research conducted by the team of authors, an attempt was made to compare the vibroacoustic signals originating from the commercial knocking sensor with a three-axis acceleration sensor. These signals were subject to a quick Fourier transform in the purpose of analysing the amplitude spectra.

1. Introduction

Searching for alternative sources of supplying the combustion engines is a natural phenomenon in relation to the limited resources of petroleum and to the more and more demanding standards regarding the exhaust gas emissions [1]. Popularity of the specific fuel in a given region depends on the geographic, political, and economic conditions. Additionally, the distribution network of a given fuel is not of lesser importance, enabling the access to the filling-up facilities. The geopolitical system is divided into two poles of gaseous fuels - CNG and LPG. In many European countries, a propane – butane mixture is very popular. This is a result of a low fuel price, wide distribution network, and a relatively cheap method of storing and transport [2].

LPG is a liquid fuel used basically as a replacement for a basic fuel in engines with spark ignition. For the last few years, the tests have been conducted connected with the effect of the LPG addition to the engines with compression-ignition [3-6]. It results from the attempts to replace a part of a diesel oil dose with a cheaper fuel, in order to cause the cost reduction.

Significant physicochemical differences of these two fuels induce both advantages and problems related to the use of the diesel oil and LPG [7] during simultaneous powering of the combustion engine. The diesel oil has a property of compression-ignition, thus being an initiator of the combustion process. These fuels have a different flame speed, and therefore a different speed of pressure increase. The current diesel oil supply systems enable a very flexible adjustment regarding the number of doses and their percentage distribution. LPG, however, is a gaseous fuel, fed to the intake manifold and creating the mixture of fuel and air during the compression stroke, which in engines with a high compression ratio results in susceptibility to the phenomenon of a knocking combustion, which is disadvantageous during combustion of such a mixture. A high compression ratio causes the increase in the probability of occurrence of this phenomenon as a result of high temperature at the end of the compression phase. The phenomenon of a knocking combustion has a degenerative effect on the engine parts [8], causing its permanent damage.

In the recent years, in Poland and in Europe, the growing interest in installations enabling injection of an LPG addition to the engines with compression-ignition can be observed. The control systems of such installations are based to a large extent on the low amount of information about the current parameters of the engine operation. Reduction or turning off the LPG addition takes place when a specific temperature of exhaust gasses is exceeded or when the information comes from the additional sensor of knocking combustion. Based on the little amount of information, it is difficult to adjust the maximal participation of LPG that will not induce the knocking combustion phenomenon.

The knocking combustion can be observed by means of registering the vibrations of the engine body, and in particular the specific frequency bands of these vibrations. The reason why the authors discuss these issues is the fact of the growing interest in using the information included in generated vibrations and in the acoustic emission [9, 10], as well as the reduction of the acoustic emission as one of the environmental threats. Another emission of vibrations and noise in a dual-fuel engine is generated by the pressure change in the combustion chamber, which correlates with the measured parameters, among others, the torque. This phenomenon seems to be interesting if an attempt to optimise the selected control parameters is taken into consideration, where the optimisation criterion is minimising the vibrations and noise generated by the combustion engine. Within the framework of the research conducted by the team of authors, an attempt was made to compare the vibroacoustic signals originating from the commercial knocking sensor with a three-axis acceleration sensor. These signals were subject to a quick Fourier transform in the purpose of analysing the amplitude spectra.

2. Testing stand

The testing stand employed in this paper is a SEAT Cordoba6L vehicle, equipped with the modified dual-fuel engine based on the unit marked 1.4 TDI AMF.

The engine of the tested vehicle was adapted to the dual-fuel operation by means of mounting an additional LPG installation with injectors located in the intake manifold tubes. The gas is injected sequentially, supplying all cylinders one by one (Fig.1). Additionally, the compression ratio was decreased from 19.5 to the value of 18.5, by means of using a thicker gasket and increasing the space of the combustion chamber by means of the mounting sleeves for the installation of the glow plugs with the pressure sensor (Fig.2).

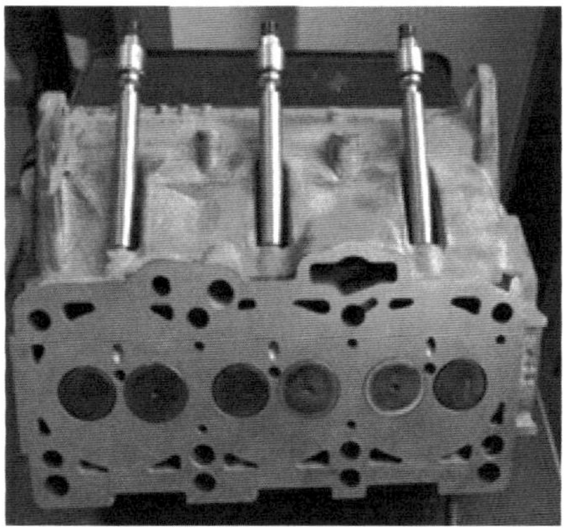

Figure 1. Glow plug mounting sleeves PSG.

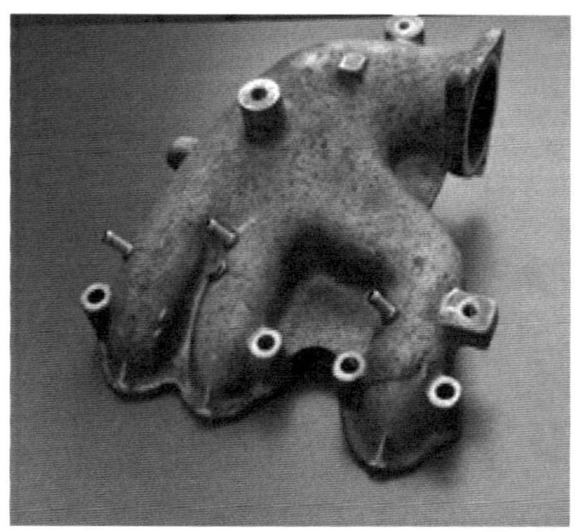

Figure 2. LPG injection nozzle.

During the tests, the head vibrations (the B&K and knock sensors) were registered among other parameters, as well as the excess oxygen ratio (λ), the air temperatures in the intake and exhaust manifolds, fuel doses, amount of air, pressure in the intake manifold, pressure inside the combustion chamber (PSG), rotational velocity, engine load, torque, and many others. During the tests, the LabVIEW environment was used to register all measured signals. For the registration of vibrations, a knock sensor (Bosch 0 261 231 004) (Fig.3) was employed, and for the registration of accelerations - a three-axis vibration sensor (B&K, 4504 type), (Fig. 3). The signals were sampled with the frequency of 51.2 kHz, and every measurement lasted 5 seconds. The measurement cards NI 9234 were used for the measurement-taking (Fig. 4). The whole vehicle situated on the chassis dynamometer is shown in the last picture of this chapter (Fig. 5). The tests were performed for different LPG shares and for various rotational velocities.

Figure 3. Installation of sensors: a combustion knock sensor and a three-axis acceleration sensor.

Figure 4. Measurement setup.

Figure 5. Tested vehicle on the chassis dynamometer.

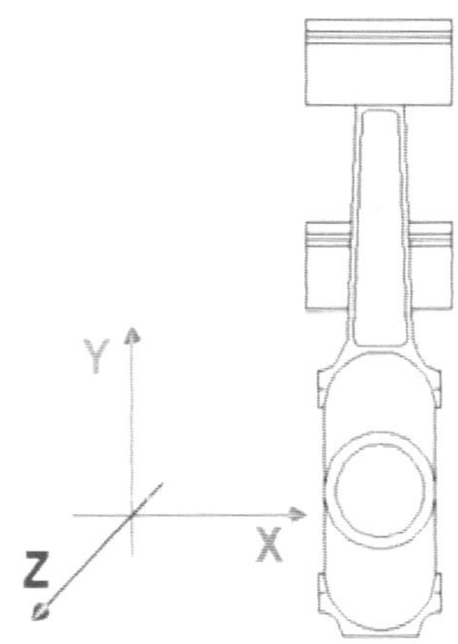

Figure 6. Acceleration axis layout of the three-axis sensor

3. Result analysis

Analysing the obtained measurements of signals connected with the dual-fuel engine's vibroacoustic activity, the amplitude-frequency analysis was conducted in the first step. In the first stage, the low frequency band, in which changes related to the ignition and combustion of the air-fuel mixture processes can be seen, was the main focus. The result is shown in Figures 7-18

IOP Conf. Series: Materials Science and Engineering **148** (2016) 012057 doi:10.1088/1757-899X/148/1/012057

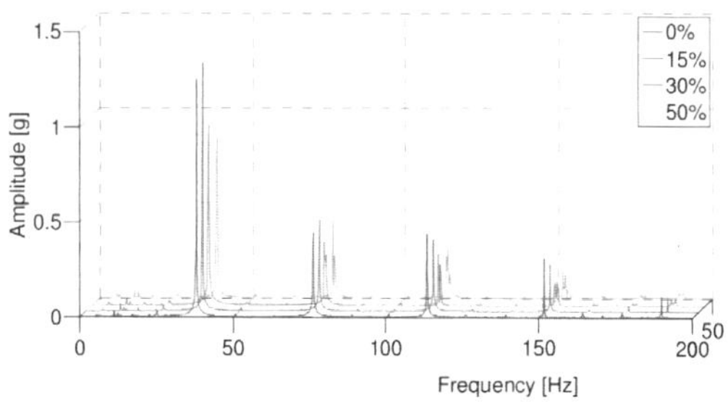

Figure 7. Amplitude spectrum, 1500 RPM, LPG share (0-50%), X-axis vibration sensor.

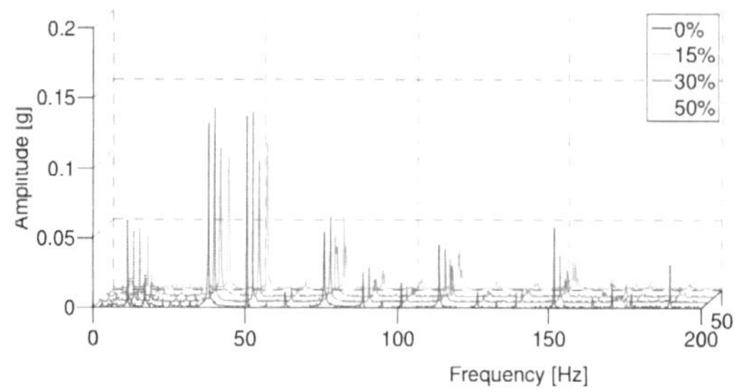

Figure 8. Amplitude spectrum, 1500 RPM, LPG share (0-50%), Y-axis vibration sensor.

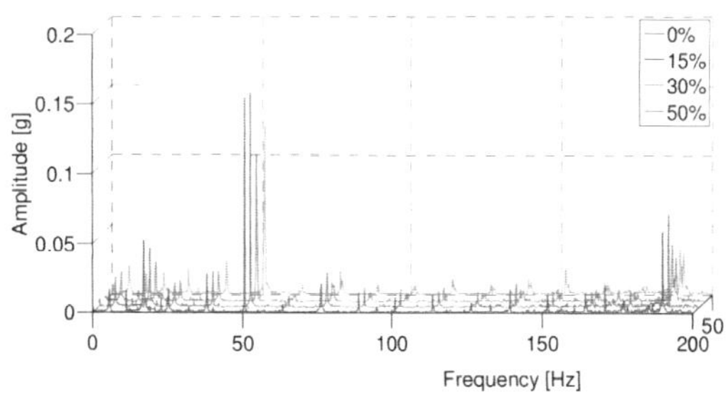

Figure 9. Amplitude spectrum, 1500 RPM, LPG share (0-50%), Z-axis vibration sensor.

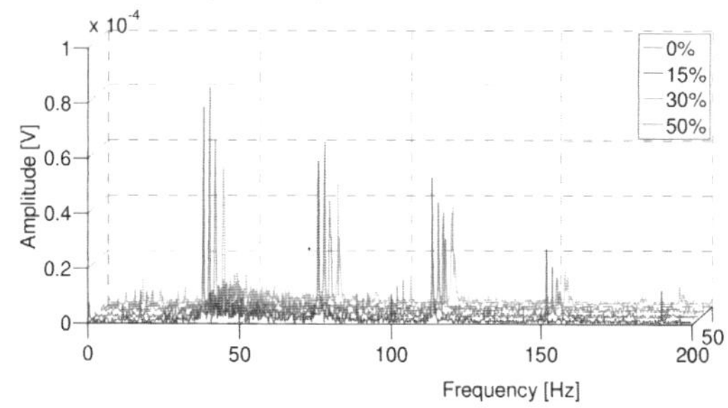

Figure 10. Amplitude spectrum, 1500 RPM, LPG share (0-50%), knock sensor.

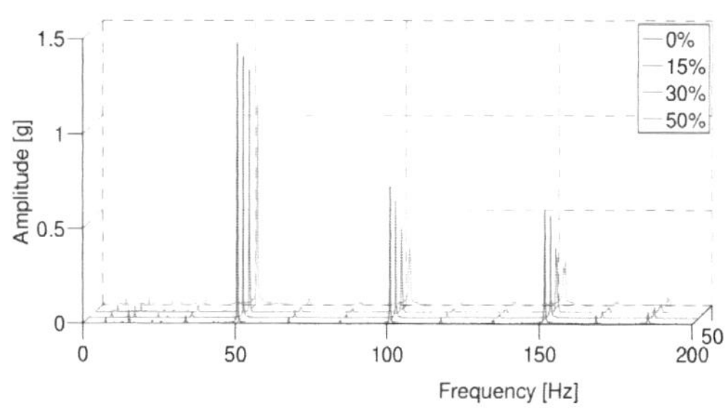

Figure 11. Amplitude spectrum, 2000 RPM, LPG share (0-50%), X-axis vibration sensor.

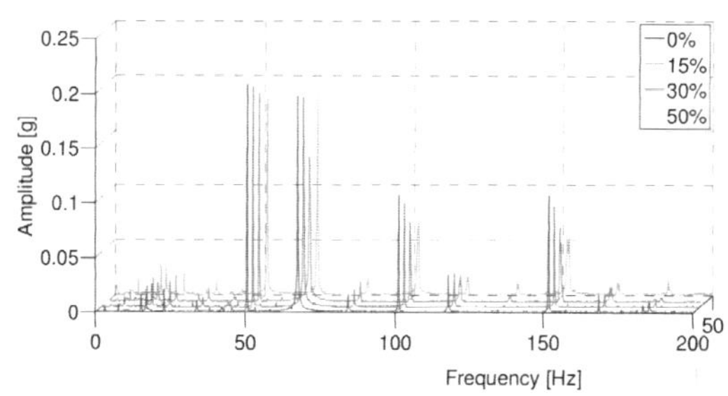

Figure 12. Amplitude spectrum, 2000 RPM, LPG share (0-50%), Y-axis vibration sensor.

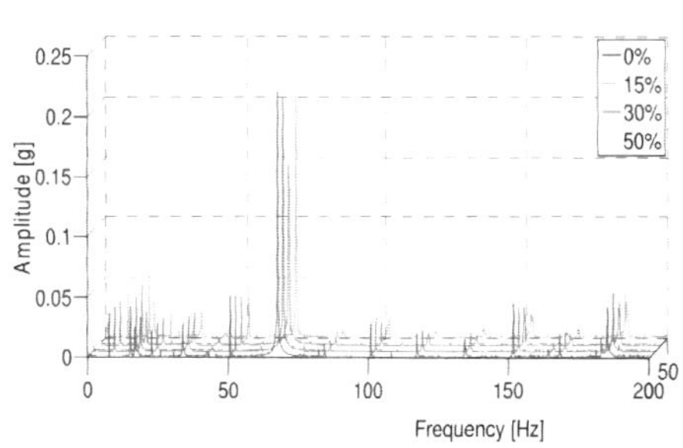

Figure 13. Amplitude spectrum, 2000 RPM, LPG share(0-50%), Z-axis vibration sensor.

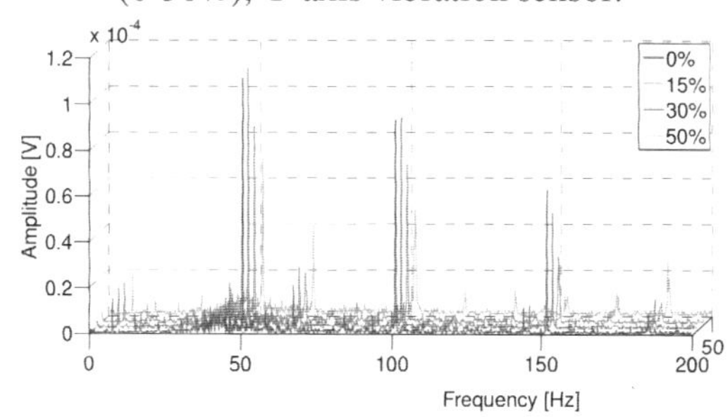

Figure 14. Amplitude spectrum, 2000 RPM, LPG share (0-50%), knock sensor.

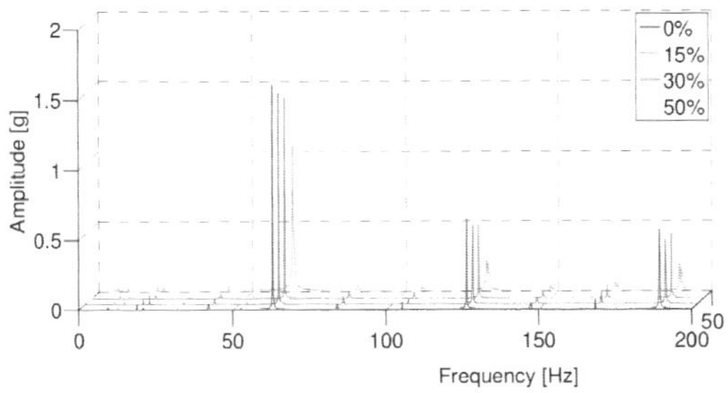

Figure 15. Amplitude spectrum, 2500 RPM, LPG share (0-50%), X-axis vibration sensor.

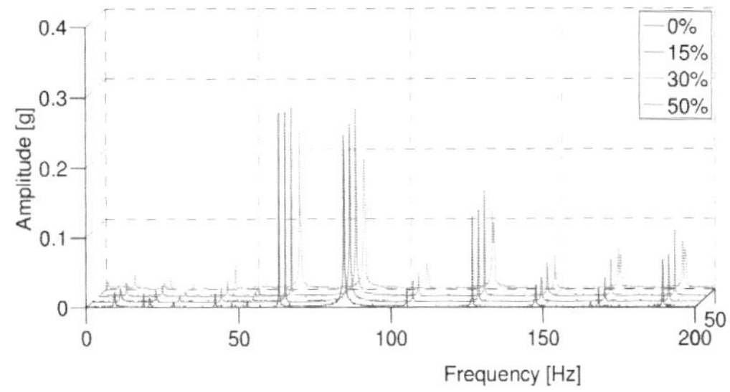

Figure 16. Amplitude spectrum, 2500 RPM, LPG share (0-50%), Y-axis vibration sensor.

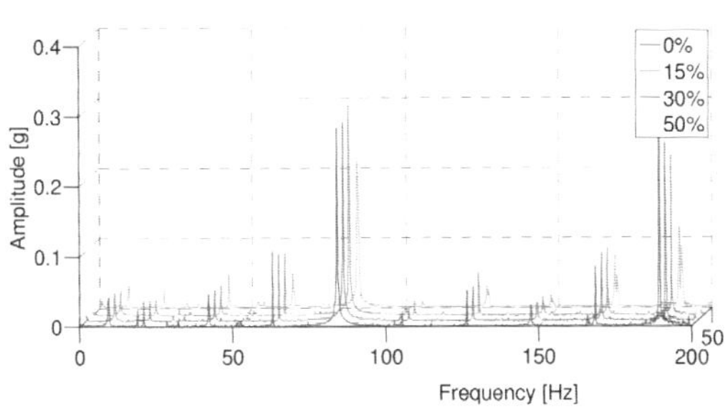

Figure 17. Amplitude spectrum, 2500 RPM, LPG share (0-50%), Z-axis vibration sensor.

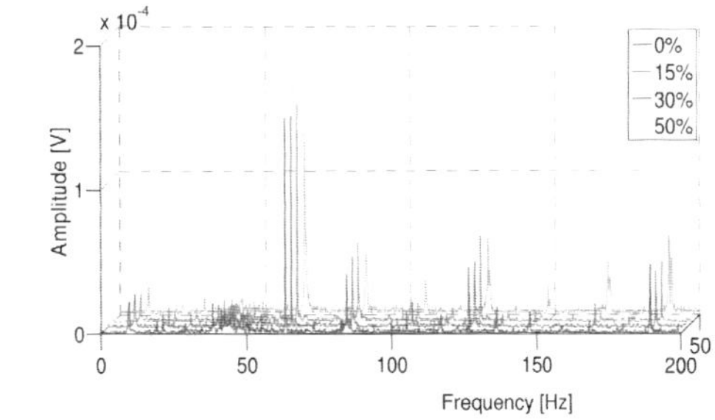

Figure 18. Amplitude spectrum, 2500 RPM, LPG share (0-50%), knock sensor.

The harmonics of the frequency corresponding to the mixture ignition indicate the decrease in vibration amplitude with the growing LPG share, apart from the last measurement for the greatest participation of the gaseous fuel, which was examined, i.e. 50% LPG. Both the spectra for the individual vibration directions of the acceleration sensor, as well as the spectrum of the signal from the knock sensor, do not indicate unequivocally the occurrence of knocking combustion. The next step consisted in accomplishing spectra for the broader frequency band. According to relation 1, the estimated frequency corresponding to the occurrence of knock combustion amounts to approximately 7 kHz. This relation was developed on the basis of the empirical data for the engines with the spark ignition, in which the resonance frequency of knock is described by the relation:

$$f[kHz] = \frac{1800}{3{,}14 \cdot piston\ diameter\ [mm]} \tag{1}$$

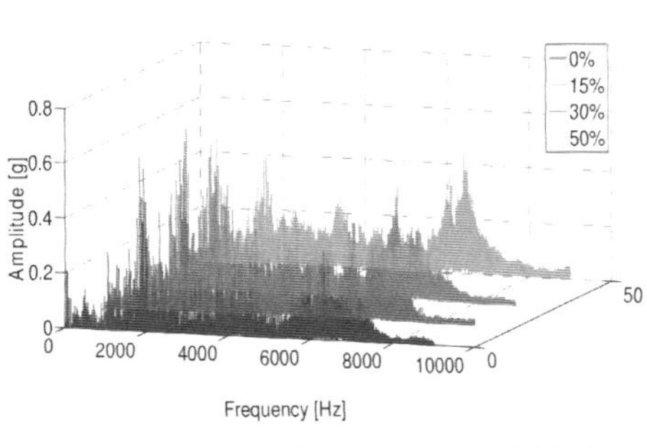

Figure 19. Amplitude spectrum, 2000 RPM, LPG share (0-50%), Y-axis vibration sensor.

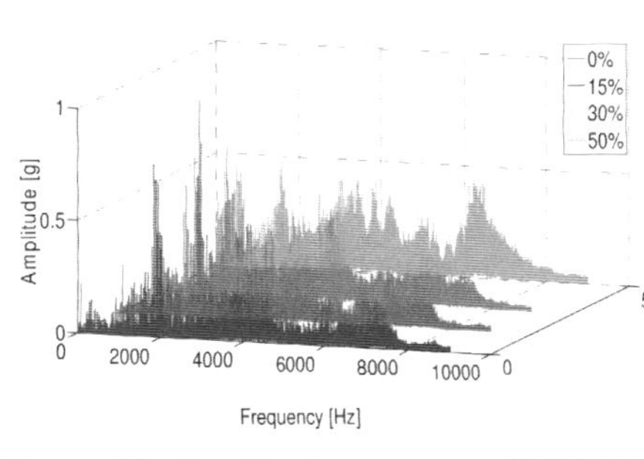

Figure 20. Amplitude spectrum, 2000 RPM, LPG share (0-50%), Z-axis vibration sensor.

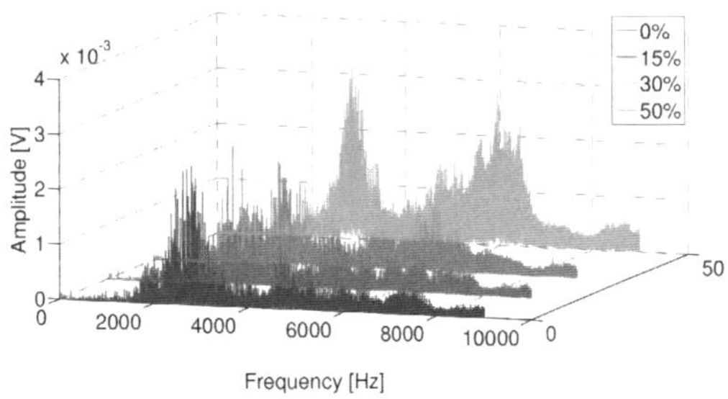

Figure 21. Amplitude spectrum, 2000 RPM, LPG share (0-50%), knock sensor.

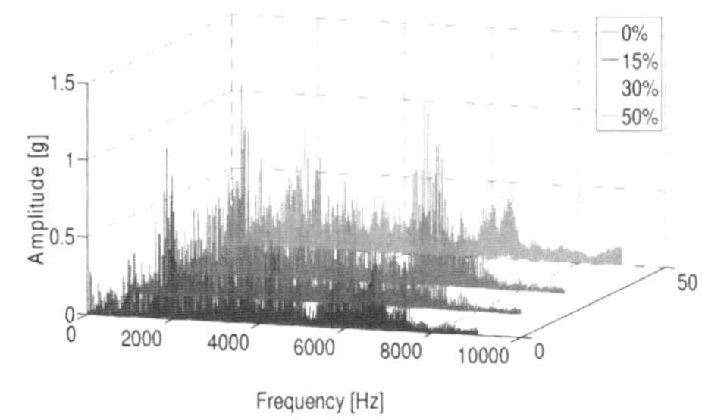

Figure 22. Amplitude spectrum, 2500 RPM, LPG share (0-50%), Y-axis vibration sensor.

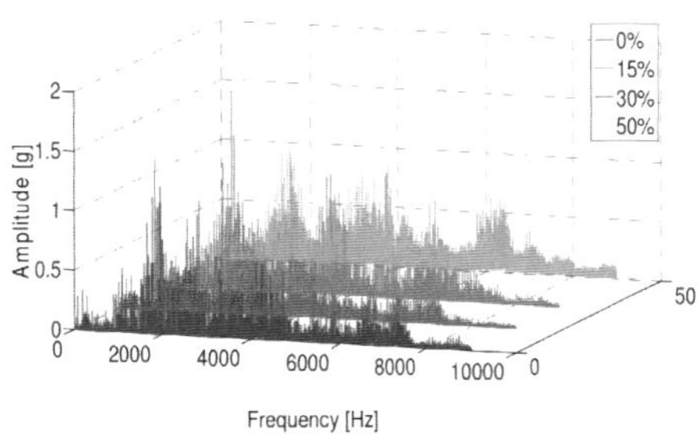

Figure 23. Amplitude spectrum, 2500 RPM, LPG share (0-50%), Z-axis vibration sensor.

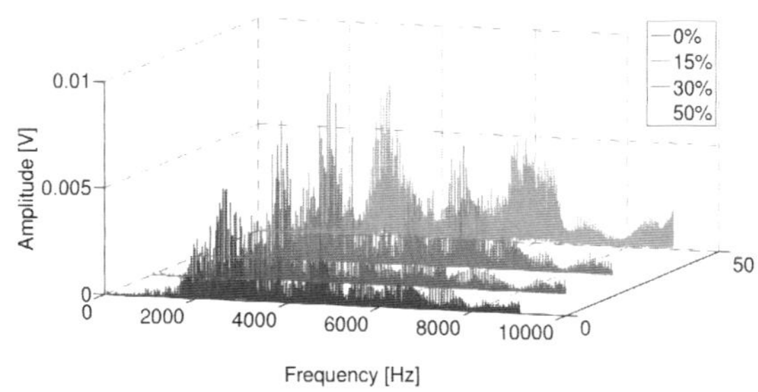

Figure 24. Amplitude spectrum, 2500 RPM, LPG share (0-50%), knock sensor.

In the spectra for the broader frequency band, a greater vibroacoustic activity in the range of 5-7 kHz can be observed, especially for bigger LPG shares, i.e. 30% and 50%. The results of this analysis indicate the symptoms of knocking, unequivocal statement of the occurrence of this phenomenon, however, is not possible. The greatest amplitudes, however, can be observed on the signal from the Z-axis and the signal from the knock sensor.

The next step was determining spectra of the proportionality constant to the vibration energy, understood as a product of the square of the amplitude and the frequency. The results are shown in Figures 26-30. As can be observed, with all measurement directions, as well as on the combustion knock sensor, the increase in amplitudes of the determined constant was registered. Selection of information proved to be a very interesting task, which was carried out by the sensor in the direction of "Z" Fig. 26 and 29 As can be observed, a proper setting of the measurement direction can serve as a filter for information regarding the occurrence of vibrations from the range of knocking combustion. This information is more unequivocal in the context of the knocking combustion occurrence.

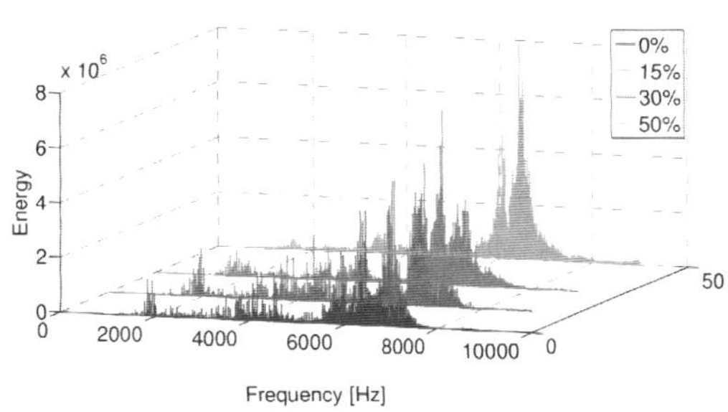

Figure 25. Amplitude spectrum of vibration energy, 2000 RPM, LPG share (0-50%), Y-axis vibration sensor.

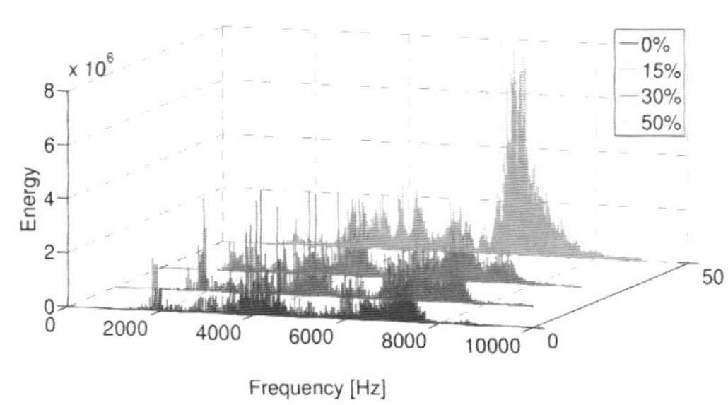

Figure 26. Amplitude spectrum of vibration energy, 2000 RPM, LPG share (0-50%), Z-axis. vibration sensor.

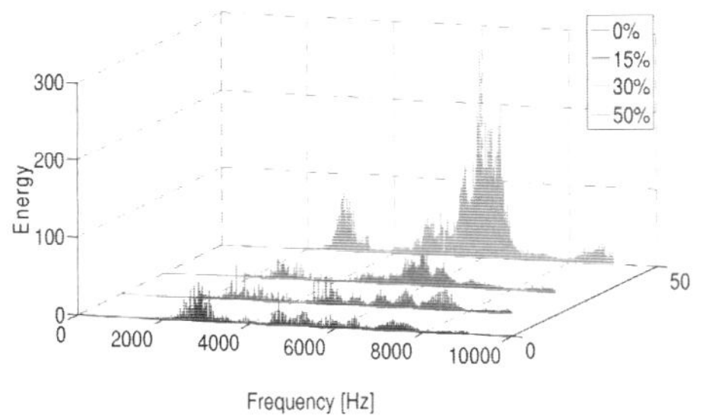

Figure 27. Amplitude spectrumof vibration energy, 2000 RPM, LPG share (0-50%), knock sensor.

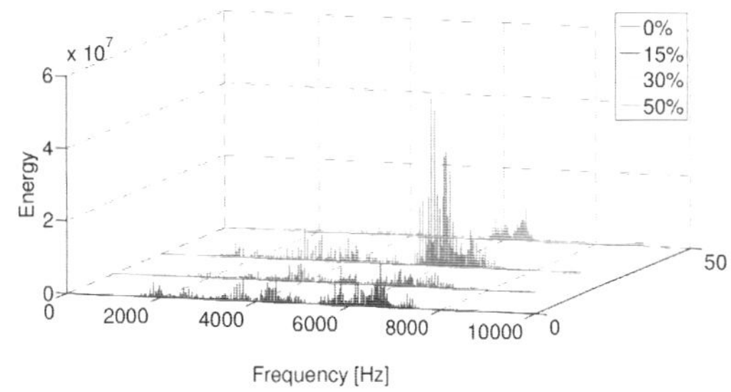

Figure 28. Amplitude spectrumof vibration energy, 2500 RPM, LPG share (0-50%), Y-axis vibration sensor.

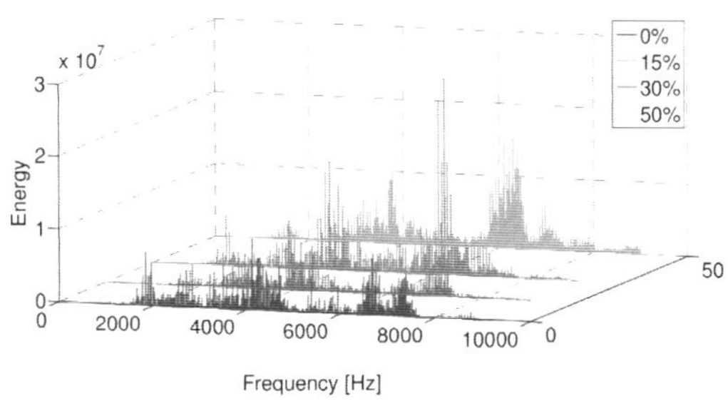

Figure 29. Amplitude spectrumof vibration energy, 2500 RPM, LPG share (0-50%), Z-axis vibration sensor.

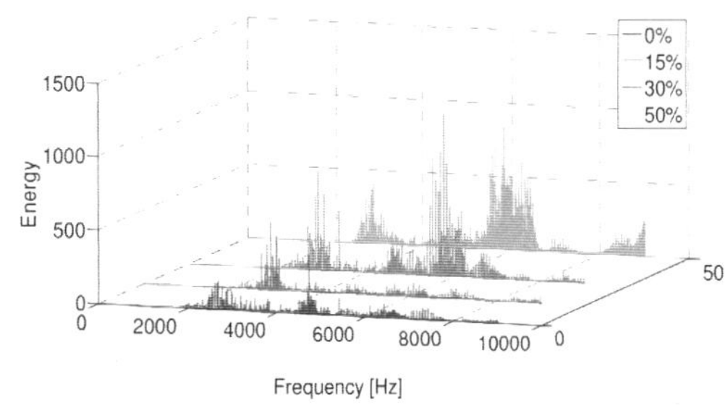

Figure 30. Amplitude spectrumof vibration energy, 2500 RPM, LPG share (0-50%), knock sensor.

The last stage presented in this paper is the analysis of the knocking combustion occurrence with the use of the summation of amplitudes of the subsequent spectral lines of the vibration energy ratio so as to illustrate the increase in vibration energy in specific frequency ranges. The plots of the summed up values of the energy lines are shown in the plots below.

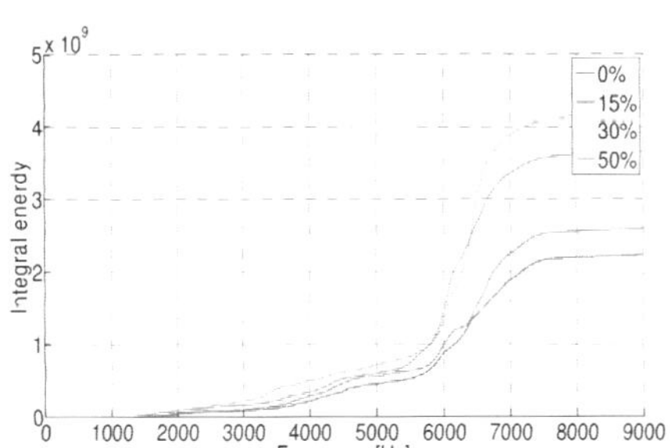

Figure 31. The integral spectrum of the vibration energy, 2000 RPM, LPG share (0-50%), Y-axis vibration sensor.

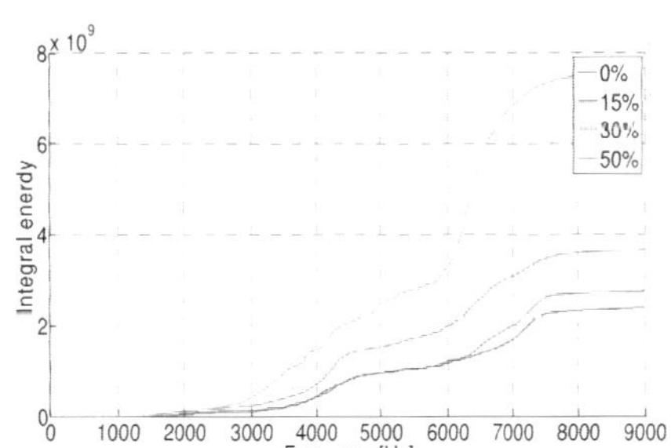

Figure 32. The integral spectrum of the vibration energy, 2000 RPM, LPG share (0-50%), Z-axis vibration sensor.

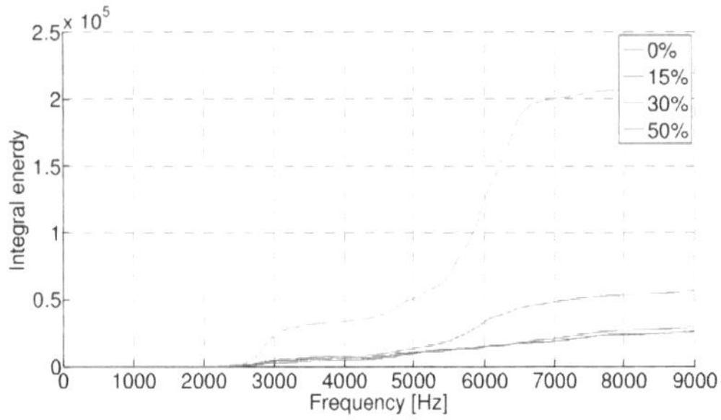

Figure 33. The integral spectrum of the vibration energy, 2000 RPM, LPG share (0-50%), knock sensor.

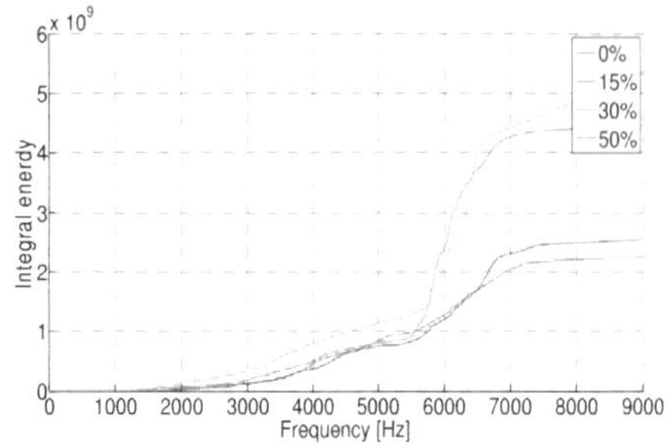

Figure 34. The integral spectrum of the vibration energy, 2500 RPM, LPG share (0-50%), Y-axis vibration sensor.

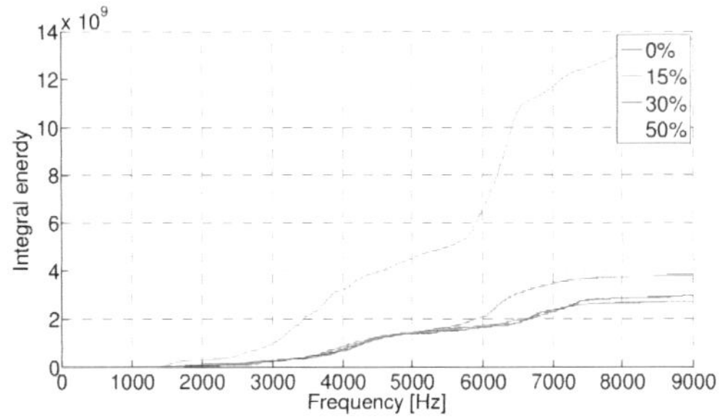

Figure 35. The integral spectrum of the vibration energy, 2500 RPM, LPG share (0-50%), Z-axis vibration sensor.

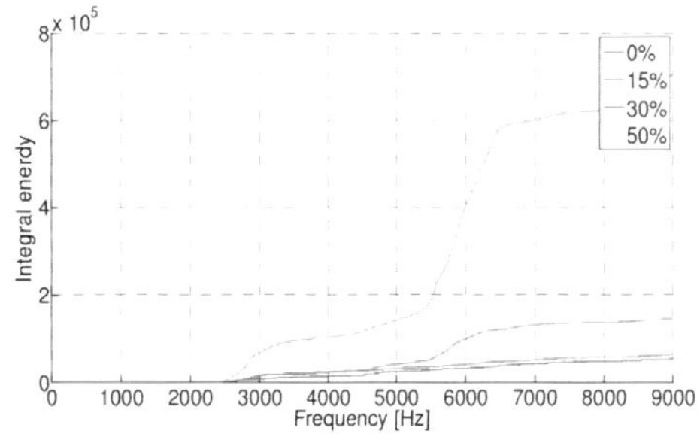

Figure 36. The integral spectrum of the vibration energy, 2500 RPM, LPG share (0-50%), knock sensor.

The curves shown in Figures 31-36 indicate, in which frequency ranges the highest increase in vibration energy can be seen, this increase is connected with the rapid increase in pressure caused by knocking combustion. The rapid pressure growth in the engine with the compression-ignition (significantly higher pressure and faster increase than in the engine with the spark ignition, for which the phenomenon of knocking combustion is much better investigated) makes it difficult to unequivocally indicate the knocking combustion occurrence. The best method consists in measuring the combustion pressure in the chamber during the engine operation. Unfortunately, these measurements are also flawed with significant disorders and they require more precise analyses, which is proved by the research results carried out so far.

In order to assess, if knocking combustion took place, the velocity of pressure increase is determined [11], whose task is defining the limit, when the combustion is normal, and when knocking combustion starts. The analysis presented in [11] does not take into account many factors influencing the velocity of the pressure increase, hence the need of additional tests and analyses.

4. Conclusions

The analysis of the vibration accelerations in the low frequency band does not enable the recognition of the occurrence of the knocking combustion in the process of combusting the mixture of fuel and air. In this band, other information can be found, related to the combustion process, which is not used in the presented task. Moving the analysis to the higher frequency band enables observations of the range, in which the amplitudes of the vibration accelerations grow, which can be connected with the occurrence of the knocking combustion, the unequivocal proof, however, causes interpretation problems.

Introduction of the proportionality constant to the energy of vibration significantly improves the interpretation possibilities, especially if it is subject to additional transformations.

A very important realisation is a phenomenon of information filtering dependent upon the direction of registering the vibration accelerations. Certain procedures require additional processing like filtering signals, cutting out bands, or different type of information preselection. In the discussed case of selecting

the direction of registration, the band related to knocking combustion was vividly marked at one of the measurement directions, which allows for additional conclusion-drawing, that does not include a series of preliminary signal processing.
A credible proof of the occurrence of knocking combustion in the given frequency ranges are observations of pressure inside the combustion chamber.

5. Abbreviations

LPG – liquefied petroleum gas
SOI – start of injection
CI – compression engine
PSG – pressure sensor glow plug
B&K – Bruel&Kjaer
NI – National Instruments
f – frequency
RPM –revolutions per minute

References

[1] Regulation (EC) No 715/2007 of the European Parliament and of the Council of 20 June 2007 on type approval of motor vehicles with respect to emissions from light passenger and commercial vehicles (Euro 5 and Euro 6) and on access to vehicle repair and maintenance information.

[2] Raslavicius L, Kersys A, Mockus S, Kersiene N and Starevicius M 2014 *Liquefied petroleum gas (LPG) as a medium-term option in the transition to sustainable fuels and transport*, Renew. Sustain. Energy Rev. **32** (2014) 513–525.

[3] Tira H S, Herreros J M, Tsolakis A and Wyszynski M L 2012 *Characteristics of LPG–diesel dual fuelled engine operated rapeseed methyl ester and gas-to-liquid diesel fuels* Energy **47** (2012) 620–629.

[4] Wagemakers A M L M and Leermakers C A J 2012 *Review on the Effects of Dual-Fuel Operation, Using Diesel and Gaseous Fuels, on Emissions and Performance*. SAE 2012-01-0869:2012.

[5] Murthy K, Madhwesh N and Shrinivasa R 2012 *Influence of injection timing on the performance of dual fuel compression ignition engine with exhaust gas recirculation* Int. J. Eng. Res. Dev. 1 (11) (2012) 37–42.

[6] Mikulski M, Wierzbicki S and Piętak 2012 *A Numerical Studies on Controlling Gaseous Fuel Combustion by Managing the Combustion Process of Diesel Pilot Dose in a Dual-Fuel Engine Chemical and Process Engineering* **36** (2), 225-238

[7] Chang Chih-Chung, Jiunn-Guang Lo and Jia-Lin Wang 2001 *Assessment of reducing ozone forming potential for vehicles using liquefied petroleum gas as an alternative fuel* Atmospheric Environment 35.35 (2001): 6201-6211.

[8] Mohsen M A, Basem A. Rabee A and Hegab H 2013 *Effect of adding oxygen to the intake air on a dual-fuel engine performance, emissions, and knock tendency* Energy **61** (2013) 612–620.

[9] Radkowski S 2007 *Use of vibroacoustical signal in detecting early stages of failures*. Eksploatacja i Niezawodność (Maintenance and Reliability) 3.35 (2007): 11-18.

[10] Zawisza M 2015 *Symptomy wibroakustyczne jako uzupełnienie systemu OBD silników wysokoprężnych* Warszawa – Radom 2015

[11] Różycki A 2014 *Zjawisko stuku w tłokowym silniku spalinowym* Logistyka 6/2014 pp 9179-9189

A concept of the assessment of Electric Vehicles' Operational Safety (EVOS)

S Łukjanow and W Zieliński

Automotive Industry Institute (PIMOT), Warsaw, Poland

E-mail: ble@pimot.org.pl

Abstract. The Electric Vehicles' Operational Safety (EVOS) is becoming an important issue due to the fact of the popularization of environment-friendly electrically driven vehicles. The expansion of new types of electric vehicles releasing by automotive companies may lead to yet unknown safety-related problems. The paper presents a three-level concept of examining and assessing the EVOS developed at PIMOT. The proposed criteria may be utilized at research works on electric vehicles, at the production of such vehicles and their components, and at the selection of vehicles, especially by transport companies.

1. Introduction

Increasingly common use and growing trend towards the popularization of electric vehicles induce the necessity to tackle the problem of safety of electrically driven vehicles of new types. Their use may be accompanied by new safety-related phenomena that have not been sufficiently explored so far and have not been addressed in the UN ECE Regulations and EU Directives in force.

This paper presents a three-level concept of examining and assessing the Electric Vehicles' Operational Safety (EVOS), developed at PIMOT:

- 1^{O} - The EVOS 1 assessment is carried out in respect of criteria related to mechanical vibrations of vehicle body, internal noise, electromagnetic compatibility (EMC), and protection against electric shock.
- 2^{O} - The EVOS 2 assessment is an expanded version of the EVOS 1, where criteria concerning the vehicle's safety equipment are added. At the EVOS 2 assessment, a "safety systems factor" is introduced to take into account the current technological progress in this field and the safety equipment provided in the electric vehicles under consideration.
- 3° - The granting of EVOS classes: there are 5 different EVOS classes, which may be granted to a specific EV type and to individual sets of equipment offered as options to vehicle users, based on the EVOS 2 assessment.

The safety examination criteria applicable to electric vehicles according to the normative documents in force and to the EVOS system have been illustrated in figure 1.

The most important aspect of using the EVOS system should be the improvement in the safety level that might be achieved by showing which electric vehicles are better from others in terms of EVOS.

The criteria prepared may be utilized at research works on electric vehicles, at the production of such vehicles and their components, and at the selection of vehicles, especially by transport companies. The implementation of the EVOS criteria having been prepared should help to improve the road traffic safety

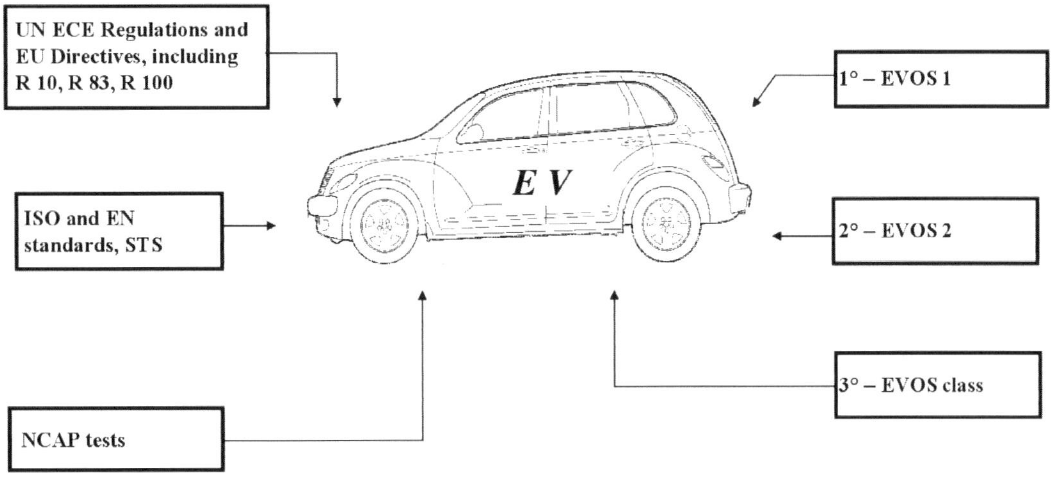

Figure 1. Electric vehicle safety examination and assessment systems with taking into account the EVOS system.

by eliminating or lessening the hazards revealed in the tests. Within the next works, it would be reasonable to develop an EVOS assessment procedure for vehicles with hybrid drive systems, which have already become a considerable segment of the market.

2. Technical requirements related to the EVOS

2.1. Identification of electric vehicles

The electric vehicle submitted for testing is identified on the grounds of the data provided in an excerpt from the type-approval certificate (for new vehicles) or the data provided in the vehicle registration certificate (for used vehicles).

2.2. Requirements concerning mechanical vibrations

The requirements concerning the mechanical vibrations of an electric vehicle are based on determining the levels of discomfort according to standards ISO 2631-1 and BS 6841; they have been specified in table 1 below, with the corresponding scores proposed for the assessment having been given in the third column of the table.

Table 1

Acceleration, RMS (m/s2)	Discomfort scale	Score (points)
Less than 0.315	Not uncomfortable	6
0.315-0.63	A little uncomfortable	5
0.5-1.0	Fairly uncomfortable	4
0.8-1.6	Uncomfortable	3
1.25-2.5	Very uncomfortable	2
Greater than 2	Extremely uncomfortable	1

Instead of specifying the measured RMS acceleration values translated into discomfort feelings, the intensities of subjective human's sensations may be optionally defined on the grounds of measured vibrations of the electric vehicle under test, as presented in table 2 below; the corresponding scores proposed for the assessment have been given in the third column of the table. An analysis of the impact of vibrations on vehicle occupants has been presented, inter alia, in publication [16].

Table 2

Relative acceleration values (m/s2)	Subjective sensation felt	Score (points)
< 0.001	Imperceptible vibration	6
< 0.01	Barely perceptible vibration	5
< 0.1	Distinctly perceptible vibration	4
< 1	Unpleasant subjective sensation of low intensity	3
< 10	Unpleasant subjective sensation	2
> 10	Unpleasant subjective sensation of very high intensity	1

2.3. Requirements concerning internal noise

The requirements concerning internal noise in an electric vehicle are not mandatory and no such requirements have been laid down in the UN ECE Regulations and EU Directives.

The internal noise in a motor vehicle may be measured in accordance with Polish Standard PN-90/S-04052. This standard was established for motor vehicles with internal combustion engines (ICE) and it is still in force.

There is no separate standard that would be applicable to electric vehicles in this respect. Pursuant to the said Polish Standard, the maximum acceptable level of internal noise in passenger cars is 79 dB(A). For motor vehicles with a single row of seats, the measurements are carried out for a single measuring point situated at driver's seat. For passenger cars with two rows of seats, the noise is additionally measured at one more measuring point. The measurements are carried out when the vehicle is accelerated from an initial speed as specified in the standards to 120 km/h or 90 % of the vehicle speed corresponding to the maximum-power engine speed, whichever is lower.

In the tests carried out by automotive magazines, the measurements are carried out at constant vehicle speeds of 50 km/h, 90 km/h, and 130 km/h. The background noise level should be lower by at least 10 dB than the noise values measured.

The following scores have been proposed for the assessment of internal noise:

76 dB - 1 point
74 dB - 2 points
72 dB - 3 points
70 dB - 4 points
68 dB - 5 points
66 dB - 6 points

The requirements adopted here are lower in comparison with those of the Standard referred to above, in consideration of the time elapsed from the Standard publication date, the progress made in the car noise damping technology, and the fact that the noise generated by electric vehicles is lower than that emitted by vehicles with internal combustion engines.

An analysis of the sources of noise emitted by transport facilities, especially motor vehicles, has been described, inter alia, in publication [1].

2.4. Requirements concerning electromagnetic compatibility (EMC)

The EMC requirements are based on the provisions of UNECE Regulation No. 10 and the corresponding EU Directive 2004/104/EC (current version). For an electric vehicle, measurements are carried out in a test site with the lowest possible electromagnetic background emission or in a semi-anechoic chamber. The assessment is performed by comparing broadband emissions. The emission limits for a 3 m distance between the measuring antenna and the vehicle have been shown in figure 2. The levels represented by line L1 are defined by the requirements laid down in the normative documents referred to above. The levels represented by lines L2 and L3 are lowered by 10 dB and 20 dB, respectively, in relation to level L1 and show tightened requirements.

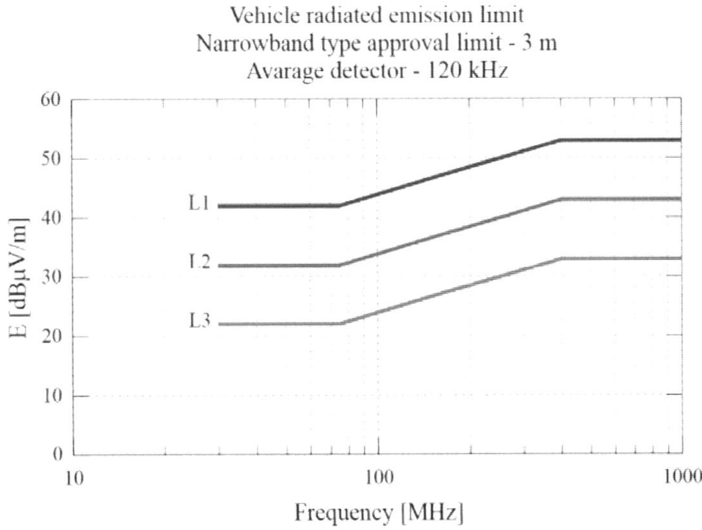

Figure 2. Proposed reference levels of broadband electromagnetic emission for vehicles (for a 3 m distance between the measuring antenna and the vehicle).

For an antenna-vehicle separation distance of 10 m, the levels represented by lines L2 and L3 are adopted as lowered by 8 dB and 16 dB, respectively, in relation to level L1.
The concept of assessing the electromagnetic emission levels with taking lines L1, L2, and L3 as a reference has been presented in publication [5].
The scores proposed for the assessment have been given in table 3 below.

Table 3

Electromagnetic emissions	Score (points)
Below L3	6
Between L3 and L2	4.5
Between L2 and L1	3
Exceeding L1 but by no more than 2 dB	1.5*
Higher than the above	0.0

At the broadband emission assessment method adopted, some small exceedances (by up to 2 dB) may be accepted in consideration of the fact that the tests may also be carried out on used vehicles, retrofitted with electronic systems or devices of varying quality. The asterisk (*) means that this score may only be applied to used vehicles.

2.5. Requirements concerning electrical safety

These requirements concern the assessment of safety related to the possibility of electric shock to a person or persons present inside the vehicle or outside it when it is not in motion as well as e.g. during the battery charging process. The electrical installation of an electric vehicle must meet the requirements of UN ECE Regulation No. 100 and the corresponding EU Directive.
For the preliminary assessment, three major factors have been chosen from many factors that have an impact on the electrical safety related to the use of electrical equipment and power supply systems in electric vehicles:

- degree of protection of electrical enclosures (IP Code);
- working voltage in electric drive and control systems;
- insulation resistance.

The meaning of individual digits in the IP Code has been explained in table 4 below.

Table 4

IP Code breakdown

Symbol	**Meaning**	**Remarks**
IP letters	Protection against accidental contact with human body parts and ingress of solid objects and water	"IP"stands for "International Protection"
1st digit: 0 to 6	Degrees of protection against contact with human body parts and ingress of solid foreign matter, e.g. dust	In the case of a large number of electricity receivers, not all digits may be present in the IP Code
2nd digit: 0 to 8	Degrees of protection against ingress of water	

If only one digit is present in the IP Code, then letter X is put in the place for which the degree of protection is not specified.

Explanation of individual digits in the IP Code

Protection symbol	**Meaning**	**Remarks**
Protection against accidental contact with human body parts and ingress of solid foreign matter		
IP 0X	No protection against contact with human body parts and against ingress of solid foreign matter	(Corresponding graphic symbols as defined below)
IP 1X	Protection against ingress of solid foreign matter $\geq$ 50 mm dia.	„Protected from touch by hand"
IP 2X	Protection against ingress of solid foreign matter $\geq$ 12.5 mm dia.	„Protected from touch by finger"
IP 3X	Protection against ingress of solid foreign matter $\geq$ 2.5 mm dia.	„Protected from touch by tools"
IP 4X	Protection against ingress of solid foreign matter $\geq$ 1 mm dia.	„Protected from touch by wires"
IP 5X	Protection against harmful dust deposition inside	„Protected from ingress of dust"
IP 6X	Total protection against dust ingress	„Dust tight"
Protection against ingress of water		
IP X0	No protection against ingress of water	
IP X1	Protection against water drops falling vertically	"Drip proof"
IP X2	Protection against water drops falling at an angle of up to 15 ° from vertical	„Rainproof"
IP X3	Protection against water spray received at an angle of up to 60 ° from vertical	„Semi splash-proof"
IP X4	Protection against water spray received from any direction	"Splash-proof"
IP X5	Protection against low-pressure water jets	„Hose-proof"
IP X6	Protection against high pressure water jets received from any direction	„Semi watertight"
IP X7	Protection against short-term water immersion	„Watertight"
IP X8	Protection against long-term water immersion	„Pressure watertight"

The following scores have been proposed for electric vehicle drive and control systems:

IP 42 - 2 points
IP 44 - 3 points
IP 54 - 4 points
IP 56 - 5 points
IP 66 - 6 points

As regards the assessment of working voltage, the scores as specified in table 5 have been proposed.

Table 5

Working voltage (V)	Score (points)
< 60	6
60-100	5
100-200	4
200-300	3
300-400	2
400-500	1

The scores proposed for the assessment of insulation resistance have been given in table 6.

Table 6

Insulation resistance (Ω/V)	Score (points)
500-750	6
750-1000	5
1000-1250	4
>1250	3

For the overall assessment, the average value of the above electrical safety factors is adopted. The IP Code system of classifying the degree of protection of electrical enclosures has been presented in table 4.

2.6. Requirements concerning the safety equipment of electrical vehicles

The road traffic safety hazards that accompany the motorization stimulate development of systems designed to improve the active (pre-accident) and passive (post-accident) safety of motor vehicles and their users. This is fostered by technological development, in particular by achievements in the field of sensors used to measure mechanical quantities, radars, lidars, digital cameras, positioning systems based on the GPS (Global Positioning System) technology, radio data transmission systems (especially GSM, i.e. Global System for Mobile Communication), signal transducers, microprocessors and computers with software for real-time data processing, data transmission networks (especially local network named CAN, i.e. Control Area Network), as well as precisely operating servomechanisms and other actuating devices. Based on the elements mentioned above, many mechatronic systems and devices have been developed that perform fragmentary tasks related to the monitoring and automatic control in motor vehicles [11].

There are a lot of systems, separate technical units, and modules that constitute the Vehicle Safety System (VSS) and 10 items, specified in table 7 below, have been selected from among them for the assessment. Item S10 has been defined as “others”, where other equipment, important for the improvement of vehicle’s operational safety and provided in the vehicle submitted for testing, may be covered. In this part of the EVOS assessment procedure, scores of 0 to 10 may be given for each of the items S1, S2, … to S10, which would make up to 100 points in total. For the overall EVOS assessment, these partial scores are converted into an appropriate safety systems factor Bx, added to the overall assessment score.

Table 7

Equipment symbol	Description of the system or separate technical unit	Manufacturer's symbol	Score (points)
S1	Seat belts with pretensioners	SRS	0-10
S2	Airbags	Various	0-10
S3	Braking systems: anti, anti-slip and supportive	ABS, ASR, EBO, EBS, others	0-10
S4	Vehicle travel path stabilization system	ESP	0-10
S5	Active cruise control system	ACC	0-10
S6	Obstacle detection system and system of automatic braking in traffic	Various	0-10
S7	Warning systems, monitoring driver fatigue or lane departure	Various	0-10
S8	Lighting, signalling and visibility improving systems	Various	0-10
S9	Monitoring of road incidents	e-Call	0-10
S10	Others		0-10
	Total, maximum		100

3. EVOS assessment criteria

Pursuant to the concept adopted as described in Section 1 herein, the EVOS assessment is carried out at three levels:

1° – EVOS 1 assessment;
2° – EVOS 2 assessment;
3° – Granting of an EVOS class.

3.1. EVOS 1 assessment

The EVOS 1 assessment is carried out based on results of the tests for conformity with the requirements described in subsections 2.1 to 2.5 herein. The assessment results are obtained by calculating the "overall EVOS assessment index" from the weighted average calculation formula as given below. For the four quantities adopted as indicators (W1, W2, W3, and W4), this formula takes the form:

$$W_{B1} = \frac{W_1 * u_1 + W_2 * u_2 + W_3 * u_3 + W_4 * u_4}{u_1 + u_2 + u_3 + u_4}, \quad (1)$$

where: W_{B1} – overall EVOS 1 assessment index;
W_1 – score given in the assessment of mechanical vibrations;
W_2 – score given in the assessment of internal noise;
W_3 – score given in the EMC assessment;
W_4 – score given in the assessment of electrical safety;
u_1, u_2, u_3, u_4 – weights of individual scores in the overall index.

In results of an analysis carried out, the following weights were adopted, with $\sum u = 1$:

$u_1 = 0.25$;
$u_2 = 0.20$;
$u_3 = 0.25$;
$u_4 = 0.30$.

For the above weight values, the formula (1) used for determining the W_{B1} index takes the following form:

$$W_{B1} = 0.25\ W_1 + 0.20\ W_{B2} + 0.25\ W_{B3} + 0.30\ W_4 \tag{2}$$

3.2. EVOS 2 assessment

For the EVOS 2 assessment, results of the tests and assessment of the electric vehicle to the EVOS 1 procedure are utilized. The scope of this assessment is widened by additionally taking into account the VSS provided. The overall EVOS 2 assessment index is determined from the following formula:

$$W_{B2} = W_{B1}*(B_x + 1), \tag{3}$$

where B_x is the safety systems factor for the vehicle under assessment.

When equation (1) is substituted to equation (3), the following formula is obtained for determining the overall EVOS 2 assessment index:

$$W_{B2} = \frac{W_1 * u_1 + W_2 * u_2 + W_3 * u_3 + W_4 * u_4}{u_1 + u_2 + u_3 + u_4} * (B_x + 1), \tag{4}$$

with B_x being calculated from the following equation, for the number k of the systems and separate technical units dedicated to vehicle safety as adopted in table 7:

$$B_x = \frac{\sum_{i=1}^{i=k} s_i}{k_p}, \tag{5}$$

where: $\sum_{i=1}^{i=10} s_i$ – - actual total score;

k_p - maximum possible total score.

If, according to table 7, the maximum possible total score is $k_p = 10 \times 10 = 100$ and

$$\sum_{i=1}^{i=k} s_i = s_1 + s_2 + \ldots\ldots\ldots\ldots + s_{10}, \tag{6}$$

then formula (5) may be written in the following form, convenient for calculations:

$$B_x = \frac{s_1 + s_2 + \ldots\ldots\ldots\ldots + s_{10}}{100} \tag{7}$$

In result of the assessment, every system and separate technical unit installed in the vehicle under consideration and dedicated to vehicle safety (represented in table 7 by symbols S1, S2, …, S10) is given a specific score (s_1, s_2, …, s_{10}, respectively). When the sum of these scores is put into equation (7), the safety systems factor B_x can be calculated. In graphical form, the function $W_{B2} = f(B_x)$ is represented by curves shown in figure 3. Four initial values W_{B1} = 3, 4, 5, and 6 were assumed for a preliminary analysis.

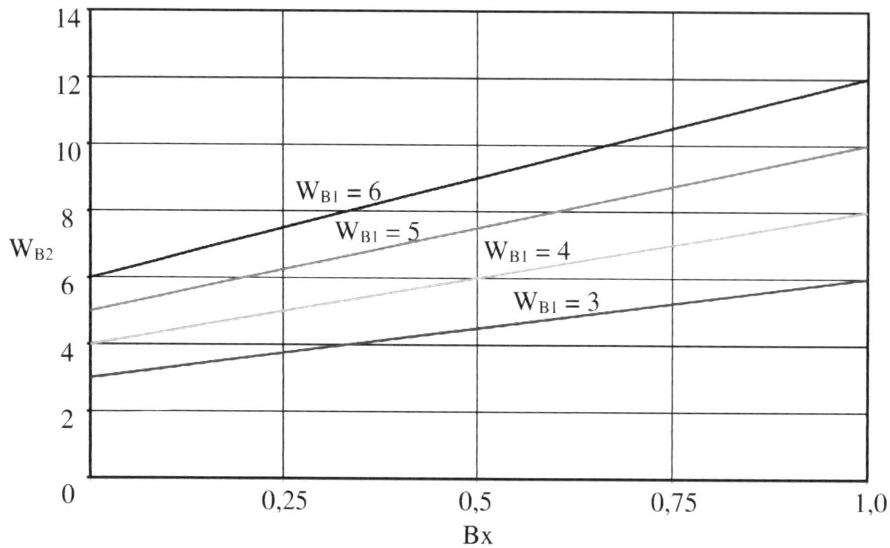

Figure 3. Curves representing the function $W_{B2} = f(B_x)$ for $W_{B1} = 3, 4, 5, 6$

3.3. Granting of an EVOS class

An EVOS class may be granted to an electric vehicle that has satisfactorily passed the EVOS 2 assessment. For individual classes, the scores as specified in table 8 below and in figure 3 have been adopted.

Table 8

EVOS class	EVOS 2 score (points)
AA	10.0 - 12.0
A	8.0 - 10.0
B	4 - 8.0
C	3 - 6.5
D	2 - 5.5

4. Recapitulation

- The objective of this work was to prepare preliminary technical requirements for the purposes of evaluation of Electric Vehicles' Operational Safety (EVOS).
- In this study, a three-level EVOS assessment concept has been presented, which covered:
  1° – EVOS 1 assessment;
  2° – EVOS 2 assessment;
  3° – Granting of an EVOS class.
- The EVOS 1 assessment is carried out in respect of the criteria described in subsections 2.1 to 2.5 herein.
- The EVOS 2 assessment is an expanded version of the EVOS 1, where criteria concerning the safety equipment of the vehicle are added. At the EVOS 2 assessment, a "safety systems factor" is introduced to take into account the current technological progress in this field and the safety equipment provided in the electric vehicles under consideration.
- As regards the granting of EVOS classes, there are 5 different classes, which may be granted to a specific EV type and individual sets of its optional equipment, based on the EVOS 2 assessment.

- Based on the results of this work, detailed EVOS criteria may be prepared, e.g. in the form of Standard Technical Specifications (STS), for the assessment of electric vehicles. The STS may constitute a basis for the certification carried out by accredited units, e.g. PIMOT. Within the next works, it would be reasonable to develop an EVOS assessment procedure for vehicles with hybrid drive systems, which have already become a considerable segment of the market.
- The criteria prepared may be utilized at research works on electric vehicles, at the production of such vehicles and their components, and at the selection of vehicles, especially by transport companies. The implementation of the EVOS criteria having been prepared should help to improve the road traffic safety by eliminating or lessening the hazards revealed in the tests.

References

[1] Chłopek Z 2002 *Pojazdy samochodowe. Ochrona środowiska naturalnego (Motor vehicles: Environmentalprotection).* WKiŁ Warszawa
[2] Chłopek Z 2012 Research on energy consumption by an electric automotive vehicle. *The Archives of Automotive Engineering – ArchiwumMotoryzacji* No 3
[3] Evaes S,Evaes J 2004 A cost comparison of fuel-cell and battery electric vehicles *Journal of Power Sources* 130
[4] Hussein I 2003 *Electric and Hybrid Vehicles Design Fundamentals.* 2003 (1st issue) 2011 (2nd issue)
[5] Łukjanow S, Kołodziejczak M and Pijanowski B 2009 Project of the evaluation and classification system of vehicles and automobile devices in aspect of electromagnetic compatibility. *Journal of KONES Powertrain and Transport,* Vol. 16 No. 1
[6] Proceedings of the International Conference *InnowacyjneRozwiązania w Przemyśle Transportowym. Pojazdy CNG iElektryczne (Innovative Solutions in the Transport Industry. CNG and Electric Vehicles)* Warszawa 8-9 November 2011.
[7] Merkisz J and Pielecha I 2006 *Alternatywne napędy pojazdów (Alternative vehicle drive systems)* Poznań University of Technology
[8] McCrone A 2011 *Price of Electric Vehicle Batteries to Fall as Manufacturing Capacity Outstrips Demand.* Bloomberg New Energy Finance 14 September 2011
[9] Olszowiec P 2002 *Gdy zabraknie prądu. Superkondensatory (When the power fails. Supercapacitors).* EnergiaGigawat June 2002.
[10] *Opracowanie i badanie samochodowego systemu bezpieczeństwa w ramach struktury inteligentnegopojazdu (Developing and testing of a vehicle safety system within the intelligent vehicle structure).* Research project No. N509 573 239 sponsored by the Ministry of Science and Higher Education, Project Manager: BogusławPijanowski M. Eng., PIMOT 2010-2013.
[11] *Opracowanie teoretycznego modelu przepływu i analizy informacji (Preparation of a theoretical model of information flow and analysis).* Research project No. N509 573 239, Task No. 3, sponsored by the Ministry of Science and Higher Education, carried out by a team headed by: Prof. Dariusz Żardecki PIMOT 2011
[12] UN ECE Regulation No. 10
[13] UN ECE Regulation No. 83
[14] UN ECE Regulation No. 100
[15] Walter F 2010 *Going Green with Electric Vehicles.* Technology and Engineering Teacher
[16] Więckowski D 2013 *Ocena drgań pionowych samochodu oddziałujących na dzieci posadowione w fotelikach (Evaluation of vertical car vibrations transmitted to children when riding in safety seats).* Wydawnictwo Naukowe PIMOT Warszawa

An assessment of consistence of exhaust gas emission test results obtained under controlled NEDC conditions

K Balawender[1], A Jaworski[1], H Kuszewski[1], K Lejda[1], A Ustrzycki[1]

[1]Rzeszow University of Technology, Faculty of Machine Engineering and Aeronautics, Department of Combustion Engines and Transport, 8 Powstancow Warszawy Av., 35-959 Rzeszow, Poland

E-mail: klejda@prz.edu.pl

Abstract. Measurements concerning emissions of pollutants contained in automobile combustion engine exhaust gases is of primary importance in view of their harmful impact on the natural environment. This paper presents results of tests aimed at determining exhaust gas pollutant emissions from a passenger car engine obtained under repeatable conditions on a chassis dynamometer. The test set-up was installed in a controlled climate chamber allowing to maintain the temperature conditions within the range from –20°C to +30°C. The analysis covered emissions of such components as CO, CO_2, NO_X, CH_4, THC, and NMHC. The purpose of the study was to assess repeatability of results obtained in a number of tests performed as per NEDC test plan. The study is an introductory stage of a wider research project concerning the effect of climate conditions and fuel type on emission of pollutants contained in exhaust gases generated by automotive vehicles.

1. Introduction

Harmful impact of transport on the environment includes emissions of gas and dust pollutants generated by combustion engines. Since many years, enormous research effort is put into studies aimed at minimisation of energy consumption and reduction of emissions of pollutants included in exhaust gases. The related progress is reflected in newly introduced emission standards and procedures applicable to combustion engine testing. Passenger cars and lightweight utility vehicles in the European Union are subject to Euro procedures which determine standards for emissions of pollutants in exhaust fumes determined in the course of drive tests under controlled conditions of chassis dynamometers[1].

Limited resources of crude oil which is a basic raw material transformed into petrol and diesel fuel encourage to look for substitution fuels. It is assumed that introduction of such substitutes on commercial fuel markets will not result in emissions of pollutants contained in exhaust gases exceeding the limits set out in currently applicable standards.

The Rzeszów University of Technology's Department of Combustion Engines and Transport also undertakes a number of research projects aimed at reduction of harmful impact of the means of automotive transport on natural environment. The main topics include combustion engine design improvement and search for alternative fuels. Current activities are focused on evaluation of emission levels characterising individual pollutants contained in passenger car exhaust gases. The related tests are performed on an engine test bench, in actual road conditions, and in controlled climate environment as defined by the European NEDC (New European Driving Cycle) testing procedure.

In view of the fact that in most cases, the effect of consecutive design and operation changes on fuel consumption and emission of pollutants is relatively small, the issue of determining precision of tests confirming substantiality seems to be of key importance. For this reason, numerous laboratories worldwide undertake studies concerning repeatability and reproducibility of results obtained in the course of pollutant emission measurements carried out on engine testing facilities [2–5]. Studies on pollutant emissions performed on chassis dynamometers represent a very complex issue. The obtained results depend, among other things, on behaviour of the operator (vehicle driver) reconstructing the drive test schedule, climate conditions, physicochemical properties of the fuel, and on precision available with the used measuring apparatus. In view of the above, the key issue is repeatability of measurements performed under conditions of repeatability, which include:

- the use of the same test method;
- performing the tests in the same laboratory;
- tests being performed by the same operator;
- performing the tests with the use of the same equipment;
- completing the tests one after another in short intervals.

This paper is a presentation of a study on repeatability of results of tests performed as per the Euro procedure used to determine emissions of pollutants contained in exhaust fumes of petrol engines driving passenger cars.

2. Testing procedure

The tests were performed on ROADSIM 48″ Chassis Dynamometer (AVL Zöellner) installed in a controlled climate chamber. The research set-up was constructed in the Automotive Ecology Laboratory run by the Department of Combustion Engine and Transport, Faculty of Mechanical Engineering and Aeronautics, Rzeszów University of Technology. Basic technical data characterising the chassis bench are summarised in table 1. A view of the tested car on the chassis on the research set-up is presented in figure 1, whereas figure 2 is a view of the control room.

Figure 1.A view of the car on the research set-up.

Table 1. Basic technical data of AVL ROADSIM 48″ Chassis Dynamometer [6].

Analysed quantity	Value
Dimensions (Length / Width / Height)	3,600 mm / 1,600 mm / 1,300 mm
Roller diameter	1,219.2 mm
Roller mass	765 kg
Rated power	153 kW
Instantaneous power	258 kW
Maximum speed	200 km/h
Inertia simulation range	454 kg ... 2,722 kg
Maximum continuous tractive force	5,987 N
Maximum instantaneous tractive force	10,096 N
Tractive force measurement error	≤0.1 %
Speed measurement error	≤0.02 km/h
Distance measurement error	0.001 %/m
Maximum axle load	2,000 kg

Figure 2. A view of the control room.

The subject of the test was a passenger car manufactured in 2003 with total mileage of about 60,000 km and other basic technical data summarised in table 2. To ensure that the resistance to motion observed on chassis dynamometer are as close as possible to those occurring in actual road conditions, an on-road coasting tests were performed within the speed range from 130 km/h to car stoppage. To measure the coasting parameters, a measuring set-up with DATRON DLS-2 optoelectronic sensor system. A view of the car in the course of the road test with the sensor mounted on it is presented in figure 3. Based on the determined coasting times, the following formula was derived for the resisting force on the car:

$$F_o = 155.11 - 1.2343\,V + 0.4685\,V^2 \tag{1}$$

where F_r is the total resisting force representing combined rolling resistance and air drag (N) and V is the driving speed (km/h).

Table 2. Technical data of the vehicle.

Parameter	Value
Length / Width / Height	4,665 mm / 1,760 mm / 1,445 mm
Wheelbase	2,670 mm
Weight	1,430 kg
Engine type	Petrol (gasoline)
Fuel System	Multi-point injection
Engine displacement	1998 cm^3
Engine power	115 kW @ 6,000 rpm
Engine torque	190 Nm @ 4,500 rpm
Number of cylinders	4
Number of valves	16
Wheel drive	Front
Number of gears (manual transmission)	5
Tire size	205/55 R16

Figure 3. A view of the car in the course of road test.

The tests were carried out in an air-conditioned chamber in which temperature of 21°C and relative humidity of 40% were maintained. According to the test procedure determined in the applicable regulation [1], the car was stabilised in the chamber before the test for a period of at least 6 hours. The test was carried out as per Type 1 test plan (verifying the average exhaust emissions after a cold start). To measure emissions, AMA i60 Exhaust Measuring System equipped with CVS i60 Exhaust Gas Dilution System (both devices by AVL) were used.

Emissions are tested over the NEDC chassis dynamometer procedure. Part one of the test is made up of four elementary urban cycles (ECE). Each elementary urban cycle comprises fifteen phases (idling, acceleration, steady speed, deceleration, etc.). Part Two of the test is made up of one extra-urban cycle (EUDC). The extra-urban cycle comprises 13 phases (idling, acceleration, steady speed, deceleration, etc.). During the test, the exhaust gases were diluted and a proportional sample collected in bags. The exhaust gases of the vehicle tested were diluted, sampled and analysed, and the total volume of the diluted exhaust were measured.

3. Research results

Results concerning content of toxic pollutants in the exhaust gas obtained in consecutive tests carried out as per Type 1 test plan are summarised in table 3 and illustrated in figure 4.

For the obtained emission measurement results, an analysis concerning accuracy of the obtained values was performed. To characterise quantitatively the accuracy of a measuring method, two fundamental terms, "trueness" and "precision", are commonly used [7, 8].

The trueness determines the closeness of the mean of a set of measurement results obtained from a large sequence of tests and an adopted reference value.

Precision is defined as the closeness of agreement between independent measurement results obtained under stipulated conditions. Precision depends on the distribution of random errors and is not referred to any actual value. Usually, precision is characterised quantitatively by means of the standard deviation of measurement results, value of which is the smaller the largest is precision of the measuring apparatus.

Repeatability is the precision under the repeatability conditions being met. To evaluate repeatability, the following quantities were determined:

- average values of measurement results x_{av};
- minimum values x_{min} and maximum values x_{max};
- the range R_x;
- the standard deviation of repeatability s_r;
- the repeatability limit r;
- the relative standard deviation of repeatability RSD_r;
- the coefficient of variation CV,
- ratio of the range to the repeatability limit.

Table 3. Emission results (bag emissions) and the limits (Euro 4 standards).

Pollutants	Emissions results (gkm^{-1})				EU Emission Standards for Passenger Cars using gasoline engines (gkm^{-1})
	Test 1	Test 2	Test 3	Test 4	
THC	0.046	0.046	0.051	0.046	0.1
NMHC	0.041	0.041	0.046	0.042	—
CH_4	0.0047	0.0048	0.0049	0.0042	—
NO_X	0.02	0.019	0.016	0.019	0.08
CO	0.439	0.441	0.448	0.428	1.0
CO_2	201.411	202.796	203.332	202.615	—

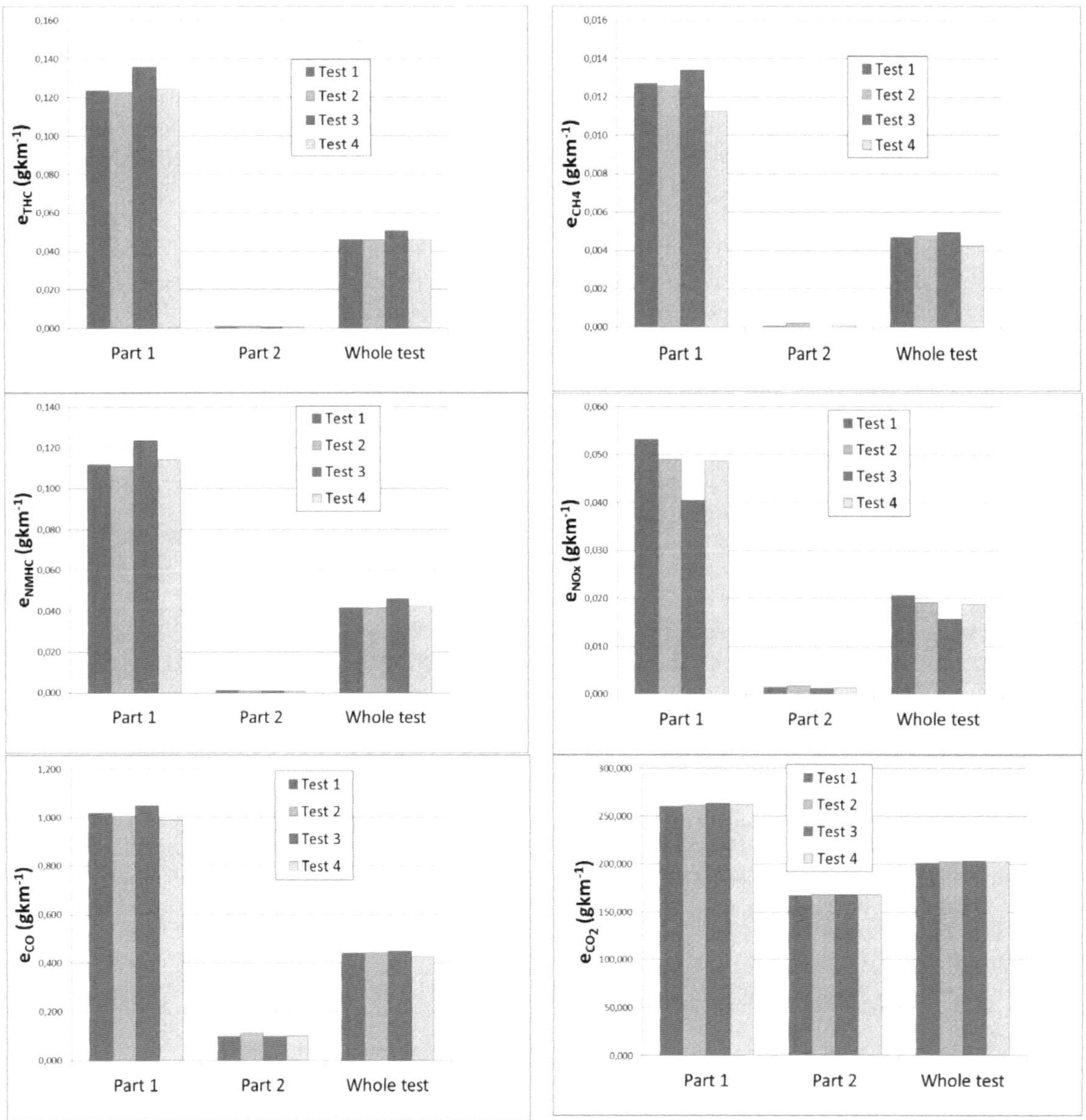

Figure 4. Results of THC, CH_4, NMHC, NO_X, CO, and CO_2 emissions for NEDC tests.

The repeatability standard deviation is a measure of spread of test results obtained under repeatability conditions defined by the formula:

$$s_r = \left[\frac{1}{n-1} \sum_{i=1}^{n} (x_i - x_{av})^2 \right]^{\frac{1}{2}} \tag{2}$$

where s_r is the repeatability standard deviation, x_i — result of i-th measurement, x_{av} — arithmetic mean of n measurement results taken into account.

The repeatability limit r is defined as a quantity which at the 95% probability level is not exceeded by the absolute value of the difference between two test results obtained under repeatability conditions.

In case of standardised methods used in laboratory practice, at least two parallel determinations are performed and the absolute value of difference between the results are compared with certain critical value known as the repeatability limit and defined by the formula [9]:

$$r = 2{,}8 \cdot s_r \tag{3}$$

where s_r is the repeatability standard deviation determined for all measurements.

It is recommended that the absolute value of difference between measurement results does not exceed the repeatability limit r. It this is the case, both results are considered correct, and the mean value of the two is presented as the final result.

The relative standard deviation repeatability RSDr is defined as:

$$\mathrm{RSD_r} = \frac{s_r}{x_{av}} \tag{4}$$

The corresponding non-dimensional relative value which also can be used to characterise non-repeatability of elements of a set is the coefficient of variation CV defined as:

$$\mathrm{CV} = \frac{s_r}{x_{av}} \cdot 100\% \tag{5}$$

Table 4. Parameters of repeatability analysis.

Parameter	Values for THC	Values for CH_4	Values for NMHC	Values for NO_X	Values for CO	Values for CO_2
x_{av} (gkm^{-1})	0.04594	0.005	0.04299	0.01857	0.43924	202.538
x_{min} (gkm^{-1})	0.04594	0.004	0.04149	0.01569	0.42776	201.411
x_{max} (gkm^{-1})	0.0506	0.005	0.04601	0.02053	0.44867	203.331
R_x (gkm^{-1})	0.00466	0.001	0.00452	0.00484	0.02091	1.92
s_r(gkm^{-1})	0.0022	0.0003	0.00206	0.00205	0.00866	0.81106
RSDr	0.04639	0.06	0.04782	0.1104	0.01971	0.004004
CV (%)	4.63915	6.59	4.78203	11.0403	1.97151	0.40045
CNR	0.046	0.066	0.048	0.11	0.020	0.004
r (gkm^{-1})	0.00615	0.00086	0.00576	0.00574	0.02425	2.27097

Guidelines concerning assessment of pollutant emissions published in works [3,4] include also a proposal to use the coefficient of non-repeatability CNR defined by means of formula:

$$\mathrm{CNR} = \frac{s_r}{|x_{av} - \mathrm{RV}|} \tag{6}$$

where RV is a reference value. According to [3], the reference value equals 0 for emissions of pollutants.

Values of the calculated repeatability measures are summarised in table 4. Plots in figure 5 represent results concerning the repeatability limits r and absolute differences Δ between two results of emission measurements under repeatability conditions for the analysed exhaust gas components. Figure 6 shows values of the coefficient of non-repeatability CNR for measurements of the mass emission of the analysed exhaust gas component throughout the test. To illustrate repeatability of measurements, ratio of the measurements range R_x to the repeatability limit r was also calculated (figure 7).

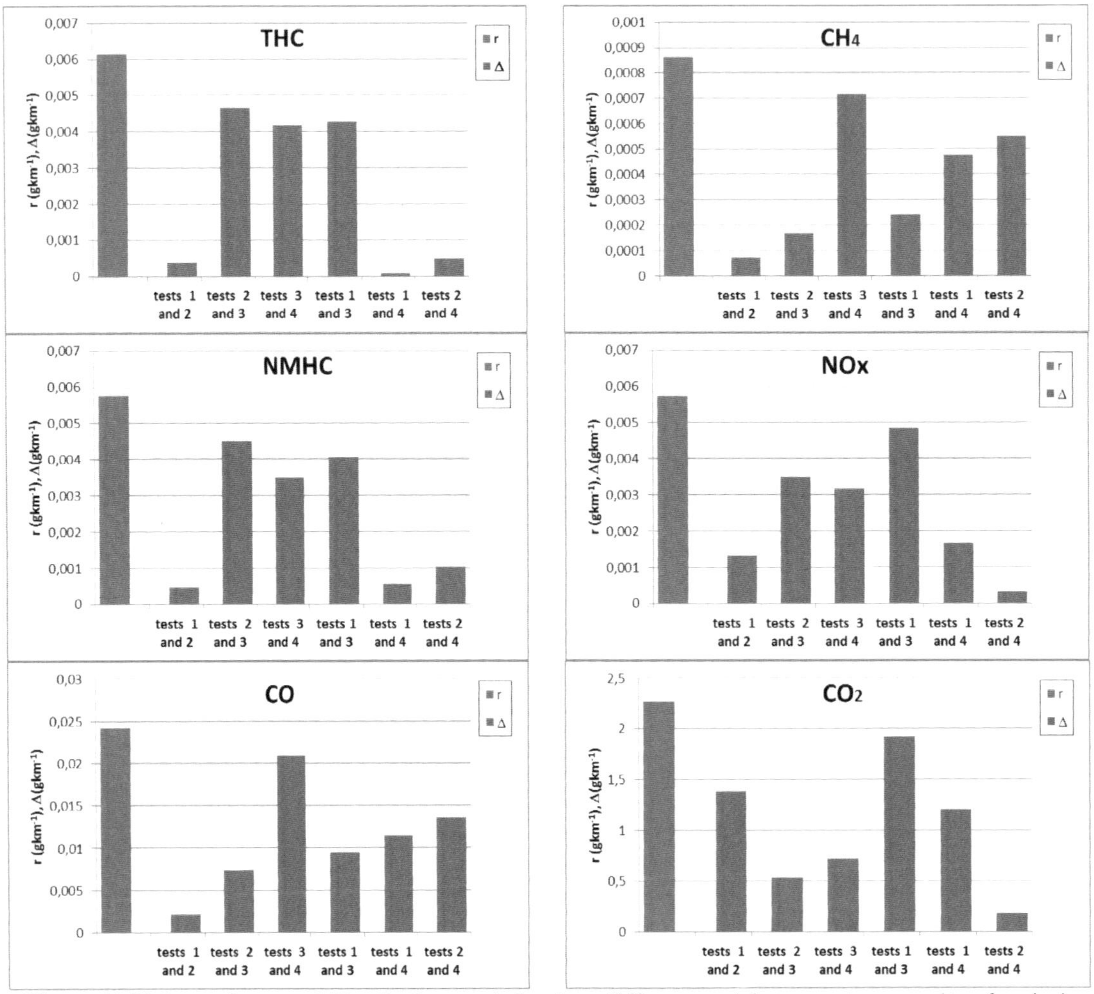

Figure 5. Values of repeatability limits r and absolute differences Δ between two results of emission measurements of the analysed exhaust gas components over the whole test.

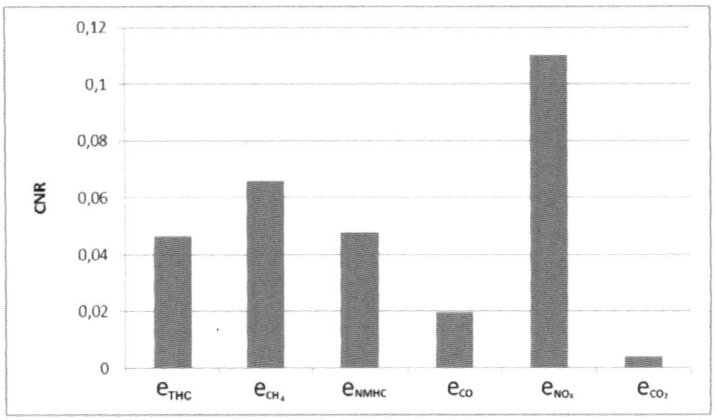

Figure 6. Values of the non-repeatability coefficient for mass emissions of the analysed exhaust gas components over the whole test.

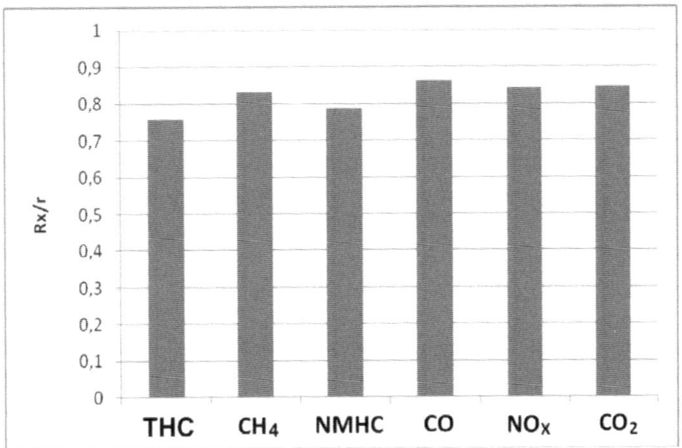

Figure 7. Values of ratio of the range *Rx* to the repeatability limit *r* of the analysed exhaust gas components over the whole test.

4. Conclusions

Analysis of the obtained results indicates that the performed measurements can be considered repeatable. Emission differences for all the measured toxic components evaluated in two measurements did not indicate any exceedances with respect to the repeatability limits *r*.

It is therefore possible under repeatability conditions to assume that on the examined research set-up, measurement results will be repeatable. The spread of pollutant emission values measured in the tests was characterised by the following values of the range R_x:

- $R_x = 0.00466$ g/km for THC ($r = 0.00615$ gkm^{-1});
- $R_x = 0.001$ g/km for CH_4 ($r = 0.00086$ gkm^{-1});
- $R_x = 0.00452$ g/km for NMHC ($r = 0.00576$ gkm^{-1});
- $R_x = 0.00484$ g/km for NO_X ($r = 0.00574$ gkm^{-1});
- $R_x = 0.02091$ g/km for CO ($r = 0.0245$ gkm^{-1});
- $R_x = 1.92$ g/km for CO_2 ($r = 2.027097$ gkm^{-1}).

The coefficient of variation CV for the analysed exhaust gas components is less than 10%, except for NO_X for which CV = 11.04% which is the evidence of repeatability of the results being at the acceptable level [2].

It should be noted that in the case when the adopted ultimate test result is the mean value of individual measurement results the number of which exceeds 2, then the repeatability limits will change and they can be calculated with the use of appropriate formulae according to [9].

Values of the coefficient of non-repeatability CNR were smallest for CO_2, both for the whole test and for its individual parts. The highest values of the coefficient were obtained for the CH_4 mass emission in the second part of the test, whereas in case of the mass emission observed in the course of the whole test, the largest values CNR value was obtained for the mass emission of NO_X.

Taking into account the range-to-repeatability limit ratio (figure 7), the largest value of the coefficient amounting to about 0.86 was obtained for CO emission. It can be however noted that values of the coefficient characterising other analysed exhaust gas components are similar and their values fall into the range from about 0.76 to about 0.86.

It can be easily seen from the graph representing requirements concerning the number of tests to be performed with respect to exhaust gas pollutants subject to limitations (figure 8) that is the result obtained from the first test is on the level equal to or less than 0.7 L for all pollutants, such result is considered final (regardless of repeatability of the measurement). In other cases (figure 8), the average from two or three tests is adopted to be the final result, whereas possible measurement results can be characterised with a large spread and thus exceed the determined repeatability limits.

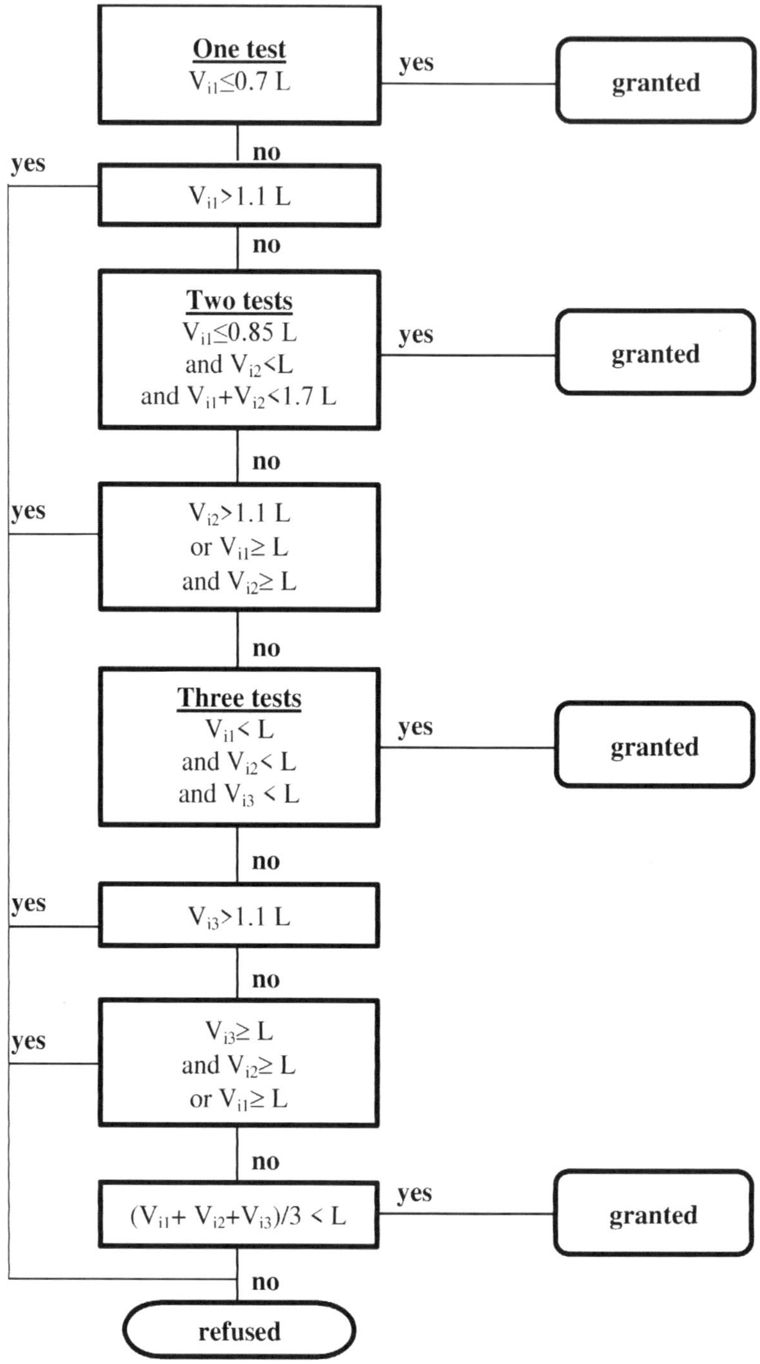

Figure 8. Flow chart for Type I Approval (Vi — emission value from *i*-th test, L — the limit value as per emission standard) [1].

To sum up, in consecutive assessments concerning exhaust gas emissions concerning e.g. dependence of pollutant levels on the vehicle operating mileage, it will be possible to consider results of any two measurements falling into the repeatability limits determined with the use of the same measuring set-up as representative ones.

References

[1] Regulation No83 of the Economic Commission for Europe of the United Nations (UNECE) – *Uniform provisions concerning the approval of vehicles with regard to the emission of pollutants according to engine fuel requirements* Official Journal of the European Union Vol 58 2015

[2] Cha-Lee Myung, Hyungmin Lee, Sangil Kwon, Sangmin Lee, Jongik Jun, Youngjae Lee, Youngmin Woo, Minho Lee, Gwi-Nam Bae and Simsoo Park 2009*Inter-laboratory correlation exerciseon a light-duty diesel passenger vehicle to verify nano-particle emission characteristics by Korea particle measurement program.*Journal of Mechanical Science and Technology 23 pp 729-738

[3] Chłopek Z. and Rostkowski J. 2015*Non-repeatability of exhaust emission test results.* Combustion Engines., 163(4), ISSN 2300-9896 pp 92-100

[4] Chłopek Z. and Szczepański T. 2015*Rating causes the non-repeatability of an internal combustion engine operating states in the sets of his work conditions.* Combustion Engines, 162(3), ISSN 2300-9896 pp 708-711

[5] Giechaskiel Barouch, Dilara Panagiota and Andersson Jon 2008 *Particle Measurement Programme (PMP) Light-Duty Inter-Laboratory Exercise: Repeatability and Reproducibility of the Particle Number Method*, Aerosol Science and Technology, 42:7, 528-543, DOI: 10.1080/02786820802220241

[6] Service Manual 2014 *AVL- ROADSIM 48"*

[7] PN-ISO 5725-1:2002 *Accuracy (trueness andprecision) of measurement methodsand results – Part 1:General principles anddefinitions*

[8] PN-ISO 5725-2:2002 *Accuracy (trueness and precision) of measurement methods and results – Part 2:Basic method for the determination of repeatability and reproducibility of a standard measurement method*

[9] PN-ISO 5725-6:2002 *Accuracy (trueness and precision) of measurement methods and results – Part 6: Use in practice accuracy values*

A comparison of exhaust emissions from vehicles fuelled with petrol, LPG and CNG

P Bielaczyc[1], A Szczotka[1] and J Woodburn[1]
[1] Automotive Research & Development Institute Ltd, BOSMAL, ul. Sarni Stok 93, 43-300 Bielsko-Biala, Poland

E-mail: Andrzej.szczotka@bosmal.com.pl

Abstract. This paper presents an analysis of THC, NMHC, CO, NO_x and CO_2 emissions during testing of two bi-fuel vehicles, fuelled with petrol and gaseous fuels, on a chassis dynamometer in the context of the Euro 6 emissions requirements. The analyses were performed on one Euro 5 bi-fuel vehicle (petrol/LPG) and one Euro 5 bi-fuel vehicle (petrol/CNG), both with SI engines equipped with MPI feeding systems operating in closed-loop control, typical three-way-catalysts and heated oxygen sensors. The vehicles had been adapted by their manufacturers for fuelling with LPG or CNG by using additional special equipment mounted onto the existing petrol fuelling system. The vehicles tested featured multipoint gas injection systems. The aim of this paper was an analysis of the impact of the gaseous fuels on the exhaust emission in comparison to the emission of the vehicles fuelled with petrol.
The tests subject to the analyses presented here were performed in the Engine Research Department of BOSMAL Automotive Research and Development Institute Ltd in Bielsko-Biala, Poland, within a research programme investigating the influence of alternative fuels on exhaust emissions from light duty vehicle vehicles with spark-ignition and compression-ignition engines.

1. Introduction
Due to necessity of limiting vehicle emissions there is a growing interest in alternative fuels like hydrogen, natural gas – methane (CNG – compressed natural gas), liquefied petroleum gas (LPG), methanol, ethanol, bio-diesel and others. Of particular interest in the above are: natural gas (methane), occurring naturally in the earth [1,2] and LPG which is colourless gas comprised of various hydrocarbons, mainly propane and butane. The driving forces of financial and ecological pressure groups are the main reason for the evaluation studies of alternative fuels (LPG and CNG in particular). Gaseous fuel distribution is well developed in many countries and these fuels are less expensive than petrol due to tax differences and subsidies in some countries.
From the ecological point of view these fuels are interesting due to impact on CO_2 emissions (greenhouse effect) - gaseous fuels contain more hydrogen and less carbon than petrol or diesel fuel (figure1), and higher carbon fractions produce more carbon dioxide CO_2 [3]. There are the additional advantages of the reduction in evaporative and refuelling emissions, particulate and NO_x emissions benefits and reduction of cancirogenous hydrocarbons emissions. Alternative fuels are suitable for SI stoichiometric or lean-burn engines with MPI or DI fuelling systems and CI engines with indirect (IDI) or direct-injection (DI) fuelling systems. Major advantages and disadvantages of LPG and CNG fuels are shown in table 1 [3-5].

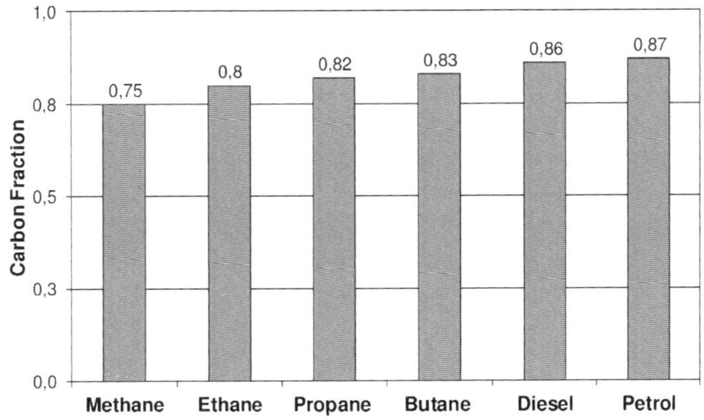

Figure 1. Carbon fractions for different fuels.

Table 1. Major advantages and disadvantages of gaseous alternative fuels

Fuel	Advantages	Disadvantages
Methane (Natural Gas)	• can be produced from a variety of feedstocks, including renewable, • very low emissions of ozone-forming hydrocarbons, particulate matters and toxic compounds, • higher calorific value than petrol on a mass basis, • potentially more complete combustion and lower cold start emissions than petrol (due to its gaseous state), • negligible sulphate and evaporative emissions.	• higher vehicle purchase cost, • lower vehicle range, • exhaust emissions of methane (a potent greenhouse gas) are relatively high, • limited refuelling infrastructure – CNG requires special refuelling stations.
Propane-Butane (LPG)	• cheaper than petrol today, • lower emissions of ozone-forming hydrocarbons and toxic compounds, lower particulate matter emissions, • most widely available clean fuel today, • excellent fuel, especially for fleet vehicles.	• cost will rise with demand, • limited supply, • its composition varies widely between countries and regions.

As a vehicular fuel, CNG exhibits significant potential for the reduction of gaseous emissions and particle emissions and improvements in energy security [6-22]; such effects have been discussed and confirmed in previous studies, in the context of both passenger cars (e.g. [13], [17]) and other vehicle types (e.g. [2]).

The aim of this study was to assess and compare the emissions performance of Euro 5 vehicles, representative of the European market, operating on petrol and CNG or LPG, for comparison to the Euro 6 limits introduced in September 2014 (figure 2). Some unregulated exhaust emissions (CO_2 and CH_4) were also measured and analysed.

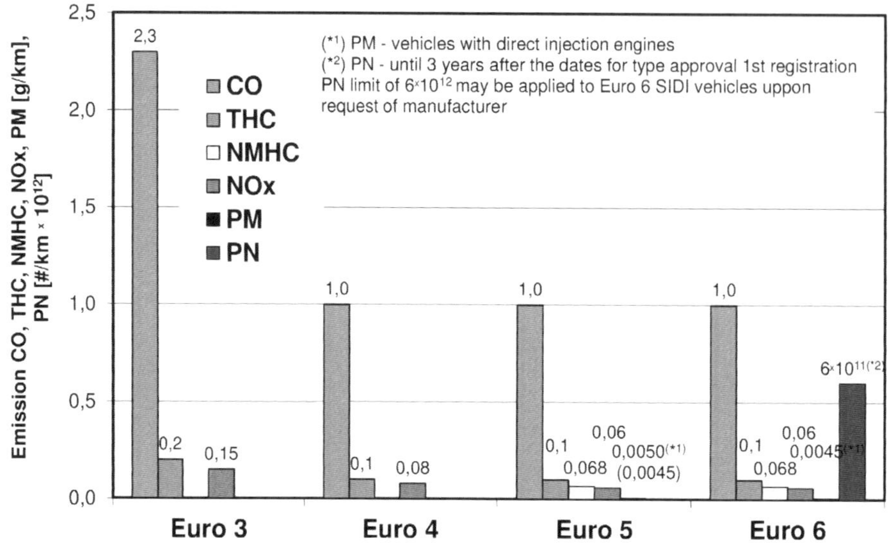

Figure 2. . Progress in European emission regulations for passenger cars fitted with spark ignition (SI) engines.

2. Research programme

The analysis of exhaust emissions from the bi-fuel vehicles with a spark-ignition engines fuelled with CNG or LPG and, later, with petrol during the NEDC cycle was carried out on a chassis dynamometer in the Emissions testing laboratory (figures 3 and 4) of BOSMAL Automotive R&D Institute in Bielsko-Biala, Poland. The tests presented in this paper were carried out within a research programme investigating the influence of alternative fuels on exhaust emissions from automotive vehicles with spark ignition and compression ignition engines. The objective of the research presented here was to determine the influence of CNG and LPG fuel usage on exhaust emissions in comparison to the emission of the vehicles fuelled with petrol.

The New European Driving Cycle (NEDC – figure 5), i.e. the current European legislative cycle, described in UNECE Regulation No. 83 [21], was selected as a representative test for this study. The test consists of two phases: the Urban Driving Cycle (UDC), followed by the high-speed Extra Urban Drive Cycle (EUDC).

Figure 3. BOSMAL Emission Testing Laboratory – internal view of the climatic chamber.

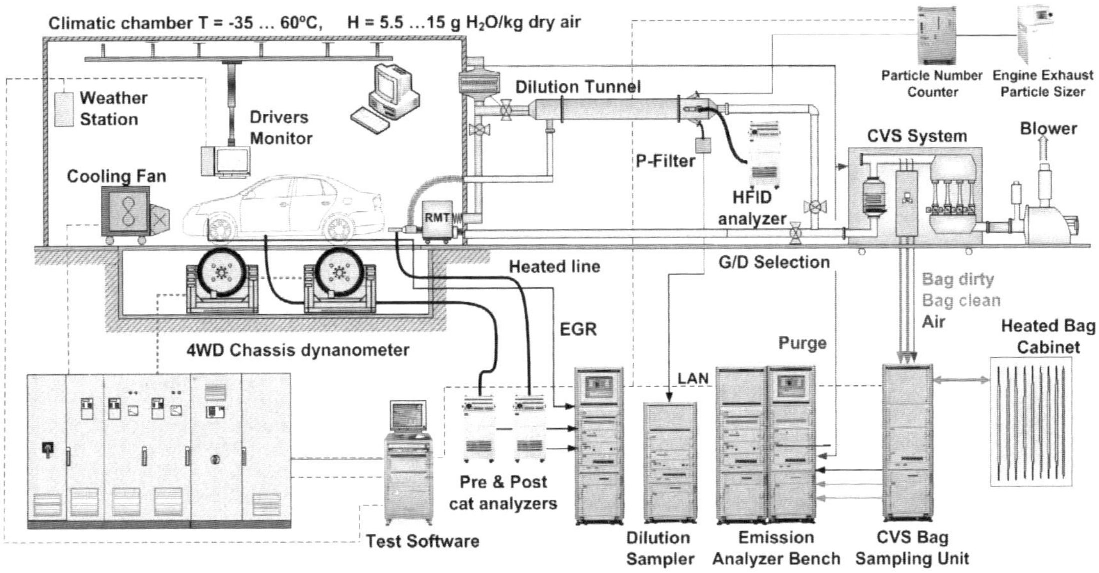

Figure 4. BOSMAL Emission Testing Laboratory – schematic overview, showing the chassis dynamometer, the climatic chamber and the various emissions measurements systems.

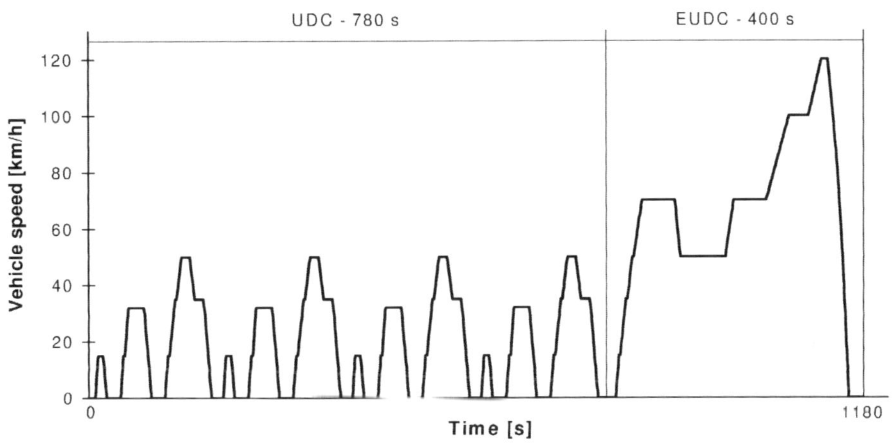

Figure 5. The New European Driving Cycle (NEDC)

The tests were conducted on one Euro 5 bi-fuel vehicle (petrol/LPG) and one Euro 5 bi-fuel vehicle (petrol/CNG), both with SI engines equipped with MPI feeding systems operating in closed-loop control, typical three-way-catalysts and heated oxygen sensors. The vehicles had been adapted by their manufacturers for fuelling with LPG or CNG by using additional special equipment mounted onto the existing petrol fuelling system. The vehicles tested featured multipoint gas injection systems. The engines always start up on petrol and are automatically switched over to CNG or LPG operation after a few seconds (if the vehicle is in CNG) or after 1-2 minutes (if the vehicle is in LPG mode). The aftertreatment systems of the cars tested consisted of a three way catalytic converters (TWC) specially adapted for bi-fuel cars (fuelled with petrol and CNG or LPG). The usage of CNG or LPG fuel also caused some changes in the engine constructions of the cars tested. The constructions of the cylinder heads were modified for CNG or LPG operation by changes made to valve and seat materials. The changes to the engines and fuelling systems, together with the installation of a larger TWC cause a slight increase in vehicles mass compared to the standard (petrol only) model.

3. Test results

A series of tests were performed on a chassis dynamometer facility on the test vehicles, fuelled with petrol and with LPG or CNG (in turn). Tests were undertaken in order to determine the influence of gaseous fuels (LPG and CNG) on emissions in comparison to Euro 6 limits and in comparison to tested petrol. Results obtained using all fuels are presented side-by-side for ease of comparison. Figures 6-11 present the average emissions (in g/km) of THC, CH_4, NMHC, CO, NO_x, and CO_2 for both phases, e.g. the UDC and EUDC, as well as for the complete NEDC (UDC+EUDC) from the test vehicles.
Hydrocarbons emissions i.e. THC, CH_4 and NMHC (figures 6-8) of vehicle A fuelled with petrol and LPG in turn were similar for both fuels. For vehicle B fuelled with petrol and CNG in turn, the total emissions of hydrocarbons (figure 6) over the entire NEDC were by 25% higher when running on CNG, which is caused by very high methane emissions (figure 7) over the UDC phase (seven fold higher than for petrol).

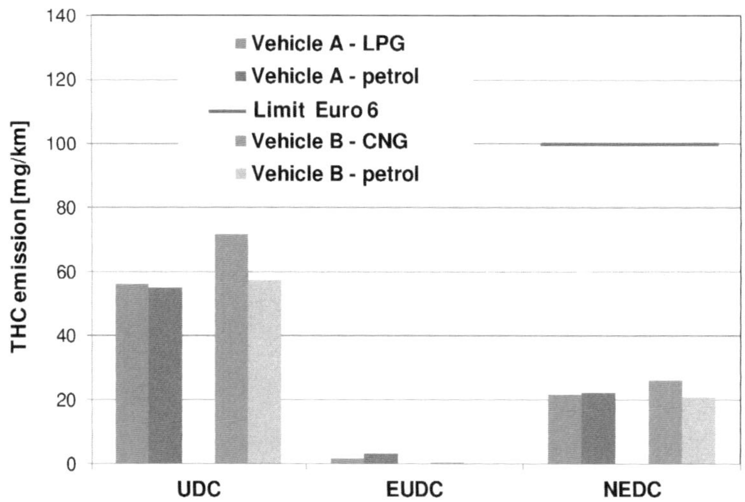

Figure 6. THC emissions over the NEDC cycle and its UDC and EUDC phases from the vehicles fuelled with petrol and gaseous fuels (LPG or CNG) in turn.

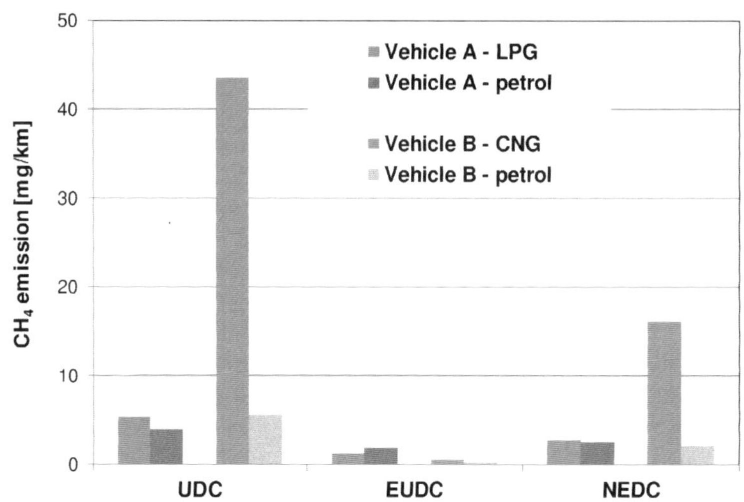

Figure 7. CH_4 emissions over the NEDC cycle and its UDC and EUDC phases from the vehicles fuelled with petrol and gaseous fuels (LPG or CNG) in turn.

The high CH_4 emissions from vehicle B fuelled with CNG were due to the fact that the main composition of the fuel was CH_4 (more than 96%), a compound which is very difficult to oxidize in a TWC. Although methane emissions are currently not subject to direct regulation in the EU, they are regulated in the US and are of concern due to methane's role as a greenhouse gas (21 times as absorptive of infrared radiation as CO_2) [23]. Furthermore, CH_4 can be considered to be indirectly regulated in the EU, since there is a THC limit and a non-methane HC limit – the difference being methane itself.

Methane, as the main (but not sole) hydrocarbon of natural gas vehicle (NGV) exhaust gas is a relatively inert component (no double carbon bonds or longer carbon chains and saturated molecular structure), which leads to it having a higher light-off temperature for conversion than other hydrocarbons (e.g. those which are unsaturated or with longer chains). Petrol does not contain any methane; however, methane is formed in petrol combustion by cracking processes.

NMHC emissions (figure 8) of both vehicles fuelled with all fuels were lower than 30% of the Euro 6 limit. NMHC emissions over the entire NEDC for vehicle B were 45% lower when running on CNG. The results show that using CNG led to higher THC emissions but lower NMHC emissions [24], relative to petrol, a phenomenon directly related to the composition of the two fuel types.

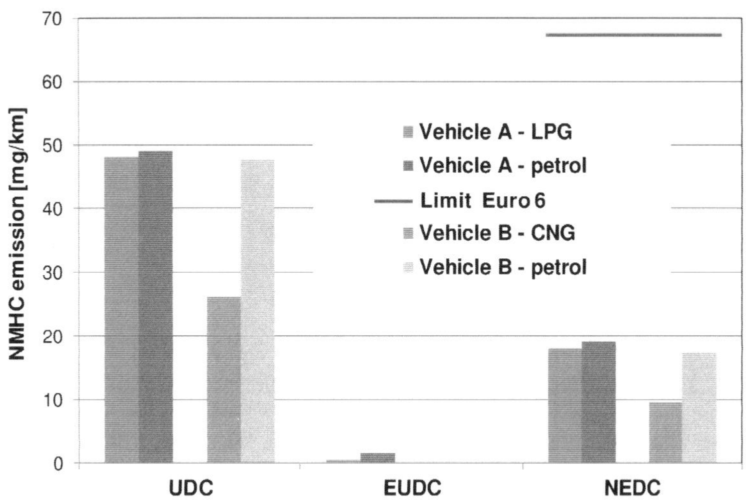

Figure 8. NMHC emissions over the NEDC cycle and its UDC and EUDC phases from the vehicles fuelled with petrol and gaseous fuels (LPG or CNG) in turn.

CO emissions of both vehicles fuelled with both fuels were lower than 50% of Euro 6 limit. For vehicle A, emissions of CO were considerably lower when running on LPG (a difference of 43% over the entire NEDC). A contrary trend was observed for vehicle B - CO emission over the entire NEDC was by 24% higher when running on CNG. It is worth pointing out that the vehicles are always started in petrol mode, even if CNG or LPG fuel was chosen as operating fuel and engines are automatically switched over to CNG or LPG operation after a few seconds (if the vehicle is in CNG) or after 1-2 minutes (if the vehicle is in LPG mode). This goes some way towards explaining the small difference in CO emission for both vehicles during the UDC phase (figure 9). During the second phase of the NEDC cycle, i.e. the EUDC phase, these differences were at times much greater, since when in CNG or LPG mode, no petrol is consumed at all by the vehicles.

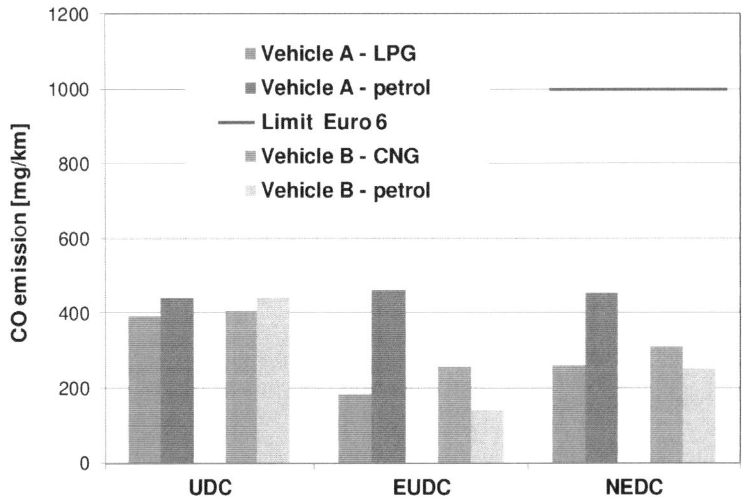

Figure 9. CO emissions over the NEDC cycle and its UDC and EUDC phases from the vehicles fuelled with petrol and gaseous fuels (LPG or CNG) in turn

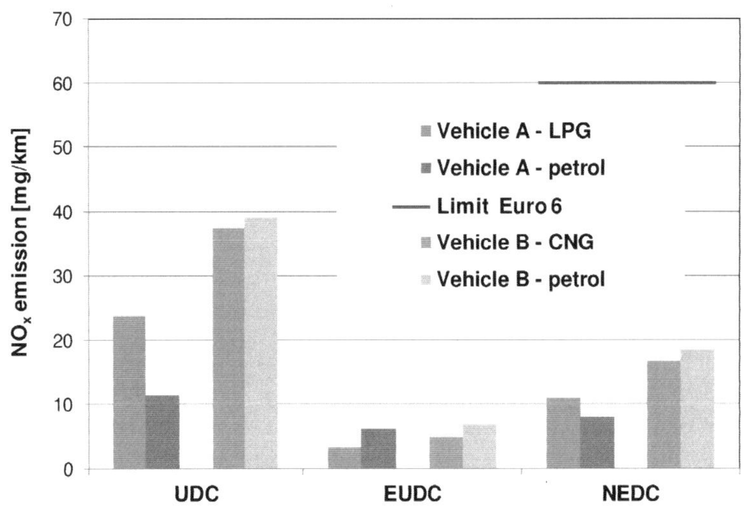

Figure 10. NO_x emissions over the NEDC cycle and its UDC and EUDC phases from the vehicles fuelled with petrol and gaseous fuels (LPG or CNG) in turn.

NO_x emissions (figure 10) from both vehicles fuelled with all fuels were lower than 30% of the Euro 6 limit. For vehicle A, emissions of NO_x were considerably higher when running on LPG (a difference of 35% over the entire NEDC), but for vehicle B they were 10% lower when running on CNG. The engine calibration and the precise chemistry of the TWC have a great impact on NO_x emissions and it is possible that for other vehicles fuelled with petrol and with LPG or CNG in turn these trends could be contrary. Because of this, the observed change may not be directly attributable in full to the fuels chemistry itself, but to the interaction of the fuel chemistry with engine operation variables (spark timing, etc.) and also TWC chemistry.

Although they are presently not subject to direct limits in automotive emissions legislation, CO_2 emissions are important with respect to global warming [23] and because CO_2 emissions and fuel economy are intimately related.

As LPG and CNG have lower carbon/hydrogen ratio (figure 1) than petrol, they produce lower CO_2 emissions per unit energy released during combustion. CO_2 emission results are shown in figure 11. It can be noted that for vehicle A for LPG these emissions were about 10% lower and for vehicle B about 25% lower for CNG fuel than for petrol in both phases, as well as in the complete NEDC cycle.

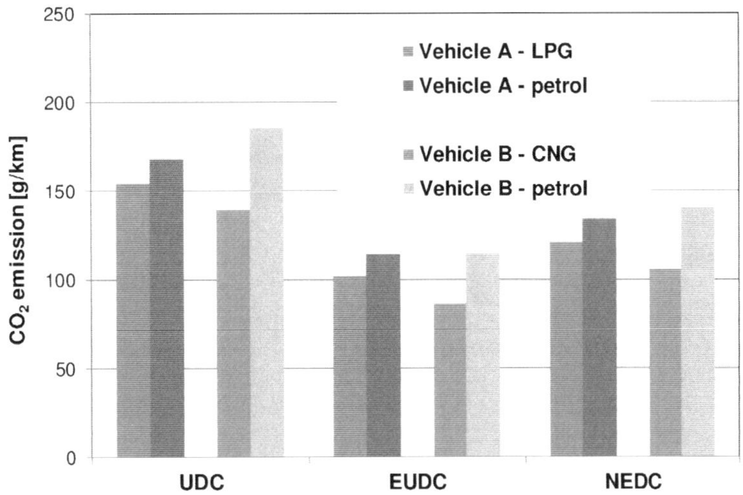

Figure 11. CO_2 emissions over the NEDC cycle and its UDC and EUDC phases from the vehicles fuelled with petrol and gaseous fuels (LPG or CNG) in turn

4. Summary/ Conclusions

The main aim of this paper was to determine the influence of CNG and LPG fuels on emissions in the context of the Euro 6 emissions requirements and in comparison to the emission of the vehicles fuelled with petrol. The analysis were performed on two Euro 5 bi-fuel light duty vehicles (one bi-fuel vehicle petrol/CNG and one bi-fuel vehicle petrol/LPG) .

On the basis of the analyses of results obtained during the NEDC emissions test, it has been found that both vehicle tested with a CNG or LPG multipoint gas injection and an integrated petrol/CNG or petrol/LPG ECU (Electronic Control Unit) already meets the Euro 6 emissions limits, without any further modifications, in particular:

- THC, NMHC, CO and NO_x emissions meet Euro 6 limits for the Light-Duty Vehicle (LDV) category.
- Observed THC emissions for both vehicles and all fuels (CNG, LPG, petrol) were even lower than the NMHC limit stipulated by these regulations, so emissions limits in this area were comfortably met. NMHC emissions were almost 7 times lower than the Euro 6 limit when running on CNG.
- THC emissions during the NEDC cycle increased when vehicle B was fuelled with CNG, in comparison to petrol, but this increase was far too small to cause problems with the Euro 6 emissions limit. THC emissions for vehicle A fuelled with LPG and petrol were similar for both fuels.
- CO emission during the NEDC cycle increased by 24% when vehicle B was fuelled with CNG, although CO emission from both fuel types was well below the limit. For vehicle A, emissions of CO were considerably lower (by 43%) when running on LPG.
- NO_x emissions from vehicle A when fuelled with LPG were about 35% higher, but from vehicle B when fuelled with CNG were about 10% lower in comparison to petrol.
- CO_2 emissions were decreased by 10% when vehicle A was fuelled with LPG and by 25% when vehicle B was fuelled with CNG

Gaseous fuels (LPG and CNG) are commonly used in SI engines because their powertrains are relatively easy to convert from liquid to gaseous fuels. These fuels are very attractive as they are cheaper than petrol or diesel. This paper has shown that even certain pre-Euro 6 technologies can meet Euro 6 emissions standards when LPG or CNG is used as a fuel. Additionally, the CO_2 emissions from vehicles operation on gaseous fuels (LPG or CNG) are low. For both these reasons, interest in producing and marketing bi-fuel passenger cars for the European market (among others) is sure to remain high, notwithstanding the current inequalities in NG distribution infrastructure and local availability.

The good performance of LPG and CNG in terms of low emissions of particulates and CO_2 confirm that LPG and CNG have significant environmental potential, even if certain emissions are sometimes higher than when running on petrol.
The emissions and performance of bi-fuel vehicles changes when operating on fuel other than gasoline (e.g. LPG or CNG). The engine and aftertreatment technology, the individual engine calibration strategy, fuel and lubricant oil properties, ambient temperatures and driving conditions are the main factors influencing tailpipe emissions.

References

[1] Suga T, Muraishi T, Brachmann T and Yatabe F 2000 *Potential of a Natural Gas Vehicle as EEV (Environmentally Enhanced Vehicle)* CEC/ SAE International Spring Fuels & Lubricants Meeting 2000 Paper No. 2000-01-1863
[2] Checkel D and Dhaliwal B 2000 *Tailpipe Emissions Comparison Between Propane and Natural Gas Forkfits* CEC/ SAE International Spring Fuels & Lubricants Meeting 2000 Paper No. 2000-01-18654
[3] Sykes R 1999 *Gas works* Engine Technology International 4/1999
[4] Watson H and Gowdie D 2000 *A Systematic Evaluation of Twelve LP Gas Mixtures for Emissions and Fuel Consumption* CEC/ SAE International Spring Fuels & Lubrificants Meeting 2000 Paper No. 2000-01-1867
[5] West J F 2000 *All aboard the clean motor* Engine Technology International 4/2000
[6] McTaggart-Cowan G P, Reynolds C C O and Bushe W K 2006 *Natural gas fuelling for heavy-duty on-road use: current trends and future direction* International Journal of Environmental Studies 2006 63 421-440
[7] Directive 2003/30/EC of the European Parliament and of the Council of 8 May 2003 on the promotion of the use of biofuels or other renewable fuels for transport Official Journal of the European Union L 123/42 17.5.2003
[8] Bhandari K, Bansal A, Shukla A and Khare M 2005 *Performance and emissions of natural gas fueled internal combustion engine: A review* Journal of Scientific & Industrial Research Vol. 64 May 2005 p. 333-338
[9] Warnecke W, Wilbrand K and Scholey H 2007 *The Drive for Energy Independence in the Next Decades – Fuels for the Future* Proceedings of the 28th International Vienna Motor Symposium (Vienna Austria)
[10] Suthawaree J, Sikder H A, Jones C E, Kato S et al. 2012 *Influence of extensive compressed natural gas (CNG) usage on air quality* Atmospheric Environment 54 (2012) 296-307
[11] Semin, R A B 2008 *A Technical Review of Compressed Natural Gas as an Automotive Fuel for Internal Combustion Engines* American J. of Engineering and Applied Sciences 1 (4) 302-311
[12] Checkel D and Dhaliwal B *Tailpipe Emissions Comparison Between Propane and Natural Gas Forklifts* SAE Paper No. 2000-01-1865 2000 doi:10.4271/2000-01-1865
[13] Allgeier T, Bischoff C and Foerster J 2004 *Natural Gas as an alternative fuel for motor vehicles* FISITA paper F2004V040 2004 FISITA World Congress
[14] Karavalakis G, Durbin T, Villela M and Miller W 2012 *Air pollutant emissions of light-duty vehicles operating on various natural gas compositions* Journal of Natural Gas Science and Engineering Vol. 4 January 2012 p. 8-16
[15] Bielaczyc P, Szczotka A and Wojnarowicz M 2010 *Analysis of the Exhaust Emissions and Performance of a Modern Vehicle Fueled with Petrol and CNG in Turn* FISITA paper F2010-A-133 FISITA 2010 World Automotive Congress Budapest 2010

[16] Bielaczyc P and Szczotka A 2012 *The potential of current European light-duty CNG-fueled vehicles to meet Euro 6 requirements* Paper 2012-SS4-403 Combustion Engines 4/2012 (151)
[17] Bielaczyc P, Woodburn J and Szczotka A 2013 *An Assessment of Regulated Emissions and CO_2 Emissions from a European Light-Duty CNG-Fueled Vehicle in the Context of Euro 6 Emissions Regulations* Applied Energy 117 2014 134-141 doi:10.1016/j.apenergy.2013.12.003
[18] Bielaczyc P, Szczotka A and Woodburn J 2015 *Regulated and Unregulated Exhaust Emissions from CNG Fueled Vehicles in Light of Euro 6 Regulations and the New WLTP/GTR 15 Test Procedure* SAE Int. J. Engines 8(3):2015 doi:10.4271/2015-01-1061 SAE 2015 World Congress Detroit
[19] Verbeek R, Ligterink N, Meulenbrugge J, Koornneef G et al. 2013 *Natural Gas in Transport - An assessment of different routes* Delft report 31.4818.38
[20] Edwards R, Hass H, Larivé J F, Lonza L et al. 2014 *Well-To-Wheels Analysis Of Future Automotive Fuels And Powertrains In The European Context* JRC85329 Report 2014 doi:10.2790/95533
[21] UNECE Regulation No. 83 – Annex 8 2011 available online: http://unece.org
[22] Jahirul M I, Masjuki H H, Saidur R, Kalam M A, Jayed M H and Wazed M A 2010 *Comparative engine performance and emissions analysis of CNG and gasoline in a retrofitted car engine* Applied Thermal Engineering 30 2010 2219-2226
[23] Seinfeld J H and Pandis S N 1998 *Atmospheric Chemistry and Physics: From Air Pollution to Global Climate Change* Wiley Interscience New York
[24] Howes P and Rideout G 1995 *Evaluation of Current Natural Gas Vehicles Technology Exhaust Emissions at 10 Various Operating Temperatures* SAE Paper 952437

Determination of operating parameters of industrial engine fuelled with post processing gases with high hydrogen content

M Brzeżański[1], M Mareczek[1], W Marek[1] and T Papuga[1]

[1]Institute of Automobiles and Internal Combustion Engines, Cracow University of Technology, 37 Jana Pawla II Avenue, 31-864 Krakow, Poland

E-mail: mbrzez@pk.edu.pl

Abstract. The results of investigations of SI engine fuelled with hydrogen and mixed n-butanol with isobutanol have been presented in article. The idea of flexible feeding system and the aim and methodology of carried out measurement have been also described. Obtained results have been compared to the results of tests carried out during flexible feeding of the same engine. The proposed control system enables not only application of different liquid and gaseous fuels but also application of the fuels which chemical composition vary within the relatively short time intervals.

1. Introduction

An important objective of the scientific and economic, leading to the rationalization of the use of different energy sources is the exploration and use of new fuels for combustion engines. There is a large group of fuels which are not suitable for use in traction vehicles for various reasons. That group of fuels, however, can be used in power generation applications. To this group belong the various types of combustibles, which are the waste product of certain industrial processes. One such type of fuels is industrial hydrogen or gas mixture containing a large proportion thereof, as well as various types of alcoholic liquid phases. This type of waste fuels are typically used at most heating purposes, but they are often irretrievably lost by burning in a torch.

Their development for low-energy seems to be a reasonable solution, the more that this type of system allows for flexible adaptation of the type of energy to current needs. This corresponds to the modern trend towards diversification of energy sources. Suitably adapted, the internal combustion piston engine, working with a power generator, can be both, a heat source and a source of electricity.

2. Waste fuels from industrial processes

In the study conducted at the Institute of Automobiles and Internal Combustion Engines Cracow University of Technology the waste fuel from the chemical industry has been taken into account. That fuel was in the form of a gas containing a high concentration of hydrogen and liquid fuels, which are a mixture of alcohols. An example of such fuel is the so-called butanol loop waste gas of the consisting of: H_2 –85%, N_2 – (6-8)%, CH_4 – (2-4)%, O_2 – (1-2)%, CO_2 – (1-2)%, or so-called octanol loop waste gas consisting of: H_2 – 80%, N_2 – (8-10)%, CH_4 – (4-6)%, O_2 – (1-2)%, CO_2 – (1-2)%. The waste liquid phase is usually comprised mainly of a mixture of alcohols: n-butanol, isobutanol, propanol, and octanol. In the conducted laboratory studies substitute fuels were used to capture the physical and chemical characteristics of listed waste fuels. As a gas fuel, industrial hydrogen was used, and as liquid fuel, a mixture of 50% n-butanol and 50% isobutanol. For comparative purposes the high methane natural gas was used.

3. Object of research and measurements

As the object of study was selected 6-cylinder supercharged gas engine type MAN E2876 LE302, of displacement Vs =12.82 dm^3, which is designed to drive a generator. The natural gas mixer feeding was the factory configuration. In order to achieve the possibility of new fuel feeding, there was designed and built the special fuel supply system similar to the common rail enabling separation of the gas line and liquid fuel line. To control the size of the gas and liquid fuel doses, ignition timing adjustment, the throttle opening and trim the emergency system, the special programmable motor controller, developed in LabView program, was constructed and applied.

This type of system allows feeding the engine with gas and liquid fuel separately, as well as the practical implementation of dual fuel supply. The controller is equipped with a knock sensor, which has clearly defined the clattering limit for each of the cylinders, in all analyzed operating conditions of the engine. The process of combustion in each cylinder is supervised by a thermocouple placed in the outlet channel of each of the cylinders, directly at the exhaust port of the cylinder head. There were examined and recorded the concentration of: carbon monoxide CO, hydrocarbons THC, nitric oxides NO_x, carbon dioxide CO_2 and oxygen O_2. In addition, the effect of that fuel feeding on other engine parameters such as torque, power, fuel consumption, total efficiency, air excess number λ and exhaust gas temperature has been investigated.

Due to the special properties of applied gaseous fuels, against possible leakage of the fuel, the special double-walled pipe system delivering natural gas or hydrogen to the engine was used. Furthermore, the test stand was equipped with a system for cooling and suctioning fumes from above the engine. The laboratory facility sensors identified concentrations of methane and hydrogen.

4. Determination of the engine working parameters

All measurements on the energy performance of engine and toxic exhaust emissions were carried out at a constant engine speed 1500 1/min and different engine load. These conditions correspond to the operation of the engine in the unit generator. Among the tested engine control parameters which have to be adapted to the type of fuel tested were: ignition timing and air excess number λ.

Both of these parameters have a significant impact on the process of combustion in the cylinder, resulting in both the value of the parameters of delivered energy and toxic exhaust emissions. The test engine temperature must not exceed 700°C because of the thermal resistance of the turbocharger (turbine), which is important in determining the maximum power of the engine fed with such type of fuel. In addition, both the value of the excess air number λ, and especially the value of ignition timing are closely related to the occurred phenomenon of knock. For this reason for each operating point of the engine, the two control parameters were adjusted individually. Simultaneously with measurements of the engine operating parameters, measurements of the concentration of selected components of the exhaust gas were carried out.

It is necessary to evaluate and analyze the combustion process in the engine cylinder. The concentration of oxygen in the engine exhaust also allowed calculating the excess air number λ, the values of which are crucial for the combustion of fuel and values of toxic exhaust emissions.

The change of operating parameters of the engine was carried out by quantitative and qualitative regulation and thereby also had an impact on the development of the combustion process and the concentration of individual exhaust gases components. For the practical implementation of the research results associated with the dual fuel fed engine, the strategy differentiating supply of chemical energy flux depending on the current availability of both gaseous fuel and liquid phase fuel was elaborated.

A reflection of this strategy is an example of the imaging method of engine power regulation shown in figure 1 where engine power 100 kW, marked on the graph as H2+MIX, was achieved by hydrogen feeding only. Further increase of the engine power up to 140 kW was obtained by increasing the dose of the liquid phase at constant hydrogen consumption. For the case of H2+MIX the power 130 kW was obtained when only hydrogen was applied. Only above of this power, the engine was additionally fed with a liquid fuel.

In both cases, the liquid phase consisted of a mixture of 50% n-butanol and 50% isobutanol. Fuel is tentatively designated as MIX. In the case of using only the liquid phase, power output was limited to 105 kW. Engine operation using only the liquid phase led to a phenomenon of diffusion of fuel to the lubricating oil. This was mainly caused by the imperfect process of atomization of the liquid phase during the formation of air-fuel mixture. Application of such a dual fuel supply system is generally associated with the necessity of obtaining a compromise between cost of equipment and accessories, and the actual possibilities of use of chemical energy fluxes of the administered fuels. In the case of the strategy used for dual fuel feeding presented in this publication, dosing of the liquid phase made possible once to extend the scope of the load from Ne = 150 kW (supplying pure hydrogen) to Ne = 190 kW (with use of dual fuel feeding). This value of load (for pure hydrogen feeding) was impossible to achieve, because increase the load above 150 kW resulted in the anomalies of combustion process.

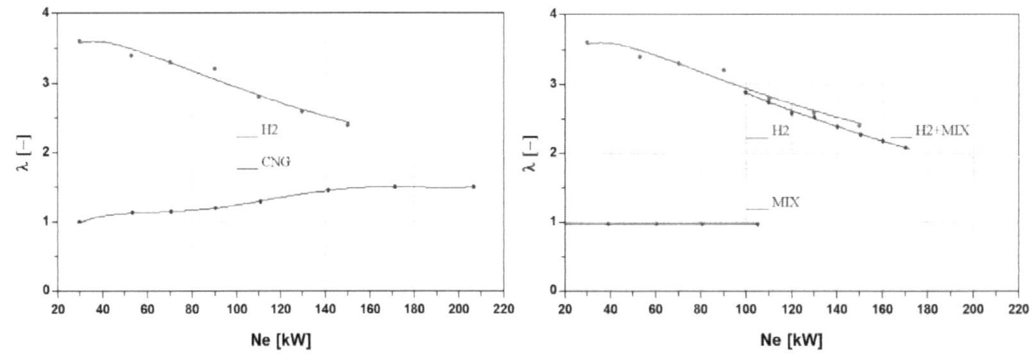

Figure 1. Comparison of air excess number λ, when the engine was fed with mixture of natural gas and hydrogen, and with a mixture of hydrogen and alcohols.

Compared to natural gas feeding, when the engine was fed with hydrogen, the concentration of carbon monoxide and hydrocarbons was negligible and resulted mainly from the combustion of lubricating oil (figure 2).

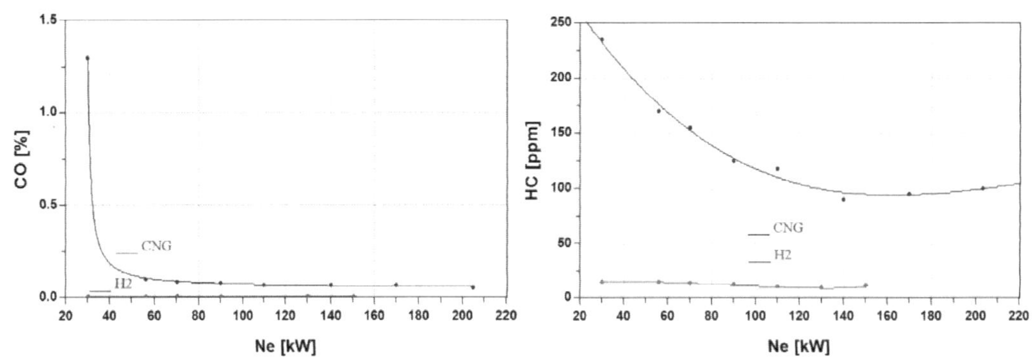

Figure 2. Comparison of the concentration of carbon monoxide and hydrocarbons for MAN engine fuelled with hydrogen or natural gas (n = 1500 1/ min).

During the engine feeding with hydrogen, slight increase in the concentration of nitrogen oxides was observed only for the state of load close to the maximum, and throughout the range of concentrations the value was very small (figure 3).

An important parameter is the engine's total efficiency. The higher value of total efficiency for engine fed with natural gas is obtained mainly due to a smaller share of the mechanical power loss in relation to the effective power. The given constant value of mechanical losses of the engine as well as the lower indicated power achieved by the hydrogen supplied engine result in a smaller value of total efficiency. In both cases, however, the value of total efficiency of the engine is high and during the natural gas feeding achieves the maximum value of approximately 42% and during the hydrogen feeding approximately 37%.

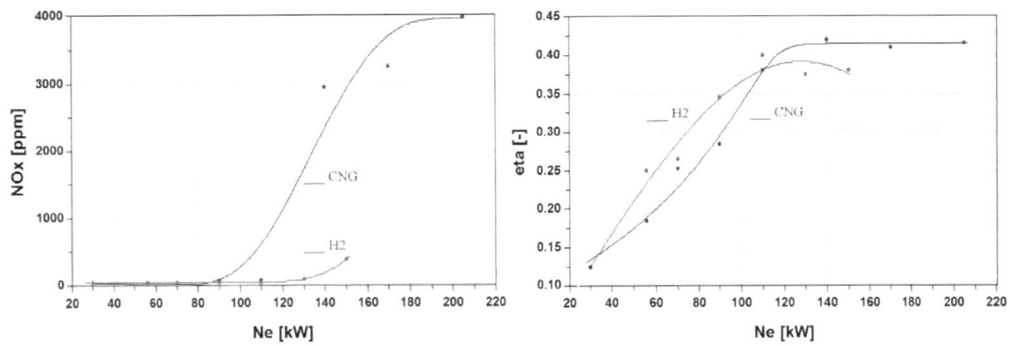

Figure 3. Comparison of the concentration of nitric oxides and total efficiency of MAN engine fuelled with natural gas or hydrogen (n = 1500 1/ min).

5. Conclusions

The study showed that the unconventional waste fuel from the chemical industry can be used to power the internal combustion engines. In this case, obtaining favorable values of operating parameters should be regarded as a success. With sophisticated engine control system in comparison to the typical mixing system, a significant increase in total efficiency has been obtained. It is also possible to use the liquid fraction of the alcohol which requires an appropriate strategy of supply. The proposed control system enables not only application of different liquid and gaseous fuels but also application of the fuels which chemical composition vary within the relatively short time intervals. Besides it is possible to apply the different optimizing procedures thanks to which one can obtain the biggest engine general efficiency and the lowest value of concentration of the individual exhaust gas toxic components.

In the domain of control systems development, the described system is unique, because is based on the generally available engineering tool, which is LabView environment and creates the possibility of individual fitting to every application in which the source of propulsion is combustion engine.

References

[1] Brzeżański M Cisek J Marek W Mareczek M and Papuga T 2013 Investigation of the supply systems of the stationary combustion engine fuelled with natural Gas *V Congress on Combustion Engines* (Bielsko Biała: PTNSS- 2013-SC-193)

[2] Brzeżański M Cisek J Marek W Mareczek M and Papuga T 2013 Investigation of the combustion engine fuelled with hydrogen and mixed n-butanol with iso-butanol *V Congress on Combustion Engines* (Bielsko Biała: PTNSS- 2013 SC 194)

[3] Brzeżański M Marek W Mareczek M and Papuga T 2013 Zastosowanie odpadowych produktów przemysłu chemicznego w energetyce *Analiza systemów energetycznych* (Kraków: chapter in a book) pp 43-56

[4] Fuchs J Leitner A Tinschmann G and Trapp C 2013 Konzept für direkt gezündete Groß-Gasmotoren *MTZ* 05

A study on the indirect urea dosing method in the Selective Catalytic Reduction system

M Brzeżański[1], **R Sala**[2]

[1]Institute of Automobiles and Internal Combustion Engines, Cracow University of Technology, Al. Jana Pawla II 37, 31-864 Krakow, Poland
E-mail:mbrzez@usk.pk.edu.pl
[2]Automotive Research & Development Institute Ltd, BOSMAL, ul. Sarni Stok 93, 43-300 Bielsko-Biala, Poland
E-mail: rafal.sala@bosmal.com.pl

Abstract. This article presents the results of studies on concept solution of dosing urea in a gas phase in a selective catalytic reduction system. The idea of the concept was to heat-up and evaporate the water urea solution before introducing it into the exhaust gas stream. The aim was to enhance the processes of urea converting into ammonia, what is the target reductant for nitrogen oxides treatment. The study was conducted on a medium-duty Euro 5 diesel engine with exhaust line consisting of DOC catalyst, DPF filter and an SCR system with a changeable setup allowing to dose the urea in liquid phase (regular solution) and to dose it in a gas phase (concept solution).

The main criteria was to assess the effect of physical state of urea dosed on the NOx conversion ratio in the SCR catalyst. In order to compare both urea dosing methods a special test procedure was developed which consisted of six test steps covering a wide temperature range of exhaust gas generated at steady state engine operation condition. Tests were conducted for different urea dosing quantities defined by the α equivalence ratio.

Based on the obtained results, a remarkable improvement in NOx reduction was found for gas urea application in comparison to the standard liquid urea dosing. Measured results indicate a high potential to increase an efficiency of the SCR catalyst by using a gas phase urea and provide the basis for further scientific research on this type of concept.

1. Introduction

Increasingly more and more stringent Europeans emission standards aim to reduce the fuel consumption and toxic compounds from automotive exhaust gasses. It poses new challenges to OEM's in terms of refining combustion engines and exhaust after-treatment systems itself. In the light of Euro 6 legislation limits and their upcoming correlation factor to Real Driving Emission figures the main attention is paid to reduce nitrogen oxides emission what necessitated application into exhaust line an additional catalyst for NOx conversion. In practice it concerns nearly all diesel applications and lean burn spark-ignition engines.

Currently there are two dominant methods of NOx reduction in the exhaust gases. The first method is selective catalytic reduction (SCR) which uses an aqueous urea solution (AdBlue) as a reductant and the second is so-called DeNOx Trap which adsorbs and stores NOx during engine lean operation $\lambda > 1$ and subsequently reduces it in short phase of gas enrichment at $\lambda \approx 0.9$. These types of catalysts are also known as NSC (NOx Storage Catalysts) or LNT (Lean NOx Traps).

The SCR is the most effective method of catalytic NOx reduction, but it requires carrying a tank on board the vehicle and dosing infrastructure, which makes the entire system more complex and extensive. The SCR catalyst is often wash-coated on the DPF substrate to integrate both catalysts (SCRF) and save the space within the vehicle's chassis and body.
DeNOx traps are more compact in structure due to the fact that regular diesel fuel is used as the reductant. DeNOx traps can be also wash-coated on a DOC substrate which reduces the size of the exhaust aftertreatment system.
The SCR and DeNOx Trap systems allow to meet always tougher NOx emission limits. An intensive research and development works carried out over last decades provided required efficiency and durability features, but still there are some technical issues to overcome.

2. Selective catalytic reduction

2.1. Introductory remarks

An SCR is nowadays the most effective method of NOx conversion into nitrogen. In mobile applications the preferred reductant is an aqueous solution of nontoxic urea, which is used as an ammonia source. The so-called AdBlue solution is injected into the exhaust system where it is decomposed to NH_3 and CO_2.
There are several complications in using urea-SCR. One of them is the need for efficient ammonia release from the urea solution and the related risks of urea deposits formation in the exhaust system. Another issue is the accurate ammonia dosage: there should be enough ammonia present on the catalyst to reduce all NOx, but at the same time there must be no excess of ammonia, to prevent its slippage from the catalyst.
Depending on an SCR catalyst coating, the biggest NOx reduction efficiency is typically achieved in a temperature window from 250°C to 450°C. Most commonly used SCR coatings developed for the Euro V emission limits were based on metal oxides: TiO_2-WO_3-V_2O_5, while for Euro VI requirements the coatings were based on zeolites: Cu-Zeolite for lower temperature range and Fe-Zeolite for higher gas temperature range.

2.2. SCR chemistry

In the SCR applications an aqueous urea solution (32.5% urea) is injected into the hot exhaust gas upstream of the SCR catalyst. The decomposition of urea into ammonia and carbon dioxide precedes the SCR reaction.
The first step is the evaporation of water from the droplets, thus leading to solid or molten urea: [4, 6]:

$$NH_2\text{–}CO\text{–}NH_2\,(\text{aqueous}) \rightarrow NH_2\text{–}CO\text{–}NH_2\,(\text{molten}) + x\;H_2O\,(\text{gas}) \qquad (2.1)$$

Molten urea will then heat up and decompose thermally according to:

$$NH_2\text{–}CO\text{–}NH_2(\text{molten}) \rightarrow NH_3(\text{gas}) + HNCO(\text{gas}) \qquad H_{298} = +186\text{ kJ} \qquad (2.2)$$

Equimolar amounts of ammonia and isocyanic acid are thus formed. Isocyanic acid is very stable in the gas phase, but hydrolyzes easily on many solid oxides reacting with water vapor originating from the combustion process:

$$HNCO(\text{gas}) + H_2O(\text{gas}) \rightarrow NH_3(\text{gas}) + CO_2(\text{gas}) \qquad \Delta H_{298} = -96\text{ kJ} \qquad (2.3)$$

The thermo-hydrolysis of urea is globally an endothermic process. The reactions (2.1) and (2.2) may also occur in the gas phase upstream of the catalyst, whereas the hydrolysis of the isocyanic acid (reaction 2.3) proceeds mainly on the SCR catalyst itself.

3. Concept of the indirect urea dosing

Studies were carried out on heavy duty diesel engine with exhaust after-treatment consisted of a DOC, DPF filter and an SCR system with the standard and development urea dosing methods.

The concept of indirect urea dosing assumed to inject the urea into a mixing module fitted upstream of SCR catalyst. The module was made of aluminium alloy in shape of gas carburettor and had an inbuilt gallery with a number of holes at its internal periphery. The working principle was that the injected liquid urea solution before reached the exhaust gas was flowing first inside the module's gallery what give an extra time for the urea evaporation and then the vapour escaped through the holes and mixed with exhaust. The mixing module induced a pressure drop which facilitated to disperse the vapour urea in an exhaust upstream of the SCR catalyst. Figure 3.1 presents the exhaust system layout with a concept urea dosing that was set up for the test purposes on the engine test bed.

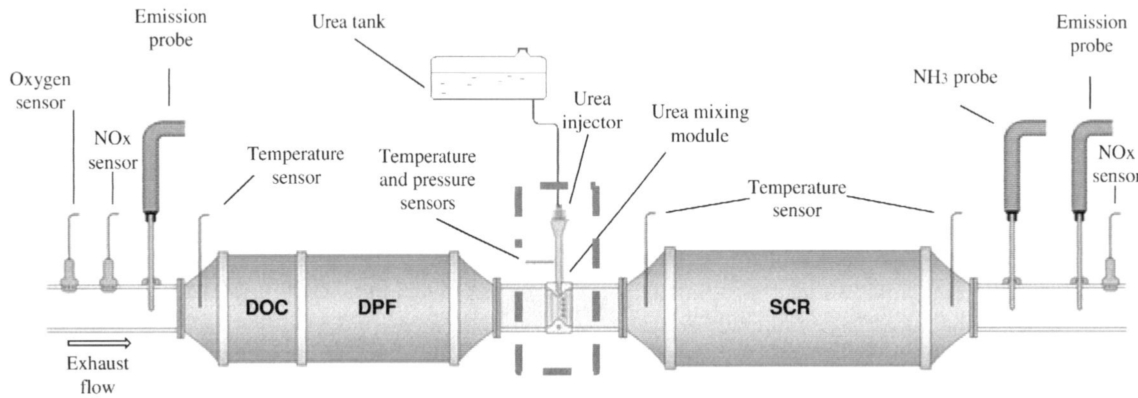

Figure 3.1. Concept of indirect urea dosing in the SCR system.

In a concept solution standard urea multi holes injector was used and dosed liquid urea into cone-shaped funnel directly connected to the mixing element via straight piece of metal tube. During the test run pressure and temperature measurements of urea dosed were taken at the inlet of the mixing module.

Figure 3.2. General view on the indirect urea dosing system: 1 – urea injector, 2 – urea temperature port, 3 – urea pressure port, 4 – tube, 5 – mixing module, 6 – DPF filter, 7 – SCR catalyst.

Figure 3.3. General view on the urea mixing module.

Standard urea dosing is shown in Figure 3.4, where the urea is injected directly into the exhaust gas. Between the injector and the SCR catalyst, a urea mixer is fitted which is made of metal blades with fixed geometry and which helps to break up urea droplets and achieve (as much as possible) even urea distribution on the SCR catalyst face.

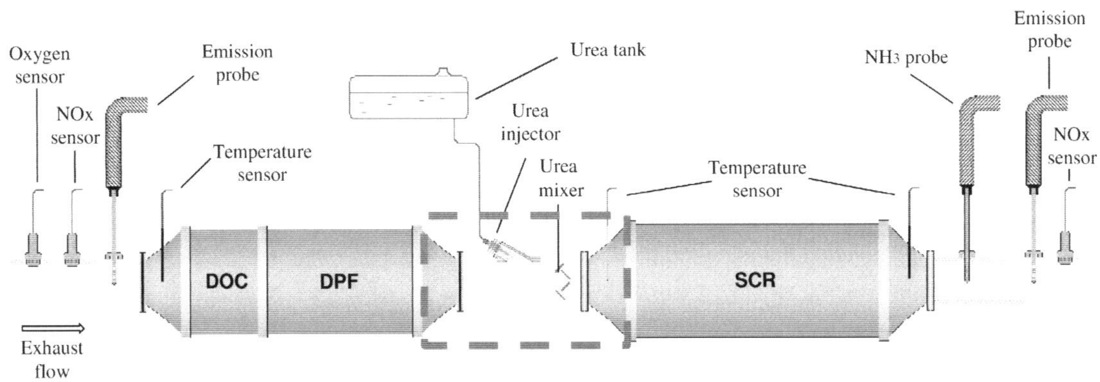

Figure 3.4. SCR system with a standard urea dosing in a liquid phase.

4. Test methodology

The study on the influence of urea dosing method on the SCR performance was carried out under steady-state engine operation. A dedicated test procedure was prepared, covering a wide range of exhaust gas temperatures and the test was run with both standard and concept urea dosing methods. As indicated by the dash line on Figures 3.1 and 3.4 urea dosing systems were changed over accordingly while the rest of exhaust layout remained unchanged. In the first approach standard urea dosing was tested and afterwards the concept dosing was put under the test.

The test itself lasted two hours and consisted of six 20-minute test steps covering exhaust gas temperature in a range from 180°C to 380°C. Test was carried out at fixed engine speed of 1900 rpm. Exhaust gas temperature was adjusted by changing the engine load in a range between 60÷400 Nm and it the gas temperature for each following step was greater by approximately 40°C than in the previous step. The engine was run under EGR valve shut-off conditions - therefore the NOx concentration in the raw exhaust was high and varied from 580 ppm to 1100 ppm. Exhaust mass flow changed in the range between 230-400 kg/h.

Above test procedure allowed to compare NOx conversion efficiency for both urea dosing methods in a function of exhaust gas temperature.

The quantity of urea dosed was expressed by a stoichiometric ratio α calculated as a quotient of the amount of ammonia molecules from the urea (NH_3IN) and the amount of nitrogen oxides molecules (NOxIN) in the elementary exhaust gas mass flow.

$$\alpha = \frac{NH3\ IN}{NOxIN} \tag{4.1}$$

Studies were carried out at three α ratios set to 0.6, 0.8 and 1.0. Exhaust gas emission was measured continuously at two points: at the engine outlet upstream of the DOC and downstream of the SCR catalyst, where the ammonia concentration was also measured. Further measurements were: gas temperature at SCR inlet and outlet, urea flow rate, temperature, pressure and a number of regular engine parameters to monitor and ensure the repeatability of the engine operation over the test time.

The quantity of urea injection corresponding to the α ratio was calculated by the control unit as a function of the actual exhaust mass flow rate and current NOx concentration measured by the portable NOx sensor fitted at the engine outlet. The urea solution was then injected at a pressure of approx. 5 bar by the multi-hole injector depending on the system configuration: directly into the exhaust gas stream (standard) or to the mixing element (concept).

Before each test was started, the entire exhaust line was carefully pre-conditioned in order to clean-up any urea deposits and carbon particles collected inside the DPF filter. It was also necessary to eliminate ammonia saturation of the SCR catalyst that could be left after previous test run and to ensure good

IOP Conf. Series: Materials Science and Engineering **148** (2016) 012062 doi:10.1088/1757-899X/148/1/012062

stability and repeatability of the initial test's conditions. The pre-conditioning was executed at high exhaust gas temperature generated by the engine running under high load conditions with the urea dosing system switched off.

5. Test results

5.1. Test results analysis

Obtained test data were processed and presented in a form of graphs with selected SCR operating parameters. The graphs directly compare the standard and concept methods of urea dosing for the test run. Test results were prepared separately for each α ratio and include the following parameters:

- NOx conversion efficiency calculated by the formula:

$$NOx\ conversion\ [\%] = \frac{NOxDOC_{IN} - NOxSCR_{OUT}}{NOxDOC_{IN}} * 100\% \quad (5.1)$$

- NH_3 concentration measured at SCR outlet.
- NOx concentration at DOC inlet and SCR outlet.
- Exhaust gas temperature and pressure at the inlet and outlet of the SCR.
- Urea solution mass flow.
- Gas urea temperature and pressure.

5.2. Test results under steady-state engine operation

The graphs below present modal analysis of test run under engine steady-state operation. The core parameter for test results evaluation was the NOx conversion efficiency as a function of the urea dosing method. Based on test data, it was stated that the concept urea system was beneficial in terms of NOx reduction over almost the entire range of the test cycle. This statement was further confirmed by the test runs with other α ratios.

The biggest benefits of concept dosing system for NOx reduction were obtained at α ratio equal to 0.8, where the difference in NOx reduction reached 28% at a gas temperature of 240°C.

The value of exhaust gas temperature measured at the SCR outlet was always greater than the one positioned at the SCR inlet. It indicates that the NOx reduction process is an exothermic reaction.

The first test was carried out for α ratio set to 0.6. During the first 130 seconds of the test at the lowest exhaust temperature of 180°C the NOx conversion for concept dosing was significantly lower compared to the standard one, but then it started to rise and had an advantage over the standard dosing for the majority of the test. Lower NOx conversion during the initial phase can be explained by the time delay of urea reaching the exhaust stream, due to the necessity of initial saturation of the mixing module with urea. The biggest difference in NOx conversion values for concept urea dosing was up to 10% within the gas temperature range 202-300°C. Ammonia slip was not detected at this α ratio.

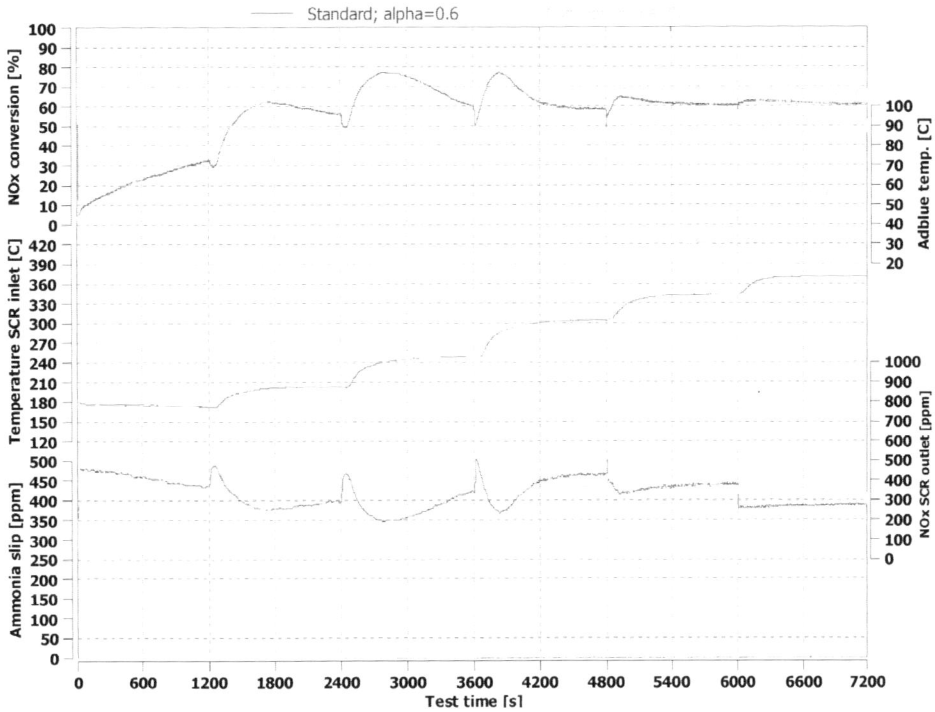

Figure 5.1. Comparison test results of SCR operating parameters for standard and concept urea dosing at $\alpha = 0.6$.

The following test with α ratio equal to 0.8 was most advantageous for concept dosing over the standard solution (Figure 5.2). The NOx conversion efficiency for the concept method was up to 28% greater than for the standard dosing , besides the initial 240s of the test where both conversion traces were quite in line. The ammonia slip was slightly higher for the concept solution.

In the last case of α ratio equal to 1.0 (Figure 5.3) the NOx conversion efficiency was lower by up to 9% for the concept dosing up to the exhaust temperature of 215°C. For the remaining part of the test the NOx conversion was greater by up to 11%. The ammonia slip was significantly higher for the concept solution.

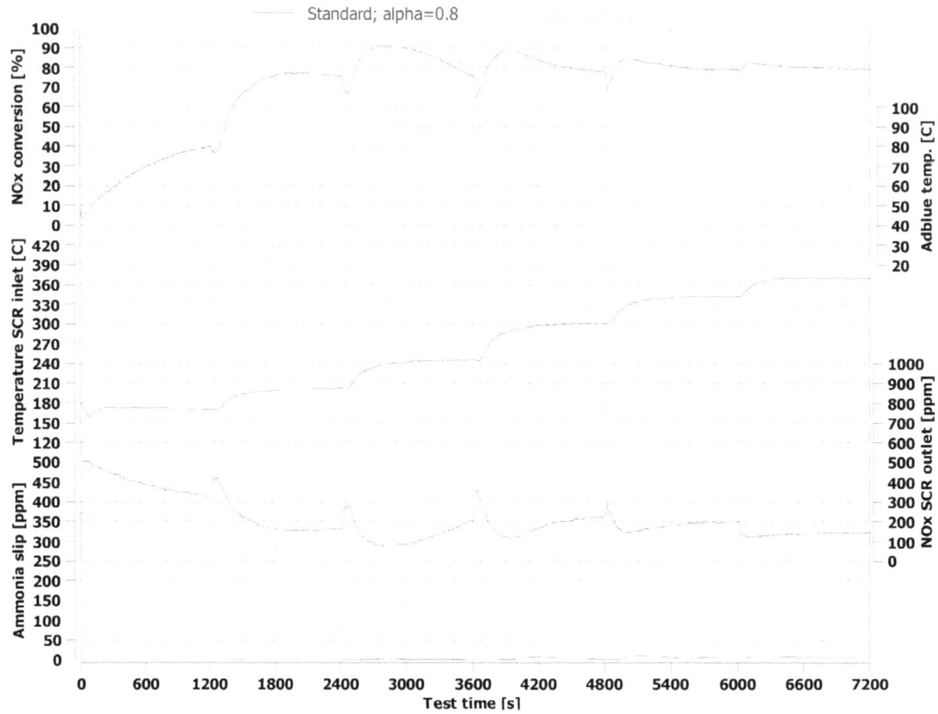

Figure 5.2. Comparison test results of SCR operating parameters for standard and concept urea dosing at α = 0.8.

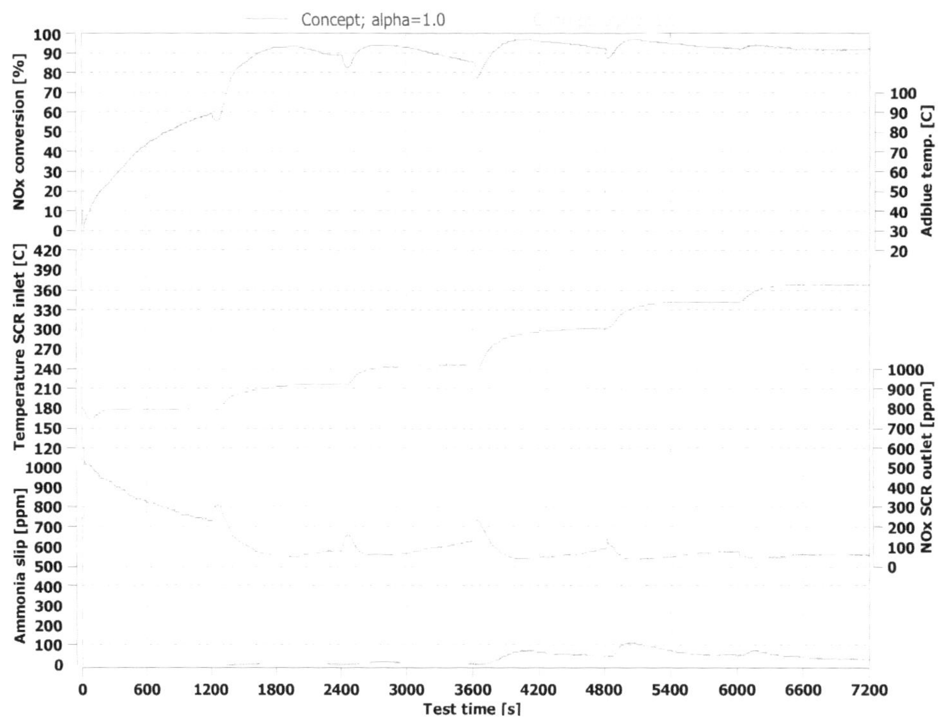

Figure 5.3. Comparison test results of SCR operating parameters for standard and concept urea dosing at α = 1.0.

Considering the concept dosing method it was noticed, that for all tested α ratios at initial phase of the test the NOx conversion rate was lower than for the standard solution. This fact can be explained by the time delay of urea reaching the exhaust.
Following the initial phase, the timing of which varied according to the α ratio, the NOx conversion for concept system was greater and the difference in NOx reduction was most evident in the middle and upper range of the exhaust gas temperature what corresponded to the test steps number 3-5.
Further improvement of concept solution should concentrate on optimization of the shape and volume of the mixing element in order to decrease the urea delay and also the exhaust back pressure.
Ammonia emission downstream of the SCR catalyst (NH_3 slip) is a critical parameter defining the operation properties of SCR system. The slip itself is difficult to control and often takes a form of a sudden release of volatile ammonia stored inside the SCR catalyst. The ability of ammonia storage within the SCR catalyst decreases with increasing temperature and vice versa – the lower the temperature, the greater the amount of ammonia that can be stored inside the catalyst. In the case of an increase in gas temperature, the ability to store ammonia is reduced, causing spontaneous ammonia release, even if the urea dosing was stopped. The ammonia slip may also occur if the urea is overdosed or if the NH_3 distribution across SCR face is uneven.
For the concept solution at dosing ratios of α=0.8 and α=1.0, ammonia concentration was higher than for the standard solution. The reason may be an uneven ammonia distribution on the SCR inlet face or possibly greater urea to ammonia conversion efficiency, which was essentially the gist of the concept solution.

6. Discussion

Test results obtained give a certain overview on a concept system performance under steady state operation. The concept has proven to be more efficient in NOx reduction for a wide gas temperature range, but also can be disadvantageous at low gas temperature. The best combination was seen for the test at α=0.8, in which the concept solution presents a significant increase in NOx reduction.
The idea of an indirect urea dosing aimed for better NOx reductant preparation before introducing it into the exhaust. Preliminary evaporation of urea by means of mixing element supposed to enhance the rate of urea decomposition into ammonia, what is currently the subject of much research.
Investigation on presented concept revealed that there are still many open questions which need to be resolved. Some critical conclusions can be pointed out, as follows:

- For the concept dosing, the NOx conversion efficiency during low gas temperature was insufficient comparing to the standard dosing. This issue needs to be further investigated and improved.
- Secondly, the method of gas urea introduction into the gas stream via the mixing unit needs to be optimized to not create an excessive increase in the exhaust backpressure.
- Finally, it would be worth carrying out a study on the properties of the material the module is made of, as they might influence the pace of thermal urea decomposition. Selection of a material may be advantageous for enhancing urea to ammonia conversion before the reductant reaches the SCR catalyst.

Further research work requires an investigation on ammonia slip (in order to decrease it). Ammonia concentration measurements on the SCR face should be carried out and optimized to achieve (as much as possible) an even NH_3 distribution.
The final point is to verify the concept for deposit formation during regular operation.

7. Summary

The exhaust after-treatment with selective catalytic reduction system is a complex but very effective way of NOx removal from combustion gases. The SCR method still requires development especially when it is operated under engine warm-up and idle conditions. Urea injection at low exhaust gas temperature is critical due to the limited ability to evaporate and chemical decomposition in the exhaust and further in the catalyst. An important in-service issue is urea deposit formation inside the SCR system, which leads to a deterioration in system functionality. Deposit formation is reversible and the solid fraction gets removed at high gas temperature.

The purpose of this study was to evaluate the effect of urea preparation on NOx conversion in an SCR catalyst. The preparation process relies on indirect urea dosing via a mixing module in order to help it to evaporate prior to introduction upstream of the SCR catalyst. An expected effect was to accelerate thermal decomposition of urea and ultimately to increase NOx conversion efficiency and possibility reduce urea consumption.

The study was conducted on a diesel engine and after-treatment with SCR system in which the urea dosing module was changeable between the standard and concept solutions. This allowed the influence of dosing method on SCR performance to be verified.

A custom made concept system was prepared and put under test. As it was a very simple solution it needed development works, but it was able to give an insight into achievable benefits.

It has been found that concept dosing method can increase NOx conversion in the SCR system compared to standard liquid urea dosing. The gain was noticed especially in the gas temperature range from 220°C to 350°C under steady-state engine operation.

The studies carried out demonstrated a certain potential for the concept and the results obtained are a great incentive for further research on this subject.

References

[1] Brzeżański M and Sala R 2013 *In-service problems of selective catalytic reduction systems for reduction of nitrogen oxides.* Combustion Engines 2013, 154(3), 969-976 ISSN 0138-0346

[2] Brzeżański M and Sala R 2013 *The influence of AdBlue dosage on the process of selective catalytic reduction of nitrogen oxides.* Combustion Engines 2013, 154(3), 1032-1037 ISSN 0138-0346

[3] Bielaczyc P 2012 *Automotive Emissions Legislation/Test Method Development over the Last 40 Years and Future Predictions for the Next 40 Years.* 3rd International Exhaust Emissions Symposium, Bielsko-Biała Poland

[4] Madia G 2002 *Measures to enhance the NOx conversion in urea-SCR systems for automotive applications.* Diss. ETH No 14595, Swiss Federal Institute of Technology, Zurich

[5] Majewski W A 2014 *Urea Dosing and Injection Systems,* www.DieselNet.com [access in 10.2014].

[6] Majewski W A 2014 *Selective Catalytic Reduction,* www.DieselNet.com [access in 10.2014].

IOP Conf. Series: Materials Science and Engineering **148** (2016) 012063 doi:10.1088/1757-899X/148/1/012063

Micro turbine engines for drones propulsion

J Dutczak[1]

[1]Institute of Automobiles and Internal Combustion Engines, Cracow University of Technology, 37 Jana Pawla II Avenue, 31-864 Krakow, Poland

E-mail: jdutczak@pk.edu.pl

Abstract. Development of micro turbine engines began from attempts of application of that propulsion source by group of enthusiasts of aviation model making. Nowadays, the domain of micro turbojet engines is treated on a par with "full size" aviation constructions. The dynamic development of these engines is caused not only by aviation modellers, but also by use of micro turbojet engines by army to propulsion of contemporary drones, i.e. Unmanned Aerial Vehicles (UAV) or Unmanned Aerial Systems (UAS). On the base of selected examples the state of art in the mentioned group of engines has been presented in the article.

1. Introduction

Construction of contemporary micro-turbine engines (MTEs) as well turboshaft as turbojet ones is the result of about twenty years of development of that idea. Interesting is that the first designs were done by a group of enthusiasts of aviation model making. There was the need of finding of real engines, but made in micro scale, for propulsion of turbojet or turboprop aircraft models. The development of that area has been the object of particular interest of army. The idea of micro turbine engines application to the propulsion of small and medium size "professional" military Unmanned Aerial Vehicles (UAVs) is at present highly developed. Some of modeller engines have been used for special purposes in the technology demonstrator UAVs like for example X-56A built by Lockheed Martin and NASA [1].

On the base of market demand many specialized firm manufacturing different kinds of micro turbine engines have been established in the world. There are several concepts of construction of mentioned engines depending on the custom demands and available know-how. The selected examples of contemporary micro turbine engines, illustrating contemporary state of art within this area are presented below.

2. Types of Micro Turbine Engines (MTEs)

It is possible to distinguish several types of MTEs taking into account their construction and destination:

- turbojet MTE applied for propulsion by use of a thrust exclusively. There are two kinds of MTEs in this case:
  - with radial compressor and radial turbine,
  - with axial compressor and axial turbine (generally still experimental ones).
- turboshaft MTE applied for propulsion by use of the mechanical torque (power) delivered to the engine shaft (spool). In this case one can distinguish two kinds of construction:
  - single spool (single shaft) engines with gear box used for propulsion of helicopter drones,
  - two spool (dual shaft) engines for turboprop and helicopter aircraft drones.

The last ones are rarely used, but seem to be very "stylish" solution being the functional copy of full size engines. Cross section of a typical micro turbojet engine with radial compressor and axial turbine is presented in the figure 1. That is the example of construction of reverse flow combustion chamber. The compressor rotor is adapted from a serial produced car turbocharger. It is the often used solution, particularly by the amateur constructors, in the aim of manufacturing costs reduction.

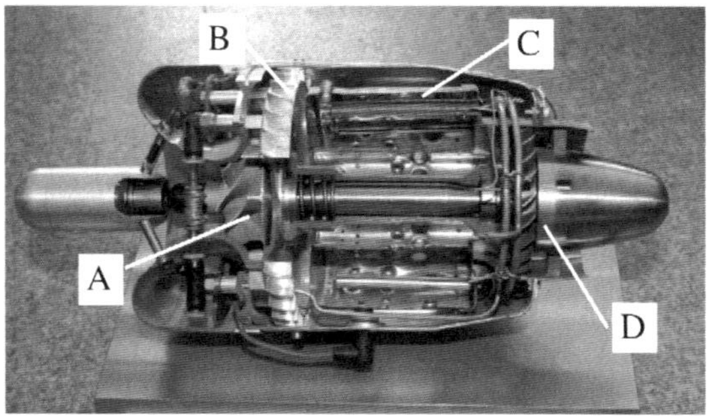

Figure 1. Cross section of a typical micro turbojet engine (MTE):
A - radial compressor rotor, B - diffuser,
C - flame tube, D - axial turbine [2].

The example of new, more advanced construction of turbojet MTE with single shaft 5-stage axial compressor and axial turbine is shown in the figure 2. Such two engines manufactured by Bladon firm, equipped with integral electric generators were applied to the battery charging of experimental Jaguar C-X75 hybrid car (2010) [3, 27].

Figure 2. Turbojet MTE with 5-stage axial compressor and axial turbine made by Bladon firm [4].

View of the similar engine shaft with 6-stage axial compressor and axial turbine is presented in the figure 3. The coin visible at the shaft gives outlook on the total scale of the engine and required precision of making. Multistage axial compressor MTEs are contemporary still experimental, however it seems this kind of engines will be dominating in the micro turbojet applications in next several years.

Figure 3. Example of the turbojet MTE single shaft with 6-stage axial compressor and axial turbine [5].

The examples of single spool (single shaft) micro turboshaft engines (MTEs) applied for propulsion of helicopter drones are presented in the figures 4 and 5.

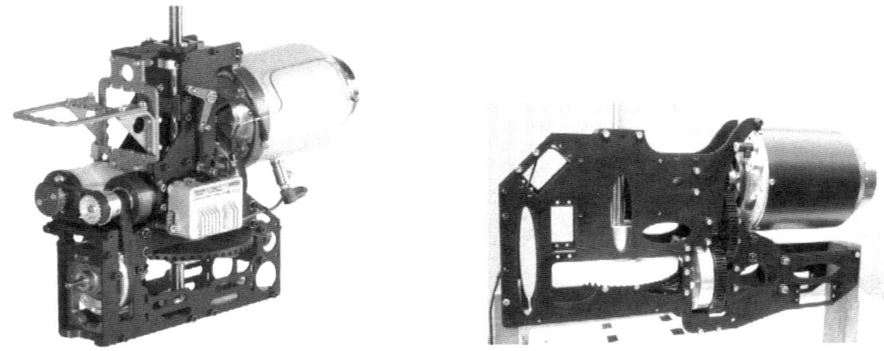

Figure 4. Single shaft MTE engine with gear box used for propulsion of helicopter drones: two different arrangements [6, 7].

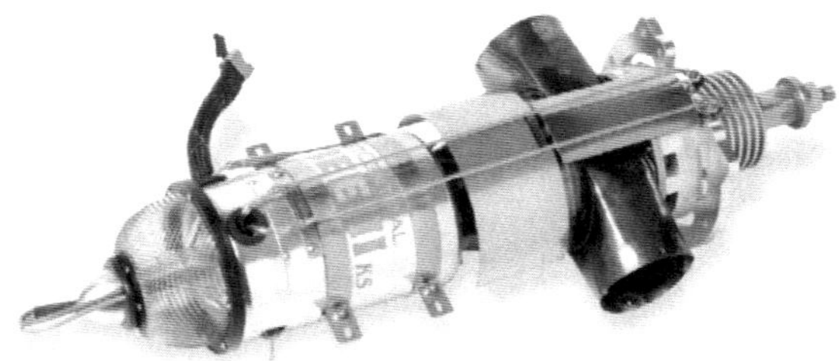

Figure 5. Single shaft engine with gear box used for propulsion of turboprop drones [8].

The dual shaft MTEs look similar to presented above single shaft arrangement, but have the free turbine (second shaft) for power delivery to the external system (helicopter gear box or propeller). The examples of dual shaft engines are presented in the figures 6 and 7.

Figure 6. Taurus dual shaft MTE engine with gear box used for propulsion of helicopter drones [9].

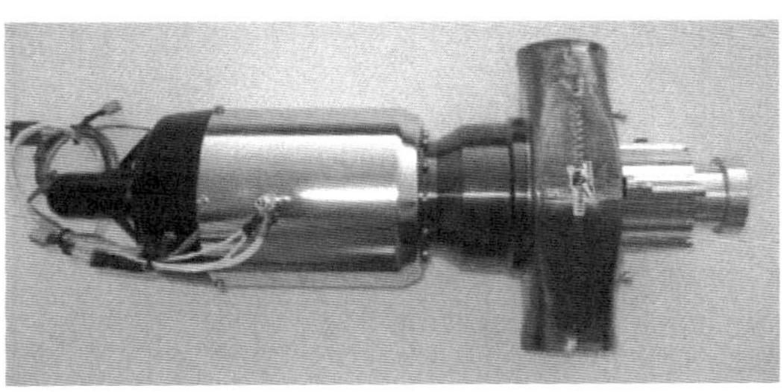

Figure 7. Taurus dual shaft MTE engine used for propulsion of turboprop drones [10].

3. Typical specifications of MTEs

3.1. Thrust range

There is no official classification for the micro turbojet engines. These engines are used mainly by modellers for propulsion of remotely piloted fixed wing aircraft and helicopters. Commonly used classification is the range of thrust generated by MTEs. Thrust values of such engines are included generally within the range of 50–250 N. The example of characteristic of thrust versus shaft rotational speed of micro turbojet engine is presented in the figure 8. The more thrust generating turbojet engines belong to the normal "adult" constructions, however sometimes for aircraft scaled models or flying mocks application of more powerful engines may be necessary.

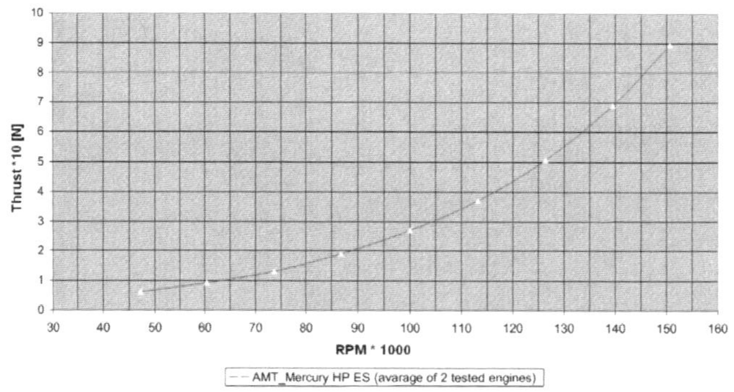

Figure 8. Thrust versus shaft rotational speed of AMT Mercury HP ES micro turbojet engine [11].

3.2. Shaft (spool) rotational speed range

The range of shaft rotational speed depends on the dimensions of an engine and required thrust value. The typical shaft RPM values are from about 35000 to 180000. The smallest MTEs have the idle speed of about 50000–95000 RPM. The biggest ones have idle speed of about 35000. The RPM of the maximum thrust (nominal RPM) are situated within the range from about 120000 to 240000 and are obtained at the maximum permissible engine speed [6, 7, 8, 9, 10, 11, 12, 16, 18, 24, 25, 26]. The ceramic shaft bearings are contemporary commonly used making possible relatively long maintenance interval up to even 50 hours. The earlier inspection is necessary in case of bearing noise or other reasons.

3.3. Compressor type

There is commonly used the single stage radial type of compressor (figure 9), as it was mentioned above. There are at least two reasons of such a concept: the simplicity of construction and possibility of adaptation, particularly in amateur designs, a standard compressor rotor taken from a car turbocharger. The professional MTEs elements are manufactured with use of advanced CAD and CNC machine tools including the contemporary tries of 3D printing application. The striving for obtaining more thrust and parallel a small diameter of an engine will be probably the impulse to wider application of multistage axial compressors. The previous technological problems of precise making of engine compressor and turbine elements are nowadays solved, so one can assume that it is only a problem of close future market demand.

Figure 9. Hawk 100R micro turbojet engine without housing; visible compressor diffuser [12].

3.4. Pressure ratio

The single stage radial compressor is sufficient to obtain the value of pressure ratio appropriate for proper work of MTE combustion chamber. The average values of pressure ratio obtained in MTEs are from about 1.2 to 4.6 depending on the engine size and power demand. The highest value of pressure ratio while maintaining a small diameter of engine will require application of multistage axial compressor.

3.5. Combustion chamber type

The reverse flow combustion chamber with flame tube made usually of Inconel alloy is the common used solution. That construction enables providing the adequate conditions for proper preparing and combustion of the air-fuel mixture. Besides the reverse flow combustion chamber makes the whole engine shorter, more compact and as the result lighter. As low as possible weight of a turbojet engine is one of the essential parameters taken into account in aviation.

3.6. Turbine type

Two types of turbines are used in MTE: radial and axial ones. The radial type is a standard at present but one can forecast the axial turbine as a close future solution. Mentioned above development of the machining techniques will enable the free choice of combination of compressor and turbine types applied in a given case.

3.7. Exhaust Gas Temperature (EGT)

The maximal exhaust gas temperature of MTEs is limited by the thermal load of turbine rotor. The threshold is usually of about 830°C. Because of need of keeping of the safety distance from that value, the real exhaust gas temperature does not exceed 800°C. It is most often the value of about 700–760°C (figure 10).

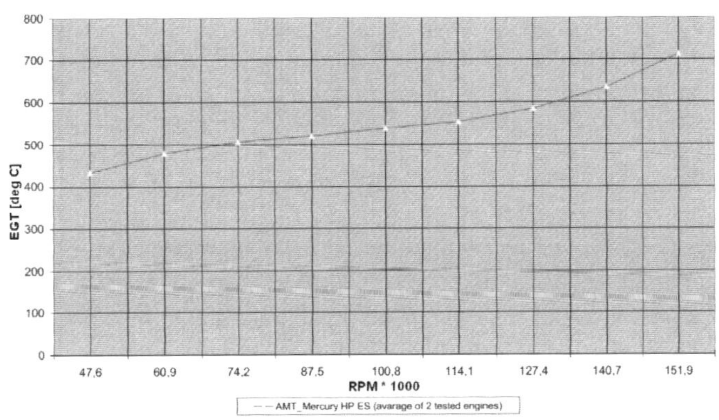

Figure 10. Exhaust gas Temperature (EGT) versus shaft rotational speed of AMT Mercury HP ES micro turbojet engine [11].

3.8. Fuel type

Micro turbojet engines are fed generally with kerosene-based fuel, i.e.: premium paraffin, kerosene, Jet-A1 or Diesel fuel. This is the reason why this kind of propulsion source is very interesting particularly for military, meeting the requirement of "single fuel in battlefield" concept. In the very begin of MTEs development, the LPG fuel was applied as the main fuel. At present LPG is used optionally as the start gas to support the engine start procedure.

3.9. Fuel consumption

Fuel consumption per minute of MTEs is relatively high when compared to the fuel consumption of piston engines. However taking into consideration the different way of power reception (mainly thrust) and the engine power to mass ratio, the values of total fuel consumption (TFC) are acceptable. The TFC on idle run is usually of about 30–65 g/min and for the maximum thrust about 260–520 g/min, depending on the engine size and construction. The example of load characteristic of the Hawk 100R micro turbojet engine is presented in the figure 11 and for comparison the load characteristic of the second end of the size scale – Hawk 260R engine is presented in the figure 12.

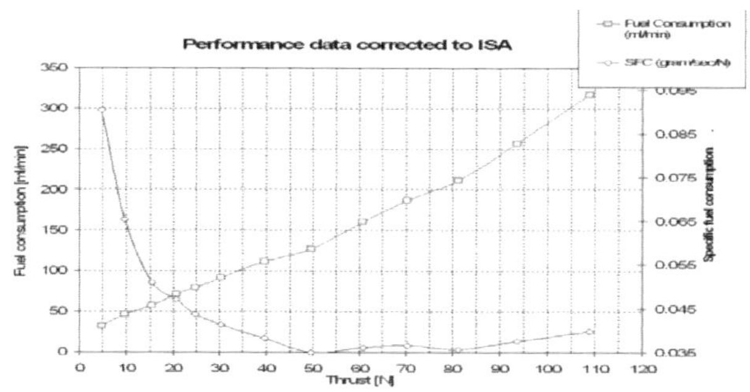

Figure 11. Load characteristic of Hawk 100R micro turbojet engine [13].

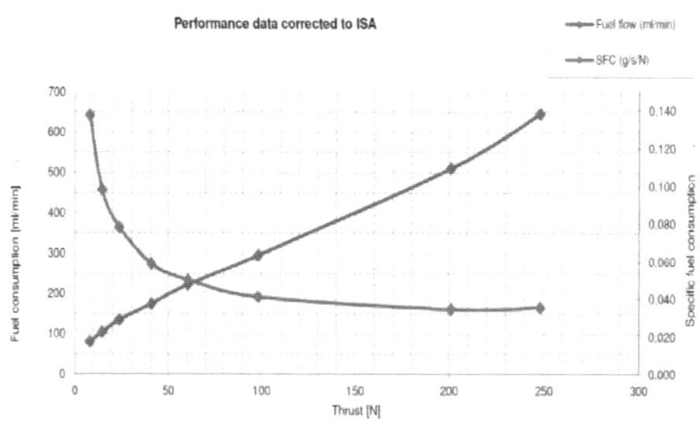

Figure 12. Load characteristic of Hawk 260R micro turbojet engine [14].

3.10. Lubrication

The bearings of micro turbojet engines are lubricated with the fuel-oil mixture. It is the open lubricating system but in the case of necessity of the engine weight diminishing, it is the only reasonable solution. Most frequently is used a two-stroke motorcycle or outboard oil but some manufacturers prefer a turbine oil.

The oil-fuel ratio depends on individual construction of a given engine and a kind of fuel applied (kerosene or diesel) and may amount to 1.5–5.0%. Generally the higher amount of lubricating oil is added to a kerosene-based lighter fuel.

3.11. Start system

There are three starting systems being contemporary in use: the air, propane-butane (Gas) and kerosene (Kero) system.

The air-start system is realized from outside of the MTE engine with use of the air stream directed into the engine air inlet on the compressor rotor (figure 13). The start procedure requires application of an external compressor but makes the engine lighter without a starter motor.

The autostart system (Gas or Kero) requires use of the integral electric starter motor (figure 14).

The propane-butane assisted starting was until recently the most widespread method. Propane-butane mixture (liquefied) has the good evaporating properties in a low ambient temperature making the easy flammable air-fuel mixture. This is the reason why gas fuel is used to support the start of MTE engines (figure 15). After reaching the temperature about 100°C auto-start system changes the fuelling to kerosene what is the basic fuel.

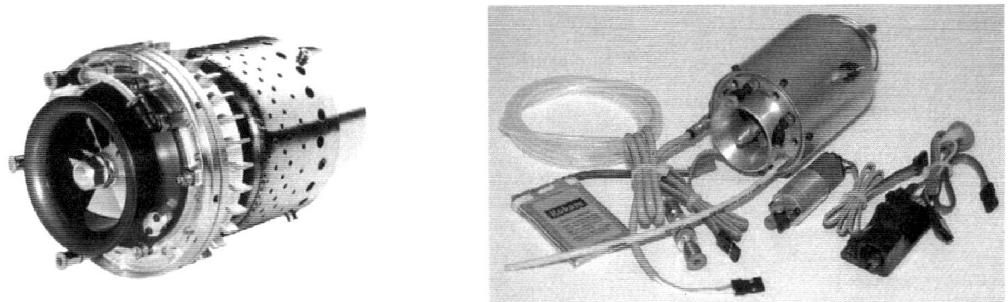

Figure 13. Air-start system: left – AMT Pegasus HP (view without engine housing) [15]; right – Lambert T15 Kolibri [16].

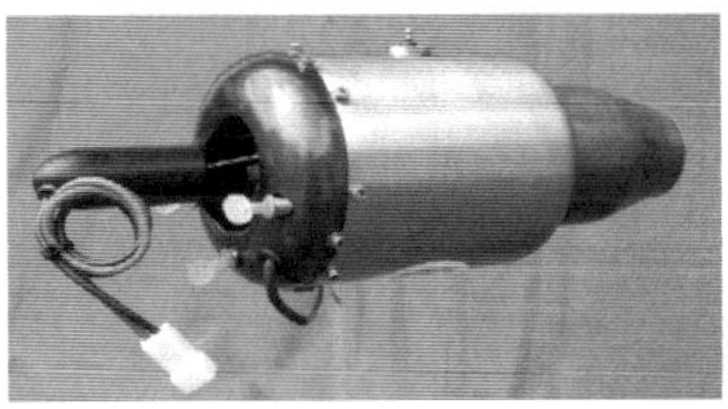

Figure 14. AMT Mercury HP electric start system (propane-butane or kerosene) [11].

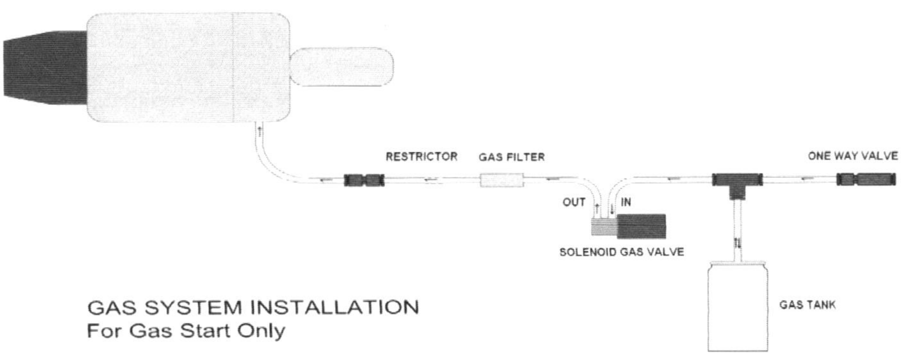

Figure 15. Gas (propane-butane) start system [17].

Kerosene auto-start system presented in the figure 16 is contemporary available due to the electronic control system (FADEC) what is the integral part of turbojet propulsion system. However the special care and attention must be paid, including a potential application of the CO_2 fire extinguisher.

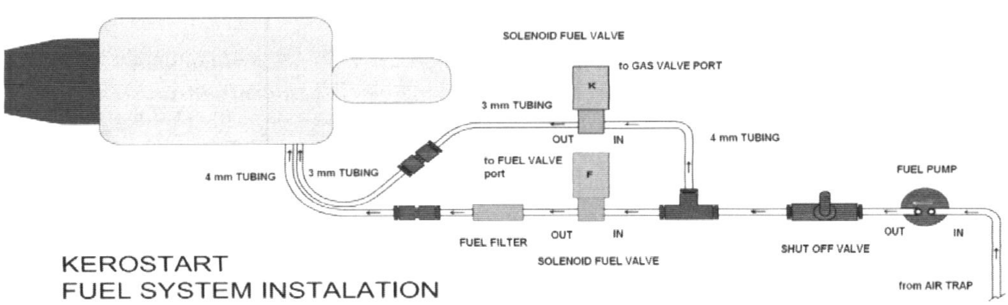

Figure 16. Kerosene auto-start system [17].

3.12. Engine control

Contemporary MTEs are equipped with electronic control system enabling start and then proper run of engines. This control unit is called FADEC (Full Authority Digital Engine Control) in the same way as in "full size" aviation. The examples of the external view of FADEC electronic control units are presented in the figure 17. The scheme of fuel supply system controlled by FADEC ECU is presented in the figure 18.

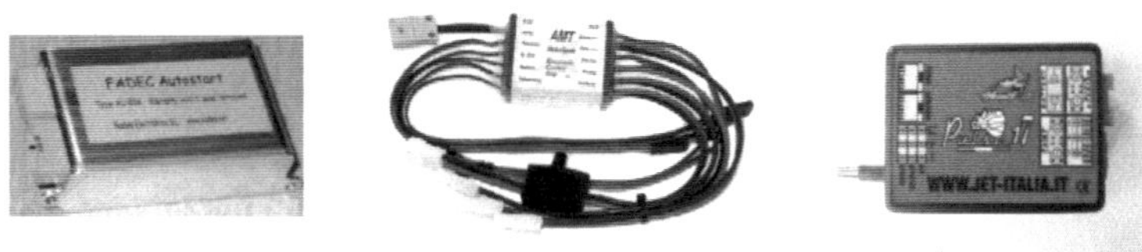

Figure 17. Examples of FADEC electronic control units [12, 11, 18].

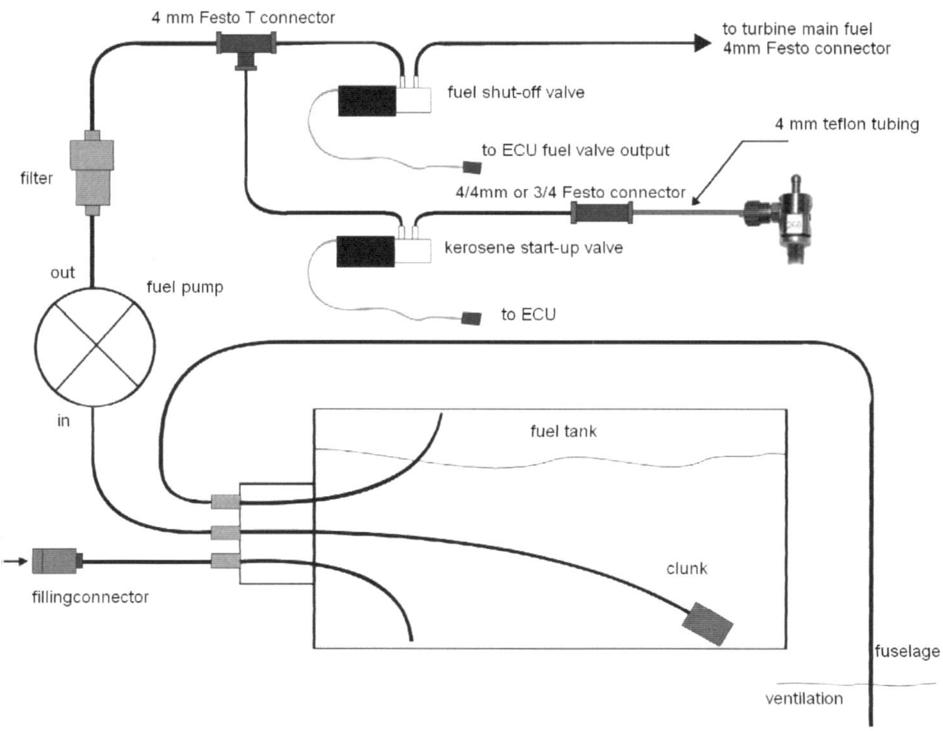

Figure 18. Scheme of fuel supply system controlled by FADEC ECU [6].

3.13. Performance (transient performance)

Transient performance is expressed by so called "spool time". It is the time of engine thrust response to the rapid "throttle" opening, measured from idle speed to maximum power (thrust) speed. As the example one can present Hawk 100R turbojet engine. For the factory settings the Hawk 100R spool time is about 4 s from idle to max power. It is however possible to apply custom regulation data and diminish the spool time to 2.5 s (figure 19). This effects in the spool time from half throttle (approximately 50 N of thrust to maximum thrust) in 0.5 s. This is a very valuable result enabling rapid manoeuvres of drone driven by such engine.

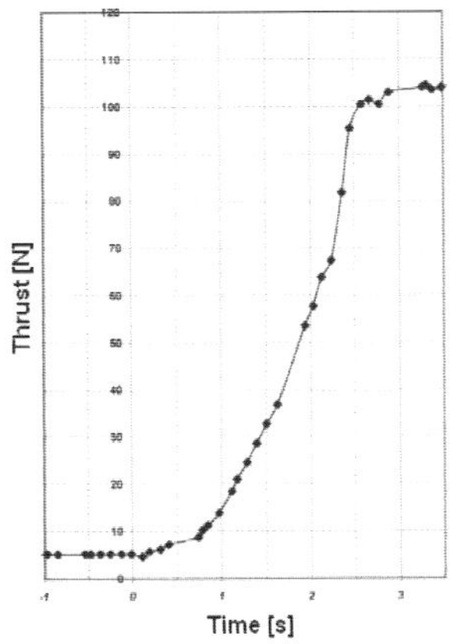

Figure 19. Hawk 100R turbojet engine transient performance – “spool time” [19].

3.14. Measures and weight

The development of technology enables to design small turbojet engines with high enough performance and durability. Particularly interesting are the smallest constructions because of the need of high precision of manufacturing.

The overall dimensions of micro turbojet engines are in the range approximately: housing diameter 90 – 120 mm, length 200 – 350 (450) mm (single-spool turboprop versions with gear box or two-spool turboprop).

The total weight with fuel supply, start and EC systems is in the range of about 1000 – 3200 g.

4. Examples of Micro Turbine Engines (MTE) specifications

Table 1. Technical specifications of chosen MTE engines [8, 20, 21, 22, 23, 28].

	Kolibri T15	Hawk 240R	Hawk 100R	Bee II Kero Start	Bee II Turbo Prop KS	Taurus 10
Idle thrust [N] / RPM	ca. 2 / 95000	8 / 39000	6 / 60000	3.6 / 55000	–	–
Maximum Thrust [N] / RPM	15 / 243000	240 / 132000	110 / 175000	70 / 185000	–	–
Maximum Power [kW] / RPM	–	–	–	–	9 / 180000	10 / 123000
Pressure ratio	2.2	1.2 – 4.6	1.2 – 3.8	–	–	–
RPM	95000 – 243000	39000 – 132000	60000 – 175000	55000 – 185000	55000 – 180000	33000 – 123000
Prop shaft RPM range	–	–	–	–	6000 – 8000	9000 (max)
Exhaust Gas Temperature [°C]	max 680	400 – 760 (800)	390 – 760	500 – 700	500 – 700	550 – 750
Fuel consumption [g/min]	ca. 50 (max)	65 – 515	30 – 260	180 (max)	135 (average)	135 – 230
Fuel type	Kerosene JetA1 Petroleum	Kerosene Jet A1 Diesel (EN590)	Premium paraffin Kerosene Jet-A1 Diesel (EN590)	Kerosene Jet A1 Diesel (EN590)	Kerosene Jet A1 Diesel (EN590)	Jet A1
Spool time (idle to max. thrust) [s]	ca. 2.5	4.5	4	3	4.2	ca. 4
Lubrication (oil-fuel mixture)	5 [%] Aeroshell 500	1.5 [%] Two stroke outboard	1.5 [%] Two stroke outboard	2.0 – 2.5 [%] Two stroke synth. / 2.5 – 5.0 [%] Turbine synthetic	2.0 – 2.5 [%] Two stroke synth. / 2.5 – 5.0 [%] Turbine synthetic	5 [%] Turbine synthetic
Engine control	FADEC	FADEC	FADEC	FADEC	FADEC	FADEC
Start system	Propane-butane External Air	Propane-butane Full auto start	Propane-butane Full auto start	Kerosene Full auto start	Kerosene Full auto start	Kerosene Full auto start
Compressor	–	Radial	Radial	Radial	Radial	–
Combustion chamber type	–	Reverse flow	Reverse flow	Annular	Annular	–
Turbine type	–	Radial	Radial	Axial	Axial	–
Number of turbines	1	1	1	1	2	2

Engine diameter [mm]	55	108	108	82	86	109
Engine length [mm]	125	346	346	232	430	450
Engine weight [g]	200	1650	1650	880	2300	3200

5. Conclusions

Micro turbine (turbojet) engines (MTE) are the relatively new propulsion sources applied mainly to aircraft drones. However the level of technical advancement of these engines is contemporary very high and is worthy of special attention. The area of MTE engines has been until now "reserved" for amateur modellers but one can observe the more and more interest in such propulsion taken by professionals (the military or aviation R&D centres). The small measures and high value of thrust or power make the MTE engines the attractive alternative for drones' propulsion. Miniaturisation of turbojet engines is possible thanks to contemporary CAD and machining systems including the advanced 3D printing. The lack of reliable technical details of micro turbojet engines is understandable because of the know-how secret particularly in the strong market competition. The newest constructions of MTE engines dedicated to turboprops or helicopters are based on the aviation designs with free turbine. These engines are the most advanced and seems to be the most "elegant" source of propulsion possessing all the best features of speed-torque characteristic of two-spool engines.

References

[1] Dutczak J 2014 *Combustion engines for propulsion of contemporary drones*, (Kraków: Silniki Spalinowe i Ekologia, Wydawnictwo Politechniki Krakowskiej) pp 63-82
[2] http://pfmrc.eu/uploads/monthly_09_2012/post-697-0-22752100-1346612586.jpg
[3] http://www.jaguar.com/about-jaguar/concept-cars/cx75.html
[4] https://www.gov.uk/government/case-studies/bladon-jets-innovative-mobile-and-low-cost-generator-takes-off
[5] http://www.gtba.co.uk/pgallery/index.php?spgmGal=Axial&spgmPic=2#spgmPicture
[6] www.cat-ing.de/turbines, Bedienungsanlaitung JetCat PHT2 ECU 6.0 2016
[7] http://www.jet-italia.it
[8] http://www.jetcentral.com.mx/products/turbo-prop-ks/
[9] http://www.pahlhelicopter.com/index_htm_files/470.png
[10] http://www.pahlhelicopter.com/index_htm_files/472.png
[11] www.amtjets.com/, Mercury HP gas turbine specifications, January 2008
[12] http://www.hawkturbine.com, Hawk Turbine RCJI 2006/2007
[13] http://www.hawkturbine.com/Fuelfigures.htm
[14] http://www.hawkturbine.com/FuelFiguresxxx.htm
[15] http://4.bp.blogspot.com/-N_U95pECvYg/T_1VkXRMz8I/AAAAAAAAAKU/T_Hu8OrhROg/s320/ac.jpg
[16] http://www.lambert-modellturbinen.de/assets/images/Kolibri2_s.jpg
[17] Jet Central, Bee II Operation and Maintenance Manual, February 2012
[18] Jet-Italia, Speed13-Pulse17-Kaiman19-Lion22-Atom25, Istruzioni e Garanzia, Rev1 2015
[19] http://www.hawkturbine.com/Performance.htm
[20] http://www.hawkturbine.com/Technicalspec.htm
[21] http://www.hawkturbine.com/Technicalspecxxx.htm
[22] http://www.jetcentral.com.mx/products/bee-ii-ks/
[23] http://www.pahlhelicopter.com/index_htm_files/Taurus%20Daten_2012
[24] http://www.frankturbine.de
[25] www.jets-munt.com/
[26] www.bf-turbines.de
[27] http://www.caranddriver.com/news/jaguar-c-x75-concept-auto-shows
[28] http://www.modellbau-peters.de/kolibri-turbine-t15-ohne-autostart.html#product_tabs_additional_tabbed

The use of gaseous fuels mixtures for SI engines propulsion

M Flekiewicz[1] and G Kubica[1]

[1]Silesian University of Technology, Faculty of Transport, Katowice, PL

E-mail: marek.flekiewicz@polsl.pl, grzegorz.kubica@polsl.pl

Abstract. Paper presents results of SI engine tests, carried on for different gaseous fuels. Carried out analysis made it possible to define correlation between fuel composition and engine operating parameters. Tests covered various gaseous mixtures: of methane and hydrogen and LPG with DME featuring different shares. The first group, considered as low carbon content fuels can be characterized by low CO_2 emissions. Flammability of hydrogen added in those mixtures realizes the function of combustion process activator. That is why hydrogen addition improves the energy conversion by about 3%.

The second group of fuels is constituted by LPG and DME mixtures. DME mixes perfectly with LPG, and differently than in case of other hydrocarbon fuels consists also of oxygen makes the stoichiometric mixture less oxygen demanding. In case of this fuel an improvement in engine volumetric and overall engine efficiency has been noticed, when compared to LPG. For the 11% DME share in the mixture an improvement of 2% in the efficiency has been noticed.

During the tests standard CNG/LPG feeding systems have been used, what underlines utility value of the research. The stand tests results have been followed by combustion process simulation including exhaust forming and charge exchange.

1. Introduction

The use of alternative fuels is one of the main solutions allowing the reduction of pollutants emissions nowadays. Those fuels allow also the increase of energy conversion efficiency in the entire life cycle as well as during the chemical energy conversion to mechanical or electrical. Among gaseous alternative fuels the most significant substitutes to conventional fuels are natural gas, LPG and DME in their quite popular applications in both vehicles and stationary applications. Carried on research on those fuels proved that it is possible to replace with them conventional fuels allowing it also to differentiate energy sources. Gaseous fuels have been normalized by means of their chemical composition and physicochemical properties, while modern propulsion and engine control systems became precise enough allowing it to achieve satisfactory ecological and economical results [1,2,3,5,6,7]. The number of dual- fuel and monofuel gas powered vehicles is therefore increasing. Considering the use of already highly accurate and dedicated sequential injection systems, the effective use of gaseous fuels energy still becomes a challenge. The problems with effective combustion process control are mainly related to high ignition temperature, low flame front speed especially in the case of natural gas [8,9,10].

One of the ways of solving this problem is increasing the flame front propagation speed, mainly by adding more reactive fuels like hydrogen or DME. The influence of the addition of those fuels to natural gas and LPG on the combustion process have been presented in many publications.

Fuel mixtures of NG and hydrogen have been widely studied in IC engines [6-11] and their results show that the engine performance can improve and the exhaust emissions can be reduced by adding a small amount of hydrogen.

2. Fuels properties and measurement set-up

Selected chemical and physical properties, of gaseous fuels have been presented on figure 1.
Fourteen of different fuels were used: gasoline, methane, LPG and fuel blends consisting of:
- methane with Hydrogen addition from 5%, to 50%, by volume,
- LPG with DME addition from 0% to 28% DME by mass, respectively.

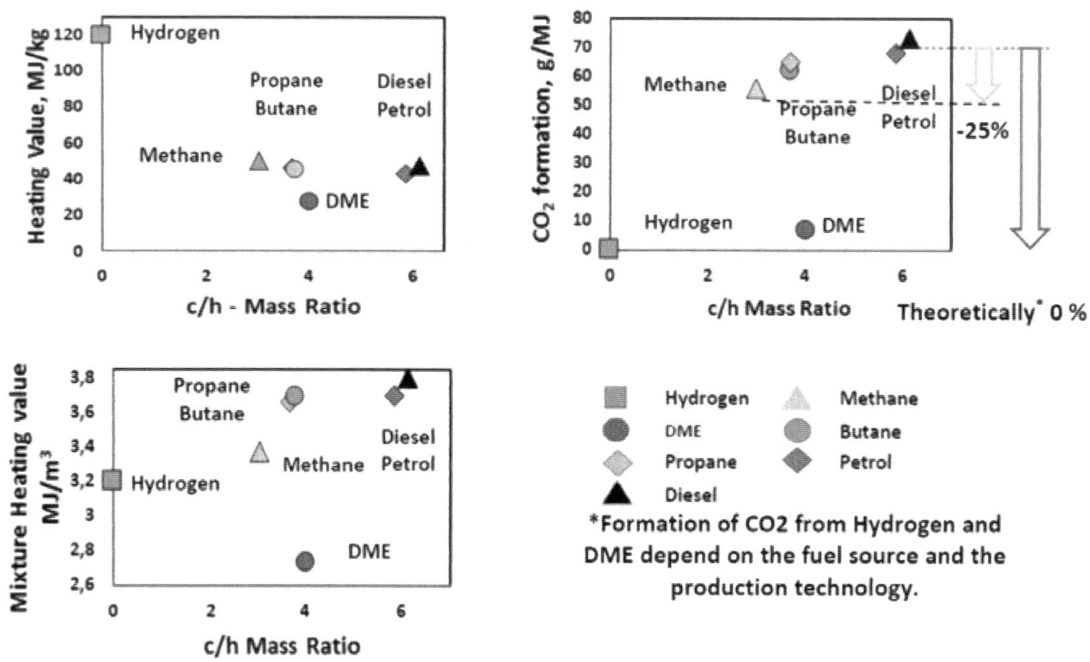

Figure 1. Main properties of selected gaseous fuels as a function of c/h ratio (at p = 100 kPa, T = 283 K, λ = 1).

The popular passenger car powered by 1.6 liter engine, naturally aspirated with a compression ratio of 9.6, port fuel injection, two valves per cylinder, flat pistons and without external EGR was used in the experiments. The experiments were performed on a BOSCH FLA 203 chassis dynamometer. Main features characterizing the engine installed on the tested vehicle have been listed in the Table 2. Engine performance has been estimated on the basis of acquired dynamic characteristics, defining the power on wheels in function of vehicle speed. Test stand has been equipped with various transducers and sensors allowing the identification of engine operating conditions. Basic measurements and control systems allowed continuous acquisition of engine operating conditions, through registrations of:

- in-cylinder pressures, crank angle, with the TDC identification,
- power on wheels, manifold pressure, inlet air temperature,
- exhaust gases temperature,
- fuel mass flow to the engine.

The in-cylinder pressure was measured by Kistler 6121 piezoelectric pressure transducers and a charge amplifier, Kistler 5011A. The signals were processed in type NI PCI-6143 board in a computer for online pressure measurements. The pressure recording system was also connected to the Kistler 2613B crank angle encoder giving the temporal resolution of the pressure recordings of 0.5 CA. The pressure measurements were recorded and stored on a computer, with recordings performed for 300 subsequent cycles in each test, and were further processed with the help of a script debugged in LabView 7.1 environment.

Table 1. The composition of mixtures prepared for tests.

Fuel	**Symbol used on figures**	**Molecular mass of fuel [kg/kmol]**	**Stoichiometric air fuel ratio A/F [kg/kg]**
LPG 50/50	LPG	51	15.76
Methane	NG	16	17.48
95% CH_4 and 5% H_2 by vol.	95NG5H	15.3	17.596
90% CH_4 and 10% H_2 by vol.	90NG10H	14.6	17.72
80% CH_4 and 20% H_2 by vol.	80NG20H	13.2	18.01
70% CH_4 and 30% H_2 by vol.	70NG30H	11.8	18.37
60% CH_4 and 40% H_2 by vol.	60NG40H	10.4	18.826
50% CH_4 and 50% H_2 by vol.	50NG50H	9	19.424
95%LPG and 5%DME by mass	5DME	50.75	15.466
89%LPG and 11%DME by mass	11DME	50.45	15.101
83%LPG and 17%DME by mass	17DME	50.15	14.73
79%LPG and 21%DME by mass	21DME	49.95	14.482
74%LPG and 26%DME by mass	26DME	49.7	14.168

Table 2. Engine characteristic.

Cylinder number and layout	4R
Maximum power	55 kW@5200 rpm
Maximum torque	128 N·m @ 2800 rpm
Displacement	1598 ccm
Bore stroke	79,0 x 81.5 mm
Compression ratio	9.6

3. Research Methodology

The research has been developed according to the predefined program which covered:

- Estimation of power on wheels of the tested vehicle in the function of vehicle speed for the all of the tested blends,
- Estimation of in cylinder pressure in function of crank angle,
- Identification of specific fuel consumption.

The research covering engine indication and specific fuel consumption measurements were carried on the idle, and for WOT at speeds of 1500, 2000, 2500 and 3500 rpm, for each of the prepared blends. During the tests no modifications in the engine control were done. Ignition timing was set up for the petrol operation, while stoichiometric air fuel ratio was continuously controlled in a closed loop mode by the means of the ECU responsible for gas fuel dosing. EGR valve remained closed.
Results recorded during stand tests were used as an input to the mathematical model in the GT-Power. The so-called "reverse run combustion simulation" uses the recorded in cylinder pressure traces as well as other required data including: fuel mass and composition, engine volumetric efficiency to calculate main combustion parameters. The model calculations are based on the equations of energy balance in a closed combustion chamber. The model calculations are supplementary to conducted measurements. Figures 5 and 6 present the influence of the hydrogen and DME content on the investigated vehicle basic parameters, describing its performance.

4. Results and Discussion

4.1. Effects of Fuel Composition on Power

One of the important performance factors is definitely power measured on wheels Pw. The measurements were done on chassis dynamometer at the WOT, for different engine speeds. For the purpose of engine output power estimation, it is necessary to take into consideration transmission efficiency. For the investigated vehicle a correction ratio of k=0.946 (according to EEC standard) has been assumed. Tests on a chassis dynamometer determined the influence of different fuels on engine overall performance. power and torque results have been presented on the chart. Generally torque and power curves have a similar shape, the only difference regards varied position of peak values for different fuels.

Figure 2 presents power on wheels for all tested blends. The power increases with rising hydrogen share. The exceptions are 5% and 10% hydrogen mixtures. Lower power output registered in these cases results from the fact that at 7% of hydrogen share in the mixture, the flame front speed is the lowest. In the case of DME enriched LPG, with a share not exceeding 17%, a small power increase has been noticed. Increasing the DME share above 17%, results in the drop of both power and when compared to propane-butane mixture.

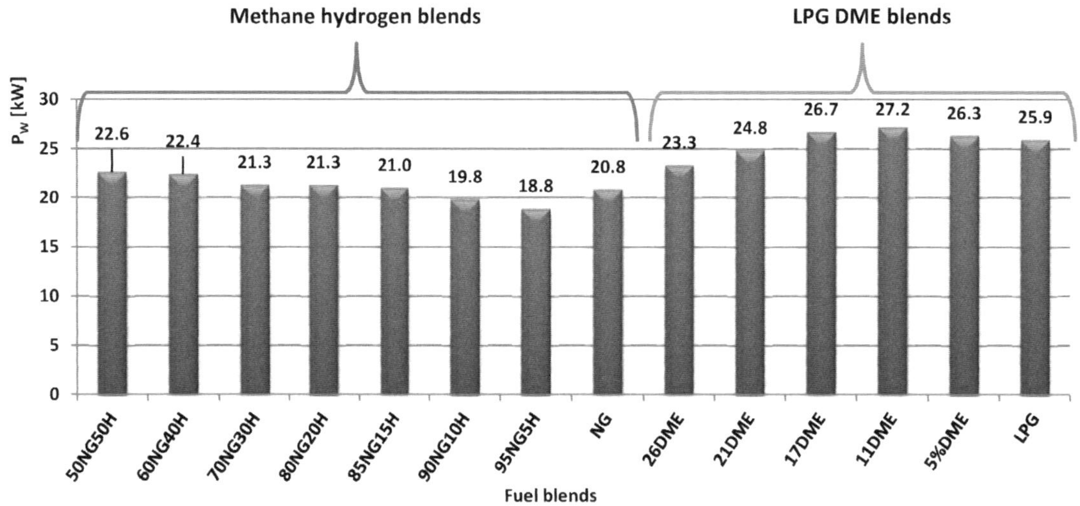

Figure 2. The variation of power at the wheels, for the hydrogen enriched methane and LPG DME blends, for rpm=2500, 100% load, λ=1.0.

4.2. In-Cylinder Pressure Traces and Heat Release

The combustion analysis has been undertaken using in-cylinder pressure traces. Figure 3 shows the p-V diagrams, while the IMEP for the methane hydrogen mixture at 2500 rpm and $\lambda = 1$ has been presented on Figure 4. In order to identify the cycle that most significantly represents the average burn characteristics of the evaluated point, the representative cycles have their IMEP, PMEP, peak pressure and location of peak pressure closest to that of the mean value of each parameter, which in turn have been calculated on the bases of 300 subsequently recorded cycles.

Increasing the hydrogen share results in the IMEP drop for mixtures featuring 15 % of hydrogen, higher IMEP values compared to methane can be noticed for shares of hydrogen over 15 %.

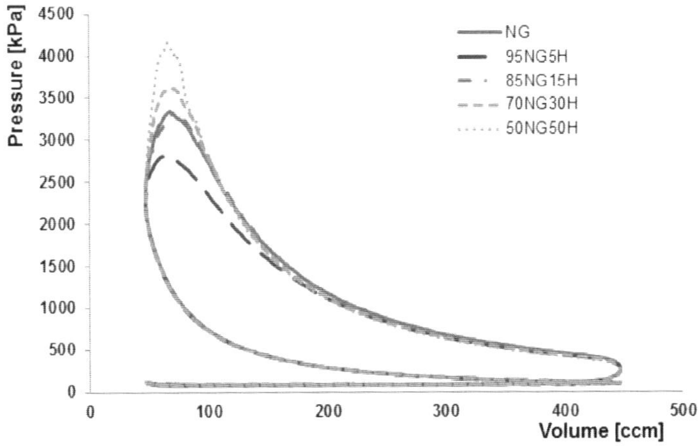

Figure 3. p-V diagram and IMEP for engine fuelled by methane enriched by hydrogen, at 2500 rpm; WOT and λ = 1.0.

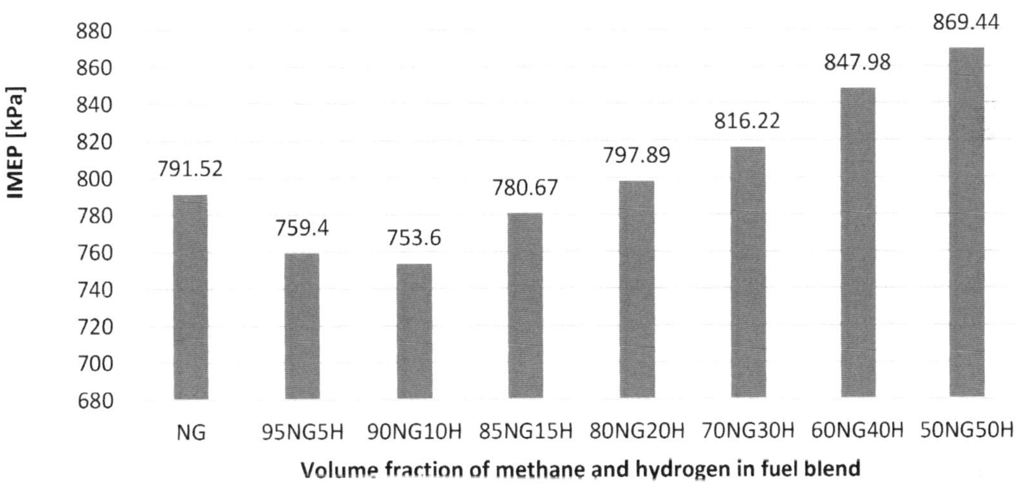

Figure 4. IMEP for engine fuelled by methane enriched by hydrogen, at 2500 rpm; WOT and λ = 1.0

Increasing the hydrogen share in the mixture accelerates the combustion process, however in the cases of hydrogen shares under 15% the amount of heat released in the combustion process is lower or equal to the one obtained for pure amount of heat released in the combustion process is lower or equal to the one obtained for pure methane.

The DME-LPG blends insignificantly increased the peak and the mean pressure values despite the fact that the engine speed and ignition timing remained unchanged. The stability of the combustion process, described by COV_IMEP does not exceed 2.5 % for the mixtures with the DME shares up to 17 % while for higher amounts of DME it reaches 4 %.

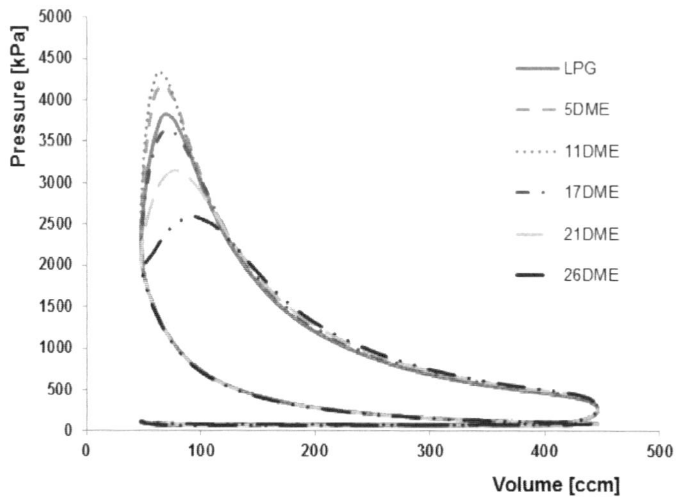

Figure 5. p-V diagram and IMEP for engine fuelled by LPG-DME mixtures, at 2500 rpm; WOT and $\lambda = 1.0$.

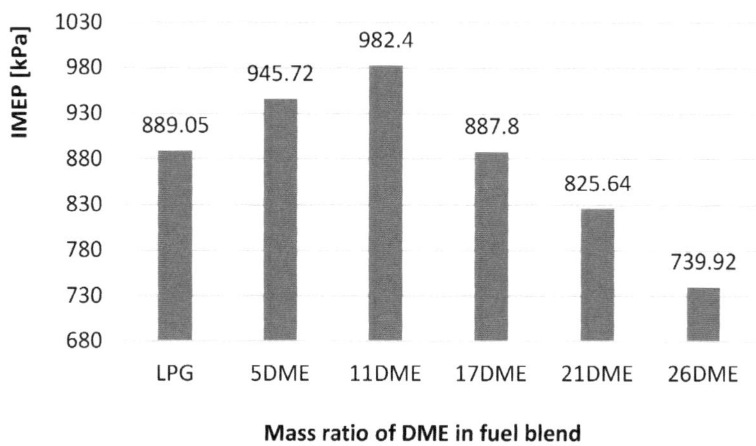

Figure 6. IMEP for engine fuelled by LPG-DME mixtures, at 2500 rpm; WOT and $\lambda = 1.0$.

In the case of the LPG-DME blend stoichiometric mixture combustion, an early growing in-cylinder pressure and a higher value of the peak pressure have been observed but only in the case of mixtures that did not exceed 17 % of the DME share, Fig. 5. With the increase of the DME fraction above 17 % the cylinder pressure curve rise retarded and the peak pressure crank-shaft angle was delayed leading to a decrease of the peak cylinder pressure along with the increasing DME fraction in the blends.
Although the flame propagation speed rises with the increase of the DME fraction in the blends, the retardation in the optimum ignition timing with the increase of the DME fraction still delays the rise of the cylinder pressure. In addition, the constant heating value of the blend with an increase of the DME fraction without the correction of the ignition angle lowers the peak value of the cylinder pressure. Lower ignition energy and temperature of DME is also the main reason for reaching of the maximum pressure values that are close to TDC. The DME mass share in the mixture, however, has a significant influence on the crankshaft angle at which the pressure reaches its maximum value. For the mixtures featuring 5 and 11 % DME content the maximum pressure was obtained faster than in the case of the LPG fuel feed. IMEP for all LPG DME mixtures has been presented on Figure 6.

4.3. Burn Rates

Figure 7 shows typical mass fraction burned (MFB) curves obtained from the GT-Power reverse run simulations.

For a given fuel, at constant λ and spark advance, as well as in reduction of the final MFB inside the cylinder. The burn rate also falls as the air fuel mixture is progressively diluted. When comparing results between the different fuels, it is evident that natural gas has the lowest burn rate and is considerably more susceptible to dilution than the hydrogen bearing fuels. This is consistent with its lower laminar flame speed and narrower flammability limits.

On the basis of MFB, presented on Figure 7 it can be noticed that mixtures featuring the DME shares of 5 % and 11 % not only initiate the combustion faster, but also the dynamics of their combustion is higher than for LPG. For those mixtures the combustion duration is also shorter, while the DME share rising over the 11 % retards the combustion initiation, and prolongs the process of combustion. This tendency is characteristic for the entire range of engine speeds.

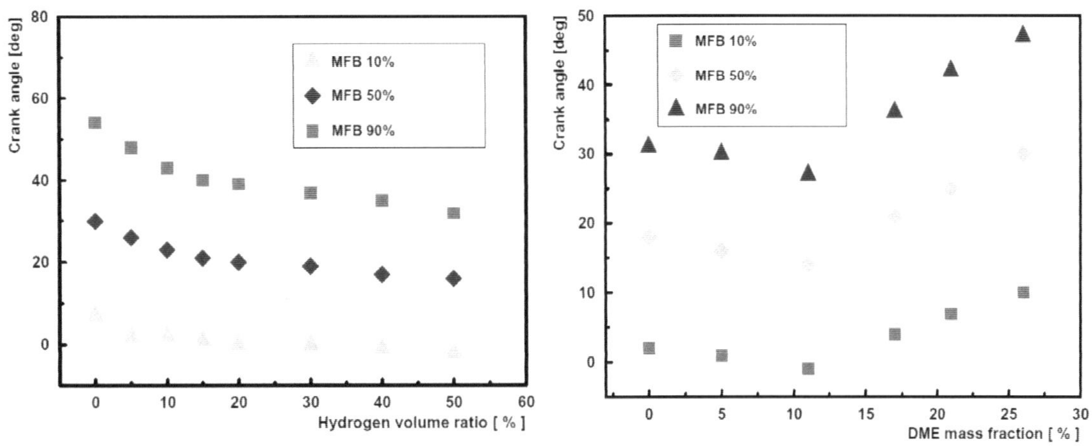

Figure 7. Variations of CA corresponding to 10, 50 and 90 % MFB for various cases of Hydrogen share and DME mass fraction 2500 rpm; load 100 %; spark advance 30°, λ = 1.0.

5. Efficiency of Energy conversion

One of the most factor describing efficiency of Energy conversion in the IC engine is definitively brake specific fuel consumption [g/kWh]. The value if this indicator in the case of described research has been calculated on the basis of the fuel consumption and engine power measurements acquired during completed stand tests. The individual fuel consumption of engine powered by methane with hydrogen addition decreases with increasing volume of hydrogen in mixture, Fig. 8.

Decreasing of the brake specific consumption affects the efficiency of the engine, a changes of efficiency are shown on Figure 9. Increasing efficiency of the engine with rising hydrogen amount in the blend does however limited, because it is connected with the risk of knocking combustion occurrence. The presence of this phenomenon can be noticed in the acquired in-cylinder pressure signal, and was audible in the engine during test.

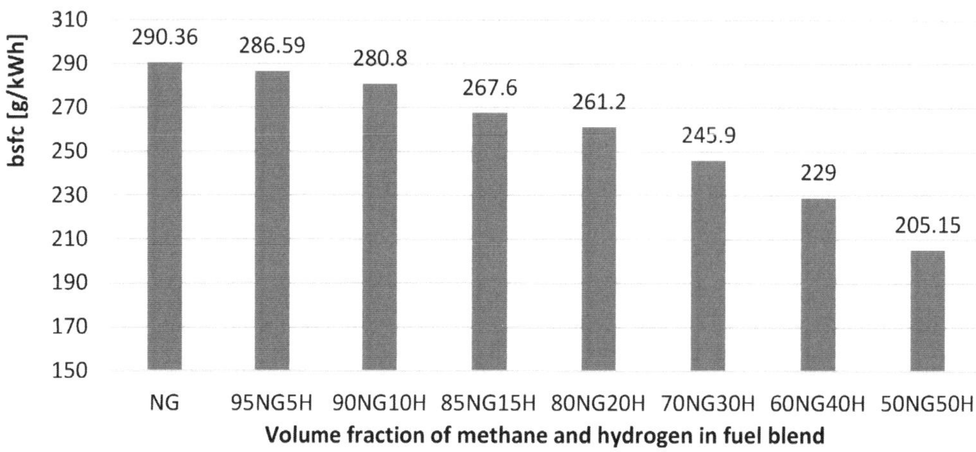

Figure 8. Brake specific fuel consumption for all methane and hydrogen mixtures.

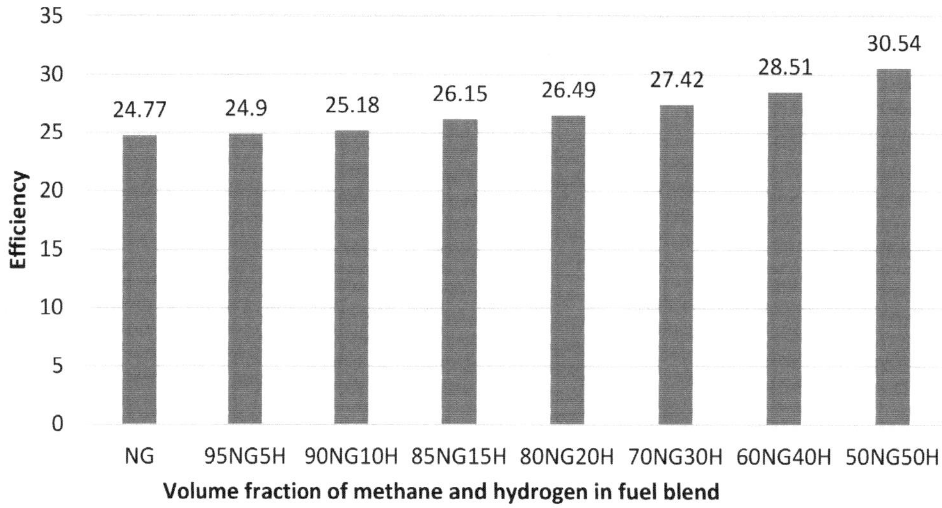

Figure 9. Engine efficiency for all methane and hydrogen mixtures.

In the case of LPG and DME mixture the bsfc and overall efficiency evolution have been presented on Figures 10 and 11. Overall efficiency values have been averaged for all recorded points, obtained for defined DME shares in the mixture, and for different engine speeds at WOT. The highest values of overall efficiency have been obtained for the mixtures featuring the DME shares from 5 to 11%. In this range, the DME share allowed it to obtain higher efficiency than in the case of LPG propulsion. The increase of DME content in the blends, over the 11% caused a slight overall efficiency drop.

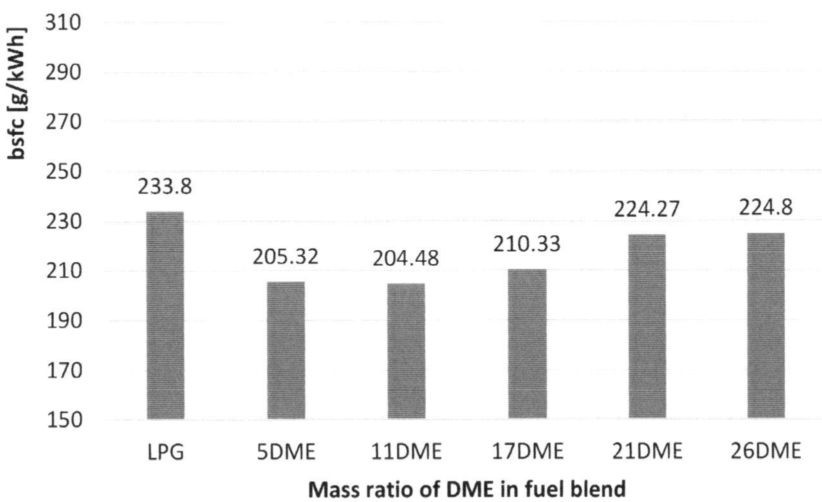

Figure 10. Brake specific fuel consumption for all LPG DME blend

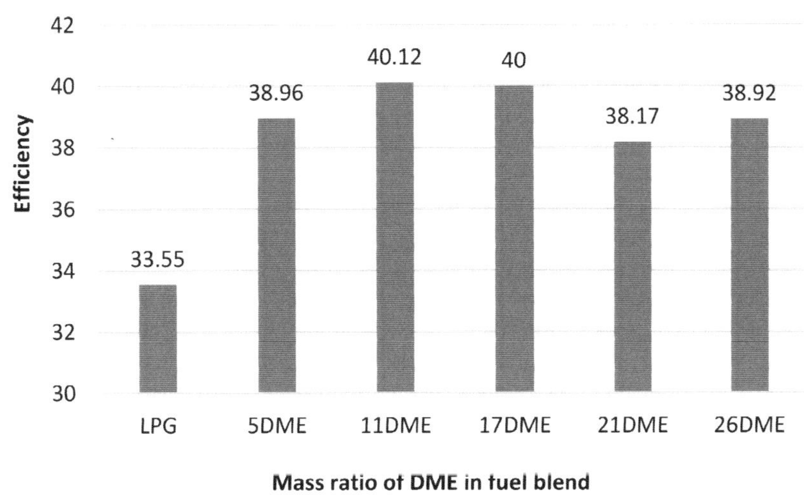

Figure 11. Engine efficiency for all LPG DME blends

6. Conclusions

Research covered stand tests of the vehicle equipped with the SI engine, fueled with methane - hydrogen and DME- LPG blends. It was possible to identify the influence of hydrogen volume rate and DME mass share in the mixtures on the overall engine performance, its efficiency and on the combustion process. Moreover, carried out tests give a possibility to understand fundamental combustion properties of HNG and LPG+DME blends, what is important for developing advanced NG and LPG based combustion engines with necessary operating strategies.

Chemical energy of the charge closed in a combustion chamber is released during combustion process the speed and energy profiles depend on the gaseous fuel composition.

General aim of adding hydrogen to methane and DME to LPG is to minimize negative parts in their combustion. Because of the differences in selected properties of methane, hydrogen, DME and LPG an improvement of the conditions for the flame initiation and shortening of the charge heating period are expected.

It has been shown that:
- Stoichiometric combustion of HNG and LPG-DME mixtures reduced ignition delay and create conductive for faster burn,
- HNG reacts faster than pure methane, allowing it to shorten combustion duration, especially in the first phase of flame development.
- LPG and DME blend reacts faster the pure LPG but only for mixtures which consist of no more than 11 % of DME, by mass ratio.
- Heat release during combustion of HNG compares to pure methane increasing for mixtures containing more than 15 % of Hydrogen. The increase of HR does not does not exceed 4.5 % Increasing of HR does not exceed 4.5 % for tested engine.
- The share of DME in the mixture of LPG DME significantly influence on changes of HR for tested engine.

Future experimental development would foresee the optimization of the emissions of the both pollutant and CO_2 along the reduction of fuel consumption for vehicle driving cycle. It is necessary to investigate widely following set of engine variables:
- Spark advance,
- Compression ratio,
- A wide spectrum of λ values,
- EGR.

References

[1] Pipitone E and Beccari S 2009 Performances Improvement of a S. I. CNG Bi-Fuel Engine by Means of Double-Fuel Injection," *SAE Technical Paper* 2009-24-0058, doi:10.4271/2009-24-0058

[2] Karim G. A 2003 Hydrogen as a spark ignition engine fuel *International Journal of Hydrogen Energy* 28 569-577, Pergamon, Elsevier

[3] Karim G A, Wierzba I and Al-Alousi Y 1996 Methane- hydrogen mixtures as fuels, ht. *J. Hydrogen Energy* Vol. 21, No. 7, pp. 625-631Copyright @ 1996 International Association for Hydrogen Energy, Elsevier Science Ltd, 0360-3199(95)001344

[4] Shrestha S, Bade O, Karim G.A. 1999 Hydrogen as an additive to methane for spark ignition engine applications *International Journal of Hydrogen Energy* 24 577-586, Pergamon, Elsevier

[5] Bauer CG and Forest TW 2001 Effect of hydrogen addition on the performance of methane-fueled vehicles. Part I: Effect on S.I. engine performance *Int. Journal of Hydrogen Energy* 26:55-70

[6] Hailin Li and Karim Ghazi A 2005 Exhaust emissions from an SI engine operating on gaseous fuel mixtures containing hydrogen *International Journal of Hydrogen Energy* 30 1491-1499, Elsevier

[7] Sierens R and Rosseel E 2001 Variable composition hydrogen/natural gas mixtures for increased engine efficiency and decreased emissions *J Eng Gas Turbines Power* 122:135-40

[8] Tang CL and Wei 2012 Shock tube measurements and kinetic investigation on the ignition delay times of methane/dimethyl ether mixtures *Energy Fuels* 26:6720–8

[9] Shrestha SOB and Karim GA 1999 Hydrogen as an additive to methane for spark ignition engine applications. *Int J Hydrogen Energy* 24:577–86

[10] Bauer CG 2002 Effect of hydrogen addition on the performance of methane-fuelled vehicles. Part II: Effect on SI engine performance. *Int J Hydrogen Energy*

[11] Flekiewicz M and Kubica G 2009 Hydrogen Enriched CNG - a Tool for Dual Fuel Engine Overall Performance Improvement SAE Technical Paper 2009-01-2681doi:10.4271/2009-01-2681

Analysis of the CO_2, NO_x emission and fuel consumption from a heavy-duty vehicle designed for carriage of timber

P Fuc[1], P Lijewski[1] and A Ziolkowski[1]

[1]Poznan University of Technology, Institute of Combustion Engines and Transport, Piotrowo str. 3, 60-965 Poznan, Poland

E-mail: andrzej.j.ziolkowski@put.poznan.pl

Abstract. The paper presents the results of measurements of the CO_2 and NO_x emission and fuel consumption recorded under actual operating conditions of a heavy-duty vehicle designed for loading and carriage of timber. The tests were performed on a specially designed test route that reflected the arrival of the vehicle to the felling site in the forest, the loading process and return to the lumberyard. The route ran through paved (asphalt) and unpaved (forest) portions. Its total length was 8.6 km. An advanced PEMS (Portable Emission Measurement System) device was used for the measurement of the exhaust emissions – SEMTECH DS by Sensors Inc. The paper analyses the CO_2 and NO_x emission and fuel consumption on all portions of the test route and presents a comparison between the forest and asphalt roads.

1. Introduction

In developed countries for the last 20 years the system of tree felling has changed significantly. First, the commonly used axes were replaced with chainsaws and then specialized machines were applied capable of cutting and processing trees for transport in the amount of up to 350 m^3 of timber (Figure 1). The method of transport of timber to the storage site has also changed Transporting timber out of the forest is realized by vehicles fitted with specialized grabber cranes for loading and unloading of the cut trees. These vehicles fall into the homologation category of heavy-duty vehicles (HDV) and must comply with the applicable emission standards [1-4].

Figure 1. Example of modern machines for shearing in the forest [5].

The group of heavy-duty vehicles varies in terms of the role of a vehicle in the transport task. We can distinguish heavy-duty vehicles used for long haulage, distribution vehicles, special vehicles and city buses and coaches. Currently, all listed vehicles are fitted with technologically advanced engines with complex aftertreatment systems that, compared to engines manufactured several years ago, are characterized by a greater volume-to-power ratio, reliability and durability. Aside from modern engines, heavy-duty vehicles are also equipped with systems that assist the driver in the reduction of the fuel consumption and extension of life of many functional components – telematic systems that control the driver's behavior and supervise the operation of the entire powertrain [6]. The mentioned equipment ensure a consequent decrease of the negative impact of heavy-duty vehicles on the natural environment.

Out of the discussed group of vehicles, the heaviest work is done by special vehicles operated in the non-road areas. In this case, the experience and behavior of the driver under given operating conditions have the greatest influence on the exhaust emissions and fuel consumption. It is noteworthy that under non-road conditions a precise evaluation of parameters is difficult. Hence, the measurements of CO_2 and NO_x were performed under actual conditions of operation using state-of-the-art technology based on portable PEMS devices. The object of the investigations was a heavy-duty vehicle designed for transport of timber and the measurements were performed on a road/non-road test route.

2. Research methodology

2.1. Characteristic of the route

The tests were performed under actual conditions of operation of a heavy-duty vehicle designed for loading and transport of timber on both, paved (asphalt) and unpaved roads. In order to accurately reflect the operating conditions of this type of vehicles the tests were performed in the following configuration (Figure 2):

a) run to the loading site:
- paved road (portion A–B),
- unpaved road (portion B–C),

b) timber loading (point C),

c) run to the unloading site:
- unpaved road (portion C–D),
- paved road (portion D–A).

The total length of the test route was 8.6 km, 5.3 km of which were unpaved forest roads.

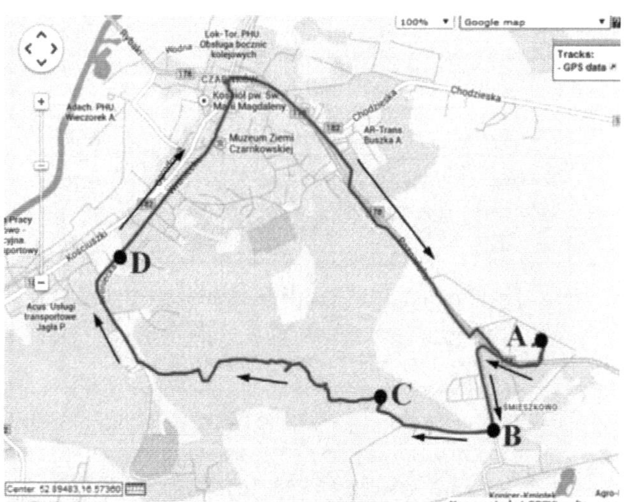

Figure 2. The test road section used in the on-road emission test for the heavy-duty vehicle.

2.2. The test vehicle
The test vehicle was a tractor-trailer system (MAN TGA 26.440) of the 6x6 axle configuration designed for transport of timber (table 1, Figure 3). The vehicle was fitted with a 324 kW (440 KM) Euro IV engine. An additional equipment of the tested object was the EPSILON E165Z95 crane for loading and unloading of timber. In the tested heavy-duty vehicle, a two-axle semi-trailer with the rear twist beam axle was applied.

Table 1. Characteristics of the test object.

Parameter	Value
Engine capacity	10.5 dm3
Number of cylinders/configuration	6 / straight
Maximum power output	324 kW @1900 rpm
Maximum torque	2100 Nm @1000-1400 rpm
Emission standard	Euro IV
Exhaust gas aftertreatment	MAN PM-KAT®
Crane type	EPSILON E165Z95 fitted with grabbers

Figure 3. A heavy-duty vehicle during the exhaust emission test under actual operating conditions.

2.3. Research equipment
For the measurement of the emission of CO_2 and NO_x, a PEMS device was used (SEMTECH by Sensors Inc.) designed for testing of exhaust emissions under actual conditions of operation (Figure 4). This is a unique combination of measurement and recording device of the following parameters [5, 7]:

a) CO and CO_2 concentration (NDIR–Non-Dispersive Infrared analyzer), NO_x = NO+NO_2 (NDUV–Non-Dispersive Ultraviolet analyzer), THC (FID–Flame Ionization Detector), O_2 (electrochemical sensor);
b) Mass flow of the exhaust gas–Pitot tube-based flow meter,
c) Ambient conditions – ambient pressure, temperature, humidity;
d) Vehicle position and speed–GPS system;
e) Data from the vehicle on-board diagnostic system.

Thanks to the mentioned measurements ensure, a comprehensive analysis of the vehicle exhaust emissions under actual operating conditions is possible allowing for the actual measurement conditions such as driving technique, ambient conditions, traffic congestion, etc.

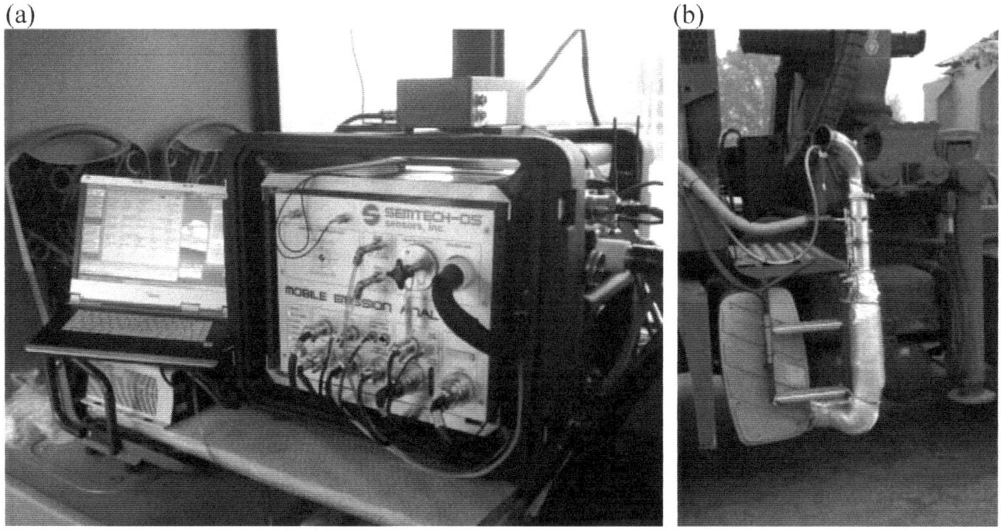

Figure 4. The SEMTECH DS analyzer (a) and the exhaust gas flow meter (b).

3. Analysis of results

3.1. Measurement conditions

The first stage of the test route covered the distance from the lumberyard to the felling site (Figure 5). The length of this portion was 2.1 km, 1.7 km of which were paved (asphalt) roads (table 2). In this road section the maximum speed of the vehicle reached 50 km/h. The next stage of the test was timber loading performed using the grabber crane fitted on the tested object. The crane has a hydraulic system powered through a power take off from the vehicle transmission. The timber loading constituted 46.8% of the entire measurement time.

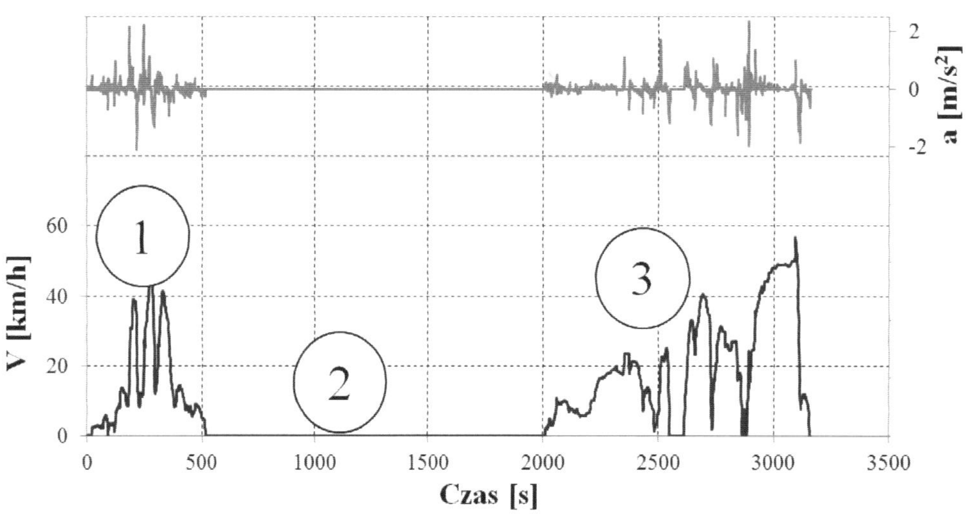

Figure 5. The vehicle speed and acceleration during the test.

Table 2. Characteristics of the parameters describing the vehicle driving profile.

Parameter	Unit	To the forest		Loading	To the lumberyard	
		Paved (asphalt) road	Unpaved (forest) road	–	Paved (asphalt) road	Unpaved (forest) road
Share	[%]	11.5	5.0	46.8	21.8	14.9
V_{max}	[km/h]	50.0	15.9	–	56.8	23.6
$V_{ave.}$	[km/h]	17.1	8.6	–	25.7	12.2
Distance	[km]	1.7	0.4	–	4.9	1.6

The last stage was the run of the vehicle with timber to the lumberyard. In this stage, paved roads prevailed in the vicinity of the town of Czarnkow. This stage revealed the greatest changes in the vehicle acceleration conditioned by the road infrastructure. The average speed on the paved roads of the vehicle was 25.7 km/h and on the forest road – 12.2 km/h. Such varied conditions of the test route had impact on the vehicle exhaust emissions (CO_2, NO_x) and fuel consumption. The measurements were performed on a sunny day without rainfall. The forest roads were passable and the all-wheel-drive system was not used (applied under difficult road conditions only).

3.2. Emission of CO_2, NO_x and fuel consumption

The greatest emission of CO_2 occurred in the last stage of the test route – the run from the felling site to the lumberyard (Figure 6). This was caused by the vehicle's greater average speed on this road section, compared to the first stage. The speeds were 18.9 and 12.8 km/h respectively. Higher vehicle speed is a result of production of greater effective power by the vehicle. Another factor influencing the CO_2 emission is the fact that the vehicle was fully loaded with timber during the run to the lumberyard (20 000 kg). When the vehicle is loaded, the average value of the engine torque increases, which increases the fuel consumption, which again is translated into higher combustion temperature inside the cylinders. Such conditions facilitate the formation of NOx, whose level depends on the temperature and pressure inside the combustion chamber. In the performed tests the emission of NO_x was close to the emission of CO_2 (Figure 7). Based on the obtained data the on-road emission of NO_x was determined (ratio of the mass of generated exhaust component to the covered distance):

a) to the loading site12.9 g/km:
- paved (asphalt) – road14.1 g/km,
- unpaved (forest) – road7.8 g/km,

b) to the unloading site –10.0 g/km:
- unpaved (forest) – 5.1 g/km,
- paved (asphalt) – 11.7 g/km.

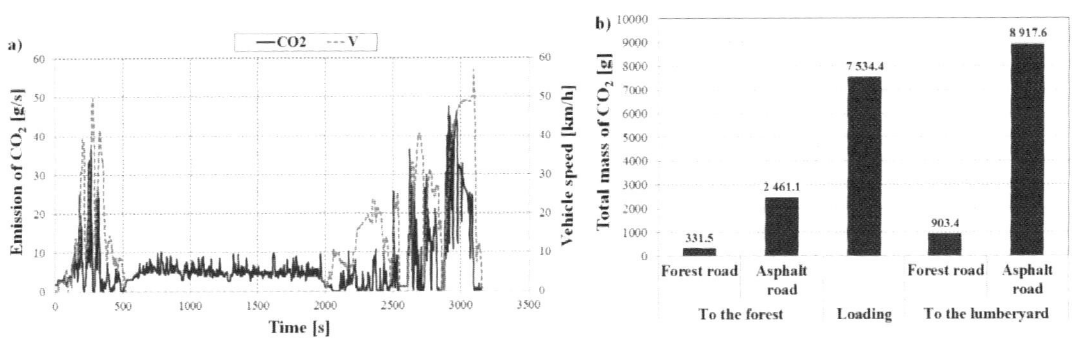

Figure 6. The emission of CO_2 and vehicle speed during the tests(a), total mass of CO_2 (b).

The vehicle, when on the way to the unloading site, had lower emission than when it was on the way to the forest. This was related to the operation of the EGR system (exhaust gas recirculation) fitted in the engine. During the run to the forest from the lumberyard, the EGR system may have supplied a lower amount of exhaust gas to the combustion chamber because it operated at low engine loads and low temperature inside the combustion chamber. During the run to the lumberyard with the trees, the engine operated at higher average load, which translated into greater mass of the recirculated exhaust gas.

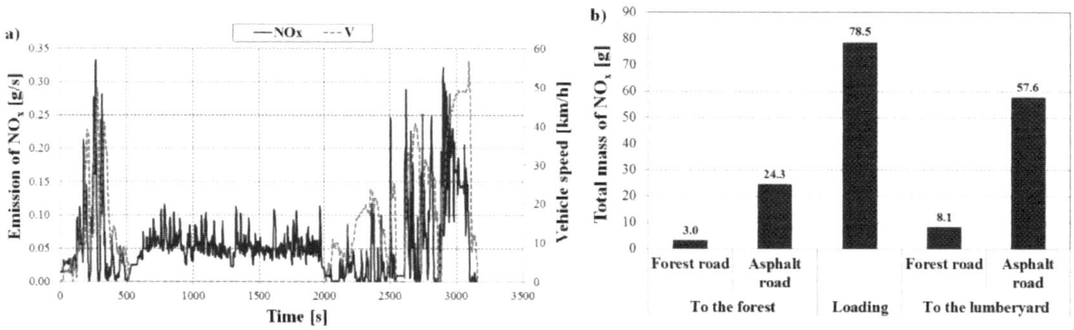

Figure 7. The emission of NO_x and vehicle speed during the tests(a), total mass of NO_x (b).

Second-by-second fuel consumption of the vehicle was determined based on the carbon balance method. This method is used mainly to determine fuel consumption based on the measured emission of CO_2, CO and THC. In order to determine the values of second-by second fuel consumption, this method was appropriately modified. A full description of the method modification can be found in earlier authors' publications [8]. For this test cycle, the second-by-second fuel consumption was determined exclusively based on the CO_2 emission rate. For heavy-duty diesel vehicles it is admissible because the values of CO and THC for this group of vehicles do not significantly influence the accuracy of determination of the carbon balance method based fuel consumption.

This is confirmed by the fact that the second-by-second fuel consumption was similar to the CO_2 emission rate (Figure 8a). Upon summing up the fuel consumption from individual stages of the test route, it was observed that the greatest amount of fuel was used by the vehicle when running from the forest to the lumberyard (Figure 8b).

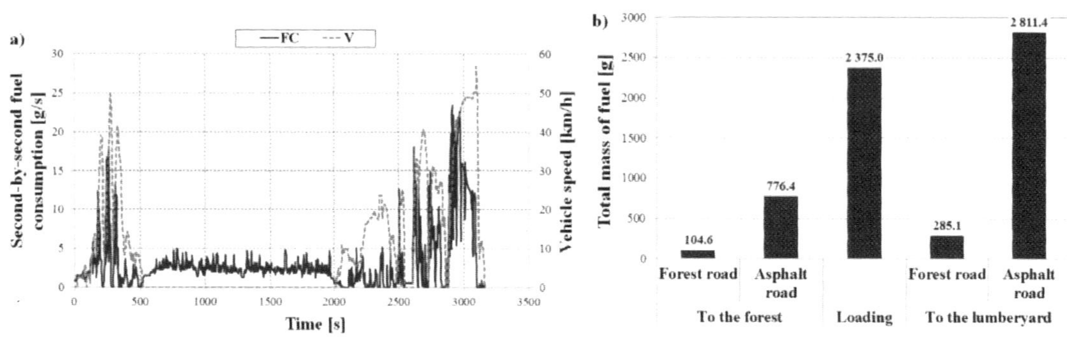

Figure 8. Second-by-second fuel consumption of a vehicle and its speed (a), total mass of fuel (b).

This confirms the above statements that higher average vehicle speed and applying a load to the vehicle translates into greater engine usable parameters (torque and power output) that higher fuel consumption is attributable to. Attention should also be drawn to increased fuel consumption when loading of the cut trees. Its maximum value was 3.2 g/s and it was greater than the maximum fuel consumption during the run to the loading site by the forest road.

Based on the second-by-second fuel consumption the fuel consumption of the vehicle was determined:

a) to the loading zone – 50.2 dm^3/100 km:
- paved (asphalt) road – 53.9 dm3/100 km,
- unpaved (forest) road – 33.1 dm3/100 km,

b) to the unloading zone – 56.7 dm^3/100 km:
- unpaved (forest) road – 21.4 dm3/100 km,
- paved (asphalt) road – 68.2 dm3/100 km.

Obtaining lower fuel consumption during the run to the loading site by the forest roads compared to the run to the lumberyard using the same roads was caused by additional manoeuvring of the vehicle on the tree-felling site to adjust the vehicle position before the loading procedure could begin. The manoeuvring was difficult because the terrain in the loading site was marshy.

4. Conclusion

The application of a portable measurement system (SEMTECH DS) in the tests of a heavy-duty vehicle used in the transport of trees in the forest allowed measuring the emission of CO_2 based on which the level of fuel consumption and NO_x emission were determined in a driving cycle reflecting day-to-day operation of this type of vehicles – paved asphalt roads and unpaved forest roads as well as loading and unloading of trees. This allowed determining the influence of the conditions of work on the exhaust emissions and energy demand of the vehicle engine. To date, only average fuel consumption was determined from a given period of vehicle operation. The paper also presented the fuel consumption of the vehicle during the tree loading procedure, whose maximum values were greater than the value recorded during the vehicle run on the forest road. The emission of NO_x during the tests may have been influenced by the EGR system that most likely, in the initial stage of the tests, supplied less exhaust gas to the cylinders due to lower engine loads and lower temperature of the exhaust gas at this stage. The tests were performed in good weather conditions and the forest roads were fully passable. It was not necessary to use the all wheel drive system. The authors plan further research using the applied method under more severe conditions. This will allow an accurate determination of the influence of heavy-duty tree-transporting vehicle working conditions on the fuel consumption.

References

[1] Bajerlein M, Rymaniak L 2014 *The Reduction of Fuel Consumption on the Example of Ecological Hybrid Buses* Applied Mechanics and Materials Vol. 518 pp 96–101.

[2] Merkisz J, Kozak M, Nijak D, Andrzejewski M, Molik P, Nowak M, Rymaniak L, Ziolkowski A 2012 *The analysis of the emission level from heavy-duty truck in a city traffic* Silniki Spalinowe / Combustion Engines Vol. 3/2012 (151) pp. 80–88.

[3] Schweier, J. & Becker, 2012 *New Holland Forage Harvester's Productivity in Short Rotatation Coppice: Evaluation of Field Studies from a German Perspective*. International Journal of Forest Engineering. No. 2 (2012), Vol. 23 pp 82–88.

[4] M. P. Walsh 2011 *Global trends in motor vehicle pollution control; a 2011 update. Part 1* Combustion Engines / Silniki Spalinowe Vol 2/2011 (145) pp. 106–117.

[5] http://www.drewno.fordaq.com/fordaq/news/thinning_finalfelling_JohnDeere_ harvester__13881.hhtm

[6] Merkisz J, Fuc P, Lijewski P, Ziolkowsk A, Rymaniak L 2013 *Impact of Application Type of Engine in Heavy-Duty Vehicle for CO_2, CO Emission and Fuel Consumption* Prace Naukowe Politechniki Warszawskiej – Transport Vol. 98. P. 437–446.

[7] Merkisz J, Fuc P, Ziolkowski A 2012 *Wpływ masy ładunku na emisję CO_2, NO_x i na zużycie paliwa pojazdu ciężarowego o masie całkowitej powyżej 12 000 kg*. Postępy nauki i techniki / Advances in Science and Technology no. 15/2012.

[8] Brebbia C, Magaril E, Khodorovsky M 2014 Energy Production and Management in the 21st Century: The Quest for Sustainable Energy (2 Volume Set). Co-authorship: P. Fuc, J. Merkisz, P. Lijewski, A. Merkisz-Guranowska, A. Ziolkowski, "The Assessment Of Exhaust System Energy Losses Based On The Measurements Performed Under Actual Traffic Conditions". WIT Transactions on Ecology and the Environment. WIT Press

Acknowledgments
The research was funded by the National Centre for Research and Development (Narodowe Centrum Badań i Rozwoju) – research project within the Applied Research Programme (contract No. PBS3/A6/26/2015).

IOP Conf. Series: Materials Science and Engineering **148** (2016) 012066 doi:10.1088/1757-899X/148/1/012066

The combustion properties analysis of various liquid fuels based on crude oil and renewables

K Grab-Rogalinski[1] and S Szwaja[1]

[1] Czestochowa University of Technology, Institute of Thermal Machinery, Armii Krajowej 21 Av., 42-200 Czestochowa, Poland

e-mail: grab@itm.pcz.czest.pl

Abstract. The paper presents results of investigation on combustion properties analysis of hydrocarbon based liquid fuels commonly used in the CI engine. The analysis was performed with aid of the CRU (Combustion Research Unit). CRU is the machine consisted of a constant volume combustion chamber equipped with one or two fuel injectors and a pressure sensor. Fuel can be injected under various both injection pressure and injection duration, also with two injector versions two stage combustion with pilot injection can be simulated, that makes it possible to introduce and modify additional parameter which is injection delay (defined as the time between pilot and main injection). On a basis of this investigation such combustion parameters as pressure increase, rate of heat release, ignition delay and combustion duration can be determined. The research was performed for the four fuels as follows: LFO, HFO, Biofuel from rape seeds and Glycerol under various injection parameters as well as combustion chamber thermodynamic conditions. Under these tests the change in such injection parameters as injection pressure, use of pilot injection, injection delay and injection duration, for main injection, were made. Moreover, fuels were tested under different conditions of load, what was determined by initial conditions (pressure and temperature) in the combustion chamber. Stored data from research allows to compare combustion parameters for fuels applied to tests and show this comparison in diagrams.

1. Introduction

The combustion properties of the fuel determine suitability of the fuel for the power applications. By power applications one can understand that the fuel can be used to supply any power machines like internal combustion engines (IC). Several basics properties of the fuel ought to be determined before it is applied in practice. The simple and effective way to determine whether a fuel is suitable for the IC engine are measurements performed on the combustion research unit (CRU).

Various types of the fuel in both liquid and gaseous phase can be used for the combustion engine. The preparation of a potential fuel to be implemented in series production requires lot of tests that allow to check whether the fuel can be used not only from the energetic point of view but also regarding to chemical properties and mechanical contamination. Due to restrictive regulations for engine exhaust gas emissions as well as the need of use renewable fuels, the basic test stands are important especially for industrial engine manufacturers and end-users. The tests should answer the question on fuel usability to the specific engine. Such approach allow to save the time and money during the engine R&D and manufacturing process.

The liquid fuels commonly used in industrial combustion engines are as follows LFO (light fuel oil), HFO (heavy fuel oil), and different types of biofuels like jetropha oil or rapeseed oil (RSO). Further, it is often mentioned on use of the glycerol as a fuel for the IC engine. All those above-mentioned fuels are in use for engines with compression ignition commonly known as diesel engines.

2. Test stand description

The CRU is an instrument based on CVCC (Constant Volume Combustion Chamber), which can measure fuel combustion parameters. The CRU makes it possible to do combustion tests on the different types of liquid and gaseous fuels for SI as well as for CI engines.

The purpose of this machine is to simulate the combustion process in the piston engine in which the dynamic aspects of its nature are neglected (piston movement, valve movement). It allows to focus only on differences that result from the fuel properties. Mainly the CRU device can by divided into three depended systems: hydraulic, pneumatic and electronic ones. The hydraulic system is responsible for providing the proper pressure in injection system. The pneumatic system is responsible for flushing and venting procedures while the fuel sample is changed from previously tested to a new one. The electronic system is responsible for steering both the hydraulic and the pneumatic systems, to communicate with a computer and to read out the measurements from CRU sensors. Test rig view is presented in figure 1.

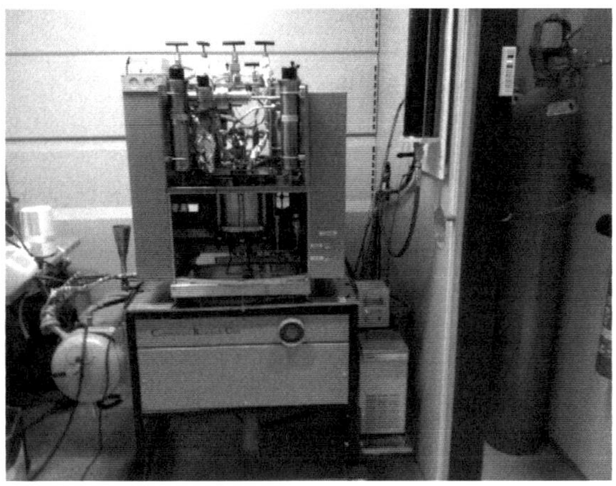

Figure. 1. CRU test rig overall view.

The CRU provides a wide range of parameters that can be adjusted during research to obtain the best results, accuracy and repeatability of the measurements. The basic parameters of the CRU unit are presented in table 1.

Table 1. Basic CRU parameters [4].

Chamber	
Fuel type	Diesel Fuel
Chamber temperature	Max 610°C
Chamber pressure	10-75bar
Chamber volume	385cm^3
Fuel injection	
Fuel pump and injection nozzle	BOSCH CR injector
	Electronic control of timing, opening period, pressure
Injection pressure	400-1600bar
Instrumentation control	
Chamber heating	2 heaters/2 sensors
Chamber pressure	Air and exhaust valve / pressure sensor
Combustion pressure sensor	Piezoelectric pressure sensor
Fuel pressure sensor	Piezoelectric pressure sensor

The CRU unit used in this research was modified to provide a possibility to use all possible fuels that could be combusted in the CI engine. For high density fuels a heating system for fuel lines were used to provide a proper kinematic viscosity during tests.

3. Tests configuration

The tests performed on the CRU were done for the four different fuels: LFO (Light Fuel Oil), HFO (Heavy Fuel Oil), RSO (Rape Seed Oil) and Glycerol. During the tests for the each fuel parameters concerning initial conditions in the combustion chamber (pressure and temperature), injection parameters (pressure and injection duration) and fuel parameters (fuel temperature – kinematic viscosity) were adjusted. In all cases during the measurements three to four injections were done for one set point. The research were done in six stages.

In first stage the kinematic viscosity as a function of temperature for HFO, RSO and Glycerol was determined by use of the rheometer. Because the LFO has low kinematic viscosity of 4,15 cSt in temperature of 20°C as Corsini et al. [2] wrote. The characteristic of this parameter against temperature wasn't determined. Afterwards, one value of kinematic viscosity for all fuels was selected to further tests. In this research, the kinematic viscosity value was $\nu \approx 23$ cSt for all fuels except LFO. The detailed measurement matrix is presented in table 2.

Table 2. Tests matrix.

Tested fuel (LFO, HFO, RSO and Glycerol)							
Initial conditions of combustion chamber		Pressure of injection p_{inj} (bar)		Injection duration inj_{time} (μs)		Kinematic viscosity ν (cSt)	
p (bar)	*T (°C)*	*55bar/550°C*	*70bar/590°C*	*55bar/550°C*	*70bar/590°C*	*55bar/550°C*	*70bar/590°C*
55	550	500		500 (1500 Glycerol)		15 (23,5 HFO)	
62,5	570	750		Optimal[a]		23 (30 HFO)	
70	590	1000		1500(2500 Glycerol)		40 (40 HFO)	

[a]The injection duration that provides 10 bar increase during the fuel sample combustion for 70bar/590°C initial conditions of the combustion chamber.

In the second stage determination of optimal injection parameters were done. For injection pressure p_{inj}=1000bar the injection duration $inj_{duration}$ was modified to achieve increase in combustion pressure, for initial conditions for the combustion chamber (70bar/590°C), of 10bar.

In the third stage for fixed injection parameters (optimal) the measurements of pressure and ROHR were performed as a function of pressure and temperature that prevailed in the CRU combustion chamber (initial conditions in combustion chamber).

In the fourth stage the pressure of injection was changed during the research. The measurements were performed for two states of initial conditions in the combustion chamber 55bar/550°C and 70bar/590°C. The time of injection was set as the optimal one determined in a second stage of this research.

In the fifth stage the injection pressure for each fuel was constant p_{inj}=1000bar and the injection duration $inj_{duration}$ was changed during the measurement according to table 1. The measurements were performed for two states of initial conditions in the combustion chamber 55bar/550°C and 70bar/590°C as well.

In the last stage the injection parameters were set optimal for each fuels and the kinematic viscosity was changed. Change in kinematic viscosity was done by change in temperature of fuel which was in the fuel line. Similarly to previously done measurements this one was also done for two states of initial conditions of combustion chamber 55bar/550°C and 70bar/590°C.

4. Results and discussion

The first measurements were focused on determination of the kinematic viscosity by use of the rheometer. The results of the measurements for HFO, RSO and Glycerol are presented in figure 2.

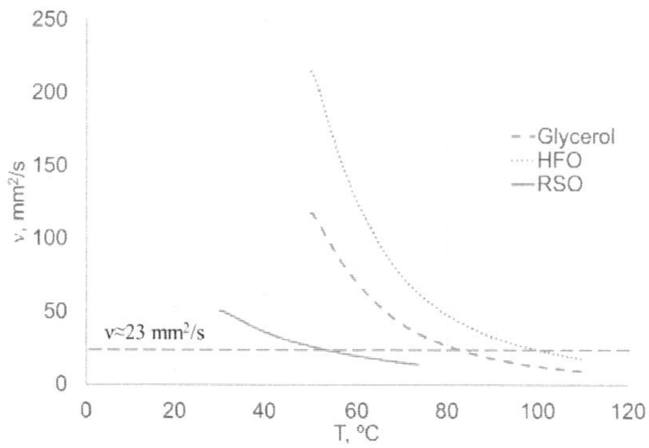

Figure 2. Kinematic viscosity vs. temperature for tested fuels (HFO, RSO and Glycerol).

As seen, these fuels are characterized by high kinematic viscosity at normal condition and use of them could be very difficult without pre-heating. The most intensive heating is required to the HFO. To obtain preliminary assumed kinematic viscosity it should be heated up to 100°C. The lowest heating up temperature is for RSO and it is about 55°C. For Glycerol the heat up temperature is approximately 82°C.

The results for optimal injection duration for these tested fuels at injection pressure of 1000bar, are presented in table 3.

Table 3. Injection duration for tested fuels.

Fuel	$Inj_{duration}$ (μs)
LFO	1000
HFO	1075
RSO	1150
Glycerol	1900

The values were obtained during the tests with 10bar increase in combustion pressure for the each fuel. As seen, for the same kinematic viscosity for HFO, RSO and Glycerol the injection duration increases what is related with lower heating value (LHV) of these fuels compared to LFO. The lowest heating value is for Glycerol LHV=16 MJ/kg [5] what means that to obtain the same energy content in an air-fuel mixture, more fuel has to be injected. It also can be seen that RSO has a lower LHV (37,6 MJ/kg [2]) compared with HFO (42,8 MJ/kg [3]) or LFO (43,2 MJ/kg [3]).

In figure 3 the comparison of combustion pressure for different initial temperature and pressure are shown. As presented, for LFO increase in combustion pressure is the same for all cases under optimal initial pressure that corresponds to proper combustion process of this fuel and good atomization and evaporation during injection. For the initial temperature change, it can be observed that with lower temperature, the pressure starts to increase later what is correlated with ignition delay (ID) which lasts longer with lower temperature and it shortens with temperature increase. In this case the ignition delay is caused by much worse evaporation in lower initial temperature.

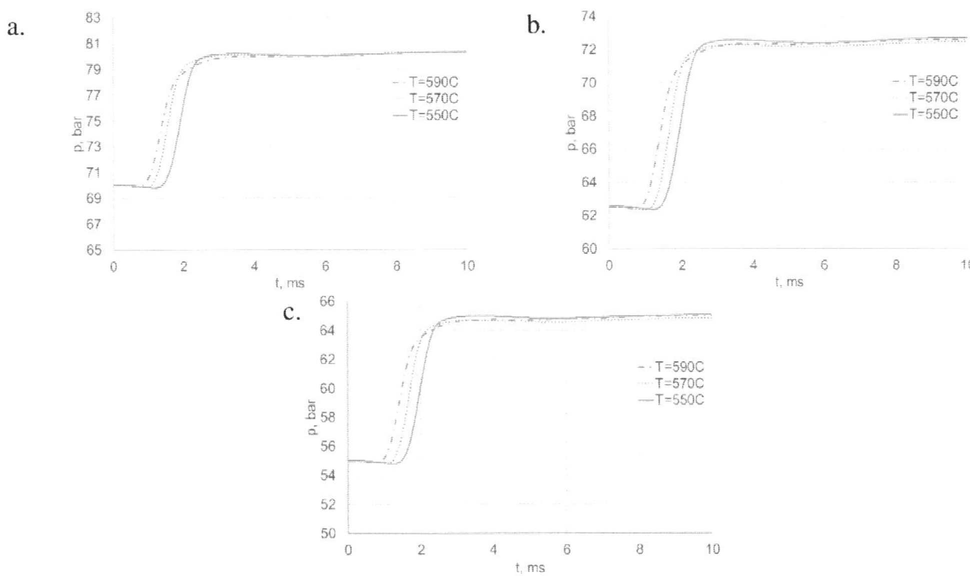

Figure. 3. Comparison of combustion pressure for various internal temperatures of combustion chamber for initial pressure a) 70bar, b) 62,5bar, c) 55bar.

The same trend can be seen for rate of pressure increase (dp/dt) where with increase of ignition delay the maximum value for dp/dt is also retarded. Figure 4 presents the rate of pressure increase

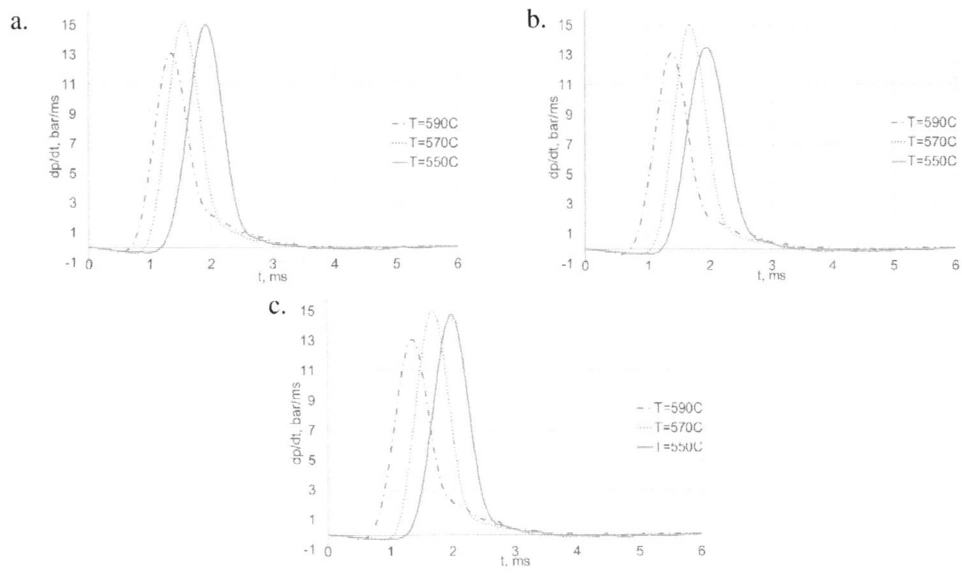

Figure. 4. Comparison of rate of pressure increase for various internal temperatures of the combustion chamber at various initial pressures a) 70bar, b) 62,5bar, c) 55bar.

For LFO, no significant change in rate of pressure increase in dependence of initial combustion chamber pressure was observed. In almost all cases the time of combustion stood not changed significantly. In this case it can be seen that with longer ignition delay the combustion at its end accelerates, as concluded from figures 3 and 4. For lower initial combustion temperature the combustion slows down. All tested fuels showed the same tendency.

Figure 5 presents ignition delay for all tested fuels as a function of initial parameters for the combustion chamber.

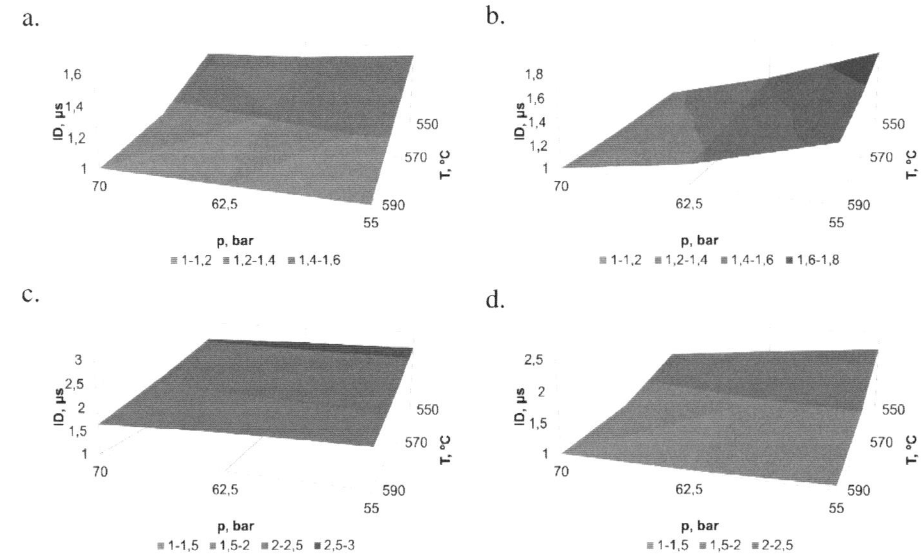

Figure. 5. Comparison of ignition delay (ID) for tested fuels, a) LFO, b) RSO, c) HFO, d) Glycerol.

For all fuels the ID has a similar tendency only for RSO more significant ID dependence of initial pressure can be observed. For LFO the major dependence on ID have an initial temperature. As can be seen, with decrease in temperature the increase in ID occurs. Furthermore, for the lowest initial combustion chamber temperature, dependence of ID and initial pressure can be observed. For HFO, Glycerol (figure 5 c and d) also for higher initial temperature the ID dependence with initial pressure can be observed. RSO however is characterized by very significant ID dependence with both initial pressure and temperature (figure 5 b). Such big difference compared to LFO can be caused by the cetane number for RSO which is 41 whereas for LFO it is above 50 according to Blin et al.[5]. HFO has a long ID which also can be related with low cetane number which is around 30 [1]. The glycerol ID is short but it is caused by pilot injection which is used to provide ignition of air-fuel mixture because pure glycerol is not able to ignite itself in the combustion engine.

As an example of the injection parameters, influence of combustion properties of HFO on combustion pressure rise is depicted in figure 6. As can be seen the combustion pressure decreases with decrease in injection pressure what is caused by two factors: worst evaporation because of larger droplets in spray and less amount of fuel injected during the injection. However for initial parameters of the combustion chamber equal to 70bar/590°C for all cases pressure increase plot is characterized by steep slope what mean that value of pressure increase rate is also high compared to initial conditions of 55bar/550°C. As can be seen in figure 6a the pressure increase slope is gentle and fuel needs more time to be burnt, what can be seen in figure 6c in which pressure increase rate has low value and lasts much longer compared to pressure release rate shown in figure 6d. This fuel is characterized with satisfied combustion properties under high initial combustion chamber conditions. For instance, LFO has good combustion properties for both low and high initial conditions. This is related with individual properties of this fuel such as the cetane number and density which affect combustion itself.

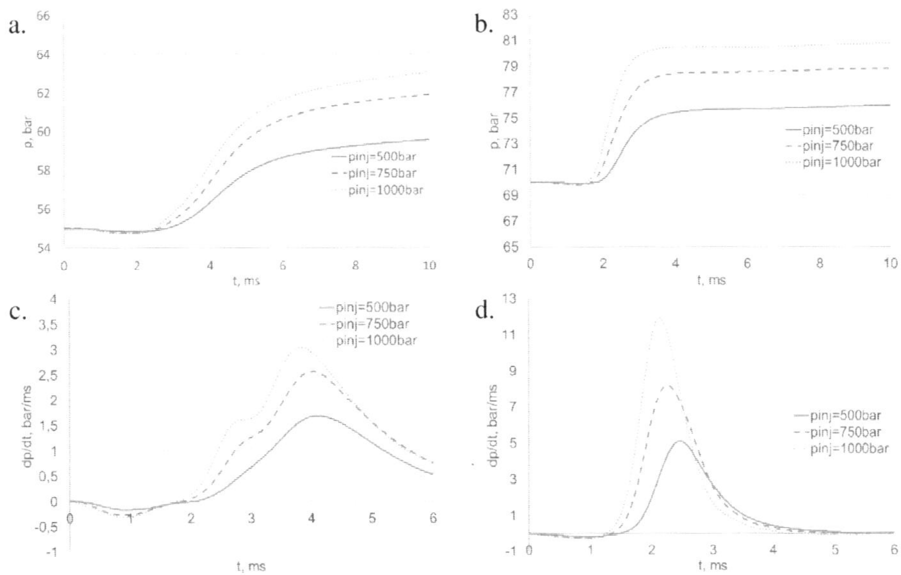

Figure. 6. Combustion pressure and rate of pressure increase for different pressure of injection and initial conditions of combustion chamber (a, c – 55 bar/550 ºC; b, d – 70 bar/590ºC; fuel – HFO).

Figure 7 presents ignition delay as a function of injection pressure for different initial conditions of the combustion chamber.

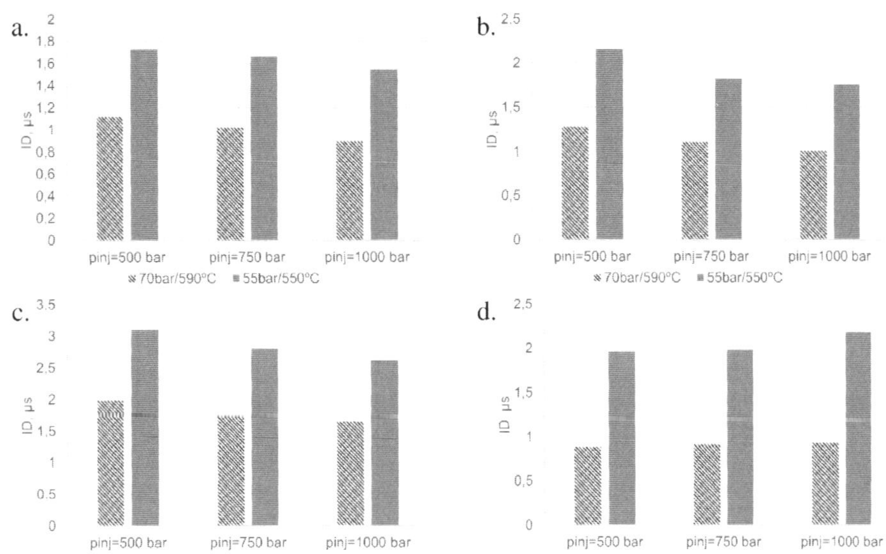

c

Figure. 7. Comparison of ID for different injection pressure and initial conditions of the combustion chamber (a) LFO, b) RSO, c) HFO, d) Glycerol).

As plotted, for LFO, RSO and HFO there is decrease in ID with increase in injection pressure what is related with better atomization of the fuel. Additionally, it can be noticed that when the initial conditions in the combustion chamber are high the ID decreases even more. Different situation is observed with glycerol, where one can notice that for low initial conditions in the combustion chamber it is slight increase in ID especially for pressure injection of 1000bar. It may be caused by that the glycerol is injected to the combustion chamber in high quantity therefore for low initial conditions it did not evaporate well because of its relatively high density and as a result the increase in ID occurred (figure 7d).

Figure 8 presents combustion pressure and rate of pressure increase for glycerol as a function of injection duration. It can be seen that for both cases of initial conditions for the combustion chamber both the combustion pressure and pressure increase rate plots as a function of injection duration start at the same moment.

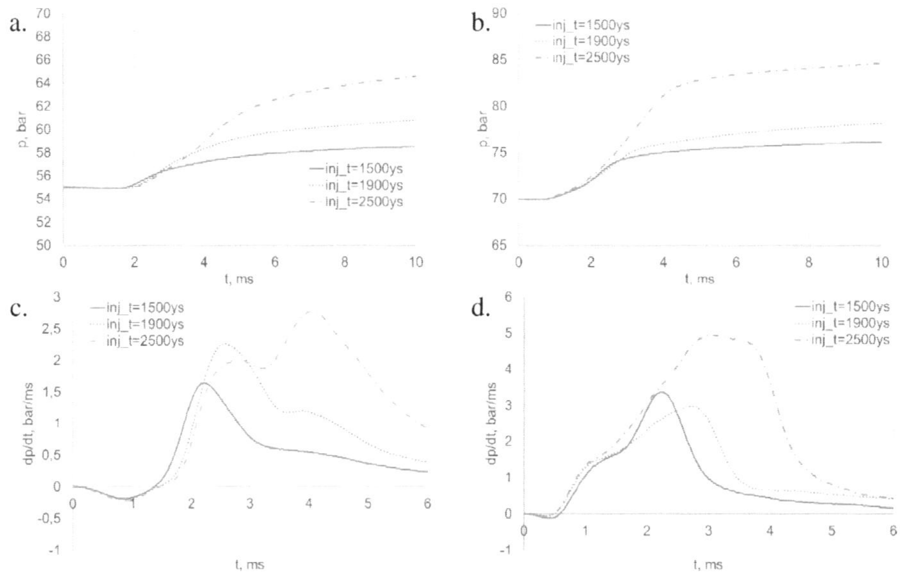

Figure 8. Combustion pressure and rate of pressure increase for different injection duration and initial conditions of combustion chamber (a, c – 55 bar/550°C; b, d – 70 bar/590°C; fuel – Glycerol).

The maximum values of the following: combustion pressure, rate of pressure increase and the time of combustion only differ from each other that is related with increasing amount of fuel injected to the combustion chamber. However, because injected fuel is the same, the changes in ID are considered as random, hence it cannot be seen any particular pattern what figure 9 presents. Most stable fuel in this case is LFO (figure 9a). As can be seen for high initial conditions ID is at the same level for different injection duration but for low initial conditions when the injection duration increases, the ID also increases what can be related to higher amounts of fuel that does not evaporate or evaporation process lasts longer before the ignition occurs.

Figure 10 presents combustion pressure, rate of pressure increase and ID as function of different kinematic viscosities for RSO in high initial conditions in the combustion chamber. As observed, in case of RSO the change in kinematic viscosity has an influence on aforementioned parameters especially combustion pressure and rate of pressure increase. With higher temperature the kinematic viscosity decreases what causes better atomization during the injection and better vaporization. As a result of this phenomenon the increase in combustion pressure occurs. The combustion for all kinematic viscosities lasts for similar time which is approximately 3ms what can be observed at figure 10b. A confirmation for this observation is figure 10c where ignition delay for all kinematic viscosities is at the same level which is approximately 1ms. For low initial conditions in combustion chamber the ID value decreases with kinematic viscosity increase which is related with better atomization of fuel, quicker evaporation and as a result advanced ignition.

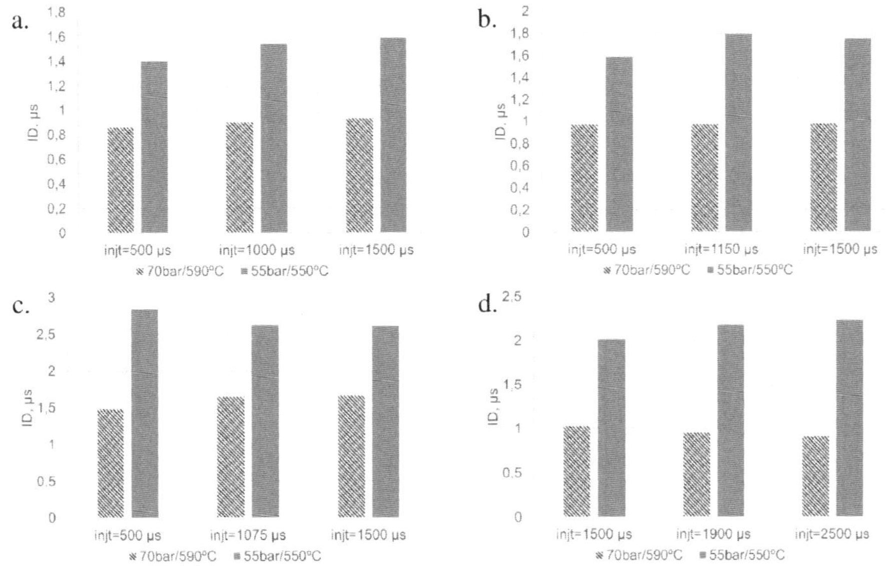

Figure 9. Comparison between ID for different initial conditions of combustion chamber a) LFO, b) RSO, c) HFO, d) Glycerol.

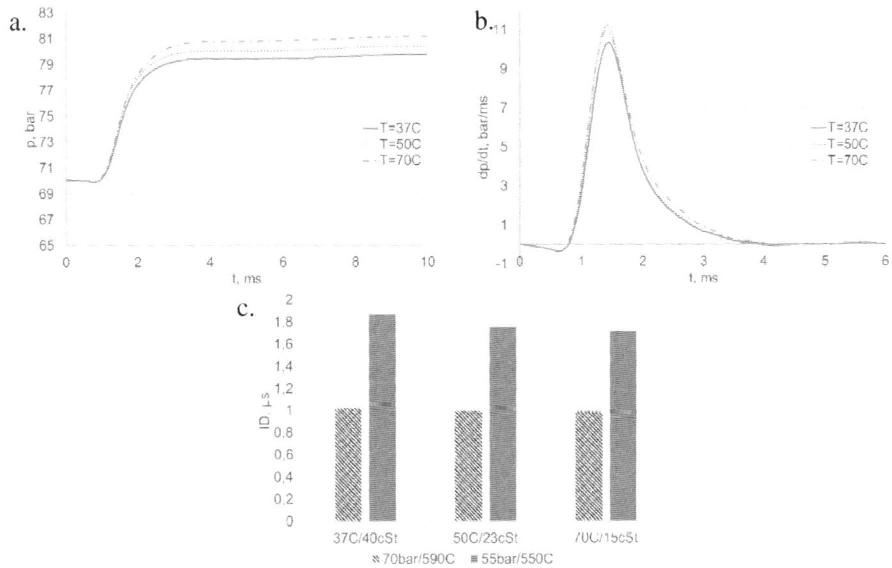

Figure 10. RSO characteristics at different kinematic viscosity a) combustion pressure, b) rate of pressure increase, c) ignition delay (ID).

5. Conclusions

The results of the research presented in this paper are a summary about possibility of use a CRU to perform fast tests for different type of fuels.

The CRU during this research was configured for combustion of the fuel that can be used in the compression ignition engine. Most advantages of this device is very fast configuration of test procedure, wide range of operational possibilities (change in combustion chamber parameters, injection parameters). This unit does not simulate real flow conditions in the engine cylinder, so there is no possibility to demonstrate evaporation process of the fuel spray which significantly affects air-fuel mixing process.

On a basis of the performed tests the following conclusions can be formed:

1) Initial temperature at the injection has crucial influence on ignition delay for all tested fuels. For RSO also initial pressure affects ID. That is caused by worse conditions for vaporization with decrease in initial temperature of combustion chamber.
2) Ignition delay depends on the cetane number of the fuel.
3) Injection pressure has bigger influence on combustion properties than injection duration especially if the ID is considered as the important parameter. That is related to the atomization of injected fuel.
4) Glycerol as a fuel for the internal combustion engine can be used only with pilot injection. Very small amount of LFO pilot dose enriched with glycerol makes the combustion less stable, slow and depleted. To avoid this problem higher amount of pilot dose should be used. However, thanks to LFO fuel used as a major pilot dose the ID is in satisfied range.
5) Kinematic viscosity of all the fuels has influence on combustion parameters. With higher kinematic viscosity the property of injection decreases which affect combustion pressure and ID, especially in low initial conditions in the combustion chamber (55bar/550°C).

Acknowledgements
This article is part of a project that has received funding from the European Union's Horizon2020 research and innovation programme under the Marie Skłodowska-Curie grant agreement No691232.
Authors would like to thank company Wärtsilä OY Finland for providing the test bench needed for realization of these research.

References

[1] Blin J, Brunschwig Ch, Chapuis A, Changotade O, Sidibe S et al. 2013 Characteristics of vegetable oils for use as fuel in stationary diesel engines - towards specifications for a standard in west africa. *Renewable and Sustainable Energy Reviews* vol. 22, pp.580-97.<10.1016/j.rser.2013.02.018>. <hal-00814899>
[2] Corsini A, Fanfarillo G, Rispoli F and Venturini P 2015 Poluttant emission in common-rail Diesel engines in extraurban cycle; rapseed oils vs Diesel fuel *Enrgy Procedia **82** (2015) 141 – 48*
[3] Demierel Y 2012 *Energy. Production, Conversion, Storage, Conservation, and Coupling* (London Springer-Verlag London Limited 2012)
[4] FUELTECH CRU materials
[5] Jiang L, Taylor RP and Agrawal AK 2012 Emissions and temperature measurements in glycerol flames Inthe 2012 Spring technical meeting of the central states section of the combustion institute, Dayton, OH, USA, (No. 030AF-0119)

IOP Conf. Series: Materials Science and Engineering **148** (2016) 012067 doi:10.1088/1757-899X/148/1/012067

Analysis of thermal stress of the piston during non-stationary heat flow in a turbocharged Diesel engine

P Gustof [1] and A Hornik[2]

[1] Silesian University of Technology, Faculty of Transport, 40-019 Katowice, Krasinskiego 8 Rd, Poland

[2] Silesian University of Technology, Faculty of Transport, 40-019 Katowice, Krasinskiego 8 Rd, Poland

E-mail: piotr.gustof@polsl.pl

Abstract. In the paper, numeric calculations of thermal stresses of the piston in a turbocharged Diesel engine in the initial phase of its work were carried out based on experimental studies and the data resulting from them. The calculations were made using a geometrical model of the piston in a five-cylinder turbocharged Diesel engine with a capacity of about 2300 cm^3, with a direct fuel injection to the combustion chamber and a power rating of 85 kW. In order to determine the thermal stress, application of own mathematical models of the heat flow in characteristic surfaces of the piston was required to show real processes occurring on the surface of the analysed component. The calculations were performed using a Geostar COSMOS/M program module. A three-dimensional geometric model of the piston was created in this program based on a real component, in order to enable the calculations and analysis of thermal stresses during non-stationary heat flow. Modelling of the thermal stresses of the piston for the engine speed $n=4250\ min^{-1}$ and engine load $\lambda=1.69$ was carried out.

1. Introduction

At present, still increasing demands both in terms of exhaust emissions and fuel economy, as well as the possibility to obtain high unit power are made in relation to modern diesel engines. Compliance with these requirements is possible, among others, due to a better understanding of the phenomena occurring inside the internal combustion engine, as well as development of various models for their subsequent multi parameter evaluation. Increasing engine power is accompanied by – apart from an increase in the mechanical loads – also an increase in thermal loads. These loads have a significant influence on such features of the engine as the exhaust blow through to the crankcase, the engine oil wear or the level of pollutants emitted by the engine to the atmosphere. Particularly a piston experiencing deformations variable in the time is exposed to these effects. Already at the stage of preliminary design work, knowledge on the thermal loads the piston will be subjected to is required. Such information can be obtained by suitable modelling of the temperature distribution in the studied cross-sections of the piston's characteristic surfaces. Design works and model tests are very expensive and prolonged. However, application of mathematical models and suitable computer techniques allows for determining the temperature distribution for different materials, sizes and shapes of the piston [12,13,14,15]. Subsequent analysis of this information helps to formulate initial design assumptions, as well as to modernize existing solutions for operation of the engine at various speeds and loads. Very good results in studies on thermal stresses of the piston are provided by application of the finite element method. However, in order to carry out any numerical calculations, some assumptions and use of empirical methods of data gathering for each of the analysed operation conditions of the engine are needed.

2. Analysis of the influence of selected parameters on the thermal loads of the piston

Thermal loads of components of a combustion chamber engine are the main factor limiting the performance and durability of Diesel engines [1]. The most loaded elements are the piston head and the surface of the engine cylinder head. Exceeding the permitted temperature of the piston causes various types of material losses and leads to an excessive increase in the diameter of the piston, resulting in its galling (Figure 1). In turn, temperature fluctuations may cause formation of cracks, leading to development of leakages and even preventing further work of the engine.

Figure 1. Traces of galling of the guide surface of a piston.

The heat effect during the engine operation causes reversible and irreversible dimensional changes in the piston. This is related closely with the suitable shape of the piston which determines the value of the clearance of the piston-cylinder assembly. The clearance of the piston in the cylinder has a significant influence on the tightness of the combustion chamber and thus on the size of the charge losses. This is particularly important for Diesel engines, where the loss of charge during compression reduces the real excess air ratio in the engine combustion chamber. Moreover, an improper clearance of the piston-cylinder assembly also has a negative impact on oil consumption, the capability to start the engine, and the noise. The piston temperature is significantly affected by the proper position of engine rings. From 60 to 80% of the heat acquired by the piston from the working medium is returned through the rings into the cylinder. Requirements for the location of the rings are incompatible. In order to reduce the thermal loads of the rings, they should be located as far away from the piston head as possible, and in order to reduce the thermal loads of the piston, they must be placed possibly high. For all engine speeds and loads, a distinct reduction in the piston temperature is achieved in all of its points by reducing the distance between the ring and the piston head. Changing the number and height of rings also causes a change in the piston temperature due to changes in the surface areas via which the heat exchange occurs.

Apart from the design conditions, also the service conditions of an engine affect the thermal loads of pistons [2,3]. The conditions of the engine operation include a change in the effective pressure. Assuming that mechanical efficiency is constant, an increase in the average effective pressure is equivalent to a higher heat emission, which leads to an increase in the piston temperature. The influence of the increase in the average effective pressure on unit heat flux for the parameter characterizing the heat load of the cylinder result from the following dependence:

$$K_c = bc_m^{0.56} \frac{p_e g_e}{D^{0.5}(\eta_v p_d)^{0.435}} \frac{T_d}{T_0} \tag{1}$$

where:

- b - coefficient of the amount of strokes (for four stroke b=2)
- η - fill coefficient
- c_m - average velocity of the piston [m/s]
- D - diameter of the cylinder [dm]
- p_d - pressure at the inlet valve [MPa]
- T_d - temperature at the inlet valve [K]
- p_e - average effective pressure [MPa]
- g_e - actual fuel consumption [g/kWh]
- T_0 - ambient temperature (T_0=293 K)

The increase in the average effective pressure of a Diesel engine is achieved by injection of a higher dose of fuel per working cycle into a cylinder. Thus, a higher amount of heat is transferred to the piston while reducing the excess air ratio. The lower the excess air ratio λ, the larger relative amount of heat may be released from the fuel in the cylinder. The temperature changes are analogous to the changes in the engine power. Both the mixture enrichment and impoverishment cause a drop in temperature of the piston. With a too rich mixture, a large part of the fuel is not burned. A lean mixture burns slowly, reducing the combustion temperature despite the fact that as a result of chronic combustion, the temperature of the exhaust gas increases. In a Diesel engine, an increase in the engine load leads to a decrease in the excess air ratio λ in the result of supplying more fuel. It leads to an increase in the amount of heat released in the engine's combustion chamber and in the temperature of the elements surrounding it, including the piston (Figure 2) [6].

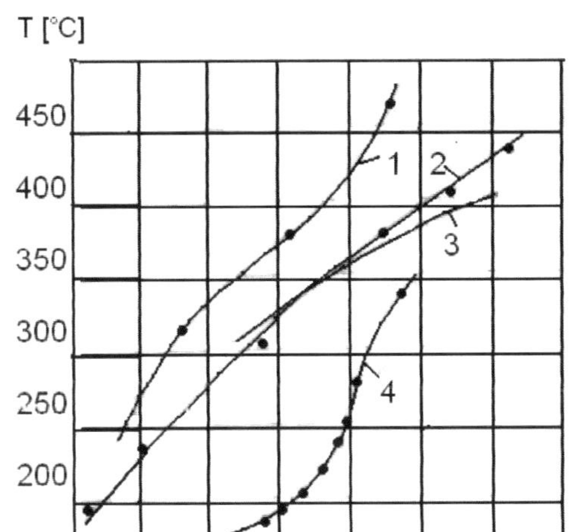

Figure 2. Temperature dependence of the piston *vs.* the excess air ratio for a self-ignition engine: 1 - engine with a pre-chamber; 2, 3 and 4 - direct injection engines.

For comparison, Figures 3 and 4 present modelled piston temperature distributions in a turbocharged Diesel engine with a capacity of about 2300 cm^3 with a direct fuel injection to the combustion chamber and engine power of 85 kW, for two loads corresponding to λ = 3.08 and λ = 1.66 at the engine speed of 2000 min^{-1} after 10 and 30 seconds of its work from the start up. Further information about boundary conditions and modelling of the temperature distribution in the piston can be found in the literature [8,9].

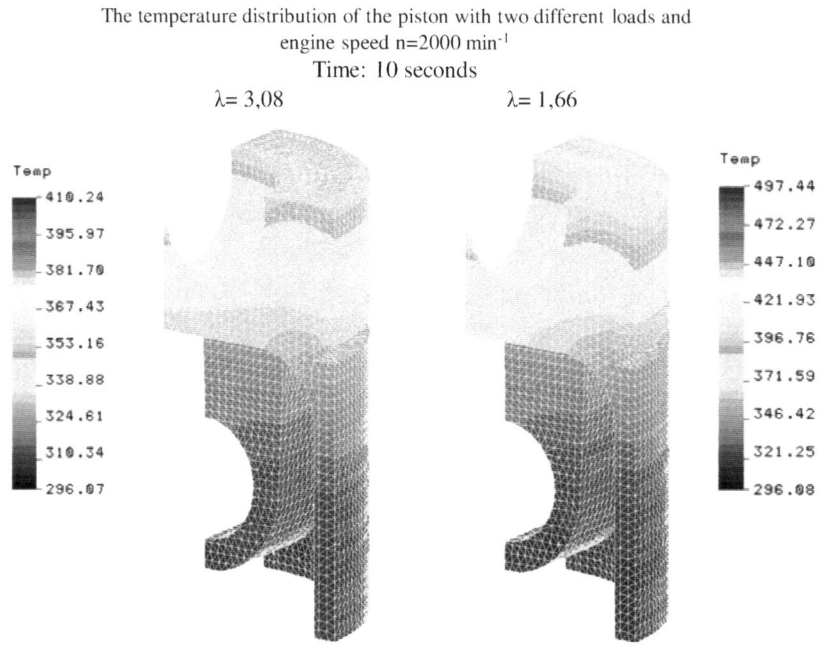

Figure 3. The temperature distribution in the piston for two loads of a turbocharged Diesel engine after 10 seconds.

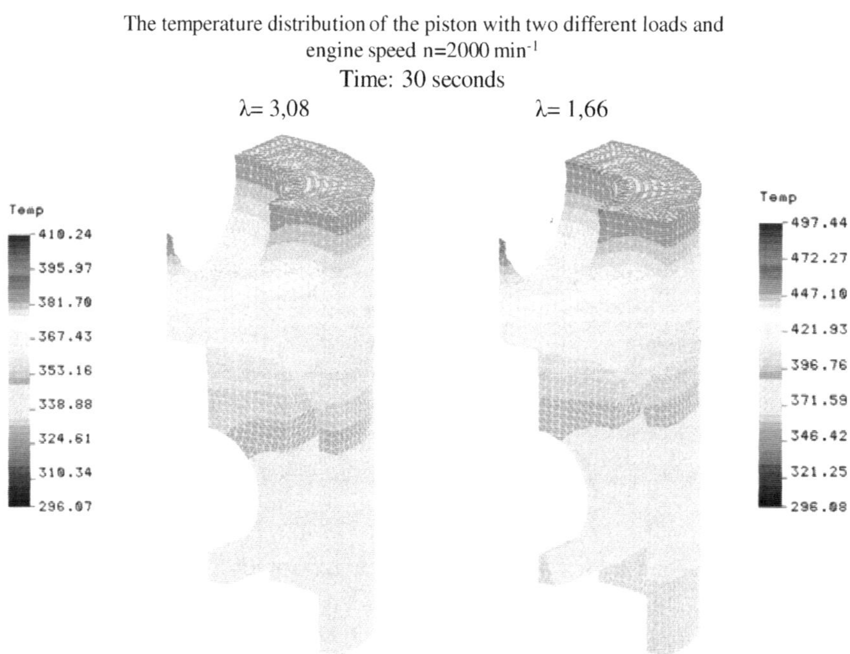

Figure 4. The temperature distribution in the piston for two loads of a turbocharged Diesel engine after 30 seconds.

IOP Conf. Series: Materials Science and Engineering **148** (2016) 012067 doi:10.1088/1757-899X/148/1/012067

Both the calculations and the research show that the temperature distribution as a function of time has a similar course for both loads. However, for a lower load at the same rotational speed of the engine, the temperature values are lower. Another parameter affecting the thermal load on the piston is the engine speed. The influence of this parameter is quite complicated. On the one hand, an increase in the engine speed is equivalent to an increase in the frequency of combustion in the cylinder, increasing the amount of the heat in the combustion chamber. On the other hand, it affects the change in the engine filling efficiency, and thus the course of the combustion process itself. In a Diesel engine, the amount of fuel being provided into the cylinder does not depend on the engine filling efficiency with air, because for all rotational speeds, the engine sucks in a different but maximum value of the air mass. The amount of the injected fuel depends on the curve of dosing characteristic of the injection pump and on the excess air ratio which depends both on the amount of intake air as well as injected fuel. Figure 5 shows the course of the calculated maximum values of the piston head surface temperature during a 60-second operation of a turbocharged Diesel engine, calculated counted from the engine start for its two rotational speeds $n = 2000\ min^{-1}$ and $n = 4250\ min^{-1}$.

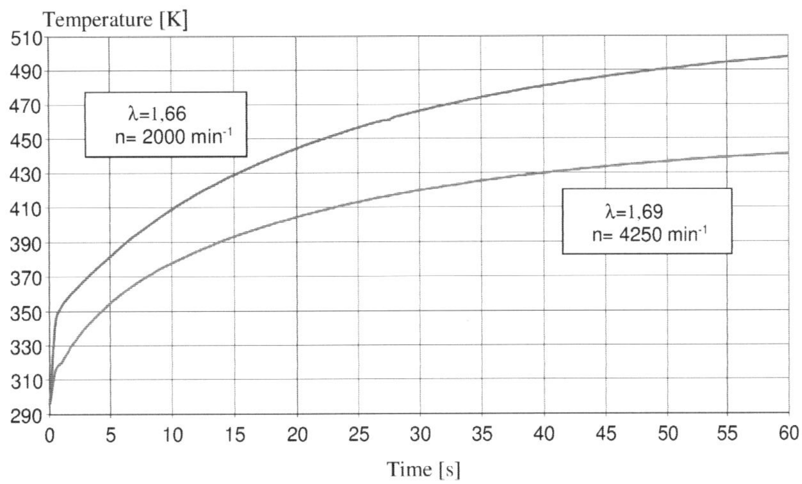

Figure 5. The course of maximum temperature on the surface of the piston head for two engine speeds.

Based on the calculations, it was found that an almost twofold increase in the rotational speed of a turbocharged Diesel engine for the same load causes lower thermal loads of the piston head [10]. This is due to a higher rate of heat exchange between the piston and its environment, in comparison with the amount of heat generated in the combustion chamber of the engine.

3. Modelling of thermal stresses of an engine piston

After starting the engine, the piston heats up until it reaches a state of equilibrium, which results from the balance between the heat taken from hot gases in the combustion chamber and transformed into useful work, and the heat transferred to the environment by, among others, the coolant and the combustion gases. Determination of thermal stresses of the piston by means of modelling required to assume equations and mathematical expressions for the calculations, describing the process of heat exchange in such a way in order for the model to reflect the actual processes occurring on the characteristic surfaces. This model was created for the piston on the basis of the differential equation of the heat flow in solids.

$$\frac{\partial T}{\partial t} = a\nabla^2 T + \frac{1\partial\lambda}{c_p\rho\partial T}\left[\left(\frac{\partial T}{\partial x}\right)^2 + \left(\frac{\partial T}{\partial y}\right)^2 + \left(\frac{\partial T}{\partial z}\right)^2\right] + \frac{q_v}{\rho c_p} \tag{2}$$

where:

a - coefficient of temperature compensation
c_p - specific heat capacity at constant pressure [J/kg K]
ρ - density [kg/m^3]
T - temperature [K]
q_v - volumetric efficiency of the internal heat source [W/m^3]
λ - thermal conductivity [W/mK]

The numerical calculations of the stresses were performed by means of the COSMOS/M program based on the knowledge of the temperature distribution for assumed operating conditions of the studied engine. In the program an actual three-dimensional discrete geometric model of the piston was created based on the real component (Figure 6) [7].

a) b)

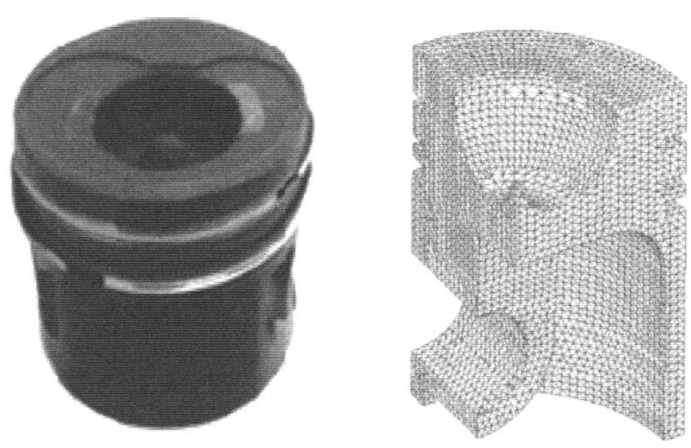

Figure 6. Geometric model of the piston: a) real component b) discrete model.

In the model, 16 characteristic the surfaces of the piston were distinguished, for which the temperature distributions and specific values of type III boundary conditions (Fourier conditions) were determined. These conditions determine the temperature of the medium surrounding the piston and the heat transfer coefficients of the characteristic surfaces. These surfaces are shown in Figure 7 [4].

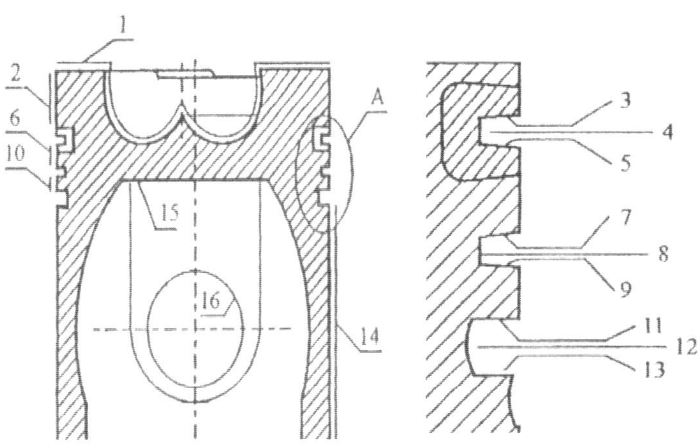

Surface of:
A
1 – piston head
2 – side of the piston over I ring
3 – the top surface of the ring groove I
4 – the groove bottom of the ring I
5 – the lower surface of the ring groove I
6 – between the rings I and II
7 – the top surface of ring groove II
8 – the groove bottom of the ring II
9 – the lower surface of ring groove II
10 – between the rings II and III
11 – the top surface of the ring groove III
12 – the groove bottom of the ring III
13 – the lower surface of the ring groove III
14 – the guide surface of the piston
15 – the internal surface of the piston
16 – the contact surface of the piston bolt

Figure 7. Characteristic of surfaces of the piston.

For numerical calculations of the thermal stresses of the piston in the state of a non-stationary heat flow, the material data of the AK 12 silumin were used. Due to the possibility to use the temperature curves in the COSMOS/M program, variable value of the thermal conductivity coefficient of the material as a function of temperature was accepted (fig. 8), among others.

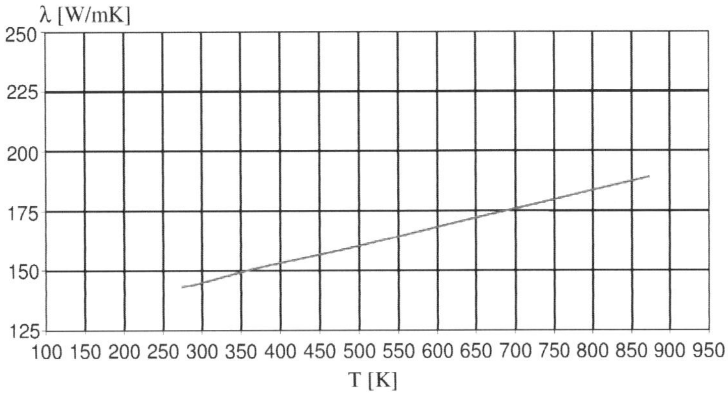

Figure 8. The graph of the thermal conductivity coefficient of the AK 12 alloy as a function of temperature

For individual piston surfaces, conditions of heat exchange equivalent to those in the combustion chamber for each cycle of the engine were assumed in the calculations. Based on the recorded indicator diagrams and the calculated total heat transfer coefficient, the temperature of the working medium surrounding the combustion chamber and the values of the heat transfer coefficient for the engine load of $\lambda = 1.69$ and speed of $n = 4250\ min^{-1}$ were determined [5]. Figures 9 and 10 show the exemplary values of thermal stresses of the piston during a non-stationary heat flow, corresponding to 10 and 20 seconds of engine operation measured from its start-up.

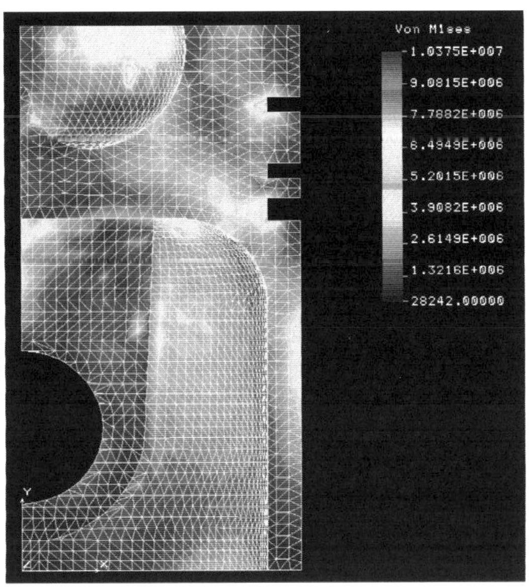

Figure 9. Thermal stresses of the piston after 10 seconds - front view.

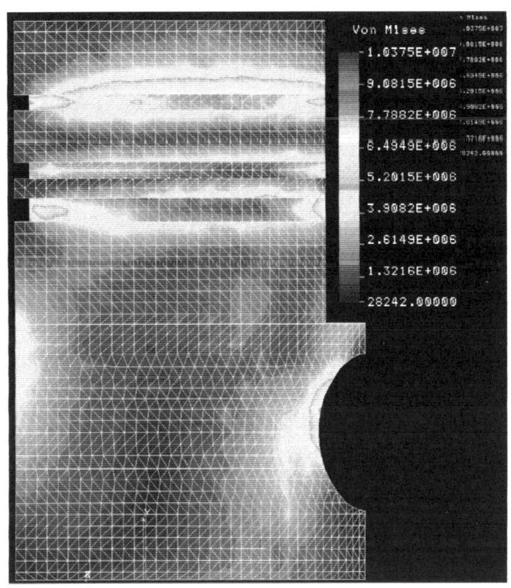

Figure 9. Thermal stresses of the piston after 10 seconds - back view.

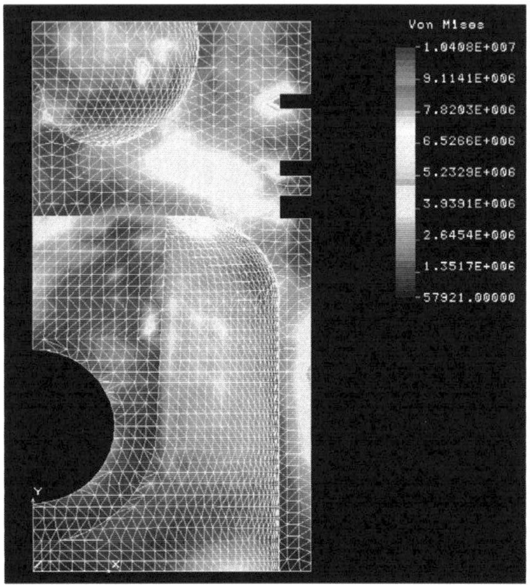

Figure 10. Thermal stresses of the piston after 20 seconds - front view.

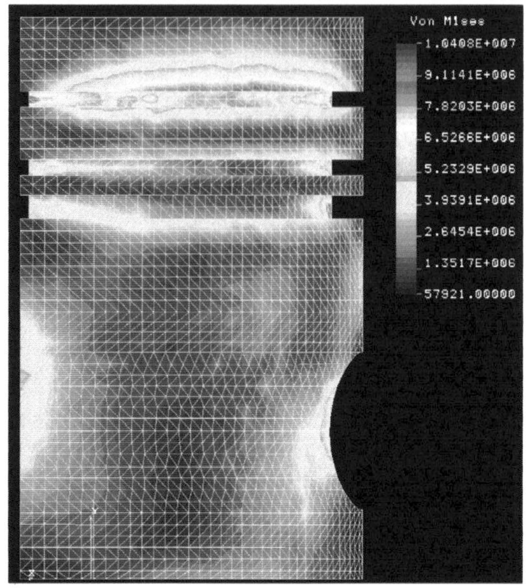

Figure 10. Thermal stresses of the piston after 20 seconds - back view.

The calculations carried out indicate that thermal stresses of the piston were increased during the warm-up of the engine. After 10 seconds of the engine operation, the highest thermal stresses occurred only on a small surface of the bottom groove of the third ring. However, after 20 seconds, the highest thermal stresses of the piston could be observed on the upper surface of the groove of the first ring and they included also the bottom side surface of the piston above this ring.

4. Conclusions

Based on the preliminary results of the calculations, it can be concluded that maximum values of thermal stresses of the piston are found mainly in the ring portion and on the surface of the piston head. On the other hand, the lowest values are found in the guide part. The obtained maximum values of thermal stresses for the engine speed of $n = 4250\ min^{-1}$ and $\lambda = 1.69$ did not exceed 12 MPa. In the state of a non-stationary heat flow, distributions of thermal stresses on the individual surfaces of the piston different in the time were obtained. According to the Authors, they represent, together with mechanical stresses, important factors in the design and subsequent operation of the piston for a specified engine kind and type. However, the calculations should be verified experimentally, to enable obtaining results of numerical calculations which would correspond to the real thermal stresses in the engine during warming up in the future [11].

References

[1] Wiśniewski S 1972 *Thermal loads piston engines*, Publisher of Communications and Connectivity, Warsaw 1972
[2] Jaskólski J, Budzik G, 1998 The influence of exploatation factors on the heat load of the piston in a diesel engine, *Technical Magazine* 5M /1998, Publisher Cracow University of Technology
[3] Jaskólski J, Budzik G 1999 Application overview finite element method for the determination of temperature fields in the piston, *Technical Magazine* 2M /1999. Publisher Cracow University of Technology
[4] Kwaśniowski S, Sroka Z, Zabłocki W. 1999 Modelling of thermal loads on components of internal combustion engines, Annexe Publishing the Technical University of Wroclaw, Wroclaw 1999

[5] Gustof P 2001 Determination of the course of the temperature load for the full working cycle in the cylinder of supercharged Diesel engine, *Scientific Papers of Silesian University of Technology Series: Transportation*. 43 Gliwice 2001
[6] Kozaczewski W 2004 The design of the piston-cylinder internal combustion engines, *Publishing Transport and Communications* Warsaw 2004
[7] Gustof P, Hornik A and Jędrusik D 2008 Determination of values and temperature distribution in the piston of supercharged Diesel engine, *Publisher Cracow University of Technology* No. 7-M / 2008 Krakow
[8] Gustof P, Hornik A 2009 The influence of the engine load on value and temperature distribution in the piston of the turbocharged Diesel engine, *Journal of Achievements in Materials and Manufacturing Engineering* Volume 35 Issue 2 August 2009
[9] Gustof P and Hornik A, 2010 Nonstationary heat flow in the piston of the turbocharged engine *Transport Problems* Volume 5 Notebook 1 Gliwice 2010
[10] Hornik A 2011 The influence of the engine speed on the temperature distribution in the piston of the turbocharged diesel engine *Transport Problems* Volume 6 Issue 3 Gliwice 2011
[11] Hornik A and Gustof P, 2012 The concept of experimental verification of modeled thermal loads of selected elements of combustion chamber in supercharged Diesel engine *Scientific Papers of Silesian University of Technology* No. Kol.1878 Katowice, 2012
[12] Bhagat A and Jibhakate Y 2012 Thermal Analysis And Optimization Of I.C. Engine Piston Using finite Element Method *International Journal of Modern Engineering Research*, Vol.2, Issue.4, July-Aug 2012 pp-2919-2921
[13] Zhang H, Lin Z, Xing J 2013 Temperature field analysis to gasoline engine piston and structure optimization *Journal of Theoretical and Applied Information Technology* 20th February 2013. Vol. 48 No.2
[14] Munyao E, He J, Zhiyuan Y and Yi Z 2014 Simulation of Thermal-Mechanical Strength for Marine Engine Piston Using FEA *Journal of Engineering Research and Applications*, Vol. 4, Issue 3(Version 1), March 2014 pp.319-323
[15] Babu K, Mahesh G and Gowd G 2014 Modelling and thermal analysis of SI Engine piston using fem *International Journal of Mechanical and Engineering and Robotics Research*, Vol. 3, No. 1, January 2014

The problems of piston skirt microgeometry in combustion engines

A Iskra[1], **M Babiak**[2] **and E Wróblewski**[3]
[1] Professor in the Faculty of Machines and Transport Poznan University of Technology, PL
[2] Doctor in the Faculty of Machines and Transport Poznan University of Technology, PL
[3] PhD Student in the Faculty of Machines and Transport Poznan University of Technology, PL

E-mail: emil.z.wroblewski@doctorate.put.poznan.pl

Abstract. Geometry of the slot between piston bearing surface and cylinder bore affects the friction losses of the IC engine to the far extent. It appears that these losses depend more on the area covered with oil than the thickness of oil layer separating collaborating parts. Barrel-shaped or stepwise piston bearing surface is the way to reduce the oil covered area. Turns out that the referred to friction losses contributes more to area covered by the oil film than the film thickness of the separation elements cooperating. The method to reduce the area covered by the oil film is a modification of the bearing surface of the piston by adjusting the profile. This paper presents the results of simulation leading to the reduction in friction losses and abrasive wear of piston bearing surface and cylinder bore. Covering the piston bearing surface with a thin layer of graphite one can get an extremely advantageous tribological properties of the piston assembly which means the expected parameters of oil film and in a case of film rupture-an ignorable abrasive wear of the graphite layer and/or cylinder bore.

1. Introduction

The advancement of modern engines is chiefly aimed at environment protection and low fuel consumption [3]. To meet the above-mentioned criteria, engines are designed based on the idea of downsizing characterized by a reduction of their displacement and increasing their effective pressure by the application of a variety of supercharging systems. The consequence of these changes is that engines operate at high thermal and mechanical loads. This is an adverse phenomenon for the piston-crankshaft assembly, particularly for pistons exposed to high temperatures and pressures inside the cylinder [4,6]. Modern engines are also expected to provide the highest overall efficiency. The way to ensure that is to reduce the friction loss in the piston-cylinder pair [7,15].
The methods to reduce the friction loss in the piston-cylinder pair are a subject of many publications that describe all types of design solutions, new materials or the application of surface layers of anti friction properties on the piston skirt [1,2].
The geometry of the gap between the piston skirt and the cylinder liner greatly affects the friction loss inside the engine. This means that the said friction loss is much more affected by the area covered by the oil film separating the mating elements than its thickness. A way to reduce the area covered by the oil film is the application of a stepped profile of the piston skirt [10,11,12]. According to theoretical considerations based on hydrodynamic theory of lubrication, the stepped shape of the gap reduces the friction loss.

The piston skirt microgeometry can be formed by the application of surface layers made from materials of good friction properties. The desired geometry of the profile can be achieved by applying thin coats on the cylindrical or conical surface of the piston skirt [5,8,14].

Modern coats are evenly applied on the surface of the piston skirt to modify the conditions of boundary friction. This, however, does not significantly affect the conditions of occurrence of liquid friction in the oil film. The most widespread coating technology is that based on graphite composites.

The results of investigations on the application of modifying layers applied on the piston skirt have confirmed a reduction in the friction loss in the piston-cylinder pair. A question arises, however, whether the microgeometry of the applied coat is optimal.

The paper presents the problems of microgeometry on the piston skirt of a combustion engine. The authors also present the results of a simulation research leading to a reduction of the friction loss and the improvement of the oil film parameters for the modified profile of the piston skirt.

2. The piston skirt profile

Piston skirt is to ensure axial displacement of the piston inside the cylinder and a transfer of lateral forces on the cylinder liner.

In reality, however, the cooperation of the piston skirt with the cylinder occurs on a much smaller area, which results in a reduction of stresses between these surfaces. Taking this fact into account, it is easily observable that the piston skirt operates in much worse conditions than those of generously lubricated sliding bearings. The lubrication of the piston skirt is much scanter. Ensuring the required durability of the piston skirt under such conditions is a very difficult task requiring a careful selection of materials and design solutions [13].

The distribution of temperatures on the piston circumference is uneven. The presence of the kingpin hubs and parts connecting the hubs with the piston makes the cross-section of this spot greater than the rest of the circumference. Parts of greater cross-section have lower thermal resistance, which makes the temperature on the piston skirt in these regions much higher than on the plane perpendicular to the kingpin axis. This results in greater thermal deformations towards the kingpin axis, which is why the piston skirt must be oval with a high ovality axis perpendicular to the axis of the kingpin and a low ovality axis alongside the kingpin to ensure that under operating temperatures the piston skirt becomes cylinder-shaped or close to it.

The operating force acting on the piston influences not only the friction conditions between the piston and the cylinder liner but also deforms the piston skirt, which is why, when cold, the guiding element should be shaped in such a way as to ensure cooperation with the cylinder liner on the largest possible area upon thermal deformation under operating temperatures and under the influence of the lateral forces.

The above-mentioned factors render the geometry of the piston skirt very complex, particularly in engines of high power ratio. When the piston reaches its operating temperatures, the play between the piston and the cylinder liner changes. Changes in the play also occur after any change of the operating conditions following the stabilization of the thermal equilibrium.

Such play variations are disadvantageous for a variety of reasons. One of them is the difficulty to ensure optimum play in terms of friction and lubrication under a sufficiently wide range of conditions. An increase in the play under lower operating temperatures results in an increase in dynamic forces exerted by the piston on the cylinder, vibration of the cylinder sleeve as well as engine noise and may lead to cavitation damage of cylinder liners in water-cooled engines.

The shape of the piston skirt in the axial cross-section, perpendicularly to the kingpin axis, and in the transverse cross-section should be designed in such a way as to ensure the maximum area of cooperation between the piston skirt and the cylinder liner under operating temperatures and under forces exerted on the piston skirt.

A modification of the geometry of the piston skirt consists in choosing such a shape of the surface that will ensure the continuity of the oil film during the piston reciprocating motion with a possibly low value of friction loss in the piston-cylinder pair.

The adopted microgeometry of the piston skirt is a result of a continuation of former research. The authors assume an application of a layer of graphite of the shape of letter H on the piston skirt. Thus

obtained stepped profile is characterized by good tribological properties under liquid friction and owing to the application of graphite, one can expect good properties under boundary friction.
In order to ascertain the friction loss in the piston-cylinder pair, 6 variants of piston microgeometry have been developed as shown in Figure 1-2.

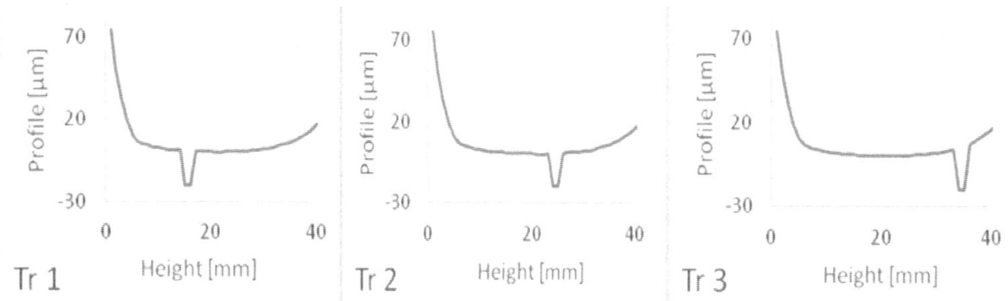

Figure 1. Tr1, Tr2, Tr3 variants of the piston bearing surface profile.

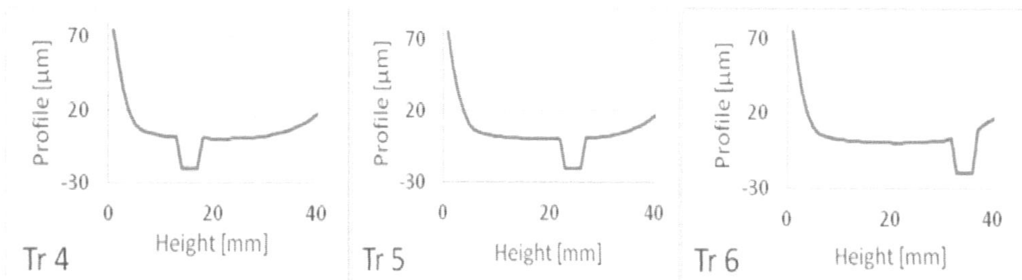

Figure 2. Tr4, Tr5, Tr6 variants of the piston bearing surface profile.

The horizontal bar of the letter H allows extending the area of high hydrodynamic pressure at the same time obtaining the effect of reduction of the oil film internal friction forces. The problem consists in determining the thickness proportion of the H bar guaranteeing high oil film bearing capacity at a limited internal friction force. Because the correctness of the computer model was validated experimentally in earlier research, the results confirm the positive results of the modification of the piston skirt microgeometry of a specified thickness of the H bar, which proves the validity of the replacement of classic barrel-shaped profile of the piston skirt with the stepped one by the application of a layer of graphite of a specified thickness.

3. Results of the simulation

The calculations of the parameters of the oil film were performed for different microgeometries of the piston skirt, starting from the barrel-shaped profile of the reference pistons and ending with the stepped piston skirt profile [9]. The thickness of the H bar is variable as well as its height in the bottom or top position against the symmetrical one. In the form of diagrams, figures 3-6 show the results of the simulation of friction loss for barrel-shaped reference pistons and the pistons of a modified piston skirt. Based on the performed analyses one can observe that the developed variants of microgeometry of the piston skirt have given the expected reduction of the friction loss in the piston-cylinder mating pair, compared to the reference pistons.

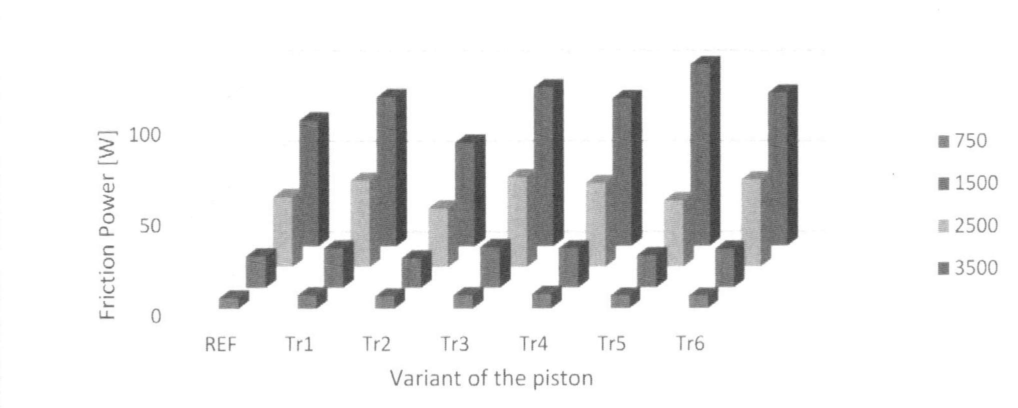

Figure 3. Friction loss in the piston skirt-cylinder liner pair as a function of engine speed and oil temperature of 100 °C.

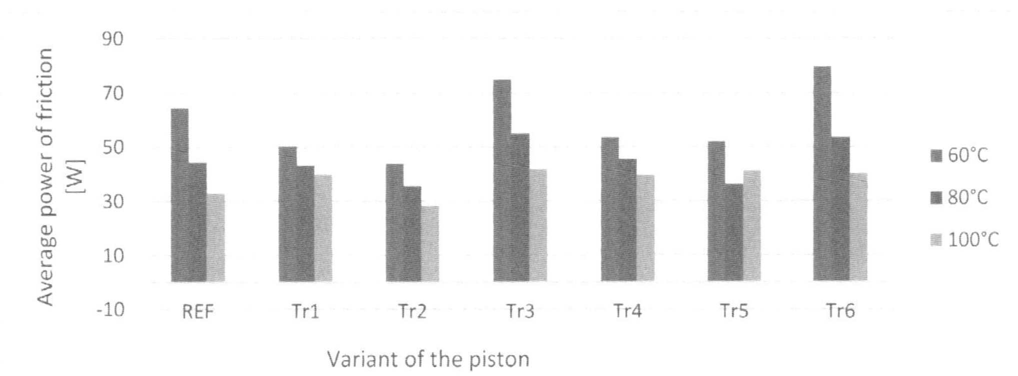

Figure 4. Results of average simulation of the friction loss in the piston skirt-cylinder liner pair for different oil temperatures.

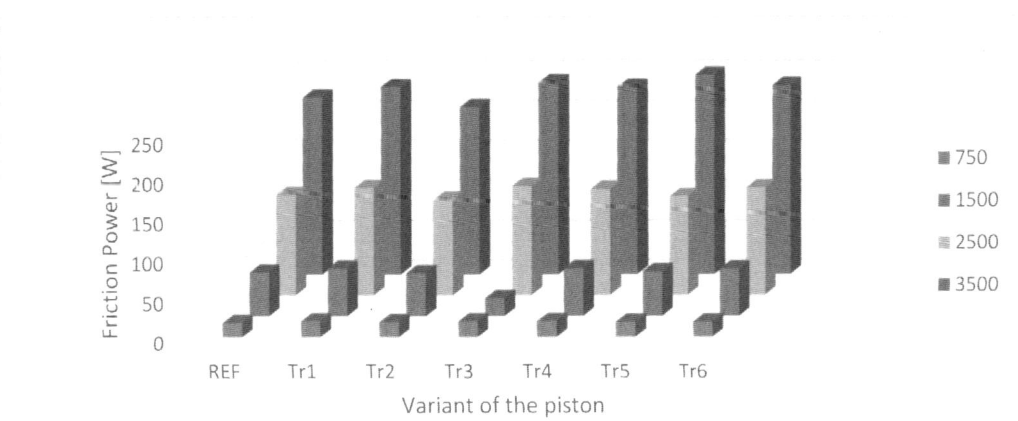

Figure 5. Friction loss in the pair (piston with a set of rings-cylinder liner) as a function of engine speed and oil temperature of 100 °C.

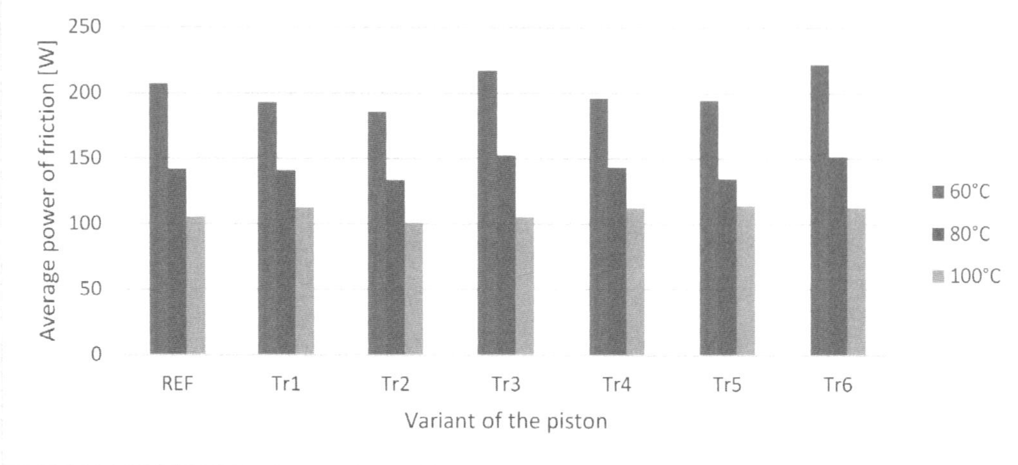

Figure 6. Results of average simulation of friction loss in the pair (piston with a set of rings-cylinder liner) for different oil temperatures.

As it turns out, the most advantageous solution is the symmetrical position of the horizontal H bar, as the friction loss decreases by approx. 10% compared to pistons of a standard profile. One could also observe that the reduction in the friction loss is greater for higher engine speeds. This is an important observation because, due to increased friction loss converted into heat when increasing the engine speed, the engine elements warm up faster and the oil viscosity drops, which may lead to boundary friction. This phenomenon may reduce the friction loss assuming that boundary friction in the piston skirt - cylinder liner pair does not occur.

From the equations of hydrodynamic lubrication theory we know that in order to form a hydrodynamic oil film one needs a narrowing oil gap. As results from the performed simulations, the microgeometry of the piston skirt significantly influences the friction loss in the piston - cylinder pair. The obtained results are explained by the graphs (figures 7-9) presenting the coverage of the piston skirt with oil film. The graphs have been made for the reference piston and the TR2 and TR5 variants for the engine speed of 2500 rpm with the following parameters:

- Nr [W] - friction loss in the piston skirt - cylinder liner pair,
- Nc [W] - total friction loss of the piston with a set of rings.

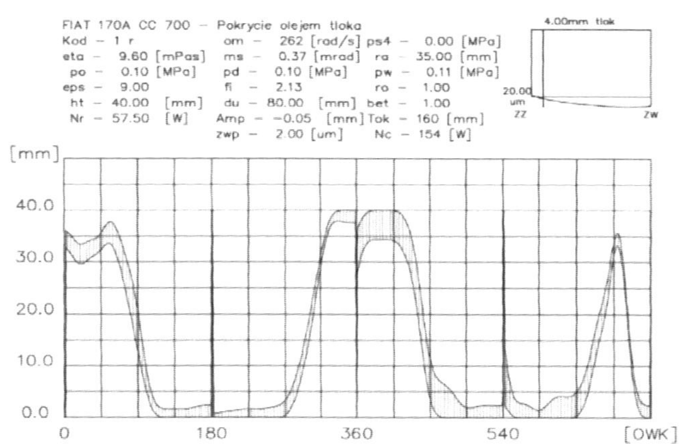

Figure 7. Oil film height on the reference piston skirt.

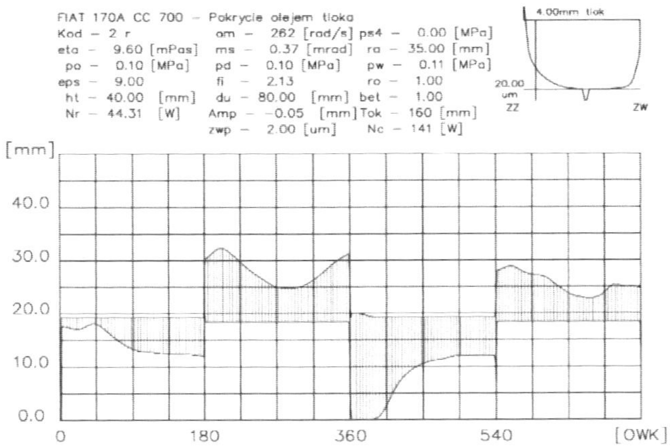

Figure 8. Oil film height on the TR2 piston skirt.

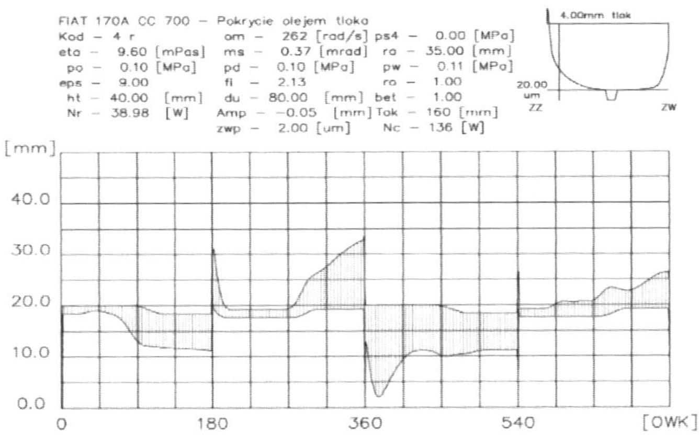

Figure 9. Oil film height on the TR5 piston skirt.

Comparing the oil films on the piston skirts one can observe that intense oil film results in an increased friction loss. If the differences in the oil film level on the piston skirt are small, even minimum variations are impactful on the friction loss. The oil film thickness is greater for the stepped profile.

4. Conclusions

- Replacing the barrel-shaped piston skirt with the stepped one may yield a reduction of the friction loss of approx. 10%.
- The stepped surface can be formed by applying coats such as graphite.
- The coats are particularly needed when lack of continuous oil film occurs i.e. the engine is off and its temperature equals the ambient temperature.
- Replacing the barrel-shaped piston skirt with the stepped one reduces the cost of the sliding surface processing of the piston and allows preserving greater tolerance at constant cost of the piston production.

5. References

[1] Baberg A, Freidhager M, Mergel H, Schmidt K 2012 Aspekte der Kolbenmaterialwahl bei Dieselmotoren MTZ 12/2012
[2] Backhaus R 2009 Kolben aus Stahl fur PkW-Dieselmotoren MTZ 12/2009
[3] Blümm M, Baberg A, Dörnenburg F, Leitzmann D 2016 Innovative Schaftbeschichtungen für Otto- und Dieselmotorkolben MTZ 02/2016
[4] Deuss T, Ehnis H, Rose R, Künzel R 2010 Reibleistungsmessungen am Befeuerten Dieselmotor-Potenziale der Kolbengruppe MTZ 5/2010
[5] Deuss T, Ehnis H, Rose R and Künzel R 2011 Reibleistungsmessungen am Befeuerten Dieselmotor-Einfluss von Kolbenschaft-beschichtungen MTZ, 4/2011

[6] Dörnenburg F, Lades K, Kenningley S 2010 Neue Technik für Höhere Warmfestigkeit von Aluminiumkolben MTZ 04/2010
[7] Fahr M, Hanke W, Klimesch Ch and Rehl A 2011 Reibungsreduzierung bei Kolbensystemen im Ottomotor MTZ 07-08/2011
[8] Golloch R 2001 Untersuchungen zur Tribologie eines Dieselmotors im Bereich Kolbenring/Zylinderlaufbuchse VDI Verlag GmbH Reihe 12 Nr 473 Düsseldorf 2001
[9] Iskra A 1995 Studium Konstrukcji i funkcjonalności pierścieni w grupie tłokowo-cylindrowej Wydawnictwo Politechniki Poznańskiej Poznań 1995
[10] Iskra A, Babiak M and Wróblewski E 2015 Wpływ schodkowego profilu powierzchni nośnej tłoka na straty tarcia silnika spalinowego Technika Transportu Szynowego 12/2015 Instytut Naukowo-Wydawniczy „TTS" sp. z o.o. Radom 2015 pp 644-647
[11] Iskra A, Babiak M and Wróblewski E 2015 Geometric shape of the support surface of the piston Journal of KONES section Vol 22 No 4 pp 95-101 European Science Society of Powertrain and Transport Publication Warsaw 2015
[12] Iskra A, Krzymień P, Wróblewski E 2015 Geometry of the improved layer over a piston skirt Combustion Engines 162(3) pp 192-196 PTNSS Bielsko-Biała 2015
[13] Kozaczewski W 2004 Konstrukcja grupy tłokowo-cylindrowej silników spalinowych WKŁ Warszawa 2004
[14] Kuberczyk R, Berner J and Bargende M 2009 Wirkungsgradunterschiede zwischen Otto- und Dieselmotor MTZ 01/2009
[15] Werkmann M, Bitter A and Künzel R 2011 Einfluss der Schaftsteifigkeit auf das Kolbengeräusch MTZ 07-08/2011

Analysis of optimisation method for a two-stroke piston ring using the Finite Element Method and the Simulated Annealing Method

M Kaliszewski[1] and P Mazuro[2]
[1] Former student of Warsaw University of Technology
[2] Assistant professor in Institute of Heat Engineering in WUT

E-mail: markaliszewski@gmail.com, pawel.mazuro@itc.pw.edu.pl

Abstract. Simulated Annealing Method of optimisation for the sealing piston ring geometry is tested. The aim of optimisation is to develop ring geometry which would exert demanded pressure on a cylinder just while being bended to fit the cylinder. Method of FEM analysis of an arbitrary piston ring geometry is applied in an ANSYS software. The demanded pressure function (basing on formulae presented by A. Iskra) as well as objective function are introduced. Geometry definition constructed by polynomials in radial coordinate system is delivered and discussed. Possible application of Simulated Annealing Method in a piston ring optimisation task is proposed and visualised. Difficulties leading to possible lack of convergence of optimisation are presented. An example of an unsuccessful optimisation performed in APDL is discussed. Possible line of further optimisation improvement is proposed.

1. Introduction

Piston ring discussed in the paper is designed in order to seal combustion space in a two-stroke barrel engine developed in The Faculty of Power and Aeronautical Engineering of Warsaw University of Technology. The aim of the paper is to analyse possible use of Simulated Annealing Method for the purpose of optimisation for the mentioned piston ring. Since optimisation did not bring a positive result, the outcome data is presented and some of improvement methods are suggested.

The radius of a cylinder associated to the ring is ϕ55mm. It was decided to use conical shape piston ring of 15' to 30' inclination and a rectangular gap. Its conical shape role is to allow for faster lapping process. Titanium alloy was chosen for a ring material. Ring properties are briefly described in table 1.

Table 1. Ring properties

Width	g	2.5 mm
Depth	h	1 mm
Deflected ring diameter	D	55 mm
Young's modulus	E	625 GPa
Poisson ratio	υ	0.21
Bending strength	k	1.6 GPa

Hence the piston is allowed to rotate freely in a cylinder due to its assembly there is no need for constraining piston ring to its piston. In a side wall of the cylinders the inlet and outer holes are present. Their presence significantly influences the rings wearing. Especially their butt ends are exposed to collide with holes edges, which is the reason why ring pressure on a cylinder wall has to be reduced at its ends [1].

Nowadays, two leading methods of determining demanded undeflected ring shapes dominate in the literature. First of them is purely analytical method described by Kozaczewski [2]. It is based on Euler–Bernoulli beam theory assuming small deflections of a ring and possibility of pressure distribution transformation into a Fourier series. Another one is a discrete method presented by Iskra [3]. The method basically approximates ring deflection by its discretisation into the set of finite number of infinitely stiff beams attached to each other by deflectable pins. The pins stiffness is computed basing on Young modulus and cross-section moment of inertia of a ring.

2. Aim of optimization

The main goal of optimization is to determine free shape of a piston ring, which would exert demanded pressure on a cylinder just while being bended to fit the cylinder. For the two-stroke piston engine function of demanded rings pressure is a constant value (p_N = 0.2 MPa) along a whole circumference of a ring excepting its ends. During 5° nearest to ring ends, pressure should drop to the value between 0 - 20% of a nominal pressure (p_E = 0÷0.04 MPa) as it is described in a figure 1. However, it is not possible to obtain this value since such pressure distribution does not fulfil static equilibrium condition because there occurs a resulting force heading in the right direction. For that reason it was decided to use another function based on a Fourier Series [3]. It may be defined by the equations (1-4) and is described in a figure 2. In fact, the mentioned equations still do not fulfil equilibrium condition perfectly due to a presence of a gap between ring ends, however unbalanced force drops by 98.4% (from R_x=0.89N to R_x=0.014N).

$$p_{opt}(\theta) = p_0 + \sum_{n=1}^{N} \left(p_n \cos(n\theta)\right) \tag{1}$$

$$p_0 = p_N \tag{2}$$

$$p_1 = 0 \tag{3}$$

$$p_n = 2(p_N - p_E)(N-1)^{-1}[1-(i-1)N^{-1}] \tag{4}$$

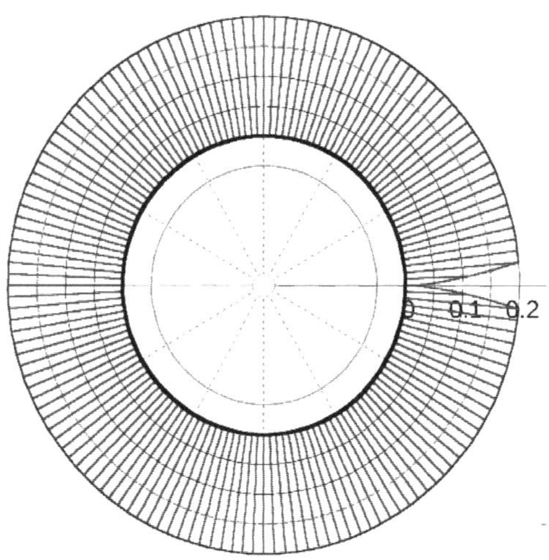

Figure 1. Demanded pressure distribution by formula suggested by Iskra [3].

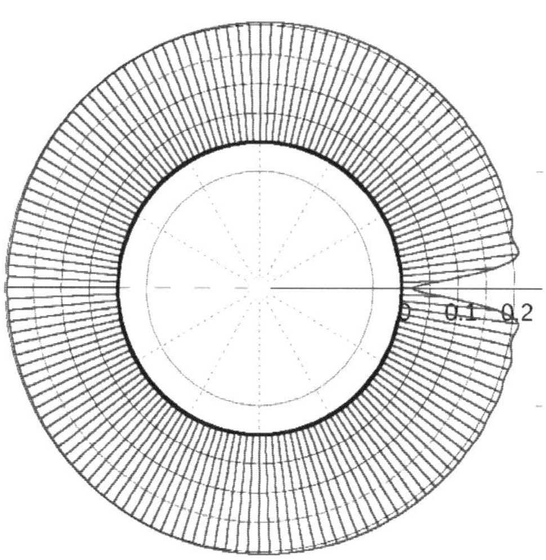

Figure 2. Pressure distribution for N = 44.

2.1. *Objective function*

The objective function is divided into two subfunctions. First of them is defined by (5) and it is a measure of how close the pressure along a circumference ($p(\theta)$) is to desired pressure. Second of them is defined by (6) and it indicates how close the pressure at the ring end is close to a demanded value. The general objective function is defined simply by adding both together.

$$F_N = \int_{-\theta_k}^{\theta_k} \left|p(\theta) - p_{opt}(\theta)\right| d\theta \cdot \left(\int_{-\theta_k}^{\theta_k} \left|p_{opt}(\theta)\right| d\theta\right)^{-1} \quad (5)$$

$$F_E = \left|p(\theta_k) - p_E\right| \cdot \left(p_E\right)^{-1} \quad (6)$$

Pressure distribution for an arbitrary geometry of the ring is determined via FEM contact simulation with no friction and no separation. Contact simulation as general algorithm was performed in ANSYS software by APDL. Due to the simple geometry of a ring it was modelled as 2D case in plane stress state with a symmetry boundary condition applied to the centre of a ring. All elements of a ring were PLANE182 quadrilateral elements as shown in a figure 3. Contact simulation was chosen to be no separation contact with penetration tolerance of constant value of 10^{-5} due to its speed and acceptable accuracy. Also an option of large displacement solution with maximum number of 10 substeps was chosen. In this kind of simulation it is possible to obtain as well positive as negative values of contact pressures. It is not physical, however it makes simulation more sensitive than separation contact simulation since for the same geometry negative pressures returns higher objective function.

Contact pressure distribution is being computed using APDL commands via reading nodal forces and dividing them by associated areas computed by equations (7-8). In those equations l_i and l_{i+1} stands for length along cylinder circumference of nearest node elements after being deflected by a contact simulation. For first and last node l_i and l_{i+1} were substituted by zeros respectively, because they are associated with only one element.

$$p_i = F_i \cdot A_i^{-1} \quad (7)$$

$$A_i = \left(l_i - l_{i+1}\right) \cdot h \cdot 2^{-1} \quad (8)$$

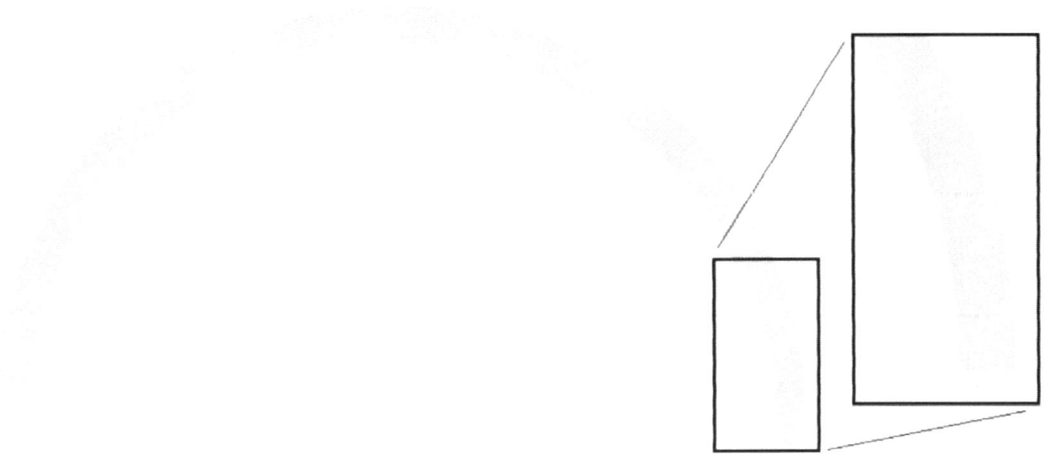

Figure 3. Typical ring mesh

2.2. *Simulated Annealing Method*

It was decided to use Simulated Annealing Method due to its insensitivity to get trapped in local minima. Simulated Annealing algorithm used in optimization is presented in the list below (prepared basing on [4] and [5]) and visualised in figure 4. In a scheme $\overline{x_0}$ stands for actual most optimal geometry, $\overline{x_1}$ stands for a new randomly generated geometry. $F(\overline{x_0})$ and $F(\overline{x_1})$ are their objective functions and

$p(\mathrm{F}(\overline{x_0}), \mathrm{F}(\overline{x_1}), T)$ is a probability function which is given by an equation (9) while r is a random number generated from uniform distribution of $\langle 0,1 \rangle$ set.

1) Generate first input geometry $\overline{x_0}$
2) Determine its objective function $F(\overline{x_0})$
3) Generate another random geometry $\overline{x_1}$ in neighbourhood of $\overline{x_0}$ according to Gaussian distribution and find its objective function $F(\overline{x_1})$
4) Check $F(\overline{x_1}) < F(\overline{x_0})$ condition:
    - if positive, substitute $\overline{x_0}$ with $\overline{x_1}$ and $F(\overline{x_0})$ with $F(\overline{x_1})$
    - if negative, check condition $p(F(\overline{x_0}), F(\overline{x_1}), T) > r$:
        - if positive, substitute $\overline{x_0}$ with $\overline{x_1}$ and $F(\overline{x_0})$ with $F(\overline{x_1})$
        - if negative, do nothing
5) Repeat commands 3-4 N_1 times
6) Lower the temperature according to the schedule
7) Repeat commands 3-6 N_2 times

$$p(\mathrm{F}(\overline{x_0}), \mathrm{F}(\overline{x_1}), T) = \exp\left(-\left(\mathrm{F}(\overline{x_1}) - \mathrm{F}(\overline{x_0})\right) \cdot T^{-1}\right) \tag{9}$$

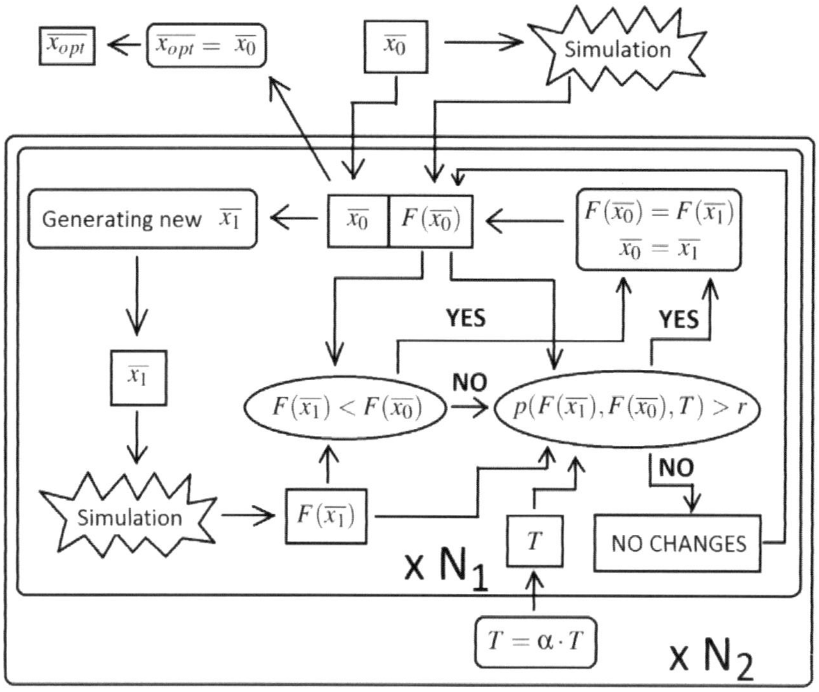

Figure 3. Simulated Annealing Method scheme

3. Problems during optimisation and results

During optimisation process design and its implementation several problems had occurred and optimisation did not bring expected results.

3.1. Local minima

It is possible to prove by mechanics of structures methods that there is only one unique geometry which would possibly bring demanded pressure distribution [2]. Due to bending deflection analysis of a ring [3] it is expected for radial coordinate value to consequently grow as long as angular coordinate get further from ring's centre. For such geometry the objective function will obtain value near to 0 (it cannot bring 0 value since defined p_{opt} does not fully fulfil equilibrium equations) which is the first deep minimum. The second deep minimum is a geometry of constant radius of $r = 27.5mm$ which will obviously result in $F = 2$. Taking into account a fact that even small geometry disturbances results in huge objective function growth (maximum objective function spotted in a simulation was over 1 million) they are very deep and very far from each other. It is also expected to spot many other local minima.

3.2. Defining geometry of the ring

Tested in the paper method of defining geometry consists of 31 polynomials of 5th order in polar coordinate system as shown in equation (10). Each of them is defined at interval of $\Delta\theta = 6°$ except last three of them which are defined at the interval of $\Delta\theta = 4°$. Their coefficients are computed from boundary conditions for r, r', r'' in 32 nodes. Common nodes for two consecutive polynomials results in automatically fulfilled continuity conditions up to the second derivative.

$$\mathrm{r}(\theta) = \mathrm{a}\theta^5 + \mathrm{b}\theta^4 + \mathrm{c}\theta^3 + \mathrm{d}\theta^2 + \mathrm{e}\theta + f \qquad (10)$$

The basic problem of such geometry definition is its complexity. Full geometry is defined by 94 variables (number of sum of polynomial coefficients is decreased by 2 because $\mathrm{r}(0)$ and $\mathrm{r}'(0)$ are already assumed to be 0 due to symmetry of a ring). Any of those variables could change in two ways – they can rise or drop – but only one of those directions leads geometry toward the optimal geometry (which is unique as it was previously noted). For that reason probability of getting closer to optimal solution is concluded to be not larger than $p = 2^{-94}$. Therefore only one of $1.98 \cdot 10^{28}$ simulations has even a chance of getting closer to a proper solution disputing that there are probably needed a few steps to obtain an optimal geometry. Even assuming that every optimisation step takes only one second (which could be obtained probably on the few fastest computers in the world, if any) it would take far too much computing time to be reasonable.

3.3. Results for 5th order polynomial piecewise function

Function presented in a previous paragraph was tested. It was tested in a few configurations of geometry definition, however only one of them was chosen as the most representative and is presented in the article. Initial geometry was very roughly estimated basing on Smoliński program [6]. Initial geometry as well as assumed standard deviations of geometry generation are presented in a figure 5 and table 2, respectively. Temperature function was set to start from $T_0 = 492956$ and its drop to be $\alpha = 0.6$. Temperature have dropped 29 times until ended at $N_2 = 30$ and T= 0.39. For last temperature probability function in case of objective function growth of 1 takes value of only $p = 0.0041$. Temperature function is presented graphically at figure 6. At each temperature the simulation was performed $N_1 = 200$ times.

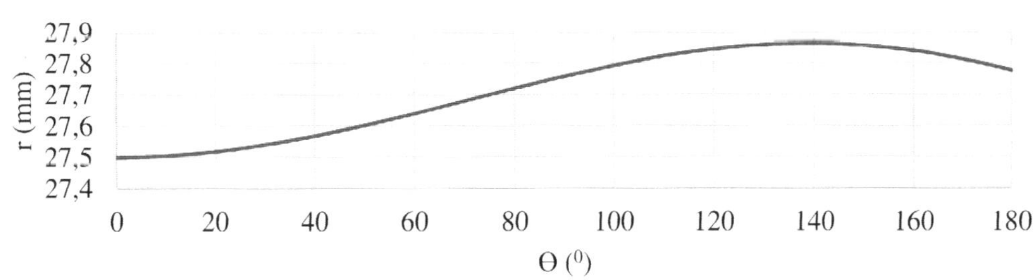

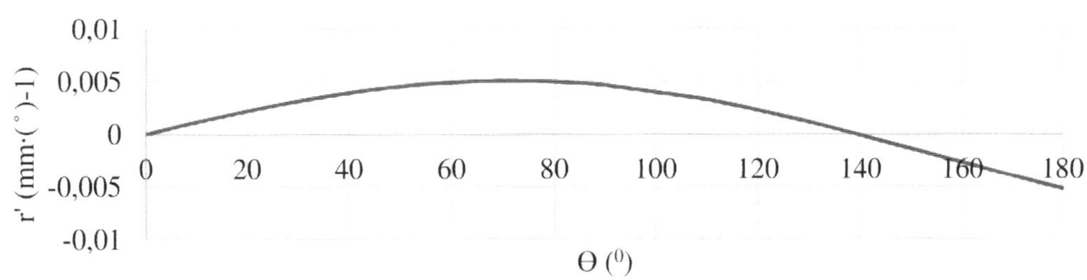

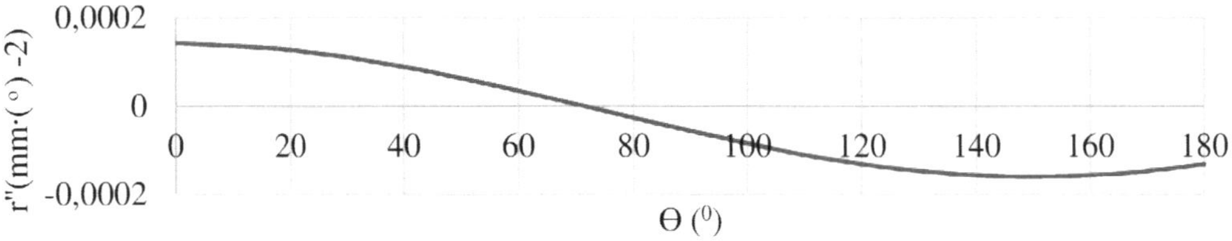

Figure 5. Initial geometry functions r, r', r''(θ).

Table 2. Standard deviation values

σ_r	0.003 mm
$\sigma_{r'}$	$0.0002\ \mathrm{mm} \cdot (°)^{-1}$
$\sigma_{r''}$	$0.0000009\ \mathrm{mm} \cdot (°)^{-2}$

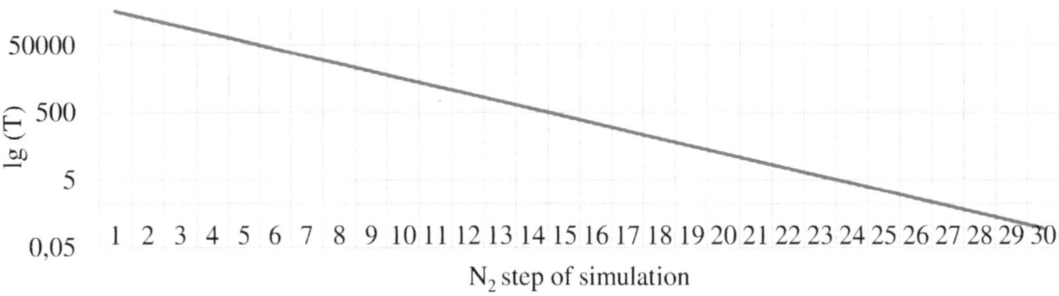

Figure 6. Temperature function

Optimization process is presented in figure 7. As it is typical for Simulated Annealing Method in the beginning the current most optimal geometry almost every time changes its value to a new value whether it is lower or not. As the temperature comes down, geometry changes less rapidly and only in case of the better values of objective function. The lowest objective function was not obtained at the end of optimisation – probably it could be improved by increasing standard deviation values. However, even the lowest of objective function values obtained cannot be accepted since its value is as much as 23603 which corresponds to summary 2360300% of pressure divergence from assumed distribution. Pressure distribution for geometry corresponding to the lowest value of objective function is presented in a figure 8.

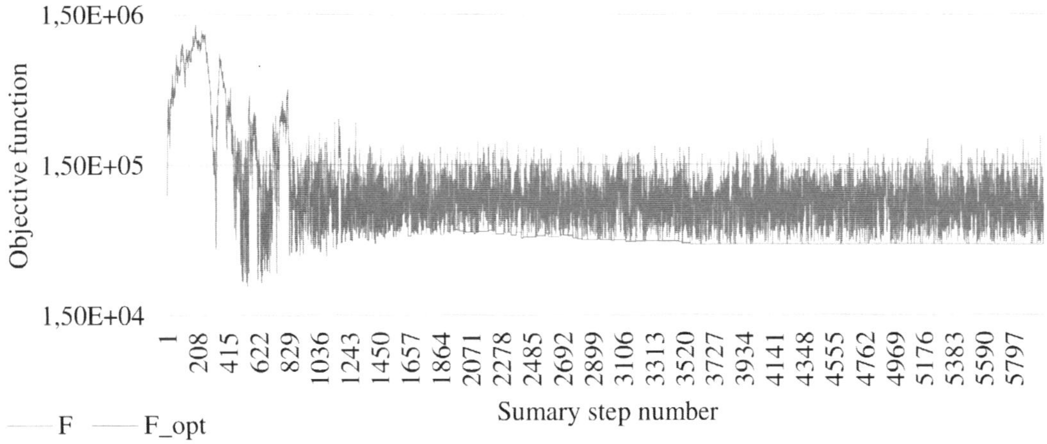

Figure 7. Optimisation progress

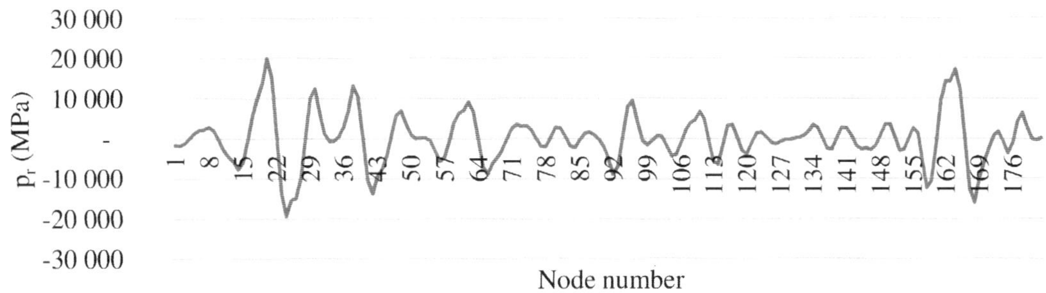

Figure 8. Pressure distribution for lowest objective function obtained

4. Conclusions

Optimisation by Simulated Annealing method was performed fully in APDL. Nevertheless it has not converged to acceptable solution inter alia due to geometry complexity which has led to a problem with a probability of finding optimal geometry. It has been observed that defining method of the geometry and standard deviations values of new geometry generation are among the most important issues for a Simulated Annealing Optimisation. The Authors' suggestion for further implementation of the method is degradation of geometry definition to a simpler one. For example two defining polynomials of 5th order would result in only 7 independent variables (assuming continuity of geometry up to the 2nd derivative). In such a case probability of finding geometry closer to an optimal one is approximately $p = 2^{-7} = 7.81 \cdot 10^{-3}$ which is a great improvement comparing to a tested case.

References

[1] Wajand J A and Wajand J T 2005 *Tłokowe silniki spalinowe średnio- i szybkoobrotowe* (Warszawa: Wydawnictwo Naukowo Techniczne)

[2] Kozaczewski W 2004 *Konstrukcja grupy tłokowo-cylindrowej silników spalinowych* (Warszawa: Wydawnictwa Komunikacji i Łączności)

[3] Iskra A 1996 *Studium konstrukcji i funkcjonalności pierścieni w grupie tłokowo-cylindrowej* (Poznań: Wydawnictwo Politechniki Poznańskiej)

[4] Yang X-S 2010 *Engineering Optimization: An Introduction with Metaheuristic Applications* (United Kington: John Wiley and Sons)

[5] Bertsimas D and Tsitsiklis J 1993 Simulated annealing *Statistical Science* **8** 10-5

[6] Smoliński R 2013 *Projekt i optymalizacja wielopierścieniowego uszczelnienia tłoka* (Master's Thesis in Faculty of Power and Aeronautical Engineering, Warsaw University of Technology)

The study on injection parameters of selected alternative fuels used in diesel engines

K Balawender[1], H Kuszewski[1], K Lejda[1] and K Lew[1]

[1] Rzeszow University of Technology, Faculty of Mechanical Engineering and Aeronautics, Department of Combustion Engines and Transport, Av. Powstancow Warszawy 8, 35-959 Rzeszow, Poland

E-mail: klejda@prz.edu.pl

Abstract. The paper presents selected results concerning fuel charging and spraying process for selected alternative fuels, including regular diesel fuel, rape oil, FAME, blends of these fuels in various proportions, and blends of rape oil with diesel fuel. Examination of the process included the fuel charge measurements. To this end, a set-up for examination of Common Rail-type injection systems was used constructed on the basis of Bosch EPS-815 test bench, from which the high-pressure pump drive system was adopted. For tests concerning the spraying process, a visualisation chamber with constant volume was utilised. The fuel spray development was registered with the use of VisioScope (AVL).

1. Introduction

The course of fuel injection process in diesel engines has a decisive effect on the combustion process and as a consequence, on other engine parameters, both operating and environment-related [3,5,10]. Values of the parameters are affected by numerous factors, from design and precision of manufacturing the injector as such, to physicochemical properties of fuels which can be subject to significant variation, even within a short period of time, due to temperature changes resulting in changeable density and viscosity [2–5].

Bearing in mind complexity of fuel spraying processes it is rather unlikely that the theory of fuel spraying will make any significant progress in the nearest future. It can be assumed therefore that the dimensional analysis will be still used in this field of research.

A large potential as far as the fuel spraying process assessment is concerned is currently offered by visualisation tests consisting in direct watching the involved phenomena. Tests of the type are in general classified as vision- and laser-based. Within each of the categories, different recording techniques are distinguished, characterised with different degree of complexity and possibility of combining several methods into one [5,7,9]. Visualisation studies allow to watch the injection process and its irregularities, facilitating the design the injection apparatus and combustion system in a way making the processes occurring in the engine more effective which ultimately leads to improvement of operating and environmental parameters of the combustion engine. Indeed, the use of the visualisation set-up does not allow to reproduce identically the conditions prevailing in the engine combustion chamber, but under certain assumptions it can be stated that there exists a relationship between tests results concerning structure of sprays obtained in extra-engine environment and actual shape of spray in a combustion

chamber. An additional benefit derived from extra-engine tests is the possibility to carry them out at a relatively low cost, especially in studies requiring a large number of engine-based tests.
The paper presents results of tests concerning fuel charging and spraying process for selected alternative fuels.

2. Research set-up and methodology

To determine charge quantities and geometries of sprays, a test set-up for examination of Common Rail-type injection system was used constructed on the base of Bosch EPS 815 test bench from which the high-pressure pump drive was utilised. Such set-up allows to test a number of different liquid fuel types [1, 4, 5]. A view and a schematic diagram of the set-up is shown in Figure 1. Basic component systems of the set-up include:

- low-pressure circuit,
- high-pressure circuit,
- controller,
- supervising computer.

Figure 1. An overview of the set-up used for visualisation studies

The low-pressure system was mounted in a mobile structure together with components of the fuel cooling system and flowmeters (Figure 2).
The high-pressure system was situated on the test bench which was used to drive the high-pressure pump. The system comprises: high-pressure pump, fuel rail, injectors (mounted in the visualisation chamber), and high-pressure lines.
The pump transfers fuel to the rail where the fuel pressure value is measured used further to control pressure modulators in the pump and the rail. From the rail, the fuel is fed to injector which injects it to a visualisation chamber equipped with a stroboscopic lighting system and designed specifically to be included in the set-up. The process of spray development of the fuel injected into the chamber was recorded by means of VisioScope visualisation system (AVL). To record high-speed courses, a data acquisition system based on measuring card AT-MIO-16E-1 (National Instruments) was used [5].
The VisioScope system combined with a high-speed courses acquisition system controlled by crank angle encoder 365C (AVL), allows to record the spray development with angular resolution in pump shaft rotation (PSR) of 0.1 deg. Pulses from the encoder were also used as signals triggering the

electronic injector controller. In this case, angular resolution was 720 pulses per full rotation which ensured high precision of injection timing [1, 5].

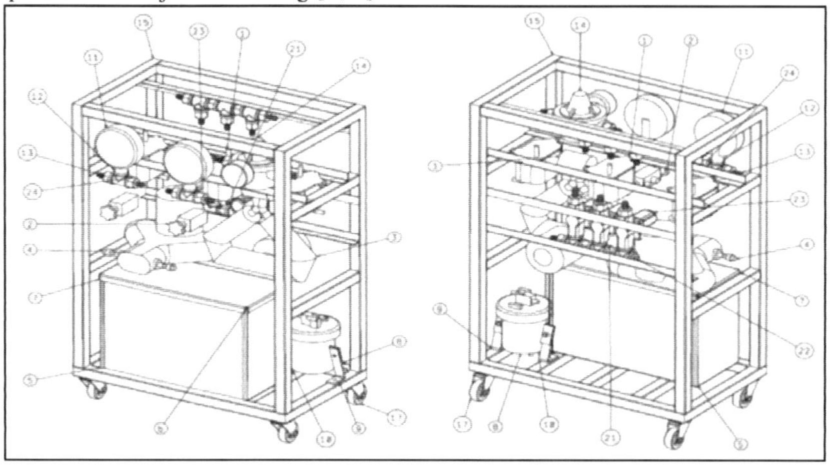

Figure 2. Arrangement of the injection system low-pressure components: 1 — fuel cooler; 2 — regulation valve; 3 — CMF 025 flowmeter; 4 — CMF 010 flowmeter; 5 — fuel tank; 6 — fuel pump; 7 — tank cover; 8 — fuel filter; 9, 10 — fuel filter brackets; 11 — manometer; 12 — T-connection; 13 — elastic line terminal; 14 — water pressure reducer with manometer; 15 — frame; 17 — mobile frame wheel; 21 — union piece, 22 — plug; 23 — solenoid valve; 24 — elastic line terminal

A view of the test chamber used to determine the injected fuel charge is presented in Figure 3, whereas the visualisation chamber is shown in Figure 4.

Figure 3. A view of injectors mounted in test chambers [1]

The research set-up, in combination with the visualisation system and a high-speed courses measuring system, allowed to carry out examination of the injection process as well as measure and record the following parameters:

- injected fuel charge,
- overflow quantity from injector,
- fuel density,

- fuel temperature at different points of the system,
- pressure in the fuel rail,
- parameters of injector controlling signal,
- high-speed courses of:
  - pressure downstream the fuel rail and in injector line upstream the injector,
  - injector needle lift (with the use of injector with needle lift sensor),
  - injector controlling signals,
  - injector voltage and current;
- spray macrostructure (by recording the process of its development) which allows to determine:
  - actual injection start and end time,
  - spray tip penetration and spray cone angle.

Figure 4. A view of the visualisation chamber with stroboscopic lighting system mounted on the test bench [4,5]

The program of tests realised on the research set-up equipped with visualisation chamber and a motion-picture camera for recording development of the spray was carried out in the following sequence:

- filling the system with the tested fuel,
- mounting the injector (W219 Bosch-type injector 0445120219) in the test chamber:
  - heating the injector and the fuel up to temperature (40 ± 2) °C,
  - checking fuel output for selected measuring points,
  - removing the injector from the test chamber;
- mounting the same injector in the visualisation chamber, and then carrying out the fuel spraying tests for the following settings: pump rotation speed 800 rpm corresponding to engine speed of 1600 rpm, injection pressure 100 MPa, and injection duration times 1800 μs and 2600 μs,
- removing the tested fuel from the system after completion of the test.

According to the assumed research program, the following fuels were subject to analysis: ONH (diesel fuel); B100 (rapeseed methyl ester); B50 (a blend of diesel fuel with an addition of 50% FAME), OR40 (a blend of diesel fuel with an addition of 40% of rape oil); and OR80 (a blend containing 80% of rape oil and 20% of diesel fuel).

Recorded motion pictures illustrating the courses of spray development allowed to determine their macrostructure, in this case represented by the spray tip penetration and spray cone angle.

3. Research results

As a result of the performed tests, a number of data files were obtained from which the average fuel charge value was calculated at each of the measuring points. Figure 5 shows results of measurements of the injected fuel charge as a function of injection duration for the pressure in fuel rail amounting to 100 MPa. The results indicate that the relationship of fuel charge changes versus the injection duration is a proportional one.
Visualisation tests were carried out in accordance with the adopted programme. For selected fuels, motion pictures were recorded illustrating the course of spray development of fuel injected to the visualisation chamber based on which, with the use of VisioScope software (AVL), spray cone angles and spray tip penetration were estimated. The injection start was determined as the angle corresponding to the image preceding this in which a fuel jet flowing out from any of the spraying nozzle could be observed. The spray tip penetration was determined at the moment corresponding to rotation of the pump shaft by 4 deg counting from the actual injection start. At such point, the tip penetration was determined for each spray separately and the average spray tip penetration value was calculated (Figures. 9 and 10). In a similar way, the spray cone angle was determined (Figures. 11 and 12). Selected photographs visualising the fuel spraying are shown in Figures. 6–8.

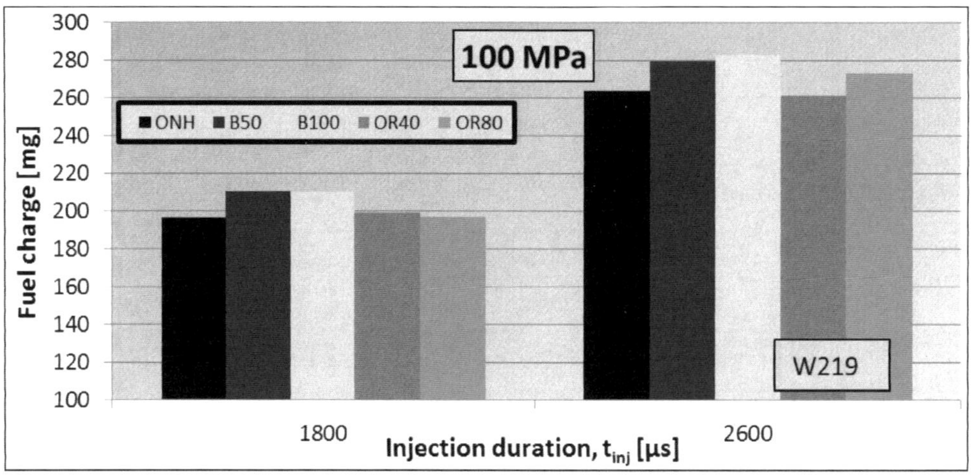

Figure 5. A summary of W219 injector deliveries depending on the injection duration for the injection pressure of 100 MPa

After visualisation tests concerning in recording views of sprayed fuel tests for all adopted measuring points and after analysis of the obtained results it has been found that the injection duration has no significant effect on the spray cone angle.
The presented results concerning the determined tip penetrations and cone angles represent average values over for all sprays (number of holes in the spray nozzle).

Figure 6. Spraying of ONH fuel (3 deg after injection start)

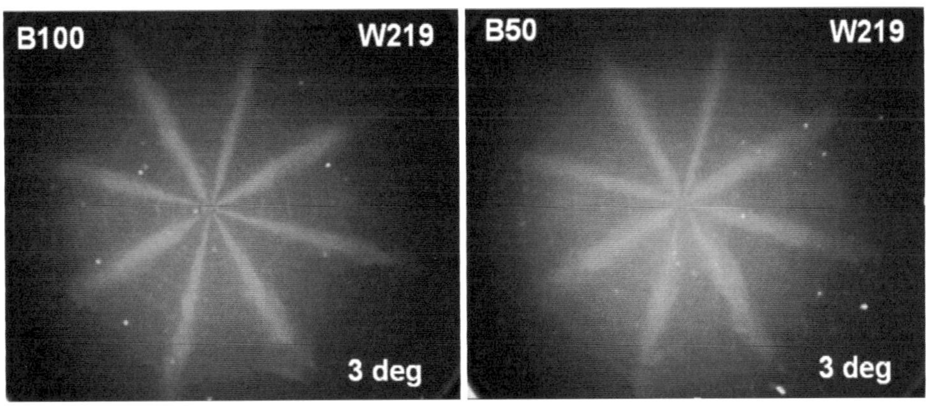

Fig. 7. Spraying of B100 and B50 fuel (3 deg after injection start)

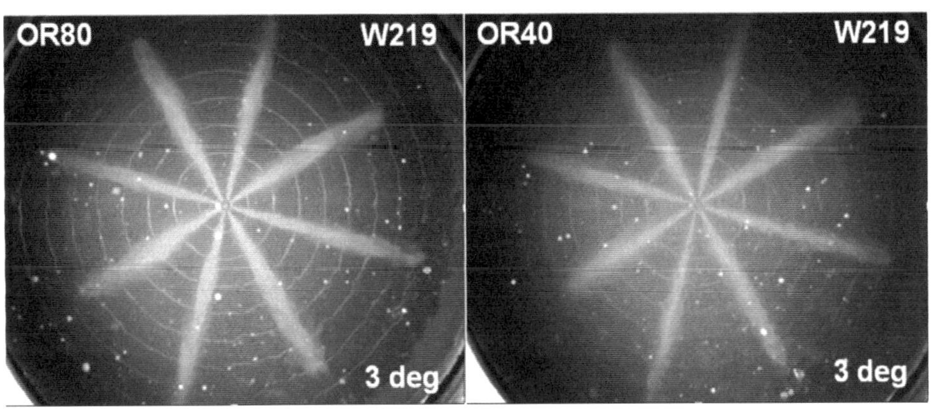

Figure 8. Spraying of OR80 and OR40 fuel (3 deg after injection start)

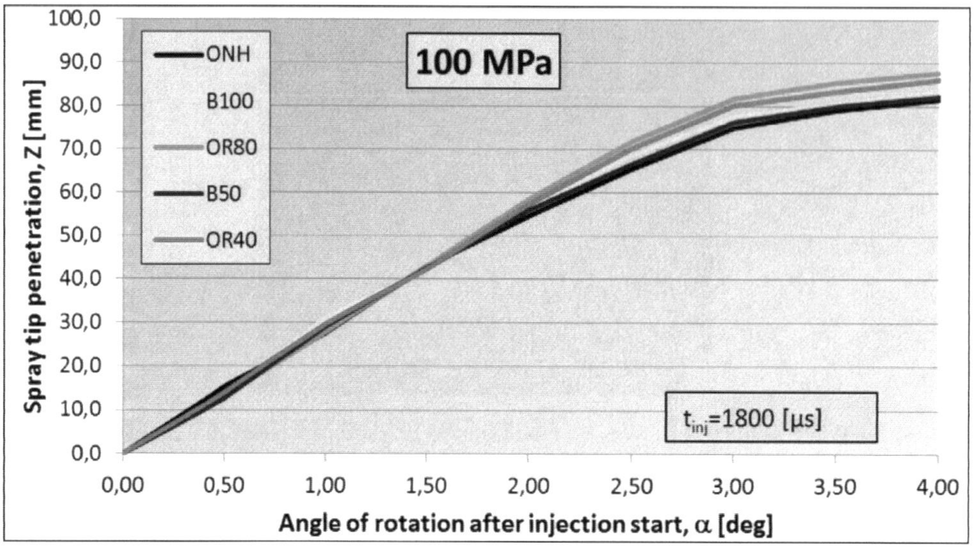

Figure 9. A comparison of average spray tip penetrations for examined fuels for W219 injector (p_{inj} = 100 MPa, t_{inj} = 1800 μs)

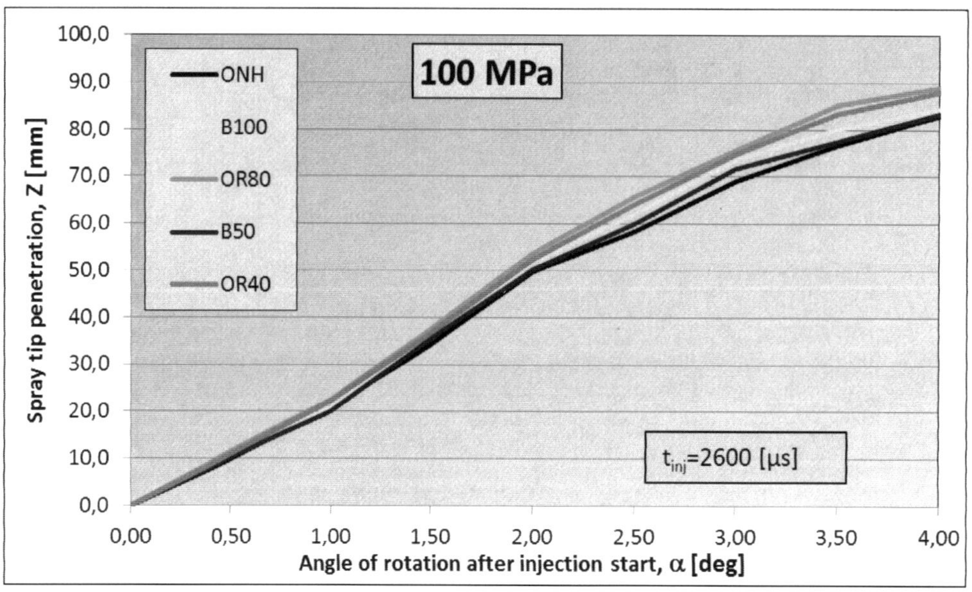

Figure 10. A comparison of average spray tip penetrations for examined fuels for W219 injector (p_{inj} =100 MPa, t_{inj} = 2600 μs)

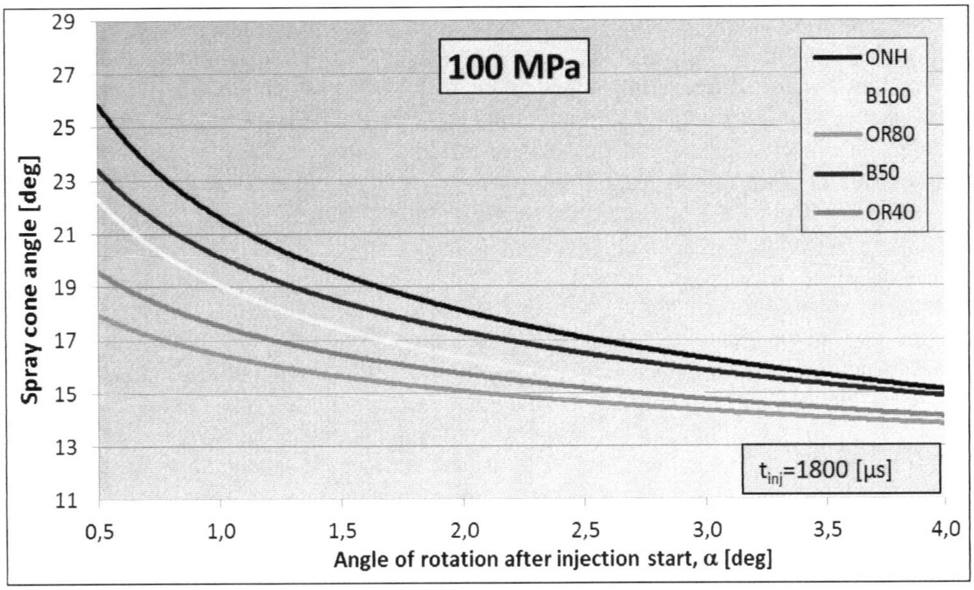

Figure 11. A comparison of average spray cone angles for W219 injector (p_{inj} =100 MPa, t_{inj} = 1800 µs)

After visualisation tests consisting in recording patterns of sprays at all measuring points and after analysis of the obtained data it has been found that the injection duration has no significant effect on the spray cone angle.

The presented results of determined tip penetrations and cone angles represent values averaged over all sprays generated by the injector.

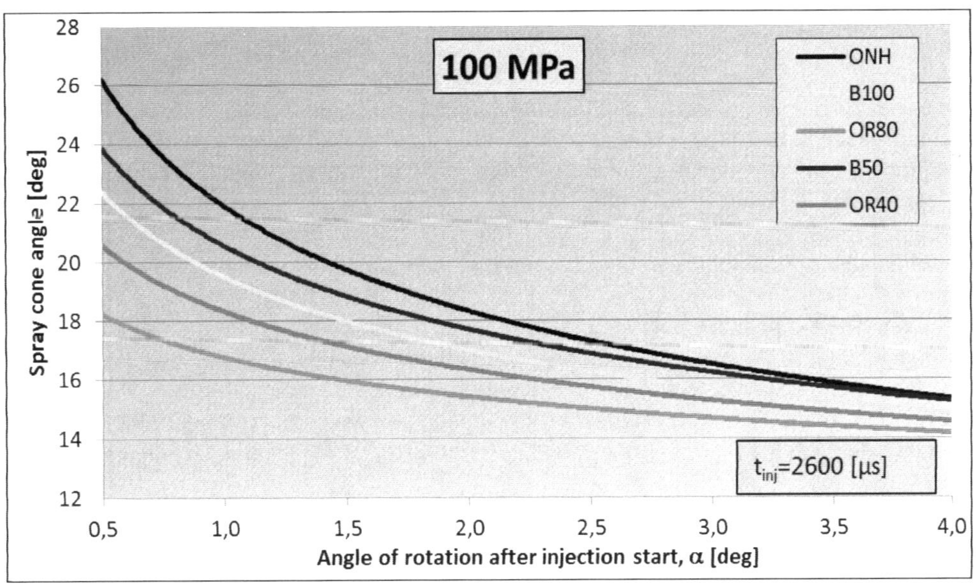

Figure 12. A comparison of average spray cone angles for examined fuels for W219 injector (p_{inj} =100 MPa, t_{inj} = 2600 µs)

Figures 9 and 10 present values of spray tip penetrations determined for the tested fuels. It follows from the graphs that the higher viscosity of fuel, the larger is the spray tip penetrations. Figs. 11 and 12, on which averaged values of spray cone angles characterising sprayed fuel for injection pressure 100 MPa and injection durations of 1800 μs and 2600 μs are presented, respectively, show a distinct differences in spray cone angles for different fuels. The main factor resulting in such differences is different viscosity of the fuels. Higher fuel viscosity disrupts the secondary disintegration of droplets [7, 10] – the spray becomes more compact and this in turn results in lower value of cone angle. For the fuel characterised with the largest viscosity (OR80), the cone angle was about 18 deg, while for the fuel with the lowest viscosity (ONH), the observed cone angle value was about 26 deg.

4. Conclusions

The undertaken visualisation studies were aimed at comparing geometries of sprays (the spray cone angle and the spray tip penetrations) for fuels characterised with various physicochemical properties and to propose a research methodology for examining the spraying process with the use of visualisation system.

Among different parameters characterising diesel engine fuels, those decisive for the jet spray cone angle and spray tip penetration include viscosity, surface tension and density (with increasing values of these quantities, the spray tip penetration and spray cone angle decreases), which was confirmed by the obtained research results. Fuels characterised with higher viscosity form more compact sprays which is a result of disrupted secondary disintegration of droplets (large drops are formed). As a result, this may lead to reaching the cylinder liner and piston head walls by sprays and consequential penetration of unburned fuel to the crank chamber. This in turn deteriorates lubrication properties of engine oil. Improper parameters of the spray macrostructure have a negative effect on the engine operation process resulting in, among other things, increased emissions of toxic exhaust gas components. Improvement of the process of spraying fuels characterised with large viscosity can be achieved by using injectors with smaller holes and a larger number of holes and increasing temperature of the injected fuel.

Although visualisation tests do not reproduce accurately conditions prevailing in the engine combustion chamber, it can be however claimed that, under determined assumptions, there exists certain correlation between structure of sprays in the extra-engine set-up and their actual shapes in the combustion chamber. Moreover, an essential advantage of extra-engine tests is the possibility to carry them out at relatively low cost.

References

[1] Balawender K, Lejda K and Ustrzycki A 2011 Stanowisko badawcze wtryskowych układów Common Rail zasilanych paliwami różnego typu. Międzynarodowa Konferencja Naukowa w Kijowie, ВІСНИК №23, s. 37-43, Kijów

[2] Jaworski A, Kuszewski H, Lew K and Ustrzycki A 2012 Ocena przydatności paliw zastępczych do silników o ZS autobusów miejskich na podstawie wybranych parametrów wtrysku. Systemy i Środki Transportu Samochodowego. Monografia pod redakcją naukową Kazimierza Lejdy, TRANSPORT'12, Nr 3, s. 143-148, Rzeszów

[3] Lejda K 2004 Selected problems of fuel supply in high-speed Diesel engines. Published by Meta Lvov (pp. 225), ISBN 966-7947-11-2, Lvov

[4] Lejda K, Balawender K, Jaworski A, Kuszewski H, Ustrzycki A, Woś P 2012 Sprawozdanie z projektu badawczego własnego nr N N509 256637 nt. Analiza determinantów dokładności procesu wtrysku paliwa w wysokociśnieniowych systemach zasilania spalinowych napędów środków transportu. Rzeszów

[5] Lew K 2014 Opracowanie kryteriów doboru paliw do specyfiki napędów wykorzystywanych w komunikacji miejskiej. Rozprawa doktorska, Poznań

[6] Orzechowski Z, Prywer J 1994 Rozpylanie cieczy w urządzeniach energetycznych. Wydawnictwa Naukowo-Techniczne, Warszawa

[7] Orzechowski Z, Prywer J 2008 Wytwarzanie i zastosowanie rozpylonej cieczy. Wydawnictwa Naukowo-Techniczne, Warszawa

[8] Pfeil J, Velji A, Spicher U 2010 High speed photography – possibilities and barriers. Combustion Engines, Nr 4/2010 (143), s. 15-22.
[9] Wisłocki K 2004 Studium wykorzystania badań optycznych do analizy procesów wtrysku i spalania w silnikach o zapłonie samoczynnym. Rozprawa nr 387, Wydawnictwo Politechniki Poznańskiej, Poznań
[10] Zabłocki M 1976 Wtrysk i spalanie paliwa w silnikach wysokoprężnych. Wydawnictwo Komunikacji i Łączności, Warszawa.

IOP Conf. Series: Materials Science and Engineering **148** (2016) 012071 doi:10.1088/1757-899X/148/1/012071

The characteristics of the combustion process occurring under real operating conditions of traction

R Longwic[1] and P Sander[2]
[1]Lublin University of Technology, Department of Vehicles, PL
[2]Lublin University of Technology, Department of Vehicles, PL

E-mail: r.longwic@pollub.pl

Abstract. The authors deal with the issues of the Diesel engine under dynamic conditions. The conditions of the dynamic operation of the engine have most frequently been mapped by the method of free acceleration of the engine caused by the change of the position of the fuel dose lever. The article presents the results of indication of the traction Diesel engine under real operating conditions. This allows for the use of a mobile system to indicate the AVL engine built in the vehicle in research. We analysed a number of thermodynamic parameters of the combustion process in various dynamic states, typical for the process of actual operation of the engine, such as working in start-up conditions and immediately after, working in conditions of acceleration and coasting. Formulated conclusions significantly expand the area of knowledge concerning the functioning of the internal combustion engine in dynamic conditions.

1. Introduction

The topic concerning the specificity of the internal combustion engine work in dynamic conditions has been discussed since the beginning of the previous century [5]. This is done due to the obvious facts which indicate that the work of the internal combustion engine of the traction vehicle proceeds mainly in dynamic conditions [6,11]. Nevertheless, no coherent and explicit methodology of testing the internal combustion engine in dynamic conditions has been developed until the present day due to the lack of normative regulations concerning this type of research. Although an interest in combustion engine work in dynamic conditions has considerably increased recently, it is mainly because of some legal regulations concerning the emission of exhaust toxic constituents. Such regulations enforce the control of emission in dynamic conditions.

Therefore, some studies which are concerned with testing the emission of fuel toxic constituents in dynamic conditions [1,2] and modelling engine work in these conditions [3,4] can be found in technical literature. However, there are few papers in which both the specificity of the internal combustion engine work in dynamic conditions and the reasons for the observed differences in the course of work processes in relation to comparable static conditions would be explained in an experimental way. In the present paper, the results of testing the engine with the dispenser fuel injection system have been presented. The engine installed in a vehicle was subjected to indication in different movement conditions. The conditions of vehicle movement were mapped on a chassis test

bench. The conditions of vehicle movement were selected in such a way that they could consequently create both static and dynamic conditions of engine work.
The aim of the research is not the mutual quantitative reference of the obtained parameters of combustion process in the conditions of static and dynamic engine work but the demonstration of the tendencies concerning changes in the course of combustion process parameters for different engine operating conditions resulting from different variants of vehicle movement. As it was mentioned, the chosen characteristic variants of vehicle movement were those which determine engine work in static and dynamic conditions.

2. The specificity of the course of the combustion process in the conditions of diesel engine dynamic work

Determined physical processes can be divided into static and dynamic processes [4,11].
A static process is the process in which:

$$\frac{dx}{dt}=0 \tag{1}$$

where:
x - selected physical parameter describing the physical process,
t – the duration of the process.
By analogy, a dynamic process is the process in which:

$$\frac{dx}{dt}\neq 0 \tag{2}$$

The dynamic process of engine work will occur when one of the equations is not satisfied:

$$\frac{dM_o}{dt}=0 \tag{3}$$

$$\frac{dn}{dt}=0 \tag{4}$$

$$\frac{d\dot{Q}}{dt}=0 \tag{5}$$

where:
M_o – usable engine torque,
n – engine crankshaft rotational speed,
Q – total thermal flux flowing through the engine (the cooling system with all possible ways of exchanging heat with the environment along with the exhaust system).

Analyzing the man-machine-environment arrangement, three groups of reasons that can cause changes in M_o, n and Q can be distinguished:

- the change of engine torque loading the engine or the change of the torque of internal losses,
- the change in the amount of fuel delivered,
- the change in air supply.

Periodicity or the lack of periodicity of the changes mentioned above will influence the occurrence of steady-state or transient engine work conditions.
Due to the impact of controllers (rotational speed, load, ignition angle etc.) and man on engine work, a dynamic process of engine work can proceed in three variants:

- M_o=const and n≠const,

- $M_o \neq$const and n=const,
- $M_o \neq$const and $n \neq$const

In papers [6, 7, 8, 9, 10, 11] the parameters of engine work were compared: in the conditions of free acceleration of the engine caused by the step change in the position of the lever controlling the fuel dose, and in static conditions which were mapped with the use of external operating characteristics. It has been shown that it is possible to compare the parameters of engine work in such conditions due to the adequacy of engine loading torque, engine crankshaft rotational speed and the dose of injected fuel.

It was stated that in the conditions of dynamic engine work during the combustion process:

- there is a slight increase in the values of mean indicated pressure,
- self-ignition lag angle increases,
- maximum velocity of pressure growth rises.

As far as the processes of pressing and injection are concerned, the acceleration of injection start angle and a slight increase in maximum injection pressures were confirmed. Those phenomena were mainly caused by the increase in residual pressures in the high pressure conduits. The engine with a classic rotational injection pump was subject to tests. However, the process of engine free acceleration does not occur in regular work of the traction engine. The research in those conditions has a lot of advantages, such as the possibility of repetitive realization of dynamic conditions and the mentioned above comparison of the measured parameters to those of the external operating characteristics or in other words to static conditions.

In the present paper, the authors wanted to confirm the hypothesis about a similar (to the one described above) tendency to change parameters describing the combustion process that proceeds in dynamic conditions in relation to static conditions. Engine work conditions were equivalent to the typical cases of vehicle operation. The engine was equipped with dispenser fuel injection system. The analysis focused on the impact of engine work conditions on: mean indicated pressures, the maximum speed of the pressure growth, the amount of produced heat and the course of producing heat was conducted. As far as the injection process is concerned, the strategy of injection realized by the engine controller was illustrated.

3. Empirical research

3.1. The research stand, the research object

The research was conducted in the Laboratory of the Department of Car Vehicles at the University of Technology in Lublin. The object of tests was Fiat Qubo with a self-ignition engine fulfilling all emission standards specified by Euro 5 and a five speed gearbox. The main technical parameters of the tested car engine are shown in Table 1.

The road conditions were mapped by simulating driving on the chassis test bench for vehicles with biaxial drive of DF4FS-HLS type. The chassis test bench consisted of:

- bench-mounted (first) roller set with an eddy current brake and a hydraulic pump;
- mobile (second) roller set with an eddy current brake, a hydraulic pump and a gear motor for overhauling;
- control panel (dashboard);
- hydraulic system control;
- axial blower for cooling the vehicle;
- PC with the test bench software.

Table 1. The main technical parameters of 1.3 Multijet engine of the tested vehicle - Fiat Qubo.

Number of cylinders	4
Cylinder diameter (mm)	69.6
Piston stroke(mm)	82
Total engine cubic capacity (cm^3)	1248
Maximum power (kW CEE)	55
Maximum power (KM CEE)	75
Functioning at maximum power (rpm)	4000
Maximum torque (Nm)	190
Rotational Speed at maximum torque (rpm)	1500
Revolutions at neutral gear (rpm)	850 ± 20
Compression ratio	16.8 : 1

The system for indicating the engine Indi Micro 602 made by AVL with a built-in signal amplifier, cooperating with four analogue input channels and two digital inputs was used in tests. The signals recorded by the AVL Indimicro system are the following:

- the course of pressure inside the cylinder – it was recorded by means of a piezoelectric sensor AVL GH13P installed in the glow plug seating of the first cylinder;
- the signal of engine crankshaft position;
- injection parameters were analysed on the basis of the analogue signal of controlling the piezoelectric injector after converting it into the digital signal.

Picture 2 presents the test stand used during the research.

Figure 1. Test stand: 1. Test car Fiat Qubo with the 1.3 Multijet engine 2. Computer with the AVL software 3. AVL Indimicro 602 system for indicating the engine 4. DF4FS-HLS chassis test bench 5. AVL IndiCom V2.7 software.

The engine installed in the vehicle was subject to indication in selected driving conditions which map static and dynamic conditions. Driving conditions were realized on the chassis test bench. The vehicle was loaded with simulated values of rolling resistance force.

The following cases of vehicle work were analysed:

- Case 1 - stop – engine neutral gear, conditions of static engine work;
- Case 2 - driving in the fourth gear at the speed of 50 km/h – conditions of static engine work, constant load from rolling resistance forces;
- Case 3 - starting the vehicle - "cold" engine start, conditions of dynamic engine work;
- Case 4 - starting the vehicle - "hot" engine start, conditions of dynamic engine work;
- Case 5 - acceleration of the vehicle in the first gear- rapid acceleration of the engine, conditions of dynamic engine work.
-

100 successive engine work cycles were recorded. The tests for each case were repeated ten times.

3.2. Analysis of the obtained results
In the pictures from 2 to 7, the obtained results of tests and calculations are shown. The course of producing heat and the amount of produced heat were analysed for the selected engine work cycles. The courses of engine rotational speed and mean indicated pressure were also analysed for the successive engine work cycles. Mean, minimum and maximum values and standard deviations of mean indicated pressure, engine rotational speed, maximum combustion pressure and maximum speed of pressure growth were calculated from all cycles of engine work. The conducted research refers to static conditions of the engine work (Cases 1 and 2) and dynamic conditions (Cases 3, 4 , 5). As for the starting process (Cases 3 and 4), engine work cycles were analysed directly after the occurrence of self-ignition. Consequently, a successful starting process is meant to lead to engine idling. However, in the initial cycles after the occurrence of combustion, the rotation speed is slightly higher than the one for engine idling. After about 20 engine work cycles, the rotation speed starts to stabilize on the level of idling rotational speed. Significant differences concern "cold" and "hot" engine start. In the case of "cold" engine start, in the second engine work cycle, the biggest increase in mean indicated pressure to the value of about 0,418 MPa occurs. At the same time, the period of kinetic combustion is almost unnoticeable. This probably results from the activity of glow plugs. In the case of "hot" engine start, the biggest increase in mean indicated pressure occurs in the first cycle, achieving the value of 0,250 MPa. The periods of kinetic and diffusive combustion are sharply visible here. Maximum speeds of pressure growth are higher in the conditions of engine start in relation to engine idle running. In the first case, they exceed 0,2 MPa/°CA; in the second case they amount to around 0,187 Mpa/°CA. Engine work in static conditions (Case 2) and engine work in the conditions of acceleration (Case 5) were subject to detailed analysis. It was stated that mean values of maximum speed of pressure growth are higher in dynamic conditions in relation to static conditions. Obviously, in the studied conditions of engine work, fuel doses are different. However, in static conditions, the maximum speed of pressure growth was 0,498 MPa/°CA for engine rotational speed of about 1583 rpm. At similar engine rotational speed of 1586.5 rpm in dynamic conditions, the maximum speed of pressure growth was 0.325 MPa/°CA. The observation presented above results from the fact that in static conditions, the maximum rate of heat release during the process of kinetic combustion was 55 kJ/m^3·°CA and in dynamic conditions 40kJ/m^3·°CA.

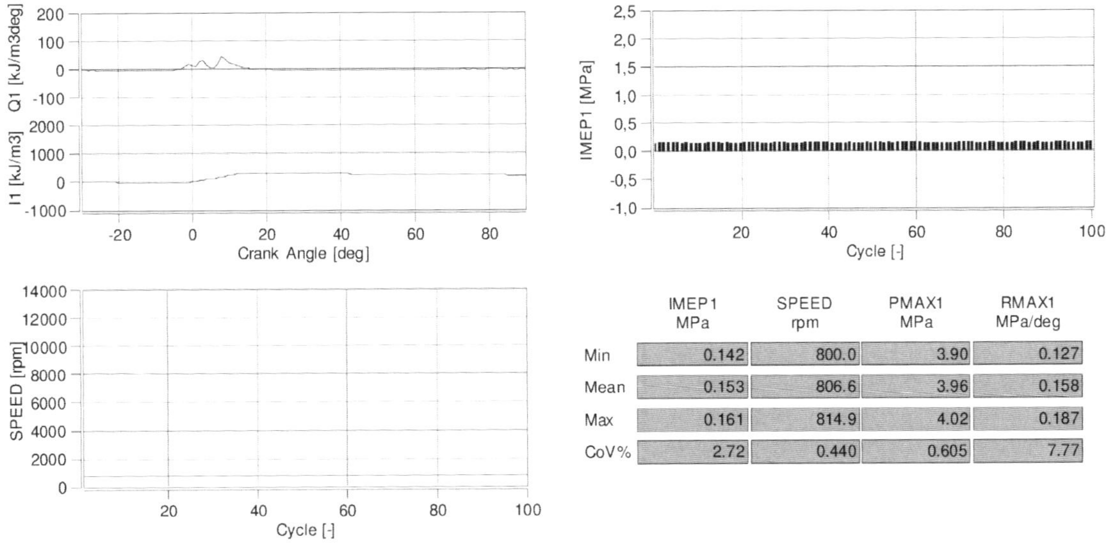

	IMEP1 MPa	SPEED rpm	PMAX1 MPa	RMAX1 MPa/deg
Min	0.142	800.0	3.90	0.127
Mean	0.153	806.6	3.96	0.158
Max	0.161	814.9	4.02	0.187
CoV%	2.72	0.440	0.605	7.77

Figure 2 . Case 1 – engine idle running, the courses of: heat production (Q1), the amount of produced heat (I1) for the first cycle of engine work and mean indicated pressure (IMEP1) and engine rotational speed (SPEED) for the tested cycles and mean, minimum and maximum values of mean indicated pressure, engine rotational speed, maximum combustion pressure (PMAX1) and maximum speed of pressure growth (RMAX1).

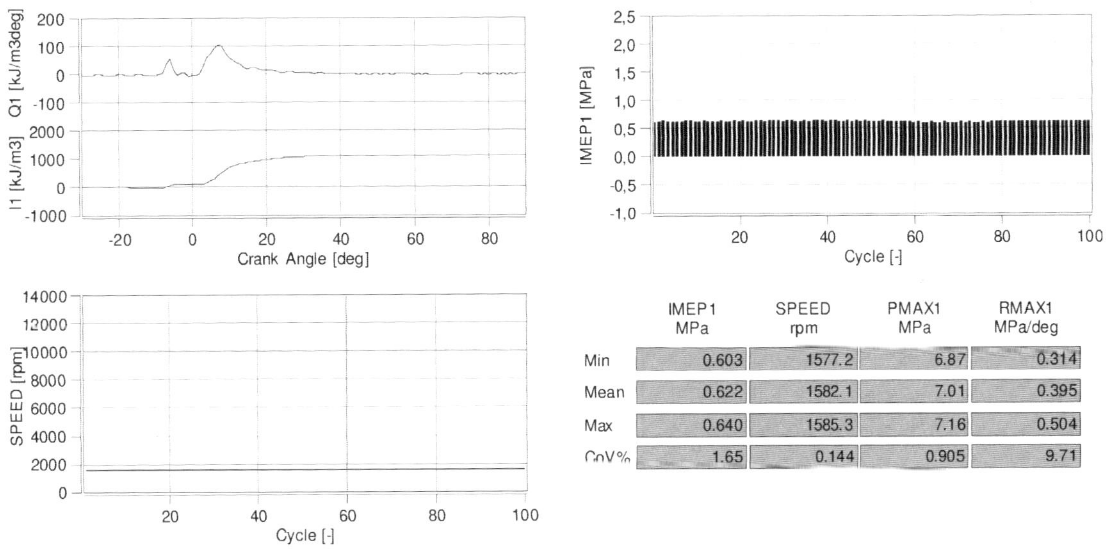

	IMEP1 MPa	SPEED rpm	PMAX1 MPa	RMAX1 MPa/deg
Min	0.603	1577.2	6.87	0.314
Mean	0.622	1582.1	7.01	0.395
Max	0.640	1585.3	7.16	0.504
CoV%	1.65	0.144	0.905	9.71

Figure 3. Case 2 – conditions of static engine work, the courses of: heat production (Q1), the amount of produced heat (I1) for the first cycle of engine work and mean indicated pressure (IMEP1) and engine rotational speed (SPEED) for the tested cycles and mean, minimum and maximum values of mean indicated pressure, engine rotational speed, maximum combustion pressure (PMAX1) and maximum speed of pressure growth (RMAX1).

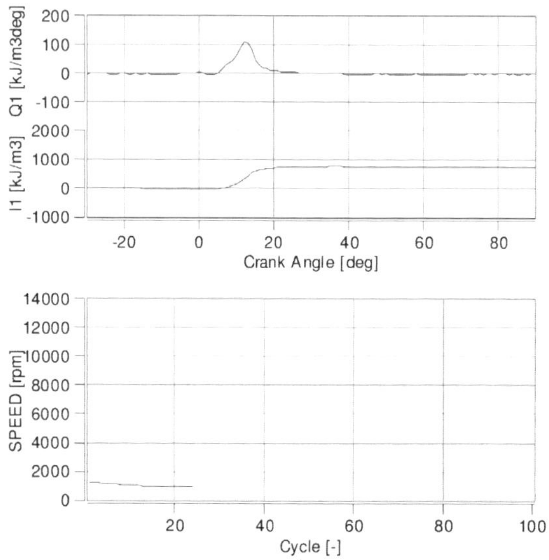

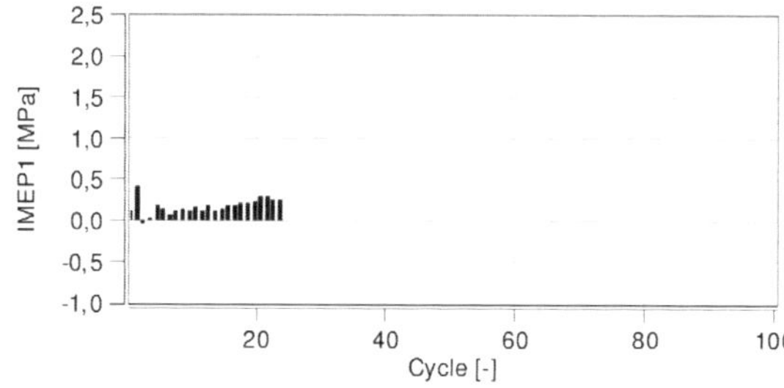

	IMEP1 MPa	SPEED rpm	PMAX1 MPa	RMAX1 MPa/deg
Min	-0.028	952.5	3.80	0.114
Mean	0.177	1059.7	4.07	0.146
Max	0.418	1285.9	4.84	0.273
CoV%	51.44	9.65	8.80	33.95

Figure 4. Case 3 – 'cold' engine start, the courses of: heat production (Q1), the amount of produced heat (I1) for the second cycle of engine work and mean indicated pressure (IMEP1) and engine rotational speed (SPEED) for the tested cycles and mean, minimum and maximum values of mean indicated pressure, engine rotational speed, maximum combustion pressure (PMAX1) and maximum speed of pressure growth (RMAX1).

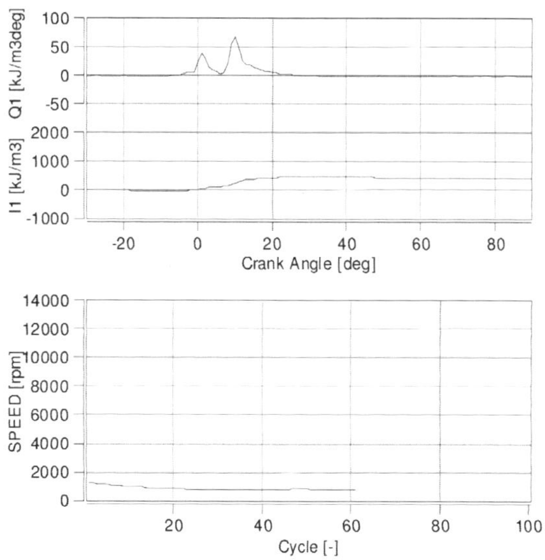

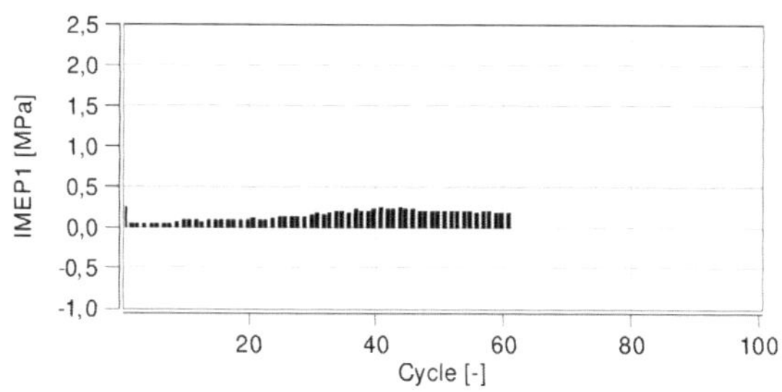

	IMEP1 MPa	SPEED rpm	PMAX1 MPa	RMAX1 MPa/deg
Min	0.046	795.9	3.34	0.096
Mean	0.158	878.2	3.88	0.152
Max	0.250	1221.0	4.18	0.202
CoV%	38.81	12.74	5.78	19.23

Figure 5. Case 4 – 'hot' engine start, the courses of: heat production (Q1), the amount of produced heat (I1) for the first cycle of engine work and mean indicated pressure (IMEP1) and engine rotational speed (SPEED) for the tested cycles and mean, minimum and maximum values of mean indicated pressure, engine rotational speed, maximum combustion pressure (PMAX1) and maximum speed of pressure growth (RMAX1)

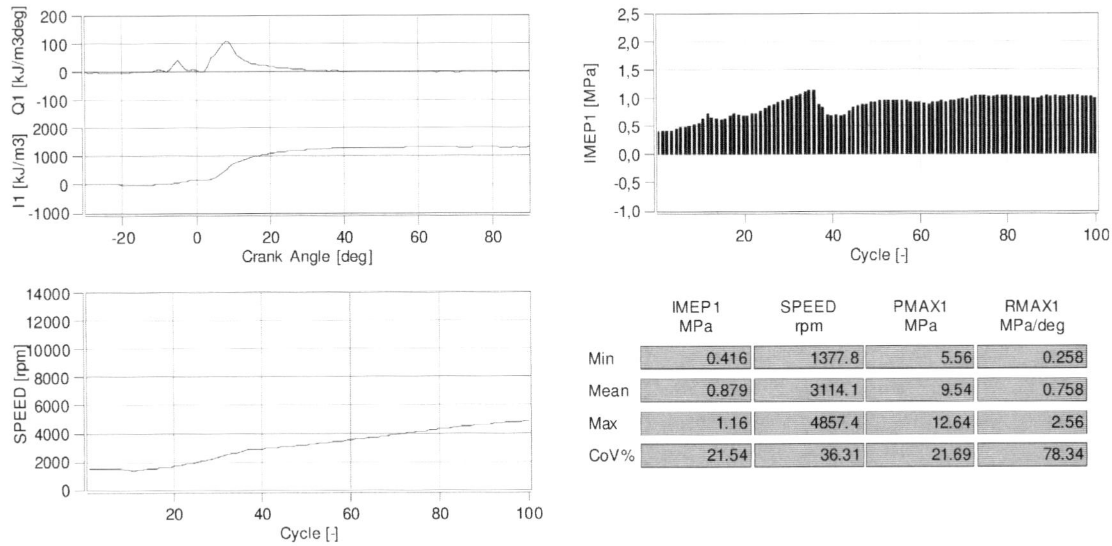

	IMEP1 MPa	SPEED rpm	PMAX1 MPa	RMAX1 MPa/deg
Min	0.416	1377.8	5.56	0.258
Mean	0.879	3114.1	9.54	0.758
Max	1.16	4857.4	12.64	2.56
CoV%	21.54	36.31	21.69	78.34

Figure 6. Case 5 – engine rapid acceleration, the courses of: heat production (Q1), the amount of produced heat (I1) for the eighteenth cycle of engine work and mean indicated pressure (IMEP1) and engine rotational speed (SPEED) for the tested cycles and mean, minimum and maximum values of mean indicated pressure, engine rotational speed, maximum combustion pressure (PMAX1) and maximum speed of pressure growth (RMAX1)

The aforementioned higher mean values of the maximum speed of pressure growth in dynamic conditions are caused by the fact that for high engine rotational speeds, the maximum speed of pressure growth in the conditions of acceleration increases rapidly up to 2,56 MPa/°CA. The period of kinetic combustion prevails while the course of heat release in the period of diffusion combustion is subject to substantial fluctuations - see Figure 8. In comparison, in the conditions of acceleration, at the speed of approximately 1586.5 rpm, 5% of heat is produced at angle -5,25°CA, 10% at angle-2,45°CA and 90% at angle 24,65°CA. At the rotational speed of around 4785,3 rpm, 5% of heat is produced at angle -2,8°CA, 10% at angle 2,15°CA and 90% at angle 42,15°CA. The strategy of realizing the injection process has a major impact on the observed phenomena - Figure 9.

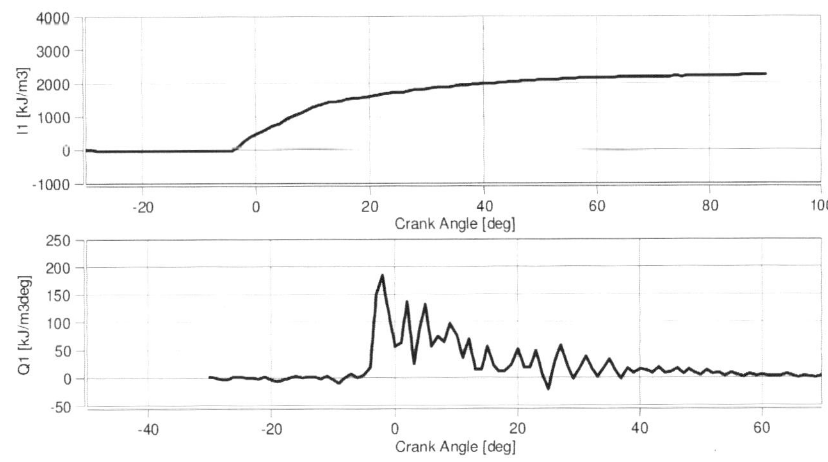

Figure 8. Case 5 – the course of heat production and the amount of produced heat in the conditions of acceleration at the speed of about 4785,3 rpm.

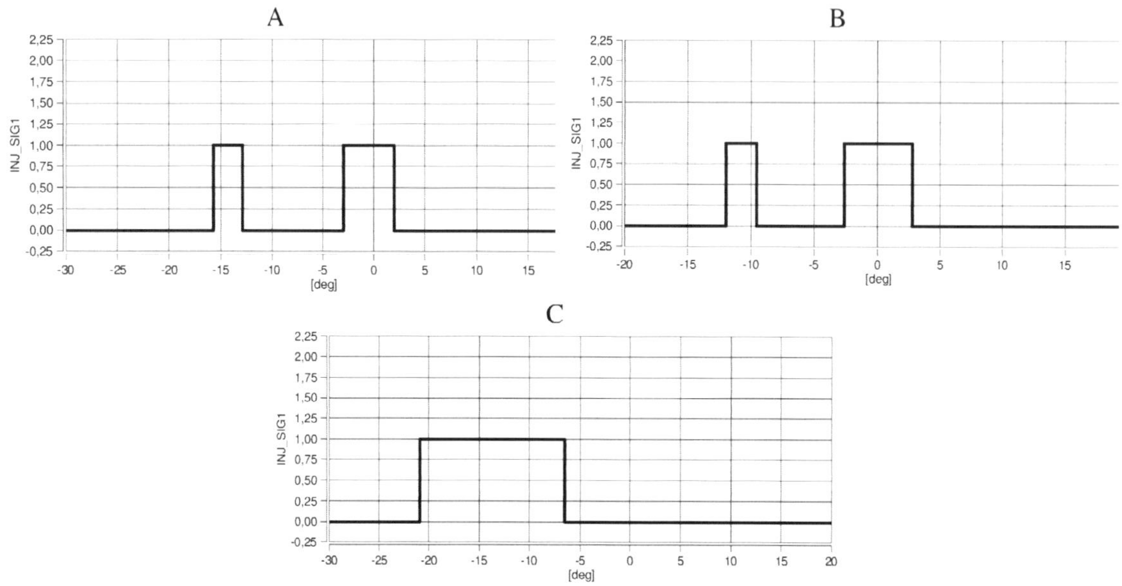

Figure 9. Digital route of the signal controlling the opening of injectors: A – static conditions (Case 2), B – acceleration (Case 5), engine rotational speed 1586,5 rpm, C – acceleration (Case 6), engine rotational speed 4785,3 rpm.

In the conditions of acceleration, the initial injection is realized closer to the TDC (top dead centre) than in static conditions. This makes it possible to shorten the period of self-ignition delay and to reduce the maximum speed of pressure growth. At high values of engine rotational speed, one injection before the TDC is realized. It is caused by insufficient amount of time for preparing the fuel for self-ignition - there are no physical conditions for realizing the initial injection.

4. Conclusions

The purpose of the experiment was to show significant differences in specific cases of engine work in static and dynamic conditions. The current state of knowledge concerning the specificity of engine work in dynamic conditions was confirmed. In the conditions of dynamic engine work, higher maximum speeds of pressure growth occur [6]. This is noticeable particularly for the process of acceleration and high values of engine rotational speed. It is possible to observe here significant speeds of heat production during the kinetic combustion and fluctuations in the course of heat production during the diffusion combustion. Large values of maximum speed of pressure growth contribute to the increase in the mechanical load of engine structural nodes. They also influence on the toxic constituents of exhaust emission. However, this issue has not been discussed in this paper. In some areas of engine work, it is possible to influence in an effective way the observed negative phenomena occurring in dynamic states by adopting the right injection strategy, mainly in the form of one or two initial injections. And yet, such steps have some limitations which result from insufficient amount of time to prepare the fuel for self- ignition at high values of engine rotational speed.

5. References

[1] Armas O., Ballesteros R. 2008 Diesel emissions from an emulsified fuel during engine transient operation, SAE Paper No. 2008-01-2430
[2] Chłopek Z 1999 Modelowanie procesów emisji spalin w warunkach eksploatacji trakcyjnej silników spalinowych, Oficyna Wydawnicza Politechniki Warszawskiej

[3] Giakoumis E 2007 Cylinder wall insulation effects on the first- and second law balances of a turbocharged diesel engine operating under transient load conditions, Energy Conversion and Management 48 Science Direct
[4] Giakoumis E and Rakopoulos C 2004 Parametric study of transient turbocharged diesel engine operation from the second-law perspective, SAE Paper No. 2004-01-1679
[5] Jante A 1933 Spezifishe Schnellaufzahlen ATZ 24
[6] Longwic R 2011 Charakterystyka działania silniki o zapłonie samoczynnym w warunkach swobodnego rozpędzania Monografia Politechnika Lubelska
[7] Longwic R, Litak G, Asok Sen and Górski K 2008 Analisys of cycle-to-cycle pressure oscillation in a diesel engine. Mechanical Systems and Signal Processing
[8] Longwic R, Litak G, Asok K and Sen 2009 Reccurence plots for diesel engine variability tests. Zeitschrift Naturforsch. Sectio A-A Journal of Physical Sciences, 64 a, 96-102
[9] Litak G and Longwic R 2009 Analisys of repeatability of Diesel engine acceleration. Applied Thermal Engineering 29, 3574-3578
[10] Longwic R, Sen A, K, Górski K, Lotko W and Litak G 2011 Cycle-to-Cycle Variation of the Combustion Process in a Diesel Engine Powered by Different Fuels. Journal of Vibroengineering VOLUME 13 ISSUE 1
[11] Longwic R 2007 Dynamic Aspects of Work of the Diesel Engine SAE Paper No. 2007-01-4210

Some aspects of the CI engine modification aimed at operation on LPG with the application of spark ignition

J Kaparuk[1], S Luft[2], T Skrzek[3], M Wojtyniak[4]

[1] PhD Student at the Faculty of Mechanical Engineering at University of Technology and Humanities in Radom
[2] Professor at the Faculty of Mechanical Engineering at University of Technology and Humanities in Radom
s.luft@uthrad.pl
[3,4] Doctor in the Faculty of Mechanical Engineering at University of Technology and Humanities in Radom

Abstract. A lot of investigation on modification of the compression ignition engine aimed at operation on LPG with the application of spark ignition has been carried out in the Laboratory of Vehicles and Combustion Engines at Kazimierz Pulaski University of Technology and Humanities in Radom. This paper presents results of investigation on establishment of the proper ignition advance angle in the modified engine. Within the framework of this investigation it was assessed the effect of this regulation on basic engine operating parameters, exhaust emission as well as basic combustion parameters.

1. Introduction

LPG enjoys great popularity in Poland due to its low price comparing to gasoline and diesel oil. It has also a well-developed distribution infrastructure. In recent years, it has been observed a trend toward modification of older generation CI engines aimed at operation on LPG. This modification consisted in reduction of the compression ratio and spark ignition application. Such kinds of modifications in CI engine constructions started in the last years of the XXth century. Description and discussion of investigations undertaken in this area are presented, among others, in [2]. Comparing to the investigations on dual-fuel engines operating on diesel oil and LPG, the scale of these investigations is much smaller. Very often the presented results do not refer to the effects resulting from the applied modifications, mainly to the following changes:

- operating engine parameters (effective power, fuel consumption and engine overall efficiency),
- parameters influencing natural environment (exhaust emissions),
- combustion parameters (that determine engine durability).

Investigation on the above-mentioned topics were carried out in the Laboratory of Vehicles and Combustion Engines at Kazimierz Pulaski University of Technology and Humanities in Radom.
The first stage of the investigation was focused on the establishment of the compression ratio of the modified engine. The obtained results were presented at the conference KONMOT 2014 and published in [1]. A further investigation concerned the establishment of the proper ignition advance angle of the modified engine. Within the framework of this investigation it was assessed the effect of this regulation on basic engine operating parameters, exhaust emission and combustion parameters.

IOP Conf. Series: Materials Science and Engineering **148** (2016) 012072 doi:10.1088/1757-899X/148/1/012072

2. Description of the test bed

The investigations were carried out with the use of a single cylinder 1HC102 research engine. In the standard version, this is a compression ignition engine fuelled with diesel oil. In the modified version, this is a spark ignition engine with a compression ratio of 9 – the value which was established earlier. This engine was coupled with an electrorotational brake Vibrometr 3 WB 15. The test bed was equipped with systems providing measurements of:

- engine torque
- engine speed
- hourly fuel consumption
- exhaust emissions
- combustion pressures.

General views of the test bed and the engine control room are presented in (figure 1 and 2).

Figure 1. Test bed: 1– 1HC 102 engine, 2 – electrorotational brake, 3 – intake manifold, 4 – air reservoir, 5 – laminar flow meter ∅20, 6 – ignition system, 7 – catalytic converter, 8 – cooling system of the cylinder pressure sensor, 9 – crankshaft rotation angle transducer, 10 – camshaft position sensor, 11 – throttle assembly, 12 – coolant temperature sensor, 13 – engine oil temperature sensor.

Figure 2. The engine control room: 1 – engine speed measurement system, 2 – engine torque measurement system, 3 – multi-channel temperature meters WRT-9, 4 – engine starting system, 5 – measurement system of the cylinder pressure and of signals from the crankshaft rotation angle transducer, 6 – PC-computer with a data acquisition card, 7 – potentiometer controlling the throttle opening, 8 – amplifier of the signals from the cylinder pressure sensor and from the crankshaft rotation angle transducer, 9 – oscilloscope, 10 – stabilized laboratory power supply.

Table 1. Comparison of characteristic features of the investigated engine.

Engine type	**1 HC102**	**1 HC102**
Way of ignition	CI	SI
Displacement [dm^3]	0,980	0,980
Compression ratio	17	9
Maximum engine speed [rpm]	2200	2200

3. Investigation procedure

In the first stage of the investigation, regulation characteristics of maximum load Mo_{max} [Nm] versus the ignition advance angle α (C.A. BTDC) were prepared and engine overall efficiency obtained in these conditions. Characteristics were prepared at three values of the engine speed: n_1=1200 rpm, n_2=1700 rpm, n_3=2200 rpm. The obtained results provided a basis for further investigations taking into consideration the beneficial value of ignition advance angle at each of the selected engine speeds. The following stage of the investigation consisted in preparation of load characteristics of the main exhaust emissions from the modified engine at each of the selected engine speeds. The final stage of the investigation consisted in preparation of load characteristics of the basic combustion parameters (maximum pressure P_{max} , mean rate of pressure rise $\frac{\Delta P}{\Delta \alpha}$ mean and maximum rate of pressure rise $\frac{\Delta P}{\Delta \alpha}$max). The present paper presents results obtained at the engine speed of 2200 rpm only. The relationships obtained for the modified engine were compared with those obtained for the standard engine operation. Results of the investigation and comparisons are presented in chapter 5.

4. Selection of the beneficial ignition advance angle regulation taking into consideration the developed torque Mo [Nm] and engine overall efficiency ηo

Selection of the beneficial ignition advance angle regulation taking into consideration the developed torque Mo [Nm] and engine overall efficiency η_o was done analyzing the regulation characteristics presented in (figure 3 and 4).

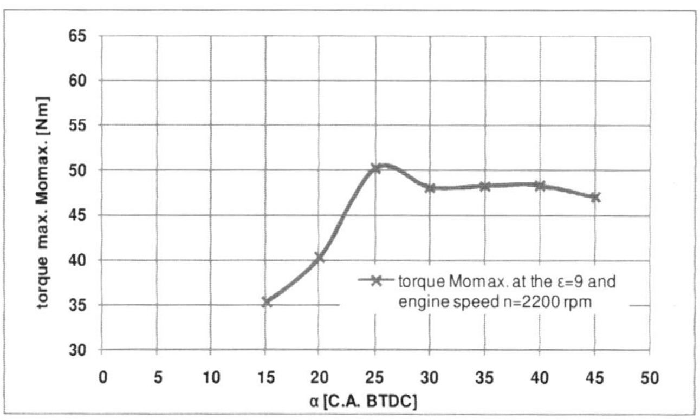

Figure 3. Characteristics of the maximum torque Mo_{max} versus the ignition advance angle α, at the engine speed n=2200 rpm.

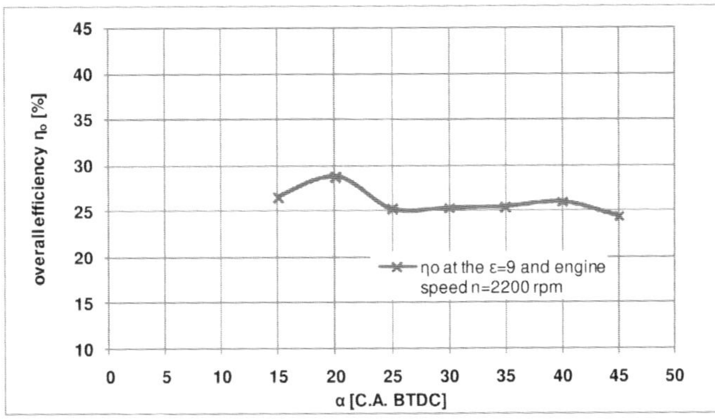

Figure 4. Regulation characteristics of the overall efficiency of the modified engine operating on LPG with the application of spark ignition for the established compression ratio $\varepsilon = 9$, at the engine speed n=2200 rpm.

It results from these characteristics that, taking into consideration the developed torque Mo [Nm], the most beneficial regulation of the ignition advance angle is α =25 C.A. BTDC (at the engine speed n=2200 rpm). The established ignition advance angle was set in further investigation on the main exhaust emissions and combustion parameters.

5. Comparison of basic operating parameters, main exhaust emissions and combustion parameters for the modified and the standard engines

In order to carry out the necessary comparisons, load characteristics of the above-mentioned parameters were prepared. In the case of the engine operating on diesel oil, the standard ignition advance angle α =20 C.A. BTDC was maintained. In the modified engine, there was applied the established earlier ignition advance angle equal to 25 C.A. BTDC.

5.1. Comparison of the engine overall efficiency η_o characteristics

The overall efficiency η_o characteristics of the investigated engines were prepared on the basis of measurements of the engine torque Mo[Nm], the hourly fuel consumption G_e[kg/h] and heating values of the applied fuels [kJ/kg]. The characteristics are presented in (figure 5).

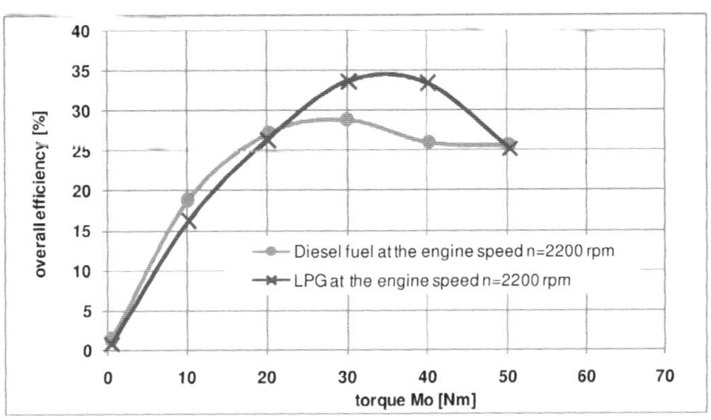

Figure 5. Comparison of the load characteristics of the overall efficiency $\eta_o = f(M_o)$ for the standard engine operating on diesel oil, at the constant ignition advance angle α =20 C.A. BTDC and the compression ratio $\varepsilon = 17$ and for the modified engine operating on LPG with the application of spark ignition, at the constant ignition advance angle α =25 C.A. BTDC and the established compression ratio $\varepsilon = 9$, at the engine speed n=2200 rpm.

Analysis of the characteristics presented in figure 5 shows that the modified engine delivers higher overall efficiency comparing to the standard operating engine, especially at loads close to the maximum ratings. It is worth mentioning the fact that both engine versions deliver the same maximum load rating of Mo=50 [Nm]. Achievement of such result was possible, to a large degree, due to the proper regulation of the ignition advance angle α for the modified engine.

5.2. Comparison of characteristics relating to the main exhaust emissions

The engine exhaust emissions were measured under load characteristics for both engine versions. The obtained characteristics are presented in (figure 6, 7 and 8). It should be noticed, that during investigation on the modified engine, the air – LPG mixture was kept at the stoichiometric ratio ($\lambda \cong 1$).

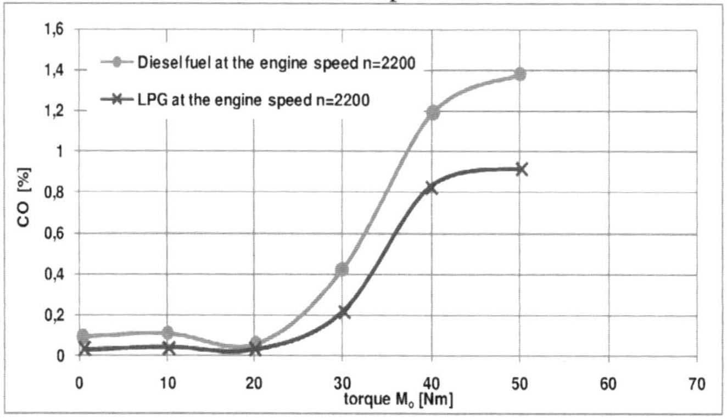

Figure 6. Comparison of the load characteristics of the CO emissions for the standard engine operating on diesel oil, at the constant ignition advance angle α =20 C.A. BTDC and the compression ratio ε = 17 and for the modified engine operating on LPG with the application of spark ignition, at the constant ignition advance angle α =25 C.A. BTDC and the established compression ratio ε = 9, at the engine speed n=2200 rpm.

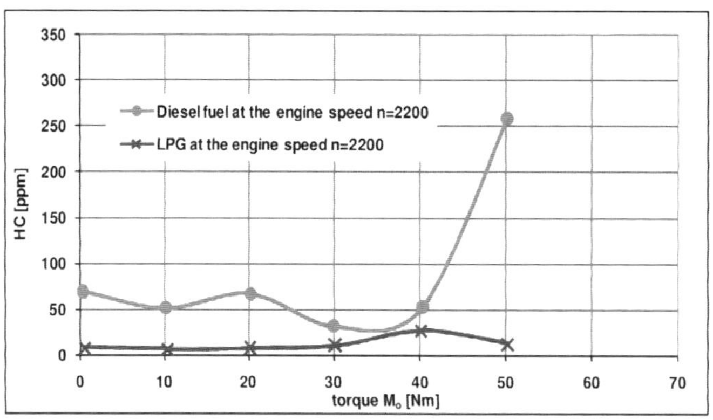

Figure 7. Comparison of the load characteristics of the HC emissions for the standard engine operating on diesel oil, at the constant ignition advance angle α =20 C.A. BTDC and the compression ratio ε = 17 and for the modified engine operating on LPG with the application of spark ignition, at the constant ignition advance angle α =25 C.A. BTDC and the established compression ratio ε = 9, at the engine speed n=2200 rpm.

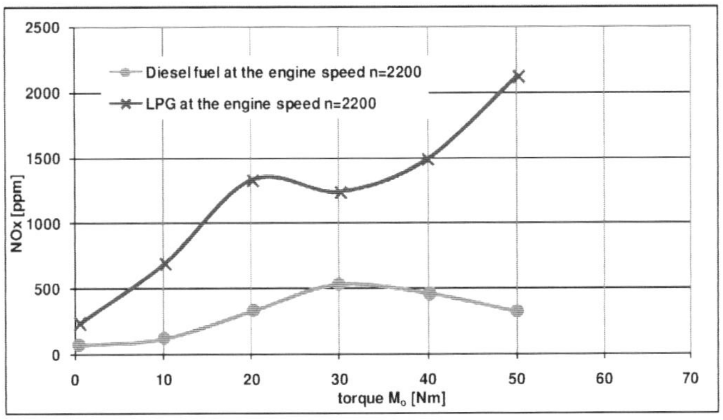

Figure 8. Comparison of the load characteristics of the NO_x emissions for the standard engine operating on diesel oil, at the constant ignition advance angle α =20 C.A. BTDC and the compression ratio ε = 17 and for the modified engine operating on LPG with the application of spark ignition, at the constant ignition advance angle α =25 C.A. BTDC and the established compression ratio ε = 9, at the engine speed n=2200 rpm.

It results from these characteristics (obtained at the engine speed n=2200 rpm and the ignition advance angle regulation α =25 C.A. BTDC) that:

- CO emission from the modified engine is lower than from the standard operating engine, particularly at loads close to the maximum ratings,
- HC emission from the modified engine is also lower than from the standard operating engine particularly at loads close to the maximum ratings as well,
- NOx emission from the modified engine is also higher than from the standard operating engine.

According to the authors, the positive effect reduction of CO and HC emissions of the modified engine is due to the following facts:

- the standard CI engine was equipped with an old style injection pump (of the inline type) operating at low injection pressure (ca. 17 MPa)
- the modified engine operated on a gaseous fuel (LPG). Fuel of this type, as a rule, has beneficial combustion characteristics.

The air–fuel equivalence ratio of the modified engine was kept at the stoichiometric condition for a SI engine $\lambda \approx 1$. In the standard CI engine, this ratio had a value varying from $\lambda \approx 8 \div 10$ (at the minimum load) to $\lambda \approx 1{,}3$ at the maximum load. At this stage of investigation, it is difficult to give a clear answer to the question of why a higher NOx emission is observed in the modified engine.
However, it should be mentioned that the investigated engine was not fitted with a catalytic converter. It is expected that application of a catalytic converter will result in lower emissions, especially in view of the fact that the engine is operating at stoichiometric conditions (λ=1).

5.3. Comparison of characteristics relating to the basic combustion parameters

During the investigation on both engine versions, measurements of the combustion pressure versus crankshaft rotation P[MPa] = f(α) [C.A.] were taken. The registered courses for consecutive load levels M[Nm] enabled to prepare the characteristics of basic combustion parameters. The characteristics relate to the following parameters:

- maximum combustion pressure P_{max} [MPa] = f(M) [Nm],
- mean rate of pressure rise $\frac{\Delta P}{\Delta \alpha}$ mean $[\frac{MPa}{{}^\circ CA}]$ = f (M) [Nm],
- maximum rate of pressure rise $\frac{\Delta P}{\Delta \alpha}$max. $[\frac{MPa}{{}^\circ CA}]$= f (M) [Nm].

These characteristics are presented in (figure 9, 10 and 11).

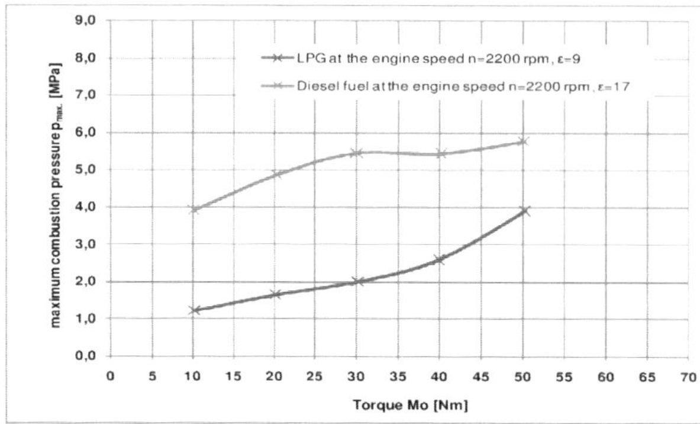

Figure 9. Characteristics of the maximum combustion pressure P_{max} [MPa] versus the engine torque M_o [Nm], at the engine speed n=2200 rpm

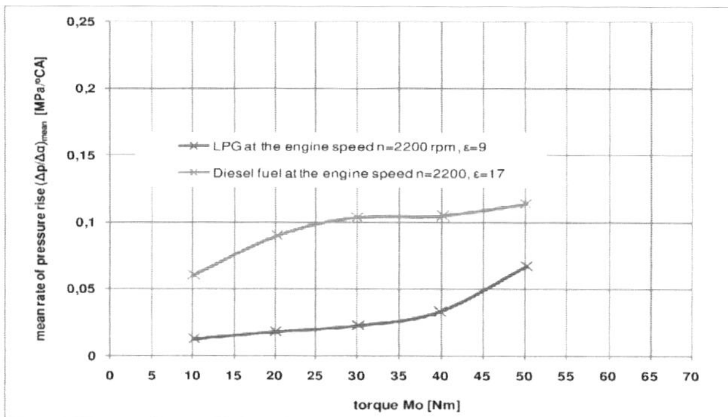

Figure 10. Characteristics of the mean rate of pressure rise $\frac{\Delta P}{\Delta\propto}$mean [$\frac{MPa}{°CA}$] versus the engine torque M_o [Nm], at the engine speed n=2200 rpm

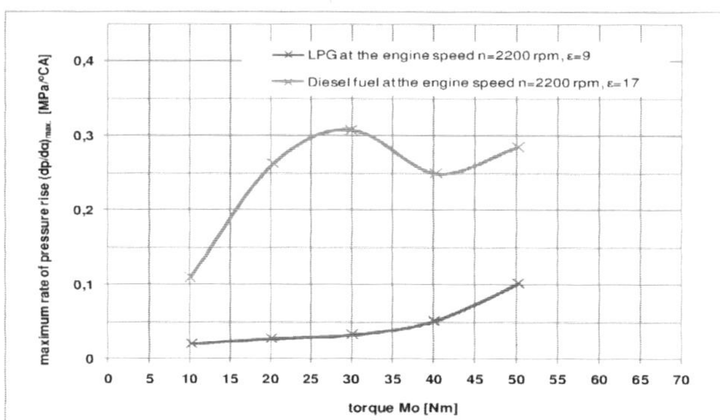

Figure 11. Characteristics of maximum rate of pressure rise $\frac{\Delta P}{\Delta\propto}$max.$\left[\frac{MPa}{°CA}\right]$ versus the engine torque M_o [Nm], at the engine speed n=2200 rpm

The presented characteristics of basic combustion parameters reveal that, at the established engine speed n=2200 rpm and at the established regulation of the ignition advance angle α =25 C.A. BTDC, the investigated combustion parameters (maximum combustion pressure P_{max}, mean rate of pressure rise $\frac{\Delta P}{\Delta\propto}$ mean and maximum rate of pressure rise $\frac{\Delta P}{\Delta\propto}$max) of the modified engine are lower, over the full range of load, than corresponding parameters of the standard operating engine. It should be mentioned, that this feature is beneficial regarding engine durability.

6. Summary – conclusions

Modification of a compression ignition engine should be preceded by investigation aimed at establishment of the compression ratio ε and the ignition advance angle α.

Proper establishment of the above-mentioned parameters enabled to obtain higher engine overall efficiency η_o of the 1HC102 engine maintaining the developed torque Mo_{max}.

Operation of the modified 1HC102 engine resulted in lower CO and HC emissions and higher NOx emission. It should be noticed, that fitting the engine with a catalytic converter would enable a further decrease of these emissions.

The process of combustion in the modified engine, in comparison to the standard operating engine, is characterized by lower values of the following parameters:

- maximum combustion pressure P_{max},
- mean rate of pressure rise $\frac{\Delta P}{\Delta \alpha}$ mean,
- maximum rate of pressure rise $\frac{\Delta P}{\Delta \alpha}$max.

This feature is beneficial regarding engine durability.

The above-presented observations enable a statement that the analysed modification of older generation CI engines would bring about a number of benefits (increased engine performances, higher engine overall efficiency, lower exhaust emissions) provided that the compression ratio ε and the ignition advance angle α will be established in a proper way.

Further improvement of the analysed operating parameters of the modified engine may be obtained in result of an accurate design of the combustion chamber.

References

[1] Kaparuk J and Luft S 2014 Wpływ stopnia sprężania na wybrane parametry pracy silnika o ZI zasilanego paliwami LPG – badania wstępne. Wydawnictwo Politechniki Krakowskiej Monografia

[2] Kaleemuddin S and Rao G.A.P 2010 *Conversion of Diesel Engine into Spark Ignition Engine to work with CNG and LPG for meeting new emission norms.* Thermal Science, Vol. 14, No. 4, pp. 913 ÷ 922. Laboratory for Thermal Engineering and Energy Research, VINČA Institute of Nuclear Sciences in Serbia, Belgrade

[3] Bardziński W and Żółtowski A 2003 *Silnik 6C107 zasilany gazem propan- butan.* Journal of KONES Internal Combustion Engines, vol. 10, s. 3 ÷ 4.

[4] Prace wykonane przez NGV AUTOGAS Sp. z o.o. i SPNG IBMER w latach 1992 - 2001 oraz INiG (2002-2006). Strona internetowa: http://www.ngvautogas.com.pl/index. php/referencje, dostępność: 12.09.2015.

[5] Gis W, Menes E and Waśkiewicz J 2011 *Paliwa gazowe w miejskiej komunikacji autobusowej w Polsce.* Transport Samochodowy 2-2011, s. 71 ÷ 95, Instytut Transportu Samochodowego, Warszawa

[6] Wojciechowski A, Chłopek Z, Gis W, Krupiński M, Menes E, Merkisz J, Waśkiewicz J, Żółtowski A 2010 *Alternative Powertrains City Busse. International Conference on Electric Vehicles.* Warsaw University of Technology

[7] Kaparuk J and Luft S 2014 *Wpływ stopnia sprężania na wybrane parametry pracy silnika o ZI zasilanego paliwem LPG - badania wstępne.* Praca zbiorowa pod red. Władysława Mitiańca. Silniki Spalinowe i Ekologia. Opracowanie monograficzne. Wydawnictwo Politechniki Krakowskiej, Kraków

[8] Kaparuk J and Luft S 2014 Wpływ modernizacji silnika o ZS w celu zasilania paliwem LPG na wybrane parametry jego pracy- badania wstępne. Logistyka 6/2014, Instytut Logistyki i Magazynowania w Poznaniu, Poznań.

Pneumatic injection system using a hot exhaust gases, developed in Institute of Automobiles and Internal Combustion Engines of Cracow University of Technology

W Marek[1] **and K Śliwiński**[1]

[1]Cracow University of Technology, Institute of Automobiles and Combustion Engines, al. Jana Pawła II 37, 31-864 Krakow, Poland

E-mail: wmarek@pk.edu.pl

Abstract. The article concerns research carried out by the Krakow University of Technology on the concept of a pneumatic fuel injection spark ignition engines. In this artkule an example of an application of this type of power to the Wankel's engine, together with a description of its design and operating principles and the benefits of its use. The work was carried out over many years by Prof. Stanislaw Jarnuszkiewicz despite the development of many patents but not widely used in engines. Authors who were involved in the team-work of the team of Prof. Jarnuszkiewicz, after conducting exploratory studies, believed that this solution has development potential and this will be presented in future articles.

1. Introduction

40 Years "Konmot" conference gives an excellent opportunity to summarize these achievements a especially that the authors of this article are derived directly from the scientific Institute of Automobiles and Internal Combustion Engines, with is the main organizer of the conference. For this particular reason, the authors proposed to publish a series of articles on the development of the unique concept of a pneumatic fuel injection for spark ignition engines, which concept was design was designed and developed by prof. Stanislaw Jarnuszkiewicz and his team. The Authors of the publication were for many years part of this team, and this way they try to commemorate a Professor and carry on to remember his achievements and carry on with his ideas.

In the introduction leading to the habilitation monograph, which was created during the first and second "KONMOT" conference. Professor indicate in his publication that ... There were already performed some tests at this time to use direct fuel injection on an industrial scale, in order to eliminate… ... large defects of two-stroke engines ... [1]. The two-stroke engines in spite of faults, due to the simplicity of design, the two stroke engines were always the point of interest for Professor. Simultaneously he was gifted designer with great intuition and the ability to deploy scientific problem into practice.

These attempts until today have not passed the test on a larger scale, because of the coast of an injection systems which ware also complicated and required professional services. These particular reasons mention above undermined obviously their attractiveness.

In the course of this work, Professor made a number of patents relating to the use on a pneumatic direct injection system, in internal combustion engines - the first in 1956.
Nevertheless, however the concept of pneumatic fuel injection using the exhaust gases proposed by Professor, still has growth potential, according to the authors' opinion. In the next series of publications, will be presented the results of studies indicating areas of possible applications of this system.

2. Hybrid engines

The dynamic development of pneumatic injection system according to the concept of Professor, occurred in the years immediately following the first oil crisis in the early seventies. There was established the cooperation with companies such as Volvo-Penta and Audi-NSU. Due to the marketing requirements, the pneumatic injection system has been called, the pumpless fuel injection system. Due to the unique characteristics of this injection system which ensured to engines with spark ignition, according to the classification of P. Schweitzer [2], identified them as hybrid engines. On the basis of this classification so called hybrid engines, an engines which combine the characteristics of spark ignition and compression-ignition engines, and their objective is to combine the advantages of both groups of engines. With the help of the injection system the stratification of charge relatively easily was obtained and qualitative power control, and the possibility of the use of liquid fuels with different fractional composition and exhaust gas recirculation. At present, after years of work in this field, each of these features can be subjected to sophisticated regulation by means of mechatronic equipment. However, there was no such possibility forty years ago, but that did not deter Professor to examine eleven prototype of engines.
As a result of stage control tests, carried out at the Cracow University of Technology in cooperation with the companies mentioned above, there were obtained in comparison with produced at that time Swedish and German low power engines, reduce fuel consumption up to 40%, with the use of gasoline and diesel. Achieved a reduction up to 90% of the content of toxic components of exhaust gas, carbon monoxide and hydrocarbons. Moreover they reached, a non-knocking combustion of liquid fuels with low octane number at constant compression ratio, and a wide range of qualitative power control. Examples of the characteristics will be presented later in this article.

3. Example of application with selected research results

The principle system developed by professor describe the use of the exhaust gases from the first cylinder with a relatively high pressure to spray the dose of fuel and inject it to the other cylinder and reveres.
Based on this principal simultaneously Professor has developed and patented injection system for two-stroke two-cylinder, three-cylinder, four-cylinder and six-cylinder engines [3]. Has also developed an adaptation of the four-stroke two-, four- and six-cylinder engines, as well as for rotary piston engine (Wankel) [4] [5]. Undoubtedly the most interesting example of application, was modified Wankel engine. Figure 1 shows the functional diagram of the solution, taken from the patent description.
As we know, over the years, the engine of this type was in the next stage of development, however, has not gained the expected level of the quantitative application. Especially recognition of accuracy of the diagnosis deserves more and proposed and documented test results, quoted in the article, the technical solution to the problem of oversize environmental burden of this type of engine. As an example, one can cite the following description [1]. Wankel type engines with rotary piston have in comparison with reciprocating engines, some differences of combustion, in addition to about 1.7 times larger faces the combustion chamber at the same volume. Therefore, in the exhaust gas of these engines there are relatively large amounts of unburned hydrocarbons and carbon monoxide with an increased specific fuel consumption. Based on previous research performed on Wankel's combustion engines, one can outline the directions to reduce the adverse effects occurring in them, for define efforts to the reduction of exhaust emissions, which include:

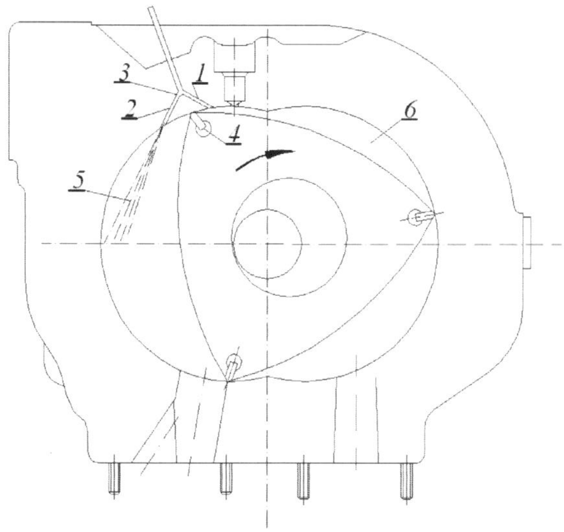

Figure 1. The functional diagram of the modified Wankel engine.

- removing particles or vapors of fuel, from the layer of boundary sidewall and from the narrowed cylinder space,
- concentration of combustible mixtures, mainly in the ahead portion of the combustion space and filling, the keeping up part of combustion space, mostly clean air, i.e. creating a stratification of charge, allowing the engine to operate also in the field of a large excess of air.

This assumption can be achieved bringing the gaseous fuel (fuel vapor) to the combustion chamber, wherein the stream of blown gas of low puncture strength, doesn't reach to the walls of the chamber, and concentrate in the central portion. The direction and phase of blow of fuel vapor should be selected so, that they focus on the overtaking part of the combustion space… end of citation.
The principle of operation of the device executing the above command is as follows (description according to Figure 1): in so-called cold bow side of engine body, done two channels 1 and 2. The ducts join together in the mixing cell 3, to which is applied the fuel in controlled quantities. The moment in which the front sealing strip 4 reveals the outlet channel 2, takes place the blow hot exhaust gas from the combustion chamber 6 through the passageway 1, the cell 3 and the orifice 2 to the charge of air in the space 5. The hot exhaust gases lead to pushing the dose of the fuel from cell 3, to atomize and evaporating, and then blows to the air charge. The fuel supply to the mixing cells occurs after exposure by the sealing strip, the channel 1, that is, during the duration of low pressure in the channels. This cycle is repeated synchronously with the cycles of the engine. The study was conducted on a prototype engine with rotary piston type Wankel - NSU KKM-613/VI.
Basic technical data:

- one rotary piston,
- chamber volume: 498 cm^3
- compression ratio: 8.6: 1
- power output: 40 kW/6000 1/min
- maximum torque: 71 Nm/2500 1/min
- gasoline supply: Solex 18/32 HHD

Changing the way the engine power has required the development of a special dispenser. Functional diagram of the unit and installation in the Wankel engine, is shown in Fig. 2.

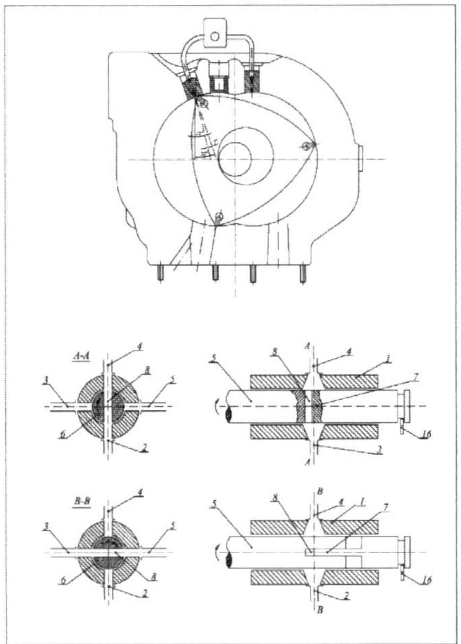

Figure 2. Functional diagram of the injection unit and installation in the Wankel engine.

The Fuel dispenser comprises a housing 1, wherein the channels 2 are formed, supplying fuel, and fuel outlet channels 4 and also inlet gas channels 3 and outlet gas channels 5, for outlet gas which makes the injection of fuel. In the body 1 the rotational shaft 6 is set, having an accommodating space 8, the volume of which varies according to the position of the strip 7, actuated by the slider 16. During the rotation of the shaft 6, the space 8 is contacted with inlet channels 2 and outlet channels 4, for fuel discharging. This space has been filled with fuel, and excess of fuel flows by channel 4. After returning to a position in which the space 8 is contacted with inlet channels and through channels 5 for discharging the gas, is followed by gas extrusion one dose of fuel interposition contained in the space 8, to the injection system. The value of the dose depends on the volume of the space 8.
The operating indicators of the modified engine in relation to the carburettor supply, are provided in Figure 3.

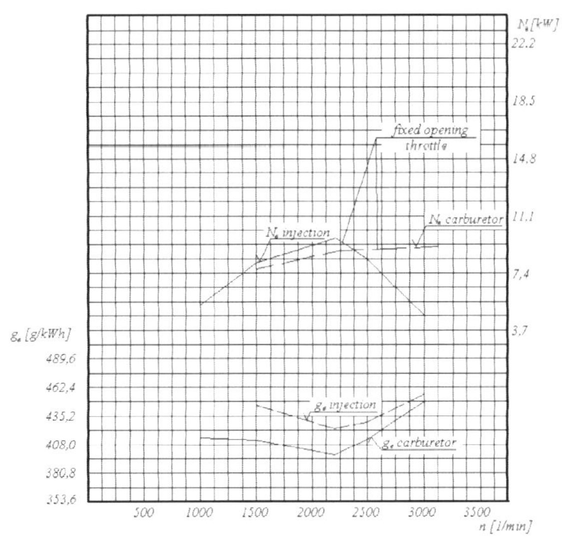

Figure 3. The operating indicators of the modified engine in relation to the carburettor supply.

Visible limited range of crankshaft speed, due to the excessive length of the injection duct. This was due to the possibility of adapting the system of injection to the accessible location, limited possibilities of assembly.
It has been found that in applied injection pipe length, the engine less than 3000 1/min obtained, and in this field has a lower specific fuel consumption while maintaining comparable power.
Because on the test engine was not possible to shorten the gas duct, Professor has developed and patented the modified fuel injection system without pump, for engines of this type. Diagram of the device shown in Figure 4.

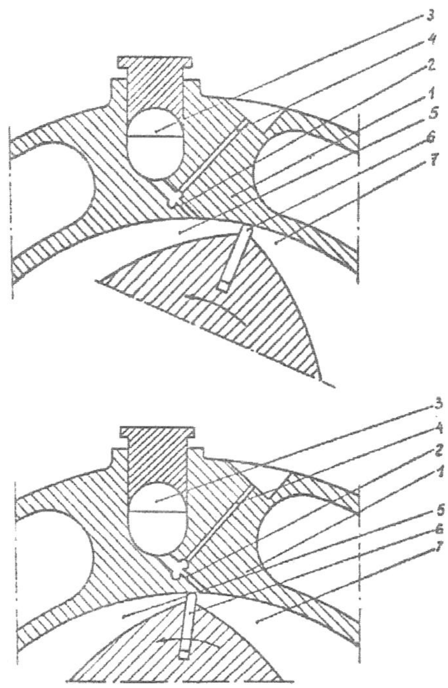

Figure 4. Modified fuel injection system for Wankel engines [1].

Unlike the previous system, the working space 5 and 7 of the engine is connected, to only one channel 2, which is located at the periphery of a trochoidal track, at the same location as the injection orifice previous version of system. This channel by this other part, is connected to the reservoir tank 3. Into this reservoir a certain amount of fuel is delivered (by channel 4), in the period, in which there is low pressure in the working chamber. During rotation of the rotating piston into the cartridge 3 flows a charge during the compression phase, then during the period of combustion and the start of expansion, flowing part of hot combustion gases under high pressure. Previously delivered dose of the fuel evaporated. At the moment, when the corner sealing strip of the piston 6 passes inlet channel 2, space of the cartridge 3 will be connected to the space 7 working chamber in which there is a low pressure (the end of the intake phase). The mixture of gases and vapors placed into the cartridge under high pressure, discharges by channel 2, into charge of air into the chamber 7. Implementation of this modification has brought next improvements in operating indicators of a modified Wankel engine. Obtained at that time results, are provided in Figure 5.
The specific fuel consumption was reduced by 15% and the content of carbon monoxide in the exhaust gas decreased repeatedly.

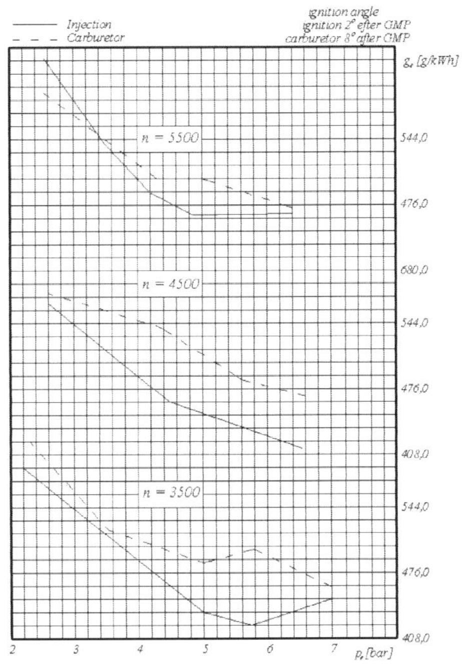

Figure 5. The operating indicators of the modified engine with a modified injection system, in relation to the carburettor supply.

4. Summary

Positive results and relatively easy to realize application requirements, created by Professor Jarnuszkiewicz, supply and combustion systems, led him to a deeper study in the description of phenomena accompanying the process of fuel injection using hot exhaust gases. One of these directions was to study, the concept of binding insensitivity of modified engine at fuel tendency to knock, with the occurrence of chemical reactions such a cracking and pyrolysis, in the space of an injection cartridge. However, further research is conducted both by the professor and his colleagues have not confirmed this concept [6] [7]. Professor all its research activities tied to the Cracow University of Technology. In the last period of its activity at our University (until 2004) he conducted research on the issue of pneumatic injection to the multi-cylinder four-stroke engines. Despite the passage of years, his achievements of application in the form of granted patents, and completed construction of low power internal combustion engines, inspires admiration and respect, and is an excellent example for future generations of researchers and engineers.

References

[1] Jarnuszkiewicz S 1981 *Wybrane zagadnienia silników hybrydowych* Politechnika Krakowska.
[2] Schweitzer P 1963 *Hybrid Motoren* MTZ.
[3] Jarnuszkiewicz S 1956 *Silnik dwusuwowy z bezpompowym wtryskiem paliwa* Patent PRL 40368.
[4] Jarnuszkiewicz S 1977 *Silnik spalinowy czterosuwowy wielocylindrowy z bezpompowym wtryskiem paliwa z zapłonem iskrowym* Patent PRL 97947.
[5] Jarnuszkiewicz S 1969 *Silnik spalinowy z wirującym tłokiem* Patent PRL 58920.
[6] Mitianiec W Śliwiński K 2003 *Analysis of thermodynamic parameters in pneumatic fuel injection system by use of hot exhaust gases in four-stroke engine* Politechnika Krakowska, Teka Komisji Naukowo-Problemowej Motoryzacji / Polska Akademia Nauk Oddział w Krakowie.
[7] Marek W 2008 *The analysis of fuel vaporization in a distributorless pneumatic injection system of a two stroke spark ignition engine* Combustion Engines.

IOP Conf. Series: Materials Science and Engineering **148** (2016) 012074 doi:10.1088/1757-899X/148/1/012074

Investigation of work parameters of SI engine dedicated to energetics aggregates with pneumatic injection system

W Marek[1] and K Śliwiński[1]

[1]Cracow University of Technology, Institute of Automobiles and Combustion Engines, al. Jana Pawła II 37, 31-864 Krakow, Poland

E-mail: wmarek@pk.edu.pl

Abstract. The article presents the possibilities of alternative fuel combustion in the engine four-stroke spark ignition engines. Power of the motor was carried out pneumatic fuel injection system using a hot gas developed by Prof. Stanislaw Jarnuszkiewicz. Presented made the position of the measuring system with the power and results. The engine experimental at the time of the study was powered by a blend of alcohol and gasoline. The main aim of the study was the question of control fuel dosage, taking into account the energy needs of forcing the engine load. During the tests carried load characteristics control the motor using the power control quality. Another issue was the elimination of penetration of fuel to the engine lubrication system, a problem occurred in the initial study on the issue of the pneumatic fuel injection using the hot exhaust gases. In summary we present the findings of this phase of the study.

1. Introduction

More recently can noted the growing interest of large industrial corporations and specialized government agencies issue of improving the efficiency of energy supply diversification. It is expected, therefore, that in the next few years will increase interest in the widest possible and universal, to use of internal combustion engines for energy purposes. The carrier of chemical energy in the working process of these engines, will be the fuels of varying fractional and group composition, and with diversified character of the phase. Therefore, it is clear that at the current stage of development of industrial civilization, piston internal combustion engine will remain for a long time the main drive source in many sectors of the economy.

This publication is another in the series [1], which will be presented the results of studies to indicating areas of possible uses, proposed by Professor Jarnuszkiewicz, the concepts of pneumatic injection by means of hot exhaust gases. In this publication will be presented one of the ways to use of waste liquid fuels, in an internal combustion engine, for the conversion of chemical energy to useful work for the production of electricity and useful heat energy in the process of recovery of waste heat from the exhaust system and engine cooling system [7]. As an example of broad and widespread use of waste fuels, can be given advanced projects in this area in the Nordic countries, despite the well-developed infrastructure of the existing energy supply from power plants.

2. Purpose and methodology of research

The main objective which was to achieve in the case of a multi-fuel engine, was the issue of controlling the dosage of fuel in the spark-ignition engines, taking into account of the energy requirements of the engine load inducement system. It not without significance was also the problem of minimizing toxic exhaust emissions, but it was treated more as a limitation, not a decision criterion. In the research work we identified another limitation, associated with the passage of the liquid phase of fuel dosing, to the lubricating oil. Was studied and recorded the impact of the fuel used for the engine, onto parameters such as torque and power, fuel consumption, the total efficiency of the engine, excess air coefficient, and exhaust gas temperature. Scheme of the engine load, was realized as the regulatory load characteristic of the engine, and was done at a fixed rotational speed of the engine shaft, using the qualitative power control.

3. Test bed investigation of the SI engine

Research of engine fed with a mixture of alcohol and the gasoline separately, were carried out on a special test bed in the Laboratory of Combustion Engines, Cracow University of Technology, whose essential elements are as follows:
– two-cylinder, four-stroke spark ignition engine,
– engine brake generator MEZ,
– AVL fuel mass flow meter , AVL Type 4210,
– The measuring systems for determining: exhaust gas temperature, ambient pressure and humidity of the intake air by the engine, the excess air ratio.

As a research object was prepared a two-cylinder four-stroke spark-ignition engine 126 000 A1, equipped in the pneumatic-fuel injection system using the hot exhaust gases, according to the concept of Professor Jarnuszkiewicz, which in the factory configuration, was carburetor equipped.

In this configuration, the engine was characterized by the following working indicators [2]: rated power on petrol 17.7 kW/4500 1/min., the maximum torque of 42 Nm/ 3000 1/min, the minimum specific fuel consumption of 300 g/kWh. Embodiment of multi-fuel supply system of such an engine, require the design and building a pneumatic injection system and the supply fuel system, which the configuration similar to a conventional, low pressure injection supply system [3] [4]. Pneumatic injection system has been designed based on the concept of Professor Jarnuszkiewicz, described in an article on his achievements, in connection with the jubilee of Conference Konmot 2016 [1]. The functional model of designed and mounted on engine, prototype system of pneumatic injection, was shown in Figure 1.

Figure 1. The view of the functional model of the pneumatic injection using hot exhaust gases for liquid fuels.

IOP Conf. Series: Materials Science and Engineering **148** (2016) 012074 doi:10.1088/1757-899X/148/1/012074

To control: the value of fuel dose and phase of its dosing, the angle of ignition timing, degree of throttle opening and the composition of the combustible mixture (excess air ratio), and also the value of dose of exhaust gases to blowing the dose of fuel, was designed and used a special motor controller. As the fuel was used waste mixture of iso-butanol and n-butanol alcohol, as a post-processing fuels of the chemical industry - marked ZAK and X. Both waste fuels come from different technology threads and characterized by the same calorific value, but different physical and chemical properties. For comparative purposes, was used motor gasoline.
Due to the prototype nature of fuel supply, and combustion system, the power control range was limited to medium loads.

4. Results and analysis of investigations

For the practical implementation of the research goal, associated with defined above as the multifuel engine supply, was developed a research program of implementing the strategy of diversified supply flow of chemical energy of the dispensed fuel. A reflection of this strategy is an example of the imaging method of regulating engine power, shown in Figure 2, when the engine is fed with a lean mixture.

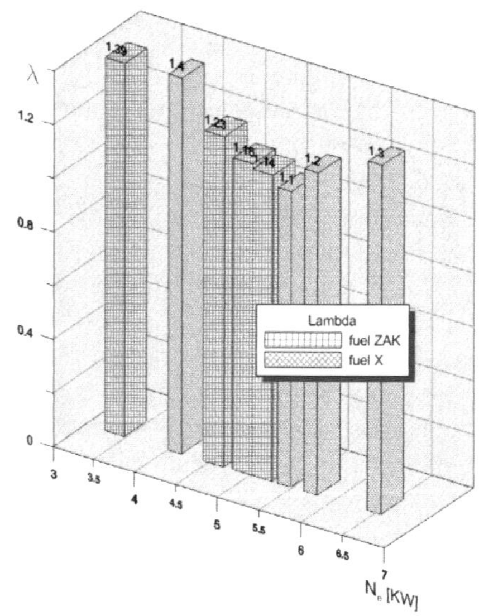

Figure 2. Characteristics of the excess air ratio of test engine, fueled with the post-processing fuels.

In the case marked on the chart as the ZAK, the engine power of 5 kW, was obtained by powered the engine only as so marked fuel.
Further increase in engine power to 6.5 kW, was obtained by application of the fuel, marked as X. This was due to the beneficial effect of fuel X physicochemical properties on the process of creating a combustible mixture, and repeatability of the heat release process. Comparatively, in the case of petrol supply to obtain the same load value, require a gradual enrichment of the mixture, up to the value of the excess air ratio of 0.83 - Figure 3.
Engine operation mode as so configured supply system, for feeding by post-processing fuels, did not cause the phenomenon of penetration of fuel into the lubricating oil. This was related mainly to the perfect process of spraying a liquid phase, when creating the fuel-air mixture. The problems associated with penetration of the waste liquid fuel to the lubricating oil are present in the case of conventional power supply systems of injection, due to insufficient atomization and evaporation of fuel in the intake and in the cylinder [5]. As we know, the process of preparing the fuel for combustion is one of the main factors influencing the efficiency of conversion of chemical energy to work for the engine [6].

The practical realization of this demand to improve the quality of spraying in the case such as that presented here supply system, is associated generally with the need to achieve a compromise between

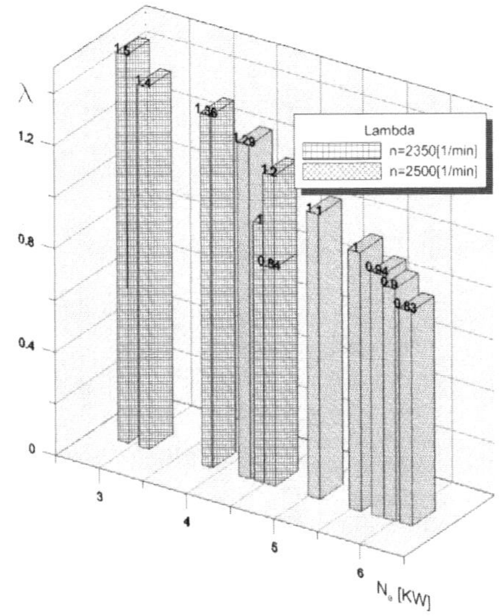

Figure 3. Characteristics of the excess air ratio of test engine, fuelled with petrol.

the cost of apparatus and equipment, and the real possibilities of maneuvering streams of chemical energy of the available fuel types. In other words, it seems rational to carry out each time of discernment concerning the ability to provide the suitable for the assumed of the engine energy parameters, fuel availability, and only then nominate the appropriate types of hardware and fuel dosage control system. Very eloquent, to verification of the scheme of the strategy of qualitative power control as given above (to constantly full throttle), is to visualize the course of the changes of the total efficiency, depending on the load, placed on the Figure 4.

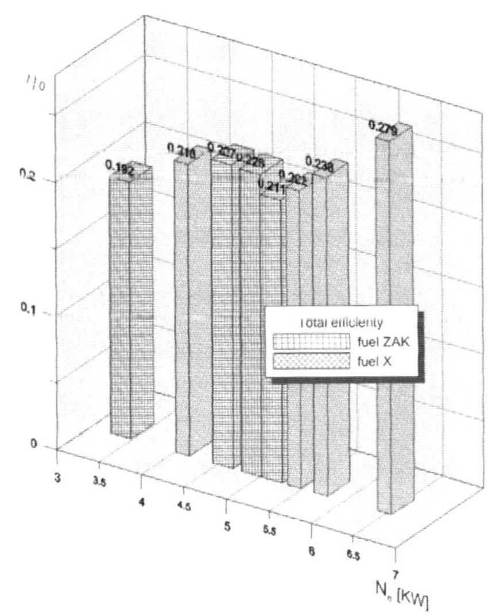

Figure 4. The total efficiency of the test engine fuelled with the post-processing fuels.

With already discussed above reasons, the total efficiency of the engine fueled ZAK, does not exceed 20%. In the case of fuel X supply, can be notice a marked increase in total efficiency of up to approx.

28%, while the occurrence of extension, of disposable load range. Comparatively, in the case of petrol supply, the value of similar total efficiency, was achieved at a lower load - Figure 5.

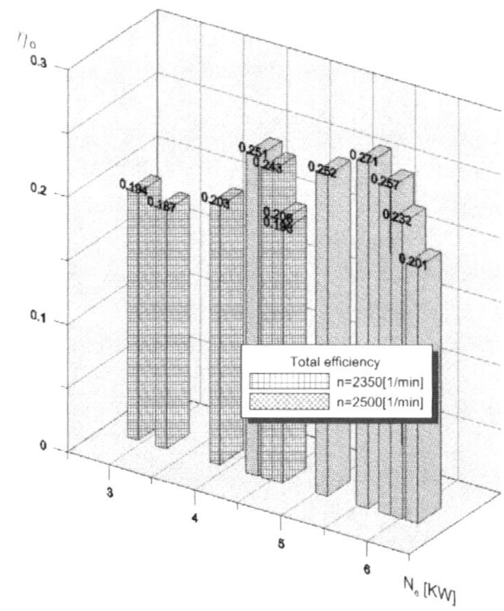

Figure 5. The total efficiency of the test engine fuelled with the petrol.

In the case of supply the engine with the post-processing fuels, may notice a steady increase in the total efficiency for the load increases - Figure 4. However, in the case of petrol supply, this sequence have not so favorable course - Figure 5.

Differences in in the process of creating a combustible mixture, have also caused the difference of the heat release in the cylinder, and the value of temperature of the combustion process. While the combustion of gasoline obtained the maximum exhaust gas temperature of 650 ° C - Figure 6, this during powering the engine with a lean mixture of post-processing fuels, the exhaust gas temperature, was reached a significantly lower value - Figure 7.

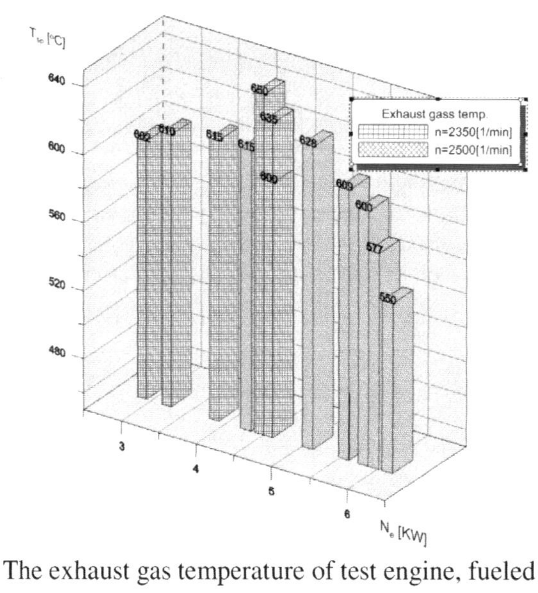

Figure 6. The exhaust gas temperature of test engine, fueled with petrol.

IOP Conf. Series: Materials Science and Engineering **148** (2016) 012074 doi:10.1088/1757-899X/148/1/012074

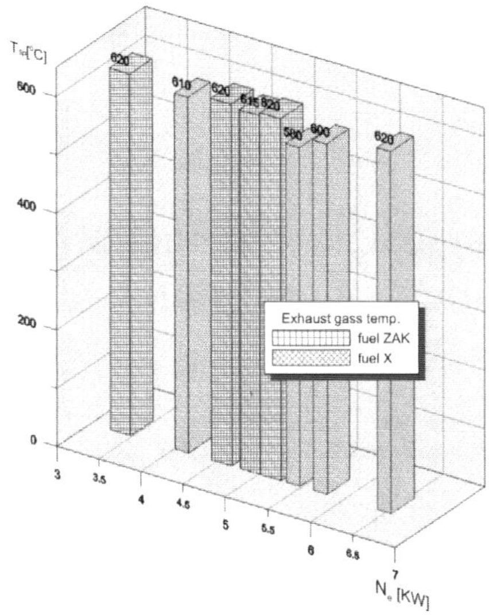

Figure 7. The exhaust gas temperature of test engine fuelled with the post-processing fuels.

5. Summary

The investigation results of the engine 126 000 A1 powered by waste fuels on alcohol base, allow to formulate the following most important conclusions:

1. The supply system designed and manufactured according to the concept of Professor Jarnuszkiewicz, allows to obtain multifuel usefulness, for the spark ignition four-stroke engine,
2. When engine was dual fuel supply, through an appropriate strategy of fuel dosage, was obtained possibility of extending disposable load range when powered by a lean mixture,
3. The results of the research allowed to define the strategy of power and control. It does not revealed the phenomenon of penetration of the liquid alcohol fuel, into a lubricating oil.
4. A preferred course of the total efficiency, and the exhaust gas temperature as a function of the load, demonstrating the effectiveness of the present fuel supply system, in the art of managing the post-processing liquid fuels.

References

[1] Marek W and Śliwiński K 2016 *Pneumatic fuel injection system using a hot exhaust gases, developed in Institute of Automobiles and Internal Combustion Engines of Cracow University of Technology* Konmot 2016.

[2] Fiat *Technical Specifications of engine*: 126 A1 000.

[3] Jarnuszkiewicz M and Śliwiński K 1992 *Odmienny sposób tworzenia mieszanki palnej w silniku wielopaliwowym* Politechnika Krakowska.

[4] Marek W 1994 *Sterowanie procesem tworzenia mieszaniny spalinowo-paliwowej przy wtrysku paliwa za pomocą gorących spalin - wybrane zagadnienia z prac własnych* Politechnika Krakowska.

[5] Brzeżański M, Mareczek M, Marek W and Papuga T 2013 *Using of waste products from the chemical industry to the energy production* Cracow University of Technology.

[6] Lejda K, Kuszewski H and Lew K 2015 *The influence of physicochemical fuel properties on operating parameters in diesel engine* Combustion Engines.

[7] Wojciechowski K T Merkisz J, Fuć P Lijewski P and Zybała R 2008 *Test bed study of waste heat recovery from the exhaust of a spark ignition engine* Proceedings of APISCEU The Ninth Asia-Pacific International Symposium on Combustion and Energy Utilization.

Investigation of ecological parameters of four-stroke SI engine, with pneumatic fuel injection system

W Marek[1] and K Śliwiński[1]

[1]Cracow University of Technology, Institute of Automobiles and Combustion Engines, al. Jana Pawła II 37, 31-864 Krakow, Poland

E-mail: wmarek@pk.edu.pl

Abstract. The publication presents the results of tests to determine the impact of using waste fuels, alcohol, to power the engine, on the ecological parameters of the combustion engine. Alternatively fuelled with a mixture of iso- and n-butanol, indicated with "X" and "END, and gasoline and a mixture of fuel and alcohol. The object of the study was a four-stroke engine with spark ignition designed to work with a generator. Motor power was held by the modified system of pneumatic injection using hot exhaust gases developed by Prof. Stanislaw Jarnuszkiewicz, controlled by modern mechatronic systems. Tests were conducted at a constant speed for the intended use of the engine. The subject of the research was to determine the control parameters such as ignition timing, mixture composition and the degree of exhaust gas recirculation on the ecological parameters of the engine. Tests were carried out using partially quality power control. In summary we present the findings of this phase of the study.

1. Introduction

Exploration and use of new fuels for combustion engines, is an important objective of scientific and economic reasons, leading to the rationalization of the use of different energy sources. Some of unconventional fuels are not suitable from various reasons, for use in traction vehicles, but can be used in power applications. For this type of fuel can include various types of combustible substances, which is a waste product of some industrial processes. One such fuel is butyl alcohol. Typically, this kind of waste fuels is most used for heating, but often irretrievably lost by burning for open space. Their development for low-energy systems seems to be a reasonable solution, the more that this type of systems allows for flexible adaptation the type of energy to current needs. This corresponds, to the modern trend, to diversify energy sources. Suitably adapted, the piston internal combustion engine, cooperating with a power generator, can be both - a heat source and a source of electricity. This publication is a next of cycle [1] [2], in which will be presented the results of studies indicating areas of possible uses, proposed by Professor Jarnuszkiewicz concept of the pneumatic injection by means of hot exhaust gases. In this publication will be presented, the results of research of ecological parameters of using liquid waste fuels, in an internal combustion engine for the conversion of chemical energy to useful work, for the production of electricity and useful heat energy.

2. Purpose and methodology of research

The main aim of the research was to determine by laboratory measurements, the impact of supply the alcohol waste fuel, on ecological parameters of the combustion engine. It became necessary to recognize, the new problems arising from the use of pneumatic injection system according to the concept of Professor Jarnuszkiewicz, as adapted to the modern mechatronic systems [3]. All measurements on engine test bench, of the toxic exhaust components, was carried out at a constant engine speed.

These conditions correspond to the operation of the engine in the power generator. The study tested the specificity of the pneumatic injection system, and therefore the selection of the rotational speed was not bound, as in typical solutions, by the operating parameters of the electrical machine. The criterion was a task - use a qualitative adjustment of power, corresponding to the energy demands of the load, with minimum values of rotation of the crankshaft of the engine. Under these conditions, the converters of mechanical energy into electrical energy, are used in the form of inverters. Among the control parameters of the test engine, which had to be matched to the type of examined fuel were primarily:

- ignition timing,
- coefficient of excess air,
- recirculation value of exhaust injection dose.

These parameters have a very significant impact on the process of combustion in the cylinder, which then results both, the value of generated energy parameters and toxic exhaust emissions. Especially in the case of qualitative power control, these parameters directly influence on the location in time, and duration of the combustion process. This influence is reflected very intensively in this case in particular, on the emission of toxic components of exhaust gases. For this reason, for the examined fuel, and for each operating point of the engine load, these control parameters were individually set. Measurements of the concentration of toxic exhaust components, was conducted by the measuring apparatus, in accordance with the applicable current standards, in which is equipped, the Engine Laboratory of Cracow University of Technology. Was studied and recorded the value of concentration: carbon monoxide CO, hydrocarbons THC, NO nitrogen oxide, and carbon dioxide CO_2 and oxygen O_2. In addition, was studied and recorded, the impact of the examined fuel, on the other engine parameters such as torque and power, the specific fuel consumption, the total efficiency, the coefficient of air excess and the exhaust gas temperature [2].

3. Test bed investigation of the SI engine

Research of engine fed with a mixture of alcohol and the gasoline separately, were carried out on a special test bed in the Laboratory of Combustion Engines, Cracow University of Technology, whose essential elements are as follows:

– two-cylinder, four-stroke spark ignition engine,
– engine brake generator MEZ,
– AVL fuel mass flow meter , AVL Type 4210,
– the measuring systems for determining: exhaust gas temperature, ambient pressure and humidity of the intake air by the engine, the excess air ratio.

As a research object was prepared a two-cylinder four-stroke spark-ignition engine 126 000 A1, equipped in the pneumatic-fuel injection system using the hot exhaust gases, according to the concept of Professor Jarnuszkiewicz, which in the factory configuration, was carburettor equipped.

In this configuration, the engine was characterized by the following working indicators [4]: rated power on petrol 17.7 kW/4500 1/min., the maximum torque of 42 Nm/ 3000 1/min, the minimum specific fuel consumption of 300 g/kWh. Embodiment of multi-fuel supply system of such an engine, require the design and building a pneumatic injection system and the supply fuel system, which the configuration similar to a conventional, low pressure injection supply system [5] [6]. Pneumatic injection system has been designed based on the concept of Professor Jarnuszkiewicz, described in an article on his achievements, in connection with the jubilee of Conference Konmot 2016 [1]. The functional model of designed and mounted on engine, prototype system of pneumatic injection, was shown in Figure 1.

Figure 1. The view of the functional model of the pneumatic injection using hot exhaust gases for liquid fuels

To control: the value of fuel dose and phase of its dosing, the angle of ignition timing, degree of throttle opening and the composition of the combustible mixture (excess air ratio), and also the value of dose of exhaust gases to blowing the dose of fuel, was designed and used a special engine controller. As the fuel was used waste mixture of iso-butanol and n-butanol alcohol, as a post-processing fuels of the chemical industry - marked ZAK and X. Both waste fuels come from different technology threads and characterized by the same calorific value, but different physical and chemical properties. For comparative purposes, was used motor gasoline.
Due to the prototype nature of fuel supply, and combustion system, the power control range was limited to medium loads.

4. Results and analysis of investigations

The energy benefits from the use of lean fuel-air mixture in the cylinder charge, have been known for a long time. Slightly worse on this background presents it's the issue of achieving, a satisfactory toxic exhaust emissions, especially when it comes to nitrogen oxides. However, in the case of dual fuel supply using the alcohol liquid phase, can be a significant reduction of the concentration of the toxic compounds in the exhaust gas. It was illustrated on Figure 2 and 3.

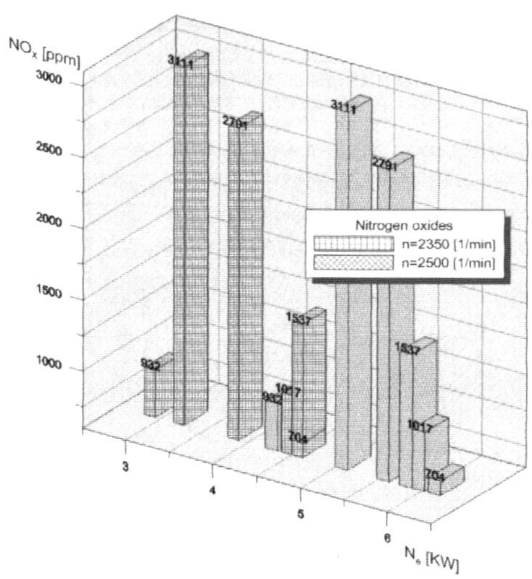

Figure 2. The concentration of nitric oxides NO_x in the exhaust gas of engine fuelled with petrol.

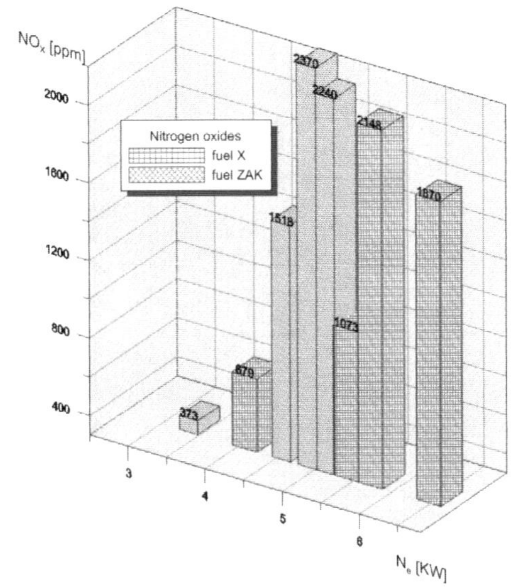

Figure 3. The concentration of nitric oxides NO_x in the exhaust gas of engine fuelled with the post-processing fuels.

In the case of petrol fuelling, the concentration of oxides of nitrogen to grow strongly while increasing the load of the engine, but by increasing the value of mixture enrichment and increase stage of recirculation while increasing a load, its concentration is significantly reduced. In the case of alcohol fuel supply, as the load increases we observed a sharp increase in the concentration of nitrogen oxides. This is facilitated by increasing the temperature of fuel combustion, as well as greater availability of oxygen, because as the load increases, the engine was powered by a mixture with ever increasing air excess. However, the adjusting of the value of exhaust gas recirculation, caused that the maximum concentration of NOx values, in case of alcohol fuel supply, it were smaller in relation to the petrol. This different strategy of fuel control dosage, was also highlighted, in the concentration of carbon monoxide in the exhaust gas. This is exemplified in Figure 4 and 5.

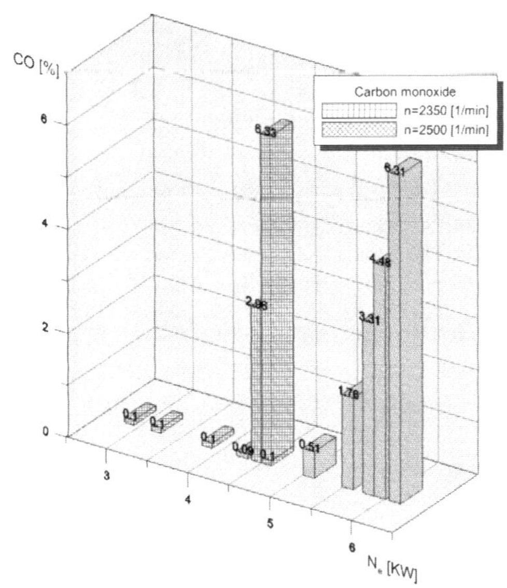

Figure 4. The concentration of carbon monoxide CO in the exhaust gas of engine fuelled with petrol.

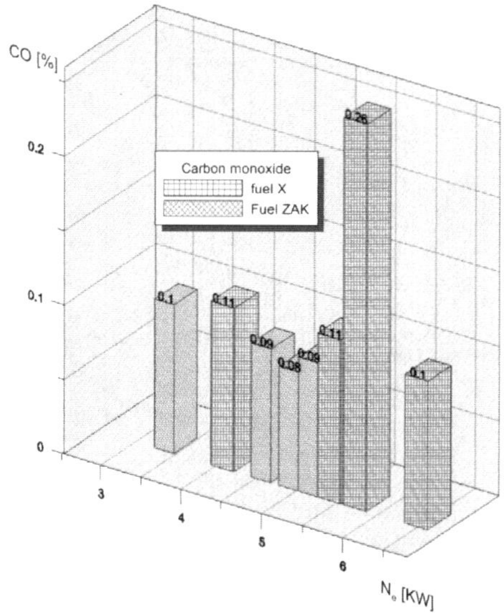

Figure 5. The concentration of carbon monoxide CO in the exhaust gas of engine fuelled with the post-processing fuels.

In the case of a gasoline supply, to increasing the load on the engine, it was necessary to enrich the air fuel mixture. This strategy allows for an increased load, without the occurrence of the phenomenon of knock, while also decreasing the concentration of NOx. The negative result, was a large carbon monoxide content in the exhaust gases - Figure 4. The strategy used for the alcohol fuel - was used a lean mixture in the whole load range - in the case of carbon monoxide concentration, will provided very positive results. It has been reduced several times - Figure 5.

The concentration of hydrocarbons in the exhaust gas, was dependent on the value of the air excess ratio - Figure 6 and 7.

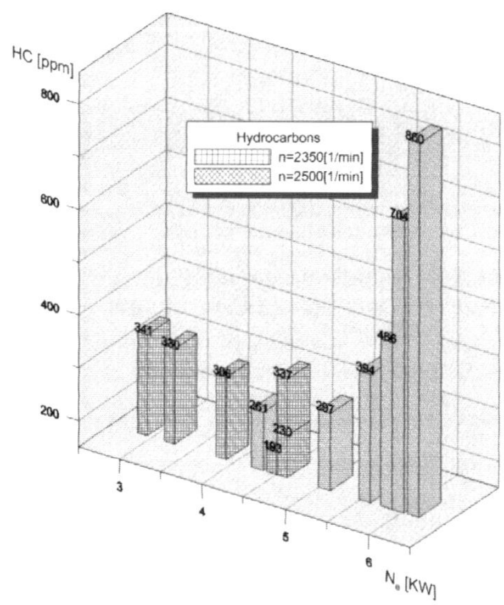

Figure 6. The concentration of hydrocarbons HC in the exhaust gas of engine fuelled with petrol.

In the case of petrol supply, an increase in the concentration level of this exhaust component (Figure 6), was mainly due, to a significant enrichment of the mixture to $\lambda = 0.83$, as the load increases.

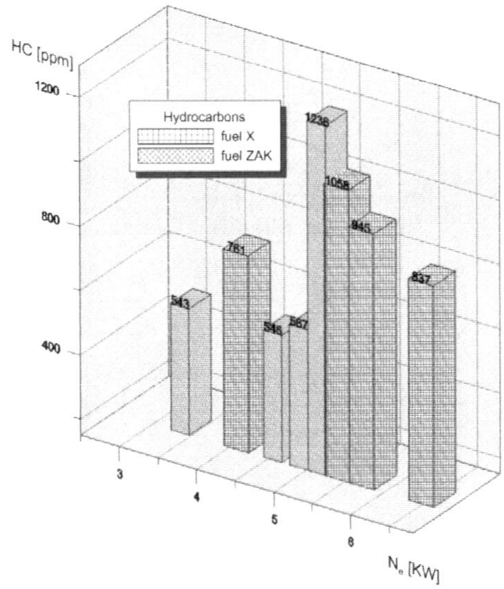

Figure 7. The concentration of hydrocarbons HC in the exhaust gas of engine fuelled with the post-processing fuels.

In the case of the alcohol fuel supply, was observed a significant spread of concentrations - Figure 7, and the higher value in relation to the petrol supply. This is due to the phenomenon of misfire and the reduction of combustion speed, during supply of engine by lean mixture of the value $\lambda \approx 1,4$.
The figure 8 presents the impact of load of the engine fuelled by alcohol, on CO_2 concentration in the exhaust gases.

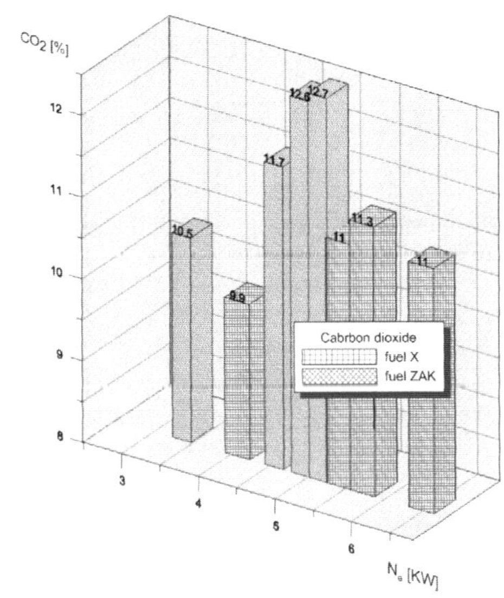

Figure 8. The concentration of carbon dioxide CO_2 in the exhaust gas of engine fuelled with the post-processing fuels.

The concentration of carbon dioxide is dependent on the composition of the mixture, also from the concentration of carbon monoxide CO and hydrocarbons HC in the exhaust gases. Because the composition of the mixture, during alcohol fuel supply, was practically solid, it also affected for stabilizing effect on the level of concentration of that component in the exhaust gases, in the final phase of the load increases. In the case of petrol supply - Figure 9, the mixture was steadily enriched, and this is reflected in the course of variability in the CO2 concentration, in accordance with the typical regulatory characteristics of the composition of the mixture.

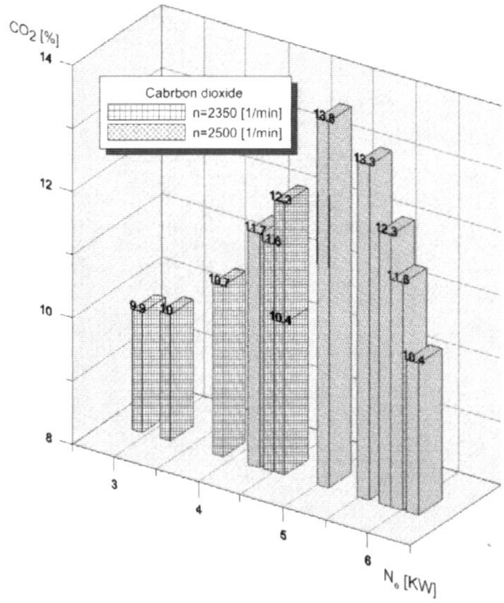

Figure 9. The concentration of carbon dioxide in the exhaust gas of engine fuelled with the petrol.

Comparing the values of CO_2 concentrations for both fuels, for the corresponding load values, found distinctive differences in reported values, due to their different chemical structure.

5. Summary

The investigation results of the ecological parameters of the engine 126 000 A1 powered by waste fuels on alcohol base, allow to formulate the following most important conclusions:

1. Due to the relatively low concentration of carbon monoxide in the exhaust gases, with the alcohol fuel supply - relative to the petrol supply, the limitation of the control of lean mixture composition, is to prevent the occurrence of the misfire,
2. Reduction of nitrogen oxides concentration, will require more precisely determine the value of exhaust gas recirculation injection dose,
3. Was achieved positive results, which testify to the fact that the combustion of alcoholic waste fuels in the internal combustion engine with pneumatic fuel injection, is reasonable for both energy and ecological issues.

References

[1] Marek W, Śliwiński K 2016 *Pneumatic fuel injection system using a hot exhaust gases, developed in Institute of Automobiles and Internal Combustion Engines of Cracow University of Technology* Konmot 2016.
[2] Marek W and Śliwiński K 2016 *Investigation of working parameters of four-stroke SI engine with pneumatic fuel injection system* Konmot 2016
[3] Juda Z and Marek W 2013 *Method of the fuel dosage management in the direct-injection system of the internal-combustion engine in the cogeneration unit* Combustion Engines, 2013
[4] Fiat *Technical Specifications of engine*: 126 A1 000
[5] Jarnuszkiewicz M and Śliwiński K 1992 *Odmienny sposób tworzenia mieszanki palnej w silniku wielopaliwowym* Politechnika Krakowska,
[6] Marek W 1994 *Sterowanie procesem tworzenia mieszaniny spalinowo-paliwowej przy wtrysku paliwa za pomocą gorących spalin - wybrane zagadnienia z prac własnych* Politechnika Krakowska,

Impact of the injection dose of exhaust gases, on work parameters of combustion engine

W Marek[1] **and K Śliwiński**[1]

[1]Cracow University of Technology, Institute of Automobiles and Combustion Engines, al. Jana Pawła II 37, 31-864 Krakow, Poland

E-mail: wmarek@pk.edu.pl

Abstract. This article is another one from the series in which were presented research results indicated the possible areas of application of the pneumatic injection using hot combustion gases proposed by Professor Jarnuszkiewicz. This publication present the results of the control system of exhaust gas recirculation. The main aim of this research was to determine the effect of exhaust gas recirculation to the operating parameters of the internal combustion engine on the basis of laboratory measurements. All measurements were performed at a constant engine speed. These conditions correspond to the operation of the motor operating an electrical generator. The study was conducted on the four-stroke two-cylinder engine with spark ignition. The study were specifically tested on the air injection system and therefore the selection of the rotational speed was not bound, as in conventional versions of operating parameters of the electrical machine. During the measurement there were applied criterion which used power control corresponding to the requirements of load power, at minimal values of engine speed. Recirculation value determined by the following recurrent position control valve of the injection doses inflator gas for pneumatic injection system. They were studied and recorded, the impact of dose of gases recirculation to the operating and ecological engine parameters such as power, torque, specific fuel consumption, efficiency, air fuel ratio, exhaust gas temperature and nitrogen oxides and hydrocarbons.

1. Introduction

Combustion engines are used not only as driving source for various vehicles but also as industrial machine drive. These engines found application in various branches of industry e.g. building, power industry, mining, gastronomy etc. Combustion engines work successfully all over the world in various latitudes and climatic conditions, practically in every field of economy and are characterized by high reliability. In dependence on destination of the engines driven mechanism and its working conditions, the engine may be fuelled with liquid or gaseous fuel. Working in these conditions combustion engines, may be used as the main driving source, additional source (used in case of failure of the main source or in chided temporally in order to obtain on assumed power aim) or may work in cogeneration with another source e.g. electric engine or heat pump. Two or more individual processes compounded, each of them aiming at producing one useful product, is called a compound process. Association of some thermal processes aims shortening the thermodynamic changes process what leads to reduction of power losses. Application of compound processes leads to reduction of exploitation costs and decrease in cumulated power consumption, hence in consequence reduction of harmful (toxic) waste products. The main part of the consumed energy is used for production of the main product. Municipal power plants and professional ones of high power are the basic cogeneration systems. However, present costs of power production force creation of newer alternative sources of power production. Small systems are created based upon reciprocate engines or gas turbines – so called CHP systems. Power of these systems oscillates from some kilowatts to a dozen or so megawatts. They permit application of compound production in small arrangements everywhere where demand for heat and electric power is rather low. Development of this type of systems is connected with high costs of electric and heat energy from the existing network. Application of small compound system gives the possibility of effective use of chemical energy of the fuel where high demand for heat and electric energy occurs, and investing in a big thermal electric power plants is not economical or impossible out of technical reasons. At the same time should be satisfied with the government and interest groups to diversify sources of energy supply [1]. It seems appropriate to propose small cogeneration systems that provide heat with the greatest possible efficiency using a variety of gaseous and liquid fuels. Conversion efficiency is considered as a factor in the optimization process. Small four-stroke internal combustion engines with spark ignition, is distinguished from other internal combustion engines relatively low total efficiency of about 25%. However, they are perfect as a base for multifuel engines [2]. If we assume that the production of electricity at maximum efficiency, will be not the primary task of a cogeneration system, it is possible to create a new application cogeneration system with multifuel engine, designed for optimized heat production from various fuels, with a significantly reduced (to larger engines) emissions of toxic exhaust gases components. This publication is a next of cycle [3] [4] [5], in which will be presented the results of studies indicating areas of possible uses, proposed by Professor Jarnuszkiewicz concept of the pneumatic injection by means of hot exhaust gases. In this publication will be presented, the results of research the regulation system of exhaust gases recirculation, in an internal four-stroke combustion engine for the control of conversion a chemical energy of fuel, to an engine work

2. Purpose and methodology of research

The main aim of the research was to determine by laboratory measurements, the impact of exhaust gases recirculation, on working parameters of the combustion engine. All measurements on engine test bench, was carried out at a constant engine speed. These conditions correspond to the operation of the engine in the power generator. The study tested the specificity of the pneumatic injection system, and therefore the selection of the rotational speed was not bound, as in typical solutions, by the operating parameters of the electrical machine. The criterion was a task - use an exhaust gases recirculation adjustment of power, corresponding to the energy demands of the load, with minimum values of rotation of the crankshaft of the engine. Under these conditions, the converters of mechanical energy into electrical energy, are used in the form of inverters. Among the control parameters of the test engine, which had to be matched to the type of examined fuel were primarily:

- ignition timing,
- coefficient of excess air,
- recirculation value of exhaust injection dose.

These parameters have a very significant impact on the process of combustion in the cylinder, which then results primarily, for the value of generated energy parameters. Especially in the case of exhaust gases recirculation, these parameters directly influence on the location in time, and duration of the combustion process. For this reason, for the examined fuel, and for each operating point of the engine load, these control parameters were individually set. Was studied and recorded, the impact of the examined fuel on the engine parameters such as torque and power, the specific fuel consumption, the total efficiency, the coefficient of air excess and the exhaust gas temperature.

3. Test bed investigation of the SI engine

Research of engine, were carried out on a special test bed in the Laboratory of Combustion Engines, Cracow University of Technology, whose essential elements are as follows:

- two-cylinder, four-stroke spark ignition engine,
- engine brake generator MEZ,
- AVL fuel mass flow meter , AVL Type 4210,
- the measuring systems for determining: exhaust gas temperature, ambient pressure and humidity of the intake air by the engine, the excess air ratio.

As a research object was prepared a two-cylinder four-stroke spark-ignition engine 126 000 A1, equipped in the pneumatic-fuel injection system using the hot exhaust gases, according to the concept of Professor Jarnuszkiewicz, which in the factory configuration, was carburettor equipped.

In this configuration, the engine was characterized by the following working indicators [6]: rated power on petrol 17.7 kW/4500 1/min., the maximum torque of 42 Nm/ 3000 1/min, the minimum specific fuel consumption of 300 g/kWh. Embodiment of this supply system of such an engine, require the design and building a pneumatic injection system and the supply fuel system, which the configuration similar to a conventional, low pressure injection supply system [2] [7]. Pneumatic injection system has been designed based on the concept of Professor Jarnuszkiewicz, described in an article on his achievements, in connection with the jubilee of Conference Konmot 2016 [3]. The functional model of designed and mounted on engine, prototype system of pneumatic injection, was shown in Figure 1.

To control: the value of fuel dose and phase of its dosing, the angle of ignition timing, degree of throttle opening and the composition of the combustible mixture (excess air ratio), and also the value of dose of exhaust gases to blowing the dose of fuel, was designed and used a special engine controller. Value of recirculation was determined by next reproducible positions of the control valve, an injection dose of the blowing gases into the pneumatic injection system. As the fuel was used engine petrol. Due to the prototype nature of fuel supply, and combustion system, the power control range was limited to medium loads.

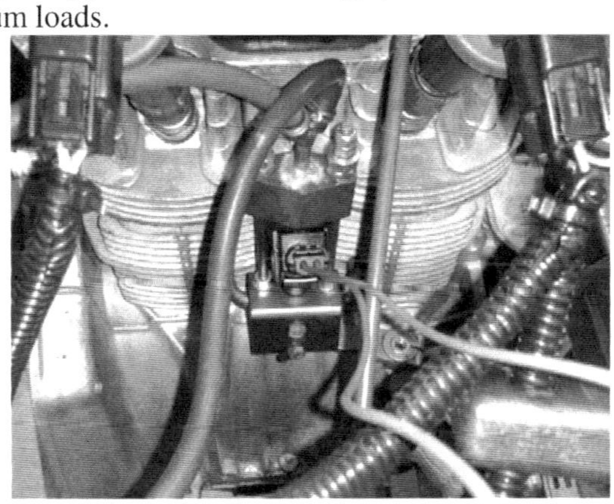

Figure 1. The view of the functional model of the pneumatic injection using hot exhaust gases for liquid fuels.

4. Results and analysis of investigations

Exhaust gas recirculation is generally used to modify the combustion process to reduce the intensity of the formation of nitrogen oxides. In the pneumatic injection system, using exhaust gases, naturally we are dealing with a kind of internal exhaust gas recirculation, by the accompanying of fuel injection, exhaust gases blown dose. Dose of the value of exhaust gas at a constant engine speed, depends mainly on the value of the free cross section of flow conduit of pneumatic fuel injection. The study used nine valve positions and the corresponding of them, value of the cross-section flow, from a completely free flow (point 0) to a position with a minimum cross-section (point 8), which provide yet correct fuel injection. The results are shown in the following figures. The fuel delivery strategy, implemented an assumption of power adjustment corresponding to the energy demands of the load, with minimum values of rotation of the crankshaft of the engine. For this case, it was 6 kW. Regardless of the degree of recirculation, the dosage was practically on the same level – figure 2.

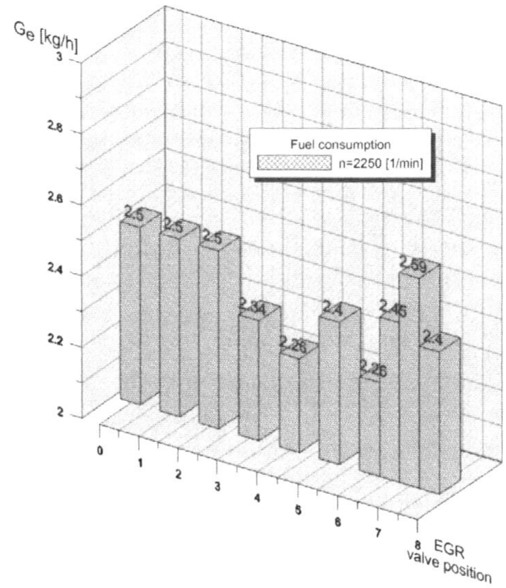

Figure 2. The fuel consumption, for the required level of engine load.

Also the load on the engine remained at a similar to required level, only at a very low value of recirculation, a load was increasing to 7 kW. This was related with favourable conditions for the process of heat evolution in cylinder – Figure 3.

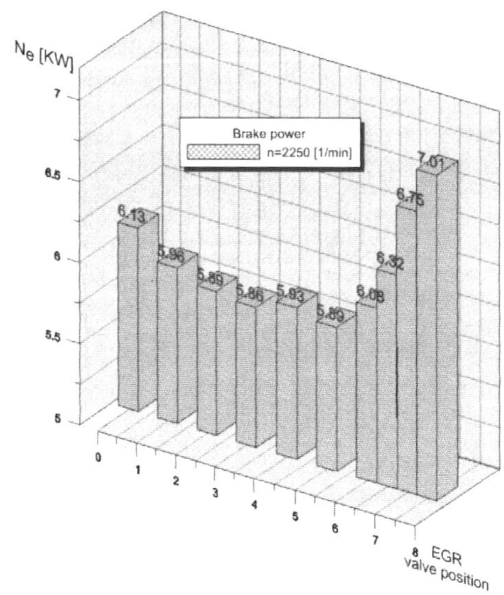

Figure 3. The value of engine load.

Inspection results to determine the excess air ratio, have reflected the characteristic gradual depletion of combustible mixture (from rich to lean), as far as reducing the value of recirculation – Figure 4.

This is due to the gradual reduction of the dose of exhaust gas supplied to the cylinder in the intake phase of the fresh air charge.

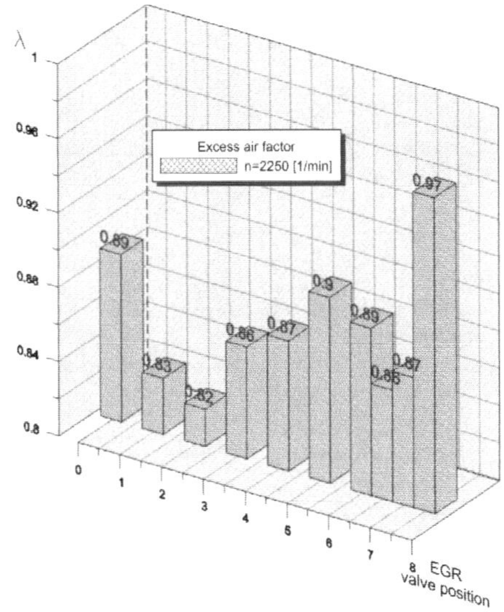

Figure 4. The checking value of the excess air ratio.

The observed an increase in load of engine at a very low value of recirculation, and at a fixed dose of fuel, was resulted in a positive increase in the total efficiency – figure 5.

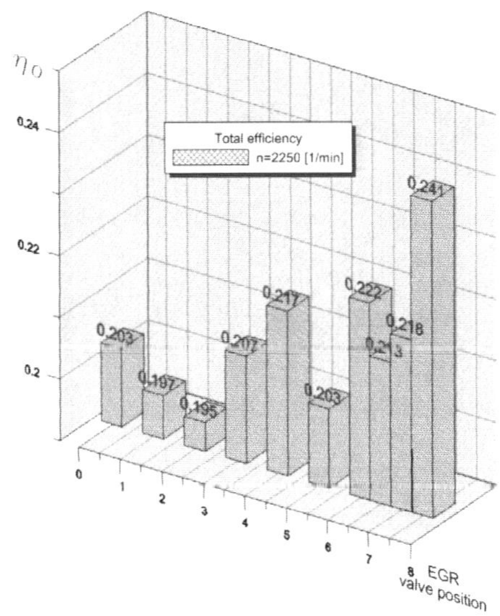

Figure 5. Comparison of the total efficiency.

The previously mentioned conditions was also observed in the case of a gradual increase in exhaust gas temperature. As to decrease the value of the recirculation rate and the increase in the value of air excess, the process of heat release occurred efficiently, at higher combustion temperatures - Figure 6. The variation of the concentration of nitrogen oxides is strongly associated with the value of the load. The increase in load accompanying drastically reduced recirculation, caused a significant increase in the concentration of NOx – Figure 7.

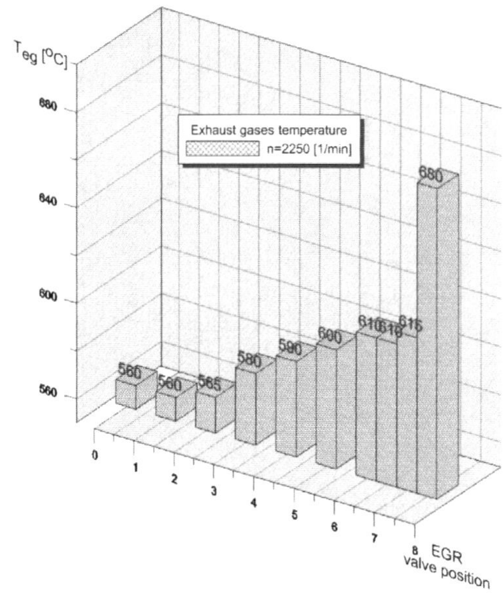

Figure 6. Comparison of the value of exhaust gas temperature.

On the other hand, the variation of the concentration of hydrocarbons, is strongly associated with the value of recirculation. The decrease in the value of recirculation has a positive effect on the process of heat release and on working cycle's repeatability. This is a typical observation, reduce the concentration of hydrocarbons, as a consequence the reduction of the degree of recirculation.

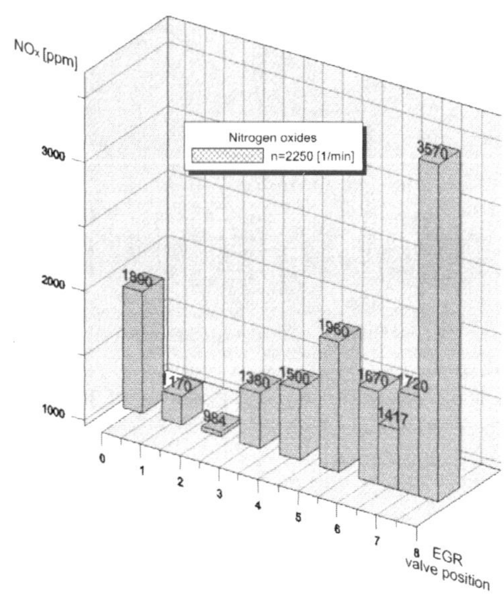

Figure 7. Comparison of the value of concentration of nitric oxides NO_x in the exhaust gas

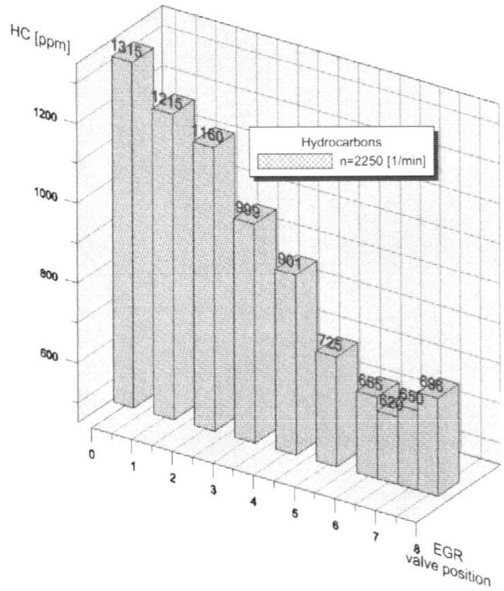

Figure 8. Comparison of the value of concentration of hydrocarbons HC in the exhaust gas.

5. Summary

The investigation results of impact of the injection dose of exhaust gases, on working parameters of combustion engine 126 000 A1 powered by petrol, allow to formulate the following most important conclusions:

1. The study revealed a significant effect of recirculation exhaust gas injection dose, on all working parameters of an engine, fitted in the pneumatic injection system using the hot exhaust gases,
2. In the case of variability in the concentrations of hydrocarbons and nitrogen oxides in the exhaust gas, their course is typical for the impact of value of recirculation, on the working process of the internal combustion engines,
3. Disclosed is also, a significant impact of the volatility of the recirculation value, on the volatility of air excess ratio. This phenomenon is very important, because it concerns a system of direct injection into the cylinder. We found that, although the value of charge of fresh air in the cylinder also depends on the exhaust gas injection dose. This is related probably to the fact, that the present injection system of the four-stroke engine, is performed at the start of the charging phase of the cylinder.
4. Choosing a strategy for controlling the dosage of fuel, in relation to the injecting exhaust gas dose, should strive for a compromise between maximizing the values of engine energy parameters and minimization of environmental performance.

References

[1] Markussen M V Østergård H Schmidt J E Oleskowicz-Popiel P and Pugesgaard S 2009 *Energy and emergy evaluation of Danish organic dairy farms: potentials for energy self-sufficiency by production of biogas, bioethanol or biodiesel* 1st Nordic Organic Conference.

[2] Jarnuszkiewicz M and Śliwiński K 1992 *Odmienny sposób tworzenia mieszanki palnej w silniku wielopaliwowym* Politechnika Krakowska.

[3] Marek W and Śliwiński K 2016 *Pneumatic fuel injection system using a hot exhaust gases, developed in Institute of Automobiles and Internal Combustion Engines of Cracow University of Technology* Konmot 2016.

[4] Marek W, Śliwiński K 2016 *Investigation of working parameters of four-stroke SI engine, with pneumatic fuel injection system* Konmot 2016

[5] Marek W and Śliwiński K 2016 *Investigation of ecological parameters of four-stroke SI engine, with pneumatic fuel injection system* Konmot 2016.
[6] Fiat *Technical Specifications of engine*: 126 A1 000.
[7] Marek W 1994 *Sterowanie procesem tworzenia mieszaniny spalinowo-paliwowej przy wtrysku paliwa za pomocą gorących spalin - wybrane zagadnienia z prac własnych* Politechnika Krakowska.

IOP Conf. Series: Materials Science and Engineering **148** (2016) 012077 doi:10.1088/1757-899X/148/1/012077

Development of the methodology of exhaust emissions measurement under RDE (Real Driving Emissions) conditions for non-road mobile machinery (NRMM) vehicles

J Merkisz[1], P Lijewski[1], P Fuc[1], M Siedlecki[1] and A Ziolkowski[1]

[1]Poznan University of Technology, Institute of Combustion Engines and Transport, Piotrowo 3, 60-965 Poznan

E-mail: maciej.s.siedlecki@doctorate.put.poznan.pl

The paper analyzes the exhaust emissions from farm vehicles based on research performed under field conditions (RDE) according to the NTE procedure. This analysis has shown that it is hard to meet the NTE requirements under field conditions (engine operation in the NTE zone for at least 30 seconds). Due to a very high variability of the engine conditions, the share of a valid number of NTE windows in the field test is small throughout the entire test. For this reason, a modification of the measurement and exhaust emissions calculation methodology has been proposed for farm vehicles of the NRMM group. A test has been developed composed of the following phases: trip to the operation site (paved roads) and field operations (including u-turns and maneuvering). The range of the operation time share in individual test phases has been determined. A change in the method of calculating the real exhaust emissions has also been implemented in relation to the NTE procedure.

1. Introduction

In the beginning of the twenty-first century, one of the major challenges of humanity is to reduce the negative effects of the development of civilization. An undesirable aspect of this development is the degradation of the environment, and for this reason measures to reduce this degradation are increasingly gaining in importance. Undoubtedly, widely understood transport has a large negative impact on the environment, namely transport powered by internal combustion engines. For this reason, the development of modern internal combustion engines is based around ecological issues.

Engines for off-road applications (non-road) represent a large share of engines in addition to those used in road vehicles. Engines used in vehicles for non-road applications NRMM (Non-Road Mobile Machinery) belong in this group. Engines of these vehicles are characterized by specific working conditions among all non-road engines that do not allow for them to be classified as traction engines. The group of NRMM vehicles includes machinery, tractors and agricultural machinery, and special purpose machines. The main problem for NRMM vehicles is the emission of PM (Particulate Matter) and NOx (Nitric Oxides). The scale of the problem can be illustrated by the situation in Germany, where the share of PM emissions coming from the engines of off-road application vehicles remained unchanged for years, at around 50% [1].

An important aspect of reducing emission of toxic fumes is the methodology of research. Currently, research on emission of toxic compounds from NRMM vehicles is performed mainly in the laboratory. These measurements apply only to the engine itself, and not the entire vehicle - and therefore the results of such studies do not give full information about emissions and fuel consumption during real operation. For the ecological assessment of the vehicles tests under real operating conditions are the most important, because only in this way it is possible to obtain complete information about the actual emission of harmful exhaust has components. The disadvantage of such research is the high cost of the measuring equipment and the need to adapt the vehicle for the needs of the test. However, it generates reliable results that are impossible to obtain in a laboratory, during tests on an engine or chassis dynamometer [2]. Currently, the measurements of emissions are carried out under real operating conditions of vehicles using mobile devices such PEMS (Portable Emission Measurement System).

2. Legal conditions for the type approval of NRMM vehicles with regard to emissions

Guidelines for testing emission of harmful exhaust compounds from NRMM vehicles are contained in the 97/68/EC Directive. The directive developed later [2] introduced changes and additions, which were mainly due to technical progress in the construction of engines and the increasing degree of standardization of regulations on a global scale. Hence, the introduced provisions were often common agreements between the European Commission and the US Environmental Protection Agency [3]. The desire to harmonize the regulations is an action desirable not only in Europe but throughout the whole world. The last directive on emissions of toxic gases from motor NRMM vehicles was published in December 2012 [4]. The Directive pays particular attention to the problem of reducing the NO_x emissions. This is connected with the necessity of using an SCR. An important part of the directive, which should be noted, is the statement about the need to include the regulations from the Directive 595/2009/EC relating to the engines of road HDVs in the 97/68/EC Directive. This provision appeared with the introduction of the procedures for type approval testing under real operating conditions. Therefore, it is expected that the engines of NRMM vehicles solutions will be accepted that have been well established already in the tests for HDV engines.

Currently in force worldwide approval static test for non-vehicle applications of combustion engines is the emission test ISO 8178 (Figure 1) developed by the ISO (International Standard Organization) [4]. This is an 11-phase test performed on an engine dynamometer. Based on the test results the average emissions of the individual components of emissions are determined. The characteristic factors of the share in each phase of the test are selected depending on the intended application of the test engine. For NRMM vehicles a basic static test is an ISO 8178-C1.

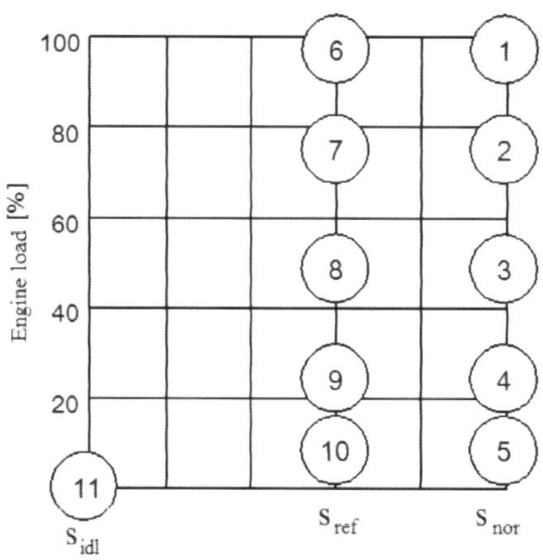

Figure 1. Schematic of an 11-phase ISO 8178 test [5]

Since coming into effect of the Stage III limit the emission measurements for NRMM vehicles are also carried out in the dynamic test NRTC (Non-Road Transient Cycle, Figure 2). In addition, the name of the ISO 8178 test changed to NRSC (Non-Road Stationary Cycle). NRTC test was developed jointly by the European and American legislation. The NRTC test is performed twice on each tested engine: once for cold and once for hot start. The final result is a weighted average, with the share of 0.1 for the test with a cold engine start (in the United States - 0.05) [5]. In the NRTC test the speed of the engine crankshaft is determined according to:

$$s = \frac{\% s_{nor} (s_{ref} - s_{idl})}{100} + s_{idl} \tag{1}$$

where: s_{nor} – normalized engine speed,
s_{idl} – engine speed at idle
s_{ref} – reference speed calculated with the formula:

$$s_{ref} = s_{50} + 0{,}95(s_{70} - s_{50}) \tag{2}$$

where: s_{50} – minimum engine speed, at which the engine reaches 50% of its nominal power,
s_{70} – maximum engine speed, at which the engine reaches 70% of its nominal power.

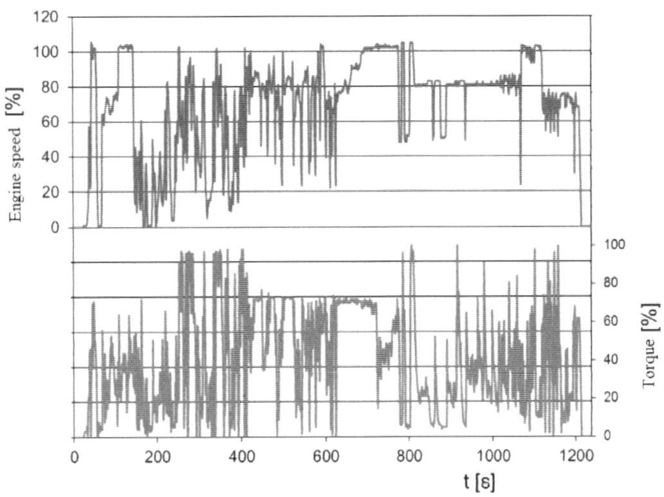

Figure 2. The NRTC test [5].

The engine torque in the NRTC test is determined according to the relationship:

$$T_o = \frac{\% T_{o\,nor} \cdot T_{o\,max}}{100} \tag{3}$$

where: $T_{o\,nor}$ – normalized engine torque,
$T_{o\,max}$ – maximum engine torque.

These tests, however, do not reflect actual operating conditions of many machines and equipment belonging to the NRMM group. For this reason, an attempt was made to develop procedures for measuring emissions of toxic compounds in real operating conditions, the RDE (Real Driving Emission), using mobile analyzers of exhaust gases belonging to the PEMS category (Portable Emission Measurement System). The European legislation assumes that the procedures developed to test vehicles of HDV category (Heavy Duty Vehicles) laid down by Directive UE582/2011 will be adapted to test NRMM vehicles. In the case of US regulations it is assumed that the research in actual operating conditions will be performed in the NTE test (Not-To-Exceed), the same test that applies for

heavy duty vehicles. And so starting from 2011 an NTE test is performed for NRMM vehicles with an engine power greater than 130 kW, and from 2012 for engines with power of 56-130 kW.
The US Environmental Protection Agency has also introduced regulations limiting emissions during the operation of motor vehicles with off-road applications. Under these provisions, NTE test (Not-To-Exceed) were introduced along with emission limits set in this test, as an additional tool to control the emission of toxic exhaust compounds (NTE is also the name of the norms for this testing procedure) [5]. Measurements in this test are performed under real operating conditions of engines. The NTE was originally introduced for engines of HDV road vehicles, and since 2011 it also applies to motor vehicles with off-road applications with power greater than 130 kW, from 2012 for engines with power between 56-130 kW, and from 2013 for engines with less than 56 kW of power. It should be noted, however, that the NTE test was adopted for the NRMM vehicles in the same form as it exists for HDVs.
Measurements in the NTE test are not associated with a particular drive cycle of the vehicle or the operating points of the engine, there is no specified test run and time constraints (Figure 3). These studies cover a range of engine operating conditions which exists within the NTE control area (the NTE zone), including steady state and in transient conditions. Emissions of toxic compounds are averaged with the engine cycle lasting a minimum of 30 seconds, called the measuring window (30 sec window). In the NTE test control an area bounded by the values of load and speed of the crankshaft was established on the engine operation map:

- the minimum speed of the engine crankshaft A is set analogous to the speed designated in the ESC test (European Stationary Cycle, likewise with speed B),
- engine load equal to or greater than 30% of the maximum engine torque,
- all operating points for which the engine reaches less than 30% of the maximum power are excluded from the NTE area,
- the engine manufacturer may request that speeds and loads for which the specific fuel consumption is greater than 5% of the minimum be excluded from the NTE area when they can expect that the engine will not work when operating in those areas. This does not apply to engines cooperating with the automatic transmission with a certain number of gear ratios and vehicles with manual transmission.

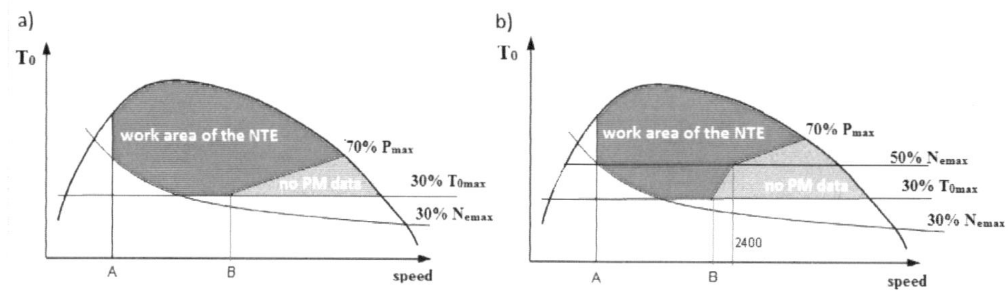

Figure 3. NTE test for engines of NRMM vehicles, a) – for engines with a maximum speed of 2400 rpm, b) – for engines with maximum speed of up to 2400 rpm [5].

In accordance with the requirements set by the EPA all NRMM engines of vehicles that meet the Tier 4 norms, they must also meet the requirements of NTE. For engines of NRMM vehicles with power greater than 130 kW, these provisions apply from 2011, for engines with power of 56-130 kW from 2012, while for engines of less than 56 kW from 2013. Limits on emission of toxic compounds in an NTE test has been established as 1.25 of the permitted emissions of a given compound in Tier 4 norms. Only for engines that have NO_x emissions of less than 2.5 g/kWh and PM emissions of less than 0.07 g/kWh the chosen multiplier is 1.5. The NTE rules apply for type approval tests as well as for the entire lifespan of the engine [5].

3. Research methodology

The study used a John Deere farm tractor 6630. Technical data was given in Table 1. A picture of the vehicle with mounted test apparatus is shown in Figure 4.

Table 1. Basic technical specifications of the tested vehicle.

Name	John Deere 6630
Number of cylinders / displacement	6 cylinder inline / 6,8 dm^3
Maximum power	110 kW at 2100 rpm
Curb weight	5230 kg
Charging	VGT Turbocharger
Cooling	Liquid with forced circulation
Injection system	Common Rail
Number of valves	24
Emission norm	Stage IIIB

Figure 4. A picture of the tractor and measuring equipment during testing.

The authors used a portable exhaust emissions analyzer (SEMTECH DS by SENSORS, Figure 5) to measure the concentration of the exhaust components. The analyzer measures the concentration of the components and simultaneously measures the flow rate of the gas. The exhaust gas is introduced into the analyzer through a probe maintaining the temperature of 191°C. The PM is filtered out (compression ignition engines) and the exhaust gas is directed to the flame-ionizing detector (FID) where the concentration of HC is measured. The exhaust gas is then chilled to the temperature of 4°C and the measurement of the concentration of NO_x, (NDUV analyzer), CO, CO_2 (NDIR analyzer) and O_2 follows in the listed order. It is also possible to collect data from the vehicle diagnostic system. Aside from the exhaust emission values, signals from the on-board diagnostic system were recorded (engine speed, load, etc.) for the purpose of comparison. The lack of rules concerning the connectors and protocols of data transmission means that during the research of NRMM vehicles it is necessary to use the service diagnostic equipment to measure parameters of the engine.

Tractor sowing was performed on a rectangular field, located near Poznan. The route was characterized by a long, monotonous ride parallel to the sides of the field and U-turns (Figure 6). Before and after turning the tractor stopped and raised or lowered the cultivator.

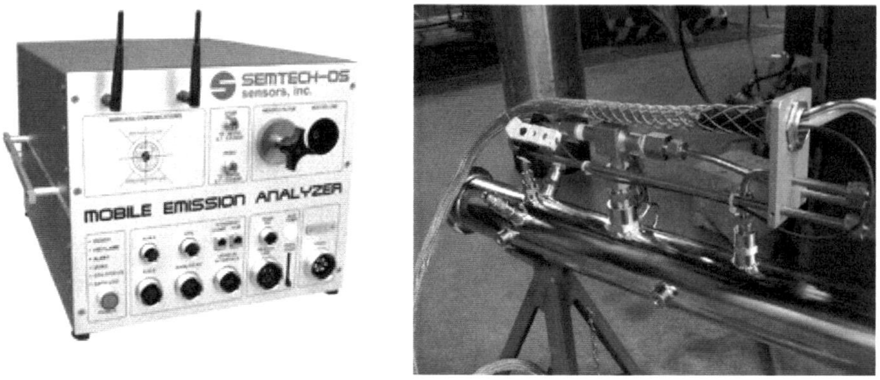

Figure 5. SEMTECH DS-analyzer and an exhaust gas mass flowmeter from SENSORS Inc.

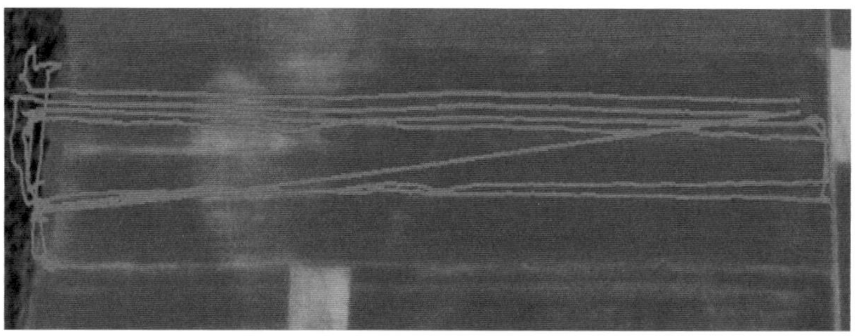

Figure 6. The route of the tractor during field work [developed based on: www.gpsvisualizer.com].

A very important aspect when making measurements in the field is the terrain. In the first cycle of the research presented, in the middle of the field length, a significant change in altitude was recorded, which amounted to a maximum of 7 meters. This trough had a length of several meters and then returned to the normal level, which varied in the range of 80-82 m (Figure 7). Changing the height of the terrain affects the emissions of CO_2, which in the first part decreased and then increased. The engine operating points changed accordingly, stepping beyond the bounds of engine operating area for the NTE test.

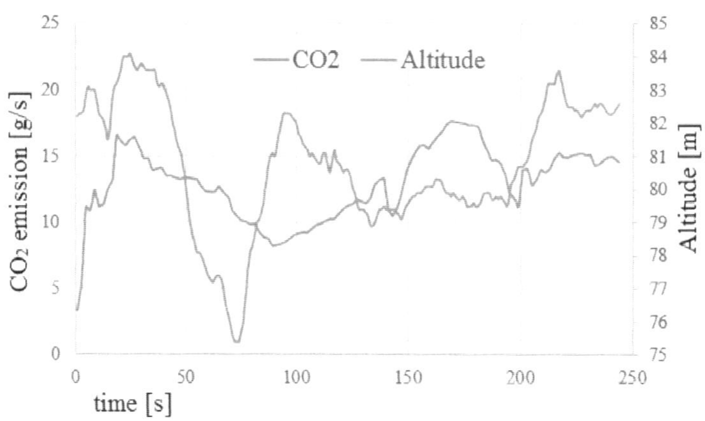

Figure 7. Changes in altitude and carbon dioxide emissions during field work in one measurement cycle.

4. The results of rde tests for nrmm vehicles

The performed research allowed to collect data on the performance of engines under real operating conditions. Particularly useful in the described analysis are: the rotational speed of the crankshaft and the engine torque. Electronic control and diagnostics systems used in modern NRMM vehicles provide some data and parameters on the operating conditions of the engine. This data is used for the needs of vehicle servicing. The type and amount of information that can be read depend on the manufacturer and also on the vehicle. These are individual solutions – each manufacturer selects the range of available service information according to their own criteria. Work is currently underway to standardize data exchange protocols, diagnostic connectors, as it has already been done for road vehicles. For the analyzes carried out for this article only some of the parameters relating to the working conditions of engines were relevant.

During the research the farm tractor was performing field work with a cultivator. Two measuring cycles were registered. In the first the tractor was performing light work of cultivation, representing a large group of works for seedbed preparation and cultivation performed in agriculture. The second measurement cycle concerned the passage of the tractor on asphalt roads from the farm to the farmland. In the first cycle the tractor engine worked in a narrow range of crankshaft speed and torque. The most commonly used operating range is 1700-1800 rpm and 200-500 Nm, it represented 68% of total work time (Figure 8). Significant shares were also for the speed range 1800-1900 rpm and a load of 200-400 Nm and the engine idle operation area 900-1100 rpm – these accounted for 11 and 7% of the time share respectively. In the field test cycle, a significant work time share of 6%, was the idle area or 900-1000 rpm.

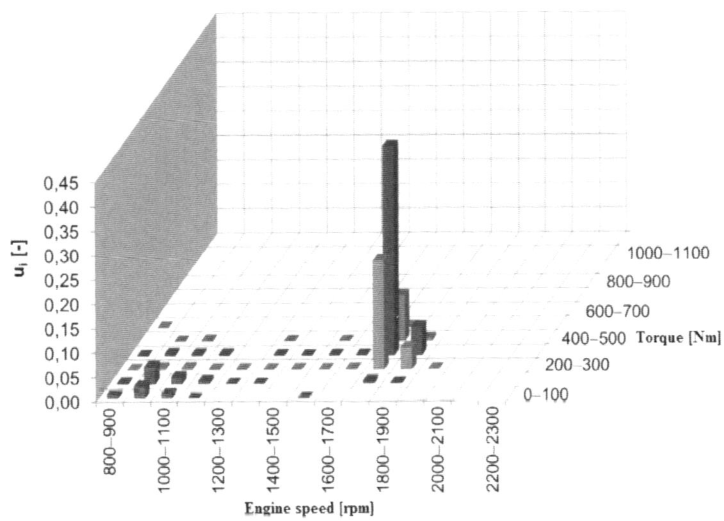

Figure 8. Engine time density characteristics of a tractor during field cycle

During the drive on the road of the tractor almost the entire range of crankshaft speed and engine load has been used (Fig. 9). The resulting characteristics of the participation of speed and load of the tractor is similar to the characteristics of the engines of rail vehicles [6]. However, it is possible to distinguish two main operating areas that were most frequently used. The first is the engine speed range of 800-1300 rpm and load 0-210 Nm - its share in the cycle was 35% (1 in Fig. 9). The second is the engine speed range of 1300-1900 rpm and a load of 140-550 Nm. The share of this area was 57% (2 in Figure 9). During the drive on a road the work time share of load exceeding 75% of the maximum was very small. In addition, the large share of the engine idling time - 15%, and thus a share higher than during field work, should be noted.

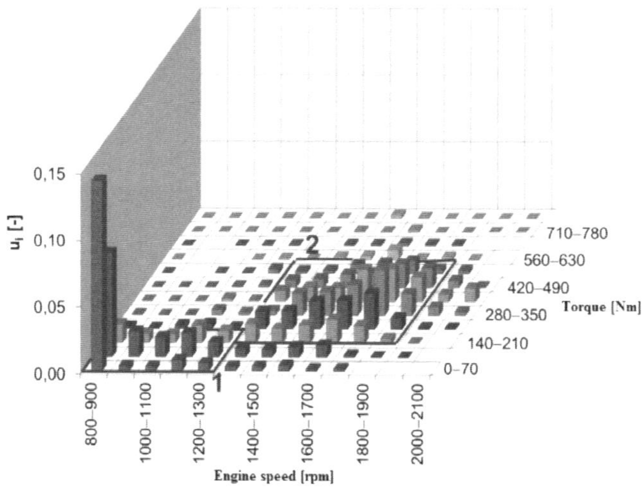

Figure 9. Engine time density characteristics of a tractor during) drive cycle.

Points of engine operation recorded during testing in actual operating conditions were compared to the NTE test. For the tractor area of the engine work in the first test cycle is only partially located in the area of the NTE test (Figure 10). Seven important NTE test windows were registered in it, which represents 25% of the time the engine operation. During the passage of the tractor on the road 48% of the work time share the engine worked in the area of the NTE test and recorded 17 major windows, with a total length of 1,015 s, which accounted for 22% of the total work time of the engine (Figure 11). This small part is the result of the frequent changes in speed and load of the engine. The required condition of maintaining constant operating parameters of the engine for 30 s under NTE test was difficult to meet.

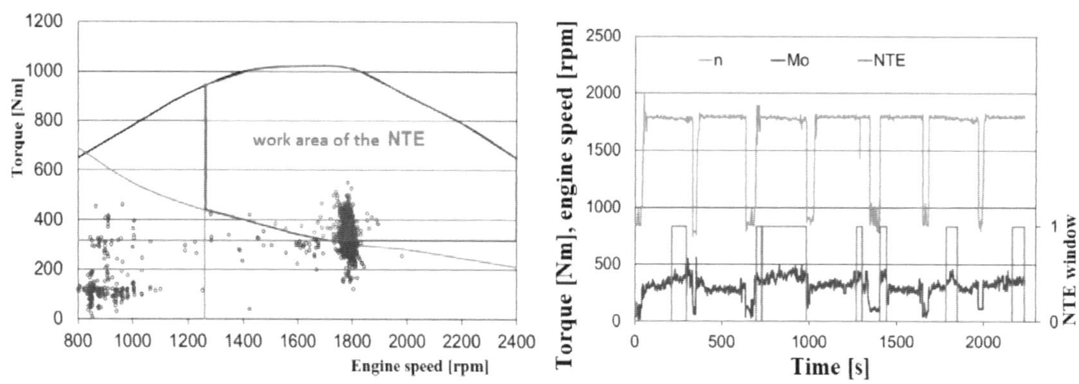

Figure 10. Operating parameters in real operating conditions and the NTE test of a tractor engine (1 measuring cycle).

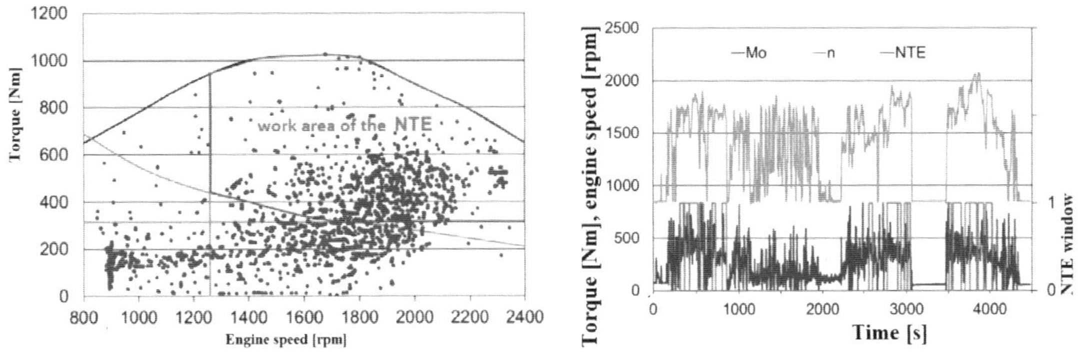

Figure 11. Operating parameters in real operating conditions and the NTE test of a tractor engine (road cycle).

5. Assumptions of rde measurement methodology for agricultural vehicles

The RDE measurements in real operating conditions made using PEMS equipment for the tractor demonstrated that the NTE method to assess emissions is not representative of the conditions of the engines in operating agricultural vehicles. This was also confirmed by previous research done by the authors presented in publications [6, 7]. Based the information acquired, assumptions for research methodology have been developed, that enables accurate analysis of emissions during RDE measurements. The methodology consists of the following elements: establishing a test route, selection and preparation of the vehicle for testing, assembly and preparation of measurement equipment for the test, measurement procedure and calculation of emissions.

I. Selection of the test cycle

The test cycle should include performing measurements on both the roads and in the field. It should start in the place where the equipment was assembled. Subsequently, it should run along asphalt roads constituting the access to the place of work in the field, and return to the starting point. The share of this phase should not exceed 25% of the total engine work time. The main part of the measurement (75% of the time share) should be done in the field. The maximum slope of a field should not exceed 3%. The measurement should be at least 7200 seconds long. The length of the cultivated field should be enough to allow for the registration of parameters for at least 10 minutes, without the need for maneuvering. The vehicle must defeat four such episodes in one direction during the measurement. This is to increase the share of the work time of the internal combustion engine load characteristics – at constant speed of the engine crankshaft but variable torque. Making turns in the field causes a large variation in operating parameters of the engine, making it difficult for subsequent analysis using the NTE method.

II. Selection and preparation of the vehicle for testing.

The study should use agricultural vehicles which are equipped with internal combustion engines that enable communication between the measurement equipment PEMS and their diagnostic systems. This is very important because it enables the measurement or calculation of the torque generated by the engine during measurements. This is used to calculate the power output and the work that the engine generated during the measurements. In the case where the diagnostics network only enables the measurement of the parameter of Engine Load, which is the ratio of the current torque to the maximum torque at a given speed of the engine crankshaft, it is necessary to obtain the full power characteristics over the entire operating area of the engine from its manufacturer. The second parameter determining the choice of the vehicle should be the number of operating hours, which is a measure of the work time of the machine and is calculated as the difference between the time the work is completed and the it commenced. It is assumed that the agricultural vehicle must have a minimum of 500 operating hours total for it to be subjected to the RDE tests. When performing the measurements it should be equipped with proper cultivating machinery.

The preparation of an agricultural vehicle for RDE testing should mainly concern the installation of measuring instruments. To place the PEMS apparatus on the vehicle, in many cases it is necessary to prepare an additional supporting structure mounted to the vehicle. It is associated with the large dimensions and weight of the PEMS research equipment. The designed support should be characterized by appropriate endurance parameters, to not cause changes in the stability of the vehicle and to not adversely affect the safety of road traffic.

III. The assembly and preparation of the measurement equipment for the test

The PEMS apparatus should be installed in the vehicle in such a manner so as not to affect its stability and to not pose any threats when conducting measurements. Exhaust flow meter EFM should be tightly connected to the exhaust system of the internal combustion engine by means of a rigid, smooth tube with the lowest possible number of curving. It is important to remember that there must be a straight section of pipe with a minimum length equal to double its diameter before the EFM flowmeter. This is to reduce the pressure pulsations of exhaust gases. Due to the measurement including particulate matter the length of the pipe segment connecting the exhaust system with a flow meter should be as short as possible. The test apparatus must be powered from an external source mounted on the vehicle. This may be a battery, or a generator. It is not permissible to power the measuring equipment directly from the vehicle, because it would affect the work load of the engine. The vehicle must also be equipped with a GPS position sensor and a temperature, pressure and humidity sensors for the ambient air.

After the assembly of research equipment on the vehicle, its stability and the quality of the mounting should be checked – as during sudden longitudinal and transverse movements the measuring equipment should not move. Once checked it should be run in order to allow it to warm up and stabilize the working conditions. After stabilizing the temperature (device showing the ready signal or readiness signaled in the control software) it is suggested to wait 10-15 minutes to permanently stabilize the operating conditions of the apparatus.

IV. Procedure for measurement and calculation of emissions

Before the measurements the zeroing and calibration procedure of the research equipment must be performed. It should be carried out after the stabilization of the apparatus (including those suggested additional few minutes). But the wait time cannot exceed one hour. To do this, follow the instructions of the PEMS equipment. First, the airtightness of the heated conduits should be tested, both those for the measurement of gaseous compounds and particulate matter. Then, a zeroing of the device should be performed – there are two methods involving either zeroing with the use of ambient air or a calibrating gas (synthetic air, nitrogen). In the case of ambient air, zeroing procedure should take place in an area where there is no excess emission of toxic compounds. When calibrating with ambient air the equipment requires the control device to be zeroed at least once per hour in the course of the measurement, and in the case of using calibration gases the reference gas canister should be placed inside the vehicle and connected to the apparatus. Then the procedure is performed using a calibration gas mixture recommended by the manufacturer. The whole process should be recorded. After completion of the measurement a control calibration of the instrument should be performed according to the methodology described before.

After completion of the zeroing and calibration of the instrument it is necessary to flush and reset the EFM flowmeter. Before the measurement of particulate matter a filter should be placed through which the diluted exhaust sample will flow. Filters should be conditioned prior to measurement, weighed, and stored and transported in a special instrument protecting against weather conditions. After finishing the measurements the filters should also be put back into it, and then weighed again.

Measurements should be performed at ambient temperatures above -7°C. Before the measurement the engine of the vehicle should be started and given sufficient time for the temperature of the coolant to reach 70°C. This should not last longer than 5 minutes. If this time is exceeded the measurements should begin. During the warm-up of the apparatus it should not be provided the exhaust gas sample, neither dedicated to the gaseous compounds not the particulate matter measuring devices. After completion of the measurement, the engine of the vehicle should be turned off and only after that the process of control zeroing and calibration.

The calculation of pollutant emissions should begin with the designation of the area specified in the NTE method. Then the measuring windows should be designated and if they accounted for over 50% of the total work time the emissions can be calculated according to the guidelines of the NTE (presented in Chapter 2). If the required validity of the measuring windows is not satisfied the work which the engine has generated in the test needs to be determined. If it's greater than the work generated in a NRTC test the measurement results can be considered valid and unit emissions can be determined as proposed by the European procedure, the draft of which is close to the Regulation EU 582/2011 dedicated for HDVs.

6. Conclusions

The analysis of emissions performed in this article is in accordance with the rules of the NTE procedure, performed on the basis of RDE measurements has shown that the NTE procedure under typical operating conditions of agricultural vehicles does not meet its target. For this purpose, it is necessary to modify or replace it with another. On the basis of the tests performed and previous experience the authors suggested research methodology, which takes into account:

a) the choice of test route,
b) selection and preparation of the vehicle for testing,
c) assembly and preparation of the measurement equipment for the test,
d) Procedure for measurement and calculation of emissions.

Especially important in this method is to determine a test cycle that takes into account operation on asphalt roads and work in the field. It was assumed that its duration must be at least 7,200 seconds and 75% of the operating time is to dedicated to the work in the field. The characteristics of the field was also defined – one passage on the field is to take 10 minutes without making turns. Four runs in each direction should be performed during the measurements. As a result, the share of valid measurement windows in the NTE method is expected to increase. If it does not exceed 50% of the work time then the engine work performed in the measurements should be determined. If it is larger than the work generated in the NRTC test the calculations should be carried out according to the initial proposal the European method dedicated for NRMM vehicles proposed by the European Commission.

References

[1] Commission Directive 2010/26/EU of 31 March 2010 amending Directive 97/68/EC of the European Parliament and of the Council on the approximation of the laws of the Member States relating to measures against the emission of gaseous and particulate pollutants from internal combustion engines to be installed in non-road mobile machinery

[2] Chandan M Boopathi S John .and Arun C 2014 System Strategy To Meet Tier4 Final Emission Regulations In a Locomotive Application For a Large Bore High Speed Diesel Engine *Proc. ASME 2014 Internal Combustion Engine Division Fall Technical Conf.* (USA, Columbus)

[3] Analysis of Diesel Emissions in the U.S.-Mexico Border Region 2007 report prepared by: Industrial Economics, Incorporated Cambridge and Ross & Associates Environmental Consulting, Ltd. prepared for: U.S. EPA Office of International Affairs

[4] Friedrich A 2008 Emission Legislation for Off-Highway Application in Germany. *Proc. Int. CTI Forum Emission Reduction Systems for Off-highway Applications* (Gemrany, Stuttgart)

[5] Worldwide Emissions Standards, Heavy Duty and Off-Highway Vehicles. Delphi brochure 2015/2016

[6] Lijewski P Merkisz J Fuc P Siedlecki M and Ziolkowski A 2015 The Measurement of Particulate Matter from Construction Machinery under Actual Operating Conditions *SAE Technical Paper 2015-01-2810, 2015, doi:10.4271/2015-01-2810*

[7] Lijewski P Merkisz J and Fuc P 2013 The Analysis of the Operating Conditions of Farm Machinery Engines in Regard to Exhaust Emissions Legislation. *Applied Engineering in Agriculture, Vol. 29(4), 2013*

Acknowledgments

The research was funded by the National Centre for Research and Development (Narodowe Centrum Badań i Rozwoju) – research project within the Applied Research Programme (contract No. PBS3/A6/26/2015)

Conversion of the exhaust emission results obtained from combustion engines of heavy-duty vehicles

J Merkisz and J Pielecha

Poznan University of Technology, 3 Piotrowo Street, 60-965 Poznan, Poland

e-mail: jerzy.merkisz@put.poznan.pl

Abstract. The use of internal combustion engines as the drive for heavy-duty vehicles forces these engines to be tested on an engine dynamometer. Thus, these engines operate under forced conditions, which are significantly different from their actual application. To assess the ecology of such vehicles (or more accurately the engine alone) the emission of pollution per unit of work done by the engine must be determined. However, obtaining the results of unit emissions (expressed in grams of the compound per a unit of performed work) does not give the grounds for determining the mass of pollutants on a given stretch of the road travelled by the vehicle. Therefore, there is a need to change the emission value expressed in units referenced to the engine work into a value of road emissions. The paper presents a methodology of determining pollutant emissions of heavy-duty road vehicles on the basis of the unit emissions, as well as additional parameters determined on the basis of the algorithm presented in the article. A solution was obtained that can be used not only for heavy-duty vehicles, but was also extended to allow use for buses.

1. Introduction

The increasing number of vehicles in the world and the environmental pollution lead to rise in the requirements in terms of exhaust emissions. The problems tackled in this article applies to very important aspects in the ecological sense, as a part of the trend of environmental protection. The experiment conducted uses the available research potential in the form of mobile gas analysers, which can measure harmful gaseous exhaust components. Use of data from on-board diagnostics system for measuring concentrations of toxic compounds for vehicles powered by natural gas enables the ecological assessment of means of transport during operation. Determination of the exhaust emission components from any vehicle or engine is possible in several ways [3, 4]:

- based on engine dynamometer test – determines the concentration of the exhaust gas components; taking into account – in accordance with the standards – conversion methods with the values for the individual components, and using the measured values of engine power, it becomes possible to determine specific emissions of the given exhaust component,
- on the basis of tests on a chassis dynamometer – emissions in a particular test are determined, and also based on the standards and regulations it is possible to determine the road emissions of the given exhaust component,
- on the basis of road tests – the concentration of gaseous components and the weight and number of particles are determined, and taking into account the exhaust mass flow also the road emission of these components.

Recent research on pollutant emissions under real traffic conditions performed with the use of mobile devices [2, 7, 15] reflects the state of ecological vehicles very well. Most attention is paid to the possibility of using such tests to calibrate the engines [1], in such a way that allows to reduce emissions not only during the engine tests, but also in the entire range of engine operation [6]. Comparative studies conducted in laboratories [2] indicate that the emission limits during the operation of vehicles with petrol engines are met, while also indicating that vehicles with diesel engines far exceed the permissible emission limits of nitrogen oxides [8, 11, 12]. Attention is drawn to significant emissions of particulates predominantly in the range of nanoparticles for engines fed with alternative fuels (e.g. natural gas) [9, 14] and the dependence of the test results on the terrain topography [10]. The results of such studies are currently not separate, but are confirmed by articles which relate to several years of research [5] and comprehensive summaries of research on vehicles in real traffic conditions [13]. However, experience in measuring emissions from vehicles shows that it is possible to determine a mutual relationship between the obtained values of specific emissions and road emissions, which will be complemented by other additional features, characterizing the test engine or characteristics relating to the vehicle.

2. Aim and methodology

The aim of this study was to determine the relation between road pollutant emissions and unit emissions, as well as the generalization and expansion of the obtained results to other categories of vehicles (e.g. urban buses). Comparative studies of the presented relations have been performed for a heavy-duty vehicle. The vehicle has an emission class of Euro IV and it was fitted with exhaust aftertreatment systems. The test route was divided into three main parts: the first – urban, the second – rural, and the third – motorway driving and return in the opposite direction. The character of the tested section of road enabled the mapping of the conditions of everyday truck traffic with a maximum mass exceeding 16,000 kg (i.e. long haul) with the following characteristics:

- urban conditions – vehicle speed in the range 0–30 km/h,
- rural conditions – vehicle speed in the range 30–60 km/h,
- motorway conditions – vehicle speed in the range 60–90 km/h.

Dividing the entire test route and all the measurement data obtained on it relations have been obtained, which show that 28% of the duration of the test was performed in urban conditions, 35% - rural conditions and the remainder 37% is motorway driving (Figure 1).

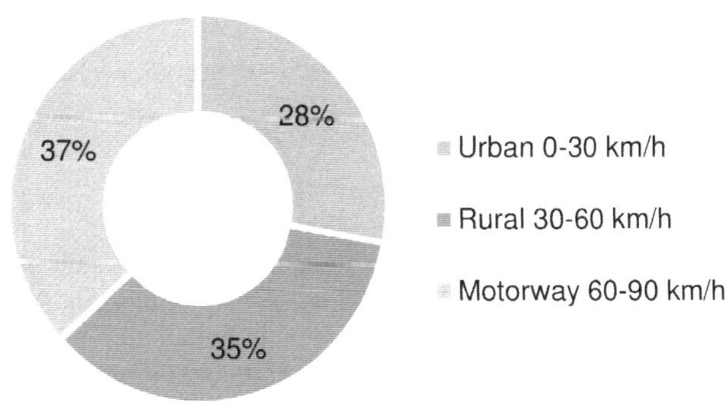

Figure 1. Characteristics of the incidence of different traffic conditions for the vehicle.

Emission measurements were done in real driving conditions; this approach requires the installation of the gas sampling apparatus on the vehicle in such a way that ensures its normal operation. Therefore a gas sampling system was prepared, which together with the system measuring the flow rate of the exhaust gas also performed partial sampling of the flue gas for the analysers for the measurement.

For measuring the concentration of harmful substances in the exhaust gas the mobile Semtech DS Company Sensors (Sensors Emission Technology) analyser was used. It facilitated the measurement of harmful compounds – carbon dioxide, carbon monoxide, hydrocarbons and nitrogen oxides. Further data directly transmitted from the vehicle's diagnostic system was sent to the central unit of the analyser and a GPS was used. Information on the results from mobile gas analysers in conjunction with the data recorded with the on-board diagnostic systems confirms the desirability of taking the assessment of emissions in real traffic conditions with the use of the measuring apparatus.

3. Analysis of the emission results

The analysis of road emissions and unit emissions was done for carbon dioxide, carbon monoxide, hydrocarbons and nitrogen oxides. Average values of emission of carbon dioxide in the various driving conditions are: 1968 g/km in urban conditions, 908 g/km in rural conditions and 681 g/km in motorway driving (Figure 2a). Analysis of the unit emissions of carbon dioxide allowed for determining the extent of variation: the value ranged from 100 to about 800 g/kWh. Average unit emission values of carbon dioxide in different driving conditions are: 570 g/kWh in urban conditions, 308 g/kWh in rural conditions and 261 g/kWh during motorway driving (Figure 2b).

The instantaneous values of carbon monoxide emission in the various stages of the test exceed 50 g/km, however, the end values in the individual stages (the ratio of the total weight of the emitted compound and the total distance covered) produced significantly smaller values. Average values for the emissions of carbon monoxide in various driving conditions are: 5.03 g/km in urban conditions, 2.16 g/km in rural conditions, and 1.16 g/km in motorway driving. Emission in motorway driving conditions was about 4 times less than in urban driving (Figure 3a). Average unit value of carbon monoxide emissions in the various driving conditions are: 1.46 g/kWh in urban conditions, 0.73 g/kWh in rural conditions, and 0.44 g/kWh during motorway driving (Figure 3b). The resulting unit emissions of carbon monoxide are below the limit, which for these vehicles is 4 g/kWh.

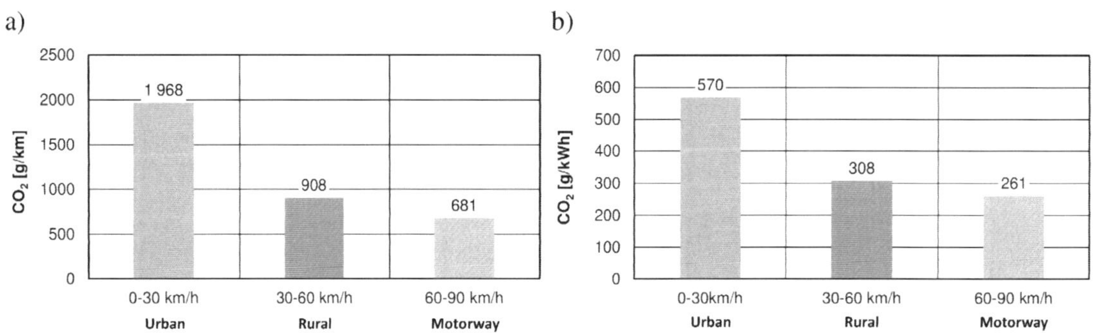

Figure 2. Characteristics of carbon dioxide emissions: a) road emissions, b) unit emissions.

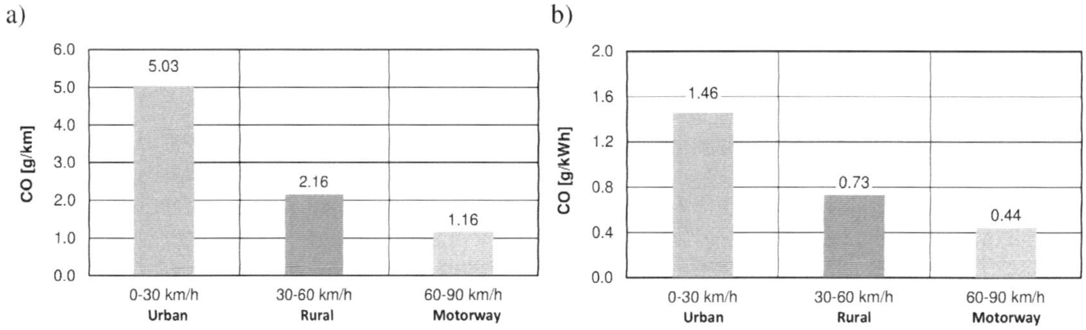

Figure 3. Characteristics of carbon monoxide emissions: a) road emissions, b) unit emissions.

The instantaneous values of emission of hydrocarbons in various stages of the test exceed 30 g/km, but the end values in individual stages produced significantly smaller values. Average values of carbon monoxide emissions in the various driving conditions are: 3.31 g/km in urban conditions, 1.28 g/km in rural conditions and 1 g/km in motorway driving. Emission in motorway driving conditions were more than 3 times less than in urban driving (Figure 4a). While the average unit value of hydrocarbon emissions in different driving conditions are: 0.96 g/kWh in urban conditions, 0.44 g/kWh in rural conditions and 0.38 g/kWh in motorway driving (Figure 4b). The obtained values of hydrocarbons emissions are lower than the permissible limit (which for these vehicles is 0.55 g/kWh), but only for driving in rural and motorway conditions.

The instantaneous values of nitrogen oxides emission in the various stages of the vehicle exceed 100 g/km, but the end values in the individual stages produced smaller results. Average values for carbon monoxide emissions in the various driving conditions include: 14.36 g/km in urban conditions, 5.37 g/km in rural conditions, and 2.05 g/km in motorway driving. Emission of nitrogen oxides in highway driving conditions was approximately 7 times lower than in urban driving (Figure 5a). While the average unit value of nitrogen oxide emissions in the various driving conditions are: 4.16 g/kWh in urban conditions, 1.82 g/kWh in rural conditions, and 0.787 g/kWh in motorway driving (Figure 5b). The values of hydrocarbons emission are lower than the permissible limit (which for these vehicles is 2 g/kWh), but only for driving in rural and motorway conditions.

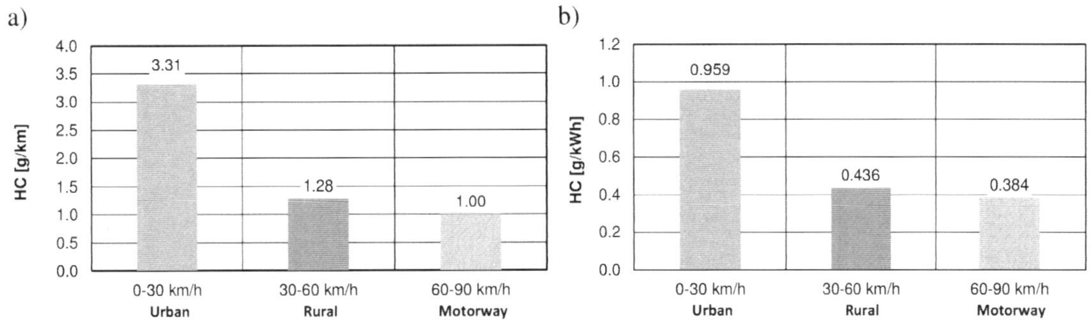

Figure 4. Characteristics of hydrocarbons emissions: a) road emissions, b) unit emissions.

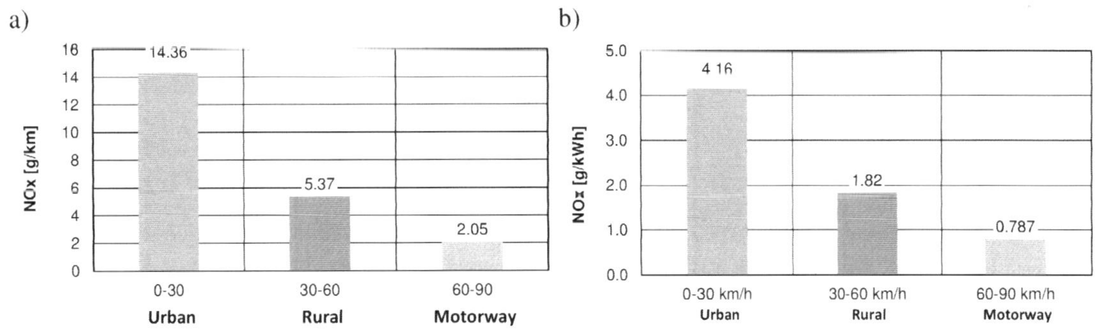

Figure 5. Characteristics of nitrogen oxides emissions: a) road emissions, b) unit emissions.

The summary of the obtained research results are the average emissions generated for the entire test route, which are equal to (Figure 6):

- for road emissions: carbon dioxide – 848 g/km, carbon monoxide – 1.77 g/km, hydrocarbons – 1.26 g/km and nitrogen oxides – 4.06 g/km,
- for unit emissions: carbon dioxide – 305 g/kWh, carbon monoxide – 0.64 g/kWh, hydrocarbons – 0.45 g/kWh, nitrogen oxides – 1.46 g/kWh.

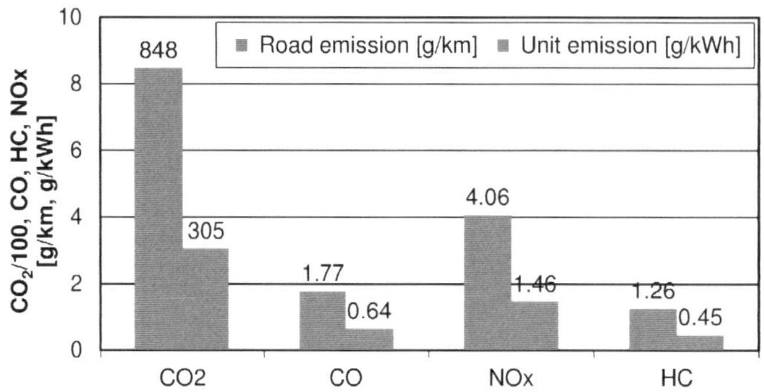

Figure 6. The obtained average values of road emissions and unit emissions in the performed tests (the values for the entire route).

Expressing the obtained emission values of the road (b_d) and unit (e_j) emissions throughout the study (in various stages and also taking into account the end values), a ratio in the form of road emissions to unit emissions (b_d/e_j) of the given exhaust emission was determined (Figure 7). A characteristic feature of the results is a constant value of this ratio for the various test route stages and their similarity as to the nature and value. Determining the relationship between the considered parameters (the road and unit emissions), the equation $b_d = 3.164\ e_j$ was defined (Figure 8) on the basis of which the value of unit emissions can be determined knowing the value of road emissions (or vice versa).

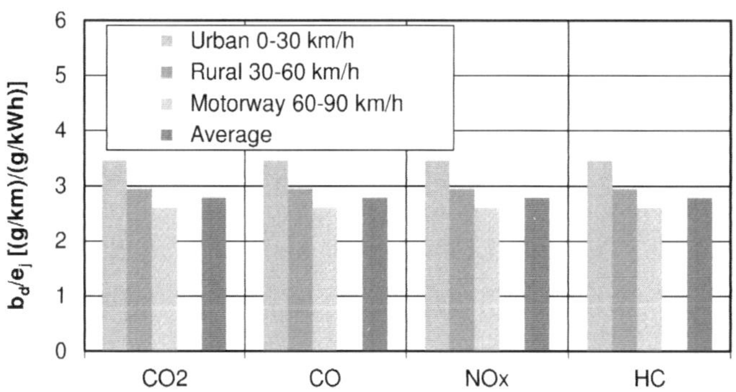

Figure 7. The ratio of the values of road and unit emissions in various test route stages.

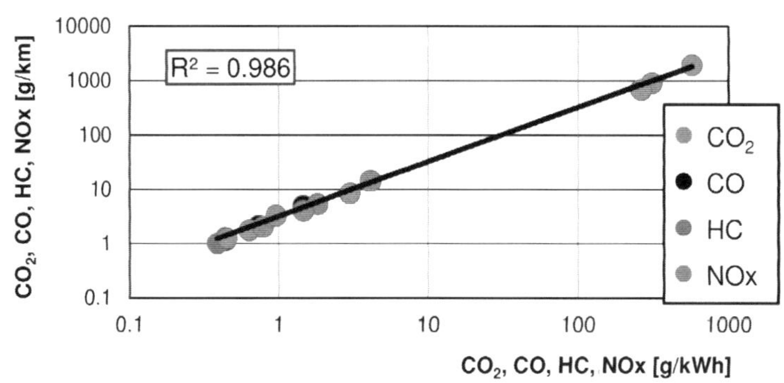

Figure 8. Dependence of road emissions as a function of emission of pollutants in various stages of research.

This relationship is valid for the conducted research (one case), but the resulting high value of the determination coefficient (0.986), hints on the nature of similarity between the work performed by the vehicle and the distance travelled. By plotting this dependence (Figure 9), showing the impact of the distance travelled and the work done by the vehicle engine a linear relationship was obtained between the two quantities. The determination coefficient obtained with this relation is 0.993, therefore, it can be considered that the values of the road and unit emissions of pollutants are related.
The obtained results cannot be applied to tests of all types of heavy-duty vehicles. Therefore, the attempt to generalize these results to a larger group of heavy-duty vehicles. 12 tests were performed in various driving conditions, yielding the data presented in Table 1. These values are obtained for urban buses with different types of engines (spark ignition and Diesel), varying maximum power and displacement, as well as different sizes (total weight range of 11,000–26,000 kg) and emission class.

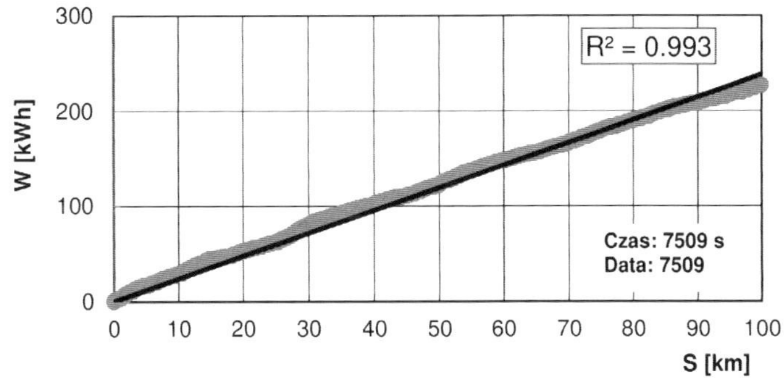

Figure 9. Relation between road emissions as a function of unit emission of pollutants in various stages of research.

Table 1. Example of data used to determine the relationship between the road emissions (b_d) and unit emissions (e_j) of heavy-duty vehicles.

Test no.	b_d/e_j (g/km)/(g/kWh)	N_e (kW)	V_s (dm^3)	m (kg)
1	2.56	230	9.2	25,000
2	2.01	192	6.9	24,000
3	2.95	213	9.2	26,000
4	3.16	412	15.6	35,000
5	1.89	192	6.9	20,000
6	1.51	230	9.2	20,000
7	1,31	230	9.2	18,000
8	2.67	209	9.2	26,000
9	2.61	221	9.2	26,000
10	2.31	221	9.2	25,000
11	1.30	176	10.2	13,000
12	1.39	123	10.2	11,000

The relationships between different values shown in Table 1 and the ratio of b_d/e_j, which reflects the relationship between road emissions and unit emissions of pollutants was investigated. The following relations were investigated:

- the impact of the maximum vehicle power on the ratio b_d/e_j:

$$b_d / e_j = f(N_e) \quad (1)$$

- the impact of the engine displacement on the ratio b_d/e_j:

$$b_d/e_j = f(V_s) \tag{2}$$

- the impact of the total mass of the vehicle on the ratio b_d/e_j:

$$b_d/e_j = f(m) \tag{3}$$

In the first step the relations between the ratio b_d/e_j and the maximum power (Figure 10), the engine displacement (Figure 11) and the mass of the vehicle (Figure 12) were prepared. Maximum power affects the value of the work done by the engine, so the unit emission of pollutants will depend on the vehicle engine power.

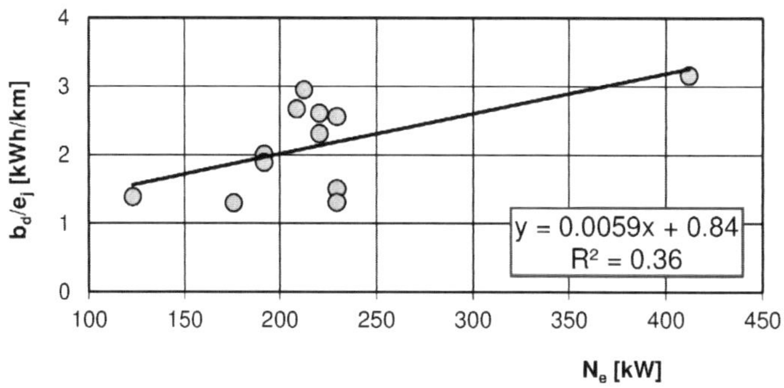

Figure 10. The relation between the ratio of road to unit emissions and the maximum engine power of the vehicle.

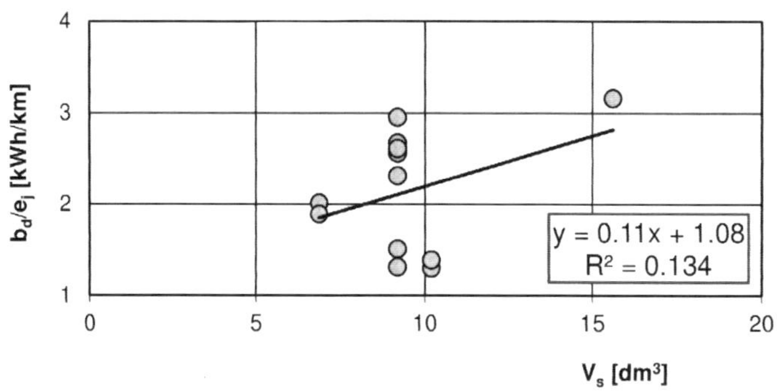

Figure 11. The relation between the ratio of road to unit emissions and the engine displacement of the heavy-duty vehicle.

Equation (1) is characterized by a low value of the coefficient of determination (0.36), but the analysis of Figure 10 shows that this factor must be taken into account in further analysis (it has a clear impact). The impact of engine displacement on emissions can be assessed under the assumption that with increasing displacement the value of the road emission of pollutants (b_d) increases, while the respective value of unit emission may be unchanged. Generally speaking, an increase in engine displacement affects the value of the road emission, but not necessarily the value of the unit emission (Figure 11). The situation is similar with equation (2), where the resulting coefficient of determination value of 0.82, leads to the conclusion that this value can be used in the further analysis. Vehicle mass is a value with a significant impact on the value of the ratio b_d/e_j (Figure 12).

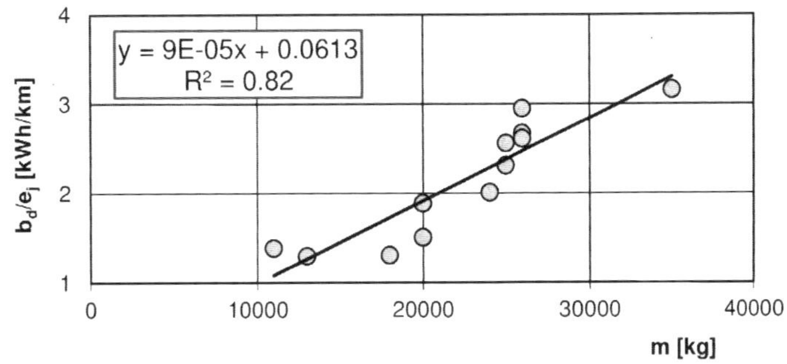

Figure 12. The relation between the ratio of road to unit emissions and the total vehicle mass.

Further steps were to use the obtained relations which was to result in a generalized equation, that could be used as a conversion method between the road emissions of pollutants and the unit emissions. The influence of the parameters previously taken for analysis was determined:

- the impact of the ratio of total vehicle mass and the engine displacement on the value of b_d/e_j:

$$b_d/e_j = f(m \cdot V_s) \tag{4}$$

- the impact of the ratio of total vehicle mass and the maximum engine power on the value of b_d/e_j:

$$b_d/e_j = f(m/N_e) \tag{5}$$

Equation (4) shown in Figure 13, has a lower coefficient of determination (0.57), than the ratio $b_d/e_j = f(m)$, but higher than for the ratio $b_d/e_j = f(V_s)$. There is a significant range of the results of the product $(m \cdot V)$, which leads to the conclusion that other parameters presented in Table 1 should also be taken into account.

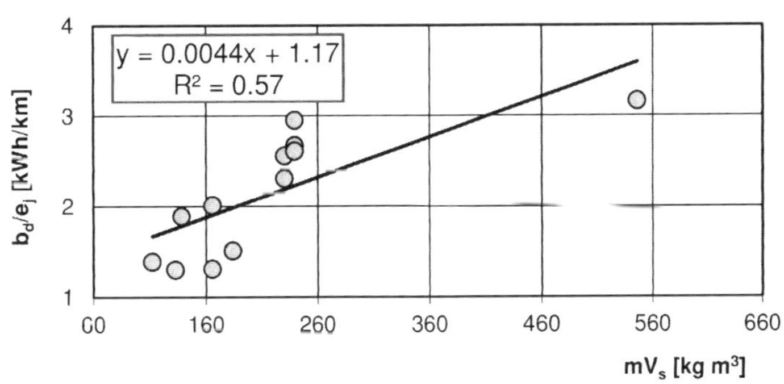

Figure 13. The relation between the ratio of road to unit emissions and the ratio of total vehicle mass to the engine displacement.

For equation (5) presented in Figure 14 a coefficient of determination equal to 0.354 was obtained – a value close to the value of the coefficient of determination obtained for the relation $b_d/e_j = f(N_e)$, and less than for the relation $b_d/e_j = f(m)$. The results presented in Figures 13 and 14, allow for using these values to determine the relationship, which combines both of these values. The generalized solution can be written as:

$$b_d/e_j = f(m \cdot V_s/N_e) \tag{6}$$

Equation (6) has been shown graphically in Figure 15, which has a coefficient of determination equal to 0.81, which is a very high value, given the small range of individual deviating points.

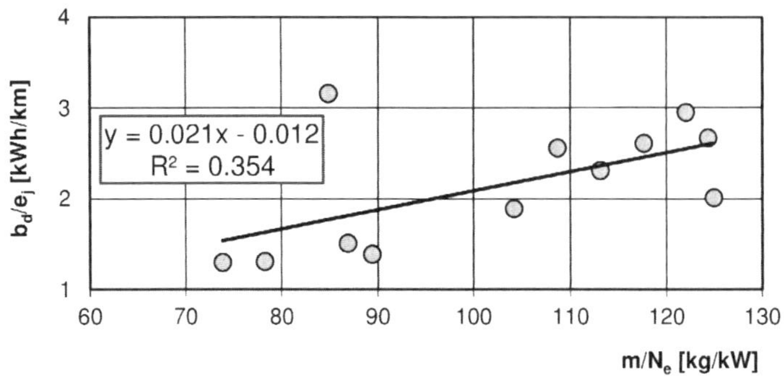

Figure 14. The relation between the ratio of road to unit emissions and the ratio of total vehicle mass to the maximum engine power.

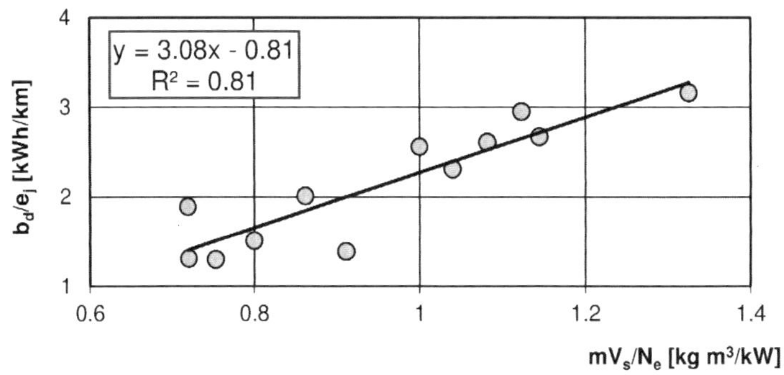

Figure 15. The relation between the ratio of road to unit emissions and the ratio of total vehicle mass to the maximum engine power.

The relationship shown in Figure 15 can be expressed as:

$$b_d / e_j = 3.08 \frac{m \cdot V_s}{N_e} - 0.81 \quad (7)$$

and then convert to an expression that describes the relationship $b_d = f(e_j)$:

$$b_d = \left(3.08 \frac{m \cdot V_s}{N_e} - 0.81 \right) e_j \quad (8)$$

Equation (8) does not have a physical significance, and the units given are only used for the conversion of road emissions and unit emissions. The equation presented above is not final and may require further work, which will result in an increase in the number of parameters resulting in an increase in the accuracy of the data obtained. This procedure should be carried out for heavy-duty vehicles and buses with different operating parameters (including different vehicle mileage). However, this relationship can be used to convert the road and unit emissions, by utilizing the factors described earlier.

4. Conclusions

The method that reflects the real emissions of harmful compounds into the environment consists of studies of vehicles that are conducted in real driving conditions. Road emission testing of vehicles are considered innovative and necessary around the world. The main advantage of these tests is the measurement of the emissions and fuel consumption of the vehicle in real driving conditions, rather than on the test bench, with well specified and controlled conditions. The most critical in terms of emissions and fuel consumption are the conditions for transient operation of the vehicle (engine) as well as driving vehicles at maximum speeds. These elements of the test drive determine the load of the drive unit and the share of transient phases in the total time of engine operation. In addition, points of operation of the engine, i.e. the load range and speed of the engine crankshaft during actual operation, are significantly different from the operating points of the engine during the certification test. Moreover, in the case of real driving research, any route of travel and load of the test vehicle can be chosen. For these reasons, it seems necessary to develop an algorithm that allows for the conversion of road emission values and unit emissions.

References

[1] Bergmann D 2013 Developing the Technology Innovation Process for Further Emissions Reduction, 6th Integer Diesel Emissions Conference and Diesel Exhaust Fluid Forum, Atlanta
[2] Bougher T, Khalek I, Trevitz S and Akard M 2010 Verification of a Gaseous Portable Emissions Measurement System with a Laboratory System Using the Code of Federal Regulations Part 1065, *SAE Paper Series* 2010-01-1069
[3] Commission Regulation (EU) No 133/2014 of 31 January 2014 amending, for the purposes of adapting to technical progress as regards emission limits, Directive 2007/46/EC of the European Parliament and of the Council, Regulation (EC) No 595/2009 of the European Parliament and of the Council and Commission Regulation (EU) No 582/2011
[4] Commission Regulation (EU) No 582/2011 of 25 May 2011 implementing and amending Regulation (EC) No 595/2009 of the European Parliament and of the Council with respect to emissions from heavy duty vehicles (Euro VI) and amending Annexes I and III to Directive 2007/46/EC of the European Parliament and of the Council
[5] Franco V, Kousoulidou M, Muntean M, Ntziachristos L, Hausberger S and Dilara P 2013 Road Vehicle Emission Factors Development: A Review, *Atmospheric Environment* **70** pp 84-97
[6] Johnson K, Durbin T, Cocker D, Miller J, Agama R and Moynahan N, 2008 On-Road Evaluation of a PEMS for Measuring Gaseous In-Use Emissions from a Heavy-Duty Diesel Vehicle, *SAE Technical Paper Series* 2008-01-1300
[7] Mamakos A, Bonnel P, Perujo A and Carriero M 2013 Assessment of portable emission Measurement Systems (PEMS) for Heavy-Duty Diesel Engines with Respect to Particulate Matter, *Journal of Aerosol Science* **57** 5 pp 4-70.
[8] MerkiszJ., Jacyna M, Merkisz-Guranowska A and Pielecha J 2014 Exhaust emissions from modes of transport under actual traffic conditions, *Energy Quest*, Ekaterinburg
[9] Merkisz J and Pielecha J 2015 Nanoparticles emissions from combustion engines, *Springer Tracts on Transportation and Traffic* **8** p 139
[10] Merkisz J and Pielecha J 2015 Real driving emissions – vehicle tests in variable terrain, *Journal of KONES* **22** (1) pp 217-225
[11] Merkisz J, Pielecha J and Fuc P 2015 LDV and HDV Vehicle Exhaust Emission Indexes in PEMS-Based RDE Tests, *Fortschritt-Berichte VDI Verlag* **12** 1(783) pp 240–265
[12] Merkisz J, Pielecha J, Fuc P and Lijewski P 2012 The Analysis of the PEMS Measurements of the Exhaust Emissions from City Buses using Different Research Procedures, VPPC 2012, Vehicle Power and Propulsion Conference, Seoul
[13] Merkisz J, Pielecha J and Radzimirski S 2014 New Trends in Emission Control in the European Union, *Springer Tracts on Transportation and Traffic* **1** p 170
[14] Myung C and Park S 2012 Exhaust Nanoparticle Emissions from Internal Combustion Engines: A Review, *International Journal of Automotive Technology* **13** (1) pp 9-22
[15] Villafuerte P, del Parque A, Bonnel P, Carriero M., Forni F and Riccobono F 2015 EU-PEMS PM Pilot Program: Testing, Data Analysis and Results, *Sustainable Transport Unit*, Report EUR 27326 EN, Joint Research Centre p 57

IOP Conf. Series: Materials Science and Engineering **148** (2016) 012079 doi:10.1088/1757-899X/148/1/012079

Characterisation of solid particles emitted from diesel and petrol engines as a contribution to the determination of the origin of carbonaceous particles in urban aerosol

M Michalik[1], M Brzeżański[2], W Wilczyńska-Michalik[3], K Fisior[4], B Klimas[4], L Samek[5] and B Pietras[3]

[1]Institute of Geological Sciences, Jagiellonian University, 2a Oleandry Street, 30-063 Krakow, Poland
[2]Institute of Automobiles and Internal Combustion Engines, Cracow University of Technology, 37 Jana Pawla II Avenue, 31-864 Krakow, Poland
[3] Institute of Geography, Pedagogical University, 2 Podchorążych Street, 30-084 Krakow, Poland
[4] Faculty of Chemistry, Jagiellonian University, 3 Ingardena Street, 30-060 Krakow, Poland
[5] Faculty of Physics and Applied Computer Science, AGH – University of Science and Technology, 30 Mickiewicza Avenue, 30-059 Krakow, Poland

E-mail: marek.michalik@uj.edu.pl

Abstract. Solid particles emitted from diesel and petrol engines were studied using a scanning electron microscope fitted with an energy dispersive spectrometer. The soot emitted from different engines under different operating conditions differed in particle size, and the form and size of aggregates. Identification of the soot particles emitted from diesel or petrol engines in urban aerosol based on their size and morphology was found to be impossible.

1. Introduction

The particulate matter (PM) emitted into the atmosphere together with engine exhaust gases is very difficult to define in a simple way. The complex composition of PM causes problems in determining its specific physical and chemical characteristics. Single particles are usually an irregular mixture of chemical compounds. They exhibit different shapes and sizes. The results from their study vary depending on the methodology of the measurement. In addition to the difficult determination of highly diverse chemical compositions, a major research challenge is determining the size of the individual particles. Separation of given individual size-fraction is usually made using a filter, with a pore size larger than the particle test fraction.

In engine measurements of solid particles the total content of organic or inorganic matter, both solid or liquid, which accumulates on the filter with absolute retention of 99% of solids and pore-size ≥ 0.3 µm is usually determined after passing through an exhaust gas air diluter at a temperature of 52 ± 3°C.

The environmental hazard of particles emitted together with the engine exhaust gases is caused by their chemical activity and very small size. The atmospheric residence time of small particles of size less than 100 nm is long, and they easily penetrate into the human body via the respiratory system [1]. Respirable particles containing heavy metals, sulphur and nitrogen compounds, and heavy

hydrocarbons like polycyclic aromatic hydrocarbons (PAHs) are dangerous for human health. Exposure to engine-originated PM increases human morbidity and mortality [2].
Particle emissions from road vehicles are studied extensively worldwide because of their dominant contribution towards the total airborne particle number concentrations found in the atmosphere in numerous cities [3].
The concentration of air pollution in Krakow has been high for over fifty years [4, 5]. It is possible to note variations in the concentration of different components of air pollution, as well as variations in the dominant sources of pollution from a historical perspective. Recently, household, vehicular exhaust and industrial emissions are considered as meaningful anthropogenic sources of air pollution. The contribution of different sources is evaluated using modelling methods. Analytical data related to the share of various sources of PM in the atmosphere are limited.
Single particle analysis of aerosol in Krakow indicates that carbonaceous particles are the most common when taking into account the total number of particles, especially in PM2.5 fraction of airborne dust [6, 7]. The morphology of carbonaceous particles varies significantly. Both single particles and aggregates of different shape and size are present.
Carbonaceous particles are considered to be more harmful than mineral ones due to their small size and composition [8]. Proper assessment of their sources is crucial to undertake reasonable efforts to improve the situation in highly polluted cities. The aim of this study was to compare the morphology and composition of solid particles emitted from diesel and petrol engines with carbonaceous particles present in aerosol. Solid particles from different engines operating in different conditions were collected. The results might be important for the elaboration of a method for the determination of sources of air pollution. The results of detailed studies of exhaust particles are also useful in the modelling of the dispersion of pollution emitted from vehicles, the determination of health effects, prediction of the soiling of building facades, determination of the deterioration mechanism of building materials, the influence on chemical reaction in the atmosphere, and the modification of climatic conditions.

2. Sampling

Investigations of PM emission from different diesel and spark ignition (SI) engines were conducted on the engine test bed of the Institute of Automobiles and Internal Combustion Engines of Cracow University of Technology.

Solid particle samples were collected on a polycarbonate membrane using a personal air sampling pump, GilAir Plus. The volume of exhausts sampled was 2 dm^3 for the diesel engine and 10 dm^3 for the petrol engines. The engines' operating conditions are listed in tables 1–3.

Table 1. Operating conditions of the Volkswagen 1.9 TDI diesel engine (1900 cm^3 displacement).

	Sample D1	Sample D2	Sample D3
Engine speed (rpm)	2000	835	2000
Torque (Nm)	115	0	200
Exhaust gas opacity (°B)	8	0.35	0.3
EGR[a] (%)	60	60	0

[a] Exhaust Gas Recirculation

Table 2. Operating conditions of the Fiat type 170A1 – conventional SI engine (900 cm^3 displacement).

	Sample B1 (warming-up phase)	Sample B2 (thermal stabilised phase)
Cylinder head coolant temperature (°C)	30–75	87
Torque (Nm)	30–40	40

Table 3. Operating conditions of the direct injected research SI engine (different injection advance angle).

	Sample B10	Sample B11	Sample B12	Sample B13	Sample B14	Sample B15
Exhaust gas temperature (°C)	500	500	500	760	900	890
Air excess number (λ)	0.875	0.875	0.875	1	1	1
Engine speed (rpm)	2480	2480	2480	2600	2600	2600
Injection advance angle (° c.a.)	340	180	90	340	180	90

Samples of the aerosols were collected on Nuclepore filters over 24 hr periods. Three size fractions (<2.5 μm, 2.5–10 μm and >10 μm) were collected by an NILU Stucked Unit connected with a flowmeter and pump. The flow rate of air was equal to 30 l per min.

3. Methods

The exhaust particles and aerosols were analysed using a field emission scanning electron microscope ([SEM], HITACHI S-4700) equipped with an energy dispersive spectrometer ([EDS], NORAN) with a Nordlys detector. Both observation and chemical analyses were performed at the voltage of 15 kV. The standardless method was used for quantification of the chemical components.
Fragments of polycarbonate membranes with deposited particles were mounted on graphite holders using conductive carbon discs, and then coated with a gold conductive layer.

4. Results

Carbonaceous particles (soot) dominate in the collected engine exhaust solid particles' samples. Their characteristics (particle or aggregate size, form of aggregates, number of particles in aggregate) are presented below.

4.1. Diesel engine - Volkswagen 1.9 TDI

The size of single particles emitted from the diesel engine observed in SEM vary from 50 nm to 100 nm in samples D2 and D3 (figures 1, 2). Particles below 50 nm occur rarely. Determination of the single particle size in sample D1 was not possible because of the presence of a complex network of elongated aggregates composed of numerous particles (figures 3, 4). Elongated chain-like aggregates occur commonly in sample D2 (figure 5); however, isometric aggregates are also present. Their size varies from 10 μm in the case of the elongated aggregates to <100 nm for the isometric ones composed of several particles. The isometric aggregates dominate in sample D3 (figure 6). Elongated aggregates are also present in sample D3, but they are shorter compared with sample D2. Their size is between 8 μm and 200 nm. Comparison of the solid particles' samples and operating conditions of the engine suggest that exhaust gas recirculation (EGR) is an important factor. Two samples collected at EGR=60% contain elongated chain-like aggregates.

4.2. Conventional petrol engine - Fiat type 170A1

The characteristics of the solid particles collected during the warming-up phase (sample B1) and during the thermal stabilised phase (sample B2) are similar. Single particles or small aggregates composed of several particles are common; larger aggregates containing numerous particles also occur. The single particles vary in size from <20 nm to 50 nm (figures 7–10). Larger particles occur rarely.

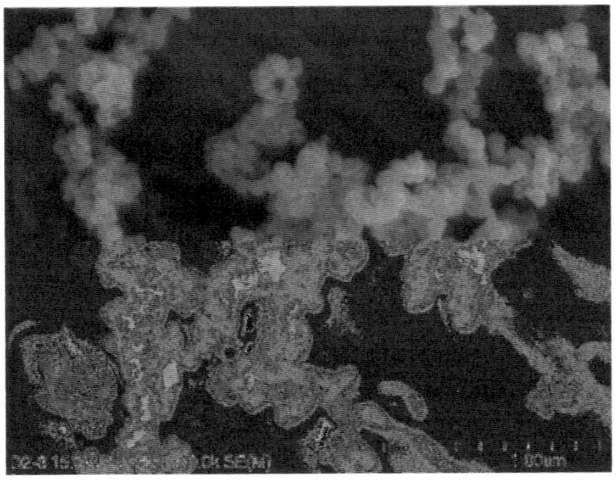

Figure 1. Soot particles in chain-like aggregate - Sample D2.

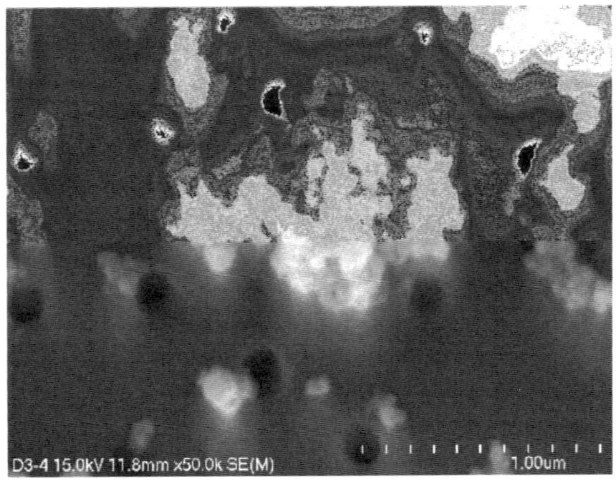

Figure 2. Soot particles in aggregate - Sample D3.

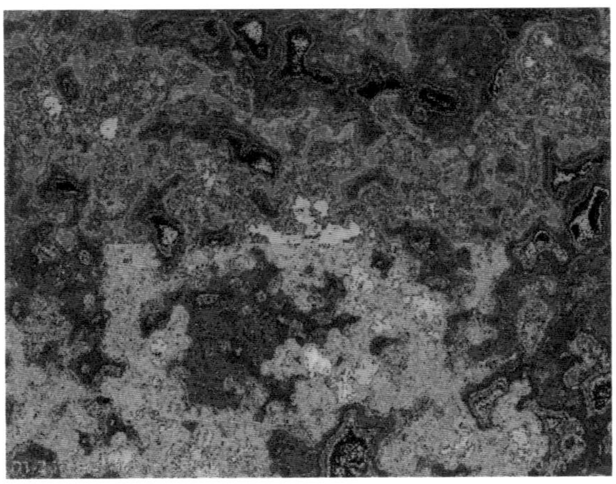

Figure 3. Complex network of soot particles' aggregates - Sample D1.

Figure 4. Densely packed network of soot particles' aggregates - Sample D1.

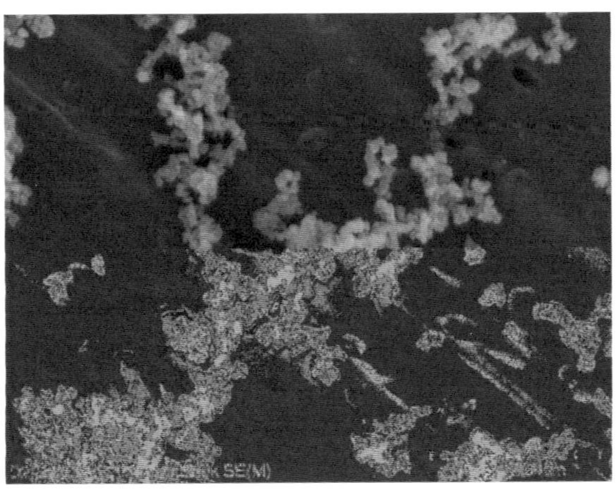

Figure 5. Chain-like aggregates of particles - Sample D2.

Figure 6. Isometric aggregate of soot particles - Sample D3.

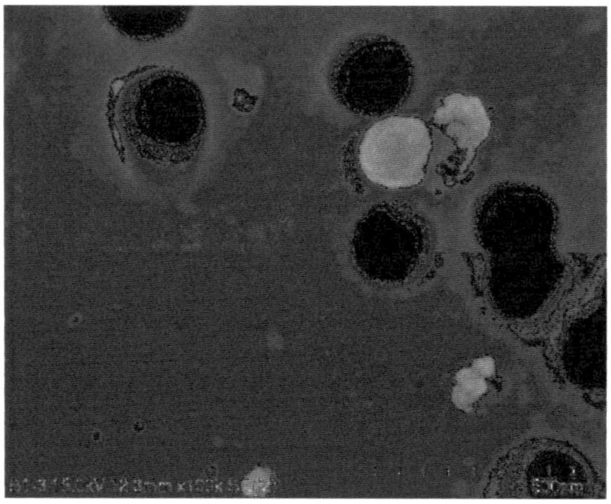

Figure 7. Single particles of soot - Sample B1.

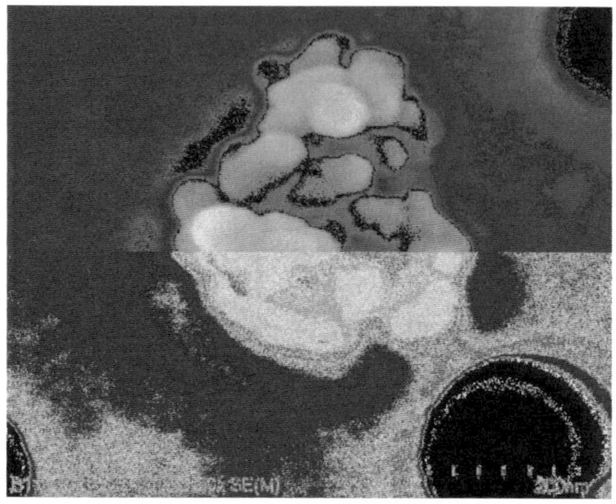

Figure 8. Aggregate of soot particles – Sample B1.

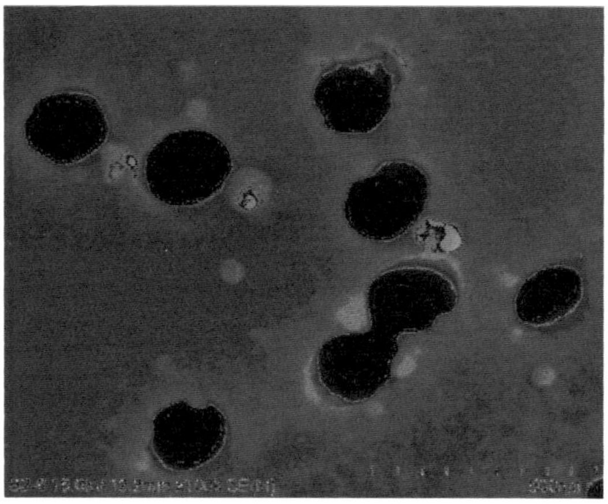

Figure 9. Single particles of soot - Sample B2.

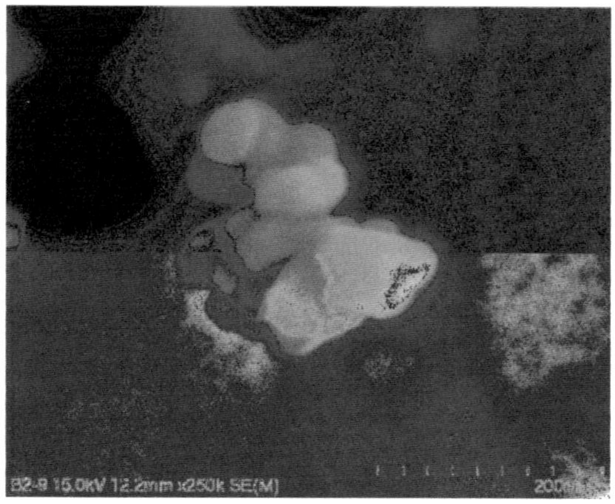

Figure 10. Aggregate of soot particles - Sample B2.

4.3. Direct injected research SI engine

The single particles or small aggregates are common in all the samples, but a large number of particles occur in the form of very complex aggregates. The aggregates are elongated and branched. Their length is usually up to 70 μm; some chain-like aggregates reach 350 μm (figures 11–14). Aggregates forming irregular ring-like patterns are less common. The size of the single particles varies from 25 nm to 50 nm (figures 15, 16).

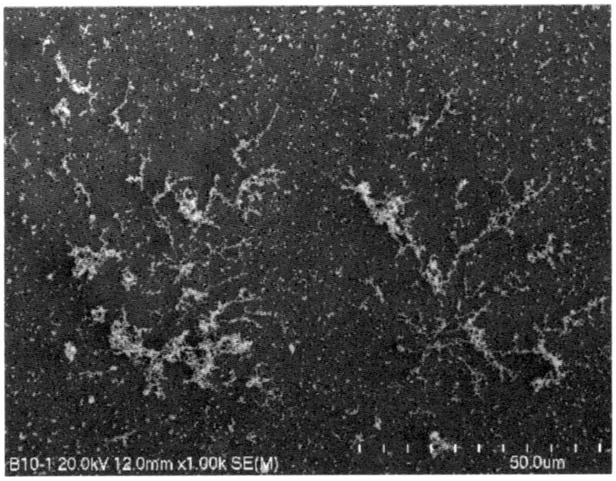

Figure 11. Single particles and complex aggregates of soot - Sample B10.

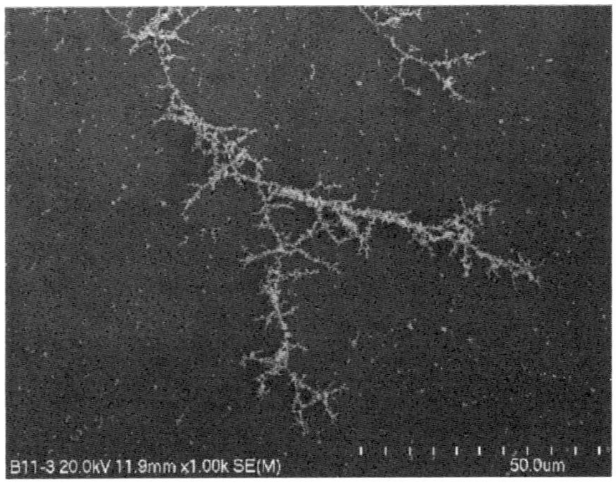

Figure 12. Branched chain-like aggregates of soot particles - Sample B11.

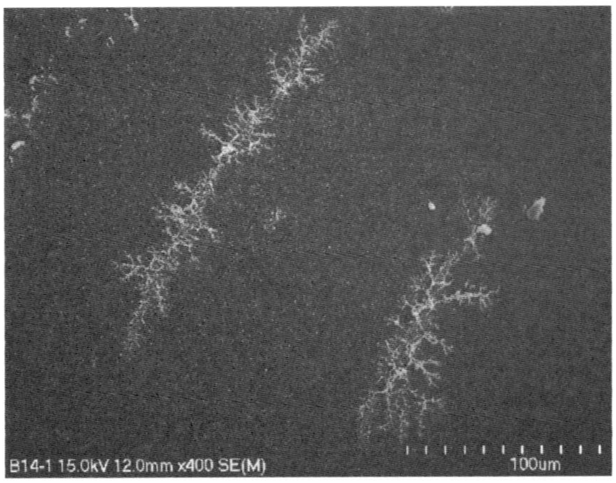

Figure 13. Chain-like aggregates of soot particles - Sample B14.

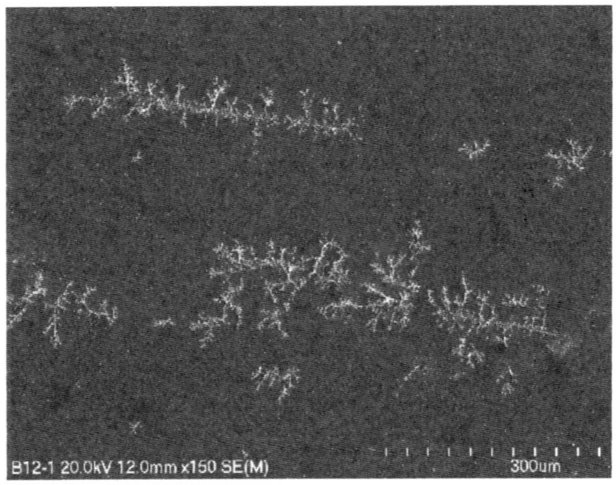

Figure 14. Chain-like aggregates of soot particles - Sample B12.

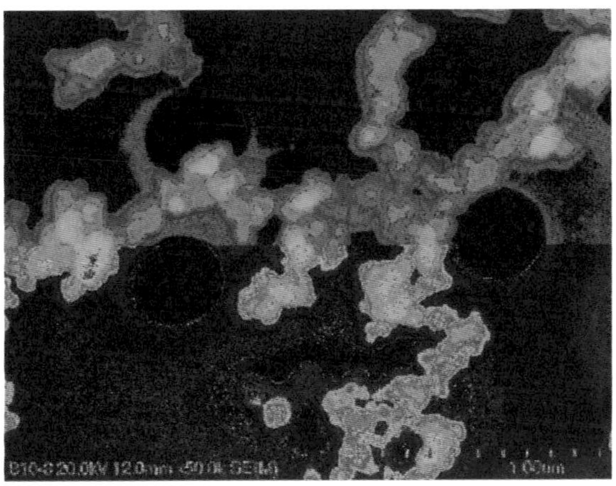

Figure 15. Soot particles in chain-like aggregate - Sample B10.

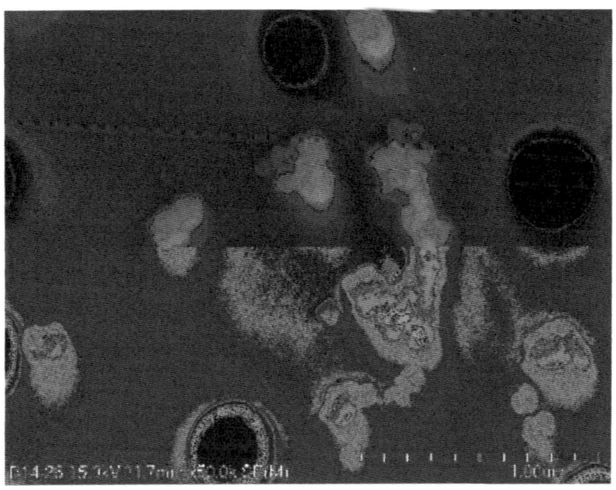

Figure 16. Single soot particles and small aggregates - Sample B14

4.4. Metal containing particles

Metal containing particles occur rarely in the analysed samples. Their size and morphology vary within a broad range. Cr-containing particles were noted in samples collected from the diesel and conventional petrol engines (figures 17 and 18). Other particles (e.g. containing Si, Al, Cl, K, Ca) were noted only as unique examples.

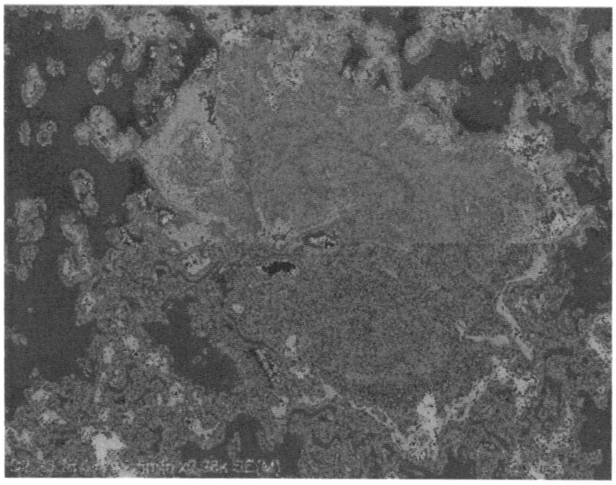

Figure 17. Cr-rich particle - Sample D2.

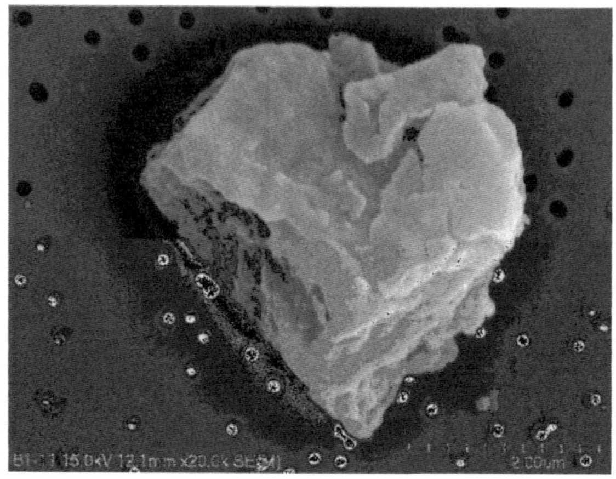

Figure 18. Cr-rich particle - Sample B1.

5. Discussion and conclusions

The soot emitted from different engines differs in particle size, and the form and size of aggregates. Detailed discussion of the relationships between operating conditions and soot characteristics needs more systematic study using a larger number of samples.

Identification of the soot particles emitted from diesel or petrol engines in urban aerosol based on their size and morphology is impossible. In urban aerosol, numerous carbonaceous particles similar to those identified in engine exhaust occur (figures 19–24) [6, 7]. Soot particles emitted from different sources (e.g. coal-powered household stoves, power plant boilers, biomass and waste combustion) could be similar in terms of the size and form of aggregates [6]. The opinion that soot particles emitted from engines are smaller in size compared with soot from other sources is not supported by our results.

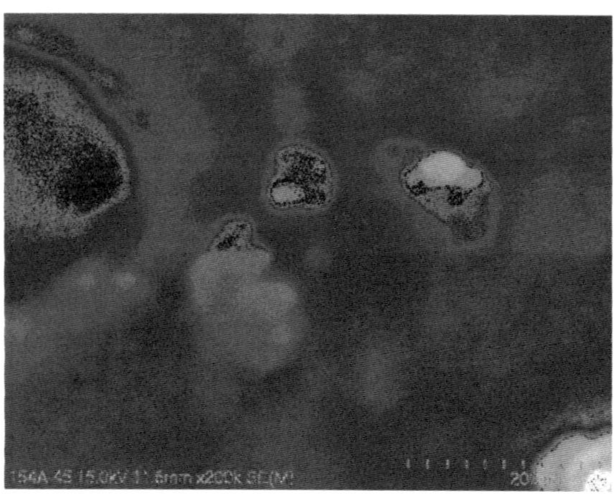

Figure 19. Single soot particles - aerosol in Krakow.

Figure 20. Single soot particles and aggregates - aerosol in Krakow.

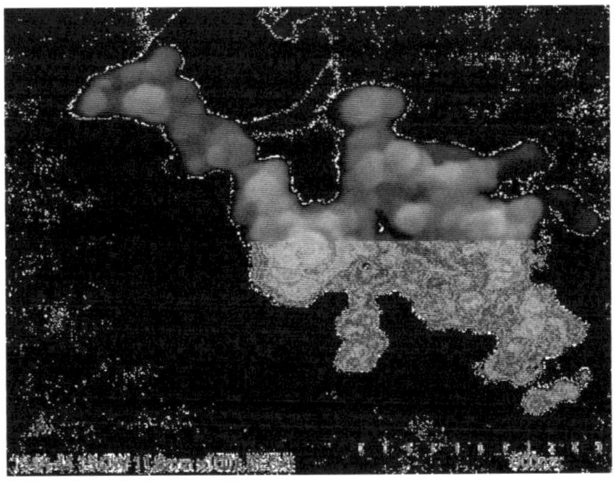

Figure 21. Soot aggregate - aerosol in Krakow.

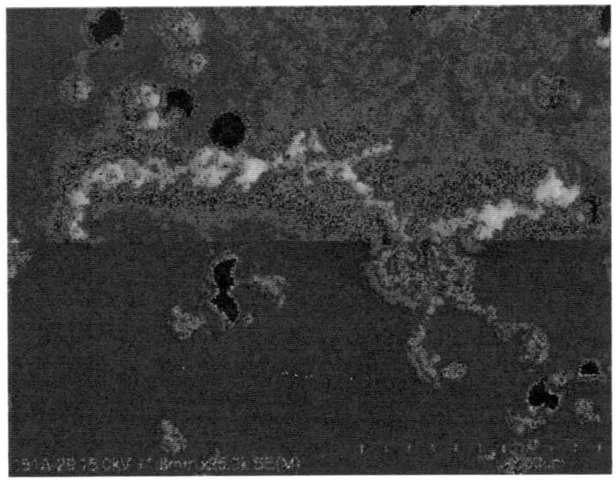

Figure 22. Chain-like soot particles' aggregate - aerosol in Krakow.

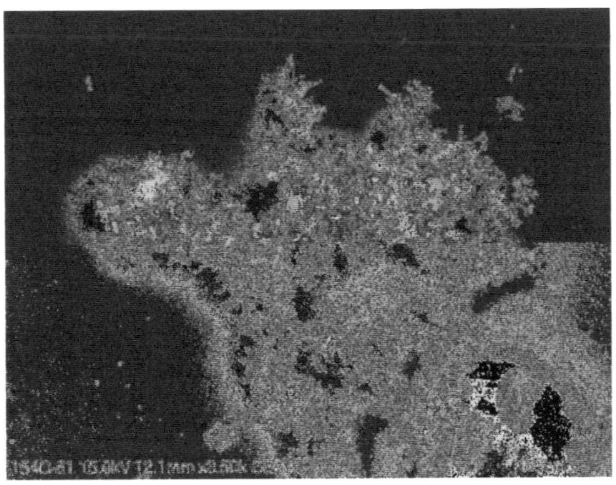

Figure 23. Loosely packed isometric aggregate of soot particles - aerosol in Krakow.

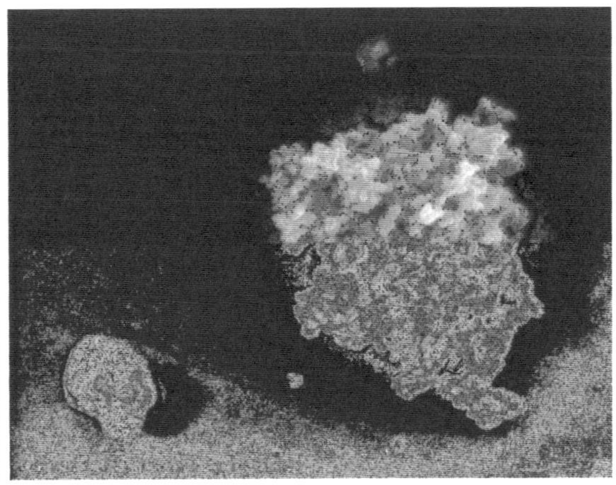

Figure 24. Densely packed aggregate of soot particles - aerosol in Krakow.

Determination of the contribution of different sources of carbonaceous compounds in urban aerosols needs the application of detailed chemical analyses in aerosol studies [e.g. 9, 10], coupled with systematic analyses of the material from all possible sources.

References

[1] Broday D M and Rosenzweig R 2011 Deposition of fractal-like soot aggregates in the human respiratory tract *Journal of Aerosol Science* **42** pp 372–386.

[2] Lawal A O, Davids L M and Marnewick J L 2016 Diesel exhaust particles and endothelial cells dysfunction: An update. *Toxicology in Vitro* **32** pp 92–104.

[3] Kumar P, Pirjola L, Ketzel M and Harrison R M 2013 Nanoparticle emissions from 11 non-vehicle exhaust sources. A review. *Atmospheric Environment* **67** pp 252-277.

[4] Ocena jakości powietrza w województwie małopolskim w 2015 roku. 2016 Wojewódzki Inspektorat Ochrony Środowiska w Krakowie.

[5] Raport o stanie środowiska w województwie małopolskim w 2014 roku. 2015 Wojewódzki Inspektorat Ochrony Środowiska w Krakowie.

[6] Wilczyńska-Michalik W and Michalik M 2015 Skład i pochodzenie cząstek pyłów w powietrzu atmosferycznym w Krakowie. *Aura* **3** pp 12-16.

[7] Wilczyńska-Michalik W, Pietras B, Samek L, Furman L, Łatkiewicz A, Rzeźnikiewicz K and Michalik M 2015 Submikronowe pyły w powietrzu atmosferycznym w Krakowie. *Aura* **8** pp 4-8.
[8] Lelieveld J, Evans J S, Fnais M, Giannadaki D and Pozzer A 2015 The contribution of outdoor air pollution sources to premature mortality on a global scale. *Nature* **525**, 3 pp 67-371.
[9] Simoneit B R T, Bi X.-H, Oros D R, Medeiros P M, Sheng G-Y and Fu J-M 2007 Phenols and hydroxy-PAHs (arylphenols) as tracers for coal smoke particulate matter e source tests and ambient aerosol assessments. *Environmental Science and Technology* **41**, 7294-7302.
[10] Górka M, Rybicki M, Simoneit B R T and Marynowski L 2014 Determination of multiple organic matter sources in aerosol PM10 from Wroc1aw, Poland using molecular and stable carbon isotope compositions. *Atmospheric Environment* **89** pp 739-748.

IOP Conf. Series: Materials Science and Engineering **148** (2016) 012080 doi:10.1088/1757-899X/148/1/012080

Assessment of total efficiency in adiabatic engines

W Mitianiec

Cracow University of Technology, Al. Jana Pawla II 37, 31-864 Krakow

e-mail: wmitanie@usk.pk.edu.pl

Abstract. The paper presents influence of ceramic coating in all surfaces of the combustion chamber of SI four-stroke engine on working parameters mainly on heat balance and total efficiency. Three cases of engine were considered: standard without ceramic coating, fully adiabatic combustion chamber and engine with different thickness of ceramic coating. Consideration of adiabatic or semi-adiabatic engine was connected with mathematical modelling of heat transfer from the cylinder gas to the cooling medium. This model takes into account changeable convection coefficient based on the experimental formulas of Woschni, heat conductivity of multi-layer walls and also small effect of radiation in SI engines. The simulation model was elaborated with full heat transfer to the cooling medium and unsteady gas flow in the engine intake and exhaust systems. The computer program taking into account 0D model of engine processes in the cylinder and 1D model of gas flow was elaborated for determination of many basic engine thermodynamic parameters for Suzuki DR-Z400S 400 cc SI engine. The paper presents calculation results of influence of the ceramic coating thickness on indicated pressure, specific fuel consumption, cooling and exhaust heat losses. Next it were presented comparisons of effective power, heat losses in the cooling and exhaust systems, total efficiency in function of engine rotational speed and also comparison of temperature inside the cylinder for standard, semi-adiabatic and full adiabatic engine. On the basis of the achieved results it was found higher total efficiency of adiabatic engines at 2500 rpm from 27% for standard engine to 37% for full adiabatic engine.

1. Introduction

Measurement of heat transfer in internal combustion engines encounters great difficulties due to unsteady heat flow from the changeable temperature of the hot gas in the cylinder to the cooling medium, in most cases water. The biggest temperature of the cylinder charge takes place at the end of the combustion process at initial period of expansion stroke. On the other hand the lowest temperature of the charge occurs during intake stroke. Sometimes temperature of the cooling medium is higher than temperature of the cylinder charge. From technical point of view there is problem with measurement of changeable temperature in the cylinder and therefore also assessment of the heat flows. Increasing of total efficiency of IC engines requires a decreasing of thermal losses to the cooling system, decreasing of exhaust losses (lower energy flowing into outflow system) and decreasing of friction losses (increasing of mechanical efficiency). The most interesting is lowering of heat transfer from the gas inside the cylinder to a cooling medium (water or air). In order to achieve higher thermal efficiency the adiabatic processes should occur in engine. During last many years much research works were done for increasing of engine efficiency by using materials with lower conductivity in order to decrease heat transfer. The problem was presented for example by Kamo [9] for diesel engine in 1987. His study concerned the problem of utilization high exhaust energy in adiabatic engine. His work helped the Cummins Engine Company, Inc. and the U S. Army to develop

an adiabatic turbo-compound engine during the last nine years [10] by using exhaust gas energy. The researchers from Ocean & Ship Foundation (Japan) [4] by insulation of the diesel engine fuelled by CNG and applying of recovering turbine improved engine efficiency from 42% to 57.5%. Another researcher Khoshravan [5] carried out the CFD calculations of heat transfer in the diesel engine with ceramic coating on the cylinder liner. The calculations were next verified by experimental work on single research engine. Many researchers used adiabatic engines for checking of influence of higher temperature on combustion of biofuels. Such work was done by scientists from BIT Rajshahi and Jadavpur University [6] on semi-adiabatic diesel engine fuelled by vegetable oil. They found that exhaust temperature, CO emission and smoke have increased, but engine thermal efficiency and NOx has decreased. The adiabatic engines both SI and CI indicate higher temperature inside the cylinder which influences on exhaust gas emission. The researcher Gosai and Nagarsheth [7] during their research work on diesel adiabatic engine found a few percent decrease of molar concentration of CO2, HC and CO were decreased in comparison to standard diesel engine. Also smoke density was reduced about 39.5% due to zirconium coating on the cylinder liner. In this case also thermal efficiency was increased. Predicting of work parameters of four-stroke adiabatic engine was carried out by Reddy and Pandurangadu [8]. They also found higher exhaust gas losses due to higher combustion temperature, but they achieved higher thermal efficiency. Influence of different ceramic materials such as SiO2, Al2O3, ZrO2, SiC, Si3N4, which were used as coating surfaces with thickness 0.5 mm in combustion chamber in diesel engine was presented by Turkish scientists in the work [11]. They confirmed about 3% increase of engine power at high rotational speed and about 10% lower heat transfer rate.
This paper concerns a determination of influence thermal barrier on engine parameters and heat balance which is caused by ceramic coating on inner surfaces of the combustion chamber (cylinder liner, piston and cylinder head). Mathematical model of heat transfer to the cooling medium by convection, conductivity and radiation enables calculation of engine thermodynamic processes and prediction of cooling and exhaust heat losses. Theoretical studies of adiabatic engines help to develop a new design of such engine. The study was carried out only for SI four stroke engine, where all previous works were done on diesel engines.

2. Heat transfer in piston engines

The thermodynamic model of heat exchange in piston internal combustion engines takes into account changeable temperature of the cylinder charge and constant temperature of the cooling medium. In this paper the water is considered as cooling medium, because most of the modern engines are currently equipped with liquid cooling system. In order to increase total efficiency of the engines the difference between higher and lower heat sources should reach higher value. Usually higher temperature T_{max} of the combustion process influences on higher total efficiency.

$$\eta_0 = \frac{T_{\max} - T_{\min}}{T_{\max}} = 1 - \frac{T_{\min}}{T_{\max}} \tag{1}$$

Heat transfer in the piston engines is carried out through three phenomena: convection, radiation and conductivity. Usually calculation is carried out for assessment of heat transfer in IC engines. Temperature of the charge in the cylinder changes during piston motion from value near 3000 K to 350 K in SI engine, thus the heat stream to the cooling medium changes rapidly and fluctuation of heat transfer can be defined in a function of crank rotation. Diagram showing change of temperature in the cylinder walls and cooling medium is presented in Fig. 1. The cylinder wall consists two layers: the first with lower conductivity and thickness g_c in mostly cases as ceramic coating and second layer with bigger thickness g_t made from cast iron with higher conductivity. The surface area of heat exchange on the cylinder liner changes according to the piston movement. However the surfaces of heat transfer through the piston and cylinder head do not changes. The heat transfer depends on convection process in the cylinder and cooling medium and cannot be controlled by engine geometry and materials, but the heat flow resistance can be reduced by applying of wall material with different conductivity. Heat transfer as a result of radiation is only due the temperature difference between the gas and walls and emissivity of the surfaces and the cylinder gas.

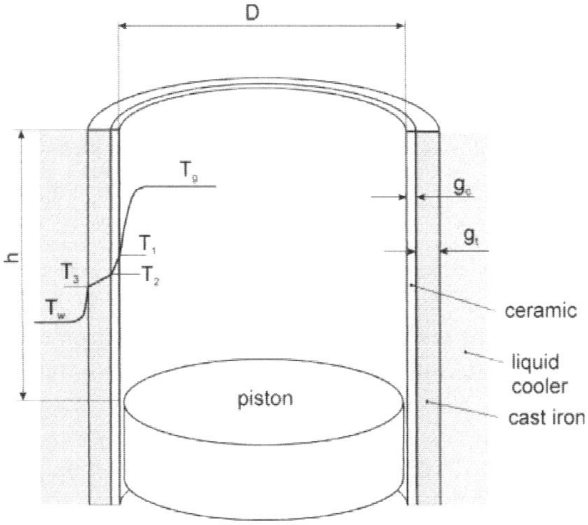

Figure 1. Schematic temperature course during heat transfer in IC Engines with ceramic coating on the cylinder liner.

In order to decrease the heat flow stream from the cylinder wall to water it is considered application of the special coating with the material of lower conductivity, usually a ceramic coating. The surface heat flow rate $\dot{q}$ through the wall with thickness l and wall conductivity λ can be expressed as follows:

$$\dot{q} = -\lambda \frac{dT}{dl} \quad [\mathrm{W/m^2}] \tag{2}$$

For constant wall conductivity λ this expression can be written as a function of the temperature difference ΔT, thickness g of the wall or the heat transfer resistance R_λ:

$$\dot{q} = -\lambda \frac{\Delta T}{g} = -\frac{\Delta T}{R_\lambda} \tag{3}$$

Real surface heat stream from the gas inside the cylinder to the liquid cooling medium through the wall depends on thermal resistance during convection and conductivity of the wall layers. Total heat flow resistance for flat walls can be written as follows:

$$R_t = R_{\alpha 1} + R_{\lambda 1} + R_{\lambda 2} + R_{\alpha 2} = \frac{1}{\alpha_1} + \frac{g_c}{\lambda_1} + \frac{g_t}{\lambda_2} + \frac{1}{\alpha_2} \tag{4}$$

where: $R_{\alpha 1}$ - convection resistance in the cylinder,

$R_{\alpha 2}$ - convection resistance in the cooling medium,

$R_{\lambda 1}$ - conductivity resistance in the ceramic layer,

$R_{\lambda 2}$ - conductivity resistance in the cast iron layer,

α_1, α_2 - convection coefficient of gas near the walls and cooling medium, respectively,

λ_1, λ_2 - conductivity coefficient of the ceramic and cast iron layers,

g_c, g_t - thickness of ceramic coating and cylinder sleeve, respectively.

In reality heat flow occurs through the cylindrical walls and for this case the flow resistance is not a linear function of wall thickness but for multi-layer walls is expressed as follows:

$$R_\lambda = \frac{r_i}{\lambda_i} \ln \frac{r_{i+1}}{r_i} \tag{5}$$

where: r_i, r_{i+1} - inner and outer radius of the layer walls, respectively,

λ_i - conductivity of layer material.

Convection coefficient α_1 inside the cylinder (heat flow from the gas to the inner cylinder wall) can be calculated using Woschni experimental formulas for intake, compression, expansion and exhaust processes given in different sources [1, 3]. On the other hand inflow of the heat to the liquid medium is a linear function of convection coefficient α_2 and can be obtained from Nusselt number, which is a function of Reynolds and Prandtl numbers. However, the simple expression as a function of water velocity given by Heywood [1] was used. Convection coefficient for heat outflow from the bottom surface of the piston head is calculated using Grashoff and Prandtl numbers for laminar flow from the crankcase side. Temperature of the gas inside the crankcase is almost constant (about 350 K) and gas flow near the piston walls is laminar.
Knowing all heat transfer resistances from the cylinder gas to the external cooling medium the surface heat flow rate $\dot{q}$ (without radiation) can be determined from (1). The changeable temperature of the cylinder gas is obtained from simulation process of engine work with taking into account unsteady gas flow in inlet and outlet systems. At such assumptions temperature of internal cylinder wall T_1 and temperature T_2 can be determined from the following expression:

$$T_1 = T_g - \frac{\dot{q}}{\alpha_1} \qquad\qquad T_3 = T_w + \frac{\dot{q}}{\alpha_2}\frac{r_1}{r_3} \tag{6}$$

where: $r_1 = 0.5D$ and $r_3 = r_1 + g_c + g_t$

Radiation heat between cylinder gas containing also soot and walls is determined from the following formula given by Horlock and Winterbone [2]:

$$\dot{q}_r = \frac{F_b}{F_1}\sigma_0\varepsilon_z\left[T_g^{\ 4} - T_1^4\right] \tag{7}$$

where $\sigma_0 = 5.67$ is Boltzman constant, ε_z is emissivity of central body, F_b and F_1 are central and surroundings surfaces, respectively.

3. Research object

The simulation research was carried out on SI four stroke engine SUZUKI DR-Z400S, which geometrical parameters are presented in Table 1.

Table 1. Geometry of simulation engine based on SUZUKI DR-Z400S

Geometrical parameters	Data
Cylinder diameter	90 mm
Stroke	62,6 mm
Compression ratio	8
Number of valves	2+2
Opening and closing of inlet valve	20° CA BTDC/ 45° CA ABDC
Opening and closing of exhaust valve	40° CA BBDC / 19° CA ATDC

In order to perform the simulation test of influence of ceramic coating on the combustion chamber surfaces on engine work parameters the calculation computer program was made by author, which takes into account all thermodynamic processes taking place during one engine cycle. The combustion processes was modelled by using Viebe function and gas flow in engine ducts was treated as an unsteady flow with taking into account the waveform of pressure. For comparison of all considered cases of the tests the coefficients in Viebe function were assumed with the same value. However heat release depends also on there initial temperature of the charge in the cylinder. Validation of the assumed combustion model for all calculation models is required in the future. In this work the Viebe coefficients a and m were taken from the experimental work done before on the conventional engine Suzuki by Soczowka [12] with using of the following exponential Vibe formula:

$$x_b = 1 - \exp\left[a \left(\frac{\varphi - \varphi_b}{\Delta\varphi_b} \right)^m \right] \tag{8}$$

where: x_b - fuel burnt ratio,

φ - current angle of crank rotation,

φ_b - crank angle of beginning of fuel burning,

$\Delta\varphi_b$ - duration of crank angle rotation of full combustion process.

The following values of coefficients were taken from calibration of the standard engine: a= -6.908 and m=3. At higher temperature at the end of compression process in the adiabatic engine and semi-adiabatic engine one can expect knocking problem. The knocking phenomenon can be reduced by decreasing of compression ratio, however, which leads to lower engine total efficiency.

The purpose of this work a special attention was paid to heat transfer from the cylinder gas to the water as a cooling medium with constant temperature 357 K flowing around the cylinder with velocity 0.8 m/s. Input data containing many constant parameters were read by the program from the text table. The results of calculation were transferred to the output files giving many thermodynamic parameters as a function of crank angle rotation for given engine speed, constant ignition point and combustion duration. Convergence calculations were provided by the iterative cycle work until the indicated pressure difference achieved an assumed value. The calculations were made for changing a thickness of the ceramic coating with constant low conductivity 5 W/m and for cylinder sleeve made from cast iron with constant conductivity 45 W/m. The cylinder head and piston were treated as an aluminium alloy with conductivity equal 180 W/m. For comparison of calculation results only changing of ceramic coating was considered. The purpose of the calculations was to determine the effect of adiabatic process on engine work parameters, especially total engine efficiency. Three cases of four stroke engine were compared: standard engine (without ceramic coating), full adiabatic engine (without cooling effect) and semi adiabatic (with changeable ceramic coating).

4. Influence of ceramic thickness on engine work parameters

The first stage of calculation of was determination of influence of thickness of the ceramic coating on four-stroke engine work parameters, which were carried out at 3000 rpm and air excess ratio λ=1 for whole range of thickness from 0 to 2 mm. The thickness equal 0 mm of the coating means that only the cast iron cylinder sleeve is the thermal barrier for conductivity. Figure 2 presents variation of mean indicated pressure and specific fuel consumption as a function of thickness of ceramic coating on all surfaces of combustion chamber (cylinder walls, cylinder head and piston head), wherein conductivity of the ceramic coating were constant and amount 5 W/m. It was found increasing of mean indicated pressure only 0.2 bar (near 2%) with increasing of the ceramic coating thickness from 0 to 2 mm. On the other hand increasing of this value causes a decreasing of specific fuel consumption about 3%. Figure 3 presents courses of cooling heat losses, exhaust heat losses and engine total efficiency for engine with combustion chamber covered with different thickness of the ceramic coating at constant rotational speed 3000 rpm.

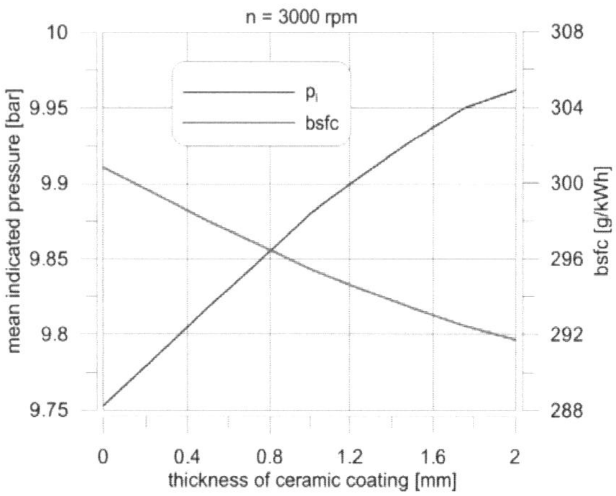

Figure 2. Influence of ceramic coating thickness on mean indicated pressure and specific fuel consumption.

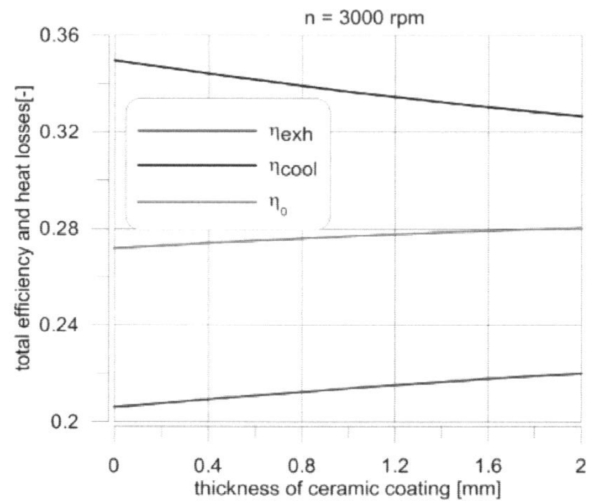

Figure 3. Exhaust and cooling heat losses and total efficiency versus thickness of ceramic coating.

After increasing of ceramic coating thickness the heat transfer to the cooling liquid is decreased but at the expense of increasing the amount of the exhaust heat. On the other hand the increasing of total efficiency is seen from 27% to 28%. It was observed lowering of engine power for three considered cases at 3000 rpm (Fig.4). In standard engine (without ceramic coating) the cooling heat losses amount about 35% of total heat released from fuel during combustion process.

5. Adiabatic engine versus standard engine

Further simulation studies were carried out on three types engines: 1) fully adiabatic, 2) with thickness of the ceramic coating equal 1 mm and 3) standard engine at wide throttle opening (λ=1) and different rotational speed. Figure 4 presents effective power of engines on speed characteristic.

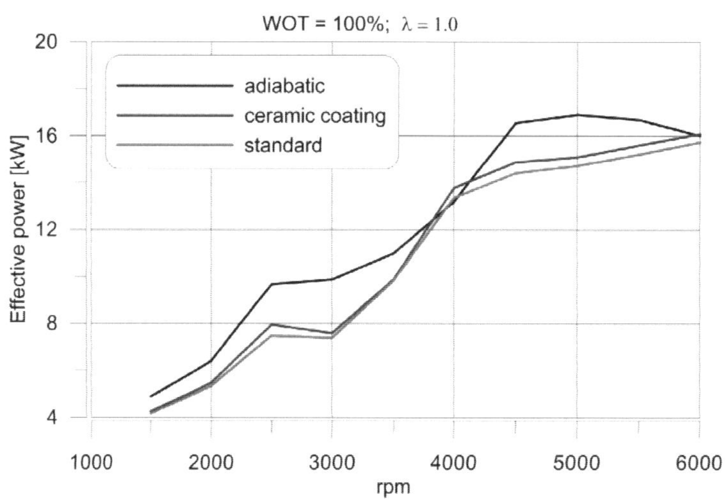

Figure 4. Comparison of effective power for three engines cases.

The adiabatic engine indicates higher power in whole range of rotational speed. The highest difference between standard and adiabatic engine amounts 2.5 kW at 3000 and 4500 rpm (increase about 33% and 17%, respectively). The cooling heat losses can be reduced by covering of the combustion chamber with layer of material of low conductivity. It was found a decreasing of cooling heat losses with increasing of rotational speed (Fig. 5). In the adiabatic engine these losses theoretically amount zero.

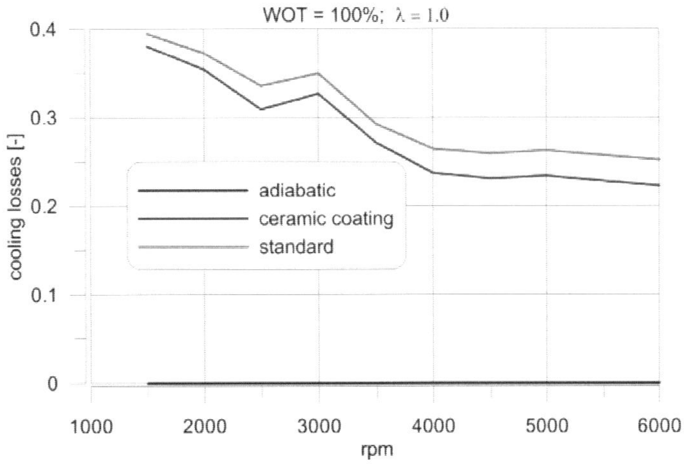

Figure 5. Comparing of cooling heat losses for three engine types.

In the adiabatic engine these losses theoretically amount zero. The engine with ceramic coating of the combustion chamber indicates decreasing of the cooling heat about 3 %. It is not as much as was expected. The decreasing of the cooling heat influences on higher amount of the exhaust heat at the increase of rotational speed. In the engine with full adiabatic processes of the working cycle the exhaust heat losses amount almost 50% of total delivered energy from fuel (Fig. 6). In the standard engine these losses amount only 28% at 4000 rpm. The engine with semi-adiabatic processes (ceramic thickness 1 mm) indicates about 8% increase of energy in the exhaust system. Increasing of exhaust energy can be used in cogeneration systems but in the engine the first step is using this energy for driving the turbocharger for obtaining a higher engine power.

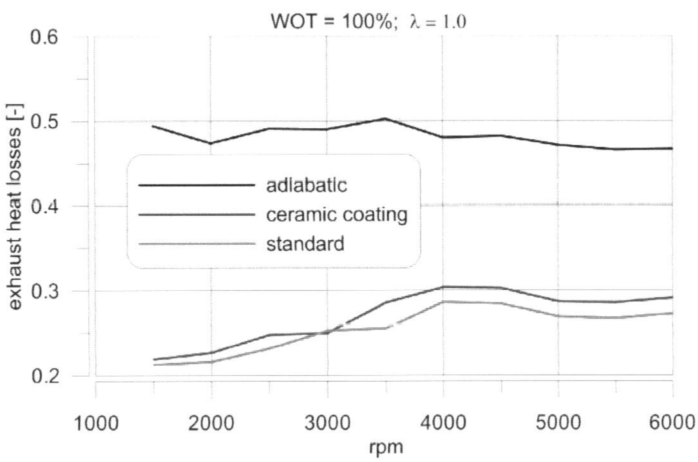

Figure 6. Comparing of exhaust heat losses for three engine cases.

The engine work is assessed by determination of total efficiency. The computer program calculates the mechanical work and total energy delivered with gasoline by knowing its caloric value. Figure 7 presents courses of total efficiency for three considered engine cases. The adiabatic engine has a greatest efficiency at lower rotational speeds compared to the standard engine approximately 35% but at higher rotational speeds only 13%. Covering of the combustion chamber surfaces with a thin ceramic coating (1 mm) will increase the total efficiency about 4% in comparison to the standard engine. The adiabatic processes in SI engine cause an increase of internal energy, because there is any

IOP Conf. Series: Materials Science and Engineering **148** (2016) 012080 doi:10.1088/1757-899X/148/1/012080

heat transfer to the cooling medium. In such engine temperature of the cylinder gas reaches value about 3000 K, which is about 250 k higher than in the standard engine (Fig. 8).

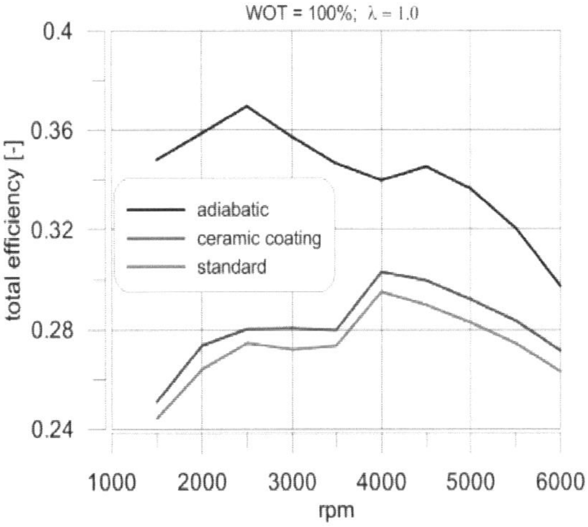

Figure 7. Variation of total efficiency for three engines cases.

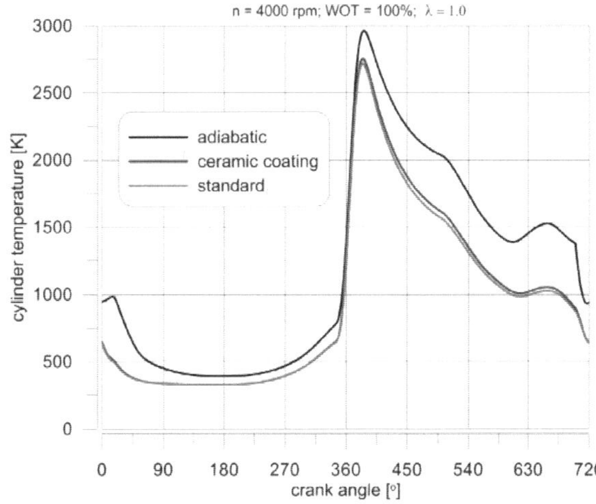

Figure 8. Comparison of cylinder gas temperature for three engine cases at 4000 rpm and WOT.

During all engine processes this temperature is always higher even during intake stroke. This phenomenon has a big influence on thermal stresses of all engine elements, especially on the exhaust valves. Such high exhaust temperature is too high for all turbocharger's elements.

6. Summary

The presented work shows a certain aspects of applying the ceramic coating for adiabatic processes in SI engines, which was considered mostly in diesel engines in the past. On the basis of heat transfer mathematical model the simulation of work cycle of three SI engine cases was made for different initial conditions. One of the effects of the work is elaboration of computer program for analyzing the adiabatic engine work. The achieved results of the numerical studies give the following conclusions:

1. The full adiabatic engine without heat transfer to the cooling medium indicates higher engine power than standard engine from 17% at higher rotational speed to 33% at lower rotational speed.
2. Increasing of the ceramic thickness coating on the combustion chamber walls gives a linear increase of mean indicated pressure, but the increment is only 2% at thickness 2 mm. Technology of ceramic covering is expensive and so low yield energy does not compensate the cost of this technology.
3. The adiabatic engines influences on higher temperature of the cylinder charge and high exhaust heat losses. However the exhaust energy can be used to drive a turbocharger and also utilized in cogeneration systems.
4. Almost a half of delivered energy from fuel is discharged with exhaust gases for the adiabatic engine, when in semi-adiabatic only 28% of total energy is discharged to the outflow.
5. The calculations indicated a lowering of cooling heat losses with increasing of rotational speed both for standard and semi-adiabatic engine.
6. Total yield of energy is possible only for full adiabatic engine with ceramic coating with a conductivity close zero.

Future work concerning the adiabatic SI engine should be conducted on experimental test in order to calibrate the parameters of combustion process especially to determine the Vibe coefficients and possibility of knock combustion due to higher temperature at the end of compression process.

References

[1] Heywood J B 1988 *Internal Combustion Engines Fundamentals* (New York: Mc Graw-Hill)
[2] Szargut J 1975 Termodynamika (Warszawa: PWN)
[3] Horlock J H and Winterbone D E 1986 The thermodynamics and Gas Dynamice of Internal-Combustion Engines Vol. II (Oxford: Clarendon Press)
[4] Kawamura H and Akama M 2003 *Development of an Adiabatic Engine Installed Energy Recover Turbines and Converters of CNG Fuel* SAE 2003-01-2265
[5] Khoshravan E 1991 *Heat transfer Studies in an Adiabatic Diesel Engines* SAE 912502
[6] Beg E A Rahman M S Bose P K and Ghosh B B 2002 *Performance Studies on a Semi-Adiabatic Diesel Engine Using Vegetable Oil as Fuel* SAE 2002-01-2692
[7] Gosai D C and Nagarsheth H J *Performance and Exhaust Emission Studies of an Adiabatic Engine with Optimum Cooling* Procedia Technology 14 (2014) 413-421
[8] Reddy S S K and Pandurangadu V *Theoretical & Experimental Investigations on the Performance of a four stroke adiabatic DI diesel engine* Int. J. of Adv. In Eng. & Tech. Vol. 6
[9] Kamo R 1987 *The adiabatic engine for advanced automotive applications* (New York: Springer Science+Business Media)
[10] Kamo R and Bryzik W 2008 *Cummins/TACOM advanced adiabatic engine* Proc. Of the 12th Autom. Mat. Conf. Eng. and Sc. Vol 5 Issue 5/6 doi: 10.1002/9780470320211.ch4
[11] Ciniviz M Salman M S Canli E Köse H and Solmaz Ö 2012 *Ceramic Coating Applications and Research Fields for Internal Combustion Engines* (http://cdn.intechopen.com/pdfs-wm/29756.pdf InTech Europe)
[12] Soczowka J 2008 *Adaptive system of controlling the ratio of compression with spark ignition engine* Doctor Thesis Cracow Univ. of Techn.

Influence of increased exhaust gas recirculation ratio on the thermodynamic processes in CI DI engine

I Pielecha, M Skowron, W Bueschke, W Cieślik and K Wisłocki

Institute of Combustion Engines and Transport, Poznan University of Technology, Piotrowo 3 street, Poznan, Poland

ireneusz.pielecha@put.poznan.pl

Abstract. The aim of the conducted research was the thermodynamic identification of combustion process involving up to 50–60% recirculated exhaust gasses in compression ignition engine. The values of the respective indicators obtained for the high share of exhaust gasses in the cylinder were compared to the values obtained in the engine working without EGR. The research was conducted on the single-cylinder AVL 5804 engine equipped with recirculated gas cooling system. The conditions of combustion process were determined using indicators of engine operation and measurements of fast-varying cylinder pressure. The evaluation of the influence of different share of exhaust gases in the cylinder on the combustion course and heat release was conducted. As a result of the conducted research the possibility of utilizing the high share of exhaust gasses (50–60%) with simultaneous ecological benefits, not only in relation to nitrogen oxides, was demonstrated.

1. Introduction and analysis of the literature

Requirements for meeting emission standards give rise to the need to seek new structural solutions in the construction of modern combustion engines and engine exhaust aftertreatment systems. Structural methods to meet the stringent energy and emissions requirements (with regard to the emission of CO_2) by propulsion units are known as the concept of downsizing [1] or hybridization of propulsion systems [2]. Among these solutions the most important is the role of the modern combustion and exhaust gas recirculation systems [3–5]. Exhaust gas recirculation, which is supplying of exhaust gases into the combustion chamber, is one of the methods to reduce nitrogen oxides emissions, especially for the medium loads. In certain scopes of engine operation it also allows to reduce fuel consumption without significant deterioration of the operating indicators. The efficiency of the exhaust gas recirculation systems, however, depends to a large extent on the way they are accomplished. Supplying a given amount of exhaust gases into the operating cylinder can be achieved in two ways:

- using the internal recirculation (IGR—internal gas residual), which involves retaining in the cylinder a part of the exhaust gases from the previous cycle by appropriate control of variable valve timing;
- by the use of an external recirculation system (EGR—exhaust gas recirculation), returning the part of the exhaust gases leaving the cylinder (cylinders) from the previous engine cycle.

Despite numerous possibilities of EGR systems application, still a major concern is the method of their accomplishment and the control over the amount of recirculated exhaust gases. Figure 1 shows the division of the cylinder exhaust gas supply systems into internal systems of exhaust gas recirculation and external systems supplying the exhaust gases into the cylinder. The former can be accomplished in two ways: by variable valve timing and by variable valve lift. Both methods of regulation involve retaining in the cylinder a part of the exhaust gas from the previous engine operation cycle. The external exhaust gas recirculation systems can be divided into high-pressure (1, figure 1), low-pressure (2) and mixed systems. The external exhaust gas recirculation can take place before the exhaust aftertreatment system—by the so-called "dirty EGR" (2a on figure 1) and after the aftertreatment system—by the so-called "clean EGR" (2b on figure 1).

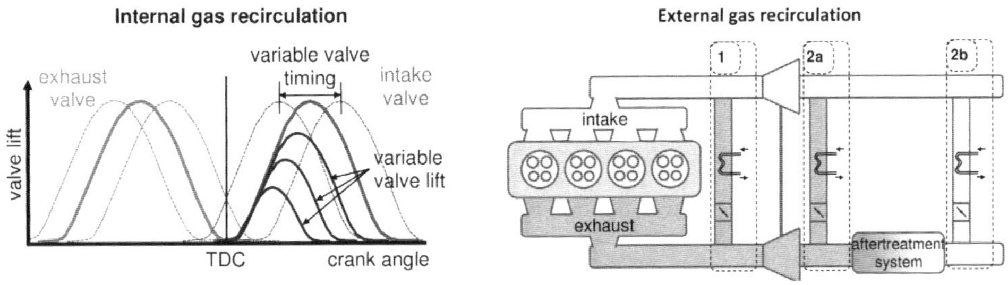

Figure 1. Systematics of the cylinder exhaust gas supply systems [3, 4, 6].

From the analysis of literature [7–10], it ensures that one of the most beneficial solutions for controlling thermodynamic processes in CI engines is low-temperature combustion (LTC), which causes the reduction of emissions of nitrogen oxides and particulate matter (soot). Such a process is accomplished with participation of significant share of recirculated exhaust gases reducing the maximum combustion temperature. This process can be supported by fuel dose splitting and a reduction of the compression ratio, although in the LTC combustion system it is also possible to use single doses of injected fuel. Obtaining the LTC type combustion requires the appropriate control of the injection time in the close to the piston top dead centre (TDC) while applying a large quantity of recirculated gases (50–70%) [11, 12]. Such a large share of the exhaust gases in a charge, what also mean significant share of chemical inert molecules instead active molecules of oxygen, leads to increased self-ignition delays (due to the longer time of mixing), resulting in an increased homogeneity of the mixture.

An important aspect of the exhaust gas recirculation systems use is their impact on the volumetric efficiency of fresh air in the cylinders. As it was indicated in [13], that the introduction of 25% share of EGR reduces the value of fresh air volumetric efficiency by about 14%. Within certain scope of partial loads, an increase in thermal efficiency and reduction of specific fuel consumption were observed. The discussed decrease in the volumetric efficiency of fresh air can be compensated by changing the supercharging parameters, as a result of which there will be a further increase in thermal efficiency. The vast majority of published research findings on the EGR systems include works with the maximum participation of the exhaust gases of 25–30%, which raised the need to complement our knowledge on the matter in terms of a larger share of the recirculated exhaust gases.

The studies conducted up to now by [14, 15] as well as by Knight [16] indicate only the influence of EGR amount on the exhaust gas emission. However, the influence of the recirculated exhaust gases on the thermodynamic parameters of the combustion engine in selected operating points is omitted. In the cited works the research involved one selected point of engine operation (n = 1500 rpm, P_{inj} = 85 MPa) taking into account, however, its variable supercharging. It was proved that for large EGR share of 50–60% it is possible to obtain reduced emissions of NO_x and soot; those conclusions confirmed the expectations of the authors of these works.

The thermodynamic analysis of the combustion process was conducted only in few studies limited to the systems operating with the share of the recirculated exhaust gases of about 36% [17]. The authors of this study conducted the research on the thermodynamic characteristics of the combustion process in the CI engine with significant share of recirculated exhaust gases of up to 60% [18].

2. The research problem

The basic research problem formulated by the authors was the explanation of the significance of the influence of the recirculated exhaust gases amount on the change in the basic parameters and indicators of the thermodynamic engine cycle. It was decided to demonstrate the extent to which the increase in the share of exhaust gases in the closed cylinder affects the variation of the compression-ignition engine operation on the basis of fluctuations in the charge pressure during combustion, the variability of mean indicated pressure and temperature of the charge and exhaust gas. Also the changes in the heat release ratio and the angular position of the stipulated centre of combustion (CoC, position of the point of 50% heat release) were assessed. It was also expected that the possibility and the scope of reduction in the nitrogen oxides and particulate matter (soot) emissions for the low temperature combustion will be explained.

3. Research methodology

3.1. Research method

In order to explain the issues and research problems defined in section 2 it was decided that the method of experimental test bench would be utilized. To this purpose, the engine test bench with electrical dynamometer brake was appropriately adapted to enable maintaining the constant engine speed. On the test bench was installed a tested object which was a direct injection compression-ignition engine.

3.2. Research object

The study was carried out on a single-cylinder AVL 5804 test engine equipped with the system of cooling recirculated exhaust gases. Engine technical specification is shown in table 1. The asynchronous electric dynamometer brake AMK ASYN of DW13-170-4-AOW type was used. The EGR system was equipped with two solenoid valves controlled by the pulse-width modulation (PWM), which enabled repeatable setting of the valves position on the air inlet and exhaust gas outlet. The system of two valves made it possible to obtain the exhaust recirculation ratio of up to 60%. The amount of recirculated exhaust gases was determined by volume on basis of carbon dioxide volumetric concentration measurement in the air in the intake channel and in the engine outlet according to the equation:

$$EGR = (CO_{2\text{-ex}} - CO_{2\text{-in}})\, CO_{2\text{-ex}}^{-1} \cdot 100\% \quad (1)$$

where: $CO_{2\text{-ex}}$ – the percentage of carbon dioxide in exhaust gases, $CO_{2\text{-in}}$ – percentage of CO_2 in the engine outlet channel.

Table 1. Characteristics of the AVL 5804 test bench.

Parameter	Value
Engine capacity	510.7 cm^3
Piston stroke	90 mm
Cylinder diameter	85 mm
Compression ratio	16.2
Injector type	piezoelectric, common rail, 8-hole, d = 0.117 mm
Cylinder head gasket thickness	2 mm (required to obtain expected compression ratio)

3.3. Measuring equipment

The test stand (figure 2) was equipped with control and measuring apparatus including: (a) a system for fast-varying processes measurement AVL IndiCom 621 enabling measurement of pressure in a cylinder (P_{cyl}) with the use of pressure sensor AVL GH14D with the measuring range 0–250 bar and sensitivity of 18.84 pC/bar, (b) system for data acquisition AVL Concerto, (c) system for control of the fuel injection process enabling the control of the injection time and injection angle with resolution of $\Delta\alpha$ = 0.5 CA and fuel pressure up to 200 MPa developed by Mechatronika Poland, (d) the system controlling settings of the solenoid valves for exhaust gas recirculation with the use of PWM signals with resolution of 1% developed by Mechatronika, e) the system of measuring gaseous components—Horiba Mexa 7100D for type approval testing, measuring concentration of CO, THC (HFID), NO_x, CO_2, (f) the system measuring particle matter (soot)—Micro Soot Sensor by AVL.

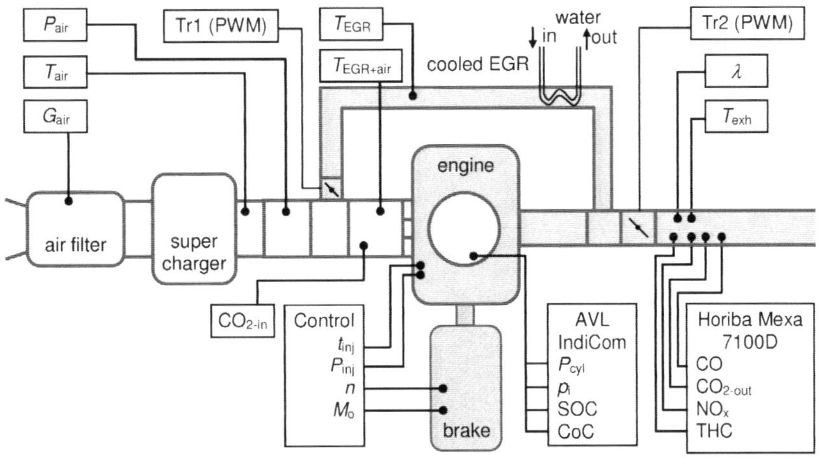

Figure 2. Diagram of the test bench with testing apparatus and characteristic parameters of the process.

3.4. The methodology and scope of the research

The tests were conducted for two measurement points for which, at varying amount of recirculated exhaust gases, were defined: (1) indicators of the combustion engine operation, (2) thermodynamic indicators of the combustion process, (3) emissions of toxic components of exhaust gases.

The indicators of combustion engine operation included (figure 3): (1) maximum pressure in the cylinder—P_{mx} and the phase (angle of CA) of its occurrence AP_{mx}, (2) mean indicated pressure—designated as p_i; (3) the maximum pressure rise after self-ignition R_{mx} and the phase (angle of CA) of its occurrence AR_{mx}.

There were also determined: (1) exhaust gas temperature—T_{exh}, (2) exhaust gas temperature at the valve of recirculation before the inlet channel—T_{EGR}, (3) temperature of the charge at the inlet (air and exhaust gas mixture)—$T_{EGR+air}$.

Thermodynamic indicators of engine operation included: (1) the start of combustion (SOC)—defined as the beginning of the positive heat release value during compression, (2) angle of combustion of 5% of a fuel dose—MBF5 (Mass Burnt Fuel), (3) angle of occurrence of the centre of combustion (CoC) which is combustion of 50% of fuel dose—MBF50, (4) angle of the end of combustion (EoC) defined as combustion of 90% of fuel dose—MBF90.

Specific emission of fuel components (in g kWh^{-1}) was defined, with the use of engine operation indicators, on the basis of concentration measurements of CO, HC, NO_x in ppm and carbon mass C in mg m^{-3} as:

$$e_C = C\,\rho_{exh}\,(G_e + G_{air})\,p_i^{-1} \tag{2}$$

$$e_{CO} = 0.000966\ CO\,(G_e + G_{air}))\,p_i^{-1} \tag{3}$$

$$e_{HC} = 0.001578\ HC\,(G_e + G_{air}))\,p_i^{-1} \tag{4}$$

$$e_{NOx} = 0.001519\ NO_x\,(G_e + G_{air}))\,p_i^{-1} \tag{5}$$

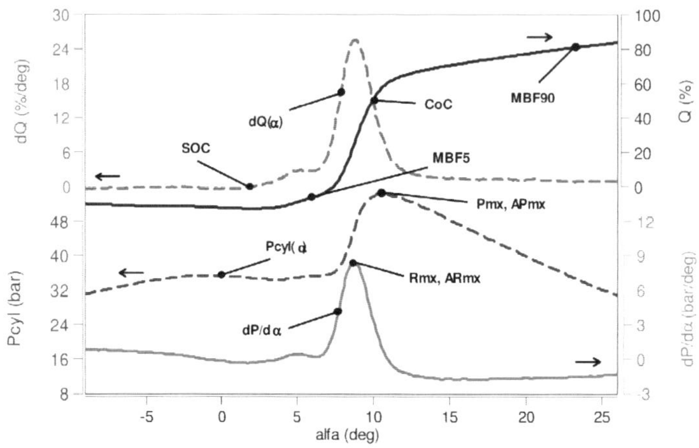

Figure 3. Selection of indicators analysed during tests on diesel fuel combustion with significant share of exhaust gases.

Using the results of indicator tests determined were also the rate of heat release – $dQ/d\alpha$ and accumulated heat Q as:

$$dQ/da = \frac{\kappa}{\kappa - 1}\left(\frac{P_\alpha + P_{\alpha+1}}{2}\right)(V_{\alpha+1} - V_\alpha) + \frac{1}{\kappa - 1}\left(\frac{V_\alpha + V_{\alpha+1}}{2}\right)(P_{\alpha+1} - P_\alpha) \tag{6}$$

where the indexes α and $\alpha + 1$ are the CA current and the subsequent values of pressure in cylinder (P) or their corresponding volume of cylinder (V).

The tests were carried out for two values of engine speed: 1000 rpm and 1500 rpm for different diesel fuel doses: $q_o = 12$ and $q_o = 23$ mg inj^{-1}. The tests were carried out so as to maintain relatively constant angle of combustion of 50% of fuel dose – the angle is called the centre of combustion (CoC). In the analysed operating conditions this angle amounted to, respectively: 9 and 11° of CA after TDC. Without taking into account this condition, the fuel burning would be much delayed and the correct operation of the engine would be impossible. Conditions of conducting measurements are presented in table 2.

Table 2. Measuring points – the research plan.

No.	n (rpm)	q_o (mg inj^{-1})	CoC (deg aTDC)	P_{inj} (MPa)	P_{in} (hPa)	Share of exhaust gases (%)
1	1000	12	9	44	1024	0; 9; 19; 30; 45; 51
2	1500	23	11	120	1219	0; 9; 21; 30; 42; 50; 60

4. Estimation of indicators of engine operation for different shares of exhaust gases

4.1. General indicators

Before proceeding to data analysis, the characteristics of pressure in cylinder were smoothed with the use of low-pass filter of 4 kHz. Supplying exhaust gases for the subsequent cycle of engine operation has dominating influence on the composition of the exhaust gases and indicators of engine operation. The analysis of the tests results was conducted in terms of general, thermodynamic and ecological indicators.

The dependency determining the influence of the amount of recirculated exhaust gases in the engine cylinder on the general indicators of its operation is shown in figure 4. For smaller fuel dose ($q_o = 12$ mg inj^{-1}) the highest value of the maximum combustion pressure P_{mx} occurs at 20% share of exhaust gases and amounts to 54.6 bar. Introduction of 10% share of EGR caused a decrease of P_{mx} to 53.2 bar (a change by approx. 3%), without any influence on the value of the mean indicated pressure—p_i. Its higher value increases from 2.9 bar to 3.7 bar along with the increasing share of recirculated exhaust gases (up to the value of 50%). It means a change of the p_i value by more than 25%. It can be concluded, therefore, that an increase in the share of exhaust gases to 50% causes an increase in the mean indicated pressure by about 25%.

(a) (b)

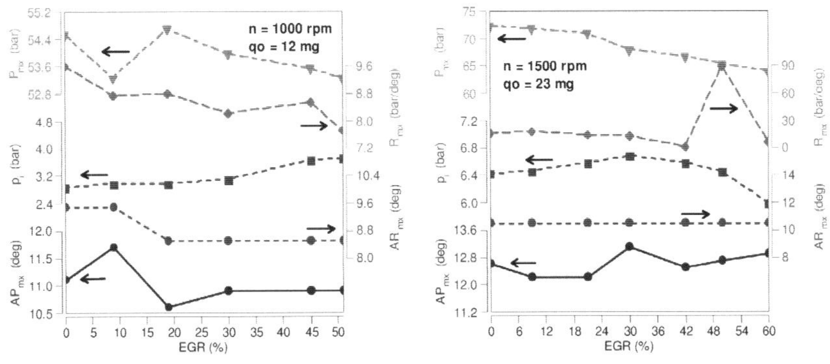

Figure 4. The main indicators of engine operation in relation to the share of recirculated exhaust gases in the engine: (a) at n = 1000 rpm and q_o = 12 mg inj^{-1}, (b) at n = 1500 rpm and q_o = 23 mg inj^{-1}.

At the same time, along with the increase of the EGR share, the pressure increase rate decreases which results in the decrease of the maximum pressure to the value of 53.4 bar for the share of recirculated exhaust gases ranging from 20% to 50%. Simultaneously the changes of the occurrence angle of the maximum pressure AP_{mx} and the angle of the maximum pressure increase rate AR_{mx} were observed for the crankshaft speed 1.5 deg.

Use of a larger dose and higher engine speed result in a change in the relationship between these parameters. Along with the increase of the EGR share, the value of the maximum combustion pressure P_{mx} and the value of the pressure increment R_{mx} decrease. The highest value of the mean indicated pressure of 6.7 bar and the occurrence angle of the maximum pressure AP_{mx} are observed for 30% share of EGR. A change in the share of recirculated exhaust gases does not affect the angle of the maximum increase rate occurrence of the pressure AR_{mx} and amounts to 10.5 deg CA.

The results of the analysis concerning the impact of the recirculated exhaust gases share on the excess air ratio and exhaust gas temperature are shown in figure 5. With the increasing share of exhaust gases in the cylinder, the excess air ratio decreases irrespective of the applied fuel dose. During combustion of a large fuel dose the global excess air ratio achieved is below 1.0. These are conditions rendering combustion of diesel oil difficult.

During a smaller fuel dose combustion, the exhaust gases temperature decreases and the temperature of recirculated exhaust gases increases along with the increase of their share in the cylinder charge. This increase is associated with an increased flow rate of the exhaust gases. An increased share of exhaust gases with higher temperature results in an increase of the mixture of air and exhaust gases temperature ($T_{EGR+air}$) irrespective of the applied fuel dose. The highest exhaust gases temperature occurs at 30% share of EGR. For this point of operation was obtained the largest value of the mean indicated pressure, which is shown in figure 5b.

(a) (b)

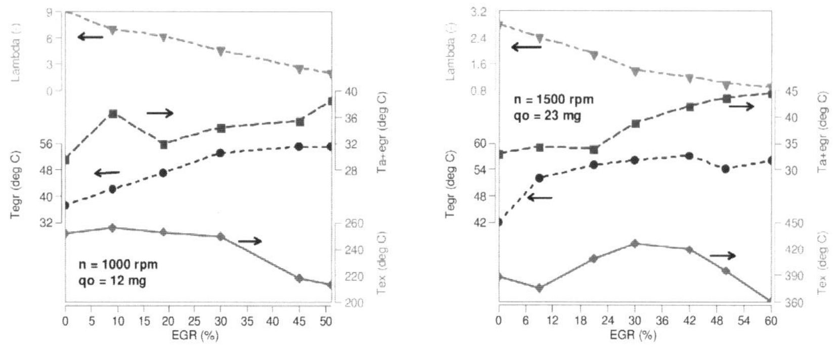

Figure 5. Changes in the excess air ratio and temperatures of: exhaust gases, recirculated exhaust gases and mixture of air and exhaust gases in relation to the share of recirculated exhaust gases in the engine: (a) at n = 1000 rpm and q_o = 12 mg inj^{-1}, (b) at n = 1500 rpm and q_o = 23 mg inj^{-1}.

4.2. Thermodynamic indicators

The research on combustion process with significant share of recirculated exhaust gases was conducted mainly based on the analysis of thermodynamic parameters of this process. The changes in those parameters during engine operation for two different fuel doses are presented in figure 6 where the angle dependencies of the start (MBF5), centre (MBF50) and end of combustion (MBF90) and the start of ignition (SOC) are shown. Independent of the fuel dose amount , an increase in the share of recirculated exhaust gases results in the delayed start of combustion (SOC). The value (angle) of 50% combustion of fuel dose (CoC) was maintained on the assumed level according to the fuel dose: 10°CA (for reduced dose) and 12°CA for combustion of a larger dose. The variation for both strategies was below 1°CA which should be considered an appropriate value. Maintaining such characteristics of the combustion process for an increased share of the recirculated exhaust gases requires acceleration of the start of combustion, which causes elongation of the whole process. Such dependencies were confirmed for both fuel doses.

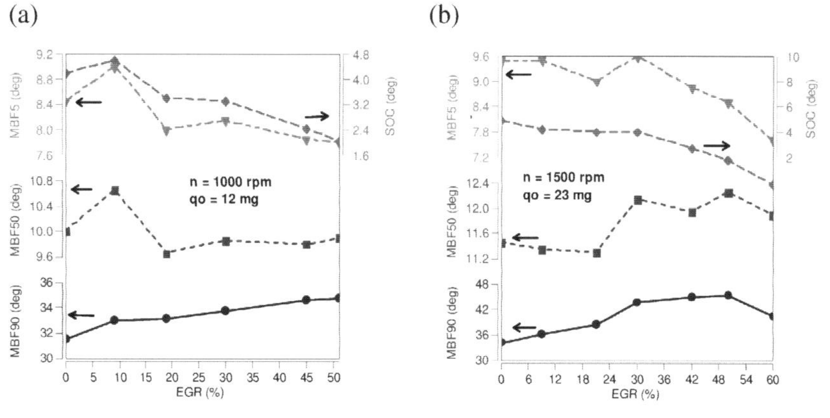

Figure 6. Thermodynamic indicators of engine operation in relation to the recirculated exhaust gases share: (a) at n = 1000 rpm and q_o = 12 mg inj^{-1}, (b) at n = 1500 rpm and q_o = 23 mg inj^{-1}.

During the tests the analysis of the cylinder pressure characteristics was conducted, which is presented in figure 7. The use of recirculated exhaust gases for engine speed of n = 1000 rpm causes initially (for

10% EGR) significant delays in pressure increase after self-ignition. Further increase of the exhaust gases share in the cylinder only slightly deviates from the characteristics of pressure at the lack of exhaust gases. The analysis of this operation indicates point insignificant decrease in the maximum pressure in the cylinder for increased exhaust gases share (which was also confirmed in figure 4a). The second measurement at the engine speed of n = 1500 rpm, depending on the exhaust gases share in the cylinder, was characterised by a higher variability compared to the first measuring point. Here are observed both, changes in the pressure value before combustion and maximum pressure changes. An increase in the exhaust gases share resulted in prolonging the start of combustion, which corresponds to the almost flat cylinder pressure characteristics for the range from 4–8° CA after TDC. The maximum value change of P_{cyl} at EGR = 0% and 60% achieved the value of 7 bar.

(a) (b)

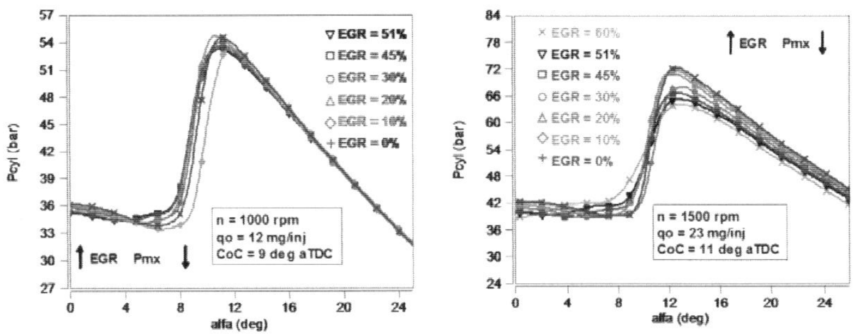

Figure 7. Analysis of the maximum cylinder pressure changes for engine speed of: (a) n = 1000 rpm, (b) n = 1500 rpm.

Introduction of exhaust gases recirculation resulted in an increase in the heat release rate in the cylinder (figure 8). Increase in the share of exhaust gases during small fuel dose combustion leads to insignificant limitation in the heat release rate for small and medium exhaust gases share. During fuel combustion with maximum exhaust gases share the limitation reaches the value of 20%. The occurrence angle of the maximum heat release rate does not undergo significant changes and amounts to 8.5–10° CA after TDC.

(a) (b)

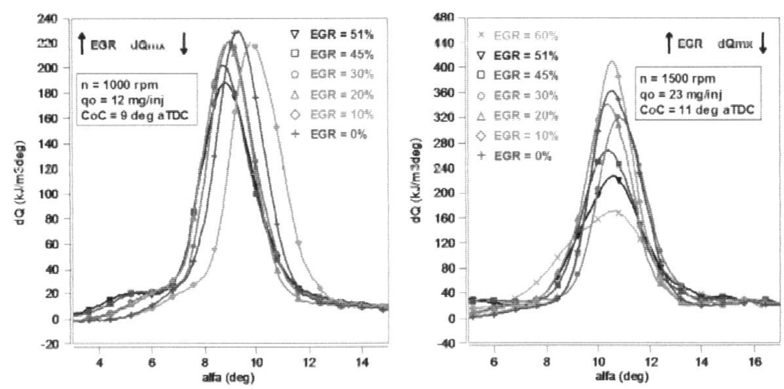

Figure 8. Analysis of the heat release rate values for the speeds of: (a) n = 1000 rpm, (b) n = 1500 rpm.

Combustion of a large fuel dose with a recirculated exhaust gases share leads to significant changes in the heat release rate. The changes in the heat release rate obtain values exceeding 50%. It means a two-fold increase in the values of changes compared to the small fuel dose combustion.

5. Summary

The conducted research confirmed that the variable amount of exhaust gases in the cylinder results in changes in the engine operation parameters with significant variation of the trends for particular components. It forces the necessity to find the optimum solution for a particular application.

A noticeable drop in the maximum pressure in cylinder for an increased exhaust gases share, especially during larger fuel dose combustion, does not result in a decrease in the mean indicated pressure. The P_{mx} parameter achieves maximum values at 20% share of exhaust gases, and the mean indicated pressure value for this range stays at the constant level compared to other measuring points. Only for the 30–45% share of exhaust gases an increase in this indicator might be observed. The exhaust gases share amount has no essential influence on acceleration or the time delay of the maximum pressure in cylinder occurrence in relation to the crankshaft angle (for the assumed value of CoC).

On the ecological analysis basis it might be concluded that combustion of fuel with a large share or recirculated exhaust gases is efficient only when the excess air ratio at large exhaust gases share (over 40%) is much higher than 1.0. Such a situation was obtained only for a small fuel dose combustion. During a large fuel dose combustion the increase in the engine supercharging value with air would be necessary.

Symbols and abbreviations

AP_{mx}	crankshaft angle of maximum cylinder pressure
AR_{mx}	crankshaft angle of maximum rise of cylinder pressure
CoC	centre of combustion (MBF50)
CoV	coefficient of variation
dQ	heat release rate
dQ_{mx}	maximum of heat release rate
EGR	exhaust gas recirculation
ID	ignition delay
IGR	internal gas residual
n	engine speed
p_i	indicated mean effective pressure
LTC	low temperature combustion
MBF	mass burnt fuel
P_{cyl}	cylinder pressure
P_{mx}	maximum cylinder pressure
P_{inj}	injection pressure
R_{mx}	maximum rise of cylinder pressure
SOC	start of combustion
T_r	throttle
t	time characteristic for the combustion phase
$T_{EGR+air}$	mixture temperature (air and EGR)
T_{EGR}	temperature of recirculated gases
T_{exh}	exhaust gases temperature
q_o	fuel dose
Q	heat release
λ	air excess ratio

References

[1] Pielecha I, Cieślik W, Borowski P, Czajka J and Bueschke W 2014 *Comb. Eng.* **159** p 12–25
[2] Pielecha I, Cieślik W, Borowski P, Czajka J and Bueschke W 2014 *Comb. Eng.* **158** p 23–35
[3] Berg T, Thiele O, Seefeldt S and Vanhaelst R 2013 *Mag. MTZ* **6** p 472–7

[4] Bourhis G, Chauvin J, Gautrot X and de Francqueville L 2013 *SAE Paper* 2013-01-0256
[5] Sroka ZJ 2012 *J. Therm. Anal. Calorim.* **110** p 51–8
[6] Cavina N, Siviero C and Suglia R 2004 *SAE Paper* 2004-01-2943
[7] Anselmi P, Kashdan J, Bression G, Ferrero-Lesur E, Thirouard B and Walter B 2010 *SAE Paper* 2010-01-2162
[8] Bittle J, Knight B and Jacobs T 2011 *SAE Paper* 2011-01-1350
[9] Brijesh P, Chowdhury A and Sreedhara S 2013 *SAE Paper* 2013-01-0321
[10] Kashdan J and Thirouard B 2009. *SAE Int. J. Engines* **2** p 1857–72
[11] Asad U, Zheng M, Ting D and Tjong J 2015 *ASME J. Eng. Gas Turbines Power* **137**
[12] Pielecha I, Borowski P, Czajka J, Wisłocki K and Kaźmierowski J 2015 *J. Therm. Anal. Calorim.* **119** p 695–703
[13] Hussain J, Palaniradja K, Alagumurthi N and Manimaran R 2012 *Alexandria Eng. J.* **51** p 241–7
[14] Asad U, Divekar P, Zheng M and Tjong J 2013 *SAE Paper* 2013-01-0283
[15] Divekar P, Asad U, Han X, Chen X and Zheng M 2014 *J. Eng. Gas Turbines Power* **136**
[16] Knight B, Bittle J and Jacobs T 2011 *SAE Paper* 2011-01-1354
[17] Maiboom A, Tauzia X and Hétet J 2008 *Energy* **33** p 22–34
[18] Czajka J, Wisłocki K, Pielecha I and Borowski P 2014 *Comb. Eng.* **157** p 45–59

Acknowledgements
The research presented in this paper was performed within the European Research Project PowerFul FP7, grant agreement No. 234032, POWERtrain for Future Light-duty vehicles.

The impact of air-fuel mixture composition on SI engine performance during natural gas and producer gas combustion

G Przybyła[1], S Postrzednik[1] and Z Żmudka[1]
[1]Silesian University of Technology, Institute of Thermal Technology, Konarskiego 22, 44-100 Gliwice, Poland
E-mail: grzegorz.przybyla@polsl.pl

Abstract. The paper summarizers results of experimental tests of SI engine fuelled with gaseous fuels such as, natural gas and three mixtures of producer gas substitute that simulated real producer gas composition. The engine was operated under full open throttle and charged with different air-fuel mixture composition (changed value of air excess ratio). The spark timing was adjusted to obtain maximum brake torque (MBT) for each fuel and air-fuel mixture. This paper reports engine indicated performance based on in-cylinder, cycle resolved pressure measurements. The engine performance utilizing producer gas in terms of indicated efficiency is increased by about 2 percentage points when compared to fuelling with natural gas. The engine power de-rating when producer gas is utilized instead the natural gas, varies from 24% to 28,6% under stoichiometric combustion conditions. For lean burn (λ=1.5) the difference are lower and varies from 22% to 24.5%.

1. Introduction

Running out of fossil fuels and increase in climate change are the main reason to seek alternative methods for more efficient use of fuel energy and increase in application of renewable resources. That is the resources which do not emit greenhouse gases and contribute to major reduction of emissions (considering fuel sustainability and carbon life cycle). Among all the types of renewable fuels, biomass is considered the most important with respect to the Polish climate and geographical conditions [1-3] Liquid fuels derived from biomass processing such as vegetable oils (i.e. fatty acid methyl esters - FAME) and alcohols are currently common in use as a blend component or additive to conventional fuels [4-6]. These fuels play an important role and can be considered as a possible alternative for the internal combustion engines. When considering biomass processing technologies, gasification process is one of the possible options for biogas production often called producer gas (PG) or syngas. The producer gas is characterized by its low calorific value and it can be used as a fuel in internal combustion engines and in process heat generation (boilers).

1.1. Producer gas utilization in SI engine

The number of studies on the use of synthesis gas from biomass gasification or pyrolysis in SI engines is quite limited considering the studies on eg. the use of biogas from fermentation processes. The authors of [7] presents the results of SI engine fuelled with natural gas and a gas having a composition corresponding to gas from the biomass gasification (CO = 17% CO_2 = 20% H_2 = 30%, CH_4 = 3%, N_2 = 30%). Tests were carried out in a range of variable composition of natural gas, which means it was diluted with CO_2, N_2 and air. For different values of the natural gas dilution with carbon dioxide, nitrogen and air parameters characterizing the combustion process were evaluated. The coefficient of variation of IMEP remained at favourable levels during the combustion of syngas in relation to diluted natural gas. Comparing the results for the same excess air ratio (λ = 1.25), the highest value COV of IMEP = 30% was observed in diluted gas by carbon dioxide, and the lowest value COV of IMEP ~ 0,7% was observed in the synthesis gas. In addition, indicated efficiency and utility range of the excess air ratio were higher during the combustion of the syngas. Both in terms of the stoichiometric and lean mixtures the combustion of the synthesis gas is accompanied with significantly higher content of CO in the exhaust gas in relation to the use of natural gas. On the other hand, the content of unburned hydrocarbons was several tens of times higher than during the combustion of natural gas. During the combustion of stoichiometric mixtures the content of NOx in the exhaust gas was significantly higher than during the combustion of natural gas. Realizing the combustion of lean mixtures the NOx content was at a similar level for both fuels. Although the work contains many valuable results and observations unambiguous comparison of emissions is impossible. Values given by the study's authors represent the molar proportions of individual components at the same excess air ratio. It should be stressed that natural gas and syngas have a different compositions, hence the content of ingredients such as CO_2 and N_2 in the engine exhaust is different for both fuels. One of the studies that is characterized by a wide range of research and unequivocal comparison of the parameters is the work [8]. The authors present SI engine research results fuelled with gas from the gasification of sewage sludge and a mixture of methane - the gas from the gasification of sewage sludge (CO = 16% CO_2 = 15% H_2 = 13%, CH_4 = 3%, N_2 = 53%). The study was conducted during the combustion of stoichiometric mixtures and for combinations of lean mixtures. The engine test were conducted at constant rotational speed, the load was a result of the intake pressure, which was set at constant value of 80 kPa. As shown in the presented report, the gas from the gasification of sewage sludge is troublesome fuel to power internal combustion engines. The combustion of fuel in the SI engine is unstable and often accompanied by the phenomenon of the misfire, consisting in the absence of combustion of a combustible mixture in some work cycles. Therefore, the authors determined the effect of the additive gas to the gas from the gasification of biomass on the correct operation of an internal combustion engine including the aspect of energy and environmental impacts. The optimum value of the volume fraction of methane in the mixture of the gas from the gasification of sewage sludge has been set at 40%. The addition of methane in to gas from gasification of sewage sludge significantly improves the stability of the internal combustion SI engine operating in the lean mixtures. Research in the combustion of lean methane-gas from the gasification of sewage sludge with the air shows no significant difference in the performance characteristics of the engine in relation to the energy supply of pure methane. Emissions of harmful substances (such as HC, NOx and CO) while using mixtures of synthesis gas - methane with variable ratio of excess air was at a similar level as it was when using pure methane. The test results of an installation consisting of a downdraft gasifier and SI engine shown in [9] indicate a higher efficiency of electricity generation during the combustion of the synthesis gas with respect to gasoline. In addition, presented in this paper results indicate that the molar ratio of CO and NOx in the exhaust gas of the engine fed with the synthesis gas was lower in relation to the petrol respectively 30-96% and 54-84% (depending on the engine load). The authors do not specify at what excess air ratio the tests were conducted. These values are ambiguous shares molar compared to the operating parameters of the engine during petrol and gas from the gasification, as well as energy efficiency of the system. The provided values of molar fractions and efficiency are ambiguous compared to the working parameters of the engine during combustion of fuel or syngas. Another example of research of gasification installation combined with SI engine is publication [10]. The tested object is based on commercial installations produced by the Canadian company Ankur code-named WBG-120. The internal combustion engine used in the system is a spark ignition engine type Cummins G-855-G driving the three-phase electricity generator type Stamford UC 27E (125 kVA). Average gas composition obtained by a gasification process of olive trees seeds was: CO = 10.7%, CO_2 = 4.6%, H_2 = 24.1%, CH_4 = 4.2%, N_2 = 55.1%, O_2 = 1.3%. During the research the engine worked at a constant speed (1500 rev/min) and full load. As a result the efficiency of electricity generation was 16.1%. The authors did not examine the effect of control parameters on engine efficiency and emission of harmful substances. One suspects that the low efficiency of power generation is a result of lack of control parameters optimization of the internal combustion engine.

The present work reports on the indicated performance of SI engine fuelled with a natural gas and producer gas substitute (clean and dry gas mixture obtained by mixing the gaseous compounds stored in high pressure cylinders). The stability limits in combustion process of lean air-fuel mixtures was investigated. As an stability indicator of combustion process the COV of IMEP was used.

2. The fuel properties and experimental test bench characteristic

Gaseous fuels are attractive for use in internal combustion engine (ICE) because of their wide ignition limits and ability to form homogeneous mixtures. Moreover, gaseous fuels usually have high hydrogen to carbon ratio, and thus relatively low CO_2 emissions are possible when they are used in SI engines. Natural gas is readily available from deep underground natural rock formations, or as a petroleum-based fuels, while producer gas can be obtained from renewable sources using biomass. When biomass is used sustainably to displace fossil fuels, the net impact is a lower CO_2 level in the atmosphere. The volumetric calorific value of producer gas is usually more than 5 times lower compared with natural gas. The LHV of producer gas varies with its composition that is heavily dependent on the gasification process. The main variable that determines the producer gas composition is the relative air flow direction. Gasifiers with a parallel flow of biomass feedstock and air impost higher technological requirements on the biomass feedstock compare to those with a counter- flow arrangement of biomass and air feeds. However, the producer gas from parallel flow gasifiers contains less undesired liquid residuals, such as tar and water. For application, where small and medium power outputs are needed, the most advanced technologies offer gasifiers with a parallel downdraft biomass and air delivery [11-13].

2.1. Properties and production method of producer gas

The major components of producer gas are nitrogen (N_2), carbon dioxide (CO_2), hydrogen (H_2), carbon monoxide (CO), methane (CH_4), and water (H_2O). Occasionally minute amounts of higher hydrocarbons such as ethane (C_2H_6) can be traced. At the exit from a gasifier the producer gas is usually at elevated, 500 °C to 800 °C, than ambient temperature. Therefore, the producer gas energy content exceeds its lower caloric heating value, by the enthalpy change between these temperatures. It follows that efficiency of the gasification process can be asset based on hot or cooled gasification products. In both methods, the energy content of producer gas is referenced with respect to the biomass energy input. The process efficiency based on hot products can as high as 90%, while the cold products efficiencies are in 40 to 70% range. Mole fractions of combustible components in a dry producer gas vary and depend on the gasification process operational variables; 15 to 40% for CO, 10 to 35% for H_2, and 2 to 5% for CH_4. Similarly the producer gas LHV can usually range from 4 to 10 MJ/Nm^3 50 [14]. Literature reporting on producer gas properties and the gasification process efficiencies can be recognised as extensive [15-16].

There are various significant parameters that can be used to characterize the usefulness of a gaseous fuels to powering the internal combustion engine. Different fuels need different amount of air to create expected air-gas mixtures which can be burned in an engine. Thus, the calorific value ($e_{d,v}$) of an air-gas mixture, the flammability limits, and the flame speed, are the next important parameters to, consider [17].

The properties of fuels and air fuel-mixtures used during the experimental tests are presented in Table 1. The ed,v were computed using equation 1 and assuming the values of following data T_l = 298.15 K, p_l = 101.325 kPa, relative air humidity, φ = 45% to determine molar humidity ratio, the residual exhaust gases fraction δ_{ex} = 5% as a molar fraction in fresh air, the air excess ratio λ = 1. The influence of the air excess ratio on the calorific value of air-gas mixture for fuel described in Table 1 is shown on Figure 2.

Table 1. Properties of tested fuels.

Fuel	Mixture description (used on Figures)	Composition (by volume)	LHVm *MJ/kg*	LHV *MJ/m³*	e_{dv} *MJ/m³*	$V_{a,min}$, *m³/m³*
Natural Gas	GZ50	CH_4 = 98,5% CO_2 = 0,1% N_2 = 1% other: C_2H_6, C_3H_8, C_4H_{10}	48.84	35.3	2.91	9.4
Producer gas	HC1	H_2 = 18,2% CO = 18,5% CH_4 = 3% CO_2 = 12% N_2 = 48,3%	4.85	5.38	2.19	1.16
Producer gas	HC0.4	H_2 = 10,5% CO = 27,5% CH_4 = 2% CO_2 = 8,5% N_2 = 51,5%	4.52	5.32	2.24	1.1
Producer gas	HC0.2	H_2 = 4,8% CO = 23,2% CH_4 = 2,8% CO_2 = 10,7% N_2 = 58,5%	3.55	4.45	2.03	0.93

The calorific value of air-gas mixture can be calculated using following equation:

$$e_{d,v} = \frac{p_1}{(MR)T_1} \frac{LHV}{\left[\frac{1}{M_f} + \lambda n'_{a,\min} (1 + X_{za} + \delta_{ex}) \right]} \quad (1)$$

where:

p_1, Pa; T_1, K – in-cylinder thermodynamic parameters after filling process, (MR) – universal gas constant; LHV, J/kg – Lower heating value; M_f, kg/kmol – fuel atomic weight, λ – air excess ratio; X_{za}, $kmol_{H2O}$/ $kmol_a$ – molar humidity ratio.

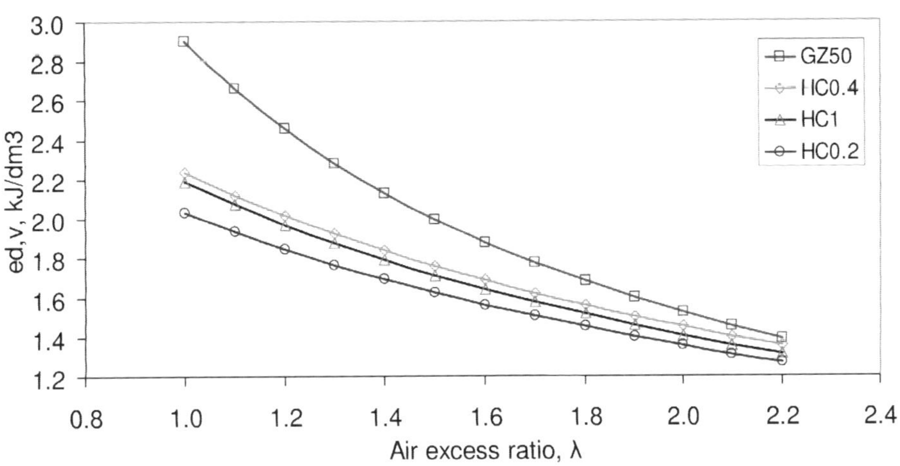

Figure 1. Calorific value of air – fuel mixture in a range of air excess ratio changes.

The LHVs of the simulated producer gases in this investigation varied from 4.45 MJ/m^3 to 5.38 MJ/m^3, compared against natural gas containing 35.3 MJ/m^3. The energy density per unit mixture volume of the producer gas is about 30% lower than that for a natural gas and air mixture at an air excess ratio $\lambda = 1$ despite the calorific value of the producer gas is 1/7.5 of natural gas. However, as the air excess ratio increase up to 2.2 the charge energy density differences become smaller and are equal 8.6% for HC0.2, 5.8% for HC1 and 2.2% for HC0.4 producer gas mixture. This shows that producer gas fuelling of the lean burn SI natural gas engine should not lead to significant power de-rating of the engine when operating with lean mixtures.

2.2. *Experimental test rig*

An overview of the engine test bench and measuring equipment is presented in Figure 2. The main components of the experimental set up include:

- Three cylinders SI engine with a capacity of 796 cm3 and compression ratio equal to 9.3. The engine is without turbocharging and was originally powered by petrol. For the purpose of experiment and possibility of gaseous fuel application, the control and power supply system of the engine have been modified.
- Electric motor with the power take-off system, capable of operating in two modes, the motor and generator. The main purpose of this system is to start the engine and then to apply load on the selected point of the operating cycle.
- High-pressure cylinders with gas mixtures and dual stage gas regulators.
- Measuring devices for flow rate, temperature and pressure evaluation including: rotameters, manometers and thermocouples.

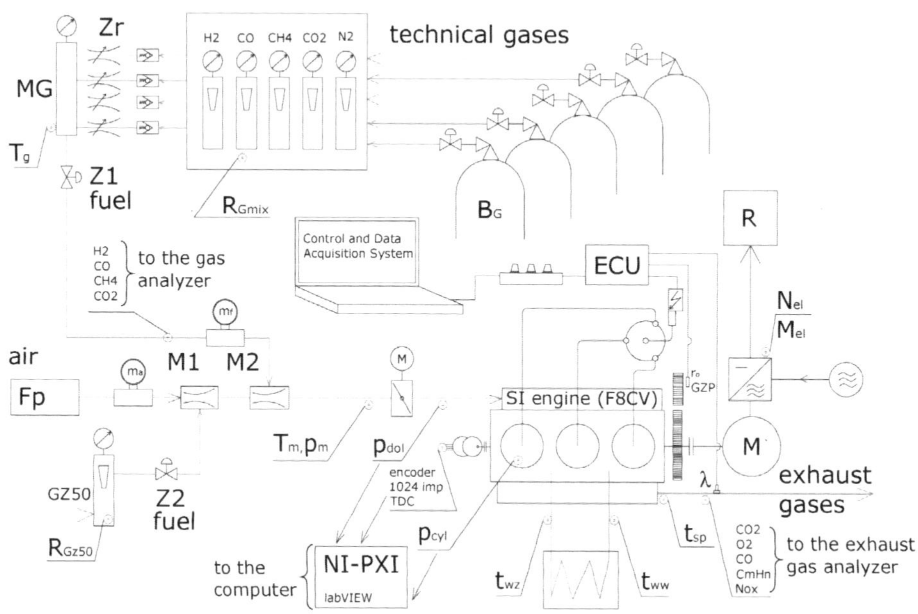

where: Zr, Z1, Z2 – control valves, MG – gas mixer, M1, M2 – air-fuel mixers, m – mass flow meter, B_G – gas cylinders, R_{Gmix}, R_{Gz50} – rotameters, R – braking resistor, M – electric motor

Figure 2. The experimental test rig.

The engine test bench has a cooling system for engine lubrication and cooling liquids. Within the cooling system two plate heat exchangers connected to the valves have been employed. Controlling thermostat is located in the primary circuit of the engine cooling system. The settings of the control

valves on the secondary side of the heat exchangers allow to adjust the amount of the removed heat, which is then transmitted to the local central heating system. In this way, the stabilization of the engine temperature in the range of 80 ± 5°C can be assured.
During the experiment, the operation of the engine was controlled using Electronic Control Unit (ECU). This type of device driver has the capability to program and monitor engine operational parameters, i.e. pre-programmed ignition timing maps and mixture composition. The composition of the mixture was controlled using the signal from lambda sensor, which can operate in a closed loop mode with the controller. During the tests the natural gas (GZ50) was collected from local gas grid, while the producer gas substitute was stored at high pressure cylinders. The gas fuel was supplied to the engine through the gas mixer. Two mixers were selected to fuelling the engine by air-gas mixture, separately for natural gas and for producer gas. The volumetric flow of each gaseous component (CO, CO_2, CH_4, N_2) have been manually adjusted by special valve to obtain demanded producer gas mixture composition and value of air excess ratio during combustion process. Value of air excess ratio was controlled by wideband lambda sensor. The volumetric flows of the gases was measured by rotameters equipped in temperature and pressure sensor. Additionally the producer gas flow before air-fuel mixer was measured using Coriolis mass flow meter. The producer gas substitute composition was controlled using gas analyzer.
The pressure measurements in the first cylinder were performed using piezoelectric pressure transducer. This type of transducer through so called charge amplifier generates an analogue (voltage) pressure signal which is then sampled at a sufficient frequency by the data acquisition system. In addition, the absolute pressure within the intake manifold was recorded with piezoresistive absolute pressure transducer. At both measurement ducts the pressure signal was sampled at predefined crank angle using encoder. Measurements were carried out with a resolution of 1024 measurement points per revolution of the crankshaft. The encoder was also equipped with a position marker device for indicating the position of a piston in a cylinder. Each sets of measurement consisted of 100 consecutive engine cycles.

3. Results and discussion

In this study the first point of interest in evaluating the validity of the producer gas fuel as a suitable fuel for SI engine was to obtain stable combustion at a given intake air to fuel ratio. Once stable combustion was achieved the fuel flow was changed (and the ignition timing was adjusted) therefore varying the mixture equivalence ratio to determine an affective lean operating limit. The upper limit value of COV of IMEP on 5% was agreed as an indicator of combustion process stability. The tested engine fuelled with natural gas and producer gas HC0.2 mixture composition reached the stable combustion for an air excess ratio of $\lambda = 1.5$. When the producer gas mixture included more hydrogen the lean burn limit was higher. For the producer gas substitute HC0.4 mixture stable combustion was obtained for an air excess ratio of $\lambda = 1.65$, while for HC1 the limit value of air excess ratio was $\lambda = 1.95$. The MBT ignition timing for both fuels and for all tested mixtures in a range of air excess ratio are presented in Table 2.

Table 1. Optimal ignition timing for tested gaseous mixtures.

Fuel	Engine speed $\mathbf{r_o}$ = 1500 rpm, full open throttle					
	Air excess ratio λ					
	1	**1,25**	**1,5**	**1,65**	**1,85**	**1,95**
	MBT Ignition timing α_z, CA before TDC					
GZ50	**21**	**26**	**44**	P_{lim}	P_{lim}	P_{lim}
HC1	**21**	**30**	**34**	-	**59**	**64**
HC04	**24**	**31**	**42**	**51**	P_{lim}	P_{lim}
HC02	**33**	**41**	**49**	P_{lim}	P_{lim}	P_{lim}

where:- – outside the plan of experiment, P_{lim} – unstable combustion or misfire

Indicated efficiency of the engine is an significant parameter to report when discussing the indicated performance of an engine. In Figure 3 and 4 the indicated efficiency of the SI engine fuelled with natural gas and producer gas substitute (three mixtures) as a function of air excess ratio is presented. The influence of fuel type and air excess ratio on the IMEP is presented on Fig. 5 and Fig. 6. IMEP values are always of prime interest in performance evaluation. The IMEPs for natural gas range between 6.6 bar (for λ=1.5) to 9.1 bar (for λ=1). When the producer gas is utilized the IMEP varied between 3.8 bar (for λ=1.95 and fuel composition HC1) to 6.9 bar (for λ=1 and fuel composition HC1) what is the maximum value obtained for producer gas. It is somewhat surprising because when the calorific value of air fuel mixture is compared at a range of air excess ratio (Fig. 1) the $e_{d.v}$ of producer gas composition HC1 is lower than for HC0.4. But for the air excess ratio λ=1 and λ=1.25 the utilization of the fuel composition HC1 brings higher engine indicated efficiency than fuel HC0.4 (see Fig. 4).

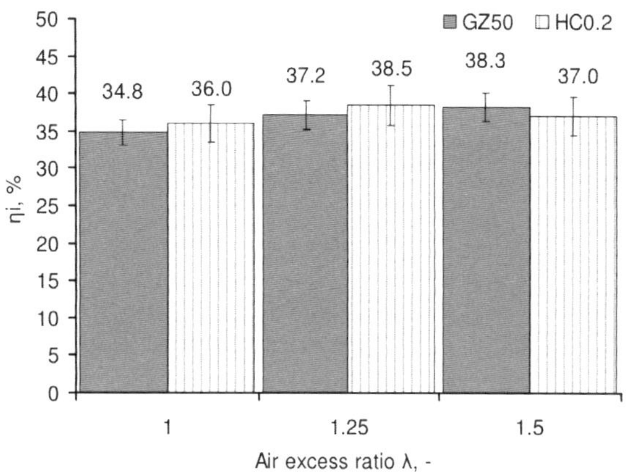

Figure 3. Indicated efficiency vs. air excess ratio utilizing natural gas and PG mixture HC0.2.

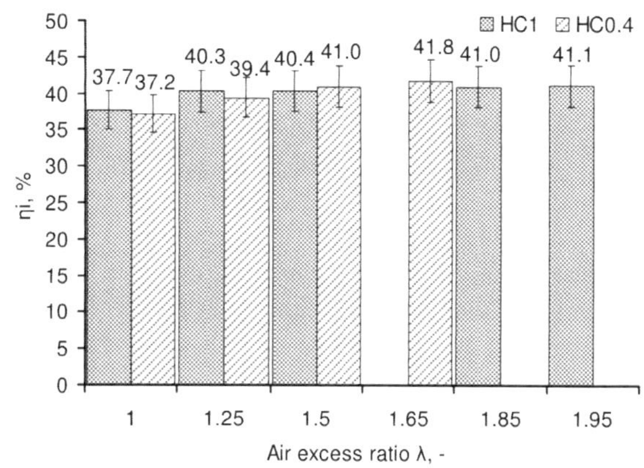

Figure 4. Indicated efficiency vs. air excess ratio utilizing PG mixture HC1 and HC0.4.

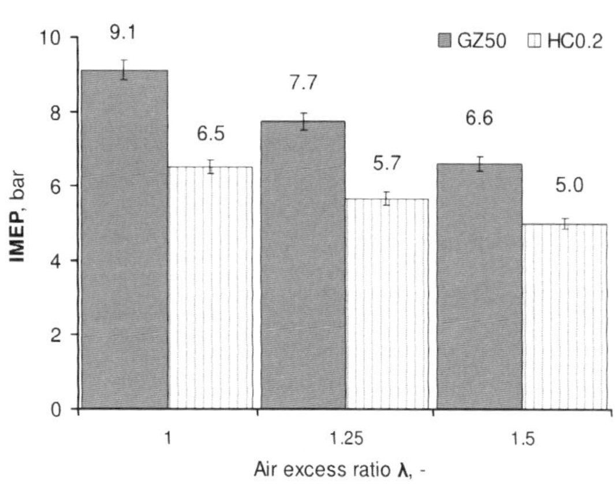

Figure 5. IMEP vs. air excess ratio utilizing natural gas and PG mixture HC0.2.

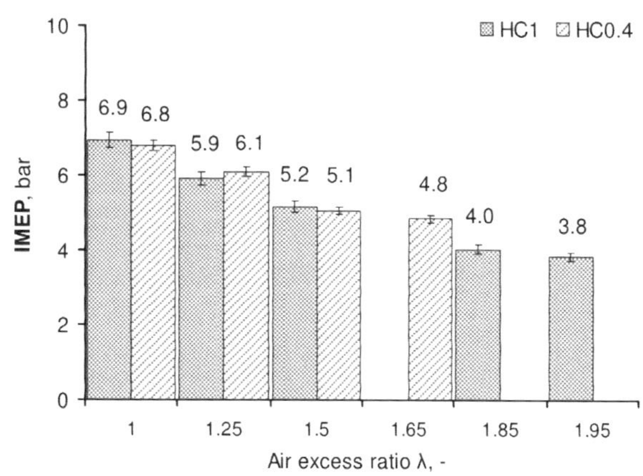

Figure 6. IMEP vs. air excess ratio utilizing natural gas and PG mixture HC1 and HC0.4.

The similar impact of air excess ratio change on the indicated efficiency value for all tested fuels was observed. For most tested fuels as the air excess ratio is increased and the indicated efficiency increased too. Exception is producer gas mixture HC0.2 (with the highest content of carbon monoxide from all used mixtures), for this fuel the maximum of indicated efficiency was achieved when the engine was operated with air excess ratio λ=1.25. However, for all tested fuels combusted with this

value of air excess ratio (λ=1.25) the highest increase of indicated efficiency was observed comparing the values obtained for stoichiometric air fuel mixture. In this range of air excess ratio (λ=1 to λ=1.25) the indicated efficiency increases averagely of about 2 percentage points. The increase of indicated efficiency for lean burn condition is mainly associated with lower level of in cylinder temperature what is directly related with lower heat losses during combustion period. The tested engine reached highest level of indicated efficiency when producer gas mixture HC0.4 has been burned with air excess ratio λ=1.65. It should be noted here that as the air excess ratio is increased the indicated mean effective pressure of the engine is decreased and it is the effect of calorific value change of air-fuel mixture (Fig. 1). Because the friction work of the engine internal parts is the same in all tested conditions covered by the experiment, the engine efficiency (based on the crankshaft power output) can be constant or can decrease with increase of air excess ratio.
In Fig. 7 and Fig. 8 the Coefficient of Variation (COV) of the IMEP values are depicted. In the constant air excess ratio one can see that the variation in combustion stability between the tested fuels is relatively small as the span of variations is 0.6 percentage point for λ=1.

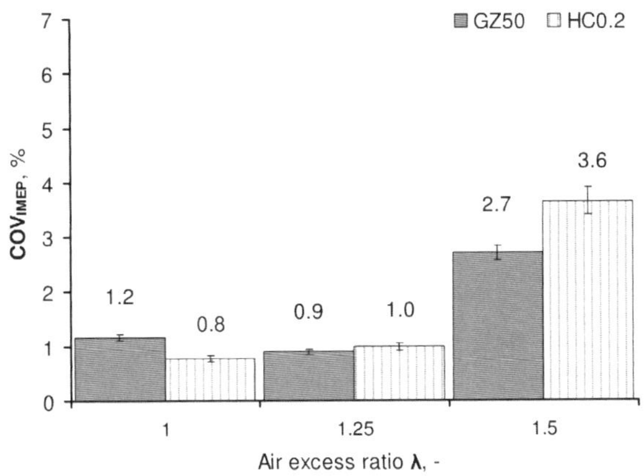

Figure 7. COV of IMEP vs. air excess ratio utilizing natural gas and PG mixture HC0.2.

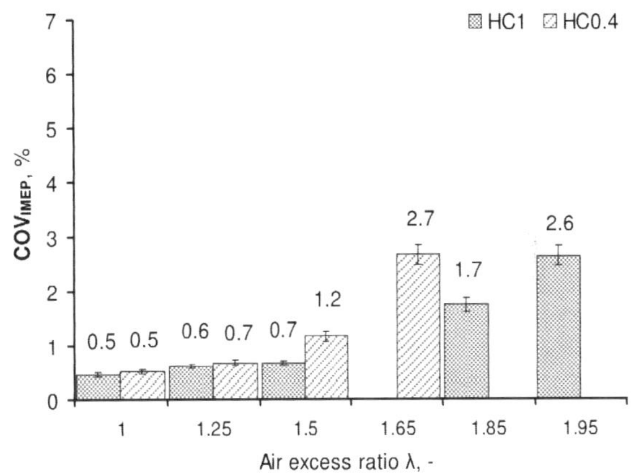

Figure 8. Indicated COV of IMEP vs. air excess ratio utilizing PG mixture HC1 and HC0.4.

The observed values of the COV for both fuels are acceptable and compare well to those reported in literature [7-10] For natural gas and producer gas composition HC0.2 (see Fig. 7) a COV of IMEP range of 1.2% (for λ=1) to 2.7% (for λ=1.5) for GZ50 and 0.8% (for λ=1) to 3.6% (for λ=1.5) for HC0.2 fuel.
For fuel composition HC1 a COV of IMEP range of 0.46 % (for λ=1) to 2.6% (for λ=1.95) what was the lowest value of COV obtained during all engine tests. Fuel HC1 includes most hydrogen in the composition comparing to the other producer gas mixtures. In this case the role of hydrogen in producer gas mixture is visible. The hydrogen can extend the flammability limits of the fuel and stabilising the combustion process of producer gas.

4. Conclusions

In this work the influence of air-fuel mixture change and a simulated biomass gas composition change that resembles in its composition so called producer gas for SI engine performance has been investigated. The engine performance was evaluated based on indicated performance parameters derived from in-cylinder cycle-resolved pressure traces.
It is shown that the simulated biomass gas can be stable burned in SI engine. The useful range of air excess ratio change is wider for producer gas mixtures with higher H_2/CO ratio. In particular the results indicate the following:

- The engine performance in terms of indicated efficiency is increased by about 2 percentage points when compared to fuelling with natural gas.
- The engine power de-rating when producer gas is utilized instead the natural gas, varies from 24% for producer gas HC1 to 28,6% for producer gas HC0.2 under stoichiometric combustion conditions. For lean burn (λ=1.5) the difference are lower and varies from 22% for producer gas HC1 to 24.5% for producer gas HC0.2.
- To obtain maximum value of IMEP the ignition timing should be advanced for the engine fuelled with producer gas if compare with the values for natural gas. The ignition timing must be more advancing for producer gas with low H_2/CO ratio and for lean mixture combustion.
- For most of engine working conditions the COV of IMEP has lower values when the engine is fuelled with simulated producer gas. The lowest values is obtained for stoichiometric combustion.

References

[1] Berggren M., Ljunggren E., Johnsson F., *Biomass co-firing potentials for electricity generation in Poland-Matching supply and co-firing opportunities*, (*Biomass and Bioenergy* vol 32), (Elsevier 2008), pp 865 – 879
[2] Igliński B., Iglińska A., Kujawski W., Buczkowski R., Cichosz M., *Bioenergy in Poland*, (*Renewable and Sustainable Energy Reviews* vol 15), (Elsevier 2011), pp 2999– 3007
[3] Leszczynski, S.; Brzychczyk, P.; Sekula, R., *Review of biomass as a source of energy for Poland*, (*Energy Sources*, vol 19), pp 845-50, Oct. 1997; ISSN: 0090-8312
[4] McKendry P., *Energy production from biomass (part 2): conversion technologies*, (*Bioresource Technology* vol 83), (Elsevier 2002), pp 47-54
[5] Mancaruso E., Sequino L., Vaglieco B. M., *First and second generation biodiesels spray characterization in a diesel engine*, (*Fuel* vol 90), (Elsevier 2011),pp 2870–2883
[6] Silva M. J., Souza S. M. N., Chaves L., Rosa H. A., Secco D., Santos R. F., Baricatti R. A., Nogueira C. E. C., *Comparative analysis of engine generator performance using diesel oil and biodiesels available in Parana´ State, Brazil*, (*Renewable and Sustainable Energy Reviews* vol 17) (Elsevier 2013), pp 278–282
[7] Yamasaki Y., Tomatsu G., Nagata Y., Kaneko S.: *Development of a Small Size Gas Engine System with Biomass Gas (Combustion Characteristics of The Wood Chip Pyrolysis Gas)*, SAE Technical paper 716 2007-01-3612.
[8] Szwaja S., Kovacs V.B., Bereczky A., Penninger A.: Sewage *sludge producer gas enriched with methane as a fuel to a spark igniter engine.* (*Fuel Processing Technology*, vol. 110), (Elsevier 2013), pp 160–166
[9] Shah A., Srinivasan R., To S.D, Columbus E.P.: *Performance and emissions of a spark-ignited engine driven generator on biomass based syngas*, (*Bioresource Technology* vol. 101) (Elsevier 2010), pp 4656–721 4661
[10] Margarits N.K., Grammelis P., Vera D., Jurado F.: A*ssessment of operational results of downdraft biomass gasifier coupled with a gas engine, Transport Research Arena – Europe 2012*, (Proc*edia – Social and Behavioral Sciences* vol 48) (Elsevier 2012), pp 857–867
[11] Bridgwater AV. *Renewable fuels and chemicals by thermal processing of biomass.* (*Chemical Engineering Journal* vol 91) (Elsevier 2003), pp 87-102
[12] Beenackers AACM. *Biomass gasification in moving beds, a review of European Technologies.* (*Renewable Energy* vol 16) (Elsevier 1999), pp 1180-886
[13] Kirubakaran V.; Sivaramakrishnan V.; Nalini R.; Sekar T.; Premalatha M.; Subramanian P. *A review on gasification of biomass.* (*Renewable and Sustainable Energy Reviews* vol 13). (Elsevier 2009), pp 179-86
[14] Jun H.; Heejoon K. T*he reduction and control technology of tar during biomass gasification/pyrolysis. An overview.* (*Journal of Renewable and Sustainable Energy Reviews* vol 12) (Elsevier 2008), pp 397-416

[15] Jun Wander P.R.; Altafini C.R.; Barreto R.M. *Assessment of a small sawdust gasification unit.* (*Biomass and Bioenergy* vol 27). (Elsevier 2004), pp 467–476

[16] Lv P.; Yuan Z.; Ma L.; Wu C.; Chen Y.; Zhu J. *Hydrogen-rich gas production from biomass air and oxygen/steam gasification in a downdraft gasifier.* (*Renewable Energy* vol 32). (Elsevier 2007), pp 2173–2185

[17] Przybyła G., Ziółkowski Ł., *Analysis of energy conversion process of SI engine fuelled with LCV gas.* Proceedings of the Ninth Asia-Pacific International Symposium on Combustion and Energy Utilization, Beijing, 2 - 6 November, 2008. (World Publishing Corporation 2008), pp 113 – 118

Acknowledgments
This work is supported by the Silesian University of Technology from the statutory research source of Institute of Thermal Technology.

Numerical investigation of CAI Combustion in the Opposed-Piston Engine with Direct and Indirect Water Injection

R Pyszczek[1], P Mazuro and A Teodorczyk

[1] Corresponding author. Warsaw University of Technology, Institute of Heat Engineering, Nowowiejska Street 21/25, 00-665 Warsaw, Poland

E-mail: rafal.pyszczek@itc.pw.edu.pl

Abstract. This paper is focused on the CAI combustion control in a turbocharged 2-stroke Opposed-Piston (OP) engine. The barrel type OP engine arrangement is of particular interest for the authors because of its robust design, high mechanical efficiency and relatively easy incorporation of a Variable Compression Ratio (VCR). The other advantage of such design is that combustion chamber is formed between two moving pistons – there is no additional cylinder head to be cooled which directly results in an increased thermal efficiency. Furthermore, engine operation in a Controlled Auto-Ignition (CAI) mode at high compression ratios (CR) raises a possibility of reaching even higher efficiencies and very low emissions. In order to control CAI combustion such measures as VCR and water injection were considered for indirect ignition timing control. Numerical simulations of the scavenging and combustion processes were performed with the 3D CFD multipurpose AVL Fire solver. Numerous cases were calculated with different engine compression ratios and different amounts of directly and indirectly injected water. The influence of the VCR and water injection on the ignition timing and engine performance was determined and their application in the real engine was discussed.

1. Introduction

The possibility of achieving high thermal efficiency brings Opposed-Piston (OP) engines back into interest of research centers. An advantage of such design is that combustion chamber is formed between moving pistons. The main thermal benefits arise from the geometrical shape of the combustion chamber. As heat loses are proportional to the area *A* of the reactor and heat generation is proportional to the volume *V*, the factor *A/V* is simple but very efficient estimator of internal combustion engine efficiency and in OP engines it is twice lower than in conventional SI [1]. In the OP engine designed as a 2-stroke the poppet valves become redundant and with adapting an uniflow scavenging, the volumetric, trapping and scavenging efficiency become comparable with a 4-stroke. Because of the lack of the cylinder head, a scavenging process need to be handled differently than in a conventional 4-stroke engine. In the OP engine usually intake and exhaust ports in the cylinder liner are responsible for the scavenging. When pistons reach their Bottom Dead Centre (BDC), ports are opened and charge is exchanged due to pressure difference between the intake and exhaust manifolds.

There are several types of OP engine configurations (cranckless, single crankshaft, multiple crankshaft, rotary or barrel) [2]. The barrel type OP engine is of particular interest for the authors because of its robust design, high mechanical efficiency and relatively easy incorporation of a Variable Compression Ratio (VCR) [3]. In the barrel type OP engine cylinders' axes are parallel to the drive shaft axis. Linear to rotary motion is changed through the special plate with connecting rods mounted on ball bearings. An 8-cylinder barrel type OP piston engine concept is presented in **Błąd! Nie można odnaleźć źródła odwołania.**.

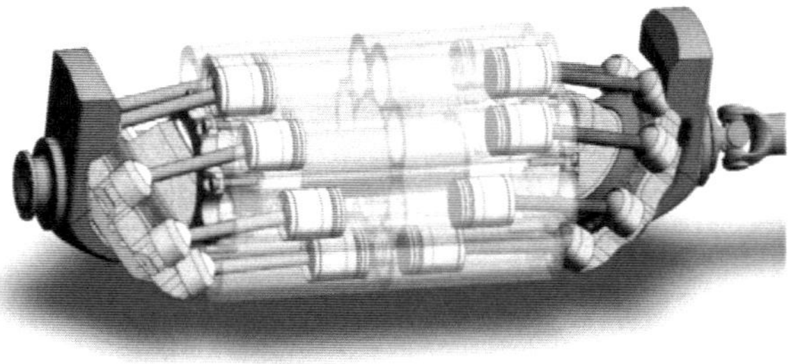

Figure 1. An 8-cylinder barrel-type OP engine concept.

Another measure for increasing engines efficiency is Homogeneous Charge Compression Ignition (HCCI)/Controlled Auto-Ignition (CAI) combustion mode. It is an unique form of combustion based on charge auto-ignition at desired crank angle. It has been demonstrated that a CAI gasoline engine can achieve fuel economy levels comparable to those of a Compression Ignition (CI) engines, while producing engine-out NOx emissions that are as low as tail-pipe NOx emissions from a conventional SI engine equipped with a three-way catalyst [4]. Although the idea of HCCI/CAI combustion is desirable, it is also very challenging to implement in an internal combustion engine due to the absence of direct ignition timing control (i.e. spark). In order to guarantee correct combustion timing, closed-loop combustion control is necessary. This type of control is supposed to vicariously influence the ignition timing via different measures (i.e. VCR or water injection).

This paper is focused on the CAI combustion control in a turbocharged 2-stroke barrel-type opposed-piston engine which is currently under development at Warsaw University of Technology. The engine will be equipped with an online VCR and water injection for indirect ignition timing control. The water injection is also supposed to prevent knock by charge cooling through evaporation and extend engine operation area. Water will be injected either directly into the cylinder or indirectly into the intake manifold. The turbocharging will increase the engine power output. Such a complex CAI combustion control system is challenging to implement in the real engine. Therefore, numerical simulations can become a useful tool for better understanding of the combustion process and can shorten the time span for engine development. In this study 3D CFD numerical simulations of the scavenging, injection and combustion processes were performed in order to determine possible engine operating points in CAI combustion mode.

2. HCCI/CAI fundamentals and modeling

In a conventional SI engine combustion starts when an electrical spark is used to initiate a flame kernel that develops across the combustion chamber in a turbulent manner. During HCCI/CAI combustion auto-ignition takes place when the conditions are right in terms of temperature and pressure for the particular mixture of fuel and air. However, the type of ignition and further combustion differ between HCCI and CAI leading to different benefits of these two types of combustion and different strategies of the ignition and combustion control. In the HCCI mode premixed, homogeneous mixture ignites at multiple sites and invokes the simultaneous reactive envelopment of the entire fuel/air mixture without a flame front. The heat release process is much faster than the conventional SI combustion and is most likely to result in undesired knock. In order to limit the heat release rate and prevent knock HCCI engines usually work with charge highly diluted either by the excess of air or by EGR. In cases where charge stratification occurs, the mixture will ignite at a single spot, where it firstly reach the ignition condition and the flame will propagate across the combustion chamber. Although this Controlled Auto-Ignition (CAI) operation is similar to conventional SI combustion, the conditions (pressure and temperature) in the chamber at the moment of ignition would lead to undesired knock in case of stoichiometric mixture. Hence, engines in CAI mode, similarly as in HCCI mode, work with lean mixtures.

In recent years various experimental and numerical studies of HCCI/CAI combustion have been carried out [4–8]. Researchers investigated mainly conventional 4-stroke engines proving that with VCR or intake charge heating it is possible to achieve CAI combustion and indirectly control ignition timing.

Engine operation in this mode is usually limited to low and medium loads due to reduced equivalence ratios for avoiding knock. However, with proper internal or external EGR, thermal and fuel stratification even a stoichiometric charge can be combusted with acceptable pressure-rise rates. Furthermore, supercharging is one way to dramatically increase IMEP for HCCI and extend its operation area.

In order to properly predict ignition and combustion in the engine running in CAI/HCCI mode it is necessary to account for the charge stratification and wall interaction, what leads to multi-dimensional CFD modeling. Furthermore, HCCI/CAI ignition of heavy hydrocarbons is a two-stage process involving a low temperature cycle followed by a high temperature cycle [4] so the fuel chemical kinetics plays a dominate role during the whole combustion process. With rapid growth of the HCCI/CAI engine research, the study on chemical kinetics mechanism of different surrogate fuels which can properly describe combustion characteristics in these new concept engines has become very active. The attention was focused on reduced mechanisms which can be applied directly to the engine combustion simulation using multidimensional CFD codes. In recent years several mechanisms were developed for iso-octane which exhibits similar trends to gasoline over the HCCI operating range [9–12]. The most recent one is by Liu et al. [12] and consists of 32 species and 111 reactions. Authors validated their skeletal model against experimental data in various reactors under diverse operating conditions, with emphasis on laminar flame speed and important species evolution. The predicted results showed good agreement with experiments. Also, the performance of the mechanism was compared to that of other reduced mechanisms available in the literature, showing its superiority in a HCCI relevant conditions.

3. Engine setup

This study concerns numerical simulations of CAI combustion in the Opposed-Piston engine being currently under development at Warsaw University of Technology. The investigated research engine is a 2-cylinder, 2-stroke, barrel-type OP engine with cylinders' axes parallel to the shaft axis. The basic engine parameters are given in Table 1.

Table 1. Engine parameters.

Engine type	2-stroke, OP, turbocharged
Number of cylinders	2
Bore	0.055 m
Displacement volume	~1800 ccm
Engine speed	1500 rpm
Compression ratio	VCR 8÷18
Fuel	Gasoline
Ignition	SI/CAI
Fuel injection	Direct Injection (DI)
Water injection	Direct/Manifold (DWI/MWI)

The biggest challenges associated with CAI combustion are ignition timing control and operation outside the knock area. As for the knock, it is usually related to excessive heat release rate and can be avoided in CAI engines running with reduced equivalence ratios. However, it is desirable to reduce heat release rate with different methods in order to increase equivalence ratio and extend the engine operation area to higher loads. The investigated engine will be equipped with the following solutions that will allow CAI combustion control:

- Variable Compression Ratio for indirect ignition timing control
- Variable Port Timing for scavenging and EGR control
- Direct fuel injection for fuel stratification and limiting the heat release rate
- Manifold Water Injection for indirect ignition timing control and limiting knock
- Direct Water Injection for ignition control through direct internal EGR cooling
- Turbocharging for increased engines IMEP and wider operation area

4. Simulation setup

4.1. Numerical mesh and boundary conditions

Numerical simulations of combustion in this study were performed using the AVL Fire 3D CFD solver based on Finite Volume Method (FVM) discretization. Firstly, numerical mesh was prepared. Simulations were performed for one cylinder and half of the intake and exhaust manifolds. Two types of meshes were defined – steady mesh for intake/exhaust manifolds and moving mesh for cylinder volume. Movement of the cylinder mesh was handled by changing positions of the nodes representing intake-side and exhaust-side pistons separately according to the given piston displacement curves. Simultaneously, positions of the nodes between pistons were interpolated every time step according to the pistons displacement. Meshes of the cylinder and intake/exhaust manifolds were connected with *Arbitrary Interface* at the contact of the cylinder wall and intake/exhaust ports. Total number of mesh elements for the entire model was of 1 200 000, while mesh for the cylinder itself consisted of 200 000 elements. The mesh of the cylinder used for simulations together with applied temperature boundary conditions is presented in Figure 2. In this study temperature on the intake side piston wall was assumed to be T_{in}=800 K, temperature on the exhaust side piston T_{ex}=880 K and on the cylinder wall a temperature distribution was defined in range of T_c=400–800 K. Such high temperatures were assumed due to the high engine thermal load.

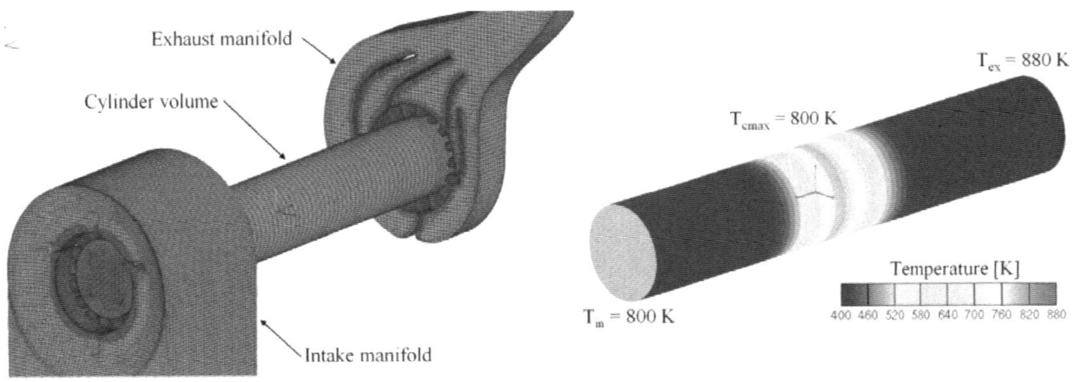

Figure 2. Numerical mesh and temperature boundary conditions.

4.2. Combustion modeling

There are various combustion models available in commercial CFD solvers for premixed or non-premixed combustion. In case of piston engines these models are usually developed for SI and CI combustion [1]. Although combustion models predict Auto-Ignition, they are not be suitable for low temperature combustion of CAI. In order to account for modeling both low and high temperature combustion of transportation fuel surrogates, such as iso-octane it is necessary to use reduced kinetic schemes. In this study one of the most recent reduced mechanisms developed by Liu et al. [12] for iso-octane HCCI/CAI combustion was used. Additionally, a submechanism for NOx formation was taken from [10] and included in the Liu scheme. Final kinetic scheme for iso-octane oxidation used for CAI combustion simulations in this study consisted of 37 species and 124 reactions.

Turbulence is a key issue in CFD modeling of processes taking place combustion engines. It has to be taken into account in combustion simulations. In this work k-ζ-f turbulence model was used for turbulence modeling. It is a four-equation model developed by Hanjalic et al. [13] which solves equations for turbulent kinetic energy k, turbulent kinetic energy dissipation rate ε, normalized velocity scale ζ and explicit relaxation function f. The model details are given in [1] and [13].

4.3. Operating conditions in simulations

In previous sections several solutions for CAI combustion control in 2-stroke OP engine were named and described. In order to limit the number of investigated cases in this study only two measures relevant for CAI combustion were considered – VCR and water injection. A constant boost pressure of 3 bars (absolute) was assumed for the simulations to increase engine power output. Reduced equivalence ratios were consider to limit the maximum pressure rise and avoid knock. Operating conditions for this study are summarized in Table 2.

Table 2. Engine operating conditions

Boost pressure	3 bar (absolute)
Intake temperature	410 K (at 3 bar boost)
Equivalence ratio	0.4÷1.0
Compression Ratio	8÷12
Fuel in simulations	iso-octane
Fuel direct injection	45÷120 mg/cycle
Start of fuel injection	-120°CA ATDC
Fuel temperature	300 K
Water Injection	0÷130 mg/cycle
Start of Direct Water Injection	-160°CA ATDC
Water temperature	300 K

5. Results and discussion

5.1. Operation targets for investigated cases

Simulations were performed for different compression ratios, equivalence ratios and different amounts of directly or indirectly injected water. For each case two engine cycles were calculated. For the first cycle approximate initial conditions in the cylinder before the scavenging were assumed. At the end of the first cycle realistic conditions in the cylinder were obtained for the second cycle scavenging and combustion. Only results from the second cycle are presented and compared in this study.

In order to compare the results and draw conclusion specific targets for the combustion results need to be assumed. In this study following parameters are considered:

- Crank angle of 50% accumulated heat release (CA50) was considered as a measure of combustion timing [4]. Target for CA50 in this study was ±1.0°CA ATDC
- Maximum pressure rise dP/dCA was assumed as the limiting parameter for the knock, since the knock itself was not modeled in this work. Target for maximum dP/dCA was of 20 bar/CA

In CAI combustion high temperature heat release is followed by the low temperature heat release. Low temperature combustion usually starts earlier in the cycle and is responsible for 5% to 10% of the total heat released depending on the fuel [4]. For the iso-octane reduced scheme used in this study low temperature combustion delivers ~5% of the total heat released. The crank angle at which 5% of the heat is released (CA5) can be considered as the high temperature ignition. However, in this study CA5 timing was not analyzed and focus was paid only on the CA50 timing adjustment.

5.2. CAI without water injection

Firstly, engine operation in CAI mode without water injection was investigated. In order to meet the targets for CA50 and dP/dCA it was necessary to work with the lowest possible compression ratio of CR=8. Results for this operating point are given in Figure 3

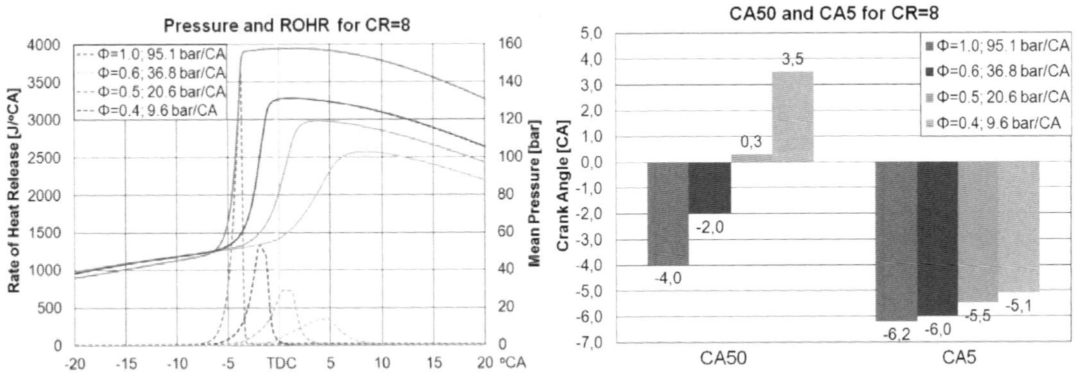

Clearly, the engine cannot work with stoichiometric mixture under these conditions due to excessive heat release resulting in high maximum pressure rise and probably knock. Hence, equivalence ratio need to be lowered to the level of ϕ=0.5 at which dP/dCA=20.9 bar/CA and CA50=0.3°CA ATDC. It can be seen that CA50 timing differs with the equivalence ratio significantly. It is due to different combustion duration. On the other hand CA5 timing differs with the equivalence ratio only slightly meaning that equivalence ratio has small impact on the high temperature ignition timing of the mixture.

In order to have a benchmark for CAI combustion performance, an additional case of conventional SI combustion with indirect gasoline injection (intake manifold injection MI) in the investigated engine was calculated for compression ratio of CR=9 and equivalence ratio of ϕ=1.0. To make the case comparable with the CAI cases, the boost pressure and intake temperature were adjusted for similar power output. Also ignition timing was adjusted for CA50 to occur at TDC.

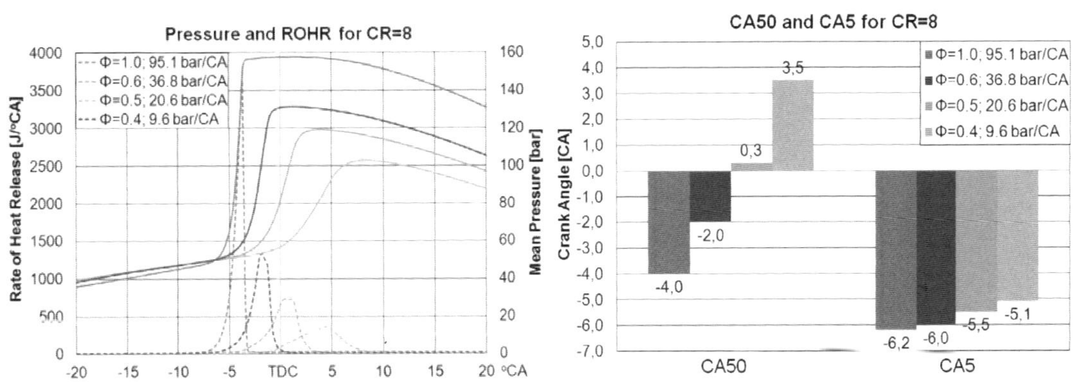

Figure 3. Results of pressure trace, ROHR, CA50 and CA5 for CR=8.

5.3. CAI with water injection

The possibility of adjusting compression ratio is of particular importance for CAI combustion control. Raising the compression ratio is expected to result in an increased engine efficiency and lower fuel consumption. In previous section CAI combustion was obtained for the fixed compression ratio of CR=8. If one would increase the compression ratio, simultaneously auto-ignition would occur too early and affect the engine operation. The other solution for indirect ignition control proposed by the authors is water injection. The water injected into the intake manifold would cool the intake charge through the evaporation and retard the ignition. Additionally, it could increase the charge thermal stratification and limit maximum heat release and knock. The water can be also injected directly into the cylinder during the scavenging process. In that case evaporation takes place inside the cylinder and have more direct impact on the cylinder walls but can result in uneven water distribution.

In order to investigate simultaneous application of the VCR and water injection in CAI engine a number of cases was calculated including compression ratios of CR=8, 9, 10, 11 and 12. Considered amount of injected water was ranging in 0÷130 mg/cycle. In case of Manifold Water Injection (MWI) water evaporation calculations were carried out separately and their results were applied as initial conditions in the intake manifold for CFD simulations. For Direct Water Injection (DWI) water was injected into the cylinder with a pintle-type injector during the scavenging process. The injection timing should be as early as possible for droplets to have more time for evaporation. On the other hand injection should not be too early in order for water not to reach the exhaust. The best timing for the start of DWI in investigated cases was determined to be -160°CA ATDC.

Results of the temperature field at the cross-section of the cylinder and water droplets for different amount of the injected water just after the intake ports closing are compared in Figure 4. Thermal stratification developed during the scavenging is clearly visible. Differences between Direct and Manifold Water Injection are also visible. In case of MWI and 30 mg/cycle of water all droplets evaporated before the end of the scavenging. For 60, 90 and 120 mg/cycle intake temperature was too low for all droplets to evaporate. Droplets which entered the cylinder are distributed in the whole volume and are slightly more concentrated on the intake side. Their evaporation during the compression stroke increases the thermal stratification. High temperatures remains at the exhaust side of the combustion chamber. The Direct Water Injection behavior is different. During the injection water droplets are drifted by the scavenging air momentum to the exhaust side of the cylinder. The internal EGR and exhaust side piston are cooled by the injected water, while intake side of the cylinder remains unaffected, what means that thermal stratification is reduced. Direct Water Injection might be effective in ignition timing control. However, its influence on the heat release rate and knock is expected to be limited because of the uneven water distribution in the cylinder.

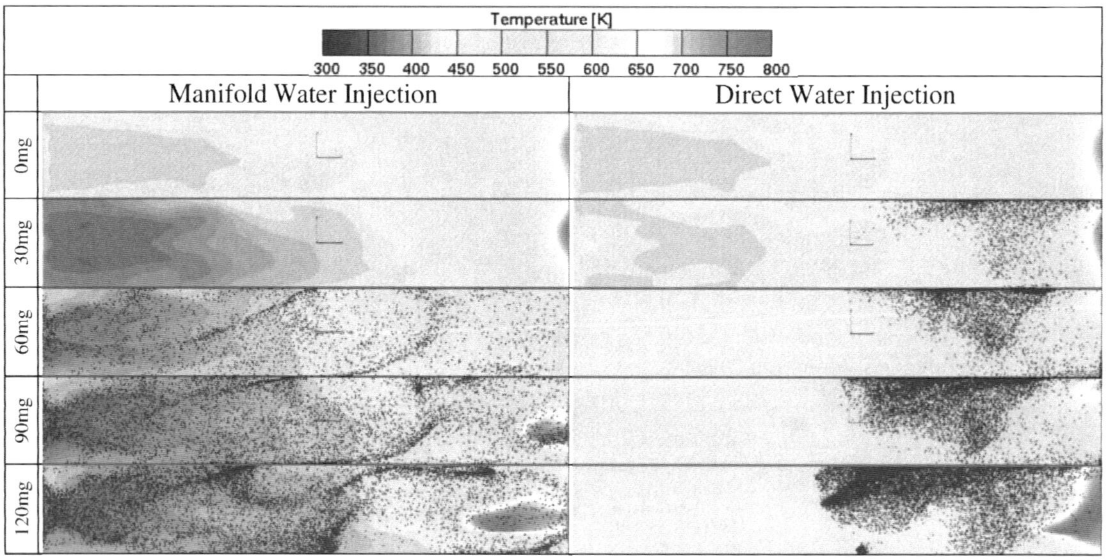

Figure 4. Temperature fields and water droplets after the scavenging at -120°CA ATDC. Manifold Water Injection on the left and Direct Water Injection on the right

Table 3. Summary of the final operating points

Case	Combustion mode	Equivalence ratio [-]	Absolute boost pressure [bar]	Fuel mass [mg]	Water mass [mg]	Water to Fuel ratio	dP/dCA [bar/CA]	CA50 [CA]	EGR [-]
CR9 SI	SI	1.0	1.2	61.7	-	-	15.8	0.4	5.5%
CR8 CAI	CAI	0.5	3.0	58.5	-	-	20.6	0.3	5.1%
CR9 DWI	CAI with	0.4	3.0	46.8	5	0.10	18.8	-1.1	5.5%
CR10 DWI		0.45	3.0	48.5	70	1.44	16.7	0.5	10.6%

CR11 DWI	DWI	0.5	3.0	54.0	100	1.85	18.3	-0.5	10.2%
CR12 DWI		0.5	3.0	53.4	130	2.43	16.1	0.7	11.0%
CR9 MWI	CAI with MWI	0.5	3.0	61.5	30	0.49	18.4	-0.4	5.9%
CR10 MWI		0.5	3.0	60.1	60	1.00	20.7	-0.9	9.8%
CR11 MWI		0.5	3.0	58.0	90	1.54	19.5	-0.4	8.3%
CR12 MWI		0.5	3.0	57.1	120	2.10	13.3	-0.3	9.4%

Equivalence ratio adjustment for higher compression ratios with water injection was performed similarly as for CR=8 in section 5.2. Of course, combination of compression ratio and water injection required calculating considerable number of cases. Firstly, cases with equivalence ratio of ϕ=0.5 were calculated in order to determine which combination of the VCR and water injection is the closest to defined limits. Then, final adjustments were done to fit the targets. Finally defined operating points for each investigated compression ratio are given in Table 3 together with the benchmark SI combustion case. It is interesting to compare Direct and Manifold Water Injection for the same compression ratios. For CR=9 only 5 mg/cycle of DWI was enough to meet the CA50 target. However, water injection cooled down only the surroundings of the exhaust piston wall and it was necessary to lower the equivalence ratio to the ϕ=0.4 to fit the dP/dCA limit. On the other hand MWI of 30 mg/cycle for CR=9 helped to meet the CA50 target, but the water was evenly distributed and it was possible to maintain equivalence ratio of ϕ=0.5 for the dP/dCA limit. For the higher compression ratios the trend was different and more water needed to be injected directly than into the manifold to obtain defined targets. It can be explained by the fact that DWI had more impact on the exhaust side of the cylinder volume and was less effective in cooling down the intake side of the cylinder volume, while for the MWI water was more evenly distributed in the whole cylinder volume.

5.4. Ignition positioning

Water injection for CAI combustion influence not only ignition timing, but also ignition positioning. The charge will ignite at the position where ignition conditions are firstly achieved. Probably wall temperatures are of importance for ignition in CAI combustion mode. Results of ignition positioning together with remaining water droplets are given for final cases in Figure 5.

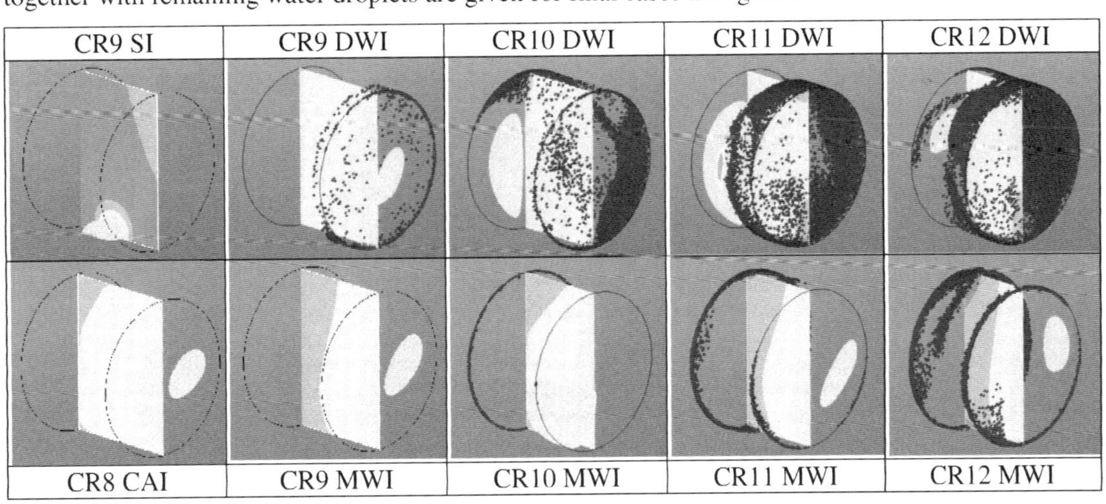

Figure 5. Ignition positioning in investigated cases. Exhaust-side piston in the front, intake-side piston in the back.

In case of SI combustion spark ignition at the cylinder wall is visible. For CAI combustion with Manifold Water Injection ignition always occurred at the exhaust-side piston wall. It was because of the high wall temperature and surrounding of the high-temperature internal EGR. Remaining water droplets are visible for higher compression ratios, for which more water was injected. As for Direct Water Injection, it is visible that the water was dragged to the exhaust-side piston and most of it evaporated in this area during the compression stroke. In the effect surroundings of the exhaust-side piston were cooled down, so that the ignition relocated to the cylinder wall for CR10 DWI, CR11 DWI and CR12 DWI cases. More water droplets remain unevaporated than in MWI cases due to less time available for evaporation and increased water concentration at the exhaust-side of the chamber.

These results suggest that it might be beneficial to simultaneously inject water directly into the cylinder and indirectly into the intake manifold in order to retard the ignition even more for higher compression ratios. Water injected into the manifold would be more efficient in cooling down the intake-side piston and intake-side of the cylinder volume, while water injected into the cylinder would directly affect exhaust-side piston as well as high temperature EGR and its surroundings.

5.5. Performance and emissions for final operating points

In order to determine what was the effect of an increased compression ratio and water injection on the engine performance such parameters as IMEP, indicated efficiency and emissions are compared for the final operating points from Table 3 and presented in Figure 6. If it comes to IMEP (which also corresponds to engine power output), it is related to the engines load and cannot be directly compared between different compression ratios. However, it can be seen that generally with Manifold Water Injection IMEP was increased. It was due to lower intake temperatures, which allowed to trap more air in the cylinder, as well as maintain equivalence ratio at the level of ϕ=0.5. On the other hand, Direct Water Injection decreased IMEP, because of the worse scavenging (more EGR trapped in the cylinder). Additionally, for CR9 and CR10 cases equivalence ratio needed to be adjusted for dP/dCA limit, which also contributed to the lower power output. The lowest results were obtained for CR9 case with Direct Water Injection, while the highest results were for CR12 case and Manifold Water Injection.

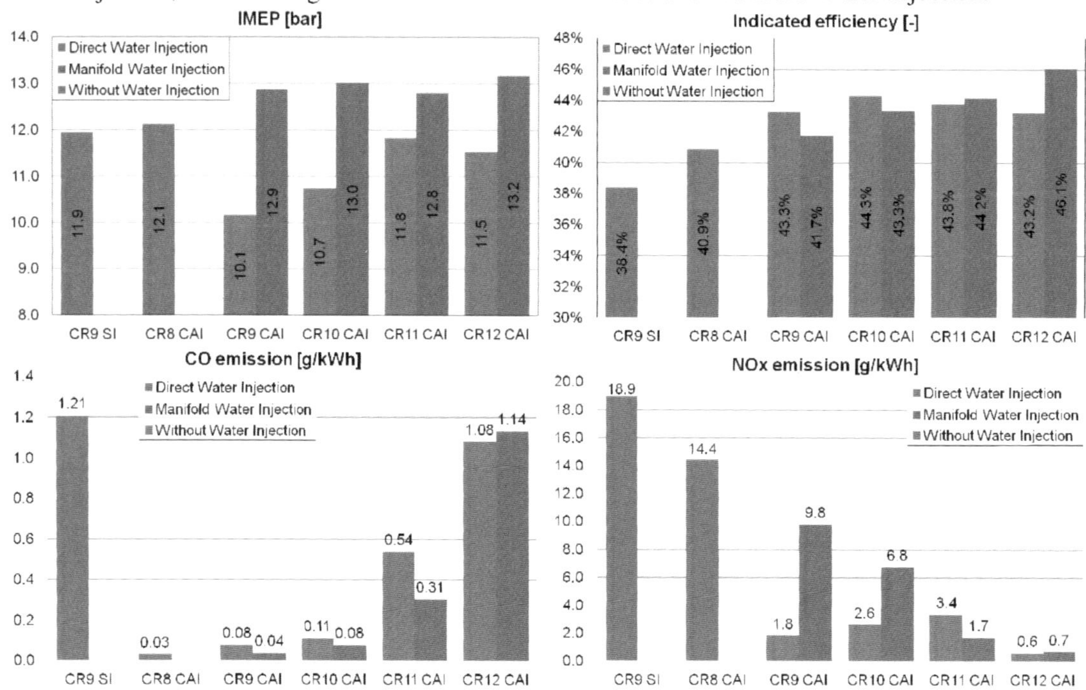

Figure 6. IMEP, indicated efficiency, CO and NOx emission

The engines efficiency is probably of more importance than power output, because it directly translates into the specific fuel consumption. Performed simulations prove that switching from SI mode to CAI mode results in efficiency increase. Then, in CAI mode with Manifold Water Injection efficiency was further increased with the compression ratio reaching 46.1% for CR=12. Although this trend is not clear with Direct Water Injection, for this mode the efficiency was generally higher for higher compression ratios. Differences between cases with DWI might be related to uneven water distribution that affected heat transfer coefficient and heat losses to the cylinder wall. Finally, the maximum efficiency of 46.1% obtained for CAI combustion is a significant improvement in regard to the benchmark SI combustion efficiency of 38.4%.

The most important result nowadays are emissions. For considered operation points switching from SI to CAI combustion decreased the NOx emission by over 20%. It was mainly due to lower equivalence ratio which resulted in lower maximum local temperatures. As for the CAI with Manifold Water Injection NOx emission was also decreased with every compression ratio increase. The more water was injected, the lower maximum local temperatures and local oxygen concentrations were, what favored decrease in NOx formation. In case of Direct Water Injection NOx reduction was even more effective for compression ratios CR=9 and CR=10. However, it was caused mainly by the reduced equivalence ratio in these two cases. For CR=11 MWI was more effective in NOx reduction than DWI, while for CR=12 similar, very low NOx emissions were obtained for MWI and DWI.

When it comes to CO emission, switching from stoichiometric SI combustion to CAI mode resulted in significant CO concentration reduction in exhaust gases. It was possible due to air excess which allowed for CO to CO_2 oxidation at temperatures above 1500 K. CO emission remained low for CR9 and CR10 cases (despite the water injection method). With more water injected in CR11 and CR12 cases CO emission was increased probably due to lower combustion temperatures that inhibited CO oxidation. CR10 and CR11 cases can be considered as a trade-off between CO and NOx emission.

6. Conclusions

In this study CAI combustion in a 2-stroke barrel type OP engine with direct gasoline injection was investigated. The VCR and water injection were considered for the ignition timing control. 3D CFD numerical simulations were performed in conjunction with a reduced kinetic scheme for iso-octane oxidation in order to account for both low and high temperature heat release. To avoid excessive heat release and knock in case of CAI combustion equivalence ratio was adjusted to the level of ϕ=0.4÷0.5 for investigated cases. Possible CAI engine operating points were determined and performance results were compared with conventional SI engine operation. Simulations proved that with VCR and water injection it may be possible to indirectly control the ignition timing in the OP engine working in CAI combustion mode. Furthermore, in defined operating points engine efficiency was improved in CAI mode. As for emissions, NOx concentration in exhaust gases was significantly reduced with water injection thanks to lowered combustion temperatures while CO emission can also be reduced with CAI mode as long as temperatures after the combustion process are sufficient for CO to CO_2 oxidation.

Acknowledgments

Presented work is a part of the research which received funding from the Polish-Norwegian Research Program operated by the National Centre of Research and Development under the Norwegian Financial Mechanism 2009-20014 in the frame of Project Contract No Pol-Nor/199058/94

References

[1] AVL Fire Manual v2014
[2] Flint M, 2010 Opposed Piston Engines: Evolution, Use, and Future Applications. *Warrendale: SAE International*
[3] Mazuro P, Rychter T, Teodorczyk A Piston engines with cylinder axis parallel to drive shaft axis - classification and review. *Journal of KONES Powertrain and Transport*. Vol.13, No. 3
[4] Zhao H 2007 HCCI and CAI engines for the automotive industry, *Woodhead Publishing*
[5] Oakley, A, Zhao, H, Ma, T, and Ladommatos N 2001 Experimental studies on controlled auto-ignition (CAI) combustion of gasoline in a 4-stroke engine. *SAE* paper 2001-01-1030
[6] Cairns A, Blaxill H 2005 The Effects of Internal and External Exhaust Gas Recirculation on Gasoline Controlled Auto-Ignition. *SAE* paper 2005-01-0133
[7] Sjöberg M, Dec J E 2005 Potential of Thermal Stratification and Combustion Retard for Reducing

Pressure-Rise Rates in HCCI Engines, Based on Multi-Zone Modeling and Experiments. *SAE* paper 2005-01-0113

[8] Christensen M, Johansson B, Amnéus P, Mauss F 1998 Supercharged Homogeneous Charge Compression Ignition. *SAE* paper 980787

[9] Tanaka S, Ayala F, Keck J C 2003 A reduced chemical kinetic model for HCCI combustion of primary reference fuels in a rapid compression machine. *Combust Flame*;133:467–81

[10] Jia M, Xie M 2006 A chemical kinetics model of iso-octane oxidation for HCCI engines. *Fuel* 85(17–18):2593–604

[11] Tsurushima T 2009 A new skeletal PRF kinetic model for HCCI combustion. *Proc Combust Inst*;32(2):2835–41

[12] Liu Y, Jia M, Xie M, Pang B 2013 Improvement on a skeletal chemical kinetic model of iso-octane for internal combustion engine by using a practical methodology. *Fuel*;103:884–891

[13] Hanjalić K, Popovac M, Hadziabdić M 2004 A robust near-wall elliptic-relaxation eddy-viscosity turbulence model for CFD. *Int Journal of Heat and Fluid Flow,* 25(6), pp. 1047–1051

IOP Conf. Series: Materials Science and Engineering **148** (2016) 012084 doi:10.1088/1757-899X/148/1/012084

Statistics of indicated pressure in combustion engine.

L J Sitnik and M Andrych-Zalewska

Wroclaw University of Since and Technology, Faculty of Mechanical Engineering
Wybrzeze Stanislawa Wyspianskiego 27, 50-370 Wroclaw (Poland)

E-mail: lech.sitni@pwr.edu.pl; monika.andrych@pwr.edu.pl

Abstract: The paper presents the classic form of pressure waveforms in burn chamber of diesel engine but based on strict analytical basis for amending the displacement volume. The pressure measurement results are obtained in the engine running on an engine dynamometer stand. The study was conducted by a 13-phase ESC test (European Stationary Cycle). In each test phase are archived 90 waveforms of pressure. As a result of extensive statistical analysis was found that while the engine is idling distribution of 90 value of pressure at any value of the angle of rotation of the crankshaft can be described uniform distribution. In the each point of characteristic of the engine corresponding to the individual phases of the ESC test, 90 of the pressure for any value of the angle of rotation of the crankshaft can be described as normal distribution. These relationships are verified using tests: Shapiro-Wilk, Jarque-Bera, Lilliefors, Anderson-Darling. In the following part, with each value of the crank angle, are obtain values of descriptive statistics for the pressure data. In its essence, are obtained a new way to approach the issue of pressure waveform analysis in the burn chamber of engine. The new method can be used to further analysis, especially the combustion process in the engine. It was found, e.g. a very large variances of pressure near the transition from compression to expansion stroke. This lack of stationarity of the process can be important both because of the emissions of exhaust gases and fuel consumption of the engine.

1. Introduction

The challenge facing the automotive industry, mainly reduce toxic exhaust emissions. This limitation must be achieved while reducing energy consumption. This means the need to raise the efficiency of drives.

Progress is seen in hybrid drives. Although this type of drive is unquestionable past and future are:

- The density of accumulated energy is too low - which means the need for unproductive by carrying extra weight in vehicles, and it must reflect negatively on the energy efficiency of vehicles.
- Casa charging is too long - at the same time prevents efficiently recover energy, e.g. In the braking process.
- There are no reports as to where the estimates have come up energy, especially electricity, in quantities enabling; meet the demand of modern transportation.

Until these problems are resolved intentional is still perfecting the classic power sources, especially combustion engines. The issue of improving the internal combustion engine is part of this work. It deals with questions related to the theoretical foundations of internal combustion engines associated with the course indicated pressure. Presents the classic form of indicator diagram engine but based on strict analytical basis for amending the displacement volume. This description has been verified measurement results course of the pressure in the combustion engine running on an engine dynamometer. The study was conducted by a 13-phase test ESC. In each test phase Archived 90 waveforms indicated pressure in the later part of the work at each of the tested values of the angle of rotation of the crankshaft, values of descriptive statistics. Thus the data obtained, which in its essence, a new way to approach the issue of analysis indicator diagram can be used to further works, especially on the combustion process in the engine.

2. Pressure waveforms in the burn chamber of the engine

An indicator diagram showing the development of the internal combustion engine cylinder pressure as a function of engine crank angle p = p (α) - called indicator diagram of an open - Fig. 1. If the pressure is shown as a function of the volume of the gases contained in the cylinder between the piston and cylinder head the p = p (V) - a function that is called the graph of the indicator closed - Fig. 2.

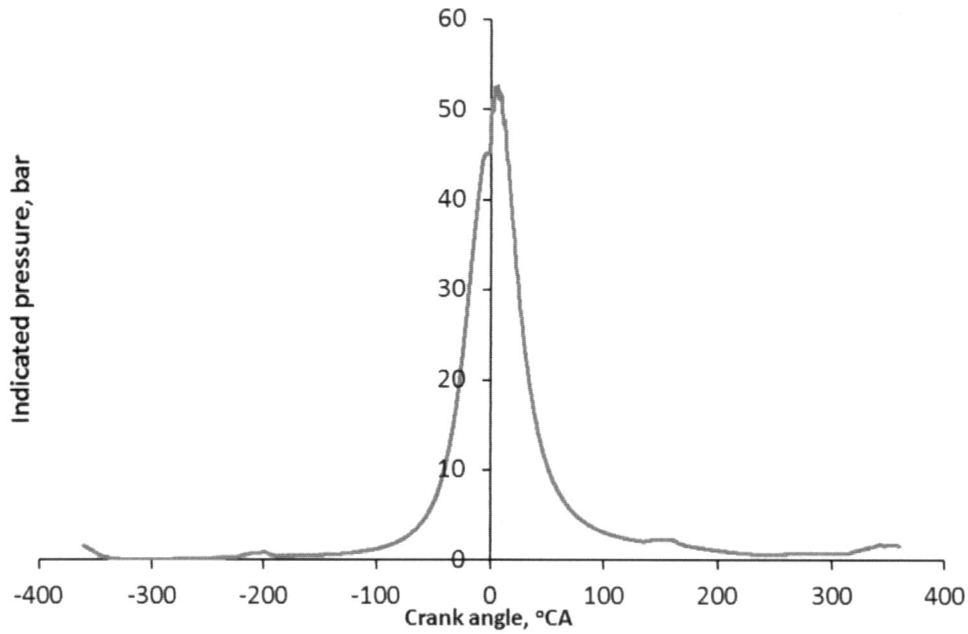

Figure 1. Open indicator diagram test engine

The transition from open to the graph chart contained requires knowledge of changes in the volume above the piston as a function of crank angle. Commonly known from the literature, however, is the basis of all the work, there is a dependency strict. Close relationship here is as follows describing calculation of fuel consumption [14].

The size of the geometric test engine necessary to determine the path of the piston is shown later in this article.

This relation was used to determine the piston travel in function of the angle of rotation of the crown and therefore also the determination of the relationship Vrms = Vrms (α), and consequently, p = (Vrms).

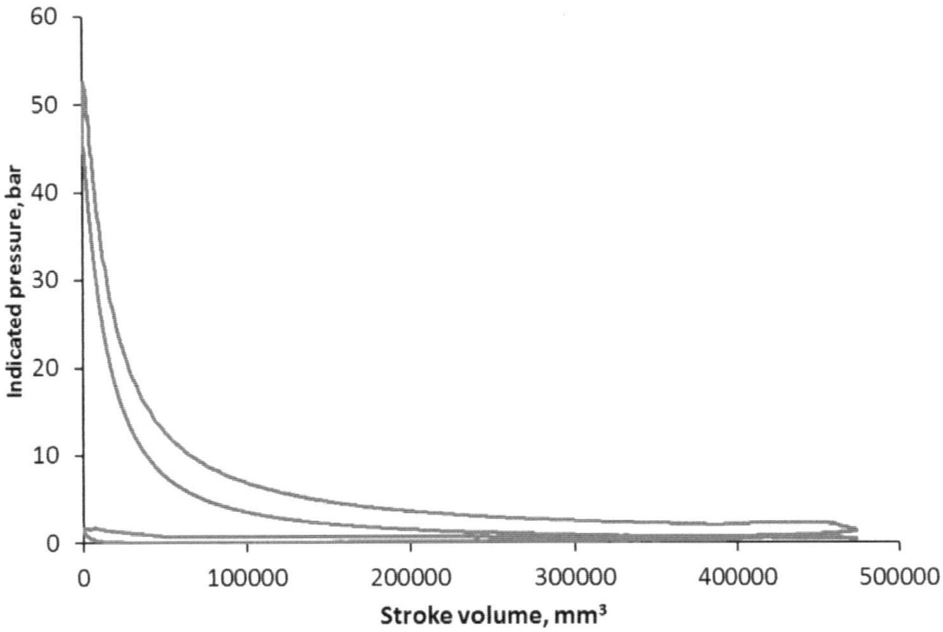

Figure 2. Closed indicator diagram test engine.

An indicator diagram determined using the apparatus, the indication consisting of:

- a pressure sensor arranged in the combustion chamber,
- charge amplifier,
- a sensor of the angular position of the engine crankshaft, and
- DVR.

Indicated engine is thus brought virtually to the pressure prevailing in the working cylinder of the internal combustion engine depending on the angle of rotation of the crankshaft during one cycle.

The real engine cycle differs quite significantly from the theoretical cycle. The engine exchanged the working medium after each cycle. Due to chemical reaction changes the number of moles of the thermodynamic medium is expanded and thus the factor have different physical properties compared to compressed medium, As a result of leakage of the combustion chamber of the thermodynamic medium amount of change no heat is supplied to the thermodynamic medium from the outside, but they result in the combustion of the fuel in the cylinder, – conduct heat input is not consistent with the assumptions p = const. and V = const. There is a loss of incomplete and incomplete combustion of fuel, – compression and expansion factor is not done isentropic; in the initial period of the compression heat is supplied from the hot cylinder wall to the agent at a later stage during the expansion and compression heat is removed from the refrigerant to the cooler walls of the cylinder and its head,• no heat is dissipated from the refrigerant after it has completed its work, but the agent exiting the cylinder carries away a large quantity of thermal energy. All of this belies the thermodynamic assumptions underlying the creation known from the literature indicator diagrams and their subsequent analysis. By joining doa analysis obtained indicator diagrams, especially in terms of optimizing processes in the engine runs you should be aware of this.

3. Engine and his research

Parameters of the tested diesel engine

- Capacity: 1896 cm^3
- Bore: 79.5 mm
- Stroke: 95.5 mm

- Compression ratio: 19.5
- Arrangement of cylinders: in line
- Number of cylinders: 4
- Number of valves: 8
- Injection system: pump injectors,
- The maximum power of engine; 100 kW.

The study was conducted using an engine dynamometer with the appropriate gear. Pressure measurement performed using the sensor.

4. Points characteristics of the ESC test

Engine test was carried out using engine dyno. Engine loaded by the ESC test. The detailed procedure determine the parameters of the test in clinical engine described in [15]. The values of engine speed and load corresponding shown in Figure 3. The size of the circles in the figure correspond to the share of the load in the total load in the assay.

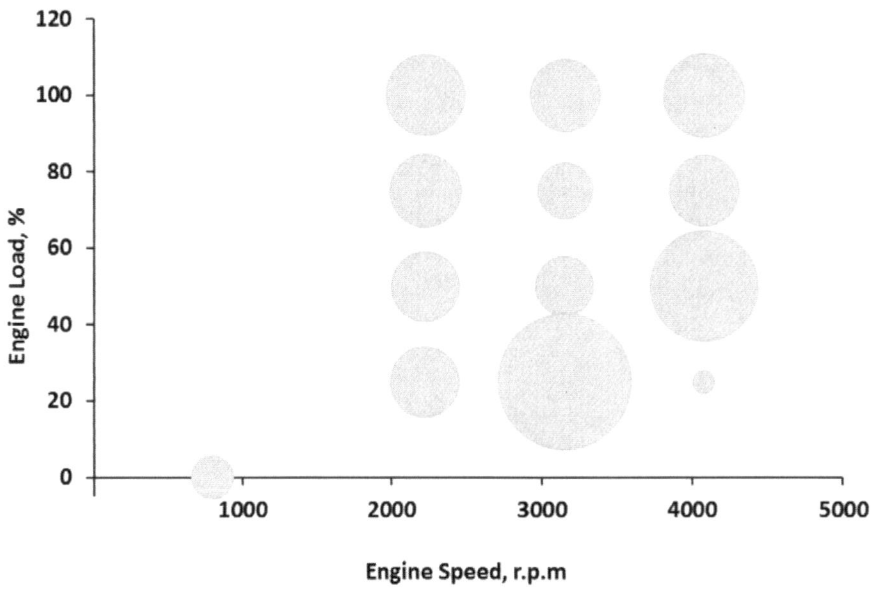

Figure 3. Points motor characteristics corresponding to the requirements of the ESC test.

5. Averaged pressure Indicated

As a result of analysis (not presented here) we found that while the engine is idling (phase 1 test ESC), the statistical distribution of 90 of the pressure at each value of the crank angle, can be described statistically uniform distribution. While in the other phases of the test, 90 such values can describe a normal distribution. The verification tests were performed using the Shapiro-Wilk, Jarque-Bera, Lilliefors, Anderson-Darling.

Measuring the angle of rotation of the crankshaft was carried out with an accuracy of 0.1 degrees. As a result, during one cycle of the engine 7200 obtained the angle of rotation of the crankshaft and the same, the corresponding values of pressure. In each phase, the ESC test measurement was repeated 90 times. In each phase of the test obtained so 7200X90 = 648 000 of the crank angle and the corresponding pressure.

It must be emphasized that the pressure measured at one point in the combustion chamber. Location of measuring point due to the construction of the engine and was not in any way chosen

IOP Conf. Series: Materials Science and Engineering **148** (2016) 012084 doi:10.1088/1757-899X/148/1/012084

Presentation of the whole experimental material is, in this work, for obvious reasons, impossible. To illustrate the results obtained are used merely an example of data 13, part of the test. This phase corresponds to the load of 50% at a speed of 4075 rpm.

At each value of the angle of rotation of the crankshaft 90 of pressure values were available. It was possible, therefore, to carry out statistical analysis of these values and setting the parameters of descriptive statistics. Selected results from these statistics are shown in Table 1.

Table 1. An example of the results of descriptive statistics 90 pressure at selected values of the crank angle in the 13th stage of the test ESC.

CA [deg]	Average	Standard error	Median	Mode	tandard deviation	Sample variance	Kurtosis	Inclination	Range	Minimum	Maksimum	Sum	Counter	Trust level (95%0
-10	41.35	0.02	41.33	41.29	0.15	0.02	0.38	0.20	0.77	41.00	41.77	3721.92	90.00	0.03
-9	42.36	0.02	42.32	42.32	0.20	0.04	0.49	0.47	1.04	41.84	42.88	3812.05	90.00	0.04
-8	43.41	0.07	43.21	43.12	0.67	0.44	6.93	2.56	3.74	42.66	46.40	3907.34	90.00	0.14
-7	44.33	0.12	43.88	43.75	1.12	1.24	6.40	2.50	6.02	43.43	49.45	3989.55	90.00	0.23
-6	44.68	0.09	44.34	44.31	0.90	0.80	4.40	2.21	4.30	43.91	48.20	4021.54	90.00	0.19
-5	45.01	0.12	44.53	44.53	1.18	1.40	4.63	2.29	5.51	44.13	49.63	4051.32	90.00	0.25
-4	45.11	0.14	44.55	44.55	1.31	1.71	4.20	2.20	6.04	44.04	50.08	4059.50	90.00	0.27
-3	45.12	0.17	44.43	44.43	138	2.49	4.29	2.20	7.42	43.97	51.39	4061.08	90.00	0.33
-2	45.14	0.20	44.22	44.13	1.94	3.75	3.85	2.07	9.45	43.71	53.16	4062.44	90.00	0.41
-1	45.41	0.25	44.19	44.00	2.38	5.65	1.18	1.56	8.94	43.53	52.46	4087.27	90.00	0.50
0	47.66	0.31	47.04	46.(8	258	8.85	-0.88	0.52	10.98	43.38	54.36	4289.47	90.00	0.62
1	50.92	0.21	50.74	52.55	1.95	3.81	-0.10	-0.23	9.49	45.15	54.64	4582.50	90.00	0.41
2	50.30	0.22	50.19	52.28	2.12	4.49	0.48	-0.09	10.97	44.33	55.31	4527.27	90.00	0.44
3	51.04	0.26	51.08	50.06	2.44	5.95	0.01	-0.23	11.49	44.38	55.87	4593.34	90.00	0.51
4	52.45	0.24	52.10	52.08	2.28	5.20	-0.47	0.26	11.01	47.23	58.24	4720.63	90.00	0.48
5	51.79	0.28	51.68	51.22	2.66	7.07	-0.08	0.08	13.32	46.05	59.37	4661.54	90.00	0.56
6	52.53	0.22	52.59	51.61	2.13	4.55	-0.05	-0.43	10.13	46.01	56.14	4727.47	90.00	0.45
7	51.89	0.24	51.99	50.64	2.24	5.04	1.43	0.26	14.64	45.31	59.95	4670.11	90.00	0.47
8	51.21	0.27	51.46	50.84	2.60	6.77	-0.16	-0.35	12.87	43.58	56.44	4608.84	90.00	0.54
9	51.68	0.27	51.38	50.82	252	6.35	-0.33	0.48	11.98	46.52	58.49	4650.97	90.00	0.53
10	50.41	0.20	50.53	51.22	1.94	3.76	-0.55	0.09	8.11	46.85	54.95	4536.71	90.00	0.41

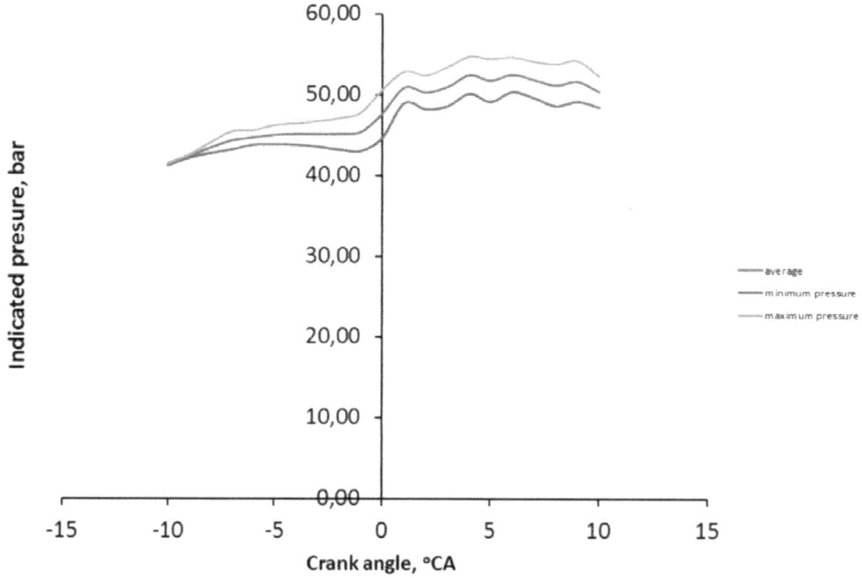

Figure 4. Averaged pressure indicated and range of variability of this pressure for selected values of the crank angle in a ambient of TDC in the 13th phase of the ESC engine test.

Mileage averaged pressure changes and changes the scope of this variation. For each value of the angle of rotation of the crankshaft is registered 90 pressure values, which are derived from 90 individual engine cycles. From a statistical point of view, these 90 pressure values are part of the general population. These 90 values can describe a normal distribution (out of phase idling). Determine can be further such parameters as kurtosis and skewness of the statistical distribution. An example thereof is given in Table 1 and Figure 5 are illustrated.

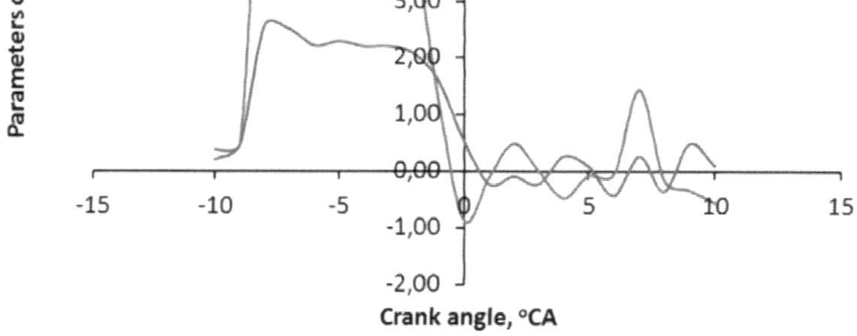

Figure 5. Kurtosis and skewness of pressure distributions for selected values of the angle of rotation of the engine crankshaft in the 13th stage of the test ESC.

It is characteristic that the value of kurtosis and skewness do not change their character before TDC piston (during the build-up of pressure). After TDC the pressure is maintained for a certain period in a certain, fairly constant level, and the kurtosis and skewness oscillate about zero. If we analyze all the data relating to the 13th phase of the ESC test shown above observations are confirmed - Figures 6 and 7.

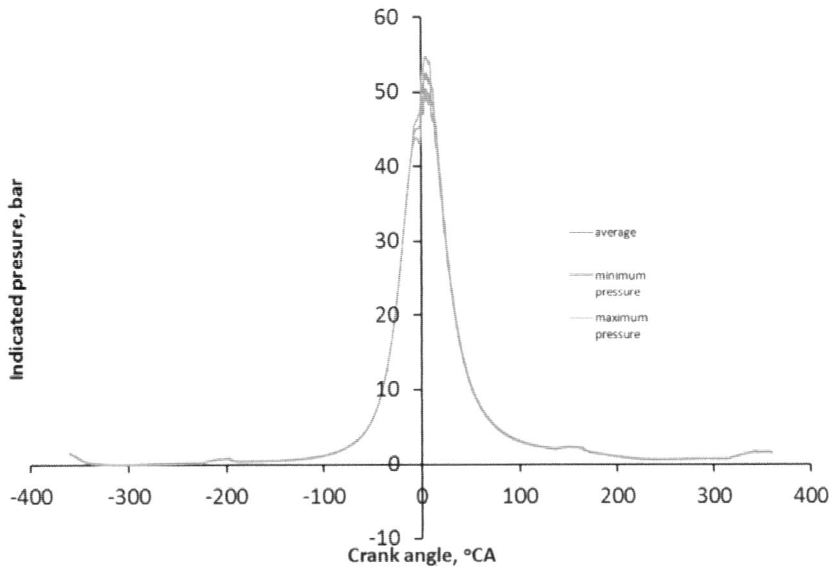

Figure 6. Averaged pressure Indicated and the scope of variation as a function of angle of rotation of the crankshaft engine to be tested in the 13-test phase ESC.

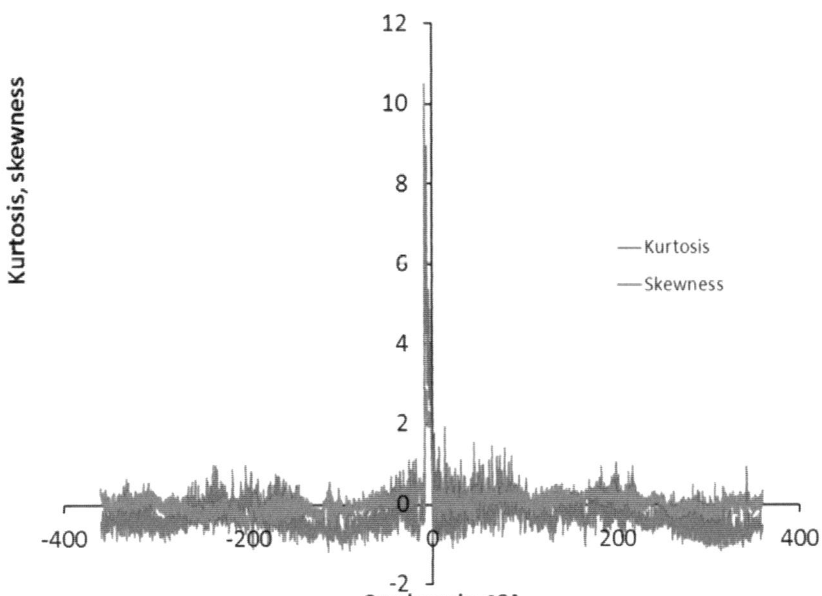

Figure 7. Kurtosis and skewness pressure distributions as a function of the angle of rotation of the engine crankshaft in the 13th stage of the test ESC.

But this is not the correctness at all points of the characteristics of the engine. In the drawings 8 and 9 show the analyzed data with respect to all the test phases ESC.

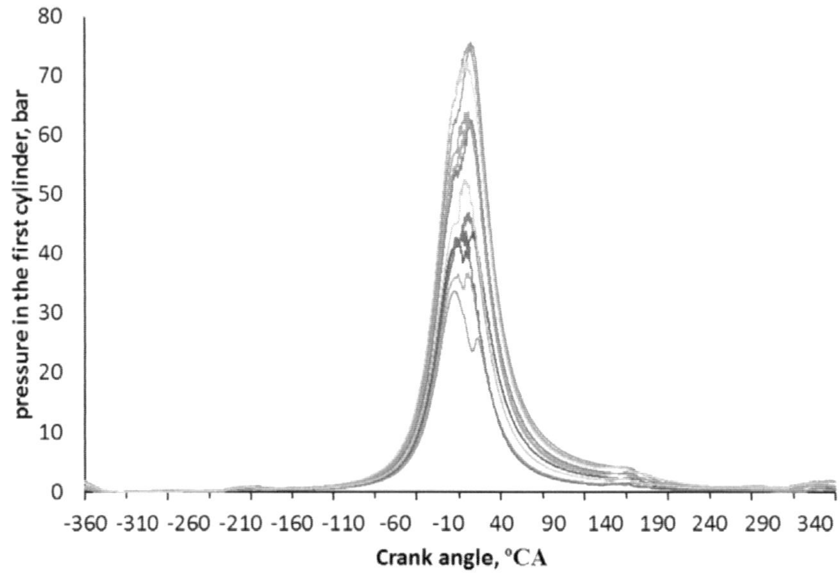

Figure 8. Indicated averaged pressure as a function of the crank angle of the engine under test 13 then ESC test phases.

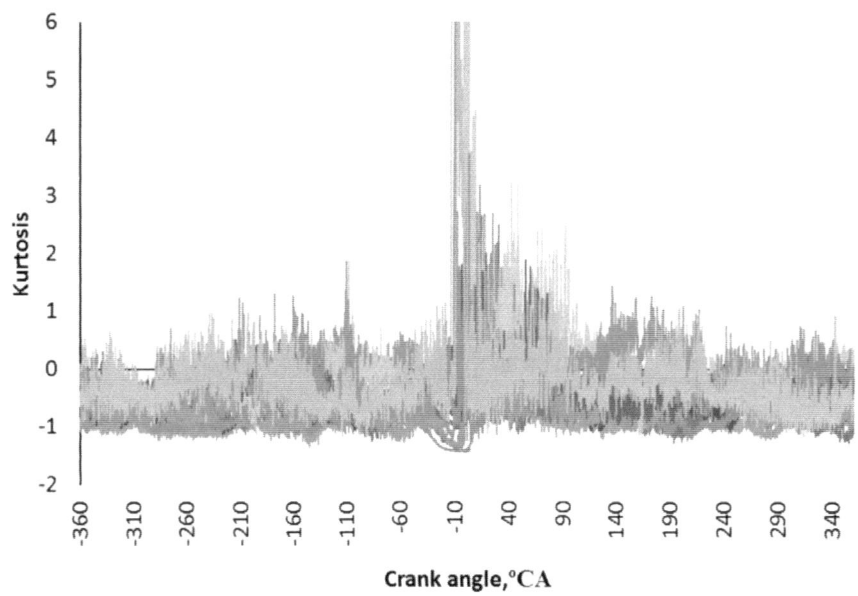

Figure 9. Kurtosis as a function of the angle of rotation of the crankshaft engine to be tested in 13 ESC test phases.

The data presented indicate that the processes in the vicinity of TDC between compression and expansion strokes are random processes with clearly non-stationarity. This lack of stationarity as a function of crank angle for the realization of a substantial part of the fuel energy conversion is not raised so far in the literature. Lack of stationarity has no basis in existing views on the implementation of the combustion process. This lack of stationarity is undoubted but also affects the transition temperature, and in the wake of the permanently reactions during the combustion and expansion of gases. This results in effects on the economy and engine emissions of toxic exhaust gas components. Progress in engine development requires an explanation of this phenomenon, his mastery and control it.

6. Conclusion

Was studied diesel engine on the engine test bench. Studies carried out in stationary conditions. The motor was loaded by the ESC test. In each phase of the test recorded 90 engine cycles. Was measured a crank angle and the corresponding combustion chamber pressure. Was obtained a database for statistical analysis. It was found that in addition to idle the engine, all the values indicated pressure for each crank angle can be described normal distribution. The parameters of this distribution change significantly with TDC neighborhood of the motor at the transition from compression stroke to a power stroke.
The data presented indicate that the processes in the vicinity of TDC between compression and expansion strokes are random processes with clearly non-stationarity. This lack of stationarity as a function of crank angle for the realization of a substantial part of the fuel energy conversion is not raised so far in the literature. Lack of stationarity has no basis in existing views on the implementation of the combustion process. This lack of stationarity is undoubted but also affects the transition temperature, and in the wake of the permanently reactions during the combustion and expansion of gases. This results in effects on the economy and engine emissions of toxic exhaust gas components. Progress in engine development requires an explanation of this phenomenon, his mastery and control it.
Based on the results presented in this work can draw the following conclusions:

1. Statistical analysis of waveforms processes in the engine can be a tool to assess the correctness of their implementation.
2. There was a lack of stationarity of processes in the engine.
3. Lack of stationarity of the processes is not well described in the literature.
4. Lack of analysis of instability can lead to erroneous process control in the engine, and thus also be an obstacle in achieving satisfactory performance of their work.

References

[1] Bernhardt M and Kruczyński S 1982 *Praca silnika w warunkach nieustalonych,* Technika Motoryzacyjna nr 4
[2] Chłopek Z 2001 *On selected methods of numerical differentiation on the example of cylindrical pressure course differentiation*, Journal of Kones, Warszawa-Gdynia
[3] Chłopek Z 2001 *The subject of task explicitness in numerical differentiation*, Journal of Kones, Warszawa-Gdynia
[4] Heywood J.B 1988 *Internal Combustion Engines Fundamentals*, Mc Graw Hill Co. N. Y.
[5] Jante A 1933 *Spezifishe Schnellaufzahlen.* ATZ 24
[6] Kostin A K, Ermiekbajew K B 1988 *Ekspłuatacionnyje rieżimy transportnych dizieliej*, Ałma-Ata, Nauka
[7] Litak G, Kamiński T, Czarnigowski J, Żukowski D and Wendeker M 2007 *Cycle-to-cycle oscillations of heat release in a spark ignition engine*, Wydawnictwo Springer, Meccanica 42
[8] Longwic R, Litak G and Asok K 2009 *Reccurence plots for diesel engine variability tests*, Zeitschrift Naturforsch, Sectio A-A Journal of Physical Sciences, 64 a, 96-102
[9] Longwic R, Litak G, Asok K. Sen and Górski K 2008 *Analisys of cycle-to-cycle pressure oscillation in a diesel engine*, Mechanical Systems and Signal Processing
[10] Lotko W, Longwic R and Górski K 2010 *Nieustalone stany pracy silnika wysokoprężnego zasilanego olejem napędowym z eterem etylo – tert butylowym,* Monography WNT Warszawa
[11] Luft S 2003 *Podstawy budowy silników* WKiŁ, Warszawa
[12]. Longowic R 2011 *Charakterystyka działania silnika o zapłonie samoczynnym w warunkach swobodnego rozpędu,* Lublin
[13] Chłopek Z and Piaseczny L 2004 *Badania procesów szybkozmiennych zachodzących w silniku spalinowym,* Zeszyty Naukowe Akademii Marynarki Wojennej rok XLV nr 2
[14] Sitnik L 2010 Some aspect of theory of cumulated fuel consumption *Journal of KONES Powertrain and Transport* Vol. 17, No. 4 Warsaw
[15] Sitnik L 2004 *Ekopaliwa silnikowe (Ecological fuels)* Wroclaw University of Technology ISBN:83-7085-767-1, s. 336

Transient states analysis of CI engine injectors with the use of optical methods

M Skowron, I Pielecha and K Wisłocki

Institute of Combustion Engines and Transport, Poznan University of Technology, Piotrowo 3 street, Poznan, Poland

maciej.skowron@put.poznan.pl

Abstract. The main aim of research was to define real injection time delay against the time of electrical control signal for piezoelectric and electromagnetic diesel fuel injectors. The second objective of this work was the evaluation of influence of typical injection parameters on this delay. The analysis was focused on the occurrence of appropriate control signals recorded with the use of fast-varying data acquisition system and compared with the data recorded by high speed camera. The tests were conducted in constant volume chamber for different injector types used in combustion engines with direct injection of diesel fuel. The tests were performed under variable conditions: different fuel pressure, air back-pressure and injection duration time.

1. Introduction

High pressure injection of liquid fuel realized by common rail system requires high precision in fuel quantity delivered into combustion chamber by injector. Currently there are two main types of injectors commonly used for both compression ignition and spark ignition engine: electromagnetic and piezoelectric. Although operational fuel pressure in gasoline injection systems is much lower in comparison to diesel fuel systems, but injector opening times are almost the same for both types of construction. That fact makes the requirements similar for CI and SI injectors. These devices should be characterized by possibly fast response to proper control signals for both opening and closing operation. Payri et al. [1] investigated the influence of the electromagnetic and piezoelectric technologies on injection process for indirect-acting piezoelectric and solenoid injectors. In that aspect hydraulic characteristics were prepared for both designs and then compared with each other. In this research the fuel rate indicator based on Bosch method was applied. The achieved results showed faster actuation of piezo-injectors compared the solenoid ones. What it means is that their hydraulic delay is smaller. The differences were increasing inversely proportional to the injection pressure. Second observation from that test was the higher flow speed of fuel for piezo-injector compared to the other design with similar hole diameter.

Similar research was conducted by Yu et al. [2], but in that case fuels selected for investigations were kerosene and diesel fuel. Based only on control signals and variation of fuel flow rate from injector nozzle, the injection delay against the time of electrical opening signal was obtained. The values of this parameter were constant independently of injection pressure (in this case 60 and 100 MPa) and the delay times were respectively 0.45 ms for piezo- and 0.55 ms for solenoid injector. Additionally, faster flow from injector nozzle for lower injection pressures was observed. That fact gives significant benefits to combustion engine operation such as better atomization and evaporation of fuel. The said authors used

optical methods only to evaluate the rate of fuel spray development during injection but they did not use these methods to estimate the hydraulic delay of fuel injection.
Salvador et al. [3] determined injection delay as a difference between the start of injector energizing and start of fuel spray outflow from injector, which was measured in diagnostic system for estimation of fuel flow rate. This research was conducted for injection pressure range 30—180 MPa and for different fuel temperatures. The results obtained indicated the influence of both parameters on the injection delay. Shorter delay time was confirmed for higher fuel temperature and pressure.
Similar results were obtained by Armas et al. [4]. In fuel pressure range 40—120 MPa and for time of energizing of 0.5—2 ms obtained injection delay was contained within the range of 0.2—0.3 ms. Lower values were reached for piezo-injectors, however while the energizing time was increasing, the delay time was equalizing to the same value—of around 0.25 ms—for both injector types. The optical tests of piezoelectric injectors were conducted by Magno et al. [5] but those tests referred to the influence of injector wear on fuel flow rate. They proved that brand-new injector had a faster response then the used one. The used injector had also different profile of the divided fuel dose: a bigger pilot and smaller main dose. The injection delay against the start of electrical opening impulse was not analysed in this research. Some prototypes of direct acting piezoelectric injector were presented by Delphi [6] and Continental [7]. The study on these designs was focused on the influence of voltage on fuel flow rate [3, 8, 9]. Payri et al. [9] in their studies used a prototype injector with different rate-shaping models to evaluate the compression ignition indexes of diesel fuel.
Literature studies carried out in the framework of this work have shown that the available knowledge and explanations of the injector control in the aspect of injection delay and its influence on the spray development is insufficient. There are some works available focusing on electrical parameters of control signals and hydraulic parameters for piezo- and solenoid injectors. However, there are no studies focusing on optical investigations in the aspect of injection delay where both electrical delay and hydraulic delay were combined. The authors have already made some preliminary studies in that field [10]. For optical observations the recording speed of 50 kfps was used. In this paper some additional results are presented with accuracy improved by enhancing recording speed up to 250 kfps and for both types of injectors used in CI engines.

2. The research problem

The tests conducted are mainly focused on hydraulic delay and are based on determination of fuel outflow rate from the injector. However, such studies don't deliver enough data to define precisely the start of injection. Unfortunately, the methods are not sensitive enough to record pressure increase in the research chamber from the very beginning. Free of that disadvantage are the optical methods, the application of which allows observing the occurrence of the first fuel drops flowing from the injector nozzle. For that reason the authors decided to use the optical observation of fuel spray creation for their studies on injection delay.
The main research problems of these studies were:

1) determination of the time delay of electrical signal against starting impulse of the process,
2) defining the hydraulic delay of fuel spray creation after starting impulse of the process which can be determined with recording speed of up to 250 kfps.

The methodology applied in this research is presented below in chapter 3.

3. Research methodology

3.1. Test stand

The research was conducted on the test bench in the Institute of Combustion Engines and Transport of Poznan University of Technology. This test bench consists of (its scheme is shown in figure 1):

- A Constant Volume Chamber (CVC) for simulating in-cylinder conditions (back-pressure) typical for injection in internal combustion engine; air pressure range was kept up to 10 MPa; the chamber has optical access and two 500 W halogen lamps to illuminate fuel drops.

- A sequencer to generate control signals (5 V, TTL) for the injection system and the signals powering the diode; resolution of this device is ±1 ns.
- A common rail stand for fuel supply; it consists of high pressure fuel pump Bosch CP 4.1, a control system to achieve fuel pressure in rail up to 200 MPa and another control unit to transform signals from the sequencer into proper electric signal for the injectors to start and finish the injection process.
- Current clamps (Chauvin Arnoux E3N with measuring resolution of 10 mV/A) and passive high-impedance oscilloscope probe M007 by Pico Technology. Both were used to measure electrical signals before the injector.
- An acquisition system which consists of two main elements: the first one is AVL IndiModul to record electric signals (measured by the clamps and the probe) and control signals from the sequencer; the resolution of 500 kHz for signals during this test was selected; the other element is the high speed camera HSS5 by LaVision with recording speed of 250 kfps.

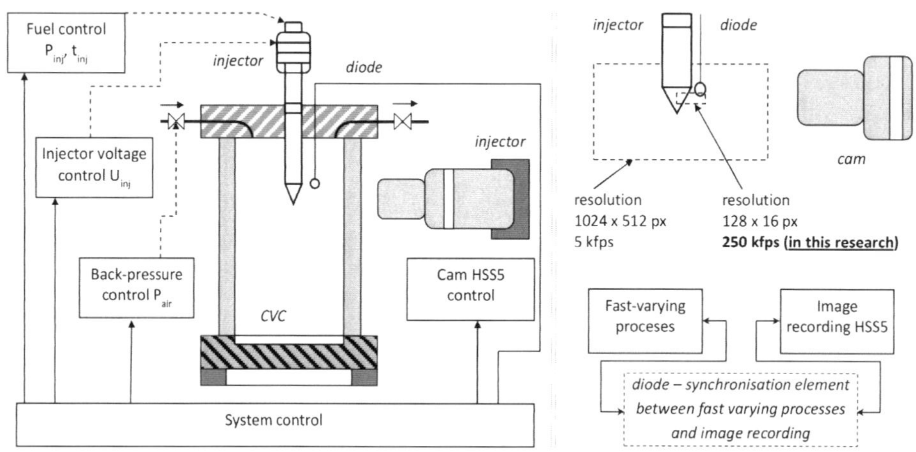

Figure 1. Research stand.

3.2. *Test objects and measuring range*

For the tests the two common-rail injectors of different types were chosen. The first one was a piezoelectric injector by Bosch and the other one was of electromagnetic type, also by Bosch (their parameters are shown in table 1).

Table 1. Characteristics of the injectors.

Parameter	Injector type:	Electromagnetic	Piezoelectric
Static flow		520 cm^3/min	705 cm^3/min
Spray angle		162 deg	170 deg
No. of holes		9	8
Injection pressure		25—180 MPa	up to 200 MPa

The investigations were conducted for values of parameters typical for normal operation of internal combustion engine. The research plan was created with the aim to determine the influence of each parameter on the injection delay time. As the variables of the process were chosen: the injection pressure, the back-pressure and the injection duration. It should be noticed that the last parameter has to get higher value for solenoid injector because of lower fuel flow rate. The range of parameters variation is shown in table 2.

IOP Conf. Series: Materials Science and Engineering **148** (2016) 012085 doi:10.1088/1757-899X/148/1/012085

Table 2. Points of measurements.

Parameter	Value
Injection pressure	30—120 MPa; ΔP = 30 MPa
Back-pressure in CVC	2.5—4 MPa; ΔP = 0.5 MPa
Injection duration	0.2; 0.4; 0.6 for piezo-injector 0.3; 0;5; 0.7; for solenoid injector
No. of iterations	2 for each point

3.3. Electrical signal analysis

During the tests, the electric parameters were recorded (figure 2). This data was used to determine the time delay between the starting impulse and the proper electrical signal (measured in the injector with the use of AVL Concerto software). The starting impulse was selected as the start of TTL control signal for the injector; the finish point of this analysis was set as the time when the response on current clamps was observed. The difference between those two points was signed as *t-e* and describes the hardware delay in the injection system. In this analysis the time *t-d* was defined as the time from the start of the impulse (for injector) to the diode response. The meaning of this time will be explained further in this chapter.

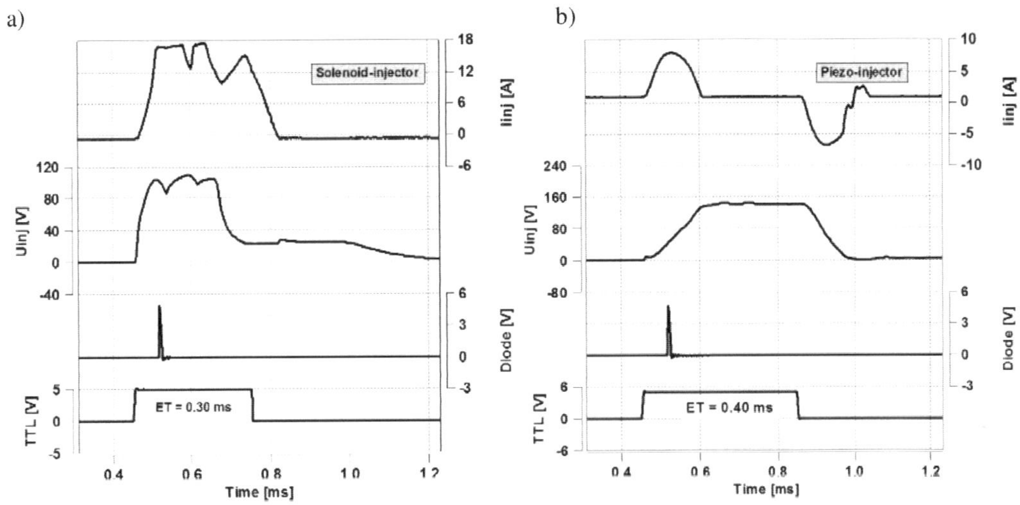

Figure 2. Electric parameters recorded during injector operation:
a) solenoid injector – P_{inj} = 30 MPa, P_{back} = 25 bar, ET (energized time) = 0.3 ms,
b) piezo-injector – P_{inj} = 30 MPa, P_{back} = 25 bar, ET = 0.4 ms.

3.4. Analysis of optical tests

Pictures from optical tests were analysed with the usage of DaVis software from LaVision. The methodology of picture analysis is explained in figure 3. The recording speed was set on 250 kfps with the spatial resolution of 128 × 16 pixels.

At the beginning, the background was subtracted from the raw images to achieve a clear picture of the fuel spray development. The first picture analysed to determine the injection delay was the picture with a diode visible on left side (third photo in figure 3). The diode flash time was set at the value of 4 μs, so it was possible to observe the diode on one frame only. The next step of the analysis was to find the frame where the pixels near the injector nozzle changed their illumination level. The change in illumination means occurrence of fuel drops. That time was described as *t-o*.

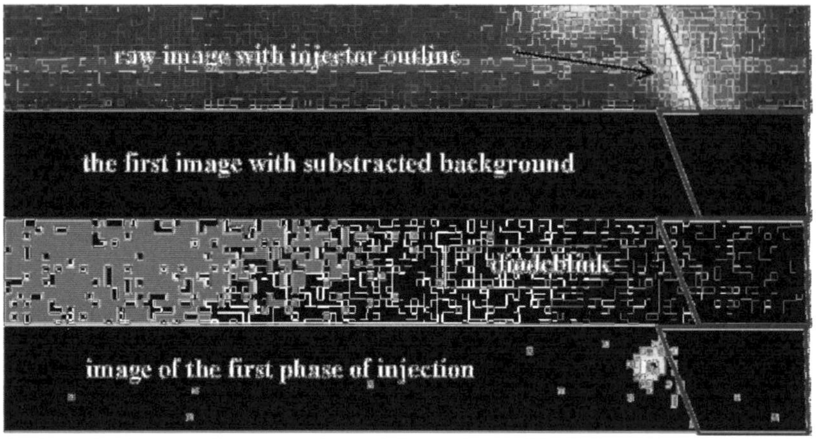

Figure 3. Optical research algorithm of fuel injection based on data from solenoid injector: ET = 0.5 ms, P_{inj} = 120 MPa, P_{back} = 40 MPa.

3.5. Determination of the injection delay

The total delay time from the input signal to the beginning of the injection was defined here as the time difference between the time of the electrical signal from sequencer starting the energizing process and the time of the image with first fuel droplets occurring in the injector nozzle. The diode was used only to synchronize the results achieved from camera and from IndiModule; it reflects the resolution of one recorded frame. In this way, the calculation of fuel injection delay consists of the two main parts (equation 1): electrical and hydraulic delays. As the electric delay the authors understand the value of *t-e* presented in chapter 3.3 (figure 2) and it describes only the delay of the hardware. The hydraulic part (*t-h*, equation 2) consists of *t-d* described in chapter 3.3 and of *t-o* described in chapter 3.4. The hydraulic delay time describes mainly the injector and less the physical parameters of the fuel, such as e.g. viscosity. Equation 1 shows the total delay time according to the presented methodology:

$$t = t\text{-}e + t\text{-}h \quad (1)$$

$$t\text{-}h = t\text{-}d + t\text{-}o \quad (2)$$

The above equations are illustrated in figure 4, whose specific characteristics of the recorded parameters were generated in AVL Concerto. Vertical lines in the figure 4 were assigned to proper frames on the basis of the optical tests.

4. Investigation of the results for piezo- and solenoid injectors

The research results were gathered into groups and presented in form of the charts. The demanded injection time was selected as constant. The results are presented in 3 groups: for short, medium and for long injection time. This type of presentation allows comparing the influence of each parameter on the fuel injection delay. For both injectors the hardware response delay is considered as constant and it is 0.012 ms for piezo- and 0.016 for solenoid injector. As the diode light signal occurs on the same frame after the start of each measurement, it could be stated that the measurements are stable and repeatable.

Figure 5 presents short energizing time for both types of injectors, as was already mentioned in the research methodology. This time amounts to 0.2 ms and 0.3 ms for piezo- and solenoid injector, respectively. On the chart for piezo-injector, the trend to speed up the injection with higher pressure in the chamber was presented. Only for P_{inj} = 60 MPa this tendency is different, but it could be influenced by random circumstances.

The solenoid injector was working in a more repeatable manner and the back-pressure did not make any significant difference in the delay time. In that case the injection pressure was the factor which speeded up the injection, as it was already observed for piezo-injector. What's worth mentioning is the fact, that the results for solenoid injector from P_{inj} over 60 MPa in most cases differ by ±1 frame only, so in terms of the measurement accuracy they should be treated as no difference. Piezo-injector at low injection pressures obtained lower delay than electro-injector, but with the increase in pressure that was reversed. It should be noted that a piezo-injector is an older design: the electromagnetic injector replaced the piezo-injector in the manufacturer designs.

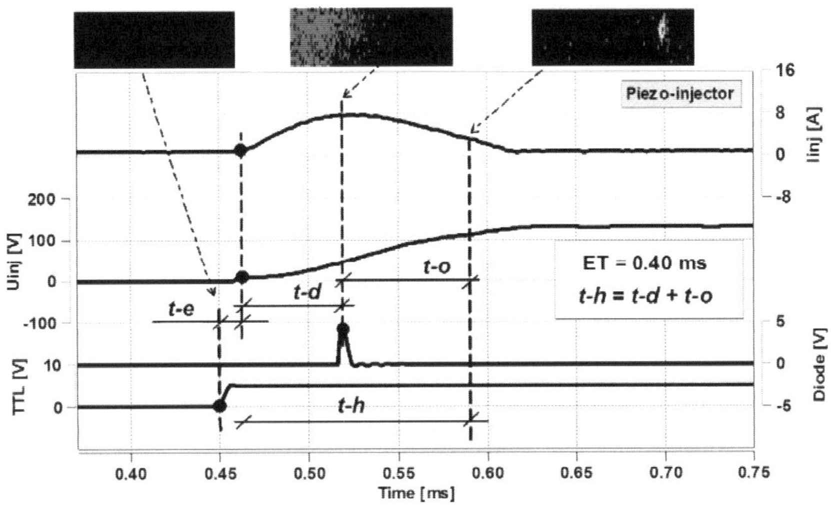

Figure 4. Determination of injector opening delay time based on the recorded data.

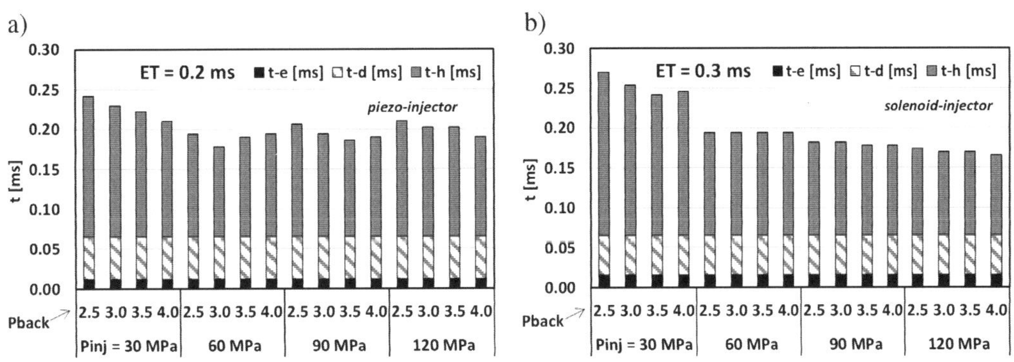

Figure 5. Injection delay for short injection duration; a) piezo-injector, b) solenoid injector.

The results of the injection delay for medium and long injection durations are presented in figure 6 and figure 7, respectively. The piezo-injector operation could be evaluated as more random. On the basis of the results comparison it could be stated, that air pressure (back-pressure) has most significant influence on the delay and the influence of fuel pressure is marginal. Very similar results were achieved for solenoid injectors, but in this case the greatest influence on the injection delay had the fuel pressure and the influence of air pressure was negligible.

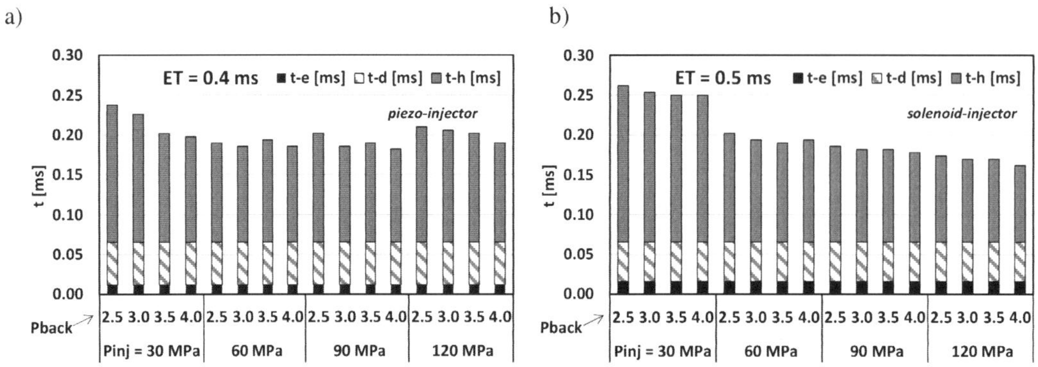

Figure 6. Injection delay for medium injection duration; a) piezo-injector, b) solenoid injector.

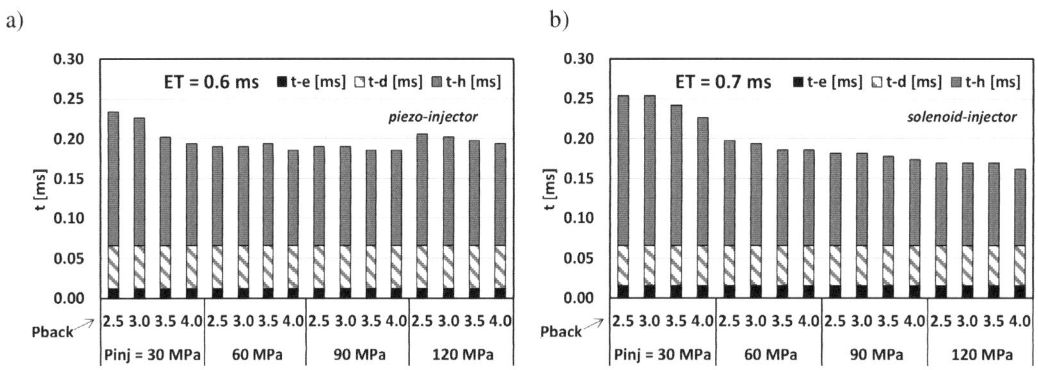

Figure 7. Injection delay for long injection duration; a) piezoinjector, b) solenoid injector.

The above presented results show that these two types of injectors have different properties in terms of their reaction on the electrical opening signal. The piezo-injector is more sensitive to the back-pressure and the fuel pressure has marginal impact on its performance. A contrary situation is noticed for solenoid injector – the biggest impact on the injection delay has the injection pressure. Smaller, but still noticeable impact on this parameter was observed for back-pressure. The injection duration has no impact on the injection delay. According to electrical signals charts such conclusion is correct: the signal from the sequencer continued when the drops of fuel occurred (so if the amplification caused by the appearance of the signal occurred, it lasted for all the measuring points).

5. Summary

In this paper, the study on time delay between the electrical signal for the start of the injection and the beginning of the fuel outflow from the injector nozzle were performed and described. For that purpose the measuring system for electrical signals recording of the injector energizing device was implemented and synchronized with optical observations recorded with the use of high speed camera. The use of the light impulse from the diode made it possible to synchronize all recorded data and to determine electric and hydraulic injection delays on this basis.

The comparison of two injector types was performed in conditions typical for normal operation of an internal combustion engine. In the above conditions, the injection delay time was determined as 0.17—0.24 ms for piezoelectric injectors and as 0.16—0.26 ms for solenoid injectors.

The analysis of these tests allows drawing the following conclusions:

a) Electric delay is constant regardless the injector type and its impact on total delay time is marginal; this parameter describes only the hardware system which controls the injection process.
b) Hydraulic delay is much more variable and it depends on the injection process conditions such as back-pressure and injection pressure; the piezoelectric injector demonstrated more sensitivity to the back-pressure; the reaction time of the solenoid injector was mainly dependent on the fuel injection pressure and less dependent of the air back-pressure.

The total delay time obtained during these tests indicates that solenoid injector seems to be faster than the piezo-injector. Additionally, the solenoid type of injectors offers the possibility to control the injection process more precisely. However, this research does not include fuel dose division, where also the closing time delay is crucial. Further investigations should consider these aspects for better understanding of the injection control process.

Symbols and abbreviations

CI	compressed ignition engine
CVC	Constant Volume Chamber
fps	frames per second
HSS	High Speed Star (cam)
I_{inj}	injector current
P_{air}, P_{back}	back pressure
P_{inj}	injection pressure
SI	spark ignition engine
t	time (overall)
t_{inj}, ET	injection duration, energizing time
$t\text{-}d$	time from beginning point to diode power supply signal starts
$t\text{-}e$	electric delay
$t\text{-}h$	hydraulic delay
$t\text{-}o$	time from diode to fuel drops occur
TTL	transistor-transistor-logic
$\underline{U}_{inj}$	injector voltage
ΔP	delta pressure

References

[1] Payri R, Salvador FJ, Gimeno J and De la Morena J 2011 *Appl. Energy* **88**4 p 1068–1074
[2] Yu W, Yang W, Tay K, Mohan B, Zhao F and Zhang Y 2016 *Exp. Therm. Fluid Sci.* **76** p 12–23
[3] Salvador FJ, Gimeno J, Carreres M and Crialesi-Esposito M 2016 *Energ. Convers. Manage* **114** p 364–375
[4] Armas O, Martínez-Martínez S, Mata C and Pacheco C 2016 *Fuel* **167** p 199–207
[5] Magno A, Mancaruso E and Vaglieco BM 2014 *Energy* **64** p 842–852
[6] Schöppe D, Zűlch S, Hardy M, Geurts D, Jorach RW and Baker N 2008 *MTZ worldwide* **69** 10 p 32–38
[7] Ferrari A and Mittica A2012 *Appl. Energy* **99** p 471–483
[8] Macian V, Payri R, Ruiz S, Bardi M and Plazas AH 2014 *Appl. Energy* **118** p 100–113
[9] Payri R, Gimeno J, Bardi M and Plazas AH 2013 *Appl. Energy* **106** p 152–62
[10] Skowron M and Pielecha I 2015 *Comb. Engines* **162** 3 p 327–334

IOP Conf. Series: Materials Science and Engineering **148** (2016) 012086 doi:10.1088/1757-899X/148/1/012086

Assessment of the effect of gaseous fuel delivery mode on thermal efficiency and fuel losses during the valve overlap period in a dual-fuel compression ignition engine

T Skrzek

Doctor in the Faculty of Mechanical Engineering at University Technology and Humanities in Radom

E-mail: t.skrzek@uthrad.pl

Abstract. The paper describes the effect of dual fuelling of single cylinder AVL test CI engine with the use of two ways of gas delivery to the engine manifold. The engine was fuelled diesel oil and propane. For all the tests, gas consumption was maintained at the same level. In the first mode the gas was delivered by injector located under inlet valve. In the second method, there was used a mixer fitted to the intake manifold. The paper compares the results of thermal efficiency and emissions of propane in the exhaust for both fuelling modes. Research clearly show how important it is to synchronize the injector opening time of the intake stroke. This is especially important for supercharged engines in which there is a valve overlap.

1. Introduction

For many years in the Institute of Vehicle and Machine Exploitation at Kazimierz Pułaski University of Technology and Humanities in Radom there are caring out investigations on dual fuelling of compression ignition engines. [3,4,5,7]. A growing interest in this type of fuelling comes from the possibility utilizing alternative fuels. The dual-fuel concept assumes the highest possible share of the alternative fuel in the total fuel consumption. Most often, the alternative fuel is delivered to the engine via a separate fuel system which provides the fuel of low cetane number to the intake manifold (figure 1).

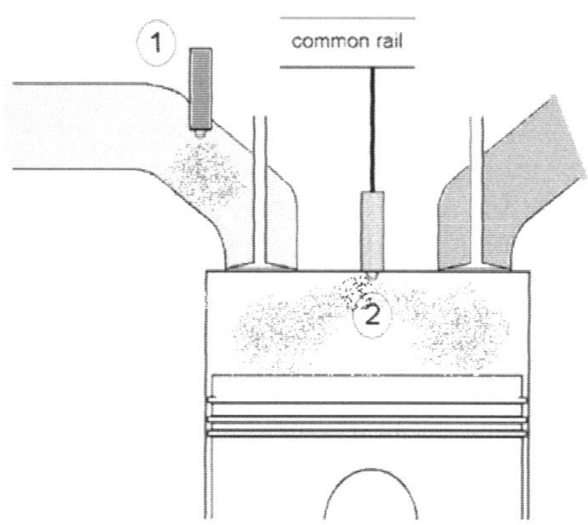

Figure 1. An overall concept of the compression ignition dual-fuel engine with alternative fuel injection into the intake manifold
1 – gaseous fuel injector
2 – diesel oil injector

Previous investigation on dual fuelling revealed many factors determining engine operational parameters. It was showed that diesel oil injection parameters are the most important, regarding engine performance and exhaust emissions. The dual fuel engine demands much more complex control of combustion process. Both the pilot diesel oil quantity and injection angle as well as the pilot dose division affect the process [8] and they should vary in a broad range depending on, engine load and speed. Modern fuel supply systems of compression ignition engines provide broad possibilities to control the above mentioned parameters. In effect, the combustion may be more efficient. Moreover, an increase of the alternative fuel share in the total fuel consumption is one of the main aim of the realization the concept. The top priority of this concept is to obtain the highest possible engine overall efficiency and, at the same time, reduction the toxic exhaust emissions during the process of combustion [2]. Possibility to control diesel oil injection parameters enables a proper selection of them for dual fuel operation, maintaining engine performance on the level of a standard engine operation. However, it should be noticed that despite the broad possibilities, offered by diesel oil supply system, regarding the selection of injection parameters, in the case of dual fuelling it is very difficult to obtain the hydrocarbon emission at the level comparable to that of a standard engine operation. Usually, this may be due to a few reasons. An increased hydrocarbon emission may result from incomplete and imperfect combustion, flame quenching near the combustion chamber walls, chronic combustion of a lean mixture and fuel blow-by resulting from the valve overlap during the operation on gaseous fuel delivered into the intake manifold [6,9]. The last of the above-mentioned factors contributing to the increase of hydrocarbon emission was a reason why engine test attempted to explain the question: to what extent fuel losses provided inlet stroke during the valve overlap period may be reduced in effect of injection parameters adjustment? Two most commonly used methods of gaseous fuel delivery were applied in the investigation. The first one is gas injection near the inlet valve. In the second method, there was used a mixer mounted on the intake manifold.

The valve timing system is responsible for effective working medium exchange [10]. Both, in SI and CI engines, it occurs a simultaneous opening of inlet and exhaust valves, so-called valve overlap. It occurs at the end of the exhaust stroke when the exhaust valve is not closed yet and the inlet valve is just being open, to effectively remove exhaust gas from the cylinder and cool the combustion chamber it is important particularly for turbocharged engines. The cooling effect in the combustion chamber provides more efficient cylinder filling, improves the engine overall efficiency and power as well as affects the exhaust emission. Generally, most modern engines are equipped with valve operation control systems where opening and closing phases vary in a broad range. A dual-fuel engine suffers from the problem of gaseous fuel losses into the exhaust port during delivery into the intake manifold, what creates both economic and ecological problems. It is, therefore, desirable to solve this problem via a direct injection of the gaseous fuel into the cylinder. However, such solution is expensive and is usually applied in big, low speed marine engines [1]. The investigations described in the paper indicate that there is possible to improve the way of gaseous fuel delivery into the intake manifold in order to reduce fuel losses and in effect – to improve the engine overall efficiency.

2. Investigation procedure

The investigations were carried out with the use of AVL 5402 single cylinder compression ignition research engine with a displacement of 511 cm^3 and compression ratio of 17,5. The engine was fueled with diesel oil and propane as a dual fuel engine (figure 1, 2).

Figure 2. AVL test bed: a) general view of the test bed, b) view of the control room.

In the investigation it was applied the most commonly used gas system. Gaseous fuel was delivered in two modes. In the first one, a propane injector was placed near the inlet valve and its operation was synchronized with intake stroke (figure 3a).

In the second method, there was used a mixer mounted on the intake manifold, and the gaseous fuel was supplied in continuous way (figure 3b).

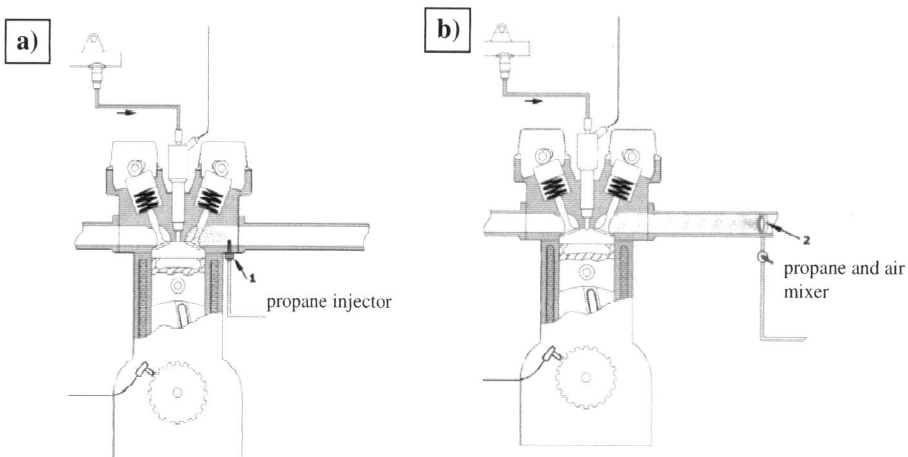

Figure 3. Propane delivery modes: a) application of an injector – gas injection near the inlet valve, b) application of a mixer on the intake manifold.

In the first method, the injector opening time was constant regardless of the boost pressure and was about 16,5 ms. This enabled to maintain a steady propane (kg/h) consumption for all tests. The injector opening time was chosen on a base of earlier tests. Results of them indicated that this time should be long enough to obtain a possible high share of propane in the combustible mixture and, at the same time, it should be short enough to start and finish gaseous fuel delivery in a period shorter than the inlet valve opening period. Taking into consideration the research needs, it was applied a variable gas injection start angle. This parameter varied during the complete cycle, i.e. 720° C.A. For the reason of an essential influence of the start of injection on engine performances and emissions, mainly during the valve overlap period, it was decided to take measurements every 10° C.A. ranging from 225° to 515° C.A. (figure 4). Since any important changes were observed over the remaining area, measurements were taken every 90° C.A. Investigations were carried out in the engine operating both in a naturally aspirated and turbocharged modes. Two boost pressure values (400 and 600 mbar) were applied.

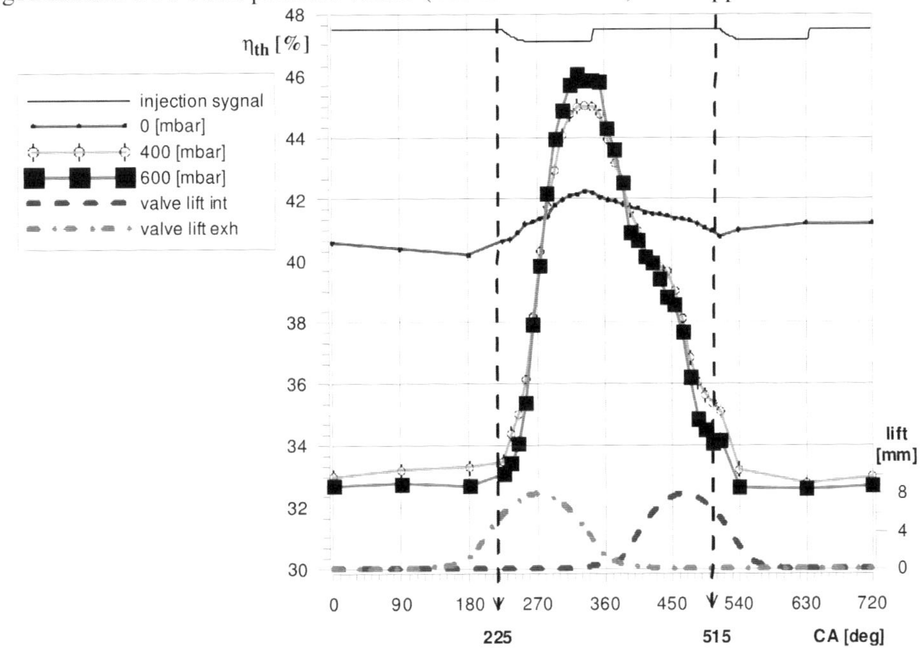

Figure 4. Thermal efficiency versus the angle of propane injector opening for three boost pressure values: 0, 400, 600 [mbar].

As conserving of the propane hourly consumption at various boost pressures causes that the value of air excess ratio varies significantly, it was decided not to compare, in this work, combustion parameters and exhaust emissions.

3. Results and discussion

Results obtained with application of the first gas delivery method are presented in figures 4 and 5. To illustrate the operating range of the injector, it was analysed the course of the steering current which enables to determine approximately both start and end of propane injection. Area with the most apparent changes of thermal efficiency η_{th} starts with gas injection at 225° C.A., when the exhaust valve is visibly open. After 16,5 ms, the injector finishes gas delivery, just before the inlet valve opening. From the point of view of the obtained thermal efficiency, this is the less favourable moment. It happens this way because the gaseous fuel charge, waiting ahead of the inlet valve, penetrates into the exhaust system in effect of the valve overlap. In result, apparent decrease of thermal efficiency η_{th} and, at the same time, equally apparent increase of propane emission are observed (figure 5). Slight delay of the gaseous fuel injection start angle results in a rapid increase of thermal efficiency. This is observed mainly during the turbocharged operation, when an increase of the pressure difference between inlet and exhaust manifolds favours fuel escape. During the naturally aspirated operation, the pressure differences (although significantly lower) also favours fuel escape during the valve overlap period. In this case, it is rather observed sucking out the fuel charge from the intake manifold towards the exhaust manifold. The higher the speed of exhaust gas leaving the cylinder (resulting from higher engine speed) the more fuel escapes (Kadenacy effect) [10]. Thus, the gaseous fuel injection near the inlet valve should be closely synchronized with valve timing. Both, the beginning and end of injection are essential in this case. However, it should be noticed that the registered electrical signal of the injector operation does not correspond to the real start and end moments of fuel injection. It should be taken into account the time necessary to open the injector and the time required for the gas to pass the distance between the injector and the inlet valve. Careful analysis of the changes in thermal efficiency and propane concentration in the exhaust gas may indicate the most effective synchronization of the gas injector opening with valve timing. It results from the investigation, that the synchronization moment depends slightly on the boost pressure as well. Analysing the obtained results, one may conclude that that the beginning of the opening injector can start even before the opening intake valve, only during the naturally aspirated operation. Slight delay of the start of gaseous fuel injection is advantageous for higher boost pressures (the lowest propane emission – figure 5). The highest efficiency and, at the same time, the lowest propane emission are obtained at the moment when the injector opens during the valve overlap period what indicates a delay between the start of the electrical signal and the time necessary for the gas to get near the inlet valve. This means that during the valve overlap period, when the combustion chamber is being flushed, the air charge does not contain propane yet. Unfortunately, the most favourable injector operation area lies in the narrow range of 320° to 360° C.A. Delivery of gaseous fuel a few degrees beyond this range causes a rapid decrease of thermal efficiency resulting from the charge escape into the exhaust manifold. Figure 5 presents a result of the selection of the most favourable gas injection timing when considering propane concentration in the exhaust gas.

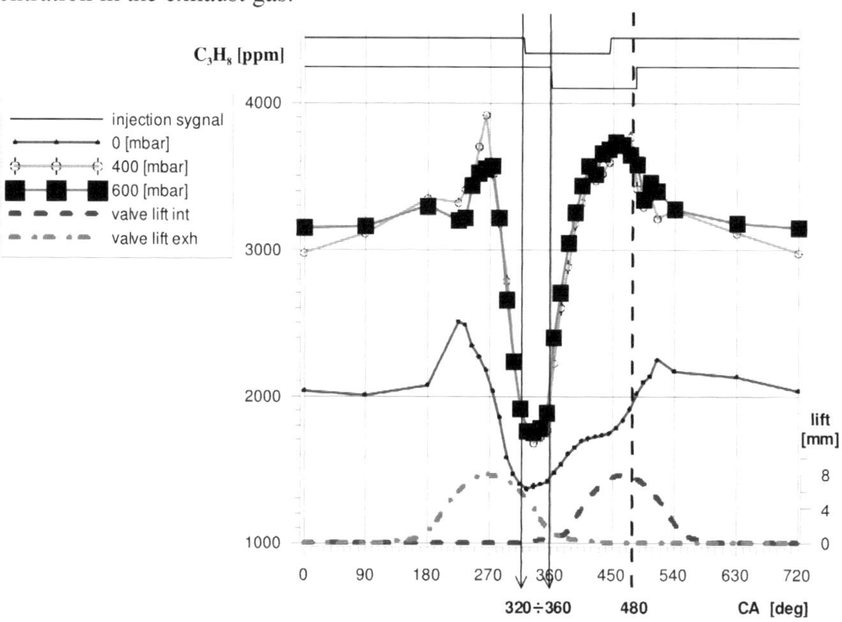

Figure 5. Propane volumetric concentrations in the exhaust gas obtained for various start angles of propane injector opening, for three various boost pressure versus C.A.

The continuous line indicates beginning of the favourable injector operating area at 320° C.A. indicating at the same time the lower limit. The dashed line indicates the end of this area, when the injector should finish gas delivery, what occurs at 480° C.A., indicating at the same the upper limit. It is turning out that setting the injector closing angle is equally important as setting the injector opening angle. Because, delaying the injection time beyond the upper limit results in an effect similar as in the case of too early injection. There occurs here another phenomenon consisting in fuel backflow from the cylinder into the intake manifold. Characteristics presented in figure 6 were obtained with the use of gas exchange and combustion analysis software (GCA) for fuel mass calculation (application dedicated for the single-cylinder research engine used in the present investigation).

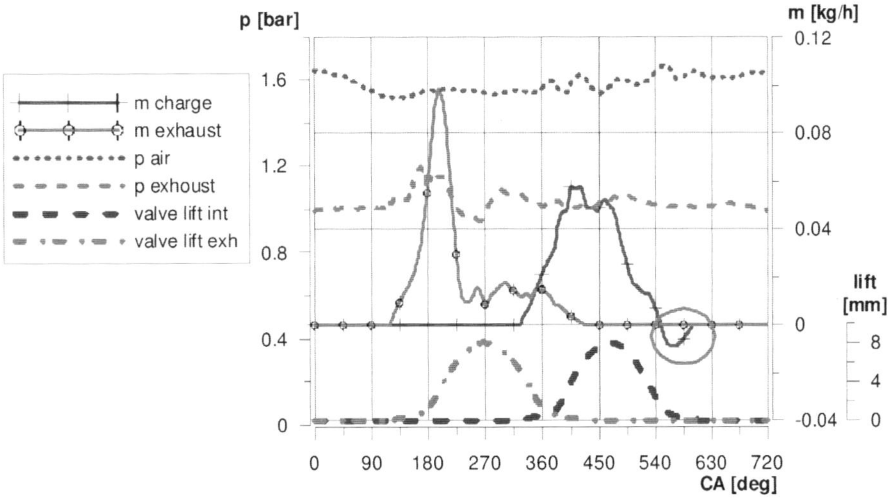

Figure 6. Exhaust mass and charge (gas and air mixture) flows versus crank angle for the boost pressure of 600 mbar.

In this phenomenon, toward the end of the filling cycle, the pressure in the cylinder may be higher than in the intake manifold forcing a part of charge to flow back into the intake manifold just before closing the inlet valve (area marked in circle). Thus, it is favourable to synchronize the time of gaseous fuel delivery with the beginning of filling the cylinder with fresh air. However, the injection must be finished before the fuel flow back into the intake manifold. In this way, in the early phase of refuelling, gas concentration in the mixture is high. Closing the gas injector during cylinder refuelling will result in further supplying it with air only. In this case, the charge is not homogeneous and may contain less fuel in the upper part of the cylinder. So, the gaseous fuel share in the flowing back mixture is low. In result, gaseous fuel losses during the cylinder flushing are small, what is confirmed by investigation. Taking this into account, one may come to a conclusion that a gaseous fuel, left near the inlet valve in result of too late injector closing, i.e. at the end of the intake stroke or after complete injector closing, escapes most intensively into the exhaust manifold during the valve overlap. So, the best solution is when a cylinder is flushed with air only.

Another, common method of gas delivery during a dual-fuel operation consists in gas injection with the use of single injector or a couple of them placed in some distance from the intake manifold, usually behind or in front of the supercharger. Application of this method results in diffusion of gaseous fuel in the air sucked by the engine. However, it does not mean that the mixture is homogeneous. Such prepared mixture is distributed into the individual engine cylinders. A similar effect may be obtained with the use of a mixer. Certainly, this is not the right method, taking into consideration obtaining the highest possible thermal efficiency, due to the inevitable fuel charge escape during the valve overlap. Moreover, gas delivered in this way fills the intake system of a significant capacity resulting from the need to use an intercooler. So, during the engine braking the energy supplied to the intake system is lost. This is due to cutting off diesel oil supply necessary to initiate the combustion. The above considerations raise the following question: which of the applied gas delivery methods will result in a better use of energy? To answer this question, the results obtained with the application of both methods were compared. Figure 7 presents calculated thermal efficiency values obtained for both delivery methods. It should be noticed, that for the first method there are presented the best results i.e. the ones obtained in the most favourable synchronization of the gas injector opening time with valve timing. In the second method, all engine operational parameters i.e. engine speed, boost pressure, hourly consumption of both fuels and diesel oil injection parameters were reproduced. Gaseous fuel was delivered in a continuous manner using a mixer mounted at the entrance to the intake manifold.

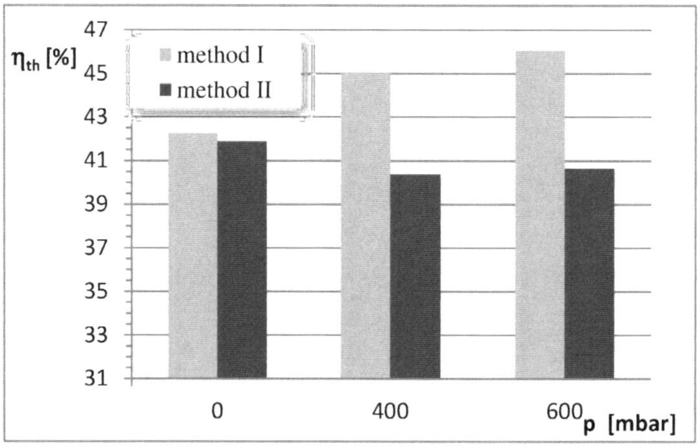

Figure 7. Comparison of thermal efficiency of the investigated engine for both applied gas delivery methods.

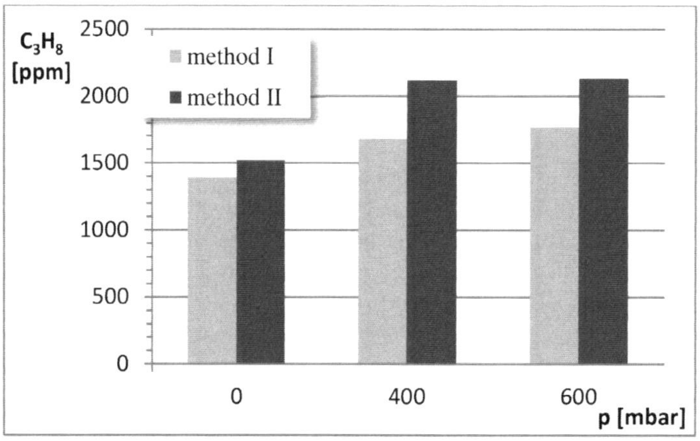

Figure 8. Comparison of propane volumetric concentration in the exhaust gas of the investigated engine for both applied gas delivery methods.

Analysis of the above presented results unequivocally reveals that application of the first method gives better results. In particular, a clear difference is observed in the case of turbocharging, when a higher pressure difference between inlet and exhaust manifolds favours fuel escape during the valve overlap. Benefits resulting from application of the first method are more apparent in the graph of thermal efficiency (figure 4). A graph representing propane volumetric concentration in the exhaust gas clearly indicates the superiority of this method as well. However, it should be noticed that together with propane also more air flows towards the exhaust system resulting in mixture dilution, what may cause an incorrect interpretation of the results regarding the volumetric concentration of the exhaust components.

Summary

On the basis of the carried out investigation, considering the overall efficiency, it may be clearly stated that the first method of gaseous fuel delivery shows superiority over the second one. The gas injector close to the inlet valve and application of the full injection timing control both: the start of injection and the injection duration, essentially influence improvement of the engine thermal efficiency, what clearly limits the fuel escape during the valve overlap. However, it should be considered the need to use a separate system controlling the operation of injectors, applying an intentionally prepared map including at least three basic parameters: engine load, boost pressure and engine speed. It should be added that providing the correct mode of gaseous fuel injection control by keeping at the same time a constant ratio of both fuels (the gaseous fuel to the standard fuel) may be problematic, particularly at high engine speed. Injection time of the gaseous fuel depends on the engine speed. So, the injection must happen within a specific range, beyond which fuel escape is observed. This is related to the need to provide a correct gaseous fuel flow in a shorter time, what may be realized by an increase of the injection pressure or by delivering the gaseous fuel in the liquid state. Therefore, comparing the both gas delivery methods, taking into consideration their complexity, one can see the superiority of the second method. However, the investigation results indicate unequivocally the need to apply the first method which, despite the objective difficulties, shows a high potential.

References

[1] Chen S and Beck N 2001 *Gas Engine Combustion Principles and Applications* SAE Technical Paper 2001-01-2489, doi:10.4271/2001-01-2489
[2] Konigsson F and Stalhammar P and Angström H 2011 *Combustion Modes in a Diesel-CNG Dual Fuel Engine* SAE Technical Paper 2011-01-1962, doi:10.4271/2011-01-1962
[3] Kowalewicz A and Wołoszyn R 2011 *Comparison of performance end emission of turbocharged C.I. engine fuelled either with diesel fuel or CNG and diesel fuel* Combustion Engines, No. 3/2011 (146),
[4] Luft S 2007 *Analysys of combustion process in dual fuel compression ignition engine fuelled with LPG in the liquid phase* Journal of Kones, Powertrain and Transport, Vol. 14, No.3,
[5] Luft S 2008 *Atempt to compare basic combustion parameters of a dual-fuel compression ignition engine for various main fuels and their delivery modes.* Czasopismo Techniczne z. 11. Mechanika z. 7-M,
[6] Mustafi N and Raine R 2008 *A Study of the Emissions of a Dual Fuel Engine Operating with Alternative Gaseous Fuels* SAE Technical Paper 2008-01-1394, doi:10.4271/2008-01-1394
[7] Różycki A 2008 *Granica spalania stukowego w dwupaliwowym silniku o zapłonie samoczynnym* Czasopismo Techniczne z. 11. Mechanika z. 7-M,
[8] Skrzek T 2015 *Wpływ podziału dawki oraz kąta początku wtrysku dawki inicjującej na parametry pracy dwupaliwowego silnika o zapłonie samoczynnym* Combustion Engines, No. 3/2015 (162) - PTNSS–2015–3480
[9] Stelmasiak Z 2013 *Dwupaliwowe silniki o zapłonie samoczynnym* Bielsko-Biała – Radom: ITE
[10] Wajand A 1988 *Silniki o zapłonie samoczynnym* Wydawnictwa Naukowo–Techniczne, Warszawa

Utilization of waste glycerin to fuelling of spark ignition engines

Z Stelmasiak[1] and D Pietras

[1] Corresponding author. Akademia Techniczno-Humanistyczna, Katedra Silników Spalinowych i Pojazdów, 43-309 Bielsko-Biała, ul. Willowa 2

E-mail: zstelmasiak@ath.bielsko.pl

Abstract. The paper discusses a possibilities of usage a simple alcohols to fuelling of spark ignition engines. Methanol and blends of methanol with glycerin, being a waste product from production of bio-components to fuels based on rapeseed oil, have been used in course of the investigations. The main objective of the research was to determine possibilities of utilization of glycerin to blending of engine fuels. The investigations have been performed using the Fiat 1100 MPI engine. Parameters obtained with the engine powered by pure methanol and by methanol-glycerin mixtures with 10÷30%$_{Vol}$ content of glycerin were compared to parameters of the engine fuelled conventionally with the E95 gasoline. The investigations have shown increase of overall efficiency of the engine run on pure methanol with 2.5÷5.0%, and run on the mixture having 10% addition of glycerin with 2.0÷7.8%. Simultaneously, fuelling of the engine with the investigated alcohols results in reduced concentration of toxic components in exhaust gases like: CO, THC and NO_x , as well as the greenhouse gas CO_2.

1. Introduction

Necessity of usage of biofuels in Poland results from implementation of the Polish National Indicative Targets (NCW in short), imposing gradual increase in share of renewable fuels in total volume of the engine fuels [3, 4, 17]. According to decree of the Polish Council of Ministers on the NCW of the 20.07.2013, energetic share of biofuels in the year 2018 should reach the level of 8.5%, Figure 1. This requires numerous changes in agricultural production, new technologies of production of biofuels and changed fuelling systems of the engines [1, 2, 11, 17].

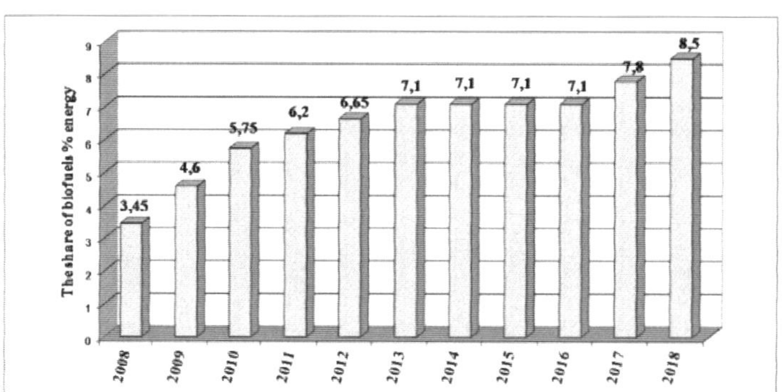

Figure 1. Anticipated share of biofuels in Poland, according to the Polish National Indicative Targets [17].

Achievement of the NCW targets can be attained by fuelling of some percentage of engines with pure biofuels or blends of biofuels with gasoline and Diesel oil [5, 6, 8, 9, 18]. Maximal content of biofuels in mixtures with petroleum fuels depends on a possibility of blending and creation of stable, homogeneous mixtures (in changing ambient conditions and long time intervals), as well as on the octane number (RON) and the cetane number (LC) of the bio-additives [13, 14,17]. For this reason, simple alcohols are added to gasoline: methanol and ethanol mainly, or esters of the alcohols: ethyl-tert-butyl ester and methyl-tert-butyl ester. In turn, in fuels to Diesel engines are used esters of unsaturated fatty acids; in Polish conditions produced from rapeseed oil. The by-product in production of these esters is glycerin in quantity of about 100 kg per 1 tone of processed rapeseed oil [12]. Considering average yearly consumption of Diesel oil, quantity of waste glycerin can be assumed as 100 thousand tons per year. The glycerin can be used in pharmaceutical, agricultural and food industries, but consumption capacity of these industries is much lower. Due to this, surplus in production of glycerin generates a big logistic issue, and some other methods of utilization of glycerin should be explored.

Additionally, another important issue, which should be also taken into considerations, is impact of combustion of alcohols on the natural environment, inclusive of greenhouse effect. The simple alcohols can be produced from a biomass, reproducible source of energy, possible to obtain in large quantities. In nature the biomass undergoes natural decomposition, resulting in production of carbon dioxide (CO_2) and methane (CH_4) – the both gases are classified among so called greenhouse gases. However, due to natural processes of decomposition of the biomass, emission of CO_2 and CH_4 in such case is not counted among harmful emissions. Utilization of the biomass and transforming it into alcohols, and next combustion of such alcohols in the engines results in emission of CO_2 to atmosphere, which in process of photosynthesis is used to production of a new biomass. In result, usage of the biomass can be considered, in aspect of CO_2, as zero-emission one, which additionally reduces natural emission of CH_4 from processes of biodegradation.

Carbon dioxide and water are the effects of combustion of methyl and ethyl alcohols and glycerin, while this process occurs according to the following summary reactions:

$$CH_3OH + \frac{3}{2}O_2 \rightarrow CO_2 + 2H_2O \tag{1}$$

$$C_2H_5OH + 3O_2 \rightarrow 2CO_2 + 3H_2O \tag{2}$$

$$C_3H_5(OH)_3 + \frac{7}{2}O_2 \rightarrow 3CO_2 + 4H_2O \tag{3}$$

Mass fraction of carbon atoms in molecule of alcohol is lower, comparing to traditional fuels, and amounts respectively to 0.375 for methyl alcohol, 0.520 for ethyl alcohol, and 0.391for glycerin, while in case of gasoline and Diesel oil this ratio amounts approximately to 0.845÷0.855. It results in lower emission of CO_2 comparing to combustion of gasoline.

Methanol and ethanol, due to their excellent properties, high octane number, high heat of vaporization and high combustion rate mainly, can be successfully used both to spark ignition engines (as pure fuels or additives to traditional fuels) and to Diesel engines, as additives burnt simultaneously with Diesel oil. Methanol and ethanol have been used for more than eighty years to production of fuels to aviation and sports engines. So far, the glycerin, due to its high viscosity, lower calorific value, and infinitesimal production volumes, hasn't been used to fuelling of engines. However, big market supply of glycerin, connected with production of biofuels, has resulted in interest of many research centres, what can be evidenced by growing number of publications printed recently in scientific literature [7, 9, 10, 12].

The paper presents results of research work on effects of methanol and methanol-glycerin mixtures on parameters of automotive spark ignition engine of the Fiat 1100 MPI type. Utilization of waste glycerin should facilitate achievement of the NCW targets and increase pool of biofuels produced from domestic feedstock.

2. The engine and the engine test bend

The tests were performed on four cylinder, spark ignition engine with multipoint fuel injection of the Fiat 1100 MPI type. Technical data of the engine are presented in the Table 1.

Table 1. Technical data of the engine

Type of the engine	Fiat 1100 MPI
Bore x stroke	70 x 72 mm
Swept capacity	1108 cm^3
Compression ratio	9.6
Max output power/rotations	40 kW/5000 rpm
Max torque/rotations	88 Nm/3000 rpm

On the engine test bend it has been installed a duplex feeding system for fuelling with gasoline and alcohol. Each from these two systems was equipped with individual fuel pump, pressure stabilization system and system to measurement of fuel consumption. The fuelling system enabled controlling of transient fuel consumption, what considerably facilitated selection of engine adjustment when the fuels were changed. Size of injected doses of the fuel was selected depending on type of the fuel, engine load and rotation speed. View of the engine test bed is shown in the Fig. 2.
The engine test bed has incorporated a system to automatic acquisition of measured data to the Excel data sheet. Recorded values were used directly from computer files in programs to processing of the measured data.

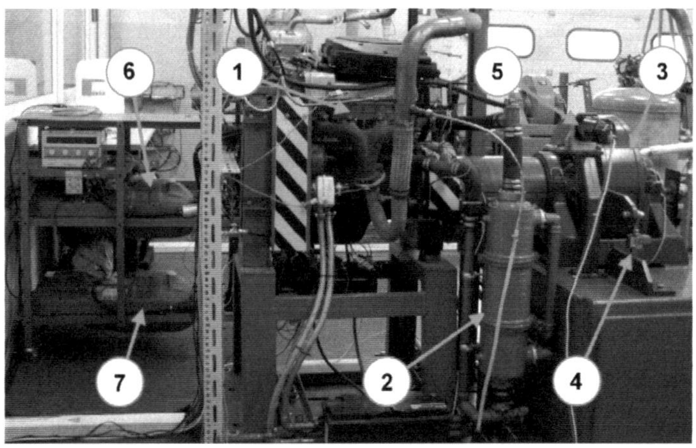

Figure 2. Test bend: 1 – the Fiat 1100 MPI engine, 2 – radiator of cooling fluid, 3 – Eddy-current brake made by Schenck, 4 – transducer of engine load, 5 – sensor of rotational speed, 6 – alcohol tank, 7 – gasoline tank.

3. Results of the investigations

Before start of the tests on the engine dyno it has been performed attempts to mix technical glycerin with Diesel oil, with gasoline, and with methyl alcohol. Assessment of a possibility of the mixing was performed by observation of the mixtures after about 15 minutes from intensive mixing and after about two months of immovable storage. In the Figure 3 is presented a view of the mixtures after about two months from time of the mixing.
Glycerin in ambient conditions is not mixable both with gasoline and Diesel oil. Process of the mixing doesn't depend on content of glycerin; shares of the glycerin within interval of $5 \div 50\%_{Vol}$ were used in course of the investigations. Anyhow, after intensive mixing it can be seen a segregation of the compounds, and glycerin as a heavier compound, in result of gravitation starts to accumulate on bottom of vessel, Figure 3a. After short time, about 15÷20 minutes, it has been observed complete segregation of glycerin, gasoline and Diesel oil fractions.

Glycerin is easily soluble in methanol and ethanol, regardless of percentage share of glycerin in the mixture. After gentle shaking are created homogenous mixtures, maintaining its stable form for a long time, Fig. 3b. From engine applications point of view such situation is beneficial because none operational problems could occur in case of fuelling of engines with mixtures of various alcohols.

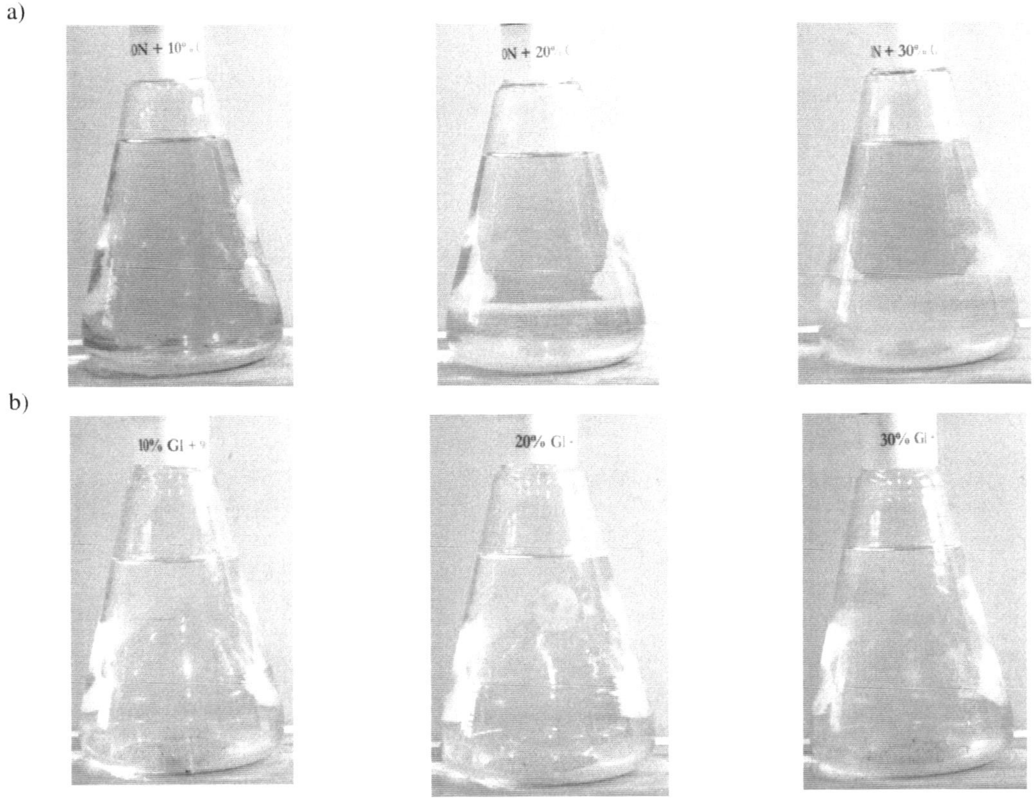

Figure 3. View of the mixtures after two months from mixing: a) Diesel oil – glycerin mixture, b) methanol – glycerin mixture.

On the base of mixing tests of glycerin with different fuels, three mixtures of alcohol and glycerin having specified below volumetric composition have been selected to further comparative tests:

- Met-Gl 10% - 90%$_{Vol}$ methanol + 10%$_{Vol}$ technical glycerin,
- Met-Gl 20% - 80%$_{Vol}$ methanol + 20%$_{Vol}$ technical glycerin,
- Met-Gl 30% - 70%$_{Vol}$ methanol + 30%$_{Vol}$ technical glycerin.

In course of the tests it has been used 5 different fuels: commercial gasoline as a reference fuel, marked in the diagrams as E95, technical methanol marked as Met 100%, and three methanol-glycerin mixtures marked as 10%, 20% and 30% respectively, whereas the percentages denote volumetric content of glycerin in the mixtures. Properties of implemented fuels are specified in the Table 2.

Table 2. Properties of fuels used in course of the tests.

Fuel	Gasoline E95	Methanol	Met-Gl 10%	Met-Gl 20%	Met-Gl 30%
Density [kg/dm^3]	0.74	0.796	0.842	0.889	0.935
Calorific value [MJ/kg]	42.5	19.5	15.9	16.2	16.6
Octane number	95	115	-	-	-

Usage of methanol to fuelling of spark ignition engine results in growth of overall efficiency of the engine, Fig. 4. Improvement of the efficiency is reported in complete range of engine loads and for all tested rotational speeds. Especially significant changes of the overall efficiency were observed for a higher engine loads. Higher combustion rate of methanol is the reason for the observed changes, what was confirmed by earlier studies [6, 13].

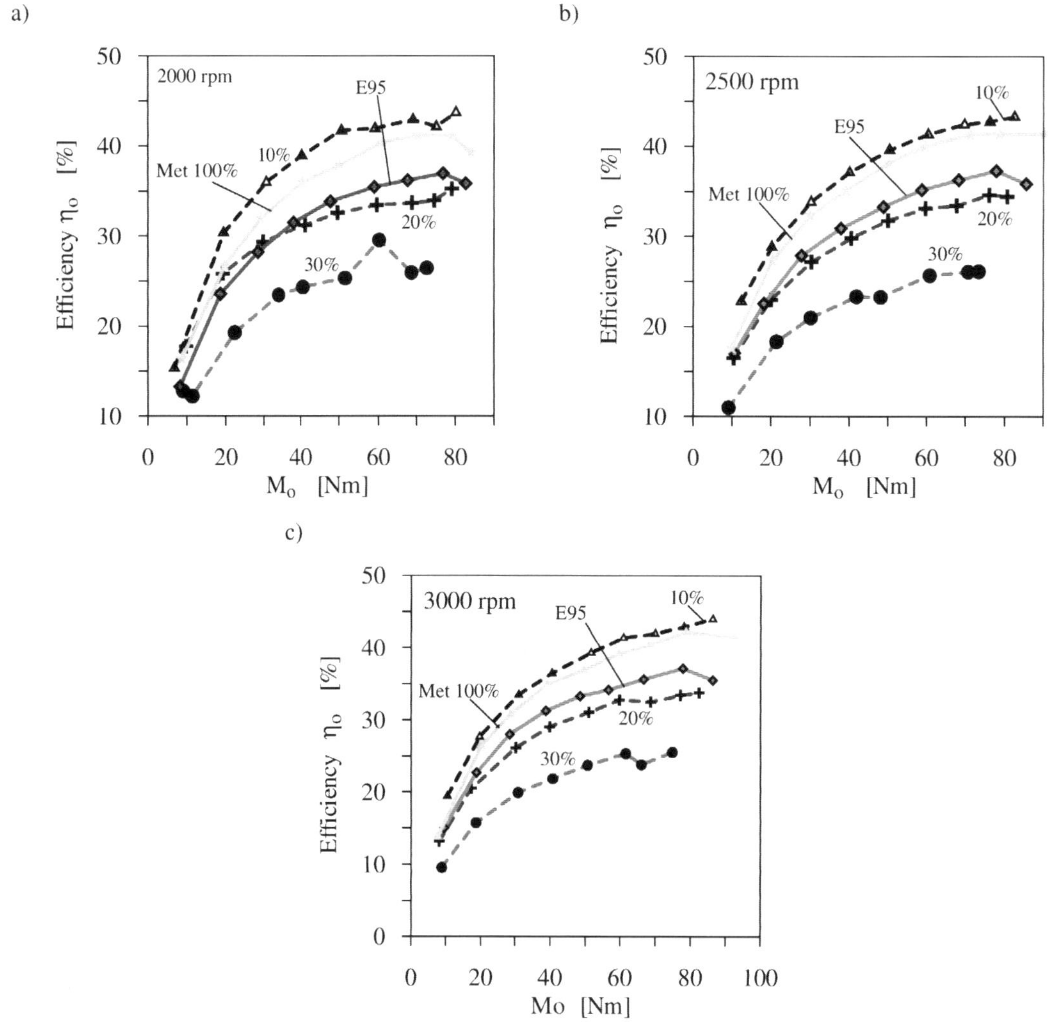

Figure 4. Comparison of overall efficiency of the engine run on gasoline E95, methanol Met 100% and gasoline-methanol mixtures with different content of glycerin: efficiencies of gasoline-methanol mixtures are marked with dashed lines.

10% additive of glycerin in methanol results in further increase of the overall efficiency seen in complete range of change of engine loads and rotational speeds, Figure 4. More big contents of glycerin, 20% and 30%, resulted in reduction of the overall efficiency, and as consequence, the overall efficiency is lower than in case of fuelling with gasoline. Especially significant changes in the overall efficiency were observed in case of 30% additive of glycerin.

Absolute change of the overall efficiency in measuring points was determined as difference in efficiency of the engine run on methanol and mixtures with 10% glycerin, and gasoline, at the same engine load and rotational speed:

$$\Delta\eta_o = \eta_{Met100\%} - \eta_{E95} \tag{4}$$

$$\Delta\eta_o = \eta_{Met-Gl10\%} - \eta_{E95} \tag{5}$$

where: $\Delta\eta_o$ – change of overall efficiency of the engine [%],
$\eta_{Met100\%}$ – efficiency of the engine run on pure methanol [%],
$\eta_{Met-Gl\ 10\%}$ – efficiency of the engine run on mixture of methanol and glycerin [%],
η_{E95} – efficiency of the engine run on gasoline [%].

Course of values of absolute change of the efficiency in case of the engine run on alcohols is shown in the Fig. 5. Engine run on pure methanol results in increase of the overall efficiency at rotational speed of 2000 rpm within range of 2.5÷5.0% of the absolute values, depending on engine load, comparing to the engine run on gasoline. Similar growths of the efficiency were also obtained for higher rotational speeds. Differences in the overall efficiency are growing together with increase of engine load. It is worth to be emphasized that even at minimal engine loads it has been obtained increase of the efficiency with more than 2% in spite of reduced temperature of the charge, what is connected with higher heat of vaporization of methanol and worsened conditions of combustion.
Usage of methanol mixture with 10% additive of glycerin resulted in further increase of the overall efficiency. For the engine speed of 2000 rpm, differences in the efficiency, comparing to gasoline fuelling, amount to 2.0÷7.8% of the absolute values. The highest differences are seen in range of medium engine loads and decrease at maximal loads. For higher rotational speeds it has been observed a little bit lower differences of the efficiency, comparing to gasoline fuelling. It is worth to underline that 10% additive of glycerin to methanol resulted in increase of the overall efficiency, comparing to fuelling with pure methanol, in complete range of change of engine load and rotational speed, Figure 4.

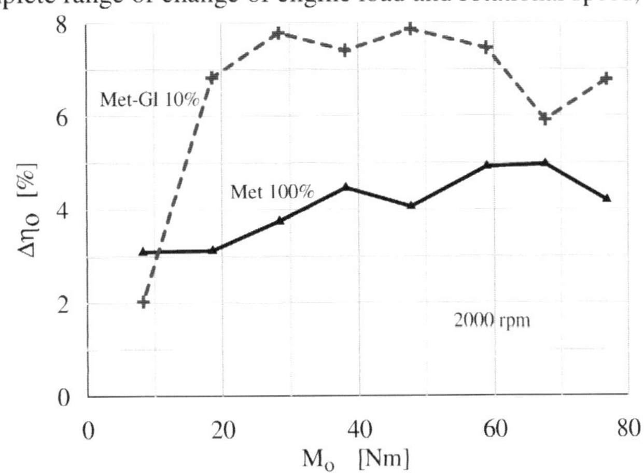

Figure 5. Absolute change of overall efficiency of the engine run on methanol Met 100% and methanol-glycerin mixture with 10% content of glycerin: rotational speed 2000 rpm.

Effect of alcohol fuels on overall efficiency of the engine can be also evaluated by relative change of the efficiency, described from the following formula:

$$\delta\eta_o = \frac{\eta_{Met100\%} - \eta_{E95}}{\eta_{E95}} \cdot 100\% \tag{6}$$

Course of absolute value of change of the overall efficiency in function of engine load presented in the Figure 6 is pointing at considerable improvement of engine efficiency in area of medium and maximal engine load, especially for the mixture with 10% additive of glycerin. Relative changes of the efficiency above 10% should contribute to reduction of operational consumption of energy and improvement of ecological features of the engine (significant reduction of carbon dioxide emission). Analysis of the Fig.

6 is also pointing at considerable increase of the efficiency in area of lower engine loads. Relative changes of the efficiency in such operational conditions amount to 12÷20% in case of the methanol, and to 15÷28% in case of mixture with glycerin. Such changes can have an effect on operational fuel consumption, because percentage of engine operation at low loads is considerable, especially in conditions of heavy traffic urban driving.

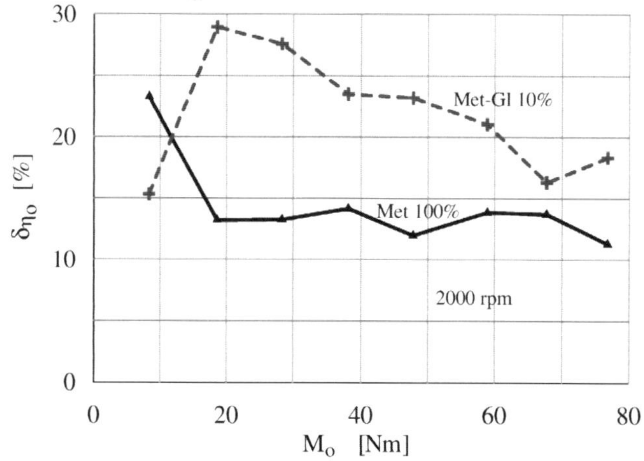

Figure 6. Relative change of overall efficiency of the engine run on methanol Met 100% and methanol-glycerin mixture with 10% content of glycerine.

Quantitative changes of the efficiency caused by 10% additive of glycerin in relation to the efficiency in case of methanol are presented in the Figure 7. For the rotational speed 2000 rpm it have been obtained increase of the overall efficiency, comparing to the run on pure methanol, in range of 1.0÷4.0% of the absolute values. For higher rotational speeds, changes in the efficiency were lower and amounted to 0.5÷1.5%, what also needs to be classified as a beneficial phenomenon.

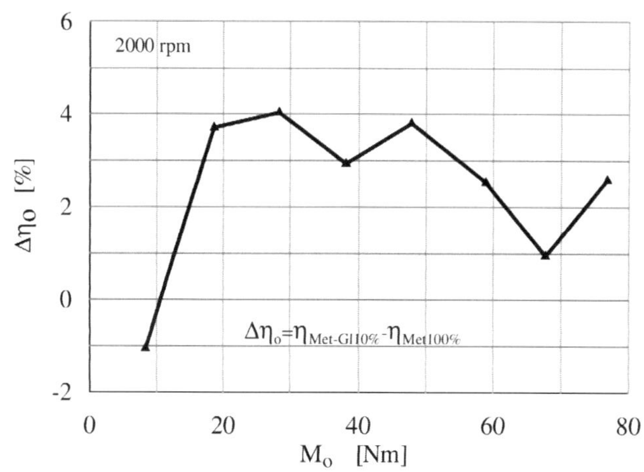

Figure 7. Absolute change of overall efficiency of the engine run on methanol-glycerin mixture with 10% content of glycerin with respect to the engine run on pure methanol

Use of alcohols to fuelling of the engine has also significant effect on emission of toxic components of exhaust gases, what can be testified by change of concentration of carbon oxide CO, summary hydrocarbons THC and nitrogen oxides NO_x shown in the Figure 8. Usage of both methanol and its mixture with 10% addition of glycerin results in reduction of carbon oxide and hydrocarbons concentration in exhaust gases of the engine, observed in complete range of changes of engine load and rotational speed, Fig. 8a and 8b. Especially important is reduction of THC: in case of methanol 2÷3 fold and in case of mixture with glycerin 1.2÷2.0 fold. However, it should be noticed that additive of up-to10% of glycerin results in tendency to lower emissions of CO and higher emission of THC, comparing to the engine run on pure methanol.

Combustion of methanol results in reduction of concentration of nitrogen oxides in exhaust gases, Figure 8c. Root cause of such phenomenon is lower temperature of flame and shorter combustion time of methanol-air mixtures [6]. Especially significant differences are present in area of high engine loads, when the biggest quantities of nitrogen oxides are generated. The changes described here are especially significant due to difficulties in reduction of nitrogen oxides in reduction type catalytic converters. Simultaneously it should be noted that 10% additive of glycerin, in case of medium and maximal engine loads, leads to increase of NO_x emission, comparing to fuelling with pure methanol. This issue needs additional investigations in range of dynamics of combustion process.

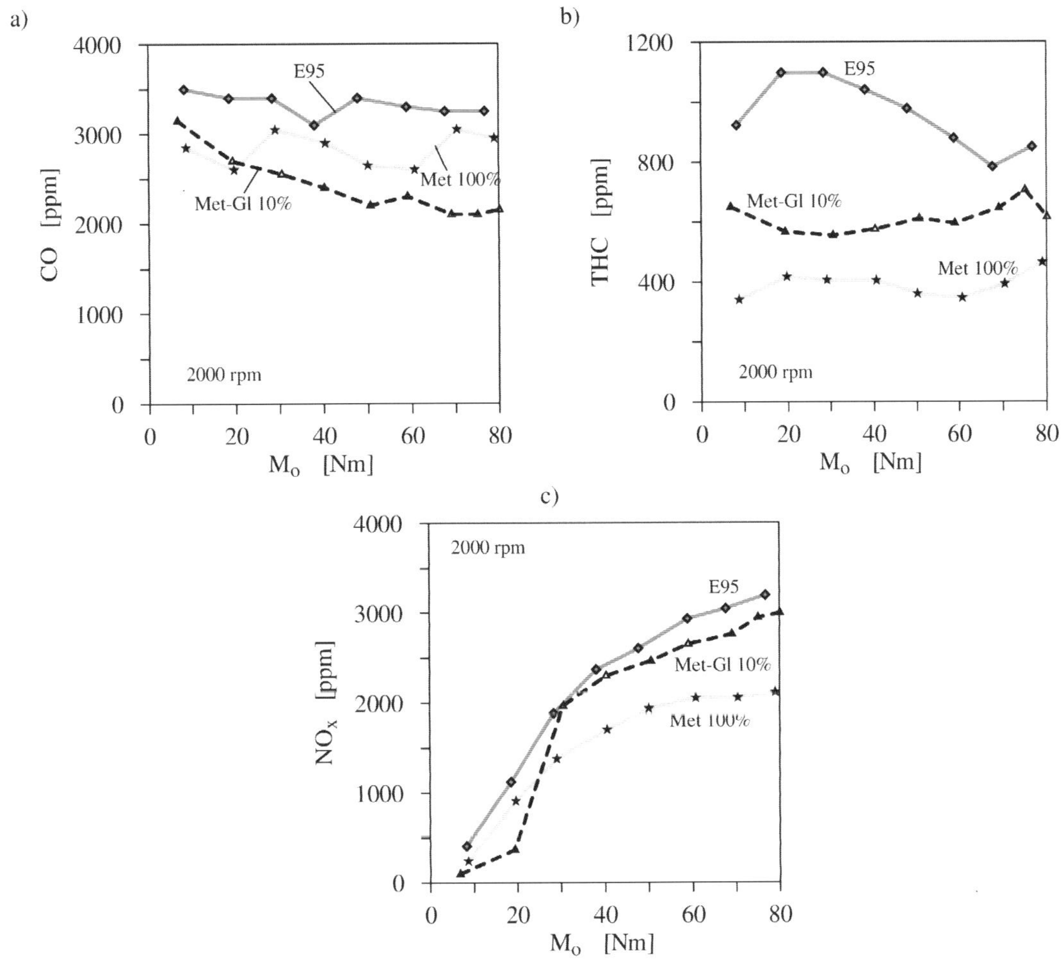

Figure 8. Changes in concentration of toxic components of exhaust gases of the engine run on Met 100% and methanol-glycerin mixture with 10% content of glycerin

Bigger contents of methanol (20% and 30% additives) result in increased concentration of carbon oxide and summary hydrocarbons, and reduced concentrations of nitrogen oxides in exhaust gases, Fig. 9a÷9c. Direction of change of these components in exhaust gases points at worsening of combustion parameters of the charge, what is connected with worsening of conditions of atomization and vaporization of the mixture due to higher viscosity of glycerin comparing to methanol. It can be also confirmed by fact of reduction of overall efficiency of the engine, described earlier.

Higher additions of glycerin result in reduction of temperature of exhaust gases behind the exhaust valve, what also can be indicative of worsening of combustion conditions of flammable mixture at high contents of glycerin, Figure 9d.

It seems that distinctive worsening of overall efficiency of the engine and increased emission of carbon oxide and hydrocarbons restrict possibility of adding glycerin in quantities lower than 10%, calculated volumetrically. However, in light of required regulations in area of the Polish National Indicative Targets such addition should be considered as significant, what should substantially increase possibility of obtaining of national bio-components to fuels.

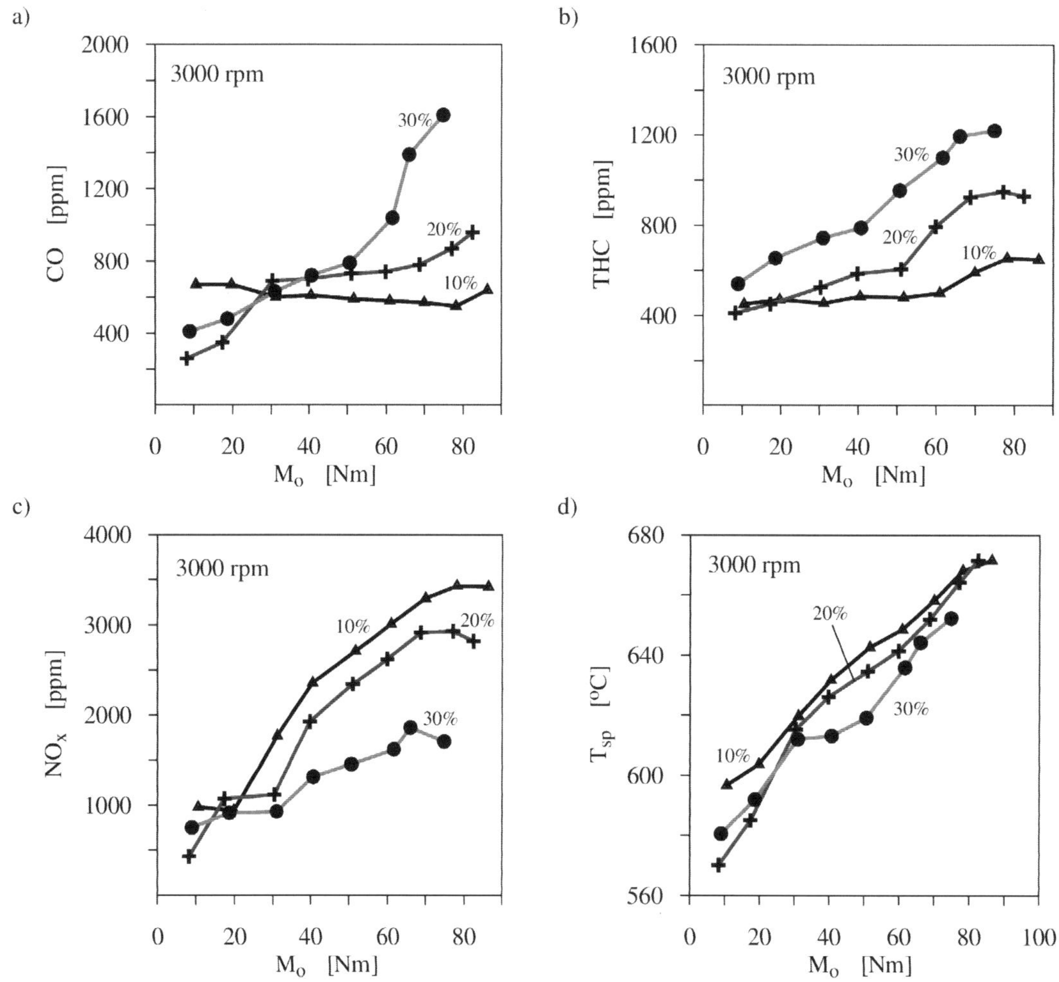

Figure 9. Changes in concentration of exhaust gas components, CO, THC and NO_x, and temperature T_{sp} of exhaust gas in function of engine load for various content of glycerin: engine rotational speed 3000 rpm.

4. Summary

Based on performed investigations it is possible to draw the following conclusions of a general nature:

- Use of simple alcohols to fuelling of spark ignitron engines leads to improvement of operational economy and ecological properties.
- In case of the engine powered by pure methanol it has been obtained increase of overall efficiency of the engine, comparing to gasoline fuelling, in range of 2.5÷5.0% of relative values and about 12÷20% of absolute values. 10% additive of glycerin resulted in further increase of overall efficiency of the engine. Absolute increases of the efficiency in comparison to gasoline fueling

amounted to 2.0÷7.8%, while relative ones increased with 15÷28%. The highest increase of the efficiency was obtained in area of average and maximal engine loads.

- Distinct improvement of the overall efficiency observed in area of all investigated engine loads and rotational speeds can contribute to reduction of operational consumption of energy, especially in urban conditions with heavy traffic. It can lead to considerable reduction of emission of greenhouse gases (CO_2) because share of carbon atoms in molecule of glycerin and methanol is lower comparing to gasoline and amounts respectively to 0.375 for methanol and 0.391 for glycerin, while in case of average gasoline amounts to 0.855.
- Usage both of methanol and its mixture with 10% additive of glycerin results in reduction of concentration of carbon oxide and hydrocarbons in complete range of changes of engine load and rotational speed. Especially important is reduction of THC: for methanol 2÷3 fold, and for mixture with glycerin 1.2÷2.0 fold.
- Combustion of methanol results in decreased concentration of nitrogen oxides in exhaust gases. Especially big differences are seen in area of high engine loads, when the biggest quantities of nitrogen oxides are generated. Due to difficulties in reduction of nitrogen oxides in reduction type catalytic converters, the changes described here are very important. Addition of small quantities of glycerin can lead, in area of medium and high engine loads, to increased emission of NO_x comparing to fuelling with pure methanol.
- Advantageous effects of addition of glycerin to methanol, in terms of improvement of overall efficiency of the engine and reduction of concentration of toxic components of exhaust gases, suggests that glycerin can be used in quantities below 10% to production of mixtures with methanol and ethanol, destined to spark ignition engines. This allows usage of waste glycerin from production of bioesters from rapeseed oil, what significantly can increase pool of Polish national bio-components to engine fuels.

References

[1]. Baczewski K and Kołdoński T 2005 Paliwa do silników o zapłonie iskrowym. WKiŁ, W-wa
[2]. Borychowski M 2012 Produkcja i zużycie biopaliw płynnych w Polsce i na świecie – szanse, zagrożenia, kontrowersje. *Roczniki Ekonomiczne Kujawsko–Pomorskiej Szkoły Wyższej w Bydgoszczy* nr 5, s. 39-59.
[3]. *Biuletyn Urzędu regulacji Energetyki* nr 2/2010.
[4]. Gmyrek J 2010 Bioetanol w realizacji NCW w PKN ORLEN S.A. *Konferencja „FUEL,S ZOOM – BIOETANOL"* Kraków 27-28 kwietnia 2010 r.
[5]. Kowalewicz A 2004 Emission Characteristics of Compression Ignition Engines Fuelled with RME/DF and Ethanol. *Journal of KONES* vol. 11/2004 No 1-2 2004.
[6]. Kowalewicz A and Pajączek Z 2004 Eco – Diesel engine with additional injection of ethanol, *Archives Combustionis* vol. 23/2003 No. 3-4
[7]. Lábaj J, Barta D and Lenhard R 2008 CFD simulation of glycerol combustion in diesel engine. *Prague : Institute of Thermomechanics AS CR*, ISBN 978-80-87012-14-7 pp. 35–50.
[8]. Larisch J and Stelmasiak Z 2013 Dual Fuelling SI Engine by Mixing Alcohol and Gasoline. *Combustion Engines* No. 3/2013.
[9]. McNeil J, Day P and Sirovski F 2012 Glycerine from biodiesel: The perfect diesel fuel. *Process Safety and Environmental Protection - ELSEVIER.* Vol 90 Tom Part B No 3,
[10]. McNeil J 2011 Efficient Combustion of Glycerol and Other Low Cetane Fuels in the Diesel Engine. *The Institution of Diesel and Gas Turbine Engineers publication Power Eng.*
[11]. Pańczyszyn T 2010 Forum roślin energetycznych. Produkcja biopaliw płynnych – potencjał surowcowy. Poznań 12.02.2010.
[12].Rychlik A and Kibalczyc Ł 2015 Zastosowanie gliceryny do zasilania tłokowych silników wysokoprężnych dużej mocy *Combustion Engines* No 3/2015.
[13]. Semikow J 2012 Studium dwupaliwowego zasilania silnika o zapłonie iskrowym benzyną i alkoholem. *Praca doktorska* ATH Bielsko-Biała.
[14]. Stelmasiak Z, Larisch J, Semikow J 2009 Some aspects of dual fuelling SI engine with gasoline and alcohol *12th EAEC European Congress Bratislava EAEC 2009* June 29-July 1
[15]. Stelmasiak Z, Semikow J 2010 The possibilities of improvement of spark ignition engine efficiency trough dual fueling of methanol and gasoline *Combustion Engines*, No 3/2010, pp. 59-67

[16]. Stelmasiak Z 2011 A New Concept of Dual Fuelled SI Engines Run on Gasoline and Alcohol, *The Archives of Transport*, nr 2/2011, pp. 73-85
[17]. Stelmasiak Z 2014 Applications of alcohols to dual-fuel feeding the spark-ignition and self-ignition engines. *Polish Maritime Research* nr 3/2014, s. 86-94.
[18]. Westcott P. C 2007 Ethanol Expansion in the United States: How Will the Agricultural Sector Adjust?. *Economic Research Service.*

Research on the 2nd generation biofuel BIOXDIESEL in aspects of emission of toxic substances in exhaust gases

M S Struś[1], W Poprawski[1] and M Rewolte[2]

[1] Wrocław University of Science and Technology, Faculty of Mechanical and Power Engineering, Wybrzeże Wyspiańskiego 27, 50-370 Wrocław, Poland; tel.+48713204965,
[2] Proagroenergia Przedsiębiorstwo Innowacyjno-Wdrożeniowe Sp z o.o. Chwałowice, ul. Szkolna 49, 55-220 Jelcz Laskowice, Poland, tel.+48 501486121

E-mail: mieczyslaw.strus@pwr.edu.pl

Abstract. This paper presents results of research of Diesel engines emission of toxic substances in exhaust gases fuelled with a second generation biofuel BIOXDIESEL, which is a blend of Fatty Acid Ethyl Esters obtained from waste resources such waste vegetable and animal fats, bioethanol and standard Diesel fuel. Presented results are very promising, showing that the emission of toxic substances in exhaust gases are significantly reduced when fuelling with BIOXDIESEL fuel in comparison with standard Diesel fuel.

1. Introduction

Manufacturers of on and off highway vehicles, agriculture and construction machinery as well as power generators find many ways to reduce emission of toxic substances in exhaust gases. They use advanced systems of accurate fuel injection, catalytic converters, particulate matter filters, engines supercharging to selective catalytic reduction. They also design fuel feeding systems to enable application of biofuels.

One of the ways to reduce negative impact of combustion engines on environment is using biofuels, which are made from waste feedstock, such as animal and vegetable resources.

Current Polish and European Union law allow addition of up to 7% of biocomponents containing Fatty Acid Methyl Esters (FAME) to the fuel used for Diesel engines.

The main components of the BIOXDIESEL fuel are based on the Fatty Acid Ethyl Esters (FAEE), which are manufactured from waste resources like animal and vegetable fats, which cannot be further processed in food industry.

2. Description of fuel used for testing

For testing purposes the experimental fuel BIOXDIESEL has been blended. It is three component mixture 70% of biocomponents containing Fatty Acid Ethyl Esters (FAEE) manufactured from waste animal and vegetable fats, rapeseed oil, bioethanol with 30% of standard Diesel fuel. The properties of BIOXDIESEL fuel are meeting the requirements of the PN-EN 590:2013. Currently there are no legal or standard requirements regarding FAEE, this is why the properties of the ethyl esters are compared with PN-EN 14214 standard, which is designed for methyl esters.

The Fatty Acid Ethyl Esters (FAEE) are manufactured in transesterification process with bioethanol in presence of alkali catalyst. The feedstock for the process has been carefully selected from wide range of resources [1].

Table 1 presents selected properties of the BIOXDIESEL fuel compared to the PN-EN 590:2013 and PN-EN-14214 standards.

Table. 1. Selected properties of BIOXDIESEL requirements of the PN-EN 590:2013 and PN-EN 14214+A1 standard.

Parameter	Units	Properties of BIOXDIESEL fuel	PN-EN 590:2013 requirement		PN-EN 14214+A1 requirement	
			min	max	min	max
Cold Filter Plugging Point	[°C]	-17	From April 16 to September 30: **0°C** From March 1 to April 15 and October 1 Novermber 15: **-10 °C** From November 16 to end of February: **-20 °C**		From April 16 to September 30: **0°C** From March 1 to April 15 and October 1 Novermber 15: **-10 °C** From November 16 to end of February: **-20 °C**	
Cetane number		51,9	51		51	
Density at 15°C	kg/m^3	868	820	845	860	900
Flash point	°C	24	55		101	
Viscocity at 40°C		2,33	2	4.5	3,5	5
Calorific value	MJ/kg	38,5	42,8		38	

3. Description of engines used for testing

The efficiency of engines fuelled with BIOXDIESEL fuel has been a subject of previous research. The results presented in [2] and [3] show that the development of the BIOXDIESEL fuel may be crucial in terms of the reduction of impact of exhaust gases on environment.

For testing purposes the 1.3 SCE turbocharged Diesel engine was used. This is modern 4 cylinder turbocharged engine equipped with Common Rail fuel injection system, displacement 1248 ccm, 69,6 bore. It generates maximum 51 kW (70 BHP) at 4000 RPM. Maximal torque generated by the engine reaches 180 Nm at 1750 RPM. This engine is also equipped with exhaust gases recirculation (EGR) system controlled by the ECU. The pressure in the Common Rail system is 1400 bars. This engine is commonly used for small and compact passenger vehicles and small vans.

4. Description of testing procedure

The emission tests were carried out in 20 points which were derived from the EUDC testing cycle on the engine test bed. The EUDC test procedure reflects the road driving conditions, combining urban and highway duty cycle. The measurement were taken on the engine installed on test bed equipped with the exhaust gases analysis equipment. Additionally there were also performed another load tests at 1500, 2500 and 3500 RPM engine crankshaft speed at various loads.

In order to compare the two fuels (BIOXDIESEL and standard Diesel) the following quantities were measured during the tests:

- Concentration of CO (carbon oxide) in exhaust gases;
- Concentration of HC (hydrocarbons) in exhaust gases;
- Concentration of NOx (nitrogen oxides) in exhaust gases;
- Smoke (with an opacimeter);
- Soot concentration.

During the testing performance of the engine fuelled with BIOXDIESEL and standard Diesel fuelled was measured.

In table 2 the measurement points of the EUDC test cycle are presented. There are also presented vehicle speed and corresponding gear at which such engine load is achieved. The points also present the constant driving speed parts of the test.

Table. 2. List of measurement points according to EUDC testing procedure

Measurem ent point	Engine revolution [RPM]	Engine Torque [Nm]	Corresponding vehicle speed/gear [km/h /gear]
1	820	45	
2	850	20	
3	1050	15	
4	1250	20	
5	1400	40	
6	1550	4	35/III
7	1590	10	50/IV
8	1650	37	
9	1730	11	70/V
10	1740	26	
11	1800	38	
12	1920	3	15/I
13	2050	26	
14	2200	4	50/III
15	2260	4	32/II
16	2400	55	
17	2500	34	100/V
18	2700	73	
19	2930	80	
20	3020	55	120/V

5. Results of engine performance and emission measurement

The emission measurement was performed according to the EUDC procedure, the points listed in table 2. Below, figure 1 presents comparison of engine efficiency, obtained during fuelling with standard Diesel fuel and BIOXDIESEL. The results show, that majority of measurement point of EUDC test the efficiency measured with experimental biofuel is bigger in comparison with standard Diesel fuel. Figure 2 presents comparison of the value of fuel dose. The values are very similar, when measured in the points of the EUDC test. This shows, that the engine was very accurately controlled by the ECU. Even though the calorific value of the BIOXDIESEL fuel is about 10% lower in comparison with standard Diesel fuel the unit fuel consumption and hourly fuel consumption are very similar for both fuels (figures 3 and 4). The accuracy and repeatability of the engine control can be seen on the comparison of the position of the EGR valve (figure 5). Both values follow each other for most of the points of the EUCD test.

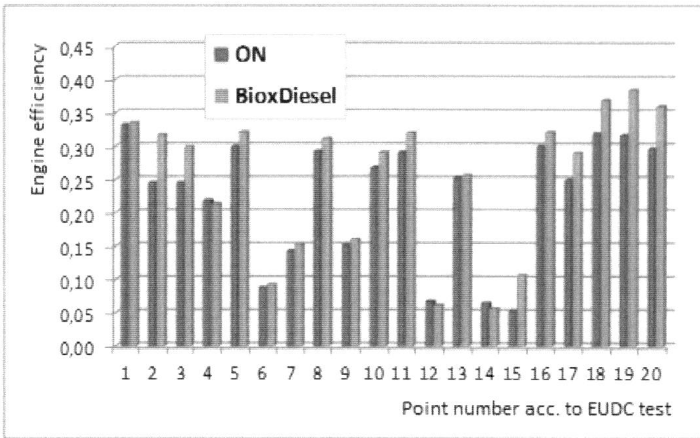

Figure. 1. Comparison of engine efficiency for BIOXDIESEL and standard Diesel fuel.

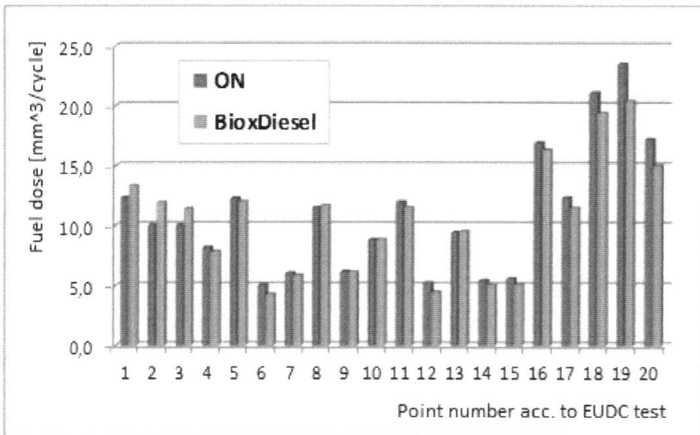

Figure. 2. Comparison of the fuel dose BIOXDIESEL and standard Diesel fuel.

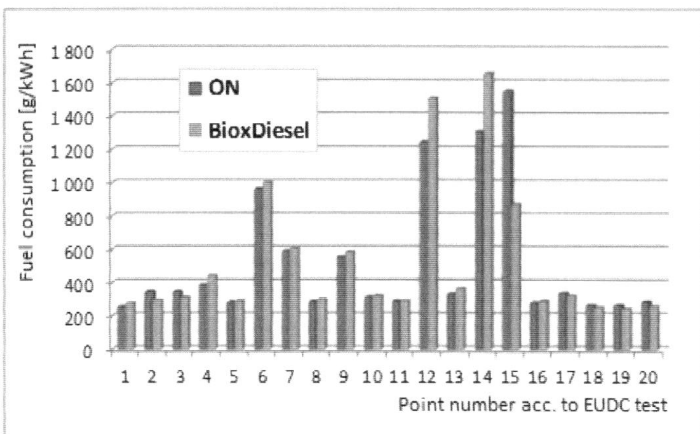

Figure. 3. Comparison of the unit fuel consumption for BIOXDIESEL and standard Diesel fuel.

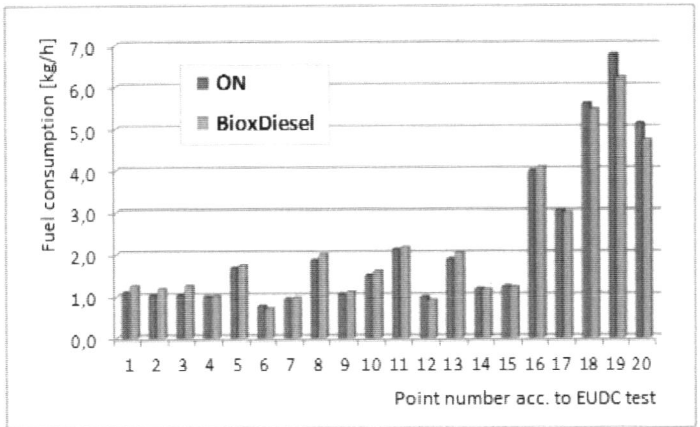

Figure. 4. Comparison of hourly fuel consumption for BIOXDIESEL and standard Diesel fuel.

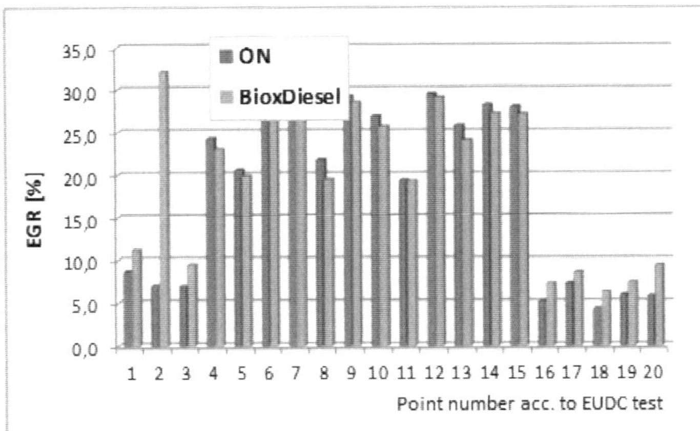

Figure. 5. Comparison of position of the EGR valve for BIOXDIESEL and standard Diesel fuel.

The results of emission tests show the advantage of the BIOXDIESEL fuel over standard Diesel. The emission of carbon oxide and hydrocarbons is much lower over the test for experimental biofuel (figures 6 and 7). Such observation may prove better evaporation of the BIOXDIESEL over standard Diesel fuel.

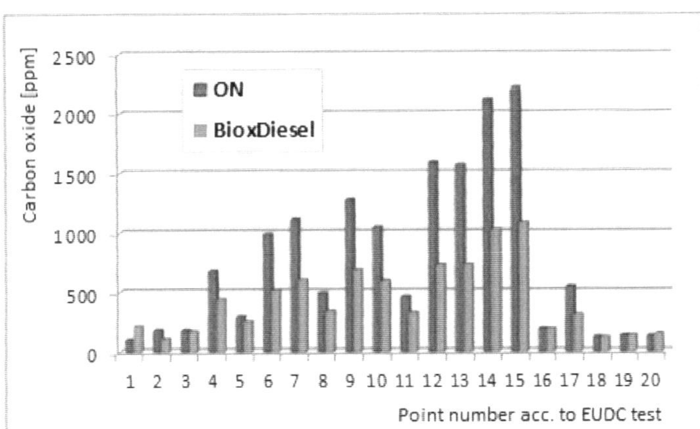

Figure. 6. Comparison of emission of carbon oxide in exhaust gases for BIOXDIESEL and standard Diesel fuel.

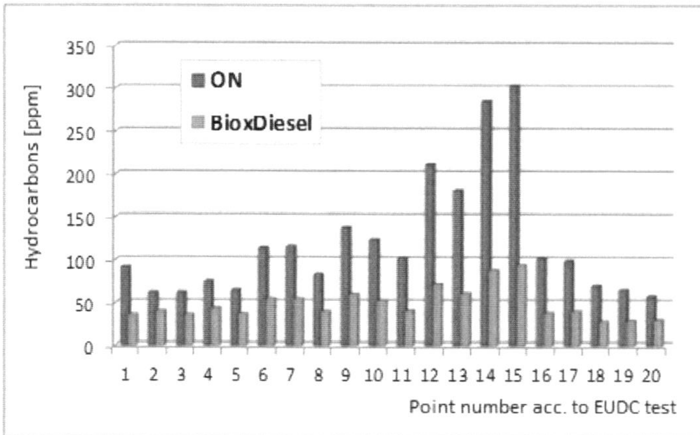

Figure. 7. Comparison of emission of hydrocarbons in exhaust gases for BIOXDIESEL and standard Diesel fuel.

Figure 8 presents emission measurement of nitrogen oxides during fuelling with BIOXDIESEL follows the emission measured on Diesel fuel. But in the points, when the engine works with higher load (points 18, 19 and 20, which correspond to highway driving) one can observe significantly lower emission when fuelling with BIOXDIESEL.
Similarly, when comparing emission of soot and smoke (figures 9 and 10) it is not possible to clearly point which fuel used for testing has its advantages in terms of lower emission. Similarly to NOx, smoke emission at the points with higher engine load caused lower emission of soot and smoke.
Comparison of the Lambda coefficient show again (figure 11), that the engine was very accurately controlled. The burning efficiency for both fuels was comparable.

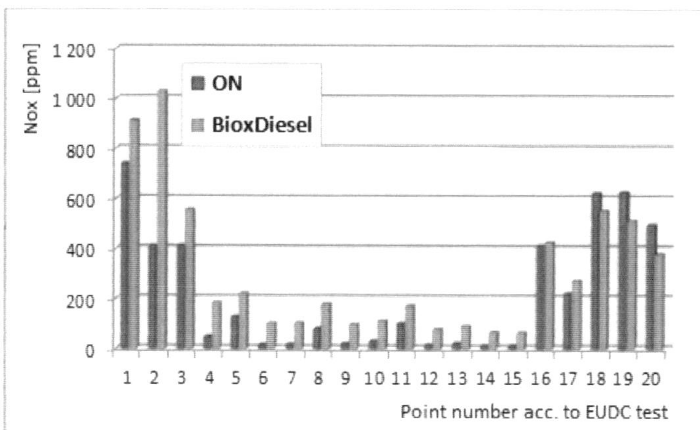

Figure. 8. Comparison of emission of nitrogen oxides in exhaust gases for BIOXDIESEL and standard Diesel fuel.

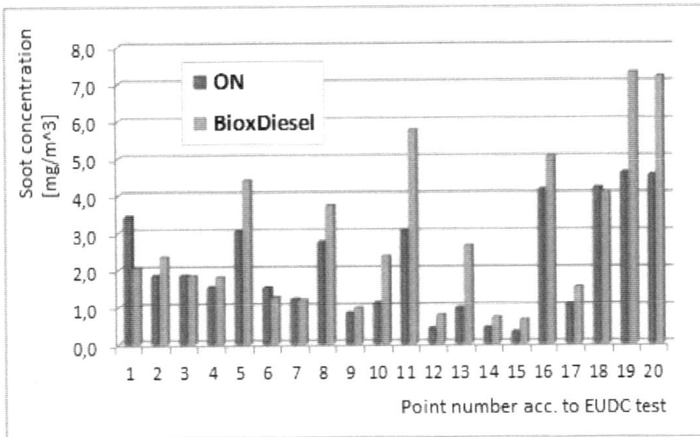

Figure. 9. Comparison of soot concentration in exhaust gases for BIOXDIESEL and standard Diesel fuel.

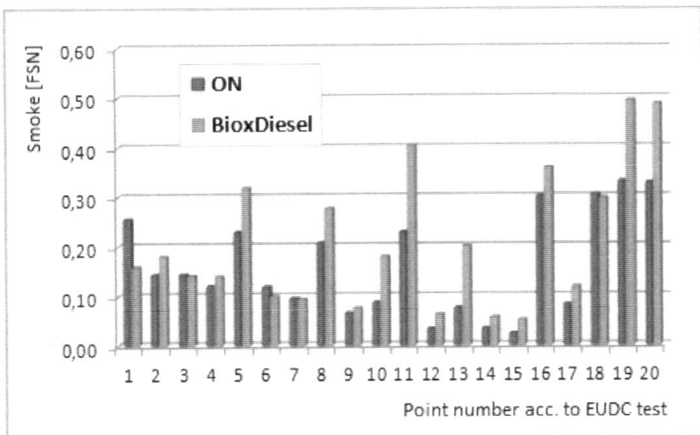

Figure. 10. Comparison of emission smoke exhaust gases for BIOXDIESEL and standard Diesel fuel.

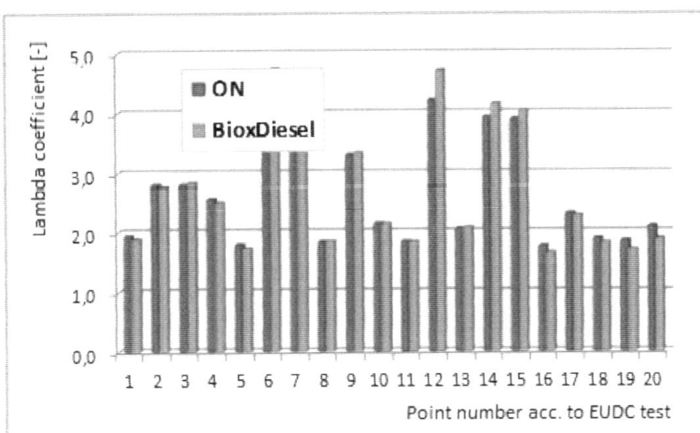

Figure. 11. Comparison of the value of LAMBDA coefficient for BIOXDIESEL and standard Diesel fuel.

6. Conclusions

Presented results of emission measurement show, that the BIOXDIESEL is very advanced fuel, which can replace standard Diesel fuel without negative impact on environment. The BIOXDIESEL fuel shows its advantage over standard fuel in terms of emission of carbon monoxide and hydrocarbons. This is achieved by very good evaporation properties of the BIOXDIESEL over standard Diesel fuel.
The measurement of nitrogen oxides (NOx) in exhaust gases also proved, that the BIOXDIESEL has an advantage over Diesel fuel for the points of the test cycle, which correspond to the highway driving with higher speed and load. The similar situation was for measurement of soot and smoke in exhaust gases.
Another big advantage of the BIOXDIESEL fuel is higher engine efficiency when fuelled with BIOXDIESEL in comparison with standard Diesel fuel. The differences are the biggest when the load was the biggest. Similarly, the fuel consumption of BIOXDIESEL fuel was lower.
Obtained results of the research on the BIOXDIESEL fuel proves that it is advanced fuel and may be used alternatively for all Diesel engines, however the finite verification can be done after currently realised vehicle durability testing on the distance of 600'000 – 1'000'000 km in the period of 8 – 12 years.

References

[1] Struś M S, Poprawski W and Rewolte M 2015"Efficiency of feedstock selection for the second generation BIOXDIESEL biofuel for Diesel engines," *Combustion Engines,*, 3/2015

[2] Struś M S 2012 *Ocena wpływu biopaliw na wybrane właściwości eksploatacyjne silników*, Oficyna Wydawnicza Politechniki Wrocławskiej Wroclaw

[3] Struś M S 2014 Efektywność silników spalinowych o zapłonie samoczynnym zasilanych paliwem bioxdiesel z komponentami całkowicie odnawialnymi *Inżynieria Maszyn,*, vol. 2, no. 19, pp. 108-115

Research on the influence of ozone dissolved in the fuel-water emulsion on the parameters of the CI engine

M K Wojs[1], P Orliński[1], W Kamela[1] and P Kruczyński[2]

[1] Combustion Engines Department, Warsaw University of Technology, Poland
[2] BU Power Systems Polska Sp. z o.o., Warsaw, Poland

E-mail: mwojs@simr.pw.edu.pl

Abstract. The article presents the results of empirical research on the impact of ozone dissolved in fuel-water emulsion on combustion process and concentration of toxic substances in CI engine. The effect of ozone presence in the emulsion and its influence on main engine characteristics (power, torque, fuel consumption) and selected parameters that characterize combustion process (levels of pressures and temperatures in combustion chamber, period of combustion delay, heat release rate, fuel burnt rate) is shown. The change in concentration of toxic components in exhausts gases when engine is fueled with ozonized emulsion was also identified. The empirical research and their analysis showed significant differences in the combustion process when fuel-water emulsion containing ozone was used. These differences include: increased power and efficiency of the engine that are accompanied by reduction in time of combustion delay and beneficial effects of ozone on HC, PM, CO and NO_X emissions.

1. Introduction

The more rigorous standards on toxic substances emission in CI engines, as well as growing requirements concerning their effectiveness have led to the search for new technical solutions. In the case of CI engines, the solutions which will shorten the delay time of fuel self-ignition are being searched for. They will enable the proper control of combustion process, which – in turn will have a positive influence on operating parameters of the engine and toxic substances emission in exhaust gas [2]. One of the most common methods is using new, alternative fuels or fuel additives, which can be applied without significantly interfering with the engine design. One such fuel type could be, among others, a fuel-water emulsion [1, 3]. Water as an additive to the combustion chamber during its working cycle has been applied for many years now, in many different forms e.g. water injection, or the abovementioned emulsion [5]. In order to boost the diesel oil and water mixing capacity and to lengthen the durability of such a mixture, additives in the form of emulators are used. They decrease fuel surface tension.

Applying the emulsion may have a beneficial effect on the chosen engine parameters: it will lower NO_X and solid particles emissions, it will also decrease specific fuel consumption. However, fuel-water emulsions have numerous disadvantages. Its durability is largely influenced by ambient temperature and the possible foaming. Water content may also hinder fuel injection.

Another solution, which could have a positive effect on engine performance and reducing toxic substances emissions in exhaust gas, is to saturate the fuel consumed by the working engine with ozone [4, 5]. Ozone is an allotrope of oxygen consisting of triatomic molecules. It dissolves in water and oils. It is thermally unstable and breaks down to oxygen even in room temperature. It has strong oxidizing properties, which contributes to precipitating fuel oxidation. It may trigger the simplest alkanes oxidation in temperatures lower than the oxygen. Furthermore, the presence of ozone in fuel supplied to the engine may accelerate fuel self-ignition and increase its cetane number [6, 7].

2. Research methodology

In this article, the chosen parameters of PERKINS 1104C-E44T – a CI, turbocharged engine were presented and analyzed, based on averaging the courses of working medium pressures in the combustion chamber, obtained as a result of empirical research. During the research, the engine was powered with fuel-water emulsion and the emulsion that underwent ozonization. The emulsion was created by mixing diesel oil and water (demineralized). The basic parameters of the mixture are shown in Table 1.

Table 1. Parameters of fuel-water emulsion

Parameter	Value
Water phase	15 [%]
Calorific value	36,55 [MJ/kg]
Density in 15°C	0,8681 [kg/m^3]
Ozonization time	30 [min]

The ozonization process was conducted using a set of three ozone generators tanks together with powering compressors. The diagram of the test bench is shown in Figure 1.

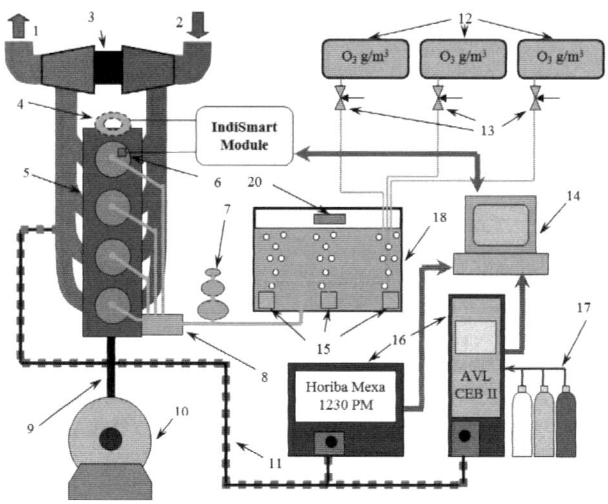

Figure 1. Diagram of the test bench: 1 – exhaust outlet, 2 – air inlet, 3 – turbocharger, 4 – engine crankshaft angle indicator, 5 – diesel engine, 6 – cylinder pressure indicator, 7 – fuel consumption measurement unit, 8 – fuel pump with injectors, 9 – propeller shaft, 10 – dyno brake SCHENCK W450, 11 – heated gas lines, 12 – regulated ozone flow generators, 13 – valves with flow indicators, 14 – PC computer with data acquisition, 15 – oxidizing stones, 16 – exhaust gas analyzers, 17 – main gases for analyzers, 18 – liquid fuel tank, 20 – O_3 concentration sensor

The fuel ozonization process was conducted in a research tank by diffusers placed symmetrically at the bottom. Three ozone generators worked at full capacity 35g/h and ozone was obtained from air (0.6 m^3/h flow) using dielectric barrier discharge method. They operated until the possible ozone saturation of the studied fuel was achieved.

3. Defining combustion process parameters

The delay time of fuel self-ignition was defined empirically based on the experimentally determined the injection start (using a sprayer needle lift sensor) and combustion start (owing to the created indicatory graph) in accordance with the methodology described in detail in the publication [8].
In order to define the rate of released heat, based on experimentally measured courses of working fluid pressures in the combustion chamber, a methodology proposed by AVL was used [10].
A simplified process is used to calculate the heat release rate by computing the energy the cylinder pressure effectively delivers to the gas. Surface losses are disregarded. Therefore, the heat release rate calculated is lower than the actual energy released. The calculation is based on the first law of thermodynamics. The heat release rate is calculated using the change of cylinder pressure in the following equation:

$$HRR_i = \frac{K}{m-1}[m \cdot p_i \cdot (V_{i+n} - V_{i-n}) + V_i \cdot (p_{1+n} - p_{1-n})] \qquad (1)$$

Explanation of terms:
n = interval (1 deg. crank angle)
m = polytropic coefficient
p = cylinder pressure (bar)
V = volume (m^3)
K = constant (100… due to unit conversion)
It is also used to calculate other thermodynamics parameters, e.g. combustion start, ignition delay and maximum cylinder pressure rise.

4. Results

Figure 2 presents model courses of working medium pressures in the combustion chamber with 100% engine load and crankshaft rotational speed of 1400 and 2200 rev/min. During the research, the engine was powered with fuel-water emulsion (E) and with emulsion that underwent ozonization process (EO).
Based on the defined pressures, in accordance with the methodology described above, the most important combustion parameters, such as heat release rate and degree of burnout dose, were determined (Figure 3 and 4). The self-ignition delay time was also determined (Figure 5). Furthermore, two additional combustion parameters were defined: the angle of crankshaft revolution corresponding to 50% and 90% fuel dose burn, in the case of powering the engine with the studied fuels (Figure 6)

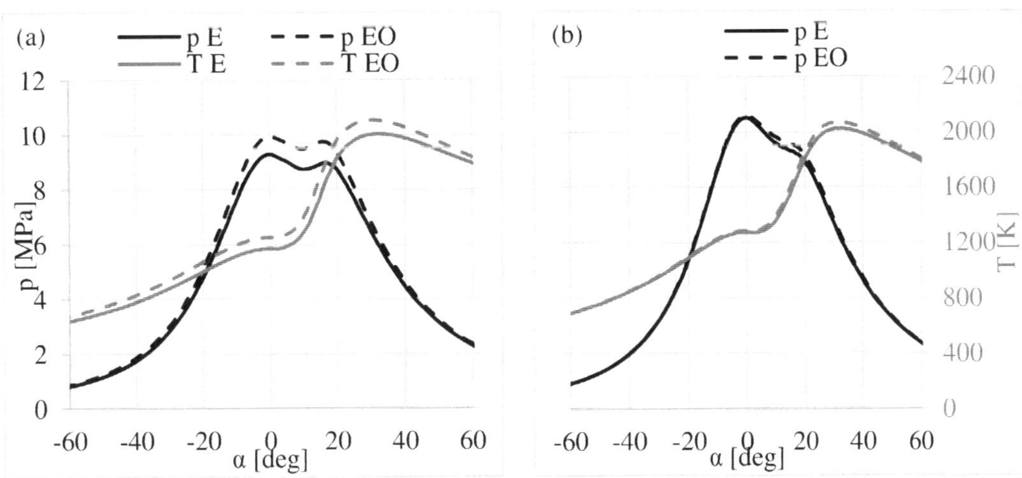

Figure 2. The dependence of pressure – p and temperature – T of the working medium in the combustion chamber for 100% maximum load for the emulsions – E and emulsion ozonated – EO, (a) at n = 1400 rev / min, (b) at n = 2200 rev / min

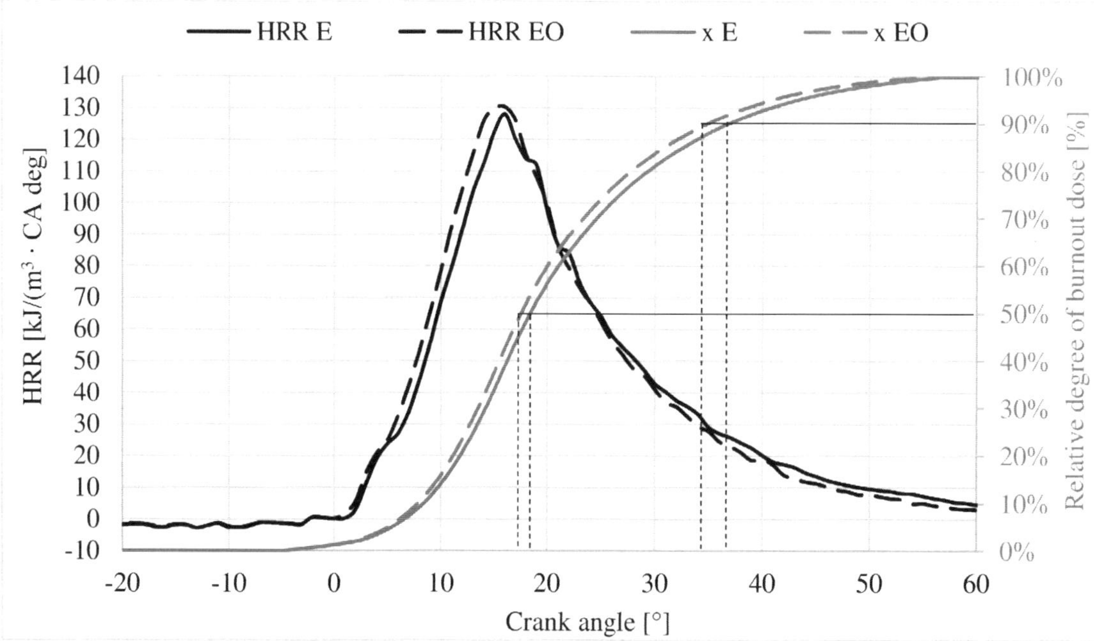

Figure 3. The dependence of heat release rate – HRR and relative degree of burnout dose – x of the working medium in the combustion chamber for 100% maximum load at n = 1400 rev / min and the emulsions – E and emulsion ozonated – EO

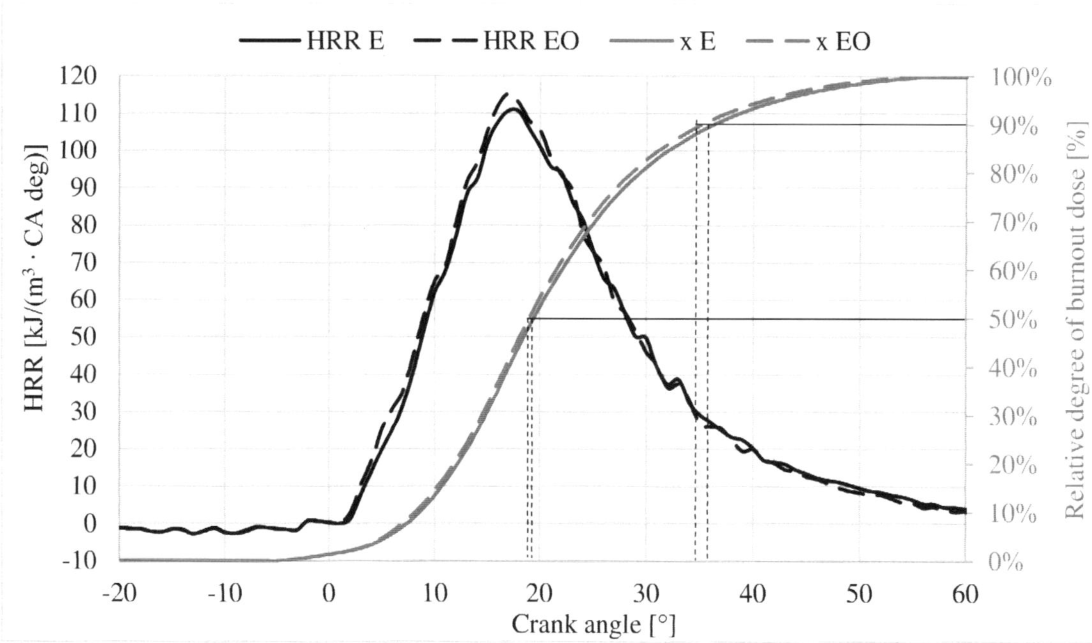

Figure 4. The dependence of heat release rate – HRR and relative degree of burnout dose – x of the working medium in the combustion chamber for 100% maximum load at n = 2200 rev / min and the emulsions – E and emulsion ozonated – EO

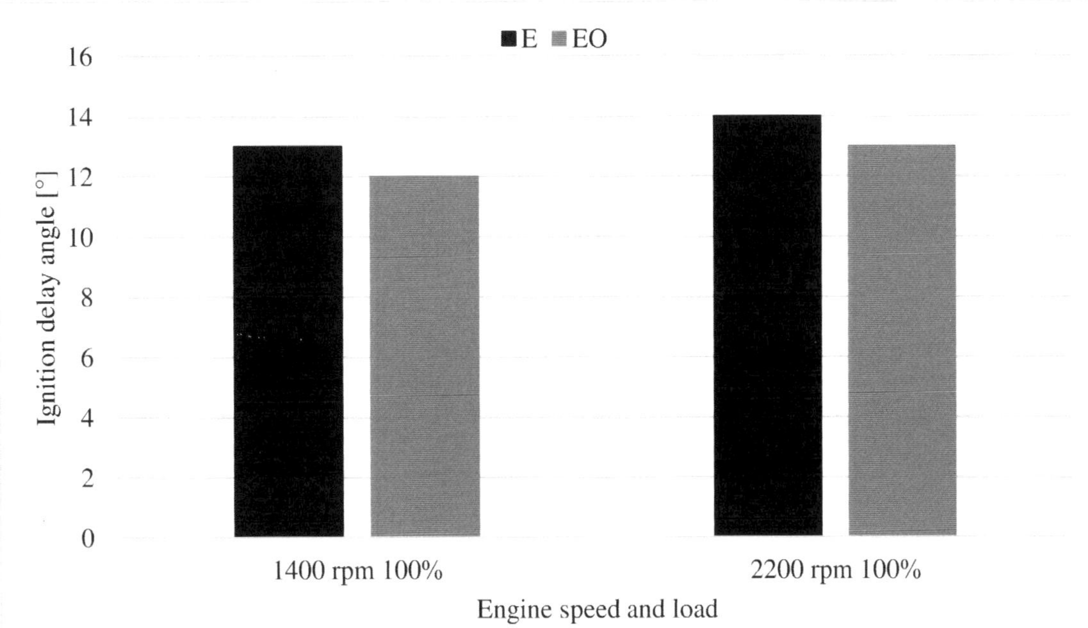

Figure 5. Crankshaft revolution angle corresponding to self-ignition delay phase

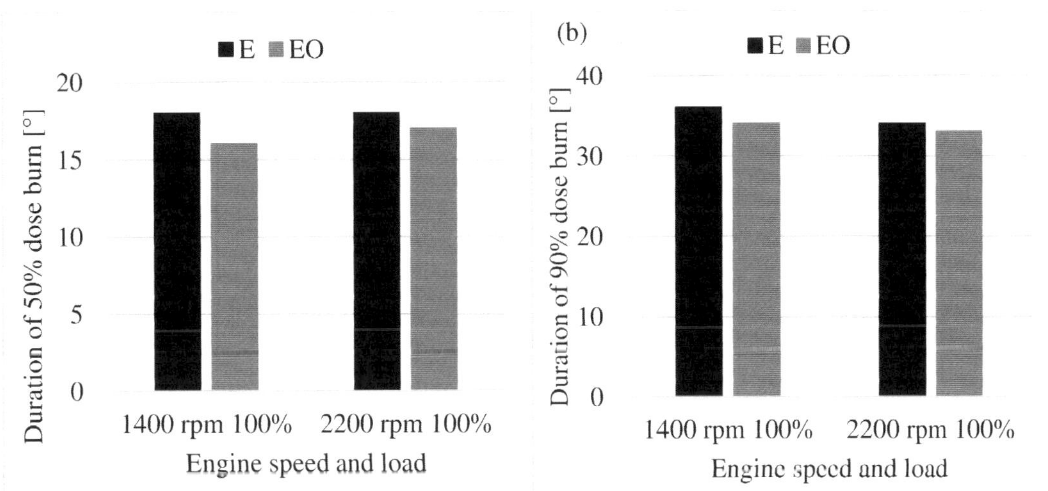

Figure 6. Crankshaft revolution angle corresponding to: (a) 50% fuel dose burn, (b) 90% fuel dose burn

During empirical research, the external characteristics of the engine was defined. It presented the basic parameters (power, torque and engine efficiency) – Figures 7 and 8. Moreover, the concentration of toxic exhaust gas components were determined (HC, PM, CO, NO_X) with performance load at crankshaft rotational speed of 1400 and 2200 rev/min.

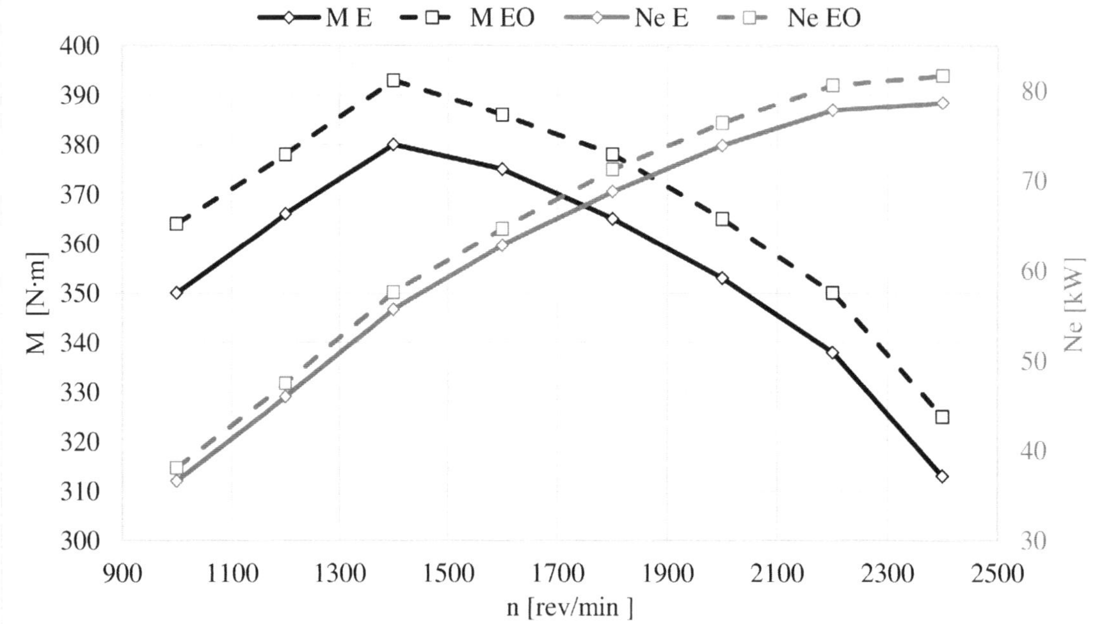

Figure 7. The dependence of torque – M and power – N_e of rotational speed for the emulsions – E and emulsion ozonated – EO

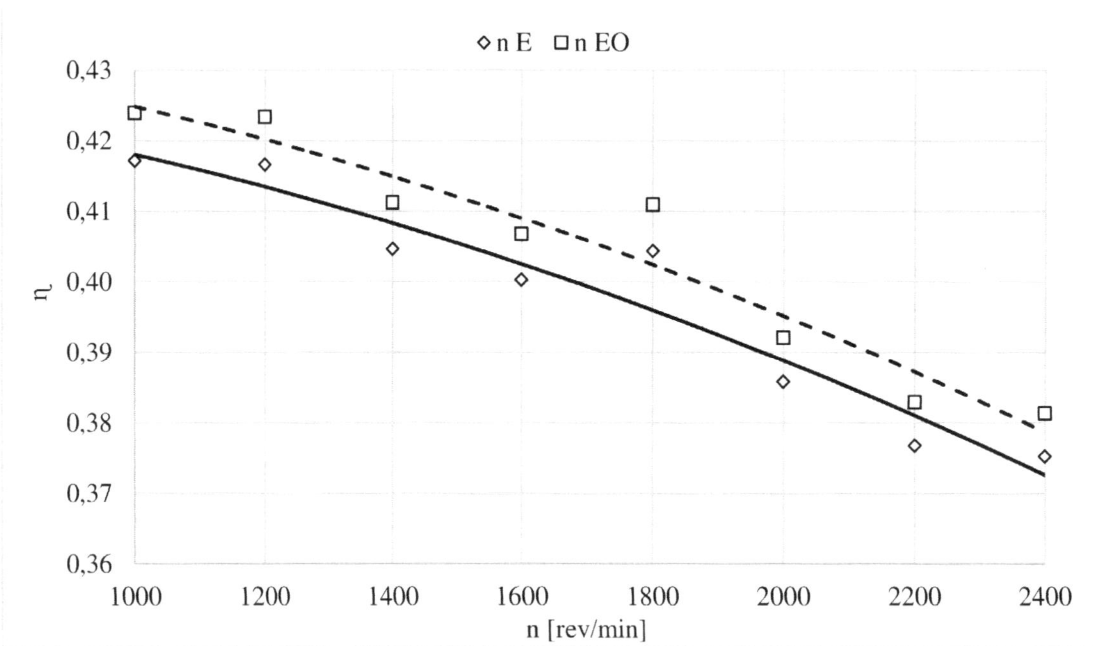

Figure 8. The dependence of efficiency – η of rotational speed for the emulsions – E and emulsion ozonated – EO

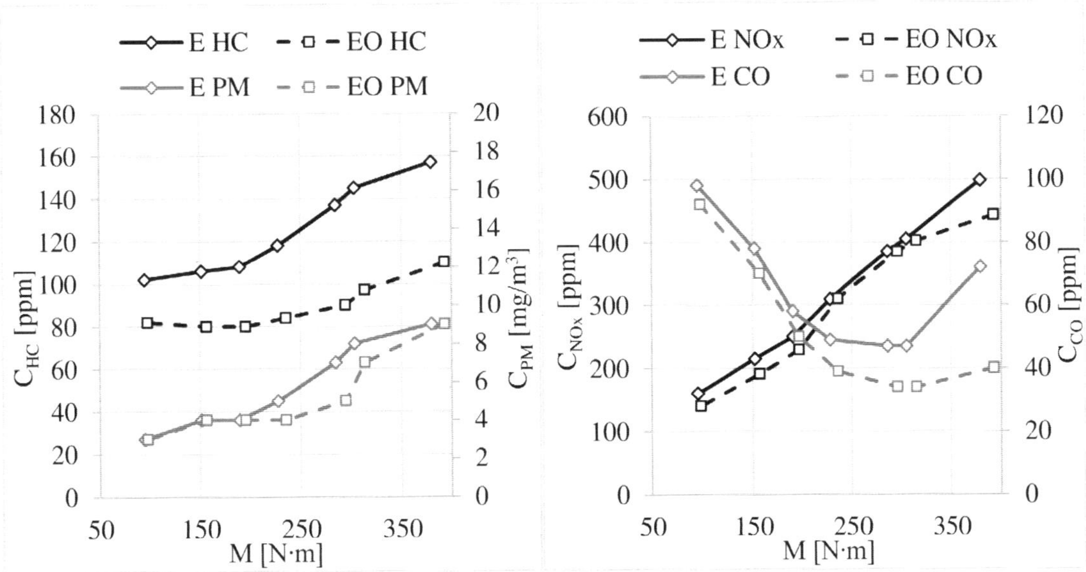

Figure 9. The dependence of HC and PM concentration (a), NO_X and CO (b) of torque for the emulsions – E and emulsion ozonated – EO, 1400 rpm

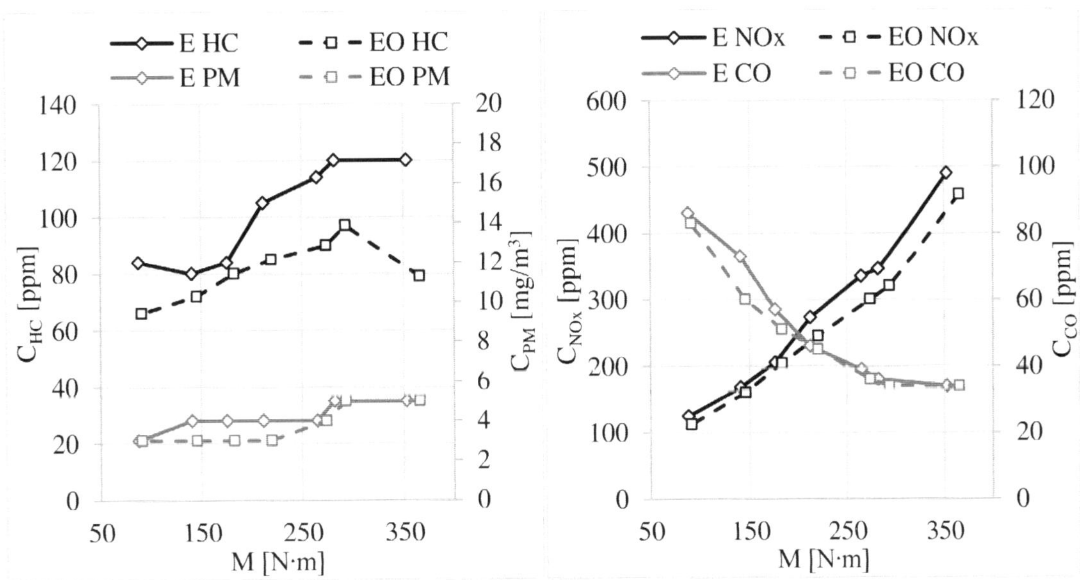

Figure 10. The dependence of HC and PM concentration (a), NO_X and CO (b) of torque for the emulsions – E and emulsion ozonated – EO, 2200 rpm

5. Conclusions

Under the same conditions of engine operation, it was observed that using the ozonated emulsion causes:

- earlier combustion start maximally by approximately 2 degrees of crankshaft revolutions, resulting from shortening of self-ignition delay – Figure 2-5,
- 8% increase in maximum value of working medium pressure and temperature, at the rotational speed of 1400 rev/min and with powering the engine with a maximum fuel dose – Figure 2a,

- an approximately 2% increase in maximum value of working medium pressure and temperature, at the rotational speed of 2200 rev/min while powering the engine with a maximum fuel dose – Figure 2a,
- an approximately 2% increase in maximum heat release rate – Figure 3-4,
- a maximally 6% increase in engine power and torque in all range of operation, in the case of powering it with the emulsion which underwent ozonization process, compared to pure emulsion – Figure 7,
- an approximately 1% increase in general efficiency of the engine in all range of the studied engine operation, powered by the ozonated emulsion – Figure 8,
- maximally an approximate 40% decrease in HC concentration at both crankshaft rotational speeds – Figure 9a and 10a,
- maximally an approximate 10% decrease in PM concentration for ozonated emulsion under high engine loads – Figure 9a and 10a,
- maximally an approximate 10% decrease in NOx concentration for ozonated emulsion under high engine loads – Figure 9b and 10b,
- maximally an approximate 30% decrease in CO concentration at 1400 crankshaft rotational speed for high loads and an approximate 10% decrease at 220 crankshaft rotational speed for low engine loads – Figure 9b and 10b,
- in the empirical research results analysis presented above, possible measuring errors were considered (connected with measuring crankshaft torque, fuel consumption).

Preliminary study of the influence of fuel-water emulsion indicate a positive influence of ozone dissolved in fuel on combustion and effective parameters of engine operation and toxicity.

References

[1] Kruczyński S, Danilczyk W and Kamela W 2009 Water-fuel micro emulsions for supplying self-ignition engines, *Combustion Engines* **2009- SC2** 353-359
[2] Kruczyński S 2002 *Wybrane metody ograniczania emisji tlenków azotu w silnikach o zapłonie samoczynnym* (Warszawa-Radom: Wydawnictwo ITE PIB)
[3] Draft Technical Report 2002 *Impacts of Lubrizol's PuriNOx Water/Diesel Emulsion on Exhaust Emission from Heavy-Duty Engines* (United States Environmental Protection Agency, EPA420-P-02-007)
[4] Wilk M 2007 *Badanie wpływu ozonu na proces spalania gazu* (Kraków: Akademia Górniczo-Hutnicza im. Stanisława Staszica)
[5] Tacibana T, Hirata K, Nishida H and Osada H 1991 *Effect of Oznone on Combustion in Compression Ignition engines*
[6] Depcik C, Mangus M, and Ragone C 2014 Ozone-Assisted Combustion-Part I: Literature Review and Kinetic Study Using Detailed n-Heptane Kinetic Mechanism *Gas Turbines Power* **136(9)**
[7] Stoeck T and Prajwowski K 2012 Wpływ dodatku ozonu na parametry pracy i toksyczność spalin silnika o zapłonie samoczynnym, *Autobusy TEST* **10**
[8] Ambrozik A 2010 *Analiza cykli pracy czterosuwowych silników spalinowych* (Kielce: Wyd. Politechniki Świętokrzyskiej)
[9] Orliński P 2013 Wybrane zagadnienia procesu spalania paliwa pochodzenia roślinnego w silnikach o zapłonie samoczynnym (Radom: Instytut Naukowo – Wydawniczy: SPATIUM)
[10] AVL Concerto 2010, Instructions for Software Version 4.2

Influence of sliding surface roughness and oil temperature on piston ring pack operation of an automotive IC engine

A Wolff

Faculty of Transport, Warsaw University of Technology
75 Koszykowa Street, PL-00-662 Warsaw, Poland

E-mail: wolff@wt.pw.edu.pl

Abstract. In the paper a comprehensive model of a piston ring pack motion on an oil film has been presented. The local oil film thickness can be compared to height of the combined roughness of sliding surfaces of piston rings and cylinder liner. Equations describing the mixed lubrication problem based on the empirical mathematical model formulated in works of Patir, Cheng and Greenwood, Tripp have been combined and used in this paper. The developed model takes the following phenomena into account: hydrodynamic and contact forces, spring and gas forces acting on piston rings. The rings motion concerning low and high temperature of cylinder surface has been compared. These results concern cases of hydrodynamic and mixed lubrication. Changes of oil wetted area and contact zone of piston rings have been shown. In addition the oil film thickness distribution along cylinder liner and all the forces acting on piston rings have been analysed and discussed. The results have been presented in form of relevant diagrams. The developed model and software can be utilized for optimization of piston rings design.

1. Introduction

Piston rings are important part of internal combustion engines. Commonly a set of piston rings is used to form a dynamic gas seal between the piston and cylinder wall [4,5]. The sliding motion of the piston forms a thin oil film between the ring land and cylinder wall, which lubricates the sliding components [1,3,8]. The hydrodynamic force generated by this thin oil film is opposed by a combination of the gas pressure acting on the back side of each ring and the ring stiffness. Due to the dynamic nature of these forces, each individual ring is periodically compressed and extended as the piston runs through its cycle. The problem of studying this interaction is further complicated by the high temperatures involved, as these result in low oil viscosity and subsequently very low oil film thickness. The oil film is typically thick enough to expect the existence of mixed lubrication, so this phenomenon should also be taken into account [1,2,6,7,9]. The use of modern oil of low viscosity, working at a high temperature causes the existence of a very thin oil film thickness comparable to the value of the liner surface roughness. In such conditions, the possibility of direct contact between the ring and cylinder liner surface exists. Therefore the numerical simulation of these processes, which take place in a typical piston ring pack operation, is important from practical point of view.

2. Modelling of piston ring pack operation

2.1. Developed sub-models

A combined model of piston rings operation has been developed. It consists of two main models: a) model of gas flow through the labyrinth seal piston-rings-cylinder (PRC), b) model of oil flow in the lubrication gap between the ring and cylinder liner. The two mentioned models are coupled. In addition, sub-models of the following mechanical phenomena have been used: a contact of rough surfaces, an axial movement of rings within piston grooves and an elastic torsional deformation of piston rings (Figure 1). But this time the effect of ring twist has not been taken into account.

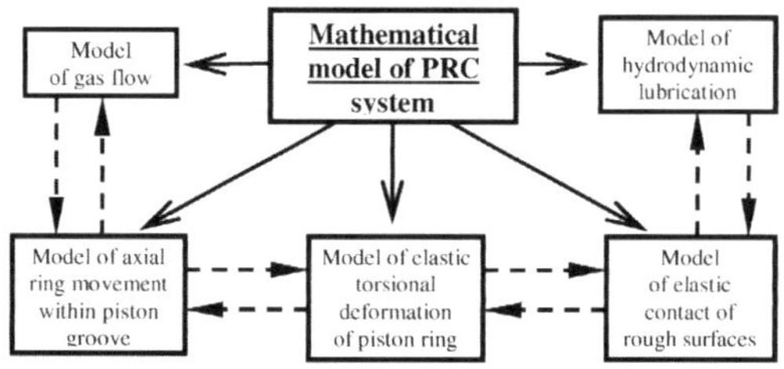

Figure 1. Developed comprehensive model of the system: piston-ring-cylinder (PRC).

All the sub-models are described in detail in publications [10,11,13-16] of the author. In this paper only the sub-model of mixed lubrication [2,6,7] is shortly presented.

2.2. Model of mixed lubrication
Two main cases of oil flow in the system piston ring – cylinder liner are presented in Figure 2.

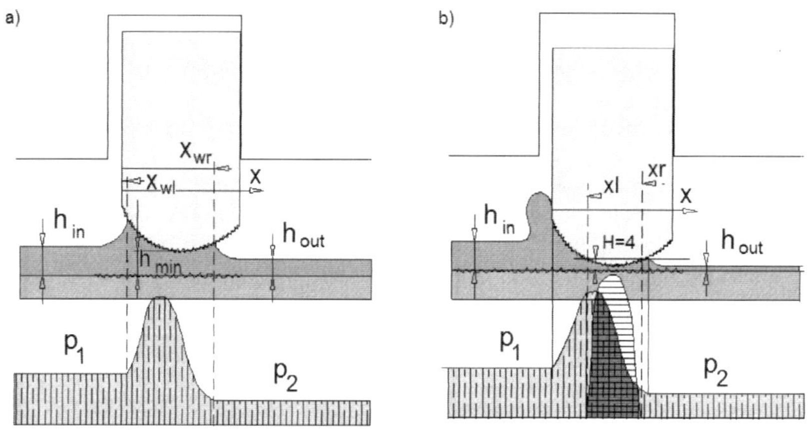

Figure 2. Scheme of gap between the ring face and cylinder liner in the case of:
a) hydrodynamic, **b)** mixed friction.

A one dimensional form of the modified Reynolds equation developed by Patir and Cheng [6,7] has been used to calculate hydrodynamic forces in the case of rough gap surfaces. This equation is applicable to any general roughness structure and takes the following form:

$$\frac{\partial}{\partial x}\left(\phi_x \frac{h^3}{12\mu}\frac{d\bar{p}}{dx}\right) = \frac{U}{2}\frac{d\bar{h}_T}{dx} + \frac{U}{2}\sigma\frac{d\phi_S}{dx} + \frac{d\bar{h}_T}{dt} \tag{1}$$

where: t - time; x – coordinate along cylinder liner; h – nominal oil film thickness; h_T – average gap (ring-cylinder); p – hydrodynamic pressure; U – axial ring velocity; μ – dynamic oil viscosity; $v=\partial h_T/\partial t$ – radial ring velocity, σ – composite root mean square roughness of sliding surfaces.

The significance and mathematical description of empirical coefficients ϕ_x , ϕ_S and boundary conditions of equation (1) are presented in [6,7] and also in [10].

The effects of interacting asperities of piston ring and cylinder liner surfaces were modelled using the mathematical model developed by Greenwood and Tripp [2]. In this case the asperity contact force per unit circumference is given by

$$F_A = 16\sqrt{\frac{2}{15}}\pi(\eta\beta\sigma)^2 E'\sqrt{\left(\frac{\sigma}{\beta}\right)}\int_{x_l}^{x_r} F_{5/2}\left(\frac{h}{\sigma}\right)dx \tag{2}$$

where the integration limits x_l and x_r define a continuous interval, $x_l \le x \le x_r$ in which $\frac{h}{\sigma} \le 4$

η - asperity density; β - asperity radius of curvature;

$\sigma = \sqrt{\sigma_1^2 + \sigma_2^2}$ composite roughness of sliding surfaces.

The model is also described in detail in publication [10] of the author of this article.

2.3. Viscosity as a function of oil temperature

The viscosity of the oil used for lubrication is a key factor influencing oil film thickness. Automotive engines have a relatively high temperature of oil film and due to that a relatively low oil viscosity (Figure 3). This explains the very thin oil film left by the periodically moving ring pack. In the analysis presented, it was assumed that the oil film temperature is equal to the liner temperature.

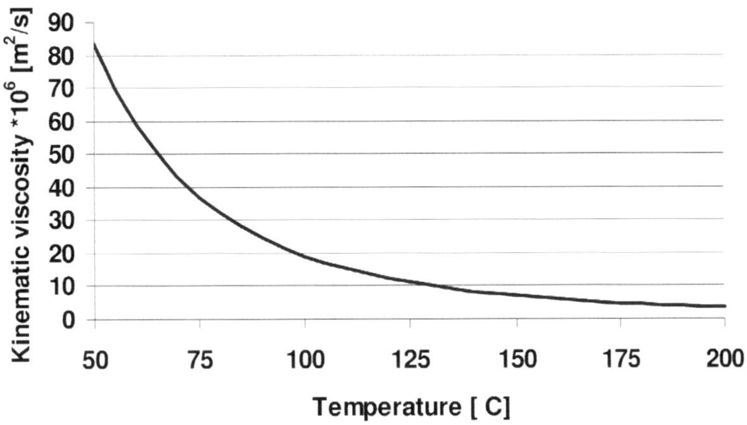

Figure 3. Oil viscosity versus oil temperature.

The variation of oil viscosity with temperature is determined by the commonly used Vogel equation:

$$\mu_0(T) = k \exp\left(\frac{\theta_1}{\theta_2 + T}\right) \tag{3}$$

where: k = 0,0352 cSt; θ_1 = 1658,88 °C; θ_2 = 163,54 °C for the SA-10W50A multigrade oil and temperature T [°C] .

2.4. Experimental verification of the developed model

A verification of the simulation model has been done by the author for a two- and four-stroke marine engine [12,13,16]. The experimental verification of the model of gas flow through the labyrinth seal of piston rings was carried out using measurements of unsteady gas pressure in the cylinder, between the piston rings and under piston performed by piezoelectric sensors mounted in the piston. A satisfactory qualitative and quantitative compatibility of the analyzed pressure variations has been achieved. The maximal relative differences between measured and calculated pressure values have not exceeded 15% [12,16]. On the other hand, the experimental verification of the hydrodynamic model of piston rings involved measurement results of scraped oil volumes by a gland-box of a two-stroke marine engine. Unfortunately, similar measurements for piston ring packs of tested engines have not been carried out. Examination of scraped oil volumes by the ring pack (of the gland-box of marine internal combustion engine) proves a satisfactory quantitative agreement between numerical and experimental results. The maximal relative differences between measured and calculated values have not exceeded 10% [12,16].

3. Computational results

3.1. Main data of chosen engine

The simulation investigations have been done for a four-stroke spark ignition engine of a middle class passenger car. The main data of the engine is presented in Table 1.

Table 1. Main engine parameters

Cylinder diameter	D_C = 80 mm
Piston diameter	D_P = 79,92 mm
Piston stroke	S = 79.5 mm
Engine rotational speed	n = 3400 rpm

The type of ring set considered is common in car engines. It consists of three rings: a compression ring, a scraper ring and a two-lip oil ring. The package includes conventional straight ring end gaps. The surface geometry of the piston ring package, with vertical dimensions magnified by factor of 1000 relative to the horizontal ones, is depicted in Figure 4. All the rings are barrel shaped.

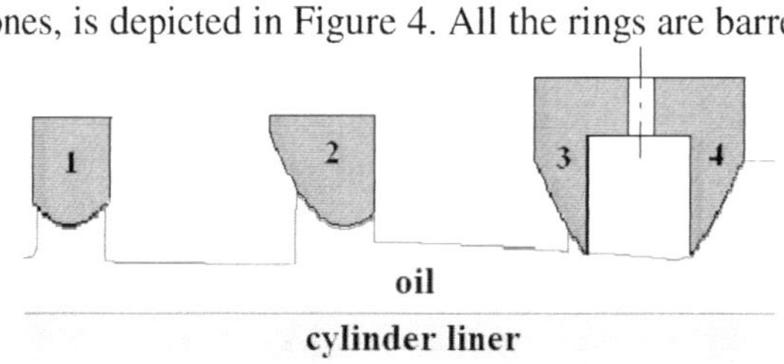

Figure 4. Ring pack geometry under consideration (1-compression ring, 2 – scraper ring, 3 - upper lip of the oil ring and 4 – lower lip of the oil ring).

3.2. Calculation results for assumed oil temperature

An automotive internal combustion engine works where a wide range of loads and the resulting thermal conditions exist. In the paper two characteristic thermal states of engine operation have been analysed and compared. It has been assumed that the first characteristic engine load state is engine warm-up, when the cylinder liner temperature is low. The second characteristic engine state corresponds to the nominal thermal load, when the liner temperature is much higher.

Two calculations have been performed. The first one assumed that the oil temperature equals 93°C. Under this assumption, the oil film is thick enough in comparison with the mean value of the surface roughness. Therefore, the surfaces of rings and cylinder liner can be treated as smooth surfaces. The second calculation has been done under the assumption that the oil temperature equals 200°C. In this case, the possibility of mixed lubrication is expected and taken into account. The main parameters of the rough structure of the liner and ring surface are presented in table 2. An important calculation result is the oil film thickness distribution on the cylinder liner. The motion of the ring pack scraping and distributing oil on the cylinder liner leaves the oil film profile shown in Fig. 5 after a few cycles of operation. A comparison of results concerning low (T_{oil}=93°C) and high (T_{oil}=200°C) oil temperature is presented. An uneven oil film distribution along the cylinder liner can be clearly seen.

Table 2. Surface roughness parameters

Surface data	Value
Asperity density	$\eta = 1 \cdot 10^6$ [m^{-2}]
Asperity radius of curvature	$\beta = 0,2$ [μm]
RMS roughness of cylinder liner	$\sigma_1 = 0,22$ [μm]
RMS roughness of rings sliding surface	$\sigma_2 = 0,044$ [μm]

The low oil film thickness near the piston top dead centre (TDC) and peaks of accumulated oil near the leading ring lips should be noticed. The difference of the oil film thickness along the area of ring pack motion is very important. In the case of lower oil temperature, the minimum film thickness is higher than 1 μm in contrary to the case of higher temperature, where the oil film thickness decreases to 0,3 ÷ 0,4 μm. This value is comparable with the root mean square (RMS) roughness of the cylinder liner that equals 0,22 μm. The very low local film thickness values, in both cases, near the top dead centre can be explained by the existence of high gas pressure values existing in this area during the compression and working phases of engine operation.

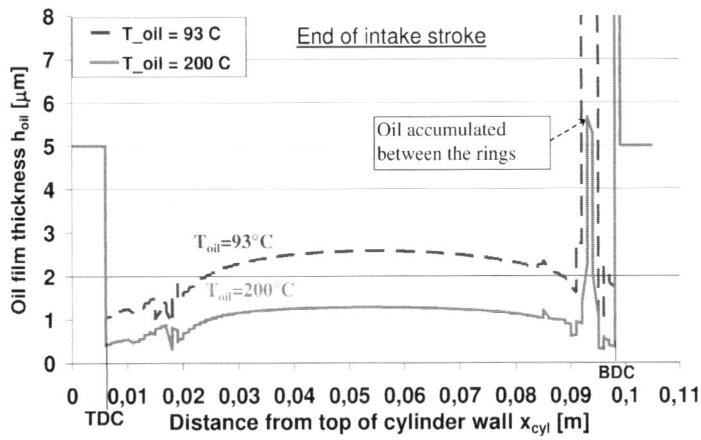

Figure 5. Comparison of the oil film thickness distributions left by the ring pack along cylinder wall for the oil temperature of 93°C and 200°C.

In the following figures variations in some physical parameters as functions of the crankshaft rotation angle, beginning from the piston bottom dead centre (BDC) of the four-stroke engine operation (0°) are shown. In this case the end of compression phase is at 180° of crank angle (piston top dead centre - TDC).

The hydrodynamic friction forces as functions of the crankshaft rotation were calculated and presented in Figure 6. These forces significantly depend on piston velocity. For this reason the highest values of hydrodynamic friction forces should be noticed at crank angles, where the maximum piston velocity is reached. These forces could be neglected in the piston motion phases corresponding to low velocity near the reverse points.

In the case of a higher oil temperature (Figure 6b) the variation of hydrodynamic friction forces has a slightly different character. Changes in the amplitude of these forces, and thus the absolute extreme values are significantly lower (approximately 3 times for the compression ring).

Gas and ring elastic tensing forces must be compensated by hydrodynamic forces generated in the ring-liner gap and, additionally, in the mixed lubrication cases expected, by elastic contact forces.

a) b)

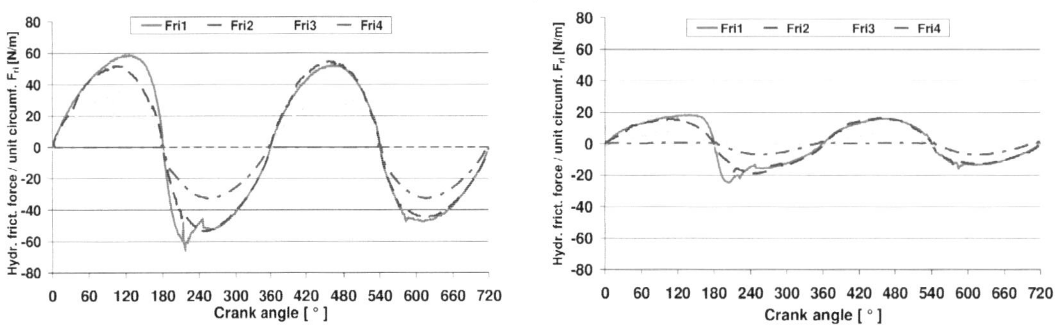

Figure 6. Variation in hydrodynamic friction force F_{fri_i} for each piston ring lip (i =1,2,3,4) versus crank angle. Results for oil temperature: a) 93°C, b) 200°C.

In Figure 7a variations in radial components of elastic contact forces of piston rings with cylinder liner are visible. The force values are referenced to the unit of ring circumference. These forces occur in the case of a high oil temperature (T_{oil}=200°C) near the piston top and bottom dead centre. The values of elastic contact forces are much lower than hydrodynamic forces acting on rings [10,13]. In the case of higher oil temperature and lower oil film thickness additionally tangential components of contact forces occur near piston reverse points (TDC and BDC - Figure 7b).

a) b)

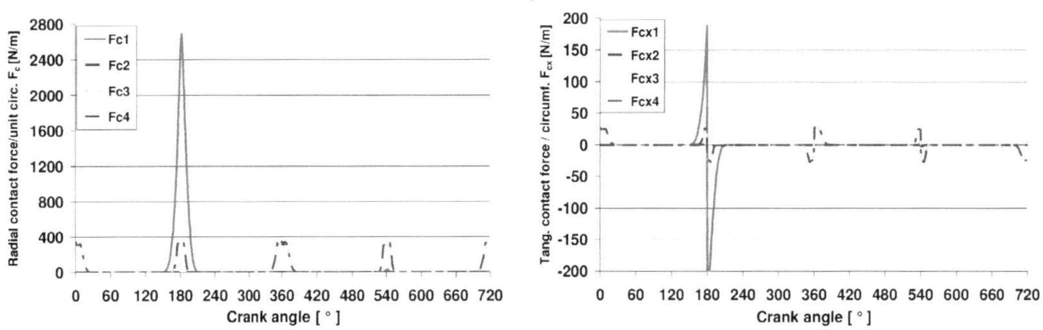

Figure 7. Variation in: a) radial F_{c_i} and b) tangential F_{cx_i} component of contact force for each piston ring lip (i =1,2,3,4) versus crank angle. Results for rough surfaces and oil temperature of 200°C.

One important computational problem is the definition of the boundaries of the ring wetted area (see Figure 8). After a series of piston operation cycles, each ring scrapes and accumulates excessive oil, leaving behind an oil film not sufficient for full lubrication. The variation of wetted area has an essential influence on the hydrodynamic bearing force of the ring and the resulting radial ring velocity. Oil film thickness, piston velocity, ring stiffness and ring surface roughness parameters have significant influence on changes of wetted area boundaries and areas of direct contact of the ring and cylinder liner surfaces in the case of mixed lubrication. Most often rings are only partially wetted in phases of high piston velocity. Rings are fully wetted in the piston motion phases corresponding to low velocity near the reverse points. Figure 8a depicts changes of the wetted area boundaries of the compression ring for low oil temperature. In this case the full lubrication can be observed. Oil is mostly concentrated in the central part of the ring profile. In the rear part the hydrodynamic pressure drops and this area is ventilated by the gas outside the ring. In the case of very thin oil film, the wetted area is also reduced in the upstream part of the ring surface. Additionally, in Fig. 8b corresponding to the case of high oil temperature and mixed lubrication, boundaries of the contact area are shown. Due to very low oil film thickness and assumed contact limit of $h/\sigma = 4$ [2,6,7], an arrival of elastic contact area is noticeable. In both cases (Figures 8a and 8b) the shapes of the wetted areas look very similar. However, in the second case of high oil temperature, the wetted area is a little bit smaller.

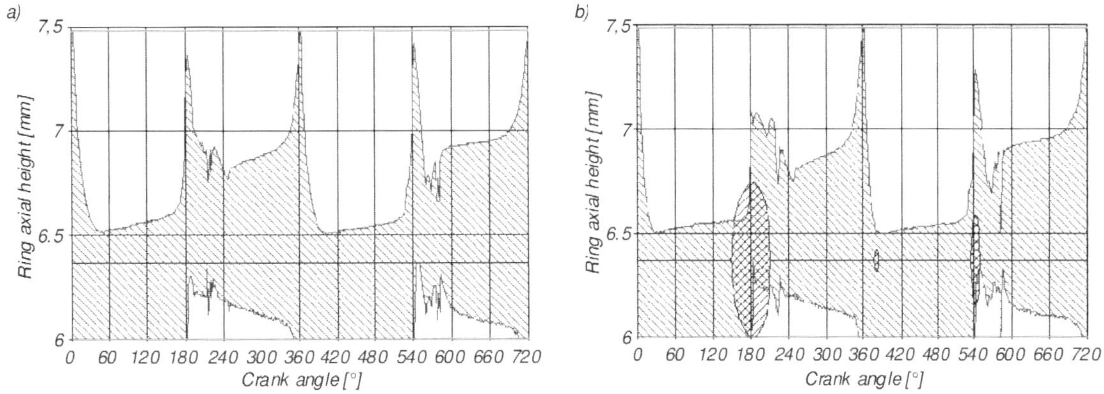

Figure 8. Oil wetted area (\\\\\\) and contact zone (//////) of the 1st ring (compression ring) versus crank angle. Results for the oil temperature: a) 93°C, b) 200°C.

The oil wetted area and contact zone of the 2nd (scraper) ring, but for the higher surface roughness ($\sigma_1 = \sigma_2 = 0{,}4$ μm), has been presented by the author in article [10].

The two-land oil ring shows the same pattern (Figures 9a and 9b) of wetted area distribution in both cases. It can be seen that depending on the direction of the piston motion, only the leading ring land is wetted with starvation on the trailing land. It should be noted that the ring twist effect has been neglected. The hydrodynamic lubrication case is presented in Figure 9a, and the mixed lubrication case in Figure 9b. In addition, a small area of contact in the central ring part is observed in the case of high oil temperature.

The calculations showed that at higher oil temperatures, its viscosity is so low that it causes the movement of the ring in the distance of the cylinder liner compared to the size of medium asperities characterizing this surface. In this case, the model of the oil flow in the lubricating gap described by the Reynolds equation is replaced by the model developed by Patir and Cheng [6,7] (equation (4)). In this case, there are also mechanical interactions between the mating rough surfaces. It is therefore necessary to take these forces into account, which allows the model of Greenwood and Tripp [2].

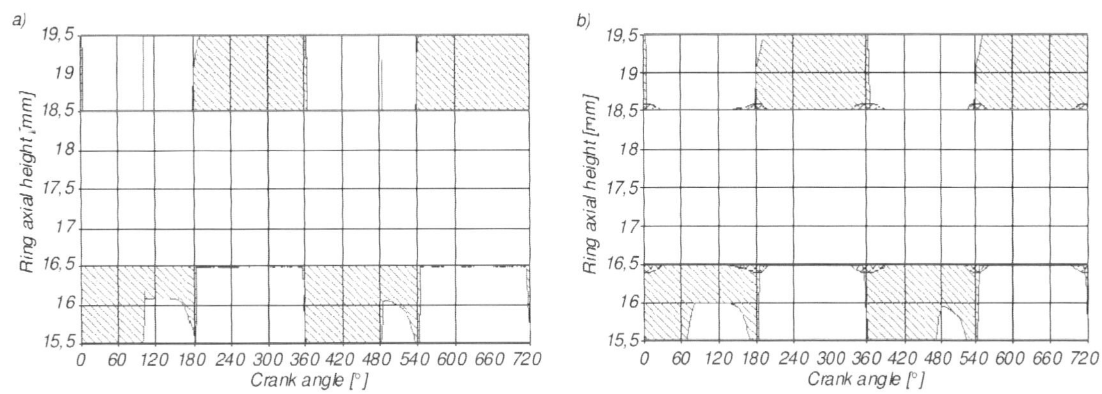

Figure 9. Oil wetted area (\\\\\\) and contact zone (//////) of the 3rd ring (oil ring with 2 lands) versus crank angle. Results for the oil temperature: a) 93°C, b) 200°C.

3.3. Calculation results for the chosen surface roughness of cylinder liner

Simulation investigations were carried out for four selected surfaces of cylinder liner characterized by the following root mean square (RMS) roughness: σ_1 = 0,1 μm, σ_1 = 0,22 μm, σ_1 = 0,3 μm and σ_1 = 0,4 μm. Every time a constant value of σ_2 = 0,044 μm was assumed for piston rings. In addition, for comparison calculations concerning perfectly smooth surfaces of cylinder liner and piston rings (σ_1 = 0 μm and σ_2 = 0 μm) were performed.

Due to the presence of relatively thin oil layers the contact between the rough surface of the cylinder and piston rings occurred (mixed friction). It was important to obtain a response of the models of Patir and Cheng [6,7] and Greenwood and Tripp [2] to the change of the surface roughness. However, the main objective of the study was to evaluate the influence of the surface roughness of cylinder liner on friction losses. The computational results are shown in Figs. 10 and 11. The first one depicts the total power loss (due to hydrodynamic and mixed lubrication) versus crank angle for all of the aforementioned surface roughness of the cylinder liner.

In turn, Fig. 11 shows average values of previously presented friction power losses associated with the selected values of surface roughness. Generally it can be concluded that the smallest friction power loss occurs in the case of perfectly smooth surface. The higher RMS of cylinder surface roughness is applied the more friction power loss is detected.

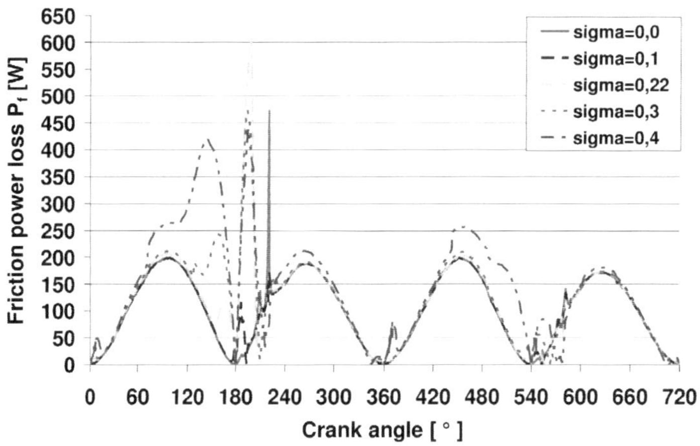

Figure 10. Total power loss (due to hydrodynamic and mixed lubrication) versus crank angle. Results for the chosen RMS surface roughness of cylinder liner: σ_1 = 0 μm, σ_1 = 0,1 μm, σ_1 = 0,22 μm, σ_1 = 0,3 μm, σ_1 = 0,4 μm

A significant increase of friction power loss at the end of the compression stroke and at the beginning of expansion can be seen (Figure 10), i.e. when a high gas pressure in cylinder exists and an elastic contact of piston rings with surface roughness of cylinder liner occurs. In this range of operating conditions an increase of wear intensity of the cylinder liner can be predicted. It is noteworthy that at piston reverse points the axial velocity of piston rings decreases to zero, which means the simultaneous drop of friction power loss to zero value.

Although the model of Patir and Cheng [6,7] and Greenwood and Tripp [2] allows to perform a number of interesting analysis, it is difficult to directly choose the optimum surface roughness. For this purpose it would be necessary to examine more tribological factors such as the microstructure of surface treatment, etc. It would be also good to collect an extensive experimental data.

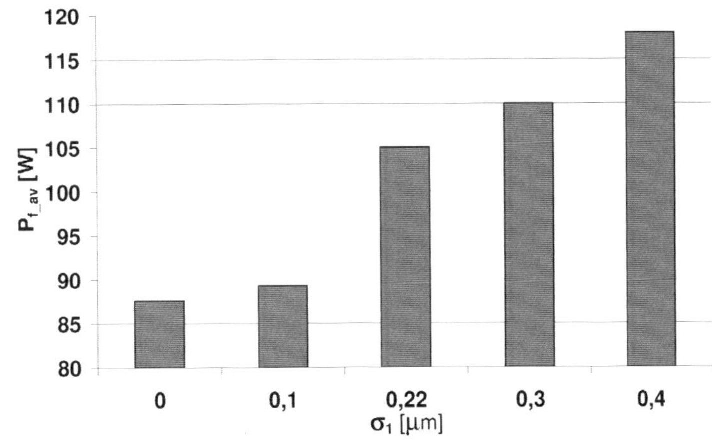

Figure 11. Average value of total power loss P_{fri_av} (due to hydrodynamic and mixed lubrication) versus RMS of surface roughness of cylinder liner.

4. Conclusions

The major conclusions that may be drawn from the results are as follows:

1. The developed mathematical model and simulation programme give a lot of practical information that would be more complicated and expensive to obtain using experimental methods;
2. Elastic radial contact forces due to surface roughness occur in the area of piston TDC and BDC (Figure 7). These forces are relevant to high gas pressure and low oil viscosity caused by high temperature. However, the values of elastic contact forces are much lower than hydrodynamic forces acting on piston rings [10,13];
3. The hydrodynamic forces are generated by relatively low pressure acting on a large surface in contrast to high local contact pressure concentrated on a very small area of elastic contact. Due to that fact, the elastic contact seems to be responsible for wear process despite the low speed of the piston close to TDC and BDC;
4. The most severe lubrication conditions occur close to TDC towards the end of the compression and beginning of expansion stroke, especially for the first (compression) ring (Figure 7). Due to high temperature, the oil viscosity is very low and, consequently, the oil film along the cylinder liner is very thin: 0,3 ÷ 0,4 μm at TDC (Fig. 5). In this case it is essential that a model of mixed lubrication accounting for the surface roughness is used (RMS = 0,22 μm for the cylinder liner of analyzed engine);
5. The higher RMS of cylinder surface roughness is applied the more friction power loss is detected (Figure 11);
6. Although the model of Patir and Cheng [6,7] and Greenwood and Tripp [2] allows to perform a number of analysis concerning mixed lubrication, it would be difficult to choose the optimum surface roughness. For this purpose it would be necessary to examine more tribological factors such as the microstructure of surface treatment, etc. It should be also good to collect an extensive experimental data.

References

[1] Dowson D 1993 *Piston Assemblies: Background and Lubrication analysis*, Engine Tribology. Amsterdam, New York: Taylor C. M. (editor), Elsevier Science.

[2] Greenwood J Tripp J H 1971 *The contact of Two Nominally Flat Rough Surfaces*. Proc Inst. Mech. Eng. 185: 625-633.

[3] Iskra A 2001 *Parametry filmu olejowego w węzłach mechanizmu tłokowo – korbowego silnika spalinowego*. Poznań: Wydawnictwo Politechniki Poznańskiej.

[4] Koszałka G, Niewczas A, Guzik M 2008 *Predicted and Actual Effect of Cylinder Liner Wear on the Blowby in a Truck Diesel Engine*. SAE Paper No. 2008-01-1717.

[5] Koszałka G 2010 *Application of the piston-rings-cylinder kit model in the evaluation of operational changes in blowby flow rate*. Eksploatacja i Niezawodność – Maintenance and Reliability 4: 72-81.

[6] Patir N, Cheng H S 1978 *An Average Flow Model for Determining Effects of Three-Dimensional Roughness on Partial Hydrodynamic Lubrication*. Transactions of ASME; 100.

[7] Patir N, Cheng H S 1979 *Application of Average Flow Model to Lubrication between Rough Sliding Surfaces*. Transactions of ASME; 101.

[8] Serdecki W 2002 *Badania współpracy elementów układu tłokowo – cylindrowego silnika spalinowego*. Poznań: Wydawnictwo Politechniki Poznańskiej.

[9] Tian T 2002 *Dynamic Behaviors of Piston Rings and Their Practical Impact-Part II: Oil Transport, Friction*, and Wear of Ring/Liner Interface and the Effects of Piston and Ring Dynamics. Proc. Inst. Mech. Eng., Part J: Journal of Engineering Tribology; 216: 229-247.

[10] Wolff A, Piechna J 2005 *Numerical simulation of piston ring pack operation in the case of mixed lubrication*. The Archive of Mech. Engineering; LII, No. 3: 157-190.

[11] Wolff A, Piechna J 2007 *Numerical simulation of piston ring pack operation with regard to ring twist effects*. The Archive of Mech. Engineering; LIV, No. 1: 65-99.

[12] Wolff A 2009 *Experimental verification of the model of piston ring pack operation of an internal combustion engine*. The Archive of Mechanical Engineering; LVI, No. 1: 73-90.

[13] Wolff A 2009 *Numerical analysis of piston ring pack operation*. Silniki spalinowe / Combustion Engines; 2: 128-141.

[14] Wolff A 2011 *Numerical analysis of piston ring pack operation of a marine two-stroke engine*. Silniki spalinowe / Combustion Engines; 3(146), (article on CD enclosed to the periodical).

[15] Wolff A 2012 *Influence of engine load on piston ring pack operation of a marine two-stroke engine*. Journal of KONES Powertrain and Transport; Vol. 19, No. 2: 557-569.

[16] Wolff A 2014 *Simulation based study of the system piston-ring-cylinder of a marine two-stroke engine*, Tribology Transactions, Vol. 57, Issue 4, pp. 653-667.

Acknowledgments

The author expresses his gratitude to *Wärtsilä's* R&D centre in Winterthur (Switzerland) for having the opportunity to work on projects concerning mathematical modeling and numerical simulation of tribological systems of piston rings during several research periods at this company.

The influence of fuel type on the cooling system heat exchanger parameters in heavy-duty engines

B Worsztynowicz
AGH University of Science and Technology

E-mail: worsztyn@agh.edu.pl

Abstract. The paper discuses the problem of selection of cooling systems for heavy-duty engines fitted in city buses. Aside from diesel engines, engine manufacturers also have in their portfolio engines fueled with natural gas, whose design is based on that of a conventional diesel engine. Based on the parameters of the engines from this type-series (the same displacement and rated power) an analysis has been performed of the influence of the applied fuel on the heat flows directed to the radiators and charge air coolers, hence, their size and space necessary for their proper installation. A replacement of a diesel engine with a natural gas fueled engine of the same operating parameters results in an increased amount of heat released to the coolant and a reduced heat from the engine charging system. This forces a selection of different heat exchangers that require more space for installation. A universal cooling module for different engines is not an optimal solution.

1. Introduction

Natural gas is an alternative source of energy for combustion engines, exhibiting good calorific and emission-related parameters. Due to its low energy density (has to be accumulated under high pressure) and a limited distribution network natural gas is best for stationary engines or engines used in municipal service and public transit vehicles. The above is the main reason for the engine manufacturers to launch a new engine powered by natural gas whose design is based on that of a conventional diesel engine and whose operating parameters are comparable. The process of natural gas combustion, however, is characterized by a lower flame propagation rate leading to a greater engine thermal load, hence the heat flows rejected to the cooling and exhaust systems are greater.

2. Heavy – duty engine cooling system

Cooling systems of modern heavy – duty engines, similarly to other engine systems, are increasingly sophisticated, contain an increasing number of elements and in order to optimize the engine operating parameters the operation is continuously monitored by a variety of sensors. All the mentioned facts are due to the necessity of complying with the exhaust emission standards. Apart from the cylinders and the cylinder head, where the coolant chills the fuel injectors, the valve seats and valve guides, the cooling circuit also includes other components such as the engine oil cooler, the fuel cooler, the exhaust gas cooler in the EGR system, the heater core, the Urea heater (AdBlue) and the AdBlue injector cooler, the air compressor in the pneumatic system, let alone the supercharging pressure, EGR or DPF regeneration actuators (Figure 1). Sometimes, the engine coolant also cools the transmission or retarder fluid.

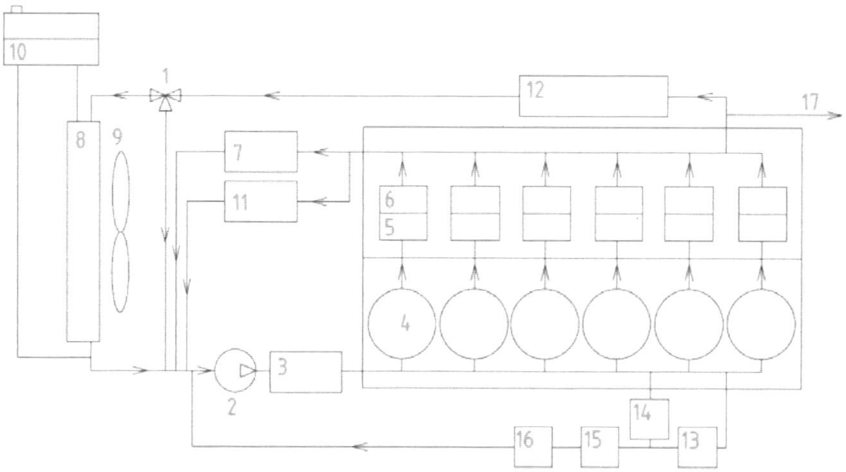

Figure 1. Diagram of the cooling system of a heavy – duty Euro 6 compliant engine [1]

1 – coolant thermostat, 2 –coolant pump, 3 – oil – water heat exchanger, 4 – cylinder, 5 – lower coolant level in cylinder head (fuel injectors and valve seats), 6 – upper coolant level in cylinder head (valve guides), 7 – heater core, 8 – radiator, 9 – fan, 10 – coolant expansion tank, 11 – urea heater , 12 – exhaust gas recirculation cooler, 13 – air compressor, 14 – fuel cooler, 15 – exhaust gas recirculation positioned, 16 – transmission oil – water cooler (optional), 17 – retarder (optional)

An increasing number of engine components requiring constant temperature (hence cooling) results in an increased heat rejection to the cooling system [1]. The temperature difference of the coolant obtained in the heat exchanger must be maintained at a constant level of approx. 6 – 8 K, because an excess temperature gradient may lead to the damage of engine components in contact with the coolant. Reception of excess heat is realized through increased coolant pump rate. The need to cool many engine components at the same time results in an extended coolant flow distance, hence greater major and minor head loss, directly affecting the pump load (its efficiency, Fig. 2).

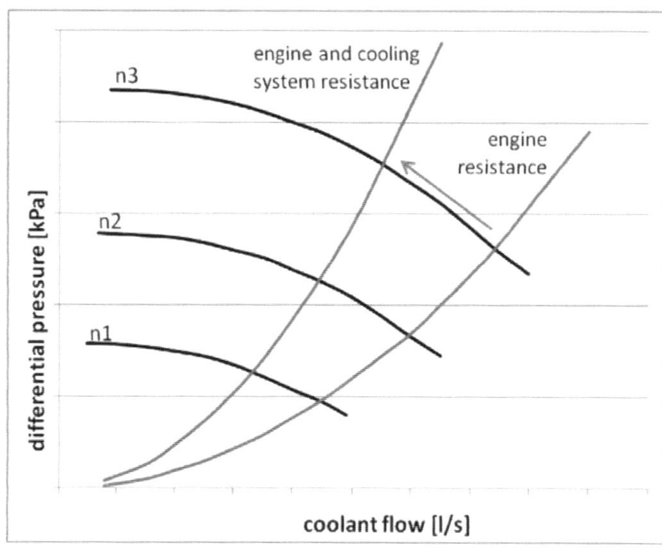

Figure 2. Characteristics of the coolant pump cooperation with the cooling system and the engine.

3. Engine characteristics

The engines selected for the analysis were EURO VI complaint, dedicated for city buses. From the point of view of natural gas used as fuel, this application is optimal. The analysed engine has six cylinders in a straight configuration, double overhead camshafts and four valves per cylinder. The engine block and cylinder head are made from cast iron and the crankshaft and the connecting rods are forged. In both fuel versions, oil-cooled aluminium pistons, fixed geometry turbochargers and cooled EGR systems were applied. The base diesel engine is a *Common Rail* direct injected engine fitted with the exhaust aftertreatment system composed of: an oxidation catalytic converter, a diesel particular filter and a reduction catalytic converter with an AdBlue system. The natural gas fuelled version is an indirect injected engine fitted with a three-way catalytic converter.

Both engines produce a torque of 1200 Nm in the same range of engine speeds (from 1200 rpm to 1600 rpm). The maximum power of 220 kW produced by the diesel engine is reached at 2200 rpm and 2000 rpm by the natural gas fuelled engine. As per manufacturer's specifications, minimum specific fuel consumption for the diesel engine is on the level of 190 g/kWh and for the natural gas fuelled engine - approx. 185 g/kWh.

The natural gas combustion process is characterized by a lower flame propagation rate compared to other fuels, the consequence of which is the extension of the combustion towards the power stroke and a greater thermal load of the engine, which is why the adaptation of a diesel engine to natural gas fuelling requires an increased efficiency of the cooling system. For the analysed engines, the amount of heat rejected to the cooling system is greater by approx. 13 % for the natural gas fuelled engine at a simultaneous reduction of the maximum admissible coolant temperature by 10 K (table 1). The lower difference of temperature between coolant and cooling air influence the reduction of the specific heat rejection of heat exchanger, consequently an application of greater space of heat exchange is necessary. The extension of the combustion process increases the exhaust gas temperature at the moment of exhaust valve opening, which, in the case of turbocharged engines, may improve the turbocharging efficiency.

Table 1. Selected engine data at ambient temperature of 25 ^{0}C

		Diesel engine	Natural gas fueled engine
Heat rejection to coolant	kW	152	172
Coolant temperature	^{0}C	105	95
Charge air temperature downstream of the turbocharger	^{0}C	175	180
Charge air temperature upstream of the engine	^{0}C	41	37
Exhaust gas temperature	^{0}C	420	540

4. Characteristics of the cooling system

Buses have their engines installed in the back of the vehicle and, consequently, the cooling modules are also in the back and are fitted sideways, which, compared to passenger vehicles and trucks, seriously inhibits the speed-induced airflow through the exchangers. Depending on the solution, the cooling module may be mounted on the level of the engine or, in the vertical configuration, above it. City buses frequently use the vertical mounting system, which is why the fan drive is rarely based on a belt transmission. Hydraulic fan drive is applied instead, or, as has recently become popular, several smaller fans with electric motors are used.

The analysis of the influence of the type of engine (diesel/natural gas) on the parameters and size of the cooling module was performed in several stages. In all calculations, a constant value of the coolant volumetric flow was assumed corresponding to the maximum admissible pressure drop in the cooling system for a given engine. Given the European climate, the calculations were performed for the ambient

temperatures of 45 ^{0}C for which the admissible temperature of the charge air is increased by 20 K respectively.
The calculations were performed in GT-SUITE using the characteristics of the BSPL heat exchanger cores.

4.1. Constant cooling air flow rate
Initially, the operation of the fan was not allowed for in the calculations. A constant value of the cooling air mass flow was assumed instead and the size of the heat exchangers for the coolant and the charge air were adjusted to obtain the cooling parameters defined in the engine characteristics. Because the diesel engine is the base unit, it was the engine for which the value of the airflow rate was adjusted. Figure 3 presents the calculation models shown in the same scale and table 2 contains the minimum theoretical size of the frontal area of the heat exchanger cores (the thickness was not modified) and respective temperature values obtained in the calculations.

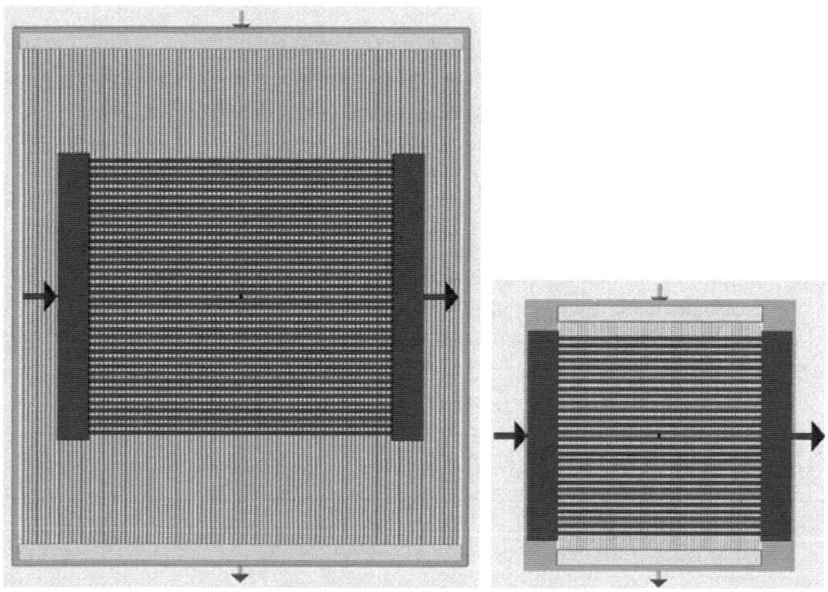

Figure 3. Calculation models of the cooling modules – NG engine on the left, diesel engine on the right.

Table 2. Calculation results for the cooling air flow of 4.5 kg/s (ambient temperature of 45 ^{0}C).

		Diesel engine	Natural gas fueled engine
Radiator core	mm	650x600	1400 x 1300
Coolant temperature downstream of the engine	^{0}C	104.3	94.9
Charge air cooler core	mm	600x600	900 x 800
Charge Air Temperature upstream of the engine	^{0}C	59.3	56.3

Obtaining the temperature values of the mediums defined in the characteristics for both engines, assuming the same cooling air flow rate makes the frontal area of the radiator core of the natural gas fuelled engine more than 4,5 times larger compared to the diesel engine. A further increase in the radiator size does not result in a measurable reduction of the temperature.

4.2. Constant size of the heat exchangers

The second stage of the analysis was performed for the cooling module composed of the coolant radiator, the charge air cooler, the fan with a hydraulic drive and the cooler of the fan hydraulic fluid. In this case, the size of the exchangers was constant and the required parameters were obtained through adjustment of the fan speed.

Figure 4 presents the calculation model of the cooling module, and table 3 includes the sizes of the basic components of the cooling module, calculated temperature values, the matching fan speeds and the resulting fan power demand.

Figure 4. Calculation model of the cooling module fitted with a axial fan.

Table 3. Calculation results for different fan speeds (ambient temperature of 45 ^{0}C)

		Diesel engine	Natural gas fueled engine
Radiator core	mm	800 x 750	
Coolant temperature downstream of the engine	^{0}C	100.4	93.1
Charge air cooler core	mm	650 x 640	
Charge air temperature upstream of the engine	^{0}C	58.9	49.8
Axial fan diameter	mm	702	
Axial fan speed	rpm	2000	3000
Cooling air flow	kg/s	4.64	6.31
Axial fan power consumption	kW	13.9	47.4

The components of the cooling module have been combined in such a fashion as to obtain (through a change of the axial fan speed) the parameters defined in the characteristics of both engines. As a result, for the diesel engine the fan speed of approx. 2000 rpm is sufficient while the fan speed for the natural gas fuelled engine must be higher (3000 rpm). For the fan speed of 3000 rpm, the power demand is approx. 3.5 times greater compared to 2000 rpm, which results in a significant engine power loss. A standard assumption is that the fan power demand (at its maximum speed) should not exceed 10% of the engine power output. An increase in the fan speed also generates a greater noise.

4.3. Dedicated cooling systems

In actual conditions of the cooling system operation, additional pressure drops of the cooling air need to be allowed for. The pressure drops result from the flow resistance generated at the shutters of the cooling module and depend on its design. Another source of pressure drops is the shape of the channel through which the air flows. The operation of cooling systems is also influenced by other elements of the engine compartment, whose actual impact is hard to predict at the calculation stage. Additional safety margin should be added accordingly.

Figure 5 presents the calculation models of the cooling modules, selected individually for each of the engines. During the calculations, the author assumed pressure drops resulting from the fitting specificity and the safety margins for the radiator and the charge air cooler. Table 4 includes the dimensions of the basic components of the cooling module, calculated temperature values, the axial fan speeds and the resulting power demands.

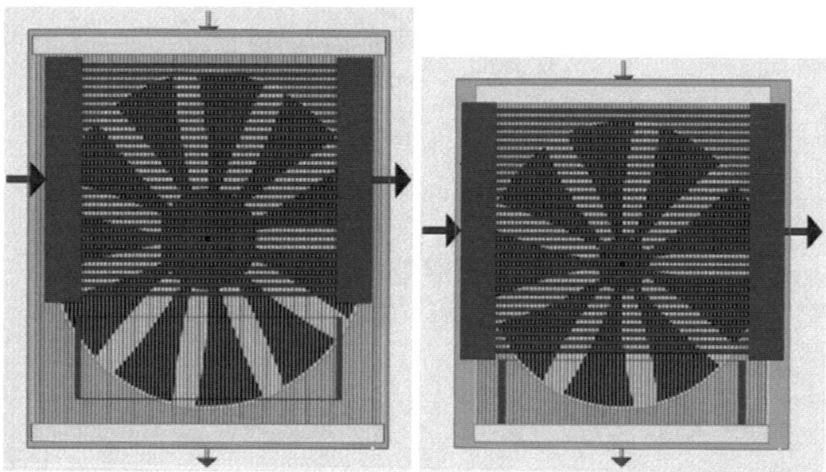

Figure 5. Calculation models of the cooling modules fitted with a radiator fan – NG-fuelled engine on the left, diesel engine on the right.

Table 4. Calculation results for the cooling modules (ambient temperature of 45 ^{0}C).

		Diesel Engine	Natural gas fueled engine
Radiator core	mm	800 x 750	900 x 900
Coolant temperature downstream of the engine	^{0}C	103.3	93.5
Charge air cooler core	mm	650 x 640	650 x 600
Charge air temperature upstream of the engine	^{0}C	58.3	53.5
Axial fan diameter	mm	702/ 9 blades	814/ 11 blades
Axial fan speed	rpm	2300	2000
Cooling air flow	kg/s	4.95	6.56
Axial fan power consumption	kW	21.3	28.6

A greater heat rejection (13%) of the natural gas fueled engine to the cooling system compared to the diesel engine, results in a larger frontal area of the radiator core by approx. 35 %. The fan drive power demand also increased by approx. 34 %.

5. Conclusions

Natural gas fueled engines are increasingly used in city buses. A replacement of an engine fitted in a given bus with an engine of a newer emission standard or of different operating parameters does not typically force a modification of the engine compartment. This may result in a very close fitting of the engine components that may have a thermal impact on one another, which may disturb their proper operation. A replacement of a diesel engine with a natural gas fueled engine of the same operating parameters results in an increased amount of heat rejected to the coolant and a lower heat from the charging system to the charge air cooler, which is a result of lower charge air mass flow to engine. Consequently, a selection of heat exchangers of different size is necessary, which requires more space for the fitting of the cooling module. A universal cooling system for different engines is not an optimum solution in terms of typical operation and may lead to significant loss of power due to energy needed for the fan drive (tab. 3).

References

[1] Worsztynowicz B 2014 Thermal balance of havy – duty combustion engines in the aspect of exhaust emissions standards *Silniki spalinowe i ekologia - praca zbiorowa, opracowanie monograficzne.* **978-83-7242-763-2**
[2] Mercedes-Benz – factory data
[3] GT-SUITE

IOP Conf. Series: Materials Science and Engineering **148** (2016) 012092 doi:10.1088/1757-899X/148/1/012092

Enhancing the properties of Fischer-Tropsch fuel produced from syngas over Co/SiO2 catalyst: Lubricity and Calorific Value

O Doustdar, M L Wyszynski*, H Mahmoudi and A Tsolakis
Department of Mechanical Engineering, School of Engineering, College of Engineering and Physical Sciences,The University of Birmingham, UK, B15 2TT
*Corresponding author: M.L.Wyszynski@bham.ac.uk

Abstract. Bio-fuel produced from renewable sources is considered the most viable alternatives for the replacement of mineral diesel fuel in compression ignition engines. There are several options for biomass derived fuels production involving chemical, biological and thermochemical processes. One of the best options is Fischer Tropsch Synthesis, which has an extensive history of gasoline and diesel production from coal and natural gas. FTS fuel could be one of the best solutions to the fuel emission due to its high quality. FTS experiments were carried out in 16 different operation conditions. Mini structured vertical downdraft fixed bed reactor was used for the FTS. Instead of Biomass gasification, a simulated N_2 -rich syngas cylinder of, 33% H_2 and 50% N_2 was used. FT fuels products were analyzed in GCMS to find the hydrocarbon distributions of FT fuel. Calorific value and lubricity of liquid FT product were measured and compared with commercial diesel fuel. Lubricity has become an important quality, particularly for biodiesel, due to higher pressures in new diesel fuel injection (DFI) technology which demands better lubrication from the fuel and calorific value which is amount of energy released in combustion paly very important role in CI engines. Results show that prepared FT fuel has desirable properties and it complies with standard values. FT samples lubricities as measured by ASTM D6079 standard vary from 286μm (HFRR scar diameter) to 417μm which are less than limit of 520μm. Net Calorific value for FT fuels vary from 9.89 MJ/kg to 43.29 MJ/kg, with six of the samples less than EN 14213 limit of 35MJ/kg. Effect of reaction condition on FT fuel properties was investigated which illustrates that in higher pressure Fischer-Tropsch reaction condition liquid product has better properties.

1. Introduction

In previous years, growing concerns about the fast depletion of fossil energy resources and need to reduce greenhouse gas emissions make renewable energy sources much more attractive. The Energy concerns in the world and economic growth in the developing countries are on the drastically rise [1-3]so the energy demands in the field of transportation and manufacturing have increased[1, 2]. Rudolph Diesel, the inventor of the first diesel engine, used vegetable oil (groundnut oil) as fuel in 1900 [4, 5]. His motivation was to improve efficiency and also to down size the price and costs of engines, compared to huge steam engines, so that smaller companies would be able to afford the machinery [6].

During a demonstration he said, "The use of vegetable oils for engine fuels may seem insignificant today, but such oils may become, in the course of time, as important as petroleum and the coal tar products of the present time." [7]. Ethanol fuel from sugar cane was implemented in Brazil after the 1970's oil crisis, however the advantages began to disappear due to the enhancement of ethanol fuel costs in 1989 [8]. Over the past few decade, there has been a surge in global warming and climate change which has brought about by fossil fuels, mainly petroleum-based liquid fuels, natural gas and coal [9]. This has renewed interest in biofuels from vegetable oils as alternative fuel more than 100 years after the time of Rudolph Diesel's predictions [4-6].There are several options for biomass derived fuels production involving chemical, biological and thermochemical processes. Two of the most promising fuels appear to be biodiesel or synthetic fuels such, as Fischer-Tropsch diesel [10]. This is because other potential fuels, such as Ethanol, Methanol and LPG, do not perform as well in modern engines. In addition, in the near future biomass is expected to play an important role and it would be one of the important renewable energy sources [3]. An option for production of renewable fuels from gasified biomass is the Fischer-Tropsch Synthesis (FTS). The FTS is considered as an efficient solution to the problem of finding appropriate substitutes for liquid fossil fuels [11, 12]. Fischer-Tropsch Synthesis is a technology that has an extensive history of gasoline and diesel production from coal and natural gas. The fuel produced by using FT synthesis could be one of the best solutions to the fuels emission due to its high quality. FTS technology devised nearly 90 years ago by Franz Fischer and Hans Tropsch in Germany. Equation 1 represents the main reaction in FTS in which Syngas (Hydrogen and Carbon monoxide) is converted to long chain Hydrocarbons over cobalt catalysts during the reaction [13].

$$n\,(CO + 2H_2) \dashrightarrow (\,\text{--}\, CH_2 \,\text{--}\,)_n + n\,(H_2O) \qquad (1)$$

FTS is a surface polymerization reaction in which the reaction between CO and H_2 takes place on the surface of the cobalt catalysts. This process is described as a Carbide formation on the surface of the catalysts, which is discovered, by Fischer and Tropsch nine decades ago. They assumed that the carbide carbons are decomposed by hydrogen with regeneration of the catalysts metal and can be transformed to hydrocarbons. CH_2 entities are formed on the surface of the catalyst and arranged in a row on the surface, which implies the carbide mechanism. Another assumption is freely movement of CH_X units along the catalyst [13, 14]. Two other reactions occur after a hydrocarbon molecule is released from the catalyst surface and reabsorbed to follow other reaction paths. Methane formation as an unwanted product in FTS is shown in the equation 2 and 3 in which are considered as an irreversible reaction. The other important reaction which occurs in FTS process is water gas shift reaction that produces water as co product; this reaction play vital role in reactors which reaction take place over cobalt catalysts and produce carbon dioxide as an undesirable product [15].

$$CO + 3H_2 \dashrightarrow CH_4 + H_2O \qquad (2)$$

$$CO + H_2O \dashrightarrow CO_2 + H_2 \qquad (3)$$

Catalysts, reaction conditions, type of reactor, H_2/CO ratio influence on the type of products in FTS [13]. FTS process over cobalt catalyst at normal pressure and temperature of 200-300°C produces linear α olefins C_nH_{2n} as the main products and the small amount of nonlinear products which contain mono methyl branch compounds and also it was considered that by increasing the residence time in reactor, which leads to secondary reactions with the same skeleton of both products, subsequent hydrogenation of olefins forms paraffin C_nH_{2n+2} . High pressure Fischer Tropsch Synthesis over cobalt catalyst produces less olefin in favour of Alkane C_nH_{2n+2} content due to increase of the molecular weight[14].

2. Experiments Methodology

2.1. Fixed Bed Reactor system and operating procedures

Figure 1 shows the schematic diagram of the experimental set-up designed and build by H. Mahmoudi [1] that was conducted in a fixed bed reactor. The experimental work of this project concentrated on developing a miniaturized version of the plant that could achieve the preliminary investigation regarding the F-T process, before scale up to pilot plant and building of a pilot scale bio-fuel generator. For this purpose, a small scale F-T bio-diesel generator via the Fischer-Tropsch synthesis process was designed, built and commissioned in a laboratory of the School of Mechanical Engineering of the University of Birmingham. A mini structured downdraft fixed-bed reactor was employed in order to find the optimum reaction conditions for maximum production of synthetic fuel, as well as to convert the syngas into the synthetic bio-fuel [1]. Experiments were carried out in 16 different operation conditions that indicated in Table 1. The reactor was fixed in a tube furnace in order to provide the heat temperature and to provide the uniform wall temperature; cast iron jacket was installed between the furnace and reactor. Instead of Biomass gasification, a simulated rich syngas bottle of 17% CO, 33% H_2 and 50% N_2 was used [1, 2, 16]. Two liquid/gas separators were used to separate liquid from gaseous products. A Gas Chromatogram Flame Ionization detector (GC-FID) (HP5890) was employed to analyse the $HC_1 - HC_8$ online. The heavier liquids products were analyzed offline by using Gas Chromatogram Mass Spectrometry (GC-MS) (Perkin Elmer Clarus 600). Utilization of nitrogen-rich syngas (with 50% of N_2 (volumetric percentage)) leads to reduction in production cost of diesel oil by eliminating the need for application of gas recycling loop after production of syngas by air partial oxidation. Waste generated heat of FT reaction could be removed effectively and temperature runaway probability can be minimized by nitrogen gas [1].

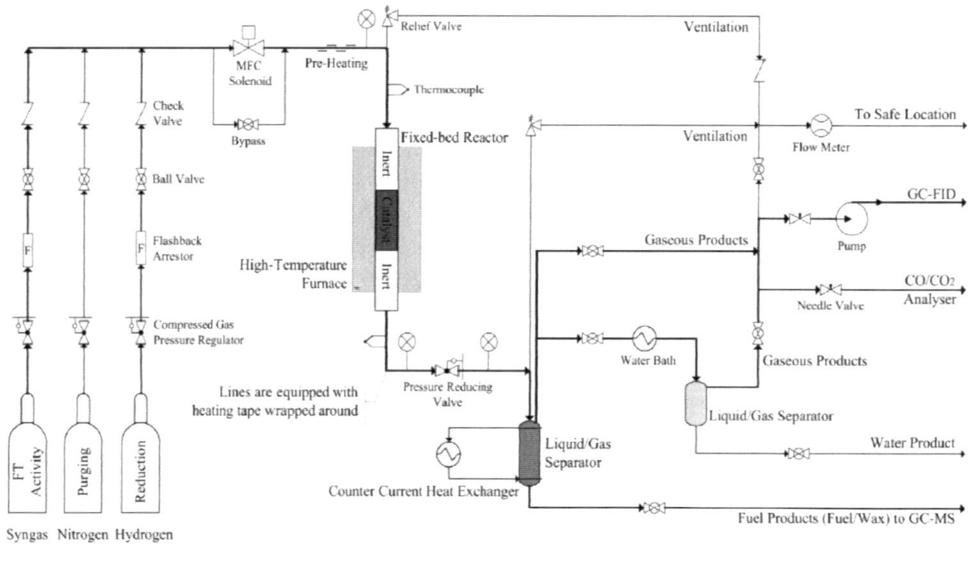

Figure 1. Schematic diagram of small scale bio-diesel generator via Fischer-Tropsch Synthesis (FTS) process and utilized equipment along with Process Path Flow(PPF) implementation conversion of synthesis gas into the liquid heavy by hydrocarbons (Shown with thicker lines) [1].

2.2. F-T Fuel Characterization

Many researchers state that the most important characteristics for fuel application are density, viscosity, heating value and cetane number due to their strong control. over emissions characteristics and engine efficiency, ultimately indicating the quality of the fuel [8]. Literature was examined to find key physicochemical properties for biodiesel and synthetic diesel produced by either transesterification or Fischer-Tropsch. In this paper Lubricity and Heat of Combustion of FT fuel was investigated.

2.2.1. F-T Fuel Composition-Gas Chromatography Analysis. Gas Chromatogram Mass Spectrometry (GC-MS) (Perkin Elmer Clarus 600) used in this study for liquid hydrocarbon products analysis which were collected in a cold trap cooled externally at 283 K using a counter current heat exchanger. The product distributions were analyzed off-line employing a DBI column combined with Gas

Chromatogram Mass Spectrometry (GC-MS). A 1 μL portion of each sample containing the mixture of gasoline, diesel and waxes was injected into the GC with a split ratio of 34:1. Non-reactive inert helium gas was used to carry the gases' samples through the instrument. The injection port was heated to 573 K; the oven temperature was increased at 8 K/min from 303 K to 493 K and held at this temperature for 5 minutes. For liquid samples, the qualitative analysis was performed to identify the constituents (elements of functional groups) [1]. 16 samples of FT fuels were analyzed and total values and percentage of paraffin, iso-paraffin, olefins and alcohols within each of the gasoline (C_7-C_{11}), diesel (C_{12}-C_{22}) and waxes (C_{23+}) were calculated.

2.2.2. Lubricity. Lubricity has become an important quality, particularly for biodiesel, due to higher pressures in new diesel fuel injection (DFI) technology which demands better lubrication from the fuel [17]. Lubricity is simply the ability of the fuel to reduce friction between moving parts within the engine to help it run smoothly. Supposedly, as much as 30% of mechanical energy is consumed by friction [18, 19]. This makes fuel lubrication vital to reduce scarring of components inside the engine. If the fuel does not contain enough lubricating ingredients, it is considered as a "dry fuel" for its incapacity of lubricating the components like fuel delivery and injection system, cylinder liners, etc. [18]. To asses lubricity a tribological test can be performed which measures the size of the wear mark in an HFRR (high frequency reciprocating rig) test [19]. The smaller the wear scare the better the lubricating properties of the fuel, and they must comply with ASTM D6079 maximum value of 520μm.

Assessment of the lubrication properties of FT fuel was carried out on HFRR (PCS Instrument Ltd) according to the EN ISO 12516-1:2006 standard [20]. This is shown schematically in Figure 2a. The test specimens comprised of 6mm diameter steel ball and steel disc. Fuel temperature maintained at $60^{\circ C}$ and the volume of the sample used set at 2ml. A humidity and temperature controlled cabinet was employed to provide the laboratory air conditions defined by the standard as shown in Figure 2b. These conditions are defined according to the ISO standard defining air conditions are defined according to the ISO standard defining air conditions for testing diesel fuel lubricity [20]. During the test the disc was fully submerged in the tested samples at a reciprocation frequency of 50 Hz lasting 75 min. Optical microscopy (with a 100x magnification lens) was used to assess the size of the wear scar on the upper specimen again to the procedure defined by the ISO standard (including a correction factor to account for standard 1.4 kPa water vapour pressure). All of the lubricity experiments were repeated twice and if a difference in corrected wear scar diameters obtained from the same fuel was higher than 20 μm [21].

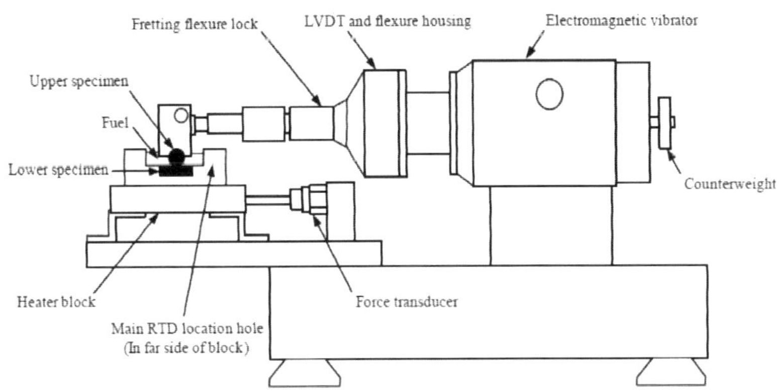

Figure 2a. Schematic diagram of high frequency reciprocating rig (HFRR) [22].

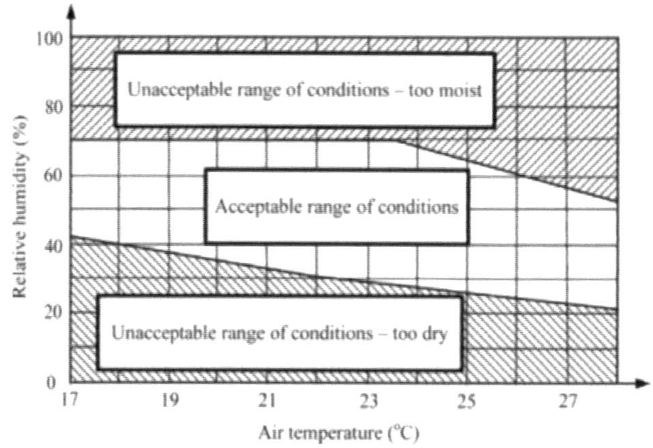

Figure 2b. Laboratory air conditions [23].

2.2.3. Heat of Combustion(Heating Value, Gross Calorific Value). One of the very important factor in the fuel economy and power deliverability is the calorific value of a fuel. The presence of oxygen in fuel improves combustion properties and emissions but reduces the calorific value. A fuel heating value is a gauge of the amount of thermal energy it releases during its burning and is an influential factor in the fuel economy and power deliverability [24]. There are several values to consider when concentration on the available energy within a fuel, the lower heating value (LHV-net) and the higher heating value (HHV-gross) included. These differ, as the lower HV does not include energy in the combustion of water vapour, whereas the higher does. The net heating value is the appropriate quantity for comparing fuels as the engine exhaust water in gas phase [25]. A diesel engine desires fuels of higher calorific value because it facilitates the heat release during combustion and improves engine performance during combustion [4, 9].As was stated earlier a high heating value is desired, though a minimum value is not specified in the biodiesel standards ASTM D6751 but is prescribed in EN 14213 at 35 MJ/kg [26].

IKA C200 oxygen bomb calorimeter according to the ASTM D240 standard was carried out to measure heating value of the FT fuels. The use of gross or net calorific value varies with type of industry. Engine manufacturers use net calorific value and UK boiler manufacturers use gross [27].Heat of combustion is determined in this test method by burning a weighed sample in an oxygen bomb calorimeter under controlled conditions. The heat of combustion is computed from temperature observations before, during and after combustion with proper allowance for thermochemical and heat transfer corrections. Gross calorific value is gathered from instrument. If the percentage of hydrogen in fuel sample is not known, the net heat of combustion may be calculated as follows[28]:

$$Q_n = 10.025 + (0.7195)Q_g \tag{4}$$

where:

Q_n Q_n = Net heat of combustion at constant pressure, MJ/kg,

Q_g Q_g = Gross heat of combustion at constant volume, MJ/kg.

3. Results and Discussion

3.1. F-T Fuel Composition-Gas Chromatography Analysis

Hydrocarbon distribution of FT synthesis on liquid phase of products measured and analysed by PerkinElmer Clarus 60 Gas Chromatography Mass Spectrometry. GC analysis is performed for identifying the hydrocarbon compounds. Table 1 indicates the total values and percentage of paraffin, iso paraffin, olefins and alcohols within each of the gasoline (C_1-C_{12}), diesel (C_{12}-C_{22}) and waxes (C_{23+}) which were calculated by analysing the result of GCMS results.

Table 1. F-T Fuel Characterization.

FT Sample	Fischer Tropsch Experiment Reaction Conditions			W, Liquid Product (%)			Composition of products in Diesel Fraction (%)			FT Characterization		
	Temperature (K)	Pressure (bar)	(NL/gramWHSV	W (C_7-C_{11})	W (C_{12}-C_{22})	W (C_{22+})	Paraffin	Olefin	Alcohol	Gross Calorific Value (MJ/kg)	Net Calorific Value (MJ/kg) % H unknown	Lubricity (μm)
FTS 1	503	10	1.8	17.66	81.35	0.99	97.4	2.18	0.42	18.8	13.54	389
FTS 2	503	15	2.4	21.06	77.55	1.39	92.83	4.63	2.53	56.95	40.99	342
FTS 3	503	20	3	17.52	81.53	0.95	92.27	3.39	4.34	51.45	37.03	403
FTS 4	503	25	3.6	20.69	78.19	1.12	90.42	4.5	5.08	48.33	34.78	360
FTS 5	518	10	2.4	26.64	71.96	1.4	98.11	1.1	0.79	49.37	35.53	296
FTS 6	518	15	1.8	21.12	78.08	0.08	98.68	0.81	0.51	29.15	20.98	346
FTS 7	518	20	3.6	21.98	76.53	1.49	95.42	2.47	2.11	50.9	36.64	339
FTS 8	518	25	3	14.91	81.7	3.39	96.78	2.04	1.18	51.8	37.28	417
FTS 9	528	10	3	32.76	66.76	0.48	97.11	1.82	1.07	36.2	26.06	286
FTS 10	528	15	3.6	24.35	70.89	4.76	97.7	1.84	0.46	54.13	38.96	294
FTS 11	528	20	1.8	23.58	75.14	1.28	98.35	1.22	0.43	43.76	31.5	336
FTS 12	528	25	2.4	21.56	76.78	1.66	92.83	3.59	3.58	50.62	36.43	340
FTS 13	543	10	3.6	24.85	73.09	2.06	97.45	2	0.55	13.75	9.89	320
FTS 14	543	15	3	25.035	73.93	1.04	98.96	0.57	0.48	38.16	27.47	330
FTS 15	543	20	2.4	17.62	80.12	2.26	96.35	2.73	0.92	60.14	43.29	372
FTS 16	543	25	1.8	24.59	73.84	1.57	98.09	1.4	0.51	57	41.02	325

3.2. Lubricity

The tribological lubricity results for 16 Fischer Tropsch samples are gathered in Table 1. Wear Scar Diameter varies between 286 μm to 417μm. From figure 3a, it can be seen that the change of FT reaction conditions effect on the lubricity of the fuels. Increasing the pressure affects the poor lubricity property and in high temperature lubricity is better. WHSV does not have significant effect on lubricity property. By validating the result with Mahmoudi [1] which stated that optimum condition for fuel production via FTS is high pressure reaction condition, it can conclude that lubricity in higher pressure indicates poor quality.

Hydrocarbon distribution of F-T synthesis on liquid products which were measure by GCMS as a function of the lubricity in figure 3b represents that decrease in production of the olefins and gasoline products (C_7-C_{11}) could increase the lubricity by controlling the initiation, growth of carbon chain and enhancing the chain termination. From figure 3b it can clearly see that in heavier hydrocarbons by increasing the percentage of the heavier hydrocarbons in product lubricity increases. So by increasing value of the Diesel in liquid product lubricity become poor. In contrast presence of the light hydrocarbons improves lubricity of the F-T fuel. Figure 4 clearly states that all fuel samples are below the ASTM D6079 Maximum limit (520μm) with comparison of the Ultra-Low Sulphur Diesel fuel which was tested in the same condition by HFRR. This contradicts Alleman et al. [29] and Norton et al. [30] who stated that the lack of aromatic content and Sulphur in FT diesel produces unacceptable lubricity fuels. Sulphur compounds in diesel fuels create some natural lubricity which is counterproductive to current aims to reduce the Sulphur content and pollutants in fuels. Additives could be a solution to this and it is thought that biodiesels from transesterification (FAME) can have superior lubricity than other alternative fuels. Best Lubricant fuel is FT 9 which is in low pressure of

10bar and temperature of 528K, however poor lubricant fuel is FT 8 with reaction condition of high pressure 25bar and low temperature 518K.

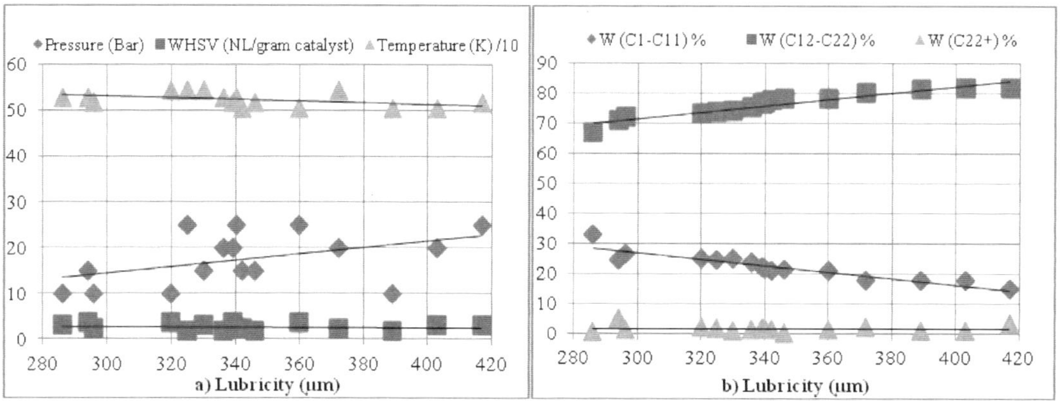

Figure3a. Relation between Lubricity and FTS Reaction Conditions (Pressure, Temperature and WHSV) **Figure 3b.** Relation between Lubricity and FT Fuel mass fraction.

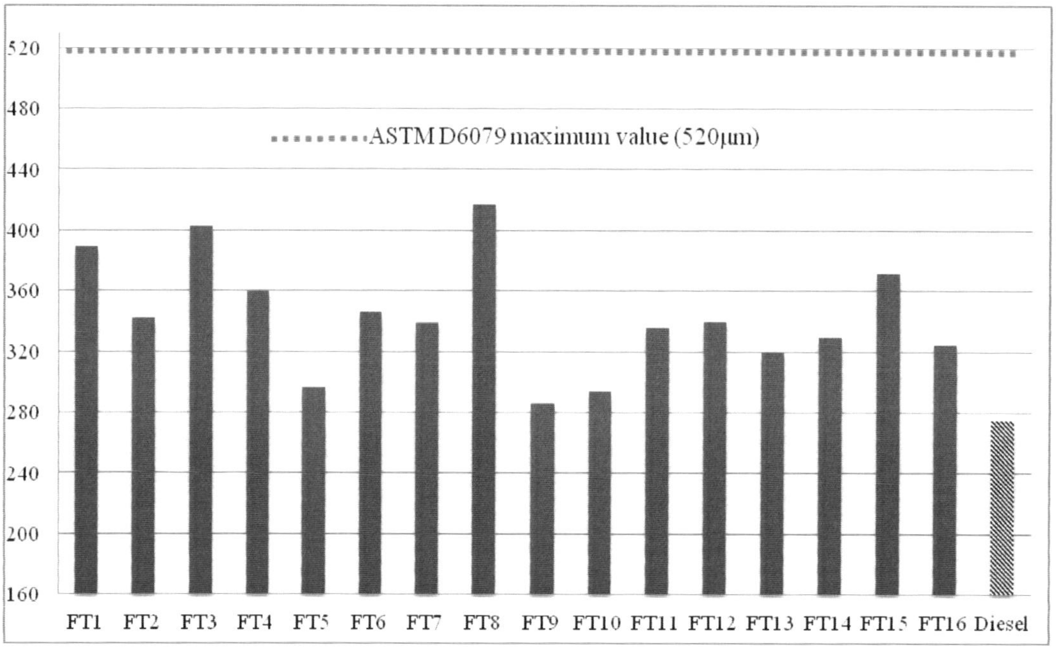

Figure 4. FT Fuel Lubricity (μm).

3.3. Heat of Combustion

Heating value is the amount of heating energy released by the combustion of a unit value of the fuels. Figure 5a indicates effect of FT reaction condition on Calorific value of fuels. It is desirable to have higher calorific value around 45 MJ/kg for diesel fuel. So in high pressure reaction condition FT fuels have appropriate heating value. Temperature and WHSV do not have any significant effect on Calorific value. Figure 5b shows effect of presence of hydrocarbon distributions on calorific value. Presence of light hydrocarbons causes decrease in calorific value but heavier hydrocarbons does not have any significant effect on amount of heat of combustion. Figure 6 shows comparison between the heating values of the Fischer-Tropsch Fuels produced in School of Mechanical Engineering at

University of Birmingham and also ULSD fuel were tested in IKA C200 oxygen bomb calorimeter. Six of these samples are in below limit of 35 MJ/Kg which needed some additives to increase their heat of combustion. FT15 in high pressure 20bar and high temperature 543 K is the fuel with highest net calorific value of 53.29MJ/kg.

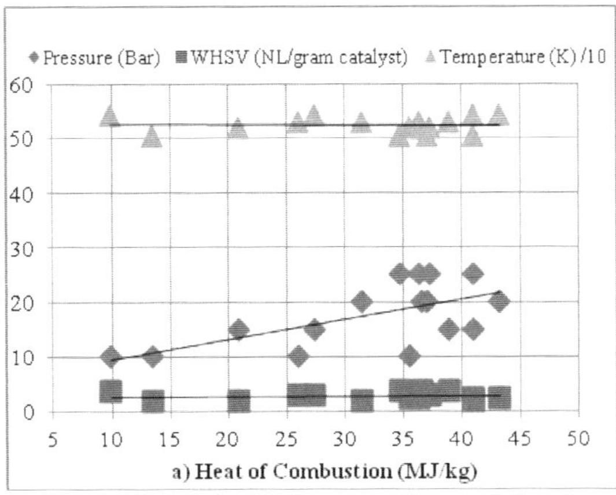

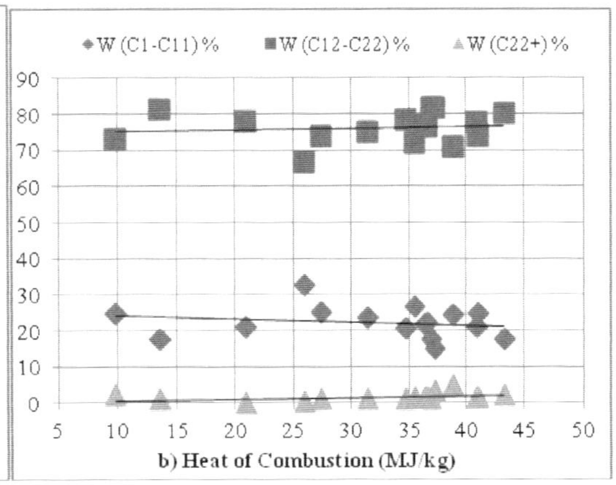

Figure 5a. Relation between Heat of Combustion and FTS Reaction Conditions (Pressure, Temperature and WHSV).

Figure 5b. Relation between Heat of Combustion and FT Fuel mass fraction.

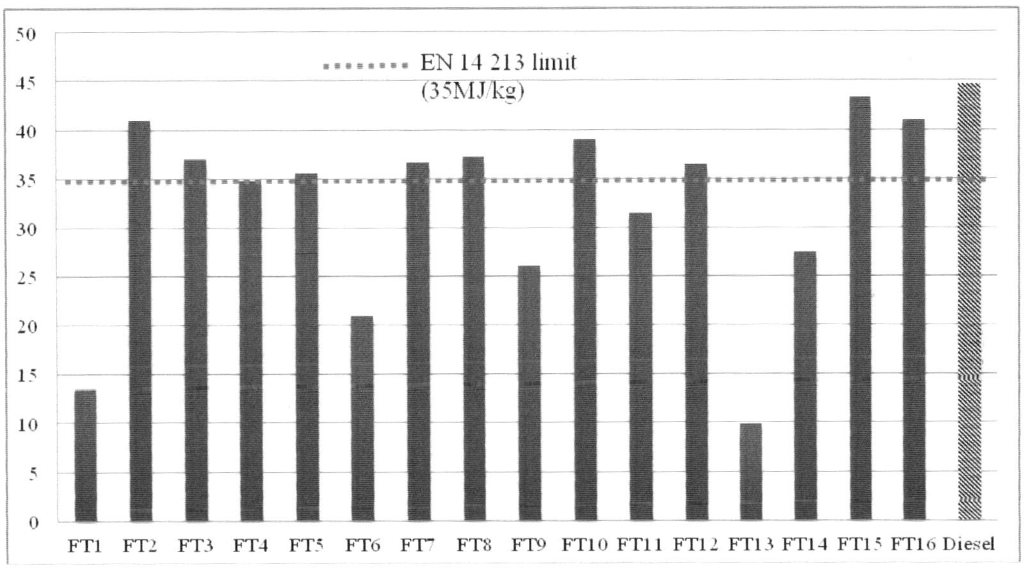

Figure 6. FT Fuel Net Calorific Value (MJ/kg).

4. Conclusions

In this paper, Fischer Tropsch Synthesis as one of the best options for biomass derived fuels production is used which has extensive history of gasoline and diesel production from coal and natural gas. FTS fuel could be one of the best solutions to the fuel emission due to its high quality. FT fuel produced in fixed bed reactor over Cobalt catalyst and it has been characterized in GCMS to determine its hydrocarbon distributions. Lubricity and calorific value measured in HFRR and IKAC200 respectively. Results show that prepared fuels in School of Mechanical Engineering at University of Birmingham have good lubricities which are more than Diesel fuel but it comply the ASTM D6079 Standard 520μm. Calorific value of only six samples are below of the limit. And result from graphs show that high pressure reaction condition improves calorific value of fuels. Because all of the samples have good lubricity so high pressure Fischer Tropsch Synthesis reaction condition produce more desirable fuels. Also Mahmoudi [1] in his PhD thesis used Taguchi method to find an optimum reaction condition which is operating at Temperature 528 K, Pressure 20-25 bar and WHSV 3-3.6NL/gram catalyst. Similarly in high pressure FTS condition, FT fuel has desirable lubricity and calorific value.

Acknowledgments
The Department of Mechanical Engineering at the University of Birmingham (UK) is gratefully acknowledged for the PhD Scholarship to Mr. Omid Doustdar.

References

[1] Mahmoudi H 2015 *Perfomance of cobalt-based eggshell catalyst in low temperature Fischer tropsch synthesis process to produce long-chain hydrocarbons from synthesis gas utilizing fixed-bed reactor technology* **PhD Thesis** School of Mechanical Engineering The University of Birmingham,

[2] Moazami N, Mahmoudi H, Rahbar K, Panahifar P, Tsolakis A and Wyszynski M L 2015 Catalytic performance of cobalt–silica catalyst for Fischer–Tropsch synthesis: Effects of reaction rates on efficiency of liquid synthesis *Chemical Engineering Science* **134**: pp 374-384

[3] Tristantini D, Lögdberg S, Gevert B, Borg Ø and Holmen 2007 A The effect of synthesis gas composition on the Fischer–Tropsch synthesis over Co/γ-Al2O3 and Co–Re/γ-Al2O3 catalysts *Fuel Processing Technology* **88**(7): pp 643-649

[4] Arbab M I, Masjuki H H, Varman M, Kalam M A, Imtenan S and Sajjad H 2013 Fuel properties, engine performance and emission characteristic of common biodiesels as a renewable and sustainable source of fuel *Renewable and Sustainable Energy Reviews* **22**: pp 133-147

[5] Murugesan A, Umarani C, Subramanian R and Nedunchezhian N 2009 Bio-diesel as an alternative fuel for diesel engines—A review *Renewable and Sustainable Energy Reviews* **13**(3): pp. 653-662

[6] Sadeghinezhad E, Kazi S N, Sadeghinejad F, Badarudin A, Mehrali M, Sadri R, and Reza Safaei M, A comprehensive literature review of bio-fuel performance in internal combustion engine and relevant costs involvement.*Renewable and Sustainable Energy Reviews*, 2014. **30**: pp 29-44.

[7] Midland A D, Biodiesel Technical Information. *ADM Biodiesel Technical Services*.

[8] Sadeghinezhad E, Kazi S N, Badarudin A, Oon C S, Zubir M N M and Mehrali M 2013 A comprehensive review of bio-diesel as alternative fuel for compression ignition engines.*Renewable and Sustainable Energy Reviews* **28**: pp 410-424

[9] Ashraful A M, Masjuki H H, Kalam M A, Rizwanul Fattah I M, Imtenan S, Shahir S A and Mobarak H M 2014 Production and comparison of fuel properties, engine performance, and emission characteristics of biodiesel from various non-edible vegetable oils: A review *Energy Conversion and Management* **80**: pp 202-228

[10] Torregrosa A J, Broatch A, Plá B and Mónico L F 2013 Impact of Fischer–Tropsch and biodiesel fuels on trade-offs between pollutant emissions and combustion noise in diesel engines *Biomass and Bioenergy* **52**: pp 22-33

[11] Maniatis K and Millich E 1998 Energy from biomass and waste: the contribution of utility scale biomass gasification plants *Biomass and Bioenergy* **15**(3): pp 195-200

[12] Zeng S, Du Y, Su H and Zhang Y 2011 Promotion effect of single or mixed rare earths on cobalt-based catalysts for Fischer–Tropsch synthesis *Catalysis Communications* **13**(1): pp 6-9

[13] Rafiq M H, Jakobsen H A, Schmid R and Hustad J E 2011 Experimental studies and modeling of a fixed bed reactor for Fischer–Tropsch synthesis using biosyngas *Fuel Processing Technology*, **92**(5) pp 893-907

[14] Henrici-Olivé G and Olive S 1984 Mechanism of the Fischer-Tropsch synthesis: Origin of oxygenates *Journal of Molecular Catalysis* **24** (1): pp 7-13
[15] Guettel R and Turek T 2008 Reactors for Fischer-Tropsch Synthesis. Institute of Chemical Process Engineering Clausthal University of Technology.*Chem. Eng. Technol*, 2008
[16] Moazami N, Mahmoudi H, Panahifar P, Rahbar K, Tsolakis A and Wyszynski M L 2015 Mathematical Modeling and Performance Study of Fischer-tropsch Synthesis of Liquid Fuel over Cobalt-silica *Energy Procedia* **75**: p. 62-71
[17] Knotthe J V G and Jurgen Krahl J The Biodiesel Handbook. 2005*: AOCS Press*. 303
[18] Hazrat M A, Rasul M G and Khan M M K 2015 Lubricity Improvement of the Ultra-low Sulfur Diesel Fuel with the Biodiesel.*Energy Procedia* **75**: pp 111-117
[19] Maru M M, Trommer R M, Cavalcanti K F, Figueiredo E S, Silva R F and Achete C A 2014 The Stribeck curve as a suitable characterization method of the lubricity of biodiesel and diesel blends *Energy* **69**: pp 673-681
[20] BSI, BS EN ISO 12156-1:2006 Diesel Fuel, Assessment of lubricity using the high frequency reciprocationg rig (HFRR) Part 1. Test Method
[21] Magı´n Lapuerta R G and John R. Agudelo 2009 Lubricity of Ethanol-Biodiesel-Diesel Fuel Blends *Energy & Fuels* **24**: p 6
[22] Sukjit E 2013 Synergistic Effects of Alcohol Based Renewable Fuels: Fuel properties and Emissions **PhD Thesis** *School of Mechanical Engineering The University of Birmingham*
[23] Eslami F 2013 Properties, Perfomance and emissions of biofuels in blends with gasoline **PhD Thesis** *School of Engineering, The University of Birmingham*
[24] Gandure J, Ketlogetswe C and Temu A 2014 Fuel properties of biodiesel produced from selected plant kernel oils indigenous to Botswana: A comparative analysis *Renewable Energy* **68**: p. 414-420.
[25] Bezergianni S and Dimitriadis A 2013 Comparison between different types of renewable diesel.*Renewable and Sustainable Energy Reviews* **21**: pp 110-116
[26] Ali O M, Mamat R, Abdullah N R and Abdullah A A 2016 Analysis of blended fuel properties and engine performance with palm biodiesel–diesel blended fuel *Renewable Energy* **86**: pp 59-67
[27] Piaszyk J 2012 Animal Fat (Tallow) as Fuel for Stationary Internal Combustion Engines **PhD Thesis** *School of Mechanical Engineering, The University of Birmingham*
[28] Standard A N 2007 Standard Test Method for Heat of Combustion of Liquid Hydrocarbon Fuels by Bomb Calorimeter / ASTM D240-02. 2007 *ASTM International United States*
[29] Alleman T L and McCormick R L 2003 Fischer-Tropsch Diesel Fuels - Properties and Exhaust Emissions: A Literature Review *SAE InternationalTechnical Paper* 2003-01-0763 DOI: 10.4271/2003-01-0763
[30] Norton P, Vertin K, Bailey B, Clark N N, Lyons D W, Goguen S and Eberhardt J 1998 Emissions from Trucks using Fischer-Tropsch Diesel Fuel *SAE International Technical Paper* 982526 DOI: 10.4271/982526